Handbook of
Vegetable Pests

Handbook of Vegetable Pests

John L. Capinera

Professor and Chairman
Department of Entomology & Nematology
University of Florida

ACADEMIC PRESS

A Harcourt Science and Technology Company

San Diego   San Francisco   New York   Boston   London   Sydney   Tokyo

Academic Press
A Harcourt Science and Technology Company
525 B Street, Suite 1900, San Diego, California 92101-4495, USA
http://www.academicpress.com

Academic Press
Harcourt Place, 32 Jamestown Road, London NW1 7BY, UK
http://www.academicpress.com

Library of Congress Catalog Card Number: 2001086233

International Standard Book Number: 0-12-158861-0

PRINTED IN THE UNITED STATES OF AMERICA
01 02 03 04 05 06 EB 9 8 7 6 5 4 3 2 1

This book is dedicated to my family, Marsha and Erica, who assisted me in many ways. I extend to them my love and appreciation.

Contents

Class Insecta—Insects
Order Coleoptera—Beetles, Weevils, White Grubs and Wireworms

Contents

Appendices

Preface

Vegetables are an important element in the culture and economy of North America. Consumption of vegetables is often a daily ritual, and according to nutritionists, vegetables should be eaten several times per day. Fresh vegetables have increased in importance in the diet of many people, displacing canned and frozen vegetables as the principal form of consumption. Better transportation systems and higher incomes have helped to increase the consumption rate of fresh commodities. Vegetable growing is a popular hobby, and for many consumers it serves to guarantee both low-cost and high-quality (fresh) food. Vegetable production in the commercial sector is immensely important. Not only are the farm-gate value and retail value of vegetables measured in billions of dollars, but manufacture and sale of components of vegetable production and processing (seed, fertilizer, pesticides, irrigation, planting and harvesting equipment, labor, etc.) add billions of dollars to economies throughout the United States and Canada.

Much has been written about vegetable culture, but surprisingly few efforts have been made to compile a comprehensive treatment of vegetable pests for North America. Pests often are an important constraint on vegetable production. Those tasty crops bred to please humans also entice pests! Effective management of vegetable pests is sometimes easy and sometimes difficult, but is assuredly easier when pertinent information is readily available. Although there are several reference books that include vegetable pests, each suffers in some area. Several years ago I set out to attempt to rectify some of the shortcomings of existing references, and this book is the result of that labor.

The *Handbook of Vegetable Pests* was produced as a reference for professionals in agriculture. It should be a handy reference for entomologists, of course, but horticulturists, cooperative extension service personnel, crop consultants, members of the agrochemical industry, and others will find it useful. I have attempted to minimize the use of scientific terms (undoubtedly to the chagrin of some entomological colleagues), and otherwise simplified the discipline specificity of the book to make it easier for specialists in related disciplines to use as a reference without needing additional reference books to decipher the terms. Some producers and dedicated amateurs may find use for this *Handbook*, but it was not really written for them.

The key elements of the *Handbook* are the **introduction**, which provides an overview of vegetable crops grown in North America and the major groups of pests that cause injury; **identification guides**, which list the major and minor pests known to attack each crop plant family and provide some distinguishing characteristics for the pests; **pest profiles**, which describe in some detail the appearance, life history, and management of the pests; **appendices**, which provide some simple keys to important groups of pests, additional sources of information on vegetable pests, cross-listings of plant names, and journal abbreviations and titles; **glossary**, which defines many of the scientific terms used in the *Handbook*; **references**, a complete list of the journals, books and other publications cited in the pest profile section; and **index**, which will assist in finding entries in the *Handbook*.

Anyone with a basic knowledge of entomology will find it easy to use the *Handbook*. The major requirement for effective use of the *Handbook* is knowledge of the pest's identity. Lacking a specific identification, the user can narrow down the pest's identity by using the identification guides or keys, or by comparing with pictures and descriptions of similar insects. The pests are organized by major taxon (usually order and family) to make scanning and comparisons easier.

I have avoided recommending specific insecticides because registrations change frequently. Instead, I have emphasized timing, target area, and intervals of application. Also, in many cases I have reported specific formulations or methods of application that have been found effective. However, the availability of insecticide formulations and application techniques may change as insecticide registrations are lost. For example, availability of granular fomulations of some insecticides seems to be decreasing even though they may have the preferred application technology for some uses. Remember that the instructions on the insecticide label must always be followed.

Acknowledgments

Much of the entomological literature published in the past 125 years that pertains to North American vegetable pests was reviewed during preparation of this *Handbook*. This was a complex task, and undoubtedly there are numerous oversights and errors in interpretation. I take full responsibility for such errors and hope that readers will be so kind to point them out so that one day these may be corrected. I have attempted to correct as many mistakes as possible by obtaining reviews from knowledgeable colleagues, and I am indebted to them for their efforts:

Guy Boivin, Agriculture Canada, Horticulture Research and Development Centre (Quebec)—Selected Cabbage, Carrot, and Onion Insects.

Paul Choate, University of Florida—Beetles, Earwigs, Flies, and Hemipterans.

Whitney Cranshaw, Colorado State University—Numerous Midwestern and Western Insects.

Joe Eger, Dow AgroChemical (Tampa)—Pentatomids.

Sanford Eigenbrode, University of Idaho—Selected Pea Pests and Crucfer Flea Beetle.

Joe Funderburk, University of Florida—Thrips; Anthomyiid Flies.

Susan Halbert, Florida Department of Agriculture, DPI—Aphids.

Marjorie Hoy, University of Florida—Mites.

Dave Horton, United States Department of Agriculture, ARS (Yakima)—Selected Potato Insects.

Gary Leibee, University of Florida—Leaf Miners and Selected Cabbage Insects.

Everett Mitchell, United States Department of Agriculture, ARS (Gainesville)—Noctuids.

Tom Mowry, University of Idaho—Potato Aphid.

Steve Naranjo, United States Department of Agriculture, ARS (Phoenix)—Whiteflies; Rootworms.

David Oi, United States Department of Agriculture, ARS (Gainesville)—Red Imported Fire Ant.

Frank Peairs, Colorado State University—Numerous Corn and Bean Insects.

Jorge Pena, University of Florida—Selected Tomato, Cucurbit, and Bean Pests.

Bob Pfadt, University of Wyoming—Grasshoppers and Mormon Cricket.

David Riley, University of Georgia—Thrips, Whiteflies, Selected Crucifer, and Cucurbit Insects.

David Robinson, United States Department of Agriculture, APHIS (Philadelphia)—Slugs and Snails.

Dak Seal, University of Florida—Wireworms, Thrips, and Cornsilk Fly.

Harold Toba, United States Department of Agriculture, ARS (Yakima)—Selected Wireworms and Green Peach Aphid.

Tom Mowry, University of Idaho—Potato Aphid.

Tom Walker, University of Florida—Field and Mole Crickets.

Mike Weiss, University of Idaho—Crucifer Flea Beetle.

Cal Welborne, Florida Department of Agriculture, DPI—Mites.

Steve Valles, United States Department of Agriculture, ARS (Gainesville)—Beetles.

David Williams, United States Department of Agriculture, ARS (Gainesville)—Red Imported Fire Ant.

Goeff Zehnder, Auburn University—Selected Potato and Cucurbit Pests.

Scientific names are surprisingly dynamic, and common names—particularly plant common names—are a constant source of confusion because each plant tends to acquire many common names. The insect pest names used in this *Handbook* usually are based on The Entomological Society of America's Common Names of Insects and Related Arthropods (1997). The names of parasitic Hymenoptera and Diptera follow K. V. Krombein, P. D. Hurd, D. R. Smith and B. D. Burks'

Catalog of Hymenoptera in America North of Mexico (1979), and P. H. Arnaud, Jr.'s publication, A Host-Parasite Catalog of North American Tachinidae (Diptera), published in 1979 as USDA Miscellaneous Publication 1319, respectively, except where I know of more recent systematic revisions. I have attempted to standardize the botanical terminology, based on Gray's Manual of Botany (1950; 8th edition), authored by M. L. Fernald, and Food and Field Crops of the United States (2nd edition; 1998) by G. M. Markle, J. J. Baron and B. A. Schneider.

Illustrations are based principally on previous publications, though all have been modified to some extent to improve their appearance. The major sources of illustrations are various USDA bulletins, circulars, and technical bulletins, though many illustrations are original. I am indebted to the following institutions and individuals for permission to reproduce illustrations:

Arkansas Agricultural Experiment Station—Spotted Cucumber and Black Blister Beetles.

Dick Baranowski, University of Florida (How to Know the Bugs)—Squash, Harlequin, Southern Green Stink, and Negro Bugs.

Carl Childers, University of Florida—Several Thrips species.

Colorado Cooperative Extension Service—Numerous Caterpillar Drawings from Publication 535A and Grasshopper drawings from Publication 584S.

Connecticut Agricultural Experiment Station—Asiatic Garden Beetle.

Florida Agricultural Experiment Station—Pepper Weevil.

Florida Department of Agriculture and Consumer Services, DPI—Colorado Potato Beetle; Banded, Southern and Northern Corn Rootworms; Yellow Margined Leaf Beetle; Mole Crickets.

Gene Gerberg (Belkin's Fundamentals of Entomology)—Bugs, Thrips, Aphid, Leafhopper, Weevil, Scarab, Maggot, Earwig, Negro Bug, and Sucking Mouthparts.

O. E. Heie, Skive Seminarium—Lettuce Aphid.

Hawaii Agricultural Research Center (William's Handbook of Sugarcane Pests)—Black Cutworm, Chinese Rose Beetle, Click Beetle, Millipede, and Slug.

Hawaii Agricultural Research Station—Sweetpotato Leafminer and Vine Borer.

Illinois Agricultural Experiment Station—Black, Dingy, and Bronzed Cutworms; Sod Webworm; Wheat Wireworm; Seedcorn Beetle.

Illinois Natural History Survey—Alfalfa and Rapid Plant Bugs.

Iowa Agricultural Experiment Station—False Chinch Bug.

Massachusetts Agricultural Experiment Station—Grass Thrips, Carrot Plant Bug.

Charles Papp, California Department of Food and Agriculture (Handbook of Agricultural Pests)—Bird Cherry-Oat, Cowpea, Foxglove, Green Peach, Pea, Bean, Melon, Cabbage, Turnip, and Potato Aphid; Tomato Russet, Twospotted and Strawberry Mites; Gray Garden, Tawny Garden, Greenhouse, *Laevicaulis*, and other Slugs; Decollate, *Bradybaena*, *Helix* and other Snails.

New York Agricultural Experiment Station—Cabbage Root Maggot and Hop Vine Borer.

North Carolina Agricultural Research Service, North Carolina State University—Corn Earworm and Tobacco Budworm Pupae, Adults, and Larval Spines from Bulletin 196.

Pennsylvania Cooperative Extension Service—Tomato Hornworm.

Puerto Rico Agricultural Experiment Station—Bean Leafskeletonizer.

Tom Riley, Louisiana State University—Selected Wireworms.

University of California (Agricultural Experiment Station)—Fire Ant (Hilgardia)—Artichoke Plum Moth, Green Peach Aphid, and Garden Centipede.

Jim Tsai, University of Florida—Corn Delphacid.

Utah Agricultural Experiment Station—Beet Leafhopper, Pale Western Cutworm, and Tomato Psyllid.

Vernon Vickery, Lyman Entomological Museum, McGill University—Mormon Cricket.

Photographs were kindly provided by numerous individuals, and they are acknowledged in the picture captions. Most that were not original were taken by Drs. Jim Castner and Paul Choate while employed as photographers for University of Florida.

Special thanks are due Dr. Steve White and Syngenta (Novartis) Crop Protection for financial support that allowed inclusion of the color photographs.

Introduction

NORTH AMERICAN VEGETABLE CROPS

Production of vegetable crops is a fundamental industry, a popular hobby, and an important element of our culture. The farm-gate value of North American vegetable crops exceeds $10 billion annually and fuels several other industries such as equipment manufacture, transportation, and retail and restaurant food sales worth many times the value of the raw materials. Vegetable production also has immense value as a form of recreation and relaxation. Surveys of recreational habits consistently reveal gardening to be the number one American hobby. Also, food is an important element of all cultures, and to be able to access diverse and freshly harvested produce, and perhaps to share it with family and friends, remains a desirable and cherished behavior among North Americans. Because the United States and Canada are populated primarily by people whose ancestors migrated from another country in relatively recent times, there are many specialty crops desired by one ethnic group or another that are not yet part of the mainstream of commerce. Thus, local or home garden production of specialty vegetables flourishes in many ethnic communities. For example, oriental vegetables are too small a component of vegetable commerce to be tabulated separately by statisticians concerned with vegetable production. However, broccoli was rare outside Italian-American communities until the 1950s and has since become one of the most popular vegetables. Other changes are imminent as North American consumers become more cosmopolitan in their dietary habits.

What Are the Major Vegetable Crops?

Numerous types of vegetables are available to the people of the United States and Canada. Availability changed radically during the 20th century. Early in that century a great deal of vegetable production was local, people ate what was seasonally available, and in the winter months they ate only what stored well in root cellars or could be preserved through canning. With the improvement in transportation in the middle of the century, southern vegetable production areas could keep northern consumers supplied with fresh vegetables through most of the winter season. Now not only is fresh produce grown in southern states, but increasing quantities are being imported from Mexico, Central America, and even southern South America.

Complete agreement is lacking on what plants should be called vegetables. In government statistics, potato and sweet potato tend to be grouped with field or staple crops such as wheat, barley, and rice. However, most people think of potato and sweet potato as vegetables. Dry beans such as pinto beans and navy beans also usually are treated as field crops, whereas the same species grown for edible pods as bush beans and pole beans are considered vegetables; even dry beans are thought of as vegetables by many. Strawberry is treated as a vegetable in some statistical compilations, but most people think of strawberry as a fruit. Also, plants such as parsley and coriander might be considered as herbs rather than vegetables because they are used mostly for seasoning.

What vegetables do Americans and Canadians prefer to eat? Despite increased importation of vegetables from other countries, examination of the commercial production data (area planted and value of crops) provides a good indication of what is popular. Following is a summary of fresh market vegetable-production data for the most important vegetable crops grown in the United States for the year 1997:

Crop	Area planted		Value
	Acres	Hectares	$ (millions)
Artichoke	9,100	3,683	67.6
Asparagus	77,640	31,420	181.2
Beans, lima	5,500	2,226	4.9
Beans, snap	88,660	35,880	156.4
Broccoli	134,300	52,367	495.5
Brussels sprouts	4,200	1,700	26.8
Cabbage	82,550	33,407	279.3
Cantaloupe	118,450	47,936	417.8
Carrot	99,030	40,076	441.2
Cauliflower	47,600	19,263	197.9
Celery	27,150	10,987	273.4
Corn, sweet	241,000	97,530	398.3
Cucumber	58,250	25,573	185.2
Eggplant	2,600	1,052	17.6
Escarole and endive	3,300	1,335	13.1
Garlic	37,000	14,974	261.5
Honeydew melon	30,700	12,424	109.3
Lettuce	279,150	112,969	1,603.6
Onion	165,580	67,009	648.4
Pepper	67,600	27,357	58.7
Potato	1,325,500	536,420	2,759.5
Potato, sweet	83,500	33,790	205.8
Spinach	16,800	6,799	58.7
Tomato	129,180	52,279	1,246.8
Watermelon	204,300	82,678	309.2

The relative importance of processed (canned and frozen) vegetables is evident from the following list of crops, areas planted, and crop values:

Crop	Area planted		Value
	Acres	Hectares	$ (millions)
Beans, lima	54,090	21,890	35.0
Beans, snap	204,450	82,751	129.7
Beet	7,780	3,148	8.1
Cabbage	5,870	2,376	8.2
Carrot	23,850	9,652	37.4
Corn, sweet	476,930	193,009	247.8
Cucumber	107,780	43,617	146.0
Pea	292,600	118,412	136.9
Spinach	20,090	8,130	15.7
Tomato	293,700	118,857	605.3

A survey of the important vegetable crops of Canada shows a similar trend among vegetable production, though the inclusion of rutabaga and the absence of broccoli are significant distinctions and reflect cultural differences rather than horticultural limitations. The Canadian data (Howard *et al.*, 1994) also include greenhouse production of cucumber and tomato—a factor that inflates the value of these crops—but not the area planted. Greenhouse production of vegetables, long an important industry in Europe, as yet is a small industry in the United States. The Canadian data follow:

Crop	Area planted		Value
	Acres	Hectares	($ millions)
Asparagus	4,250	1,700	7.6
Bean	19,500	7,800	18.0
Beet	2,250	900	4.7
Cabbage	12,000	4,800	27.6
Carrot	19,250	7,700	61.0
Cauliflower	7,500	3,000	21.0
Celery	2,000	800	13.5
Corn, sweet	89,250	35,700	60.6
Cucumber	7,250	2,900	58.0
Lettuce	7,250	2,900	35.2
Onion	11,250	4,500	34.4
Parsnip	1,000	400	1.4
Pea	48,750	19,500	23.4
Pepper	5,250	2,100	13.8
Potato	314,000	125,600	361.6
Radish	2,000	800	4.0
Rutabaga	6,500	2,600	18.0
Spinach	1,290	500	1.9
Tomato	28,750	11,500	129.7

Most vegetable crops can be grown in all areas of the United States and Canada, though some crops clearly are favored by long seasons and hot summers, and others by cooler summers. Home gardeners are remarkable in their ability to nurture vegetables through all sorts of climatic difficulties, but commercial producers tend to gravitate to the ideal climatic conditions for a particular vegetable. Thus, at least in commercial vegetable production, crops are concentrated in a few areas. The leading fresh market vegetable producing states are:

State	Percentage of national production	
	Area harvested	Value
California	43.7	53.4
Florida	10.0	15.1
Georgia	6.5	6.6
Arizona	6.4	3.8
Texas	4.4	3.1

The southern or warm-weather states tend to dominate the fresh market vegetable industry because vegetables can be grown only in these locations during the winter months. In contrast, northern states contain a high proportion of the processed vegetable industry, though California leads in processed vegetables as well. The leading producers of processed vegetables in the United States are:

State	Percentage of national production	
	Area harvested	Value
California	20.9	43.7
Wisconsin	17.3	9.8
Minnesota	14.8	7.8
Washington	11.2	7.9
Oregon	6.9	5.4

In Canada, commercial vegetable production similarly is concentrated in a few provinces:

Province	Percentage of national production
Ontario	35.6
Quebec	29.8
British Columbia	15.5
Prince Edward Island	7.2
Manitoba	6.5

Not evident from these data are the historical trends in consumption. Although a detailed description of dietary changes is beyond the scope of this treatment, it is evident that salad vegetables such as lettuce, cucumber, and tomato have increased in importance. Some vegetables such as potato and sweet corn remain popular, though in the case of sweet corn there has been a shift from canned and frozen to fresh produce. Many of the cool-weather vegetables such as Brussels sprouts, cabbage, pea, rutabaga and turnip, and spinach have decreased in popularity.

Characteristics of the Major Vegetable Crops

Artichoke (Family Compositae)

Artichoke, more correctly known as globe artichoke, is a thistle-like plant grown for the edible blossom bud. It is one of only a few vegetable plants grown in North America as a perennial, with new growth arising from the roots annually. Like some other perennials, it can be grown with some success as an annual crop by planting roots, but this is not a common practice. Although artichoke can be grown over a broad geographic area, it is not cold-hardy. Commercial production is limited to the California coast, where the cool, moist climate favors its growth. It is not a popular vegetable and is considered by many to be a "luxury" vegetable. The origin of artichoke appears to be the western Mediterranean region of Europe. A similar plant grown for the leaf stalks is cardoon; this plant is not grown commercially in North America. The key pest of artichoke is artichoke plume moth, *Platyptilia carduidactyla* (Riley), though such aphids as artichoke aphid, *Capitophorus elaeagni* (del Guercio), and bean aphid, *Aphis fabae* Scopoli, can be quite damaging at times.

Asparagus (Family Liliaceae)

The asparagus is a hardy perennial plant and if once established remains productive for many years. It is grown in all except the warmest regions of North America and is winter-hardy in most climates. It is usually grown from crowns (roots), but seeds may be used. Seeds are spread freely by birds, and it is common to find asparagus growing wild along roadsides, fences, and irrigation ditches. California, Washington, and Michigan are leaders in asparagus production, though commercial production is also important in eastern Canada. Asparagus is popular in home gardens in northern areas, probably because it is one of the first crops available for harvest in the spring. It has been an important vegetable since ancient times, and it originated in the Mediterranean region. The spears or stems as they first push up from the soil are harvested; once they begin to branch they become tough and inedible. The most serious pest is asparagus aphid, *Brachycorynella asparagi* (Mordvilko), though its economic impact is mostly a western phenomenon. In eastern North America asparagus beetle, *Crioceris asparagi* (Linnaeus), and spotted asparagus beetle, *C. duodecimpunctata* (Linnaeus), can be nuisances.

Bean and Related Crops (Family Leguminosae) (Bush Bean, Chickpea, Cowpea, Dry Bean, Faba Bean, Lentil, Lima Bean, Pea, Snap Bean, etc.)

The legumes are known for their ability to harbor nitrifying bacteria; nitrogen enhances soil productivity. The cultivated legume vegetable crops are not particularly efficient as sources of nitrogen for plant growth, however, so fertilization is still required. Most of the leguminous vegetable crops are warm-weather crops and are killed by light frosts. Pea is a notable exception, thriving under early season and cool-weather conditions, though it is killed by heavy frost. Although not often grown in the United States, and principally a home garden crop in Canada, faba bean is commonly grown in the northern regions of Europe. The legume vegetables are annual crops. The most commonly grown bean crops, snap bean and lima bean, are native to Central America. Pea likely originated in Asia, cowpea in Africa. In the United States, snap beans are quite popular, in both fresh and processed form. Florida and California lead in production of fresh market beans and Minnesota and Wisconsin in processing pea. Cowpea production occurs principally in the southeastern states, where historically it was quite important because it remained productive through the long summer months when most other vegetable crops failed. In Canada, pea is a more important crop than bean, and Ontario and Quebec lead in production of these crops. The legumes are cultivated for their seeds or seed pods. The seed-attacking maggots, *Delia* spp., can be important pests under cool-weather conditions. The key pest of bean in most areas is Mexican bean beetle, *Epilachna varivestris* Mulsant. Pea is quite susceptible to infestation by pea aphid,

Acyrthosiphon pisum Shinji. Cowpea is plagued by cowpea weevils, *Callosobruchus* spp. Locally, a number of other pests can be important, particularly thrips, leaf miners, leafhoppers, and flea beetles.

Beet and Related Crops (Family Chenopodiaceae) (Beet, Chard, Spinach, Swiss Chard)

The beet apparently originated in the Mediterranean region, spinach in Iran. Beet and its relatives are biennial crops, requiring more than one year but less than two to complete their natural life cycle. They are grown as annuals when cultivated as vegetables. Although beet originally was grown entirely for its foliage, cultivars of beets with edible below-ground portions (the edible ''root'' is mostly thickened stem material) became popular in Europe beginning about 1800. Beet, once popular in North America, has declined greatly in popularity. Chard has never been an important crop and remains relatively obscure. Spinach is not very popular, but commercial production is stable. Most commercial cultivation of beets in the United States occurs in the Great Lakes area, but spinach production is more western in distribution, from Arkansas and Texas to California. In Canada, these crops are grown primarily in Ontario, Quebec, and British Columbia. In home gardens, these crops thrive nearly everywhere. The principal pests are green peach aphid, *Myzus persicae* (Sulzer), and beet and spinach leafminer, *Pegomya* spp. In the west, beet leafhopper, *Circulifer tenellus* (Baker), can be very damaging.

Cabbage and Related Crops (Family Cruciferae) (Broccoli, Brussels Sprouts, Cabbage, Cauliflower, Chinese Cabbage, Collards, Kale, Kohlrabi, Mustard, Radish, Rutabaga, Turnip)

The cruciferous vegetables, often called ''cole'' crops, are grown for their leaves. These are cool-season crops; they tolerate light freezes and even brief heavy freezes, but prolonged deep freezes are fatal. Though naturally biennials, they are grown as annuals. Cabbage and its many forms originated along the shores of Europe; mustard and radish are from Asia. Some are popular foods in North America; others are of regional significance. Perhaps the most interesting vegetable in this group is broccoli, which has become popular only since the 1950s. Cauliflower is a moderately important crop. Cabbage and turnip, once popular in the United States, have declined in importance though cabbage remains a significant crop. Rutabaga is an important dietary element in Canada, but not in the United States. Collards and, to a lesser degree, kale, are popular vegetables in the southeastern states. In the United States, commercial production of broc-

coli, cauliflower, and Brussels sprouts is concentrated in California. California and Florida are the primary producers of radish. Cabbage is produced widely. In Canada, Ontario and Quebec are the important producers of these crops. In home gardens, these crops are produced throughout North America. The principal pests of cabbage and its closest relatives include the root maggots, *Delia* spp.; cabbage aphid, *Brevicoryne brassicae* (Linnaeus); diamondback moth, *Plutella xylostella* (Linnaeus); cabbage looper, *Trichoplusia ni* (Hübner); and imported cabbageworm, *Pieris rapae* (Linnaeus). Mustard and radish tend to be plagued more by green peach aphid, *Myzus persicae* (Sulzer).

Carrot and Related Crops (Family Umbelliferae) (Carrot, Celery, Celeriac, Chervil, Coriander, Fennel, Parsley, Parsnip)

The umbelliferous vegetables are biennial, but are grown as annuals. They require cool weather to develop properly. Most survive in heavy frost but are killed by prolonged freezing weather. The major crops of this group, carrot and celery, originated in the Mediterranean region. Carrot and parsnip are grown for their root; celery, celeriac, and fennel for the swollen stem bases; and parsley and coriander for their foliage. These crops are grown widely in the United States, in both northern and southern locales, but California dominates commercial production. In Canada, commercial production occurs predominantly in Quebec and Ontario. Carrot and celery have long been popular vegetables. Parsnip once was an important crop because it could be stored well during the winter months and even be left in the soil during freezing weather. While Canadians and Europeans retain a fondness for this crop, Americans rarely consume it. Parsley retains its utility in the American diet, but coriander is rapidly growing in popularity as Mexican and oriental recipes gain broader acceptance. Fennel is similarly growing in popularity, but as yet is a minor crop awaiting discovery by the American palate. Chervil is rarely grown in North America, and different types are grown for their root and foliage. In some areas, carrot weevil, *Listronotus oregonensis* (LeConte), and carrot rust fly, *Psila rosae* (Fabricius), are important pests, and leafhoppers sometimes transmit aster yellows disease. American serpentine leafminer, *Liriomyza trifolii* (Burgess), is often the most serious threat to commercially grown celery.

Lettuce and Related Crops (Family Compositae) (Celtuce, Chicory, Endive, Escarole, Lettuce, Radicchio)

The lettuce is an immensely popular vegetable. Among vegetables grown in the United States, it surpasses all other vegetables except for potato in area of land devoted to production and in crop value. Canada lacks the warm-weather production areas found in the United States so it is not as important a commercial crop there. It is popular among Canadian consumers and much lettuce is imported from the United States. California and Arizona dominate commercial production in the United States; Quebec is the major producer in Canada. Lettuce apparently originated in Europe or Asia and has been grown for over 2000 years. Lettuce is grown everywhere in North America for home consumption and despite the apparent concentration of lettuce production in a few states or provinces, specialty lettuces are grown around many large metropolitan areas for local consumption. Lettuce and most related crops are grown for their leaves, though in the case of celtuce the stem is eaten. Lettuce and related crops are cool-season annuals. Although killed by heavy frost, these crops are also susceptible to disruption by excessive heat. Hot weather causes lettuce to flower and become bitter tasting. Several insects are important pests of lettuce. Aster leafhopper, *Macrosteles quadrilineatus* Forbes, is an important vector of aster yellows in some production areas. Several species of aphids may be damaging, though green peach aphid, *Myzus persicae* (Sulzer), generally is most important. Numerous caterpillars such as corn earworm, *Helicoverpa zea* (Boddie), and cabbage looper, *Trichoplusia ni* (Hübner), threaten the lettuce crop.

Okra (Family Malvaceae)

The okra is thought to be native to Africa and is an important crop in tropical countries. It is also an important element of southern cooking because it is one of the few vegetables that remain productive throughout the long summer of the southeast. It is an annual plant and is killed by light frost. Okra is grown for the seed pod, which, like snap bean, is harvested before it matures. It is unusually tall for a vegetable crop, often attaining a height of two meters. It is a relatively minor vegetable crop from a national perspective and so production statistics are infrequent. Commercial production in the United States occurs in the southeast from South Carolina to Texas. The pods are subject to attack by several pests, with the most damaging being red imported fire ant, *Solenopsis invicta* Buren; southern green stink bug, *Nezara viridula* (Linnaeus); and leaf-footed bugs, *Leptoglossus* spp.

Onion and Related Plants (family Alliaceae) (Chive, Garlic, Leek, Onion, Shallot)

Onion and its relatives are biennial or perennial plants, but cultivated as annuals. These have long been important crops, with use of onion documented for nearly 5000 years. Their origin is thought to be Asia. Onion is grown principally for the below-ground leaf bases, which form a bulb, but tops are also edible. They are tolerant of cool weather, but also thrive under hot conditions. Onion is cultivated widely in North America, though commercial production of the sweet varieties tends to be concentrated in southern areas whereas the pungent varieties are cultivated at more northern latitudes. In the United States, leading producers are California, Washington, Texas, Colorado, New York, and Georgia. California also leads the nation in garlic production. Ontario and Quebec produce most of the onions grown in Canada. The key pests of onion are onion thrips, *Thrips tabaci* Lindeman, and onion maggot, *Delia antiqua* (Meigen).

Rhubarb (Family Polygonaceae)

The rhubarb is one of the few perennial vegetable plants cultivated as a perennial. Its origin is northern Asia, and its use can be traced back about 5000 years. Rhubarb thrives where summers are cool. The stalks or leaf petioles are used as food, though this vegetable is infrequently consumed. Most commercial production occurs in northern states such as Michigan, Oregon, and Washington, but it remains principally a home garden crop. There are a few important pests of rhubarb, with rhubarb curculio, *Lixus concavus* (Say), perhaps the most serious.

Squash and Related Crops (Family Cucurbitaceae) (Cucumber, Pumpkin, Squash, Cantaloupe, Watermelon, and Other Melons)

The cucurbit crops are important vegetables, though they vary in economic importance. They are annual plants and are warmth-loving crops. All are cultivated for their fruit. Light frost will kill cucurbits, and even cool weather will permanently disrupt growth. Squash and pumpkin originated in Central and South America; cucumber, melons, and watermelon are from Africa or perhaps Asia. The summer squashes and cucumbers are stable in production or increasing slightly in importance, though squash is considered to be a minor vegetable crop and statistics are scant. The winter squashes are of decreasing significance in diet and commerce. Pumpkin is grown mainly as an ornamental plant, though some is processed for food. All types of melons are increasing in popularity and economic significance. Cucumbers are produced in many states, but Georgia and Florida lead in fresh market production and Florida in cucumbers for pickles. Texas, Florida, and Georgia are the leading producers of watermelon, but commercial production is widespread. Cantaloupe and other melon production is centered in California and Arizona. Cucurbit crops have some serious insect pests. Home garden plants are plagued by squash vine borer, *Melittia cucurbitae* (Harris); squash bug, *Anasa tristis* (De Geer); and in the southeastern states by pickleworm, *Diaphania nitidalis* (Stoll). Commercial crops are affected by whitefly and aphid plant virus vectors, *Diabrotica* spp. and *Acalymma* sp. cucumber beetles in the north central states, and pickleworm in the southeast.

Sweet Corn (Family Graminae)

Corn, commonly known as maize outside the United States, apparently was domesticated in Mexico and perhaps is descended from a similar grain, teosinte. Corn originally was cultured because it was productive, a good source of carbohydrates and other nutrients, and the grain stored well. Sweet corn is a recent innovation that was first developed in the mid-1700s, but lacks the storage characteristics of the older types, or grain corn. Corn is grown for the seeds, which are clustered in a structure called the "ear." Sweet corn is a popular vegetable, though canned corn is less popular than it once was due to the availability of frozen corn. The more recent availability of fresh corn that does not quickly lose its sweetness (supersweet cultivars) has increased demand for whole-ear corn. Sweet corn is cultured widely in North America. In the United States, Florida, California, New York, and Georgia lead in fresh sweet corn production, whereas Minnesota and Wisconsin lead in processed sweet corn. In Canada, the provinces producing most of the sweet corn are Ontario and Quebec. As might be expected from a crop that originated in North America, many pests feed on corn. Thus, native species such as corn earworm, *Helicoverpa zea* (Boddie); fall armyworm, *Spodoptera frugiperda* (J.E. Smith); and the *Diabrotica* rootworms are the most serious pests. However, some introduced insects also have become frequent pests, including European corn borer, *Ostrinia nubilalis* (Hübner); Japanese beetle, *Popillia japonica* Newman; and corn leaf aphid, *Rhopalosiphum maidis* (Fitch).

Sweet Potato (Family Convolvulaceae)

Sweet potato is an immensely important crop in some parts of the world, but not in North America, where acreage and consumption are declining. Sweet potato probably originated in Mexico and is well adapted to tropical growing conditions. Some moist-fleshed varieties of sweet potato are called yams, but yams are a separate species normally found in Polynesia, and infrequently seen in North America. A perennial crop, sweet potato is normally grown as an annual. It is cultivated for its tuber. Sweet potato cannot tolerate prolonged cool weather and perishes if exposed to light frost. Most of the domestically produced sweet potatoes are grown in North Carolina and Louisiana. Sweet-potato weevil, *Cylas formicarius* (Fabricius), is the most damaging pest of this crop, but anything that damages the tubers, including wireworms and flea beetle and cucumber beetle larvae, is economically threatening.

Tomato and Related Plants (Family Solanaceae) (Eggplant, Pepper, Potato, Tomatillo, Tomato)

The solanaceous crops are among the most popular vegetable crops in North America. Potato ranks as the most valuable vegetable crop grown in the United States, with tomato third and pepper fifth in value. In Canada, potato and tomato are considered first and second in importance, but warm-season crops such as pepper do not thrive in Canada. Among home gardeners, tomato is the most popular crop. Potato is grown for its tuber, the other crops for their fruit. The origins of the solanaceous crops are diverse; eggplant originated in India, potato and tomato in Peru, tomatillo and pepper in Mexico or Guatemala. Tomato, tomatillo, pepper, and eggplant are warm-season perennials that are cultivated as annuals. Potato is a cool-season perennial cultured as an annual. Potato in most of North America is at risk from Colorado potato beetle, *Leptinotarsa decemlineata* (Say), but several aphids also are commonly damaging, particularly green peach aphid, *Myzus persicae* (Sulzer). Tomato is affected by silverleaf whitefly, *Bemisia argentifolii* Bellows and Perring; corn earworm, *Helicoverpa zea* (Boddie); and many other pests. In the west, beet leafhopper, *Circulifer tenellus* (Baker), can be damaging. Home-garden production, but not commercial production, is at risk from tobacco and tomato hornworms, *Manduca* spp. In southern production areas, pepper weevil, *Anthonomus eugenii* Cano, is the key pest of pepper.

INSECTS AND INSECT RELATIVES

With over 90,000 insects inhabiting the United States and Canada, it might seem to be an impossible task to identify one specimen among so many possibilities. However, a surprisingly small number of insects injure plants, and even fewer damage vegetable crop plants. About 300 species are documented to be pests, though a few more are capable of damage or feed on vegetables only rarely.

Insects are members of the group of animals collectively classified as *Insecta*. Insecta is a major class of animals, much like other major groups such as *Reptilia* (reptiles), *Mammalia* (mammals), *Aves* (birds), *Pices* (fish), and *Gastropoda* (snails and slugs). *Insecta* is subdivided into major groups called orders, and further subdivided into smaller groups of like organisms called families. Occasionally, further subdivisions are noted, such as subfamilies, but the most important designations, other than the order and family of insects, are the genus and species names. The genus and species designation (sometimes along with the original describer or species "author") is also called the scientific name and is used to provide universal recognition. Common names often vary among different regions of the world, and insects even acquire different names depending on the crop attacked, but scientific names are universally accepted and are changed rarely.

Frequency of Pests Among the Major Insect Orders

The occurrence of vegetable pests among the orders of insects is more or less proportional to the total number of species in each order. The approximate number of species in each major order inhabiting the United States and Canada and the number of species within the corresponding order that are considered in this book are:

Insect order	Number of species in the United States and Canada	Number treated in this book
Coleoptera (beetles)	24,000	98
Hymenoptera (ants, bees, wasps)	18,000	2
Diptera (flies)	17,000	26
Lepidoptera (moths and butterflies)	12,000	89
Homoptera (aphids, leafhoppers, etc.)	7,000	38
Hemiptera (bugs)	3,500	25
Orthoptera (grasshoppers and crickets)	1,000	15
Thysanoptera (thrips)	700	6
Dermaptera (earwigs)	20	2

The order Hymenoptera and, to a lesser degree, the order Diptera contain many beneficial species (parasitoids and pollinators), accounting for their underrepresentation as vegetable pests.

Noninsect Vegetable Pests

There are several other groups of invertebrate animals that act like insects, inflicting the same type of injury to plants. Some even resemble insects in general appearance. Therefore, they are included in this treatment of "insect" pests. Best known among these other invertebrates are the snails, slugs, mites, millipedes, sowbugs, and pillbugs.

Order Coleoptera — Beetles

Coleoptera is the largest order of insects, containing about 40% of the known species in the class Insecta. The most distinctive feature is the structure of the wings. The front pair of wings, called elytra, are normally thickened and hard, providing protection for the thinner, membranous hind pair of wings. The hind wings are longer than the front wings and fold beneath the front wings when the insect is not in flight. Beetles have chewing mouthparts, with well-developed mandibles. The immatures undergo complete metamorphosis, displaying egg, larval, and pupal stages before attaining adulthood. Their behavior and ecology are diverse. Larvae are soft bodied and usually bear three pairs of legs on the thorax, and no prolegs on the abdomen. Larvae often develop in protected habitats such as within stems and soil, but some are found feeding on foliage. About 24,000 species are known in the United States and Canada.

Family Bruchidae—Seed Beetles and Weevils

The adult bruchids have a short, broad downturned snout bearing mouthparts at the tip. These beetles are small, measuring no more than 6.5 mm long, and generally less than 5 mm. Their body tapers toward the head. The antennae are slightly expanded at the tips or serrate. The elytra extend only about two-thirds the length of the abdomen, exposing the terminal segments. The larvae live within the seeds of legumes. They possess legs only during the first instar; thereafter they are legless, plump, and humpbacked. There are over 130 species of bruchids known in the United States and Canada.

Family Carabidae—Ground Beetles

Although the pest species are somber brown or black, carabids are sometimes brightly colored and metallic. They tend to be slightly flattened and long legged, with the elytra marked with long grooves and ridges. They vary widely in size. The antennae are thread-like. Nearly all carabids are beneficial, with both the larval and adult stages feeding on other insects. Larvae generally inhabit the soil. They have a thin, elongate body with a large head. About 2300 species are known in the United States and Canada.

Family Chrysomelidae—Leaf Beetles

This is a large and diverse family. Many are brightly colored. They range in length from 1.5 to 13.0 mm. The body tends to be oval and strongly convex. The antennae and legs are moderately long to short. The antennae are thread-like or slightly expanded apically, but never clubbed. About 1500 species of chrysomelids occur in the United States and Canada. As an aid to identification, two of the easily recognizable subfamilies are treated separately below.

Subfamily Alticinae—Flea Beetles

The flea beetles are a distinctive group of leaf beetles. They derive their name from the impressive leaping behavior of adults, which is made possible by muscular, greatly enlarged hind femora. These dark-colored beetles are small, usually less than 5 mm in length, sometimes considerably smaller. They may be metallic or even brightly colored on occasion. They tend to leave small round holes in leaves when they feed, sometimes consuming only one surface of the leaf and leaving the other intact. Most larvae live below ground, feeding on the roots of plants.

Subfamily Cassidinae—Tortoise Beetles

The tortoise beetles are distinguished by the flaring of the pronotum and elytra, which largely cover the head and legs when viewed from above. This protected, armored, or turtle-like appearance is the basis for the common name of this group of leaf beetles. Many species are brightly colored, often orange or gold. Some species can be confused with lady beetles, family Coccinellidae. However, in lady beetles the head is clearly visible. Also, in lady beetles there are three tarsal segments on each leg, whereas tortoise beetles appear to have four tarsal segments. Larvae are surface feeders, found on leaves alongside adults. They bear numerous branched spines and often carry debris or fecal material on their backs.

Other Subfamilies—Leaf Beetles

A few other subfamilies of Chrysomelidae contain pests, but they are not so readily distinguished as are the aforementioned subfamilies. For example, subfamily Criocerinae contains the asparagus-feeding species, Galucerinae the corn rootworms and cucumber beetles, and Chrysomelinae contains Colorado potato beetle. As with the other chrysomelids, the adults feed above-ground on plant tissue; larval feeding behavior is variable. The adults are often brightly colored and moderate in size, often measuring 4–12 mm long. The antennae are usually moderately long in length.

Family Coccinellidae—Lady Beetles

This family is best known for its beneficial, predatory members. However, a small number are plant feeders. Most lady beetle species are brightly colored and spotted, but their general pattern is misleading; not only are some important predators spotless, but the spot pattern varies greatly even within a species, redu-

cing its diagnostic value. The legs are short or moderately long, and bear only three tarsal segments. The antennae are clubbed. The body is strongly convex, almost hemispherical in shape. There are about 400 species known from the United States and Canada. The larvae of plant-feeding species bear large branched spines.

Family Curculionidae—Billbugs, Curculios and Weevils

The members of this family are distinguished by their elongated snout, which bears mouthparts at the tip. The elbowed and clubbed antennae are attached at about the mid-point of the snout. The body form varies among species, but it is usually elongate-oval. These beetles tend to be moderate in size, ranging from 3–12 mm, but sometimes considerably larger. They usually are dark, with the elytra attaining the tip of the abdomen. This latter character is useful for distinguishing curculionids from bruchids. The plump larvae lack legs, and normally are found burrowing within plant tissue. Over 2600 species are known from the United States and Canada.

Family Elateridae—Click Beetles and Wireworms

The click beetles are elongate, parallel-sided, and usually relatively flattened in appearance. The pronotum is relatively large, and as wide as the thorax. They bear a structural mechanism that allows them to flex rapidly, producing a clicking noise. The head is usually hidden below the pronotum. The antennae are usually serrate, but not clubbed. The elytra bear ridges and grooves. These beetles usually are obscurely colored, often brown and black. The beetles are usually moderate or large in size, measuring 12–30 mm in length. Although the adults are phytophagous, it is the larval stage, known as wireworms, that is most destructive. The larvae are slender (hence the name wireworm), round in cross-section, hard-bodied, and shiny. They feed on the roots of plants. About 900 species are known in the United States and Canada.

Family Meloidae—Blister Beetles

These elongate, narrow-bodied beetles are unusual because their elytra are soft and flexible. The pronotum is narrower than the head and thorax. The tips of the elytra tend to diverge. The legs and antennae are long and thread-like. Blister beetles are moderate to large in size, measuring 12–25 mm long. The adults feed on foliage and blossoms, but the larvae live below-ground where they feed on insects; the common species feed on the egg pods of grasshoppers. About 310 species are known in the United States and Canada.

Family Nitidulidae—Sap Beetles

These beetles tend to be small to moderate and less than 12 mm long. They are oval and often distinguished by short elytra, which expose the terminal abdominal segments. Also, the antennae are distinctly clubbed. Both larvae and adults feed on decaying fruit. About 185 species occur in the United States and Canada.

Family Scarabaeidae—Scarab Beetles and White Grubs

The most distinctive feature of scarab beetles is the shape of the antennae. The three terminal segments are expanded into plate-like or finger-like structures that can be closed to form a club. Many species have long and spiny legs. Scarab beetles vary widely in size, from two to over 100 mm. The May and June beetles are oval and strongly convex. Others, however, are more flattened, with a wide prothorax and an abdomen that tapers acutely toward the posterior end. Most species are drab in color, usually brown or black. Some, however, such as Japanese beetle, are brightly colored and metallic. The adults feed above-ground on foliage, blossoms, and fruit. Larvae are plump and C-shaped in general form, usually living below-ground on roots of plants. About 1400 species of scarabs live in the United States and Canada.

Family Tenebrionidae—Darkling Beetles and False Wireworms

The darkling beetles typically inhabit arid environments, particularly the western regions of North America. Adults tend to be long-legged with a hump-backed appearance. They generally are black. The antennal structure is variable. Darkling beetles are easily confused with carabid beetles, but can be distinguished by their tarsal structure. The number of tarsal segments in darkling beetles is five, five, and four on the front, middle, and hind legs, respectively. In contrast, carabid beetles have five tarsal segments on all legs. Darkling beetles vary from 2–35 mm long. About 1000 species occur in the United States and Canada. Although adults feed on plants, larvae are more destructive due to their tendency to attack seedlings. The larvae are called false wireworms, because they are narrow and elongate in form, resembling wireworms in general appearance.

Order Dermaptera — Earwigs

Dermaptera is a small group of beetle-like insects. They may be winged or wingless, though the forewings are short when present. They are omnivorous in feeding behavior, often feeding on other small animals. Their most distinctive feature is the pincer-like cerci found at the tip of the abdomen.

Order Diptera—Flies and Maggots

The flies are notable in possessing only a single pair of wings. The wings are membranous. The second pair is reduced to a pair of small-knobbed structures, called halteres, that function to provide balance. Flies tend to be small and fragile. The mouthparts are variable, but usually function to sponge up liquids or they are needle-like and are used to pierce and suck-up liquids. The antennae are variable. They undergo complete metamorphosis, displaying egg, larval, and pupal stages in addition to the adult or fly form. The larvae generally are worm-like and legless, and are called maggots. Maggots are usually quite featureless, but the anterior end is equipped with hook-like structures that are used to tear host tissue. Pupation occurs within the old larval exoskeleton (outer covering of the maggot) and is called a puparium to distinguish it from the pupa found in most other insects, and which lacks the larval covering. About 17,000 species of flies are known in the United States and Canada.

Agromyzidae—Leaf-Mining Flies

These small black or grayish flies usually measure less than 2 mm long and are often marked with yellow. The wings lack color. Leafminers are best known for the winding mines or tunnels created by larvae as they feed between the upper and lower surfaces of leaf tissue. The typical mine is narrow, growing wider as the larva increases in size, but some species cause irregular, blotchy mines. About 190 species of Agromyzidae are known in the United States and Canada.

Drosophilidae—Pomace Flies

These small fruit flies are also known as vinegar flies, because they are attracted to fermenting fruit and vegetables. They are small in size, rarely exceeding 3 mm long. The wings are colorless, broad, and have few veins. The antenna bears a hair (brista) that is generally plumose. About 120 species of Drosophilidae are known in the United States and Canada.

Otididae—Picture-winged Flies

These medium-sized flies attain a length of up to 12 mm. They have wings that are marked with broad bands of color. The fly's body is often shiny or metallic in appearance. Larvae are found in fruits and roots, or on decaying tissue. Many species are not capable of injuring healthy plants, but are incorrectly implicated as being damaging because they are found in rotten tissue. About 125 species occur in the United States and Canada.

Psilidae—Rust Flies

These small- to medium-sized flies measure 3–8 mm long. They tend to be rather slender and bear long antennae. Their wings are not colored. The only pest species is carrot rust fly, *Psila rosae* (Fabricius), which is covered with a dense coat of hairs. Larvae burrow in roots. Only 34 species are known in the United States and Canada.

Syrphidae—Flower Flies

The adults of many species in this common family of flies are brightly colored, often black and yellow, and are sometimes mistaken for bees. Most species measure 10–20 mm long and their wings are not marked. The larvae of some species are predatory, and are often found feeding within colonies of aphids. A few species, however, feed on the below-ground portions of plants. The number of syrphid species inhabiting the United States and Canada is about 875.

Tephritidae—Fruit Flies

These medium-sized flies often attain a length of 10–12 mm. Their wings are spotted or banded. Their larvae usually develop within fruit, and some species are severe fruit pests, especially in tropical areas. A few species mine leaves. About 280 species occur in the United States and Canada.

Order Hemiptera (Heteroptera)—Bugs

The Hemiptera are often called the "true bugs" to differentiate them from other orders of insects that are casually termed bugs. The spelling of insect families in the order Hemiptera consists of two words, with "bugs" a separate word (e.g., burrower bugs and ambush bugs), whereas insects from other orders combine "bugs" into a single word (e.g., lightningbugs which are beetles, and lovebugs which are flies). The Hemiptera are distinguished from other groups of four-winged insects by the presence of long piercing-sucking mouthparts in the form of a "beak" that originates at the front of the head. In addition, the basal portion of the front pair of wings (the corium) is thickened and usually without veins, whereas the apical portion is thin, membranous, and normally bears veins. The hind wings are entirely membranous and also bear veins. The wings normally are held flat over the back when the insects are at rest, not angled or roof-like as in the order Homoptera. Insects in the order Hemiptera can be confused with insects in the order Homoptera, which also possess tubular piercing-sucking mouthparts. In Homoptera, however, the mouthparts originate at the back of the head rather than the front of the head, and the entire front wing is

membranous. Many Hemiptera have scent glands that produce strong odors. Some species possess short wings, and some are wingless. Not all Hemiptera feed on plants; many are predatory, feeding on other insects. Metamorphosis in Hemiptera is incomplete, with the adult stage preceded by the egg and nymphal stage, but the pupal stage lacking. The immatures, or nymphs, greatly resemble the adults in form, though lacking fully developed wings. Their food and habitat is the same as in the adult stage.

Family Coreidae—Squash and Leaffooted Bugs

These bugs tend to be fairly large, usually 10–30 mm, and dark colored. If disturbed, they tend to produce a foul-smelling odor. The antennae and mouthparts each bear four segments. The legs are long, and in some species they are flattened. The membranous portion of the front wings bear six or more veins. They tend to be dull colored, usually brown or black. Only a few of about 120 species occurring in the United States and Canada are destructive to vegetables.

Family Cydnidae—Burrower Bugs

Burrower bugs are oval, resembling stink bugs in general body form, though they lack the elongate lateral lobes of the pronotum usually present in true stink bugs, family Pentatomidae. The scutellum of burrower bugs is enlarged, though not as large as is found in the negro bugs, family Thyreocoridae. Perhaps the most distinctive feature is the numerous spines on the tibiae, which likely aid in burrowing through the soil, its major habitat. There are about 35 species in North America. Most are small, less than 8 mm in length, and blackish. They are nocturnal, and uncommonly observed except in lights. Only one species, *Pangaeus bilineatus* (Say), is known as a pest.

Family Lygaeidae—Seed Bugs

Lygaeid bugs resemble coreids, but usually are small, often 4–12 mm long. Sometimes they are colorful. As suggested by its common name, most species are seed feeders, though some suck plant sap or are predatory. The species damaging vegetables are less than 5 mm in length. The antennae and mouthparts (beak) each bear four segments. The membranous portion of the front wing bears only 4–5 veins, a character that is useful for distinguishing them from the Coreidae. In some species the femora of the front legs are enlarged. About 300 species are known in the United States and Canada.

Family Miridae—Plant Bugs

These common bugs tend to be narrow-bodied and elongate. They are moderate in size, measuring 4–

10 mm long. They are soft-bodied relative to most other bugs. The antennae and mouthparts (beak) each consist of four segments. The veins at the tips of the front wings are few in number, forming semi-circular loops, originating and ending in the thickened basal portion of the wing (the corium), rather than divergent or parallel veins terminating at the wing margin. Also, the apical portion of the corium is triangular, and separated from the rest of the corium by a groove. Mirids are normally plant feeders, but some feed opportunistically on insects.

Family Pentatomidae—Stink Bugs

The stink bugs derive their name from disagreeable odors, which are produced when the bugs are handled. They are also known as shield bugs, a reflection of their overall shape, and a more distinguishing character than their odor. These bugs are moderate to large in size, usually 6–23 mm long. Stink bugs possess a triangular structure behind the prothorax and at the base of the wings known as a scutellum; it reaches to the mid-point of the abdomen or beyond. The number of mouthpart (beak) segments is four and the number of antennal segments is five. Stink bugs may be plant feeders, especially seed feeders, or predators. The predatory species tend to have a thick beak and the phytophagous species a thin beak. There are about 250 species in the United States and Canada.

Family Thyreocoridae—Negro Bugs

These small oval bugs resemble stink bugs, and are often confused with beetles. They measure only 3–6 mm long and possess a greatly enlarged scutellum that covers most of the abdominal segments and wings, superficially resembling the elytra found on beetles. They usually are black. The antennae consist of five segments. Only about 30 species are known in the United States and Canada.

Family Tingidae—Lacebugs

The lacebugs are probably the most distinctive family of the Hemiptera. The adults possess ornate, lacy wings consisting of many small cells. In some species the pronotum is expanded and similarly ornate. The immatures are spiny. These are small insects, usually 3–5 mm long. The antennae and mouthparts (beak) each consist of four segments. Most species feed on trees, and about 160 species are known in the United States and Canada.

Order Homoptera—Aphids, Plant- and Leafhoppers, Psyllids and Whiteflies

The order Homoptera consists of a diverse assemblage of insects that are difficult to characterize, because superficially they are so different in general appearance. They are closely related to Hemiptera, and sometimes treated as members of that order. The mouthparts of Homoptera are the piercing-sucking type, but instead of arising at the front of the head as in the Hemiptera, they arise basally, at the back of the head. Sometimes, the mouthparts appear to originate between the base of the front pair of legs. Normally, they have four wings of relatively uniform texture throughout, though in some the front wings are slightly thickened. When at rest, the wings are usually held angled or roof-like over the body, not flat over the back as is typically found in Hemiptera. The antennae may be short or bristle-like, or long and thread-like. Metamorphosis is incomplete, with the egg, nymphal, and adult stages present. Nearly 7000 species are known in the United States and Canada.

Family Aleyrodidae—Whiteflies

The whiteflies are small but distinctive insects. They rarely exceed 2–3 mm long and usually are entirely white. Veins are not apparent on the wings because they are covered with white scales. The antennae are usually fairly long, thread-like, and consist of seven segments. The whiteflies are tropical species, and most abundant in warm climates and in greenhouses. Because they reproduce rapidly and attain very high densities, and can transmit plant viruses, they can be quite damaging to vegetable crops. The nymphal stage is flattened, often not really resembling an insect and certainly not the adult. The nymph often produces waxy filamentous secretions; it generally is sedentary. About 100 species are known in the United States and Canada.

Family Aphididae—Aphids

The aphids are small to moderate in size, normally measuring 2.0–3.5 mm long. They undergo complex life cycles, often alternating between winged and wingless generations and between perennial and annual host plants. The aphids usually occur in colonies, sometimes attaining very high densities. Most bear a pair of tubular structures called cornicles near the tip of the abdomen. They secrete from the anus sugary secretions that cause stickiness and discoloration on foliage. The thread-like antennae are moderate in length, and consist of 3–6 segments. Some species produce waxy secretions that obscure their general appearance, causing them to resemble tufts of cotton or other inanimate objects. Occasionally, they live below-ground on roots, though generally they feed on leaves and young stem tissue. They are important as direct pests of vegetable crops due to the plant sap they ingest, but they also transmit plant viruses very effectively, which greatly exacerbates their economic impact. Nearly 1400 species are known in the United States and Canada.

Family Cicadellidae—Leafhoppers
Family Delphacidae—Planthoppers

These small- to medium-sized insects rarely exceed 12 mm in length and are narrow-bodied in shape. They generally have a sharply or bluntly pointed head. The wings are normally fully formed, extending the length of the abdomen, but occasionally they are short-winged. The front wings are slightly thickened. The antennae are thread-like, originating between the eyes in Cicadellidae and beneath the eyes in Delphacidae. These insects produce sound, but it is barely audible to humans. The leafhoppers and planthoppers are second only to aphids as important plant disease vectors. Leafhoppers are extremely diverse in North America, with over 2500 species known in the United States and Canada. Planthoppers, on the other hand, are few in number; only about 145 species are known from the same geographic area.

Pseudococcidae—Mealybugs

The mealybugs are oval, flattened insects that secrete waxy filamentous material over their body, sometimes making their identity difficult to determine. Many species also produce filaments around the periphery of the body, including particularly long filaments in the anal region that resemble ''tails.'' About 280 species are known in the United States and Canada, but are usually known as pests of weedy plants and ornamentals in greenhouses. Only pink hibiscus mealybug, *Maconnellicoccus hirsutus* (Green) is known to cause serious injury to vegetable crops in the field.

Psyllidae—Psyllids

The psyllids are small insects, measuring 2–5 mm long. They resemble aphids superficially, and cicadas upon close examination. In adults the antennae are long, consisting of 9–11 segments. The hind legs are stout and capable of producing long jumps. Their leaping ability is an important diagnostic character. The body of nymphs is flattened, with a short fringe of filaments along the periphery. This stage is sedentary. In the United States and Canada, about 260 species are known to occur. Few species are important to crops.

Order Hymenoptera—Ants, Bees, Sawflies and Wasps

The order Hymenoptera is diverse in appearance and biology. Hymenopterans bear two pairs of membranous wings. Most have chewing mouthparts. Often a significant constriction occurs between the thorax and abdomen, producing a thread-like ''waist.'' A distinct ovipositor is present in many females. The order Hymenoptera contains many social insects such as ants and bees, which live in colonies, but also many solitary species. Some hymenopterans such as cicada killer wasps attain a large size, sometimes over 100 mm. In contrast, hymenopterans such as egg and thrips parasitoids are among the smallest insects, measuring about 0.2 mm long. Most members of this order are not plant pests, rather serving to pollinate plants, scavenge detritus, recycle nutrients, or parasitize other insects. Note that although only a single species of ant is treated as a plant pest in this book, many other species are indirectly detrimental, because they tend aphids, affecting their location and abundance, and thus indirectly influence the amount of damage to crops. The order Hymenoptera has complete metamorphosis, displaying egg, larval, pupal and adult stages. About 18,000 species are known in the United States and Canada, but some authorities believe that many species await discovery.

Family Argidae—Argid Sawflies

There are several families of sawflies, but most sawflies feed only on trees and shrubs. Only one species, in the family Argidae, attacks vegetable crops, and it is limited to feeding on sweet potato. The adults are broad-bodied, lacking the constriction or narrowing of the abdomen found in most Hymenoptera. Larvae of sawflies are elongate-cylindrical in form and resemble caterpillars, but bear six or more pairs of prolegs along their abdomen, more than what is found on caterpillars. The prolegs of sawflies also lack the terminal hooks or crochets that occur on caterpillar prolegs. There are only about 60 species of argid sawflies in the United States and Canada, though the number of sawflies in the several sawfly families totals nearly 1000 species in this geographic area. Adult argid sawflies measure 8–15 mm long and bear antennae with only three segments; the third segment is disproportionately long.

Family Formicidae—Ants

The ants are social insects, generally living in large colonies. They also display division of labor, which is reflected in the appearance of different ''castes'' within a colony, which differ in their appearance and behavior. Thus, we find some ants are specialized for reproduction, but others for food collection (workers) or colony defense (soldiers). Ants measure 1–20 mm long. The anterior portion of their abdomen is markedly constricted, but the first 1–2 segments bear bumps. The antennae are elbowed, an important diagnostic feature, and consist of 6–13 segments. Few ants can be considered to be direct pests of vegetables, and only red imported fire ant, *Solenopsis invicta* Buren, is regularly damaging. However, because ants attend aphids, they influence the abundance of these other plant pests, and exacerbate damage. This is somewhat offset by the predatory behavior of ants, as they can be voracious predators of caterpillars and other insect pests.

Order Lepidoptera—Moths, Butterflies and Caterpillars

This large group contains many destructive pests. The most important diagnostic character is the presence of flattened scales on the wings and often on the other parts of the adult's body. Adult lepidopterans have four wings, and are often rather heavy-bodied, and bear medium-sized to long antennae. The antennae vary in appearance; in butterflies the tip is usually expanded to form a knob, whereas in moths the apical expansion is lacking. Adults also have long, tubular mouthparts that uncoil to collect nectar from flowers. This structure is not capable of piercing plant tissue. Lepidopterans undergo complete metamorphosis, possessing egg, larval, pupal and adult stages. The immature stage, or caterpillar, is cylindrical and bears three pairs of thoracic legs and 2–5 pairs of abdominal prolegs. The prolegs bear small hooks, called crochets, at their tips. The head capsule is well-developed and hardened. The head bears chewing mouthparts and small, obscure antennae. The caterpillar may be naked or covered with short or long hairs. Larvae usually feed on plant foliage, but sometimes on other plant parts. About 12,000 species are known in the United States and Europe.

Family Arctiidae—Tiger Moths and Woolly Caterpillars

The Arctiids are heavy-bodied, medium-sized moths, usually measuring 25–30 mm in wingspan. They are often strikingly marked. The front wings are unusually long and pointed. The caterpillars bear thick and long coats of hairs, giving rise to the "woolly" designation. Few species are vegetable pests, and those that are tend to be associated primarily with weeds. There are about 265 species of arctiids in the United States and Canada.

Family Gelechiidae—Gelechid Moths and Leafminers

These moths are small, measuring only 8–25 mm in wingspan. The wings usually bear a fringe of long hairs. The larvae of vegetable-infesting species are primarily leafminers, though potato tuberworm larvae also enter potato tubers. Approximately 630 species of gelechids are known in the United States and Canada.

Family Hesperiidae—Skipper Butterflies

The skippers are not usually considered to be plant pests, and only one species affects vegetable crops, bean leafroller, *Urbanus proteus* (Linnaeus). Skippers are small to moderately sized, heavy-bodied insects. Many species, including bean leafroller, have elongate extensions or "tails" on their hind wings. They tend to be brownish, with antennae bearing thickened and hooked tips. Larvae generally feed within rolled leaves. Larvae have an enlarged head and the thorax immediately behind the head is constricted. About 300 species of hesperiids are known in the United States and Canada.

Family Lycaenidae—Hairstreak, Copper, and Blue Butterflies

These small brightly colored butterflies are not often thought of as pests. However, gray hairstreak, *Strymon melinus* (Hbner), feeds on legumes. The hairstreaks generally bear 2–3 narrow extensions or "tails," trailing from the back of the hind wing. The larvae generally are short and broad, with the head largely hidden by the thorax. About 140 members of this family occur in the United States and Canada.

Family Lyonetiidae—Lyonetiid Moths

These minute moths have a wingspan of about 4–11 mm. The wings tend to be narrow and pointed, but bear a long fringe of hairs. The larvae are leaf miners, and among vegetables only sweet potato is affected. About 125 species of lyonetid moths occur in the United States and Canada.

Family Noctuidae—Noctuid Moths, Armyworms, Cutworms, Loopers, and Stalk Borers

This is one of the largest and most important groups of vegetable pests. The moths are mostly nocturnal and obscurely marked. They are moderate in size, measuring 20–45 mm in wingspan, and are heavy bodied. The front wings are rather narrow, the hind wings broad. The hind wings are usually paler in color than the forewings. The larvae are variable in appearance, but generally dull-colored and lacking long hairs or stout bristles. Many larvae are nocturnal, seeking shelter during the day in soil and beneath debris. Cutworm and armyworm larvae normally bear five pairs of abdominal prolegs. Armyworms are so named because of their tendency to migrate in groups, but this is a transient property that is usually associated with high densities, and even some insects, called cutworms take on such migratory tendencies at times. Looper larvae bear only 3–4 pairs of abdominal prolegs, allowing them to arch their back as they crawl. Most larvae are leaf feeders or stem borers. Nearly 3000 species of noctuids occur in the United States and Canada.

Family Oecophoridae—Oecophorid Moths

The oecophorid moths are small, measuring 9–27 mm in wingspan. The front wings are usually narrow, either pointed or rounded. Only parsnip webworm, *Depressaria pastinacella* (Duponchel), among the many oecophorids found in North America causes damage to vegetable crops. About 225 species are known in the United States and Canada.

Family Papilionidae—Swallowtail Butterflies and Celeryworms

The swallowtail butterflies are large insects, often 60–100 mm in wingspan. They usually are brightly colored, and have elongate extensions or "tails" trailing from the hind wings. Few attack vegetables, with only umbelliferous crops being damaged. The larvae are large, heavy bodied, and colorful. When disturbed, larvae evert colorful, scented glands from behind the head. Only 35 species of swallowtails occur in the United States and Canada.

Family Pieridae—White, Sulfur and Orange Butterflies, and Cabbageworms

The pierid butterflies are among the most common species in gardens, meadows, and roadsides. They are medium in size, usually 40–60 mm in wingspan. The white or yellowish color of their wings is normally marked with black spots or borders. The larvae of several species affect crucifers and are often called "cabbageworms." About 60 species inhabit the United States and Canada.

Family Pyralidae—Snout Moths, Borers, Budworms, Leaftiers, and Webworms

This family is second only to Noctuidae in importance among the Lepidoptera. The adults are small to medium in size, with a wingspan of 10–40 mm. The head appears to protrude forward, providing the basis for the "snout moth" designation, though it is enlarged mouthparts (labial palps) that account for the extension. Because of their manner of holding the wings when at rest, pyralid moths generally have a distinctive triangular appearance. The feeding behavior of larvae is quite diverse, though stem borers, leaf tiers and webworms predominate. The larvae are not easily distinguished from noctuid larvae and some other caterpillars, but many pyralids have a light spot or a ring on the body at the base of body hairs. Nearly 1400 species of Pyralidae are known in the United States and Canada.

Family Sphingidae—Hawk Moths and Hornworms

The hawk moths are among the largest moths, with a wingspan sometimes exceeding 160 mm. The front wings are longer than the hind wings, sometimes twice as long. Hawk moths are strong fliers, often hovering like hummingbirds and sometimes mistaken for these small birds. Although sometimes active during daylight, hawk moths are most commonly observed near dusk, usually hovering at flowers or darting from flower to flower. The larvae are large and heavy bodied, and usually bear a single prominent tapered appendage or "horn" at the posterior tip. This horn is the basis for the common name of larvae, and is incorrectly considered as harmful to humans. Hornworms feed on a variety of plants, but only a few injure vegetables. Approximately 125 species of hawk moth are known in the United States and Canada.

Family Tortricidae—Leafrollers and Borers

The tortricid moths generally are small, but range from 10–35 mm in wingspan. They are obscure moths, usually mottled brown or gray. The larvae often roll leaves or bore into plant tissue, and many are main plant pests. Over 1000 species of tortricids are known in the United States and Canada. However, among vegetable crops, only pea is affected by a tortricid; pea moth, *Cydia nigricana* (Fabricius), burrows into pea pods and feeds on seeds.

Order Orthoptera—Grasshoppers and Crickets

The order Orthoptera consists principally of grasshoppers, crickets, and katydids. Most are medium (20–50 mm body length) but some are large (up to 80 mm) in size. They tend to be relatively thin and elongate in body form, and possess long legs, particularly the hind legs. Most species have two pairs of wings; however, some have abbreviated wings and a few are wingless. The front wings are narrow and slightly thickened, but not very hard. They serve mostly to protect the hind wings from damage rather than for flight. The hind wings are large but thin and membranous. The hind legs are enlarged and can be used for leaping, though leaping is a defense reaction and grasshoppers normally move about by walking. The mouthparts of grasshoppers are the chewing type. The antennae are variable in length, often quite long, and usually thread-like. Grasshoppers undergo incomplete metamorphosis, with egg, nymphal and adult stages, but not a pupal stage. The immatures, or nymphs, greatly resemble the adults, with nymphs differing primarily owing to small body size and poorly developed wings. Nymphs occur in the same habitat as adults and, like adults, feed principally on plant foliage. Some orthopterans produce sound. About 1000 species of Orthoptera are known in the United States and Canada.

Family Acrididae—Grasshoppers

This family of Orthoptera is also known as "short-horned grasshoppers" due to the length of the antennae. Acridid antennae tend to be about one-half the length of the body or less, which serves to differentiate this group from the Tettigoniidae. Females lack the large ovipositors found in many other groups of orthopterans. The Acrididae is the largest and most destructive group of orthopterans, though they typically reach their greatest abundance in arid grassland environments, where vegetables are not grown. They are quite dispersive, however, with some species migrating long distances and causing great crop destruction. Many species are indiscriminate feeders, accepting a broad range of host plants. About 550 species live in the United States and Canada.

Family Gryllidae—Field Crickets

The crickets are heavier-bodied than grasshoppers and have longer antennae. The ovipositor of females is conspicuous. Sound production is an important aspect of cricket biology. Among the many types of crickets, only field crickets are crop pests, and that is only infrequently. About 40 species of field crickets are known in the United States and Canada.

Family Gryllotalpidae—Mole Crickets

The mole crickets are among the most distinctive orthopterans. These medium-sized insects, about 20–40 mm long, have front legs that are modified for digging through soil. Their antennae are relatively short. The female's ovipositor is not pronounced. As pests, they are limited to the southern states. Only seven species are known in the United States and Canada.

Family Tettigoniidae—Shieldbacked Crickets and Katydids

This family is best known from the large green leaf-mimics known as katydids. The katydids are not vegetable pests, but a few related insects known as shieldbacked crickets can cause damage. Within the Rocky Mountain region, some species can be very destructive. The large (40–60 mm), hump-backed, short-winged, shieldbacked crickets resemble field crickets more than katydids, though they can be green. The antennae of tettigoniids are longer than the body. The female's ovipositor is large. Although flightless, shieldbacked crickets are quite dispersive and may migrate into crop-growing areas. About 250 species of tettigoniids occur in the United States and Canada.

Order Thysanoptera—Thrips

These very small insects usually measure only 1–2 mm long. They have two pairs of wings, consisting largely of fringe hairs. The antennae are medium in length and thread-like. The mouthparts pierce plant tissue and remove plant sap, though the mouthparts are often described as rasping rather than piercing-sucking. These insects exhibit a form of incomplete development in which the first two instars feed actively, but are inactive. Tarsal claws usually are not present. The nymphal instars generally resemble the adults in body form, the principal exceptions being the smaller size and undeveloped wings of nymphs. Thrips can be found on foliage and in blossoms. Despite their small size they can be quite destructive owing to their ability to transmit plant viruses. About 700 thrips species inhabit the United States and Canada.

Noninsect Invertebrate Pests

Other classes of animals resemble insects, or inflict similar damage to vegetable crops, four basic groups of such pests:

1. Class Acari (phylum Arthropoda)—Mites. These animals are smaller than most insects, and usually bear eight legs.
2. Class Collembola—Springtails, and class Symphyla—Symphylans (both phylum Arthropoda). These small soil-dwelling animals have been considered to be insects in the past and share many morphological characteristics with insects.
3. Class Isopoda—Pillbugs and Sowbugs, and class Diplopoda—Millipedes (both phylum Arthropoda). These many-legged animals can be distinguished by their many-segmented bodies and numerous legs.
4. Class Gastropoda (phylum Mollusca)—Slugs and Snails. These slimy animals may or may not bear

shells, but the lack of body segmentation and legs allows for easy recognition.

PEST MANAGEMENT PHILOSOPHY AND PRACTICES

What Is a Pest?

A pest is not a biological phenomenon; it is an anthropomorphic designation. For example, we consider termites as beneficial organisms when they live in forests, converting dead trees into soil organic matter. The same insects are pests when they feed on wood-associated human structures. In both cases, the termite behavior is the same, but in one case we place no value on the wood being consumed, and in the other we assign high economic value.

Sometimes the mere presence of insects causes concern or alarm. When simple insect occurrence or very minor feeding is the basis for designating an insect as a pest, the insect is said to be an *aesthetic* or *cosmetic pest*. However, when insects decrease the value of a commodity they are said to be *economic pests*. There is no absolute distinction between aesthetic and economic injury, especially with respect to vegetable crops. A home gardener may consider a dimple on a tomato fruit caused by the feeding of a stink bug to be an insignificant blemish, an aesthetic injury, but the same type of blemish can cause a produce buyer to downgrade the value of a crop, causing significant economic loss to a tomato farmer. Similarly, a few holes in the leaves of cabbage or lettuce plants caused by flea beetles is of no significance early in the life of a crop, because the affected foliage is not harvested. The same type of injury, should it be more frequent or appear late in the development of the crop, could constitute injury either by reducing the growth rate of the plant or by affecting the appearance of the harvested commodity.

It is very useful to differentiate between direct and indirect pests. *Direct pests* are those that attack the portion of the vegetable plant that is harvested for food. *Indirect pests* attack some other portion of the plant that is not harvested. A stink bug feeding on a tomato fruit is a direct pest; a leaf miner burrowing in tomato foliage and a wireworm feeding on tomato roots are indirect pests. Ultimately, the presence of large numbers of indirect pests can be damaging, but on an individual insect basis indirect pests are less injurious than direct pests. Knowing the identity of pests, and their feeding behavior, allows us to distinguish between a severe and modest threat to a vegetable crop, and determines our course of action in dealing with the threat.

In commercial vegetable production, economic considerations are very important. Commercial vegetable producers normally are aware of the "economic injury level," the point at which pest suppression is economically feasible. There is little value of spending more money on production costs such as pest suppression than can recovered in increased yield. Therefore, a grower may not be very concerned about indirect pests, relative to direct pests. Similarly, growers may be unconcerned about direct pests that are few in number unless they burrow into the produce and are not easily detectable; such cryptic damage can dramatically lower the value of a crop, because consumers are intolerant of even slight insect contamination of food products. Commercial growers often make decisions that affect large acreages, and have a huge financial investment at stake. Under such circumstances, it is appropriate to assume a worst-case scenario, and to act decisively to protect crops from insect damage. This approach is viewed as a form of crop insurance, and in many cases, it is a good investment. However, in many cases good crop monitoring can establish precisely how abundant and damaging insects might be, and because they often are not numerous enough to cause injury, significant financial savings can be realized by treating crops for pests only when treatment is actually needed. It is important to bear in mind that unnecessary treatments for pests also are a financial loss. Such unnecessary treatments also encourage the development of insecticide resistance in pest populations. In the home garden environment there is greater latitude for risk, and it is often possible to abstain from pest suppression actions unless pests are observed. In either the commercial or home garden situation, we can make sound decisions about the need for pest control only when we have information about pest identity, biology, and damage potential.

How Damaging Are Pests?

Despite our best efforts to prevent injury, vegetable crops do sustain damage by insects and other pests. The level of loss varies among crops, locations, and years. Also, some pests are consistently injurious, others rarely so. Crop losses were estimated by Metcalf and Metcalf (1993), using various sources, to average about 12.5% for commercially grown vegetable crops in the United States. Undoubtedly, entomologists could quibble about the validity of any estimate, particularly because losses vary so much from region to region and from year to year, but the loss clearly is significant. The estimated crop losses and the economic value of the insect damage (based on 1988 figures) are:

Crop	Loss crop (%)	Loss $ (millions)
Asparagus	15	22.8
Beans, snap	12	11.8
Broccoli	17	49.7
Cabbage	17	3.8
Carrots	2	6.9
Cauliflower	17	34.4
Celery	14	32.3
Cucumber	20	26.0
Lettuce	7	73.1
Melons	10	9.7
Onion	18	76.7
Pea	10	7.3
Potato	12	230.2
Sweet corn	14	50.5
Sweet potato	5	8.0
Tomato	7	98.5
Watermelon	15	11.3

These losses do not include the cost of preventing insects from causing even greater injury. The financial cost of managing vegetable pests is difficult to estimate, but the value of insecticides alone is measured in hundreds of millions of dollars.

Not surprisingly, the cost of preventing injury varies considerably among locations and crops, and from year to year. As an example, following are some estimates of insect management costs relative to some other crop production and marketing costs for tomatoes grown in central Florida during the spring and autumn growing seasons (based on 1995–1996; Smith and Taylor, 1996):

	Average cost during spring and autumn cropping seasons			
	Spring		Autumn	
	$/acre	$/hectare	$/acre	$/hectare
Operating costs				
Transplants	224	560	224	560
Fertilizer and lime	326	815	326	815
Scouting	35	87	35	87
Fumigant	562	1,405	562	1,405
Fungicide	175	437	267	667
Herbicide	37	92	37	92
Insecticide	381	952	488	1,220
Labor	347	867	387	967
Machinery	306	765	360	900
Interest	121	302	696	1,740
Total operating cost	$ 3,187	$ 7,967	$ 3,530	$ 8,825
Fixed costs				
Land rent	300	750	300	750
Machinery	212	530	255	637
Management	715	1,787	787	1,967
Overhead	894	2,235	984	2,460
Total fixed cost	$ 2,122	$ 5,305	$ 2,328	$ 5,820

(continues)

	Average cost during spring and autumn cropping seasons			
	Spring		Autumn	
	$/acre	$/hectare	$/acre	$/hectare
Total preharvest cost	$ 5,309	$ 13,272	$ 5,858	$ 14,645
Harvest and marketing costs				
Harvest and haul	980	2,450	962	2,405
Packing	2,590	6,475	2,312	5,780
Containers	1,050	2,625	937	2,342
Marketing	280	700	250	625
Total harvest and marketing costs	$ 4,900	$ 12,250	$ 4,462	$ 11,155
Total costs (per acre/hectare)	$ 10,209	$ 25,522	$ 10,320	$ 25,800

The cost of insecticide use in the central Florida tomato production region during 1995–1996 was high, $ 380/acre ($ 950/ha) in the spring production period and $ 488/acre ($ 1220/ha) in the autumn period. This does not include application costs because machinery costs are often spread over several practices; for example, fungicides and insecticides are often applied simultaneously. Similarly, fumigation practices are directed primarily to disease and nematode management, though weed and insect suppression also accrues. Crop scouting costs cannot be assigned to any particular pest group and are shown separately.

Note that insect-control costs are appreciably higher in the autumn period. This is because of better survival of pests during the summer inter-crop period than the winter inter-crop period, resulting in greater abundance and greater risk of damage. Also noteworthy is that even in the relatively insecticide-intensive Florida cropping system, insecticide costs represent no more than 13.8% of the operating costs, 8.3% of total preharvest costs, and 4.7% of total tomato production costs.

Farmers tend to focus their cost-cutting efforts on the higher priced elements of the crop-production system, often labor and packing operations. The prevalent attitude among most farmers, as long as insecticides are available, effective and affordable, is that their use will minimize an element of risk at relatively low cost. However, if all the pest-related costs are aggregated, the total cost is greatly enhanced, and the true cost of "crop insurance" can be better appreciated. In the Florida tomato production example cited above, the impact of pest management practices is estimated to cost about 49% of operating costs, 29% of total preharvest costs, and 17% of total costs.

The use of insecticide is expensive in many crops other than tomatoes. For the period 1995–1996, insecticide costs on some other vegetable crops grown in Florida were:

Crop	Region of Florida	$/acre	$/hectare
Beans, snap	South	108	270
Cabbage	North	113	282
Celery	South	352	880
	Central	201	502
Corn, Sweet	South	211	527
	Central	122	305
Cucumber	South	172	430
Eggplant	Central	457	1,142
Pepper	South	427	1,067
	Central	387	1,769
Potato	Central	119	297
	North	24	60
Squash	South	67	167
Watermelon	South	131	327
	Central	48	120
	North	9	23

As is evident from this example, insecticide costs tend to be higher in southern Florida than in central and northern Florida (Smith and Taylor, 1996). This is due to the better survival and greater abundance of insects in the subtropical regions of southern Florida. Such variation in insect abundance, and costs associated with insect suppression, are not restricted to Florida, nor are they due only to weather-related factors. The presence of certain prevailing wind patterns, culture of alternate crop hosts, and the abundance of weeds or plant disease also influence the severity of pest damage.

Sometimes insect problems are so severe or expensive that farmers abandon culture of certain crops in a particular area. Home gardeners similarly avoid culture of certain crops if they regularly encounter difficulty. Both farmers and home gardeners sometimes lack the sophistication, motivation, and economic resources to manage pests with the latest and most effective technology. On the other hand, if the garden plot is enough small and the gardener adequately motivated, extraordinary efforts are sometimes applied. The culture of vegetables under row-cover material or screening is a good example of strong desire to overcome pest pressure, or sometimes a strong desire to avoid insecticide use.

The Types of Insect Injury

Insects cause several different types of injury to vegetables, with some species capable of causing more than one type of damage. Recognition of the nature of injury is often an important diagnostic feature in the process of determining pest identity. Because the mouthparts of insects are principally useful for chew-ing, or for piercing and sucking liquid plant contents, the principal forms of injury reflect these major behaviors.

The most common form of injury occurs when insects live externally on a plant, biting off portions of tissue and swallowing the small pieces. Such insects, known generally as *chewing insects*, are usually *foliage feeders* or *root feeders*. There are numerous variations in behavior among chewing insects. Some insects, particularly young stages of insects, feed only on the leaf surface or avoid consumption of leaf veins. This selective feeding results in destruction of leaves, though the veins remain and sometimes the outline of the leaf is still quite apparent. This damage is called *skeletonizing*, because the "skeleton," or veins, are left intact. *Leaf rollers* or *leaftiers* web together leaves to provide shelter for the insect. They may remain within the shelter to feed, and venture forth only to return to the shelter when not actively feeding. Leaves may also provide shelter for *leaf miners*, which feed on the internal tissues of leaves, while leaving the outermost layers intact. Leaf miners normally spend their entire larval life within a single leaf, leaving only when it is time to pupate. Some insects are *stem borers*, chewing into leaf petioles, stalks, or roots. In many respects stem borers are similar to leaf miners, differing mostly in the site of feeding. However, they sometimes move considerable distances within the stem, and tend to be more damaging than leaf miners. By burrowing through root, stalk, and petiole tissues, borers disrupt the transport of water and nutrients for large sections of the plant, or even the entire plant. This not only inhibits normal growth, but sometimes leads to complete loss in productivity or death of the plant. Perhaps the most damaging of the chewing insect pests are *blossom* or *fruit feeders*, insects that feed on or within the reproductive tissues of the plant. The fruit is often the most valuable portion of the plant, and even a small amount of feeding can cause rejection for cosmetic (aesthetic) reasons, or can lead to invasion by plant pathogens.

In addition to the direct injury caused by chewing mouthparts, chewing insects sometimes cause damage indirectly. Chewing insects sometimes vector plant diseases, though they are much less efficient than piercing-sucking insects, or open a potential site of infection for plant pathogens borne in the soil, water, air, or insect fecal material.

Many insects have long tubular mouthparts that allow them to pierce the conducting elements of plants, or individual cells, and to extract the liquid contents. These insects, called *piercing-sucking insects*, may be found on leaves, stems, or roots, but most often are associated with the terminal or youngest leaf and stem tissue. Feeding by piercing-sucking insects often results in discoloration and deformity, apparently because insects not only extract liquids, but secrete digestive enzymes and toxins while feeding. A common response is *chlorosis*, or yellowing of leaf tissue; this may occur generally, locally, or even in individual cells. Premature *leaf drop* sometimes occurs following feeding by piercing-sucking insects. *Leaf curling* or *cupping* is a common form of deformity, usually associated with feeding by aphids. *Galls*, or swellings caused by an increase in plant-cell size or number, often envelop feeding insects and provide them with both protection and enhanced nutrition. Piercing-sucking insects often cause *stunting*, or decrease in stem and leaf elongation. Reduced stem elongation but normal lateral growth results in dense, bushy vegetation called *witch's broom*. Perhaps the most important form of injury caused by piercing-sucking insects is *pathogen* (disease) *transmission*. Piercing-sucking insects are particularly adept at transmitting diseases caused by viruses and similar organisms.

Pest Management Philosophy

There are several ways to approach the management of vegetable pests. In some instances profit is the major concern, though in other cases maximum yield or pesticide-free vegetables are desired. In some cropping systems, or for some particular pests, efforts are made to prevent pest populations from establishing by the preventative application of insecticide. On the other hand, often crops are monitored and are not sprayed until the first appearance of a particular pest. In still other situations, small or moderate numbers of pests are tolerated, but the density is carefully monitored and suppression initiated when densities reach some threshold of abundance.

The philosophy behind commercial growers' pest management practices is quite variable, despite the common objective of making a profit. The difference in approach to pest management stems from the uncertainty of production and profit. When a particular vegetable is in relatively short supply, the price is high and very good profits are made. Under such circumstances, the cost of pest management practices seem insignificant and maximization of yield pays handsome dividends. On the other hand, when a vegetable in in good supply relative to demand, the price received by the grower is low, perhaps barely covering costs of production. Under these circumstances, farmers who cut costs where they can, such as unnecessary insecticide use, make greater profit. Thus, a farmer's decision to spend money on pest management practices is tempered by the future outlook of supply and

profit. One thing is certain, however, any vegetable crop must be high in quality and free of blemishes if it is to be marketed at a profit. Thus, it is not surprising that producers invest heavily in insecticides.

Home gardeners are less sensitive to appearance of the produce. They are often motivated as much by the quality of the produce, which often means freshness and absence of pesticide residues, as they are due to possibility of saving money by growing their own vegetables. Because profit is not a key element, home gardeners can engage in practices that would be cost-prohibitive on a commercial scale. Home gardeners also tend to have less ready access to highly effective but very toxic pesticides. Commercial producers of organically grown vegetables are faced with the difficult challenge of producing blemish-free produce with a very limited array of "organic" insecticides.

It is always desirable to prevent pest populations from developing, because this eliminates the need for corrective action and the risk of crop loss. Certain actions can be taken, such as growing a crop in an isolated location, cultivating a pest-resistant cultivar, or modifying the date of planting to avoid the major flight period of a pest. Such practices often virtually eliminate problems with certain pests. Unfortunately, there are many potential pests for any vegetable crop, and such extraordinary efforts can only be directed to one or a few pests per crop. Therefore, such manipulations usually are directed only at the most severe, regularly occurring pests that must be eliminated if a crop is to be grown successfully. That leaves other pests to be managed in a curative or corrective mode when, and if, they appear. Insecticides are very useful for this purpose, though they are not necessarily the only option.

Residual, broad-spectrum insecticides, if used frequently, can prevent invasive insects from establishing, as well as can eliminate most pest problems that do exist. By using such materials on a regular basis, growers can avoid concern about regularly occurring pests as well as sporadic pests. Historically, such broad-spectrum insecticides were readily available, economic, unfailingly effective, and perceived as relatively safe to people and the natural environment. Although conditions and perceptions relative to the characteristics of insecticides have changed, it is easy to see how the use of insecticides would have great appeal to producers, and how insecticide use would become an integral part of commercial vegetable production.

Increasingly, growers cannot depend on availability of broad-spectrum insecticides to rid their fields of pest problems. There are three primary constraints on use of insecticides: cost, effectiveness, and availability. The cost of insecticides and their application can be quite high. Insecticides are applied to some crops at 2–3 day intervals during the periods of growth when the crop is particularly vulnerable to damage. This can result in considerable cost to the producer. If material and application costs increase faster than the revenue farmers receive for their produce, it is no longer cost-effective to grow the crop. Also, insecticides are not always effective. Some vegetable crops are attacked by insects that are inherently difficult to control. It is also possible to exacerbate the problem caused by certain pests, such as mites, leaf miners, and whiteflies by excessive use of insecticides. The principal causes for increased problems with pests following insecticide application are insensitivity of the target pest to the insecticide (i.e., resistance), and destruction of potentially effective natural enemies by the insecticide. Lastly, availability of pesticides is decreasing. The high cost of developing new products discourages insecticide manufacturers from developing new products. Concern over the health of farm workers, consumers, and non-target wildlife, such as birds and fish, also has led to regulatory constraints on availability and use of insecticides. Some insecticides are quite hazardous to fish and birds, but others have few deleterious effects, disrupting the metabolism or behavior of insects but not other animals.

The use of more selective pesticides, or the use of pesticides more selectively, is increasingly encouraged. Selectivity implies targeting only pests that need to be eliminated, and use of insecticides only when they are actually needed. To accomplish the objective of shifting from broad-spectrum to selective pest suppression materials, it is imperative that the identity, biology, and damage potential of pests be known.

Vegetable producers are also encouraged to use alternatives to pesticides, or a combination of tactics, rather than depending solely on insecticides for their pest suppression needs. They attempt to integrate pest management practices such as using cultural practices which limit pest potential, with careful monitoring of pest abundance, and application of selective insecticides only when it is necessary.

Pest Management Practices

There are numerous types of practices that can be implemented to protect a crop from pests, or to eliminate pests that have invaded. These practices can be organized into four broad categories: biological control, cultural manipulations, physical manipulations, and insecticides. They differ greatly in effectiveness, ease of implementation, cost of implementation, time interval required before becoming effective, and relia-

bility. No single practice is effective for all pest problems, though insecticides come closest to being universally suitable. Unfortunately, the benefits of insecticide use are sometimes offset by health or environmental hazards.

Biological Control. The use of natural enemies to suppress pest insects is one of the oldest and most effective approaches known. Although using natural organisms to our advantage is highly desirable, it is often difficult to implement and the outcome is difficult to predict. Biological control tends to work best when the pest species is an invader from another region, and is maintained at a low or moderate level of abundance in its native land by natural enemies. In this case we import the native enemies and then culture and release them where they are needed. Once established, if the native enemies are well-adapted to the new environment they provide permanent, no-cost suppression. This approach is called *classical biological control*. However, sometimes natural enemies cannot persist at a high enough level in the new environment to provide effective suppression, or effective enemies cannot be located. In this case we must culture and release natural enemies that have some promise of providing suppression, in large enough numbers to overwhelm the biotic potential of the pest, and drive their population down. This often requires regular, timely release of natural enemies, and is called *augmentive biological control*, because we are augmenting the natural population of natural enemies with supplemental biological control agents. The final major approach to using biological control organisms is to modify the environment or otherwise preserve and favor existing natural enemies. This is called *conservation biological control*, and usually involves preserving some habitat or food resource, including alternate host insects, or protecting the beneficial insects from the deleterious effects of pesticides. The conservation approach is especially appealing when the insect pest is a native species and no source of exotic natural enemies is apparent, or when the economics of producing the commodity do not favor mass culture and release of natural enemies—often an expensive undertaking.

Cultural Manipulations. Modification of planting, crop maintenance, and harvesting practices can sometimes affect pests. For example, a common practice is to delay planting so as not to have the plant present when the insect pest becomes active, or to plant early so the crop plants are well established and able to withstand some root feeding or defoliation and remain healthy. This is most effective when insects occur synchronously and for a relatively brief period of time, and when working with a crop that will not suffer from such manipulations. Similarly, tillage of the soil and destruction of crop residue often reduces the over-wintering survival of insects. Other cultural practices that often are manipulated in an effort to manage pests are crop irrigation, rotation between dissimilar crops, crop isolation from sources of pests, fallow or crop-free periods, and cultivation of crop varieties that are less attractive or unsuitable for pests. Cultural manipulations are often the most cost-effective approach to pest management, but it is difficult to manage an entire complex of pests using these techniques.

Physical Manipulations. Physical manipulations sometimes are useful, especially for small-scale or home gardenvegetable production. For example, metal and tar paper barriers around the base of seedlings can provide protection against the feeding of cutworms and oviposition by root maggots, respectively. Also, reflective or colored mulch sometimes will repel certain flying insects, including vectors of important plant viruses. Screen and floating row covers often can be used to protect insects from reaching susceptible crops; however, allowance may need to be made to provide entry by pollinators. Occasionally, traps can be used for collection and destruction of pests. Traps can be baited with food-based or chemical lures, they may be attractive owing to an insect's natural orientation to a particular color, or can take advantage of the pest's tendency to seek a certain type of shelter. Some large insects can be observed readily and simply collected by hand, not necessitating any type of attractive device. Physical manipulations tend to be material-intensive, and sometimes labor-intensive, so their use is most common in high value crops or in small plots.

Insecticides. The most common approach to vegetable crop protection is to use insecticides. Most insecticides are derived from synthetic organic chemicals, though some are derived from naturally occurring minerals or plants. Insecticides disrupt the physiology of the insect. Most of the currently available, *synthetic organic insecticides*, particularly the broad-spectrum insecticides, disrupt the nervous system of insects. Some are quite specific to insects, but many are biocides, general poisons that can affect fish, birds, and mammals if they are exposed sufficiently to high levels. The *botanical insecticides* are favored by organic gardeners, because they are perceived to be ''natural.'' Botanicals degrade quickly in the environment, but some are quite toxic to humans and should be handled as carefully as synthetic insecticides. A few products are derived from, or consist of, insect-pathogenic microorganisms. These *microbial insecticides* tend to be very specific, and safe to most non-target organisms. The best example of this is the bacterium, *Bacillus thuringiensis*. Increasing in popularity is the use of *soaps* (detergents) and *oils* (both mineral and vegetable). Although their effectiveness is generally limited to small organisms, they pose few hazards to humans and other animals.

Insecticides are formulated in many different ways. This flexibility as well as long shelf life help account for the popularity of insecticides. Most insecticides are mixed with water and applied as a spray, but sometimes dust or granule formulations are most efficient, or aerosol applicators are easier to use. Least frequent of the application methods is bait formulations, wherein the toxicant is mixed with an attractive and edible material.

Pest Identification

GENERAL CONSIDERATIONS

When an unknown pest is found on vegetables, it is important to reduce the number of potential identities to only a few by assigning the unknown to a small group, usually an insect order and family. Each order and family contains only a fraction of the approximately 300–400 species that are capable of causing damage to vegetables. You can greatly simplify identification by knowing the order, and perhaps the family. Usually only a few characteristics such as wing number and leg or mouthpart type, are needed to assign an unknown pest to order and family. The "quick guide" in this section and a key in Appendix A will aid you in this process.

WHY IDENTIFICATION IS SO IMPORTANT?

There is a great wealth of information about vegetable pests. Hundreds of careers have been devoted to determining which species are damaging, the nature of their damage, the life cycle of the pest, and how the pest is best managed to eliminate or minimize injury to plants. It is easy to overestimate the damage potential of some large, leaf-feeding pests, and similarly easy to underestimate the damage caused by small and sap-sucking species. Failure to recognize the presence of certain pests, such as plant virus vectors early in the season, can prove a grievous error, resulting in severe crop injury. On the other hand, sometimes vegetable plants are treated with an insecticide, because someone has observed lady beetles or another beneficial insect, and mistaken them for pests. Therefore, the collective knowledge and experience of generations of entomologists are available once you identify the pest—but are useless unless correct identification is made of the pest at hand.

APPROACHES TO IDENTIFICATION

There are three guiding principles to identification that, if followed, enhance correct identification. Remember that some species can be identified only by authorities, and some cannot be distinguished based only on appearance, but most serious pests are readily distinguished with but little effort. You may need to use some low-level magnification such as a 10 × hand lens, and such equipment is inexpensive and easy to use.

1. Although there are numerous pests, most are easily assigned to groups based on easy-to-discern characters such as the number of legs or wings and the type of mouthparts. With a little effort it is easy to distinguish such common groups as aphids, caterpillars, whiteflies, and wireworms.
2. Behavior is as important as appearance in distinguishing among pests. How or where a pest feeds is critical knowledge. The presence of silk webbing, the positioning of eggs, and other aspects of pest biology often are key elements in distinguishing the pests. When attempting to identify an unknown pest, try to observe as much as possible about pest behavior and plant damage.
3. Details are important; look carefully at the pest and its damage, and do not trust your memory. Collect the pest you want to identify, do not just kill it. No one who is unfamiliar with the pest's identity can anticipate what characters will be critical in distinguishing it from its relatives. It is entirely too easy not to notice the number of wings or the color of the legs, or not to remember to count the stripes or spots.

Compare the written descriptions with the physical evidence carefully. Do not ignore aspects of the description because you are unfamiliar with the terminology. Entomologists, like members of all scientific disciplines, use technical words to describe the physical appearance and behavior of their subjects of study. In this book the use of terminology is minimized, and terms are described in the glossary and illustrated with diagrams.

If a pest poses a major economic threat to a crop and you are uncertain of its identity, there are several available sources of assistance. The principal source in the United States is the Cooperative Extension System associated with the Land-Grant University found in each state. A cooperative extension service office is

Quick guide to orders of vegetable-feeding insects

Order of insects	Common name	Example of adult	Front wings	Hind wings
Coleoptera	Beetles, weevils; grubs		Thickened, hard, meeting in straight lines on back	Membranous, folding beneath front wings
Dermaptera	Earwigs		Short, thickened (if present)	Large, membranous, folding fan-like under front wings (if present)
Diptera	Files, maggots		Membranous	Absent
Hemiptera	Plant, seed, stink, lace and leaf-footed bugs		Thickened basally, membraous distally	Membranous

normally found in each county, and the employees are commonly called county agents. The county agents often can assist with identification, and if they cannot they will forward the pest to the state university for identification. Research centers and experiment stations are other good sources of expertise, in both Canada and the United States. Crop consultants pro-vide identification and pest management recommendations on a for-fee basis and usually can be found in important agricultural areas. You can access considerable additional information to aid in identification if you have access to the Internet. Suggested sources of supplementary information and assistance are found in Appendix B.

Quick guide to orders of vegetable-feeding insects   *(Continued)*

Antennae	Mouthparts	Other features	Example of immature	Immature features
Short to moderately long, thread-like or clubbed	Chewing			Usually with distinct head capsule and three pairs of thoracic legs
Moderately long, thread-like	Chewing	Large forceps-like cerci		Resemble adults except for reduced wings
Short, variable in from	Lapping-sucking	Small balancing organs (halteres) in place of hind wings		Head reduced; legs absent
Usually moderately long four to five segments	Piercing-sucking, originating at front of head	Sometimes resembling beetles		Resemble adults except for reduced wings

(Continues)

Quick guide to the important orders of vegetable-feeding insects   (Continued)

Order of insects	Common name	Example of adult	Front wings	Hind wings
Homoptera	Aphids, leafhoppers, whiteflies, psyllids		Membranous	Membranous
Hymenoptera	Sawflies, ants		Membranous	Membranous, smaller than front wings
Lepidoptera	Moths, butterflies, caterpillars, cutworms, loopers, webworms		Broad, with scales	Broad, with scales
Orthopetera	Grasshoppers, crickets		Thickened (if present), many veins	Large, membranous, folding fan-like under front wings (ifn present)
Thysanoptera	Thrips		Very slender, with fringe of long hairs	Very slender, with fringe of long hairs

Quick guide to the important orders of vegetable-feeding insects (Continued)

Antennae	Mouthparts	Other features	Example of immature	Immature features
Usually short	Piercing-sucking, originating at back of head	Extremely diverse, sometimes with waxy or thread-like axudates		Often resemble adults, but sometimes flattened
Thread-like	Chewing			Leg number variable, sawflies with seven pairs of prolegs in addition to thoracic legs
Usually moderately long, thread-like, feathery, or clubbed	Coiled sucking tube			Two to five pairs of prolegs in addition to thoracic legs
Moderately long to long, thread-like	Chewing	Hind legs well developed		Resemble adults except for reduced wings
Short	Minute, piercing-sucking	Minute in size		Resemble adults except for reduced wings

Quick guide to the adults of important families of vegetable-feeding beetles

Family	Common name	Example of adult	Antennae	Other features
Bruchidae	Seed beetles and weevils		Moderately long, weakly clubbed or branched	Tip of abdomen exposed, not covered by elytra
Chrysomelidae	Flea beetles		Moderately long, thread-like	Small; often metallic
	Tortoise beetles		Short to moderately long, thread-like	Pronotum and elytra expanded, flaring over legs; colorful
	Leaf beetles		Moderately long	Elongate body form; colorful
Curculionidae	Weevils		Short to moderately long, elbowed and clubbed	Elongate head forming shout
Coccinellidae	Lady beetles		Short, clubbed	Conves; colorful, spotted

(Continues)

Quick guide to the adults of important families of vegetable-feeding beetles   (*Continued*)

Family	Common name	Example of adult	Antennae	Other features
Elateridae	Click beetles		Short, branched	Elongate
Meloidae	Blister beetles		Moderately long, thread-like	Colorful or metallic
Scarabaeidae	Scarab and June beetles		Short, branched or clubbed	Convex; legs spiny

Quick guide to the adults of important families of vegetable-feeding flies

Family	Common name	Example of adult	Antennae	Other features
Agromyzidae	Leafminers		Short	Small; wings transparent
Anthomyiidae	Root maggots and leafminers		Short	Medium size; wings transparent; legs medium length
Otididae	Picture-wing flies		Short to moderately long	Patterned wings

(Continues)

Quick guide to the adults of important families of vegetable-feeding flies   (Continued)

Family	Common name	Example of adult	Antennae	Other features
Tephritidae	Fruit flies		Short	Patterned wings

Quick guide to the adults of important families of vegetable-feeding bugs

Family	Common name	Example of adult	Antennae	Other features
Coreidae	Squash and leaffooted bugs		Moderately long, four segments	Front wing veins highly branched
Lygaeidae	Seed bugs		Moderately long, four segments	Small size
Miridae	Plant bugs		Moderately long, four segments	Slender body
Pentatomidae	Stink bugs		Moderately long, five segments	Shield-shaped body

Quick guide to the adults of important families of vegetable-feeding Homoptera

Family	Common name	Example of adult	Antennae	Other features
Aleyrodidae	Whiteflies		Thread-like, seven segments	Opaque white wings
Aphididae	Aphids		Thread-like, three to six segments	Wings transparent; cornicles usually present
Cicadellidae	Leafhoppers		Small	Front wings thickened and opaque
Psyllidae	Psyllids		Long, nine to eleven segments	Wings transparent; cornicles absent

Quick guide to the adults of important families of vegetable-feeding moths and butterflies

Family	Common name	Example of adult	Wings	Other features
Arctiidae	Tiger moths		Front wings much longer than hind wings	Nocturnal; robust body, moderately large
Gelechiidae	Leafminer moths		Fringed	Nocturnal; small size

(Continues)

Quick guide to the adults of important families of vegetable-feeding moths and butterflies   (Continued)

Family	Common name	Example of adult	Wings	Other features
Noctuidae	Armyworms, cutworms, and loopers		Muted colors, usually brownish	Nocturnal; robust body, moderately large
Pyralidae	Borers, webworms		Wings form triangle when at rest	Nocturnal; palps protrude markedly from head, small to medium size
Sphingidae	Hawk or sphinx moths		Front wings long and pointed	Nocturnal; large size
Papilionidae	Swallowtails		Hind wing with protruding lobe or "tail"	Diurnal; large size
Pieridae	White and sulfur butterflies		White or yellow with black spots	Diurnal; moderate size

Quick guide to the adults of important families of vegetable-feeding grasshoppers and crickets

Family	Common name	Example of adult	Antennae	Other features
Acrididae	Grasshoppers		About one-half length of body	Medium to large body size; usually colorful
Gryllidae	Field crickets		Length of body or longer	Medium body size; color black
Gryllotalpidae	Mole crickets		About length of pronotum	Front legs enlarged for digging; color brownish
Tettigoniidae	Mormon crickets		Length of body or longer	Large body size; color variable

Quick guide to major noninsect groups of pests affecting vegetables

Class of Noninsects	Common name	Example of adult	Front wings	Hind wings
Acari	Mites		None	None
Collembola	Springtails		None	None
Diplopoda	Millipedes		None	None
Gastropoda	Slugs, snails		None	None
Isopoda	Pillbugs, sowbugs		None	None
Symphyla	Symphylans		None	None

Quick guide to major noninsect groups of pests affecting vegetables (Continued)

Aetennae	Mouthparts	Other features	Immature features
Minute	Piercing-sucking	Usually 4 pairs of legs	Resemble adults
Short to moderately long	Chewing	Three pairs of legs	Resemble adults
Minute	Chewing	Two pairs of legs per body segment	Resemble adults
Extrudable	Hidden	Produce mucus; shell present or absent	Resemble adults
Minute	Chewing	Seven pairs of legs	Resemble adults
Moderately long	Chewing	12 pairs of legs	Resemble adults

Guides to Pests Arranged by Vegetable Crops

Following are some guides to assist in identification of common vegetable pests. The guides are based on characters that are easy to observe, and thus consist of obvious attributes such as the portion of the plant that is affected, general morphological characteristics such as the presence or absence of legs or wings, the mouth parts or feeding habits (chewing or sucking insects), and the general category of pest (commonly orders or families). To use the guides effectively the reader must have some basic knowledge of pests, but with only a little experience, it is relatively easy to narrow down a candidate pest to only a few possibilities. Once you have some idea of the identity, the individual descriptions and keys should aid in positive identification.

Some pests are commonly associated with certain crops and often cause damage. The identification guides are designed to assist you in identifying these common pests. However, there are other pests that are found only occasionally or infrequently. These are not integrated into the main identification guides, but follow the main list as "other pests." This arrangement or prioritization of pests is not perfect, because at a certain location or time the minor pests or "other pests" may assume significance. However, you are unlikely to find the "other" pests regularly damaging.

The crops in the pest guides are arranged by plant family rather than by individual crop. This reduces redundancy when listing pests. The clustering of crops works quite well because insects usually make their food selection based on plant chemisty, and plants in the same family share similar chemistry. Though it may not be obvious that crops such as Brussels sprouts and radish are related, insects detect their similar chemistry and thus these crops share many common pests. If you are uncertain of crop-plant relationships, consult Appendix C.

GUIDE TO COMMON PESTS AFFECTING ARTICHOKE

Pests Feeding on Buds or Blossoms, or Boring into Stems

Borers: Artichoke plume moth
Leaf beetles: Spotted and western spotted cucumber
Other bugs: Leaffooted
Stink bugs: Southern green

Pests Feeding Externally on Leaves or Stems

Pests with Chewing Mouthparts
Leaf beetles: Spotted and western spotted cucumber
Loopers: Cabbage looper
Other caterpillars: Artichoke plume moth
Webworms: Celery leaftier
Pests with Piercing-Sucking Mouthparts
Aphids: Artichoke, bean, cowpea, green peach
Other bugs: Leaffooted
Mites: Twospotted spider

Other Pests Occasionally Found on Artichoke

Armyworms and cutworms: *Black and granulate cutworm, corn earworm*
Caterpillars: *Saltmarsh*
Grasshoppers: *Differential, migratory, redlegged, two-striped*
Leafhoppers: *Western potato*
Loopers: *Alfalfa*
Maggots: *Seedcorn*

Plant bugs: Tarnished
Slugs
Weevils: Vegetable

GUIDE TO COMMON PESTS AFFECTING ASPARAGUS

Pests Boring into Roots

Leafminers: Asparagus
Symphylans: Garden

Pests Producing Small Mines in Stems

Leafminers: Asparagus

Pests Feeding Externally on Leaves (Cladophylls) or Stems

Pests with Chewing Mouthparts
    Armyworms and cutworms: Redbacked cutworm
    Caterpillars: Saltmarsh
    Leaf beetles: Asparagus, spotted asparagus
    Scarab beetles: Japanese
Pests with Piercing-Sucking Mouthparts
    Aphids: Asparagus
    Thrips: Bean, western flower, onion

Pests Boring into Berries

Leaf beetles: Spotted asparagus

Other Pests Occasionally Found on Asparagus

Aphids: Bean, cowpea, green peach, melon, potato
Armyworms and cutworms: Beet, fall, southern, yellowstriped and western yellowstriped armyworm; corn earworm; zebra caterpillar; black, spotted, sweet potato, variegated, and velvet cutworm
Blister beetles: Black, immaculate, spotted, striped
Borers: stalk
Grasshoppers: Differential, migratory, twostriped
Loopers: Soybean
Mites: Bulb
Other caterpillars: Yellow woollybear
Plant bugs: Alfalfa, pale legume, tarnished, western tarnished; garden fleahopper
Scarab beetles: Chinese rose
Stink bugs: Green, harlequin, onespotted, Say
Wireworms: False

GUIDE TO COMMON PESTS AFFECTING BEAN AND RELATED CROPS

(Bean, Chickpea, Cowpea, Faba Bean, Lentil, Lima Bean, Pea, and Others)

Pests Boring into Roots or Stems

Pests with Legs
    Borers: Lesser cornstalk
    Flea beetles: Palestriped
    Leaf beetles: Bean; banded, spotted, striped and western striped cucumber; grape colaspis
    Scarab beetles: Green June
    Symphylans: Garden
Pests without Legs
    Maggots: Bean seed, seedcorn
    Weevils: Pea leaf, vegetable; whitefringed beetle

Pests Producing Small Mines in Leaves

Leafminers: American serpentine, pea, vegetable

Pests Feeding Externally on Leaves or Stems

Pests with Chewing Mouthparts
Caterpillars
    Armyworms and cutworms: Armyworm, beet armyworm, and zebra caterpillar
    Leafrollers: Bean
    Loopers: Alfalfa, cabbage, and soybean looper; bean leafskeletonizer; green cloverworm
    Other caterpillars: Saltmarsh caterpillar, yellow woollybear, gray hairstreak beetles
    Flea beetles: Palestriped, potato, western potato
    Leaf beetles: Bean; banded, spotted, striped, and western striped cucumber
    Other beetles: Mexican bean
    Weevils: Pea leaf, vegetable, whitefringed beetle
Pests with Piercing-Sucking Mouthparts
    Aphids: Bean, cowpea, green peach, pea
    Leafhoppers: Beet, potato, western potato
    Mites: Broad; strawberry, tumid, and twospotted spider
    Other bugs: Leaffooted
    Plant bugs: Garden fleahopper, tarnished, western tarnished
    Stink bugs: Brown, green, onespotted, Say, southern green
    Thrips: Bean, melon, onion, western flower
    Whiteflies: Silverleaf, sweetpotato, greenhouse

Pests Feeding on Flowers, Seeds, or Seedpods

Armyworms and cutworms: Corn earworm, western bean cutworm

Borers: Pea moth, limabean pod, European corn, gray hairstreak

Loopers: Bean leafskeletonizer

Stink bugs: Brown, green, onespotted, Say, southern green

Other bugs: Leaffootted, tarnished plant

Weevils: Bean, broadbean, cowpea, pea, and southern cowpea weevil; cowpea curculio

Other Pests Occasionally Found on Bean and Related Crops

Ants: Red imported fire

Aphids: Bean root, blue alfalfa, buckthorn, foxglove, melon, potato

Armyworms and cutworms: Bertha, fall, southern, sweetpotato, velvet, yellowstriped and western yellowstriped armyworms; army, black, clover, darksided, dingy, glassy, granulate, pale western, redbacked, spotted, and variegated cutworms; tobacco budworm

Blister beetles: Black, immaculate, spotted, striped

Borers: Stalk

Crickets: Fall, spring and southeastern field; Mormon

Earwig: European

Flea beetles: Redheaded, smartweed, tobacco, tuber, western black

Flies: Melon

Grasshoppers: American, differential, migratory, redlegged, twostriped

Loopers: Bilobed and plantain

Maggots: Radish root

Millipedes: Garden

Other caterpillars: Alfalfa caterpillar, banded woollybear

Pillbug: Common

Plant bugs: Rapid, superb, legume, pale legume

Scarab beetles: Asiatic, oriental, spring rose, Japanese, white grubs

Slugs

Snails

Springtails: Garden

Stink bugs: Harlequin

Thrips: Tobacco

Webworms: Alfalfa, beet, garden; celery leaftier

Wireworms: Eastern field, false, Great Basin, Pacific Coast, sugarbeet, wheat

GUIDE TO COMMON PESTS AFFECTING BEET AND RELATED CROPS

(Beet, Chard, Spinach, Swiss Chard)

Pests Feeding on Roots

Aphids: Sugarbeet root

Maggot: Sugarbeet root

Scarab beetles: Green June

Symphylans: Garden

Pests Producing Mines in Leaves

Leafminer: Beet, spinach

Pests Feeding Externally on Leaves or Stems

Pests with Chewing Mouthparts

Armyworms and cutworms: Beet and southern armyworm; clover and variegated cutworm; zebra caterpillar

Blister beetles: Black, immaculate, spotted, striped

Flea beetles: Palestriped, potato, tuber

Other caterpillars: Saltmarsh

Webworms: Alfalfa, beet, garden, Hawaiian beet, southern beet, spotted beet

Pests with Piercing-Sucking Mouthparts

Aphids: Bean, green peach

Leafhoppers: Beet

Other Pests Occasionally Found on Beet and Related Crops

Aphids: Buckthorn, foxglove, melon, potato

Armyworms and cutworms: Armyworm, bertha, fall armyworm, glassy, pale western, redbacked, sweet potato, velvet, yellowstriped, and western yellow striped armyworm; corn earworm; army, black, granulate, and spotted cutworm

Borer: European corn, lesser cornstalk, stalk

Crickets: Fall, southeastern, and spring field; shortwinged, southern and tawny mole; Mormon

Earwig: European

Flea beetles: Eggplant, hop, redheaded, spinach, smartweed, sweet potato, threespotted, western black, western potato, yellownecked

Flies: European crane

Grasshoppers: Differential, migratory, redlegged, twostriped

Leaf beetles: Banded cucumber, grape colaspis, spotted and western spotted cucumber

Leafhopper: Western potato

Leafminer: American serpentine, pea
Loopers: Alfalfa, cabbage, celery
Maggots: Seedcorn
Mites: Bulb, broad, twospotted spider
Other bugs: False chinch
Other caterpillars: Banded and yellow woollybear; whitelined sphinx
Plant bugs: Alfalfa, garden fleahopper, tarnished, western tarnished, pale legume
Scarab beetles: Asiatic garden, carrot, Japanese, oriental; white grubs
Slugs
Snails
Springtails: Garden
Stink bugs: Harlequin
Thrips: Bean, onion, tobacco
Webworms: Celery and false celery leaftier
Weevil: Vegetable
Wireworms: Eastern field, Gulf, Pacific Coast, southern potato, sugarbeet, tobacco

GUIDE TO COMMON PESTS AFFECTING CABBAGE AND RELATED CROPS

(Broccoli, Brussels Sprouts, Cabbage, Cauliflower, Chinese Cabbage, Collards, Kale, Mustard, Radish, Turnip, and Others)

Pests Boring into Roots or Stems

Pests with Legs
　Flea beetles: Cabbage, crucifer, horseradish, striped, western black, western striped
　Scarab beetles: Green June, white grubs
Pests without Legs
　Maggots: Cabbage, radish, seedcorn, turnip root
　Weevils: Cabbage curculio, vegetable, whitefringed beetle

Insects Producing Small Mines in Leaves

Pests with Legs
　Flea beetles: Horseradish and Zimmermann's
　Webworms: Cabbage, oriental cabbage
Pests without Legs
　Leafminers: Cabbage, pea

Pests Feeding Externally on Leaves or Stems

Pests with Chewing Mouthparts Caterpillars
　Armyworms and cutworms: Zebra caterpillar

Cabbageworms: Cross-striped, imported, purplebacked, southern, mustard, and southern white
Loopers: Alfalfa, cabbage
Other caterpillars: Cabbage budworm, diamondback moth
Webworms: Cabbage and oriental cabbage beetles
Flea beetles: Cabbage, crucifer, horseradish, striped, western black, western striped, Zimmermann's
Leaf beetles: Red turnip, yellowmargined
Weevils: Vegetable
Pests with Piercing-Sucking Mouthparts
　Aphids: Cabbage, green peach, turnip
　Stink bugs: Harlequin
　Thrips: Onion
　Whiteflies: Silverleaf, sweetpotato

Pests Feeding on Flowers, Seeds, or Seedpods

Cabbageworms: Southern
Weevils: Cabbage seedpod

Other Pests Occasionally Found on Cabbage and Related Crops

Ant: Red imported fire
Aphids: Buckthorn, potato
Armyworms and cutworms: Beet, bertha, fall, southern, yellowstriped and western yellowstriped armyworm; corn earworm; army, black, clover, darksided, dingy, glassy, granulate, redbacked, spotted, and variegated cutworm
Blister beetles: Black, immaculate, spotted, striped
Borers: European corn, lesser cornstalk, stalk
Crickets: Fall, spring, and southeastern field; shortwinged, southern, and tawny mole; Mormon
Earwig: European, ringlegged
Flea beetles: Hop, palestriped, potato, redheaded, southern tobacco, smartweed, tobacco, tuber, western potato
Flies: European crane, small fruit
Grasshoppers: Differential, eastern lubber, migratory, twostriped
Leaf beetles: Banded and spotted cucumber
Leafhoppers: Aster, western potato
Loopers: Bean leafskeletonizer; bilobed, celery, plantain, soybean
Millipedes: Garden
Other bugs: False chinch
Other caterpillars: Saltmarsh, whitelined sphinx, yellow woollybear
Pill bugs: Common
Plant bugs: Garden fleahopper, pale legume, tarnished, western tarnished

Scarab beetles: Asiatic garden, Chinese rose, Japanese
Slugs
Springtails: Garden
Snails
Stink bugs: Brown, green, Say, southern green
Thrips: Bean, melon
Webworms: Celery and false celery leaftier; alfalfa, beet, garden
Whiteflies: Greenhouse
Wireworms: Corn, eastern field, false, Gulf, Oregon, Pacific Coast, potato, southern potato, sugarbeet, tobacco, wheat

GUIDE TO COMMON PESTS AFFECTING CARROT AND RELATED PLANTS

(Carrot, Celery, Celeriac, Chervil, Coriander, Fennel, Parsley, Parsnip)

Pests Boring into Stems or Feeding on Roots

Aphids: Carrot root
Flies: Carrot rust
Scarab beetles: Green June
Symphylans: Garden
Weevils: Carrot, Texas carrot, vegetable

Pests Producing Small Mines in Leaves

Leafminers: American serpentine, parsnip, and pea

Pests Feeding Externally on Leaves or Stems

Pests with Chewing Mouthparts
   Armyworms and cutworms: Beet, fall, and southern armyworm; black and granulate cutworm
   Flea beetles: Potato, western potato
   Loopers: Alfalfa, cabbage, celery
   Other caterpillars: Black and anise swallowtails, saltmarsh caterpillar
   Webworms: Celery and false celery leaftier
   Weevils: Vegetable
Pests with Piercing-Sucking Mouthparts
   Aphids: Coriander, green peach, honeysuckle, melon, willow carrot
   Leafhopper: Aster
   Mites: Twospotted spider

Plant bugs: Carrot, pale legume, tarnished, western tarnished plant; garden fleahopper
Stink bugs: Southern green

Pests Feeding on Flowers or Seeds

Plant bugs: Carrot, tarnished

Other Pests Occasionally Found on Carrot and Related Crops

Aphids: Bean, bean root, buckthorn, cowpea, foxglove, potato
Armyworms and cutworms: Yellowstriped and western yellowstriped armyworm; army, dingy, pale western, and spotted cutworm; zebra caterpillar
Blister beetles: Black, immaculate, spotted, striped
Crickets: Fall, spring, and southeastern field; shortwinged, southern and tawny mole; Mormon
Earwigs: European
Flea beetles: Palestriped, redheaded, smartweed
Grasshoppers: Differential, eastern lubber, migratory, redlegged, twostriped
Leafhoppers: Western potato
Leaf beetles: Banded cucumber
Looper: Bean leafskeletonizer, plantain, soybean
Maggots: Seedcorn
Millipedes: Garden
Mites: Bulb
Other bugs: False chinch, leaffooted, little negro
Other caterpillars: Yellow woollybear
Plant bugs: Rapid, superb
Scarab beetles: Asiatic garden, carrot, white grubs
Slugs
Symphylans: Garden
Thrips: Onion, tobacco
Webworms: Alfalfa, beet, garden
Weevils: Whitefringed beetle
Whiteflies: silverleaf, sweetpotato
Wireworms: Eastern field, Great Basin, Gulf, southern potato, tobacco, wheat

GUIDE TO COMMON PESTS AFFECTING LETTUCE AND RELATED CROPS

(Celtuce, Chicory, Endive, Escarole, Lettuce, Radicchio)

Pests Feeding on Roots

Aphids: Lettuce root
Scarab beetles: Green June, white grubs
Symphylans: Garden

Pests Producing Mines in Leaves

Leafminers: American serpentine, pea, vegetable

Pests Feeding Externally on Leaves or Stems

Pests with Chewing Mouthparts

Armyworms and cutworms: Armyworm, beet, yellowstriped and western yellowstriped armyworm; corn earworm; zebra caterpillar; black, granulate, and variegated cutworm

Flea Beetles: Palestriped

Grasshoppers: American, differential, eastern lubber, migratory, redlegged, twostriped

Loopers: Alfalfa, cabbage, celery

Other caterpillars: Saltmarsh

Slugs

Snails

Pests with Piercing-Sucking Mouthparts

Aphids: Green peach, lettuce, potato

Leafhoppers: Aster, potato, western potato

Plant bugs: Garden fleahopper; tarnished, western tarnished, and pale legume

Thrips: Western flower

Whiteflies: Greenhouse, silverleaf, sweetpotato

Other Pests Occasionally Found on Lettuce and Related Crops

Aphids: Bean, cowpea, foxglove

Armyworms and cutworms: Bertha armyworm; clover, dingy, glassy, redbacked, and spotted cutworm; tobacco budworm

Crickets: Fall, spring, and southeastern field; shortwinged, southern, and tawny mole; Mormon

Earwigs: European, redlegged

Flea beetles: Palestriped, potato, redheaded, smartweed, tuber, western black, western potato

Flies: European crane

Leaf beetles: Banded and spotted cucumber

Loopers: Bilobed, soybean

Maggots: Seedcorn

Millipedes: Garden

Other bugs: False chinch

Other caterpillars: Whitelined sphinx

Pillbug: Common

Springtail: Garden

Stink bugs: Harlequin, Say

Thrips: Bean, melon and western flower

Webworm: Alfalfa, beet, garden; celery and false celery leaftier

Weevils: Vegetable

Wireworms: Corn, eastern field, false, Great Basin, Oregon, Pacific Coast, sugarbeet

GUIDE TO IDENTIFICATION OF COMMON INSECT PESTS AFFECTING OKRA

Pests Feeding on Roots or Lower Stem

Leaf beetles: Banded and striped cucumber

Pests Producing Mines in Leaves

Leafminers: Vegetable

Pests Boring into Stems

Borers: European corn

Pests Feeding Externally on Leaves

Pests with Chewing Mouthparts

Armyworms and cutworms: Corn earworm; yellowstriped and southern armyworm

Leaf beetles: Spotted cucumber

Loopers: Okra caterpillar

Other caterpillars: Gray hairstreak

Scarab beetles: Japanese

Pests with Piercing-Sucking Mouthparts

Aphids: Green peach, melon

Plant bugs: Garden fleahopper

Mites: Twospotted and tumid spider

Thrips: Melon

Whiteflies: Silverleaf, sweetpotato

Pests Feeding on Blossoms or Fruits

Ants: Red imported fire

Armyworms and cutworms: Corn earworm

Bugs: Leaffooted

Other caterpillars: Gray hairstreak

Stink bugs: Brown, green, southern green

Other Pests Occasionally Found on Okra Plants

Armyworms and cutworms: Black cutworm

Blister beetles: Black

Flea beetles: Palestriped, redheaded, smartweed

Grasshoppers: American

Leaf beetles: Grape colaspis, striped and western striped cucumber

Leafhoppers: Potato, western potato

Leafminers: Pea

Loopers: Soybean

Scarab beetles: Chinese rose, Japanese, white grubs

Stink bugs: Harlequin
Weevils: Whitefringed beetle
Whiteflies: Greenhouse

GUIDE TO COMMON PESTS AFFECTING ONION AND RELATED PLANTS

(Chive, Garlic, Leek, Onion, Shallot)

Pests Boring into Bulbs or Roots

Flies: Onion and lesser bulb
Maggots: Onion, seedcorn
Mites: Bulb

Pests Producing Small Mines in Leaves

Leafminers: American serpentine, pea

Pests Feeding Externally on Leaves or Stems

Pests with Chewing Mouthparts
Armyworms and cutworms: Beet armyworm
Other caterpillars: Saltmarsh
Pests with Piercing-Sucking Mouthparts
Thrips: Onion, tobacco, western flower

Other Pests Occasionally Found on Onion and Related Plants

Aphid: Bean
Armyworms and cutworms: Armyworm, fall, yellow-striped, and western yellowstriped armyworm; army, black, clover, darksided, dingy, granulate, pale western, redbacked, spotted, and variegated cutworm
Blister beetles: Black, immaculate
Borers: Potato stem
Crickets: Shortwinged, southern and tawny mole; Mormon
Flea beetles: Palestriped
Grasshoppers: Differential, eastern lubber, migratory, twostriped
Leaf beetles: Banded cucumber
Loopers: Alfalfa
Maggots: Bean seed
Plant bugs: Pale legume, tarnished, western tarnished,
Scarab beetles: Asiatic garden, oriental, white grubs
Springtails: Garden
Stink bugs: Onespotted
Symphylans: Garden
Thrips: Bean, melon,
Webworms: Alfalfa, beet, garden
Weevils: Vegetable
Wireworms: Eastern field, false, Great Basin

GUIDE TO COMMON PESTS AFFECTING RHUBARB

Pests Feeding on Roots

Leaf beetles: Spotted cucumber beetle
Weevils: Rhubarb curculio

Pests Living within Stems and Stalks

Borers: Potato stem, stalk
Weevils: Rhubarb curculio

Pests Feeding Externally on Leaves or Stems

Pests with Chewing Mouthparts
Armyworms and cutworms: Spotted and variegated cutworm
Flea beetles: Hop, potato
Scarab beetles: Japanese
Pests with Piercing-Sucking Mouthparts
Aphids: Bean, green peach, potato
Leafhopper: Potato

Other Pests Occasionally Found on Rhubarb

Armyworms and cutworms: Bertha, southern, yellowstriped, and western yellowstriped armyworm; army cutworm
Borers: European corn
Earwig: European
Flea beetles: Palestriped
Looper: Alfalfa
Maggots: Seedcorn
Other caterpillars: Yellow woollybear
Scarab beetles: Asiatic garden, oriental, white grubs
Symphylan: Garden
Webworms: Alfalfa, beet, garden

GUIDE TO COMMON PESTS AFFECTING SQUASH AND RELATED CROPS

(Cucumber, Pumpkin, Squash, Cantaloupe, Watermelon, and Other Melons)

Pests Boring into Roots, Stems, Blossoms, or Fruit

Pests with Legs
Borers: Pickleworm; melonworm; squash and southwestern squash vine
Leaf beetles: Banded, spotted, striped, and western striped cucumber

Pests without Legs
Flies: Melon; small fruit

Pests Producing Small Mines in Leaves

Leafminers: American serpentine, pea, vegetable

Pests Feeding Externally on Leaves, Stems, Blossoms, or Fruit

Pests with Chewing Mouthparts
Armyworms and cutworms: Southern armyworm; granulate cutworm
Leaf beetles: Banded, spotted, striped, and western striped cucumber; western corn rootworm
Other beetles: Squash
Other caterpillars: Melonworm
Pests With Piercing-Sucking Mouthparts
Aphids: Green peach, melon
Leafhoppers: Potato; western potato
Mite: Broad; twospotted spider
Other bugs: Leaffooted; squash and horned squash
Thrips: Melon, western flower
Whiteflies: Silverleaf, sweet potato, greenhouse

Other Pests Occasionally Found on Squash and Related Crops

Ants: Red imported fire
Aphids: Bean, buckthorn, potato, rice root
Armyworms and cutworms: Beet, fall, yellowstriped and western yellowstriped armyworm; black, darksided, dingy, redbacked, and variegated cutworm; corn earworm, tobacco budworm
Blister beetles: Black, immaculate
Borers: Lesser cornstalk, stalk
Crickets: Fall, spring, and southeastern field; shortwinged, southern, and tawny mole
Earwig: European
Flea beetles: Palestriped, potato, tuber, western potato
Flies: Oriental fruit
Grasshoppers: American, differential, eastern lubber, migratory, twostriped
Leaf beetles: Grape colaspis
Leafhoppers: Beet
Loopers: Alfalfa, cabbage, soybean
Maggots: Bean seed, seedcorn
Millipedes: Garden
Other beetles: Whitefringed
Other bugs: False chinch
Other caterpillars: Whitelined sphinx, yellow woollybear
Pillbugs: Common

Plant bugs: Garden fleahopper, pale legume, tarnished, western tarnished
Sap beetles: Dusky and fourspotted
Scarab beetles: Carrot, Chinese rose, green June, white grubs
Slugs
Snails
Springtails: Garden
Stink bugs: Brown, green, onespotted, southern green, harlequin
Symphylans: Garden
Thrips: Onion, tobacco
Webworms: Alfalfa, beet, garden
Wireworms: Eastern, false, field, Gulf, Pacific Coast, southern potato, sugarbeet, tobacco, wheat

GUIDE TO COMMON PESTS AFFECTING SWEET CORN

Pests Feeding on Roots

Aphids: Corn root
Flea beetles: Corn, desert corn, redheaded, toothed
Leaf beetles: Northern and western corn rootworms; spotted and banded cucumber
Scarab beetles: Green June, white grubs
Webworms: Sod and root
Weevils: Whitefringed beetle
Wireworms: Corn, eastern field, false, Great Basin, Gulf, Oregon, Pacific Coast, southern potato, sugarbeet, tobacco, wheat

Pests Feeding on Seed or Seedling

Armyworms and cutworms: Black, darksided, granulate, spotted, and variegated cutworm
Borers: Hop vine, potato stem
Flea beetles: Corn, desert corn, redheaded, toothed
Maggots: Bean seed, seedcorn
Other beetles: Seedcorn, slender seedcorn
Webworms: Sod and root
Weevils: Maize and southern corn billbug

Pests Boring into Stems or Taproot

Borers: European corn, lesser cornstalk, southern cornstalk, southwestern corn, stalk, sugarcane
Weevils: Maize and southern corn billbug

Pests Producing Small Mines in Leaves

Borer: Stalk
Leafminer: Corn blotch leafminer

Pests Feeding Externally on Leaves, Tassel, or Stalk

Pests with Chewing Mouthparts

Armyworms and cutworms: Armyworm, fall armyworm, corn earworm

Borers: European corn, southern cornstalk, southwestern corn, stalk, sugarcane

Grasshoppers: American, differential, migratory, twostriped, redlegged

Leaf beetles: Grape colaspis, western corn rootworm

Scarab beetles: Japanese

Webworms: Sod and root

Weevils: Maize and southern corn billbug

Pests with Piercing-Sucking Mouthparts

Aphids: Corn leaf, bird cherry-oat

Mites: Twospotted spider, Banks grass

Other bugs: Chinch

Leaf and planthoppers: Corn delphacid, corn leafhopper

Thrips: Grass

Pests Feeding on Ears or Silk

Armyworms and cutworms: Corn earworm, fall armyworm, western bean cutworm

Borers: European, southwestern corn

Earwigs: European

Flies: Cornsilk

Leaf beetles: Northern and western corn rootworm; spotted and banded cucumber

Other beetles: Dusky and fourspotted sap

Scarab beetles: Green June, Japanese

Other Pests Occasionally Found on Corn

Ant: Red imported fire

Aphids: Bean, green peach, potato

Armyworms and cutworms: Army, bertha, bronzed, glassy, and pale western cutworm; beet, sweet potato, velvet, yellowstriped, and western yellowstriped armyworm; zebra caterpillar

Blister beetles: Black, striped

Crickets: Mormon

Flies: European crane

Leaf beetles: Grape colaspis

Leafhoppers: Aster and western potato leafhopper

Loopers: Soybean

Other caterpillars: Banded and yellow woollybear; saltmarsh

Millipedes: Garden

Plant bugs: Tarnished

Scarab beetles: Asiatic, carrot, Chinese rose

Symphylans: Garden

Stink bugs: Brown, green, onespotted, southern green

Slugs

Thrips: Bean, tobacco

Webworms: Alfalfa, beet, garden

Weevils: Whitefringed beetle

GUIDE TO COMMON PESTS AFFECTING SWEET POTATO

Pests Boring into Vines or Roots

Borers: Sweet potato vine

Flea beetles: Sweet potato

Leaf beetles: Banded and spotted cucumber; sweet potato

Scarab beetles: Spring rose

Wireworms: Corn, Gulf, southern potato, tobacco

Weevils: Sweetpotato and West Indian sweetpotato; whitefringed beetle

Pests Producing Small Mines in Leaves

Leafminers: Morningglory, sweetpotato

Pests Feeding Externally on Leaves or Stems

Pests with Chewing Mouthparts

Armyworms and cutworms: Fall, southern, and yellowstriped armyworm; black, darksided, and granulate cutworm

Flea beetles: Sweetpotato

Leaf beetles: Sweetpotato

Other caterpillars: Sweetpotato hornworm

Tortoise beetles: Argus, blacklegged, golden, mottled, striped

Pests with Piercing-Sucking Mouthparts

Leafhoppers: Potato, western potato

Whiteflies: Silverleaf, sweetpotato

Other Pests Occasionally Found on Sweet Potato

Ant: Red imported fire

Aphids: Green peach, foxglove, melon, potato

Armyworms and cutworms: Beet, sweetpotato, and velvet armyworm; army, dingy, and variegated cutworm; corn earworm

Borers: Lesser cornstalk

Blister beetles: Black, striped

Crickets: Fall, spring and southeastern field; shortwinged, southern, and tawny mole

Earwing: Ringlegged

Flea beetles: Elongate, palestriped, redheaded, smart-weed

Grasshopper: Differential, migratory, twostriped

Loopers: Cabbage, soybean

Mites: Tumid, twospotted spider

Maggots: Seedcorn

Other bugs: Little negro

Other caterpillars: Yellow woollybear

Plant bugs: Garden fleahopper, pale legume, tarnished, western tarnished

Sawflies: Sweet potato

Scarab beetles: Asiatic garden, carrot, Chinese rose, white grubs

Stinkbugs: Southern green

Thrips: Onion

Weevils: Vegetable

Whiteflies: Greenhouse

GUIDE TO COMMON PESTS AFFECTING TOMATO AND RELATED PLANTS

(Eggplant, Pepper, Potato, Tomatillo, Tomato)

Pests Feeding on Roots, Tubers or Lower Stem

Borers: Potato tuberworm

Crickets: Shortwinged, southern, and tawny mole

Armyworms and cutworms: Black, granulate, and variegated cutworm

Flea beetles: Eggplant, potato, tuber, western potato

Maggots: Seedcorn

Symphylans: Garden

Scarab beetles: Green June, white grubs

Weevils: Vegetable weevil, whitefringed beetle

Wireworms: Corn, eastern field, Great Basin, southern potato, sugarbeet, tobacco

Pests Producing Mines in Leaves

Leafminers: American serpentine, vegetable

Other caterpillars: Eggplant leafminer, potato tuberworm, tomato pinworm

Pests Boring into Stems

Borers: European corn, potato stalk, potato stem, and stalk borer; potato tuberworm; tomato pinworm

Pests Feeding Externally on Leaves or Upper Stems

Pests with Chewing Mouthparts

Armyworms and cutworms: Beet, fall, southern, yellowstriped and western yellowstriped army-worm; tobacco budworm; corn earworm; variegated cutworm

Flea beetles: Eggplant, potato, tuber, western potato, tobacco, southern tobacco

Leaf beetles: Colorado potato

Loopers: Alfalfa, cabbage

Other caterpillars: Tobacco and tomato horn-worm; saltmarsh

Tortoise beetles: Eggplant

Weevils: Vegetable; whitefringed beetle

Pests with Piercing-Sucking Mouthparts

Aphids: Buckthorn, foxglove, green peach, potato

Lacebugs: Eggplant

Leafhoppers: Aster, beet, potato, western potato

Mites: Broad, tomato russet, twospotted spider

Other bugs: Leaffooted

Plant bugs: Garden fleahopper, tarnished, western tarnished

Psyllids: Potato

Stinkbugs: Southern green

Thrips: Melon, onion, western flower

Whiteflies: Greenhouse, silverleaf, sweetpotato

Pests Feeding on Blossoms or Fruits

Armyworms and cutworms: Beet, yellowstriped and western yellowstriped armyworm; corn ear-worm; variegated cutworm

Borers: Corn earworm, tobacco budworm, tomato pinworm

Flies: Small fruit

Maggots: Pepper

Other bugs: Leaffooted

Other caterpillars: Tobacco and tomato hornworm

Sap beetles: Dusky

Scarab beetles: Green June

Stink bugs: Brown, consperse, green, Say, southern green, onespotted

Weevils: Pepper

Other Pests Occasionally Found on Tomato and Related Plants

Ant: Red imported fire

Aphids: Bean root, rice root, melon

Armyworms and cutworms: Army, bertha, bronzed, clover, darksided, dingy, pale western, redbacked, and spotted cutworm; zebra caterpillar

Blister beetles: Black, immaculate, spotted, striped

Borer: Lesser cornstalk

Crickets: Fall, spring, and southeastern field; short-winged, southern, and tawny mole; Mormon

Earwigs: European, ringlegged

Flea beetles: Hop, palestriped
Flies: European crane, Mediterranean fruit
Grasshoppers: American, differential, migratory, two-striped, redlegged
Leaf beetles: Grape colaspis; spotted, striped, and western striped cucumber
Leafminers: Pea
Loopers: Soybean
Millipedes: Garden
Mites: Bulb
Other bugs: Little negro

Other caterpillars: Whitelined sphinx, yellow woollybear
Pillbugs: Common
Plant bugs: Alfalfa, pale legume, rapid, superb
Scarab beetles: Carrot, Chinese rose, Japanese
Slugs
Springtails: Garden
Stink bugs: Harlequin
Symphylans: Garden
Thrips: Bean, tobacco
Webworms: Alfalfa, beet, garden
Wireworms: False, Gulf, Oregon, Pacific Coast, wheat

Class Insecta—Insects

Order Coleoptera—Beetles, Weevils, White Grubs and Wireworms

FAMILY BRUCHIDAE—SEED BEETLES AND WEEVILS

Bean Weevil
Acanthoscelides obtectus (Say)
(Coleoptera: Bruchidae)

Natural History

Distribution.   Bean weevil is believed to have originated in Central America. It is now widespread in the United States and southern Canada, most of Central and South America, Africa, southern Europe, and New Zealand.

Host Plants.   This insect is a pest of several legumes including bean, chickpea, cowpea, faba bean, lentil, lima bean, mung bean, pea, and soybean. Many cultivars of bean, *Phaseolus vulgaris*, are susceptible, including kidney bean, navy bean, pink bean, red bean, wax bean, and white bean. Not all legumes are equally suitable, with faba bean, lima bean, pea, and soybean often cited as being relatively poor hosts.

Natural Enemies.   Surprisingly little is known about the parasitoids of bean weevil. Two species known from North America are *Dinarmus laticeps* (Ashmead) (Hymenoptera: Pteromalidae) and *Eupelmus cyaniceps* Ashmead (Hymenoptera: Eupelmidae).

Life Cycle and Description.   A life cycle of bean weevil can be completed in just 21 days during summer months in California, but also may extend up to six months under less ideal conditions. At the latitude of Washington D.C., as many as six generations have been reared annually.

Egg.   The eggs of bean weevil are white, and measure about 0.5–0.8 mm long and 0.2–0.4 mm wide. They are elliptical in shape with one end broader than the other. Eggs are deposited in small clusters of 3–30

on legume seeds, with fecundity of 40–60 eggs per female. They are not flattened against the seed, as is the case of the cowpea weevils, *Callosobruchus* spp. Also, they are not tightly glued to the seed, and can easily be removed. Duration of the egg stage is highly variable and determined principally by temperature. When reared under warm condition (30°C), eggs may hatch in five days, whereas during the winter nearly 30 days may be required.

Larva.   There are four instars. On hatching, the young larva is whitish with a black head, and measures about 0.5 mm long. It is cylindrical in shape with the posterior end markedly tapered. The young larva is equipped with well-developed thoracic legs and bears stout bristles along the back. Bristles are also located near the thoracic legs. In the second and subsequent instars the legs are reduced. After attaining the fourth instar, the larva remains white, but measures about 4 mm long, is very stout in appearance, and has a small head with stubby, fleshy thoracic legs. Head capsule widths are about 0.15, 0.22, 0.38, and 0.46 mm, respectively, for instars 1–4. Duration of the larval stage may be as little as 12 days, or as long as six months, depending not only on temperature but also on humidity and type of food. When reared at 27°C, duration of the instars was reported to be about six, seven, seven, and six days, respectively. Larvae feed entirely within the seed, and near the completion of larval development they burrow to the seed surface where they transform to pupae.

Pupa.   The pupa greatly resembles the adult in general appearance, but the wings are reduced in size. The body initially is white but darkens as the pupa completes its development. The pupa measures about 2.5–4.0 mm long. Duration of the pupal stage is about

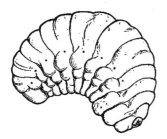

Bean weevil larva.

8–20 days, but probably can be even longer under cool conditions.

Adult. The adult is oval in shape, brownish gray, and covered with short, golden brown pubescence. The head is small. Antennae are serrate, with the basal four segments and the distal segment red, and intermediate segments black. Antennal segments tend to become larger toward the tip. The thorax is quite narrow behind the head, but widens rapidly. Elytra are dark with irregular yellowish-white transverse lines. The tips of the elytra are broadly rounded. The tip of the abdomen is exposed, and reddish when viewed from above. The beetle measures 2.6–3.8 mm long. The adult cuts a circular hole in the epidermis of the bean seed and escapes to seek mates.

The biology of bean weevil was given by Garman (1917) and Larson and Fisher (1938). Some elements of developmental biology were provided by Howe and Currie (1964). Description of the larval instars was given by Pfaffenberger (1985).

Damage

The bean weevil is principally a pest of stored beans. It breeds continuously in stored beans, reducing them to dust. Up to 28 weevils have been reared

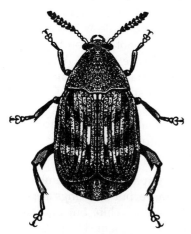

Adult bean weevil.

from a single seed. However, the adults also fly to the field and oviposit on bean pods that are green but close to maturity. In the field, the female normally gnaws a hole through the pod wall and oviposits within. The hatching larvae then move within the pod to seek an appropriate feeding site, and burrow into the seed. The larva creates a cavity in the seed, reducing its nutrient content, germination potential, and commercial value. Damage potential is much greater in southern states.

Management

Sanitation is very important. Bean weevils overwinter in stored beans, and production areas should be located distant from storage. Also, insect-free seed must be planted or problems will persist. (See the section on cowpea weevil, *Callosobruchus maculatus* (Fabricius), for discussion of disinfestation of seed.) Wild beans have been identified with resistance to bean weevil, but this resistance has not yet been incorporated into commercially available varieties (Schoonhoven *et al.*, 1983; Cardona *et al.*, 1989).

Broadbean Weevil
Bruchus rufimanus (Boheman)
(Coleoptera: Bruchidae)

Natural History

Distribution. Broadbean weevil is widespread in Europe, Asia, and northern Africa, and is established in Australia and North America. It was first observed in North America near San Francisco, California, in 1888. Despite spreading southward in California, it has not become a national problem, largely owing to the limited culture of faba beans elsewhere in North America. Thus, it is a pest only in California, although it is occasionally transported to eastern states and to Canada along with dried faba beans.

Host Plants. This insect attacks only faba bean (Middlekauff, 1951). Although there are reports of this species attacking pea, this probably stems from the similarity of adults with pea weevil, *Bruchus pisorum* (Linnaeus). Broadbean weevil adults emerge from dried seeds but cannot reinfest dried beans, so they are not a threat to beans in storage.

Natural Enemies. Several wasp parasitoids were reported to be associated with broadbean weevil by Chittenden (1912e), but only one species, *Triaspis stictostiba* Martin (Hymenoptera: Braconidae), was listed by Krombein *et al.* (1979). Additional parasitoid species were imported from Europe and released, but did not establish successfully. An interesting natural enemy is a predatory mite, *Pediculoides ventricosus*

Newport. Although it is not known to inflict high levels of mortality, the gravid females of this mite are greatly swollen and easily observed within the cells caused by the feeding of the weevil larvae.

Life Cycle and Description. There is a single generation per year, with the adult stage overwintering. In California, egg laying occurs principally in March and April. The larval period typically extends from March to mid-October. Pupae occur from August to late October. Adults are found from August until the following June.

Egg. The whitish eggs are deposited singly on the outside of the bean pods, although each pod may receive several eggs. Oviposition occurs at night. There is no relationship between the number of eggs deposited and the availability of food. Sometimes, many more eggs are deposited than can develop within a pod. Females prefer larger pods. The egg is elliptical in shape, and measures 0.55–0.60 mm long and 0.25–0.28 mm wide. One end of the egg is more pointed than the other. The chorion lacks significant sculpturing and appears quite smooth. Duration of the egg stage averages 12.5 days, with a range of 9–18 days.

Larva. The hatching larva chews a hole through the underside of the egg, burrows directly into the pod, and then into the seed. Initially the young larva is pale yellow except for the blackish brown head and mouthparts. As the larva matures, it becomes cream colored, but retains the dark head. At maturity, it measures 4.5–5.5 mm long. As the larva approaches maturity, it creates a cavity just beneath the seed covering. Therefore, when the weevil is ready to emerge, it has only to break through a thin covering. Duration of the larval stage is about 90 days, with a range of 65–105 days.

Pupa. The pupa greatly resembles the adult beetle in form. Initially it is light-yellow or cream-colored, but gradually turns darker until it attains a light brown. It measures about 5 mm long. Mean duration (range) of the pupal stage is 10 (7–16) days.

Adult. Emergence of adults from seeds in the autumn is more likely in warm weather or if the bean seeds are handled. Adults emerging from seed seek sheltered overwintering sites such as under bark of trees, and remain in reproductive diapause until spring. Increasing day length and consumption of plant pollen terminate reproductive diapause (Tran *et al.*, 1993). During cool weather, many adults remain within the seeds up to eight months. Adults may emerge from beans planted as seed, thus resulting in infestation of new, and sometimes remote, fields. The adult is blackish, with white marking on the elytra and tip of the abdomen. The white markings form an irregular white line across the distal portion of the elytra, giving the beetle a mottled appearance. The four basal segments of the antennae are reddish brown, with the remaining segments black. The forelegs are reddish brown and black, while the middle and hind pairs are black.

Broadbean weevil is easily confused with pea weevil, as the adults are very similar in appearance. In pea weevil, the terminal abdominal segment, when viewed from above, bears a pair of distinct black spots; in broadbean weevil these spots are lacking or poorly defined. In pea weevil, the posterior femora are equipped with sharp spines; in broadbean weevil spines are absent from the hind femora or blunt when they occur.

The biology of broadbean weevil was best summarized by Campbell (1920), although Chittenden (1912e) provided a few additional details.

Damage

Larvae feed within growing bean seed, pupate, and the adults emerge from the seed. Adults may emerge soon after completion of development, or remain in the seed for several months. The weight of infested seeds is reduced and germination potential decreased, but the most important consequence of infestation is the marked reduction of seed value. Even low infestation levels preclude sale of the product for human consumption. The amount of bean tissue consumed is surprisingly low, with each weevil consuming only about 3% of the bean seed's weight. However, several weevils may develop in a single seed.

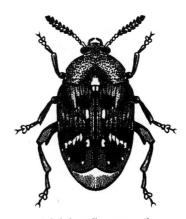

Adult broadbean weevil.

Management

Insecticides. Chemical suppression of beetles in the field is not very satisfactory unless residual insecticides are used. It is essential that insecticides be present when adults invade the field so that they can be killed before oviposition (Middlekauff, 1951). Sanitation is frequently stressed, and if carefully implemented, greatly reduces the need for insecticides. As infestations often originate with beans, it is important that clean, insect-free seed be planted. Chemical fumigation of seed is very effective and economically practical.

Cultural Practices. Beetles within the bean seeds are killed by soaking in a water bath at 65°C for 20 minutes. Also, if seed is held for at least a full year before planting—a period that does not reduce germination—the weevils contained within the seeds will emerge and perish before planting. Seed in storage, spilled seed, and culled seed that is destined to be fed to livestock are the principal sources of infestation, so care must be taken to eliminate these sources or fumigate the seed.

Cowpea Weevil
Callosobruchus maculatus (Fabricius)

Southern Cowpea Weevil
Callosobruchus chinensis (Linnaeus)
(Coleoptera: Bruchidae)

Natural History

Distribution. Cowpea weevil and southern cowpea weevil species are found throughout the United States, Central America, South America, Africa, southern Asia, and Australia. Their origin is uncertain, but their distribution is closely linked with that of their principal host plant, cowpea. This plant is reported to have originated in central or western Africa, and so the beetles may have a similar origin. In North America, they are a serious pest only in the southern states. Although both species occasionally are detected in the central provinces of Canada, they are usually associated with imported legume seed, and are not considered to be damaging.

Host Plants. These species develop successfully on a number of legume seeds including chickpea, cowpea, faba bean, lentil, pea, and soybean. However, cowpea is most suitable, and faba bean and soybean are relatively poor hosts. These insects are primarily pests of stored seed, with little damage occurring in the field.

Natural Enemies. Several parasitoids are known, including *Anisopteromalus calandrae* (Howard), *Choetos-pila elegans* Westwood, *Dinarmus laticeps* (Ashmead), and *Lariophagus texanus* Crawford (all Hymenoptera: Pteromalidae), but their incidence generally is low. An egg parasitoid, *Uscana semifumipennis* Girault (Hymenoptera: Trichogrammatidae) is an important mortality factor in some locations (Paddock and Reinhard, 1919).

Life Cycle and Description. These species complete several generations per year. An entire generation can be completed in 30–40 days. In the southern United States, the number of generations is commonly reported to be 5–8. The life cycles of these two species of *Callosobruchus* are very similar, and here they are treated as one except for where significant differences are known.

Egg. The eggs are oval, and broader at one end. The female glues the egg to the seed, and the side that attaches to the seed is flattened. The egg is 0.4–0.8 mm long and 0.3–0.5 mm wide. The duration of this stage is only 3–4 days under ideal conditions, but may extend to over 30 days during cold weather. The young larva chews a hole through the egg chorion directly into the seed, or through the pod wall and then into the seed. Females of cowpea weevil commonly produce 100 or more eggs during their life span, and southern cowpea weevil about 50, although some strains produce fewer eggs. Access by adults to water or sugar water increases fecundity. Females also produce more eggs when they are not constrained by availability of suitable oviposition sites.

Larva. There are four instars. First instars are whitish, short and thick in appearance, with small thoracic legs. Long bristles are present dorsally, and shorter bristles near the legs. Latter instars are similar, but lack the long bristles, and the thoracic legs are even less apparent. They attain a length of about 3.5–4.5 mm at maturity. The body of larvae of southern cowpea weevil is expanded behind the head, giving a humpbacked appearance when viewed in profile. Head capsule widths are about 0.14, 0.24, 0.33, and 0.48 mm for instars 1–4, respectively. Duration of the instars is normally three, three, four, and five days, respectively, for southern cowpea weevil at 30°C, but it varies greatly with temperature and host suitability. For cowpea weevil, the length of each instar is about a day longer. During the summer in California, the larval development period is reported as 17–22 days. Optimal temperature for development of both species is 30–32°C, with little or no development at temperatures below 20°C. Low humidities (< 70%) also are limiting, especially for eggs and young larvae (Howe and Currie, 1964).

Pupa. The pupa is white and stout in shape, resembling the adult in appearance. The length is 3.2–5.0 mm. The duration of the pupal stage of cowpea weevil was reported to average 6.4 days at 25°C by Howe and Currie (1964). The adult remains at the pupation site for sometime before emerging, however, so the period from transformation to a pupa to emergence of the adult from the seed totals about 10.2 days at 25°C.

Adult. The adults are oval in general shape, with cowpea weevil somewhat more elongate than southern cowpea weevil. In cowpea weevil, the general color is whitish to light brown below and brownish marked with dark brown or black above. The color pattern is rather variable, however. The head and antennae are dark brown or black, with a dark spot on the thorax behind the head. Elytra are broadly rounded at their tips. Each elytra bears a large black, quadrate spot along the leading edge; the tip is also darkened. The legs are reddish to blackish. The tip of the abdomen is exposed, and is readily visible from above. It is blackish, but is usually marked with a longitudinal gray line dorsally.

Cowpea weevil exhibits density-related polymorphic morphology, behavior, and physiology. Adults developing from insects reared under high density conditions have less pubescence on their body, allowing more black color to be visible, resulting in an overall darker color. High density beetles are also more oval in body shape, with the tip of the abdomen extending far beyond the elytra. More important, high-density beetles tend not to fly, but produce more eggs and deposit them early in adult life. Presumably, this polymorphism is an adaptation to varying resource abundance (Utida, 1972). (See color figure 134.)

In southern cowpea weevil, general body color is reddish brown, marked with black. The thorax is reddish, with a white spot at the center of the hind margin. The elytra are marked with a dark spot at the tip, and another at the leading edge, a pattern very similar to that of cowpea weevil. However, the tip of the abdomen does not markedly protrude, as in cowpea weevil. Also, the reddish antennae are much more serrated, particularly in the male. As in the case with cowpea weevil, the color patttern can be variable.

The mature beetle chews a hole in the seed coat and escapes from the seed. The exit hole may be a complete circle or a portion may remain attached and function as a door hinge. It is not necessary for the adults to feed before they mate and commence oviposition. In fact, there is little evidence that consumption of more than nectar occurs and even this is infrequent. However, the adults seemingly benefit from access to water and sugar water, as those provided with such nutrients live longer and produce more eggs. Oviposition may begin within one day of the emergence of adult. Duration of the adult stage is estimated as 7–18 days, with an average longevity of about 15 days in warm weather, but these values are lengthened considerably during cool weather. In the field, oviposition generally occurs on over-ripened pods that have split along the suture. They prefer to deposit directly on bean seed, but if this is not available, they will deposit eggs on pods. They will also oviposit on immature pods. Messina (1984) estimated that 20% of eggs deposited on green pods successfully develop into adults. In storage, they deposit eggs freely on all surfaces of dry beans. They avoid damaged seeds when selecting oviposition sites. Females of cowpea weevil produce a sex pheromone (Lextrait *et al.*, 1995) and oviposition-deterring pheromone (Wasserman, 1981; Credland and Wright, 1990).

A good general description of these insects was given by Garman (1917) and Back (1922). Paddock and Reinhard (1919) and Larson and Fisher (1938) provided detailed information on cowpea weevil, and Chittenden (1912f) described southern cowpea weevil. The developmental biology of both species was given by Howe and Currie (1964).

Adult cowpea weevil.

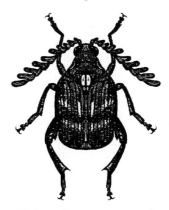

Adult southern cowpea weevil.

Damage

Larvae feed within the seed. A single larva often consumes only a small portion of the embryo, and may not disrupt seedling germination. However, attack by more than one insect per seed is possible, and is more damaging. Seeds may become infested either in the field or in storage, with most damage resulting from the latter. Thus, these insects are not usually considered pests of fresh-market crops. These bruchids, and particularly cowpea weevil, are major pests of cowpea and some other legumes where seed is not treated with insecticide or is stored in a manner that allows invasion of insects. The presence of even a single insect in a seed causes significant weight loss in seeds (Ofuya and Bamigbola, 1991), and even low levels of contamination are unacceptable in American commerce. The problem is severe enough that American dried-bean production is centered in northern climates not because these legumes cannot be grown in the southern states, but because northern locations largely avoid problems with damaging populations of bruchids.

Management

Sampling. Blacklight traps can be used to sample adults (Keever and Cline, 1983).

Insecticides. A great deal of effort has been made in less-developed countries in evaluating the natural materials that have insecticidal or repellent properties for protection of cowpea under primitive storage conditions. A variety of effective techniques can protect this vital protein resource. Natural materials such as ash, sand, vegetable oil, and fruit, cashew, ginger, and neem extracts provide protection (Su, 1976, 1978 and 1991; Singh *et al.*, 1978; Echendu, 1991; Cockfield, 1992). The mode of action of oil, one of the most effective suppression techniques, seems to be in reducing egg respiration and ability of the egg to release toxic metabolites (Don–Pedro, 1989).

Chemical fumigation is the standard practice to protect stored seeds. However, it is also possible to disinfest seed through modification of storage atmosphere; displacement of oxygen by nitrogen is effective (Storey, 1978).

Host-Plant Resistance. Although some legumes are quite resistant to these insects, some others seem uniformly susceptible. Cowpea, probably the most important host, displays measurable but low-level resistance (Fitzner *et al.*, 1985). Sources of pod and seed resistance have been identified (e.g., Schalk 1973; Talekar and Lin, 1981 and 1992; Rusoke and Fatunla, 1987), but commercially acceptable cultivars with high levels of resistance are not generally available. Rough-

seeded varieties are less preferred for oviposition by adults (Nwanze *et al.*, 1975).

Cultural Practices. Disinfestation of seed is critical both for storage of food and preservation of viable seed stock. Seeds can be disinfested with solar heaters, which take advantage of solar radiation to produce high temperatures that are lethal to larvae, pupae, and adults. Solar heaters can be constructed with some clear plastic, and achieve temperatures of 57°C for 1 h or 65°C for 5 min, which provides effective disinfestation (Kitch *et al.*, 1992).

The spatial relationship of seed storage to production areas is also significant. In addition to planting insect-free seed, it is immensely valuable to have crop-producing areas distant from storage. This is important, because dispersal of beetles from storage is a major source of field infestation. If on-farm storage is desired, storage bins should be cleaned of all old seed and the walls treated with insecticide.

Pea Weevil
Bruchus pisorum (Linnaeus)
(Coleoptera: Bruchidae)

Natural History

Distribution. Pea weevil was first observed near Philadelphia, Pennsylvania in the 1740s, and then noted in nearby states in the 1750s. It had spread across the country to Washington and Oregon in the 1890s. The origin of pea weevil is uncertain, but peas are not native to North America, and the beetles were likely introduced along with the crop. The origin of peas is believed to be the mountainous middle-eastern region from Ethiopia to Afghanistan, which may be the ultimate source of these insects, but they likely arrived via Europe. At present, pea weevil is distributed throughout the United States and southern Canada, and other temperate areas of the world including Asia, Europe, North Africa, and southern Australia.

Host Plants. Pea weevil larvae have a very restricted host range, feeding only on peas. Other legumes, including weeds, will not support growth and reproduction of this insect. Adults, particularly females, feed on pollen. If pea pollen is not available, weevils feed on pollen from Canada goldenrod, *Solidago canadensis*; yarrow, *Achillea millefolium*; dogfennel, *Anthemis cotula*; field mustard, *Brassica campestris*; grasses, and many other plants.

Natural Enemies. Several native parasitoids were found parasitizing eggs or larvae of pea weevil, including *Uscana* sp. (Hymenoptera: Trichogrammatidae), *Dinarmus laticeps* (Ashmead), *Microdontomerus* anthonomi (Crawford), *Anisopteromalus* sp., and *Eupteromalus leguminis* (Gahan) (all Hymenoptera: Pteromalidae), and *Eupelmus amicus Girault* (Hymenoptera: Eupelmidae). However, none were observed to be particularly effective in suppressing pea weevil abundance (Larson *et al.*, 1938; Annis and O'Keeffe, 1987). Therefore, attempts were made to introduce parasitoids from Europe. Despite repeated attempts at introduction in several areas of the United States and Canada, and short-term persistence of parasitoids, no significant benefit has accrued in pea weevil management (Clausen, 1978).

In addition to hymenopteran parasitoids, a few other natural enemies have been noted. A mite, *Atomus* sp. (Acari: Erythraeidae), was observed attacking pea weevil eggs in Idaho. Predators such as rove beetles (Coleoptera: Staphylinidae) and birds also feed on weevils, but the weevil's protected feeding site reduces the ability of general predators to access the larvae.

Life Cycle and Description.   There is usually only a single generation annually, with weevils overwintering in the adult stage. A few beetles in Oregon and Idaho were observed producing eggs in the autumn; some eggs were fertile and resulted in a complete second generation. Beetles invade pea fields in the spring and deposit eggs on green pods. The weevil completes its life cycle in the pea, the adults emerging in the summer or autumn to seek overwintering habitat. The mean (range) duration of the egg-adult developmental period in Idaho is about 57 (46–71) days.

Egg.   The egg is oval, measuring about 1.5 mm long and 0.6 mm wide. The eggs, which are yellowish to orange, are deposited singly or in small clusters of about two on the surface of the pea pod. Sometimes 30 or more eggs may be found on a single pod. The egg normally hatches in 8–9 days, with a range of 5–14 days.

Larva.   The young larva, on hatching, bores directly from the egg into the pod at the point of egg attachment. Occasionally, larvae construct mines in the tissue of the pea pod before entering. At this stage, the larva bears six thoracic legs and moves freely and also possesses a spiny thoracic plate. Both types of structures enable the larva to easily penetrate the pod and gain access to the pea seed. However, after burrowing into a pea seed, the whitish larva molts and loses these structures, assuming a legless form that is typical of weevil larvae. Duration of the larval stage is about 30–50 days, during which time it grows from about 1.5 mm to 6–7 mm long. There are four instars, with mean head capsule widths of 0.14, 0.35, 0.64, and

Pea weevil larva.

0.91 mm, respectively. Duration of each instar is about 12, 6–8, 12–14 and 10–14 days for instars 1–4, respectively. At larval maturity the insect has consumed most, or all, of the seed's contents, leaving a very thin layer of the seed coat in one area of the seed. It is through this thin area that the beetle will exit as an adult. More than one beetle larva may enter a pea seed, but only one survives to emerge.

Pupa.   After completing its larval development, the weevil pupates within the seed. The pupa closely resembles the adult in form and size, except that it is cream-colored. Duration of the pupal stage is about 15 days (range 8–27 days).

Adult.   The adult normally chews through the seed coat and escapes from the pea seed immediately after completing its development. Under cool and dry conditions, however, the beetle may remain within the seed during the winter and emerge in the spring. The beetle does not attack dry seeds, so it cannot multiply in stored seed. It may persist for 18–24 months within peas, so it is readily transported with dried peas. The adult weevil is brownish, but bears patches of white, gray, and black. There is a triangular white patch on the prothorax, and white patches form an irregular band beyond the middle of the elytra. The antennae and legs are black. The beetle is oval in shape and measures about 4.5–6.0 mm long and 2.0 mm wide. Most beetles seek shelter under bark of trees,

Pea weevil pupa.

or within crevices of fence posts, bales of hay, and organic debris on the soil surface. Pea weevil can tolerate winter temperatures of $-16°$ to $-19°C$, if not protected by snow cover or other shelter. When overwintering in protected locations, however, temperature as low as $-23°C$ is tolerated. Beetles become active at the time peas begin to bloom. Flights into fields occur when the temperature exceeds $21°C$, and may continue for a period of two months, usually in May and June. Beetles usually alight on reaching a pea field, so the field edges are most heavily infested. Adults do not feed on the pea plant except to consume pollen. The female must ingest pollen to develop a normal complement of eggs, and usually feeds for 4–5 days on pea pollen before beginning oviposition. She may deposit up to 50 eggs per day with an average fecundity of about 460 eggs (range 235–740 eggs). Few eggs are deposited at temperature below $21°C$, and maximum production occurs above $27°C$. Beetles take flight readily, and also feign death, when disturbed.

Pea weevil is easily confused with broadbean weevil, *Bruchus rufimanus* Boheman, as adults are very similar in appearance. In pea weevil, the terminal abdominal segment, when viewed from above, bears a pair of distinct black spots; in broadbean weevil these spots are lacking or poorly defined. In pea weevil, the posterior femora are equipped with sharp spines, and in broadbean weevil spines are absent from the hind femora or blunt.

A complete summary of pea weevil life history, based on work in the Pacific Northwest, was given by Larson *et al.* (1938). Brindley *et al.* (1946) provided a few additional observations. Information of physical ecology was given by Smith (1992).

Damage

Damage to peas is caused by larvae feeding within the pea seeds. Each larva destroys only a single pea,

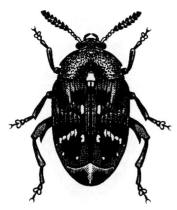

Adult pea weevil.

but contamination of peas by even a few larvae makes peas highly undesirable for human consumption. Although a lesser problem when grown for livestock food, pea weevil infestation lowers the weight and value of peas fed to livestock. Peas damaged by weevils display reduced germination potential, so they are undesirable for seed.

Management

Sampling. Sampling protocols for pea weevil eggs have been developed, including sequential sampling methods (Smith and Hepworth, 1992). Weevils selectively oviposit on the most exposed pods, and plants along field edges are more heavily infested. Normally, pods are picked and examined visually for eggs. Because the eggs are small and difficult to detect, the presence and absence are usually tabulated rather than attempting to count the eggs. Adults are best sampled by sweeping peas with a sweep net.

Insecticides. The principal form of crop protection is foliar insecticides applied when the crop is in bloom. Growers sometimes treat only the perimeter (about 50 m) of the field with insecticide because beetles typically land at field margins where they encounter a lethal dose of insecticide, but this is only appropriate for large fields. Persistent insecticides are more effective (Horne and Bailey, 1991). Fumigation of infested peas destined for seed is also recommended.

Cultural Practices. Several cultural techniques help to reduce incidence of pea weevil infestation. Destruction of crop residues that may contain weevils is perhaps the most important consideration. Infested seed should not be planted unless it is fumigated to kill overwintering beetles. Careful harvesting of peas is important, because if pea pods are shattered during harvest, a great number of seeds containing weevils may remain on the soil surface and produce beetles in the spring. It is helpful to plow under the residue of crops containing pea weevils, but they must be buried to a depth of at least 20 cm to provide marked reduction in emergence. Shattered peas also may germinate and produce crops of volunteer peas. Such volunteer plants, if not destroyed, may produce large numbers of insects.

Peas are also sometimes harvested for hay, and if they contain fairly mature pods the beetles contained within may serve as a source of inoculum. Thus, pea hay should be harvested while pods are very small, or fed to livestock soon after harvest, if peas for human consumption are also grown nearby.

Early planting, especially when accompanied by early harvesting, is desirable for pea weevil suppression. Such crops show a lower incidence of infestation, and larvae in early-maturing seeds are more juvenile, and thus more likely to be killed by the harvesting operation.

Host-Plant Resistance. Resistance is not well developed among commercially acceptable cultivars of pea. However, several sources of genetic resistance, which function mostly in deterring oviposition, have been identified for breeding purposes (Pesho *et al.*, 1977).

FAMILY CARABIDAE—GROUND BEETLES

Seedcorn Beetles
Stenolophus comma (Fabricius)
Stenolophus lecontei (Chaudoir)

Slender Seedcorn Beetle
Clivina impressifrons (LeConte)
(Coleoptera: Carabidae)

Natural History

Distribution. The seedcorn and slender seedcorn beetles are commonly found in the United States in the midwest and northeast, and occur west through the Great Plains and south to South Carolina. In Canada, they are present in eastern provinces, west to at least Ontario. They are native insects. The early accounts of the *Stenolophus* spp. are unreliable because the species were often confused; despite early reports suggesting that *S. lecontei* was the most important of the seedcorn beetles, it now appears that *S. comma* is the most serious pest.

Host Plants. Carabid beetles are known principally as predators of other insects. Animal matter also is the preferred food source of the seedcorn and slender seedcorn beetles. Eggs, larvae, and probably pupae of soil-dwelling insects are readily consumed. For example, Wyman *et al.* (1976) documented the reduction of cabbage maggot, *Delia radicum* (Linnaeus), populations when seedcorn beetles were numerous in crucifer plots.

As suggested by the common names, on occasion these beetles will injure germinating seeds, particularly corn seed. However, beetles also feed on weed seeds. Adults of *S. comma* feed readily on black nightshade, *Solanum nigrum*; crabgrass, *Digitaria sanguinalis*; foxtail, *Setaria glauca*; lambsquarters, *Chenopodium album*; and purslane, *Portulaca oleraceae*, and to a lesser degree on other seeds (Pausch, 1979).

Natural Enemies. The natural enemies of these insects are poorly known. The mite *Ovacarus clivinae*

(Stannard and Vaishampayan) develops inside slender seedcorn beetle, where it is associated with the reproductive system (Stannard and Vaishampayan, 1971).

Life Cycle and Description. These beetles are quite similar in biology. Adults overwinter and can be found throughout the year. Maximum aboveground activity in South Dakota was observed to occur during early summer, followed by almost total absence in late summer, and then reappearance in the autumn. Eggs are most abundant in the spring, larvae and then pupae occur in late spring and early summer, and new adults are produced by mid-summer. Only a single generation is produced annually. Development of the insects from egg to adult requires about 50 days.

Egg. The eggs are laid singly in horizontal tunnels dug about 2 cm deep in the soil. The eggs are white and oval in shape, measuring about 1 mm long and 0.5 mm wide. Mean duration of the egg stage is about 5.4 days (range of 3–12 days), when reared at about 22°C.

Larva. The larvae are yellowish, with a dark-brown head and prothoracic plate. The larvae display three instars, and all instars can be found during the summer months. They are normally found in burrows in the soil at depths of 2–3 cm. Mean duration (range) is reported 10.5 (4–19), 9.3 (5–16), and 12.8 (9–21) days for instars 1–3, respectively, when reared at about 22°C.

Pupa. Pupation occurs within the soil in small cells, each measuring about 13 mm in length and 5 mm in width. The pupa is dark brown or black. Mean duration of the pupal stage is 9.4 days, with a range of 7–13 days, when reared at 22°C.

Adult. The adults are small oblong beetles measuring 5–8 mm long. Adults of the *Stenolophus* spp. are dark below and yellowish brown or reddish brown above, but with the elytra blackened except for the margin. Slender seedcorn beetle differs in that it is entirely reddish brown and has especially enlarged and flattened front legs suitable for tunneling through soil. Adults are commonly found in moist soil, with a high content of organic matter. In the late autumn, beetles burrow beneath soil, rocks, and logs, where they remain in a small cell until the soil warms to about 5°C in the spring. Adults may fly during the daylight hours in the spring, but are more active in the evening during the summer months. Apparently, copulation occurs below-ground in small chambers constructed

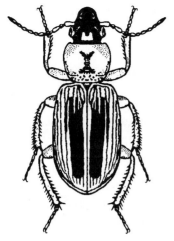

Adult seedcorn beetle, *Stenolophus coma*.

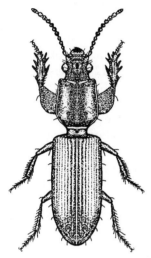

Adult slender seedcorn beetle.

by the adults during the spring. Oviposition commences within 14–15 days of emergence, and continues for 5–10 weeks.

The biology of *Stenolophus comma* was described by Kirk (1975); both *S. comma* and *S. lecontei* were considered by Pausch (1979). Slender seedcorn beetle was treated by Phillips (1909) and Pausch and Pausch (1980). These species were included in the keys to adult beetles by Downie and Arnett (1996).

Damage

Damage to germinating seeds caused by seedcorn beetles occurs principally during cool and wet weather in spring. Rarely is an ungerminated seed attacked. The contents of the seed are often consumed, leaving only the hull or seed coat. The result of attack is often a poor stand, as seedlings attacked in this manner may perish. However, if only a small portion of the seed is consumed, the seedling may recover and grow normally.

Management

The adult populations can be sampled with pitfall and blacklight traps. Rows of corn in which plants are missing should be examined carefully because beetles can often be found in association with the poorly germinating seeds. The populations of seed corn beetles can be controlled by application of insecticides to the seed, or in liquid or granular form to the soil at planting or shortly after planting (Daniels, 1977). Resistance to insecticides is evident in some populations (Sechriest *et al.*, 1971).

FAMILY CHRYSOMELIDAE, SUBFAMILY ALTICINAE—FLEA BEETLES

Cabbage Flea Beetle
Phyllotreta albionica (LeConte)
(Coleoptera: Chrysomelidae)

Natural History

Distribution. This native flea beetle is widely distributed in the western United States and Canada. It is found only as far east as Saskatchewan, Colorado, and New Mexico.

Host Plants. Cabbage flea beetle feeds principally on cruciferous plants. Vegetable crops attacked by this beetle include broccoli, Brussels sprouts, cabbage, cauliflower, kale, radish, rutabaga and turnip, but radish and turnip are most preferred. It also attacks alfalfa, sugarbeet, and tomato, but these events are unusual, or are cases of misidentification. Cabbage flea beetle feeds readily on cruciferous weeds.

Natural Enemies. Natural enemies are not well known, but *Microctonus epitricis* (Viereck) (Hymenoptera: Braconidae) has been reared from cabbage flea beetle.

Life Cycle and Description. There is a single generation of this insect in Canada, but its biology has not been studied in detail elsewhere. Adults overwinter in leaf litter along fencerows and in woodlots. They become active in March and April and begin feeding on crucifers. Mating soon ensues, and egg laying commences after a pre-oviposition period of about 5–10 days. Adults live for about three months, continually feeding and depositing more eggs.

Egg. The eggs are deposited beneath the host plants in the soil at a depth of about 2.5–5.0 cm. Clusters of 15–20 eggs are common, and females deposit eggs over a three-week period. Fecundity is estimated at about 60 eggs per female, but this is likely an underestimate owing to laboratory rearing conditions. Eggs hatch in 15–20 days.

Larva. Larvae are whitish, and are found feeding on the roots at depths of 5–15 cm. Larval development requires about four weeks, and can be found during May–July.

Pupa. As larvae are about to pupate, they characteristically move closer to the soil surface and create a small pupal chamber in the soil. The non-feeding prepupal period lasts about 10–12 days, and is followed by a pupal period of about 11 days.

Adult. The adults of the summer generation emerge from the soil in August and begin feeding. Cabbage flea beetle is a shining blackish species, measuring 1.5–1.9 mm long. This beetle is easily confused with crucifer flea beetle, *Phyllotreta cruciferae* (Goeze), a more damaging species. However, cabbage flea beetle has a bronze luster, whereas crucifer flea beetle has a blue luster. Also, the fifth antennal segment is elongate in the female, and elongate and broad in the male cabbage flea beetle. The legs of cabbage flea beetle are brown except for the femora, which are black. The hind femora are enlarged. The antennae are black basally and reddish black toward the tip. The summer adults feed for about six weeks before seeking overwintering shelter.

The biology of this flea beetle was given by Chittenden (1927), Burgess (1977, 1982), and Campbell *et al.* (1989).

Damage

Damage is caused principally by the adults in the form of small holes in the leaf surface. These beetles

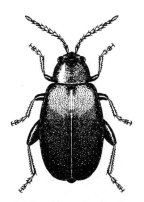

Adult cabbage flea beetle.

are rarely abundant enough to damage anything but seedling plants. The larvae feed on the roots of crucifers, but they are not usually considered a problem, probably in view of their low numbers.

Management

Management of this insect is similar to other crucifer-feeding flea beetles. Management has been discussed in detail in the section on crucifer flea beetle.

Corn Flea Beetle
Chaetocnema pulicaria Melsheimer

Toothed Flea Beetle
Chaetocnema denticulata (Illiger)
(Coleoptera: Chrysomelidae)

Natural History

Distribution. Corn flea beetle and toothed flea beetle are found primarily in the northeastern and midwestern United States, but their range extends as far west as the Rocky Mountains. The northern limit of these native species seems to be Massachusetts, New York, and southern Ontario. Survival is affected by winter weather, and damage resulting from flea beetle feeding is noticed most commonly following mild winters.

Host Plants. The adults and larvae of corn flea beetle and toothed flea beetle feed and develop on a wide variety of cultivated and wild grasses and sedges. In addition to corn, other crops attacked include barley, chufa, oats, orchard grass, wheat, and timothy grass. Examples of weeds suitable for these flea beetles are crabgrass, *Digitaria* spp.; barnyardgrass, *Echinochloa crusgalli*; witchgrass, *Panicum capillare*; straw-colored sedge, *Cyperus strigosus*; and yellow bristlegrass, *Setaria glauca*.

Natural Enemies. The natural enemies of corn flea beetle and toothed flea beetle are poorly known. Several investigators of corn flea beetle biology noted an absence of parasitoids. A wasp was found attacking toothed flea beetle eggs in Virginia, and identified as *Patasson pullicrura* (Girault) (Hymenoptera: Mymaridae) (Poos, 1955).

Life Cycle and Description. Corn flea beetle and toothed flea beetle are very similar in appearance and biology. Toothed flea beetle is much less common in corn, and so is less well studied. Except where noted below, their biology is believed to be virtually the same.

Two generations of these flea beetles are reported annually on corn in Connecticut and Virginia, with adults from the spring generation occurring in late June and early July, and those from the second brood appearing in mid-August. Poos (1955) suggested that in Virginia, corn flea beetle might complete a generation on wild grasses and sedges before moving to corn, leaving some doubt about the actual number of generations. The time required for completion of a generation (egg to adult) is estimated to average 30 days.

Overwintering of corn flea beetle and toothed flea beetle occurs in the adult stage in soil. Beetles are most commonly recovered from the upper 2–3 cm of grass sod planted in the vicinity of corn. Bluegrass, *Poa* sp., and other grasses have been found to be suitable habitats. The beetles become active when the soil surface warms to about 15–20°C.

Egg. Poos (1955) speculated that eggs were deposited at the base of corn or wild grasses in the manner of other flea beetles. However, he was unable to locate eggs in the field, and obtained them by confining beetles in the laboratory. The eggs of corn flea beetle are oval and white in color. They measure about 0.2 mm wide and 0.4 mm long. Development time of the egg stage averages 4–6 days, with a range of 2–10 days, depending principally on temperature. The eggs of toothed flea beetle are slightly larger, measuring about 0.6 mm long, and are pale yellow.

Larva. Poos found that larvae were also difficult to locate, but a few could be found in the soil surrounding corn and other grasses. Larval development requires about 16 days (range 10–23 days), not including an additional two days (range 1–5 days) as a pre-pupa. Larvae attain a maximum length of about 4.5 mm before pupation. The pupal stage requires about five days (range 3–7 days).

Adult. The adult corn flea beetle is shining black with a slight greenish, bluish or bronze luster. They measure about 1.5–1.8 mm long. Elytra are marked with rows of closely spaced punctures.

The adult toothed flea beetle is bronze in color, and measures 2.3–2.5 mm long. The elytra of this beetle also are marked with rows of punctures, but the punctures are not spaced closely. The head also bears punctures, a feature lacking from corn flea beetle. Although these two flea beetle species are not especially difficult to distinguish, there are several other flea beetle species that may occur in corn, usually in small numbers.

The biology of corn flea beetle and toothed flea beetle were given by Poos and Elliott (1936), Elliott and Poos (1940), and Poos (1955).

Adult corn flea beetle.

Damage

The adults skeletonize leaves of seedling corn in the spring, sometimes completely defoliating fields. They feed on the lower surface of foliage in narrow linear strips, usually restricting their feeding to the first three leaves.

The principal form of injury, however, is through transmission of Stewart's bacterial wilt, a disease caused by the bacterium *Erwinia stewartii*. Stewart's wilt is transmitted when beetles feed on corn, and defecate in or near the feeding site. The bacterium, which is harbored in the digestive tract of the beetles, thus gains entry to the plant through the wound. Infected beetles remain infected and are capable of transmitting the disease for the duration of their life. The disease is harbored by the beetle during the winter months, but the bacteria also occur in wild grasses, including many species that do not express symptoms of infection (Poos, 1939). The studies conducted in Connecticut demonstrated that the incidence of beetles containing bacteria generally increased from about 40% to 70% during the season, although there was some evidence of a mid-season reduction in disease levels (Heichel *et al.*, 1977). The incidence of disease in beetles is quite variable. Although both species may exhibit high levels of infection, corn flea beetle is much more abundant, and therefore is much more important as a vector. Elliott and Poos (1940) provided a long list of insects that were found contaminated with *E. stewartii*. Although the list is lengthy, incidence of infection was slight except for the aforementioned flea beetles.

The symptoms of Stewart's wilt infection in sweet corn includes wilting, pale green linear streaking, stunting, and death. Surviving plants may tassel prematurely and produce deformed ears. About three weeks are required before symptoms of infection are evident. Early season infection, or wilt phase of the disease, is most damaging to sweet corn. Field corn varieties are generally resistant to early infection, but suffer from late season infection. Stewart's wilt is a limiting factor in sweet corn production in the northeastern United States in some years.

Management

Sampling. A system to aid prediction of the severity of Stewart's wilt, based on winter temperature, was developed in Pennsylvania (Castor *et al.*, 1975). Mean monthly temperatures during December, January, and February, exceeding 3°C, favor adult overwintering survival. Beetle numbers can also be monitored during the growing season. Adams and Los (1986) found that yellow sticky traps hung close to the soil were most efficient in capturing corn flea beetles. Beetles can be counted on plants, and a threshold of six beetles per 100 plants is sometimes used to initiate control (Adams and Los, 1986). Hoffmann *et al.* (1995) studied the distribution of beetles between and within plants. Beetles tend to aggregate; about 50% are found on the uppermost of fully emerged leaf.

Insecticides. Systemic insecticides are commonly applied in-furrow at planting, and after emergence, to protect the young corn from feeding injury and transmission of Stewart's wilt (Ayers *et al.*, 1979; Munkvold *et al.*, 1996). For example, Ayers *et al.* (1979) reported that Stewart's wilt infection was reduced from about 34% in untreated corn plants to only 2–4% when effective insecticide was used. The insecticide carbofuran seems to interfere directly with the bacterium, imparting resistance to the plant (Sands *et al.*, 1979).

Cultural Practices. Sweet corn varieties differ in their susceptibility to Stewart's wilt. Some varieties are quite resistant, but if high, overwintering populations of corn flea beetles are forecast, insecticide treatment is also advised (Ayers *et al.*, 1979). Early maturing varieties tend to be more susceptible to injury.

Crucifer Flea Beetle
Phyllotreta cruciferae (Goeze)
Coleoptera: Chrysomelidae

Natural History

Distribution. Crucifer flea beetle was first found in North America in 1921, in British Columbia. It moved steadily eastward, and was recovered from Canada's Prairie Provinces in the 1940s, and Ontario and Quebec by 1954. A second introduction on the east coast is also possible, as this insect was found in Pennsylvania in 1943 and was well established in Delaware by 1951 (Westdal and Romanow, 1972). It is now widely distributed in southern Canada and most of the northern United States. Crucifer flea beetle is common in prairie and other open environments, and is rare in forested areas (Burgess, 1982). Crucifer flea beetle also is found in Europe, Africa, and Asia. The expansion of acreage planted to rape seed both in Europe and North America has facilitated the spread of this insect and enhanced its status as a crop pest.

Host Plants. Crucifer flea beetle feeds principally on plants in the family Cruciferae, though other plant families containing mustard oils (Capparidaceae, Tropaeolaceae, Limnanthaceae) have been reported to be attacked (Feeny *et al.*, 1970). Broccoli, Brussels sprouts, cabbage, cauliflower, Chinese cabbage, horseradish, kohlrabi, radish, rutabaga, and turnip are the vegetable crops commonly damaged by this flea beetle. Preference among crucifer crops is sometimes reported, but the preferences are rarely consistent owing to changes in leaf age and leaf type (Palaniswamy and Lamb, 1992). Crucifer flea beetle also damages rape, and sweet alyssum as well as several cruciferous weeds such as tansymustard, *Descurainia* sp.; wild mustard, *Brassica kaber*; stinkweed, *Thlasi arvense*; pepperweed, *Lepidium densiflorum*; yellow rocket, *Barbarea vulgaris*; and hoary cress, *Cardaria draba*. Reports of this insect feeding on non-cruciferous crop plants such as beets apparently stems from misidentification; these insects are easily confused with related species.

Natural Enemies. There are few effective natural enemies of crucifer flea beetle in North America. A parasitoid, *Microctonus vittatae* Muesebeck (Hymenoptera: Braconidae), attacks adult *P. cruciferae*, placing about two-thirds of its eggs in the host's head. Although several eggs may be deposited in each beetle, only a single wasp survives, emerging 16–19 days after parasitism (Wylie and Loan, 1984). The level of parasitism by *M. vittatae* may be 30–50% (Wylie, 1982), but crucifer flea beetle is less preferred than striped flea beetle, *Phyllophaga striolata* (Fabricius), for oviposition (Wylie, 1984). Parasitized-flea beetles emerge earlier from overwintering sites than unparasitized beetles (Wylie, 1982). Parasitism by nematodes, particularly by the allantonematid *Howardula* sp., is generally low, and the nematodes are not very pathogenic. In Europe, several other nematodes attack crucifer flea beetle (Morris, 1987). General predators such as lacewings (Neuroptera: Chrysopidae), soft-wing flower beetles (Coleoptera: Melyridae), and big-eyed bugs (Hemiptera: Lygaeidae), and omnivores such as field crickets (Orthoptera: Gryllidae), are occasionally observed feeding on adults, but their impact is undetermined (Gerber and Osgood, 1975; Burgess, 1980; Burgess and Hinks, 1987).

Life Cycle and Description. There is generally only one or possibly two generations annually, although development from egg to adult can be completed in just seven weeks. Overwintering occurs as an adult in soil and leaf litter. Fence rows, windbreaks, and other forms of shelter are thought to favor overwintering sites, although this does not seem to be active dispersal to such sites. Sometimes large aggregations of overwintering beetles are found; densities as high as two million beetles per hectare have been observed overwintering in a grove of trees (Burgess and Spurr, 1984; Turnock et al., 1987). Adults diapause under short day length (8 h) conditions, and readily survive five months of refrigeration. Peak emergence of overwintering adults was reported occurring by early May in Ontario (Kinoshita et al., 1979), with summer generation adults most abundant in late June and then again in late July. The two summer generations overlap considerably, and could easily be a protracted single generation. These authors also reported only a single summer generation during a cool summer.

Egg. The eggs are laid in the spring, singly or in groups of 3–4 in the soil, usually near the base of food plants. The temperature threshold for oviposition is 16.7°C. The egg measures 0.38–0.46 mm long and 0.18–0.25 mm wide, and is yellow in color. Eggs hatch after about 11–13 days.

Larva. There are three instars. Larvae are white except for a brown head and anal plate. Head capsule width measurements for the instars are 0.13, 0.17, and 0.26 mm, respectively. Body lengths are approximately 0.9, 4.5, and 6.7 mm, respectively. Development time is about five, three, and four days, respectively, at 25°C. Larvae feed on the root hairs of plants for 25–30 days at 20°C, and then form a small earthen pupal cell in the soil. The prepupal period, during which feeding ceases and the larval body shortens and thickens, lasts 3–6 days.

Pupa. The pupa is white, measuring about 2.4 mm long. Pupation lasts 7–9 days, followed by emergence of a whitish adult that darkens completely in about two days.

Adult. The adult measures about 2.2 mm long and is metallic blue-black except for the tarsi and antennae, which may be partly amber. The elytra, and to a lesser degree the head and thorax, bear small punctures. The hind femora are enlarged. The adults disperse principally by jumping, and are usually trapped within

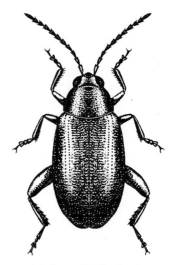

Adult crucifer flea beetle.

20 cm of the soil surface (Vincent and Stewart, 1983). They will fly throughout the season, however, and in the earlier part of the year they are more likely to be found flying at greater heights (1–2 m) (Lamb, 1983).

The biology of crucifer flea beetle was given by Westdal and Romanow (1972), Burgess (1977), and Kinoshita et al. (1979). Rearing procedures were provided by Kinoshita et al. (1979). Evidence for a male aggregation pheromone is presented by Peng and Weiss (1992).

Damage

The adult injury is in the form of small holes in the foliage. They do not eat completely through the leaf, but leave the lower epidermis intact. However, the remaining tissue soon dies, dries, and falls from the plant, producing a hole. When beetles are abundant, all leaf tissue may be riddled with holes, resulting in drying of adjacent tissue and death of emerging seedlings. Severe injury typically occurs in spring when the weather is hot and dry. Defoliation of young seedlings and small broccoli transplants in Manitoba at levels of less than 10% caused stand reductions of 5% (Soroka and Pritchard, 1987). Older plants, while rarely killed, may suffer reduced growth. Crucifer species differ in susceptibility to attack and ability to tolerate defoliation (Bodnaryk and Lamb, 1991a). Beetles also may feed on the young florets of broccoli, greatly reducing yield.

The larvae feed on roots. Although root hairs are eaten, roots are sometimes scarred or contain burrows, reducing the marketability of root crops such as radish and rutabaga (Kinoshita et al., 1978).

The ability of crucifer flea beetle to transmit *Xanthomonas campestris* pv. *campestris*, the causal agent of

Color Figures

There are three major components to the color figure section:

1. Damage by vegetable pests.

Insect damage is often characteristic of the pest causing the damage, and in some cases is quite diagnostic. A representative sampling of insect damage is shown in Figures 1–36.

2. The damaging forms of vegetable pests.

Figures 37–198 show some damaging forms of vegetable pests. These stages can do considerable injury to vegetables if they are numerous. These figures are more or less grouped by appearance, with armyworms and cutworms first through slugs and snails; otherwise there is no order to presentation.

3. The nondamaging forms of vegetable pests.

Not every stage of an insect is damaging, but often these nondamaging stages are indicative of potential damage. For example, cutworm moths cause no damage, but the presence of large numbers in light traps should stimulate careful monitoring of crops. In addition to some nondamaging adult forms (all moths and butterflies), representative egg and pupal forms of selected pests are shown in Figures 199–270.

1 Early stages of defoliation by the southern armyworm. Found on page 422 of the text.	**2** Severe defoliation or skeletonizing by the pickleworm. Found on page 485 of the text.	**3** Small holes and pits in leaves caused by palestriped flea beetle adults. Found on page 70 of the text.
4 Pitting of an asparagus spear (stem) by asparagus beetle adults. Found on page 91 of the text.	**5** Removal of leaf tissue from the lower bean-leaf surface by the Mexican bean beetle. Found on page 144 of the text.	**6** Removal of leaf tissue from a corn leaf by western corn rootworm adults. Found on page 113 of the text.
7 Serpentine leaf mine in a tomato leaf formed by a *Liriomyza* leafminer. Found on pages 196 and 202 of the text.	**8** Blotch leaf mine in corn caused by corn blotch leafminer larvae. Found on page 201 of the text.	**9** Webbing and defoliation caused by alfalfa webworm larvae. Found on page 457 of the text.
10 Discoloration or silverleaf in squash caused by the silverleaf whitefly. Found on page 283 of the text.	**11** Necrosis of squash leaf tissue caused by the squash bug. Found on page 244 of the text.	**12** Burning of leaf margins (hopperburn) caused by the potato leafhopper. Found on page 338 of the text.
13 Speckling along bean leaf veins caused by onion thrips feeding. Found on page 541 of the text.	**14** Speckling of onion foliage caused by onion thrips feeding. Found on page 541 of the text.	**15** Curling of onion foliage following heavy feeding by onion thrips. Found on page 541 of the text.
16 Speckling of a bean leaf caused by the garden fleahopper. Found on page 255 of the text.	**17** Leaf drop in beans caused by extensive feeding by melon thrips. Found on page 538 of the text.	**18** Leaf curl of a potato caused by potato psyllid feeding. (Photo by Whitney Cranshaw). Found on page 344 of the text.

19
Leaf curl of a beet caused by bean aphid feeding.
Found on page 289 of the text.

20
Leaf cupping of Brussels sprouts caused by the cabbage aphid.
Found on page 295 of the text.

21
Rolling of a bean leaf caused by the bean leafroller.
Found on page 364 of the text.

22
Blossom tissue removal of okra caused by the red imported fire ant.
Found on page 348 of the text.

23
Scaring or rindworm injury of a cantaloupe caused by the melonworm.
Found on page 484 of the text.

24
Internal fruit damage in a pepper caused by pepper weevil larvae.
Found on page 129 of the text.

25
Frass produced by a pickleworm larva and extruded from a tunnel in a pumpkin.
Found on page 486 of the text.

26
Cavity within a cucumber formed by a young pickleworm.
Found on page 486 of the text.

27
A tunnel within a squash vine formed by a squash vine borer larva.
Found on page 500 of the text.

28
Exit holes in squash blossoms following feeding by a pickleworm larvae.
Found on page 486 of the text.

29
Damage to a corn ear tip caused by grasshoppers.
Found on pages 513 and 524 of the text.

30
Fruit drop in pepper caused by the pepper weevil.
Found on page 129 of the text.

31
Kernel damage in corn caused by a western bean cutworm larva.
Found on page 439 of the text.

32
Pod injury in green beans caused by gray hairstreak larvae.
Found on page 366 of the text.

33
Wilting and leaf dieback in an onion caused by below-ground feeding of onion maggots.
Found on page 213 of the text.

34
Root clipping in corn (left) caused by western corn rootworm larvae.
Found on page 113 of the text.

35
Toppling (goose-necking) of cornstalks due to loss of roots from rootworm feeding.
Found on page 113 of the text.

36
Destruction of sweet potato tubers by sweetpotato weevil larvae.
Found on page 134 of the text.

37	38	39
Alfalfa looper larva *Autographa californica* (Lepidoptera: Noctuidae) Found on page 370 of the text.	Army cutworm larva *Euxoa auxiliaris* (Lepidoptera: Noctuidae) Found on page 372 of the text.	Armyworm larva *Pseudaletia unipunctata* (Lepidoptera: Noctuidae) Found on page 375 of the text.
40	41	42
Beet armyworm, young larva *Spodoptera exigua* (Lepidoptera: Noctuidae) Found on page 378 of the text.	Beet armyworm, mature larva *Spodoptera exigua* (Lepidoptera: Noctuidae) Found on page 378 of the text.	Bilobed looper larva *Megalographa biloba* (Lepidoptera: Noctuidae) Found on page 382 of the text.
43	44	45
Black cutworm larva *Agrotis ipsilon* (Lepidoptera: Noctuidae) Found on page 384 of the text.	Bronzed cutworm larva *Nephelodes minians* (Lepidoptera: Noctuidae) Found on page 387 of the text.	Cabbage looper larva *Trichoplusia ni* (Lepidoptera: Noctuidae) Found on page 390 of the text.
46	47	48
Celery looper larva *Anagrapha falcifera* (Lepidoptera: Noctuidae) Found on page 392 of the text.	Clover cutworm larva *Discestra trifolii* (Lepidoptera: Noctuidae) Found on page 394 of the text.	Corn earworm larva *Helicoverpa zea* (Lepidoptera: Noctuidae) Found on page 396 of the text.
49	50	51
Darksided cutworm larva *Euxoa messoria* (Lepidoptera: Noctuidae) Found on page 400 of the text.	Dingy cutworm larva *Feltia sp.* (Lepidoptera: Noctuidae) Found on page 402 of the text.	Fall armyworm larva *Spodoptera frugiperda* (Lepidoptera: Noctuidae) Found on page 405 of the text.
52	53	54
Glassy cutworm larva *Apamea devastator* (Lepidoptera: Noctuidae) Found on page 410 of the text.	Granulate cutworm larva *Agrotis subterranea* (Lepidoptera: Noctuidae) Found on page 411 of the text.	Okra caterpillar larva *Anomis erosa* (Lepidoptera: Noctuidae) Found on page 413 of the text.

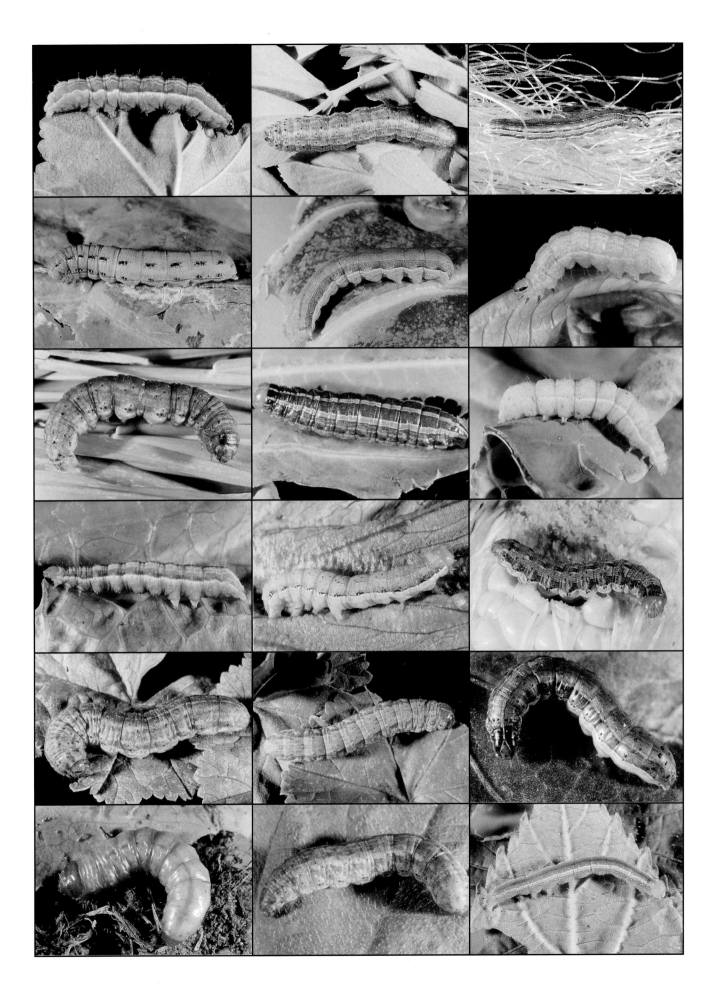

55 Pale western cutworm larva *Agrotis orthogonia* (Lepidoptera: Noctuidae) Found on page 415 of the text.	56 Southern armyworm larva *Spodoptera eridania* (Lepidoptera: Noctuidae) Found on page 422 of the text.	57 Soybean looper larva *Pseudoplusia includens* (Lepidoptera: Noctuidae) Found on page 424 of the text.
58 Spotted cutworm larva *Xestia* sp. (Lepidoptera: Noctuidae) Found on page 426 of the text.	59 Striped grass looper larva *Mocis latipes* (Lepidoptera: Noctuidae) Found on page 431 of the text.	60 Sweetpotato armyworm larva *Spodoptera dolichos* (Lepidoptera: Noctuidae) Found on page 432 of the text.
61 Tobacco budworm larva *Heliothis virescens* (Lepidoptera: Noctuidae) Found on page 434 of the text.	62 Velvet armyworm larva *Spodoptera latifascia* (Lepidoptera: Noctuidae) Found on page 432 of the text.	63 Variegated cutworm larva *Peridroma saucia* (Lepidoptera: Noctuidae) Found on page 437 of the text.
64 Western bean cutworm larva *Loxagrotis albicosta* (Lepidoptera: Noctuidae) Found on page 439 of the text.	65 Yellowstriped armyworm larva *Spodoptera ornithogalli* (Lepidoptera: Noctuidae) Found on page 441 of the text.	66 Zebra caterpillar larva *Melanchra picta* (Lepidoptera: Noctuidae) Found on page 443 of the text.
67 Alfalfa webworm larva *Loxostege cerealis* (Lepidoptera: Pyralidae) Found on page 458 of the text.	68 Beet webworm larva *Loxostege sticticalis* (Lepidoptera: Pyralidae) Found on page 460 of the text.	69 Cabbage webworm larva *Hellula rogatalis* (Lepidoptera: Pyralidae) Found on page 463 of the text.

70
Cross-striped cabbageworm
  larva
Evergestis rimosalis
(Lepidoptera: Pyralidae)
Found on page 466 of the text.
	71
	Diamondback moth larva (right)
	  and pupa (left)
	Plutella xylostella
	(Lepidoptera: Pyralidae)
	Found on page 468 of the text.
		72
		Southern beet webworm larva
		Herpetogramma bipunctalis
		(Lepidoptera: Pyralidae)
		Found on page 477 of the text.

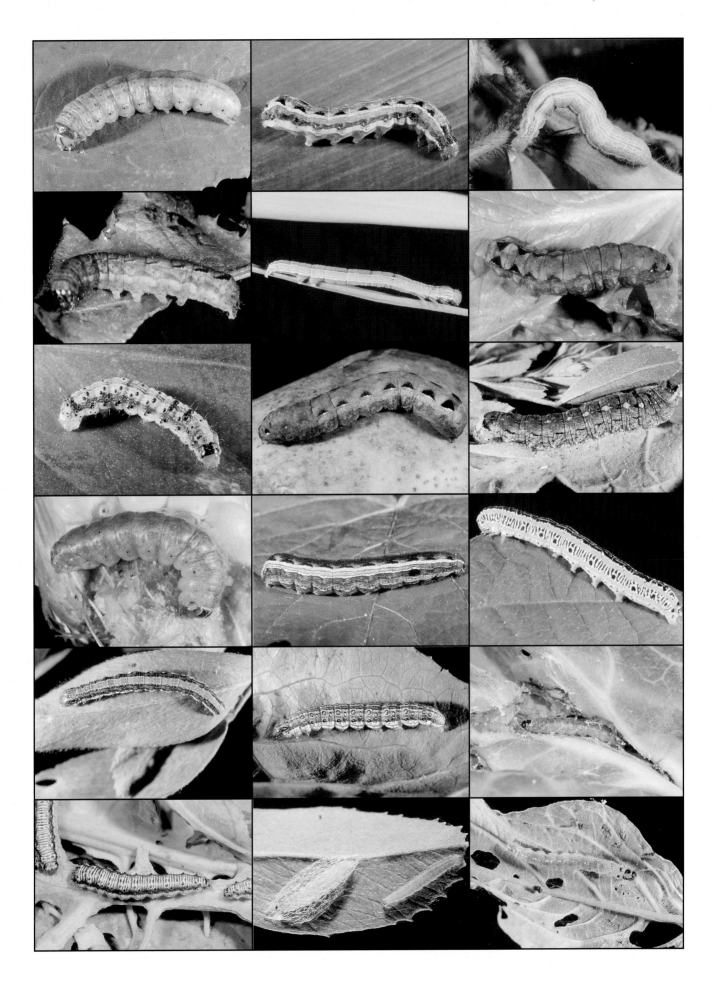

73
Hawaiian beet webworm larva
Spoladea recurvalis
(Lepidoptera: Pyralidae)
Found on page 477 of the text.

74
Pickleworm, mature larva
Diaphania nitidalis
(Lepidoptera: Pyralidae)
Found on page 486 of the text.

75
Pickleworm, immature larva
Diaphania nitidalis
(Lepidoptera: Pyralidae)
Found on page 486 of the text.

76
Melonworm larva
Diaphania hyalinata
(Lepidoptera: Pyralidae)
Found on page 484 of the text.

77
European corn borer larva
Ostrinia nubilalis
(Lepidoptera: Pyralidae)
Found on page 472 of the text.

78
Lesser cornstalk borer larva
Elasmopalpus lignosellus
(Lepidoptera: Pyralidae)
Found on page 480 of the text.

79
Limabean pod borer larva
Etiella zinckenella
(Lepidoptera: Pyralidae)
(Photo by Jim Castner)
Found on page 482 of the text.

80
Southwestern corn borer larva
Diatraea grandiosella
(Lepidoptera: Pyralidae)
Found on page 494 of the text.

81
Sugarcane borer larva
Diatrea saccharalis
(Lepidoptera: Pyralidae)
Found on page 497 of the text.

82
Tobacco hornworm larva
Manduca sexta
(Lepidoptera: Sphingidae)
Found on page 504 of the text.

83
Tomato hornworm larva
Manduca quinquemaculata
(Lepidoptera: Sphingidae)
(Photo by Paul Choate)
Found on page 504 of the text.

84
Whitelined sphinx larva
Hyles lineata
(Lepidoptera: Sphingidae)
Found on page 506 of the text.

85
Sweetpotato hornworm larva,
  green form
Agrius cingulatus
(Lepidoptera: Sphingidae)
Found on page 502 of the text.

86
Sweetpotato hornworm larva,
  brown form
Agrius cingulatus
(Lepidoptera: Sphingidae)
Found on page 502 of the text.

87
Saltmarsh caterpillar larva
Estigmene acrea
(Lepidoptera: Arctiidae)
Found on page 355 of the text.

88
Yellow woollybear larvae
Spilosoma virginica
(Lepidoptera: Arctiidae)
Found on page 357 of the text.

89
Squash vine borer larva
Melittia curcubitae
(Lepidoptera: Sesiidae)
Found on page 500 of the text.

90
Tomato pinworm larva
Keiferia lycopersicella
(Lepidoptera: Gelechiidae)
(Photo by Jim Castner)
Found on page 363 of the text.

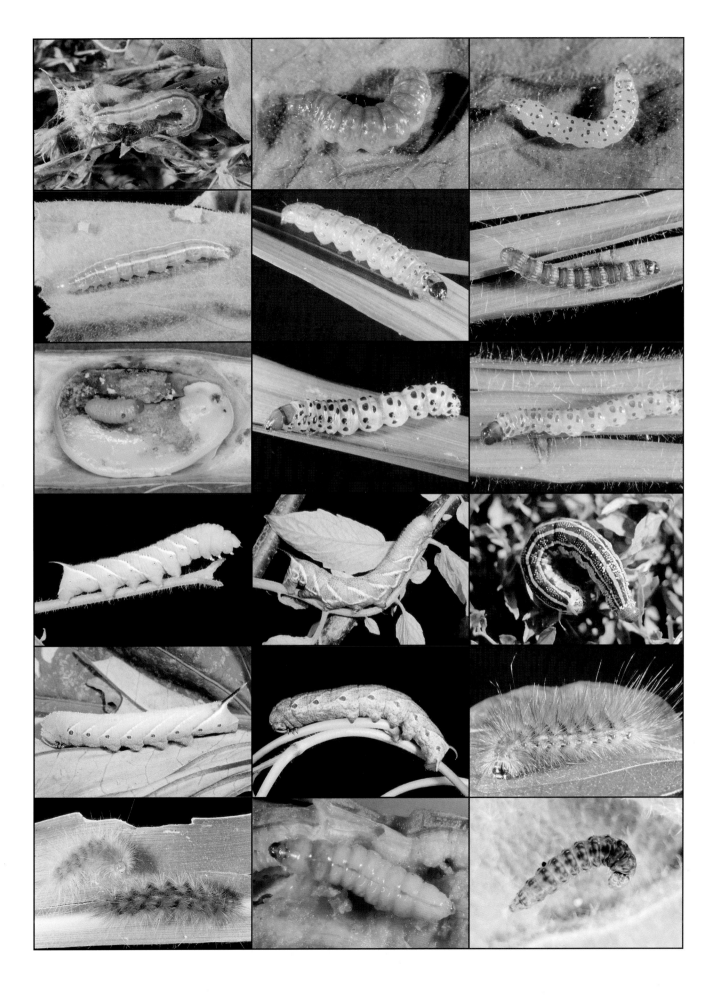

91 Morningglory leafminer larva *Bedellia somnulentella* (Lepidoptera: Lyonetiidae) Found on page 368 of the text.	92 Alfalfa caterpillar larva *Colias eurytheme* (Lepidoptera: Pieridae) Found on page 448 of the text.	93 Imported cabbageworm larva *Pieris rapae* (Lepidoptera: Pieridae) Found on page 450 of the text.
94 Southern cabbageworm larva *Pontia protodice* (Lepidoptera: Pieridae) (Photo by Jaret Daniels) Found on page 454 of the text.	95 Southern white larva *Ascia monuste* (Lepidoptera: Pieridae) (Photo by Jaret Daniels) Found on page 455 of the text.	96 Bean leafroller larva *Urbanus proteus* (Lepidoptera: Hesperiidae) Found on page 365 of the text.
97 Gray hairstreak larva *Strymon melinus* (Lepidoptera: Lycaenidae) Found on page 366 of the text.	98 Black swallowtail larva *Papilio polyxenes* (Lepidoptera: Papilionidae) Found on page 446 of the text.	99 Black blister beetle adult *Epicauta pensylvanica* (Coleoptera: Meloidae) Found on page 163 of the text.
100 Striped blister beetle adult *Epicauta vittata* (Coleoptera: Meloidae) Found on page 168 of the text.	101 Immaculate blister beetle adult *Epicauta immaculata* (Coleoptera: Meloidae) Found on page 165 of the text.	102 Spotted blister beetle adult *Epicauta maculata* (Coleoptera: Meloidae) Found on page 166 of the text.
103 Spotted asparagus beetle adult *Crioceris duodecimpunctata* (Coleoptera: Chrysomelidae) Found on page 106 of the text.	104 Asparagus beetle adult *Crioceris asparagi* (Coleoptera: Chrysomelidae) (Photo by Paul Choate) Found on page 91 of the text.	105 Asparagus beetle larva *Crioceris asparagi* (Coleoptera: Chrysomelidae) Found on page 90 of the text.
106 Colorado potato beetle adult *Leptinotarsa decemlineata* (Coleoptera: Chrysomelidae) Found on page 97 of the text.	107 Colorado potato beetle larva *Leptinotarsa decemlineata* (Coleoptera: Chrysomelidae) Found on page 97 of the text.	108 Western corn rootworm adult *Diabrotica virgifera* (Coleoptera: Chrysomelidae) Found on page 115 of the text.

109 Western corn rootworm larva *Diabrotica virgifera* (Coleoptera: Chrysomelidae) Found on page 114 of the text.	110 Spotted cucumber beetle adult *Diabrotica undecimpunctata* (Coleoptera: Chrysomelidae) Found on page 108 of the text.	111 Striped cucumber beetle adult *Acalymma vittatum* (Coleoptera: Chrysomelidae) Found on page 111 of the text.
112 Banded cucumber beetle adult *Diabrotica balteata* (Coleoptera: Chrysomelidae) Found on page 92 of the text.	113 Golden tortoise beetle adult *Charidotella bicolor* (Coleoptera: Chrysomelidae) Found on page 86 of the text.	114 Golden tortoise beetle larva *Charidotella bicolor* (Coleoptera: Chrysomelidae) Found on page 86 of the text.
115 Tobacco flea beetle adult *Epitrix hirtipennis* (Coleoptera: Chrysomelidae) (Photo by Jim Castner) Found on page 80 of the text.	116 Palestriped flea beetle adults *Systena blanda* (Coleoptera: Chrysomelidae) Found on page 70 of the text.	117 Threespotted flea beetle adult *Disonycha triangularis* (Coleoptera: Chrysomelidae) Found on page 75 of the text.
118 Yellowmargined leaf beetle larva *Microtheca ochroloma* (Coleoptera: Chrysomelidae) Found on page 117 of the text.	119 Yellowmargined leaf beetle adults *Microtheca ochroloma* (Coleoptera: Chrysomelidae) Found on page 117 of the text.	120 Green June beetle adult *Cotinus nitida* (Coleoptera: Scarabaeidae) Found on page 178 of the text.
121 June beetle (white grub adult) *Phyllophaga* sp. (Coleoptera: Scarabaeidae) Found on page 189 of the text.	122 White grub (June beetle larva) *Phyllophaga* sp. (Coleoptera: Scarabaeidae) Found on page 188 of the text.	123 Carrot beetle adult *Bothynus gibbosus* (Coleoptera: Scarabaeidae) Found on page 175 of the text.
124 Rose chafer adult *Macrodactylus subspinosus* (Coleoptera: Scarabaeidae) Found on page 185 of the text.	125 Japanese beetle adult *Popillia japonica* (Coleoptera: Scarabaeidae) Found on page 281 of the text.	126 Corn wireworm larva *Melanotus communis* (Coleoptera: Elateridae) Found on page 149 of the text.

127 Sweetpotato weevil adult *Cylas formicarius* (Coleoptera: Curculionidae) (Photo by Catharine Mannion) Found on page 134 of the text.	128 Vegetable weevil larva *Listroderes difficilis* (Coleoptera: Curculionidae) Found on page 136 of the text.	129 Mexican bean beetle adult *Epilachna varivestris* (Coleoptera: Coccinellidae) Found on page 145 of the text.
130 Mexican bean beetle larvae *Epilachna varivestris* (Coleoptera: Coccinellidae) Found on page 145 of the text.	131 Pepper weevil adult *Anthonomus eugenii* (Coleoptera: Curculionidae) Found on page 129 of the text.	132 Whitefringed beetle adult *Naupactus* sp. (Coleoptera: Curculionidae) Found on page 143 of the text.

133
West Indian sugarcane rootstalk
  borer weevil adult
Diaprepes abreviatus
(Coleoptera: Curculionidae)
Found on page 139 of the text.
	134
	Cowpea weevil adult
	Callosobruchus maculatus
	(Coleoptera: Bruchidae)
	(Photo by Jim Castner)
	Found on page 53 of the text.
		135
		Squash bug nymph (left) and
		  adult (right)
		Anasa tristis
		(Hemiptera: Coreidae)
		Found on page 245 of the text.

136 Leaffooted bug adult *Leptoglossus oppositus* (Hemiptera: Coreidae) Found on page 243 of the text.	137 Leaffooted bug adult *Leptoglossus phyllopus* (Hemiptera: Coreidae) Found on page 243 of the text.	138 Garden fleahopper, adult female *Halticus bractatus* (Hemiptera: Miridae) Found on page 256 of the text.
139 Garden fleahopper, adult male *Halticus bractatus* (Hemiptera: Miridae) Found on page 256 of the text.	140 Tarnished plant bug adult *Lygus lineolaris* (Hemiptera: Miridae) (Photo by USDA, ARS) Found on page 260 of the text.	141 Alfalfa plant bug nymph *Adelphocoris lineolatus* (Hemiptera: Miridae) (Photo by USDA, ARS) Found on page 253 of the text.
142 Chinch bug adult (left) and nymph (right) *Blissus leucopterus* (Hemiptera: Lygaeidae) (Photo by Jim Castner) Found on page 249 of the text.	143 Harlequin bug adult *Murgantia histrionica* (Hemiptera: Pentatomidae) (Photo by Jim Castner) Found on page 268 of the text.	144 Brown stink bug adult *Euschistus servus* (Hemiptera: Pentatomidae) Found on page 263 of the text.

145 Southern green stink bug adult *Nezara viridula* (Hemiptera: Pentatomidae) Found on page 274 of the text.	146 Say stink bug adult *Chlorochroa sayi* (Hemiptera: Pentatomidae) Found on page 272 of the text.	147 Potato aphids *Macrosiphum euphorbiae* (Homoptera: Aphididae) (Photo by Jim Castner) Found on page 321 of the text.
148 Pea aphids *Acyrthosiphon pisum* (Homoptera: Aphididae) Found on page 317 of the text.	149 Asparagus aphids *Brachycorynella asparagi* (Homoptera: Aphididae) Found on page 287 of the text.	150 Bird cherry-oat aphids *Rhopalosiphum padi* (Homoptera: Aphididae) (Photo by Paul Choate) Found on page 292 of the text.
151 Bean aphids *Aphis fabae* (Homoptera: Aphididae) Found on page 289 of the text.	152 Cowpea aphids *Aphis craccivora* (Homoptera: Aphididae) (Photo by Paul Choate) Found on page 303 of the text.	153 Green peach aphid *Myzus persicae* (Homoptera: Aphididae) (Photo by Paul Choate) Found on page 307 of the text.
154 Corn leaf aphids *Rhopalosiphum maidis* (Homoptera: Aphididae) Found on page 300 of the text.	155 Sugarbeet root aphids *Pemphigus* sp. (Homoptera: Aphididae) Found on page 325 of the text.	156 Turnip aphids *Lipaphis erysimi* (Homoptera: Aphididae) Found on page 327 of the text..
157 Cabbage aphids *Brevicoryne brassicae* (Homoptera: Aphididae) Found on page 295 of the text.	158 Melon aphids *Aphis gossypii* (Homoptera: Aphididae) Found on page 314 of the text.	159 Shortwinged mole cricket adult *Scapteriscus abbreviatus* (Orthoptera: Gryllotalpidae) (Photo by Paul Choate) Found on page 530 of the text.
160 Tawny mole cricket adult *Scapteriscus vicinus* (Orthoptera: Gryllotalpidae) (Photo by Paul Choate) Found on page 530 of the text.	161 Southern mole cricket adult *Scapteriscus borellii* (Orthoptera: Gryllotalpidae) (Photo by Paul Choate) Found on page 530 of the text.	162 Mormon cricket, adult male *Anabrus simplex* (Orthoptera: Tettigoniidae) Found on page 533 of the text.

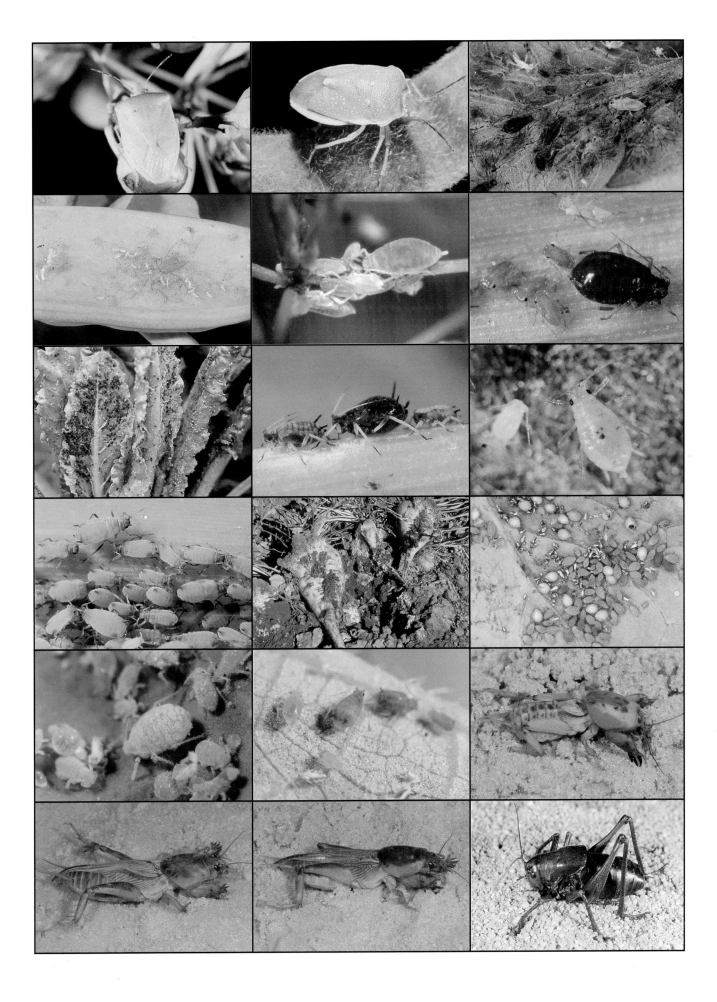

163
Redlegged grasshopper,
  adult male
Melanoplus femurrubrum
(Orthoptera: Acrididae)
Found on page 523 of the text.
	164
	Differential grasshopper,
	  adult male
	Melanoplus differentialis
	(Orthoptera: Acrididae)
	Found on page 514 of the text.
		165
		Migratory grasshopper,
		  adult male
		Melanoplus sanguinipes
		(Orthoptera: Acrididae)
		Found on page 519 of the text.

166
Twostriped grasshopper,
  adult male
Melanoplus bivittatus
(Orthoptera: Acrididae)
Found on page 525 of the text.
	167
	American grasshopper,
	  adult male
	Schistocerca americana
	(Orthoptera: Acrididae)
	Found on page 512 of the text.
		168
		American grasshopper nymphs
		Schistocerca americana
		(Orthoptera: Acrididae)
		Found on page 511 of the text.

169
Eastern lubber grasshopper,
  adult female
Romalea microptera
(Orthoptera: Acrididae)
Found on page 516 of the text.
	170
	Eastern lubber grasshopper
	  nymph
	Romalea microptera
	(Orthoptera: Acrididae)
	Found on page 515 of the text.
		171
		Field cricket, short-winged
		  female adult
		Gryllus sp.
		(Orthoptera: Gryllidae)
		(Photo by Paul Choate)
		Found on page 530 of the text.

172 Twospotted spider mites *Tetranychus urticae* (Acari) (Photo by Jim Castner) Found on page 556 of the text.	173 Springtail adult (Collembola) (Photo by Jim Castner) Found on page 561 of the text.	174 Silverleaf whitefly adults *Bemisia argentifolii* (Homoptera: Aleyrodidae) (Photo by Jim Castner) Found on page 284 of the text.

175
Silverleaf whitefly nymphs (left
  and right) and pupa (center)
Bemisia argentifolii
(Homoptera: Aleyrodidae)
(Photo by Jim Castner)
Found on page 283 of the text.
	176
	American serpentine leafminer
	  adult
	Liriomyza trifolii
	(Diptera: Agromyzidae)
	(Photo by Jeff Brushwein)
	Found on page 197 of the text.
		177
		Spinach leafminer larva
		Pegomya hyoscyami
		(Diptera: Anthomyiidae)
		Found on page 208 of the text.

178 Ringlegged earwig adult *Euborellia annulipes* (Dermaptera) Found on page 195 of the text.	179 Red imported fire ant worker *Solenopsis invicta* (Hymenoptera: Formicidae) Found on page 348 of the text.	180 Melon thrips *Thrips palmi* (Thysanoptera) Found on page 538 of the text.

181 Onion thrips *Thrips tabaci* (Thysanoptera) (Photo by Paul Choate) Found on page 541 of the text.	182 Flower thrips *Frankliniella* sp. (Thysanoptera) (Photo by Jim Castner) Found on page 547 of the text.	183 Potato psyllid nymph (left) and adult (right) *Paratrioza cockerelli* (Homoptera: Psyllidae) (Photo by Whitney Cranshaw) Found on page 345 of the text.
184 Sowbug *Porcellio* sp. (Isopoda) (Photo by Jim Castner) Found on page 565 of the text.	185 Pillbugs *Armadillidium* sp. (Isopoda) (Photo by Jim Castner) Found on page 565 of the text.	186 Veronicellid slug *Leidyula floridana* (Mollusca: Gastropoda: Veronicellidae) Found on page 567 of the text.

187
Gray garden slug
Deroceras reticulatum
(Mollusca: Gastropoda:
    Agriolimacidae)
Found on page 567 of the text.
	188
	Marsh slug
	Deroceras laeve
	(Mollusca: Gastropoda:
	    Agriolimacidae)
	Found on page 567 of the text.
		189
		Tawny garden slug
		Limax flavus
		(Mollusca: Gastropoda:
		    Limacidae)
		Found on page 567 of the text.

190
Brown garden snail
Helix aspersa
(Mollusca: Gastropoda:
    Helicidae)
Found on page 570 of the text.
	191
	Brown garden snail shells
	Helix aspersa
	(Mollusca: Gastropoda:
	    Helicidae)
	(Photo by Paul Choate)
	Found on page 570 of the text.
		192
		Giant African snail shells
		Achatina fulica
		(Mollusca: Gastropoda:
		    Achatinidae)
		(Photo by Paul Choate)
		Found on page 570 of the text.

193
Singing snail shells
Helix aperta
(Mollusca: Gastropoda:
    Helicidae)
(Photo by Paul Choate)
Found on page 570 of the text.
	194
	Roman snail shells
	Helix pomatia
	(Mollusca: Gastropoda:
	    Helicidae)
	(Photo by Paul Choate)
	Found on page 570 of the text.
		195
		Bradybaenid snail shells
		Bradybaena similaris
		(Mollusca: Gastropoda:
		    Bradybaenidae)
		(Photo by Paul Choate)
		Found on page 570 of the text.

196
Brown-lipped snail shells
Cepaea nemoralis
(Mollusca: Gastropoda:
    Helicidae)
(Photo by Paul Choate)
Found on page 570 of the text.
	197
	White-lipped snail shells
	Cepaea hortensis
	(Mollusca: Gastropoda:
	    Helicidae)
	(Photo by Paul Choate)
	Found on page 570 of the text.
		198
		Decollate snail shells
		Rumina decollata
		(Mollusca: Gastropoda:
		    Subulinidae)
		(Photo by Paul Choate)
		Found on page 570 of the text.

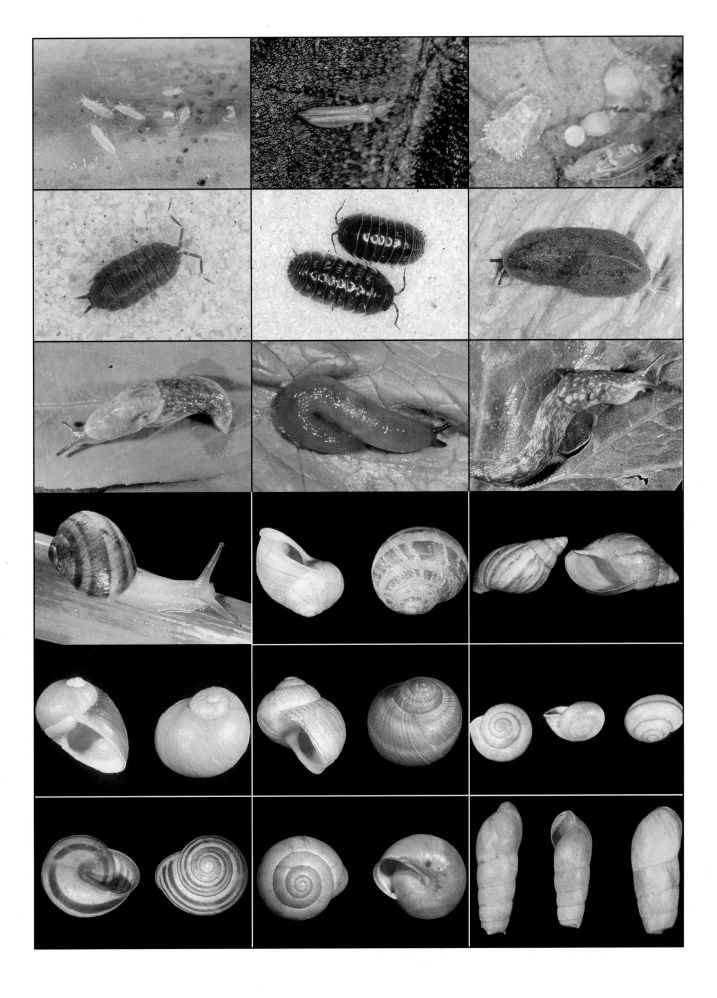

199 Alfalfa caterpillar adult *Colias eurytheme* (Lepidoptera: Pieridae) Found on page 449 of the text.	200 Imported cabbageworm adult *Pieris rapae* (Lepidoptera: Pieridae) (Photo by Jim Castner) Found on page 451 of the text.	201 Southern white adult *Ascia monuste* (Lepidoptera: Pieridae) (Photo by Jaret Daniels) Found on page 455 of the text.
202 Bean leafroller adult *Urbanus proteus* (Lepidoptera: Hesperiidae) (Photo by Jim Castner) Found on page 365 of the text.	203 Gray hairstreak adult *Strymon melinus* (Lepidoptera: Lycaenidae) Found on page 367 of the text.	204 Black swallowtail adult *Papilio polyxenes* (Lepidoptera: Papilionidae) Found on page 446 of the text.
205 Alfalfa webworm adult *Loxostege cerealis* (Lepidoptera: Pyralidae) Found on page 459 of the text.	206 Beet webworm adult *Loxostege sticticalis* (Lepidoptera: Pyralidae) Found on page 461 of the text.	207 Cross-striped cabbageworm adult *Evergestis rimosalis* (Lepidoptera: Pyralidae) Found on page 466 of the text.
208 European corn borer, adult male *Ostrinia nubilalis* (Lepidoptera: Pyralidae) Found on page 472 of the text.	209 European corn borer, adult female *Ostrinia nubilalis* (Lepidoptera: Pyralidae) Found on page 472 of the text.	210 Hawaiian beet webworm adult *Spoladea recurvalis* (Lepidoptera: Pyralidae) Found on page 478 of the text.
211 Melonworm adult *Diaphania hyalinata* (Lepidoptera: Pyralidae) Found on page 484 of the text.	212 Pickleworm adult *Diaphania nitidalis* (Lepidoptera: Pyralidae) Found on page 486 of the text.	213 Sod webworm adult *Crambus* sp. (Lepidoptera: Pyralidae) Found on page 490 of the text.
214 Southern beet webworm adult *Herpetogramma bipunctalis* (Lepidoptera: Pyralidae) Found on page 478 of the text.	215 Southwestern corn borer adult *Diatraea grandiosella* (Lepidoptera: Pyralidae) Found on page 494 of the text.	216 Saltmarsh capterpillar, adult female *Estigmene acrea* (Lepidoptera: Arctiidae) Found on page 356 of the text.

217 Tobacco hornworm adult *Manduca sexta* (Lepidoptera: Sphingidae) Found on page 505 of the text.	218 Tomato hornworm adult *Manduca quinquemaculata* (Lepidoptera: Sphingidae) Found on page 505 of the text.	219 Whitelined sphinx adult *Hyles lineata* (Lepidoptera: Sphingidae) Found on page 507 of the text.
220 Army cutworm adult, striped form *Euxoa auxiliaris* (Lepidoptera: Noctuidae) Found on page 372 of the text.	221 Army cutworm adult, brown form *Euxoa auxiliaris* (Lepidoptera: Noctuidae) Found on page 372 of the text.	222 Alfalfa looper adult *Autographa californica* (Lepidoptera: Noctuidae) Found on page 370 of the text.
223 Alfalfa looper adult *Autographa californica* (Lepidoptera: Noctuidae) Found on page 370 of the text.	224 Armyworm adult *Pseudaletia unipunctata* (Lepidoptera: Noctuidae) Found on page 375 of the text.	225 Beet armyworm adult *Spodoptera exigua* (Lepidoptera: Noctuidae) Found on page 379 of the text.
226 Bertha armyworm adult *Mamestra configurata* (Lepidoptera: Noctuidae) Found on page 381 of the text.	227 Bilobed looper adult *Megalographa biloba* (Lepidoptera: Noctuidae) Found on page 383 of the text.	228 Black cutworm adult *Agrotis ipsilon* (Lepidoptera: Noctuidae) Found on page 385 of the text.
229 Bronzed cutworm adult *Nephelodes minians* (Lepidoptera: Noctuidae) Found on page 387 of the text.	230 Cabbage looper adult *Trichoplusia ni* (Lepidoptera: Noctuidae) Found on page 390 of the text.	231 Cabbage looper adult *Trichoplusia ni* (Lepidoptera: Noctuidae) Found on page 390 of the text.
232 Celery looper adult *Anagrapha falcifera* (Lepidoptera: Noctuidae) Found on page 393 of the text.	233 Celery looper adult *Anagrapha falcifera* (Lepidoptera: Noctuidae) Found on page 393 of the text.	234 Clover cutworm adult *Discestra trifolii* (Lepidoptera: Noctuidae) Found on page 394 of the text.

235 Corn earworm adult *Helicoverpa zea* (Lepidoptera: Noctuidae) Found on page 397 of the text.	236 Darksided cutworm adult *Euxoa messoria* (Lepidoptera: Noctuidae) Found on page 401 of the text.	237 Dingy cutworm adult *Feltia* sp. (Lepidoptera: Noctuidae) Found on page 403 of the text..
238 Fall armyworm, adult male *Spodoptera frugiperda* (Lepidoptera: Noctuidae) Found on page 406 of the text.	239 Fall armyworm, adult male *Spodoptera frugiperda* (Lepidoptera: Noctuidae) Found on page 406 of the text.	240 Fall armyworm, adult female *Spodoptera frugiperda* (Lepidoptera: Noctuidae) Found on page 406 of the text.
241 Glassy cutworm adult *Apamea devastator* (Lepidoptera: Noctuidae) Found on page 410 of the text.	242 Granulate cutworm adult *Agrotis subterranea* (Lepidoptera: Noctuidae) Found on page 412 of the text.	243 Granulate cutworm adult *Agrotis subterranea* (Lepidoptera: Noctuidae) Found on page 412 of the text.
244 Green cloverworm adult *Plathypena scabra* (Lepidoptera: Noctuidae) Found on page 409 of the text.	245 Okra caterpillar adult *Anomis erosa* (Lepidoptera: Noctuidae) Found on page 414 of the text.	246 Pale western cutworm adult *Agrotis orthogonia* (Lepidoptera: Noctuidae) Found on page 415 of the text.
247 Redbacked cutworm adult *Euxoa ochrogaster* (Lepidoptera: Noctuidae) Found on page 421 of the text.	248 Spotted cutworm adult *Xestia* sp. (Lepidoptera: Noctuidae) Found on page 427 of the text.	249 Sweetpotato armyworm adult *Spodoptera dolichos* (Lepidoptera: Noctuidae) Found on page 433 of the text.
250 Variegated cutworm adult *Peridroma saucia* (Lepidoptera: Noctuidae) Found on page 438 of the text.	251 Western bean cutworm adult *Loxagrotis albicosta* (Leidoptera: Noctuidae) Found on page 439 of the text.	252 Yellowstriped armyworm adult *Spodoptera ornithogalli* (Lepidoptera: Noctuidae) Found on page 442 of the text.

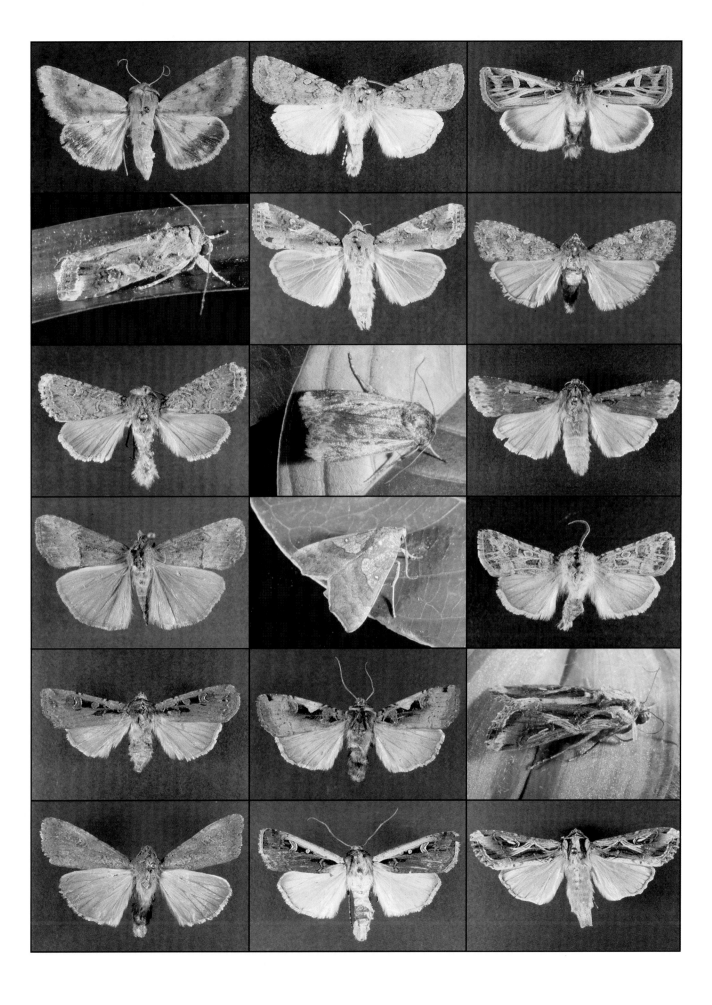

253 Zebra caterpillar adult *Melanchra picta* (Lepidoptera: Noctuidae) Found on page 443 of the text.	254 Squash vine borer adult *Melittia cucurbitae* (Lepidoptera: Sesiidae) Found on page 500 of the text.	255 Western bean cutworm egg cluster *Loxagrotis albicosta* (Lepidoptera: Noctuidae) Found on page 439 of the text.
256 Yellow woollybear egg cluster *Spilosoma virginica* (Lepidoptera: Noctuidae) Found on page 357 of the text.	257 Melonworm eggs and newly hatched larva *Diaphania hyalinata* (Lepidoptera: Pyralidae) (Photo by Rita Duncan) Found on page 484 of the text.	258 Alfalfa webworm egg clusters *Loxostege cerealis* (Lepidoptera: Pyralidae) Found on page 458 of the text.
259 Mexican bean beetle egg cluster *Epilachna varivestris* (Coleoptera: Coccinellidae) Found on page 145 of the text.	260 Asparagus beetle egg cluster *Crioceris asparagi* (Coleoptera: Chrysomelidae) Found on page 90 of the text.	261 West Indian sugarcane rootstalk borer weevil egg cluster *Diaprepes abbreviatus* (Coleoptera: Curculionidae) Found on page 138 of the text.
262 Squash bug egg cluster *Anasa tristis* (Hemiptera: Coreidae) Found on page 245 of the text.	263 Beet leafminer egg cluster *Pegomya betae* (Diptera: Anthomiidae) Found on page 207 of the text.	264 Mormon cricket eggs *Anabrus simplex* (Orthoptera: Tettigoniidae) Found on page 533 of the text.
265 Corn earworm pupa *Helicoverpa zea* (Lepidoptera: Noctuidae) Found on page 397 of the text.	266 Bilobed looper pupa *Megalographa biloba* (Lepidoptera: Noctuidae) Found on page 382 of the text.	267 Imported cabbageworm pupa *Pieris rapae* (Lepidoptera: Pieridae) Found on page 451 of the text.
268 Tobacco hornworm pupa *Manduca sexta* (Lepidoptera: Sphingidae) Found on page 504 of the text.	269 Sweetpotato hornworm pupa *Agrius cingulatus* (Lepidoptera: Sphingidae) Found on page 502 of the text.	270 Mexican bean beetle pupa *Epilachna varivestris* (Coleoptera: Coccinellidae) Found on page 145 of the text.

black rot of crucifers, was investigated by Shelton and Hunter (1985). Although the potential to transmit this bacterium mechanically was demonstrated, there was little evidence that the beetle was an efficient or important vector of this disease.

Management

Sampling. Allyl isothiocyanate is attractive to crucifer-feeding flea beetles, especially prior to host plant emergence, and can be used as a lure for various traps (Burgess and Wiens, 1980, Pivnick *et al.*, 1992). Adults also are collected by sweep net, and the other stages separated from soil by flotation (Burgess, 1977). Yellow water pan or sticky traps are sometimes useful for monitoring beetle populations, but day-to-day variation in trap catches reduces the value of such observations (Lamb, 1983).

Insecticides. Multiple foliar applications to young plants, or granular applications to soil, may be necessary for good control of flea beetles. Increases in stand density and yield commonly result from protection of seedlings with granular insecticides (Reed and Byers, 1981). Insecticides are commonly incorporated into planting water to afford protection of young transplants (Kinoshita *et al.*, 1978). There is some evidence of incipient resistance in some Canadian populations (Turnock and Turnbull, 1994). Neem products, at high concentrations, induce flea beetle mortality and provide short-term feeding deterrent effects (Palaniswamy and Wise, 1994).

Biological Control. Although some *Phyllotreta* spp. are susceptible to the steinernematid nematode *Steinernema carpocapsae* under laboratory conditions, this does not appear as a viable option for crucifer flea beetle management. Morris (1987) studied susceptibility in the field, and concluded that the beetle larvae may be too small for ready entry by infective juvenile nematodes.

Cultural Practices. Various cultural manipulations have been investigated for avoidance of crucifer flea beetles or their damage. In the studies conducted in Virginia, intercropping collards with nonhost plants resulted in lower beetle numbers and damage levels relative to collards monoculture, but collards yield was not increased owing to plant competition in polycultures (Latheef *et al.*, 1984). Similar results were observed in North Dakota with canola-field pea intercrops (Weiss *et al.*, 1994). However, in California, slight benefit was noted from similar polycultural systems (Gliessman and Altieri, 1982). Larger plot size tends to result in higher beetle densities owing to random dispersal from smaller plots (Kareiva, 1985). Dispersal rates are higher from dicultures than monocultures (Elmstrom *et al.*, 1988).

Mustard, a highly preferred food plant, can be interplanted or planted as a border to reduce beetle densities on other susceptible crops (Kloen and Altieri, 1990). Companion herbs, often suggested to confer benefit to adjacent crops, differ in their effect on crucifer flea beetle. Catnip, southernwood, tansy, and wormwood reduced beetle densities, whereas hyssop and santolina did not affect beetle numbers on collards (Latheef and Ortiz, 1984).

The role of weeds in flea beetle management is variable. Traditionally, weeds have been viewed as alternate hosts that favor survival of beetles when crops are not available. If the crop is not yet available as a food source, destruction of weeds can deprive flea beetles of food, leading to their demise. However, weeds also confer a benefit by attracting beetles from the crop plants, thereby reducing beetle density and damage to crops (Altieri and Gliessman, 1983), especially if the weeds are more preferred than crops. Reduced tillage practices, which leave more weeds, thereby may result in reduced flea beetle damage (Reed and Byers, 1981; Milbrath *et al.*, 1995).

Early season plantings tend to have more severe flea beetle problems, so delayed planting is recommended if constraints such as season length and market conditions will allow this practice (Milbrath *et al.*, 1995). Direct seeding of broccoli, rather than the use of transplants, necessitates the use of granular insecticides to protect the young seedlings (Soroka and Pritchard, 1987). Plants growing from small seeds are less tolerant to flea beetle defoliation than those growing from large seeds (Bodnaryk and Lamb, 1991b).

Crucifer cultivars differ in the level of waxy bloom present on the leaf surface. Low-wax or glossy lines offer promise for reduction in caterpillar numbers, owing both to less preferred oviposition by adults and lower survival of young larvae on glossy host plants. However, such glossy varieties tend to support higher numbers of flea beetles (Stoner, 1990, 1992). Wild crucifers, with high densities of leaf trichomes, are resistant to feeding by crucifer flea beetles, but this character has not yet been adequately incorporated into commercial cultivars (Palaniswamy and Bodnaryk, 1994).

Desert Corn Flea Beetle
Chaetocnema ectypa Horn
(Coleoptera: Chrysomelidae)

Natural History

Distribution. Desert corn flea beetle is western in distribution. It is generally considered damaging only in the southwestern states of New Mexico, Arizona, and California. However, it has apparently spread north to Canada, where Beirne (1971) reported it from British Colombia starting about 1949, and where it occasionally causes severe damage to corn.

Host Plants. Desert corn flea beetle is known principally as a pest of corn, but it readily attacks other grass crops such as barley, sorghum, sudangrass, sugarcane, and wheat. Occasionally, it has been reported to feed on non-grass crops such as alfalfa, bean, cantaloupe, and sugarbeet. Native and weedy grasses are suitable hosts, and include such species as wild barley, *Hordeum leporinum*; saltgrass, *Distichlis spicata*; Johnsongrass, *Sorghum halepense*; and hairgrass dropseed, *Sporobolus airoides*. Wildermuth (1917) reported that the adults could be found feeding on the aboveground plant parts, and larvae on the roots, whenever the plants were actively growing.

Natural Enemies. Natural enemies are poorly known. Wildermuth (1917) found some general predators such as ground beetle larvae (Coleoptera: Carabidae), and parasites such as a mite, *Pymotes* sp. (Acari: Tarsonemidae) and a wasp, *Neurepyris* sp. (Hymenoptera: Bethylidae) attacking corn flea beetle in Arizona, but their importance is uncertain.

Life Cycle and Description. Desert corn flea beetle has been reported to have three generations regularly in Arizona, with part of the population undergoing a fourth generation. First generation adults, which result from the reproduction of the overwintering beetles, appear in early June, followed by the second generation adults in mid-July and the third in mid-September. Generations overlap considerably and all stages are present through most of the year. The time required for a complete generation is about 46 days, though this may vary from 31 to 79 days depending on weather.

Egg. The eggs are about 0.35 mm long and 0.15 mm wide and whitish. They are unusual among flea beetle eggs in that they are bean shaped. Eggs are deposited at or near the surface of the soil, sometimes being attached below-ground to the plant stem. They usually are deposited singly, but groups of up to six eggs have been observed. The eggs require, on average, 5.8 days to hatch. However, the hatching time varies from as little as three days in the summer to 15 days in the spring.

Larva. The larvae initially are only about 0.8 mm in length, but eventually grow to about 4 mm. They are whitish, although the head capsule and anal plate are straw-colored. The larval development time is about 32 days, but it may vary from 20–47 days depending on temperature. The mature larva, as it pre-

Desert corn flea beetle egg.

Desert corn flea beetle larva.

pares for pupation, creates a small cell in the soil. The larva becomes thicker in form and shortens to about one-half its former length prior to pupation. The prepupal period typically is about 3–4 days.

Pupa. The pupa resembles an adult beetle in appearance, but initially it is white and gradually turns brown. The pupa measures about 1.5–2.0 mm long. Duration of the pupal stage averages 5.6 days (range 3–12 days). Pupation occurs near the soil surface under wet conditions, but as deep as 10 cm in dry soil.

Adult. The adult flea beetles are shining black, measuring 1.5–2.0 mm long. The adult bears parallel rows of small punctures on the elytra. The hind femora are greatly enlarged. Beetles are mature and copulation commences about 4–6 days after emergence. Egg deposition begins about 4–5 days after copulation. Adults commonly live a month or more, and produce 40–80 eggs. Desert corn flea beetle apparently overwinters above-ground, among organic debris and weedy grasses. The latter is important as a food source when the weather warms. However, the number of adults observed above-ground during the winter months is very low, so the possibility of some beetles overwintering in the soil should not be ruled out.

Damage

As is the case with most flea beetles, adults feed on foliage, while larvae attack the roots. Adults tend to create long narrow feeding strips, caused by removing epidermis between the parallel veins of grasses. They do not usually eat completely through the leaf. The

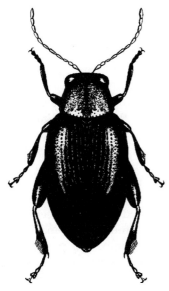

Adult desert corn flea beetle.

time and site of feeding varies. During cool weather, the beetles are most active during the heat of the day, but in hot weather feeding is limited to cooler periods. Beetles often work their way down to the growing point of the grasses and feed on the most tender young tissue.

The larvae consume rootlets completely, but create grooves in, or burrow within, larger roots. Sometimes the entire root system is tunneled, or the roots entirely severed. Early planted corn is especially susceptible to injury.

Management

This insect is not considered as a major pest, and planting-time applications of insecticide to the soil, or foliar applications post-planting, usually prevent or remedy problems. However, elimination of dense stands of weedy grasses such as Johnsongrass and saltgrass will deprive the insects of alternate hosts and do much to alleviate problems. Damage by larvae can be minimized with adequate irrigation.

Eggplant Flea Beetle
Epitrix fuscula (Crotch)
(Coleoptera: Chrysomelidae)

Natural History

Distribution. Eggplant flea beetle is widely distributed in the United States. It is found from the east coast to the west coast except in northern areas. It is found only as far north as New Jersey, Illinois,

Nebraska, and Utah. This native species is not known as a pest in Canada.

Host Plants. Eggplant flea beetle feeds on solanaceous plants such as eggplant and potato. The host range of eggplant flea beetle is poorly known, but likely is equivalent to that of potato flea beetle, *Epitrix cucumeris* (Harris). Jewett (1929) reported eggplant flea beetle to be more abundant on potato in Kentucky than potato flea beetle.

Natural Enemies. The natural enemies of eggplant flea beetle are poorly studied, but should be equivalent or identical to those of potato flea beetle.

Life Cycle and Description. There are two generations annually in Kentucky, with overwintering occurring in the adult stage. Overwintering occurs in the soil. A complete generation, egg to adult, is estimated to require 30–45 days.

Adults emerge in the spring, usually April in Indiana and early May in Kentucky, and soon begin egg production.

Egg. The eggs are deposited in the soil beneath host plants, usually at a depth of 1–2 cm. They are laid singly, but sometimes small clusters are produced. The egg is elliptical in shape and whitish. The egg measures 0.40–0.45 mm in long and 0.21–0.23 mm wide. Eggs hatch in 6–8 days.

Larva. Upon hatching, the larva measures about 1 mm long, but eventually attains a length of 3.5–4.5 mm. The larva is whitish with a brown head and brownish anal plate. The larva bears anal prolegs. The larvae feed on roots and rootlets, but also burrow into the tissues, creating surface scars and tunnels. Their development time varies considerably, depending on the weather, but generally averages about 20 days (range 17–25 days). There are three instars, with development times of about 4, 9.5, and 7.1 days, respectively.

Pupa. During the last 1–4 days of the larval stage, the larva shortens, thickens, and forms a pupal cell in the soil. Pupation occurs at depths ranging from less than 1.0 cm to 7.5 cm. Initially, the pupa, which resembles the adult in form, is white. It darkens considerably, however, before emergence. Duration of pupation averages about 6 days (range 4–8 days). The adult may remain in the pupal cell for 2–3 days before escaping.

Adult. The adult is rather oval and measures about 2 mm long. The body is black and densely covered with short hairs. The antennae and legs are reddish-yellow, except for the black femora. The hind femora

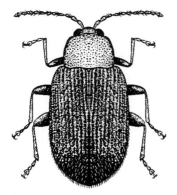

Adult eggplant flea beetle.

are enlarged. The pronotum is densely marked with large, deep punctures, and the elytra with rows of fine punctures. This beetle is quite similar to potato flea beetle, but eggplant flea beetle can be distinguished using the following characters: the elytra of eggplant flea beetle are more densely hairy, the transverse depression at the posterior of the pronotum is not pronounced, and the femora of the front and middle legs are completely black. Overwintering and summer adults can live for more than two months. Feeding commences soon after emergence from the soil, and egg production begins two weeks later. In Kentucky, overwintering beetles produced eggs from mid-May to mid-June, whereas first brood adults produced eggs from mid-July to mid-August. Second brood beetles do not produce eggs. As the weather becomes cold, beetles aggregate under organic litter and eventually dig into the soil and enter diapause until spring.

The biology of eggplant flea beetle was given by Jewett (1929).

Damage

The adults chew small holes in the leaves. They feed on either the upper- or lower-leaf surface, but the latter is preferred. Adjacent tissue, while not eaten, often dies. The leaves acquire a rusty or burned appearance when heavily damaged. When attacked early in the growth of the plant, stunting results.

The larvae feed belowground on rootlets, roots, and tubers. Severe root pruning may occur. Damage to potato tubers, which is typically in the form of pitting or roughening, resembles damage by potato flea beetle and tuber flea beetle.

Management

Management of eggplant flea beetle, which has not been well studied, should be equivalent to potato flea beetle.

Hop Flea Beetle
Psylliodes punctulata (Melsheimer)
(Coleoptera: Chrysomelidae)

Natural History

Distribution.   Hop flea beetle is found throughout northern areas of the United States and southern regions of Canada, from the Atlantic to the Pacific coasts. However, it is known as a serious pest principally in the western portion of its range—British Columbia, Washington, and Oregon.

Host Plants.   This insect has a wide host range. Vegetables attacked include beet, cabbage, cauliflower, cucumber, mustard, spinach, potato, radish, rhubarb, tomato, turnip, and probably others. Other crops damaged include canola, hops, sugarbeet, and strawberry. Many weeds are known to be suitable hosts, including Canada thistle, *Cirsium arvense*; dock, *Rumex* sp.; flixweed, *Descurainia sophia*; pepperweed, *Lepidium* spp. ; lambsquarters, *Chenopodium album*; nettle, *Urtica* sp.; wild buckwheat, *Polygonum convolvulus*; and tumbleweed, *Amaranthus albus*.

Natural Enemies.   General predators such as birds and ground beetle (Coleoptera: Carabidae) larvae are known to feed on the adults and larvae, respectively. Wylie (1984) observed parasitism of hop flea beetles by *Microctonus punctulatae* Loan and Wylie and *Townesilitus psylliodis* Loan (both Hymenoptera: Braconidae); the latter seems to attack only hop flea beetle.

Life Cycle and Description.   After overwintering as adults, beetles become active early in the spring, often while snow remains on the ground. Overwintering occurs in grass and grain stubble, and in leaf litter and other debris found in woodlots. They are usually the first flea beetles found in pastures, fencerows, and roadsides among dry grasses and weeds, seeking green plants. They are fairly active fliers in the spring—a habit that seems to diminish later in the season. Although there are reports of two generations, it is now believed that only one generation occurs annually in Canada. The overwintering adults disappear in July after initiating the summer generation. Adults from the summer generation, which are destined to overwinter, begin to appear in August. Overwintering beetles do not mate until the following spring. About two months are required for a generation to be completed.

Egg. Mating commences as soon as the beetles become active in the spring, and continues until the decline of the overwintering generation, usually in July. Eggs are deposited in the soil, normally at a depth of 3.5–5.0 cm. The eggs are elliptical in shape, yellow in color, and measure about 0.30 mm long

Hop flea beetle larva.

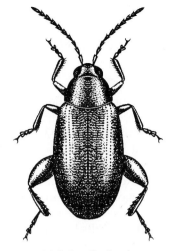

Adult hop flea beetle.

and 0.15 mm wide. They are laid singly or in small clusters and require 2–3 weeks for hatching.

Larva. Young larvae are whitish or gray, and measure about 0.5 mm long. They have a dark head and anal plate. The larvae eventually attain a length of about 5 mm, and require about 35 days to complete development of the stage. The larvae are quite active in the soil. Most are found at depths of 7.5–20.0 cm, but deeper depths are attained by some. They feed on the rootlets of crops and weeds.

Pupa. The larvae pupate in the soil and reportedly do not prepare a pupal cell. However, the larva shortens and thickens, remaining immobile for about two weeks before pupation. The pupa greatly resembles the adult beetle in form. Initially it is white, but gradually darkens over the two-week period the insect remains in this stage.

Adult. Upon emergence from the soil, the beetle is blue-black, but soon acquires a shining bronze-black appearance. It measures 1.5–2.5 mm long, with the males consistently smaller than females. The elytra are well marked with rows of punctures. The antennae are brown and lighter in color basally. The legs are reddish yellow except for the femora, which are darker. The hind femora are enlarged. Hop flea beetle

can be distinguished from the other common pest flea beetle species by the antennae. A characteristic of the genus *Psylliodes* is 10 antennal segments, whereas *Chaetocnema, Disonycha, Epitrix, Phyllotreta,* and *Systena* have 11 segments.

A fairly complete account of the biology was provided by Parker (1910). Useful observations were provided by Burgess (1977) and Wylie (1979).

Damage

The adults are the damaging stage of this insect. Adults inflict typical flea beetle injury, making irregular round feeding sites in which they eat all except one layer of leaf epidermis—in this case the lower surface. The lower epidermis at the site of feeding eventually dies, however, leaving a small round hole where the beetle fed. At high densities, beetles riddle the leaves with tiny holes, effectively eating all except the veins and causing severe damage to small plants. Leaf petioles are eaten occasionally. Except for hops, mature crops are not usually injured. Although larvae feed on roots, they are not reported to be severe pests of crops. Thus, weed hosts must figure prominently in the biology of these insects.

Management

Protection of seedlings is accomplished with systemic insecticides applied at planting or foliar insecticides applied early in plant development. Row covers would protect young plants from dispersing beetles. Parker (1910) observed that later in the season the beetles were reluctant to fly or to leap. Therefore, large plants such as hop vines could be protected by placing a barrier of adhesive around the base of the plant to entangle crawling beetles.

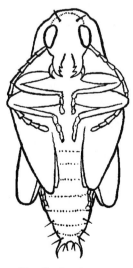

Hop flea beetle pupa.

Horseradish Flea Beetle
Phyllotreta armoraciae (Koch)
(Coleoptera: Chrysomelidae)

Horseradish flea beetle larva.

Natural History

Distribution. Horseradish flea beetle is an introduced species, probably from Europe. It was first found in North America in 1893 in the vicinity of Chicago, Illinois. It now is widely distributed in the northern United States and southern Canada, but principally east of the Rocky Mountains. It also has been collected from Oregon and Idaho, but in these locations it seems not to be a serious pest.

Host Plants. The adults have been collected from several cruciferous plants, but damage by adults and feeding by larvae is largely confined to horseradish. Other plants from which adults have been collected are radish, turnip, and marsh yellow cress, *Rorippa islandica*. The latter species occurs commonly in moist habitats, and it is considered a preferred host.

Natural Enemies. This insect is poorly studied and its natural enemies undetermined.

Life Cycle and Description. There is a single generation per year. The adult is the overwintering stage. Development from egg to the adult stage requires about 75–90 days.

Egg. The eggs are deposited from April or May until early August in Wisconsin. They are most often deposited on the petioles of young leaves, often where the petiole meets the root. Some eggs are also deposited on or in the soil. Eggs are elliptical in shape and orange. The average length is 0.57 mm (range 0.43–0.84 mm), and the width is 0.33 mm (range 0.26–0.47 mm). The eggs are usually deposited in small clusters, and individual eggs are sometimes aligned side by side. They are not securely attached to the plant. Egg production has not been exhaustively studied, but over 400 eggs have been produced by a single female. Eggs hatch in 7–14 days.

Larva. Upon hatching, the larva is whitish with a brown head and anal plate, and measures about 1.3 mm long. The larva becomes yellowish white as it grows, eventually attaining a length of about 4.8 mm. Larvae feed in the petioles and midribs of leaves and occasionally in the roots. This mining may cause the death of the foliage. Its development requires 50–60 days.

Pupa. At maturity, the larva leaves the plant tissue and descends to the soil. There is a subterranean prepupal period of 2–7 days followed by pupation. The pupa is white and is the same size as the adult beetle. Pupation usually lasts 10–13 days.

Adult. The adults are fairly distinctive, relative to other *Phyllotreta* spp. Newly emerged beetles begin appearing in July. The blackish beetles bear broad, straw-yellow or cream-colored stripes, one on each elytron. The stripes are considerably wider than those of other striped species, resulting in a relatively narrow dark line down the center of the back. The hind femora are enlarged. The legs are yellowish red distally, and darkened basally. The antennae lack segments that are greatly expanded, and are colored brown except for the basal 3–4 segments, which are lighter or yellowish. The beetles measure 2.5–3.4 mm long and average slightly larger than most other *Phyllotreta* spp. The adults seem to be long lived, being captured during nearly all months of the year. The

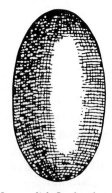

Horseradish flea beetle egg.

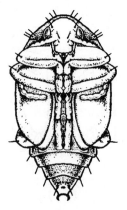

Horseradish flea beetle pupa.

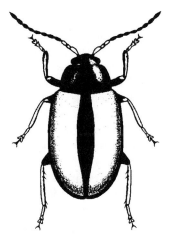

Adult horseradish flea beetle.

winter is passed by the adults in dry and sheltered locations.

The biology of horseradish flea beetle was given by Chittenden and Howard (1917).

Damage

The larvae mine the petioles and large veins of leaves, and occasionally enter roots to feed. Young leaves may wilt and die when severely mined. They feed on foliage also, sometimes making numerous small holes, but also making pits in the thicker tissue. Damage to foliage results in uneven growth and decreased root yield. Adults tend to be more damaging early in the season, and larvae later in the season.

Management

Crops should be monitored early in the year for feeding injury, because adult feeding damage often suggests additional larval damage later in the season. Crop rotation is recommended for horseradish flea beetle control, but this is impractical for home-garden production of horseradish.

Palestriped Flea Beetle
Systena blanda Melsheimer

Elongate Flea Beetle
Systena elongata (Fabricius)
(Coleoptera: Chrysomelidae)

Natural history

Distribution. Palestriped flea beetle is widely distributed in the United States. It is found from New York to Florida on the east coast and west to California. This native pest is rare or absent in the northern states from New England to Washington. Its pest status seems to be limited to the central United States from Virginia to Colorado. In Canada, this flea beetle is known from Alberta and Ontario, but is so infrequent as not to be considered a pest.

A closely related species, elongate flea beetle, is found over the same geographic area. However, it is most abundant in the southeastern United States. In Canada, it is known only from Manitoba. Rarely it has been known to reach damaging levels of abundance.

Host Plants. A very broad range of plants is eaten by the adults of *S. blanda*. Elongate flea beetle, although less well-known, seems to have a very similar host range. Vegetable crops reported to be attacked include beet, cabbage, carrots, cucumber, eggplant, lima bean, lettuce, melon, okra, onion, parsnip, potato, pepper, pumpkin, radish, rhubarb, snap bean, Swiss chard, sweet potato, tomato, turnip, and watermelon. Other crops such as alfalfa, apple, clover, corn, cotton, peanut, pear, soybean, strawberry, sugarbeet, and sunflower also are reported as hosts. Numerous common weeds such as field bindweed, *Convolvulus arvensis*; kochia, *Kochia scoparia*; lambsquarters, *Chenopodium album*; purslane, *Portulaca oleracea*; plantain, *Plantago* spp. ; nightshade, *Solanum* spp. ; povertyweed, *Franseria discolor.*; ragweed, *Ambrosia* spp.; redroot pigweed, *Amaranthus retroflexus*; and sorrel, *Rumex* spp. also support palestriped flea beetle adults. Suitable larval hosts are known to include plantain, ragweed, and lambsquarter (Hawley, 1922; Underhill, 1928), but there are probably many others. Despite the apparent wide host range, sugarbeet and bean are most frequently observed to be damaged. Weeds, especially bindweed and povertyweed, are often preferentially consumed, if they are available.

Natural Enemies. These insects are poorly studied, so the natural enemies are not well known. The parasitoids *Microctonus epitricis* Viereck (Hymenoptera: Braconidae) and *Howardula* sp. (Nematoda: Allantonematidae) were recovered from palestriped flea beetle in Colorado (Capinera, 1979b). The wasp *M. epitricis* is often found associated with *Phyllotreta* and *Epitrix* spp. flea beetles (Loan 1967b); in Colorado it parasitized over 50% of palestriped flea beetles in the August collection, but generally the incidence of parasitism was negligible. Parasitized palestriped flea beetles also ate less foliage than unparasitized insects. Flea beetles, on average, supported 2.4 reproductive female nematodes per beetle. Infection by nematodes shortened beetle longevity by about three days. Nematodes probably increase the rate of larval mortality, but this has not been studied in palestriped beetle. Elsey (1977a,b), Elsey and Pitts (1976) and Poinar

(1979) should be consulted for information on *Howardula* spp.

Life Cycle and Description.   The number of generations per year for palestriped flea beetle is one in Canada and New York and two in Virginia. In California, the number is not known, but beetles are active from January through September. The adults of palestriped flea beetle are active from mid-June until September in New York, whereas in Virginia the first generation of adults occurs from May through July, and the second generation from mid-July until September. Oviposition occurs from late July to late September in New York, where only a single generation occurs annually. In Virginia, where two generations occur, eggs are deposited from mid-May until mid-August for the first generation, and from late July until early September for the second generation. The insects overwinter as larvae in all locations.

Egg.   The egg of palestriped flea beetle is oval, and yellow. Eggs measure about 0.4 mm in width and 0.8 mm long. Eggs are deposited in the soil, adjacent to larval food plants, at a depth of 3.0–7.5 mm. The female may deposit the eggs singly, or in groups of up to eight, but 1–3 eggs per cluster seems to be normal for a single location. Beetles select several oviposition sites for each oviposition event. Thus, 12–18 eggs are normally deposited per event, with 8 to 14 events observed during the life of each beetle. Total egg production per beetle averages 68–153, with fecundity of the spring generation tending to be lower. Incubation period of the eggs is 11–23 days.

Larva.   The larvae of palestriped flea beetles are elongate and slender and equipped with stout and black spines. Their color is white or yellowish, except for the reddish brown head. The abdomen bears an anal proleg, like most other flea beetle larvae, but also bears a large fleshy process dorsally at the tip of the abdomen. At hatching, young larvae measure about 1 mm long but attain a length of 8–11 mm at maturity. There are three instars, each requiring about six days for completion. Larvae consume the fibrous roots and the surface of larger roots. Occasionally, they tunnel into roots. Near the completion of the larval stage, larvae prepare a small cell in the soil, become shorter in length and thicker in diameter, and molt into the pupa. This nonfeeding prepupal stage lasts 3–4 days.

Pupa.   The pupae are white and resemble the adult in form. Pupae measure 3.2–3.6 mm long. Duration of the pupal stage is about 8–21 days.

Adult.   The adult palestriped flea beetles measure 3.0–4.3 mm long. The most pronounced feature, and

the basis for the common name, is the pair of pale yellow stripes along the back, one on each elytron. The stripe is broad, usually measuring one-third the width of the elytron. Although the background color of the elytron is usually dark, often black, in some specimens it is much paler, approaching yellow. In the latter case, the yellow band may not be evident. The thorax and abdomen are usually black or dark brown. The head is orange or reddish-brown. The hind femora are enlarged. (See color figures 3 and 116.)

Elongate flea beetle is poorly known. The immature stages have not been described. The adult is nearly identical to palestriped flea beetle. Palestriped flea beetle is quite variable in appearance, so there are no simple diagnostic characters to distinguish it from elongate flea beetle. In fact, further study may show that they are different forms of the same species. Blatchley (1910) and Smith (1970) suggested that these two species could be separated by the punctations on the gena (face), head, and pronotum; *S. blanda* is smooth to moderately punctate, whereas *S. elongata* is coarsely punctate.

The large size of *Systena* spp. is a good diagnostic character. Among the common crop-feeding flea beetles only *Systena* and *Disonycha* spp. exceed 3.5 mm in length.

The most comprehensive treatment of palestriped flea beetle was given by Underhill (1928). He distinguished *S. blanda*, palestriped flea beetle, from *S. taeniata* Say, which he called the banded-flea beetle. Although these are now considered one species, Underhill (1928) suggested that the former overwinters as larvae while the latter passes the winter as adults. The biology of elongate flea beetle was reviewed by Smith (1970).

Damage

As is the case with most flea beetles, palestriped flea beetle larvae feed on roots. There are surprisingly few

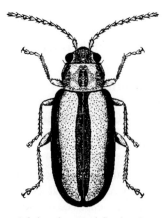

Adult palestriped flea beetle.

reports of larval feeding injury, possibly because weeds are preferred for oviposition. A notable exception is serious damage to young onions in Iowa, where not only roots but young bulbs were consumed (Drake and Harris, 1931). Sweet potato tubers also have been reported as injured.

The adults of palestriped flea beetle have been noted as serious pests, especially of seedling plants. Adults normally feed on the upper surface of foliage, making an irregular round feeding site. Beetles consume foliage at a rate of 0.3 sq cm per day (Capinera, 1978a). The lower epidermis, while left intact by the beetle, soon dies, forming a translucent membrane. Eventually, this membrane dries and drops from the leaf, leaving a hole. Leaves are often riddled with holes, with little more than veins remaining, where beetle density is high and availability of foliage is low. Such seedlings may perish, especially if water or temperature conditions are stressful for the plant. Beetles often aggregate, which leads to some plants being severely damaged, while others escape with minimal injury.

Management

Sampling. Beetles can be sampled with a sweep net or with traps. Yellow water-pan traps are very effective at capturing beetles relative to blue, green, black, red, orange, or white traps. Water-pan traps also are more effective than sticky traps (Capinera and Walmsley, 1978). If sticky traps are used, those positioned close to the soil (0.6 m above the soil) will catch more beetles than those that are elevated (1.2 m) (Capinera, 1980).

Insecticides. Granular insecticide formulations, applied at planting time, often are used to protect seedlings from flea beetle larvae. Systemic materials also confer protection to seedlings, or aqueous materials may be applied as seedlings emerge. In most cases, it is only the seedlings that require protection.

Potato Flea Beetle
Epitrix cucumeris **(Harris)**

Western Potato Flea Beetle
Epitrix subcrinita **LeConte**
(Coleoptera: Chrysomelidae)

Natural History

Distribution. Potato flea beetle occurs widely in North America, east of the Rocky Mountains. In the United States, it is found from South Carolina to Colorado, and all northward states. It is considered as a pest mostly in northern states. In Canada, potato flea beetle is found from Nova Scotia to Alberta. It is destructive as far west as Manitoba, but is considered to be a severe pest in the Maritime Provinces.

Western potato flea beetle occurs on the Pacific coast from northern California to British Columbia, and in the intermountain area of the United States. In the intermountain area, the western portions of Montana, Wyoming, Colorado, and New Mexico support western flea beetle, as well as northern Arizona and Nevada, and all of Utah and Idaho.

Host Plants. Potato flea beetle feeds principally on plants in the family Solanaceae. Vegetable crops attacked by both larvae and adults include eggplant, potato, tomato, and pepper. Other plants reported to be fed upon occasionally by adults include alfalfa, bean, beet, cantaloupe, celery, cucumber, lettuce, pumpkin, radish, squash, sunflower, tobacco, and turnip. Solanaceous weeds such as jimsonweed, *Datura stramonium*; ground cherry, *Physalis pubescens*; horsenettle, *Solanum carolinense*; and black nightshade, *Solanum nigrum* are also suitable hosts, and non-solanaceous plants such as redroot pigweed, *Amaranthus retroflexus*; kochia, *Kochia scoparia*; and lambsquarters, *Chenopodium album* may be fed upon. Western potato flea beetle apparently has the same host range.

Natural Enemies. An allantonematid nematode, probably *Howardula* sp., was reported from potato flea beetle in Washington, as was occasional infection by a fungus (Hanson 1933). In Ontario, Loan (1967b) recovered *Microctonus epitricis* (Viereck) (Hymenoptera: Braconidae) from potato flea beetle. The importance of these biotic agents in potato flea beetle control has not been determined.

Life Cycle and Description. Potato flea beetle has only one generation per year in northern locations such as Washington and all of Canada. However, two generations occur in southern portions of the potato flea beetle's range such as Kentucky and Delaware. A generation can be completed in 30–50 days. Adults overwinter at the soil surface beneath plant litter and undergrowth, or in the soil to depths of 15–25 cm, often in or near potato fields. In the spring, after feeding on newly emerged crop plants or weeds, female beetles deposit eggs in the soil at the base of host plants.

In Colorado, the overwintering beetles are active from late May until August, with eggs produced by the overwintering individuals present from June until August, larvae from June until September, and pupae from July until September. The adults produced from this summer generation are present from late July until October. Thus, with the two overlapping broods of adults, beetles are present throughout the summer months. There is functionally only one generation per year in Colorado, but a small portion of the beetles produce eggs and initiate a second generation (Hoerner and Gillette, 1928).

Egg. The eggs of potato flea beetle are elliptical, and pearly white in color. They measure about 0.2 mm wide and 0.5 mm long. They are usually buried by the beetles in the soil at a depth of 12–20 mm, but occasionally deeper, sometimes to a depth of 50 mm. They are laid singly, but several eggs may be placed in the vicinity of a single plant. Females produce, on average about 100 eggs during their life span, but some produce more than 200 eggs. Females may produce eggs over a period of about two months. Moisture is an important criterion for egg deposition and hatching. Females seek moist areas for egg deposition, and lack of moisture retards hatching. Temperature also influences hatching, with embryonic development being completed in 6–8 days during warm weather, but 11–12 days during cool weather.

Larva. The larvae are about 1 mm long and white when they first hatch from the egg. As they mature, the head and thoracic plate and to a lesser degree the anal plate, become brown, while the remainder of the body remains pale. The larva eventually attains a length of about 4–5 mm. The terminal abdominal segment bears an anal proleg. Larvae normally complete their development in about 20–25 days, but the larval development period may range from 13–45 days, depending on weather.

Pupa. The larvae pupate in the soil after preparing a small pupal chamber from soil particles. Pupation occurs near the food plant, close to the area of feeding. The pupa measures about 1.5–2.0 mm long and initially it is white, but eventually darkens. Pupal development requires about 6–10 days (range 3–22 days).

Adult. The adult potato flea beetle is fairly light in color after emerging. However, by the time it has dug to the soil surface it is fully darkened and cannot readily be distinguished from individuals that have long been emerged, including those that have overwintered. Adults can be found feeding until the plants have been killed by frost. Beetles measure 1.5–2.0 mm in length and are black. The antennae and legs are orange, except that a portion of the fore and middle femora are black, and the hind femora are completely black. The hind femora are enlarged. The pronotum is marked with fine punctures, the features of which can be used to distinguish this insect from the closely related tuber flea beetle, *Epitrix tuberis* Gentner. The fine punctures in the center of the pronotum of potato flea beetle are separated by a distance exceeding the diameter of the punctures. In contrast, the punctures at the center of the pronotum of tuber flea beetle are relatively coarse, and are separated by a distance less than the diameter of the punctures (Gentner, 1943). The pronotum of tuber flea beetle also has a deep transverse depression near the hind margin. Both the pronotum and elytra bear numerous fine hairs, but their density is less than with the similar eggplant flea beetle (see eggplant flea beetle for comparative description).

Western potato flea beetle has been poorly studied, but its biology in Washington seems similar to that of potato flea beetle (Hanson, 1933). It seems to be two-brooded in Oregon. It is similar in appearance to potato flea beetle, but is bronze rather than black, and lacks the transverse depression on the prothorax.

A good account of potato flea beetle was given by Johannsen (1913), based on the work conducted in Maine. However, a more comprehensive treatment of biology, including rearing methods, was given by Hoerner and Gillette (1928) from Colorado. This and most other life-history research from the western United States was conducted before a systematic revision of the *Epitrix* flea beetles resulted in recognition of tuber flea beetle as a separate species (Gentner, 1944). Thus, some studies may contain mixed populations, and despite Gentner's attempt to identify the true species involved in previous research on "potato flea beetle," the biology of potato flea beetle remains somewhat uncertain. The work of Hill and Tate (1942) in Nebraska is particularly suspect, because tuber flea beetle appears to be present in this area. Beirne (1971) and Campbell *et al.* (1989) provided a useful perspective on potato flea beetle in Canada.

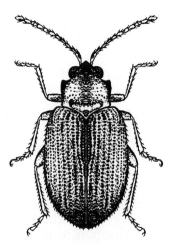

Adult potato flea beetle.

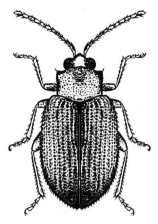

Adult western potato flea beetle.

Damage

The damage by adult potato flea beetles is typical flea beetle injury. Small pits are eaten in either the upper or lower surface of the foliage. Although the beetles may not eat completely through the foliage, the remaining epidermis soon dries and dies. Dead leaf tissue eventually falls away, leaving a hole. Leaves on heavily infested plants may be riddled with holes, although the veins usually remain. Small plants may be killed. Once potato plants become established, however, they can withstand considerable flea beetle feeding. Senanayake *et al.* (1993) estimated that densities of 100 beetles per plant would not depress potato yield of healthy plants. If potatoes were previously stressed by Colorado potato beetle feeding, then densities as low as 5–25 flea beetles per plant would depress yield. In eastern Canada, flea beetle populations regularly reduce potato yield by 15% unless corrective actions are taken. Damage by adults is greater under hot and dry conditions.

The larvae are root feeders, but occasionally cause considerable damage by feeding on potato tubers. By feeding on the tuber surface, larvae produce tracks or scars that materially damage the commercial value of the potato tubers. Occasionally they burrow directly into the tuber to a depth of up to 3 cm. The pits eventually are filled with corky material that blackens. Although tuber feeding is noted wherever potato flea beetle occurs, it is not as serious a problem as it is with tuber flea beetle. In the case of potato flea beetle, adult damage is the predominant form of damage.

Potato flea beetle has been implicated in transmission of several plant diseases. Tuber feeding also enhanced infection of tubers by Rhizoctonia and potato scab fungi (Hanson, 1933). Potato flea beetle also has been shown to be capable of transmitting ring-rot bacteria (Christie *et al.*, 1993), and probably many other diseases.

Damage to plants by western potato flea beetle is similar to potato flea beetle damage, although much less frequent. Also, western potato flea beetle larvae are less prone to feed on the tubers of potatoes, feeding instead primarily on the fibrous roots.

Management

Sampling. Senanayake and Holliday (1988) compared various sampling procedures for potato flea beetle monitoring. Use of a sweep net was efficient early in the season, but declined as plants matured. Visual samples were not highly satisfactory because beetles were often feeding on the underside of foliage, and difficult to observe. Whole plant bag sampling was determined unwieldy for routine sampling despite its high degree of precision. Stewart and Thompson (1989) determined the spatial distribution of potato flea beetles on potato and recommended that 10–20 plants be sampled when beetle densities were four per plant or higher, but that sample size be increased to 50–60 plants when beetle densities were only one per plant.

Because the beetles are small, mobile, and often feed on the underside of foliage, plant damage may be a better indicator of insect density than beetle counts. The relationship of feeding puncture abundance to yield has been studied, but owing to variability in plant response, the damage threshold remains uncertain (Howard *et al.*, 1994).

Insecticides. Foliar applications are made for suppression of adults, usually 1–2 weeks after adults first appear. Commonly, several applications are necessary to protect young plants from overwintering beetles, and the first or spring generation adults, due to protracted emergence. Granular formulations are applied at planting or after, if there is reason to expect damage by larvae. Systemic insecticides have been applied at planting, using various formulations; although some systemic treatments provide excellent seedling protection against both adults and larvae, phytotoxicity is a potential hazard (Chalfant *et al.*, 1979b). Insecticide resistance has been detectable since the 1950s (Kring, 1958), but populations remain manageable (Ritcey *et al.*, 1982). In many areas, excellent control of flea beetles occurs as part of Colorado potato beetle chemical suppression. As more selective management techniques are developed for Colorado potato beetle, flea beetle problems may be increased.

Cultural Practices. Glandular trichomes deter feeding by potato flea beetle (Tingey and Sinden, 1982), but this character is not yet incorporated into horticulturally acceptable varieties.

Row covers can prevent attack of plants by adult flea beetles, and subsequent damage by larvae, if the crop is grown on land not previously infested by potato flea beetle. If the beetles were abundant in the previous season, however, the adults may be overwintering in the soil and can emerge under the row cover. Thus, there is considerable benefit from crop rotation with a non-solanaceous crop.

Destruction of solanaceous weeds is beneficial. Because adults do not fly readily, planting of crops at some distance from previous crops or weed hosts is considered beneficial. Wolfenbarger (1940), working in New York, studied the distribution of potato flea beetle injury in potato fields with respect to weedy, uncultivated areas. Damage to tubers was greatest within 25–30 m of uncultivated areas containing solanaceous weeds. Further, the number of beetles overwintering in soil of uncultivated areas was estimated about 470,000 per hectare, whereas it was only 55,000 per hectare in potato fields.

Redheaded Flea Beetle
Systena frontalis (Fabricius)

Smartweed Flea Beetle
Systena hudsonias (Förster)
(Coleoptera: Chrysomelidae)

Natural History

Distribution. Redheaded flea beetle is distributed widely in the eastern United States and Canada. Although it has been collected in Montana, it is rarely found west of Manitoba, Kansas, and eastern Texas.

Smartweed flea beetle occupies a similar range. It is distributed widely in the eastern states and provinces, but is found only as far west as South Dakota, Colorado, and New Mexico.

Host Plants. The adults of both redheaded and smartweed flea beetle have been observed to feed on various vegetables, including bean, beet, cabbage, corn, eggplant, lettuce, okra, parsley, potato, and sweet potato. Other crops such as alfalfa, apple, clover, cranberry, currant, gooseberry, grape, horseradish, raspberry, soybean, sugarbeet, sunflower, strawberry, and numerous woody ornamental shrubs are also attacked. As might be expected of insects with such a wide host range, numerous weed hosts have been reported, including smartweed, *Polygonum pensylvanicum*; lambsquarters, *Chenopodium album*; giant ragweed, *Ambrosia trifida*; plantain, *Plantago major*; beggartick, *Bidens frondosa*; redroot pigweed, *Amaranthus retroflexus*; Canada lettuce, *Lactuca canadensis*; Canada thistle,

Cirsium arvense; and giant foxtail, *Setaria faberii*. Of the aforementioned weeds, the first four are thought to be preferred by redheaded flea beetle. Redheaded flea beetle larvae have been observed developing on corn, and both species attack sugarbeet.

Life Cycle and Description. Apparently, there is only a single generation per year. Adults are present from mid-summer until winter. Eggs are present at all times except spring. The biology of smartweed flea beetle is almost unknown. Following is a description of redheaded flea beetle.

Egg. The eggs are elliptical, and slightly more rounded at one end. They measure 0.70–0.85 mm long. Eggs are deposited in the soil beneath suitable food plants, usually at a depth of 1.5–5.0 cm. This is the overwintering stage, and eggs must receive a period of chilling before they will hatch.

Larva. Larvae are whitish with a brownish head capsule, and bear numerous spines. There is a prominent anal tubercle bearing a tuft of setae. Larvae develop from May to July, feeding on plant roots. Larvae display three instars and reach a length of about 8 mm during development, which requires about 30 days.

Pupa. Pupation occurs in the soil.

Adult. Adults measure 3.5–5.5 mm long. They are dark reddish brown, with an orange-red head. Head color, the basis for the common name, is diagnostic, although related species such as *S. elongata* (Fabricius) may have a reddish brown head. The hind femora are enlarged. The large size of *Systena* spp. is a good diagnostic character, and among the common crop-feeding flea beetles only *Systena* and *Disonycha* spp. exceed 3.5 mm. The adults are active from June through September or October in the North, and through November in the South, depending on weather. There is a report of adults overwintering in Indiana, but this has not been confirmed. Eggs are deposited beginning in July, but remain in diapause until the following spring.

The adults of smartweed flea beetle are distinguished from redheaded flea beetle only by the lack of an orange-red head. In all other respects, they are virtually identical (Smith, 1970). Further study may demonstrate that smartweed flea beetle is merely a color variant of redheaded flea beetle.

The biology of redheaded flea beetle is not well studied. Hawley (1922), Smith (1970), and Jacques and Peters (1971) provided the most complete information on life history.

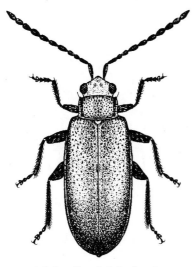

Adult redheaded flea beetle.

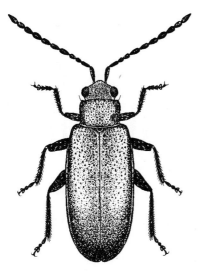

Adult smartweed flea beetle.

Damage

The adults feed on the epidermis of foliage, usually the upper surface. They create elongate holes, often leaving the lower epidermis intact as a transparent membrane. They may aggregate on certain plants, and often prefer weeds to crops. Jacques and Peters (1971) noted that the adults were abundant on corn only in the absence of preferred weeds. Corn silks were observed to be preferred over leaf tissue, and beetles were observed to leave corn fields after silks were mature. Hawley (1922) noted that bean plants usually recovered from defoliation; the exception was during dry weather, when permanent damage or death could occur.

The larvae feed on, and burrow within, plant roots. Jacques and Peters (1971) found redheaded flea beetle larvae interspersed with corn rootworm larvae among corn roots in Iowa, but flea beetle larvae were much less numerous, and less damaging than northern corn rootworm. Riley (1983) documented severe damage to germinating soybean seeds and seedlings by redheaded flea beetle in Mississippi. The larvae scored the seed surface, and burrowed through the seed, stem, and roots.

Management

Weed management is a key to effective management of both redheaded flea beetle and smartweed flea beetle, as these insects consistently are reported as pests only where favored weeds are abundant. Chemical treatments applied for corn rootworm larvae are adequate for redheaded flea beetle larvae, though seldom needed. Adult damage is easily prevented with foliar applications of insecticide if the crop is carefully monitored.

Spinach Flea Beetle
Disonycha xanthomelas (Dalman)

Three-spotted Flea Beetle
Disonycha triangularis (Say)

Yellow-Necked Flea Beetle
Disonycha mellicollis (Say)
(Coleoptera: Chrysomelidae)

Natural History

Distribution. Spinach flea beetle and yellow-necked flea beetle occur throughout the United States and Canada, east of the Rocky Mountains. Three-spotted flea beetle occupies the same area as the aforementioned species, but also occurs in British Columbia, Idaho, and Utah. These insects are native to North America.

Host Plants. These flea beetles are known principally as pests of the family Chenopodiaceae—beet, spinach, and Swiss chard. Therefore, it is not surprising that they damage sugarbeet, but they also are known to damage cabbage, canola, horseradish, lettuce, and radish on rare occasion. Weeds are their principal host, principally chickweed, *Stellaria media*; purslane, *Portulaca* spp.; lambsquarters, *Chenopodium album*; and pigweed, *Amaranthus* spp.

Natural Enemies. Little is known concerning the natural enemies of these beetles, although Chittenden (1899) reported that *Medina barbata* (Coquillett) (Diptera: Tachinidae) was a parasitoid of the adult spinach flea beetle, and Loan (1967a,b) found threespotted flea beetle to be attacked by *Microctonus disonychae* (Loan) (Hymenoptera: Braconidae).

Life Cycle and Description. These insects are poorly known, but spinach flea beetle has two generations annually in Maryland, with adults present throughout the year. The life cycle is reported to require 30–60 days. Apparently, these beetles overwinter in the adult stage and have been observed under loose bark of trees and other sheltered locations during the winter months. From the report of Beirne (1971) that adults are present in May and June and again in September and October, we might surmise that only one generation occurs in Canada. The life cycle of the other *Disonycha* flea beetles seems to be similar.

Egg. The eggs of spinach flea beetle are laid in clusters of about 4–30 eggs, attached on end to leaves, stems, and sometimes on soil. They measure 1.25–1.50 mm long and 0.40–0.57 mm wide, and are orange in color. The eggs of threespotted flea beetle seem to be undescribed, but they likely are similar to spinach flea beetle. Yellownecked flea beetle is known to deposit clusters of eggs in a manner similar to spinach flea beetle, but they are somewhat unusual in that they are red in color. Eggs hatch in 4–10 days, the larvae escaping by chewing a slit in the side of the egg.

Larva. The larvae are normally grayish, though sometimes purplish when feeding on beet foliage. The head is darker, and the body is well-equipped with short and stout spines. Initially measuring about 1.8 mm long, the larvae reach 8–9 mm at maturity. Larvae require 10–30 days to complete their development.

Pupa. Larvae drop to the soil to form a small pupation cell. Pupae resemble adults, but are grayish in color. Pupation requires 10–14 days.

Adult. Adults of *Disonycha* are fairly large for flea beetles, measuring 5–6 mm long. Among the common crop-infesting flea beetles, only *Systena* spp. approach *Disonycha* in size. Their elytra are shiny black, sometimes tinged with blue or green. The prothorax is yellow or red. Threespotted flea beetle can be distinguished from the other species by the three black spots on the thorax. (See color figure 117.) Yellownecked flea beetle can be distinguished from spinach flea beetle by the color of the femora: in the former the femora are entirely yellow, whereas in the latter they are partly blue or green. Females produce an estimated 100–300 eggs.

The biology of spinach flea beetle was provided by Chittenden (1899), that of threespotted flea beetle

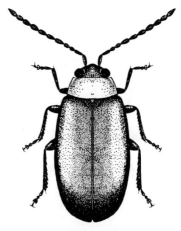

Adult spinach flea beetle.

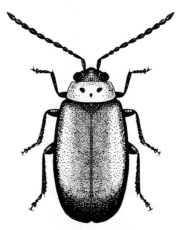

Adult threespotted flea beetle.

by Maxson (1948), and of yellownecked flea beetle by Chittenden (1912b).

Damage

The larvae of *Disonycha* flea beetles differ from the more common genera in that they feed on foliage rather than roots. They tend initially to be gregarious, but this habit dissipates as they mature. They usually feed on the underside of the leaves, initially feeding only partially through the leaf tissue, but eventually creating holes. Adults similarly feed on the leaf tissue, skeletonizing the foliage. Generally, these flea beetles are not considered serious pests.

Management

Because weeds are often important in the biology of these insects, an important element of insect control is weed suppression. Should the insects require suppression, foliar insecticides provide quick relief. The beetles do not seem to overwinter in the soil, so covering crops with netting or row cover material will prevent damage in the spring.

Striped Flea Beetle
Phyllotreta striolata (Fabricius)

Western Striped Flea Beetle
Phyllotreta ramosa (Crotch)
(Coleoptera: Chrysomelidae)

Natural history

Distribution. Striped flea beetle, *Phyllotreta striolata* (Fabricius), is common in Europe and Asia, and gained access to North America before 1700 (Bain and LeSage, 1998). It is now widely distributed in Canada and the United States; however, it is not abundant in the Rocky Mountain region and west coast. It is well known from Canada's Prairie Provinces, where crucifer oil-seed crops are grown extensively, but is not as abundant, or damaging, as crucifer flea beetle, *Phyllotreta cruciferae* (Goeze). In the United States, striped flea beetle is most common in the northeast and midwest.

Western striped flea beetle, *Phyllotreta ramosa* (Crotch), apparently is native to North America. Its distribution is restricted to the western United States, principally Washington, Oregon, and California. A report of this insect from British Columbia is suspect.

Host Plants. Striped flea beetle is a crucifer-feeding species, and is often called the striped cabbage flea beetle. As is the case with other Cruciferae-feeding insects, it will also feed on mustard oil-containing plants from the families Capparidaceae and Tropaeolaceae. It commonly attacks such vegetables as Brussels sprouts, cabbage, cauliflower, collards, kale, mustard, radish, rutabaga, turnip, and watercress. It is reported to have attacked non-cruciferous garden vegetables, but these are erroneous, with the insects actually feeding on cruciferous weeds found amongst the vegetables. Weeds serving as suitable hosts include black mustard, *Brassica nigra*; flixweed, *Descurainia sophia*; pepperweed, *Lepidium densiflorum*; shepherdspurse, *Capsella bursa-pastoris*; yellow rocket, *Barbarea vulgaris*; and other common cruciferous weeds. Information on host preference was provided by Feeny *et al.* (1970) and Tahvanainen (1983). Western striped flea beetle has similar host preferences.

Natural Enemies. A parasitoid, *Microtonus vittatae* Muesebeck (Hymenoptera: Braconidae) attacks the adults of several crucifer-feeding flea beetles, including striped flea beetle. The biology of this parasite was given by Wylie (1982, 1984) and Wylie and Loan (1984). General predators such as the shield bug, *Podisus maculiventris* (Hemiptera: Pentatomidae) also sometimes feed on beetles (Culliney, 1986).

Life Cycle and Description. In Canada, striped flea beetle has only one generation per year, though a complete life cycle requires only about one month. Overwintering occurs in the adult stage in the soil and leaf litter (Burgess and Spurr, 1984; Turnock *et al.*, 1987). Adults become active in early spring, which is March in the southern states, April in the midwest and southern Canada, and as late as June in northern Canada. In Saskatchewan, abundance of overwintering adults is high through mid- to late-June, with summer generation beetles becoming abundant in late July. Similar population trends have been observed in New York. Beetles are more active fliers early in the season, and fly at greater heights (1–2 m) than later in the season (Lamb, 1983). They become active about two weeks earlier than crucifer flea beetle, with which they commonly coexist. A second generation is reported from New York and North Carolina, but this is uncertain. The life cycle of western striped flea beetle is unknown, but likely is quite similar to that of striped flea beetle.

Egg. Soon after their emergence, beetles begin to deposit eggs in the soil adjacent to host plants. In Manitoba, egg-laying females were collected from April until early August. Eggs hatch in about five days.

Larva. There are three instars. The larvae measure about 1.25 mm long at hatching, but eventually attain a length of about 4.9 mm. They are white or yellowish white, with a brownish head and anal plate. The larvae bear an anal proleg.

Pupa. Pupation occurs in the soil within a small cell formed from soil particles. Pupae resemble the adults in form but lack fully developed wings.

Adult. Adults are about 2.0–2.4 mm long. The hind femora are enlarged. The head and pronotum are black, and the elytra are brownish black, each with a slight metallic luster. The elytra usually bear an irregular yellow band running nearly the length of each elytron. However, the band is sometimes discontinuous, resulting in two yellow spots on each elytron. The proportion of the population with the aforementioned discontinuous banding pattern is higher in southern portions of the beetle's range. The adults emerge and are very numerous in July and August in northern states.

Striped flea beetle and western striped flea beetle are very similar in appearance. Males can be easily differentiated, however, by the size of the fifth antennal segment. In western striped flea beetle, antennal segments four to six are nearly the same size. In striped flea beetle, the fifth segment is wider and longer than adjacent segments, nearly twice the length of the sixth segment. The other common striped flea beetle that may get confused with these species is Zimmermann's flea beetle. However, male *P. zimmermanni* have their fifth antennal segment even more enlarged, about three times the length of the sixth segment.

Information on striped flea beetle biology was given by Dugas (1938), Burgess (1977), and Wylie (1979). Rearing techniques were provided by Burgess and Wiens (1976). Information on western striped beetle can be found in Chittenden (1927) and Smith (1985).

Damage

Striped flea beetle damages both the above-ground and below-ground portions of crucifers. Adults are damaging to seedlings in the spring, but even mature plants can be severely damaged when beetles are particularly abundant. Dugas (1938), for example, reported that the foliage from an entire field of mature turnips in Louisiana was so damaged that it appeared

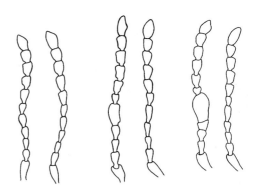

Antennae of some *Phyllotreta* flea beetles: (left to right) western striped, striped, Zimmermann's.

burned. Although the beetles produce only small holes in the foliage, when excessive feeding occurs the adjacent tissue dies, leading to a bronzed or burned appearance. Even modest levels of feeding reduce the value of foliage crops such as mustard if the plants are nearing maturity. Beetles will sometimes feed on the young florets of broccoli, greatly reducing yield.

Larvae feed on both rootlets and the principal portions of the plant root. Larvae often feed along the surface of the main root, leaving shallow trenches on the root surface. On plants where the root is harvested, such as turnip, larval feeding destroys the crop's marketability. Even in crops where the root is not harvested, such as cabbage, flea beetle larvae can be very damaging, because their root pruning activities stunts the plants and may cause seedling death.

Management

These insects are very similar to crucifer flea beetle, and often coexist simultaneously. The management considerations discussed under crucifer flea beetle are also applicable to the striped and western striped flea beetles.

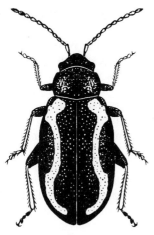

Adult striped flea beetle.

Sweetpotato Flea Beetle
Chaetocnema confinis Crotch
(Coleoptera: Chrysomelidae)

Natural History

Distribution. In the United States, this insect occurs practically everywhere sweet potato is cultivated, including California. However, it is considered a pest east of the Rocky Mountains only, and is not known to be damaging in Canada.

Host Plants. Among vegetable crops, only sweet potato and corn are attacked, and corn only rarely. However, several other crops are sometimes consumed, including clover, oats, raspberry, rye, sugar-

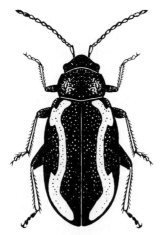

Adult western striped flea beetle.

beet, timothy, and wheat. Weeds of the family Convolvulaceae, particularly bindweed, *Convolvulus* spp., are the preferred hosts.

Natural Enemies. No natural enemies are known from this poorly studied insect.

Life Cycle and Description. The life cycle of this insect is poorly known. It appears that there is only one generation annually in the north, while several generations may occur per year in the south. The complete life cycle is estimated to require 4–8 weeks for its completion. Adults are found throughout the season, from April through October, but they are damaging to sweet potato foliage principally early in the season.

Egg. The egg is deposited in the soil. It is white and elliptical, measuring about 0.2 mm long.

Larva. The larvae are white, and likely resemble larvae of corn flea beetle, *Chaetocnema pulicara* Melsheimer, or other members of the genus *Chaetocnema*. Larval development requires about three weeks. The larvae attain a length of about 4.8 mm at maturity.

Pupa. Pupation occurs in the soil. Pupae are white, resembling the adult in form.

Adult. The adult is small, measuring only about 1.5–1.8 mm long. Its color is black, but has a bronze cast. The elytra are marked with rows of punctures. The legs are reddish yellow, and the hind femora are enlarged. The adult is the overwintering stage, and is found under debris and in wood lots near cultivated areas. The adult becomes active early in the year, often in May, and seeks suitable host plants for feeding. Beetles are reported to abandon sweet potato later in the season, as soon as bindweed is available.

There is not a good source of biological information on this insect. Elements of biology can be found in Smith (1910) and Sorensen and Baker (1983).

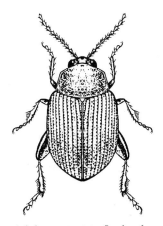

Adult sweetpotato flea beetle.

Damage

Damage by adults is not the typical round-hole injury found in most flea beetles, but rather like that of other *Chaetocnema* spp. Specifically, they feed in long, narrow strips on the foliage. Often the feeding strips are parallel to the major veins, but as damage levels increase the feeding becomes more general. It is only the young plants that are injured severely.

The larvae are not usually found feeding on sweet potato; the adults apparently prefer to oviposit near bindweed. When feeding on sweet potato, the damage is principally due to consumption of fibrous roots, and larvae do not bore within the roots. Occasionally they etch the skin of the tuber, however, forming winding channels on the surface.

Management

In fields that have a history of flea beetle infestation, pre- and post-planting applications of granular insecticides are recommended (Chalfant *et al.*, 1979a). If systemic materials are not applied, then foliar applications may be necessary to prevent injury to the vines. Some sweet potato varieties are resistant to root injury (Cuthbert and Davis, 1970). Resistant and nonresistant cultivars can be interplanted, with some lessening of injury among interplanted susceptible stock (Schalk *et al.*, 1992). Reduction in larval injury was reported by application of the entomopathogenic nematode, *Steinernema carpocapsae* (Nematoda: Steinermatidae) (Schalk *et al.*, 1993).

Tobacco Flea Beetle
Epitrix hirtipennis (Melsheimer)

Southern Tobacco Flea Beetle
Epitrix fasciata Blatchley
(Coleoptera: Chrysomelidae)

Natural History

Distribution. Tobacco flea beetle, *Epitrix hirtipennis* (Melsheimer), is distributed widely in the United States. It is found most commonly in the southeast, but also as far north as Maryland and Michigan, and west to southern Colorado and California. It is infrequent in the northwest and Great Plains, but apparently occurs in Hawaii and Mexico. In Canada, it is known from Ontario and Quebec.

Southern tobacco flea beetle, *Epitrix fasciata* Blatchley, has limited distribution in the United States, and is known only along the Gulf Coast, from Florida to Texas. It occurs widely in Central America and the Caribbean, and as far south as Argentina.

Host Plants. Tobacco flea beetle adults feed readily upon tobacco and other plants in the family Solanaceae, but sometimes attack other plants as well. Vegetable crops most frequently attacked are eggplant, potato, and tomato, but cabbage, cowpea, pepper, snap beans, and turnip are consumed occasionally. Weeds commonly serving as hosts are nightshade, *Solanum* spp.; jimsonweed, *Datura stramonium*; groundcherry, *Physalis heterophylla*; pokeweed, *Phytolacca americana*; burdock, *Arctium minus*; cocklebur, *Xanthium* spp.; and many others. The larvae develop successfully on many solanaceous plants, but do not survive on non-solanaceous plants. In studies conducted in Virginia, tobacco, potato, and jimsonweed were particularly good hosts for larvae (Glass, 1943).

The little information available on southern tobacco flea beetle suggests that the host range is nearly identical to that of tobacco flea beetle (White and Barber, 1974).

Natural Enemies. Tobacco flea beetle is preyed upon to a limited extent by general predators such as bigeyed bug, *Geocoris punctipes* (Say) (Hemiptera: Lygaeidae) (Dominick, 1943). The principal parasitoid seems to be *Microctonus epitricis* (Vierick) (Hymenoptera: Braconidae), which causes parasitism rates of up to 25% (Dominick and Wene, 1941). This parasitoid also attacks southern tobacco flea beetle.

A nematode, *Howardula dominicki* Elsey (Nematoda: Allantonematidae), parasitizes up to 70% of larvae and 50% of adults of tobacco flea beetle. Host larvae are killed, and adult females are made sterile by the nematodes. Male beetles are important in nematode dissemination. The biology of this nematode was provided by Elsey and Pitts (1976), and Elsey (1977b,c).

Life Cycle and Description. The number of annual generations of tobacco flea beetle varies, with three reported from Kentucky, 3–4 in Virginia, and 4–5 from Florida. In Kentucky, Jewett (1926) observed first generation adults, resulting from reproduction by overwintering beetles, in mid-June. This was followed by a second generation in late July, and a third in September. Not all insects undergo the third generation, however, some commence overwintering after only two generations. In Florida, Chamberlin *et al.* (1924) reported four well-defined generations, but there may be additional generations. Overwintering occurs in the adult stage, under plant debris, and often in weedy or wooded areas adjacent to crop fields. In the south, the beetles may remain active throughout the winter.

Egg. Tobacco flea beetle eggs are elongate and slightly pointed at one end. Initially, the egg is white, but gradually changes to lemon yellow as it matures. The egg measures about 0.41 mm (range 0.36–

0.48 mm) long and 0.18 mm (range 0.16–0.26 mm) wide. Eggs are deposited in the soil near the base of the host plant, frequently in clusters of 5–6 eggs. Moist areas are preferred for oviposition. Eggs hatch in 6–8 days. Overwintering females deposit, on an average, 2.2 eggs per day, and total production is estimated at about 200 eggs per female. Later generations seem to live for a shorter period of time, and the number of eggs produced per female is about 100.

Larva. The larvae have three instars, and grow in size from less than 1 mm at hatching to about 4.2 mm at maturity. The larva is whitish, except for the head, which is yellow or yellow-brown. The terminal abdominal segment bears an anal proleg. Larvae feed principally on the fine roots of solanaceous plants, but occasionally larger roots are girdled or tunneled. Except when the soil is very dry, larvae also feed near the soil surface. Larval development time averages 16–20 days, depending on generation and weather. Jewett (1926) determined the duration of the three instars to be 4.6, 3.9, and 5.4 days, respectively. Average head capsule width measurements for the instars is 0.12, 0.17, and 0.21 mm, respectively (Martin and Herzog, 1987). Larvae are cannibalistic.

Pupa. At maturity, larvae prepare a small cell in the soil and after 1–2 days transform into the pupal stage. The pupa closely resembles the adult in form, measures about 1.75 mm long, and is whitish. Pupation occurs within the upper 1–3 cm of soil, and lasts 4–5 days.

Adult. The adults measure only about 1.4–2.2 mm long. They are reddish-yellow, with a brown abdomen and a brown irregular transverse patch or band crossing, or nearly crossing, the yellow elytra. The legs are also reddish-yellow except for the darker femora of the hind legs. The hind femora are enlarged. As is the case with all *Epitrix* beetles, the entire body bears a coat of short hairs. Adults live for several weeks, resulting in overlapping generations and almost continuous egg production. The pre-oviposition period is 2–3 weeks.

Tobacco flea beetle, *E. hirtipennis*, and southern tobacco flea beetle, *E. fasciata*, are quite similar in appearance. *E. hirtipennis* is slightly larger and narrower, measuring 1.6–2.2 mm long and 1.9–2.0 times as long as wide. *E. fasciata* is 1.4–1.7 mm long and 1.7–1.8 times as long as wide. There is a broad transverse band across the elytra of both species, but although it is continuous in *E. hirtipennis* it is generally interrupted in *E. fasciata* (White and Barber, 1974).

The biology of the tobacco flea beetle was given by Metcalf and Underhill (1919), Jewett (1926), and Dominick (1943). Simple rearing techniques were pro-

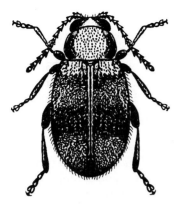

Adult tobacco flea beetle.

vided by Jewett (1926), and Martin and Herzog (1987). Most research focuses on the relationship of tobacco flea beetle with tobacco, despite its common association with vegetable crops. The biology of southern tobacco flea beetle is poorly known, but is presumed similar to tobacco flea beetle. (See color figure 115.)

Damage

The nature of damage caused by these species is typical of flea beetle injury. Adults eat small holes partly or completely through the leaves, usually feeding from the underside. In the former instance, the adjacent remaining tissue usually dies and drops from the plant, leaving a somewhat circular hole. Foliage, especially of young plants, may be riddled with these small holes. In extreme cases, only the veins may remain, seriously disrupting the physiology of the plant. The result is usually stunted plants, although seedlings sometimes are killed. Feeding by as few as five beetles for a three-week period early in the life of a plant may result in significant yield reduction in tobacco, and presumably other crops (Semtner, 1984a). The beetles are more active and damaging during warm and sunny weather. They often feed selectively on shaded, or lower sections of the plant.

The larvae feed on the roots, particularly the rootlets. Even a single larva feeding on a germinating seedling may cause enough damage leading to death of the plant. The discoloration of the seedling stem base, wilting, and rapid death of the plant caused by larval feeding greatly resembles seedling "damping off" due to plant parasitic fungi (Gentile and Stoner, 1968a).

Management

Sampling. Various approaches for monitoring abundance of tobacco flea beetle have been investigated. Soil cores can be collected for estimation of larval, pupal, and adult numbers. Larvae can be separated from soil with heat extraction by using Berlese funnels. Pupae, being nonmotile, must be separated

by flotation or visual examination. Soil samples can be held in containers for adult emergence, but field emergence traps are more usual. Duke and Lambert (1987) recommended sample sizes of about 50, 12, and 30 soil cores for these three life stages, respectively. Adults can also be monitored with colored sticky traps; yellow is the most attractive color (Dominick, 1971).

Insecticides. Foliar and soil-applied insecticides are commonly used for tobacco flea beetle suppression. Systemic insecticides are particularly useful for the young and rapidly growing seedlings because the leaf tissue is expanding very rapidly and multiple applications of non-systemic material would be necessary to protect the foliage from adult feeding damage.

Cultural Practices. Modification in planting date has been examined for reduction in flea beetle impact, but without much success. Late-planted crops may escape the attack of overwintering beetles, but plants then are relatively small and susceptible to injury by first brood adults (Semtner, 1984b). Also, crop yields or crop value are often adversely affected by late planting, which can outweigh the benefits of avoiding flea beetle injury.

Host-plant nutrition affects flea beetles differently, depending on the nutrient (Semtner *et al.*, 1980). High nitrogen levels sometimes seem to affect flea beetles, but the pattern is not consistent. Plants with higher levels of phosphorus are less preferred by beetles, whereas higher levels of potassium favor adult abundance.

Among other methods of cultural control that have been studied are row covers, burning, and host plant resistance. Row covers work quite well for most pests, but as the beetles may overwinter in the field, they can emerge under the protective covering. Thus, this practice has limitations if the field was infested previously. Burning of crop residue and field edges, where overwintering may occur, may kill some beetles, but it is no longer recommended routinely because the benefits are slight. Host-plant resistance has been sought for some crops, such as *Lycopersicon* spp. (tomato and its relatives), but without much success (Gentile and Stoner, 1968b). Although some selections had glandular hairs that reduced feeding by adults, even wild species were susceptible to larvae.

Tuber Flea Beetle
Epitrix tuberis Gentner
(Coleoptera: Chrysomelidae)

Natural History

Distribution. The tuber flea beetle is western in distribution, although some eastward movement is evident in Canada, as it has only appeared recently in Alberta. It is highly damaging in British Columbia, where it is considered by some as be the most serious insect pest of potato, and in Washington and Oregon. Gentner (1944) found that in addition to the far western populations there was a disjunct population in Colorado and western Nebraska. Gentner speculated that this beetle might be native to eastern Colorado, and have spread to other states.

Host Plants.   Tuber flea beetle feeds on a variety of solanaceous crops and weeds. The preferred host seems to be potato. Other crops that have been observed to be attacked, when potato was not available, include bean, cabbage, cucumber, lettuce, pepper, radish, spinach, swiss chard, and tomato. Several weeds, such as buffalo bur, *Solanum rostratum*; ground cherry, *Physalis lanceolata*; marsh elder, *Iva xanthifolia*; kochia, *Kochia scoparia*; dandelion, *Taraxacum officinale*; wild mustard, *Brassica kaber*; tansymustard, *Descurainia pinnata*; and lambsquarters, *Chenopodium album*; also are consumed by adults.

Hill (1946) conducted plant suitability experiments in Nebraska, assessing adult longevity, egg production, and larval survival on various host plants. Potato foliage was the most suitable food—egg production was highest and mortality lowest. Tomato also was a fairly suitable host, although not as favorable as potato. Buffalo bur was quite suitable for adult survival, but egg production was reduced considerably relative to potato. Plants that were decidedly less suitable included bean, ground cherry, marsh elder, and kochia. However, it is worth noting that even the less suitable hosts considerably extended longevity of adults as compared to the absence of food, and some egg production resulted from beetles fed these plants. No larval development was detected on marsh elder or kochia, but the tests were not exhaustive. Thus, though not optimal hosts, various weeds certainly would assist survival of beetles in the absence of potato and tomato.

Natural Enemies.   Natural enemies seem to be rare or nonexistent (Neilson and Finlayson, 1953).

Life History and Description.   There are 1–2 generations annually, though two is most common. Most of the second brood beetles enter diapause, but a few produce eggs and initiate a third generation. In some locations, only a single generation develops annually due to adverse weather. A complete generation requires about six weeks. Overwintering occurs at the adult stage within the soil or under plant debris, in or near potato fields, with adults first emerging in May or June (Vernon and Thomson, 1991).

Egg.   Egg deposition occurs for about a month, with an average of 90 eggs produced per female. However, Hill (1946) reported good survival for 40–60 days, with fecundity of up to 250 eggs per female. The elliptical eggs, which measure 0.5 mm long and 0.2 mm wide, are deposited singly in the soil adjacent to food plants. Eggs hatch in about 10 days.

Larva.   The larvae feed below-ground on the roots of plants. The larva is elongate in form, and bears three pairs of true legs. It is white, though its head capsule is brown. It measures only about 1 mm at hatching, but grows to about 5.3 mm long and 0.8 mm wide at maturity. First-generation larvae are present from early June to mid-July. Larval development requires about three weeks. Second-generation larvae are present from about mid-July to mid-August. Third-generation larvae are present from about mid-August until October.

Pupa.   Pupation occurs in the soil. The pupa is white and measures about 2.5 mm long and 1.5 mm wide. The duration of the pupal stage is about two weeks.

Adult.   The adult is a small, shiny black beetle, measuring 1.5–2.0 mm long. Its antennae and legs are orange, and the elytra bear rows of fine punctures. The thorax and elytra are covered with a fine layer of hair. The hind femora are enlarged. The pre-oviposition period after adult emergence from overwintering is 5–8 days. Oviposition can occur over a period of 12–54 days, but averages about 37 days.

The biology of tuber flea beetle was given by Hill (1946), Neilson and Finlayson (1953), Beirne (1971), and Howard *et al.* (1994).

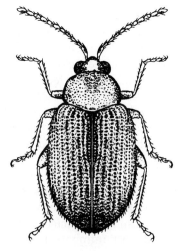

Adult tuber flea beetle.

Damage

Damage by tuber flea beetles is typical of flea beetle injury. Adults eat small holes in foliage and larvae feed on the below-ground parts of plants. (See the discussion on Potato Flea Beetle for a more detailed description of Flea Beetle Damage.) Tuber flea beetle apparently has a high rate of reproduction with excellent survival or both, because low spring populations often result in considerable damage by autumn (Vernon et al., 1990). Second-generation larvae are particularly damaging, because they feed on the surface of potato tubers, in the same manner as potato flea beetle, reducing their commercial value. Growers attempt to maintain a first-generation beetle density of less than 0.05 beetles per meter of row (or 1 beetle per 10 sweeps), a density that produces minimal damage to tubers by the second generation (Vernon and Thomson, 1993). Damage is often greatest along the edges of fields. Tuber flea beetle is considered much more damaging to potato tubers than to potato flea beetle.

As is the case with many flea beetle species, activity and damage are higher under hot and dry conditions. Overwintering survival is poor under exceptionally cold conditions, reducing damage during the subsequent summer. Organic soils are more favorable for flea beetle population growth than to mineral soil (Vernon and MacKenzie, 1991a).

Management

Sampling. Tuber flea beetles prefer to feed and mate on the upper surface of foliage, and are easily monitored visually. This is the only practical method for newly emerged or small potato plants, as other methods are too destructive. As plants mature, other techniques, such as use of a sweep net, are more appropriate (Vernon et al., 1990). Flea beetle densities tend to be higher along the margins of potato fields (Cusson et al., 1990), suggesting that although overwintering may occur within a field, most beetles may overwinter in weedy or untilled areas. This aggregated distribution must be considered when deciding on the need for control.

Insecticides. Potato growers in British Columbia apply insecticides from plant emergence to maturity to prevent damage by these flea beetles (Vernon and MacKenzie, 1991a). Systemic and nonsystemic granular insecticides are commonly applied to prevent damage. Foliar insecticides also are effective, but it may be necessary to apply insecticides every week (Vernon and MacKenzie, 1991b). If the overwintering population can be reduced adequately, the need for late season pest suppression is reduced. Insecticide resistance has been a problem in British Columbia (Finlayson et al., 1972), but now many insecticides are effective.

Cultural Practices. Flea beetle densities are higher in potato fields where potatoes have been grown previously (Cusson et al., 1990), suggesting that there is considerable value in rotating crops. If potatoes are to be grown for the first time in a field, there is little likelihood that flea beetles are present; thus, treatment of the edge rows with a granular systemic insecticide should provide a chemical "barrier," protecting the remainder of the crop from invading beetles. Early harvesting of potatoes can help avoid most of the damage caused by the summer generation of larvae.

In some areas potatoes can be planted either early or late in the growing season. In such locations, beetles typically attain moderate densities in the early season plantings and then disperse to later plantings, where significant damage occurs. It may be advisable to treat early season plantings to avoid damage in adjacent late season plantings if the early plantings are smaller in acreage. In any event, growers should be aware of the inoculum potential of early crops.

Western Black Flea Beetle
Phyllotreta pusilla Horn
(Coleoptera: Chrysomelidae)

Natural History

Distribution. Western black flea beetle is distributed widely in the western regions of the United States and Canada. It is particularly damaging along the western edge of the Great Plains, from Saskatchewan in the north to New Mexico in the south.

Host Plants. This insect has also been called the western cabbage flea beetle, which is a fair indication of its host preference. However, unlike many crucifer-feeding flea beetles, it will feed readily on other crops. Western black flea beetle is most damaging on mustard, radish, and turnip; but broccoli, cabbage, cauliflower, horseradish, kohlrabi, rutabaga, and watercress also are attacked. Among other vegetable crops, damage also has been reported on carrot, corn, beet, bean, lettuce, pea, potato, and tomato. Other economic plants injured include sugarbeet and nasturtium. Common weed hosts include pepperweed, Lepidium spp. and tansymustard, Descurainia pinnata. Feeding by larvae seems to be restricted to crucifers.

Natural Enemies. The principal natural enemies seem to be wasp and nematode parasites. Microctonus pussillae Muesebeck (Hymenoptera: Braconidae) is found throughout the summer months, but parasitism in excess of 16% has not been observed. A closely related parasitoid, *M. punctulatae* Loan and Wylie, also readily parasitizes this flea beetle under laboratory conditions, but levels of parasitism in the field are unknown (Wylie, 1984). Nematodes, probably *Howardula* sp. (Nematoda: Allantonematidae) have been observed to be numerous in western cabbage flea beetle, but the only reported effect seems to reduce egg production by the beetles (Chittenden and Marsh, 1920). Birds also will consume flea beetles.

Life Cycle and Description.　The number of generations per year is about three in Colorado, where the season of activity ranges from April until September. The duration of the life cycle is about 30 days. Adults overwinter in soil or under organic debris in northern climates. In warmer climates, beetles may remain active throughout the year.

Egg.　The eggs are deposited in soil near plants, with females each producing up to 250 eggs over a ten-week period. The oval eggs are light yellow, and measure about 0.5 mm long. Eggs hatch in 5–10 days.

Larva.　Young larvae feed on root hairs, while older larvae also feed on large roots. Larvae are whitish, except the head and anal plate that are brown. The larva bears an anal proleg, and eventually attains a length of about 5 mm. Duration of the larval stage is about 20–25 days.

Pupa.　Mature larvae construct small pupal chambers in the soil, and spend two to five days just before to pupation. Duration of pupation is about 10 days.

Adult.　Beetles are blackish with a shining surface, and have a metallic copper or bronze cast. The hind femora are enlarged. They measure 1.5–2.9 mm long.

The biology of western black flea beetle was outlined by Chittenden and Marsh (1920).

Adult western black flea beetle.

Damage

The principal form of damage is leaf feeding by adults. Beetles chew small holes in foliage, which can kill or impair the growth of seedling plants. Larvae also feed on the roots, reducing growth rates of plants. A more complete description of flea beetle injury can be found in the section on crucifer flea beetle, *Phyllotreta cruciferae* (Goeze).

Management

Management practices for this insect are similar to those recommended for crucifer flea beetle. The principal difference to consider is the wider host range of western black flea beetle. Not only may noncrucifer crops be damaged, but a wider variety of weeds will support beetle survival.

Zimmermann's Flea Beetle
Phyllotreta zimmermanni (Crotch)
(Coleoptera: Chrysomelidae)

Natural History

Distribution. Zimmermann's flea beetle is widespread in North America, and is known from nearly all states and provinces. However, it is most frequently found in the northeastern and midwestern states. Its abundance is reported to have diminished with the introduction from Europe of crucifer flea beetle, *Phyllotreta cruciferae* (Goeze), and striped flea beetle, *Phyllotreta striolata* (Fabricius), presumably owing to competition.

Host Plants.　This beetle feeds principally on crucifers, including such vegetables as broccoli, cabbage, collards, mustard, radish, turnip, and watercress. The adults sometimes feed on other plants, however, and have occasionally been observed feeding on alfalfa, clover, horseradish, peach, sunflower, tobacco, and wheat. Weeds serving as suitable hosts are black mustard, *Brassica nigra*; Virginia pepperweed, *Lepidium virginicum*; shepherdspurse, *Capsella bursa-pastoris*; yellow rocket, *Barbarea vulgaris*; and especially common hop, *Humulus lupulus*; and pepperweed, *Lepidium* spp. Larval hosts are reported as watercress, pepperweed, and infrequently rock cress, *Arabis* sp.

Natural Enemies.　A common flea beetle parasitoid, *Microctonus vittatae* Muesebeck (Hymenoptera: Braconidae), attacks Zimmermann's flea beetle and attains levels of parasitism of 60–70% (Loan, 1967b). The parasitoid is subject to hyperparasitism by *Mesochorus phyllotreta* Jourdheuil (Hymenoptera: Ichneumonidae), which diminishes the effectiveness of the primary parasitoid in regulating flea beetle abundance. A nematode, *Howardula* sp. (Nematoda: Allantonematidae) also was reported to parasitize this beetle (Elsey, 1977a).

Life Cycle and Description. Apparently there is only a single generation per year, though the complete life cycle requires about 30 days. Adults are the over-wintering stage, and become active early in the spring, which is March in most of the United States, but April or May in Canada.

Egg. Beetles deposit their eggs on the upper surface of foliage along the midvein. They were reported by Riley (1884) to be white and elongate, and measure only 0.02 mm long, but this measurement is likely an error; 0.2 mm is more probable.

Larva. The larvae are dark orange, with darker spots, and a head that is nearly black. They attain a length of about 5 mm at maturity. The larvae burrow into the foliage and feed as leaf miners between the upper and lower epidermis. This feeding habit is unusual among crop-feeding flea beetles, although several weed-feeding species also are leaf miners. In Missouri, the larvae are very abundant during May and June (Riley, 1884). The larvae may leave their mine and form another in a more suitable location, including another leaf.

Pupa. Pupation occurs in the soil. The pupa seems to be undescribed, but likely is very similar to that of crucifer flea beetle, *Phyllotreta cruciferae* (Goeze).

Adult. Beetles are black or blackish, with a straw-yellow irregular band running the length of each elytron. They greatly resemble striped flea beetle and western striped flea beetle. These flea beetles are most easily distinguished by the size of the fifth antennal segment of males. In western striped flea beetle, *Phyllotreta ramosa* (Crotch), antennal segments 4–6 are nearly the same size. In striped flea beetle, the fifth segment is wider and longer than adjacent segments, about twice the length of the sixth segment. In males of Zimmermann's flea beetle, the fifth antennal segment is even more enlarged, about three times the length of the sixth segment. Zimmermann's flea beetle measures 2.0–3.0 mm long and averages larger than the other two species.

The biology of Zimmermann's flea beetle was outlined by Riley (1884), Chittenden (1927), and Smith (1985).

Damage

Although larvae are leaf miners, they generally confine to their feeding activity to weeds, and so cause no

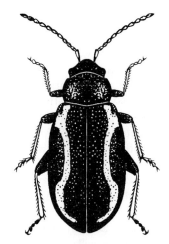

Adult Zimmermann's flea beetle.

economic loss to crops. Adults can sometimes become abundant, and in early spring they can be damaging to seedling plants. Adults eat small holes in the foliage, sometimes completely skeletonizing leaves, leading to the death of young plants.

Management

Management of adult flea beetles was discussed in detail under crucifer flea beetle, which should serve as a guide for Zimmermann's flea beetle control. Unlike crucifer flea beetle, however, larval control is not an issue.

FAMILY CHRYSOMELIDAE, SUBFAMILY CASSIDINAE—TORTOISE BEETLES

Argus Tortoise Beetle
    Chelymorpha cassidea **(Linnaeus)**

Blacklegged Tortoise Beetle
    Jonthonota nigripes **(Olivier)**

Golden Tortoise Beetle
    Charidotella bicolor **(Fabricius)**

Mottled Tortoise Beetle
    Deloyala guttata **(Olivier)**

Striped Tortoise Beetle
    Agroiconota bivittata **(Say)**
        (Coleoptera: Chrysomelidae)

Natural History

Distribution. These native species are distributed widely in eastern North America, west to Iowa and

Texas. Most also have a broader range, with *D. guttata* found in Mexico and South America, *A. bivittata* in Arizona and Mexico, and *C. cassidea* and *J. nigripes* in western states and Mexico.

Host Plants. The aforementioned species of tortoise beetles are associated with sweet potato and related species such as morningglory, *Ipomoea* spp., and bindweed, *Convolvulus* spp. Argus tortoise beetle is reported from other plants such as cabbage; corn; milkweed, *Asclepias* sp.; sunflower, *Helianthus* sp.; and dodder, *Cuscuta*. However, most records indicate Convolvulaceae as the host, and other records are suspect.

Natural Enemies. Several wasp parasitoids are known, including *Emersonella niveipes* Girault (Hymenoptera: Eulophidae) and *Brachymeria russelli* Burks (Hymenoptera: Chalcididae) from argus tortoise beetle, and *Tetrastichus cassidus* Burks (Hymenoptera: Eulophidae) from golden tortoise beetle and mottled tortoise beetle. Fly parasitoids include *Eucelatoriopsis dimmocki* (Aldrich) from striped tortoise beetle and golden tortoise beetle, and *Eribella exilis* (Coquillett) and *Eucelatoriopsis dimmocki* from argus tortoise beetle (all Diptera: Tachinidae).

Numerous predators are reported to feed on tortoise beetle larvae, including lady beetles such as *Coccinella* spp. and *Coleomegilla* spp. (Coleoptera: Coccinellidae) but especially insects with piercing-sucking mouthparts, such as damsel bugs (Hemiptera: Nabidae), shield bugs (Hemiptera: Pentatomidae), and assassin bugs (Hemiptera: Reduviidae). The "shield" carried by larvae (see description of larvae below) is somewhat effective against small predators, but large predators, especially those with long piercing-sucking mouthparts, are not deterred (Olmstead and Denno, 1992, 1993).

Life Cycle and Description. Very little biological information is available on these species, probably reflecting their slight economic importance. In the northern states, there usually is only one generation annually. In New Jersey, the beetles first appear in May or June, commence feeding on weeds, and deposit eggs soon thereafter. A new population of adults is evident in July. New adults feed briefly before entering diapause until the following spring. In argus tortoise beetle a few beetles deposit eggs in August, which successfully develop into a second generation. Development time from egg to adult requires about 40 days.

Egg. The eggs of most species are attached singly to the underside of leaves or on stems, and usually are white. Eggs are oval and flattened, but sometimes decorated with sharp angles or spines. They measure

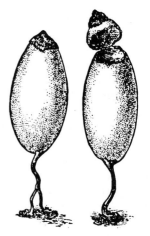

Argus tortoise beetle eggs.

only about 1 mm in length and hatch in 5–10 days. In argus tortoise beetle, the eggs are elongate-spherical and attached to the leaf by a stalk or pedicel. Each egg of argus tortoise beetle bears a small pointed cap. Eggs are deposited in clusters of about 20 eggs. The eggs of this species are slightly larger than the others, measuring about 1.6 mm long. Fecundity is not well documented, but Rausher (1984) reported oviposition of 50–90 eggs by argus tortoise beetle in a two-week period.

Larva. The larvae are broad and flattened and adorned with branched spines. Their thoracic legs are short and thick, and unlike many chrysomelids they lack an anal proleg. The color of the larva varies among and within species. Most are yellowish, but larval color may be reddish brown in golden tortoise beetle (see color figure 114), and greenish in mottled tortoise beetle. There are three instars. Larvae display the habit of carrying their cast skins and fecal material attached to spines arising from the posterior end of their body, a structure called an "anal fork." The anal fork is movable, and is usually used to hold the debris over the back of the body, forming a "shield" which deters predation. Striped tortoise beetle differs from the other species in not carrying debris on the anal fork. Larvae of tortoise beetles mature in 14–21 days.

Pupa. When mature, the larva attaches itself to the leaf by its anal end and pupates. Pupae are brownish, except that in mottled tortoise beetle the pupa is greenish. As is the case with larvae, the pupae bear spines. The fecal material and other debris carried by the larval stage may also remain attached to the pupa. The pupa measures 5–8 mm long. Duration of the pupal stage is usually 7–14 days.

Adult. The adult beetles are distinctive in that the margins of the prothorax and elytra are expanded,

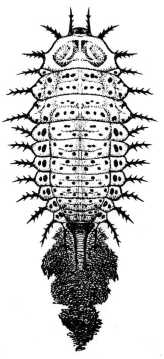

Argus tortoise beetle larva.

Argus tortoise beetle pupa.

name suggests, particularly golden-metallic (see color figure 113). When the adult first emerges, however, the beetles are orangish with six spots, but this soon fades and the golden color is acquired. Mottled tortoise beetle usually bears considerable black pigment, especially at the base of the pronotum and dorsally on the elytra. The periphery of the beetle is mostly transparent, however. The black pattern on the beetle often resembles a shield, but is highly variable. Striped tortoise beetle, perhaps the most distinctive of the sweet potato-feeding species, is yellow with two black stripes on each elytron.

Complete description of these interesting beetles is lacking. Riley (1870b), Smith (1910), and Barber (1916) gave useful observations, and Chittenden (1924) described argus tortoise beetle. Downie and Arnett (1996), and Riley (1986) provided keys to the genera and brief descriptions of the species.

Damage

Both larvae and adults feed on the foliage of sweet potato. The typical form of injury is creation of numer-

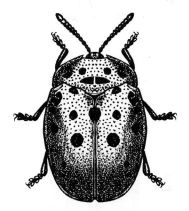

Adult argus tortoise beetle.

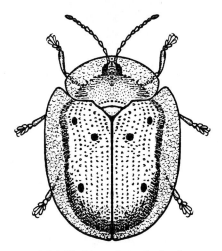

Adult blacklegged tortoise beetle.

largely concealing the head and appendages. The expanded margins usually are not heavily pigmented. The beetles are fairly small, measuring 5.0–7.0 mm long. Argus tortoise beetle is a notable exception, measuring 8.0–11.5 mm. The beetles vary in color, but invariably are brightly colored, often metallic, and are sometimes called "goldbugs." Argus tortoise beetle is reddish or yellowish, and usually it has six black spots on the pronotum and on each elytron, plus one spot dorsally that overlaps each elytron. Blacklegged tortoise beetle is red to yellowish, and bears several (usually three) conspicuous black spots and several inconspicuous small spots on each elytron. A distinctive feature of this species is the black legs, the basis for its common name. Golden tortoise beetle is, as its

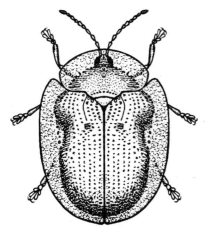

Adult golden tortoise beetle.

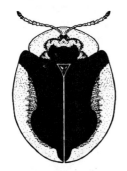

Adult mottled tortoise beetle.

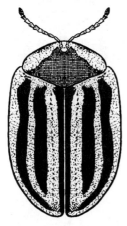

Adult striped tortoise beetle.

ous small- or medium-sized irregular holes. The leaf may be completely riddled with holes, but the major veins are left intact. Both stages usually inhabit the lower surface, but eat entirely through the foliage. The tortoise beetles are rarely abundant enough to be considered damaging.

Management

The larvae and adults are easily controlled with foliar insecticides, but this action is rarely warranted.

Eggplant Tortoise Beetle
Gratiana pallidula (Boheman)
(Coleoptera: Chrysomelidae)

Natural History

Distribution. The eggplant tortoise beetle is found widely in the United States, from Maryland to Florida in the east, and west to California. A native species, eggplant tortoise beetle is most common in southern states, and is not known as damaging in northern states and Canada. A strikingly similar insect, *G. lutescens* (Boheman), is common in South America and interbreeds successfully with eggplant flea beetle, so the status of these species is uncertain (Siebert, 1975).

Host Plants. This species feeds on plants in the family Solanaceae. In addition to eggplant, potato, and tomato, eggplant tortoise beetle has been collected from horse nettle, *Solanum carolinense*; silverleaf nettle, *Solanum elaeagnifolium*; and nightshade, *Solanum* sp.

Natural Enemies. An unspecified egg parasitoid was noted by Jones (1916b). A pupal parasitoid *Spilochalcis sanguiniventris* (Cresson) (Hymenoptera: Chalcididae) also is known, but its significance is uncertain.

Life Cycle and Description. The number of generations per year is about five in Louisiana. A complete life cycle requires about 30 days. The insects are active from May to October.

Egg. The egg is elongate oval, measuring about 1.24 mm long (range 1.16–1.30 mm) and 0.67 mm wide (range 0.62–0.76 mm). The egg is whitish initially, but turns brown with age. Eggs generally are deposited on the underside of foliage, but sometimes elsewhere. Eggs are deposited singly or in small clusters of 2–4, and covered with 1–2 layers of brownish adhesive. Fecundity is estimated at 250 eggs or more. Duration of the egg stage is generally 4–5 days.

Larva. The larvae are light green or yellowish green. The surface of the body is equipped with about 16 pairs of branched spines occurring laterally. The tip of the abdomen also bears a pair of spines called the "anal fork," which collects the shed cuticles and excrement. The anal fork bearing debris may be held over the body, presumably providing camouflage or physical protection from natural enemies. There are five instars. Larval head capsule widths are about 0.46, 0.55, 0.67, 0.82, and 0.97 mm, respectively. Mean development time of larvae is about 17 days (range 12–20 days). Larvae eventually attain a length of about 5.0–5.5 mm.

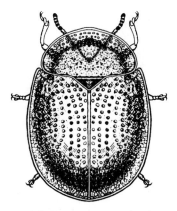

Adult eggplant tortoise beetle.

Pupa. Pupation occurs on the foliage, where the pupa is attached by the anal region. The pupa measures about 5.4 mm long, and retains most of the branched spines present on the larva. Duration of the pupal stage is 4–7 days.

Adult. The adult is oval, and measures about 5.0–6.0 mm long. It is greenish or yellow-green, with yellowish-brown appendages. The elytra bear several rows of coarse punctures. Adults normally live 2–3 months.

The biology of eggplant tortoise beetle was described by Jones (1916b).

Damage

The larvae and adults occur on the foliage of solanceous species, where they feed and create small, somewhat circular holes. Unlike feeding by most flea beetles, with which tortoise beetle injury could be confused, eggplant tortoise beetle feeds completely through the foliage. This insect is rarely considered damaging, and is usually restricted to eggplant.

Management

Foliar insecticides are effective against both larval and adult stages of this insect.

FAMILY CHRYSOMELIDAE, SEVERAL SUBFAMILIES—LEAF BEETLES

Asparagus Beetle
Crioceris asparagi (Linnaeus)
(Coleoptera: Chrysomelidae)

Natural History

Distribution. Asparagus beetle is found throughout North America wherever asparagus is grown. This insect is of European origin, and was first observed in the western hemisphere at Long Island, New York, about 1856. It spread rapidly in the northeast and midwestern states and eastern Canada, and reached California in 1904, Oregon and Washington in the 1910s and 1920s, and British Columbia in 1926. This species is invariably more abundant than the co-occurring spotted asparagus beetle, *Crioceris duodecimpunctata* (Linneaus).

Host Plants. Asparagus beetle feeds only on asparagus. Asparagus has escaped cultivation and grows wild throughout the northern United States and southern Canada wherever there is adequate moisture. These wild plants, as well as homegarden plants, are the principal food resources of asparagus beetle. Regular harvesting, and especially insecticide application, generally keep asparagus beetle from attaining high densities on commercial crops. Nevertheless, in northern climates it occasionally is damaging. California, a major producer of commercial asparagus, is relatively free of asparagus beetle problems; hot weather likely is the basis for few problems in California and southern states.

Natural Enemies. Several general predators and parasitoids are known from asparagus beetles. The lady beetles *Coleomegilla maculata* Mulsant and *Hippodamia convergens* Guerin-Meneville, *Coccinella transversoguttata* Brown, and *Coccinella novemnotata* Herbst (all Coleoptera: Coccindellidae) feed on eggs and larvae. Larvae also are attacked by the stink bugs, *Perillus bioculatus* Fabricius and *Stiretrus anchorago* (Fabricius) (both Hemiptera: Pentatomidae) and the assassin bugs, *Sinea* sp., *Pselliopus* sp., and *Arilus cristatus* (Linnaeus) (all Hemiptera: Reduviidae).

The most important mortality factor is *Tetrastichus asparagi* Crawford (Hymenoptera: Eulophidae), which feeds on young eggs and parasitizes older eggs. The wasp uses her ovipositor to puncture the egg and then, in young eggs, applies her mouthparts to the puncture. She then sucks the contents from the chorion, causing collapse of the egg, but the egg remnants remain attached to the plant. In Massachusetts, about 50% of eggs are fed upon by *T. asparagi*, and a similar proportion of the survivors is parasitized (Capinera and Lilly, 1975a). Although parasitized larvae complete their development, they perish after forming their pupal cell. A mean of 4.75 parasitoids develops in each parasitized beetle larva. The parasitoids, which are specific to *C. asparagi*, are well-synchronized with their hosts, and emerge in time to feed on and parasitize the subsequent generation (Capinera and Lilly, 1975b; van Alphen, 1980). There are three generations annually. The biology of this parasitoid was given by Russell and Johnston (1912), and Johnston (1915).

There have been attempts to relocate parasitoids and to increase parasitism rates. *T. asparagi* was shipped from Ohio to the west coast in the 1930s and successfully established in Washington, but not in California (Clausen, 1978). Additional species of parasitoids were imported from Europe, and one, *Lemophagus cioceritor* Aubert (Hymenoptera: Ichneumonidae) is likely established in Ontario (Hendrickson *et al.*, 1991).

Life Cycle and Description.   The number of generations is probably three throughout this insect's range, though there is some speculation that there may be more generations in mild climates. In Ontario, Taylor and Harcourt (1975, 1978) reported three generations, with peak egg-laying in early June, July, and August. This was found consistent with observations from Massachusetts, though only a portion of the adults from the second generation produced eggs. Thus, the third generation was very small (Capinera and Lilly 1975a), giving the impression of only two generations. This species overwinters as adults.

Egg.   The adults commence mating and oviposition soon after emerging from overwintering quarters. Eggs are laid on end in short rows, usually numbering 3–10 in each cluster, and deposited on foliage, stems, and flower buds over the canopy of the plant (see color figure 260). The eggs are elongate-oval and dark brown in color, and measure about 1.2–1.3 mm long and 0.5 mm wide. Mean duration of the egg stage is 3.25 days at 26°C, with a range of 3–8 days in normal temperature.

Larva.   The larvae are brown or dark gray, and measure about 1.5 mm long at hatching, but attain a length of about 8 mm at maturity. The head and legs

Asparagus beetle larva.

are black, the body is rather plump, and the tip of the abdomen bears anal prolegs, which help to anchor the larva to the plant while it feeds. There are four instars. Mean development times are about 1.6, 1.7, 2.2, and 6 days for instars 1–4, respectively, at 26°C. The rate of development increases from about 10°C, the lowest temperature at which development occurs, until the upper limit of development is attained at about 34°C. At 32–34°C, larval development is completed in about 7.9 days. When survival rates at various temperatures are considered in addition to development rates, optimal temperature for larval asparagus beetle is about 30–32°C. Taylor and Harcourt (1978) used temperature relations to predict accurately the development of beetle populations under field conditions in Ontario. (See color figure 105.)

Pupa.   Mature larvae drop to the soil and construct pupal cells from oral secretions and soil particles. The pupa resembles the adult beetle except for the poorly developed wings and the yellowish coloration, and measures about 6 mm in length. Mean duration of the pupal stage is 5.7 days at 26°C, decreasing to 4.0 days at 30–32°C.

Asparagus beetle eggs on blossom bud.

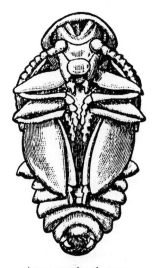

Asparagus beetle pupa.

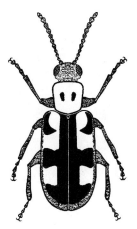

Adult asparagus beetle.

Adult. The adult is a brightly colored insect, measuring about 5–6 mm long. The head, legs, antennae, and ventral surfaces are bluish black. The thorax is reddish, and usually marked with two black dots. The elytra are highly variable in color. The tip and outer margin (leading edge) are orange, and the remainder is bluish black, but marked centrally with up to three yellowish white spots. The beetles are reluctant to fly when disturbed, usually dodging around to the opposite side of the stem. They also drop readily from the plant and feign death if they or the substrate are disturbed (Capinera, 1976). Adults are reported to overwinter in hollow asparagus stems, under debris in crop fields, and under loose bark of trees. (See color figures 4 and 104.)

The biology of asparagus beetle was given by Chittenden (1917) and Dingler (1934). Developmental biology was reported by Taylor and Harcourt (1978).

Damage

The larvae and adults feed on foliage during the summer months, reducing photosynthetic capacity of the plants and affecting subsequent crop yield. This damage is relatively infrequent, however, compared to direct damage by adults early in the season, when they feed directly on emerging asparagus spears. The pits and gouges caused by adult feeding render the spears of no commercial value. Adults also deposit eggs on spears, and on occasion eggs are serious contaminants because they cannot be removed by washing. Asparagus beetles are a regular problem in homegardens and on volunteer asparagus, and beetles may disperse to commercial crops occasionally. Because of the nature of the damage, growers in northern asparagus-producing areas consider asparagus beetle a serious threat.

Management

Monitoring of adult populations is recommended early in the season. Visual observation is adequate. The distribution of all stages is clumped (Taylor and Harcourt, 1975).

Foliar insecticides can be applied to destroy adults and prevent egg laying on spears (McClanahan, 1975a). Spears are harvested almost daily, so use of an insecticide with little residual activity is essential. Beetles will often congregate on plants with an abundance of foliage, so it is advisable to leave a small portion of the crop as a decoy, while other areas are harvested. Destruction of volunteer plants, which often spring up along roadsides and irrigation ditches, can prevent population increases.

Banded Cucumber Beetle
Diabrotica balteata LeConte
(Coleoptera: Chrysomelidae)

Natural History

Distribution. Banded cucumber beetle is a tropical insect. Before 1910, its distribution in the United States was limited to southern Arizona and Texas (Saba, 1970). Now it is distributed all over the southern United States from North Carolina to southern California, and south through Mexico and Central America (Krysan, 1986). Although, it has spread rapidly in the United States, its intolerance to freezing temperatures probably limits its northward distribution to its current status. Saba (1970) observed poor survival of all stages at 0°C, and no survival below 0°C.

Host Plants. Adults feed on a wide range of plants, but seem to prefer plants from the family Cucurbitaceae, Rosaceae, Leguminoseae, and Crucifereae (Saba, 1970). Vegetable crops damaged include cucumber, squash, beet, bean, pea, sweet potato, okra, corn, lettuce, onion, and various cabbages (Chittenden, 1912a; Saba, 1970). Despite the common name, bean and soybean are especially favored and in tropical countries this insect is commonly viewed as a bean pest (Cardona *et al.*, 1982).

Natural Enemies. Except for nematodes, the natural enemies of banded cucumber beetle are poorly known. The mermithid nematode, *Filipjevimermis leipsandra* (Poinar and Welch), affects this species, as well as spotted cucumber beetle and several flea beetle species (Creighton and Fassuliotis, 1980). In banded cucumber beetle, natural infection levels in South Carolina peaked in late summer, but levels in the range of 20–40% were common from May through October. *F. leipsandra* reproduces parthenogenetically and has a life cycle duration of 65–70 days under favorable conditions (Cuthbert, 1968). The heterorhabditid nematode *Heterorhabditis heliothidis* (Khan, Brooks, and Hirschman), has also been isolated from field populations of banded cucumber beetle (Creighton and Fassuliotis, 1985), but it is uncertain whether this nematode is an important natural mortality factor. Ants have been shown as important egg predators in the tropics (Risch, 1981) and likely are important elsewhere.

Life Cycle and Description. Banded cucumber beetle does not enter diapause (Saba, 1970). It remains active as long as the weather remains favorable, with up to 6–7 generations per year reported in Louisiana (Pitre and Kantack, 1962) and Texas (Krysan and Branson, 1983). Under optimal conditions, a life cycle can be completed in 45 days.

Egg. The eggs are yellow, oval in form, and measure about 0.6 mm long and 0.35 mm wide. They are deposited in cracks in the soil, and 5–9 days normally are required before hatching (Marsh, 1912a).

Larva. The three instars have mean head capsule widths measuring about 0.24, 0.35, and 0.51 mm, respectively. The body length during these instars is reported to be about 2.3, 4.5, and 8.9 mm. Larval color is somewhat variable; initially it is white, but may also take on a pale-yellow color, depending on the food source. Development time is temperature dependent, but the range is about 4–8, 3–11, and 4–15 days for instars 1–3, respectively. Total larval development time is usually 11–17 days.

Pupa. Pupation occurs in the soil, with a duration of 4–6 days.

Adult. The adults are 5–6 mm long, greenish yellow, with a red head and black thorax. Usually there are three transverse bands across the elytra, green in color but sometimes with a bluish tint, and a thin green band running down the center of the insect's back. The banding pattern is variable, and sometimes almost absent (Chittenden, 1912a). Copulation occurs about six days after adult emergence, whereas egg deposition begins about 16 days later. Oviposition takes place at 2–3-day intervals for 2–8 weeks. Females normally deposit 2–15 egg clusters up to 100 eggs each. A total of 850 eggs may be produced by a female. Adult longevity is 17–44 days, but averages 26 days. (See color figure 112.)

Detailed information on banded cucumber beetle biology was provided by Pitre and Kantack (1992), and Saba (1970). Rearing techniques were provided

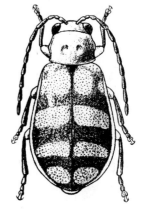

Adult banded cucumber beetle.

by Schalk (1986). Biosystematics of *Diabrotica* spp. was presented by Krysan and Branson (1983).

Damage

Banded cucumber beetle is nearly omnivorous, and in addition to numerous plants being attacked, all parts of the plant are injured. Marsh (1912a), for example, noted damage to foliage, blossoms, silk, kernels, the plant crown, and roots. Larvae feed only on the roots. The most frequent forms of serious injury are defoliation by adults and root feeding on plant seedlings by larvae. Some of the most serious injury results from larval feeding on sweet-potato roots. There are no reports of cucurbit wilt transmission by banded cucumber beetle, but this species is much less well studied than striped and spotted cucumber beetles. Banded cucumber beetle is known as a vector of virus diseases in beans (Cardona *et al.*, 1982; Gergerich *et al.*, 1986), and Latin and Reed (1985) suggested that larval feeding might increase the incidence and severity of Fusarium wilt.

Management

Sampling. A female-produced sex pheromone has been identified (Chuman *et al.*, 1987).

Insecticides. Insecticides are sometimes used to prevent damage to roots by larvae. Most often, granular insecticides are applied in a band over the row, either at time of planting or soon thereafter. Foliar insecticides are sometimes needed to prevent excessive damage to seedlings, but adults are rarely abundant enough to warrant control on large plants.

Biological Control. Two nematodes have been well studied for suppression of banded cucumber beetle: the mermithid, *F. leipsandra* and the heterorhabditid, *H. heliothidis*. Creighton and Fassuliotis (1983) induced high levels of parasitism in banded cucumber beetle larvae by *F. leipsandra* with application of nematode eggs to field microplots. This nematode also infects several other chrysomelids. Some progress toward mass production of the nematode also has been reported (Creighton and Fassuliotis, 1982), but the nematode is not available commercially. *Heterorhabditis heliothidis* was found naturally parasitizing banded cucumber beetle (Creighton and Fassuliotis, 1985), and effectively reduced beetle larval numbers in trials with potted plants. Although *H. heliothidis* is not available commercially, the related *Steinernema* nematodes are effective under experimental conditions (Poinar, 1979) and are readily available.

Fungi have been evaluated experimentally for banded cucumber beetle control (Bell *et al.*, 1972). *Beauveria bassiana* and *Metarhizium anisopliae* infected both larvae and pupae, but *Spicaria rileyi* was relatively ineffective.

Cultural Practices. The wide host range of this insect, which includes many weed species (Saba, 1970) suggests that clean cultivation or some other form of weed control would be valuable in reducing damage to seedlings. Adults can be kept from attacking seedlings through use of screening or row covers.

Host-Plant Resistance. Cultivars of sweet potato partially resistant to cucumber beetle injury are known (Schalk and Creighton, 1989), but this information seems lacking for other plants.

Bean Leaf Beetle
Cerotoma trifurcata (Förster)
(Coleoptera: Chrysomelidae)

Natural History

Distribution. Bean leaf beetle is a native species, one of the large complex of legume-feeding species found in the American tropics and subtropics. Among the *Cerotoma* species, however, only *C. trifurcata* is abundant in the United States. It occurs widely east of the Great Plains, and is known from all eastern states and the southern portions of Ontario and Quebec.

Host Plants. This species feeds on plants in the family Leguminosae. Vegetable crops attacked include cowpea, lima bean, and snap bean. Other crops attacked are alfalfa, clover, and sweet clover, but bean leaf beetle is best known as a pest of soybean. Wild legumes known to support bean leaf beetle include bush clover, *Lespedeza striata*; hog peanut, *Amphicarpaea bracteata*; tick trefoil, *Desmodium cuspidatum*; and wild pea, *Strophostyles helvola*. In the spring, before legumes are available, adults may feed for a time on plants other than legumes (Helm *et al.*, 1983), though this is rarely observed. Although the host range of

adults is well-known, larval feeding is poorly documented. Larval and adult hosts are assumed to be the same.

Natural Enemies. An important parasitoid of bean leaf beetle and several other chrysomelids is *Celatoria diabroticae* (Shimer) (Diptera: Tachinidae). This species attacks mostly first-generation adult bean leaf beetles, with incidence of parasitism peaking in mid-summer. Parasitized adult beetles produce few eggs prior to death (Herzog, 1977). *Hyalomyodes triangulifer* (Loew) (Diptera: Tachinidae), also a parasitoid of several beetles, occurs occasionally in bean leaf beetle. In Minnesota, *Medina* sp. (Diptera: Tachinidae) caused high levels of parasitism, but this species has not been reported elsewhere (Loughran and Ragsdale, 1986b). Two species of ectoparasitic mites, *Trombidium hypri* Vercammen-Grandjean, Van Driesche and Gyrisco, and *T. newelli* Welbourn and Flessel (Acari: Trombidiidae) are found feeding beneath the adult elytra, but incidence of infestation is low (Peterson *et al.*, 1992).

Life Cycle and Description. There are three generations in southern states, two in northern areas such as Nebraska, Illinois, and North Carolina, and only one generation in Minnesota, the northern limit of bean leaf beetle's range. A complete generation requires about 30 days. In southern states adults become active and begin oviposition in April or May, with a new generation of adults present in June–July, August, and September. In more northern states the overwintering adults become active in May and June, producing new adults in July–August and August–September. In Minnesota, the phenology is similar to other northern states, except that the second generation does not develop. The adult is the overwintering stage, and surviving adults from all generations may overwinter.

Egg. The eggs of bean leaf beetle are deposited in the soil around the base of adult host plants to a depth of up to 3.8 cm, usually in clusters of about twelve. The eggs are oval and taper to a pronounced point at each end. They are reddish orange. Eggs measure about 0.8 mm long and 0.35 mm wide. Mean duration of the egg stage is 9.6 days at 25°C (ranging from 15 days at 20°C to 7 days at 30°C). Females generally produce about 250–350 eggs, with individuals producing as few as 100 to as many as 1,400.

Larva. The larval stage has three instars. The larva is elongate, with well-developed thoracic legs and a fleshy anal proleg on the terminal abdominal segment. The larva is white, with a blackish head and anal plate. The thoracic plate may also be darkened. The larva grows from 1 mm at hatching to 7–10 mm at maturity. Mean duration of the larval stage is 17.6 days at 25°C,

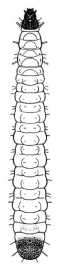

Bean leaf beetle larva.

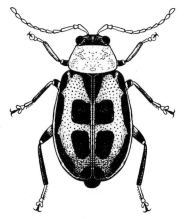

Adult bean leaf beetle.

and ranges from 23.9 days at 21°C to 12.8 days at 30°C. Development time of the three instars is reported to be 4.5, 3.4, and 7.9 days, respectively, at 26°C. The head capsule widths of the larvae are 0.17–0.19, 0.27–0.30, and 0.40–0.42 mm for instars 1–3, respectively. Only about two-thirds of the larval period is devoted to feeding; during the remainder of the time the larva is immobile in preparation for the molt to the pupal stage.

Pupa. Pupation occurs in the soil, in a small cell formed by the larva. The pupa is whitish and resembles the adult in form, though the wings are not fully developed. It measures 3.0–4.5 mm long. Mean duration of the pupal stage is 7.1 days at 25°C, and ranges from 10 days at 20°C to 5 days at 30°C. After successful molting to the adult stage, the beetle remains immobile for a few hours to a few days, depending on weather, and then digs to the soil surface.

Adult. The adult is a relatively small beetle, measuring 3.5–5.0 mm long. The beetle is yellowish to dull red and is marked with variable amounts of black. The head is black and the antennae pale. The legs are yellowish and black. In many individuals, the leading edge of the elytra bear a band of black, and there are 2–3 black spots centrally on each elytron. On other individuals, however, most of the black pigment is faded or absent. In nearly all specimens, there is a triangular spot of black on the elytra immediately behind the prothorax. The beetle feeds voraciously for a few days after emerging from the soil. During this time it mates and susequently commences oviposition. The pre-oviposition period is reported to average nine days, but this varies considerably. Availability of young seedling host plants is critical for egg development. Ovipositing females live 2–4 weeks, whereas those entering diapause persist for several months. Diapausing adults overwinter in clumps of grass, under leaf litter, and in soil. Forest litter is a favorite diapausing site (Boiteau *et al.*, 1980; Jeffords *et al.*, 1983) with emergence in the spring occurring when mean daily temperature exceeds 26°C and day length exceeds 13 h (Boiteau *et al.*, 1979).

The information on biology was provided by Isely (1930), Eddy and Nettles (1930), and Kogan *et al.* (1980). Phenology was given by Boiteau *et al.* (1980), Loughran and Ragsdale (1986a), and Smelser and Pedigo (1991). Laboratory rearing was outlined by Herzog *et al.* (1974).

Damage

The adults prefer young tissue and are most damaging to young plants. They feed on young leaves, if these are available, but also attack young pods. Typical above-ground injury involves consumption of holes of varying sizes in leaf tissues. The larger veins are usually spared. The consumption rate of foliage by adults is estimated at about 2 sq cm per day. The holes are too large to be confused with damage by flea beetles, and bean beetles eat completely through the smallest veins and the leaf blade, so they are not readily confused with Mexican bean beetle, *Epilachna varivestris* Mulsant. Damage could be mistaken for spotted cucumber beetle, *Diabrotica undecimpunctata* Mannerheim, however. The larvae of bean leaf beetle feed below-ground around the base of the seedling, causing direct injury and making it easier for soil-borne plant pathogens to gain entry into the young plant. Roots and nodules containing nitrogen-fixing bacteria also are consumed.

The adults transmit plant viruses, including pod mottle, cowpea mosaic, and southern bean mosaic viruses. Beetles are mechanical vectors of viruses, and not very efficient transmitters. They are likely the most import vectors, however, due to the abundance of the beetles (Jansen and Staples, 1970b; Patel and Pitre, 1976; Mueller and Haddox, 1980; Pitre, 1989).

Management

Sampling. Most studies of bean leaf beetle sampling have involved soybean, and the applicability of these studies to other crops is not known. Adult populations tend to be uniformly distributed to slightly aggregated. Sweep-net sampling is useful for adults in all crops, but visual observation is probably most appropriate early in the season when plants are small. Eggs are sampled from soil cores taken close to host plants, and separated from soil by washing through a series of screens. The larvae and pupae are separated from soil by brine flotation. Kogan *et al.* (1980) thoroughly discussed sampling.

Insecticides. Planting-time application of granular systemic insecticide to the seed furrow or over the row has been common to protect crops, because it prevents both root injury and foliar damage to young plants (Lentz *et al.*, 1983). Foliar insecticide applications are also effective, but do not protect against injury by larvae. A study in Alabama showed that insecticide incorporated into citrus pulp bait and applied to foliage was particularly effective for bean leaf beetle suppression (Harper, 1981), but this approach has not come into general practice.

Cultural Practices. Early plantings are highly attractive to overwintering beetles; later plantings sometimes escape injury (Witkowski and Echtenkamp, 1996). Zeiss and Pedigo (1996) examined the potential benefits of late planting and observed that beetles denied legumes early in life and forced to feed on grasses failed to reproduce, even if later supplied with soybean. Beetles provided only with alfalfa early in life initially had low fecundity, but regained a high rate of egg production once provided with soybean. Thus, area-wide delay in annual legume-crop planting has considerable benefit in reducing population increase, but it is reduced by availability of alternative legume food sources such as alfalfa.

Soil moisture and texture affect egg and larval survival, and development. The eggs are laid in moist organic soil in preference to dry and sandy soil (Marrone and Stinner, 1983a). The eggs absorb water and swell as part of their normal embryonic development. Lack of water causes egg development to cease, but both high and low amounts of water can be deleterious to eggs (Marrone and Stinner, 1983b). First instars are the larval stage that is most susceptible to mortality induced by unfavorable physical conditions in the soil. Pupae also are relatively susceptible to injury. Larvae can move several centimeters to areas of soil with higher organic matter or moisture content (Marrone and Stinner, 1983c). Young larvae are likely to burrow into root nodules under dry soil conditions. Larval survival is highest in organic and wet soils, and lowest in sandy and dry soils. Thus, fields with high organic matter content are more likely to experience damage from bean leaf beetle larvae (Marrone and Stinner, 1984).

Proximity of woodlots is implicated in outbreaks of bean leaf beetle. Availability of such suitable overwintering habitat, especially when combined with early season host availability, leads to higher likelihood of crop damage.

Colorado Potato Beetle
Leptinotarsa decemlineata (Say)
(Coleoptera: Chrysomelidae)

Natural History

Distribution. The origin of Colorado potato beetle probably is Mexico. This insect apparently dispersed northward into the Great Plains region, feeding on weeds, before introduction of extensive agriculture in the 1800s. The introduction of potato to the Great Plains provided this species with an abundant food supply, resulting in crop damage beginning about 1859 in Nebraska. Once Colorado potato beetle accepted potato as a food plant, it spread rapidly, crossing the Mississippi River by 1864, and reaching the east coast from Connecticut to Virginia by 1874. By 1878, all of eastern Canada was infested. It spread southward more slowly, not reaching Florida and Louisiana until by 1900. Colorado potato beetle is now found throughout the United States except for California, Nevada, Alaska and Hawaii. It also occurs throughout southern Canada. In recent times the Colorado potato beetle has become a pest in most of Europe, Mexico, and portions of Central America.

Host Plants. Colorado potato beetle feeds entirely on plants of the family Solanaceae. The most important vegetable host, as the common name suggests, is potato. Colorado potato beetle can sometimes damage tomato and eggplant, and rarely pepper, in the eastern United States. Among the weeds often accepted by adults, and also generally supporting larval growth, are bittersweet, *Solanum dulcamara*; buffalo bur, *S. rostratum*; horsenettle, *S. carolinense*; hairy nightshade, *S. sarachoides*; silverleaf nightshade, *S. elaeagnifolium*; *S. dimidiatum*; and henbit, *Hyoscyamus niger*; and possibly some other *Solanum* spp. (Hsiao 1988; Weber *et al.*,

1995). Buffalo bur is perhaps the original host of Colorado potato beetle, and though being a good host, potato usually is superior (Hsiao and Fraenkel, 1968). For example, Brown *et al.* (1980) compared the growth of Colorado potato beetles fed potato or hairy nightshade. Larvae fed nightshade grew more slowly, required longer to commence oviposition, and produced fewer eggs than those fed potato. Solanaceous plants that are sometimes accepted as poor hosts include petunia; tobacco; wild tobacco, *Nicotiana rustica*; jimsonweed, *Datura stramonium*; apple of Peru, *Nicandra physalodes*; ground cherry, *Physalis heterophylla*; and matrimony vine, *Lycium halimifolium*. Colorado potato beetle has repeatedly displayed the ability to adjust its host range to accept locally abundant *Solanum* species (Horton *et al.*, 1988; Mena–Covarrubias *et al.*, 1996), so its host range varies geographically and continues to expand.

Colorado potato beetles have been observed feeding on non-solanaceous plants under field conditions, and the ability of larvae to feed on these and to display short-term growth have been confirmed under laboratory conditions. Among these casual hosts are milkweed, *Asclepias syriaca*; butterfly weed, *A. tuberosa* (both family Asclepiadaceae); pea (family Leguminosae); cabbage; shepherdspurse, *Capsella bursa-pastoris* (both Cruciferae); Romaine lettuce; and bullthistle, *Cirsium vulgare* (both Compositae) (Hsiao and Fraenkel 1968). However, such feeding is rare and these plants should not be considered normal hosts. Bongers (1970) conducted extensive tests on host selection and provided an extensive review of the literature. Hsiao (1988) also addressed host specificity, including the semiochemicals that affect acceptance of plants by beetles.

Natural Enemies. Many natural enemies have been identified, but they usually are unable to keep the Colorado potato beetle population at low levels of abundance. Harcourt (1971), for example, concluded that there were no effective natural enemies in Canada and that populations were regulated by starvation and dispersal. Not every investigator is so pessimistic about natural enemies, but clearly there is a shortage of effective naturally occurring enemies. Among common predators are green lacewings, *Chrysoperla* spp. (Neuroptera: Chrysopidae); several stink bugs but especially twospotted stink bug, *Perillus bioculatus* (Fabricius) and spined soldier bug, *Podisus maculiventris* (Say) (both Hemiptera: Pentatomidae); damsel bugs, *Nabis* spp. (Hemiptera: Nabidae); many lady beetles but particularly *Coleomegilla maculata* (De Geer) and *Hippodamia convergens* Guerin-Meneville (both Coleoptera: Coccinellidae); and ground beetles, *Lebia grandis* Hentz and *Pterostichus* spp. (both Coleop-

tera: Carabidae). The aforementioned insects are general predators, feeding on a number of soft-bodied insects, and so their occurrence is not tightly linked to the abundance of Colorado potato beetle.

Parasitoids of Colorado potato beetle are more specific in their host associations. The best known is the fly *Myiopharus doryphorae* (Riley) (Diptera: Tachinidae), which characteristically builds to high densities in autumn, inflicting high levels of parasitism on the final generation of beetles. Unfortunately, despite the ability of this tachinid to cause high levels of parasitism, it has poor overwintering ability, and begins the year with low population densities and poor ability to suppress early season potato beetle populations. An exception seems to be in Colorado, where parasitism rates are high early in the season, and where Colorado potato beetles are not usually a serious pest (Horton and Capinera, 1987a). Other tachinid parasitoids of Colorado potato beetle include *Myiopharus aberrans* (Townsend), *M. australis* Reinhard, *M. macella* Reinhard, and *Winthemia* spp.

Pathogens seem to be relatively unimportant in the survival of Colorado potato beetle. The nematodes *Mesodiplogaster lheritieri* and *Pristionchus uniformis* (both Nematoda: Diplogasteridae) have been recovered from Colorado potato beetle (Poinar, 1979). A mollicute, *Spiroplasma leptinotarsae*, inhabits the gut lumen of both adults and larvae, but does not induce disease (Hackett *et al.*, 1996).

An interesting life-table study was conducted on tomato, a sub-optimal host, by Latheef and Harcourt (1974). Not surprisingly, the major mortality factor was physiological; i.e., the tomato plant was a fairly unsuitable host. However, these authors also suggested that cannibalism was the major mortality factor during the egg stage, and that rainfall affected the larvae.

Life Cycle and Description. The winter is passed in the adult stage, which emerges from the soil in the spring at the time potatoes are emerging. A complete generation can occur in about 30 days. The northernmost portion of its range usually allows only a single generation per year, but two generations occur widely, even in Canada. Southern areas such as Maryland and Virginia can support three annual generations, but not all the beetles go on to form a third generation; the third generation often develops on weeds rather than potatoes, so it is overlooked (Johnson and Ballinger, 1916). Diapause is induced by a combination of photoperiod, temperature, and host quality. Long day lengths normally promote continued reproduction, whereas short day lengths promote diapause, but diapause induction varies widely among populations. For example, populations in Washington and Utah do not reproduce when the photoperiod is less than 15 h, whereas beetles from Arizona reproduce at a 13 h photophase. Beetles from southern Texas and Mexico are relatively insensitive to photoperiod and reproduce at photoperiods of 10–18 h (Hsiao, 1988). The overwintering population may be comprised of individuals from more than one generation. In Massachusetts, egg hatch begins in late May and first generation larvae are present until July. First generation adults, the progeny of overwintered beetles, begin to emerge in July. Those emerging before August 1 normally go on to reproduce, whereas those emerging later usually enter diapause (Ferro *et al.*, 1991). Some beetles remain in diapause for more than a year.

Egg.   Overwintering adults usually feed for 5–10 days before mating and producing eggs, though some beetles mate in the autumn and can oviposit without mating in the spring. The eggs are orange and elongate oval, measuring about 1.7–1.8 mm long and 0.8 mm wide. They are deposited on end in clusters of about 5–100 eggs, but 20–60 eggs per cluster is normal. The eggs are deposited on the lower surface of foliage, and anchored with a small amount of yellowish adhesive. The eggs do not change markedly in appearance until about 12 h before hatching, when the embryo becomes visible. Mean development time of eggs is 10.7, 6.2, 3.4, and 4.6 days when held at 15°, 20°, 24°, and 30°C, respectively.

Larva.   The larvae are reddish and black, and easily observed and recognized. The larvae are very plump, with the abdomen strongly convex in shape. The larvae bear a terminal proleg at the tip of the abdomen, in addition to three pairs of thoracic legs. Young larvae are dark red with a black head, thoracic plate, and legs. Two rows of black spots occur along each side of the abdomen. The larger larvae are lighter red, with the black coloration of the thoracic shield reduced to the posterior margin. There are four instars. Head capsule widths are about 0.65, 1.09, 1.67, and 2.5 mm for instars 1–4 respectively. During these instars body length increases from 1.5–2.6, 2.8–5.3, 5.5–8.5, and 9–15 mm, respectively. Developmental thresholds vary geographically, but 8–12°C is common. Rapid development and low mortality occur between 25–33°C, with 28°C optimal. Mean development time of first instars is about 6.1, 3.7, 2.1, and 1.4 days at 15°, 20°, 24°, and 28°C, respectively. For the second instar, mean development times are about 5.0, 3.8, 2.2, and 1.6 days; for the third instar 2.8, 2.5, 2.3, and 1.7 days; and for the fourth instar 9.5, 6.6, 3.3, and 2.4 days, respectively, when reared at 15°, 20°, 24°, and 28°C. (See color figure 107.)

Colorado potato beetle larva.

Pupa.   At maturity, larvae drop to the soil and burrow to depths of 2–5 cm where they form a small cell. After about two days they develop into pupae. The pupae are oval and orangish in color. They measure about 9.2 mm long and 6.4 mm wide. The form of the adult beetle is recognizable in the pupal stage, though the wings and antennae are twisted ventrally. Mean development time of the pupal stage, exclusive of the period spent in the soil prior to actual pupation, is about 5.8 days. Ferro *et al.* (1985) report the below-ground combined prepupal, pupal and post-pupal period to average about 22.3, 14.9, 11.7, and 8.8 days at 15°, 20°, 24°, and 28°C, respectively.

Adult.   After transforming from the pupal to the adult stage, the beetle remains in the soil for 3–4 days before digging to the surface. Adults are robust in form, and oval in shape when viewed from above. The dorsal surface of adults is principally yellow, but each forewing is marked with five longitudinal black lines. The head bears a triangular black spot and the thorax is dotted with about ten irregular dark markings. The underside of the beetle and legs also are mostly dark. (See color figure 106.) Beetles produce eggs over a 4–10 week period, with most of the eggs produced during weeks 1–5. Fecundity of beetles under field conditions has been estimated about 200–500, but this is likely an underestimate. Under labora-

Colorado potato beetle pupa.

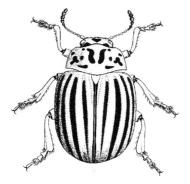

Adult Colorado potato beetle.

tory conditions, mean fecundity was 3348 eggs perfemale when fed potato, and 2094 when fed hairy nightshade (Brown *et al.*, 1980). In the autumn, newly emerged beetles feed for a time, and then dig into the soil, normally to a depth of 7–13 cm, to pass the winter. When beetles emerge in the spring they are not highly dispersive, mostly seeking out hosts by walking, though they are capable of flight. Voss and Ferro (1990) reported on dispersal behavior under field conditions in Massachusetts. Male beetles also walk upwind toward females, suggesting the presence of a pheromone that operates up to a distance of at least 50 cm (Edwards and Seabrook, 1997).

A good description of Colorado potato beetle was given by Girault and Rosenfeld (1907), general behavior and ecology by Chittenden (1907), and an overall review was published by Hare (1990). Thermal relations were presented by Walgenbach and Wyman (1984), Ferro *et al.* (1985), and Logan *et al.* (1985). Diapause in potato beetle, particularly the physiological aspects, was reviewed by de Kort (1990).

Damage

Colorado potato beetle is the major defoliator of potato throughout most of North America. Both adults and larvae feed on leaves, and when foliage has been consumed they will gnaw on stem tissue, and even tubers. First instars are responsible for about 3% of total leaf consumption, and second, third, and fourth instars for 5% 15%, and 77%, respectively. Total leaf consumption is estimated at 35–45 sq cm for larvae; adults consume foliage at a rate of 7–10 sq cm per day.

Most plant protection programs attempt to protect the early and mid-season growth of the plant. Potatoes are most susceptible to injury during bloom and shortly thereafter when tubers are rapidly expanding. Generally, defoliation should not exceed 10–25% during this period. Late season defoliation is much less

damaging (Hare, 1980; Ferro *et al.*, 1983; Zehnder *et al.*, 1995). Tomato is readily damaged by early season defoliation, and mean population densities as low as 0.5 beetles per plant reduce tomato yields (Schalk and Stoner, 1979).

Management

Sampling. Potato beetle populations initially are aggregated, but as larvae mature they disperse and aggregation is decreased but not eliminated. Sampling is usually accomplished by visual examination of plants. Harcourt (1964) recommended that entire plants be examined for above-ground life stages, and samples of 50–200 plants be examined to establish accurate population estimates. Estimates based on visual examination are highly correlated with whole plant samples (Senanayake and Holliday, 1988). Several researchers have developed sequential sampling plans for potato beetles; Martel *et al.* (1986) described a sampling protocol for Colorado potato beetle on potato, and Hamilton *et al.* (1998) proposed separate sequential sampling plans for Colorado potato beetle populations on eggplant, one for egg clusters and another for motile stages. The adults are attracted to yellow, and can be captured with traps (Zehnder and Speese, 1987), but this is not a common practice because the various life stages of this insect can so easily be detected on plants in the field.

Insecticides. In the late 1800s and early 1900s, the depredations of Colorado potato beetle were so great that this insect largely was responsible for the development of arsenical insecticides. Following the development of modern insecticides in the 1940s and 1950s, the status of Colorado potato beetle as a major pest was reduced, but it periodically resurges as a concern owing to development of insecticide resistance (Forgash 1981; Gauthier *et al.*, 1981; Heim *et al.*, 1990). Application of insecticides to protect potatoes, and to a lesser degree tomato and eggplant, is a common practice in many areas. Foliar applications are most common, but systemic insecticides are sometimes applied at planting for early season protection (McClanahan 1975b; Boiteau *et al.*, 1997). Some strains of the bacterium *Bacillus thuringiensis*, particularly *B.t. tenebrionis*, are effective against these beetles. It is usually necessary to apply *B. thuringiensis* products to the first two instars to attain high levels of suppression, though even a single application can provide enhanced potato yield (Nault and Kennedy, 1999). Some botanical insecticides are effective if applied to larvae (Zehnder 1986), but formulations may benefit from inclusion of a synergist (Zehnder and Warthen, 1988).

Cultural Practices. Cultural techniques have been studied extensively, initially because damage by this serious pest predates insecticides, and more recently because potato beetles evolved resistance to nearly all insecticides. Perhaps the most valuable cultural practice is crop rotation. Overwintered beetles initially are not highly dispersive, so moving the location of potato fields from year to year is beneficial. However, distances of at least 0.5 km are necessary to provide isolation (Hough–Goldstein and Whalen, 1996; Weisz *et al.*, 1996). Such rotation significantly delays the invasion of fields by beetles, eliminating the need for one or two early-season insecticide applications. Also, because beetles initially disperse principally by walking, trenches lined with plastic can be used to capture beetles and impede colonization of fields. Trenches with walls that slope at an angle greater than 45° capture 50% or more of the beetles, often resulting in high, but variable, reductions in beetle damage (Boiteau *et al.*, 1994).

The copper- and tin-based fungicides have useful anti-feedant properties, but are not directly toxic to beetles. Regular application of these fungicides reduces feeding and oviposition by Colorado potato beetle, particularly on less preferred hosts such as tomato (Hare *et al.*, 1983; Hare, 1984). Failure to cause mortality apparently limits the commercial acceptance of such anti-feedants.

Cultural manipulations such as application of straw mulch and compost have been studied to determine their effects on potato beetles. Application of mulch causes decreased numbers of beetles, whereas compost seemingly has no effect (Zehnder and Hough–Goldstein, 1990; Stoner *et al.*, 1996). The decrease in beetle abundance that commences within two weeks of application of mulch is attributed to increased abundance of soil-dwelling predators, and results in less defoliation and increased tuber yields (Brust, 1994). Application of nitrogen has only a slight, but negative, effect on potato beetle development rates (Jansson and Smilowitz, 1985). Intercropping and weed management also affect beetle densities. Potato monoculture is highly beneficial for potato beetle. Both beetle densities and fecundity tend to be higher under monoculture conditions (Horton and Capinera, 1987b).

Special equipment like propane flamers and crop vacuums have been used to destroy potato beetles. Flamers direct heat to the soil surface at the beginning of the season before the crop emerges, or soon after plant emergence (Pelletier *et al.*, 1995). Such treatment can kill 50–90% of the beetles, with slight damage to plants. This approach works best with young plants, and is limited by both weather constraints and cost,

so it is infrequently used. Crop vacuums can remove up to 70% of small larvae and adults, but are also limited to use early in the season, and by the considerable expense of equipment procurement (Boiteau *et al.*, 1992).

Biological Control. None of the naturally occurring beneficial organisms display the ability to regulate potato beetle populations consistently, though the tachinid, *Myiopharus doryphorae*, and the stink bugs, *Perillus bioculatus*, and *Podisus maculiventris*, have been cultured and released to suppress beetle damage (Tamaki and Butt, 1978; Tamaki *et al.*, 1983a; Biever and Chauvin, 1992; Cloutier and Bauduin, 1995; Hough-Goldstein and McPherson, 1996). Stink bugs are effective at destroying eggs and larvae, if a favorable ratio of predatory bugs to beetle egg clusters can be established—a ratio of one bug per about 100 beetle eggs can reduce injury by 80%. Efforts have been made to identify new biotic agents, the most promising of which is the egg parasitoid, *Edovum puttleri* (Hymenoptera: Eulophidae). This wasp can kill more than 80% of the eggs in an egg cluster, but the effectiveness of *E. puttleri* is limited by its requirement for warm temperatures in order to be active, and its inability to overwinter in temperate climates. Repeated release of *E. puttleri* into eggplant fields in New Jersey was reported to be an important element in producing high-quality fruit with minimal pesticide use and high financial returns (Hamilton and Lashomb, 1996). This parasitoid was described by many authors, including Grissell (1981), and Sears and Boiteau (1989).

Pathogens have been investigated extensively for Colorado potato beetle suppression and strains of *Bacillus thuringiensis*, mentioned above under ''insecticides,'' are most promising. However, the fungal pathogen *Beauveria bassiana* has been used in Europe with modest success (Roberts *et al.*, 1981). It is limited mostly by the economics of potato production, and the incompatibility of the entomopathogenic fungus with fungicides that must be applied to control foliar plant diseases. Entomopathogenic nematodes can be applied to the foliage to suppress feeding larvae or to the soil to kill larvae as they pupate (MacVean *et al.*, 1982; Toba *et al.*, 1983; Nickle *et al.*, 1994; Berry *et al.*, 1997).

Host-Plant Resistance. Although cultivars of potato and tomato differ slightly in their susceptibility to potato beetle, effective levels of plant resistance have never been available in commercial varieties. However, incorporation of *Bacillus thuringiensis* toxin into transgenic potato plants (Wierenga *et al.*, 1996) appears to be an effective basis for protection against the feeding by young larvae, at least for the short term.

Grape Colaspis
Colaspis brunnea (Fabricius)
(Coleoptera: Chrysomelidae)

Natural History

Distribution. This native insect is found widely in North America, east of the Rocky Mountains, though it is relatively infrequent in the western Great Plains. It also occurs spradically in the western states, as it also is reported from Arizona and Oregon.

Host Plants. The grape colaspis is known to feed on numerous plants, in both the larval and adult stage, though rarely does it cause much damage. The preferred hosts seem to be legumes, but it can be quite common in legume-grass mixtures. The adults have been observed to feed on such vegetables as bean, beet, cantaloupe, corn, cowpea, and potato. Other crops serving as hosts for adults include alfalfa, apple, bean, buckwheat, clover, grape, okra, pear, pecan, soybean, strawberry, sugar beet, trefoil, timothy, and watermelon. The larvae were successfully cultured on alfalfa, alsike clover, lespedeza, potato, red clover, rice, soybean, strawberry, timothy, and white clover. Suitable weed hosts include smartweed, *Polygonum* sp.; cinquefoil, *Potentilla* sp.; dock, *Rumex crispus*; and the grasses *Poa compressa*, *Paspalum* sp., and *Muhlenbergia* sp.

Natural Enemies. Few natural enemies of grape colaspis are documented. Elsey (1979) indicated that the nematodes *Howardula colaspidis* (Nematoda: Allantonematidae) and *Mikoletzyka* sp. (Nematoda: Diplogasteridae) are found in association with grape colaspis in North Carolina. Although up to 13% of the larvae were parasitized by *Howardula* and 39% by *Mikoletzyka* during certain periods of the year, there seems a little or no effect on the host. Diplogasterids typically have few effects, but *Howardula* normally affects ovarian development of female chrysomelids, reducing fecundity or inducing sterility.

Life Cycle and Description. The grape colaspis overwinters as partly grown larvae, and pupates in May and June. New adults and eggs are present beginning in June, and larvae are found throughout the summer and into the following spring. A single generation per year is reported from Iowa. In Arkansas, however, some of the new larvae do not overwinter, but go on to form adults and a second brood of larvae. The adults from the overwintering beetles disappear by mid-July, about the time the adults from the second brood begin to appear. Thus, there is a peak of adult activity in the spring, followed by another in the late summer, with larvae from both groups of adults overwintering.

Egg. The eggs are deposited in the soil near the surface unless a female descends into a soil crevice to deposit her eggs. The egg clusters contain, on average, about 35 eggs that require a mean of 6.4 days to hatch. The egg is elongate oval. The mean length (range) is 0.63 mm (0.56–0.72 mm) and mean width (range) is 0.28 mm (0.25–0.32 mm). It varies from milky white to bright yellow.

Larva. The larva is whitish with a brownish or yellowish-brown head and prothoracic shield. It tends to be thick-bodied and C-shaped, much like a white grub (Coleoptera: Scarabaeidae). The ventral surface of the body is marked with rows of stubby tubercles bearing stout spines. The larval stage usually has ten instars, and attains a length of about 3–4 mm and a width of 1.5–2.0 mm when mature. Head capsule widths are 0.22–0.28, 0.30–0.36, 0.36–0.49, 0.50–0.60, 0.60–0.74, 0.70–0.86, 0.80–0.92, 0.85–1.00, 0.95–1.10, and 1.05–1.20 mm for instars 1–10, respectively. Lindsay (1943) reported duration of the larval stage to be 102–175 days in the laboratory. He gave mean instar durations of 17.4, 6.8, 6.7, 6.0, 8.1, 8.5, 10.1, 9.5, 10.1, and 10.3 days for instars 1–10, but reported that most larvae underwent additional molts, producing up to 17 instars. Overwintering larvae are buried in the soil at depths of about 7–30 cm. The principal overwintering forms are instars 4–6. Overwintering survival is quite high, even in soils that are flooded for much of the winter. Larvae move toward the soil surface in early spring, where they complete their feeding and pupate.

Pupa. Pupation occurs in a vertical cell within the top 1 cm of soil. The pupa is white and measures about 3.5 mm long and 2.5 mm wide. It is unusually well-equipped with stout hairs and spines. Duration of the pupal stage averages 5.4 days (range 3–7 days).

Adult. The adult is oval, and measures 4–6 mm long. It is brownish yellow in color, with the ventral side darker and the legs pale-yellowish. The elytra are distinctly marked with broad ridges separated by rows of deep punctures. The adults mate soon after emergence. The pre-oviposition period is 3–5 days. Adults mate repeatedly despite their relatively short life span, which averages about eight days in males and 13 days in females. A female may deposit up to six egg masses and up to about 250 eggs. The adults are active during the daylight hours, and fly readily.

Elements of grape colaspis biology were given by Forbes (1903), Bigger (1928), and Rolston and Rouse (1965). However, the most complete treatment of grape colaspis biology was the dissertation of Lindsay

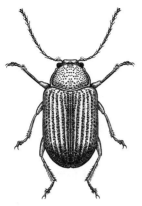

Adult grape colaspis.

(1943). Grape colaspis was included in the key to eastern beetles by Downie and Arnett (1966).

Damage

Damage usually occurs when susceptible crops are planted following grass-legume mixtures, weedy fields, and occasionally legume crops such as clover and lespedeza. Plant stand density may be reduced owing to feeding by larvae on germinating seeds. Young plants may have the roots pruned, resulting in stunting, desiccation, and even death. One common symptom of infestation in corn is foliage discoloration. The leaves turn purple due to inadequate phosphorus uptake. The adults chew small holes in the foliage, and when they are abundant, may skeletonize leaves.

Management

Tillage in the spring can be detrimental to the overwintering larvae, but this is not an entirely reliable method for larval suppression. Larvae can be controlled by application of insecticide to seed. Seed treatment is advisable if susceptible crops follow legumes or grass-legume mixtures, particularly if there is evidence of infestation during the previous year. Larvae do not tolerate dry soil very well, so frequent irrigation should be avoided.

Northern Corn Rootworm
Diabrotica barberi Smith and Lawrence
(Coleoptera: Chrysomelidae)

Natural History

Distribution.   Northern corn rootworm is a native species, and was first discovered attacking corn in Colorado. It has since spread eastward, principally to the corn belt in the midwestern states. Range expansion was caused primarily by a change in crop production practices: culture of corn continuously in the same fields. In the United States, northern corn rootworm occurs in the Great Plains region from North Dakota to Oklahoma, and east to the Atlantic Ocean. As its common name suggests, this insect is largely absent from the south, but is present in Tennessee and the northern regions of Alabama and Georgia. In Canada, northern corn rootworm is found in southern Ontario and Quebec. Northern corn rootworm and western corn rootworm, *Diabrotica virgifera* LeConte, can both be found in the same fields, but western corn rootworm has displaced northern corn rootworm, or at least reduced its abundance, in some areas. Displacement has been attributed to greater insecticide resistance in western corn rootworm, but western corn rootworm also has a much higher reproductive rate than does northern corn rootworm.

Host Plants.   Northern corn rootworm has a rather restricted host range. The larvae develop only on plants in the family Gramineae. Corn is the only crop regularly attacked by larvae, but development can also occur to a lesser extent on millet, rice, and spelt (a primitive type of wheat). Also, some survival occurs on rangeland forage grasses such as foxtail, *Setaria* spp.; wheatgrass, *Agropyron* spp.; weeping lovegrass, *Eragrostis curvula*; and Canada wildrye, *Elymus canadensis* (Branson and Ortman 1967a, 1971). Adults consume several types of food, with corn kernels, corn silk and corn tassel tissue favoring survival and reproduction, though corn leaf tissue is inadequate. Blossoms from goldenrod, *Solidago canadensis*; squash; and sunflower also are suitable (Lance and Fisher, 1987; Siegfried and Mullin, 1990). Additionally, adults feed on pollen from plants in the families Gramineae, Compositae, Leguminosae, and Cucurbitaceae.

Natural Enemies.   Several beneficial organisms have been found associated with northern corn rootworm, but they seem to be of little consequence. Predators include ground beetles (Coleoptera: Carabidae), soldier beetles (Coleoptera: Cantharidae), and mites (Acari: Laelaptidae, Rhodacaridae, and Amerosiidae). Probably the most important parasitoid is *Celatoria diabroticae* (Shimer) (Diptera: Tachinidae), a species that also attacks other *Diabrotica* spp., but its relationship with northern corn rootworm is not clearly known. Pathogens such as protozoa, fungi, and bacteria have been isolated from rootworm larvae. The soil-inhabiting fungus *Beauveria bassiana* can cause epizootics in larval populations, but this occurs infrequently.

Life Cycle and Description.   Generally there is a single generation per year, with the egg stage overwintering, but sometimes eggs pass through more than one winter before hatching occurs. Eggs hatch in late spring, larvae feed until about June, and adults begin to emerge in July. Development is not synchronous, so emergence is protracted. The temperature threshold for development of the various stages is 10–11°C. The mean period required for northern corn rootworm to develop from hatching to emergence of the adult is about 100, 47, and 30 days when reared at 15°, 21°, and 27°C (Woodson and Jackson, 1996).

Egg. The eggs are oval and whitish. Eggs are small, measuring only about 0.6 mm long and 0.4 mm wide. The surface of the eggs, when examined microscopically, are marked with a polygonal (mostly hexagonal) pattern, with each polygon containing small pits. The presence of pits serves to differentiate northern corn rootworm eggs from those of western corn rootworm, which lack pits. Eggs of spotted cucumber beetle, *Diabrotica undecimpunctata* Mannerheim, also contain polygons with pits, but the pits are proportionally larger (Atyeo *et al.*, 1964; Rowley and Peters, 1972; Krysan, 1986). Eggs are deposited in soil cracks or other breaks in the soil that allow the female to access moist soil. Thus, in irrigated fields the eggs are frequently found in the furrows (Weiss *et al.*, 1983). The female commonly deposits eggs in loose clusters of 3–10 eggs. Females select corn as an oviposition site over other crops and weeds (Boetel *et al.*, 1992). Eggs survive temperatures of 4–5°C for brief periods, but survival decreases markedly at −10°C. The temperature threshold for embryonic development is about 11°C. The number of days required for hatch of eggs in the spring is about 21, 32, and 79 days when held at 25°, 20°, and 15°C, respectively (Apple *et al.*, 1971; Woodson *et al.*, 1996). Duration of egg diapause is somewhat variable (Fisher *et al.*, 1994), and some members of the population pass through more than one winter before hatching (Landis *et al.*, 1992; Levine *et al.*, 1992).

Larva. The larva is elongate and cylindrical in shape, tapering toward the head. The body is white, bearing relatively few hairs or spines. The head capsule, and thoracic and anal plates, are yellowish brown. The three pairs of legs are brownish, and terminate in a single claw. The posterior end of the body bears a single retractile extension or tubercle. Larval development time at variable temperatures was determined by Golden and Meinke (1991) to be about 47 days. However, Woodson and Jackson (1996) reported mean total larval development periods of 73, 32, and 21.5 days at 15°, 21°, and 27°C, respectively. Mean development time of the three instars was 6.5, 6.5, and 19 days, respectively, at 21°C. Head capsule widths average 0.22, 0.33, and 0.49 mm, respectively, for instars 1–3. Larvae attain a length of about 7 mm at maturity.

Northern corn rootworm larva.

Pupa. The larvae prepare small cells in the soil for pupation. Pupation often occurs within the upper 5 cm of soil. The pupa is white, except for the reddish brown eyes. It measures about 4.5 mm long and 2.5 mm wide. In general form, it resembles the adult except that the wings are reduced in size and twisted ventrally. Also, the pupa bears a pair of hooks at the tip of the abdomen. Duration of the pupal stage is about 27, 12, and 7 days when reared at 15°, 21°, and 27°C, respectively.

Adult. The adults usually are yellowish or yellow-green, and lack the broad-black stripes or extensive black pigmentation found on the elytra of western corn rootworm. However, northern corn rootworm beetles in the northern Great Plains are sometimes striped. Northern corn rootworm beetles generally have yellowish to brownish antennae, tibiae, and tarsi, though in the eastern states they can be blackish. They measure about 6 mm long. The pre-oviposition period of beetles is about 14 days, followed by an oviposition period of 40–60 days. Adults can survive up to nearly three months, and produce clutches of 20–30 eggs at about seven-day intervals. Total fecundity averages about 185 eggs per female on a good diet, but considerably less on suboptimal diets. However, if optimal temperatures are also provided, mean fecundity can increase to 274 eggs per female (Naranjo and Sawyer, 1987).

The biology of corn rootworm was given by Forbes (1892) and Chiang (1973) and some descriptive information by Krysan *et al.* (1983). A key to the adult, *Diabrotica*, and pictures of eggs are found in Krysan (1986). Mendoza and Peters (1964) provided a key to the principal mature rootworm larvae.

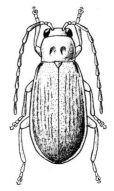

Adult northern corn rootworm.

Damage

Newly hatched larvae feed on the root hairs and outer tissue of roots, but as they increase in size and appetite they burrow into the roots, often consuming them entirely. Root damage can inhibit plant growth, but also reduces the ability of the plant to resist wind, especially when the soil is wet. Plants that topple, or lodge, owing to damaged-root systems display reduced growth potential. Lodged plants are difficult to harvest. Larval feeding also can allow entrance of Fusarium fungi, resulting in stalk rot. Feeding by adults on corn silks occasionally is enough to interfere with pollination, and beetles sometimes cause direct injury by feeding on the kernels at the tip of the ear. Foliage feeding by adults of this species is rare, though it is quite common with the related western corn rootworm. Maize chlorotic mottle virus may be transmitted by rootworm beetles (Jensen, 1985). Adult rootworm densities of one per plant or greater suggest damaging larval densities during the next year if corn is planted in the same field.

Management

Sampling. A great deal of effort has gone into development of improved sampling protocols for rootworms, though the principal concern is the vast acreage of grain corn, rather than the modest acreage of sweet corn. With grain corn, large quantities of insecticide are applied at great expense to a crop with a modest profit margin, so there is ample incentive to improve decision-making relative to insecticide application.

Sampling has been investigated for most rootworm life stages. The egg and larval sampling has been studied extensively (Weiss and Mayo, 1983; Hein et al., 1985; Fisher and Bergman 1986; Ruesink, 1986; Tollefson 1990), but remains mostly a research protocol due to the high labor requirements. Thus, most field-level sampling involves adult population monitoring, with its densities used to predict larval damage during the following year. Visual assessment of adult densities can be made by whole-plant or ear-zone counts, and sequential sampling protocols have been developed (Foster et al., 1982; McAuslane et al., 1987). However, traps are usually preferred because they can be left in the field for longer periods and are less affected by beetle movement and short-term weather phenomena. A simple and popular trap is the yellow sticky trap (Hein and Tollefson, 1984; Tollefson, 1986; Hesler and Sutter, 1993). Chemical attractants also have been investigated. Chemical arrestants such as cucurbitacin, lures including sex pheromone, and other chemicals isolated from the Cucurbitaceae such as eugenol, isoeugenol, 2-methoxy-4-propylphenol and cinnamyl alcohol have been found successful (Shaw et al., 1984; Ladd, 1984; Levine and Gray, 1994; McGovern and Ladd, 1990; Hoffman et al., 1996a).

Insecticides. Corn producers generally rely on soil-applied insecticides to protect their crops from larval rootworm damage. Either liquid or granular formulations may be applied to the root zone, and it can be applied at planting time or after the corn is partly grown. Where corn is cropped continuously, insecticide resistance and enhanced microbial degradation of insecticides have been noted (Levine and Oloumi–Sadeghi, 1991). Suppression of adults with foliar insecticides could be made to prevent damage during the subsequent year, but in many cases this is not done owing to the vagility of adults. Adult control is also needed occasionally to protect corn silks and ear tips from injury. There is considerable use of insecticide-containing bait to accomplish adult control with greatly reduced insecticide application rates in midwestern states (Lance, 1988).

Cultural Practices. Levine and Oloumi–Sadeghi (1991) provided an excellent review of how cropping practices affected corn rootworm biology. Planting synchrony is among the cultural factors affecting corn rootworm populations. Beetles depend on the availability of green silks and pollen resources that are available only briefly. Late planting of crops delays and extends the development of the insect population, and reduces survival, possibly owing to deprivation of corn roots to hatching larvae (Musick et al., 1980; Bergman and Turpin, 1984). Thus, late plantings do not require insecticide treatments for this insect, but are likely to be heavily infested in the following year owing to attraction of beetles to late-planted corn. Availability of corn flowers is important for adult survival. Beetles tend to disperse to early-flowering corn until it senesces, then disperse to later-flowering fields (Naranjo and Sawyer, 1988).

The most important cropping practice is crop rotation, which normally results in destruction of rootworms owing to the inability of larvae to survive on crops other that corn. Thus, corn routinely is rotated with a non-host such as soybean. This apparently has led to increased incidence of rootworm populations that diapause through two years before hatching, potentially reducing the effectiveness of this practice (Levine et al., 1992; Steffey et al., 1992). As yet, most areas have not experienced a serious problem with two-year life cycles, and crop rotation remains a preferred management practice.

If corn is to be planted into a field that previously supported corn, tillage and disking will have few effects on insect survival (Gray and Tollefson,

1988a,b). However, planting new rows of corn between the old rows is often desirable, because the eggs often are concentrated at the base of the old corn plants, and young larvae will experience difficulty in dispersing such distances (Chiang *et al.*, 1971). Application of irrigation water and additional nitrogen help to offset loss of roots.

Biological Control. Effective biological controls have yet to be developed for corn rootworms. Considerable research has been conducted on the use of entomopathogenic nematodes (Nematoda: Steinernematidae and Heterorhabditidae) for larval suppression. Results have not been consistent, and the cost of this approach remains prohibitive, but good progress is being made on use of these biotic agents (Thurston and Yule, 1990; Wright *et al.*, 1993; Jackson and Hesler, 1995).

Red Turnip Beetle
Entomoscelis americana Brown
(Coleoptera: Chrysomelidae)

Natural History

Distribution. Red turnip beetle occurs in the United States from Minnesota and Wisconsin west to Colorado and Washington. In Canada, it is known from the Prairie Provinces, British Colombia, and northward. It is considered to be native to western North America, and limited in its distribution by a combination of weather and host plant availability (Gerber, 1989).

Host Plants. This is a crucifer-feeding insect. Vegetable crops attacked include broccoli, Brussels sprouts, cauliflower, cabbage, kale, kohlrabi, mustard, radish, turnip, and watercress. It also feeds on rape (canola) and horseradish. As rape has assumed greater importance as an oil-seed crop, red turnip beetle has increasingly become a more abundant, and serious, pest in Canada. Cruciferous weeds are suitable hosts, and feeding has been reported on shepherdspurse, *Capsella bursa-pastoris*; pepperweed, *Lepidium* spp.; wallflower, *Erysimum parviflorum*; and the mustards *Erucastrum*, *Sisymbrium*, and *Brassica* spp. However, Gerber and Obadofin (1981a) determined that black mustard, *Brassica nigra*; flixweed, *Descurainia sophia*; and tall hedge mustard, *Sisymbrium loeselii*; were marginally suitable.

Natural Enemies. No parasites are known, and the incidence of predation by carabid beetles and microsporidan disease is slight (Gerber, 1989).

Life Cycle and Description. Red turnip beetle has an unusual life history because the adult period is interrupted by a summer aestivation. There is one gen-

eration per year, and the egg is the overwintering stage. Larval and pupal development occur in the spring, with adults emerging in June. The adults do not oviposit immediately, however, but feed for a month and then enter a period of aestivation in the soil. They re-emerge later in the summer, mate, and oviposit.

Egg. The eggs are deposited in August and September. They are elliptical, measuring about 1.5 mm long, and are red or brown. Eggs are deposited on the soil surface, or in the soil to a depth of about 1–2 cm. The eggs may be laid singly or in small clusters. They hatch in the spring, usually in March or April. The lower temperature threshold for development is about 5°C, and the upper limit about 37.5°C (Gerber and Lamb, 1982), but they hatch successfully over a broad range of temperatures (Lamb *et al.*, 1984). The eggs tolerate cold temperatures very well; soil temperatures as low as −10°C have no effect on survival (Gerber, 1981).

Larva. The larvae are wrinkled in appearance, and initially are orange with black spots, but as they mature they become black. They initially are 1–2 mm long, but eventually attain a length of about 12 mm. There are four instars, requiring 6.9, 5.1, 5.0, and 14.9 days, respectively, for their development. Larval development, which often requires about 30 days, is usually complete by late May. Larvae feed principally at night.

Pupa. Pupation occurs in the soil at a depth of 2–3 cm. The pupa is orange, and 6–10 mm long. Duration of the pupal stage is about 15 days.

Adult. The adult is distinctively marked with black and red. The ventral surfaces are primarily black, but the dorsal surface is red marked with black. The dorsum is marked with a small-black spot at the back of the head, which is contiguous with a large black spot in the center of the pronotum. The pronotum also bears a small black spot on each side of the pronotum. There also is a black stripe running about two-thirds the length of each elytron. A narrow black stripe runs along most of the center of the back, where the elytra join dorsally. Thus, the overall appearance is a red beetle with a black spot near the head and three black stripes on the elytra. Beetles measure about 10 mm long. Adults appear intermittently, first in June and July, and then again in August until late October or the onset of cold weather, when they perish. Adults are reported to invade crops by walking rather than flying.

Adult red turnip beetle.

The biology of red turnip beetle was given by Chittenden (1902), Gerber and Lamb (1982), Gerber (1981, 1989), and Gerber and Obadofin (1981a,b).

Damage

The larvae cause damage by feeding on the cotyledons, leaves, petioles, and stems of seedlings early in the growing season. Later, both larvae and adults feed on foliage, eating large, irregular holes in the leaves. This is principally a pest of homegardens rather than commercial vegetable crops. It commonly is of concern to oil-seed producers, but the impact is often slight (Gerber, 1976).

Management

Cruciferous weeds and volunteer crucifer crops can be important sources of insects in the spring, so clean cultivation is advisable if this insect previously has been abundant. Although this can be effective by denying larvae a source of food, if adults are present the destruction of weeds and volunteer cruciferous crops may drive the beetles onto the primary crop. Tillage may also destroy some of the overwintering eggs. If adults or larvae are abundant in the spring, foliar application of insecticides is an effective remedy. Row covers will also prevent invasion of vegetable gardens.

Spotted Asparagus Beetle
Crioceris duodecimpunctata (Linnaeus)
(Coleoptera: Chrysomelidae)

Natural History

Distribution. Spotted asparagus beetle is found throughout North America. This insect is of European origin, and was first observed in the western hemisphere near Baltimore, Maryland about 1881. It spread rapidly in the northeast and midwestern states and eastern Canada, reaching Idaho and Washington in the 1940s and 1950s and British Columbia in 1962. Although it is widespread, it is less abundant than asparagus beetle, *Crioceris asparagi* (Linnaeus).

Host Plants. This species is very selective in its host preference, attacking only asparagus. It is most commonly observed on wild and home-garden plants, rarely attaining high densities in commercial plantings. The larvae feed only within berries.

Natural Enemies. Little is known concerning the natural enemies of spotted asparagus beetle. Capinera (1974b) was unable to identify significant natural enemies in Massachusetts, but noted that resource availability (lack of berries for the larvae) might be a key factor in regulating insect densities. Ground beetles (Coleoptera: Carabidae) may take a toll on larvae searching the foliage for berries or droping to the ground to pupate, but this has not been measured. The asparagus beetle egg parasitoid, *Tetrastichus asparagi* Crawford (Hymenoptera: Eulophidae) has been reported in the literature as a parasite of spotted asparagus beetle, but this is an error. A related species, *Tetrastichus crioceridis* Graham, attacks spotted asparagus beetle eggs in Europe, but attempts to establish it in North America have not been successful (van Alphen, 1980; Hendrickson *et al.*, 1991).

Life Cycle and Description. Two generations occur annually, but the second generation is often small. Overwintered adults are abundant in May and June, followed by first generation adults in July and then the second generation adults in August and September.

Egg. The eggs are laid singly, usually on plants bearing berries. The elliptical eggs measure about 1.0–1.2 mm long and 0.4 mm wide, and are glued on their side to asparagus foliage. This method of attachment makes it easy to distinguish *C. duodecimpunctata* eggs from *C. asparagi* eggs; the latter are attached on end. The eggs initially are whitish, turning yellow-orange and then olive green or brown as they mature. Eggs hatch in 7–12 days and young larvae burrow into berries to begin feeding.

Larva. The larvae are reported to progress through three instars in 21–30 days, increasing in length from about 1 mm to 8.5 mm (Fink, 1913). However, careful study of the sibling species, *C. asparagi* showed four instars (Taylor and Harcourt, 1978), so Fink may have overlooked one instar. Larvae are pale-yellow or orange. The head is initially black, but at the molt to the second instar the head becomes pale. The thoracic plate is dark throughout the development. The bright color of the larvae immediately distinguishes them

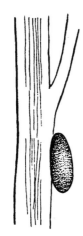

Spotted asparagus beetle egg on plant stem.

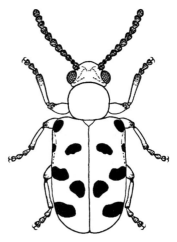

Adult spotted asparagus beetle.

Spotted asparagus beetle larva.

from the gray or brown larvae of *C. asparagi*. Larvae are normally found only within berries, but because 2–4 berries are required for complete development, they occasionally may be found searching the foliage. Also, berries sometimes fall from the plant as larvae consume the interior, forcing larvae to ascend the plant in search of additional food.

Pupa. When a larva completes its development it leaves the berry and drops to the soil to pupate. The larva prepares a small pupation chamber from oral secretions and soil particles and pupates within. The pupa is yellow and greatly resembles the adult in form, except that the wings are not fully developed. Duration of the pupal stage is 12–16 days.

Adult. The adult is largely orange, with each elytron marked with six black dots. The beetle measures 6-8 mm long. The antennae, eyes, tarsi, distal portion of the femora, and the underside of the thorax are black. When disturbed the adults are quick to take flight, and stridulate loudly if captured (Capinera, 1976). Adults emerge from overwintering in May, usually about a week later than asparagus beetle, *Crioceris asparagi*. Unlike *C. asparagi*, they do not commence oviposition soon after emergence. Oviposition is delayed until asparagus flowers are fertilized or berries are present. Adults seek hollow stems of asparagus and other sheltered locations in which to pass the winter. (See color figure 103.)

The biology of this poorly studied insect was given by Fink (1913), Chittenden (1917), and Dingler (1934).

Damage

The only serious damage is caused by adults early in the spring when they gnaw on the emerging shoots, creating small pits in the epidermis. Such damage renders the shoots worthless for commercial purposes. Later in the season the adults also feed on leaves, and remove the outer layer from stems. This latter damage generally has little effect on the growth of the plant, and subsequent yield. Larvae feed inside berries, preventing seed from forming. This is unimportant except where seed is being produced.

Management

This insect does not usually cause significant injury in commercial production. The frequent harvesting of commercial asparagus is believed to discourage feeding. In volunteer plants or homegardens the adults sometimes attain high densities, however, and deform young spears by their feeding. Insecticides are highly effective against asparagus beetles. Collection and destruction of berries has been suggested as a method of guarding against future damage by adults, but this is practical only on the scale of the homegarden.

Spotted Cucumber Beetle
Diabrotica undecimpunctata Mannerheim
(Coleoptera: Chrysomelidae)

Natural History

Distribution. Spotted cucumber beetle is also known as "southern corn rootworm," which is an indication of geographic distribution as well as an alternate favored-food plant. It is found throughout the United States, though two races are recognized. In the Rocky Mountain region and westward a variant designated as *Diabrotica undecimpunctata undecimpunctata* Mannerheim is recognized, and called the western spotted cucumber beetle. The eastern form of the spotted cucumber beetle is found east of the Rocky mountains, including eastern Canada as far west as Alberta, and is designated as *Diabrotica undecimpunctata howardi* Barber. Spotted cucumber beetle is also found in Mexico, and consists mostly of the eastern race.

Elsey (1988) studied population dynamics of spotted cucumber beetle, and also striped cucumber beetle, *Acalymma vittatum* (Fabricius), and banded cucumber beetles, *Diabrotica balteata* LeConte, affecting cucurbits in South Carolina. His data suggested that spotted cucumber beetle was the numerically dominant species during the two years of the study, though striped cucumber beetle was also quite numerous. This appears to be consistent with most observations from the southern areas of the United States, except for southern Florida and Texas, where the banded species predominates. In the northern areas of the United States and southern Canada, striped cucumber beetle is generally dominant. Spotted cucumber beetle is highly dispersive, overwintering in southern states and dispersing northward annually.

Host Plants. This insect has a very wide host range. More than 200 plants are known to be fed upon. Beetles fly considerable distances regularly in search of suitable food. Adults are especially fond of pollen and flower petals. Arant (1929) reported that though the adult will feed on the flowers of almost any plant, they preferred cucurbits, legumes, tomato, ornamental plants, and fruit crops. He also provided a long list of plants on which he had observed leaf damage, including not only the aforementioned susceptible crops but cabbage, corn, lettuce, mustard, pea, potato, and turnip. Michelbacher *et al.* (1943) presented a similar list for the western race of the species, including similar relative preferences. Artichoke is also sometimes damaged by western spotted cucumber beetle in California (Lange, 1941). Elsey (1988) indicated that zucchini squash was more attractive to beetles than was cucumber. The flower candytuft, *Umbellata* sp., was noted as especially attractive in Alabama and wild cucumber, *Micrampelis oregona*, was favored in the Pacific Northwest. Adults also feed readily on sweet substances: fruits and aphid honeydew. The larvae, however, are largely restricted to feeding on the roots of cucurbits, legumes, and grains (including corn) and grasses. Winter cover crops and weeds, especially broadleaf species, often are attractive to ovipositing females (Brust and House, 1990; Buntin *et al.*, 1994).

Natural Enemies. Several natural enemies of spotted cucumber beetles are known, the most important being *Celatoria diabroticae* (Shimer) (Diptera: Tachinidae). This fly deposits her larva within the abdomen of the adult beetle, which soon perishes. In Alabama, parasitism was highest during the late winter. Workers in Alabama, Arkansas, California, and Oregon have documented levels of parasitism reaching about 30%, but generally they discounted the importance of the parasitoid for regulating populations of cucumber beetles. Fronk's (1950) study in Virginia showed an average seasonal parasitism rate of only 3.7%. Rockwood and Chamberlin (1943), however, argued that the value of the parasites may be underestimated. Cucumber beetles also are fed upon by many bird species (Arant, 1929), and attacked by mermithid nematodes and the fungus *Beauveria* (Rockwood and Chamberlin, 1943). Fronk (1950) reported his finding several nematode species parasitizing rootworms in Virginia, with *Howardula benigna* (Cobb) (Nematoda: Allantonematidae) most abundant. He observed a seasonal average parasitism rate of 23.6%, and reported that *Howardula*-infected female beetles had poorly developed ovaries. Poinar (1979) reported experimental infection of southern corn rootworm by the mermithid nematode *Filipjevimermis leipsandra* Poinar and Welch. Various steinernematid and heterorhabditid spp. undoubtedly attack larvae in some locations.

Weather. Environmental conditions, and particularly soil moisture, have been implicated as affecting the abundance of spotted cucumber beetle. Workers in Arkansas (Isely, 1929), Alabama (Arant, 1929), and California (Smith and Michelbacher, 1949) all noted that high soil moisture was a requirement for larval survival. Smith and Michelbacher (1949) further noted that in dry years it was only the clay soils or low spots that successfully produced beetles, whereas during moist years sandy soils were more suitable.

Life Cycle and Description. The spotted cucumber beetle normally completes its life cycle in 4–6 weeks. The number of generations per year seems unusually variable, possibly because there is no true diapause in these insects and they continue to reproduce as long as the weather allows, and because they annually re-invade northern areas. However, Elsey (1988) suggested that spotted cucumber beetle populations may have a short (30–60 days) reproductive diapause. Often there are two generations per year, but Rockwood and Chamberlain (1943) reported a single generation in Oregon, and Arant (1929) and Michelbacher *et al.* (1955) indicated three in Alabama and southern California, respectively. In Iowa, Sweetman (1926) reported one generation but Drake and Harris

(1926) claimed two per year. Because the generations overlap considerably, and adults can be found continuously during warm weather, field observations could lead to an underestimate of generation number. The adult is the overwintering stage. Overwintering beetles hide in plant debris and become active when the temperature reaches about 15–20°C. In the mild weather areas of the Northwest, aggregation of beetles in protected locations or "caches" may occur during the winter, but this is the exception, not the rule. Smith and Allen (1932) indicated that beetles overwinter only in areas with mild winters, principally the southern states, dispersing northward annually. Further, they noted that northward dispersants in the spring could consist of a mixture of insects that had dispersed southward during the preceding autumn, and new adults produced at southern latitudes.

Egg. Normally, about 40 eggs are produced at each oviposition. Chalfant and Mitchell (1967) demonstrated that beetles prefer coarse soil or soil with cracks to fine, smooth soil; they also prefer wet over dry soil for oviposition. A female deposits, on average, six egg clusters during her life, for a total of 200–300 eggs. All values related to egg production are quite variable, with the values decreasing as the beetles progress from generation one, to two, to three. The yellow egg is oval, and measures about 0.7 mm long and 0.5 mm wide. Egg incubation time varies with temperature, from as little as a week to up to a month.

Larva. The three instars generally require seven, five, and four days, respectively, or about 14–24 days for complete larval development. The larva grows from about 1.8 mm long at hatching to about 12 mm long at maturity. Head capsule widths measure 0.3, 0.4, and 0.6 mm, respectively. The larva is yellowish white, with the head, thoracic, and anal plates colored brown. The mature larva constructs a small chamber in the soil and pupates within.

Pupa. The pupa is white initially, but turns yellow with age. It measures about 7.5 mm long and 4.5 mm wide. The tip of the abdomen bears a pair of stout spines, and smaller spines are found dorsally on the remaining abdominal segments. Duration of pupation is about seven days (range 4–10 days). After comple-

Spotted cucumber beetle pupa.

tion of the pupal period the beetle digs to the soil surface.

Adult. The adult is long-lived. A life span of 60 days is common during the summer, but longevity is extended considerably, up to 200 days during the winter. The adult is about 6.0–7.5 mm long. It is generally colored yellowish green with 12 black spots on the elytra and a black head. The western form of spotted cucumber beetle has black legs and abdomen, whereas the eastern form has a pale abdomen and legs partly pale. Adults usually do not begin oviposition until two to three weeks after emergence. (See color figure 110.)

Arant (1929) provided the most complete life history study. Mendoza and Peters (1964) provided characters to distinguish larvae of *D. undecimpunctata* from western corn rootworm, *D. virgifera* LeConte, and northern corn rootworm, *D. barberi* Smith and Lawrence, but Krysan (1986) observed that some specimens cannot be distinguished with certainty using this key. Techniques for rearing southern corn rootworm are summarized by Jackson (1986). Biosystematics of *Diabrotica* spp. is given by Krysan and Branson (1983).

Spotted cucumber beetle larva.

Adult spotted cucumber beetle.

Damage

The adults eat small holes in the leaves and flower petals of many plants. If provided a choice, vegetable crops rather than grain crops are usually attacked by adults. Fruit of smooth-skinned melons is susceptible to damage, especially before the skin hardens. Brewer *et al.* (1987) studied the effects of adult feeding on seedlings of zucchini squash in Louisiana. Densities of 10 beetles per plant caused significant levels of plant mortality, but their effect diminished greatly once the seedlings attained the two or three leaf stage. While beetle damage may result in complete defoliation, sometimes the feeding is limited to the pollen or fruit. Because beetles often avoid heat, the shaded portions of the fruit may be especially damaged.

Beetles also transmit cucumber mosaic virus and muskmelon necrotic spot virus (Gergerich *et al.*, 1986), and bacterial wilt of cucurbits and of corn. Transmission of cucurbit bacterial wilt by spotted cucumber beetles is a secondary, late-season spread, so these insects are considered less important vectors than striped cucumber beetles (DaCosta and Jones, 1971).

Larvae feed on the roots of plants and also bore into the base of the stems. Root pruning may result in only a discolored and stunted plant, whereas when larvae burrow into the stem they may cause the plant's death. Latin and Reed (1985) suggested that larval feeding could increase the incidence and severity of Fusarium wilt. Some injury to the surface, or rind, of fruit may occur as larvae burrow up from the soil and feed on fruit in contact with soil, and larvae are sometimes called "rindworms." Grain seedlings, particularly corn, are often damaged by larval feeding. Older plants are less susceptible to death following feeding.

Management

Sampling. A sex pheromone has been identified (Guss *et al.*, 1983), and may prove useful as a component of monitoring or suppression tactics. Traps that take advantage of the adult orientation to plant kairomones and the yellow color have been evaluated alone and in combination (Hoffman *et al.*, 1996a). Although there is some evidence for improved catches, the responses are weak relative to some other diabroticite beetles.

Insecticides. Planting time application of granular insecticides, applied in a band over the row, is often recommended for protection of plant roots if a field is expected to be heavily infested by larvae. Foliar and blossom injury is prevented by timely application of insecticides, especially to the seedling stage, because they are more susceptible to injury. There is risk of foliar insecticides interfering with pollination in insect pollinated crops such as cucurbits, though this is not a consideration in wind pollinated crops such as corn. Insecticide applications made late in the day minimize damage to pollinators. Waiting at least one day after insecticide treatment before introducing beehives is also advisable. Control of beetles helps reduce spread of cucurbit wilt, but the concern over this disease diminishes as the plants mature.

Feeding stimulants and lures have considerable potential for increasing both the effectiveness and selectivity of insecticide applications. Cucurbitacins are arrestants and feeding stimulants for spotted cucumber beetle, though they have low volatility and therefore are not valuable as lures. Volatile components of blossoms, attractive to beetles, have been identified and evaluated as lures. Various mixtures attract beetles, but phenyl acetaldehyde was found more attractive to spotted cucumber beetle than to the other common squash blossom-infesting *Diabrotica* beetles. Cinnamaldehyde alone or in a mixture with trimethoxybenzene and indole also were very attractive kairomones, and have been used experimentally in the field to capture beetles. These lures are especially good for detecting beetles when they are at low densities, and otherwise difficult to locate. Combinations of volatile attractants, cucurbitacins, and low rates of insecticides have been applied to corn cob grits to make lethal baits (Metcalf *et al.*, 1987; Lampman and Metcalf, 1987; Deheer and Tallamy, 1991; Metcalf and Metcalf, 1992).

Cultural Practices. The need for insecticides can be avoided somewhat if the timing of planting can be manipulated so that the beetles have already dispersed and deposited most of their eggs before planting. For example, Luginbill (1922a) recommended safe planting periods of April 20–May 1 for north Florida and south Georgia, May 1–May 10 for central Georgia and the southern portion of South Carolina, and May 10–May 20 for the northern regions of Georgia and South Carolina and for North Carolina.

As there is such a wide range of suitable hosts and the insects are so mobile, crop rotation is usually not considered to be a practical management approach. There are some crops, however, such as cotton, that seem to be immune to injury. Arant (1929) documented problems in corn following winter legumes, because overwintering beetles were attracted to the legumes in the spring, and deposited eggs. When corn was planted immediately after legumes, the larvae caused severe injury. Grass weeds growing in a field immediately before planting may also predispose a crop to injury, so clean cultivation is important. Buntin *et al.* (1994) studied the effects of winter cover crops on subsequent injury. Broadleaf cover crops, especially hairy vetch, induced greater insect problems than grain crops such as wheat. Fallow, weed-free fields had a low incidence of problems.

Soil moisture conditions affect the insects and their damage. Crops grown in moist soil, due either to a high water table or abnormally heavy rainfall, are more likely to be damaged by larvae, probably due to enhanced larval survival. In contrast, water-stressed, wilted cucurbit plants are more attractive to beetles, and experience more foliar damage than turgid plants.

For small plantings, screens or covers provide some protection from adults.

Trap crops have some benefit for cucumber beetle control. Pair (1997) reported that small plantings of an attractive squash crop that were treated with systemic insecticide could suppress early-season populations on nearby cantaloupe, squash, and watermelon.

Host-Plant Resistance. There are some varietal differences among cucurbits with respect to susceptibility to injury. Bitter cultivars, which have a higher cucurbitacin content, were reported to be more attractive and injured (Haynes and Jones, 1975). Consequently, non-bitter cultivars have less bacterial wilt disease (DaCosta and Jones, 1971).

Biological Control. The nematodes *Steinernema glaseri* and *S. carpocapsae* (both Nematoda: Steinernematidae) will attack both the larval and adult stages of spotted cucumber beetle (Poinar, 1979). Their use has not been perfected, however.

Striped Cucumber Beetle
Acalymma vittatum (Fabricius)

Western Striped Cucumber Beetle
Acalymma trivittatum (Mannerheim)
(Coleoptera: Chrysomelidae)

Natural History

Distribution. Striped cucumber beetle, *Acalymma vittatum* (Fabricius), is found throughout the United States and Canada east of the Rocky Mountains, and most of Central America. It is not usually considered a severe pest in the Rocky Mountain and Great Plains states, nor in the southeast. However, in the midwestern and northeastern states, and in eastern Canada, it is the most important pest of cucurbits.

Acalymma trivittatum (Mannerheim), called the western striped cucumber beetle, replaces *A. vittatum* west of the Rocky Mountains. Both *Acalymma* species are very similar in appearance, biology, and damage potential.

Host Plants. The host range consists principally of plants from four families: Cucurbitaceae, Rosaceae,

Leguminosae, and Compositae. Although apple, pear, green beans, soybeans, okra, eggplant, potato, and many other crop plants have been fed upon occasionally by striped cucumber beetle, crop injury is usually limited to cucurbit crops. Rosaceous hosts are important in the early spring, before cucurbit crops are available. At this time, beetles feed greedily on the blossoms of trees and shrubs. Alternate hosts become important again in the autumn when cucurbit crops have declined, but late in the season Compositae are favored by adults. Larvae feed on roots of cucurbits and, though damaging, are of less concern than adults.

Natural Enemies. Numerous natural enemies of striped cucumber beetle have been determined, but only a few are significant. Probably the most important parasitoid is *Celatoria setosa* (Coquillett) (Diptera: Tachinidae), which parasitizes adult cucumber beetles, and was found consistently to parasitize 10–40% of the population in Ohio (Houser and Balduf, 1925). The wasp, *Syrrhizus diabroticae* Gahan (Hymenoptera: Braconidae) also attacks adult beetles, though the incidence of parasitism was found to be less. The nematode *Howardula begnina* Cobb (Nematoda: Allantonematidae) penetrates newly hatched beetle larvae in the soil, matures in the adult beetle, and is dispersed by the female during the act of oviposition. The adult beetle may contain large numbers of nematodes within her body cavity, which undoubtedly affects her longevity and reproductive capacity, but the adults are not very pathogenic. Rates of nematode infection of 10–25% are common throughout the summer in Ohio. A number of general predators, including soldier beetles (Coleoptera: Cantharidae), ground beetles (Coleoptera: Carabidae), shield bugs (Hemiptera: Pentatomidae), damsel bugs (Hemiptera: Nabidae), and birds, as well as fungi attack striped cucumber beetle.

Life Cycle and Description. Striped cucumber beetles require about 40–60 days to complete a generation, but development time is largely a function of weather, and values given by various authors differ greatly. The number of generations varies from one in New Hampshire, to two in Ohio and Connecticut, and three in Texas. Adults overwinter and become active early in the spring, taking flight when the air temperature reaches about 18°C.

Egg. The eggs are deposited in the soil about the base of plants or along vines, usually within 1.5 cm of the soil surface. Females sometimes crawl into cracks in the soil and deposit their eggs at greater depth. Chittenden (1923) reported maximum egg production of over 1500 per female, but indicated that the

Striped cucumber beetle larva.

average was 400–500. The oval eggs are yellow or orange, and measure 0.5–0.7 mm long and 0.3–0.5 mm wide. The length of the egg stage is about 5–10 days.

Larva. The white larvae initially are about 1.5 mm long but attain a length of about 10–12 mm at maturity, and have a dark head, thoracic, and anal plates. Larvae have a proleg-like structure near the tip of the abdomen, in addition to three pairs of true legs. Total duration of the three instars is 15–45 days.

Pupa. Pupation occurs in the soil in a small chamber within 5 cm of the soil surface. The pupal stage requires about 5–7 days. The pupa resembles the adult in form except that its wings are poorly developed.

Adult. The adult is about 7 mm long. When viewed from above, the beetle is yellow with three black stripes on the elytra, a yellow thorax, and a black head. The elytra are marked with minute, but distinct, punctures. The pre-oviposition period is about seven days. Adults are the overwintering stage, with beetles found in organic debris or in the soil, often at sites distant from crop fields but adjacent to autumn food sources. Adults of striped cucumber beetle commonly are confused with adults of western corn rootworm, *Diabrotica virgifera* LeConte, which also feed freely on cucurbit blossoms. The black stripes on the elytra of striped cucumber beetle are parallel-sided, whereas in western corn rootworm the stripes tend to be a slightly wavy, especially the central stripe. (See color figure 111.)

Good accounts of cucumber beetle life history were provided by Garman (1901), Chittenden (1923), Houser and Balduf (1925), and Isely (1927).

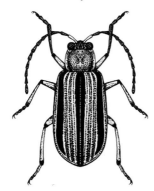

Adult striped cucumber beetle.

Damage

Striped cucumber beetle causes damage directly by its feeding and indirectly through transmission of plant diseases. In the spring, adults feed on young seedlings, often causing complete defoliation. Brewer *et al.* (1987) studied the effects of adult feeding on zucchini seedling survival in Louisiana, and noted significant levels of mortality when they attained densities of 10 beetles per plant or greater. If infestation was delayed until the second or third true leaves were formed, yields were not reduced. Beetles also selectively feed on cucurbit blossoms, if they are available, and may prevent good fruit set. They even attack the fruit, making unsightly pits in the rind. Smooth-skinned melons are especially prone to injury. Because adults seek to avoid heat, beetles may accumulate on the underside of fruit and cause damage there. Larvae feed below-ground on the roots and above-ground on the stems, usually maintaining contact with the soil. Occasionally, larvae will tunnel from the soil directly into mature fruit.

Striped cucumber beetle is an important vector of plant diseases. Latin and Reed (1985) demonstrated the ability of larvae to increase the incidence and severity of Fusarium wilt through their root feeding behavior. Adults harbor bacterial wilt, *Erwinia tracheiphila*, within their bodies during winter months and then apparently vector the disease in the spring. Disease transmission is accomplished when contaminated-fecal material comes in contact with wounds caused by beetle feeding. Although the importance of cucumber beetles in disease transmission is well-established, the role of beetles in aiding the overwintering of the bacterium is less certain (Harrison *et al.*, 1980b). Alternate, asymptomatic host plants such as weeds are likely involved. Cucumber beetles also transmit squash mosaic virus (Gergerich *et al.*, 1986).

Management

Sampling. Plant kairomones such as mixtures of indole with trimethoxybenzene and cinnamaldehyde have been used to make highly attractive sticky traps for monitoring beetle densities (Lewis *et al.*, 1990). These chemical lures are most effective when combined with visual stimuli, such as yellow-sticky traps, and positioned at the height of the plant canopy (Hoffman *et al.*, 1996a). Visual examination is often used to estimate adult density, with suppression initiated at densities of 0.5–1.0 beetles per plant. Brust and Foster (1999) suggested that a threshold of one beetle per plant was adequate to protect cantaloupe in the midwest from economic infection with bacterial wilt.

Sampling plans, including sequential plans, were discussed by Burkness and Hutchison (1997, 1998).

Insecticides. Insecticides are used routinely in commercial production of cucurbits to control striped cucumber beetle. Foliar protectants are common, but because the young, rapidly growing plants are very susceptible to injury, careful timing is important. Granular formulations are sometimes applied in a band over the plant row to protect the roots from larval tunneling. In general, striped cucumber beetles are not considered difficult to kill. Their ability to transmit disease exaggerates their impact, however, especially in the midwest. The need to control the beetles and their spread of cucurbit wilt disease diminishes as the plants reach maturity. If beehives are placed into a field to increase pollination, there should be at least a one-day interval between insecticide application and hive placement. Research in Indiana showed that insecticide applications to cantaloupe, based on thresholds of 0.5 beetles per plant before and one beetle per plant after fruit appeared, produced financial savings relative to scheduled weekly insecticide treatments (Brust et al., 1996).

As is the case with the closely related *Diabrotica* species, bitter cucurbitacins are potent arrestants and feeding stimulants for striped cucumber beetles, and can be used to enhance insecticidal control. Ferguson *et al.* (1983) reported a high correlation between cucurbitacin content of squash and feeding damage by cucumber beetles. Spotted cucumber beetle is also affected by volatile chemicals produced by squash blossoms; indole is especially implicated. Combination of cucurbitacins with blossom volatiles and insecticide has been used experimentally to formulate an effective bait that suppresses beetle abundance and disease incidence as well as foliar insecticide, but is less disruptive to pollination (Brust and Foster, 1995). A cucurbitacin bait containing insecticide is available commercially.

Cultural Practices. Late planting is sometimes suggested to minimize damage by striped cucumber beetle. Cucurbits are cold sensitive and grow much faster under warm conditions, so late planting favors the ability of the plant to outgrow feeding injury by beetles. Screen or paper covers are sometimes used with young plants. The latter not only prevents defoliation by adults but also provides some protection from cold weather early in the season.

The use of resistant cultivars or species is widely practiced, when practical. The preference of cucumber beetles for summer squash has led to the recommendation that this be planted as an attractant and trap crop to prevent damage to cucumber. Michelbacher *et al.* (1955) stated that adults avoid heat and are attracted to moisture. Thus, melon fields were frequently invaded and damaged after irrigation.

Trap crops have some benefit for cucumber beetle control. Pair (1997) reported that small plantings of an attractive squash crop that were treated with systemic insecticide could suppress early season populations on nearby cantaloupe, squash, and watermelon.

Biological Control. Undoubtedly, some degree of control of rootworm larvae can be attained with steinernematid and heterorhabditid nematodes, but this has been well-studied only for western corn rootworm, *Diabrotica virgifera* LeConte. Under laboratory conditions, striped cucumber beetle larvae are very susceptible, and the nematodes can be delivered to the root zone of plants through trickle irrigation emitters (Reed *et al.*, 1986).

Sweetpotato Leaf Beetle
Typophorus nigritus (Crotch)
(Coleoptera: Chrysomelidae)

Natural History

Distribution. This native insect is recorded from all the southeastern states from Maryland to Texas. It is also known from Central and South America and the Caribbean.

Host Plants. Sweet potato is the principal host of this insect, and the only vegetable crop to be attacked. This species has also been collected from morningglory, *Ipomoea* spp.

Natural Enemies. The natural enemies of this insect are not known.

Life Cycle and Description. A single generation occurs annually in North Carolina. Adults are present from late May to mid-July, and eggs from June until early August. Larvae, which are the overwintering stage, are present from June until the following spring. The pupal stage occurs in May. Duration of the complete life cycle is about 300–350 days.

Egg. Oviposition occurs about 14 days (range 10–22 days) after the adults emerge and commence feeding. The eggs are elongate-oval and measure 1.36–1.44 mm long and 0.56–0.64 mm wide. Eggs are yellow and are deposited in clusters of 3–25. Eggs normally are deposited in the soil, but sometimes on foliage, and females reportedly produce an average of 50 eggs. Duration of the egg stage is 6–10 days.

Larva. Upon hatching, the larva enters the belowground portions of the vine or the roots to feed. The larvae are pale yellow, with a brownish head. The larva is only about 1.5 mm at hatching, but grows to

Sweetpotato leaf beetle larva.

about 11 mm at maturity. The larval period, which includes the overwintering period, is 260–320 days.

Pupa. Pupation occurs in the soil in cells at a depth of 10–20 cm. The pupa resembles the adult in form except that the legs, antennae and wings are folded closely to the body. The pupa is pale yellow or white and measures 7–8 mm long. Duration of the pupal stage is about 18 days. Newly molted beetles remain in their pupal cells for 4–5 days before digging to the surface of the soil.

Adult. The adult is oval and metallic bluish-green to black. The antennae are reddish. The elytra are rather smooth. The beetles measure 6.0–7.5 mm long and 4–5 mm wide. Adults commence feeding on foliage about two days after emergence and live for 30–40 days. They drop from the foliage to the soil surface if disturbed, and sometimes are found hiding in leaf litter during the daylight hours. The duration of the oviposition period is about 30 days.

The biology of sweetpotato leaf beetle was presented by Brannon (1938), and additional observations could be found in King and Saunders (1984).

Damage

The adults feed on the foliage, starting with the marginal areas and working inward toward the center of the leaf. The aggregated nature of adults, especially

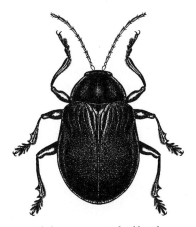

Adult sweetpotato leaf beetle.

after they have initially emerged, tends to result in localized defoliation. However, adult feeding damage is relatively minor as compared to injury by larvae. The larvae feed in stem, root and tuber tissue. Larvae can burrow deep into the tuber, creating tunnels packed with fecal material. They also may confine their feeding to the surface of the tuber, creating scars.

Management

These insects, while widespread, are infrequently numerous. Therefore, controls are rarely warranted. Foliar insecticides effectively suppress adults. Larval injury is best prevented with the application of liquid or granular insecticide to the soil.

Western Corn Rootworm
Diabrotica virgifera LeConte
(Coleoptera: Chrysomelidae)

Natural History

Distribution. Western corn rootworm is a widely spread native species, consisting of two recognized strains. The principal strain is *Diabrotica virgifera virgifera* LeConte, which occurs from Montana, Utah, and Arizona east to New York, Connecticut and Virginia. In Canada, it is known only from southern Ontario. First noticed attacking corn in Colorado in 1909, it has gradually expanded its range through the midwestern corn belt and eastern states, where it often displaces northern corn rootworm, *Diabrotica barberi* Smith and Lawrence. This species is absent from the southeastern states, where banded cucumber beetle, *Diabrotica balteata* LeConte, is present, but successfully overlaps the northern range of the related spotted-cucumber beetle, *Diabrotica undecimpunctata* Mannerheim.

In southern Oklahoma, most of Texas, and south through most of Mexico the strain *Diabrotica virgifera zeae* Krysan and Smith predominates. It is known as "Mexican corn rootworm."

Host Plants. The larval stage of this species feeds successfully only on plants in the family Gramineae. Although corn is the principal host, larvae can complete their development successfully on lovegrass, *Eragrostis* spp.; wheatgrass, *Agropyron* spp.; foxtail, *Setaria* spp.; rice; spelt (a primitive type of wheat); wheat, and perhaps barley (Branson and Ortman, 1967b; 1970). The adults feed readily on blossoms and foliage of Cucurbitaceae, and on the inflorescence of goldenrod, *Solidago canadensis*, and sunflower, *Helianthus annuus*, but these are not as satisfactory for adult longevity as is corn (Siegfried and Mullin, 1990). Although all cultivated squashes are attacked, beetles greatly prefer the cucurbitacin-rich wild species, especially the xerophytic species (Howe *et al.*, 1976). Adults also feed on pollen of weeds such as pigweed, *Amaranthus* sp., and ragweed, *Ambrosia* sp., and spores of fungi (Ludwig and Hill, 1975). In the midwestern states, adults have shown an increased tendency to consume soybean leaves, but this is a nutritionally inadequate diet. (See color figures 6, 34, and 35.)

Natural Enemies. Western corn rootworm is remarkably free of natural enemies (Chiang, 1973). Some general predators, such as ambush bugs (Hemiptera: Phymatidae), and assassin bugs (Hemiptera: Reduviidae) will feed on adults, and mites (Acari: Laelaptidae, Rhodacaridae, and Amerosiidae) feed on larvae and eggs. The fly *Celatoria diabroticae* (Shimer) (Diptera: Tachinidae) parasitizes rootworms, but like the predators, it seems relatively unimportant in population regulation.

Life Cycle and Description. There is a single generation per year. The egg is the overwintering stage. Hatching occurs in the spring, followed by a larval feeding and pupation period of 1–2 months. Under ideal temperatures of 21–24°C, rootworms can progress from the egg to the adult stage in about 30 days. Adult emergence occurs between July and August in the corn belt, but during May–June in Texas (Cocke *et al.*, 1994). Adults remain active and deposit overwintering eggs through September or October, depending on the onset of cold weather.

Egg. The eggs are deposited in the soil, about 80% found in the upper 10 cm in moist soil (Weiss *et al.*, 1983). Females use cracks in the soil, earthworm burrows, and the space between stalks and surrounding soil to gain entry to regions of the soil profile, where soil is moist (Kirk, 1981). Fecundity normally is about 400–500 eggs per female under field conditions, deposited at a rate of about 60–70 per week in loose clusters of 6–10 eggs. Higher fecundity is possible, up to an average of 850–1100 egg per female, when larvae are reared at low densities with an abundance of food (Branson and Johnson, 1973; Branson and Sutter, 1985). The eggs are oval and whitish. Eggs measure only about 0.6 mm long and 0.4 mm wide. The surface of the eggs, when examined microscopically, are marked with a polygonal (mostly hexagonal) pattern, but small pits are not found within the polygons. The absence of pits serves to differentiate western (and Mexican) corn rootworm eggs from those of northern corn rootworm, and banded and spotted cucumber beetles, which possess pits (Atyeo *et al.*, 1964; Rowley and Peters, 1972; Krysan, 1986). The eggs generally require a cold period to break diapause, and then

hatch in about 34 and 41 days when held at 25 and 20°C, respectively. The eggs tolerate prolonged exposure of −5.0° to −7.5°C, but survival decreases when eggs experience −10°C (Gustin and Wilde, 1984). Cold winters, shallow overwintering depths, and lack of crop residue and snow cover all contribute to higher egg mortality (Godfrey *et al.*, 1995). The threshold for egg development is about 11°C.

Larva. The larvae develop in the soil, feeding on the roots of grasses. They usually are found within 20 cm of their host plant and within the upper 10 cm of soil. The larval stage is elongate and cylindrical, tapering toward the head. The body is white, bearing relatively few hairs or spines. The head capsule, and thoracic and anal plates, are yellowish brown. The three pairs of legs are brownish, and terminate in a single claw. The posterior end of the body bears a single retractile extension or tubercle. The larvae of western corn rootworm can often be distinguished from those of northern corn rootworm and spotted cucumber beetle by the color pattern of the anal plate. The dark area in western corn rootworm has a central notch at the anterior margin; this notch is absent from the aforementioned co-occurring species. There are three instars, which require about 50 days for development. The mean duration (range) of the instars was given as 11 (8–17), 14 (8–20), and 27 (25–33) days for instars 1–3, respectively, when reared at 20°C (George and Hintz, 1966). However, Jackson and Elliott (1988) reported shorter development times of 8.1, 6.8, and 15 days for instars 1–3 when reared at 18°C, and 5.6, 4.9, and 11.2 days for instars 1–3 when reared at 21°C. Mean head capsule widths are 0.20, 0.32, and 0.50 mm for instars 1–3, respectively. Mean body lengths (range) are 1.5 (1.2–1.7), 3.7 (3.2–5.0), and 7.1 (6.5–8.0) mm, respectively. The larvae can disperse readily up to about 40 cm in search of food (Short and Luedtke, 1970). First instars often burrow into the roots. (See color figure 109.)

Pupa. The pupation usually occurs within 25 cm of the host plant and within the upper 10 cm of soil. The pupa is whitish and measures about 4.5 mm long and 2.5 mm in wide. Mean duration (range) of the pupal stage was reported to be 15 (14–15) days when reared at 20°C (George and Hintz, 1966). However, Jackson and Elliott (1988) found mean development only 13.5 days at 18°C and 10.1 days at 21°C. Beetles remain in the pupal cell for 1–3 days before digging to the soil surface.

Western corn rootworm larva.

Adult. The adults are yellowish to yellowish brown with black markings, and measure about 6 mm long. The beetle's head often is black, though it sometimes is pale. A common pattern on the elytra is a black stripe centrally, accompanied by another on each elytron. These normally are females. The alternate pattern commonly found in western corn rootworm is black elytra, with some yellow at the tips of the elytra and its margins. These normally are males. Mexican corn rootworm, in contrast, is green with a black head, lacking black on the elytra except for a small amount at the base. The adults of both subspecies display a fairly protracted pre-ovipositional period ranging from about 12–30 days, and then are thought to live for an additional 30–60 days in the field, though they can survive and continue to produce eggs for over 90 days under laboratory conditions. Oviposition rates peak at about 10–15 days after initiation, and then taper off. Western corn rootworm beetles are quite active, and adults disperse readily from fields with mature corn to fields that are silking or pollinating. Females are particularly vagile. Adult dispersal occurs principally during periods of no wind or low wind speed, and beetles fly mostly at heights of less than 4 m (Vanwoerkom *et al.*, 1983). (See color figure 108.)

The review of rootworms by Chiang (1973) is a helpful introduction to rootworm biology. Rearing methods for larvae generally use sprouted corn as a food source (Jackson, 1986; Branson *et al.*, 1988); adult maintenance is described by Guss *et al.* (1976). Separation of western corn rootworm larvae from northern corn rootworm and spotted cucumber beetle was described by Mendoza and Peters (1964), but Krysan (1986) reported that the characters used in this key are imperfect. A female sex pheromone has been identified (Guss, 1976). Krysan and Branson (1983) discussed the biosystematics of *Diabrotica* spp. Mexican corn rootworm was described by Krysan *et al.* (1980) and Branson *et al.* (1982). A key to the adult rootworms was presented by Krysan (1986).

Damage

The principal injury caused by rootworms is destruction of the roots by larvae. The larvae first feed on the rootlets and then burrow into the roots, often causing complete destruction of the root system. Plants with damaged roots are stunted, and tend to blow over (lodge) under conditions of high moisture and wind. Corn plants are quite tolerant to larval feeding, however, and sometimes support hundreds of larvae without great yield reduction. In the absence of adequate soil moisture, such plants may display

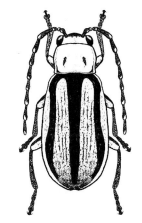

Adult female western corn rootworm.

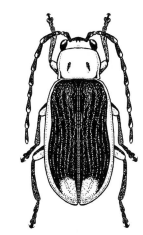

Adult male western corn rootworm.

greater yield loss, or may die (Chiang *et al.*, 1980). Root damage ratings are the preferred method of assessing larval feeding (Branson *et al.*, 1980).

The adults also cause injury, though this stage generally is considered less damaging than the larval stage. The adults feed on the surface of foliage, particularly young foliage and leaf tips, causing the leaf tissue to die and to turn brown. More importantly, they feed on the young corn silk, sometimes completely clipping the silks to the tip of the ear and interfering with pollination. Such plants have ears with kernels irregularly placed or missing, particularly near the ear tip. When beetles are abundant, they consume all the silk and feed directly on the kernels at the ear tip. They also are attracted to openings in the ear husk caused by such insects as grasshoppers and caterpillars, which allow the beetles ready access to the kernels.

Western corn rootworm beetles are implicated in disease transmission. They have been found contaminated externally and internally with *Fusarium* spp., causative agents of corn stalk rot (Gilbertson *et al.*, 1986). They also have been implicated in the transmission of maize chlorotic mottle virus (Jensen, 1985).

Management

Sampling. Extensive research has been conducted on all life stages, but crop management decisions tend to base mostly on adult abundance owing to the difficulty and expense of sampling other stages. All life stages tend to aggregate in distribution. The eggs are sampled from soil by washing and screening (Lawson and Weekman, 1966; Hein *et al.*, 1985; Ruesink, 1986). The larvae and pupae can be detected by soil sifting and visual examination, or by washing (Weiss and Mayo, 1983; Shaw *et al.*, 1984; Fisher and Bergman, 1986). The adult sampling involves visual examination of whole plants or region of the plant such as the ear zone, or traps that attract beetles using visual stimuli or volatile baits. Common traps are yellow sticky traps, or container traps baited with volatiles (Hoffmann *et al.*, 1996a) or a feeding arrestant such as cucurbitacin (Hein and Tollefson, 1984; Hesler and Sutter, 1993; Levine and Gray, 1994). Sequential sampling plans have been published (Foster *et al.*, 1982, McAuslane *et al.*, 1987).

Insecticides. Insecticides are used principally when land is planted to corn continuously. Insecticides are usually applied with the seed or over the row, during or shortly after planting. However, beetles disperse readily, and because sweet corn is very susceptible to cosmetic injury, ear tip damage resulting from adult feeding may occur. Adult suppression with foliar insecticides is easily accomplished, and commonly results from insecticide applications aimed at primary ear pests such as corn earworm, *Helicoverpa zea* (Boddie).

Considerable research has been conducted on insecticide-containing baits for beetle control. Dry baits have been formulated from corn cob grit impregnated with cucurbitacin and carbamate insecticides (Metcalf *et al.*, 1987) and starch matrices impregnated with cucurbitacin or volatile semiochemicals (Lance and Sutter, 1990; Weissling and Meinke, 1991a,b). Liquid formulations of cucurbitacin and insecticide seem preferable over dry formulations, however, due to ease of application. Also, vial traps containing cucurbitacin, volatiles, or pheromone have been used to suppress populations (Lance, 1988). However, these techniques have been limited primarily to area-wide suppression efforts.

Cultural Practices. Several cultural practices affect western corn rootworm survival. Conservation tillage (no-till) results in delayed emergence of beetles rela-tive to conventional tillage, but survival is not markedly affected (Gray and Tollefson, 1988a,b). Planting of corn between previous rows is preferable to planting within previous rows, presumably because larvae must disperse farther to find food in the first instance (Chiang *et al.*, 1971). Earlier planting dates also favor rootworm survival, so it is advisable to delay planting in heavily infested soil if insecticides are not used (Musick *et al.*, 1980; Fisher *et al.*, 1991). High moisture levels resulting from flooding can reduce adult emergence, but the level of soil saturation necessary to harm insects is detrimental to corn growth (Reidell and Sutter, 1995). Strip cropping, while deleterious to some species with a narrow host range, seems not to be a deterrent to feeding by western corn rootworm beetles, as they were more numerous on one-, two-, and four-row corn plots alternated with beans than on 16-row corn plots (Capinera *et al.*, 1985). The most important cultural practice is crop rotation. Western corn rootworm populations are suppressed or eliminated by rotation to a non-host crop such as soybean, so continuous corn cropping is discouraged. However, there are reports from Texas that Mexican corn rootworm survives rotation with sorghum (Stewart *et al.*, 1995), and there is preliminary evidence that western corn rootworm in the midwestern states is beginning to display similar behavior. It is not yet apparent whether corn rootworm has adopted a multi-year diapause, as has northern corn rootworm in some areas, or it has adapted to feeding on other hosts. Corn rootworms are not a problem on muck soils, and they do not cause appreciable injury on sandy soils unless irrigated.

Biological Control. Entomopathogenic nematodes (Nematoda: Steinernematidae and Heterorhabditidae) have been evaluated for control of western corn rootworm larvae (Wright *et al.*, 1993; Jackson and Hesler, 1995, Ellsbury *et al.*, 1996; Jackson, 1996). *Steinernema* spp. nematodes can provide good suppression, though not as effective as some insecticides. Nematodes are readily applied through irrigation systems and generally require high volumes of water for effective application. Effective suppression also require high rates of application, which can be a considerable expense.

Yellowmargined Leaf Beetle
Microtheca ochroloma Stål
(Coleoptera: Chrysomelidae)

Natural History

Distribution. This insect is a native of South America, where it is known from Argentina, Brazil,

Chile, and Uruguay. It was first found in the United States at Mobile, Alabama, in 1947. It is now distributed along the Gulf Coast from Florida to Louisiana.

Host Plants. The yellowmargined leaf beetle prefers plants in the family Cruciferae. It has been reported to damage cabbage, Chinese cabbage, collards, mustard, radish, turnip, and watercress. Occasionally, it has been found on potato. Ameen and Story (1997b) conducted choice tests with larvae and adults, and reported a preference rating of turnip > mustard > radish > collards > cabbage. It has also been collected from pepperweed, *Lepidium virginicum*; dock, *Rumex* sp.; and various clovers and vetch, but it is not clear whether the insects were feeding or just seeking shelter on these plants. Fecundity and longevity on several host plants were reported by Ameen and Story (1997a). They found that turnip was the most favorable host, with an average of 490 eggs produced per female, followed by radish (440), mustard (425), cabbage (271), and collards (199). Adult longevity varied from 68 days when fed turnip and collards to 105 days on radish. Surprisingly, there was not a significant relationship between longevity and fecundity.

Natural Enemies. This insect is inadequately studied, and no natural enemies are known.

Life Cycle and Description. The life cycle is poorly known, and no evidence of diapause has been reported. There appears to be a single generation annually, but the insect can complete its life cycle in less than a month under favorable conditions, so it is possible that more than one generation occur during mild Gulf Coast winters. In Florida, adults remain active throughout the winter months. Damage is noted in the spring and early summer, when both larvae and adults can be found feeding on crucifers. A summer aestivation from mid-June to October has been suggested.

Egg. The adults can begin copulation about six days after emergence, followed by oviposition in another 3–6 days. The eggs are bright orange, elongate, and deposited singly or in small groups on the foliage. Duration of the egg stage is 4–5 days.

Larva. The larva is yellowish brown and covered with a fine layer of hairs. The head capsule is dark brown or black. Larvae are gregarious during early instars, but eventually become solitary. There are three instars, lasting 3–4, 3–4, and 5–6 days, respectively. (See color figure 118.)

Pupa. The larva spins a loose net-like pupal case on foliage, and pupates within. Duration of the pupal stage is 5–6 days. Newly formed adults remain within the cocoon for about two days before escaping.

Adult yellowmargined leaf beetle.

Adult. The adult is about 5 mm long, and predominantly dark bronze or black. The peripheral edges of the elytra, however, are marked with a margin of yellowish or brownish, which is the basis for the common name of this insect. The elytra are also each marked with four rows of deep punctures. (See color figure 119.)

Biological information on yellowmargined leaf beetle was given by Chamberlin and Tippins (1948), Woodruff (1974), and Oliver and Chapin (1983).

Damage

The adults and larvae feed on foliage, making small holes and feeding on the leaf margin. Damage to cruciferous crops is reported in the spring. In the United States, it is a pest only in small or home garden plantings. In South America, it has been observed to damage crops grown on a commercial scale, but only before the advent of modern insecticides.

Management

Generally, this is a minor pest that does not require control except on an isolated plant or planting. If the insects are observed to be numerous, foliar insecticides can be applied to provide suppression.

FAMILY CURCULIONIDAE—BILLBUGS, CURCULIOS AND WEEVILS

Cabbage Curculio
Ceutorhynchus rapae **Gyllenhal**
(Coleoptera: Curculionidae)

Natural History

Distribution. The cabbage curculio was likely introduced accidentally to North America from

Europe in the early 1800s. The earliest confirmed specimens were collected in 1873 in Massachusetts, but several writers described injury by a similar weevil earlier in the century. It is now generally distributed in the northern portions of the United States, extending as far south as Virginia in the east and California in the west. In Canada it is found from Ontario west to British Columbia.

Host Plants.   Most cruciferous vegetable crops can be damaged by cabbage curculio, including cabbage, cauliflower, and turnip. Despite the common name, however, cabbage is one of the least preferred host plants. The beetles prefer wild crucifers such as hedge mustard, *Sisymbrium officinale*; pepperweed, *Lepidium* spp.; and shepherdspurse, *Capsella bursa-pastoris*; and then cultivated crucifers other than cabbage.

Natural Enemies.   Natural enemies of cabbage curculio are not well known, but *Euderis lividus* (Ashmead) (Hymenoptera: Eulophidae) has been reported to parasitize the larvae (Chittenden, 1900).

Life Cycle and Description.   There is one generation per year, with the adult stage overwintering. In Maryland and Missouri, adults emerge from diapause in April. Mating and oviposition occur soon after emergence. Adults feed briefly on foliage and stems before oviposition, but cause little damage.

Egg.   Females deposit eggs in stem tissue within cavities created by feeding. Some swelling of the plant tissue near the eggs is noticeable. Oviposition is complete by early May. Eggs are elliptical and gray. Egg length is 0.65–0.85 mm; and its width is 0.35–0.45 mm. Eggs hatch in 5–8 days.

Larva.   The larvae feed in stem tissue. They are whitish, with a yellowish or brownish head. The body tapers markedly toward both the anterior and posterior ends. The body is fairly wrinkled in appearance, with no evidence of legs apparent. Larvae attain a length of 5–6 mm at maturity. The larval development requires about 21 days, and most of then have abandoned their feeding sites by the end of May.

Pupa.   Larvae drop to the soil to pupate, secreting an adhesive material to form a small cell from soil particles. The larva remains in the cell for about 14 days

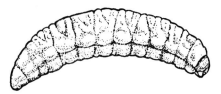

Cabbage curculio larva.

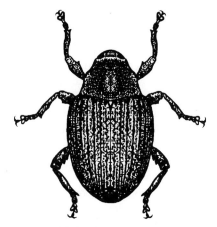

Adult cabbage curculio.

before pupation occurs. The period of pupation is an additional 21 days. The pupal cell is about 5 mm in length, only slightly larger than the white pupa. Pupation occurs at a depth of less than 2 cm. The adults emerge from their pupal cells in June and feed before entering diapause.

Adult.   The adults measure about 2.70–3.25 mm long. The base color of the beetle is black, but the body is covered with a fine layer of scales that are brownish-yellow in new adults and gray in overwintered adults. The elytra are marked with longitudinal ridges, but the most distinctive feature is the elongate snout. The snout exceeds the thorax in length, and is curved ventrally under the body.

The biology of cabbage curculio was provided by Chittenden (1900).

Damage

The cabbage curculio damages crucifers by feeding as larvae in the stems and petioles of the plant. Young tissue is preferentially selected by ovipositing females, and she may deposit her eggs basally in a small plant, or just below the flower on a mature plant. Plants may show wilted leaves or breakage when larvae are numerous. Death sometimes occurs, especially among seedlings. Curculio feeding also seems to allow plant pathogens to gain entry.

Management

Cabbage curculio is generally considered a minor pest, and has little importance to commercial plantings. Foliar applications of insecticides readily control this insect. Because of their strong preference for weeds, adults often avoid ovipositing on crops, especially cabbage. If cabbage curculio is abundant, however, it is advisable to delay destruction of cucurbitaceous weeds until oviposition has occurred. Timely destruction of weeds by tillage provides an effective trap crop and an effective non-chemical means of insect control.

Cabbage Seedpod Weevil
Ceutorhynchus assimilis (Paykull)
(Coleoptera: Curculionidae)

Natural History

Distribution. Cabbage seedpod weevil is native to Europe, and was first discovered in North America in 1931 near Vancouver, British Columbia. It soon spread to Washington, Oregon, and Idaho, becoming a serious pest of cruciferous seed crops. With the increased popularity of rape as a field crop throughout North America, there is potential for cabbage seedpod weevil to spread and assume greater importance as a vegetable pest. For example, Buntin and Raymer (1994) and Boyd and Lentz (1994) reported this insect to be abundant and damaging to oilseed rape (canola) in northern Georgia and in Tennessee, respectively.

Host Plants. This weevil seems to feed exclusively on crucifers. Vegetable crops attacked include broccoli, Brussels sprouts, cabbage, cauliflower, Chinese cabbage, kale, kohlrabi, mustard, radish, rutabaga, and turnip. It also feeds on rape and on cruciferous weeds such as tansymustard, *Descurainia* spp.; rockcress, *Arabis* spp.; shepherdspurse, *Capsella bursapastoris*; pepperweed, *Lepidium* spp.; and wild radish, *Raphanus* spp.

Natural Enemies. Numerous parasitoids are known to attack cabbage seedpod weevil, including *Bracon* sp. (Hymenoptera: Braconidae); *Eupelmella vesicularis* (Retzius) (Hymenoptera: Eupelmidae); *Eurytoma* sp. (Hymenoptera: Eurytomidae); *Necremnus duplicatus* Gahan and *Tetrastichus* sp. (both Hymenoptera: Eulophidae); *Asaphes californicus* Girault, *Habrocytus* sp., *Trichomalus fasciatus* (Thomson), and *Zatropis* sp. (all Hymenoptera: Pteromalidae); and *Megaspilus* sp. (Hymenoptera: Ceraphronidae). The most important species are *Trichomalus fasciatus* and *Necremnus duplicatus*, with the former species probably introduced from Europe accidentally at the same time as the weevil (Hanson *et al.*, 1948).

Life Cycle and Description. There is a single generation per year. Adults overwinter in soil or under debris, becoming active in the spring as the air temperatures reach about 15°C. Adults typically feed on nectar and pollen for about a month before beginning egg production. Females denied the opportunity to feed at flowers exhibited poor development of their ovaries (Ni *et al.*, 1990). Females are particularly attracted to volatile emissions of flowers (Evans and Allen-Williams, 1992). The egg development is favored by temperatures of 15° and 20°C, relative to 10° and 25°C (Ni *et al.*, 1990). Flight is limited to periods of relatively low wind speed (less than 0.5 m/s) and warm temperatures (above 22°C) (Kjaer–Pedersen, 1992).

Egg. Females deposit eggs singly in the seed pod. The normal distribution of eggs is one per pod. This is critically important in some of the wild hosts, where pods are small and only one larva can survive, but not essential in such crops as rape, where the pods provide excess food (Kozlowski *et al.*, 1983). The egg is white or cream colored, and elliptical. They measure about 0.32–0.35 mm long and 0.20–0.22 mm wide. The eggs hatch in a few days and larvae feed on developing seeds for about 2–3 weeks.

Larva. At the time of hatching, the larva is yellow, and measures 3–4 mm long. It soon turns white, with a brown head capsule, and eventually attains a length of about 5–7 mm. At maturity, larvae cut a hole in the seed pod and drop or crawl to the soil to pupate.

Pupa. The larva digs to a depth of 2–5 cm before creating a pupal chamber. Pupation requires 2–4 weeks. The pupa is white or yellow, and measures 3.2–3.7 mm long.

Adult. The adults overwinter at field margins and in woods, becoming active in March or April in Georgia, then commencing oviposition. New adults usually emerge in July, and feed before entering diapause. Adults are small, only 1.75–2.5 mm long. They are black or grayish black. When they first emerge from the pupal stage they often have a fine covering of gray scales, but the gray color diminishes with time.

The biology of cabbage seedpod weevil was provided by Hanson *et al.* (1948) and Campbell *et al.* (1989).

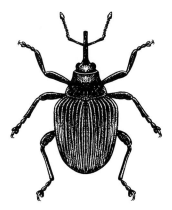

Adult cabbage seedpod weevil.

Damage

Cabbage seedpod weevil larvae feed only on the developing seed pods of crucifers, so their effect on commercial vegetable production is limited mostly to Washington and surrounding areas, where vegetable seed production is an important industry. Each larva damages three seeds on average. The larvae should not pose a threat to home garden production unless seed production is desired. As noted previously, their increased importance in recent years is owing to increased demand for oil-seed rape.

The adult feeding during spring months is similarly restricted mostly to the flower and related tissues. Adult feeding punctures often produce sap-like exudates on the pods and stems, and stamens and pistils of the flowers are consumed. Egg laying occurs in the feeding punctures. These wounds also open the pod to invasion by a gall midge, *Dasineura brassicae* (Winnertz) (Diptera: Cecidomyiidae). Because the midge ovipositor cannot penetrate the seed pod, its abundance and damage are directly related to the weevils. In studies conducted in England, rape showed strong ability to compensate for damage by weevils, but secondary invasion of rape-seed pods by gall midges caused additional damage, thereby increasing the impact of the cabbage seedpod weevil (Free *et al.*, 1983).

Feeding in the summer or autumn by pre-diapause adults occurs on stems, large leaf veins, and any pods or other thick-green plant material that is available.

Management

Insecticides. The presence of cruciferous weeds or previous crucifer crops can predispose an area to problems with seedpod weevil. If high populations are expected or detected, foliar insecticide applications are commonly recommended. To avoid destruction of pollinators, insecticides should be applied pre-bloom or post-bloom, but never during bloom.

Cultural Practices. Availability of overwintering sites may limit the abundance of weevils (Boyd and Lentz, 1994).

Host-Plant Resistance. Doucette (1947) evaluated the relative suitability of different crucifer crops for cabbage seedpod weevil. In general, mustard and radish were less susceptible to attack, whereas Chinese cabbage and turnip were most susceptible. He also documented considerable difference in susceptibility to attack among cabbage cultivars.

Pheromones. An oviposition-deterring pheromone is deposited by the female following egg-laying, presumably to ensure adequate food for her offspring (Ferguson and Williams, 1991). Eventually this may lead to the ability to deter females from oviposition, thereby reducing damage.

Carrot Weevil
Listronotus oregonensis (LeConte)

Texas Carrot Weevil
Listronotus texanus (Stockton)
(Coleoptera: Curculionidae)

Natural History

Distribution. Carrot weevil is native to North America, and has been known as a pest since the mid-1800s. It is found throughout the United States, though it is a more serious pest in northern states. It is fairly widespread in Canada and continues to increase in geographic range and severity of damage. However, it is a serious pest principally in Ontario and Quebec.

Texas carrot weevil is also a native insect, though it likely occurs in Mexico as well. Only in recent years has it become a pest, principally in the Rio Grande Valley of south Texas, but it is also known from Louisiana.

Host Plants. These insects attack crop plants in the family Umbelliferae: carrot, celery, dill, parsley, and parsnip. Carrot weevil initially was known as a pest of parsley, and for many years as the "parsley stalk weevil." However, its apparent host range (as determined by recorded damage) was expanded to include carrot by 1925, and celery by 1936. Both umbelliferous and non-umbelliferous wild plants are suitable hosts, including wild carrot, *Daucus carota*, and wild parsnip, *Pastinaca sativa* (Umbelliferae); dock, *Rumex* spp. (Polygonaceae); and plantain, *Plantago* spp. (Plantaginaceae).

Natural Enemies. Several species of ground beetles, including *Pterostichus melanarius* Illiger, *P. lucablandus* Say, *Bembidion quadrimaculatum* Say, *Clivina fossor* Linnaeus, and *Anisodactylus santaecrucis* Fabricius; (all Coleoptera: Carabidae) have been shown to feed on eggs, larvae, and pupae of carrot weevil in choice tests (Baines *et al.*, 1990). Although some are voracious, *Pterostichus* spp. consuming 10–11 weevils per day, their importance in the field is uncertain. An egg parasitoid, *Anaphes sordidatus* (Girault) (Hymenoptera: Mymaridae), inflicts considerable mortality, reaching nearly 50% parasitism and averaging 22% in Michigan (Collins and Grafius, 1986) and Ontario (Cormier *et al.*, 1996), but not functioning in a density-dependent manner (Boivin, 1993). *Anaphes* sp. also attacks Texas carrot weevil in Texas, often achieving high levels of parasitism (60–90%) (Boivin *et al.*, 1990). Entomopathogenic fungi, such as *Beauveria* *bassiana* and *Metarhizium anisophliae*, and the *Listronotus* strain of the nematode *Steinernema carpocapsae* (Nematoda: Steinernematidae) occasionally provide some suppression of carrot weevil.

Life Cycle and Description. The number of generations varies geographically, and with seasonal weather patterns. In New Jersey there are 2–3 carrot weevil generations per year (Pepper, 1942). There is considerable overlap of the generations, so all life stages may be found from May to October. Wright and Decker (1958) reported two generations from Illinois. In Massachusetts (Whitcomb, 1965), Ontario (Stevenson, 1976a), and Quebec (Boivin, 1988), there is usually only one carrot weevil generation per year, but some weevils complete a second generation if they first develop on weeds that are present early in the season. The period of time necessary for carrot weevil to complete development of the egg to adult stages is about 30–40 days, but another 10–17 days are required before eggs of the next generation are produced. Overwintering occurs in the adult stage, usually in grass sod or organic debris adjacent to crop fields. Reproductive diapause is regulated by both photoperiod and temperature (Stevenson and Boivin, 1990); for example, oviposition requires 14–16 h photophase at 20°C, but only 10–12 h at 28–30°C.

The biology of Texas carrot weevil is different because it inhabits a much warmer climate, where carrots are cultured during the winter months. All stages can be found throughout the year in southern Texas, though it is most abundant beginning in February when commercially grown carrots are abundant.

Egg. The eggs of both species are oval, and measure about 0.8 mm long and 0.5 mm wide. Although initially light-yellow, the eggs turn brown or almost black. The eggs are deposited within cavities made by the females while feeding. She places 1–10 eggs per cavity, but the average clutch size is 3–4 eggs. The number of egg cavities per plant varies widely, with up to 100 eggs being found in some plants. The eggs are normally deposited on the interior or concave sides of leaf stalks. The female seals the opening with a black anal secretion after depositing her eggs. Eggs hatch in 5–16 days, with incubation time averaging about 11 days in the spring, but only 6–8 days in the summer months. Females deposit about 150–175 eggs during their life span.

Larva. Upon hatching, larvae of both species tunnel inside the stalk, usually tunneling downwards. Some larvae exit the stalk, drop to the soil, and re-enter the plant at, or below, ground level. When plant stalks are small, larvae may move directly to the roots to feed. It may not be readily apparent that larvae are feeding in the crown or roots of plants. Four larval instars are reported by most authors, though Wright and Decker (1958) indicated five instars. Average duration of each instar is 4.1, 2.7, 3.4, and 9.2 days, respectively. Development time is affected by temperature, of course, and Simonet and Davenport (1981), and Stevenson (1986) gave temperature-related developmental data for carrot weevil. Mean head capsule width for the instars is 0.31, 0.44, 0.65, and 0.99 mm, respectively. Larvae are legless, white to pinkish brown, with yellowish brown heads. Larval development in carrot weevil requires 11–19 days, averaging 14.5 days in the spring, and about 13 days in the summer.

The developmental biology of Texas carrot weevil is similar. However, as might be expected for a species inhabiting a warmer environment, the developmental threshold is higher, 13.3°C vs. about 7°C in carrot weevil. Woodson and Edelson (1988) provided developmental data for Texas carrot weevil.

Pupa. The mature larva leaves the plant to construct a pupal chamber in the soil, usually at a depth of 3–15 cm. A prepupal period of 1–5 days precedes the pupal stage. The pupa is 5–8 mm long, creamy-white, and bears fairly prominent spines on each abdominal segment and elsewhere on the body. Duration of the pupa is 4–13 days, averaging 7–9 days. After transformation to the adult, the beetle remains in the pupal chamber for 1–5 days, and then digs to the soil surface.

Adult. Newly emerged beetles are light brown, but gradually turn darker. The dark body is covered with tan scales. A faint-striped pattern sometimes is evident on the elytra. The adults of carrot weevil measure 3.5–7.0 mm long, averaging males 6.0 mm and females 6.5 mm. Texas carrot weevil averages slightly smaller, normally 4.0–4.7 mm long. The adults are quick to drop from the plant and feign death when disturbed. Although capable of flight, they rarely disperse in this manner, traveling principally by walking.

An excellent synopsis of carrot weevil biology was given by Pepper (1942), and additional useful information was provided by Whitcomb (1965) and Martel *et al.* (1976). Information on weevil rearing was provided by Martel *et al.* (1975). Texas carrot weevil was described by Stockton (1963) and its biology was given by Edelson (1985b, 1986). Boivin (1999) provided a recent summary of biology and management considerations for both species.

Damage

Both adults and larvae damage plants. Adult feeding takes two principal forms: cavities created by

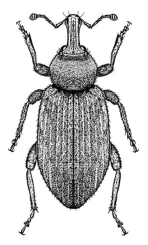

Adult carrot weevil.

gnawing into the tissue (usually accompanied by oviposition), and surface injury caused by removal of the epidermis. Females often deposit eggs on young plants, but are selective in their oviposition behavior. For example, Boivin (1988) reported that carrots were not attacked until they had attained the four-leaf stage. Larval feeding is much more injurious than adult feeding, and results from larvae burrowing through the base of the stalks or roots. On carrots, larval damage generally occurs on the upper one-third of the root, and there usually is only one weevil larva per plant. The tunneling is usually very pronounced. Fungi often invade roots that are heavily damaged by weevils. In untreated carrot fields located in areas where carrot weevils are historically damaging, crop losses of 70% may occur. In commercial fields, where insecticides are applied, losses of 10% are not uncommon. Thus, carrot weevil is considered a serious pest in some locations.

Management

Sampling. In Quebec, day-degree accumulations above a base threshold temperature of 7°C have been used to estimate timing of oviposition (Boivin 1988; 1999). Carrot-baited traps also can be used to monitor adult activity (Boivin 1985; Ghidiu and Van Vranken, 1995), though Edelson (1985b) reported that carrot-baited traps were not very effective for Texas carrot weevil. Perron (1971) compared several sampling methods and determined that root sampling was preferable to light traps, color-based sticky traps, and sweep nets for population estimation.

Insecticides. Formerly, persistent liquid or granular insecticides were applied to the furrow at planting to kill adults and larvae before they could inflict serious injury. Persistent materials are no longer readily available, so multiple applications of foliar insecticides are usually used to suppress carrot-weevil adults. Insecticide must be applied before oviposition has commenced, so timing is important. In Canada, where a single generation is the rule, 1–2 well-timed applications may be adequate. Where multiple or overlapping generations occur, more sprays are applied. A technology that has not been exploited in recent years is the use of poison baits for adult suppression. However, Pepper (1942), reported good control with dried apple pomace-based baits.

In Texas, Texas carrot weevil control may require several insecticide applications for effective suppression. Late-season applications are most important, and late planting of the crop along with late-season application affords the most economic approach to chemical suppression (Woodson *et al.*, 1989).

Biological Control. Entomopathogenic nematodes have been evaluated for management of carrot weevil under laboratory and field plot conditions. The steinernematid nematodes *Steinernema carpocapsae* and *S. bibionis* were more effective than the heterorhabditid *Heterorhabditis heliothidis*, and larvae were more susceptible than pupae or adults (Belair and Boivin, 1985; Boivin and Belair, 1989). In research conducted on organic soils in Quebec, application of *S. carpocapsae* as a soil drench or in conjunction with a bait provided moderate levels of damage suppression (Belair and Boivin, 1995).

Cultural Practices. In areas where there is only one generation per year, damage can be reduced by delaying planting until most oviposition has occurred. Organic soils tend to have more severe problems.

Sanitation and crop rotation help alleviate carrot weevil problems. Carrots left unharvested, for example, make very suitable early season oviposition sites for overwintering beetles. Crop rotation, while often recommended because beetles seem to have limited powers of flight, is often impractical because carrot and celery production tends to be concentrated in relatively small areas of organic soil, and it is difficult to achieve the necessary isolation.

Cowpea Curculio
Chalcodermus aeneus Boheman
(Coleoptera: Curculionidae)

Natural History

Distribution. This insect apparently is native to North America. It is restricted primarily to the southeastern states, from Virginia to Florida in the East, and west to Texas and Oklahoma. Occasionally it is reported in more northern states, but causes no damage there. It also is reported to occur in Central and South America.

Host Plants. Cowpea curculio feeds principally on legumes, but other plants are sometimes consumed. Cowpea, snap bean, lima bean, and pea are the vegetables injured, but cowpea is the preferred host. Other crops attacked are cotton, soybean, and strawberry. Among weed-host plants are cutleaf evening primrose, *Oenothera laciniata*; moss verbena, *Verbena tenuisecta*; wild bean, *Strophostyles umbellata* and *S. helvola*; purple cudweed, *Gnaphalium purpureum*; heartwing sorrel, *Rumex hastatulus*; sheep sorrel, *Rumex acetocella*; and spring vetch, *Vicia sativa*. Weeds and cotton are primarily attacked early in the year, before cowpea is available, but sicklepod, *Senna obtusifolia* is a host during the cowpea-cropping season. Sudbrink *et al.* (1998) provided data on the sequence of hosts in Alabama.

Natural Enemies. Several natural enemies have been reported, but most authors suggested that a fungus, *Beauveria* sp., is an important factor in overwintering survival of adults. The bacterium *Serratia marcescens* was observed to cause high mortality among larvae in South Carolina (Bell and Hamalle, 1971). Among parasitoids associated with cowpea curculio, only *Myiophasia globosa* (Townsend) (Diptera: Tachinidae) is consistently reported to be abundant. Arant (1938) also noted that unspecified ant species (Hymenoptera: Formicidae) and hot, dry weather affected larval survival. Russell (1981) observed that red imported fire ant, *Solenopsis invicta* Buren (Hymenoptera: Formicidae), significantly reduced the rate of successful pupation by cowpea curculio.

Life Cycle and Description. This insect overwinters in the adult stage, emerging in April or May to begin feeding. Oviposition usually does not occur until cowpea is available, which is often June or July. There appear to be two generations annually in Alabama, but only one in Virginia. However, the adults are long-lived, often surviving for several months, so the generations are indistinct. About 30–40 days are required for a complete generation.

Egg. The egg is oval, and white. It measures about 0.9 mm long and 0.6 mm wide. Eggs are deposited in the pod, or within the seed in the pod. The female deposits her eggs in feeding sites, with only a single egg deposited in each feeding puncture. Each female deposits, on average, about 112 eggs (range 30–280) during an oviposition period of about 45 days. Duration of the egg stage is about four (range 3–6) days.

Larva. The larva is pale yellow, though the head and prothoracic plate are yellowish brown. The larva lacks legs, but bears deep furrows around its body. The body is thickest about one-third the distance between the head and anus, tapering gradually to a fairly pointed posterior end. The body bears stiff bristles. The larva attains a length of about 7 mm at maturity. There are four instars. Duration of the larval stage was reported to require about 9.4 days in Alabama, but only about 6–7 days in Virginia. In the latter study, development times of the four instars averaged about 1, 1, 1, and 3.2 days, respectively.

Pupa. At completion of the larval stage the insect drops to the soil and burrows to a depth of about 2.5–7.5 cm. During a prepupal period of about six (range 3–14) days, the larva creates a pupal cell and molts to the pupal stage. The pupa greatly resembles the adult in shape and size, but is yellowish white. Duration of the pupal stage is about 10 (range 5–19) days. After transformation into the adult, the beetle remains in the pupal cell for 2–3 days while it hardens, and then digs to the surface to emerge.

Adult. The adult is oval and robust in appearance. It is black, with a faint bronze tint. The mouthparts are elongate, being slightly longer than the thorax, and only slightly curved. The thorax and elytra are marked with coarse punctures. The beetle measures 4.8–5.5 mm long. The adults are most active during the morning and early evening, and seek shade during the heat of the day. They feign death and drop to the soil when disturbed. Beetles rarely fly. Adults overwinter in the soil, under leaves, and other organic debris.

A good summary of cowpea curculio biology was given by Arant (1938). Additional useful observations were provided by Hetrick (1947).

Damage

Legumes are damaged by both adult and larval stages, both of which feed on seeds within the pods.

Adult cowpea curculio.

Small cavities and shallow furrows are typical forms of injury. In the spring, before pods are available, adults will feed on the lower epidermis of foliage, but this damage is insignificant.

Management

Insecticides. Insecticides are commonly applied to cowpea by conventional spraying or ultra low volume techniques to protect against injury by curculio (Dupree, 1970; Chalfant and Young, 1988), and adults are the usual target. Usually, one or more applications are made before pod development, followed by insecticide treatments at about 3–5 day intervals during pod growth. Insecticides directed to the soil beneath cowpea plants also are beneficial, because adults often aggregate during the heat of the day (Chalfant, 1973b). Treatment of mature pods is not necessarily beneficial (Chalfant *et al.*, 1982). Insecticide resistance has become a problem for some classes of insecticides (N'Guessan and Chalfant, 1990).

Cultural Practices. As the adult rarely flies, but reaches its host plant principally by walking, crop rotation is beneficial. Also, tillage and destruction of alternate crop and weed hosts, and crop residue, will destroy overwintering beetles.

Host-Plant Resistance. Considerable effort has been devoted to the identification of resistance among cowpea varieties to cowpea curculio. Pod wall (hull) thickness has been identified as a key factor in conferring resistance, and several commercially available varieties possess this desirable trait (Chalfant *et al.*, 1972; Cuthbert and Davis, 1972). However, when confronted with high curculio densities, these varieties do not impart complete protection, so they are best considered as part of a damage reduction program.

Biological Control. Entomopathogenic fungi have been evaluated for suppression of cowpea curculio. Strains of both *Metarhizium anisopliae* and *Beauveria bassiana* are effective under experimental conditions (Bell and Hamalle, 1971; Daoust and Pereira, 1986), but commercial products are not yet available.

Maize Billbug
Sphenophorus maidis Chittenden

Southern Corn Billbug
Sphenophorus callosus (Olivier)
(Coleoptera: Curculionidae)

Natural History

Distribution. There are several species with similar appearance and biology in the *Sphenophorus* billbug complex. Collectively, they occur throughout the United States and southern Canada. They are most destructive, however, in the eastern half of the continent. The most important species are maize billbug, *Sphenophorus maidis* Chittenden; and southern corn billbug, *S. callosus* (Oliver). Maize billbug is found in the midwestern and southern states. Southern corn billbug is found in the southern half of the United States, extending as far west as Arizona and north to most of the midwestern states and New England. Other billbugs occasionally mentioned as vegetable pests are bluegrass billbug, *S. parvulus* Gyllenhal; corn billbug, *S. zeae* (Walsh); claycolored billbug, *S. aequalis* Gyllenhal; cattail billbug, *S. pertinax* (Oliver); nutgrass billbug, *S. cariosus* (Olivier); and hunting billbug, *S. venatus* Chittenden. Apparently these are native species.

Host Plants. These billbugs feed mostly on grasses. Corn is the only vegetable crop damaged, and the principal cultivated host. For most species, the adult is the stage attacking corn, but in the case of maize billbug and southern corn billbug, both adults and larvae damage corn. Feeding also has been reported on barley, chufa, oat, peanut, rice, wheat, and timothy. Wild plants attacked include bentgrass, *Agrostis*; bluegrass, *Poa* sp.; jointgrass, *Paspalum* sp.; crabgrass, *Digitaria* sp.; nutgrass, *Cyperus* spp.; rushes, *Juncus* spp.; reeds, *Phragmites* sp.; and cattails, *Typha* spp.

Natural Enemies. The billbugs seem to have a few natural enemies. Ants are thought to gather the eggs. Several parasitoids have been reared from billbugs, including *Bracon sphenophori* (Muesebeck) (Hymenoptera: Braconidae) from *S. callosus*; *Bracon analcidis* Ashmead (Hymenoptera: Braconidae) and *Myiophasia* sp. (Diptera: Tachinidae) from *S. parvulus*; *Vipio belfragei* (Cresson) (Hymenoptera: Braconidae) from *S. cariosus*; and *Gambrus bituminosus* (Cushman) (Hymenoptera: Ichneumonidae) from *S. pertinax*. Severe mortality of pupae and adults has been observed in South Carolina, caused by the fungus *Beauveria bassiana*.

Life Cycle and Description. The billbugs are quite similar in form and biology, and except where noted, the following description applies mostly to maize billbug. The overwintering stage of these insects is generally the adult form, though in southern states more than one stage may be present during the winter months. There usually is a single generation, with adults ovipositing in May and June, larvae present through August, and pupation occurring in late summer and autumn. Some beetles emerge from the pupal stage and feed before overwintering, whereas others remain, as adults, in the pupal cell until spring.

Egg. The eggs are deposited singly in the stem of the host plant within a cavity created by the mouth-

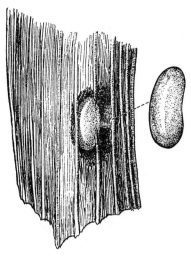

Southern corn billbug eggs.

Southern corn billbug larva.

parts of the female. The egg is white and bean-shaped. It typically measures 1.5–3.0 mm long and about 0.9 mm wide. The incubation period of the eggs is 4–15 days, usually averaging about seven days.

Larva. The larvae are short, heavy-bodied, and "C"-shaped in appearance. The body is white, but they bear a distinct brown or yellow head. They lack legs and are relatively free of setae. They feed within the stem for several weeks, 30–70 days in the case of maize billbug. If the food supply in the stem is exhausted, larvae will dig down into the soil and feed on the roots. The larvae of southern corn billbug have four instars, the duration of which averages about 4.0, 4.5, 9.7, and 12.4 days.

Pupa. Pupation may occur in either the stem of the host plant or in the soil nearby. The pupa measures about 12–15 mm long, and ranges from whitish, when first formed, to brown or black at maturity. The pupal period is usually short, about 10–14 days. The adults may evacuate the pupal cell in the autumn or may remain within the cell until spring.

Adult. The adults are elongate-oval, and vary in color from brown to black. Beetles range in size from about 5–19 mm. The head is prolonged into an elongate snout that curves strongly downward and measures about one-third the length of the body. The thorax usually bears deep pits, and the elytra distinct grooves. Adults are long-lived, surviving not only the winter but also well into the summer. They become active in the spring when the soil temperature exceeds to 18°C. Mating can occur any time after emergence, and oviposition begins about 10 days after mating. Oviposition can occur for 1–3 months, and though insectary records document an average of only about 40 eggs per female, some records suggest that up to 200 eggs can be produced by a female. The adults rarely fly and disperse instead by walking.

A synopsis of billbugs was provided by Satterthwait (1919), and good treatments of maize billbug were given by Hayes (1920), Cartwright (1929), and Kirk (1957a). Detailed treatment of southern corn billbug was presented by Metcalf (1917). Forbes (1904) pictured and described briefly several billbugs attacking corn in Illinois.

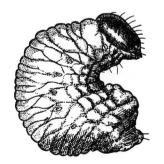

Maize billbug larva.

Adult maize billbug.

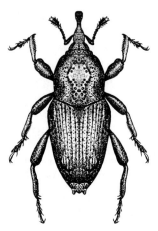

Adult southern corn billbug.

Damage

Prior to the availability of modern persistent insecticides, billbugs were the most serious pest of southern corn crops. They can cause severe reduction in stand density, though they principally attack field corn rather than sweet corn. The adults feed just above or below the soil surface, where they puncture the stem, creating large cavities in the plant tissue. When leaves later unfurl, the plant displays even rows of round holes. Extensive feeding causes distortion and death of the plant. Beetles sometimes feed on germinating seed and roots. The adults also create cavities in which they deposit their eggs. Larvae, upon hatching, burrow upward in the stem, causing wilting and excessive lateral growth. Corn rarely produces ears when supporting larvae.

Management

Low-lying and wet areas are more prone to billbug problems owing to the presence of alternate hosts such as cattails and rushes. Seedlings can be protected with liquid or granular insecticide applied at planting (Kirk, 1957b; Durant, 1974). Tillage and destruction of crop residue in the autumn can damage overwintering beetles, but the level of control attained by such practices is slight. Early planting allows the plants to gain enough size to tolerate considerable feeding, so it is sometimes recommended. Crop rotation is beneficial if crops other than grains can be grown profitably.

Pea Leaf Weevil
Sitona lineatus (Linneaus)
(Coleoptera: Curculionidae)

Natural History

Distribution. Pea leaf weevil is likely of European origin. It is a common pest in Europe, and is also known from North Africa, Israel, and Japan. It was first observed in North America on Vancouver Island, British Columbia in 1936. By 1940, it had reached Washington, and continued to spread at a rate of about 25 km per year until much of British Columbia, Idaho, and Oregon were infested. From about 1960–1980, it was confined by ecological barriers to the Pacific Northwest, but in 1984 it was detected in Virginia, where it is free to spread to the northeast and midwest. Hoebeke and Wheeler (1985) suggested that the eastern population may have originated from Bermuda.

Host Plants. Larvae develop successfully only on plants in the family Leguminosae, and legumes also are the principal food of adults. Peas and faba bean are the vegetable plants fed upon, though weevils also consume such crops as alfalfa, red clover, white clover, birdsfoot trefoil, and several types of vetch, *Vicia* spp. In North America, pea leaf weevil does not consume lentil, though this is reported a host in southern Europe. Several other legumes and other plants may be fed upon when emerging adults are denied access to favored legumes, but their damage is usually negligible (Fisher and O'Keeffe, 1979b,c).

Natural Enemies. Although parasitoids are known in Europe, little evidence has been presented in North America to suggest an important role for natural enemies in suppression of pea leaf weevil abundance. Ground beetles (Coleoptera: Carabidae) account for variable, but occasionally considerable, predation of adult pea leaf weevil in England (Hamon et al., 1990). The fungus *Beauveria bassiana* regularly occurs as a larval pathogen in the northwest, but its significance is uncertain.

Weather. Pea leaf weevil thrives under cool and moist conditions. Survival of young larvae, in particular, is decreased by warm and dry conditions. Thus, larvae hatching from eggs deposited after April fare poorly. Crops sown in the spring are less susceptible to attack, because their root growth is occurring late in the spring, at a time when warm temperatures inhibit larval survival. The opposite is true for autumn-planted crops, because their roots are vigorously attacked by larvae very early in the spring.

Life Cycle and Description. Pea leaf weevil has a single generation annually. Weevils overwinter as adults, and in the Pacific Northwest they become active in March, at a time when temperature exceeds about 14°C. Eggs are deposited in March–May, larvae occur in April–June, and pupae are found predominantly in June. New adults occur as early as June, but are more common in July, with overwintered adults persisting until the following May. Duration of the life cycle, from egg to adult emergence, is about 68 days.

Egg. The eggs are scattered singly, occasionally on plants but usually on the soil surface near the base of the plant. During dry weather, females may selectively deposit eggs on the moist underside of soil clods. Estimates of fecundity vary, but some reports suggested a mean of 1100 with a range of 350–1650 eggs per female. In Idaho, Schotzko and O'Keeffe (1988) reported that length of quiescence and adult diet greatly affected fecundity; weevils fed peas produced 30 times as many eggs as those fed alfalfa. The egg is oval, and measures about 0.36 (range 0.32–0.37) mm long and 0.29 (range 0.23–0.31) mm wide. The eggs are yellowish white when first deposited, but within 2–3 days they are changed to gray and then to shiny black. The incubation requirement for eggs in Washington is 16–18 days.

Larva. The young larvae, upon hatching, burrow down through the soil and seek plant roots. When they locate a root nodule they penetrate and feed within. Larvae may burrow to depths of about 27 cm, but most feed on the upper 15 cm. The larva is legless, but the terminal abdominal segment bears a fleshy prominence that can be extended and withdrawn, and which aids in locomotion. There are five instars, with mean (range) head capsule widths of 0.16 (0.15–0.18), 0.24 (0.18–0.29), 0.37 (0.30–0.45), 0.55 (0.45–0.63), and 0.83 (0.63–1.00) mm for instars 1–5, respectively (Hamon *et al.*, 1984). The young larva, which measures 0.9–1.1 mm long, is equipped with sparse, relatively long hairs. It is creamy white, and the small head is dark, at least peripherally. The mature larva has shorter bristles, and a relatively smaller, reddish brown head. It attains a length of about 6–7 mm before pupation. Duration of the larval stage is about 35 days.

Pupa. Pupation occurs in the upper 7.5 cm of soil. The pupa is creamy white and measures 4–5 mm long. It resembles the adult in form and the major difference is the lack of fully formed wings. Bristle-bearing elevations that give the pupa a rather spiny appearance are found on the head, thoracic, and abdominal segments. Duration of the pupal stage is 15–17 days.

Adult. The adults fly actively in early July, seeking succulent legumes. They feed until August, and then enter a period of aestivation which persists through most of the winter. Under the mild-winter conditions of the northwest, the beetles become active on warm days, crawl about, and do some feeding. Adult beha-

vior is different in Europe, where winters tend to be colder, with aestivation leading to diapause and no adult activity until spring. Fisher and O'Keefe (1979a) documented the early and late-season flights of beetles in Idaho. However, it appears that great numbers, if not most, move to overwintering sites by walking rather than by flying (Hamon *et al.*, 1987). The adult is slender, measures about 4.5 mm (range 3.6–5.4 mm) long, and is grayish-brown. The thorax is marked with three weak, light-colored stripes, one centrally and two laterally marked, that extend onto the elytra. The elytra also bear parallel striations running their length, are marked with minute punctures, and covered with a fine layer of scales. The legs are black, but reddish basally. The antennae are long and slender, brownish red in color, and markedly enlarged at the tips. In the spring, males produce an aggregation pheromone that is attractive to both sexes (Blight and Wadhams, 1987).

A valuable summary of pea leaf weevil biology, based on observations made in the northwest, was produced by Prescott and Reeher (1961). Jackson (1920) provided detailed morphology and life-history information, but the latter reflected European weather conditions, which resulted in slightly different weevil biology.

Damage

Plants are fed upon by both larvae and adults. The larvae feed within the root nodules of legumes, which are also inhabited by bacteria that provide the plant with nitrogen. Plants can also take up nitrogen directly from the soil, usually taking nitrogen from fertilizer in preference to using nitrogen from nodules. Thus, although pea leaf weevil larvae decrease the abundance of nodules on plant roots, their direct effect on crop yield is variable and often slight (George, 1962; Bardner and Fletcher, 1979). Often overlooked, however, is the potential for enhanced root disease following feeding by larvae (Lee and Upton, 1992).

The adult feeding is more injurious and easily recognizable. Adults cut notches in the shape of a half-circle along the margin of leaves. Often the cuts are in close sequence, resulting in a scalloped effect. Artificial defoliation studies in England suggested that about a 10% yield decrease was associated with four weevil-like notches cut in a young pea plant (George *et al.*, 1962). When weevils are abundant, plants may be completely defoliated. Such damage usually occurs in early spring. Adults may also disperse to nearby alfalfa or clover, defoliating these crops. If birdsfoot trefoil is invaded, usually only the flower buds and not the leaves are consumed. Some plant viruses may be transmitted by pea leaf weevil feeding (Cock-bain *et al.*, 1975), but this weevil is not a very efficient vector.

Management

Sampling. Adult populations are assessed by visual examination (in pea) or sweeping (in alfalfa and vetch). In pea, however, feeding damage is readily observed and diagnostic. Pea leaf weevil aggregation pheromone can be used to lure weevils to cone traps. Cone traps are more suitable for population monitoring than yellow sticky traps (Nielsen and Jensen, 1993). The aggregation pheromone can also be used in conjunction with neem, an anti-feedant, to concentrate the weevils in some areas and prevent feeding in others (Smart *et al.*, 1994).

Insecticides. Foliar and soil-applied granular insecticides, both systemic and nonsystemic, are often applied to alleviate damage by adults; if applied sufficiently early they may also prevent oviposition (Lee and Upton, 1992). Granular products also kill larvae that are feeding belowground, and therefore immune to foliar applications. Research in England suggested that it might not be necessary to spray entire fields for suppression of pea leaf weevil. Good control is attained by spraying strips of crop, and leaving strips untreated, because the adult weevils are mobile and move freely in the field, contacting insecticide and perishing (Ward and Morse, 1995).

Cultural Practices. Cultural practices help minimize the potential damage from pea leaf weevil. It is advisable to rotate crops, and to plant pea at a considerable distance from other legumes fed upon by adults, such as alfalfa, clover, and vetch. Adequate fertilization, and particularly irrigation, help pea plants recover from defoliation by adults. Intercropping such nonhost plants as oats with legumes serves to increase the rate of emigration from intercropped plots, and reduce the damage to legumes by adults (Baliddawa, 1984).

Biological Control. The larvae of pea leaf weevil are susceptible to infection by the entomopathogenic fungus *Metarhizium flavoviride* (Poprawski *et al.*, 1985) and nematode *Steinernema carpocapsae* (Jaworska and Ropek 1994), but practical demonstration of their efficacy is lacking.

Host-Plant Resistance. There has been relatively little effort directed toward selection of plant varieties resistant to pea leaf weevil. In Finland, evaluation of 84 varieties, in either laboratory or in field studies, showed no differences in leaf damage (Tulisalo and Markkula, 1970). However, work in Czechoslovakia suggested that feeding rate did vary among some varieties, with reduced feeding having both morphological and biochemical components (Havlickova, 1980).

Pepper Weevil
Anthonomus eugenii Cano
(Coleoptera: Curculionidae)

Natural History

Distribution. First found in the United States in Texas in 1904, pepper weevil reached California in 1923 and Florida in 1935. It is now found across the southernmost United States from Florida to California. Pepper weevil populations persist only where food plants are available throughout the year. Because transplants are shipped northward each spring, however, pepper weevil sometimes occurs in more northern locations. Pepper weevil was first observed in Hawaii in 1933 and in Puerto Rico in 1982. Although its origin is likely Mexico, it has spread throughout most of Central America and the Caribbean.

Host Plants. Pepper weevil develops only on plants in the family Solanaceae. Oviposition occurs on plants in the genera, *Capsicum* and *Solanum*, but feeding by adults extends to other Solanaceae, such as *Physalis*, *Lycopersicon*, *Datura*, *Petunia*, and *Nicotiana*. Among vegetables, all varieties of pepper are susceptible to attack, tomatillo is a moderately susceptible host, and eggplant grown in proximity to pepper may occasionally have the blossoms damaged. Several species of nightshade support pepper weevil oviposition and development, particularly black nightshade, *Solanum nigrum*, but also silverleaf nightshade, *S. elaeagnifolium*; horsenettle, *S. carolinense*; buffalo bur, *S. rostratum*; and Jerusalem cherry, *S. pseudocapsicum*.

Natural Enemies. Parasitoids of pepper weevil include *Catolaccus hunteri* Crawford (Hymenoptera: Pteromalidae) and *Bracon mellitor* Say (Hymenoptera: Braconidae). Riley and Schuster (1992) reported up to 26% parasitism of pepper weevil larvae by *C. hunteri*.

Life Cycle and Description. In California, a complete generation is reported to require about 25 days under optimal conditions, about 33 days on average. In Florida and Puerto Rico, 20 days is about average. Under insectary conditions in California, up to eight generations have been produced in a single year. In most locations, 3–5 annual generations is probably normal, though adults are long-lived and produce overlapping generations, so it is difficult to ascertain generation numbers accurately. In central Florida, adults are common from March until June, and sometimes in the autumn months, reflecting the availability of peppers. However, a few can be found throughout the year except during inclement periods, such as

December and January. Adults overwinter, but only where food is available, because diapause does not occur in this species.

Egg.   Oviposition may commence within two days of mating. Eggs are white when first deposited, but soon turn yellow. They are oval in shape and measure 0.53 mm long and 0.39 mm wide. Eggs are deposited singly beneath the surface of the bud or pod. The female creates an egg cavity with her mouthparts before depositing the egg, and seals the puncture containing the egg with a light brown fluid that hardens and darkens. Females deposit eggs at a rate of about 5–7 eggs per day, and fecundity averages 341 eggs, but is nearly 600 in some individuals. The mean incubation period is 4.3 days (range 3–5 days).

Larva.   Like other weevils in this genus, apparently the pepper weevil has three instars. The larvae are white to gray with a yellowish brown head. They lack thoracic legs and have few large hairs or bristles. They are aggressive, especially during the end of the larval period. Usually only a single larva survives within a bud, though more than one can occur within larger fruit. First instars measure about 1 mm long (range 0.8–1.5 mm). Second instars measure about 1.9 mm long (range 1.3–2.6 mm). Third instars measure about 3.3 mm (range 2.2–5.0 mm). Mean development time of the larvae is about 1.7, 2.2, and 8.4 days for instars 1–3, respectively. The figure for third instar development time contains a prepupal period of about 4.9 days, during which time the larva creates a pupal cell from anal secretions.

Pupa.   The pupal cell is brittle and found within the blossom or fruit. The pupa resembles the adult, except that the wings are not fully developed and large setae are found on the prothorax and abdomen. White when first formed, the pupa eventually becomes yellowish with brown eyes. Mean duration of the pupal stage is 4.7 days (range 3–6 days).

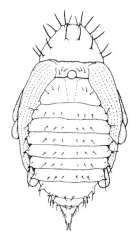

Pepper weevil pupa.

Adult.   The adult emerges from the pupal case 3–4 days after being formed. A clean, round hole marks the escape of the beetle from the bud or fruit. The beetle is oval and varies from 2.0–3.5 mm long and 1.5–1.8 mm wide. The body is strongly arched with a long, stout beak as is typical for this genus. The thorax and elytra are mostly covered with small scales. The antennae are long and markedly expanded at the tip. The femora each bear a sharp tooth. The color is dark mahogany to nearly black. Feeding begins immediately after emergence. Males produce an aggregation pheromone that attracts both sexes (Eller *et al.*, 1994). (See color figure 131.)

The most complete treatment of pepper weevil biology was given by Elmore *et al.* (1934), though Riley and King (1994) published a useful review of biology and management. Other descriptions were provided by Goff and Wilson (1937) and Gordon and Armstrong (1990). Host relations were described by Patrock and Schuster (1992).

Damage

This is the most important insect pest of pepper in the southern United States. An important form of damage is destruction of blossom buds and immature pods. On buds, adult and larval feeding cause bud drop. Adult feeding punctures appear as dark specks

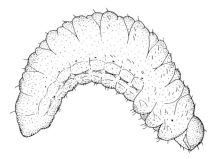

Pepper weevil larva.

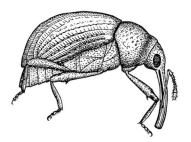

Adult pepper weevil.

on the fruit, and are not very damaging. Sometimes the fruit is deformed. Fruit drop is very common, and is perhaps the most obvious sign of infestation. Larval feeding within the mature pod is another important form of damage, causing the core to become brown, and often moldy. The stem of pods infested by larvae turn yellow, and the pod turns yellow or red at the base prematurely. In the absence of pepper blossom and fruit, adults feed on leaves and pepper stems, but cause no significant damage by these habits. Puncture of peppers by pepper weevil allows penetration of the fungus *Alternaria alternata*, an otherwise weak pathogen, and extensive fungal growth within the pepper fruit (Bruton *et al.*, 1989). (See color figures 24 and 30.)

Management

Sampling. The distribution of weevils is aggregated, especially at field margins. Adult population estimates are best obtained by visual examination and yellow sticky traps (Segarra–Carmona and Pantoja, 1988a). Traps should be placed 10–60 cm above the soil; one 375 sq cm trap captures as many weevils as are detected by inspecting 50 buds (Riley and Schuster, 1994). If visual monitoring is preferred, terminal bud sampling is effective, though more beetles are present in the morning than evening (Riley *et al.*, 1992a). Action thresholds of one adult per 400 terminal buds (Riley *et al.*, 1992b) or 1% of buds infested (Cartwright *et al.*, 1990b) have been suggested. A sequential sampling protocol was developed by Segarra–Carmona and Pantoja (1988b). Isolation of an aggregation pheromone may lead to improved sampling techniques (Eller *et al.*, 1994).

Insecticides. Insecticides are commonly applied to the foliage at short intervals once buds begin to form. Insecticidal suppression of adult weevils is feasible, but insecticides vary considerably in effectiveness, and even in the presence of chemical insecticides some loss commonly occurs (Ozaki and Genung, 1982; Schuster and Everett, 1982; Armstrong, 1994). Segarra–Carmona and Pantoja (1988b) estimated that economic damage commences with adult populations of 0.01 beetle per plant.

Cultural Practices. Cultural practices that significantly affect pepper weevil damage are limited, but compatible with other management practices including biological control. Berdegue *et al.* (1994), working in Texas, compared 35 varieties of jalapeno, bell, pimento, serrano, yellow, cayenne, chile, tabasco and cherry peppers, and found few differences in susceptibility. Sanitation can be important, if it can be implemented on an appropriate scale. Removal and

destruction of fallen fruit, for example, will result in destruction of larvae and pupae. Similarly, a crop-free period, particularly if accompanied by destruction of alternate hosts, can disrupt the life cycle. Solanaceous weeds should not be allowed to grow in proximity to pepper as the weeds may serve as an alternate host.

Potato Stalk Borer
Trichobaris trinotata (Say)
(Coleoptera: Curculionidae)

Natural History

Distribution. Potato stalk borer is found widely in eastern North America, west to the Rocky Mountains. It is considered a native insect, though it also occurs in Mexico.

Host Plants. This insect feeds only on plants in the family Solanaceae. As suggested by the common name, it feeds principally on potato. However, it can also affect eggplant. The favored weed host is horsenettle, *Solanum carolinense*; but it also feeds on buffalo bur, *Solanum rostratum*; black nightshade, *S. nigrum*; and several species of groundcherry, *Physalis* spp. Other early records of feeding on crops and weeds are suspect, and likely due to misidentification of the insect.

Natural Enemies. The natural enemies of this poorly studied insect seem to be few in number. The most important natural enemy is *Nealiolus curculionis* (Fitch) (Hymenoptera: Braconidae), but *Eurytoma tylodermatis* Ashmead (Hymenoptera: Euytomidae) is also recorded from potato stalk borer.

Life Cycle and Description. There is only one generation per year throughout the range of this insect. Adults overwinter in reproductive diapause. In the spring, usually during May, adults become active and mate, with oviposition commencing in May or June. Larvae are present in early summer, and following a brief pupal period, new adults appear, usually in July or August.

Egg. The adults usually deposit eggs within cavities in the stems and petioles, though some eggs are deposited in axils of terminal leaves. Usually only a single egg is found in each cavity, but sometimes two eggs occur at the same site. The egg is white, oval in form, and measures about 0.65 mm long and 0.39 mm wide. Mean duration of the egg stage is 7.4 days at 24°C (range 6–11 days).

Larva. The larvae are whitish to yellowish, with a pale brown head. The larva is quite elongate in form, and attains a length of 9–11 mm. There are 5–6 instars. Mean duration is about 4.1, 4.4, 6.8, 8.4, 18.4, and 16.4

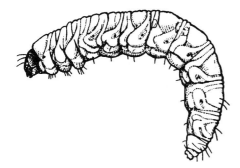

Potato stalk borer weevil larva.

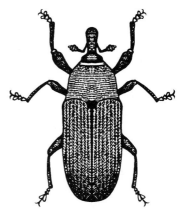

Adult potato stalk borer weevil.

days for instars 1–6, respectively. Total duration of the larval stage is 20–84 days for larvae with five instars, and 33–106 days for larvae with six instars. The larval stage lacks legs, but bears deep thoracic ridges which aid in mobility. Larvae feed within the plant, primarily within stem tissue. Prior to pupation, the larva burrows outward to the epidermis of the plant, but leaves a thin covering; this forms the point of adult escape from the plant. A quiescent period of 2–6 days precedes the molt to the pupal stage.

Pupa. Pupation occurs in the plant, within a pupal case consisting of plant particles. The pupa is oval in form, greatly resembling the adult except for poorly developed wings. It is white or yellow initially, but become dark within the final few days of this stage. Mean duration of the pupal stage is 12.1 days at 24°C (range 8–18 days).

Adult. The adult is a small ash-gray weevil, with males measuring about 4.2 mm long and 1.6 mm wide, and females 4.5 mm long and 1.7 mm wide. Close examination reveals that the true color of the weevil body is black, but its surface is covered with minute gray scales. This is most apparent at the head, which lacks scales, and therefore is black. There also are three black spots at the base of the elytra—one dorsally at the juncture of the elytra and two laterally. These three spots are the basis for the species name, *trinotata*. The

rostrum (beak) curves strongly at the base, resulting in a downward orientation in the beak. Unlike the body form of some weevils, the body of the beetle is not strongly curved. Adults mate in the spring, and oviposition commences just one day after mating. The female chews a hole to the depth of the rostrum and then oviposits within the cavity. The hole is sealed with a clear fluid secreted by the female.

This insect and its biology were described by Faville and Parrott (1899), Chittenden (1902), and Cuda and Burke (1985, 1986). A key to the species of *Trichobaris* was published by Barber (1935).

Damage

Once considered a serious pest, current agronomic practices have relegated this insect to insignificant status over most of its range. The principal form of damage is larval burrowing. Larvae gradually move from the point of hatching downward through the stem. As the larva grows it produces a correspondingly larger tunnel, causing greater injury. At some point the larva reverses direction, feeding upward again, and enlarging the tunnel. Length of the burrow may reach 30 cm. Larvae may also burrow into the roots. Adults also feed in the spring before oviposition, but apparently not in the autumn. The damage by adults consists of irregular holes eaten in the leaves.

Management

It is difficult to detect these insect, because they dwell within the stems of potato and weeds. Also, the beetles are quite small, and drop readily if the plants are disturbed. Sanitation is effective at eliminating this insect as a pest. The most important practice is to destroy the potato vines, as this is the overwintering site of weevils. Secondly, because the weevils develop on, and overwinter within, solanaceous weeds, it is useful to destroy these plants. Late-planted crops are

Potato stalk borer weevil pupa.

less injured by weevils than early plantings due to the early activities of adults in the spring. If insecticides are used, they should be applied to the foliage as the plants emerge.

Rhubarb Curculio
Lixus concavus Say
(Coleoptera: Curculionidae)

Natural History

Distribution. This native species is found throughout the eastern United States west to Idaho, Utah, and Texas. In Canada it is known from Ontario. A very similar insect, *Lixus mucidus* LeConte, may be confused with rhubarb curculio, because it shares the same geographic range, host plants, and biology. It also is native to North America.

Host Plants. These species normally are associated with such weeds as sunflower, *Helianthus* sp.; dock, *Rumex* spp.; and perhaps rosinweed, *Silphium* sp.; but both *Lixus concavus* and *Lixus mucidus* attack rhubarb. Interestingly, though the complete life cycle occurs in weeds, eggs and larvae rarely survive in rhubarb, so this vegetable is principally a food host for adults.

Natural Enemies. Population regulation is poorly known in this little-studied insect. The only parasitoid known from rhubarb curculio is *Rhaconotus fasciatus* (Ashmead) (Hymenoptera: Braconidae). The larvae apparently are cannibalistic, because although it is not uncommon to find several young larvae in the stalk of a host plant, it is rare to have more than one reach maturity in each plant.

Life Cycle and Description. There is a single generation annually. The adults overwinter, emerging in April in the vicinity of Washington D.C. Mating and oviposition occur soon thereafter. Eggs are found in April–mid-June, larvae through July or August, pupation commences in August, and adults begin emergence in September. In Canada, adults do not emerge from overwintering until June, but similarly complete a new generation by September.

Egg. The egg is oval in form and pale yellow, and is deposited in plant tissue at a depth of 2–3 mm. The eggs measure 1.5–1.9 mm long and 1.2–1.3 mm wide. They are deposited principally within the leaf petioles and flower stalk, though sometimes within the larger veins of the leaf. Apparently the weevil deposits the egg within cavities made by feeding, though there are many more feeding sites than oviposition sites. Duration of the egg stage is about eight days. The eggs fare poorly when deposited in rhubarb, apparently suffering from the flow of sap.

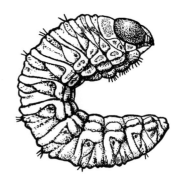

Rhubarb curculio larva.

Larva. The larva is whitish, but with a brown head that usually bears an inverted Y-shaped mark. The larva's body is curved or C-shaped, highly wrinkled, and lacks thoracic legs. Larvae bore down through the stalk into the root, usually with only a single beetle surviving in each plant. Duration of the larval stage is about 60 days, and then the larva creates a pupal cell in the base of the stem or in the roots.

Pupa. The pupal cell is 25–40 mm long and 5–7 mm wide. It is constructed just beneath the soil surface, and the larva chews an exit hole for the adults before pupating. The pupa is whitish and measures 14–15 mm long. The head bears the long snout of the adult form, though the wings are twisted beneath the ventral surface of the body. The abdominal segments are marked with short spines. Duration of the pupal stage is about eight days, though the adult remains in the pupal chamber for another 7–8 days before emerging.

Adult. The adult is 10–14 mm long, and black, but covered with fine gray hairs. Rhubarb curculio is also dusted with yellow powder though this is lacking from *L. mucidus*, and can be rubbed off. The pronotum and elytra bear punctures. After emerging in September, adults feed briefly on host plants, then seek sheltered locations for overwintering. The adults pass the winter hidden beneath debris near the host plants.

The biology was described by Webster (1889), Chittenden (1900), and Weiss (1912). These species were included in the treatment of eastern beetles by Downie and Arnett (1996) and of Canadian beetles by Campbell *et al.* (1989).

Damage

The adults of rhubarb curculio may nibble at the edge of leaves in the spring, but damage to rhubarb results principally from gnawing into the leaf stalks. The holes are oval or round, and up to 3 mm deep.

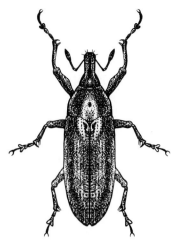

Adult rhubarb curculio.

Sticky sap secretions often exude from the wounds, collecting as glistening drops of gum. Feeding occurs mostly after most rhubarb, an early season crop, has been harvested. Larvae seldom develop in rhubarb. Nevertheless, rhubarb curculio is considered as the most injurious of the insects attacking rhubarb—a plant notably free of insect pests.

Management

Rhubarb curculio is rarely of consequence, though application of insecticides to rhubarb foliage should protect plants from injury. Because its abundance is governed mostly by the presence of favored weed hosts such as dock, elimination of such weeds will eliminate the threat of damage by weevils.

Sweetpotato Weevil
Cylas formicarius (Fabricius)
(Coleoptera: Curculionidae)

Natural History

Distribution. Sweetpotato weevil now is found in tropical regions of Africa, Asia, and North and South America. Its origin is believed to be Africa or India despite the fact that the sweetpotato plant likely evolved in northwestern South America. Thus, the association between this crop and its principal pest is relatively recent. Sweetpotato weevil was first noted in the United States in Louisiana in 1875, and then in Florida in 1878 and Texas in 1890, probably entering by way of Cuba. It is now found throughout the coastal plain of the southeast from North Carolina to Texas. It also is found in Hawaii and Puerto Rico.

Host Plants. This weevil feeds only on plants in the plant family Convolvulaceae. Although it has been found associated with several genera, its primary hosts are in the genus *Ipomoea*. Among vegetable crops only sweet potato, *I. batatas*, is a suitable host. In the United States, some cultivars of sweet potato are called yams, and these are susceptible to sweetpotato weevil. True yam, *Dioscorea* spp., belongs to the plant family Dioscoreaceae. True yams are not a host of sweetpotato weevil and are rarely cultivated except in tropical areas. Native plants can be important hosts of sweetpotato weevil. Railroad vine, *Ipomoea pes-caprae*, and morningglory, *I. panduratea*, are among the suitable wild hosts. Cockerham *et al.* (1954) gave information on host relations.

Natural Enemies. Several natural enemies are known. Wasps such as *Bracon mellitor* Say, *B. punctatus* (Muesebeck), *Metapelma spectabile* Westwood (all Hymenoptera: Braconidae), and *Euderus purpureas* Yoshimoto (Hymenoptera: Eulophidae) have been reared from sweetpotato weevil larvae in the southeastern United States. There have been no studies of parasitoid effectiveness, but these species seem to be infrequent. Among predators, ants (Hymenoptera: Formicidae) seem to be most important. Diseases, especially the fungus *Beauveria bassiana*, have been observed to inflict high levels of mortality under conditions of high humidity and high insect density, but field conditions are rarely conducive for disease epizootics. Jansson (1991) provided a complete review of natural enemies.

Life Cycle and Description. A complete life cycle requires 1–2 months, with about 35–40 days common during the summer months. The generations are indistinct, and the number occurring annually is estimated to be 5 in Texas, and about 8 in Louisiana. The adults do not undergo a period of diapause in the winter, but seek shelter and remain inactive until the weather is favorable. All stages can be found throughout the year if suitable host material is available.

Egg. The eggs are deposited in small cavities created by the female with her mouthparts in the sweetpotato root or stem. The female deposits a single egg, and seals the egg within the oviposition cavity with a plug of fecal material, making it difficult to observe. Most eggs tend to be deposited near the juncture of the stem and root (tuber). Sometimes the adult will crawl down cracks in the soil to access tubers for oviposition, in preference to depositing eggs in stem tissue. The egg is oval and creamy white. Its size is reported to average 0.65–0.79 mm long and 0.41–0.50 mm wide. There is little distinct sculpturing on the surface of the egg. Duration of the egg stage varies from about 5–6 days during the summer to about 11–12 days during colder weather. Females apparently produce 2–4 eggs per day, or 75–90 eggs during their life span of about 30 days. Under laboratory conditions, however, Jansson and Hunsberger (1991) reported mean fecundity of 122 eggs, and others reported about 50–250 eggs per female.

Larva. When the egg hatches the larva usually burrows directly into the tuber or stem of the plant. Those hatching in the stem usually burrow down into the tuber. The larva is legless, white in color, and displays three instars. The mean head capsule widths of the instars are 0.29–0.32 mm, 0.43–0.49 mm, and 0.75–0.78 mm for instars 1–3, respectively. Duration of each instar is 8–16, 12–21, and 35–56 days, respectively. Temperature is the principal factor affecting larval development rate, with larval development (not including the prepupal period) occurring in about 10 and 35 days at 30° and 24°C, respectively. The larva creates winding tunnels packed with fecal material as it feeds and grows. (See color figure 36.)

Pupa. The mature larva creates a small pupal chamber in the tuber or stem. The pupa is similar to the adult in appearance, though the head and elytra are bent ventrally. The pupa measures about 6.5 mm long. Initially the pupa is white, but with time this stage becomes grayish, with darker eyes and legs. Duration of the pupal stage averages 7–10 days, but in cool weather it may be extended up to 28 days.

Adult. Normally the adult emerges from the pupation site by chewing a hole through the exterior of the plant tissue, but sometimes it remains for a considerable period and feeds within the tuber. The adult is striking in form and color. The body, legs, and head are long and thin, giving it an ant-like appearance. The head is black, the antennae, thorax and legs orange to reddish-brown, and the abdomen and elytra are metallic blue. The snout is slightly curved and about as long as the thorax; the antennae are attached at about the mid-point on the snout. The beetle appears smooth and shiny, but close examination shows a layer of short hairs. The adult measures 5.5–8.0 mm long. Under laboratory conditions at 15°C, adults can live over 200 days if provided with food, and about 30 days if starved. In contrast, their longevity decreases to about three months if held at 30°C with food, and eight days without food (Mullen, 1981). Adults are secretive, often feeding on the lower surface of leaves, and are not readily noticed. The adult is quick to feign death if disturbed. Adults can fly, but seem to do so rarely and in short, low flights. However, because they are active mostly at night, their dispersive abilities are probably underestimated. Females feed for a day or more before becoming sexually active, but commence oviposition shortly after mating. The average pre-oviposition period is seven days. A sex pheromone produced by females has been identified and synthesized (Heath *et al.*, 1986). (See color figure 127.)

Good summaries of sweetpotato weevil biology were found in Reinhard (1923), Sherman and Tamashiro (1954), and Cockerham *et al.* (1954).

Damage

Sweetpotato weevil is often considered to be the most serious pest of sweet potato, with reports of losses ranging from 5–97% in areas where the weevil occurs. Some of the insect density and yield component relationships were documented by Sutherland (1986). Basically, he found that there was a strong positive relationship between vine damage or weevil

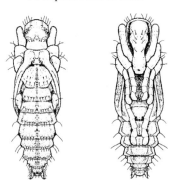

Sweetpotato weevil larva.

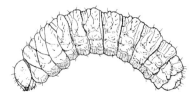

Sweetpotato weevil pupa.

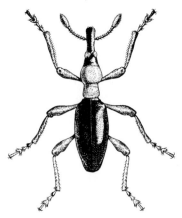

Adult sweetpotato weevil.

density, and tuber damage. However, the plants exhibited some compensatory ability, with the relationship between vine damage and yield nonlinear. Talekar (1982), in contrast, found few significant relationships between weevil numbers and damage.

A symptom of infestation by sweetpotato weevil is yellowing of the vines, but a heavy infestation is usually necessary before this is apparent. Thus, incipient problems are easily overlooked, and damage not apparent until tubers are harvested. The principal form of damage to sweet potato is mining of the tubers by larvae. The infested tuber is often riddled with cavities, spongy in appearance, and dark in color. In addition to damage caused directly by tunneling, larvae cause damage indirectly by facilitating entry of soilborne pathogens. Even low levels of feeding induce a chemical reaction that imparts a bitter taste and terpene odor to the tubers. Larvae also mine the vine of the plant, causing it to darken, crack, or collapse. The adult may feed on the tubers, creating numerous small holes that measure about the length of its head. The adult generally has limited access to the tubers, however, so damage by this stage is less severe than by larvae. Adult feeding on the foliage seldom is of consequence.

Management

Sampling. Over 90% of larvae are found in the upper 15 cm of the tubers and basal 10 cm of the vine. Early in the season larvae are found about equally in the vine and tuber, but later in the season most occur in the tubers (Jansson *et al.*, 1990a). Distribution of sweetpotato weevil in fields is aggregated (McSorley and Jansson, 1991). Pheromone traps show great promise for monitoring of adult population density (Jansson *et al.*, 1990c). Weevils respond to low concentrations of pheromone, and apparently will move up to 280 m to a pheromone source (Mason *et al.*, 1990). The sex pheromone also shows great potential for mating disruption and mass trapping (Jansson *et al.*, 1991a).

Insecticides. Planting-time applications of insecticides are sometimes made to the soil to prevent injury to slips or cuttings. Both granular or liquid formulations have been used. Systemic insecticides are preferred. Post-plant applications are sometimes made to the foliage for adult control, especially if fields are likely to be invaded from adjacent areas, but if systemic insecticide is applied some suppression of larvae developing in the vine may also occur. Because of the long duration of the plant-growth period, it is not uncommon for preplant or planting-time applications to be followed by one or more insecticide applications

to the plant or soil at mid-season. Insecticides are also applied to tubers being placed into storage to prevent reinfestation and inoculation of nearby fields.

Cultural Practices. Cultural practices are sometimes recommended to alleviate weevil problems. Isolation is frequently recommended, and it is advisable to locate new fields away from previous crops and distant from sweet potato storage facilities, because both can be a source of new infestations. However, despite the infrequency of flight by adults, dispersal can occur over considerable distances. Miyatake *et al.* (1995), for example, documented dispersal rates of 150 m per day, with dispersal more rapid in the absence of suitable hosts.

Sanitation is particularly important for weevil population management. Discarded and unharvested tubers can support large populations, and every effort should be made to remove such host material. Related to this, of course, is the destruction of alternate hosts. Control of *Ipomoea* weeds is recommended.

Soil conditions affect weevil damage potential. Because adults crawl into cracks in the soil to gain access to tubers, efforts are made to reduce soil cracking. Irrigation, mulching, and hilling are common approaches, and planting in sandy soil is preferable to soil with high clay content because cracking is less likely.

Host-Plant Resistance. Considerable research has been conducted on host-plant resistance in sweet potato, and several commercially available varieties exhibit low levels of resistance (Waddill and Conover, 1978; Barlow and Rolston, 1981; Mullen *et al.*, 1980, 1985). Unfortunately, resistance is not adequate when plants are exposed to high weevil population densities, and severe damage results. Talekar (1987) suggested that adequate resistance might not be isolated by conventional plant-breeding techniques. Son *et al.* (1991) identified chemical oviposition stimulants that may be the basis for resistance in some cultivars.

Biological Control. Entomopathogenic nematodes seem to be the organisms with the greatest potential for practical biological suppression of sweetpotato weevil (Jansson, 1991). Several strains of *Steinernema carpocapsae* (Nematoda: Steinernematidae) and *Heterorhabditis bacteriophora* (Nematoda: Heterorhabditidae) penetrate the soil and tubers, killing weevil larvae. At least in the soils of southern Florida, infective nematodes are persistent, remaining active for about four months. In some cases nematodes are more effective than insecticides at reducing damage (Jansson *et al.*, 1990b, Jansson *et al.*, 1993).

Other Methods. Other methods of suppression are sometimes used, especially for post-harvest treatment of tubers. Post-harvest treatment not only prevents damage in storage, but allows shipment of tubers to areas where sweetpotato weevil is not found but might survive. Traditionally, post-harvest treatment has been accomplished with chemical fumigants, but they have fallen out of favor. Irradiation is potentially effective, though older stages of insects are less susceptible to destruction (Dawes *et al.*, 1987; Sharp, 1995). Storage in controlled atmospheres, principally low oxygen and high carbon dioxide, is very effective for destruction of weevils, but requires good storage conditions (Delate *et al.*, 1990).

Vegetable Weevil
Listroderes difficilis Germar
(Coleoptera: Curculionidae)

Natural History

Distribution. First found in North America in 1922, vegetable weevil quickly spread from Mississippi, its point of introduction, to other southern states from South Carolina to California, and to Hawaii. Its distribution is limited to warm climates, and it is most common in Alabama, Louisiana, and Mississippi. However, vegetable weevil can withstand moderately cold weather, including brief periods of below-freezing temperature. Originally described from Brazil, vegetable weevil has also invaded Australia, New Zealand, and South Africa.

Host Plants. Vegetable weevil has a very wide host range, and is known to attack such vegetables as artichoke, bean, beet, broccoli, Brussels sprouts, cabbage, carrot, cauliflower, celery, Chinese cabbage, collards, eggplant, endive, garlic, kale, kohlrabi, leek, lettuce, mustard, onion, parsley, parsnip, pepper, potato, radish, rutabaga, salsify, southern pea, spinach, sweet potato, Swiss chard, tomato, and turnip. Despite its wide host range, it is generally known as a pest of potato and tomato (Solanaceae); cabbage, collards, and turnip (Cruciferae); and carrot (Umbelliferae). Not surprisingly, vegetable weevil is known to feed on a long list of weeds, including common chickweed, *Stellaria media*; lambsquarters, *Chenopodium* spp.; thistle, *Cirsium* spp.; dogfennel, *Eupatorium capillifolium*; pepperweed, *Lepidium* spp.; mallow, *Malva* spp.; sorrel, *Oxalis* spp.; and plantain, *Plantago* spp. In Georgia, turnip is the preferred vegetable crop, and chickweed is the most common wild host (Beckham 1953).

Natural Enemies. This insect seems relatively free of important parasitic insects, but is preyed upon by general predators, especially ants. Fire ants, *Solenopsis* spp., for example, are known to feed on both eggs and larvae of vegetable weevil. Steinernematid nematodes are known to parasitize vegetable weevil experimen-

tally (Poinar, 1979), and diplogasterid nematodes have caused natural infestations (High, 1939). *Hyalomyodes triangulifera* (Loew), a common parasitoid of Coleoptera and Lepidoptera, and *Epiplagiops littoralis* Blanchard, a little-known parasitoid (both Diptera: Tachinidae) are reported from vegetable weevil, but their effects are undetermined.

Life Cycle and Description. There is only a single generation per year, with the adults aestivating during the summer months. Typical oversummering locations for the inactive adults include beneath loose bark of trees, and under grass, weeds, and other organic debris, often at the edges of crop fields. In September, activity is resumed, and egg production begins. This beetle is parthenogenetic; only females are known, and all are capable of producing fertile eggs. Under ideal conditions, a life cycle can be completed in about 45–110 days.

Egg. Eggs are deposited over an extended period of time, usually from September until March, if weather allows adults to be active. This results in protracted periods of activity for all other life stages, also. The egg is slightly elliptical, and about 0.64 mm long and 0.54 mm wide. Initially, it is white, but turns yellow and then gray, and eventually black, as it reaches maturity. The eggs are deposited near the crown of the plant, but sometimes on leaf petioles or on the soil adjacent to the plant. Estimates of egg production are from 1–30 per day and 300–1500 per female. Duration of the egg stage is 11–33 days, but generally is 15–20 days.

Larva. Upon hatching, larvae begin feeding at the plant crown, often destroying new foliar tissue before it can develop fully. However, no part of the plant is immune to attack, and leaves, stems, and roots may be consumed. The period of larval abundance corresponds roughly to that of egg production, late autumn until early spring. The larvae initially measure about 1.7 mm long, and are creamy-white with a black head. As foliage is consumed, however, larvae acquire a yellowish or green color and attain a length of 14 mm. In larger larvae the head is yellowish-brown with dark spots. Larvae lack true legs, but use ventral and lateral ridges or protrusions on their body for locomotion. The lateral protrusions, which are pyramidal, are apparent when the larvae are viewed from above. Also present on the larvae are dark rings around the abdominal spiracles, which results in a line of dark spots along each side. There are four larval instars. Lovell (1932) gave mean head capsule widths of 0.21, 0.30, 0.44, and 0.70 mm, respectively, for the larval instars. In contrast, Beckham (1953) provided corre-

Vegetable weevil larva.

sponding measurements of 0.32, 0.48, 0.79, and 1.18 mm. The basis for this fairly significant difference in head capsule measurements is unexplained. Larval growth is completed in 23–45 days, averaging 38 days. Mature larvae drop to the ground to pupate. (See color figure 128.)

Pupa. Larvae prepare a small cell in the soil at a depth of 2.5–5.0 cm, and pupate within. The pupa measures about 7.5 mm long, is yellow-green, and resembles the adult in form. Pupation occurs in the winter and spring, with the adults not emerging until April or May. However, in the laboratory pupation may be completed in 14–16 days.

Adult. The adult emerges from the soil several days after completing the transformation from the pupal stage. It measures about 8 mm long. The adult is grayish-brown, though generally there is a light-colored "V" on the elytra. The body is also covered with tan or gray scales and scattered hairs. The snout is short and stout. The antennae are slightly clubbed. Adults are somewhat gregarious, often collecting in small groups when they seek shelter or feed. They are nocturnal, hiding during the daylight hours under plant debris or clods of soil. Despite having fully formed wings, adults rarely fly, almost always walking from place to place. Adults are active from late autumn until early summer, and then aestivate through the summer months. Adults often feign death if disturbed.

The biology of vegetable weevil was provided by Lovell (1932), High (1939), and Beckham (1953).

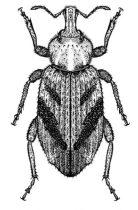

Adult vegetable weevil.

Damage

Both adults and larvae may feed upon foliage and roots, and inflict serious damage. The principal form of damage results when larvae feed on the developing tissue of plants, stunting them. Later, larvae may move onto mature foliage, consuming everything except large veins. Initially, larval feeding consists of small, round holes, but as larvae mature the holes get quite large and are irregular in shape. In crops with well-developed roots such as carrot and turnip, larvae may also move to the roots to feed, where they tunnel through the plant tissue. Adults can feed on stem tissue, sometimes cutting off the stems of young plants at the soil surface in a manner similar to cutworm damage.

Management

Insecticides. Although formerly considered an important pest, modern foliar insecticides now keep vegetable weevil under control. Botanical insecticides can be effective for suppression of this insect. Fruit-based poison bait formulations formerly were used for adult control with considerable success.

Cultural Practices. Cultural controls can be beneficial, but are rarely practiced, because insecticides applied for this, or other insects, effectively suppress vegetable weevil. Cultivation during the winter months, especially repeated cultivation, can destroy the pupal stage, which is found in the soil at this time of the year. Crop rotation can have some benefit, despite the wide host range of this insect. Repeated culture of highly preferred plants such as turnip, carrot, potato, and tomato should be avoided. Because adults almost never fly, susceptible vegetables following nonsusceptible crops such as corn are not damaged. Sanitation should be practiced, because summer-time aestivation takes place commonly under organic debris. If the soil is free of such shelter, summer survival may be poor.

West Indian Sugarcane Rootstalk Borer Weevil
Diaprepes abbreviatus (Linnaeus) (Coleoptera: Curculionidae)

Natural History

Distribution. *Diaprepes abbreviatus* is found widely in the Caribbean Islands, and was discovered in Florida in 1964. It has since then spread slowly, but occurs throughout southern and central Florida. It is thought to have been introduced with ornamental plants shipped to Florida from Puerto Rico, and therefore has the potential to spread again by this means, particularly to other warm weather regions such as southern Texas and California. In Puerto Rico these pests are known as "vaquitas."

Host Plants.   Although first reported to be a pest of sugarcane, this weevil has since been found to attack many other plants. In Florida, it is a serious pest of citrus and woody ornamentals. However, it has been known to cause serious injury to okra in the Virgin Islands, to yam, pepper and lima bean in Puerto Rico, and to potato in Florida. The adults have been found on numerous shrubs and flowers (Griffith, 1975), and sometimes trees are defoliated by adult feeding. However, adult-feeding behavior is not truly indicative of host suitability, because it gives no indication of suitability for larval development. Schroeder *et al.* (1979) evaluated over 70 plants for larval suitability, and indicated that in addition to citrus and sugarcane, larvae could develop on aloe, coralberry, croton, false aralia, waxplant, shore juniper, red cedar, liriope, and prayerplant. Eggs are deposited on ornamental plants such as ficus and cornplant (Abreu–Rodriguez and Perez–Escolar, 1983).

Natural Enemies.   Predation rates of young larvae are quite high as they drop from their point of hatch on foliage and crawl along the surface of the soil seeking a suitable site to enter the soil. Within Florida citrus groves, several species of ants account for nearly 50% destruction of young larvae (Whitcomb *et al.*, 1982). The ants, *Pheidole dentata* Mayr, *P. floridana* Emery, and *Tetramorium similimum* Roger (all Hymenoptera: Formicidae) are most effective. The ant, *Crematogaster ashmeadi* Mayr (Hymenoptera: Formicidae) also attacks weevil egg masses on foliage. In the absence of ants, the earwig *Labidura riparia* (Pallas), is an important predator (Tryon, 1986).

Several egg parasitoids of *D. abbreviatus* are known, but *Tetrastichus haitiensis* Gahan (Hymenoptera: Eulophidae) is the most important parasitoid in the Caribbean areas and it has been introduced into new infestations in the Caribbean areas and Florida (Beavers *et al.*, 1980). In Florida, performance has been erratic.

The larvae are affected by such fungi as *Metarhizium anisopliae*, *Beauveria bassiana*, *Paecilomyces lilacinus*, and *Aspergillus ochraceous*, and the nematodes *Steinernema carpocapsae* (Nematoda: Steinernematidae) and *Heterorhabditis* sp. (Nematoda: Heterorhabditidae). The nematodes cause higher levels of mortality than the fungi in Florida, often about 50% (Beavers *et al.*, 1983).

Life Cycle and Description.   A life cycle requires about a year for completion, but development time is highly variable and diapause apparently is variable. Adults can be found throughout the year, but there are two periods of peak emergence—June and September. Adult males are active for about two months, whereas females are active for four months.

Egg.   The eggs are deposited in clusters between leaves. An adhesive secreted by the female causes the leaves to stick together and glues the eggs to the leaves. Old rather than young leaves are selected as oviposition sites, presumably because adults tend to feed on young leaves, which would endanger the eggs. Egg clusters normally contain 30–100 eggs, but up to 300 they have been observed in some clusters. They are deposited in a single layer. Over the course of their life, females produce about 5000–6500 eggs. The eggs are elongate-oval with a length of about 1.2 mm and a width of 0.4 mm. Initially white in color, the eggs turn brownish before hatching. The egg hatching occurs in about 7–8 days. (See color figure 261.)

Larva.   Young larvae, upon hatching, drop from the foliage to the soil. Larvae are not effective at burrowing, rather entering the soil through cracks. They feed initially on small roots or rootlets, and do not acquire the burrowing habit as suggested by the their common name, until they attain the third or fourth instar. Larvae are whitish to yellowish-brown, with a brown head. At hatching they measure only about 1 mm long, but attain a maximum length of about 20 mm. They normally display eight instars, but individuals have been observed with as few as six instars or as many as 18. The larval growth is rapid for the first 3–4 months, but total larval development usually requires about 370 days (Beavers, 1982). The period of larval development is variable, in part, because it may include a 2- to 13-month period of diapause. The period of diapause is defined by Wolcott (1936) as a period when feeding, growth and molting do not occur; however, the larva continues to move in the soil. Wolcott speculated that this kept the larva from being infested by mites and fungi.

Quintela *et al.* (1998) studied larval development on artificial diet and citrus roots. This study suggested that larvae can have 10–11 instars, though it is not clear that all larvae display this pattern of growth. Mean head capsule widths were 0.26, 0.35, 0.47, 0.65. 0.99, 1.39, 1.81, 2.22, 2.64, 3.03, and 3.31 mm for instars 1–11, respectively. Mean duration of each instar was 7.9, 5.1, 4.0, 9.4, 3.7, 4.8, 11.9, 11.1, 37.2, 35.1, and 18.0 days, respectively. Total larval development time, about 148 days, was relatively short in this study, probably because a period of diapause was absent. Larval growth, as reflected by head capsule width and larval weight, was rapid for the first 75 days, but there was little increase thereafter.

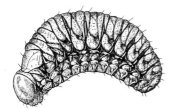

West Indian sugarcane rootstalk borer weevil larva.

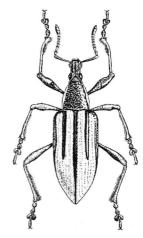

Adult West Indian sugarcane rootstalk borer weevil.

Pupa. At larval maturity a pupal cell is formed in the soil. Pupation may occur at any time during the year though spring and autumn are predominant. The pupa is waxy white in appearance and its form resembles the adult except the wings that are twisted ventrally. Duration of the pupal period is about 15 days.

Adult. The adults are grayish green to black dorsally, and usually bear light tan or orangish scales on the elytra. A striped pattern usually occurs on the elytra, where 3–7 black stripes traverse about two-thirds the length of the elytra. The ventral surface and legs are dark gray. This species has distinctly long legs and antennae, and the snout is moderately long. Adults measure 9–15 mm long. Females were reported to commence oviposition 3–7 days after they emerge from the soil by Wolcott (1936), but Beavers (1982) reported a mean pre-oviposition period of 21 days. The difference is attributable to the period of time the adults spend in the soil before emergence. Adults are capable of strong flight, but of short duration. Thus, natural spread is slow. When disturbed, adults tend to drop to the soil (Beavers and Selhime, 1978). Adults often aggregate on certain trees, and though tender young foliage is involved in attraction, there is also evidence of an aggregation pheromone (Beavers *et al.*, 1982). (See color figure 133.)

The West Indian sugarcane rootstalk borer weevil, including its color forms, was described by Pierce (1915). The most complete biology was given by Wolcott (1936). Methods of culture, including an artificial diet, were presented by Beavers (1982). A bibliography was published by Hall (1995).

Damage

Adults injure plants by feeding on foliage. Initially they feed along the edge, creating a notched appearance. Eventually, however, they may consume the entire leaf, even defoliating trees. The larvae feed below-ground on roots. They will consume rootlets and remove the outer layers of larger roots, often girdling them. The larvae also burrow into soft below-ground plant tissue such as the base of sugarcane stalks and the tubers of yam and potato. However, vegetable crops are not normally injured by this weevil. Injury occurs to vegetables mostly when planted in soil previously occupied by sugarcane or ornamental crops.

Management

Sampling. Because these beetles are not usually vegetable pests, specific monitoring practices have not been developed for use in vegetables. In citrus, adult populations are monitored by visual examination of trees for adults and leaf notching, and with screen traps that capture adults as they emerge from the soil. Visual and odor-based traps are under development (Schroeder and Beavers, 1985), but as yet they are unproven.

Insecticides. Application of insecticide to soil can reduce tunneling in tubers (Oramas *et al.*, 1990). Adult suppression is commonly practiced in citrus and ornamental crops by application of insecticide to foliage (Wong *et al.*, 1975), but oil sprays to foliage can also reduce survival of eggs (Schroeder and Green, 1983).

Biological Control. Larvae of *D. abbreviatus* are susceptible to infection by *Steinernema* and *Heterorhabditis* spp. entomopathogenic nematodes (Nematoda: Steinernematidae and Heterorhabditidae) (Figueroa and Roman, 1990a,b, Downing *et al.*, 1991, Shapiro *et al.*, 1999). Under field conditions, *Heterorhabditis bacteriophora* significantly reduced adult emergence (Downing *et al.*, 1991).

West Indian Sweetpotato Weevil
Euscepes postfasciatus (Fairmaire)
(Coleoptera: Curculionidae)

Natural History

Distribution. West Indian sweetpotato weevil is found in Central and South America, the Caribbean

Islands, and many islands in the Pacific, including Fiji, Guam, Japan, New Zealand, and Samoa. It seems a native of the western hemisphere, and probably the Caribbean region. In the United States, it is known only from Hawaii, Puerto Rico, and the Virgin Islands. However, the proximity and tropical climate of southern Florida suggest that this state could be invaded by dispersants from Cuba, Dominican Republic, Jamaica, or elsewhere. Because West Indian sweetpotato weevil survives on Bermuda, which has a mild but not tropical climate, it is possible that the potential geographic range of this insect in United States could include much of the Gulf Coast area or southeast.

Host Plants.   This weevil feeds on plants in the genus *Ipomoea* of the plant family Convolvulaceae. Among vegetable crops only sweet potato, *I. batatas*, is a suitable host. Native plants can be important hosts of sweetpotato weevil. Railroad vine, *Ipomoea pes-caprae*, and many other *Ipomoea* spp. are suitable hosts.

Natural Enemies.   Little is known concerning the natural enemies of this insect. The fungus *Beauveria bassiana* has been consistently associated with this insect, and damage by this insect is typically greater during the dry season when fungi are less effective. This relationship is merely a correlation, however, and further research is needed to determine if there is a cause-and-effect relationship. Some parasitoids have been noted. In Hawaii, *Eupelmus cushmani* (Crawford) (Hymenoptera: Eupelmidae) attacks larvae infesting the above-ground portions of the host plant. In Peru, *Eurydinoteloides* sp. and *Cerophala* sp. (both Hymenoptera: Pteromalidae) also attack larvae.

Life Cycle and Description.   The life cycle can be completed in 4–6 weeks. Thus, several generations per year can develop in the tropical climates where these insects are found.

Egg.   The female excavates a shallow cavity with her mouthparts in the sweet potato tuber or vine as a receptacle for her egg. Eggs are laid singly. The dimensions of the oval, yellowish egg given by Sherman and Tamashiro (1954) were 0.38 mm long and 0.34 mm wide. However, Raman and Alleyne (1991) reported that the egg was only 0.12 mm long and 0.10 mm wide. Both reported that the female seals the eggs within the oviposition cavity with a drop of fecal material, and that the oviposited egg appears as a raised-dark spot on the plant. The same authors suggested that this behavior does not apply to eggs laid on foliage, and that these eggs fail to hatch. The incubation period of the eggs is 7–9 days at 24°C and 11–15 days at 18°C. Females are reported to produce about 95–362 eggs during their life time, with mean fecundity estimated at 180 eggs.

Larva.   Upon hatching, larvae burrow deeper into the vine or tuber to feed. Larvae are legless, with a well-developed head. They are whitish and attain a length of about 5 mm. They greatly resemble larvae of sweetpotato weevil, *Cylas formicarius*; differences in setal patterns are pictured by Sherman and Tamashiro (1954). Mean head capsule widths for the five instars are 0.18, 0.25, 0.37, 0.56, and 0.78 mm for instars 1–5, respectively. Duration of the feeding period of the larval stage requires 25 (range 18–31) days at 24°C and 45 (range 40–53) days at 18°C (Raman and Alleyne, 1991). This does not include the nonfeeding or prepupal portion of the fifth instar, however, which is about five days. Sherman and Tamashiro (1954) indicated the number of days after oviposition in which the various instars were found as: first 7–14; second 10–15; third 12–16; fourth 13–21; fifth 17–30; and prepupal 21–32. Further, they indicate that the minimum time spent in each was 3, 2, 1, 4, 4, and 4 days for instars 1–5 and the prepupal stage, respectively. Thus, the duration of the larval stage at about 27°C, including the prepupal period, is likely about 25 days.

Pupa.   The larva prepares a small pupation cell at the final larval feeding site. The pupa is white initially, but becomes tan with age. The pupa measures about 4.0–4.9 mm long. In form, it greatly resembles the adult, except that the snout is pressed against the ventral surface and the elytra are incompletely developed and twisted ventrally. Duration of the pupal stage is 7–10 days.

West Indian sweetpotato weevil larva.

West Indian sweetpotato weevil pupa.

Adult. The adult of West Indian sweetpotato weevil is small and inconspicuous. It measures only 3.2–4.0 mm long and is reddish brown to grayish black. The tip of the elytra bear an indistinct whitish or yellowish transverse bar. The entire body, including the legs, is covered with a dense covering of short, stiff, erect bristles and scales. The adult often feeds before emerging from the pupation site. The adults feed on either the vine or the root, and have a tendency to be gregarious in the adult stage. Thus, several individuals may feed together, forming a large cavity. This species is much less apparent than the co-occurring sweetpotato weevil, *Cylas formicarius* (Fabricius). Often West Indian sweetpotato weevil feigns death when disturbed, drops from the plant, where it blends in with the soil particles, and avoids detection. Adults are long lived, often surviving more than 200 days if provided with adequate food. An interesting aspect of adult biology is that they apparently are flightless. Therefore, dispersal is dependent on walking by adults, or transport of infested tubers or other plant material containing larvae or other unapparent stage; the transport of plant material is undoubtedly the principal means of long-distance dispersal.

The biology of this species was reviewed by Sherman and Tamashiro (1954), and Raman and Alleyne (1991). Pierce (1918) provided a detailed description, and pictures the different stages.

Damage

This is the most severe weevil pest of sweet potato in the South Pacific and Caribbean areas. Successful cultivation of sweet potato on some Caribbean islands hinges largely on the successful management of this insect, which is known locally as "scarabee." It damages tubers both in the field and storage. Insect damage induces terpenoid production in sweet potatoes, making them bitter and unpalatable. Larvae infesting the tuber often burrow deep into the tuber, leaving little evidence of their presence on the surface. However, the center of the tuber may be thoroughly tunneled and packed with fecal material. Eventually, the tuber becomes blackened, and often infected with plant pathogens. Vines that are infested become wrinkled, cracked, or collapsed.

Management

Insecticides. Insecticides are often used to protect plants from injury. Planting stock sometimes is dipped in insecticide before treatment; this is particularly important if the planting stock is already infested. Insecticide is also incorporated into the planting bed and applied to foliage. Systemic insecticides are preferred because so much of the life cycle occurs inside the plant.

Cultural Practices. Several cultural practices materially affect West Indian sweetpotato weevil abundance (Sherman and Tamashiro, 1954). Foremost is sanitation. Propagation of sweet potato is often by apical cuttings or "slips," which can be infested. Plant debris is often infested and may serve as a breeding site for weevils between crops; such material should be destroyed. *Ipomoea* weeds should be removed from areas near crop fields as they serve both a source of initial infestation and provide harborage between crops. Crop rotation is particularly effective, because the adults are flightless. Storage of tubers should be distant from fields used to culture sweet potato.

Host-Plant Resistance. Considerable work has been done to assess naturally occurring host plant resistance. Varieties with thin stems are less injured by weevils. Unfortunately, they produce less satisfactory yields and their tubers tend to grow close to the surface, where they are more easily accessed by weevils. However, though numerous varieties or isolates contain some elements of resistance, considerable work remains to be done before commercially acceptable varieties with reliable resistance are available. It is likely that host-plant resistance will prove to be only a component of the management program, and cannot be relied upon to provide complete protection from attack (Raman and Alleyne, 1991).

Whitefringed Beetle
Naupactus spp.
(Coleoptera: Curculionidae)

Natural History

Distribution. The status of whitefringed beetle has long been in dispute because of disagreement over one or more species of "whitefringed beetle." Even the genus designation is disputed. There is consensus,

Adult West Indian sweetpotato weevil.

however, that whitefringed beetle invaded the southeastern United States from South America about 1936, and was first recovered from western Florida. Alabama and Mississippi are the states most infested with this insect, but it is now distributed from Virginia to Florida in the east, and to Missouri and Texas in the west. This beetle has also been found in New Mexico and California, but its distribution is limited there. The potential range of this species is believed to include most of the southern United States, from Virginia and Kentucky to southern Colorado and nearly all of California. Whitefringed beetle is considered a pest in Argentina, Brazil, Chile, and Uruguay. Australia and New Zealand also have been invaded. The species occurring in the United States apparently are *N. leucoloma* Boheman, *N. peregrinus* (Buchanan), and *N. minor* (Buchanan); the first two species are of greatest economic significance in the United States.

Host Plants. Whitefringed beetle is considered to have a very wide host range, with the number of plants eaten estimated at 385 species. Grasses, however, are considered to be relatively poor hosts. Vegetable crops attacked include bean, carrot, corn, cowpea, cucumber, lima bean, mustard, okra, potato, pea, squash, sweet potato, and watermelon. Whitefringed beetle perhaps is better known as a field crop pest, attacking alfalfa, cotton, peanut, soybean, tobacco, and velvetbean. Numerous broadleaf weeds are fed upon by adults and larvae, whereas shrubs and trees are fed upon principally by adults. Damage to young pine trees planted on converted croplands containing residual beetle populations has become a problem in recent years.

Natural Enemies. No insect parasitoids are known from the United States or South America. Nematodes, particularly steinernematids, have been observed as significant mortality factors in some locations, but apparently they are limited to heavier soil types. Several pathogens, including the fungus *Metarhizium anisopliae*, a microsporidian *Nosema* sp., and undetermined bacteria have been observed on numerous occasions to infect larvae. Also, general predators such as ground beetles (Coleoptera: Carabidae), stink bugs (Hemiptera: Pentatomidae), ants (Hymenoptera: Pentatomidae), and birds prey on adults.

Life Cycle and Description. In North America, there is generally one generation per year. In Alabama, eggs occur in mid- to late-summer, early instars tend to be found in late summer, intermediate instars in autumn and winter, and late instars in the spring months (Zehnder, 1997). In the colder regions of South America, a complete life cycle requires two years, so prolonged development may also occur among some populations in the United States, and this has been reported even among some southern populations. In all locations, the larval stage generally overwinters, but overwintering eggs have also been observed. Reproduction is parthenogenetic.

Egg. Females deposit eggs in clusters, usually numbering from 10 to 15 per egg mass. Eggs measure about 0.9 mm long and 0.6 mm wide, and are elliptical. The eggs initially are white, but after 4–5 days they turn yellow. The eggs are covered with a sticky gelatinous material that hardens, causing the eggs to adhere to the substrate. They are deposited in various locations, including on the foliage and stems of plants, and below-ground, but most commonly they are deposited at the surface of the soil adjacent to the host plant. The egg production varies only a few (15–60 per female) when larvae feed on less suitable host plants such as grasses, to numerous (1500 or more per female) when favorable host plants such as peanut or velvetbean are available. Eggs deposited during the warmer seasons hatch in 10–30 days, with an average of about 15–17 days during the summer months. During the winter months eggs may require 100 days to hatch; viability is greatly reduced during the winter. The eggs reportedly require moisture in order for hatch to occur.

Larva. The larva is legless, and yellowish-white, except for the head, which is brown. The larva lives within the soil and feeds principally on roots, though it frequently burrows into below-ground parts of plant, such as tubers. The summer, autumn, and winter months normally are spent in the larval stage, followed by pupation in the spring, but some apparently pass a second year in the larval stage.

Pupa. At maturity, the larva secretes adhesive which hardens and forms a pupal chamber. Larvae normally pupate in May–July. Pupation normally occurs within the top 5–15 cm of soil, but sometimes occurs at greater depths. The pupa measures about 12 mm long and is initially whitish but darkens as maturity approaches. Duration of the pupal stage is about 13 days, ranging from 8–15 days. Emergence is

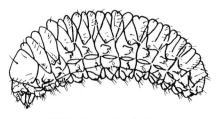

Whitefringed beetle larva.

stimulated by moist conditions—neither dry nor wet soil is favorable.

Adult. The adult is brownish-gray, with a lighter band along the outer margins of the elytra. The thorax and the sides of the head also bear light-colored bands. The body is densely covered with short hairs. The elytra are fused together and the wings reduced; thus, the beetles are unable to fly. Only females are known. Adult emergence is typically protracted, with beetles emerging from the soil in May–August. Emergence is particularly apparent after rainfall. (See color figure 132.)

The biology of whitefringed beetle was provided by Henderson and Padget (1949), Young *et al.* (1950), Warner (1975), and Zehnder (1997). Descriptions and keys to the species were found in Lanteri and Marvaldi (1995). Observations from South America were given by Berry (1947).

Damage

These insects are able to develop on a wide range of host plants, and are relatively unchecked by natural enemies. Thus, they build up to very high levels. All the larvae are long-lived; they may destroy replanted fields unless they are killed. They seem to thrive on field crops such as alfalfa, peanuts, and velvetbean, so vegetables following these crops are especially prone to injury. Lawns, woods, swamps, and old-stand vegetation are not considered as suitable habitats. Rather, well-drained and disturbed environments seem attractive.

The larvae feed on the roots, sometimes completely severing small roots. However, the principal damage results from burrowing into fleshy tissue such as the main tap root, tubers, and below-ground stems. Often damage is limited to surface scars or channels. Large sweet potato tubers are especially damaged (Zehnder, 1997). On fibrous roots, such as those found on trees, feeding may be limited to removal of the cortical tissues. Early symptoms of feeding injury include discoloration and wilting, but this may be followed by plant death.

The adults may feed on the foliage, particularly of broadleaf plants. They make irregular feeding sites, or notches, along the margins of leaves. Although they may occasionally defoliate a plant, the adult injury is not usually considered serious.

Management

Sampling. Damage potential is sometimes determined by sampling soil and sieving for larvae. Soil beneath favored plants, particularly weeds, is often selected for sampling. The frequency of leaf notching is directly related to the abundance of adults, and serves as a convenient index of damage potential.

Insecticides. Persistent soil insecticides were formerly used extensively for larval suppression, but their use has been discontinued owing to environmental concerns. Less persistent insecticides are often applied at spring planting, and kill many overwintering larvae. However, larvae that survive may produce another generation that causes damage in the autumn after insecticide dissipation. Adult control is sometimes practiced. Foliar insecticides targeted against adults have been shown to reduce grub damage to sweet potato roots (Zehnder *et al.*, 1998).

Cultural Practices. Although whitefringed beetle has a wide host range, rotation to grains and grasses is recommended because the fibrous root systems of grasses are not easily damaged, and larval growth is suboptimal. Also, planting of legume crops is discouraged, except during the winter months, because they are favored hosts.

FAMILY COCCINELLIDAE—LADY BEETLES

Mexican Bean Beetle
Epilachna varivestis Mulsant
(Coleoptera: Coccinellidae)

Natural History

Distribution. Mexican bean beetle is native to Mexico and Central America, and it likely has resided in Arizona and southern New Mexico since the introduction of cultivated beans by indigenous peoples several hundred years ago. By the late 1800s, it was

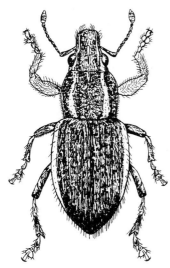

Adult whitefringed beetle.

damaging beans throughout the southwest, particularly Colorado. A major increase in damage followed the accidental transport of Mexican bean beetle to northern Alabama about 1918, apparently in shipments of alfalfa hay from Colorado and New Mexico. The beetles, once gaining access to eastern states, spread rapidly to the northeast. By 1922, Mexican bean beetle had invaded Georgia, North and South Carolina, Virginia, Tennessee, and Kentucky. It reached Ohio and Pennsylvania in 1925, Ontario in 1927, New Jersey in 1928, and Connecticut in 1929. It is now found all over the continental United States. In Canada, Mexican bean beetle is found in eastern provinces, from Ontario to New Brunswick, and also is reported from British Columbia, but it is a common pest only in Ontario.

The central Great Plains, from North Dakota to Texas, formerly provided a natural barrier in spreading of the beetles. Although this barrier was bridged through human intervention, there remains an area that is relatively inhospitable to bean beetle, so they are infrequent in this region. This insect is also infrequent in Pacific Coast states. Thus, there are two fairly discrete populations—a western population in the Rocky Mountain region including the western edge of the Great Plains, and an eastern population that inhabits most of the eastern United States west to Kansas.

Host Plants. Mexican bean beetle develops only on legumes. Other plants are occasionally reported injured, but these invariably are growing adjacent to defoliated legumes and will not support reproduction of beetles. Among vegetable crops eaten are cowpea, lima bean, and snap bean, particularly the latter two bean types. Related crops such as faba bean, lentil, and mung bean seem to be immune. Field crops that may be attacked include alfalfa, sweet clover, various dry beans, and soybean. Formerly, field crops other than dry beans were relatively unsuitable and rarely injured. However, starting in the 1970s the eastern and then midwestern states began experiencing considerable damage to soybean by Mexican bean beetle. (See color figure 5.)

The natural host appears to be tick trefoil, *Desmodium* spp.; however, in the United States, Mexican bean beetle is almost found associated with cultivated legumes. Lupine, *Lupinus* spp., were found to support adults in California, but no reproduction occurred.

Natural Enemies. Numerous predators, parasitoids, and microbial disease agents of Mexican bean beetle have been identified, but few native natural enemies are considered to be important (Howard and Landis, 1936). Among predators the soldier bug *Stiretrus anchorago* (Fabricius) and the spined soldier bug, *Podisus maculiventris* (Say) (Hemiptera: Pentato-

midae), are often cited as the most effective. Lady beetle species such as convergent lady beetle, *Hippodamia convergens* Guerin–Meneville; transverse lady beetle, *Coccinella transversoguttata* Brown; and *Coleomegilla maculata* (De Geer) (all Coleoptera: Coccinellidae) sometimes prey on the eggs or young larvae of bean beetle, and on occasion have been considered important predators.

Parasitoids have not been very effective at suppressing bean beetle. Species native to the United States have not adapted to Mexican bean beetle as a host, whereas species imported from Central and South America have failed to establish permanently. Special attention was given to a species from Mexico, *Aplomyiopsis epilachnae* (Aldrich) (Diptera: Tachinidae). It was released widely in the eastern states, but apparently is not adapted to cold winters (Landis and Howard, 1940). In recent years a parasitoid from India and Japan, *Pediobius foveolatus* (Crawford) (Hymenoptera: Eulophidae) has been cultured and released in eastern states. Annual release is necessary because the parasitoid is unable to overwinter in the United States (Schaefer *et al.*, 1983). In Maryland, nurse crops of snap beans have been planted early to attract bean beetles, with parasitoids released into these small plantings. As overwintering or reproducing beetles appear elsewhere, parasitoids naturally disperse from the nurse crops to attack the expanding bean beetle population. The nurse crop approach allows development of a large population of parasitoids with minimal effort (Stevens *et al.*, 1975a). Small gardens, such as those in urban and suburban communities, also are suitable for *Pediobius* release (Barrows and Hooker, 1981).

Microbial pathogens, especially *Nosema epilachnae* and *N. varivestris* (both Microsporida: Nosematidae), occur in bean beetles (Brooks *et al.*, 1985). These pathogens are deleterious to Mexican bean beetle; *Nosema epilachnae*, in particular, reduces longevity and fecundity in bean beetles (Brooks, 1986). However, these pathogens also infect the parasitoid, *Pediobius foveolatus*. Infection of the parasitoid occurs when the immature stage develops in its host, or when the adult ingests the pathogen (Own and Brooks, 1986).

Weather. Although natural enemies may affect bean beetle abundance, weather is also thought to play an important role in population dynamics. Hot and dry weather is thought to be detrimental to survival of all stages, but especially the egg stage. Temperatures above 35–37°C can be lethal. Mellors *et al.* (1984) are among the recent authors to give information on the effects of various temperatures for several time intervals, with important earlier studies presented by Miller (1930) and Sweetman and Fernald (1930).

Life Cycle and Description. Mexican bean beetle usually exhibits 1–3 generations annually. In the western United States there normally is one complete generation, with a small number of individuals reproducing and developing a small second generation. However, during cool summers the members of the second generation are unable to complete their development and perish. In the southeast, where three generations are more common, a few beetles deposit eggs that produce a small fourth generation. Throughout the nation, adults are the overwintering stage. Overwintered adults typically are most abundant in June, followed by first, second, and third generation (when present) beetles in July–August, August–September, and October, respectively. A life cycle may be completed in 30–40 days during the summer months, but may require 60 days during cooler weather. In recent years much of the research has concentrated on Mexican bean beetle as a soybean pest, but life history parameters, while similar, are not the same as on more suitable hosts. Larvae reared on soybean tend to have higher mortality and longer development times than on snap bean (Bernhardt and Shepard, 1978; Hammond, 1985).

Egg. The eggs are deposited on end in clusters of 40–60 eggs, usually on the underside of leaves. They are elliptical, and measure about 1.3 mm long and 0.6 mm wide. Eggs generally are yellow, but turn orange-yellow before hatching. They hatch in 5–14 days, with a mean incubation period of 5.7 days. All females from the first generation deposit eggs, but in South Carolina only 94% of the second and 60% of the third generation beetles reportedly produced eggs as more and more beetles entered reproductive diapause. (See color figure 259.)

Larva. Upon hatching, larvae are yellow, and armed with a dense covering of branched spines arrayed in six longitudinal rows. The tips of the spines, when examined closely, usually can be observed to be black. There are four instars. The mean duration of instars in a South Carolina study has been reported to be 3.9, 3.6, 3.6, and 3.6 days, respectively (Eddy and McAlister, 1927). In Colorado, its development required 4.8, 4.1, 4.9, and 5.3 days, respectively (Kabissa and Fronk, 1986). The development time has been studied extensively, and values vary somewhat with location and weather, but most of them are similar to the aforementioned studies. After attaining its full size, about 8 mm long, the larva attaches its anal end to a substrate, usually the leaf on which it fed, and pupates. (See color figure 130.)

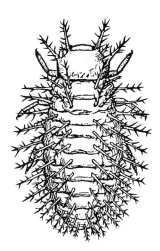

Mexican bean beetle larva.

Pupa. During the process of pupation the larval covering, which contains the spines, is pushed back toward the point of attachment to the substrate. Thus, the pupa appears to bear spines, but this is simply remnants of its earlier life, and not firmly attached. Rather, the yellow-orange pupa is quite free from projections. In some cases, particularly late in the season, the pupa is not completely yellow, but bears brown or black lines. Duration of the pupal stage averages 8.1 days in South Carolina and 9.6 days in Colorado. (See color figure 270.)

Adult. The adults are brightly colored beetles that resemble many beneficial lady beetle species, differing from the beneficial species principally in their phytophagous feeding habit. The beetle is hemispherical in shape, and bears 16 black spots. The spots are arranged in three rows, with six spots in each of the first two rows and four in the third row; thus, there are eight on each elytron. The background color is usually orange or copper, but ranges from yellow

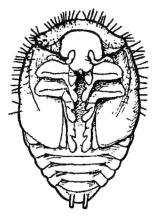

Mexican bean beetle pupa.

when freshly emerged, to reddish-brown when old. The reddish color is especially evident among the overwintered individuals. The beetles normally measure about 6–8 mm long and 4–6 mm wide, but size varies considerably depending on availability of food. In behavior, the beetle is rather sluggish, and responds to significant disturbance by dropping from the plant and by excreting small drops of blood from the articulations of their legs. This excretion reportedly is a form of a defensive chemistry that reduces predation by other insects. The pre-oviposition period of adults averages 11.5 days. Females commonly produce about 500 eggs, sometimes depositing over 1200 eggs. Mean fecundity is lowest in overwintering females, and increases in later generations. Adults overwinter under leaves and other plant debris, and under logs and stones. Aggregations of several hundred overwintering beetles are not uncommon. In some cases, overwintering may occur adjacent to larval host plants, but adults are strong fliers and often disperse several kilometers to suitable overwintering quarters, often in wooded areas. (See color figure 129.)

The biology of Mexican bean beetle was provided by many authors, but among the most complete are List (1921), Howard and English (1924), Eddy and McAlister (1927), Friend and Turner (1931), and Auclair (1959). Rearing procedures for both Mexican bean beetle and the parasitoid *Pediobius foveolatus* were given by Stevens *et al.* (1975b).

Damage

Larvae and adults feed principally on leaf tissue, but under high density conditions and when faced with starvation they also feed on blossoms, pods, and stems. Bean beetles feed on the lower surface of the foliage, removing small strips of tissue and usually leaving the upper epidermis and veins intact. The upper epidermis soon dies and becomes transparent,

leaving characteristic injury consisting of a number of small transparent spots that is reminiscent of a stained glass window. Entire leaves are quickly reduced to skeletal lace-like remains that have little photosynthetic value and usually dry and die quickly. Larvae are particularly damaging. As the larvae are not very mobile, they usually concentrate their feeding on a single leaf, inflicting complete defoliation before relocating to another leaf.

Bean plants are quite tolerant of defoliation. Bean plants can often withstand 50% leaf loss, especially if it occurs early in the season, leaving time for the plants to recover, or if the affected foliage is older and physiologically inferior. Plants are most sensitive to damage during the pod establishment and filling stages (Waddill *et al.*, 1984). On average, defoliation in excess of 10–20% usually results in decreased yield. Each larva may consume over 30 sq cm of foliage, with about 70% of the consumption occurring during the final larval stage (McAvoy and Smith, 1979). In studies of dry bean response to bean beetle defoliation, Michels and Burkhardt (1981) estimated that loss would occur with as few as 1.0–1.5 larvae per plant. In contrast, Capinera *et al.* (1987) determined that field beans could tolerate up to 19% defoliation without yield reduction, and that the number of beetles necessary to inflict this level of injury varied from 3–20 depending on the length of the adult feeding period.

Mexican bean beetle also is capable of plant disease transmission. When Jansen and Staples (1970a) allowed larval and adult bean beetles to feed on plants infected with cowpea mosaic virus, they became capable of transmitting the virus to healthy plants for a two-day period. Wang *et al.* (1994) reported slightly longer virus retention times, about four days, and suggested that beetles lost their ability to transmit virus in direct proportion to the amount of foliage consumed. Thus, Mexican bean beetles are not particularly effective vectors, but because adults fly freely they may be important in redistribution of such diseases.

Management

Sampling. The eggs and larvae are highly aggregated in distribution. Sampling is usually accomplished by visual examination of plants. Larval, pupal, and adult stages can be collected with sweep nets, however.

Insecticides. Modern insecticides have relegated Mexican bean beetle to low status in commercial bean production. They remain a serious problem, however, in homegardens and elsewhere when insecticides are not used. With all insecticides, thorough coverage of

Adult Mexican bean beetle.

foliage, particularly the lower epidermis of leaves, is necessary. Systemic insecticides applied at planting often provide good early-season protection, but when beetle densities are very high crops often benefit from later-season applications also (Webb *et al.*, 1970; Elden, 1982).

Cultural Practices. Cultural practices are of limited value. Beetles fly long distances to overwinter, so crop rotation and destruction of overwintering sites generally are not practical. It is a useful practice, however, to destroy bean plants as soon as they have been harvested, as this may disrupt development of many immature insects and inhibit development of additional generations.

Early-planted crops may be useful to attract adult beetles, where they can be destroyed by disking, insecticide application, or release of parasitoids. However, though this trap-crop approach works to lure beetles from a relatively unpreferred crop such as soybean to a preferred crop such as lima bean, it is not effective in the protection of preferred crops (Rust 1977).

Biological Control. The principal means of biological suppression of Mexican bean beetle is release of *Pediobius* wasps. (This species is discussed in the section on natural enemies). Despite extensive research demonstrating the feasibility of such releases, commercial bean producers rarely consider such an approach, tending to rely instead on chemical insecticides. Insecticides are somewhat compatible with parasitoid releases because some insecticides dissipate to non-toxic levels in 1–3 days, and parasitoid pupae are relatively immune to field applications of insecticides (Flanders *et al.*, 1984).

Host-Plant Resistance. Considerable effort has been directed toward identification of species and cultivars resistant to bean beetle oviposition or feeding. In Ohio, Wolfenbarger and Sleesman (1961) showed that *Phaseolus* spp., principally snap and to a lesser extent lima beans, were susceptible to injury. Cowpea, faba, and other beans generally were not attacked. Campbell and Brett (1966), working in North Carolina, were able to identify some commercial varieties of both snap and lima beans that displayed resistance. Resistance was reflected in reduced oviposition on resistant varieties, and decreased size and lower fecundity of adults reared on resistant plants. Insect development time was not affected. Also, some so-called vegetable-type soybean cultivars, suitable for such uses as sprouts and tofu, have been evaluated for bean beetle resistance and found to vary considerably in susceptibility to damage (Kraemer *et al.*, 1994). Raina *et al.* (1978) reported that greenhouse screening could be used to identify resistance.

Squash Beetle
Epilachna borealis (Fabricius)
(Coleoptera: Coccinellidae)

Natural History

Distribution. Squash beetle is reported to occur throughout much of the eastern United States from Massachusetts to Kansas in the north and from Florida to central Texas in the south. However, its impact as a pest is generally limited to the Atlantic Coast, from Connecticut to Georgia. It appears to be a native insect. A related and similar-appearing species, *E. tredecimnotata* (Latreille), feeds on cucurbits throughout Mexico and Central America, and has been found in western Texas, and southern New Mexico and Arizona.

Host Plants. Squash beetle larvae feed only on cucurbits, and successful larval development has been reported on various squashes, cucumber, watermelon, cantaloupe, and gourds. In addition, wild cucurbits can serve as hosts. Underhill (1923) indicated that prickly cucumber, *Echinocystis lobata;* and one-seeded bur cucumber, *Sicyos angulata*, were readily attacked in Virginia. Adults are less restrictive in their diet; in addition to cucurbits, they have been observed to feed on the blossoms and pods of lima beans and cowpeas, on lima bean foliage, and on fresh corn silks.

Natural Enemies. Because this insect is a relatively minor pest, it has not been thoroughly studied, and therefore its natural enemies are not well-known. However, general predators such as stink bugs (Hemiptera: Pentatomidae) and assassin bugs (Hemiptera: Reduviidae) are known to attack larvae. Underhill (1923) reported some evidence of attack by tachinids (Diptera: Tachinidae), and he further noted that 25–33% of eggs were sometimes consumed by predatory ladybeetles (Coleoptera: Coccinelliae) and lacewings (Neuroptera: Chrysopidae). Smith (1893) noted that squash beetle larvae frequently attack eggs, so cannibalism may be an important mortality factor.

Life Cycle and Description. Squash beetle can complete its life cycle in 25–48 days, but the average is 32 days. Two generations are produced per year in Virginia, but only one in Connecticut (Britton, 1919).

Egg. The yellow eggs are elongate, measuring about 1.8 mm long and 0.7 mm wide. They are laid on end in clusters averaging 45 eggs per mass (range 12–60 eggs). Females deposit eggs at 4–5-day intervals, and generally produce about 300–400 eggs during their life span. Eggs hatch 7–8 days after oviposition.

Larva. The larvae are yellow and have six rows of large spines running the length of the body; the spines

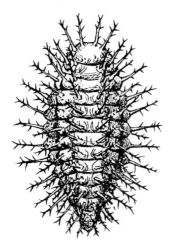

Squash beetle larva.

Adult squash beetle.

are often darker or tipped with black. Underhill (1923) indicated that the branching characteristics of the lateral row of spines could be used to distinguish the four instars. The number of branches on each spine was given as 0, 4–7, 10–12, and 16–18 for instars 1–4, respectively. The duration of the larval stage is 16–18 days, with average instar durations of 3.4, 3.2, 4.0, and 6.0 days for instars 1–4, respectively.

Pupa. Pupation occurs on the underside of cucurbit leaves, or on adjacent weeds, often in a shaded location. The pupa is about 9 mm long, yellow or orange in color, with branched brown spines attached to the old larval covering and clustered near the tip of the abdomen. Duration of the pupal stage is about seven days.

Adult. The adult is 8–10 mm long, with the females averaging larger than males. Beetles are yellow to dull red, with 12 large black spots on the elytra and four small black spots on the thorax. The spots on the elytra are arranged in three transverse rows. Adults begin production of eggs about eight days after emergence. Beetles normally abandon cucurbit fields in August or September, seeking shelter under plant debris or structures in preparation for overwintering. Cracks in the rough bark of trees are especially pre-

Squash beetle pupa.

ferred as a resting site for overwintering though they may also seek shelter under leaves and pine straw. Beetles seem not to penetrate more that 12–15 m into forested areas, preferring trees near the edge. In Virginia, they move back into the cucurbit fields in May or June. After overwintering, they may live up to several weeks without food while awaiting the availability of cucurbits. The beetles usually begin egg deposition within two weeks of feeding, so eggs from the overwintering beetles are abundant in June. The adults from this first generation begin to lay eggs in July, producing the second generation in August. By late summer, the overwintering beetles have died, but both first and second generation adults are available to overwinter.

Underhill (1923) gave a complete description of squash beetle biology, and the report of Brannon (1937) is informative.

Damage

The larvae and adults feed on the underside of cucurbit foliage, though adults may also feed on the upper surface. Larvae make a circular cut in the leaf before consuming the lower layer of epidermis within the circle. The circular cut or trenching behavior is believed to inhibit production of sticky phloem sap by the plant, which inhibits feeding (McCloud *et al.*, 1995). The plant tissue is not removed in a conventional manner, but crushed between the jaws, and the remnants left in ridges. Adults sometimes feed on the rind of the fruit. Squash beetle feeding behavior is similar to Mexican bean beetle, but these two species are unusual among the family Coccinellidae, which is normally regarded as a very beneficial group of insect predators.

Management

Insecticides. Squash beetle is relatively easy to kill with foliar insecticides, though usually this species does not warrant control.

Cultural Practices. Squash beetles are not strong fliers, so relocation of cucurbit fields away from overwintering sites is recommended. Prompt destruction of cucurbit vines in the late summer may kill many of the late-developing larvae and pupae, reducing the number of beetles that will overwinter successfully. Row covers should effectively disrupt their life cycle.

FAMILY ELATERIDAE—CLICK BEETLES AND WIREWORMS

Corn Wireworm
Melanotus communis (Gyllenhal)

Oregon Wireworm
Melanotus longulus oregonensis (LeConte) (Coleoptera: Elateridae)

Natural History

Distribution. The genus *Melanotus* is widespread in North America, but the western half of the continent has relatively few species. This is likely related to their preference for moist habitats. Corn wireworm is widespread in the eastern United States and occurs as far west as Nebraska and Texas. In Canada it is known from Ontario and Quebec. Oregon wireworm occurs in all states west of the Rocky Mountains and in British Columbia. Although these are the most common *Melanotus* spp. affecting vegetables, others including *M. depressus* (Melsheimer), *M. verberans* (LeConte), and *M. cribulosus* (LeConte) are sometimes reported to be damaging, particularly to corn. Apparently these are native species. (See color figure 126.)

Host Plants. *Melanotus* species are reported to damage vegetables such as cabbage, corn, escarole, lettuce, pepper, potato, sweet potato, and other crops such as field corn, sorghum, soybean, sugarcane, and wheat.

Natural Enemies. The natural enemies of wireworms are not well known, and generally seem to be unimportant. In Florida, *M. communis* is parasitized by the wasp *Pristocera armifera* (Say) (Hymenoptera: Bethylidae), but Hall (1982) reported that only about 4% of wireworm larvae were affected.

Life Cycle and Description. The biology of *Melanotus* is poorly studied relative to other genera of wireworms, which is surprising considering the frequency of association with corn. In Iowa, they are reported to have a five-year life cycle, with egg deposition occurring in June, egg hatching in July, and larvae requiring four additional summers to complete their development. In southern Florida, however, the life cycle apparently is reduced to 2–3 years (D. Seal, personal communication). Pupation occurs in the autumn. Although Fenton (1926) suggested that adults emerge from the soil to overwinter under the bark of trees, Hyslop (1915) documented overwintering by adults in the soil in pupal cells. Adults reportedly feed on pollen (Fenton, 1926).

Studies in Florida showed that most flight activity occurred in May and June. Adults fly in the evening hours. Females contain 50–100 eggs within their ovaries, indicating a fairly high fecundity (Cherry and Hall, 1986). A strong association of larvae with cool and moist soil is evident, though warmer soil may be tolerated if it is moist (Shepard, 1973b). Adults are reddish brown to dark brown, and measure 10–13 mm long. Larval morphology is fairly typical of wireworms. They are yellow to yellow-brown, shiny, and elongate. The head, thoracic plate, and anal plate are darker. The most distinctive feature about *Melanotus* larvae is the lack of a notch at the tip of the abdomen in combination with a flattened abdominal tip. A key to distinguish *Melanotus* from the other common vegetable-attacking genera of wireworms can be found in Appendix A.

Keys for the identification of the genera of adult Elateridae can be found in Arnett (1968), and of larvae in Wilkinson (1963) and Becker and Dogger (1991). Keys for the identification of adult *Melanotus* were provided by Quate and Thompson (1967). A key for *Melanotus* larvae of mid-western corn fields was developed by Riley and Keaster (1979). The ecology and management of wireworms was reviewed by Thomas (1940).

Damage

The larvae of *Melanotus* feed below-ground on seeds, roots, tubers, and other plant tissue. They will kill young plants and deface the surface of potato

Corn wireworm larva.

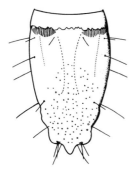

Terminal abdominal segment (*dorsal view*) of corn wireworm larva.

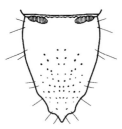

Terminal abdominal segment (*dorsal view*) of Oregon wireworm larva.

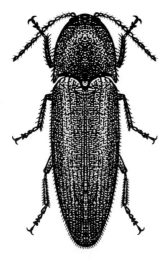

Adult corn wireworm.

and sweet potato tubers. *Melanotus* spp. prefer high moisture, and the heavier, wet portions of fields are most likely to experience damage. Damage is exacerbated by dry soil conditions, however. Larvae tend to move toward the soil surface in the spring as the soil warms, where they feed until the soil attains about 21°C; they then move deeper in the soil where it is cooler. Larvae again return to the upper layer of the soil in the autumn when the soil surface is cool, and feed until it becomes cold, then they return to a position deep in the soil until the soil warms in the spring (Fisher *et al.*, 1975).

Management

Sampling. Considerable effort has been made to develop baits for wireworm sampling. Growers are advised to assess the population densities of wireworm larvae before planting in the spring. Baiting is accomplished by burying whole wheat, corn, sorghum, or other attractive food sources in the soil at a depth of 10–15 cm and counting the number of wireworm larvae attracted (Apablaza *et al.*, 1977, Jansson and Lecrone, 1989). As few as 12 bait stations per hectare are often recommended. Soaking the bait for one day increases its attractiveness. Baiting does not work well if the soil is less than 8°C, so a sheet of transparent plastic is sometimes placed over the section of soil-containing bait; this warms the soil and increases the mobility of wireworm larvae, allowing more accurate counts (Ward and Keaster, 1977; Kirfman *et al.*, 1986). Larvae can also be separated from the soil by sifting; a screen with 6–7 meshes per centimeter is desirable. Soil sampling followed by screening is an accurate, but labor intensive, means to predict wireworm problems before planting. Spatial distribution of larvae is variable, but sometimes aggregated, requiring extensive sampling. A mobile, self-propelled sampling device has been developed (Smith *et al.*, 1981). Simmons *et al.* (1998) compared several methods of sampling and reported that corn/wheat baits were the most accurate, precise, and cost effective.

The adult populations can also be sampled. Pheromones of the tufted-apple budmoth, *Platynota idaeusalis* Walker (Lepidoptera: Tortricidae), are attractive to some *Melanotus* spp. adults (Brown and Keaster, 1983, Keaster *et al.*, 1987). This suggests that a sex pheromone of *Melanotus* spp. could be identified and synthesized for population monitoring, as has been done for Pacific Coast wireworm, *Limonius canus* LeConte (Butler *et al.*, 1975). Pitfall traps and sticky traps are useful for sampling adult populations on the soil surface and in flight, respectively (Brown and Keaster, 1986). Genung (1972) noted that click beetle catches on sticky traps and in light traps were not completely equivalent, and that both types of trap should be used in population studies. In southern Florida, blacklight traps are sometimes used to predict wireworm population densities during the next growing season.

Insecticides. Insecticide is often applied to prevent susceptible crops from being injured by wireworms. The most common approach to wireworm suppression is to apply persistent liquid or granular formulations at planting, and to incorporate the insecticide into the soil near the seeds or seedlings. Preplant applications generally, but not always, are more effective than postplant treatments (Linduska, 1979; Chalfant *et al.*, 1992b). Sometimes insecticides are applied broadcast over the entire field, but more commonly they are applied in bands over or along the rows. For long-duration crops, the planting-time application may be followed by a mid-season application. Insecticide also may be incorporated into the irrigation water (Chalfant *et al.*, 1992a). Effectiveness varies considerably among insecticides (Day and Crosby, 1972), and some insecticide resistance has been reported. Soils with high organic matter content sometimes interfere with insecticide efficacy (Campbell *et al.*, 1971).

Cultural Practices. Cultural practices are the most important approach to population management. Crop rotation and fallowing are particularly effective. Grasses are particularly attractive to *Melanotus* wireworms, and a survey of wireworm problems in Missouri corn fields established a relationship between wireworms and previous growth of sod (Belcher, 1989). The type and temporal occurrence of cover crops can affect subsequent damage by wireworms. Grass cover crops such as sorghum-sudan are highly attractive to *Melanotus* spp., and it is important to have the fields free of such grasses during peak flights (Jansson and Lecrone, 1991). *Melanotus* spp., unlike some other wireworms, will inhabit forested areas, and crops planted on recently cleared land may be damaged (Cheshire and Riley, 1988). Where crops can be flooded, six weeks or two four-week periods separated by two weeks of drying can eliminate *Melanotus* wireworms (Genung 1970).

Great Basin Wireworm
Ctenicera pruinina (Horn)

Puget Sound Wireworm
Ctenicera aeripennis aeripennis (Kirby)

Prairie Grain Wireworm
Ctenicera aeripennis destructor (Brown)

Dryland Wireworm
Ctenicera glauca (Germar)
(Coleoptera: Elateridae)

Natural History

Distribution. These native species are cool weather wireworms, and principally western in distribution. They cause damage in the United States from Oregon and Washington east to the north-central states. In Canada, *C. aeripennis* is destructive in British Columbia, the Prairie Provinces, and north into the Northwest and Yukon Territories. *C. glauca* is damaging in British Columbia and the Prairie Provinces, but *C. pruinina* is not known from Canada. Unlike some other wireworms, *Ctenicera* spp. tolerate dry soil and are damaging to dryland crops.

Host Plants. These wireworms are best known as wheat pests, the principal crop in the arid regions where they are commonly found. However, they have a wide host range. The larva, or wireworm stage, attacks many vegetables, including bean, carrot, corn, lettuce, lima bean, onion, potato, tomato, and probably many others. Additional crops injured, in addition to wheat, are barley, canola, flax, oats, rye, sugarbeet, and sunflower. *Ctenicera* spp. also feed on other insects, such as fly larvae, if provided with the opportunity (Zacharuk, 1963).

Natural Enemies. As is the case with other wireworms, *Ctenicera* spp. are relatively free of important natural enemies. Birds often feed on the adults in the spring when they emerge from the soil. The horned lark, *Otocoris alpestris alpestris*, is a species that is especially effective at gleaning adults. When the soil is tilled and larvae are exposed, a number of birds will take advantage of this temporary food source, but probably only a small proportion of larvae are ever exposed long enough to be captured by avian predators. Predatory insects such as ground beetles (Coleoptera: Carabidae) and stiletto flies (Diptera: Therevidae) are associated with Great Basin wireworm. Mortality due to the fungi *Beauveria* and *Metarhizium*, and unspecified bacteria, have been noted during rearing studies, but these pathogens seem of little significance in nature (Zacharuk, 1962a).

Life Cycle and Description. These are long-lived species, normally requiring a minimum of three years for development, and as much as 10 years under unfavorable conditions. Adults, which are also called click beetles, deposit eggs in the spring. Larvae complete their development after a minimum of three summers, and this is followed by pupation in the autumn, and overwintering by the adult. Following are brief descriptions of two of the more important species: Great Basin wireworm and prairie grain wireworm.

The eggs of Great Basin wireworm are white, and elliptical, measuring only slightly longer than wide. They are covered with viscous, sticky secretion that causes soil particles to adhere. The female normally deposits about 350 eggs over a 21-day period, often at a depth of 10–15 cm. She apparently seeks moist soil for oviposition. Eggs must absorb water from the soil in order to complete embryonic development, and they absorb water more rapidly at higher temperatures (Doane, 1969a). Duration of the egg stage is about 30 days, with hatching occurring in June. Young larvae are white, but soon turn yellow. The head, thorax, and anal plate are yellowish brown. When larvae hatch they measure about 1.5 mm long, but eventually attain a length of about 18–20 mm. As is the case with most wireworms, they are shiny in appearance and the thoracic legs are not pronounced. Larvae feed on rootlets and roots, usually molting twice per year with a range of 1–5 molts per season. Pupation occurs in the soil at a depth of 15–25 cm, in a small cell formed by the larva. The pupal stage is initially white but turns yellow with time. In form, the pupa greatly resembles the adult, but the elytra are not completely developed and are twisted ventrally. Duration of the pupal stage is only about 10 days, but beetles remain in the cell until the following spring when they dig to the surface. The adult is brown in color. It measures about 13–15 mm long. The elytra are marked with numerous parallel furrows. The beetles become active in the spring, usually May, when the soil attains a temperature of at least 13°C and air temperature reaches 18–20°C. Adults are relatively short lived, usually perishing after 3–4 weeks. The adults seem not to feed, or if they do, they feed sparingly. The adults of dryland wireworm are quite similar to Great Basin wireworm in appearance except that they average smaller in size, about 7–10 mm.

Prairie grain wireworm and Puget Sound wireworm are similar in morphology and ecology, and recently they have been treated as a single species. However, the work of Glen *et al.* (1943) suggested that they are different species. Eggs are white and oval, measuring about 0.48 mm long and 0.34 mm wide. The color is white, and eggs are coated with a sticky, viscous material that causes oil particles to adhere. Duration of embryonic development is 11.5 days at 27°C and 20 days at 21°C. Larvae are elongate and yellowish, as is typical of wireworms. Young larvae seem to feed principally on root hairs and fungal mycelia; it is only after the first two instars that larvae begin to feed on seeds. Apparently these larvae do not descend deeply into the soil during the winter months, but remain near the surface. They become active as the soil warms above 7°C. The number of larval instars varies from 9–11 and requires from one to over two years under laboratory conditions. Under field conditions, larval development time is thought to average three years. Larvae of *C. a. aeripennis* attain a length of 28 mm at maturity, but *C. a. destructor* rarely exceeds 22 mm (Wilkinson, 1963). As larvae attain maturity they form a pupal cell in the soil, usually at a depth of 5–10 cm. The pupae resemble the adult in form, although the elytra are not fully developed. Duration of the pupal stage is 2–3 weeks at 20°C. Adults remain in the pupal chambers until spring, when they dig

to the soil surface. They typically emerge in April or May. The adults have a black pronotum and bronze or greenish elytra. They measure about 14 mm long. Females mate soon after emergence. About six days after mating the female commences oviposition. She usually deposits from 400–1400 eggs over a period of about 20 days. Females that have access to food live longer and tend to produce more eggs. Doane (1963) noted that this was an unusually high rate of fecundity for a wireworm.

The biology of Great Basin wireworm was given by Lane (1931), prairie grain wireworm by Zacharuk (1962a,b), and dryland wireworm by Hyslop (1915). Keys for the identification of the genera of adult Elateridae can be found in Arnett (1968), and of larvae in Becker and Dogger (1991). Wilkinson (1963) provided an excellent key to distinguish among *Ctenicera* spp., and other wireworms found in British Columbia, and includes information on biology and economic impact. Thomas (1940) provided a review of wireworm biology and management, and King (1928a) and Glen *et al.* (1943) gave useful information on wireworm ecology and damage in Canada. A key to distinguish *Ctenicera* from the other common vegetable-attacking genera of wireworms can be found in Appendix A.

Dryland wireworm larva.

Adult dryland wireworm.

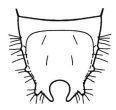

Terminal abdominal segment (*dorsal view*) of Great Basin wireworm larva.

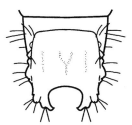

Terminal abdominal segment (*dorsal view*) of Puget Sound wireworm larva.

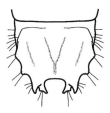

Terminal abdominal segment (*dorsal view*) of prairie grain wireworm larva.

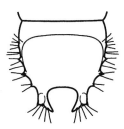

Terminal abdominal segment (*dorsal view*) of dryland wireworm larva.

Damage

The larvae feed on seeds and below-ground portions of plants, often killing several plants during the course of their development. In southern portions of their range or dry climates the damage usually occurs in the spring months, and the larvae descend deep into the soil during the summer as the soil dries and warms (Lane, 1931). In northern climates larval feeding occurs throughout the summer, with Burrage (1963) reporting maximum damage to potato tubers in Saskatchewan during June–August. Damage was positively correlated with soil temperature as long as soil moisture levels were adequate.

Management

Sampling. The statistical distribution of *Ctenicera* spp. varies with the stage sampled. Eggs and young larvae are aggregated, but the aggregation dissipates as the insects become more mature and mobile (Doane, 1977). Doane (1969b) presented a technique for separation of eggs from the soil. Population monitoring depends primarily on assessment of larval populations, and fields should be sampled before susceptible crops such as potatoes are planted. Larval populations are sampled by baiting or soil sampling.

Baiting is accomplished by burying whole wheat, corn, potato, carrot or other attractive food sources in the soil at a depth of 10–15 cm and counting the number of wireworm larvae attracted (Doane, 1981). For *Ctenicera* spp., grains seem to be more attractive than vegetable matter (Toba and Turner, 1983). Carbon dioxide emanates from such bait, and larvae follow the carbon dioxide gradient to the food source (Doane *et al.*, 1975). Soaking the bait for one day increases its attractiveness. Baiting does not work well if the soil is cold, so a sheet of black or transparent plastic is sometimes placed over the section of soil containing bait; this warms the soil and increases the mobility of wireworm larvae, allowing more accurate counts. Simmons *et al.* (1998) compared several methods of sampling and reported that corn/wheat baits were the most accurate, precise, and cost effective.

Soil sampling requires 15×15 cm soil samples, including soil to a depth of 30–40 cm, that are removed and checked for wireworms. This is normally done by sieving soil with coarse and fine screens to separate insects from the soil. Although accurate estimates of wireworm population densities are possible with soil sampling, this approach is very labor intensive; thus, baiting is more commonly used.

Insecticides. Insecticides are often applied for wireworm control when susceptible crops such as potato are planted. Insecticides may be applied broadcast before planting, but more commonly at the time of planting in the seed furrow, over the row, or broadcast. At low wireworm densities, applications limited to the rows are adequate (Toba and Turner, 1979, 1981a; Toba and Powell, 1986), but at moderate wireworm densities broadcast applications are often more effective. At high densities susceptible crops should not be planted.

Cultural Practices. *Ctenicera* spp. have a wide host range and crop rotation generally is not practical. Summer fallow sometimes can be a useful practice,

however, because if weed control is strictly practiced
during the fallow period the larvae are deprived of
food. Although the larger larvae are extremely tolerant
of starvation, sometimes surviving two years without
food, the small larvae suffer from food deprivation.
Thus, summer fallow alone is not dependable for
protection against these wireworms. Special caution
should be taken if crops follow grass sod, a favored
habitat.

Shallow cultivation is sometimes recommended
where wireworms are a problem. By cultivating shallowly, the lower layers of soil remains compact, and
the beetles are unable to burrow deeply as they prepare to oviposit. Shallow oviposition results in desiccation of the eggs.

The method of seeding also influences wireworm
damage potential. Shallow seeding is desirable if adequate moisture is available, because the warm soil
speeds germination and growth, providing the plant
the opportunity to outgrow insect feeding damage. It
also is advantageous to avoid early or late planting
because cool or dry soil inhibits plant growth and
increases the likelihood of significant injury.

Southern Potato Wireworm
Conoderus falli (Lane)

Tobacco Wireworm
Conoderus vespertinus (Fabricius)

Gulf Wireworm
Conoderus amplicollis (Gyllenhal)
(Coleoptera: Elateridae)

Natural History

Distribution. Several species of *Conoderus* damage
vegetables, especially in southeastern states and in California. Their origin is uncertain, but southern potato
wireworm and Gulf wireworm also occur in South
America and Asia, so they likely were introduced.
Southern potato wireworm and Gulf wireworm were
first detected damaging crops in Georgia and Alabama,
respectively, in 1927, and by 1955 they had spread along
the South Atlantic and Gulf Coast from North Carolina
to eastern Texas. Southern potato wireworm was first
detected in California in 1963, and Gulf wireworm in
1938. Tobacco wireworm is known from many eastern
states and provinces, but in the south it occurs as far
west as Arizona. Among other *Conoderus* spp., *C. exsul*
(Sharp) is sometimes damaging in California and other
southwestern states, and *C. rudis* (Brown) and *C. scissus*
(Schaeffer) are occasional pests in the southeast (Seal *et
al.*, 1992c; Seal and Chalfant, 1994).

Host Plants. Vegetable crops reported to be
damaged under field conditions include the below-
ground portions of cantaloupe, beet, cabbage, carrot,
celery, corn, cowpea, mustard, potato, sweet potato,
tomato, and turnip. Sweet potato is the most severely
injured crop. Other crops injured are gladiolus, peanut, strawberry, and tobacco. In studies conducted in
South Carolina, grain seeds such as oat, sorghum, and
wheat readily supported larval growth, and seeds
from non-grasses such as cowpea, lima bean, and
Amaranthus supported fair survival and growth.
Potato, sweet potato, and carrot also supported moderate survival and growth (Day *et al.*, 1971).

Natural Enemies. Day *et al.* (1971) reported that
the fungus *Metarhizium anisopliae* was commonly
found in dying southern potato wireworm larvae,
and that an unspecified nematode was a common
parasitoid. Also, birds frequently follow tractors as
fields are plowed, and feed upon exposed wireworms.
There is no data, however, to suggest that any of these
mortality factors are significant. Larvae of southern
potato wireworm, and undoubtedly other species,
are susceptible to infection by *Steinernema carpocapsae*
(Nematoda: Steinernematidae) under experimental
conditions (Poinar, 1979).

Life Cycle and Description. Generally there is
one generation of southern potato, tobacco, and Gulf
wireworm annually in the southeastern states. Most
insects overwinter as partly grown larvae and pupate
in March–May. This results in adults that oviposit
throughout the summer and produce overwintering
larvae. A small proportion of the southern wireworm
larvae complete their development in the summer and
overwinter as adults, and the adults also may produce
eggs early enough that larvae do not enter diapause
but go on to complete a second generation. In southern
Florida, a third generation is possible. The following is
a description of southern potato wireworm, but the
other *Conoderus* spp. are similar.

Egg. The eggs are spherical in shape, smooth
in texture, and white in color. They measure 0.3–
0.4 mm in diameter. When first deposited the eggs
are coated with a sticky secretion, causing soil particles
to adhere as the material dries. Eggs are deposited
near the surface, usually in the upper one centimeter.
Day *et al.* (1971) reported that beetles deposit an average of about 36 eggs, with a range of 22–63, but this
level of fecundity seems low. Mean duration of the
egg stage is about 8–20 days at 23–25°C and 7–9 days
at 27–30°C.

Larva. The larvae, also called wireworms, initially
are whitish, but as they mature they become cream or

yellowish, with the head, pronotum, and terminal abdominal segment becoming reddish orange. The number of instars is thought to be 3–4. Mature larvae measure about 17 mm long and 2 mm wide, and are smooth, shiny, and cylindrical. The caudal plate at the tip of the abdomen bears a small notch. This plate is also marked by a pair of setae near the notch, and another pair near the anterior margin of the plate. The characteristics of the caudal plate are useful diagnostic characters. Larvae are found in various grassy and weedy fields but not in heavily wooded areas. They also are infrequent in fields that previously lacked grass and broadleaf weed cover. In addition to providing larval food, plants prevent soil temperatures from becoming lethally high, and provide hiding places for adults. Larvae are principally located near the soil surface; about three-fourths are found in the upper 5 cm, and most of the rest in the 5–10 cm range. Development time of larvae in spring and early summer is typically 40–70 days, with a range of about 30–90 days. However, most larvae developing in August, and all larvae developing in September and October, overwinter in the larval stage and do not pupate until the following spring. Thus, larval development time in the overwintering larvae is 200 to over 300 days.

The shape of the notch in the caudal plate of the larvae can be used to distinguish among the common species. In *C. falli*, the notch is closed or nearly so, whereas in *C. amplicollis* and *C. vespertinus* the notch is open and V-shaped. In *C. amplicollis* the notch is relatively shallow, the depth being equal to approximately one-third the distance from the notch to the pair of setae on the dorsal plate. In contrast, the notch in *C. vespertinus* is relatively deep, being approximately equal to the distance from the notch to the pair of setae on the dorsal plate. Rabb (1963) gave a key to several *Conoderus* larvae.

Pupa.   When larvae have completed their development they twist and roll in the soil to form a pupal cell. The cell is usually within the upper 10 cm of soil. Pupae are white to yellowish, with the head and thorax becoming tan just before the emergence of the adult. In form, the pupae greatly resemble the adults except that the elytra are reduced in size and twisted ventrally. The pupae tend to be slightly larger than the adults. Duration of the pupal stage is about six days at 29°C and 18 days at 19°C.

Adult.   The adults, called click beetles, are elongate oval, and measure 6–9 mm long and 2–3 mm wide. They are uniformly dark, ranging from light to dark-brown, except that the legs are tan. The elytra are marked with numerous parallel channels, and are moderately shiny. The adults are nocturnal, and are found hiding under organic material during the daylight. The pre-oviposition period of adults is reported to be 12–30 days, and are capable of oviposition from March through September; beetles lay few eggs in the autumn or winter.

The adults of tobacco wireworm measure 7–10 mm long. The pronotum and elytra are yellowish brown, with a pair of longitudinal dark bands on the pronotum and an irregular dark band dorsally on the elytra. Adults of Gulf wireworm measure 8–9 mm long. They are dark brown with yellowish legs and antennae.

The biology of southern potato wireworm was given by Day *et al.* (1971), tobacco wireworm by Rabb (1963), and Gulf wireworm by Cockerham and Deen (1936) and Stone and Wilcox (1979). The ecology and management of wireworms were reviewed by Thomas (1940) and Glen *et al.* (1943). Chalfant and Seal (1991) provided an excellent review of wireworm problems on sweet potato. A key to distinguish *Conoderus* from the other common vegetable-attacking genera of wireworms can be found in Appendix A. Keys for the identification of the genera of adult Elateridae can be found in Arnett (1968), and of larvae in Becker and Dogger (1991).

Tobacco wireworm larva.

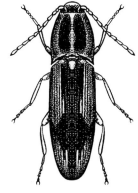

Adult tobacco wireworm.

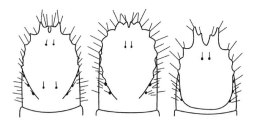

Terminal abdominal segments (*dorsal view*) of some *Conoderus* wireworm larvae: (*left to right*) southern potato, Gulf, tobacco.

Damage

The larvae, called wireworms, feed principally on seeds, roots, stems, and tubers of several plants. They seem to prefer plant parts that are high in starch content, and sometimes attack fruit where it comes into contact with soil. Larvae often scar the surface of potato and carrot, which causes slight loss in quality, but may also tunnel into these vegetables, causing serious loss. However, they can be cannibalistic and predatory on other soil-dwelling species such as rootworms (Coleoptera: Chrysomelidae). The adult stage of wireworms causes little injury.

Wireworms sometimes cause considerable damage to irrigation systems. Polyethylene and some other plastic materials may have holes completely eaten through the tubing, causing loss of pressure and uneven watering. Thick-walled tubing and harder, high-density tubing are more resistant to damage (Ota, 1973).

Management

Sampling. The adults are readily attracted to blacklight traps (Day and Reid, 1969; Stone and Wilcox, 1983). This attraction is useful for population assessment, but it is not adequate for effective suppression (Day *et al.*, 1973). Adults also can be captured on sticky traps, but the population trends ascertained by light and sticky traps are not completely equivalent (Genung, 1972). Larvae can be separated from the soil by sifting; a screen with 6–7 meshes per centimeter is desirable. Soil sampling followed by screening is accurate but labor intensive means to predict wireworm problems before planting. Spatial distribution of larvae is variable, but sometimes aggregated, requiring extensive sampling (Seal *et al.*, 1992c).

Considerable work has been done to develop baits for wireworm sampling. Growers can assess the population densities of wireworm larvae before planting in the spring (Jansson and Lecrone, 1989; Seal *et al.*, 1992a). Baiting is accomplished by burying whole wheat, corn, potato, sweet potato, carrot, or other

attractive food sources in the soil at a depth of 10–15 cm and counting the number of wireworm larvae attracted. Soaking the bait for one day increases its attractiveness. Baiting does not work well if the soil is cold or dry; when soil is cold a sheet of transparent plastic is sometimes placed over the section of soil-containing bait as this warms the soil and increases the mobility of wireworm larvae, allowing more accurate counts (Bynum and Archer, 1987). Wireworm counts of one or more per bait station suggest that wireworms perhaps will be a problem and insecticides may be used. Simmons *et al.* (1998) compared several methods of sampling and reported that corn/wheat baits were the most accurate, precise, and cost effective.

Insecticides. Insecticide is often applied to prevent susceptible crops from being injured by wireworms, principally because unless preplant sampling is done carefully, considerable damage may occur before the infestation is detected. The most common approach to wireworm suppression is to apply persistent liquid or granular formulations at planting, and to incorporate the insecticide into the soil near the seeds or seedlings. Preplant applications generally, but not always, are more effective than postplant treatments (Linduska, 1979; Chalfant *et al.*, 1992b). Sometimes insecticides are applied broadcast over the entire field, but more commonly they are applied in bands over or along the rows. For long-duration crops, the planting-time application may be followed by a mid-season application. Insecticide also may be incorporated into the irrigation water (Chalfant *et al.*, 1992a). Effectiveness varies considerably among insecticides (Day and Crosby, 1972), and some insecticide resistance has been reported.

Cultural Practices. The type and temporal occurrence of weeds and cover crops can affect subsequent damage by wireworms. Grass cover crops such as sorghum-sudan are highly attractive to *Conoderus* spp., and it is important to have the fields free from such grasses during peak flights (Jansson and Lecrone, 1991). Similarly, the rotational sequence of vegetable, or vegetable and field crops, affects wireworm abundance and species composition (Seal *et al.*, 1992b). Native grass sod is often heavily infested with wireworm larvae, and damage to vegetables might be high in years immediately following conversion of sod to crop land. On the other hand, legumes are generally poor hosts for wireworms, and little damage follows these crops.

Field flooding causes destruction of wireworms in Florida (Genung, 1970). A six-week period of flooding is needed; alternatively, two four-week periods separated by a two-week period of drying are effective.

Host-Plant Resistance. Resistance to wireworms in commercially available crop cultivars varies among crops. Cuthbert and Davis (1970) reported high levels of resistance among sweet potato cultivars, and Wiseman *et al.* (1976) demonstrated some resistance among field and sweet corn varieties. However, most crops remain susceptible to injury. Interesting work conducted in South Carolina by Schalk *et al.* (1992) demonstrated that interplanting a resistant variety of sweet potato within a susceptible variety would reduce injury by wireworms and other insects to the susceptible cultivar.

Sugarbeet Wireworm
Limonius californicus (Mannerheim)

Eastern Field Wireworm
Limonius agonus (Say)

Pacific Coast Wireworm
Limonius canus LeConte
(Coleoptera: Elateridae)

Natural History

Distribution. The *Limonius* spp. are principally western in distribution, occurring from the Rocky Mountains to the Pacific Ocean, and from British Columbia to southern California. In addition to sugarbeet wireworm and Pacific Coast wireworm, other western species that sometimes damage crops are western field wireworm, *L. infuscatus* Motschulsky; and Columbia Basin wireworm, *L. subauratus* LeConte. In the eastern United States the eastern field wireworm, *L. agonus* Say, is an occasional pest. Apparently these species are native to North America.

Host Plants. The larvae of these beetles, called wireworms, damage cantaloupe, cauliflower, corn, lettuce, lima bean, mustard, potato, tomato and other crops with tubers or a substantial root system. They are most frequently noted as pests of potato and to a lesser degree of sugarbeet and alfalfa. Weeds are suitable for larval development, and they have been observed to feed on dock, *Rumex* sp.; Johnsongrass, *Sorghum halepense*; nettle, *Urtica* sp.; pigweed, *Amaranthus retroflexus*; wild beet, *Beta* sp.; and others. *Limonius* spp. larvae are active, and aggressive toward each other, sometime displaying cannibalistic tendencies. Adults, which are called click beetles, have been observed to feed on a variety of foods, including beet roots, foliage of several plants, and pollen.

Natural Enemies. Natural enemies of wireworms are few, and often relatively ineffective. Birds are often observed to follow tractors as soil is tilled, and to consume large numbers of wireworm larvae. However, they glean only a small proportion of the larvae. Insectivorous birds also prey on adults, as do predatory ground beetles (Coleoptera: Carabidae). Perhaps the most important natural enemy is the stiletto fly *Psilocephala frontalis* Cole (Diptera: Therevidae). The larvae of this predatory species, which resemble wireworms and are sometimes mistaken for beetles, occur in the soil. They puncture the integument of wireworm larvae and feed upon them.

Life Cycle and Description. Among the *Limonius* spp. the biology of sugarbeet wireworm is well-studied. The following description applies to sugarbeet wireworm except where noted. However, the biologies of the several *Limonius* spp. seem to be similar.

Duration of the sugarbeet wireworm life cycle is highly variable, ranging from 1–5 years. In California, 70% of the insects complete their development in two years, but further north 3–4 years may be required for development. Adults emerge in the spring, lay eggs, and larvae live for various lengths of time before pupating. Pupation occurs in the autumn, with adults overwintering in the soil prior to emergence.

Egg. The egg is elongate-elliptical in shape, with both ends similar in form. The mean egg length is 0.69 mm (range 0.63–0.73 mm), and mean width is 0.50 mm (range 0.47–0.53 mm). Eggs are white and the surface appears to be smooth, but close examination shows that the surface is textured. Eggs tend to be deposited near the soil surface, but their location is affected by moisture conditions. In moist soil, most eggs are deposited within the upper 5 cm, but in drier soil the eggs are deposited more deeply. The female normally produces 200–350 eggs. The presence of vegetation does not seem to affect oviposition behavior by females. Duration of the egg stage is about 27–30 days at 21°C.

Larva. The larva is elongate-cylindrical, with the mature stage measuring 18–21 mm long and 2.5–3.0 mm wide. It is smooth, shiny and yellowish brown. The head and anal segments are flattened dorsally, and brownish. The terminal abdominal segment bears a circular notch that is nearly closed distally. The antennae and thoracic legs are short. Duration of the larval stage varies considerably—a mean of 171, 502, 857, 1233, and 1589 days for insects with a 1–5 year life cycle, respectively. The number of instars ranges from 10–13.

Pupa. Pupation occurs in the soil, in a small chamber formed by the larva. Larvae normally pupate at a depth of 15–30 cm. The pupal stage initially is white,

Sugarbeet wireworm larva.

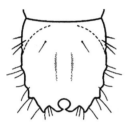

Terminal abdominal segment (*dorsal view*) of sugarbeet wireworm larva.

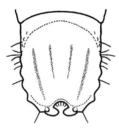

Terminal abdominal segment (*dorsal view*) of eastern field wireworm larva.

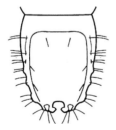

Terminal abdominal segment (*dorsal view*) of Pacific Coast wireworm larva.

but soon turns yellowish. In shape, the pupa greatly resembles the adult except that the abdomen is slightly more elongate than in the adult, and the elytra are reduced in size and twisted ventrally. Pupae measure about 11.5 mm long and 3.6 mm wide. Duration of the pupal stage is 7–9 days.

Adult. The adult is 8.5–12.0 mm long. It generally is dark brown to black, but sometimes it is orangish brown or reddish brown. The elytra are marked with numerous parallel channels. Females are more likely to have four and five year life cycles than males; males are more likely to have a one year life cycle than females. Adults mate immediately upon emerging from the soil. After a preoviposition period of about six days, females commence egg production. Duration of the oviposition period is about 25–35 days. Their length of life is given at only 30 days for males and 45 days for females; this seems short and may be an artifact of caged conditions.

Detailed biology of sugarbeet wireworm was provided by Graf (1914) and Stone (1941). A key for important *Limonius* spp. was published by Lanchester (1946). A key to distinguish *Limonius* from the other common vegetable-attacking genera of wireworms can be found in Appendix A. Keys for the identification of the genera of adult Elateridae can be found in Arnett (1968), and of larvae in Wilkinson (1963) and Becker and Dogger (1991). Wilkinson (1963) provided excellent keys to distinguish among the western wireworm species, as well as their biology and history of damage in British Columbia. Good reviews of wireworm biology and damage can be found in Thomas (1940) and Glen *et al.* (1943).

Damage

These wireworms are damaging in irrigated crops. They normally are not abundant in dryland production or in grasslands. Damage to seed or to germinating seedlings normally occurs soon after planting. Such plant material is easily destroyed by the feeding of a single wireworm larva, and results in reduction in stand. Early symptoms of infestation are poor germination, and wilting or death of seedlings. In older plants considerable root injury can occur before signs

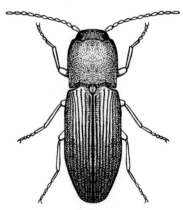

Adult sugarbeet wireworm.

of wireworm feeding are evident. In crops such as potato, however, where the appearance of the tuber is important, even superficial damage can be costly. Wireworms often tunnel deeply into potato tubers, which provides entry for soil-borne plant pathogens. Wireworms are the most serious soil pests of potato in the Pacific Northwest (Toba, 1987). Onsager (1975) noted that Pacific Coast wireworm was more damaging to potato than to southern potato wireworm, *Conoderus falli* Lane; any tubers fed upon by Pacific Coast wireworm qualify for rejection according to current standards. Small tubers are subject to more injury than are large tubers (Toba and Turner, 1981b).

Management

Sampling. Sex pheromones for Pacific Coast wireworm have been tentatively identified and may prove useful for population monitoring (Butler *et al.*, 1975). Presently, population monitoring depends on larval assessment, and fields should be sampled before susceptible crops such as potatoes are planted. Larval populations tend to be aggregated (Williams *et al.*, 1992). Larval populations are sampled by baiting or soil sampling.

Baiting is accomplished by burying whole wheat, corn, potato, carrot, or other attractive food sources in the soil at a depth of 10–15 cm and counting the number of wireworm larvae attracted. Soaking the bait for one day increases its attractiveness. Baiting does not work well if the soil is under 8°C, so a sheet of transparent plastic is sometimes placed over the section of soil-containing bait; this warms the soil and increases the mobility of wireworm larvae, allowing more accurate counts. Wireworm counts of one or more per bait station suggest that wireworms perhaps will be a problem and insecticides may be needed (Bynum and Archer, 1987). At larval densities of more than four per bait station it is not advisable to plant potatoes even if insecticides are to be used. Simmons *et al.* (1998) compared several methods of sampling and reported that corn/wheat baits were the most accurate, precise, and cost effective.

For soil sampling, 15 × 15 cm soil samples, including soil to a depth of 30–40 cm, are removed and checked for wireworms. This is normally done by sieving with coarse and fine screen to separate insects from the soil. If wireworm counts exceed 2 in 40 samples, insecticide treatment is warranted.

Insecticides. Insecticides in granular or liquid form are often applied broadcast before planting, and incorported into the soil to kill wireworms. Also, fumigation of soil is practiced in many areas with a history of wireworm problems. Alternatively, insecti-

cides can be applied with the seed at planting, or adjacent to the plants after planting. Wireworms are fairly mobile, with sugarbeet wireworms often moving 45 cm in a two-week period (Toba, 1985b); this can make band treatments risky. Broadcast applications are most desirable when wireworm densities are high, but in-row applications are adequate at low to moderate densities (Wilkinson *et al.*, 1977; Toba and Turner, 1979; 1981a).

Plant-produced toxicants such as isothiocyanates, which apparently function as biofumigant allelochemicals, have had limited testing for suppression of such insects as sugarbeet wireworm. Incorporation of rapeseed seed meal into soil demonstrated the potential to kill wireworm larvae, but available varieties of rape are too low in isothiocyanates to provide practical suppression (Elberson *et al.*, 1996).

Cultural Practices. Crop rotation can be beneficial in alleviating wireworm problems. Some crops, particularly alfalfa, tend to dry the soil, which is deleterious to wireworms. Fallowing of cropland can also disrupt wireworm populations by affecting soil moisture and by depriving larvae of food. Thus, it is beneficial to alternate such practices with production of wireworm-susceptible crops.

Biological Control. Toba *et al.* (1983) evaluated the effectiveness of *Steinernema* spp. (Nematoda: Steinernematidae) for suppression of sugarbeet wireworm in Washington. Application of nematodes to the soil surface at high rates demonstrated some infection, but high levels of inoculum are needed to provide good suppression of wireworms.

Wheat Wireworm
Agriotes mancus (Say)
(Coleoptera: Elateridae)

Natural History

Distribution. Wheat wireworm is northern in distribution, but occurs primarily in eastern North America. This native wireworm is considered to be a pest from the Atlantic Ocean to the Great Plains, and damaging as far west as eastern Saskatchewan. It is considered to be the most serious wireworm pest in Canada's Atlantic Provinces, and is well known in the midwestern states. It is not the only *Agriotes* spp. that is known to damage crops, but it is the most serious. Among other *Agriotes* spp. that sometimes damaging are dusky wireworm, *A. obscurus* (Linnaeus) and lined click beetle, *A. lineatus* (Linnaeus), both imports from Europe to the east and west coast of Canada; western wireworm, *A. sparsus* LeConte, known to damaging in the northwest; and *A. sputator* (Linnaeus), another European import that inhabits eastern Canada.

Host Plants. This species is known to feed on such vegetables as bean, cabbage, carrot, corn, cucumber, potato, tomato, turnip, and probably others. As its common name suggests, *A. mancus* is a pest of wheat, and it also damages such field crops as oats, soybean, and tobacco. It favors many grasses, and as is the case with many wireworms, crops are often damaged when planted after grass sod.

Natural Enemies. There is little information on the natural enemies of wheat wireworm. Avian predators are sometimes observed feeding on larvae exposed by tillage, but birds probably affect a small proportion of the population. Fungi are commonly associated with dying or dead wireworms, but their role in wireworm ecology is poorly known.

Life Cycle and Description. Wheat wireworm has a 3–4 year life cycle. Adults are active early in the spring, mate, and oviposit in April or May. Larvae feed through the remainder of the first year, and the entire second year, overwintering as partly grown larvae. During the third season they complete larval growth and pupate in July, forming adults in August–October. The adults remain in the soil, in their pupal cells, until the following spring. In New York, a portion of the population has an extended larval development time, producing a four-year cycle (Rawlins, 1940). Under the cooler conditons of Quebec, apparently all insects have a four-year life cycle (Lafrance, 1967).

Egg. Females are reported to produce a mean of about 100 eggs (range 21–194). Eggs of wheat wireworm are white and shiny, and deposited just beneath the surface of the soil, usually in the upper 1.5 cm, in cracks and crevices, and under stones and surface debris. They may be deposited singly or in small clusters. The eggs are spherical to oval and measure about 0.5 mm long and 0.45 mm wide. Eggs are covered with a sticky secretion that causes soil particles to adhere. Duration of the egg stage is about 20–25 days.

Larva. The larvae are pale-yellow, and elongate and cylindrical in form. Initially they measure only about 1.8 mm long, but attain a length of about 3–6, 6–20, and 20–25 mm during the first, second, and third year of growth. The terminal abdominal segment is distinctive in that it is cone-like or slightly bulbous, without a terminal notch or lateral teeth. Dorsally, however, the terminal segment bears an eye-like invagination near each side. Duration of the larval stage is thought to require nearly three years; however, a small proportion of larvae may complete their development in only two years, whereas others require four years. Larvae must have living plant material in order to develop; humus alone is inadequate. Large larvae,

Wheat wireworm larva.

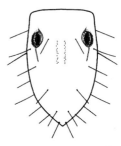

Terminal abdominal segment (*dorsal view*) of wheat wireworm larva.

but not young, are quite resistant to starvation, and survive 1–2 years without food.

Pupa. The pupa is similar to the adult in form, but with poorly developed elytra. The pupa is shiny white, and measures about 10 mm long and 3 mm wide. Pupation occurs in the soil at a depth of 10–20 cm. Duration of the pupal stage is 17–20 days.

Adult. The adult, known as a click beetle, is small, measuring only about 8 mm long and 3 mm wide. It is chestnut brown in color and covered with a short layer of yellowish hair. The legs and antennae also are yellowish. Duration of the adult stage is normally about 30 days. The peak of adult activity is June, but some adults can be found throughout the summer. Adults are susceptible to desiccation and are usually found in close proximity to the soil or in other moist locations. They are not often observed feeding, but readily ingest syrup, molasses, and other sweet substances.

A detailed description of the morphology and biology of wheat wireworm was provided by Hawkins (1936), and good life-history information was given by Rawlins (1940). A key to distinguish *Agriotes* from the other common vegetable-attacking genera of wireworms can be found in Appendix A. Keys for

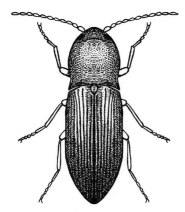

Adult wheat wireworm.

the identification of the genera of adult Elateridae can be found in Arnett (1968), and of larvae in Wilkinson (1963) and Becker and Dogger (1991). Although Wilkinson (1963) did not include *A. mancus* in his key to wireworms of British Columbia, it is very useful for identification of the other western *Agriotis* spp. Becker (1956) also provided a major review of the genus *Agriotes*. Thomas (1940) provides a review of wireworm biology and management, and King (1928a) and Glen *et al.* (1943) gave useful information on wireworm ecology and damage in Canada.

Damage

Injury by wheat wireworm is typical of wireworm damage. Larvae feed below-ground on roots, stems, stolons, and tubers. They may tunnel completely through large plant tissue such as potato tubers, which are honeycombed by their feeding passageways. In other instances the larvae limit themselves to surface feeding, which nevertheless mar the appearance and reduce commercial value. Once the larvae chew through the surface, the tissue is often invaded by fungi and other soil-borne plant pathogens.

Management

Sampling. The eggs can be separated from the soil by sifting with a 60-mesh sieve. Separation is aided by a gentle stream of water, which helps wash away the soil, including some of the soil particles that adhere to the sticky surface of the eggs. They also can be separated by flotation, using a strong solution of sodium chloride or magnesium sulfate.

Population monitoring depends primarily on assessment of larval populations, and fields should be sampled before susceptible crops, such as potatoes, are planted. Larval populations are sampled by baiting or soil sampling. Baiting is accomplished by burying whole wheat, corn, barley, flour plus molasses, or

other attractive food sources in the soil at a depth of 10–15 cm and counting the number of wireworm larvae attracted (Lafrance, 1967; Parker, 1994). Baiting does not work well if the soil is cold, so a sheet of black or transparent plastic is sometimes placed over the section of soil containing bait; this warms the soil and increases the mobility of wireworm larvae, allowing more accurate counts. Simmons *et al.* (1998) compared several methods of sampling and reported that corn/wheat baits were the most accurate, precise, and cost effective.

For soil sampling, 15 × 15 cm soil samples, including soil to a depth of 30–40 cm, are removed and checked for wireworms. This is normally done by sieving with coarse and fine screen to separate insects from the soil. Although accurate estimates of wireworm population densities are possible with soil sampling, this approach is very labor intensive; however, baiting is more commonly used.

Adults may be collected from "trap heaps," which consist of small piles of hay placed on bare soil. They are made more attractive by spraying the heap with a solution of dilute molasses. Unlike many other click beetle species, *A. mancus* cannot be readily captured in blacklight traps, apparently because the beetles disperse primarily by walking (Lafrance, 1967).

Insecticides. Insecticides are often applied for wireworm control when susceptible crops such as potato are planted. Insecticides may be applied broadcast before planting, but more commonly are applied at the time of planting in the seed furrow, over the row, or broadcast. At low wireworm densities, applications limited to the rows are adequate, but at moderate wireworm densities broadcast applications are often more effective. At high densities susceptible crops should not be planted.

Cultural Practices. Land management has a significant role in creating or alleviating problems with wheat wireworm. Land that is planted to grass sod is often heavily infested with wheat wireworm, and crops are at great risk for about 2 years following conversion from sod to crops. Also, this wireworm is usually associated with heavy soil such as silt and clay loams, and low-lying or moist sections of fields.

FAMILY MELOIDAE—BLISTER BEETLES

Black Blister Beetle
Epicauta pensylvanica **De Geer**
(Coleoptera: Meloidae)

Natural History

Distribution. *Epicauta* species and other blister beetles are native to North America, and though many species have been reported to feed on crops, few are commonly destructive. Black blister beetle, *Epicauta pensylvanica* De Geer, is usually the most common species, and occurs everywhere in the eastern United States and southern Canada west to the Rocky Mountains. Immaculate blister beetle, *E. immaculata* (Say); spotted blister beetle, *E. maculata* (Say); and striped blister beetle, *E. vittata* (Fabricius), are the other common vegetable-feeding blister beetles, and are treated separately.

Other common but less damaging *Epicauta* spp. include ashgray blister beetle, *E. fabricii* (LeConte) in eastern North America west to the Rocky Mountains but absent from the southeast; *E. cinerea* (Förster) in the northeastern and midwestern states; and margined blister beetle, *E. pestifera* Werner, which is most abundant in the northeastern and midwestern United States and Canada. The taxonomy of blister beetles is difficult and many of the old records are suspect, or host associations are reported but little information on crop damage is given, so the economic importance of many species is questionable.

Other genera of Meloidae occasionally affect crops, but incidents of damage are isolated. Among such occasional pests are *Linsleya sphaericollis* (Say) in the Rocky Mountain region and west to the Pacific Ocean; Nuttall blister beetle, *Lytta nuttalli* Say, throughout the west and east to Nebraska; and *Meloe niger* Kirby, which occurs throughout the United States and southern Canada except for the southeastern states.

Host Plants. Blister beetle adults are found on a number of vegetable crops, including asparagus, beet, broad bean, cabbage, carrot, celery, chicory, Chinese cabbage, corn, eggplant, lima bean, mustard, okra, onion, pea, pepper, potato, pumpkin, radish, snap bean, spinach, squash, sweet potato, Swiss chard, tomato, and turnip. Vegetable crops specifically known to be damaged by black blister beetle include bean, beet, cabbage, potato, onion, pea, radish, squash, pumpkin, and tomato, though blossom feeding is greatly preferred over foliage consumption. Other crops such as alfalfa, clover, soybean, and sugarbeet, as well as numerous flower crops, fruit trees, and broadleaf weeds are occasionally eaten. Adults often collect on goldenrod flowers, *Solidago* spp., in the autumn.

Natural Enemies. Surprisingly little is known concerning the natural enemies of blister beetles, reflecting their minor status as crop pests and the subterranean habits of larvae. Undoubtedly starvation of first instars is a very important factor during most seasons, and cannibalism is prevalent among larvae. Ant-like flower beetles (Coleoptera: Anthicidae), false ant-like flower beetles (Coleoptera: Pedilidae), and some plant bugs (Hemiptera: Miridae) have been implicated as mortality agents of blister beetles. The larva of the blister beetle *Epicauta atrata* (Fabricius) has also been shown as predatory on eggs of *E. pensylvanica*, and it is possible that other species within the genus are predatory (Selander, 1981; 1982). See also the discussion of natural enemies under striped blister beetle.

Life Cycle and Description. There is a single generation per year, with overwintering occurring as any of the larval instars. In Arkansas, pupation typically occurs in July or August, adults are found in August–November, eggs are found in September–November, and new larvae are present beginning in October.

Egg. The eggs of black blister beetle are elliptical, measuring about 1.3 mm long (range 1.0–1.6 mm), and 0.6–0.7 mm wide. The eggs are pearly white initially, but become darker as the embryo develops. The eggs are deposited within a cavity in the soil, in clusters of 100–200 eggs. The soil cavities are tubular structures about 20–30 mm deep and only 3–4 mm in diameter, and the entrance is closed by the female after she completes the act of oviposition. Eggs normally hatch in 14–15 days.

Larva. The larval instars are quite varied in appearance, reflecting the unusual biology of the insect. Unlike most insects, the ovipositing female does not locate a food source for her offspring, apparently depositing her eggs randomly. Thus, when young larvae hatch they must dig to the surface and disperse to find a grasshopper egg pod on which to feed. The first instar is thus quite mobile, and equipped with long legs with which to disperse. First instars explore cracks, crevices, and depressions in the soil as they search to find an egg pod. Once locating a pod, they pierce an egg and consume the liquid contents. Usually a single egg is adequate for complete development of the first instar. The number of instars is normally 6–7. Larvae are creamy white or yellowish white, with brown head capsules. After the first instar the larva moves little, and the legs become relatively smaller and smaller. The sixth instar does not feed, instead digging 2–3 cm into the soil and preparing a cell. The sixth instar may be followed by another non-feeding instar, or by the pupal stage. The sixth and seventh instar bear only minute legs, and the head capsule is reduced in size and retracted into the body. Black blister beetle consumes 21–27 eggs of *Melanoplus differentialis* Thomas during its larval development. This grasshopper, and some other *Melanoplus* spp., produce 100 or more eggs, allowing more than one blister beetle to develop. If blister beetle larvae encounter one another, however, they fight and only one survives. Many grasshoppers produce small egg pods, with less than 25 eggs, thereby limiting the ability of blister beetles to develop. The mean head capsule widths of the larval instars are 0.41, 0.57, 0.85, 1.28, 1.77, 1.55 and (if present) 1.77 mm, respectively, for instars 1–7. Mean development time is 8.5, 2.1, 2.4, 2.8, 12.5, 7.3, and 7.6 days, respectively, for a total larval development time of about 43 days unless the larva enters a period of arrested development for the winter.

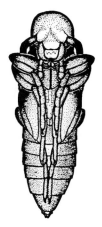

Black blister beetle pupa.

Pupa. The pupa resembles the adult beetle, though the legs and wings are folded against the underside of the body, and the color initially is creamy white, becoming black at maturity. The pupa measures about 10 mm long, and is found within a cell in the soil. Mean duration of the pupal stage is about 14 days, with a range of 11–28 days, and there appears to be no tendency for diapause in this stage.

Black blister beetle, first instar.

Adult. The adult digs to the soil surface after pupation. The adult is elongate and slender in form, black in color, and measures 7–15 mm long. As is typical with this family, the thorax is narrower than the head and abdomen, and the legs and antennae are moderately long. The body bears numerous small punctures, and is sparsely clothed with short hairs. The elytra are long, covering the abdomen but separated or divergent at the tips. The hind wings are transparent. (See color figure 99.)

Unfortunately, *E. pensylvanica* is not the only black species, though few other black *Epicauta* spp. are common or damaging. Black species of *Lytta* and *Meloe*, including those mentioned previously, occasionally are injurious, though these species are purplish or greenish black (*Lytta*) or bear abbreviated elytra and are broad rather than slender in form (*Meloe*).

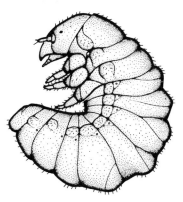

Black blister beetle, fourth instar.

Black blister beetle was treated in detail by Horsfall (1941), and also was included in the publications of Milliken (1921) and Horsfall (1943). Werner (1945) and Pinto (1991) included this species in keys to North American *Epicauta*. Horsfall (1943) is a good source of information on all the common *Epicauta* spp. and described a method of rearing blister beetles. Downie and Arnett (1996) provided keys to the eastern species of blister beetles, Werner *et al.* (1966) to those in the southwest. Information on *Lytta* spp. can be found in Selander (1960) and Church and Gerber (1977b); on *Meloe* spp. in Pinto and Selander (1970) and Mayer

Black blister beetle, seventh instar.

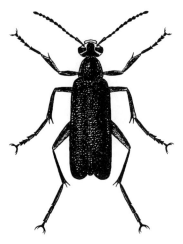

Adult black blister beetle.

and Johansen (1978); and on *Linsleya* spp. in Selander (1955) and Church and Gerber (1977a).

Damage

The adult blister beetles defoliate plants, though the pattern of injury is irregular. Beetles tend to aggregate, apparently in mating swarms, so damage can be severe in relatively small areas of a crop and absent or trivial elsewhere. Although adult black blister beetles inflict damage to vegetable crops primarily through consumption of foliage, in some instances they prefer blossoms.

Preference of black blister beetle and other blister beetle species for blossoms is most noticeable and potentially damaging in alfalfa, where beetles congregate mostly during periods of bloom. Aggregations of beetles can be incorporated into alfalfa hay when it is baled, particularly if the stems are crushed as part of the harvesting process. Crushing, or crimping, aids in the drying and preservation of hay, but also can result in death of aggregations of blister beetles, and incorporation of their bodies into the hay. Blister beetles, even dead individuals, contain a vesicating substance called cantharidin which, when ingested, damages the digestive tract of animals. Cantharidin also causes blisters to form on the skin of sensitive humans who come into contact with crushed beetles, and in formation of blisters in the mouths of livestock, particularly horses. This blistering action is the basis for the common name of the beetles. Cantharidin content varies among species, ranging from 5% in *E. immaculata* to about 1% in *E. pensylvanica* (Capinera *et al.*, 1985). Despite the relatively low toxin content in each beetle, large numbers can be incorporated into hay, and horses can ingest enough beetles to cause death (Blodgett *et al.*, 1991). Cantharidin likely serves as a feeding deterrent to most predators, thereby

protecting blister beetles and their eggs from consumption. However, some insects are attracted to cantharidin, and this compound is involved in the chemical communication among blister beetles (Young, 1984a,b; Klahn, 1987).

The damage caused by *Epicauta* spp. blister beetles is offset, at least during periods of relatively low beetle density, by the predatory behavior of blister beetle larvae. *Epicauta* spp. larvae feed on the eggs of grasshoppers, including many crop-damaging *Melanoplus* spp. During periods of grasshopper abundance the number of blister beetles tends to increase substantially. Studies of egg pod destruction in western states during a period of grasshopper abundance (Parker and Wakeland, 1957), for example, documented that 8.8% of pods were damaged by blister beetles. Although the blister beetles eventually contribute materially to the suppression of grasshopper population outbreaks, the higher numbers of blister beetles often cause greater crop injury during, and immediately after, the periods of grasshopper abundance. The larvae of blister beetles other than *Epicauta* spp., however, seem to feed principally on ground-nesting bees (Hymenoptera: Andrenidae, Halictidae, perhaps others) and the bee's nest provisions. Their abundance fluctuates less, and they provide no known agricultural benefits.

Management

Blister beetles infrequently are vegetable crop pests, though they may become quite abundant during and following long-term grasshopper population increases, particularly populations of *Melanoplus differentialis* and *M. bivittatus* (Say). Suppression of grasshoppers indirectly suppresses blister beetles by eliminating the food supply of the blister beetle larvae. Direct suppression of blister beetles usually does not occur in conjunction with chemical treatment of grasshopper populations, because the grasshoppers occur earlier in the season, when blister beetles are still in the soil. Blister beetles are easily controlled by application of common insecticides to crop foliage, and small plantings can be protected with row covers or screening. Because black blister beetle is highly attracted to alfalfa, especially during periods of bloom, large numbers of blister beetles may disperse to nearby crops following alfalfa harvest.

Immaculate Blister Beetle
Epicauta immaculata (Say)
(Coleoptera: Meloidae)

Natural History

Distribution.   This native species is found widely in the midwestern states and Great Plains region of the

United States from Kentucky west to Colorado and New Mexico.

Host Plants. Immaculate blister beetle has a fairly broad host range, and though it is especially often recorded as a pest of potatoes, it also attacks such vegetables as bean, beet, cabbage, onion, pea, pumpkin, radish, squash, and tomato. Post-bloom lima bean also is suitable, though pre-bloom plants are reportedly toxic. Other crops and flowers including alfalfa, sweet clover, hollyhock, gaillardia, and marigold can serve as hosts. Weeds including sunflower, *Helianthus* spp.; cactus blossoms, *Opuntia* sp.; wild lettuce, *Lactuca* sp.; Russian thistle, *Salsola kali*; and mullein, *Verbascum* sp.; also are consumed.

Natural Enemies. The natural enemies of immaculate blister beetle are not precisely defined, but undoubtedly are the same or similar to those affecting black blister beetle, *Epicauta pensylvanica* De Geer, and striped blister beetle, *Epicauta vittata* (Fabricius).

Life Cycle and Description. There is a single generation annually. In South Dakota, adults are present from June to August, but are most abundant in July. Eggs are present in July and August, and early instars through November. Overwintering occurs in instar six, with the final instar and pupation in May.

Egg. The eggs are deposited in a bell-shaped cavity near the surface of the soil, at a depth of perhaps 3 cm. The cavity is plugged with soil by the female after she completes oviposition. Several hundred eggs reportedly can be placed in a single cavity. The egg is elongate oval, with rounded ends. It measures 1.3–1.5 mm long and 0.5–0.6 mm wide. It is yellowish white initially, becoming darker with age. Duration of the egg stage is 11–18 days.

Larva. The young larva is yellowish white, with a brown head. It is long-legged and mobile, and after remaining in the egg cavity for 1–2 days, forages actively for grasshopper eggs. Young larvae are active on the soil surface but also crawl into cracks and crevices in search of grasshopper eggs. Upon locating a supply of food the larva commences feeding and progresses through four additional instars while continuing to feed. Duration of the first instar is 3–5 days; instars 2–4 require 1–4 days each, and the fifth is 7–13 days in duration. On average the first five instars require about 18 days. Because the larvae are not required to search for additional food, their legs are not needed and they become smaller with each molt. The fifth instar burrows deep into the soil, molts to the sixth instar, stops feeding, and overwinters. The head is reduced in this and the subsequent stage, which also is a period that is free of feeding. The final,

or seventh, instar digs closer to the surface in preparation for the molt to the pupal stage and constructs a small chamber. Duration of the seventh instar is 8–15 days.

Pupa. Pupation normally occurs at a depth of 2–5 cm. The pupa is yellowish-white, about 17 mm long, and similar in form to the adult, though the wings and legs are pulled closely to the ventral surface of the developing beetle. Duration of the pupal stage is 10–14 days.

Adult. This is one of the largest species of *Epicauta*, measuring 13–23 mm long. The background color is black, but the body is covered with short dense pubescence of various colors, imparting an overall gray, tan, or reddish brown appearance. The adult is elongate and slender and as is normal with this family, the thorax is narrower than the head and abdomen, and the legs and antennae are moderately long. The elytra are long, covering the abdomen but separated or flared at the tips. The hind wings are transparent. New adults fed voraciously for 7–10 days after emerging, and then after another week commence oviposition. Periods of feeding are interspersed by acts of oviposition. (See color figure 101.)

Immaculate blister beetle and its biology were described by Milliken (1921) and Gilbertson and Horsfall (1940). Werner (1945) and Pinto (1991) included this species in keys to North American *Epicauta*. Horsfall (1943) described a method of rearing blister beetles. Downie and Arnett (1996) provided keys to the eastern species of blister beetles, though it was derived from Werner's key.

Damage

Immaculate blister beetles are large and can cause extensive defoliation. Also, adults are quite gregarious and sometimes assemble in very large numbers.

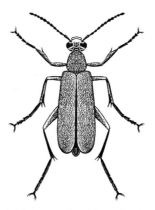

Adult immaculate blister beetle.

Such aggregations can cause severe localized injury, though many nearby areas may escape with no visible damage. In most respects, the injury caused by immaculate blister beetle is equivalent to damage by other meloids. (See the discussion of damage in the section on black blister beetle for additional information.)

Management

Blister beetles infrequently are vegetable crop pests, though they may become quite abundant during and following long-term grasshopper population increases, particularly populations of differential grasshopper, *Melanoplus differentialis* Thomas, and twostriped grasshopper, *Melanoplus bivittatus* (Say). As immaculate blister beetle is a large species and requires considerable food, the larvae develop successfully only in the presence of grasshoppers which produce large egg pods, such as differential grasshopper and twostriped grasshopper. Suppression of grasshoppers indirectly suppresses blister beetles by eliminating the food supply of the blister beetle larvae. Direct suppression of blister beetles usually does not occur in conjunction with chemical treatment of grasshopper populations because the grasshoppers occur earlier in the season, when blister beetles are still in the soil. Blister beetles are easily controlled by application of common insecticides to crop foliage, and small plantings can be protected with row covers or screening.

Spotted Blister Beetle
Epicauta maculata (Say)
(Coleoptera: Meloidae)

Natural History

Distribution. This native species occurs primarily in the Great Plains region of North America. It is abundant from southern Manitoba and Saskatchewan south through the western plains to central Mexico and Guatemala. It is rare east of Iowa, and occurs west of the Rocky Mountains only in Arizona. There are other spotted blister beetles with which it can be confused, but there is little overlap in the geographic distribution. Also see the discussion of blister beetles in the section on black blister beetle, *Epicauta pensylvanica* (De Geer).

Host Plants. Spotted blister beetle is reported from such vegetable crops as bean, beet, cabbage, potato, and spinach. Other crops affected include alfalfa, soybean, and sweet clover. Weed hosts include silver-leaf nightshade, *Solanum elaeagnifolium*; and lambsquarters, *Chenopodium album*.

Natural Enemies. Natural enemies are poorly known. For discussion of natural enemies of blister beetle, see the sections on black blister beetle, *Epicauta pensylvanica* De Geer, and striped blister beetle, *Epicauta vittata* (Fabricius).

Life Cycle and Description. The life cycle of spotted blister beetle is not as clearly defined as some of the other blister beetles. In most cases, only a single generation per year occurs, but in some cases larvae do not diapause, so the possibility exists of a second generation. The adult of this species is one of the last blister beetles to emerge and, as with other *Epicauta* spp., late instars overwinter.

Egg. The eggs are elongate oval, with rounded ends. Eggs measure about 1.2 mm long and 0.5 mm wide, and are whitish in color. The female deposits 50–100 eggs per egg mass within a cavity in the soil. The egg cavity is reported to be flared or bell-shaped, wider at the bottom than at the top, and to extend to a depth of about 2.5 cm. Duration of the egg stage is 10–21 days.

Larva. This insect is not as well-studied as some other blister beetles, so knowledge of the developmental biology is incomplete. In most respects, morphology and development should be equivalent to that of black blister beetle, *Epicauta pensylvanica*, with five instars developing on grasshopper eggs followed by two non-feeding instars. Mean duration of the first five, or feeding, instars of spotted blister beetle is 4.1, 1.6, 1.5, 2.0, and 13.6 days, respectively. The larvae are yellowish in color.

Pupa. The pupal stage resembles the adult, though the wings and legs are drawn under the body. It is yellowish white and measures about 10 mm long. Like the larval stage, it occurs in the soil.

Adult. The adult is small, measuring 6–12 mm long, and moderately elongate and slender. It is black but densely clothed with short-gray or olive-gray hairs; this pubescence imparts a grayish color to the beetles. The pubescence is absent in small to moderately sized spots, particularly on the elytra, which produces the black-spot pattern that is the basis for the common name of this species. As is typical with this family, the thorax is narrower than the head and abdomen, and the legs and antennae are moderately long. The elytra are long, covering the abdomen but separated or divergent at the tips. The hind wings are transparent. (See color figure 102.)

Spotted blister beetle and its biology were described by Milliken (1921), Gilbertson and Horsfall (1940), and Pinto (1980). Werner (1945) and Pinto (1991) included this species in keys to North American *Epicauta*.

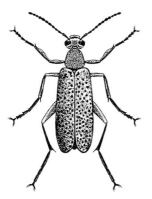

Adult spotted blister beetle.

Horsfall (1943) is a good source of information on all the common *Epicauta* spp. and he described a method of rearing blister beetles. Downie and Arnett (1996) provided keys to the eastern species of blister beetles, though their key was derived from Werner's 1945 key; Werner *et al.* (1966) provided keys to blister beetles in the southwest.

Damage

The adult blister beetles cause injury by defoliating plants, though they are relatively small beetles, and less damaging than those of the larger species because they eat less. Beetles tend to aggregate, apparently in mating swarms, so damage can be significant in relatively small areas of a crop and is absent or trivial elsewhere. Although adults of spotted blister beetle inflict damage to vegetable crops through consumption of foliage, in some instances they prefer blossoms, and larvae are beneficial due to their consumption of grasshopper eggs. For a more complete discussion of the damaging and beneficial aspects of blister beetles, see black blister beetle, *Epicauta pensylvanica*.

Management

Blister beetles are not often vegetable crop pests, though they may become quite abundant during and following long-term grasshopper population increases. Gilbertson and Horsfall (1940) reported that spotted blister beetle was generally associated with eggs of migratory grasshopper, *Melanoplus sanguinipes* (Fabricius) though certainly larger grasshopper species with larger egg pods would also be suitable hosts. Suppression of grasshoppers indirectly suppresses blister beetles by depriving the beetle larvae of grasshopper eggs. Blister beetles also are easily controlled with foliar applications of common insecticides, and small crop plantings can be protected with row covers or screening.

Striped Blister Beetle
Epicauta vittata (Fabricius)
(Coleoptera: Meloidae)

Natural History

Distribution. The striped blister beetle, *Epicauta vittata* (Fabricius), is a native species and eastern in distribution. It has been collected from all eastern states west to, and including, South Dakota, Nebraska, Kansas, and Oklahoma. In Canada, it is known from Quebec and Ontario. Populations from the southeastern coastal plain including Florida, southern Georgia, and eastern South Carolina differ in appearance from beetles found elsewhere and are called the "lemniscate race," but interbreed successfully with normal beetles. Also see the discussion of blister beetles in the section on black blister beetle, *Epicauta pensylvanica* (De Geer).

Host Plants. Vegetable crops such as bean, beet, carrot, cabbage, Chinese cabbage, corn, eggplant, melon, mustard, pea, pepper, potato, radish, spinach, squash, sweet potato, tomato, and turnip are sometimes injured. Clover and soybean can also be attacked. Pigweed, *Amaranthus* spp., is highly preferred by adults.

Natural Enemies. Little is known concerning the natural enemies of striped blister beetle, though Ingram and Douglas (1932) reported consumption by robber flies (Diptera: Asilidae) and avian predators, including meadowlark, *Sturnella neglecta* Audubon; bluebird, *Sialia sialis* (Linnaeus); and scissor-tailed flycatcher, *Muscivora forficata* (Gmelin). Selander (1981, 1982) reported predation of striped blister beetle eggs by the predatory blister beetle, *Epicauta atrata* (Fabricius). See the section on black blister beetle for further discussion on natural enemies.

Life Cycle and Description. There are 1–2 generations per year. In Arkansas, the May–June or early-emerging adults go on to produce another generation in which adults emerge in September. Maximum emergence of adults occurs in July, however, and few of the later-emerging adults contribute to the second generation. The generations overlap, resulting in adults present in the field from late-May to late-October. Overwintering occurs as instars five and six. In Florida, the season is considerably advanced, with most adults found during April–June.

Egg. The eggs of striped blister beetle are whitish and average about 1.8 mm long and 0.7 mm wide, though they vary greatly in size, with the length ranging from 1.4–2.1 mm. In shape, they are elongate oval, with rounded ends. The eggs are deposited in the soil within a tubular chamber 3–4 cm deep, normally in clusters of 100–200 eggs. The female covers the eggs after oviposition, and hatching occurs in 10–16 days. Eggs can be found from June–September in Arkansas. Females are reported to produce several hundred eggs in captivity, though this estimate may be low, because many closely related species produce 2000–3000 eggs during the life span of 20–50 days (Adams and Selander, 1979).

Larva. As is typical with blister beetles, the larval form initially has long legs and is quite mobile. The legs become reduced in size, however, as the larva develops, and instars 6–7 (if present) have small head capsules and do not feed. Newly hatched larvae are whitish but soon turn reddish brown with dark brown bands on the thorax and abdomen. The first instar must disperse from the egg mass and can locate grasshopper eggs on which to feed. Once locating a grasshopper egg pod, the larva feeds and molts at short intervals. The mean duration of the first five instars, all of which feed on grasshopper eggs, is 3.4, 1.2, 1.1, 1.6, and 11.3 days, respectively, at warm temperature. Overwintering also may occur in the fifth instar, followed by pupation in the spring. Alternatively, larvae may display a sixth and seventh instar before pupation, with the sixth instar persisting for about 230 days and the seventh for 6–14 days. Seventh instars are found in April–July at about the same time other individuals are pupating or emerging as adults. Mean larval head capsule widths (range) are 0.63 (0.53–0.82), 0.83 (0.75–0.91), 1.10 (0.93–1.20), 1.64 (1.37–1.75), 2.28 (2.18–2.37), 1.73 (1.37–2.18), and 2.42 (1.75–2.68) mm for instars 1–7, respectively.

Pupa. The pupal stage is found in the soil. The pupa greatly resembles the adult, though the wings and legs are tightly drawn to the ventral surface. Initially the pupa is whitish, but becomes darker with maturity. Duration of the pupal stage is 9–13 days. Pupae are found in May–August.

Adult. The adults measure 9–17 mm long and are black and yellow. The color pattern varies geographically, but normally two black spots occur dorsally on the head and on the thorax, and the elytra each bear two or three black stripes. Northern populations often bear only two black stripes on the elytra, whereas southern populations usually bear three black stripes. The southern Coastal Plain race, which has three stripes on each elytron, is termed the lemniscate race. Like most other blister beetles, the adult is elongate and slender, the thorax is narrower than the head and abdomen, and the legs and antennae are moderately long. The body bears numerous small punctures, and is densely clothed with short hairs. The elytra are long, covering the abdomen but separated or divergent at the tips. The hind wings are transparent.

The adults are most active during the morning and late afternoon, seeking shelter from the sun at midday. In particularly hot and arid climates, they remain inactive during the day, confining activity to the evening hours. They are easily disturbed, dropping readily from the plant and hiding or scurrying away if disturbed. Pheromonal cues are likely involved in aggregation behavior, but this is poorly defined. The pre-ovipositional interval of striped blister beetle is about 20 days, with a 10 day interval between production of egg masses. (See color figure 100.)

The adults of striped blister beetle are very similar to *Epicauta occidentalis* Werner and *E. temexa* Selander and Adams, and these species frequently have been confused in the literature. Also, "*Epicauta lemniscata*" has been used for reference to these species, and to mixed populations of these species. Adams and Selander (1979) provided a detailed description of striped blister beetle, and keys to separate this species from close relatives. The biology of striped blister beetle is included in the publications of Gilbertson and Horsfall (1940) and Horsfall (1943). Werner (1945) and Pinto (1991) included this species in keys to North American *Epicauta*. Downie and Arnett (1996) provided keys to the eastern species of blister beetles, though these are derived from Werner's. Horsfall (1943) described a method of rearing blister beetles.

Damage

Striped blister beetle is one of the most damaging of the blister beetles to vegetable crops in areas where it occurs. This is owing to its feeding preferences, which include several common crops and greater preference for foliage than other species; its propensity to feed on fruits of solanaceous plants; its relatively large size and voracious appetite; its strong tendency to aggregate into large mating and feeding swarms; and its high degree of dispersiveness, which can result in

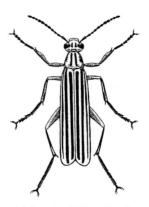

Adult striped blister beetle.

sudden appearance of large swarms of beetles. It also has been implicated in the transmission of bean pod mottle virus to soybean; this is not surprising because other beetles also are associated with spread of this disease (Patel and Pitre, 1971). In other respects, damage by striped blister beetle is nearly identical to black blister beetle, *Epicauta pensylvanica* De Geer. (See the section on black blister beetle for a thorough discussion of damage.)

Management

Insecticides. Blister beetle populations are easily controlled by foliar application of insecticides. The tendency of beetles to aggregate, however, may result in serious defoliation in small areas of a field, and little or no damage elsewhere. Thus, careful visual examination of plants is suggested, followed by spot treatment of infested areas.

Cultural Practices. Striped blister beetle is reportedly closely associated with grasshoppers that produce large egg pods, particularly twostriped grasshopper, *Melanoplus bivittatus* (Say) and differential grasshopper, *M. differentialis* Thomas. Thus, environments that favor these grasshoppers, such as an abundance of alfalfa, also favor occurrence of the blister beetles. However, young blister beetles may be present in the summer before these grasshoppers have produced egg pods, so Horsfall (1943) suggested that other, early maturing species of grasshoppers may be necessary to provide eggs to bridge the gap between the hatch of beetle larvae and the appearance of new *M. bivittatus* and *M. differentialis* eggs. This suggests that blister beetle abundance can be regulated not only by practices which affect grasshopper abundance, but that disruption of either the early season or late season species would be disruptive to striped blister beetle.

Striped blister beetle adult feeds readily on alfalfa, and can be abundant in this crop even in the absence of grasshoppers. The association of striped blister beetle with alfalfa implies that vegetable crops grown in proximity to alfalfa are at greater risk, and that the risk is elevated when alfalfa is harvested. Small crop plantings can be protected from blister beetles with row covers or screening.

FAMILY NITIDULIDAE—SAP BEETLES

Dusky Sap Beetle
Carpophilus lugubris Murray
(Coleoptera: Nitidulidae)

Natural History

Distribution. Sap beetles of the genus *Carpophilus* are found throughout the United States and southern Canada. Dusky sap beetle, a native insect, is the most important species. It is found from the Atlantic to the Pacific, and south into Central and South America, but is most numerous in eastern and midwestern states. Other *Carpophilus* spp. of occasional concern include antique sap beetle, *C. antiquus* Melsheimer; corn sap beetle, *C. dimidiatus* (Fabricius); and dried fruit beetle, *C. hemipterus* (Linnaeus).

Host Plants. Sap beetles have derived their name from their tendency to feed on sap and other sweet secretions. However, they also feed on fungi, pollen, damaged or rotting fruit, and occasionally on undamaged fruit. Among vegetables, corn is the crop most commonly injured, followed distantly by tomato. Dusky sap beetle is definitely a primary pest that feeds on uninjured corn; the other species are more likely associated with vegetable matter injured by other insects or by physical disruption.

Natural Enemies. Natural enemies of sap beetles seem to be few and their effects generally are unknown. The insidious flower bug, *Orius insidiosus* (Say), feeds on sap beetle eggs. A wasp, *Cryptoserphus abruptus* (Say) (Hymenoptera: Proctotrupidae), parasitizes beetle larvae. A nematode tentatively identified as *Hexamermis* sp. (Nematoda: Mermithidae) was found to infect up to 89% of adult beetles collected at certain sites during the spring in Illinois (Dowd *et al.*, 1995), but it was absent from most sites. Some *Carpophilus* spp. are known to be affected by *Howardula* sp. nematodes (Nematoda: Allantonematidae) and the fungus *Beauveria bassiana*. When sap beetle larvae occur in corn ears containing larvae of corn earworm, *Helicoverpa zea* (Boddie), they are often killed by the caterpillars.

Life History and Description. Overwintering occurs underground in both the pupal and adult stages, usually in association with crop debris. The adults become active early in the spring, and enter corn fields about the time tassels are produced. Several overlapping generations may occur without distinct periods of abundance. In Washington and Maryland, eggs predominate in July, larvae in late July–early August, and adults are most abundant during August and taper off in September. Total generation time requires 1–3 months.

Egg. The eggs are milky white, elongate oval, and measure about 1.2 mm long and 0.25 mm wide. Eggs are commonly deposited on corn silk, but they also can be found on the fecal material of boring insects, wet accumulations of pollen, and other food sources. Silks that are turning brown or have dried are particularly attractive to ovipositing beetles. The eggs may be deposited singly or in short chains of 3–5 eggs. Duration of the egg stage is 2–4 days at 21°C.

Larva. The larvae are white, yellowish, or pinkish, but bear a brownish head and caudal plate. The caudal plate, at the tip of the abdomen, is marked by a semicircular notch. The tip of the abdomen also bears a fleshy protuberance ventrally. The body bears sparse but moderately long hairs. Connell (1956) indicated that there are three instars, with a duration of about two, three, and eight days, respectively, followed by a prepupal period of about seven days. However, Tamaki *et al.* (1982) reported that there are four instars of dusky sap beetle with a mean duration of 4.5, 4.5, 3.5, and 9.6 days, respectively.

Pupa. After completion of the feeding period, larvae drop to the soil. They dig into a depth of 2.5–7.5 cm and prepare a small cell for pupation. Duration of the pupal stage averages 9.2 days at 21°C in warm weather, but some individuals overwinter in the pupal stage.

Adult. The adult measures 3.5–4.5 mm long, 1.8 mm wide, and is oval in form. It is dull brownish-black in color. As is common among sap beetles, the elytra are short, exposing the terminal abdominal segments. The antennae are club-shaped. Average adult longevity is reported to vary from about 130–300 days, depending on its diet, but this species is obviously a long-lived insect. Adults are active fliers. The adult males produce aggregation pheromones that are synergized by food odors (Bartelt *et al.*, 1991; Lin *et al.*, 1992). Although different sap beetles produce unique pheromones, there is considerable cross attraction (Bartelt *et al.*, 1994). Adults often overwinter either in the soil or above-ground in sheltered locations. Overwintering adults are not in diapause and become active whenever weather allows.

Some elements of dusky sap beetle biology were described by Connell (1956) and Tamaki *et al.* (1982). Description of several *Carpophilus* spp., and keys to common adult and larval sap beetles found in eastern states, were given by Connell (1956). Downie and Arnett (1996) also treated the adults of eastern species, and Hatch (1982) treated the western species. Rearing was described by Tamaki *et al.* (1982). Williams *et al.* (1983) published a bibliography of *Carpophilus*.

Damage

Sap beetles differ considerably in their tendency to attack undamaged vegetables. Among the *Carpophilus* spp., only dusky sap beetle is known to attack uninjured fruit frequently, whereas other species may be occasional contaminants of vegetable matter previously injured by caterpillars, grasshoppers, birds,

Adult dusky sap beetle.

or other factors. Research in Delaware (Connell, 1956) and North Carolina (Daugherty and Brett, 1966) indicated that the attack in early season on the corn crop by sap beetles was independent of corn earworm damage. However, a positive association of sap beetles was found later in the season with corn ears exhibiting injury by corn earworm.

The presence of young sap beetle larvae in corn kernels is of particular concern because they are hard to detect. Sap beetles are responsible for rejection of considerable quantities of corn at canneries, and also transmit fungi which produce mycotoxins.

Management

Sampling. Sap beetles can be captured in traps baited with aggregation pheromone, pheromone plus whole wheat bread dough (Bartelt *et al.*, 1994), or whole wheat bread dough alone (Peng and Williams, 1991). Various types of traps can be used with equal success, but they tend to be very efficacious if placed close to the soil surface. Dowd *et al.* (1992) described a particularly effective trap. Aggregation pheromones sometime concentrate beetles without capturing them, which can lead to localized crop damage.

Insecticides. Insecticides often are applied during the silking stage of corn production to prevent infestation by dusky sap beetle, apparently because fermenting pollen is highly attractive. Sweet corn producers usually treat at this time to prevent infestation by corn earworm and other caterpillar pests, thereby suppressing sap beetles also (Harrison, 1962).

Cultural Techniques. Cultural practices affect dusky sap beetle damage. In North Carolina, early planted corn is more heavily infested than late planted corn. Sweet corn varieties differ in susceptibility to injury, owing primarily to different survival rates among larvae rather than selective oviposition behavior by adults (Daugherty and Brett, 1966). Connell (1956) attributed increase in sap beetle damage to the loose husk found on some corn varieties, and to short husks that provide little coverage of kernels at the tips of the ears.

Sanitation is an important element of sap beetle management. Corn ears that are left unharvested, particularly those on or in the soil, support survival and overwintering of beetles. In Delaware, decomposing ears support larvae until December, when larvae leave the ears to pupate in the soil. Burying ears at a depth of 10 cm or greater inhibits sap beetle survival. Other decomposing crop debris, including lettuce and tree fruit, favors sap beetles, and they can be attracted to trees that are infected with bacterial wetwood (W. Cranshaw, personal communication). Thus, windbreaks and wood lots can be an important source of beetles.

Fourspotted Sap Beetle
Glischrochilus quadrisignatus (Say)
(Coleoptera: Nitidulidae)

Natural History

Distribution. This native insect is also sometimes called "picnic beetle" because, like most sap beetles, it is attracted to sweet substances and the odor of food. It is found throughout the northern United States and southern Canada. Fourspotted sap beetle has been found as far south as Florida, but is infrequent in southern areas. A related species, *G. fasciatus* (Oliver), is similar in appearance and range, but it is not considered to be a primary pest.

Host Plants. The adults feed principally on damaged or decaying vegetables and fruits such as cantaloupe, sweet corn, tomato, apple, peach, and pear, and also on fungi. Late-season adults also attack ripe but undamaged fruits and vegetables, particularly corn and raspberry. Sometimes they also have been known to attack onion sets, potato, and strawberry.

Natural Enemies. Mortality factors are poorly known. General predators such as lady beetles (Coleoptera: Coccinellidae) and insidious flower bug, *Orius insidiosus* (Say) (Hemiptera: Anthocoridae), occasionally feed on sap beetle larvae, but there is no evidence that they are important. Similarly, the fungus *Beauveria bassiana* is associated with sap beetles, but its significance is uncertain.

Life Cycle and Description. This beetle overwinters in the adult stage. Oviposition occurs early in the spring, and in Ontario larval development is completed in May or mid-June, with new adults appearing in July and August. This pattern is similar in such mid-

western states as Minnesota, Iowa, and Illinois. Although only a single generation occurs annually in most of the northern regions, apparently a second generation occurs in Illinois (Dowd and Nelson, 1994), and a few insects go on to produce a second generation even as far north as southern Ontario (Foott and Timmins, 1977).

Egg. The eggs are deposited in the spring, principally on decomposing ears of corn, though they also occur in the soil near food. Other oviposition sites include decaying cabbage, onions, and potato-seed pieces. Beetles dig in the soil to a depth of 15 cm to seek a suitable oviposition site. Females can produce about 300 eggs, which are deposited singly or in small clusters. Duration of the egg stage is about 3.4 days at 24°C, but about 8.3 days at 15°C. The developmental threshold is 6.5°C.

Larva. The larval stage is brownish in color, with sparse hairs. The head and caudal plate are reddish brown. The tip of the abdomen bears a fleshy protuberance ventrally. The larval stage eventually attains a length of 5–6 mm. Three instars are reported, with mean development times of about 3, 4, and 13 days when reared at 24–28°C. The developmental threshold is 5.5°C, though larvae cannot complete development at 10°C or lower. Larvae are present early in the summer, and large numbers often are associated with decaying ears of corn.

Pupa. Pupation usually occurs within a small cell in the soil in the immediate vicinity of the larval food source. The pupa initially is white, and turns tan before adult emergence. Duration of the pupal stage is usually 8–12 days, and then the adult remains below-ground for about 12 days before digging to the surface. The developmental threshold is about 10.5°C.

Adult. The adult is elongate and shiny black. The front wings bear four small orange spots—one at the base and one distally on each forewing. The front wings are truncate, exposing the tip of the abdomen. The antennae are club-shaped. The adult measures 5–6 mm long. Adults overwinter in clumps of grass, under rotting fruit and vegetables and under leaf mold and bark, but especially in the top 2.5 cm of soil. Areas with grass sod and tall weeds are preferred relative to areas with sparse vegetation or forest. Many of the females overwintering are already mated. Adults that emerge in the summer initially are tan rather than black. Adults are long lived, and at mid-summer both new adults and adults from the previous summer can be found. The pre-oviposition period of adults in the

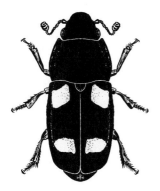

Adult fourspotted sap beetle.

spring is about 10 days. High humidity seems to be a prerequisite to oviposition. The oviposition period of females is about 55–70 days.

The biology was described by Luckmann (1963), Mussen and Chiang (1974), and Foott and Timmins (1977, 1979). Rearing was described by Mussen and Chiang (1974) and Foott and Timmins (1979). A key to adult nitidulids, including the northeastern species of *Glischrochilus*, was given by Downie and Arnett (1996). A bibliography of *Glischrochilus* was published by Miller and Williams (1981).

Damage

Fourspotted sap beetle is a significant nuisance at food processing plants, roadside fruit and vegetable stands, and when food is kept uncovered out-of-doors. It is reported to burrow readily into potato salad and bread, and to plunge into pickle jars. It readily attacks canning tomatoes that are harvested, and slightly injured. Fourspotted sap beetle attacks developing sweet corn during the silking stage, feeding down through the silk. Attack by these beetles is often followed by invasion of dusky sap beetle, *Carpophilus lugubris*. This beetle seems to be especially attracted to plants injured by European corn borer, *Ostrinia nubilalis* (Hübner), and Japanese beetle, *Popillia japonica* Neuman. It also is attracted to fields that are heavily infested by corn leaf aphid, *Rhopalosiphum maidis* (Fitch). Sap beetles may transmit the fungus *Gibberella zeae*, which is responsible for causing a plant disease known as corn ear rot.

Damage to corn is somewhat offset by destruction of young European corn borer larvae by sap beetles. Sap beetle adults are attracted to the fecal material of corn borers, enter their tunnels, and kill young borer larvae. McCoy and Brindley (1961) estimated that 8–17% of corn borer larvae were killed by fourspotted sap beetles.

Management

Sampling. Food lures can be used to assess the population density of adults. Attractants such as banana, sweet corn, and whole wheat-bread dough are most effective. The chemical components of many fruits—butyl acetate and propyl propionate—also are effective lures. Traps placed close to the soil are more effective than elevated traps (Alm *et al.*, 1986, 1989).

Insecticides. Protection of sweet corn ears during the period of growth can be attained by applying insecticides during and after silking. Insecticide residues are low at harvest time, however, and beetles frequently attack corn and tomatoes during this period. Food lures containing insecticide kill many beetles and offer the potential to reduce post-harvest infestation if adequate numbers of traps are present (Foott and Hybsky, 1976).

The attraction of beetles to vegetables during and after harvest sometimes creates problems with chemical-based suppression. It is often undesirable to treat close to harvest owing to the risk of excessive insecticide residues in the product. In processing the tomatoes harvested in Ontario, for example, beetles fly to the tomatoes that have been loaded into containers for transport to factories. The beetles burrow into cracked fruit and are difficult to detect. Because beetles usually land on the containers before crawling on the harvested fruit, however, it is possible to treat the containers with insecticides close to harvest, affect good suppression of beetles, and yet maintain low-insecticide residues on food (von Stryk and Foott, 1976).

Cultural Practices. The most important cultural practice is sanitation. All ears should be removed from the corn field after harvest, or crop residue plowed deeply to prevent development of beetles on discarded ears of corn. If beetles are found abundant in an area, it is also advisable to remove or destroy other potential oviposition sites such as old cabbage and onion crops.

FAMILY SCARABAEIDAE—SCARAB BEETLES AND WHITE GRUBS

Asiatic Garden Beetle
Maladera castanea (Arrow)
(Coleoptera: Scarabaeidae)

Natural History

Distribution. Asiatic garden beetle was first observed in North America in New Jersey during 1921, but it has since spread as far north as Massachusetts and as far south as South Carolina. Its likely origin is Japan or China.

Host Plants. This insect is reported to feed on more than 100 plants, and though much of the diet breadth is due to adult feeding, larvae also have a wide host range. Larvae normally are most abundant on turfgrass or pastureland, but sometimes they occur in high densities in vegetable and flower gardens, where they attack beet, carrot, corn, lettuce, onion, Swiss chard, strawberry, begonia, columbine, and occasionally others (Hallock, 1934). Among the adult food plants are such crop and ornamental plants as bean, beet, broccoli, cabbage, carrot, corn, eggplant, kohlrabi, parsley, parsnip, pea, pepper, potato, radish, rhubarb, spinach, sweet potato, Swiss chard, turnip, peach, cherry blackberry, sumac, ailanthus, butterfly bush, chrysanthemum, dahlia, larkspur, gaillardia, gerbera, sunflower, strawflower, phlox, viburnum, zinnia, and boxelder. The most favored vegetable crops are beet, carrot, parsnip, pepper, and turnip. Among weeds readily consumed are ragweed, *Ambrosia trifida*; burdock, *Arctium* spp.; beggartick, *Bidens frondosa*; plantain, *Plantago* spp.; smartweed, *Polygonum pensylvanicum*; and cocklebur, *Xanthium* spp. The abundance of larvae is often greater near favored adult food plants, though orange hawkweed, *Hieracium aurantiacum*, which is not a food plant, is a preferred site for oviposition. Larvae also feed on decaying organic matter.

Natural Enemies. A parasitoid, *Tiphia asericae* Allen and Jaynes (Hymenoptera: Scoliidae), is known from Asiatic garden beetle, though its importance is uncertain. Several microbial pathogens affect Asiatic garden beetle, and likely contribute to periodic population declines. In a survey conducted in Connecticut (Hanula and Andreadis, 1988), infection rates by gregarines varied from 0–63%, by the protozoan *Adelina* from 0–67%, and by the bacteria *Bacillus popilliae* and *B. lentimorbus* from 0–12.5%. Also observed in Asiatic garden beetle was the rickettsial disease *Rickettsiella popilliae*, which imparts a blue-green color to grubs.

Life Cycle and Description. There is a single generation per year in New York. Both second and third instars survive the winter, though the majority of larvae are third instars. Pupation occurs in June–July and adults are present from early July through September. Eggs are most abundant from mid-July to August, first instars in August–September, and the latter two instars until the following spring.

Egg. The eggs of Asiatic garden beetle are white, oval to nearly spherical in shape, and measure about 1 mm in diameter. They are deposited in the soil in clusters of up to 20 eggs, with the eggs held loosely together by a gelatinous secretion. Egg deposition occurs at soil depths of 1.5–10.0 cm, but mostly between 2.5–5.0 cm. The egg deposition often occurs in the vicinity of favored adult food plants, though sometimes in the vicinity of plants that provide only shelter to adults during the daylight. Mean fecundity is about 60 eggs per female, but the production of up to 180 eggs has been reported. Duration of the egg stage is about 10 days.

Larva. The larva is whitish, except for the head, which is brown. The body of the larva is the typical C-shaped form which is so common among scarab beetle larvae. The underside of the terminal abdominal segment bears a curved row of spines that has diagnostic value. There are three larval instars, with the body increasing from about 1.4 mm in length at hatching to about 19 mm at maturity. During the spring and summer feeding periods larvae feed within the top 13 cm of soil, though most are located at a depth of 5–8 cm. During the overwintering period the grubs descend to a depth of 15–30 cm. At maturity, the larva creates a small cell in the soil and pupates.

Pupa. The pupa resembles the adult in general form, except that the wings are shrunken and twisted ventrally. Also, the pupa is tan or light brown and 8–9 mm long. Pupation generally occurs in the soil at a depth of 4–10 cm. Mean duration of the pupal stage is about 10 days (range of 8–15 days).

Adult. The Asiatic garden beetle is chestnut brown in color and measures about 8–12 mm long and 4–6 mm wide. The elytra, which are well-marked with grooves, do not extend the entire length of the abdomen, leaving the terminal abdominal segments exposed. The ventral surface of the adult bears rows of short yellow hairs on each body segment. Adults are nocturnal, hiding during the day on the soil beneath any available cover, or beneath loose soil. They are active only on warm nights, or the warm part of nights. Adults are often found in great number at lights. Duration of the adult stage is about 30 days.

The biology of Asiatic garden beetle was given by Hallock (1932, 1935, 1936). Asiatic garden beetle was included in the larval keys of Ritcher (1966), and the adult keys by Downie and Arnett (1996).

Damage

Asiatic garden beetle is principally a turfgrass pest, though both larvae and adults can injure home gardens. Relative to similar immigrant scarabs, Asiatic garden beetle is generally more damaging to vegetables than oriental beetle, *Anomala orientalis* Waterhouse, but is less damaging than Japanese beetle, *Popillia japonica* Newman. The major form of damage

Adult Asiatic garden beetle.

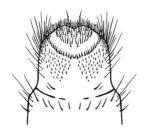

Terminal abdominal segment (*ventral view*) of Asiatic garden beetle larva.

is root feeding by larvae, but adults feed on foliage and flowers. Damage usually occurs only when large numbers of larvae are present, or in the case of crop seedlings, when plants are quite young. Thus, 2–3 grubs per grass plant affect growth, and 4–5 grubs cause enough root injury to cause plant death. There was also an interesting report of beetles that crawled into the ears of campers in Pennsylvania (Maddock and Fehn, 1958). This form of injury, though affecting hundreds of campers in this report, is not generally a frequent or widespread phenomenon.

Management

Asiatic garden beetle populations are usually censused by sampling turfgrass for grubs. The adults are highly attracted to lights, however, so light traps can be a useful tool for monitoring. Residual insecticide is often applied in either liquid or granular form to the soil for grub suppression, though adults are susceptible to insecticide applied to foliage. *Bacillus popilliae* provides partial suppression of larvae.

Carrot Beetle
Bothynus gibbosus (De Geer)
(Coleoptera: Scarabaeidae)

Natural History

Distribution. Carrot beetle is found throughout the United States and southern Canada, and in north-

ern Mexico. This native insect is not usually considered to be a pest in northern states and in Canada, but can be injurious in the southern Great Plains.

Host Plants. This insect feeds on a fairly broad range of plants, and carrot is not one of the crops most frequently damaged. Chittenden (1902) argued that it was the most serious pests of carrot in the past century, and his opinion apparently prevailed. Carrot weevil, *Listronotus oregonensis* (LeConte), and carrot rust fly, *Psila rosae* (Fabricius), have since become much more serious pests of carrot, but the name of this beetle, though inappropriate, is unlikely to change. Hayes (1917) registered his dislike with the name "carrot beetle," but failed to suggest alternatives. The name "sunflower root beetle" might be a more accurate reflection of this insect's ecology.

Vegetable crops damaged by the carrot beetle include cantaloupe, carrot, celery, corn, parsnip, pepper, potato, sweet potato, tomato, and watermelon. Other crops injured are cotton, sugarbeet, sunflower, and perhaps oats and wheat. Weed hosts are quite numerous, and include prairie sunflower, *Helianthus petiolaris*; carelessweed, *Amaranthus palmeri*; horseweed, *Conyza canadensis*; sawleaf daisy, *Prionopsis ciliata*; silverleaf nightshade, *Solanum elaeagnifolium*; and white rosinweed, *Silphium albiforum*. Most plant feeding is accomplished by adults. Larvae develop principally on decaying organic matter, and are often associated with animal manure. Occasionally, they feed on roots, however, and also are cannibalistic.

Natural Enemies. Several flesh flies (Diptera: Sarcophagidae) are parasitic on this beetle, including *Gymnoprosopa argentifrons* Townsend, *Sarcophaga sarracenioides* Aldrich, *S. cimbices* Townsend, *S. helicis* Townsend, and *S. rudis* Aldrich (Hayes, 1917; Rogers, 1974). Additionally, undetermined fungi and bacteria are associated with eggs and larvae. A bulb mite, *Rhizoglyphus echinopus* (Fumouze and Robin) (Acari: Acaridae) builds up to high densities on the overwintering adults. The importance of these natural enemies is unknown.

Life Cycle and Description. There is a single generation annually, with adults serving as the overwintering stage. Overwintering adults emerge in April–June, feed and oviposit. Larvae develop during the summer months, and produce a new brood of adults during July–October that overwinter and emerge the following spring.

Egg. The eggs are deposited in the spring, often late May–July. The egg is slightly elliptical and white. It measures 1.5–2.5 mm long. Eggs are deposited in the soil, often at a depth of several centimeters. Their deposition may coincide with the presence of a suitable host plant, but this is not a prerequisite if organic matter is abundant. The duration of the egg stage is about 11 days (range of 7–22 days).

Larva. The larvae are whitish with a brown head capsule. Larvae also acquire a bluish tint that is quite common among scarab grubs. They attain a length of about 3 cm. Larvae feed for about 52 days (range 40–70 days), passing through three instars. They then spend about seven days in a nonfeeding prepupal state while awaiting transformation to the pupal stage.

Pupa. Pupation occurs in the soil, usually during the period of August–September. Pupae measure about 15 mm long. Pupal color is whitish initially, but soon turns to brown. Duration of this stage averages 19 days (range 11–29 days).

Adult. The adults are moderately large, and measure 11–16 mm long. They are reddish-brown to black. The elytra bear parallel impressions and coarse punctures. The ventral surface of the thorax and legs, but not the abdomen, are covered with a dense layer of hairs. Adults are active and feed during the autumn months, and then enter the soil to pass the winter. The adults are nocturnal, and highly attracted to lights. They are most active between dusk and 10:00 p.m. (Rogers *et al.*, 1978). Because the beetles commonly damage other more fragile insects captured in light traps such as moths, the aforementioned authors suggested stratification of night-time sampling, with beetles being collected during early evening, but traps for moths being operated after 10:30 p.m. Overwintered adults are found abundantly on warm evenings as early as April, but return to the soil at daybreak. In South Carolina, King (1969) reported that adult emergence was greatly influenced by rainfall, but not temperature. He suggested that maximal emergence occurred following an accumulation of about 100 cm in the preceding 15 days. This apparently is not the case in Texas, however, where emergence occurs without such high levels of rainfall. Mating is reported to occur below-ground. The adults feed below-ground, and are the most damaging stage. (See color figure 123.)

The most complete biology of carrot beetle was given by Hayes (1917), but more recent studies by Bottrell *et al.* (1973) and Rogers (1974) are significant contributions.

Damage

Damage occurs on a wide variety of plants, but has been rigorously studied only on sunflower. In such a crop, where the seedlings do not emerge until mid-

Adult carrot beetle.

summer, it is the summer rather than the overwintering generation of beetles that causes injury. Damage to early season crops does occur, but it is caused by overwintering adults. Larval damage usually is insignificant. Adults feed at an average depth of 3.5 cm (range 0.6–12.9 cm), and a single adult can destroy virtually all the lateral roots of a small plant during the average feeding bout of 3.6 days duration. The primary, or tap root, is often only damaged at the surface. Sunflower varieties differ in susceptibility to injury, with perennial species less damaged (Rogers, 1980).

Management

Carrot beetles have proven to be difficult to control unless highly residual insecticides are applied to the soil. Granular and liquid formulations may be applied, but success is not always guaranteed (Bottrell *et al.*, 1970; Rogers and Howell, 1973). Fortunately, they are usually associated with wild plants rather than crops. The principal exception is damage to sunflower, which can be a problem when this crop is grown in the southern Great Plains. Rogers *et al.* (1980) reported on evaluation of resistant sunflower varieties.

Chinese Rose Beetle
Adoretus sinicus Burmeister
(Coleoptera: Scarabaeidae)

Natural History

Distribution. This beetle is native to the Orient, where it is known from the Asian mainland, Formosa, Philippines, Indonesia, and several other islands. In the United States, Chinese rose beetle is known only from Hawaii, where it was first observed at Honolulu in 1891. By 1898 it had spread throughout the Hawaiian Islands and had become a serious pest. It is frequently intercepted on plant shipments sent from Hawaii to California, and has potential to establish in warm climates such as California and Florida.

Host Plants. This species has a very wide host range. Habeck (1964) noted that about 250 species in over 50 plant families had been documented as hosts, but that the actual range might include 600–700 species. Ornamentals, trees and fruits are damaged in addition to vegetables. It is reported to be the most important defoliator of beans in Hawaii, and also damages asparagus, Chinese cabbage, broccoli, corn, cucumber, eggplant, okra, and sweet potato. Among the most highly preferred plants are strawberry, *Fragaria* sp.; Rangoon creeper, *Quisqualis indica*; plumbago, *Plumbago capensis*; seagrape, *Cocoloba uvifera*; cacao, *Theobroma cacao*; macaranga, *Macaranga grandiflora*; castorbean, *Ricinus communis*; hibiscus, *Hibiscus* spp.; and heliconia, *Heliconia* spp.

Natural Enemies. Several parasitoids and predators were imported to Hawaii for biological suppression of Chinese rose beetle. None were successfully established, but two parasitoids that were imported and released in Hawaii for suppression of Oriental beetle, *Anomala orientalis* Waterhouse, also attack Chinese rose beetle. These parasitoids, *Camposomeris marginella modesta* (Smith) (Hymenoptera: Scoliidae) and *Tiphia segregata* Crawford (Hymenoptera: Tiphiidae), were reported not to be very effective (Clausen, 1978). Nonetheless, *C.m. modesta* readily digs for, and parasitizes, grubs in open areas such as tilled fields, and Williams (1931) indicated that it was effective in agricultural fields. Unfortunately, this parasitoid is reluctant to forage in dense vegetation such as lawn grass, where Chinese rose beetle larvae also may develop. A wireworm, *Conoderus exsul* (Sharp) (Coleoptera: Elateridae), is reported to attack the larvae of Chinese rose beetle in preference to plant material. Larvae often are infected with the fungus, *Metarhizium anisopliae*, especially during the wet season.

Life Cycle and Description. This species completes a generation in 6–9 weeks during the summer, so several generations seem likely annually in Hawaii.

Egg. The eggs are deposited in soil, usually within the upper 4 cm. The female burrows in the soil and creates a small chamber before oviposition, and leaves a single egg within. Females typically create 4–6 egg chambers in close proximity. The eggs are shining white initially, becoming dull creamy white before hatching. Eggs are oval, and measure about 1.4 mm long and 0.7 mm wide. Duration of the egg stage is about 12–14 days at 24°C and 8–12 days at 28°C.

Larva. The larvae of Chinese rose beetle are not very destructive, feeding principally on decaying leaf material in the soil. In form, the larva is typical for a scarab beetle, with well-developed thoracic legs, a yellowish-brown head capsule, and a whitish body. The larva attains a length of about 35 mm prior to pupation. There are three instars. The head capsule widths of the instars measure about 1.2, 1.9, and 3.1 mm. Duration of the instars was reported to be 19.6, 16.8, and 34.3 days, respectively, during February–May. Development times were similarly reported to be 22.8, 14.5, and 44.4 days, respectively, during December–April. The final five days of the larval stage is a nonfeeding prepupal period.

Pupa. The pupa initially is yellowish white, gradually turning brownish before emergence of the adult. The pupa resembles the adult except that the elytra are reduced in size and twisted ventrally. The pupa measures about 13 mm long and 6–7 mm wide. Duration of the pupal stage is estimated at 11–17 days, averaging about 14 days.

Adult. The newly emerged adult remains in the soil and does not feed for 3–5 days. The adult is reddish brown, but as it is clothed with a layer of fine scale-like white hairs, the brownish color tends toward gray. The beetle measures 7.0–10.0 mm and is relatively long and narrow in appearance. The adult is active at night, and during the day remains secluded under tree bark or organic debris, or buried in the soil at a shallow depth. Oviposition commences about seven days after mating. Fecundity under caged conditions averaged 54 (range 22–89) eggs, with maximum oviposition of 28 eggs per week.

The biology of this insect is incompletely described. Williams (1931) and Habeck (1964) provided some information on field biology. Habeck (1963) gave a complete description of the immature stages.

Damage

The adults are the destructive stage, and feed freely on the leaves, flowers, and buds of numerous plants. The interveinal feeding pattern in dicotyledonous

Adult Chinese rose beetle.

plants results in a skeletonizing of foliage, and is fairly diagnostic. In monocotyledonous plants, however, veins as well as interveinal tissue are consumed. Photosynthetic rates of remaining plant tissue are more depressed following injury to monocot foliage, relative to dicot tissue (Furutani *et al.*, 1990). Beetles preferentially feed upon tissues high in carbohydrates (Furutani and Arita, 1990). The larvae are commonly found in lawns, gardens, and flower beds, but do not attack living-plant tissue, so they are not destructive.

Management

The adults are attracted to light traps, but females are usually captured only after they have completed oviposition. They are also attracted to stressed or beetle-damaged plants, presumably because such plant produce ethylene in a gaseous form (Arita *et al.*, 1988). Foliar insecticides are effective against adults.

Green June Beetle
Cotinis nitida (Linnaeus)
(Coleoptera: Scarabaeidae)

Natural History

Distribution. Green June beetle is found principally in the southeastern states. It occurs north to southern Connecticut and southern Illinois, and west to Kansas and Texas. This insect is native to North America.

Host Plants. This insect is well known for the feeding habits of the adult stage because the injury of this stage is very visible. The adults feed commonly on ripening fruits such as apricot, apple, blackberry, fig, grape, nectarine, pear, plum, prune, quince, raspberry, and strawberry, which explains an alternative common name of this species—"fig eater." The adults will occasionally feed on vegetables, particularly the ears of corn, and the fruits and vines of melon and tomato. They also are attracted to, and feed upon, the sap, pollen, nectar, and flowers of trees. The larval stage has a broader range of hosts, however, and undoubtedly does more damage, but it is a less visible form of injury. Larvae feed below-ground on the roots of such vegetables as bean, beet, cabbage, carrot, celery, collards, corn, eggplant, endive, lettuce, parsley, parsnip, pea, potato, turnip, and likely others. Larvae also feed on the roots of lawn grasses; such field crops as alfalfa, oat, sorghum, and tobacco; and such ornamentals as dahlia, geranium, hyacinth, rose, and violet. Larvae apparently are capable of developing on decaying organic matter, and soils which are rich in humus or manure are suitable for larval development and are attractive to ovipositing females.

Natural Enemies. Several natural enemies of this beetle are known. Vertebrate predators such as skunks, chipmunks, opossums, and moles, as well as many bird species, have been reported to feed upon green June beetle larvae. Also, *Sarcophaga* spp. (Diptera: Sarcophagidae) have been reared from pupae and adults of the beetle. The digger wasp *Discolia dubia* Say (Hymenoptera: Scoliidae) is particularly attracted to the tunnels of the beetle larvae, where it paralyzes and parasitizes the larvae. The diseases have been poorly studied, but bacterial and fungal diseases have been observed, particularly the fungus *Metarhizium anisopliae.*

Life Cycle and Description. There is a single generation per year. Larvae are present for about 10 months, from August through the winter and spring, with larval development completed in the spring and pupation occurring during May or June. In Virginia, adults are present in June–August, with egg deposition occurring during July–August. In Georgia, Beckham and Dupree (1952) and Morrill (1975) reported high levels of adult emergence anywhere from late June through August, depending on the year.

Egg. The adults apparently are attracted to sandy soil containing manure or decomposing vegetation for oviposition. The eggs are deposited in soil cavities dug by females. Clusters of 10–25 eggs are deposited at a depth of about 7–8 cm in a single cavity at the end of a tunnel, though the individual eggs are often separated by soil packed around the eggs by the females. The eggs are about 1.5 mm in diameter when first deposited, but within a day of oviposition swell to a diameter of about 3 mm as they absorb water. The egg is nearly spherical in shape and gray or dull white. Duration of the eggs stage is about 10–15 days. Females normally deposit 60–75 eggs, but sometimes up to 100 eggs, over a period of about a month. Domek and Johnson (1991) reported an oviposition rate of 2.2 eggs per day for apple–fed adults, mean longevity of 23 days, and an average fecundity of 50.8 eggs per female.

Larva. There are three instars. The young larva, or grub, at hatching measures about 6 mm long, and is creamy white. The yellowish head soon turns brown. As the larva matures, the body color remains mostly grayish white or yellowish white, but the tip of the abdomen acquires a bluish or greenish color. The legs are relatively short and yellowish in color. The body bears many short, thick bristles of yellowish color. The larva bears ridges on its dorsal surface, and uses these to move along the surface of the soil. The larva measures about 16 mm in the second instar and about

28 mm in the third. Head capsule widths are about 1.2, 2.2, and 4.1 mm in instars 1–3, respectively. Duration of the first two instars is about 15 and 25 days, respectively. The third instar is the overwintering form, thus requiring about 9–10 months for maturity. Larvae normally burrow at depths of about 15–30 cm, but move deeper during cold weather. They maintain contact with the surface, often pushing soil to the surface during the night, creating small mounds of soil ("push-ups") not unlike large ant hills. They also have the unusual habit of crawling about the soil surface at night, normally crawling on their back. Larvae of all instars can be found crawling on their back above-ground during the evening hours. Young (1995) studied seasonal movement of larvae in Mississippi and reported first instars principally in August, second in September, and third in October. During the spring and early summer larvae are at their maximum size and activity level, and cause the greatest amount of damage.

Pupa.   For pupation, the larva prepares a substantial cell of soil particles fastened together with fluid excreted by the larva. The cell is oval, measures about 16 mm long and 12 mm wide, and is found at a depth of about 20 cm. The larva transforms to the pupal stage within the cell, and resembles the adult in form except that the elytra and hind wings are wrapped ventrally. Also, the pupa is yellowish in color, acquiring a darker shade just before emerging as an adult. The pupa measures about 25 mm long and 12 mm wide. Duration of the pupal stage is about 20–30 days.

Adult.   The adult is distinctive in form and color. It is fairly large, measuring about 20–30 mm long and 12–15 mm wide. The body width is notably greatest at the base of the elytra. Dorsally, the color of the beetle is velvety green or brown, with a light band of brown or orangish yellow at the margins of the elytra. The orange color may extend to other portions of the elytra. The lower surface is metallic yellow or green. Adults are active from about dawn until noon, and males produce a buzzing sound while flying low to the ground in search of mates. Males use pheromones to locate females, and often several males can be observed competing for the opportunity to copulate with a single female. Mating is relatively brief, lasting only a few minutes, and then the females descend into the soil. The female's burrowing results in production of soil mounds similar to that produced by larvae; thus, adults can be troublesome on golf courses. (See color figure 120.)

Good summaries of green June beetle biology were given by Davis and Luginbill (1921) and Chittenden and Fink (1922). There is evidence of a female sex pheromone and an aggregation pheromone (Domek and Johnson, 1987; 1988), but they have not been characterized. For the aggregation pheromone, a beetle food source and yeasts are implicated in pheromone production (Domek and Johnson, 1990).

Damage

The adults cause injury by feeding on the leaves and fruit of various fruit and vegetable crops. Larval feeding is more important, but less apparent, because it occurs below-ground. Not only do larvae feed on the roots and succulent stems of various plants, but their tunneling habits can be quite destructive to root systems, resulting in uprooting of young plants. On lawns, and particularly on golf courses, the soil pushed upward as larvae tunnel to the surface creates small disfiguring mounds.

Management

Sampling.   Organic matter, incuding rotting or composted animal manure and vegetation, is highly

Green June beetle larva.

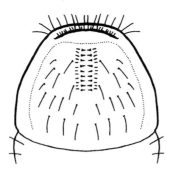

Terminal abdominal segment (*ventral view*) of green June beetle larva.

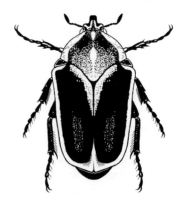

Adult green June beetle.

attractive to ovipositing females. Fields containing high levels of such soil amendments should be carefully monitored for larvae. Soil sampling normally is required to get good estimates of grub abundance, but the mounds resulting from the adult and larval habit of pushing soil to the surface can be indicative of insect densities. Also, pitfall traps can be used to capture larvae as they move over the soil surface at night. Larvae can also be brought to the surface by flooding or heavy application of irrigation water.

The adults are attracted to many volatiles. Fermenting sugar and molasses, malt extract, and isopropyl alcohol can be used as a lure for beetles (Beckham and Dupree, 1952; Landolt, 1990).

Insecticides. Insecticides can be applied to the soil to prevent injury by grubs. Application to the furrow or over the top of the row at planting is preferred for annual crops, whereas in turf broadcast application is required. Adults also can be controlled by application of insecticides, though susceptibility varies considerably among products (Thurston *et al.*, 1956).

Cultural Practices. Previously larvae were routinely captured with pitfall traps positioned at the soil surface. Such traps take advantage of the night-time movement of grubs. The simplest type of trap is a container such as a coffee can that is buried in the soil with the top level to the surface of the ground. Because larvae tend to accumulate in furrows, traps positioned in furrows tend to capture more larvae. Troughs or artificial furrows can be constructed by attaching two boards lengthwise to form a ''V'' and sinking the trough into the soil. The relatively steep, slippery sides prevent larvae from escaping once they fall into the trough. Such practices have little use in commercial crop production but may find favor with home gardeners.

Plowing is often suggested as a remedy for grub infestation. The physical disruption associated with plowing, and the potential for predation of grubs by birds, are cited as benefits. However, in the case of green June beetle, the grubs tend to be located too deep in the soil during the normal spring or autumn plowing periods to be much affected by tillage. When grubs are near the surface, plants typically are in the field and also susceptible to damage by tillage.

Biological Control. Entomopathogenic nematodes (Nematoda: Heterorhabditidae and Steinernematidae) have been evaluated for control of green June beetle larvae (Townsend *et al.*, 1994). The beetle larvae were reported to be somewhat resistant to infection because though about half of the larvae perished under optimal conditions of infection, the nematodes failed to reproduce in the cadavers.

Japanese Beetle
Popillia japonica Newman
(Coleoptera: Scarabaeidae)

Natural History

Distribution. Japanese beetle was accidentally introduced to North America with nursery stock shipped to southern New Jersey, and it was first detected during the summer of 1916. Its origin, as its name suggests, was Japan. It spread rapidly in the eastern United States, attaining Pennsylvania in 1920, Delaware in 1926, New York in 1932, Maryland in 1933, and Connecticut in 1937. During the 1940s, it spread further in the Middle Atlantic region and New England. During the 1950s, it spread to the Great Lakes region, and as far south as South Carolina. By 1975 it is was found in most states east of the Mississippi River, and in Ontario. Japanese beetle can live where the mean soil temperature in the summer is between 17.5° and 27.5°C, the mean winter soil temperature is above -9.4°C, and there is at least 25 cm of precipitation during the summer. Thus, its natural spread westward across the Great Plains is hampered by dry conditions, but it remains a threat to Pacific Coast areas, where it can thrive. Indeed, Japanese beetle has invaded California, but was successfully eradicated.

Host Plants. Japanese beetle has a very wide host range, often given as 400 species or more. The plant families Rosaceae, Malvaceae, Polygonaceae, Tiliaceae, Ulmaceae, and Vitaceae are quite susceptible. Actually, about 50 species are avidly fed upon, and another 50 species are regularly attacked. Other plants are damaged occasionally, usually during the period of great insect abundance when favored foods are exhausted. Among the most preferred hosts are trees such as birches, black walnut, crabapple, crepe myrtle, elm, horsechestnut, linden, maples, and sassafras, and fruits such as apple, apricot, blueberry, cherries, grape, nectarine, peach, plum, and raspberry. Preferred vegetables include asparagus, corn, and rhubarb, with some feeding on beet, broccoli, lima bean, mustard, okra, potato, and snap bean. Other important plants often eaten are poison ivy and rose. Vegetable crops that usually are avoided by beetles are artichoke, Brussels sprouts, cabbage, cantaloupe, carrot, cauliflower, celery, cucumber, eggplant, endive, leek, lettuce, onion, pea, parsley, pepper, pumpkin, salsify, spinach, squash, sweet potato, turnip, and watermelon. Interestingly, Japanese beetle readily consumes, but it is poisoned, by some geranium spp. and perhaps other species.

Natural Enemies. Native natural enemies, though numerous, often are not adequate to keep Japanese beetle from becoming abundant. Among the important bird predators are the grackle, *Quiscalus quiscala*; meadowlark, *Sturnella magnam*; and starling, *Sturnis vulgaris*. Mammals such as the common mole, *Scalopus aquaticus*; shrew, *Blarina brevicauda*; and skunk, *Mephitis mephitis;* similarly feed on beetles, especially the larval stage, and are often quite destructive to plants owing to their digging for the grubs. Native predatory and parasitic insects are generally lacking.

A significant effort was undertaken to introduce from Asia exotic beneficial organisms that might suppress Japanese beetle. Several flies (Diptera: Tachinidae) were introduced, including the adult parasitoid *Hyperecteina aldrici* Mesnil and the larval parasitoids *Prosena siberita* (Fabricius) and *Dexilla ventralis* (Aldrich). Among the wasps introduced, only the larval parasitoids *Tiphia popilliavora* Rohwer and *Tiphia vernalis* Rohwer (both Hymenoptera: Scoliidae) are important, with the latter species causing over 60% parasitism in some locations. However, the entomopathogenic nematode *Steinernema glaseri* (Nematoda: Steinernematidae) and the milky disease bacteria *Bacillus popilliae* and *B. lentimorbus* have proved most effective in population reduction. An excellent summary of biological control efforts was given by Fleming (1968). The effectiveness of *B. popilliae* seems to vary somewhat over time, perhaps explaining the periodic increases in Japanese beetle abundance (Dunbar and Beard, 1975). Other pathogens of Japanese beetle also are known. In a survey conducted in Connecticut, for example, gregarines infected 0–100% of grubs, the microsporidian, *Ovavesicula popilliae*, was found in 4–94% of grubs, and the rickettsial disease *Rickettsiella popilliae* and the fungus *Metarhizium anisopliae* were noted (Hanula and Andreadis, 1988). In this study, infection rates by milky disease ranged from 0–100%

Life Cycle and Description. There normally is one generation per year, but in northern areas some individuals require two years to complete their development. Adults generally emerge in May–June, remain abundant through the summer months, and then perish in September. The eggs are deposited in June–August, with the first instar developing in late summer or autumn, and larvae overwintering as second or third instars. The third instar normally is completed in the spring. Total development time is about 200 days when reared at 20°C, but only about 140 when reared at 25°C.

Egg. The eggs are usually deposited in the proximity of favored adult food plants. Grasses are the favored oviposition site, and moist soil is preferred over dry. The female digs in the soil to a depth of 5–10 cm to oviposit. She normally deposits about 3–4 eggs at a location, and repeats the digging and oviposition processes over a period of weeks until 40–60 eggs are produced. The maximum number of eggs produced is about 130. The eggs are unusual as they vary slightly in shape, ranging from spheres initially measuring 1.5 mm in diameter, to ellipsoids measuring 1.5 mm long and 1.0 mm wide. The eggs are white to yellowish white. They enlarge as the embryo matures, as is common among scarab beetle eggs, until they are about twice the original size. Enlargement is due to water absorption, and lack of water inhibits egg development or causes death of eggs. During dry periods females selectively oviposit in low, poorly drained areas and irrigated fields. Duration of the egg stage is about 21 days at 20°C, but only 8 days at 30°C. No egg hatch occurs below 15°C or above 34°C, and 30°C seems to be about optimal.

Larva. There are three instars during the larval stage. Upon hatching the young grub measures about 1.5 mm long, bears three pairs of thoracic legs, and has 10 abdominal segments. Although the larva is completely white initially, the head darkens to yellowish brown; after feeding commences the tip of the abdomen becomes grayish brown. Numerous short, dark spines occur on the dorsal surface of the abdomen. On the ventral surface of the tip of the abdomen there are several rows of spines, the configuration of which forms a ''V''; this configuration serves to distinguish this grub from other soil-dwelling beetle larvae. When dug from the soil the larva typically assumes a curled posture, sometimes referred to as C-shaped. By completion of the first instar the larva usually attains a length of about 10 mm. The second and third instars are similar in appearance, except in size. Body lengths attained by larvae are about 18.5 and 32.0 mm for instars two and three, respectively. Head capsule widths for the instars are about 1.2, 1.9, and 3.1 mm, respectively. Larvae develop at temperatures from 17.5° to 30°C. Duration of the instars is about 30, 56, and 105 days for instars 1–3, respectively at 20°C. In contrast, duration of the instars is about 17, 18, and 102 days at 25°C. Survival of grubs is favored by adequate moisture in the summer, snow cover or warm winter temperatures, and acidic soil.

The larvae move considerable distances in the soil. They will be deeper in dry than moist soil, but temperature is probably a more important determinant of location. During the summer nearly all larvae are within the top 5 cm of soil. During winter the grubs move to depths of 5–15 cm, and then return to the surface in spring. In habitats with suitable food, such as sod, larvae rarely move laterally more than 1 m during a season, whereas in fallow ground they may move several meters.

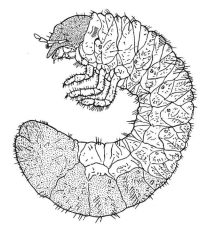

Japanese beetle larva.

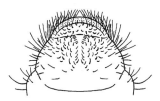

Terminal abdominal segment (*ventral view*) of Japanese beetle larva.

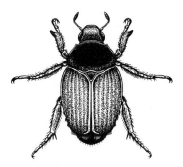

Adult Japanese beetle.

Pupa. The pupa develops within a cell in the soil. The pupa resembles the adult, though the elytra are shrunken and distorted. It measures about 14 mm long. The pupa is initially white, but gradually assumes a metallic green color. Pupae develop at temperatures of 13–35°C. Duration of the pupal stage is about 17 and seven days at 20° and 25°C, respectively.

Adult. The adult remains in its pupal cell for 2–14 days before digging to the surface and emerging. The beetle is oval, and measures 8–11 mm long and 5–7 mm wide. On average, the female is slightly larger than the male. The beetle is principally metallic green, including the legs, but the elytra are coppery brown. The elytra do not quite cover the tip of the abdomen, exposing a row of white spots on each side of the abdomen and a pair of spots on the dorsal surface of the last abdominal segment. The dorsal surface of the body is largely smooth and lacks pubescence, but the ventral surface bears short gray hairs. The head and thorax bear many punctures, and the grooves of the elytra also are punctate. Adults normally live 40–50 days, but under extreme weather conditions may live much shorter or longer. Beetles often disperse to sunny locations to feed. They also seem to be attracted to damaged or weakened plants. Adults usually mate while on foliage, and may remain in copula for several hours and mate repeatedly. Males often aggregate near recently emerged females and attempt to mate.

They sometimes grapple with one another as they attempt to mate, forming balls of beetles that consist of 25–200 males surrounding a female! Males locate virgin females by her production of sex pheromone (Tumlinson *et al.*, 1977). (See color figure 125.)

The most complete treatments of Japanese beetle were provided by Fleming (1972, 1976), but there are numerous lengthy reports on biology or management. Sim (1934) gave an account of characters to distinguish Japanese beetle larvae from similar grubs.

Damage

This is a very important insect in the northeastern United States, but less so on vegetables than some other plants. Their overwhelming abundance and the tendency of adults to aggregate are significant elements in their damage potential. Adults feed on the upper surface of foliage, eating the tissue between the veins and leaving a lace-like skeleton. Such leaves invariably perish. They also feed on the stems of succulent asparagus, and on the young silk of corn. They attack fruit, and have the curious habit of feeding readily on the fruit of peach but avoiding its foliage. Adults consume foliage at the rate of 30–40 sq mm per hour. Grubs feed on plant roots; initially the rootlets are attacked, but larger roots are consumed as the larva matures. Although the roots of grasses are preferred, roots of many other plants, including numerous vegetables, are consumed.

Management

Sampling. Adult population monitoring is usually accomplished with traps that use chemical odors as a lure. Both food-based and sex pheromone-based traps, or a combination trap, can be used successfully. The components incorporated into multilure traps usually include phenethyl propionate, eugenol, geraniol, and sex pheromone. Metcalf and Metcalf (1992) gave a good synopsis of Japanese beetle chemical attractants. Klostermeyer (1985), working in Tennessee, reported higher catches with traps baited with sex plus food lures than food lures alone. Trap design is not as important as is the composition of the lure. However, trap design and placement are important considerations, and sex attractant-containing traps placed close to the soil (30 cm height) are more effective than those positioned at 90 cm elevation (Alm *et al.*, 1994). Interestingly, traps containing only food-based lures are more effective when placed at more elevated locations (Ladd and Klein, 1982). This is not too surprising when the ecology of Japanese beetle is considered: virgin females emerge from the soil, and trees are preferred-host plants. Populations are aggregated, and Allsopp *et al.* (1992) estimated that 11–66 traps might be necessary to estimate population density accurately, depending on the density of the beetles. Smitley (1996) reported a positive correlation between adult catches in traps and larval abundance, thereby justifying the use of traps for population monitoring.

Insecticides. Persistent insecticides formerly were applied to the soil to prevent injury to roots by grubs. Insecticide resistance developed, especially on golf courses and other areas where esthetic values are important (Niemczyk *et al.*, 1975). The registrations on such persistent insecticides have since been canceled, and less persistent insecticides are used for root protection in crops or turf. Insecticides are also applied to foliage for adult control. Foliar protectants must be applied repeatedly, especially if material with short residual activity are used. Hydrated lime, though not insecticidal, is a fairly good repellent. Lime must be applied frequently, as either a dust or aqueous suspension, to be effective (Fleming *et al.*, 1934).

Biological Control. Milky spore disease formulations consisting of *Bacillus popilliae* and *B. lentimorbus* are commercially available. This product usually is recommended for control of grubs on turf, where most larvae develop, but this materially benefits vegetable production by reducing the numbers of adults present as defoliators. In recent years there has been considerable interest in the use of entomopathogenic nematodes as biological insecticides for larval control. *Steinernema* and *Heterorhabditis* spp. (Nematoda: Steinernematidae and Heterorhabditidae) can be applied to turfgrass successfully, especially if applications are followed by watering. High levels of suppression can be attained, though there are significant differences in nematode species and strain effectiveness (Klein and Georgis, 1992; Selvan *et al.*, 1994). Some work indicates that the nematodes may be more effective than insecticides (Cowles and Villani, 1994). Research on nematodes has been largely restricted to turf, though the aforementioned findings are likely applicable to crops also.

Traps. Several studies have been conducted to assess the potential of trapping to eliminate or suppress beetles, or to protect plants. In general, though large numbers of beetles have been captured in such efforts, the results are disappointing. For example, large numbers of traps containing food-based lures were placed on Nantucket Island, Massachusetts, for a three-year period. Although densities were reduced by about 50%, the beetles persisted (Hamilton *et al.*, 1971). In Kentucky, placement of 2–7 traps adjacent to host plants not only failed to protect susceptible plants, they seemed to increase damage levels.

Cultural Practices. Nonhost plants act as an impediment to movement of Japanese beetle. Sorghum, for example, which is not eaten by Japanese beetle, reduces the rate of dispersion of adults from soybean patch to soybean patch (Bohlen and Barrett, 1990). In contrast, strip cropping of corn and soybean has no effect on adult distribution (Tonhasca and Stinner, 1991), probably because both plants are suitable hosts.

Screening and row covers can often be used to prevent adults from gaining access to vegetable plants. Unfortunately, corn is probably the most highly preferred vegetable, and this plant is too large to cover conveniently.

Injury to corn can be prevented or reduced by modifying the time of planting. Early planting, in particular, is beneficial in culture of corn because the corn ears are pollinated before the adults become abundant and consume the corn silks; once the ear is pollinated the silks have no value. Where the growing season permits, planting of late-season varieties also is beneficial, because the silks are not produced until after most adults have perished.

There are numerous reports that certain plants, if eaten, are toxic to Japanese beetles. Plants reported to be toxic are geranium, *Pelargonium domesticum*; bottlebrush buckeye, *Aesculus parviflora*; and castorbean, *Ricinus communis*. There is evidence that the first two species are toxic, but there are no data to support reports of toxicity associated with castor bean. Castor bean is known to cause mammalian toxicity, which may account for this erroneous information concerning Japanese beetle. Unfortunately, though geranium and bottlebrush buckeye are, in fact, somewhat poisonous to Japanese beetle, they are rarely eaten by this insect, so impart little suppressive value.

Soil preparation can affect Japanese beetle grub populations. Because beetles favor acidic soil, application of lime to acidic soil to attain a neutral pH can result in lower grub populations (Polivka, 1960). Tillage can destroy grubs, and repeated tillage or rototilling is more disruptive than a single cultivation.

Oriental Beetle
Anomala orientalis Waterhouse
(Coleoptera: Scarabaeidae)

Natural History

Distribution. Oriental beetle is an immigrant species. It likely originated in the Philippines or Japan, but was found in Hawaii in 1908 and Connecticut in 1920. It has since spread to other northeastern states and southward, with its current distribution extending from Massachusetts and New York in the north to North Carolina in the south, and also in Ohio near Lake Erie. Since its initial introduction, its rate of spread has been surprisingly slow.

Host Plants. Oriental beetle causes relatively little injury to vegetables, though it may feed on bean, beet, onion, and rhubarb. Also, sugarcane is severely damaged in Hawaii, as well as strawberry and nursery stock in the northeastern states. Oriental beetle is principally a pest of turfgrass, where the larval stage attacks roots. Other plants occasionally injured, principally by the adults, include such flowers as dahlia, hollyhock, phlox, and rose. Most damage to economically important plants results from feeding by the larval stage.

Natural Enemies. Several natural enemies were imported into Hawaii and released, including two species which were successfully established—*Campsomeris marginella modesta* Smith (Hymenoptera: Scoliidae) and *Tiphia segregata* Crawford (Hymenoptera: Tiphiidae). The former species is considered to be quite important as a mortality agent of larvae. Although at least five species of scoliids were introduced to the northeastern states, no parasitoids were successfully established. However, a survey of microbial pathogens in Connecticut (Hanula and Andreadis, 1988) demonstrated that gregarines infected up to 100% of the population and *Bacillus popillae* was found in up to 25% of the grubs sampled. Although the effect of these pathogens is not known, they likely help to suppress population densities of oriental beetle. Birds are sometimes observed to flock to grub–infested turfgrass, where they feed extensively on larvae. Weather also seems to limit the abundance of oriental beetle, and populations diminish during abnormally dry summers.

Life Cycle and Description. Although a complete life cycle may be completed in only three months in Hawaii, there normally is but a single generation annually in the temperate climate of the northeastern states. In New England, some individuals even require a second year to complete their development. Overwintering normally occurs in the last larval instar. In Connecticut, larvae feed during April–June, and pupate in June. Pupae transform to adults in late June–August, and emerge from the soil within a few days. Following mating and oviposition, eggs begin to hatch in July and August. Larvae usually attain the third instar by late September.

Egg. The eggs are oval in shape. They typically measure about 1.2 mm wide and 1.5 mm long when first deposited, but swell slightly as they absorb water, and increase to 1.6 mm wide and 2.0 mm long. The period of incubation is about 17 days at 24°C, averaging 25–28 days under field conditions.

Larva. Upon hatching the larva digs nearly to the surface to feed on roots and dead organic matter. Larvae display three instars. The first and second instars each require about 30–40 days to complete their development. The third instar overwinters, moving from a feeding position near the soil surface to greater depth in the soil, often 25–40 cm. Larvae are about 4 mm long following hatching, and attain a length of about 4, 15, and 20–25 mm by the end of instars 1–3, respectively. Head capsule width is about 1.2, 1.9, and 2.9 mm, respectively, for the corresponding instars. The larva is similar in form to other scarab grubs—thick bodied and C-shaped. Its body is whitish, though the head is brown. The body bears numerous short hairs, with the arrangement of hairs on the ventral surface of the terminal abdominal segment important for diagnostic purposes.

Pupa. Pupation occurs in early June when larvae dig to a depth of 3–20 cm and create a small cell. The mature larva remains in the cell for 5–8 days before molting to a pupa. The duration of the pupal stage is about 15 days. The pupa is about 10 mm long and 5 mm wide, and light brown in color. The pupa generally resembles the adult though the wings are short and twisted to the ventral surface.

Adult. The adults are elongate-oval, and more tapered at the anterior end. The elytra bear pronounced ridges. The male beetles measure about 9 mm long and 4 mm wide; the females measure about 10.3 mm

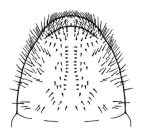

Terminal abdominal segment (*ventral view*) of oriental beetle larva.

and 4.5 mm, respectively. The beetles are highly variable in coloration, ranging from yellowish brown to black. However, the head is invariably black, and the pronotum usually bears two moderately sized irregular black areas that coalesce into a single large spot in darker forms. The elytra often bear irregular black areas that form a V-shaped pattern, but as noted previously, coloration is not consistent in this species. Adults are active during morning and afternoon as well as at night. They may be attracted to flowers, where they feed slightly, or to lights. Females produce a sex pheromone that attracts males (Zhang *et al.*, 1994). Both sexes mate repeatedly, and females commence oviposition within 1–5 days of mating. Oviposition normally continues for about seven days (range 4–20 days). Females burrow into the soil, usually to a depth of 10–20 cm, to deposit eggs.

Detailed description and biology were provided by Friend (1929). The larva of oriental beetle was included in the key by Ritcher (1966), and the adults in the keys of Downie and Arnett (1996).

Damage

Damage is caused principally by the larval stage, which feeds on the roots of plants. Adults are generally considered to be of no consequence, though occasionally they feed on foliage of vegetable crops. Normally, only turfgrass experiences severe injury, but other plants are sometimes damaged. In the northeastern states, where oriental beetle co-occurs with Japanese beetle, *Popillia japonica* Newman, and Asiatic garden beetle, *Maladera castanea* (Arrow), the oriental beetle is sometimes the most important cause of turfgrass injury (Adams, 1949; Facundo *et al.*, 1994).

Management

Population density is usually assessed by examining areas of dead or dying sod for the presence of larvae. However, populations also may be monitored with sex pheromone-baited traps (Facundo *et al.*, 1994; Alm *et al.*, 1999). Applications of insecticide or *Bacillus popillae* to turfgrass are the principal methods of oriental beetle suppression. The susceptibility of

Adult oriental beetle, various color forms.

oriental beetle grubs to *B. popillae* seems to vary over time, perhaps accounting for some of the variation in abundance (Dunbar and Beard, 1975). Compared to some other turf-damaging species, oriental beetle is relatively susceptible to most insecticides (Villani *et al.*, 1988), though there may be site-related differences in field efficacy (Baker, 1986). Oriental beetle seems to be relatively non-susceptible to entomopathogenic nematodes (Nematoda: Steinernematidae, Heterorhabditidae) (Alm *et al.*, 1992).

Rose Chafer
Macrodactylus subspinosus (Fabricius)

Western Rose Chafer
Macrodactylus uniformis Horn
(Coleoptera: Scarabaeidae)

Natural History

Distribution. Rose chafer occurs principally in eastern North America, from Maine, Quebec, and Ontario south to Virginia, Georgia, and Tennessee. However, to a lesser degree its range also extends west to Nebraska and Oklahoma, and is replaced in the southwestern states by western rose chafer. These species are native to North America.

Host Plants. Known mostly as a pest of rose, peony and grapes, rose chafer adults feed on a wide variety of plants, including such vegetables as asparagus, bean, beet, cabbage, corn, pepper, rhubarb, sweet potato, tomato, and perhaps others. Apple, blackberry, cherry, grape, peach, pear, plum and strawberry are among the fruits damaged. Flowers injured include dahlia, daisy, foxglove, geranium, hollyhock, iris, hydrangea, peony, poppy, and rose, though foxglove apparently is poisonous to beetles. Trees such as elm, magnolia, oak, sassafras, sumac and others also are attacked; even conifers are not immune to attack. Larvae attack the roots of grasses and other plants. The dietary of western rose chafer is poorly documented, but apparently is similar to rose chafer.

Natural Enemies. The natural enemies are not well documented. General predators such as birds and toads are known to feed on rose chafer beetles, and ground beetles (Coleoptera: Carabidae) are thought to consume larvae, but there appear to be no records of parasitoids. It is difficult to imagine native species without a large complement of natural enemies, so the absence of records must simply reflect lack of attention by entomologists in recent times.

Life Cycle and Description. In New Jersey the winter is passed as a partly or fully grown larva, pupation occurs in April, adults emerge in late May and

June, eggs are deposited in June–July, and larvae develop from July until cold weather drives them deep into the soil, usually in October–November.

Egg. The eggs reportedly are deposited in light, sandy soil or tilled soil, but not in dense sod or heavy, wet soil, because larvae are usually found only in the former environments. However, in preference tests adults oviposited preferentially in wet soil, and were not affected by soil texture (Allsopp *et al.*, 1992), so oviposition behavior is not fully known. The eggs are nearly spherical or oval, yellowish white, and measure about 0.8 mm in diameter. The adult deposits eggs singly along a burrow, which she digs below-ground, sometimes to a depth of 10 cm, but also very near the soil surface. Fecundity is reported to be 24–36 eggs, all of which are deposited in a single burrow. Duration of the egg stage is about 21 days.

Larva. Upon hatching, larvae may feed on organic matter in the soil but generally feed on roots. Larvae are yellowish white, with a tinge of blue toward the tip of the abdomen, well equipped with hairs, and attain a length of about 20 mm by autumn. The head is pale red. The body is C-shaped like most scarab beetles, though they are not as heavy-bodied as most white grubs. Larvae descend below the frost line in the winter, returning to the surface in the spring to feed for 2–4 weeks before pupating.

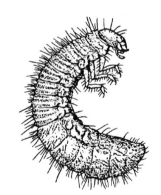

Rose chafer larva.

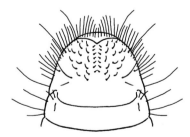

Terminal abdominal segment (*ventral view*) of rose chafer larva.

Rose chafer pupa.

Pupa. Pupation occurs in a small cell prepared by the larva near the soil surface. The pupa is yellowish white, about 15 mm long, and generally resembles the adult, though the wings are undeveloped and the legs are drawn up close to the underside of the body. Duration of the pupal stage is 2–4 weeks.

Adult. The adults of both species are quite similar in appearance, though the western rose chafer averages slightly larger in size and has longer pubescence on the elytra. The rose chafer is fairly long and slender, measuring about 8–12 mm, though the very long legs give the impression of a much larger insect. The dorsal surface of the body is reddish brown but mostly covered with yellow hairs, imparting a tan color; it is either reddish brown or blackish below. The head and legs are reddish or brown, and the spines and claws reddish brown or blackish. The antennae are expanded terminally. The legs are well armed with spines and large terminal claws. The adults tend to be abundant for about a month, usually June, during which most damage occurs. The adults are active during daylight. Beetles mate soon after emerging from the soil, feed for about 7–10 days before beginning oviposition, and oviposition is completed in a few days. Thus, adult longevity is no more than three weeks. (See color figure 124.)

The biology of rose chafer was given by Riley (1890), Smith (1891), Chittenden (1916b), and Lamson (1922).

Damage

Most damage is caused by the adults, which while preferring to feed on blossoms, also consume fruit and foliage. Leaves are often skeletonized. Light or white-colored blossoms are preferred. Damage occurs

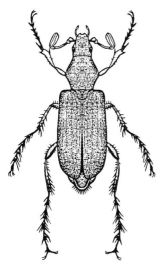

Adult rose chafer.

mostly in areas with sandy soil, with beetles infrequent elsewhere. Damage to fruit causes the greatest degree of alarm among gardeners, though flower blossoms also may be shredded. Rose chafer is not generally known as a pest in commercial crops, though commercially produced grapes sometimes suffer. The beetles also are poisonous to chickens, with mortality due to a toxin, not to mechanical injury by the beetle's long and sharp claws as was thought previously (Lamson, 1922). Beetle larvae feed on the eggs of *Melanoplus* spp. grasshoppers, but the importance of this beneficial behavior is unknown.

Management

Several chemicals have proven to be attractive to adults, and have potential for development as lures. Among the attractive chemicals are caproic acid, hexanoic acid, valeric acid, octyl butyrate, nonyl butyrate, and others. The lures are most effective when combined with a white trap (Williams and Miller, 1982; Williams *et al.*, 1990). Clean cultivation of crops is recommended both because tillage can destroy stages in the soil, particularly pupae, and because tillage deprives larvae of grasses and other preferred food. Insecticides applied to the foliage are usually recommended when adults are abundant.

Spring Rose Beetle
Strigoderma arboricola (Fabricius)
(Coleoptera: Scarabaeidae)

Natural History

Distribution.   Spring rose beetle is a native species found in the northeastern United States from New York to North Carolina, west to Colorado and Minnesota. In Canada it is known from southern Ontario.

Host Plants.   The adults feed on the blossoms of many plants. Wild and cultivated rose are favored food plants, which is the basis for its common name. Blossoms, and sometimes fruits, of other economically important plants such as blackberry, cotton, clover, coreopsis, hairy vetch, hollyhock, honeysuckle, iris, lilies, perennial pea, timothy, and peony also are consumed. Vegetable crops attacked by adults include bean, cantaloupe, corn, cowpea, and cucumber. Among weeds fed upon by adults are dewberry, *Rubus* spp.; elderberry, *Sambucus canadensis*; hoary verbena, *Verbena stricta*; horsemint, *Mondarda punctata*; plantain, *Plantago* spp.; prickly pear cactus, *Opuntia humifusa*; water lily, *Nymphaea* sp.; water willow, *Justicia americana*; wild parsnip, *Pastinaca sativa*, and others. Larval feeding habits are not well-known, but spring rose beetle larvae are reported to damage roots of corn, cotton, peanut, potato, soybean, strawberry, sweet potato, and various grasses and grain crops.

Natural Enemies.   Little is known about the natural enemies of this insect. An entomopathogenic nematode, *Steinernema glaseri* (Nematoda: Steinernematidae), is known to affect the larvae of this and other scarab beetles (Poinar, 1978).

Life Cycle and Description.   There is a single generation per year, with overwintering occurring in the last larval instar. In Virginia, adults are abundant starting in May or June, while eggs are abundant in June. In Minnesota and Ontario, adults were reported to be common in June and July, while eggs were found in July and early August.

Egg.   The eggs are deposited singly in the soil at a depth of 5–10 cm. Grayson (1946) reported a fecundity of 30–40 eggs per female, but this seems to be an underestimate. The eggs are white and oval. Size varies because the eggs absorb water as they mature; the length increases from about 1.59 to 1.86 mm, the width from 1.07 to 1.63 mm. Duration of the egg stage is reported to average about 13.5 days (range 8–21 days).

Larva.   The larva is whitish, with a light brown head. There are three instars. Duration of the instars is about 16 days for each of the first two stages and 270 for the third, or overwintering, instar. The terminal portion of the last instar, or prepupal period, is relatively short, about 11 days. Larvae occur in the soil at a depth of 12–28 cm. Under warm conditions, larval development proceeds faster. Hoffmann (1936) reported larval development of about 160 days in the laboratory.

Pupa. Pupation occurs in the soil, normally during April or May. Larvae form an earthen cell measuring about 30 mm long and 10 mm wide, and pupate within. The pupa measures 9–12 mm long and gradually changes its color from light to dark brown as it matures. Duration of the pupal stage is usually 12–24 days.

Adult. The beetles are moderate in size, measuring 8.5–12.5 mm long. They are somewhat variable in color, but the head and thorax are usually blackish green, and the elytra brownish yellow. The elytra are marked with distinct ridges. The underside of the beetles bears long hairs, and is brown or brownish gray. Adults appear to persist for about 30 days. Mating is relatively brief in duration, requiring only 2–15 minutes. Mating is often observed on flowers, a favorite food for adults. The pre-oviposition period is 7–16 days, followed by oviposition in the soil.

The biology of spring rose beetle was given by Hayes (1921), Hoffmann (1936), and Grayson (1946). The larvae were described by Ritcher (1966).

Damage

This insect is known principally as a pest of roses, as the adults feed greedily on the blossoms. However, the larvae feed on the below-ground parts of several plants, and they not infrequently are associated with sweet potato and peanuts.

Management

Historically, larvae are usually more abundant in low and heavy soil than in well-drained, sandy soil. Soils high in organic matter also are more prone to injury. Application of liquid or granular insecticide to the soil at planting-time or early in the growth of the crop has generally prevented injury. However, in Colorado injury has occurred in sandy soils, and it

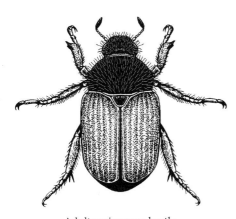

Adult spring rose beetle.

has proven difficult to attain protection with planting-time treatments, because the insecticide dissipates before eggs hatch (Peairs, pers. comm.).

White Grubs
Phyllophaga and others
(Coleoptera: Scarabaeidae)

Natural History

Distribution. The whitish, soil-dwelling larvae of scarab beetle species are called white grubs. Most are in the genus, *Phyllophaga*, but sometimes included are some members of the genera, *Anomala*, *Cyclocephala*, and *Polyphylla*. The adults are called June beetles, May beetles, chafers, or cockchafers. They are found throughout the United States and southern Canada, but are most abundant east of the Rocky Mountains. Nearly all species are native. Among the most damaging species are common June beetle, *Phyllophaga anxia* (LeConte); common cockchafer, *P. fervida* (Fabricius); northern June beetle, *P. fusca* (Froelich); fever June beetle, *P. futilis* (LeConte); southern June beetle, *P. inversa* (Horn); wheat white grub, *P. lanceolata* (Say); and tenlined June beetle, *Polyphylla decemlineata* (Say).

Host Plants. Both larvae and adults can be damaging, but they usually have markedly different feeding preferences. Larvae are found in the soil, and normally feed on the roots of grasses. Adults, however, usually feed on the leaves and flowers of deciduous trees and shrubs, particularly ash, elm, hackberry, hickory, locust, oak, poplar, walnut, and willow, but sometimes pine. Root crops such as beet, potato, and turnip are among the crops most susceptible to injury by larvae, but corn and strawberry are sometimes damaged, and the seedlings of virtually all vegetables are susceptible if grub density is high. Vegetables crops are not preferred oviposition sites, so damage usually occurs only when crops are planted into land that recently has been grass sod, pasture, or sometimes grass crops such as timothy, small grains, or sorghum. Tree seedlings, especially conifers, are damaged in both northern and southern states when planted into grass sod or pasture. Alfalfa and clover seem to be avoided even though both crops often contain some grass. Adults disperse widely, and trees adjacent to grassland, especially when few in number, are sometimes injured by defoliation.

Natural Enemies. A large number of predators, parasitoids, and pathogens of white grubs is known. Undoubtedly, many soil-inhabiting predators, especially beetles (Coleoptera), kill white grubs, but there is little documentation of this. Larvae of robber flies (Diptera: Asilidae) and beeflies (Diptera: Bombyliidae)

White grub, *Phyllophaga* sp.

are frequently found in association with grubs. Among the parasitoids are such tachinids as *Cryptomeigenia theutis* (Walker), *Eutrixa exilis* (Coquillett), *Eutrixoides jonesii* Walton, and *Microphthalma* spp. (all Diptera Tachinidae); the light flies, *Pyrgota* spp. (Diptera: Pyrgotidae); and the wasps *Pelecinus polyturator* (Drury) (Hymenoptera: Pelecinidae); and *Tiphia* spp. (Hymenoptera: Tiphiidae). Of the parasitoids, only *Tiphia* spp. are documented to be of considerable importance, sometimes accounting for up to 50% mortality among grubs. Mites (Acari: several families) are commonly associated with grubs and beetles; eggs and pupae generally are free of mites. However, few of these mite associates thought to be are parasitic or detrimental to white grubs (Poprawski and Yule, 1992). All the common groups of insect pathogens have been found associated with white grubs, especially *Metarhizium anisopliae* and *Beauveria bassiana* fungi. There is no evidence that these pathogens, and many of the other associates of white grubs, have significant impact of grub density. A review of many natural enemies of white grubs was published by Lim *et al.* (1981b).

Life Cycle and Description.   The life cycle of the various species ranges from 1 to 4 years, though in the *Phyllophaga* species it is 2–4 years, with the three-year life cycle most frequent. Duration of the life cycle is often related to latitude. In the southern United States two year life cycles are common, whereas in the north central states three-year cycles predominate, and in northern states and Canada 3–4 year cycles occur.

The general three-year life cycle starts with oviposition in the spring or early summer, followed by feeding and growth until autumn when cold weather induces a period of inactivity, and overwintering by second instar larvae. During the second year, feeding is resumed in the spring, the third instar is attained, and feeding continues until cold weather when another period of inactivity occurs. In the spring of the third year, larval growth is completed and pupa-

tion occurs during the summer. The adult remains in the soil until the following spring. Thus, the three-year life cycle occurs during a four-year calendar period.

In the typical two-year life cycle, oviposition occurs in the summer, the eggs hatch in late summer, and young larvae commence feeding, quickly attaining the third larval instar and overwinter. They feed again throughout the second summer and overwinter as mature third instars. The following spring they feed briefly, pupate, and adults emerge, mate, and oviposit. Thus, in the two-year life cycle larvae develop very quickly and the adult does not undergo a period of arrested development, as occurs in the three-year cycle. The two-year period of development occurs over three calendar years.

Following is a description of common June beetle, *P. anxia*, perhaps the most abundant and damaging *Phyllophaga* in northern areas, and known from nearly all of Canada and the United States. It serves well to illustrate the biology of white grubs and June beetles.

In Quebec, common June beetle displays the typical three-year life cycle. The adults begin to fly in May and oviposit in June. First instar grubs are abundant in July and molt into second instars in August. The second instars overwinter and molt into third instars during June of the second year. Third instars overwinter at the end of the second year, developing into prepupae during June of the third year, and pupating in July. They are fully formed by September, but remain in the soil until the following spring (Lim *et al.*, 1981a). As noted earlier, the three-year life cycle is spread over four calendar years.

Egg.   About 10 days after mating the female deposits pearly-white and elongate-oval eggs about 2.4 mm long and 1.5 mm wide. The eggs absorb water, becoming enlarged in size, and eventually measure about 3 mm long and 2 mm wide. Mean fecundity is about 55 eggs per female. Eggs hatch after about 20–30 days.

Larva.   Young grubs often feed on decaying vegetation before turning to root feeding, but they are not associated with manure. The grubs are pearly-white, with a dark head and legs. They bear a thin covering of stout hairs, and when dug from the soil, they assume a curved-body posture or C-shape. Although the grubs feed at a depth of perhaps 5–15 cm during the summer months, they descend to about 30–100 cm during the winter months. Depth of burrowing is markedly affected by soil and moisture conditions and the age of the grub. Survival of grubs is higher in light soils and in moderate moisture conditions; heavy rain is detrimental both to oviposition and survival of young grubs. Grubs pass through three instars over about a 24-month period that is spread over part of three calendar years (Lim *et al.*, 1980a). (See color figure 122.)

Pupa. The larva prepares a cell in the soil for pupation. This cell also serves as the overwintering site for the adult, which does not emerge from the soil until the following spring.

Adult. The body of adult *P. anxia* is oblong-ovate, dark brown in color, and measures 17–21 mm long. The elytra are marked with numerous fine and shallow punctures. The underside of the thorax bears a dense coat of long hairs. The antennae are 10-segmented, elbowed, and expanded at the tip. The basal segments of the legs are generally stout, the tibiae are broadened and armed with teeth or stout spurs, but the tarsal segments are elongate and narrow. (See color figure 121.)

An excellent treatment of *Phyllophaga* biology, and a taxonomic treatment of adults, was presented by Luginbill and Painter (1953). Other reliable sources of general biology, damage and management information were found in Davis (1922), Luginbill (1938), and Ritcher (1940). Occurrence of white grub species in the north central states was given by Pike *et al.* (1977) and in the southeastern states by Forschler and Gardner (1990). The publication of Ritcher (1966) was an especially comprehensive treatment of grubs. Rearing techniques were reviewed by Lim *et al.* (1980a).

Damage

White grubs are only occasional pests in vegetable production. Grubs have limited mobility; their distri-

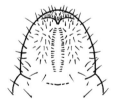

Terminal abdominal segment (*ventral view*) of white grub.

Adult white grub, also called May beetle or June beetle.

bution is largely a function of oviposition preference by females. Thus, though the roots and tubers of vegetables may be consumed, this is often due to conversion of grassland bearing partially grown grubs to vegetable production areas, and a decrease in the availability of their preferred host plants, or grasses. Typically, larvae clip the roots of plants with fibrous-roots systems, especially during the second year of larval life, resulting in the wilting and death of young plants. However, they may chew holes into larger roots and tubers, damage which may not be apparent until harvest.

Damage to trees by adults is more frequent in northern states. In such areas the transition from cold, wet weather to warm, dry weather is abrupt, causing mass emergence of adults and defoliation of nearby trees. In contrast, in the south the gradual increase in temperatures causes a protracted emergence by adults and the defoliation of trees is less apparent. Adults prefer young foliage, feeding from the edge toward the center of the leaves. They feed at night and hide during the day, so the cause of damage is often overlooked.

Management

Sampling. Estimates of grub density are often made by examining 0.3 m cubes of sod and soil. Owing to the highly aggregated distribution of grubs and eggs (Guppy and Harcourt, 1970), however, extensive sampling is required to obtain a high degree of precision, often 50–100 samples (Guppy and Harcourt, 1973). Light traps are used to monitor populations of June beetles because they are nocturnal; such traps collect mostly males. Flights normally occur between sunset and midnight, and then again at sunrise, and are initiated by a combination of temperature and photoperiod stimuli (Guppy, 1982). Examples of the results of light-traps studies can be found in Kard and Hain (1990), Forschler and Gardner (1991), and Dahl and Mahr (1991).

Insecticides. White grubs and June beetles are controlled with insecticides by application to soil or tree foliage, respectively. Soil applications predominate except perhaps for protection of foliage in fruit crops. Persistent insecticides are used in soil environments, applied preplanting (Rolston and Barlow, 1980) or at planting (Rivers *et al.*, 1977). Insecticide resistance has developed in some regions (Lim *et al.*, 1980b).

Cultural Practices. Because of the attraction of June beetles to grasses, particularly short or mowed grass, there is a considerable risk when crops are planted into land that supported grass during the previous year, or when crops are grown adjacent to grass pastures or sod. If clover is planted immediately following grass, the risk to subsequent crops is reduced, partly because grubs rarely injure clover, but also because June beetles prefer not to oviposit in clover fields. It is also important that crops be free of grass and other weeds during the flight of June beetles, or considerable oviposition may occur. However, if grubs are already in the soil, the presence of grass weeds can be advantageous because grubs will feed preferentially on the grass (Rivers *et al.*, 1977).

Plowing and disking are sometimes recommended for destruction of grubs because the soft bodies of larvae are easily damaged by tillage. Also, tillage exposes grubs to birds, which can often be seen following tractors and consuming large numbers of exposed insects.

Host-Plant Resistance. There is not much useful information on the varietal susceptibility of vegetable crops to white grubs. An exception is sweet potato, where not only has resistance been demonstrated (Rolston *et al.*, 1981), but it has been shown that interplanting a resistant cultivar with a susceptible cultivar confers a reduction in damage to the susceptible variety (Schalk *et al.*, 1991, 1992).

Biological Control. There have been few attempts to implement biological suppression of white grubs, other than the use of domestic animals such as poultry and swine to consume larvae. The spore-forming bacteria *Bacillus popillae* and *Bacillus lentimorbus* can infect and kill white grubs, but this expensive treatment is usually reserved for use on turf, where the bacteria can recycle through generation after generation of grubs. Entomopathogenic nematodes (Nematoda: Steinernematidae and Heterorhabditidae) have been evaluated for white grub suppression. Grubs are susceptible to infection and high levels of suppression have been attained by both injection and surface application, but the results are not consistent (Kard *et al.*, 1988).

FAMILY TENEBRIONIDAE—DARKING BEETLES AND FALSE WIREWORMS

False Wireworms
Blapstinus, Coniontis, Eleodes, and *Ulus* spp. (Coleoptera: Tenebrionidae)

Natural History

Distribution. Several native species of Tenebrionidae damage vegetable crops in the western United States and Canada. The affected area is west of the Mississippi River, and invariably is arid. Although numerous species may be involved, the common species and area where damage has been reported are: *Blapstinus elongatus* Casey (California), *B. fuliginosus*

Casey (California), *B. pimalis* Casey (Arizona, California), *B. rufipes* Casey (California), *B. substriatus* Champion (Montana), *Coniontis globulina* Casey (California), *C. muscula* Blaisdell (California), *C. subpubescens* LeConte (California), *Eleodes hispilabris* Say (Alberta, Washington), *E. omissa* (Say) (California), *E. tricostatus* (Say) (western United States and Canada), and *Ulus crassus* (LeConte) (California). The list of damaging species is likely longer, but species–level identifications are not usually made. In fact, these insects commonly are confused with wireworms (Coleoptera: Elateridae), a similar and much more common group of pests.

Host Plants. False wireworms are associated with grassland environments, and both larvae (called false wireworms) and adults (called darkling beetles) feed on seeds and seedlings of grasses and grain crops. Most false wireworm species, including many which are not mentioned above, are best known as wheat pests, principally because this is the crop most often grown in the arid areas inhabited by these insects. However, when irrigation is applied and vegetable crops cultured, false wireworms may also cause injury to vegetables. Among the vegetables reported injured are asparagus, cabbage, cantaloupe, corn, lettuce, lima bean, mustard, onion, pepper, potato, radish, snap bean, tomato, and watermelon.

Natural Enemies. Many parasitoids and diseases are known from false wireworms and darkling beetles, but the significance of these natural enemies is not always apparent. Flesh flies (Diptera: Sarcophagidae), particularly *Blaexoxipha eleodes* (Aldrich), are known from several *Eleodes* spp. *Eleodes* spp. also are parasitized by *Eleodiphaga* and *Sitophaga* spp. (Diptera: Tachinidae) and *Microctonus eleodis* (Viereck) (Hymenoptera: Braconidae). The fungi *Beauveria bassianna* and *Metarhizium anisopliae* occasionally are associated with false wireworm larvae (Allsopp, 1980).

Rodents are important predators of darkling beetles in grass and shrub ecosystems, but not all rodent species consume them regularly. Work in Utah indicated that though northern grasshopper mouse, *Onychomys leucogaster arcticeps*, and omnivorous deer mouse, *Peromyscus maniculatus nebrascensis*, were effective predators, Uinta ground squirrel, *Spermophilus armatus*, and Great Basin pocket mouse, *Perognathus parvus clarus*, developed an aversion to the beetles after several feedings (Parmenter and MacMahon, 1988).

Birds feed readily on larvae and pupae, and often follow tractors as the soil is tilled, and capture the exposed insects. Larvae also can be forced to the soil surface when it becomes saturated by heavy rain, and birds are frequently observed to take advantage of this temporary abundance of food. Consumption of the adult stage is more infrequent, but some species are adapted to consume these insects. Among the species which feed on darkling beetles most frequently are bronzed grackle, *Quiscalus quiscula seneus*; western crow, *Corvus brachyrhynchos hesperis*; western robin, *Planesticus migratorius propinquus*; and sage hen, *Centrocercus urophasianus* (Hyslop, 1912c)

Life Cycle and Description. The aforementioned false wireworm species have, at most, a single generation per year. Most species seem to require 2–3 years to complete their development. False wireworms generally overwinter as larvae, but some species pass the winter as adults. Some apparently overwinter in both stages.

The life cycle of *Eleodes hispilabris*, one of the most common species that affect vegetables, is fairly typical of false wireworms. Adults pass the winter under dense masses of leaf debris, in cracks in the soil, and in rodent burrows. In the spring the adults feed on young weeds. In June they commence egg laying, with eggs deposited just beneath the soil surface. The eggs hatch in July, with the larvae about one-half grown at the onset of winter. The larvae commence feeding again in the spring, and pupate in the soil in about August. They have about 11 instars, and duration of the larval stage is about 270 days. The adults emerge in the autumn, feed briefly, and then overwinter, completing the two-year developmental cycle.

The species differ somewhat in appearance, but the following description of *Eleodes letcheri* Blaisdell is typical. The egg is oval, measuring about 1.1 mm long and 0.6 mm wide. It is white, and free of sculpturing. Larvae are elongate-cylindrical, and somewhat flattened ventrally. Larvae are yellowish, but the head, mandibles, tarsi, and anal segment are brown to black. The antennae are short, enlarged or clubbed at the tip, and yellow in color. The enlarged antennae of false wireworms serves to differentiate them from wireworms, which lack expanded antennae. The thoracic legs are stout. The larvae move freely and rapidly on the soil surface, a characteristic that helps distinguish them from wireworms, which are not agile on the soil suface. The pupa generally resembles the adult, except that the elytra are shrunken. Initially white in color, the pupal stage soon turns black. The adult is a shining black beetle with punctate elytra. The legs are long and stout. The maximum width of the abdomen is considerably greater than the juncture of the thorax and abdomen, and the tip of the abdomen tapers to a point. The adults are easily distinguished from predatory ground beetles (Coleoptera: Carabidae), with which they could be confused, because the apical segments of the antennae are enlarged in tenebrionids but not in carabids.

False wireworm or darkling beetle larva, *Eleodes suturalis.*

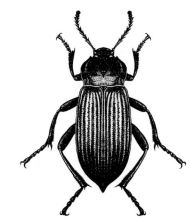

Darkling beetle or adult false wireworm, *Eleodes suturalis.*

Most adults measure about 2.5 cm long. They display a defensive posture that involves extending the hind legs and tilting the body forward so that the tip of the abdomen is greatly elevated. Release of chemical exudates usually accompanies the head–standing behavior. In some cases the exudate is forcibly expelled, but more commonly it is simply released at the tip of the abdomen as a drop, or spreads over the posterior part of the body. The exudate usually contains toluquinone, ethyl-*p*-quinone, and often other chemical components, that help deter predation.

A worldwide review of false wireworm biology was published by Allsopp (1980). Observations on several western species were provided by Wakeland (1926). Rearing was outlined by Matteson (1966) and Wright (1972).

Damage

These insect normally are pests of dryland agriculture, and particularly of fields that have recently been converted from grassland. However, as noted by Daniels (1977), irrigated crops can be invaded. Both larval and adult stages may injure vegetables, but such injury is infrequent. Larvae tend to reside in the soil near the surface and feed on roots, seeds, and the stems of seedlings. Farmers often first detect a false wireworm infestation when they observe an irregular or patchy stand of seedlings in a field; widespread or uniform damage is rare. The adults are more mobile and active above-ground, they often feed on aerial plant parts. For example, Campbell (1924) observed the beetles feeding on the stems of tomatoes, peppers and lima beans—both young seedlings sprouting from seed and transplanted plants were injured. Such injury is usually inflicted at the soil surface, and sometimes several beetles aggregate and feed on a single plant, quickly girdling it. Older plants are quite resistant to attack. False wireworms are omnivorous. In addition to feeding on seeds and seedlings, they also may feed on tubers, leaf tissue, plant detritus, and insect cadavers. Toba (1985a) studied damage to potato in Washington and reported that the surface scarring inflicted on potatoes was so minor that it would not lower the grade, and reduce the value, of potatoes.

Management

Sampling. The larval and adult populations are often assessed by direct count of the number of individuals per unit area. This count is often accompanied by soil sieving, because larvae are usually in the soil. The adults may also be censused by using pitfall traps. In irrigated fields, damage often occurs at the field margins as adults invade from nearby weedy or uncultivated areas. Therefore, field edges and adjacent areas warrant monitoring.

Insecticides. Insecticides may be applied to seeds or around the base of young plants. Phytotoxicity is a consideration with seed treatment. Residual insecticides are desirable, but not always available when protecting against injury by false wireworm. Baits containing insecticides are often effective for adult suppression, and bait application can often be limited to field margins. Daniels (1977) reported variable responses to insecticides in Texas, but complete suppression with some materials.

Cultural Practices. Soil characteristics are often linked to false wireworm problems, with damage most often occurring in dry, light, sandy or sandy loam soil (Wade, 1921). However, research in South Dakota showed that though some species (e.g., *E. hispilabris*) were most frequent in such soil types, other species (e.g., *E. tricostatus*) occur across most soil types, including heavier clay soils (Calkins and Kirk, 1975).

Order Dermaptera—Earwigs

European Earwig
Forficula auricularia Linnaeus
(Dermaptera: Forficulidae)

Natural History

Distribution. European earwig was first observed in North America at Seattle, Washington in 1907. It spread quickly and was reported from Oregon in 1909, British Columbia in 1919, and California in 1923. It reached Rhode Island in 1911, New York in 1912, and most other provinces and northern states in the 1930s and 1940s. Now it occurs south to North Carolina, Arizona and southern California, but owing to its preference for temperate climates it is unlikely to become abundant in the southeastern states. It is not very tolerant of arid environments, but survives where irrigation is practiced. European earwig is native to Europe, western Asia, and northern Africa, but also has been introduced to Australia and New Zealand.

Other earwigs are abundant in North America, but few are as numerous as European earwig, and none are as damaging. The ringlegged earwig, *Euborellia annulipes* (Lucas) (discussed separately), and the African earwig, *Euborellia cincticollis* (Gerstaecker), are probably the only other species of concern to crop producers. Ringlegged earwig is widespread, but African earwig is limited to southwestern states. Neither species is native to North America.

Host Plants. This insect is omnivorous and feeds on a wide variety of plant and animal matter. Although its predatory habits somewhat offset its phytophagous behavior, on occasion European earwig can inflict significant injury to vegetables, fruit, and flowers. Bean, beet, cabbage, celery, chard, cauliflower, cucumber, lettuce, pea, potato, rhubarb, and tomato are among the vegetable crops sometimes injured. Seedlings and plants providing the earwigs with good shelter, such as the heads of cauliflower, the stem bases of chard, and the ears of corn, are particularly likely to be eaten, and also likely to be contaminated with fecal material. Among flowers most often injured are dahlia, carnation, pinks, sweet william, and zinnia. Ripe fruits such as apple, apricot, peach, plum, pear, and strawberry are sometimes reported to be damaged.

European earwig is reported to consume aphids, spiders, caterpillar pupae, leaf beetle eggs, scale insects, spiders, and springtails as well as vegetable matter. Aphid consumption is especially frequent and well-documented (McLeod and Chant, 1952; Madge and Buxton, 1976a,b; Mueller *et al.*, 1988). In addition to the higher plants mentioned above, earwigs consume algae and fungi, and often consume vegetable and animal matter in equal proportions (Buxton and Madge, 1976a).

Natural Enemies. There are several known natural enemies, including some that were imported from Europe in an attempt to limit the destructive habits of this earwig in North America. Some authors have suggested that the most important natural enemy is the European parasitoid *Bigonicheta spinipennis* (Meigen) (Diptera: Tachinidae), which has been reported to parasitize 10–50% of the earwigs in British Columbia. Others, however, report low incidence of parasitism (Lamb and Wellington, 1975). Another fly, *Ocytata pallipes* (Fallén) (Diptera: Tachinidae), also was successfully established, but it causes little mortality. Under the cool, wet conditions of Oregon, Washington, and British Columbia, the fungi *Erynia forficulae* and *Metarhizium anisopliae* also infect earwigs (Crumb *et al.*, 1941; Ben-Ze'ev, 1986). The nematode *Mermis nigrescens* appears to be an important mortality factor in Ontario, where 10–63% of earwigs were infected during a two-year period (Wilson, 1971). However, this nematode has not been reported from earwigs elsewhere. Avian predation can be significant (Lamb, 1975).

Life Cycle and Description. One generation is completed annually, and overwintering occurs in the adult stage. In British Columbia, the eggs are deposited in late winter, hatch in May, and nymphs attain the adult stage in August. Overwintering females may also produce an additional brood; these eggs hatch in June and also mature by the end of August (Lamb and Wellington, 1975). In Washington, these events occur about one month earlier (Crumb *et al.*, 1941). Only a single brood of eggs is produced in colder climates such as in Quebec (Tourneur and Gingras, 1992).

Egg. The egg is pearly white and oval to elliptical. The egg measures 1.13 mm long and 0.85 mm wide when first deposited, but it absorbs water, swells, and nearly doubles in volume before hatching. The eggs are deposited in a cell in the soil, in a single cluster, usually within 5 cm of the surface. Mean number of eggs per cluster is reported to range from 30 to 60 eggs in the first cluster. The second cluster, if produced, contains only half as many eggs. Duration of the egg stage under winter field conditions in British Columbia averages 72.8 days (range 56–85 days). The second or spring brood of eggs requires only 20 days to hatch. They are attended by the female, who frequently moves the eggs around the cell, and apparently keeps mold from developing on the eggs (Buxton and Madge, 1974). Females guard their eggs from other earwigs and fight with any intruders.

Nymph. The nymphal stages, four in number, have the same general form as adults except that the wings increase in size with maturity. The cerci are present in all instars, growing in size with each molt. The body color darkens, gradually changing from grayish brown to dark brown, as the nymph matures. The legs are pale throughout. The wing pads are first evident in instar four. Mean head capsule width is 0.91, 1.14, 1.5, and 1.9 mm in instars 1–4, respectively. Mean body length is 4.2, 6.0, 9.0, and 9–11 mm, respectively. The number of antennal segments is 8, 10, 11, and 12 in instars 1–4. Mean duration (range) of instars under laboratory temperatures of 15–21°C is 12.0 (11–15), 10.2 (8–14), 11.2 (9–15), and 16.2 (14–19) days for instars 1–4. However, development time is considerably longer under field conditions, requiring 18–24, 14–21, 15–20, and about 21 days for the corresponding instars. Young nymphs are guarded by the mother earwig, who remains in or near the cell where the eggs are deposited until the nymph's second instar is attained.

Adult. The adult normally measures 13–14 mm long, exclusive of the pincher-like cerci (forceps), though some individuals are markedly smaller. The head measures about 2.2 mm wide. Adults, including the legs, are dark brown or reddish brown, though paler ventrally. The antennae have 14 segments. They bear a set of cerci at the tip of the abdomen. The pronounced cerci are the most distinctive feature of earwigs; in the male the cerci are strongly curved, whereas in the female they curve only slightly. They can use the cerci in defense, twisting the abdomen forward over the head or sideways to engage an enemy, often another earwig. Despite the appearance of being wingless, adults bear long hind wings folded beneath the abbreviated forewings. Although rearely observed to fly, when ready to take flight the adults usually climb and take off from an elevated object. The hind wings are opened and closed quickly, so it is difficult to observe the wings.

Earwigs are nocturnal, spending the day hidden under leaf debris, in cracks and crevices, and in other dark locations. Their night-time activity is influenced by weather. Stable temperature encourages activity, and it is favored by higher minimum temperatures but is discouraged by higher maximum temperatures. High relative humidity seems to suppress movement, whereas higher wind velocities and greater cloud cover encouraged earwig activity (Chant and McLeod, 1952). They produce an aggregation pheromone in their feces that is attractive to both sexes and to nymphs, and release quinones as defensive chemicals from abdominal glands (Walker *et al.*, 1993).

Social behavior is weakly developed in European earwig. Males and females mate in late summer or autumn, and then construct a subterranean tunnel (nest) in which they overwinter. The female drives the male from the nest at the time of oviposition. Eggs are manipulated frequently, the female apparently cleaning them to prevent growth of fungi. She will relocate the eggs in an attempt to provide optimal temperature and humidity. Although the female normally keeps the eggs in a pile, as the time for hatching approaches she spreads the eggs in a single layer. After hatching, females continue to guard the nymphs and provide them with food. Food is provided by females carrying objects into the nest, and by regurgitation. Thus, there is parental care, but no cooperative brood care (Lamb, 1976).

The most comprehensive treatment of European earwig biology was provided by Crumb *et al.* (1941), though the publications by Jones (1917) and Fulton (1924) were informative. Lamb and Wellington (1974) described methods for rearing. Keys to western earwig species were provided by Langston and Powell (1975).

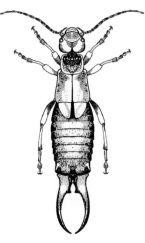

Adult male European earwig.

Eastern species were considered by Hoffmann (1987). A synopsis of European earwig, including keys to related Canadian insects, was given by Vickery and Kevan (1985).

Damage

The economic status of earwigs is subject to dispute. Undoubtedly, earwigs sometimes damage vegetable and flower crops, both by leaf consumption and fruit injury. Foliage injury is usually done in the form of several small holes. Tender foliage may be completely devoured except for major veins. However, the physical presence of earwigs as crop contaminants is perhaps even more important, because most people find their presence and odor repulsive. The annoyance associated with their presence is exacerbated by the tendency of earwigs to aggregate, often in association with human habitations; most people simply find them annoying. Their propensity to consume other insects, particularly aphids, is an important element in offsetting their reputation as a crop pest. However, augmenting the earwig population by field release, and providing them with additional shelter to enhance survival, have had mixed success in suppressing aphid populations (Carroll and Hoyt, 1984; Carroll *et al.*, 1985).

Management

Sampling. Population monitoring can be accomplished with baits and traps. Small piles of baits which can be checked during the evening, distributed in dense vegetation, often attract large numbers of earwigs. Wheat bran or oatmeal can serve as a bait. Traps take advantage of the natural tendency of earwigs to hide in crevices and dark spots, and can be used to detect presence of earwings and to estimate abundance.

Insecticides. Residual foliar insecticides and baits containing toxicant can be used to suppress earwigs. Of numerous baits evaluated, Crumb *et al.* (1941) suggested that wheat bran flakes plus toxicant and a small amount of fish oil was optimal. Fulton (1924) believed fish oil unnecessary but suggested addition of glycerin and molasses. Commercial products are rarely formulated specifically for earwigs because they rarely are a severe problem. Rather, products sold for grasshoppers, cutworms, slugs, and sowbugs are applied for earwig control. Bait is most effective if applied in the evening.

Cultural Practices. On residential property or in small gardens, persistent trapping can be used to reduce earwig abundance, though this approach is not effective if the initial earwig density is high. Boards placed on the soil are attractive to earwigs seeking shelter. Even more earwigs accumulate if there are narrow grooves or channels in the board. Moistened, rolled-up

newspaper placed in the garden in the evening and disposed of in the morning makes a convenient earwig trap for home gardens. A particularly effective collecting technique is to fill a flower pot with wood shavings and invert the pot over a short stake that has been driven into the soil. Traps can also be placed in trees because earwigs favor this habitat.

Ringlegged Earwig
Euborellia annulipes (Lucas)
(Dermaptera: Carcinophoridae)

Natural History

Distribution. First found in the United States in 1884, ringlegged earwig now is widespread in southern states. It is also known from many northern states, and the southernmost portions of British Columbia, Ontario and Quebec in Canada. Redlegged earwig occurs in Hawaii and has been transported to most other areas of the world, including both tropical and temperate climates. It likely is of European origin.

Host Plants. Ringlegged earwig is omnivorous in its feeding habits, taking both plant and animal material readily. It occurs as a minor nuisance in southern vegetable gardens and in greenhouses, where it nibbles on succulent plants such as lettuce. It is also documented to feed on roots or tubers of radish, potato, and sweet potato, and the pods of peanuts, though this is infrequent and normally of a little consequence. Ringlegged earwig is a voracious predator of insects and sowbugs. This predatory behavior probably offsets the small amount of damage done to plants. It also is highly cannibalistic.

Natural Enemies. The natural enemies of ringlegged earwig seem to be undocumented, though they are likely about the same as those attacking European earwig, *Forficula auricularia* Linneaus. Cannibalism of eggs and nymphs by adults is an important mortality factor.

Life Cycle and Description. This insect seems not to have been studied under field conditions. Three generations were observed under greenhouse conditions in Ohio—one each in the spring, autumn, and winter months. A complete generation can be completed in 61 days (Klostermeyer, 1942). Thus, under field conditions, it seems probable that at least two generations occur, one each in spring and autumn, at least in warm climates. In Illinois, adults can be found throughout the year except during winter when adults seek shelter deep in the soil. (The ringlegged earwig adult in color figure 178 is referred to on page 196.)

Egg. The eggs are nearly spherical when first deposited, and measure about 0.75 mm in diameter. As the embryo develops, however, the egg becomes more elliptical, and attains a length of about 1.25 mm. The egg is creamy white initially, becoming brown as the embryo develops. Females deposit 1–7 clutches of eggs with a mean clutch size of about 50 eggs. Total fecundity is estimated at 100–200 eggs. Duration of the egg stage is 6–17 days.

Nymph. The nymphs greatly resemble the adult in form, differing primarily in size. Wing pads are absent. The head and abdomen are dark-brown. The pronotum is considerably lighter in color, usually grayish or yellowish-brown. The legs are whitish, with a dark ring around the femur. The cerci are moderately long, and not strongly curved. Normally, five instars are found, but six are observed occasionally. Instars are difficult to distinguish and no single character is completely diagnostic. The number of antennal segments is most useful, though this is mostly effective among the early instars. The number of antennal segments is about 8, 11, 13, 14–15, 15–16, and 14–17 in instars 1–6. Head capsule width is 0.62–0.75, 0.70–0.91, 0.83–1.09, 1.04–1.56, 1.22–1.56, and 1.40–1.72 mm for instars 1–6, respectively. Body length is 3.0–4.7, 3.9–6.9, 5.7–7.7, 6.7–10.8, 8.7–13.2, and 9.8–12.9 mm, respectively, for instars 1–6. When reared at 21–23°C, Bharadwaj (1966) reported mean development times of 11.8, 10.6, 13.4, 16.3, 20.1, and 27.0 days for instars 1–6, respectively, for a total of about 99 days. However, he observed a considerably shorter mean nymphal development period, 83.6 days, when cultured at 20–29°C. Klostermeyer (1942) was able to rear the nymphal stage through to the adult in as little as 45 days, with an average of 81 days, when cultured at 18–29°C.

Adult. The adults are dark brown, and wingless. They measure 12–16 mm long, with females averaging slightly larger than the males. The legs are pale, usually with a dark band around the middle of the femur, and often tibia, of each leg. Adults generally bear 16 antennal segments. The leg bands are the basis for the common name, and are readily apparent. Cerci of the adults can be used to distinguish the sexes. In the male the cerci are more curved, with the right branch of the forceps turned sharply inward at the tip. The males also possess 10 abdominal segments, whereas females possess eight segments. (See color figure 178.)

Earwigs are nocturnal. Mating occurs 1–2 days after attainment of the adult stage. Oviposition commences 10–15 days after mating, and requires about three days to complete. The adults construct a small cell in the soil in which eggs are deposited. The female drives the male from the oviposition chamber before eggs are produced. The female protects the egg clutch from mites, fungi, and intruders, cleaning and relocating them if necessary. Maternal care decreases soon after

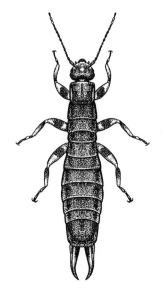

Adult female ringlegged earwig.

nymphs hatch, disappearing after about 10 days. The female cannot tolerate the presence of her progeny once she begins production of a subsequent egg clutch. Adults are long-lived, capable of living over 200 days.

The biology of redlegged earwig was given by Klostermeyer (1942), Neiswander (1944), Bharadwaj (1966), and Langston and Powell (1975). Culture was described by Bharadwaj (1966). This earwig was included in the keys by Langston and Powell (1975) and Hoffmann (1987).

Damage

These earwigs cause little direct injury to growing vegetable crops, but can feed on both the above-ground and below-ground portions of plants. More commonly, they serve as a contaminant of produce, sometimes defecating on leafy green vegetables. They also can cause injury to stored products such as potatoes and carrots, and are important contaminants of food processing plants (Gould, 1948). These earwigs are important insect predators, and are documented to feed on such diverse prey as caterpillars, beetle larvae, and leafhoppers. Unfortunately, there seems to be no quantitative data on their relative importance.

Management

These earwigs rarely warrant suppression, but are easily killed by most residual insecticides. They also take bait formulations consisting of wheat bran, molasses, and toxicant, as well as many other baits (Neiswander, 1944). For additional information on earwig damage and management, see the section on European earwig, *Forficula auricularia* Linnaeus.

*Color figure 7 is referred to on page 197.

Order Diptera—Flies and Maggots

FAMILY AGROMYZIDAE—LEAFMINER FLIES

American Serpentine Leafminer
Liriomyza trifolii (Burgess)
(Diptera: Agromyzidae)

Natural History

Distribution. This native leafminer has long been found throughout the eastern United States and Canada, northern South America, and the Caribbean. In recent years it has been introduced into California, Europe, and elsewhere, resulting in a fairly cosmopolitan distribution. There is increasing international traffic in horticultural crops, particularly flowers, which is thought to be the basis for the expanding range of this species. *Liriomyza trifolii* readily infests greenhouses, so range expansion seems likely to continue, including some northern climates. As a vegetable pest, however, its occurrence is limited principally to tropical and subtropical regions.

The taxonomy of this group was greatly confused until about 1980. Consequently, many records before this time are incorrect or unsubstantiated. Confusion with vegetable leafminer, *L. sativae* Blanchard, has been especially frequent.

Host Plants. *Liriomyza trifolii* is perhaps best known as a pest of chrysanthemums and celery, but it is highly polyphagous. For example, Stegmaier (1966b) reported 55 hosts from Florida, including bean, beet, carrot, celery, cucumber, eggplant, lettuce, melon, onion, pea, pepper, potato, squash, and tomato. Flower crops that are readily infested and which are known to facilitate spread of this pest include chrysanthemum, gerbera, gypsophila, and marigold, but there are likely many other hosts, especially among the Compositae. Numerous broad-leaved weed species support larval growth. Schuster *et al.* (1991) found that the nightshade *Solanum americanum*; Spanish needles, *Bidens alba*; and pilewort, *Erechtites hieracifolia*; were suitable weed hosts in Florida. (See color figure 7.)

Natural Enemies. In North America, parasitic wasps of the families Braconidae, Eulophidae, and Pteromalidae are important in natural control, and in the absence of insecticides, they usually keep this insect at low levels of abundance. At least 14 parasitoid species are known from Florida alone. Species of Eulophidae such as *Diglyphus begina* (Ashmead), *D. intermedius* (Girault), *D. pulchripes*, and *Chrysocharis parksi* Crawford are generally dominant in most American studies, though their relative importance varies geographically and temporally (Minkenberg and van Lenteren, 1986). Many of the parasitoids attacking *L. trifolii* also attack *L. sativae*.

Predators and diseases are not considered to be important, relative to parasitoids. However, both larvae and adults are susceptible to predation by a wide variety of general predators, particularly ants.

Life Cycle and Description. Leafminers have a relatively short life cycle. The time required for a complete life cycle in warm environments such as California and Florida is often 21–28 days, so numerous generations can occur annually in tropical climates. Leibee (1984) determined growth at a constant 25°C, and reported that about 19 days were required from oviposition to emergence of the adult. Development rates increase with temperature up to about 30°C; temperatures above 30°C are usually unfavorable and larvae experience high mortality. Minkenberg (1988) indicated that at 25°C, the egg stage required 2.7 days for development. The three active larval instars required 1.4, 1.4, and 1.8 days, respectively. The time spent in the puparium was 9.3 days. Also, there was an adult preovipostion period that averaged 1.3 days. The temperature threshold for development of the various stages is 6–10°C except that egg laying requires about 12°C. (The American serpentine leafminer adult in color figure 176 is referred to on page 198.)

Egg. The eggs are deposited in the middle or lower stratum of plant foliage. The adult seems to avoid immature leaves. The female deposits the eggs from the lower surface of the leaf, but they are inserted just below the epidermis. Eggs are oval in shape and small in size, measuring about 1.0 mm long and 0.2 mm wide. Initially they are clear but soon become creamy white.

Larva. Body length and mouth parts can be used to differentiate instars; the latter is particularly useful. For instar one, the mean and range of body and mouth parts (cephalopharyngeal skeleton) lengths are 0.39 (0.33–0.53) mm and 0.10 (0.08–0.11) mm, respectively. For instar two, the body and mouth parts measurements are 1.00 (0.55–1.21) mm and 0.17 (0.15–0.18) mm, respectively. For instar three, the body and mouth parts measurements are 1.99 (1.26–2.62) mm and 0.25 (0.22–0.31) mm, respectively. A fourth instar occurs between puparium formation and pupation; this is a nonfeeding stage and is usually ignored by authors (Parrella, 1987). The puparium is initially golden brown, but turns darker brown with time.

Adult. The adults are small, measuring less than 2 mm long, with a wing length of 1.25–1.9 mm. The head is yellow with red eyes. The thorax and abdomen are mostly grey and black though the ventral surface and legs are yellow. The wings are transparent. Key characters that serve to differentiate this species from vegetable leafminer, *Liriomyza sativae* Blanchard, are the matte, greyish-black mesonotum and the yellow hind margins of the eyes. In vegetable leafminer the mesonotum is shining black and the hind margin of the eyes is black. The small size of this species serves to distinguish it from pea leafminer, *Liriomyza huidobrensis* (Blanchard), which has a wing length of 1.7–2.25 mm. Also, the yellow femora of American serpentine leafminer help to separate it from pea leafminer, which has darker femora. (See color figure 176.)

Adult longevity is 13–18 days. Leibee (1984), working with celery as a host plant, estimated that oviposition occurred at a rate of 35–39 eggs per day, for a total fecundity of 200–400 eggs. Parella *et al.* (1983) reported similar egg production rates on tomato, but lower total fecundity, because tomato is a less suitable larval host. The female makes numerous punctures of the leaf mesophyll with her ovipositor, and uses these punctures for feeding and egg laying. The proportion of punctures receiving an egg is about 25% in chrysanthemum and celery, both favored hosts, but only about 10% in tomato, which is less suitable for larval survival and adult longevity. Although the female apparently feeds on the exuding sap at all wounds, she spends less time feeding on unfavorable hosts. Adults are weak fliers, and often are blown by the wind. Males live only 2–3 days, possibly because they cannot puncture foliage and therefore feed less than females, whereas females usually survive for about a week. Typically, they feed and oviposit during much of the daylight hours, but especially near mid-day.

A good summary of American serpentine leafminer biology was published by Minkenberg and van Lenteren (1986). Keys for the identification of agromyzid leafminers could be found in Spencer and Steyskal (1986).

Damage

The numerous punctures caused by females during the feeding and oviposition processes can result in a stippled appearance on foliage, especially at the leaf apex and margins. However, the principal form of damage is the mining of leaves by larvae, which results in destruction of leaf mesophyll. The mine becomes noticeable about 3–4 days after oviposition and becomes larger in size as the larva matures. The pattern of mining is irregular. Both leaf mining and stippling can greatly depress the level of photosynthesis in plant tissue (Parrella *et al.*, 1985). Extensive mining may cause premature leaf drop, which can result in lack of shading and sun scalding of fruit. Wounding also allows entry of bacterial and fungal diseases.

There is some disagreement about the relative importance of leaf mining, and many studies have had difficulty in demonstrating an adverse effect of leaf mines on yield. Varieties and cultural practices undoubtedly contribute to some of the inconsistency observed, but clearly crops such as tomatoes are quite resilient, capable of withstanding considerable leaf damage. It is often necessary to have an average of 1–3 mines per tomato leaf before yield reductions occur (Levins *et al.*, 1975; Schuster *et al.*, 1976). In potatoes, Wolfenbarger (1954) reported sizable yield increases with as little as 25% reduction in leafminer numbers,

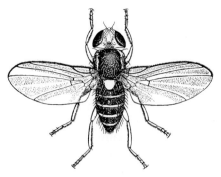

Adult American serpentine leafminer.

but leafminers were considered less damaging than green peach aphid, *Myzus persicae* (Sulzer).

Management

Sampling. Various techniques have been evaluated to assess larval and adult population densities. Counting mines in leaves is a good index of past activity, but many mines may be vacant. Counting live larvae in mines is time consuming, but more indicative of future damage. Puparia can be collected by placing trays beneath foliage to capture larvae as they evacuate mines, and the captures are highly correlated with the number of active miners (Johnson *et al.*, 1980). The adults can be captured by using adhesive applied to yellow cards or stakes. Sequential sampling plans for determining the need for treatment, based on the number of mined leaves, were developed by Wolfenbarger and Wolfenbarger (1966). Zehnder and Trumble (1985) developed similar plans based on counts of puparia and flies.

Insecticides. Chemical insecticides have long been used to protect foliage from injury, but insecticide resistance is a major problem. Insecticide susceptibility varies widely among populations, and level of susceptibility is directly related to frequency of insecticide application (Mason *et al.*, 1989). In Florida, for example, longevity of insecticide effectiveness is often about 2–4 years, and then is usually followed by severe resistance among the treated populations. Rotation among classes of insecticides is recommended to delay development of resistance. Reduction in dose level and frequency of insecticide application, as well as preservation of the susceptible genotype through nontreatment of some areas, are suggested as means to preserve insecticide susceptibility among leafminer populations (Mason *et al.*, 1989). Insect growth regulators can be more stable, but are not immune from the resistance problem. Neem is sometimes reported to be effective for leafminer suppression (Jyani *et al.*, 1995).

Some insecticides are highly disruptive to naturally occurring biological control agents, particularly parasitoids. Use of many chemical insecticides exacerbates leafminer problems by inducing extremely high leafminer populations. This usually results when insecticides are applied for lepidopterous insects, and use of more selective pest control materials such as *Bacillus thuringiensis* is recommended as it allows survival of the leafminer parasitoids.

Biological Control. Because parasitoids often provide effective suppression of leafminers in the field when disruptive insects are not used, there has been interest in release of parasitoids into crops. This occurs principally in greenhouse-grown crops, but is also applicable to field conditions. Parrella *et al.*

(1989) published information on mass rearing *Diglyphus begini* for inoculative release. *Steinernema* nematodes have also been evaluated for suppression of leaf-mining activity. One of the most complete studies, by Hara *et al.* (1993), showed that high levels of relative humidity (at least 92%) were needed to attain even moderately high (greater than 65%) levels of parasitism. Adjuvants that enhance nematode survival increase levels of leafminer mortality (Broadbent and Olthof, 1995).

Cultural Practices. Because broadleaf weeds and senescent crops may serve as sources of inoculum, destruction of weeds and deep plowing of crop residues is recommended. The adults experience difficulty in emerging from anything but a shallow layer of soil.

Cultural practices such as mulching and staking of vegetables may influence both leafminers and their natural enemies. Price and Poe (1976) reported that leafminer numbers were higher when tomatoes were grown with plastic mulch or tied to stakes. At least part of the reason seems to be due to lower parasitoid activity in plots where tomatoes were staked. However, whereas Wolfenbarger and Moore (1968) reported that aluminum mulch seemed to repel leafminers in tomato and squash plantings, Webb and Smith (1973) found this not to be the case.

Asparagus Miner
Ophiomyia simplex (Loew)
(Diptera: Agromyzidae)

Natural History

Distribution. This fly is native to Europe, where it occurs widely. It was first noticed in North America about 1861, and probably was introduced with asparagus plants. By the late 1800s and early 1900s it was found generally in New England and the Middle Atlantic States, and had been distributed to some locations on the Great Lake States. The asparagus miner was detected in California in 1905. It is now believed to occur wherever asparagus is grown.

Host Plants. Asparagus miner has a very restricted host range, feeding only on asparagus. It is rarely associated with asparagus spears (young shoots); instead, it is normally found principally on older stalks bearing "ferns." Asparagus miner attacks young plants, however, if they are old enough to bear ferns.

Natural Enemies. Little is known about the natural enemies of asparagus miner in North America. One parasitoid, *Chorebus rondanii* (Giard) (Hymenoptera: Braconidae), is known from Massachusetts, and also is found in Europe (Krombein *et al.*, 1979). Barnes (1937a) noted other parasitoid species in England which apparently have not been introduced to North America.

Life Cycle and Description. There are two generations annually throughout the range of this insect, with the pupal stage overwintering. The adults first become apparent in May, and within a few days of emergence they copulate and begin oviposition. First generation larvae are common in June, completing their development and pupating in July. By late July or early August, second generation adults have emerged and begin producing eggs. Second generation larvae begin to mature in late August and early September. The larvae form puparia in the autumn, remaining in diapause until April or May of the following year.

Egg. The female deposits eggs beneath the epidermis of stems, usually near the base of the plant. The presence of the egg in the plant tissue is not apparent. The egg is white, and elongate oval. One end tends to be more pointed than the other. The egg measures about 0.48 mm long and 0.22 mm wide. Duration of the egg stage is estimated to be 7–14 days.

Larva. The larva is whitish. It tapers slightly at both ends, and the anterior end bears black mouthhooks, while the posterior end bears a pair of black spiracles. There are three instars, and their length is

about 0.4, 2.0, and 3.5 mm, respectively, during instars 1–3. Barnes (1937a) gave characters to separate the instars. Duration of the larval stage has been poorly documented, but it seems to be 7–14 days.

Pupa. Pupation occurs within the larval mine. First-generation larvae usually pupate above-ground, whereas second-generation insect tend to pupate in the stalk or roots below-ground, often at a depth of 5 cm or more. The puparium initially is light-brown, becoming dark brown with age. It is somewhat flattened in form, and measures about 4–5 mm long. The puparium bears hooks at both ends. Duration of the pupal stage is 14–21 days during the summer, and several months during the winter.

Adult. The adult is shiny black in appearance. It measures 3–4 mm long, with a wing span of 5–6 mm. The wings are colorless. Fecundity is unknown.

The biology of the flies was given by Fink (1913) for New York, and Barnes (1937a) for England. Ferro and Gilbertson (1982) and Lampert *et al.* (1984) provided bionomics in Massachusetts and Michigan, respectively.

Damage

The larvae feed just beneath the surface of the stem, interfering with photosynthesis. Burrowing starts at about soil level and proceeds upwards, forming mines. If several larvae are present in the same stalk the feeding activity of the larvae will girdle the plant. Infested plants become yellow in color. Discoloration may occur only at the base when few flies are present, or the entire plant may be affected. Burrowing by larvae also extends downward and even includes feeding on the roots. Larvae do not seem to damage young spears. The flies do not readily oviposit on this young tissue, and if they do, the eggs and larvae are too small to be observed at harvest, so they are not culled.

The importance of asparagus leaf miner has been disputed. Eichmann (1943), for example, dismissed the possibility of injury, including the potential of interaction with crown rot disease caused by *Fusarium* spp. Although the direct effects of asparagus miner on yield are likely minimal, it is now apparent that their involvement with disease is significant. The fungi *Fusarium moniliforme* and *F. oxysporum* are associated with all life stages of asparagus leaf miner, and increased incidence of Fusarium is associated with stem mining. Commercial asparagus fields in Massachusetts were found by Gilbertson *et al.* (1985) to harbor 1.9 mines per stem and 2.9 pupae per stem. Significant decline in commercial aspragus in northeastern states has been attributed to the Fusarium and asparagus miner problem.

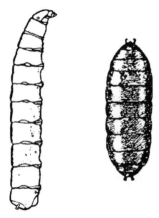

Asparagus miner larva (*left*) and pupa (*right*).

Adult asparagus miner.

Management

Sampling. Lampert *et al.* (1984) suggested that the pupal stage, particularly autumn puparia, was the best sampling unit. They also presented information on optimal sample size for fly populations with different densities. Ferro and Suchak (1980) studied methods of adult sampling and recommended yellow or lime green sticky stakes because of the relatively large number of flies captured with these techniques.

Insecticides. Insecticides applied to the stems of asparagus plants reduced abundance of asparagus miners and incidence of crown rot disease, and increased crop yield (Damicone *et al.*, 1987). Soil fumigation also enhances yield due primarily to destruction of fungi, but this approach is very expensive and useful only for the plant establishment phase of asparagus production.

Cultural Practices. Destruction of overwintering stalks is sometimes recommended because puparia are harbored in the stalks. However, this is not completely satisfactory because some flies are associated with the roots. Wild asparagus is a major source of flies, and roadsides and irrigation ditches that might harbor asparagus should be checked for plants, and efforts made to eradicate them.

Cabbage Leafminer
Liriomyza brassicae (Riley)
(Diptera: Agromyzidae)

Natural History

Distribution. The cabbage leafminer is common throughout tropical areas of North and South America, Africa, Australia, and Asia, but it also occurs in some temperate regions. In North America, cabbage leafminer has been reported from most regions of the United States, and is found as far north as Manitoba, Canada. However, it occurs as a pest principally in southern states. It likely is not native to North America.

Host Plants. Cabbage leafminer attacks many plants in the family Cruciferae, including broccoli, cabbage, cauliflower, radish, turnip, watercress, and various mustards (Stegmaier, 1967). It is also known from peas (Leguminosae), and nasturtium (Tropaeolaceae). Although it is the most common leafminer attacking crucifers, other species are sometimes found on these crops.

Natural Enemies. Stegmaier (1967) reported three species of eulophid wasps parasitizing cabbage leafminer in Florida: *Diaulinopsis callichroma* Crawford, *Chrysocharis* sp., and *Pnigalio* sp. (all Hymenoptera: Eulophidae). Oatman and Painter (1969) found seven

species of Eulophidae, as well as one species each of Pteromalidae, Cynipidae, and Braconidae attacking this leafminer in California. In the latter study, the proportion of flies parasitized was lowest in January (26%) and highest in October (84%). The eulophid *Diglyphus begini* (Ashmead) was the predominant parasitoid, and attacked the larval stage.

Life Cycle and Description. Cabbage leafminers breed continuously in tropical climates. In more temperate climates such as California, *L. brassicae* occurs throughout the year, but it is much more abundant during the summer months (Oatman and Platner, 1969).

Egg. Most feeding and egg deposition occurs beneath, and along the margins, of leaves. The shiny white egg measures 0.16 mm wide and 0.28 mm long, and hatches in about three days.

Larva. The small larva tunnels through the mesophyll tissue, initially forming an elongate mine on the lower surface of the leaf. Accumulations of fecal material are usually evident along the center of the leaf mine, though this is more evident as the larva matures and produces a wider mine. Duration of the three active instars is about 6–7 days, with their duration averaging about two, two, and three days, respectively. The mouthpart (cephalopharyngeal skeleton) length of these instars measures about 0.09, 0.18, and 0.28 mm, respectively. About half-way through the third instar the feeding behavior changes, with the larva moving from the lower leaf surface to the upper leaf surface to feed. As the larva approaches maturity it cuts a slit in the surface of the leaf and drops to the ground to form a puparium. A fourth larval instar occurs between puparium formation and pupation, but no feeding occurs so this instar is generally overlooked (Parrella, 1987).

Pupa. The yellow-brown puparium measures about 1.75 mm long and 0.8 mm wide. The time spent in the puparium is about 8–10 days. The adult emerges from a slit in the anterior region of the puparium, usually during the afternoon or evening hours. Mating occurs in the morning following adult emergence.

Adult. These small black and yellow flies are easily confused with vegetable leafminer, *L. sativae* Blanchard, but can be distinguished by using the male genitalia (Spencer, 1981). As in vegetable leafminer, the hind margin of the eyes is black, but cabbage leafminer differs in that it has less yellow color on the mesonotum. Vegetable leafminer generally has yellow on more than 25% of the mesonotum surface, whereas cabbage

---

* Color figure 8 is referred to on page 202.

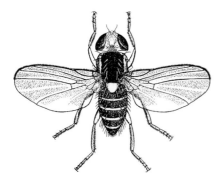

Adult cabbage leafminer.

leafminer has less than 25% yellow. The female deposits the egg singly by puncturing the lower surface of leaves with her ovipositor. Such punctures always serve as feeding sites, but sometimes also serve as egg receptacles.

The biology of this insect is not well known. Beri (1974) presented the most detailed life history information.

Damage

The principal form of plant damage is caused by larvae mining the leaf tissue. An elongate, irregular mine that widens as the larva matures is typical of this insect's feeding pattern. Feeding and oviposition punctures may damage young tissue, but this is usually insignificant. The large numbers of mines evident on many plants do not always result in serious damage. Oatman and Platner (1969), for example, reported that though mine numbers reached 178 per plant, most of the mining was confined to the old, outer leaves that drop off or are removed at harvest.

Management

Management of this insect is poorly studied. The techniques discussed under American serpentine leafminer, *L. trifolii* (Burgess), and vegetable leafminer, *L. sativae*, are probably appropriate for this insect as well.

Corn Blotch Leafminer
Agromyza parvicornis Loew
(Diptera: Agromyzidae)

Natural History

Distribution. Corn blotch leafminer is found throughout the United States and southern Canada. This species is native to North America.

Host Plants. This leafminer attacks only grasses. Its principal host is corn, but it also is sometimes found in other crops such as millet and wheat, and weedy

grasses such as barnyardgrass, *Echinochloa crusgalli*; and crabgrass, *Digitaria sanguinalis*. (See color figure 8.)

Natural Enemies. Corn blotch leafminer is heavily attacked by hymenopterous parasitoids. These beneficial insects seem to keep the leafminer at low levels of abundance. Among the parasitoids known to affect corn blotch leafminer are *Achrysocharella diastatae* (Howard) and *A. punctiventris* (Crawford); *Chrysocharis ainslei* Crawford and *C. parksi* Crawford; *Cirrospilus flavoviridis* Crawford; *Closterocerus tricinctus* (Ashmead) and *C. utahensis* Crawford; *Diglyphus begini* (Ashmead), *D. pulchripes* (Crawford), and *D. websteri* Crawford; *Notanisomorpha ainsliei* Crawford; *Zagrammosoma multilineatum* (Ashmead) (all Hymenoptera: Eulophidae); *Opius diastatae* (Ashmead), *O. succineus* Gahan, and *O. utahensis* Gahan (all Hymenoptera: Braconidae). The relative importance of the members of this diverse assemblage of wasps is unknown.

Life Cycle and Description. The number of generations is not well documented because of overlap, but a generation can be completed under warm conditions in less than 30 days. In northern states flies are present from about April to October, which allows for 4–5 generations. In the south, though some pupae apparently diapause, flies can be active throughout the year. Development time is extended during the winter months in Florida, but these insects continue to develop and reproduce.

Egg. The female inserts egg between the lower and upper epidermis. Oviposition may occur from either the lower- or upper-leaf surface. The egg is inserted with the longer axis of the egg parallel to the leaf veins. The white egg is flattened and measures about 0.45–0.50 mm long and 0.17–0.19 mm wide. It is broadly rounded at each end, but constricted slightly at the middle. The eggs are readily visible as white spots in the leaf tissue. Duration of the egg stage is 3–5 days, depending on temperature.

Larva. The larvae live their entire life within leaf tissue, and apparently they are unable to move from leaf to leaf. There likely are three instars, which require 4–10 days for total larval development. The larva attains a length of about 3 mm and a width of 1 mm. It is greenish white early in its life and becomes yellowish-white later. The body is cylindrical, but the

Corn blotch leafminer larva.

*Color figure 7 is referred to on page 203.

head region tapers to a point. The mouth hooks are darkly pigmented. At maturity, the larva drops to the ground to pupate.

Pupa. The larva normally burrows into the soil to pupate, sometimes to a depth of up to 5 cm, but usually about one cm. The puparium is reddish brown. It measures about 3 mm long and 1 mm wide. Unlike the larval stage, in the puparium its body segments are well-differentiated. Also, its anterior spiracles are prominent and protruding, and the posterior pair is less distinct. Duration of the pupal stage is normally 14–21 days, but this is also the overwintering stage.

Adult. This fly measures 3–4 mm long. It is predominantly black, though the face and legs are sometimes yellowish-brown. The wings are transparent. Females may live for 14 days or more, and produce eggs during all but the first few days. Adults are most active in the afternoon. Their reproductive capacity is not known.

The biology of corn blotch leafminer was described by Phillips (1914). It was included in the key by Spencer and Steyskal (1986).

Damage

The larvae feed between the lower and upper epidermis of the leaf. The epidermal tissue dies soon after the mesophyll are removed, leaving a thin, brown, and brittle area. Initially, larvae may feed in a straight line, and their path is directed by the parallel veins of the leaf. Later in life, however, larvae feed in an irregular pattern, leaving an large area of dead tissue that may involve the entire leaf of small grains and grasses. In corn, such injury is nominal unless numerous larvae are present.

The female also causes a small amount of injury through the production of feeding punctures. Females use their ovipositor to puncture the leaf, and then they feed upon sap that collects at the wound. Punctures

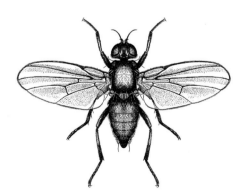

Adult corn blotch leafminer.

may be quite numerous, but injury is insignificant in all but the smallest plants.

Management

Because these flies rarely become abundant, and injury occurs on foliage rather than the ear (fruit), suppression should not be required. However, application of insecticides to foliage will eliminate these flies.

Pea Leafminer
Liriomyza huidobrensis (Blanchard)
(Diptera: Agromyzidae)

Natural History

Distribution. Pea leafminer is widely distributed in South America, and seems likely to be native to that region. In the United States, it occurs west of the Rocky Mountains but is abundant principally in California (Spencer, 1981). Reports of infestations in eastern states apparently are based on misidentification. However, this leafminer is easily transported with cut flowers and vegetables, and there is concern that shipment of produce could result in establishment of this pest in eastern states. Pea leafminer has been introduced into Europe and Israel, where it is considered to be a very serious pest (Weintraub and Horowitz, 1995; 1996).

Host Plants. This insect is highly polyphagous, though as its common name suggests, garden pea is an especially suitable host. Plants from several families are attacked, including Chenopodiaceae, Compositae, Cruciferae, Cucurbitaceae, Leguminosae, Liliaceae, Linaceae, Solanaceae, Tropaeolaceae, Umbelliferae, and Violaceae. The vegetables known to be suitable hosts include bean, beet, broccoli, cabbage, cantaloupe, carrot, celery, Chinese cabbage, cucumber, eggplant, endive, faba bean, garlic, leek, lettuce, okra, onion, parsley, pea, pepper, potato, radicchio, radish, spinach, tomato, and turnip. Several flowers, including aster, baby's breath, chrysanthemum, dahlia, and zinnia are suitable hosts, as are numerous common weeds such as lambsquarters, *Chenopodium album*; and pigweed, *Amaranthus* spp. (Lange *et al.*, 1957). (See color figure 7.)

Natural Enemies. As is the case with other *Liriomyza* species, numerous wasp parasitoids are known. Lange *et al.* (1957) reported *Diglyphus intermedius* (Girault), *D. begini* (Ashmead), *Chrysocharis ainsliei* Crawford, and *C. parksi* Crawford (all Hymenoptera: Eulophidae), *Halticoptera* sp. (Hymenoptera: Pteromalidae), and the braconid *Opius* sp. (Hymenoptera: Braconidae) from California. Parasitism was reported to reach 50–90% regularly. Insecticides interfere with parasitoids, especially *Diglyphus*. Pea leafminer has become a major pest of potatoes in Peru due to disruption of parasitoids by insecticides (Weintraub and Horowitz, 1995).

Life Cycle and Description. Pea leafminer completes its life cycle in 30–60 days, resulting in 5–6 generations per year depending on weather conditions.

Egg. The white eggs are about 0.28 mm long and 0.15 mm wide, oval in shape, and inserted into foliage. Duration of the egg stage is about three (range 1.5–4.0) days. Females produce 8–14 eggs per day.

Larva. The larvae feed in spongy mesophyll tissue and produce long, twisting mines that widen as the larvae mature. Although generally observed in leaves, mines may also be produced in stems and pods of peas. On leaves, the mine typically is initiated on the upper surface, but after feeding only a few millimeter the larva burrows to the lower leaf surface to complete its development. After passing through three instars, which normally requires 4–5 days but may take 10 days under cool conditions, the larva attains a length of about 4 mm.

Pupa. At maturity, the third instar normally cuts a slit in the leaf and drops to the ground to form a puparium in the soil. A fourth nonfeeding larval instar occurs before puparium formation, but this stage only persists for 4–5 h. The puparium measures 1.6–3.2 mm long, and is light-brown to almost black in color. The pupal stage lasts for about 7–13 days. There seems to be a relationship between puparium color and duration of the puparium; darker puparia have longer pupal periods.

Adult. The small black and yellow flies measure 1.7–2.1 mm long. The wing length is 1.7–2.25 mm, considerably larger than American serpentine leafminer, *Liriomyza trifolii* (Burgess) (1.25–1.9 mm) and vegetable leafminer, *L. sativae* Blanchard (1.25–1.9 mm). The femora are dark or at least bear dark bands in pea leafminer, whereas in American serpentine and vegetable leafminer the femora are yellow. The adults live for

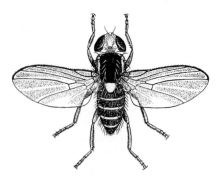

Adult pea leafminer.

12–14 days, whereas females live considerably longer than males. During this time each female may produce up to 1000 leaf punctures using their ovipositor. About 80–90% of the punctures occur on the upper leaf surface. Generally, only 10–20% of the punctures, from which the females feed, also contain eggs.

Biological information was provided by Frick (1951), Lange *et al.* (1957), and Parrella and Bethke (1984). Weintraub and Horowitz (1995) provided a useful synopsis of history, biology, and management. Keys for the identification of agromyzid leafminers can be found in Spencer and Steyskal (1986).

Damage

Pea leafminer damage results principally from reduction in photosynthetic tissue caused by larval mining. Pea leafminer characteristically, but not exclusively, mines along the leaf midrib and lateral veins. It feeds within the spongy mesophyll tissue where chloroplasts are located. Thus, pea leafminer typically is more damaging than American serpentine leafminer, *Liriomyza trifolii* (Burgess), which burrows in the palisade mesophyll. However, larval feeding is not the only source of damage. Females puncture tissue with their ovipositor to create feeding sites, at which both females and males feed. The number of feeding punctures produced by females was estimated by Prado and Cruz (1986) at about 275 per day. These feeding punctures can be dense enough to cause leaf discoloration or deformity. In the case of crops where appearance is important, as in spinach, leaf punctures alone can result in rejection of a crop. Punctures, and especially mining, also make plants more susceptible to invasion by plant pathogens.

Management

Sampling. Weintraub and Horowitz (1996) suggested using yellow-sticky cards for leafminer sampling in potato, and demonstrated that highest catches occurred with traps placed at plant height. Heinze and Chaney (1995) evaluated leafminer sampling plans in celery.

Insecticides. Pea leafminer management has not been extensively studied in the United States, probably owing to the restricted geographical distribution of the insect and susceptibility to insecticides. Parrella and Bethke (1984) suggested that *L. huidobrensis* is inherently more susceptible to insecticides than some other leafminer species, but that situation changed in the mid-1990s, when pea leafminer assumed greater importance to vegetable crops in California, presumably due to insecticide resistance. Also, about 1990 an insecticide-tolerant pea leafminer strain spread through many European countries, causing great loss to flower and vegetable crops in Europe and Israel.

Insecticides that are effective against larvae in the field generally have systemic properties. The neem-based botanical insecticides are effective when applied to the soil as a drench (Weintraub and Horowitz, 1997).

Cultural Practices. Alternatives to chemical insecticides are, as yet, few and relatively expensive. There has been some success in suppression of pea leafminer with entomopathogenic nematodes (Nematoda: Steinernematidae and Heterorhabditidae) if high humidity is maintained. Yellow plastic curtains coated with adhesive can be used to trap large numbers of flies. Some varieties of potato, particularly those with high densities of glandular trichomes, are somewhat resistant (Weintraub and Horowitz, 1995).

Vegetable Leafminer
Liriomyza sativae Blanchard
(Diptera: Agromyzidae)

Natural History

Distribution. This leaf miner is found commonly in the southern United States from Florida to California and Hawaii, and in most of Central and South America. Occasionally it is reported in more northern areas such as Ohio. Because this insect has been confused with American serpentine leafminer, *Liriomyza trifolii* (Burgess), and other related species, many location records are incorrect or unconfirmed.

Host Plants. Vegetable leafminer attacks a wide variety of plants, principally in the families Cucurbitaceae, Leguminosae, and Solanaceae. Stegmaier (1966a) reported nearly 40 hosts from 10 plant families in Florida. Among the numerous weed hosts, the nightshade, *Solanum americanum*; and Spanishneedles, *Bidens alba*; were especially suitable in Florida (Schuster *et al.*, 1991). Vegetables known as hosts include bean, eggplant, pepper, potato, squash, tomato, and watermelon. Oatman (1959) reported a similar host range in California, but also noted suitability of cucumber, beet, pea, lettuce, and many other composites. Celery is also reported to be attacked, but to a lesser extent by this leafminer species than by *L. trifolii*. In Hawaii, damage to onion foliage is a problem for the marketing of scallions (green onions) (Kawate and Coughlin, 1995). Vegetable leafminer was previously considered to be the most important agromyzid pest in North America (Spencer, 1981), but this distinction is now held by *L. trifolii*. Parrella and Keil (1984) argued that misidentification accounts for this apparent shift in pest status, and while this may be true, differential susceptibility to insecticides should not be dismissed as a significant contributing factor, or perhaps even the most important factor accounting for increased problems with *L. trifolii*.

Natural Enemies. Vegetable leafminer is attacked by many of parasitoids, with the relative importance of species varying geographically and temporally. Many species attacking *L. trifolii* apparently also attack *L. sativae*. In Hawaii, *Chrysonotomyia punctiventris* (Crawford) (Hymenoptera: Eulophidae), *Halicoptera circulis* (Walker) (Hymenoptera: Pteromalidae), and *Ganaspidium hunteri* (Crawford) (Hymenoptera: Eucoilidae) were considered to be important in watermelon (Johnson, 1987). In California and Florida, the same genera or species were found attacking vegetable leafminer on tomato or bean, but *Opius dimidiatus* (Ashmead) (Hymenoptera: Braconidae) also occurred commonly in Florida (Stegmaier, 1966a; Lema and Poe, 1979; Johnson *et al.*, 1980a). Levels of parasitism are often proportional to leafminer density, and parasitoid effectiveness is easily disrupted by application of insecticides. Steinernematid nematodes can infect *L. trifolii* larvae when the nematodes are applied in aqueous suspension and the plants are held under high humidity conditions (Broadbent and Olthof, 1995); *L. sativae* is probably equally susceptible.

Life Cycle and Description. The developmental thresholds for eggs, larvae, and pupae are estimated at about 9–12°C. The combined development time required by the egg and larval stages is about 7–9 days at warm temperature of (25–30°C). Pupal development also requires 7–9 days at this temperature. Both egg–larval and pupal development times are prolonged to about 25 days at 15°C. At the optimal temperature of 30°C, vegetableleaf miner completes its development from the egg to adult stage in about 15 days.

Egg. The white, elliptical eggs measure about 0.23 mm long and 0.13 mm wide. The eggs are inserted into plant tissue just beneath the leaf surface and hatch in about three days. Flies feed on the plant secretions caused by oviposition, and also on natural exudates. Females often make feeding punctures, particularly along the margins or tips of leaves, without depositing eggs. Females can produce 600–700 eggs over their life span, though some estimates of egg production suggest that 200–300 is more typical. Initially, females may deposit eggs at a rate of 30–40 per day, but oviposition decreases as flies age.

Larva. There are three active instars, and larvae attain a length of about 2.25 mm. Initially, the larvae are nearly colorless, becoming greenish and then yellowish as they mature. Black mouthparts are apparent in all instars, and can be used to differentiate the larvae. The average length and range of the mouthparts (cephalopharyngeal skeleton) in the three feeding instars is 0.09 (0.6–0.11), 0.15 (0.12–0.17), and 0.23 (0.19–0.25) mm, respectively. The mature larva cuts a semicircular slit in the mined leaf just before formation of the puparium. Almost invariably, the slit is cut in the upper surface of the leaf. The larva usually emerges from the mine, drops from the leaf, and burrows into the soil to a depth of only a few centimeter to form a puparium. A fourth larval instar occurs between puparium formation and pupation, but this is generally ignored by authors (Parrella, 1987).

Pupa. The reddish brown puparium measures about 1.5 mm long and 0.75 mm wide. After about nine days, the adult emerges from the puparium, mainly in the early morning hours. Both sexes emerge simultaneously. Mating initially occurs the day following adult emergence, but multiple matings by both sexes have been observed up to a month post-emergence.

Adult. The adults are mainly yellow and black. The shiny black mesonotum of *L. sativae* is used to distinguish this fly from the closely related American serpentine leafminer, *Liriomyza trifolii* (Burgess), which has a grayish black mesonotum. Also, the black hind margin of the eyes serves to distinguish this insect from *L. trifolii*, which has eyes with yellow hind margins. Females are larger and more robust than males and have an elongated abdomen. The wing length of this species is 1.25–1.7 mm, with the males averaging about 1.3 mm and the females about 1.5 mm. The small size of these flies serves to distinguish them from pea leafminer, *Liriomyza huidobrensis* (Blanchard), which has a wing length of 1.7–2.25 mm. The yellow femora of vegetable leafminer also help to distinguish this species, as the femora of pea leafminer are dark. Flies normally survive only about a month. They are less common during the cool months of the year, often reaching high, damaging levels by mid-summer. In warm climates they may breed continuously, with many overlapping generations per year.

The biology of *L. sativae* is not well documented, but important elements have been studied by Parkman *et al.* (1989), Petitt and Wietlisbach (1994), and Palumbo

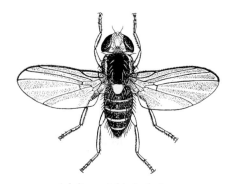

Adult vegetable leafminer.

(1995). The work by Oatman and Michelbacher (1958) probably refers to *L. sativae*. Keys for the identification of agromyzid leafminers can be found in Spencer and Steyskal (1986).

Damage

Foliage punctures inflicted by females during the acts of oviposition or feeding may cause a stippled appearance on foliage, but this damage is slight compared to the leaf mining activity of larvae. The irregular mine increases in width from about 0.25 mm to about 1.5 mm as the larva matures, and is virtually identical in appearance and impact with the mines of *L. trifolii*. Larvae often are easily visible within the mine, where they remove the mesophyll between the surfaces of the leaf. Their fecal deposits also are evident in the mines. The potential impact of the mining activity is evident from the work of Sharma *et al.* (1980), who studied the value of treating squash with insecticides in California. These researchers reported 30–60% yield increase when effective insecticides were applied, but as is often the case with leafminers, numerous insecticides were not effective.

Management

Sampling. Several methods for population assessment have been studied, and collecting puparia in trays placed beneath plants was recommended by Johnson *et al.* (1980b) as a labor-saving technique. Zehnder and Trumble (1984) used yellow-sticky traps to monitor adults, and reported that *L. sativae* flies were more active at the middle plant height of tomatoes, while *L. trifolii* was more active at low plant height. They also confirmed the value of pupal counts for prediction of adult numbers two weeks later. Yellow-sticky traps, however, have the advantage of being able to quickly detect invasion of a field by adults from surrounding areas. Sequential sampling plans were developed by Zehnder and Trumble (1985).

Insecticides. Foliar application of insecticides often is frequent. Insecticide susceptibility varies greatly both spatially and temporally. Many organophosphate and carbamate insecticides are no longer effective. Insecticides are disruptive to naturally occurring biological control agents, and leafminer outbreaks are sometimes reported to follow chemical insecticide treatment for other insects.

Cultural Practices. There is some variation among many crops in susceptibility to leaf mining. This has been noted, for example, in cultivars of tomato, cucumber, cantaloupe, and beans (Hanna *et al.*, 1987). However, the differences tend to be moderate, and not adequate for reliable protection. Nitrogen level and reflective mulches are sometimes said to influence leaf-miner populations, but the responses have not been consistent (Chalfant *et al.*, 1977; Hanna *et al.*, 1987). Placement of row covers over cantaloupe has been reported to prevent damage by leafminer (Orozco-Santos *et al.*, 1995). The same study evaluated the benefits of transparent polyethylene mulch, however, and found no reduction in leafminer populations. Sometimes crops are invaded when adjacent crops are especially suitable, as was reported by Sharma *et al.* (1980) in California, where cotton was an important source of invaders. Weeds are a source of flies (Parkman *et al.*, 1989), but also a source of parasitoids.

FAMILY ANTHOMYIIDAE — ROOT AND SEED MAGGOTS, LEAFMINER FLIES

Bean Seed Maggot
Delia florilega (Zetterstedt)
(Diptera: Anthomyiidae)

Natural History

Distribution. The distribution of bean seed maggot is not precisely known because it is often confused with seedcorn maggot, *Delia platura* (Meigen). However, it is known as a pest in the northeastern and midwestern areas of the United States, and in Canada as far west as the Prairie Provinces. It also is found in Europe, its likely source, though its date of introduction to North America is unknown.

Host Plants. Bean seed maggot has a wide host range, though it is usually confused with seedcorn maggot, so its host preferences are not well known. It attacks and develops successfully on such vegetables as cantaloupe, corn, kidney bean, pea, snap bean, squash, and probably onion, potato, and pepper. It also attacks field corn, soybean, and perhaps others. Hosts are more attractive when bacteria and yeast are present (Kim and Eckenrode, 1987). Adults feed on nectar from such flowers as dandelion, *Taraxacum officinale*, and also on aphid honeydew.

Life Cycle and Description. The temperature threshold for development is about 5°C. The optimal temperature for development, as determined by rapid development and maximal survival, is about 25–30°C. The time required for its development from the egg to the adult stage is 19.2 days at 30°C. Unlike the case with seedcorn maggot, which apparently aestivates at high temperatures, rearing temperatures of up to 35°C do not induce aestivation (Kim and Eckenrode, 1987). There likely are 2–3 generations annually in most locations, though Miller and McClanahan (1960) suggested four generations for southwestern Ontario.

Egg. The bean seed maggot egg is elongate, and white in color. It measures about 0.9 mm long and 0.3 mm wide. The egg is convex on one side and slightly concave on the other side. It is indistinguishable from that of seedcorn maggot. They are deposited near the surface of the soil in the proximity of food. The egg stage requires 1.4 days for hatching at 30°C, but development time is little influenced by temperatures in the range of 10–40°C. Oviposition potential varies considerably, and depends significantly on temperature. Mean egg production per female was 193, 99, and 47 for flies reared at 20, 25, and 30°C. The eggs are deposited throughout the life of the female. Oviposition can occur at any time of day, but occurs predominantly during the afternoon.

Larva. There are three instars, with average development time of 1.4, 1.4, and 6.1 days, respectively, for instars 1–3 at 30°C. The larva is white in color, and bears anterior spiracles with 5–8 lobes. The larva attains a length of about 7 mm at maturity. The larva feeds on the embryo of the plant or on the belowground portion of the seedling.

Pupa. The oval puparium is reddish or dark-brown, and measures about 5 mm long. Pupation occurs in the soil adjacent to the food plant. Duration of the pupal stage averages 9.7 days at 30°C.

Adult. The adults are grayish-green flies, measuring 4–5 mm long. They closely resemble seedcorn maggot, and are separated using the bristles on the legs. In males, the dorsally directed bristles of the middle metatarsus are twice as long as the width of the tarsus, whereas those of seedcorn maggot are no longer than the width of the tarsus. In females, there are four or fewer bristles on the mesothoracic tibia, whereas in seedcorn maggot the number is five or greater (Kim and Eckenrode, 1983; 1984). Mated females aged 4–6 days are capable of oviposition. In laboratory studies conducted by Kim and Eckenrode (1987), the adult females lived an average of 22.8 days, but males lived only 7.1 days when reared at 30°C.

Bean seed maggot larva, posterior view showing caudal spiracles.

*Color figure 263 is referred to on page 208.

Useful accounts of bean seed maggot biology were given by Throne and Eckenrode (1986), and Kim and Eckenrode (1987). Miles (1952), and Miller and McClanahan (1960) provided information, but these authors treated bean seed maggot and seedcorn maggot together. An interesting account of *Delia* ecology, and implications for management from a British perspective, is found in Finch (1989).

Damage

Bean seed maggot damages seeds and young plants in the same manner as seedcorn maggot. Basically, larvae burrow into developing seeds and the below-ground portions of seedlings. Stand density is often reduced, though plants sometimes recover from injury. Bean seed maggot often occur in mixed populations with seedcorn maggot. In a study conducted in Ontario, Miller and McClanahan (1960) found that about 11% of maggots were bean seed maggot, and the balance were seedcorn maggot. Bean seed maggot has also been implicated in the transmission of *Erwinia* bacteria. See the section on seedcorn maggot for a more complete discussion of damage.

Management

Management consideration discussed in the section on seedcorn maggot are also applicable for bean seed maggot. In one of the few management studies directed solely at bean seed maggot, Kim *et al.* (1985) reported that rapidly-germinating, colored-seed bean lines were more resistant to attack.

Beet Leafminer
Pegomya betae Curtis
Spinach Leafminer
Pegomya hyoscyami (Panzer)
(Diptera: Anthomyiidae)

Natural History

Distribution. There long has been confusion about whether beet leafminer and spinach leafminer are separate species. Now they are considered distinct species, but they are very similar in appearance and biology. Both leafminer species are found in Europe, northern Africa, Asia, and North America. In the United States they apparently occur widely, though absent from the southern and southwestern states. They also occur throughout southern Canada. Spinach leafminer predominates in eastern North America and beet leafminer in the west. These leafminers were accidentally introduced to North America from Europe, probably in the 1800s.

Host Plants. This species is best known for its damage to members of the plant family Chenopodiaceae, but spinach leafminer also is reported to attack a few plants in the family Solanaceae, Carophyllaceae, and perhaps others. The principal crop hosts of both leafminers are beet, spinach, sugarbeet, and Swiss chard. Alternate hosts are poorly documented, but lambsquarters, *Chenopodium album*; and redroot pigweed, *Amaranthus retroflexus*; are cited as hosts. Spinach leafminer also attacks black henbane, *Hyoscyamus niger*; deadly nightshade, *Atropa belladonna*; *Datura* sp.; and *Dianthus* sp. In studies conducted in France, beet leafminer favored *Beta vulgaris* for oviposition, and larvae developed only on this species, whereas spinach leafminer oviposited and survived on *Chenopodium* and *Datura* (D'Aguilar and Missonnier, 1957).

Natural Enemies. Several parasitoids of spinach leafminer are known from North America, including *Biosteres anthomyiae* (Ashmead), *Opius fulvicolis* Thomson, *O. middlekauffi* Fischer, and *O. nitidulator* (Nees) (all Hymenoptera: Braconidae). In addition, *Biosteres spinaciae* (Thomson) (Braconidae) has been reared from both leafminer species. Predatory bugs such as *Nabis* (Hemiptera: Nabidae) feed on larvae by puncturing the leaf epidermis to reach the insect beneath.

Life Cycle and Description. A generation requires about 30–40 days for its completion. The number of generations is reported to be 2–4 annually. Overwintering occurs in the pupal stage and commences beginning in August. Adults of the overwintering generation emerge in April or May. (The spinach leafminer larva in color figure 177 is referred to on page 209.)

Egg. The eggs are elongate-oval and measure about 0.87 mm long and 0.31 mm wide. Delicate hexagonal sculpturing is present on the surface of the eggs. The eggs are white. The eggs are glued to the lower surface of leaves, often in clusters of 2–5 but sometimes in larger groups. The eggs are positioned adjacent and parallel to one another. Females have been reported to deposit up to 70 eggs, but this has been poorly studied so the true fecundity remains unknown. Eggs hatch in 3–6 days, and larvae immediately burrow into the leaf. (See color figure 263.)

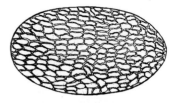

Beet leafminer egg.

Beet leafminer larva.

Larva. The larva is cylindrical in form, tapering to a point at the anterior end. At hatching, larvae are almost transparent but become whitish or yellowish with age. Larval size increases from about 1.0 mm to 1.6 mm during the first instar, to 2.0–4.0 and 8.0–9.0 mm during instars two and three, respectively. The anterior spiracles are marked with 7–8 lobes. The larval period requires about 7–8 days during the summer, but may require 10–12 days during cool weather. Larvae often feed together within a single mine, exhausting the food resource. Often only 2–3 larvae can survive on a leaf, and later–hatching larvae perish. Larger larvae are capable of leaving a mine and entering a new leaf, but it is questionable whether this happens often under natural conditions. Crowded larvae tend to produce small adults. (See color figure 177.)

Pupa. Pupation usually occurs in the soil directly beneath the plant which was fed upon by the larval stage, and often at depths of 5 cm. Pupation occasionally occurs on the plant within the larval mine. D'Aguilar and Missonnier (1957) suggested that spinach leafminer larvae were quite likely to pupate within the larval mine, especially early in the season, whereas beet leafminer always dropped to the soil for pupation. Pupae that develop within leaves do not enter diapause. The puparium is oval in form, slightly flattened, and with the anterior end more tapered. Both the anterior and posterior ends are marked with protruding spiracles. The oval puparium measures about 4.5–5.0 mm long and 1.7–2.0 mm wide. Initially yellowish, the puparium soon turns red-dish-brown to brownish black. Duration of pupation is usually 10–20 days except, of course, when overwintering.

Adult. These small, hairy flies measure about 5–7 mm long. The adult appears to be gray or grayish-brown, though long hairs clothing the body are black and the front of the head is silvery white. A dark stripe may occur dorsally on the abdomen, particularly in spinach leafminer. In beet leafminer the middle and hind tibiae and femora tend to be dark, whereas in spinach leafminer they are yellow. The thorax, abdomen, and legs are adorned with stout black hairs. The wings are transparent. Oviposition commences one or more days after copulation. On average, beet leafminer adults are larger and darker in color than spinach leafminer. Duration of adults males is about 9–12 days, and of females about 12–24 days.

Because the identity of these leafminers has been confused, there is some uncertainty about leafminer biology. Cameron (1914) and Frost (1924) provided detailed biology of spinach leafminer. Characters to

Beet leafminer puparium.

distinguish beet and spinach leafminer were given by D'Aguilar and Missonnier (1957), Chillcott (1959), and Michelsen (1980). A procedure for culture of beet leafminer was described by Rottger (1979).

Damage

The larvae feed between the lower and upper epidermis of the leaf. The form of the mine initially is long and narrow, but soon turns into an irregular blotch. Old mines become dry and brittle. Such mining is rarely of consequence to growth of beet roots, but if beet tops are desired or if leafy crops such as Swiss chard or spinach are grown, the damage can be considerable. These leafminers are mostly known as garden pests, but commercial-scale fields of spinach are sometimes damaged.

Management

The eggs are easily observed by examining the lower surface of leaves. Damage can be prevented by application of insecticides to foliage or to soil; systemic insecticides are particularly effective, but care must be taken to avoid residues on leafy vegetables. Residual soil applications are effective because larvae generally drop to the soil to pupate, thereby contacting the insecticide. This approach requires that the earliest foliage, which supports leafminer development, be discarded. An effective alternative for small-scale production is to cover the plants with screen or floating row cover material; this prevents flies from ovipositing unless they emerge from the soil beneath the cover. Crops

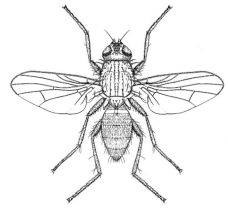

Adult beet leafminer.

grown early and late in the season often escape the principal flights of flies, thereby escaping injury.

Cabbage Maggot
Delia radicum (Linnaeus)
(Diptera: Anthomyiidae)

Natural History

Distribution. Cabbage maggot is known as a pest throughout the northern hemisphere. Not native to North America, it apparently was introduced accidentally from Europe in the early 1800s. Cabbage maggot thrives under cool conditions, and it is a pest only in the northern portions of the United States. It rarely is reported to be a pest south of about latitude 45° North, and when it is, it usually occurs at a high elevation. In Canada, it is found almost everywhere, including some of the northernmost agricultural regions. Beirne (1971) attributed greatly reduced production of cruciferous root crops in eastern Canada to this insect.

Host Plants. Cabbage maggot commonly attacks cruciferous vegetable crops, including broccoli, Brussels sprouts, cabbage, cauliflower, collards, kale, kohlrabi, mustard, radish, rutabaga, turnip, and watercress. It has been reported from noncrucifer crops on occasion, but these are misidentifications that stem from the difficulty in accurately identifying this fly. Cruciferous weeds apparently do not play a significant role in the biology of this insect; although some appear to be suitable hosts, they rarely are mentioned in the economic entomology literature.

Natural Enemies. Important natural enemies include staphylinids (Coleoptera: Staphylinidae) of several genera, particularly *Aleochara* sp.; a wasp, *Trybliographa rapae* (Westwood) (Hymenoptera: Eucoilidae); and a mite, *Trombidium* sp. (Acari: Trombidiidae). *Aleochara bilineata* is unusual in that though larvae parasitize the pupal stage of cabbage maggot, the adults feed on maggot eggs and young larvae (Royer *et al.*, 1998; Royer and Boivin, 1999). *Trybliographa* attacks the larvae, and the mite destroys the eggs. Clausen (1978) provides a useful synopsis of the aforementioned beneficial insects.

Natural control has been studied extensively in Europe, and this is relevant because many of the same biotic natural mortality factors are also found in North America. Egg predation by staphylinids and carabids may reach 90–95% annually (Hughes, 1959; Coaker and Williams, 1963). *Aleochara* are very effective predators, but become active too late in the spring to have much effect on first generation cabbage maggot. *Trybliographa* is fairly effective at high host densities, often parasitizing in excess of 50% of available

hosts (Miles, 1956; Hughes and Mitchell, 1960; Jones and Hassell, 1988). Other natural enemies of the immature cabbage maggot include numerous hymenopterous parasitoids of questionable economic importance, carabids (Coleoptera: Carabidae) and ants (Hymenoptera: Formicidae). General predators undoubtedly attack the adults, but are not considered important.

Fungi are commonly observed infecting flies. *Entomophthora muscae* and *Strongwellsea castrans* cause epizootics among adults during wet weather, and though impressive, act too late to prevent early season crop damage. *Entomophthora muscae* has been discussed further in the section on onion maggot, *Delia antiqua* (Meigen).

Weather. The spring generation tends to appear consistently, but latter generations are greatly influenced by weather. Rain and cool weather may decrease egg production and egg predation, and cause starvation of flies (Miles, 1951; 1956), but optimal egg production is associated with temperatures of 18–21°C, which corresponds well with the weather occurring during most spring generations (Miles, 1954a,b). Pupal development is particularly susceptible to delay caused by hot temperature, which normally is associated with summer generations (Finch *et al.*, 1986). Dry soil is lethal to eggs.

Life Cycle and Description. The number of generations occurring annually varies from one in the far north to three in optimal climates, though there are occasional reports of four generations. Two generations occur in southern Alberta (Swailes, 1958), and two and three in eastern and southwestern Ontario, respectively (Mukerji and Harcourt, 1970). Three generations are documented from New York (Schoene, 1916), though part of the second generation enters diapause rather than going on to form a third generation. Sandy soils in Canada's Atlantic Provinces have two generations, whereas heavy soil in the same area usually support only one generation (Beirne, 1971). In New York, the adults of the first brood were observed in May to early June, the second during mid-June to early July, and the third during late August to early October. A very similar pattern occurs in British Columbia. The generations may overlap considerably. A developmental threshold of about 6°C has been determined for most life stages (Eckenrode and Chapman, 1971a). The time required for a complete generation is estimated at 40–60 days.

Egg. The eggs normally are laid in the soil around the stem of cruciferous plants, but sometimes they are deposited directly on the stem of plants. The elongate eggs are white in color, and taper markedly at both ends, but one end is more blunt than the other. One side of the egg is flattened or slightly concave, with the opposite side convex. The eggs measure about 1.1 mm long and 0.34 mm wide. They are often laid in clumps of a few eggs, but sometimes hundreds of eggs are found at the same location—evidence that more than one female may oviposit at the same spot. Females commonly produce 300–400 eggs during their life span of 30–60 days. Eggs hatch in 3–5 days, averaging about 3.5 days at 20°C.

Larva. There are three instars. The length of the mouthparts (cephalopharyngeal skeleton) can be used to differentiate instars, with mean lengths of 0.44, 0.80, and 1.24 mm, respectively (Miles, 1952). The larvae are white and attain lengths of about 1.5, 3.7, and up to 8 mm, respectively. The mouth hooks are black. Located immediately behind the head are a pair of brownish fan–like spiracles, each of which is divided into about 12 lobes. Larvae feed externally and internally on roots, and internally on stem tissue. The larval period requires about 18–22 days under field conditions, but development time may be altered by weather. Exposure of first and second instar larvae to cool temperatures or short photoperiods seems to induce diapause in the pupal stage (Koni, 1976), but the relative importance of each factor in stimulating diapause is disputed.

Pupa. The puparium is oval, bluntly rounded at both ends, and brown in color. The average length is 5.5 mm, (range 3.5–6.5 mm). The duration of the pre-pupal stage is about 3–5 days, and the pupa requires 12–25 days during the summer. However, this is the overwintering stage, so it is prolonged for 5–8 months in the overwintering generation. Overwintering pupae require at least 22 weeks of temperature less than 6°C to complete diapause development (Collier and Finch,

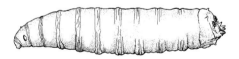

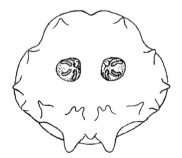

Cabbage maggot larva.

Cabbage maggot larva, posterior view showing caudal spiracles.

1983). The puparia from the summer generations are usually found in the soil immediately adjacent to the root on which the larvae last fed. Sometimes they occur within the plant tissue, including the above-ground stem tissue. Aestivation occurs in response to warm temperature above 20°C, and especially in response to hot temperature, 27–30°C (Collier and Finch, 1983). In preparation for overwintering, the larvae seem to disperse further from the plant, and deeper into the soil. Overwintering puparia may be found 12–15 cm from the plant, and commonly are found in the soil at a depth of 10 cm.

Adult. The adults are dark with gray markings. The male bears three blackish longitudinal bands on the thorax, but these markings are less distinct on the female. The flies are quite bristly, and measure 5–7 mm long. The male cabbage maggot is distinguished from its common relatives, seedcorn maggot (*Delia platura* (Meigen) and onion maggot, by the presence in cabbage maggot of a dense tuft of bristles at the base of the underside of the hind femora. Adults feed on nectar from flowering plants. If they obtain adequate food they may persist for 2–4 weeks, whereas they perish in 2–3 days if denied food. Adults are highly attracted to crucifers for oviposition (Eckenrode and Chapman, 1971b). The pre-oviposition period of adults is about six days.

The biology of cabbage maggot was given by Schoene (1916) and Fulton (1942), and information on rearing by Harris and Svec (1966) and Finch and Coaker (1969). Mass rearing technology was presented by van Keymeulen *et al.* (1981). Keys to distinguish eggs, larvae, and adults of cabbage maggot from other anthomyiids associated with crucifers were provided by Brooks (1951). An interesting account of *Delia* ecology, and implications for management from a British perspective, was found in Finch (1989).

Damage

Larvae damage crucifers by feeding on the roots and, to a much lesser extent, the stems or petioles of plants. Damage to leaf crops such as cabbage is most evident in the late spring; signs of feeding damage are initially seen as drooping or wilting of a few leaves, and then perhaps the entire plant. Delayed maturity and stunting are common responses to root maggot injury. Plant death often coincides with drought or water stress, when the injury to roots is fully expressed. When plants are small, 5–10 maggots are necessary to kill the seedling. However, later in the season densities of 100 maggots or more may be supported satisfactorily if the plant has adequate water. In

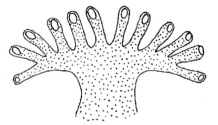

Structure of spiracles on cabbage maggot.

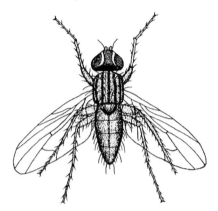

Adult cabbage maggot.

California, Zalom and Pickel (1985) found that once Brussels sprouts plants attained an age of about four weeks they were not very susceptible to injury.

Cabbage maggot larvae feed on the rootlets or feeder roots, but invariably move to the main or tap root as they mature. They scar the surface and burrow into the root. For crops that are harvested for their root, such as radish and turnip, damage results in severe crop loss, and in this case the autumn generation may be quite important. The summer generation causes little damage. Damage tends to be greater on loamy-sand soil than on sand or clay soil (Friend and Harcourt 1957), but as a general rule light soils are more problem prone (Beirne, 1971).

Although most eggs are laid on the soil, a small number are sometimes deposited on plant tissue, resulting in injury by larvae to leaflets, especially to Brussels sprout buttons. Occasionally the growing points of plants are attacked, resulting in multiple heads (Gratwick, 1992).

Management

Sampling. The adult flight periods can be monitored using cone-screen traps baited with crucifers. Baits are more effective than yellow-sticky traps (Dapsis and Ferro, 1983), though sticky trap captures are correlated with egg deposition rates (Sears and Dufault, 1986). Horizontal surfaces are more suitable than vertical for landing of flies (Finch and Collier, 1989). Dispensers that release isothiocyanates, natu-

rally occurring odors released by crucifers, can be used as lures (Finch and Skinner, 1982). Color, but not leaf pattern, also can influence host selection (Prokopy *et al.*, 1983). Bracken (1988) found that water traps baited with isothiocyanate could be used to monitor population trends in Manitoba, but that trap catches did not predict egg numbers accurately. Finch (1991) increased selectivity of yellow water-pan traps by modifying the paint pattern. Alternatively, flight activity can be predicted from thermal unit accumulations. Eckenrode and Chapman (1972), working in Wisconsin, reported that the overwintering (partly developed) generation required about 300 degree-days (above a base temperature of 43°F). Subsequent generations required about 1200 degree-days. Similar results were reported by Bracken (1988). However, accurate prediction is complicated by bimodality of emergence in both diapausing and nondiapausing populations, with emergence peaks separated by at least 20 days when flies are reared at 20°C (Biron *et al.*, 1998).

Sampling of immature stages is labor intensive. Mukerji and Harcourt (1970) estimated that at densities of about six eggs, larvae, or pupae per plant, sample sizes of about 50 plants might be required to obtain adequate precision. The optimal sampling unit was the plant root together with a 10–15 cm core of soil.

Insecticides. With the advent of chlorinated hydrocarbon insecticides, damage by cabbage maggot was greatly decreased. Long-lasting insecticides, applied to the soil at planting, protect the roots from larvae (Doane and Chapman, 1962b). This remains the principal method of plant protection in commercial crucifer production, but the insecticides have been changed over time as resistance to insecticides is developed. Loss of insecticide efficacy is due not only to selection for insecticide-resistant insects, but enhanced degradation of insecticide by soil microbes. Suett *et al.* (1993), for example, suggested that there should be at least a three-year interval in applications of the same insecticide for cabbage root maggot control.

Insecticides are typically applied as a granular formulation over the seed bed or incorporated into the soil, or as a liquid drench. Foliar applications are sometimes made to suppress adults. Foliar application of insecticides, timed according to temperature accumulations, can be superior to soil applications or calendar-based sprays (Wyman *et al.*, 1977). In studies conducted in Ontario, incorporation of sucrose bait into foliar insecticide formulations did not increase efficacy (Dufault and Sears, 1982). Seed treatment can be an effective method to provide protection to seedlings, but it does not work well for all insecticides

(Ester *et al.*, 1994). Naphthalene has been investigated in Europe for repellency to ovipositing flies; though good protection occurs for about six weeks, the cost of application is high (den Ouden *et al.*, 1984).

Cultural Practices. Modification of planting and tillage practices is often recommended for reduction in cabbage maggot damage. Delayed planting is reported to allow the young plants to escape oviposition by the spring adults. Sanitation also is quite important, as the roots and stems of crucifers left in the field can be very suitable for autumn and early spring generations of cabbage maggot. Crop residues should be deeply buried, or be pulled and allowed to dry completely. High plant densities are more attractive than low to flies, but because there are more plants on which to distribute the eggs, yield may be equivalent at both plant densities (Finch and Skinner, 1976). Introducing diversity into the landscape, as by undersowing portions of a crucifer crop with clover, will disturb the normal host orientation pattern and will reduce oviposition (Kostal and Finch, 1994). Similarly, single-row intercropping of crucifers with unrelated plants will greatly decrease the oviposition rate on crucifers (Tukahirwa and Coaker, 1982).

Physical manipulations of the crop environment also assist pest suppression. Crops covered tightly with row covers escape injury by cabbage maggot (Schoene, 1916; Millar and Isman, 1988), though it is difficult to separate the benefit of insect damage reduction from improved microenvironment under the row cover (Matthews–Gehringer and Hough–Goldstein, 1988). Even surrounding a crop with a window screen barrier, without covering the top of the crop, provides some benefit. For example, Päts and Vernon (1999) reduced cabbage maggot fly numbers by 90% in radish plantings surrounded by a 1.2 m high barrier. Presumably, the barrier interferes with fly host location. Tar paper or cardboard discs, or collars made of other weather-resistant material, placed around the stem of seedlings have long been recommended as a physical barrier to reduce the ability of females to deposit eggs at the soil-stem interface (Schoene, 1916). A precise fit is required, however, or the flies will circumvent the barrier (Matthews–Gehringer and Hough–Goldstein, 1988). Mulch that is painted blue also reduces the incidence of infestation by cabbage maggot (Liburd *et al.*, 1998).

Biological Control. Reliable biological control techniques have not yet been developed. Entomopathogenic nematodes (Nematoda: Steinernematidae and Heterorhabditidae) have been evaluated for suppression of larvae. Although heterorhabditid nematodes are attracted to cabbage maggot larvae and pupae (Lei *et al.*, 1992), under field conditions they

have not been shown to be effective (Simser, 1992). Steinernematid nematodes provide some suppression of cabbage maggot larvae in pot and field trials, but very high densities of nematodes are needed, at least 100,000 nematodes per plant (Schroeder *et al.*, 1996). This is not entirely surprising because fly larvae are less susceptible to nematodes than many other insects. In Europe, the potential of using the predatory beetle *Aleochara bilineata* (Gyllenhal) (Coleoptera: Staphylinidae) to achieve biological suppression of cabbage maggot is being studied; while technically feasible, the costs thus far are high (Finch, 1993).

Host-Plant Resistance. Despite the common name, cabbage is less attractive to cabbage maggots than some other crucifer crops. Chinese cabbage, mustard, rutabaga, and turnip tend to be more severely injured than cabbage (Radcliffe and Chapman, 1966). Doane and Chapman (1962a) reported that a larger numbers of eggs were associated with radish, rutabaga, and turnip than with mustard and cauliflower. There also is some variation within vegetable crops in resistance to attack (Ellis *et al.*, 1979); fast-growing varieties seem most injured (Swailes, 1959). Finch (1993) reported that despite considerable effort, not much progress has been made on finding cultivars resistant to cabbage maggot.

Onion Maggot
Delia antiqua (Meigen)
(Diptera: Anthomyiidae)

Natural History

Distribution. Onion maggot is found throughout the northern hemisphere, and apparently it was introduced to North America from Europe soon after European colonists arrived. It was recognized as a pest in the New England states by the early 1800s, and reached western onion-growing regions by the early 1900s. Onion maggot now is distributed widely in the United States and southern Canada. However, it apparently is absent from Florida in the southeast, and Arizona and New Mexico in the southwest. It is considered a serious pest only in northern areas.

Host Plants. Onion maggot attacks plants in the family Alliaceae (Amaryllidaceae). Onion is the principal host, but chive, garlic, leek, and shallot are sometimes attacked. Wild onion apparently is not a suitable host. It is possible to rear onion maggot larvae on abnormal hosts such as radish and turnip (Workman, 1958), but adults normally do not deposit eggs on such plants. (See color figure 33).

Onion maggot adults are particularly attracted to decaying onions, such as those infected with the soft-rotting bacteria *Erwinia carotovora* or *Fusarium* fungus

(Dindonis and Miller, 1980). Survival may be slightly higher, or development time shorter, for onion maggot larvae feeding on microbe-infected onions (Zurlini and Robinson, 1978). Older onion maggot larvae are usually fully capable of attacking and developing satisfactorily on disease-free onions (Schneider *et al.*, 1983). However, survival of young maggots is poor on mature onions unless they are wounded or infected with disease. Indeed, survival of the third generation of onion maggots is likely dependent on the availability of unhealthy onions.

Natural Enemies. Numerous insects are known, or suspected, to prey on onion maggots. Tomlin *et al.* (1985) provided a comprehensive list of natural enemies from Ontario that was likely a representative of the predator and parasitoid complex in North America. Parasitoids included *Aleochara bilineata* (Gyllenhal), *A. bipustulata* (Linnaeus), *A. curtula* (Goeze) (all Coleoptera: Staphylinidae); *Aphaereta pallipes* (Say) (Hymenoptera: Braconidae); *Spalangia rugosicollis* Latreille and *Sphegigaster* sp. (both Hymenoptera: Pteromalidae). Among the common predators were mites (Acari), rove beetles (Coleoptera: Staphylinidae), hister beetles (Coleoptera: Histeridae), sap beetles (Coleoptera: Nitidulidae), ground beetles (Coleoptera: Carabidae), and ants (Hymenoptera: Formicidae). Eggs and larvae were much more susceptible to predation than pupae, probably because of their smaller size. The authors concluded that the parasitoids *Aleochara bilineata* and *Aphaereta pallipes,* which caused 21% and 17% mortality, respectively, were among the most important mortality agents. Predation is much more difficult to measure, so the relative importance of predators is less certain. However, such predators as *Bembidion quadrimculatum* L. (Coleoptera: Carabidae) were reported to consume 25 onion maggot eggs per day (Grafius and Warner, 1989), so predators surely have an important role in onion maggot population biology. *Aleochara bilineata* is unusual in that though larvae parasitize the pupal stage of onion maggot, the adults feed on maggot eggs and young larvae (Royer *et al.*, 1998; Royer and Boivin, 1999).

The fungus *Entomophthora muscae* commonly infects onion maggot and many other flies. Dead, spore-covered flies are often found attached to elevated portions of plants following fungus outbreaks. Carruthers *et al.* (1985) conducted a study of this fungus in Michigan, and concluded that while abiotic factors such as humidity and temperature likely were important in disease cycle development, host and pathogen densities were more critical factors. Flies contact fungus, and become infected, both on host-plant foliage and in the soil. Death of adults occurs in 7–9 days after exposure to conidia (Tu and Harris, 1988). The soil is the principal overwintering reservoir for the fungus. The spring and autumn generations of onion maggot exhibited higher levels of infection.

Life Cycle and Description. The number of generations is two, or more commonly three, per year—with a proportion of each generation diapausing as puparia until the next year. In Quebec, adults from overwintering insects appear in May and initiate the first generation (Lafrance and Perron, 1959; Boivin and Benoit, 1987). Peak emergence of the first generation occurs in early July. The second generation begins emerging in mid- to late-August, with peak emergence about the first of September. Third generation insects all enter diapause, forming the overwintering population that emerges the following spring. The proportion of insects entering diapause in the first and second generations is reported to be about 25% and 60%, respectively. During exceptionally warm summers fewer insects enter diapause. Very similar patterns of emergence were observed in Pennsylvania (Eyer, 1922), and Oregon (Workman, 1958).

Egg. Oviposition occurs when temperatures attain about 21–26°C (Miles, 1955a). In Pennsylvania, maximum oviposition from the overwintering generation was observed in early June, whereas the first generation adults produced eggs in mid- to late-July, and the second generation adults produced eggs in early September (Eyer, 1922). The eggs are white and elongate, with one side convex and the other side slightly concave. One end is slightly more pointed than the other. The eggs measure about 1.2 mm long (range 1.12–1.23 mm) and 0.5 mm wide. Females tend to deposit clusters of 7–9 eggs daily, and duration of the oviposition period is about 30–60 days. In the laboratory, egg production for onion maggot colonies may average 100–250 per female, with individually-reared insects producing an average of 500 eggs (Vernon and Borden, 1979). Estimates of egg production in the field, however, are considerably less, perhaps 50 per female (Perron and Lafrance, 1961). The eggs are positioned on or near the host plant, often in soil crevices adjacent to onion plants. If the soil is wet, as occurs immediately following a heavy rain, the adults may deposit their eggs on the foliage. Ovipositing flies are highly attracted to bulbs previously infested with maggots, though low to moderate levels of damage are preferred over high levels of damage (Haussmann and Miller, 1989). Microbial activity of rotting onion bulbs enhances attractancy of onion volatiles (Dindonis and Miller, 1981). The soil temperature optimum for oviposition is about 20–22°C (Keller and Miller, 1990). The eggs hatch in about 5.5 days (range 5–7 days) in cool weather, but hatch may be shortened to 2–3 days during the warmth of the summer.

Larva. Larvae are whitish, and have three instars. The mouthparts (cephalopharyngeal skeleton) measure 0.38, 0.71, and 0.96 mm long, respectively, during instars 1–3. When they first hatch, larvae measure only about 1 mm long, but by the time they are matured larvae attain a length of about 9–10 mm (Miles, 1953). Development time is about 3, 4–5, and 9–14 days for instars 1–3, respectively. Total duration of the larval stage is estimated at 15–23 days. The anterior spiracles, which do not become visible until the second instar, bear 12 (range 10–13) finger-like lobes. The number of spiracular lobes is a useful character for differentiating onion maggot larvae from seedcorn maggot, *Delia platura* (Meigen) (Diptera: Anthonyiidae)—a species commonly found in association with onions and onion maggots. The spiracles of seedcorn maggot larvae bear only 5–8 lobes.

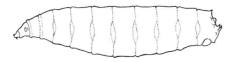
Onion maggot larva.

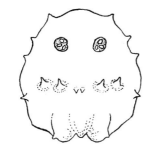
Onion maggot larva, posterior view showing caudal spiracles.

Onion maggot puparium.

Pupa. Mature larvae form a puparium in the soil. The location of the puparium may be close to the bulb, perhaps among the onion roots, but it sometimes is a considerable distance away, and at a depth of 10–12 cm. Puparium color varies from yellow-brown to dark-brown or almost black. The puparium measures 4–5 mm long. Duration of the puparium is from 15 to 19 days during the spring, 8 to 14 days during the summer, and 5 to 6 months during the winter.

Adult. The adult is greenish gray, and marked with longitudinal dark strips on the thorax. The legs are black and the wings rather colorless with black veins. The adults average about 6 mm long. Adults usually live about 30–50 days if they have adequate food, but they may attain longevities of 60–70 days. The pre-oviposition period is often about 10 days, followed by an egg-laying period of about 5 days.

The biology of onion maggot was given by Eyer (1922), Workman (1958), and Perron (1972). Methods of onion maggot rearing were provided by Allen and Askew (1970) and Blaine and McEwen (1984). Effect of radiation on onion maggot was given by McEwen *et al.* (1984). An interesting account of *Delia* ecology, and implications for management from a British perspective, was found in Finch (1989).

Damage

Onion maggots can be extremely damaging in northern onion producing areas. In New York, for example, 40–80% losses are realized regularly if insecticides are not used (Martinson *et al.*, 1989). Damage to onions often occurs in early to mid-summer; the damage that occurred earlier in the development of the onion plant is more often due to seedcorn maggot,

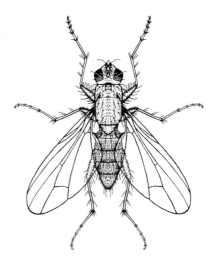
Adult onion maggot.

which may be active as early as April (Miles, 1955b). When onion maggot larvae feed on relatively small onions they may completely hollow out the young bulb, causing rapid death. A single onion maggot larva may kill several seedling onions, whereas as the onions grow larger the larvae are not compelled to disperse to other onions to find food. Later in the season the onions are more tolerant of onion maggot feeding, and there may not be above-ground signs of maggot feeding, but the bulbs can incur additional damage from secondary invaders such as soil-borne fungi and seedcorn maggot. Onions nearing maturity are most susceptible to these secondary organisms. Twelve-week-old onions are not very suitable for survival of onion maggots (Finch *et al.*, 1986) unless they are mechanically damaged or infected with disease. Mechanical damage to bulbs during harvesting can lead to attack by third generation flies of bulbs drying in the field (Eckenrode and Nyrop, 1986). Carruthers *et al.* (1984) have developed procedures for estimating plant damage.

There appears to be a close relationship between onion maggot and soft rot of onion caused by *Erwinia carotovora*. The maggot creates wounds where the bacterium can enter the onion. The surface of several stages of the insect may be contaminated with the bacterium, thus aiding dispersal. Also, the bacterium is found in the puparium, which provides protection and allows survival during inclement periods. The insect also benefits from this relationship, because onion tissue infected with soft rot bacteria are more suitable for development of larvae. The relationship is not entirely specific, however, and several other flies such as seedcorn maggot, and several otitidids (Diptera: Otitidae), are attracted to rotting onions (Harrison *et al.*, 1980).

Management

Sampling. Onion maggot flies use both olfactory and visual cues to find their host plant. This information has been exploited to develop sampling techniques. Flies will orientate up-wind in response to onion volatiles (Judd and Borden, 1989), and various trap designs employing inverted cones suspended above onion bulbs or baited with the onion chemical n-dipropyl disulfide have been devised to capture adults (Dindonis and Miller, 1980). Enzymatic yeast hydrolysate is also attractive (Miller and Haarer, 1981). Yellow traps also are recommended for capture of onion maggot flies, and sometimes yellow-sticky traps are baited with onions. Vernon *et al.* (1989), however, reported that white traps were more effective. Traps capture more flies when they are positioned within 20–30 cm of the bare soil. The adults prefer to land on horizontal surfaces. Therefore, water pan traps are more appropriate than vertical sticky traps (Finch and Collier, 1989).

Heat-unit accumulations have been used to predict developmental rates of onion maggot (Boivin and Benoit, 1987) and as an aid to estimating optimal timing of insecticide applications (Eckenrode *et al.*, 1975). Careful timing of adult control reduced the number of insecticide treatments from 7–12 annually to only two, with excellent results (Liu *et al.*, 1982). Phenological development of plants also serves as an indicator of fly flights (Boivin and Benoit, 1987).

Insecticides. In many areas, insecticides are routinely used to protect against damage by onion maggot. Application of granular or liquid insecticide formulations to the furrow at planting, or over the row after planting, are common. Treatment of seed with insecticides is also beneficial, but some materials are phytotoxic when applied in this manner (Saynor and Hill, 1977). Disruption of first generation onion maggot helps protect against damage by later generations even if the insecticide residue has dissipated (Ritcey *et al.*, 1991). However, long-lasting insecticides are more effective, and in the absence of these, adult suppression is practiced to protect against second- and third-generation onion maggots (Finch *et al.*, 1986). Insecticide applications to foliage are common, but in many areas insecticides are also applied to the harvested bulbs drying in windrows. To avoid illegal insecticide residues on onions, it is important to use insecticides that are not very persistent for treatment of dry bulbs (Frank *et al.*, 1982). Insecticide resistance is common in many onion-growing areas (Harris and Svec, 1976; Straub and Davis, 1978; Harris *et al.*, 1982; Hayden and Grafius, 1990).

Biological Control. Several fungi, including *Beauveria bassiana* and *Paecilomyces fumosoroseus*, have been evaluated for suppression of onion maggot, but the puparia and adults seem to be relatively insensitive (Majchrowicz *et al.*, 1990).

Cultural Practices. The only significant host of onion maggots is onions, and usually commercially produced crops of onions. Therefore, crop rotation is sometimes a suggested component of pest management. The potential for onion maggots to cause damage is directly related to the previous presence of onions (Martinson *et al.*, 1988). However, onion maggots disperse randomly over distances that exceed 2 km (Martinson *et al.*, 1989), so it is difficult to achieve adequate isolation.

Sanitation is an important component of onion maggot management. Damaged bulbs left in the field are an important food source for overwintering populations, much more so than are cull piles and volunteer onions (Finch and Eckenrode, 1985). Volunteer onions are, in fact, highly attractive to ovipositing flies in the spring, but few maggots survive on these plants. This suggests the possibility of using early-planted trap crops to lure females from the principal crop. Miller and Cowles (1990) suggested use of a combination of cull onions as an oviposition attractant, and application of a chemical oviposition deterrent to onion seedlings, to minimize damage. Efforts should be made to minimize damage to bulbs when tilling, or at harvest, as these injured bulbs are very suitable for onion maggot larval development. If onion culls are to be disked in the autumn, it is best to wait until third generation flies have oviposited, as this deprives flies of additional suitable oviposition sites.

Soil conditions influence onion maggot, but the relationship of soil to population biology warrants additional research. Egg hatching and larval survival are higher in moist soil. Only if larvae are exposed to saturated soil for protracted periods will high soil moisture levels be a problem. Perron (1972) reported that sites with heavy soil, rich in clay, were less suitable for onion maggot than the sites with organic soil. Onion maggots apparently prefer to oviposit on organic soils, and abundance of beneficial insects is higher in organic soils.

Crop density may affect onion maggot damage. Perron (1972) reported that high onion densities were beneficial, principally due to dilution of onion maggot injury over more onions. Parasite activity was also enhanced in high density plots.

In New York, Walters and Eckenrode (1996) demonstrated the benefit of several approaches to onion maggot suppression. Crop rotation in combination with partially resistant onion varieties and decreased rates of insecticide were effective in preventing damage.

Radish Root Maggot
Delia planipalpis (Stein)
(Diptera: Anthomyiidae)

Natural History

Distribution. Radish root maggot is a native species, and western in distribution. It is known from Alaska and Manitoba in the north, with its range extending south to California and New Mexico.

Host Plants. This insect principally attacks plants in the family Cruciferae. Vegetable crops suitable for survival of radish root maggot include Brussels sprouts, cabbage, cauliflower, Chinese cabbage, rad-

ish, and turnip. Occasionally it has been observed to develop in the pods of pea and lupine. Weed hosts include pepperweed, *Lepidium densiflorum*; and tansy mustard, *Descurainia pinnata*. However, radish is the principal host, and is the only plant known to be damaged seriously.

Natural Enemies. The natural enemies associated with cabbage maggot, *Delia radicum*, and other *Delia* spp., are likely the principal predators and parasitoids of radish root maggot. The parasitoids *Trybliographa rapae* (Westwood) (Hymenoptera: Eucoilidae) and *Aleochara bilineata* Gyllenhall (Coleoptera: Staphylinidae) are known to parasitize up to 20% of the maggot population, with the staphylinid being considerably more important (Wishart, 1957; Kelleher, 1958). Ground beetles (Coleoptera: Carabidae) are considered as important egg predators, but no data are available to be document their effects. Hyslop (1912) reported rearing parasitoids from this fly, but he did not give any indication of their importance.

Life Cycle and Description. In Manitoba, about three generations develop annually. Overwintering occurs in the puparium. Adults from overwintered pupae appear in late May to early June. This is followed by the first generation adults in late June and early July, second generation adults in early August, and the offspring of the second generation overwintering. The life cycle is not well studied.

Egg. Oviposition occurs over a 10–20-day period, producing only 40–50 eggs during their life span of about 30 days. Eggs are deposited in clusters of 1–3 on an exposed portion of the root, or on the inner surface of a leaf petiole. Thus, they differ from most other *Delia* pests, which tend to lay larger clusters on or below the soil surface. Duration of the egg stage is about two days.

Larva. The whitish larvae feed within the root and attain a length of about 7.9 mm. They possess a typical maggot form, generally cylindrical in shape but tapering from the bluntly rounded posterior end to the pointed anterior end. Duration of the three instars is about 3, 3.5, and 10 days, respectively.

Pupa. Pupation generally occurs in the soil near the host plant. The puparium is brown to brownish-black, and measures about 6.4 mm long.

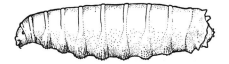

Radish root maggot larva.

Adult. The adults are very similar to seedcorn maggot flies, *Delia platura* (Meigen), in general appearance. The flies are gray or yellowish-gray in overall body color, but marked with a black dorsal stripe on the abdomen. The thorax bears a medial and two lateral stripes, but they are weak. The wings are not marked. Males measure about 5 mm long, and females 6 mm.

The only significant treatment of radish root maggot biology was provided by Kelleher (1958). Hyslop (1912) gave a description of the developmental stages. Brooks (1951) provided some valuable observations and a key to differentiate radish root maggot from similar crucifer-infesting flies.

Damage

In a study of radish-infesting flies conducted in Manitoba, Kelleher (1958) found radish root maggot to be the dominant pest species. There usually are few larvae per radish, often only one, so plants are not heavily damaged and do not exhibit obvious symptoms of attack. Although damage is not evident until the radishes are harvested, even one insect can destroy the commercial value of a radish. Because the insects feed internally and are difficult to detect, even a low incidence of infection may destroy the commercial value of a crop.

Management

Management practices developed for cabbage maggot, *Delia radicum* (Linnaeus), and seedcorn maggot, *Delia platura* (Meigen), are applicable to radish maggot. Prevention of damage is normally accomplished by application of liquid or granular insecticide to the soil at planting.

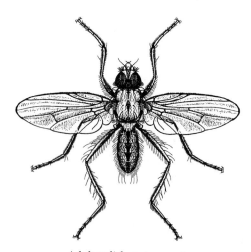

Adult radish root maggot.

Seedcorn Maggot
Delia platura (Meigen)
(Diptera: Anthomyiidae)

Natural History

Distribution. Seedcorn maggot is found in temperate areas all over the world. It was first described in Germany, but had attained North America by the mid-1800s, and like so many other invasive pests, gradually spread westward. It is now reported from throughout the United States, including Alaska and Hawaii. Seedcorn maggot is common in southern Canada, and is also reported from some far northern locations in Canada, including the Yukon and Northwest Territories. Seedcorn maggot is very difficult to distinguish from turnip maggot, *Delia floralis* (Fallén), so some reports of occurrence and damage are suspect.

Host Plants. A wide variety of plants are reported to be attacked. Among vegetables reported injured are artichoke, beet, Brussels sprouts, cabbage, cantaloupe, carrot, cauliflower, corn, cucumber, garlic, kale, lettuce, lima bean, mustard, onion, pea, potato, pumpkin, rhubarb, rutabaga, sea kale, snap bean, spinach, squash, sweet potato, tomato, and turnip. Other crops such as alfalfa, cotton, strawberry, tobacco, and wheat also are damaged. Seedcorn maggot has been reported to feed on grasshopper eggs.

Seedcorn maggot attacks seeds and seedlings, but there is some doubt about its status as a primary pest. It is often regarded as saprophagous, associated with plants that have already been damaged by other insects or plant disease (Brooks, 1951; Beirne, 1971).

Natural Enemies. The natural enemies of seedcorn maggot are largely the same as those attacking onion maggot, *Delia antiqua* (Meigen). Among the most important are the parasitoids *Aleochara bilineata* (Gyllenhal) (Coleoptera: Staphylinidae) and *Aphaereta pallipes* (Say) (Hymenoptera: Braconidae) (Tomlin *et al.*, 1985). General predators such as ants and spiders also act on seedcorn maggot. Berisford and Tsao (1974) described an epizootic of the fungus *Entomophthora muscae* (Entomophthoraceae) in seedcorn maggot. Additional information on these natural enemies is found in the section on onion maggot, *Delia antiqua* (Meigen).

Life Cycle and Description. A complete life cycle requires from 15 to 77 days, but during most of the year 16 to 21 days is adequate. However, owing to aestivation of larvae in the summer and diapause of pupae in the winter, there usually are only 2–4 generations annually. The exact number of generations is often uncertain owing to asynchronous emergence, generations that are divided into diapausing and non-diapausing populations, and variable weather. Based on thermal accumulations and flight patterns, there seem to be four generations per year in Wisconsin (Strong and Apple, 1958); this is also reported from Ontario (Beirne, 1971) and England (Miles, 1955a). Throne and Eckenrode (1985) reported four flights of adults in New York—one each in May, June, July, and August. In New York, not all overwintering flies emerged in May, so the June population consisted of both overwintering flies and progeny of the May adults. Not surprisingly, the June population tends to be both large and damaging. The number of generations per year was estimated by Reid (1940) to be three in South Carolina. No diapause was observed in South Carolina, with the insects active throughout most of the year, especially in the cooler periods. In Iowa, only two generations were observed (Funderburk *et al.*, 1984a), and Higley and Pedigo (1984) reported a summer aestivation that progressed into diapause. The threshold temperature for egg-adult development is estimated at 3.9°C (Sanborn *et al.*, 1982a), but unlike some other *Delia* spp., it has proved difficult to predict accurately the development of seedcorn maggot from temperature accumulations.

Egg. The eggs are elongate and ovoid, with one end tapering to a blunt point and the other more bluntly rounded. The egg is slightly curved, with one side being decidedly convex and the opposite being slightly concave. The pearly-white eggs measure about 0.92 mm long (range 0.90–0.95 mm) and 0.3 mm wide. The eggs are usually placed at the soil surface, singly or in a cluster of up to 10. Oviposition occurs readily at 10–27°C. Favorite oviposition sites are sprouting or decaying seeds, plant material, and organic fertilizer such as fish meal or cottonseed meal. Moist, freshly turned soil is quite favorable for egg laying. Females avoid both dry and very moist soil. Duration of the egg stage is a function of temperature. It may hatch in just one day when held at 28°C, but at a more spring-like temperature of 15°C, egg hatch requires 2–3 days, and at cool temperatures of 5–7°C about 7–9 days are required. Larvae dig down upon hatching, often burrowing to a depth of 6–8 cm to locate suitable food.

Larva. Larvae are white, and have three instars. They increase in length from about 0.7 mm at hatch to about 7 mm at maturity. The length of the mouthparts (cephalopharyngeal skeleton) is 0.33, 0.65, and 0.89 mm for the three instars, respectively. Larvae are reported to feed gregariously. First instar larvae cannot successfully attack food such as freshly cut or healed (corked) potato seed pieces; they survive and grow better if they locate food that is decaying. Larvae

Seedcorn maggot larva.

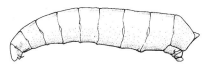

Seedcorn maggot larva, posterior view showing caudal spiracles.

may develop at rather low temperatures of 4–7°C, but optimal growth occurs at 21–23°C. Larval development time is 20–30 days at temperature of about 10°C, but 7–12 days under more optimal temperature of 21–30°C. At about 14–17°C, the duration of the first instar is 1–3 days, second is 3–5 days, and third is 5–16 days. The anterior spiracles, a character with some diagnostic value, bears 5–8 lobes. This character is particularly useful for distinguishing seedcorn maggot from onion maggot; the spiracles of the latter bear about 12 lobes.

Pupa. The oval puparium is light reddish-brown initially, but becomes quite dark before emergence of the adult. It measures about 4–5 mm long and 1.5 mm wide. Following a prepupal period of about two days pupation occurs in the soil, often at the site of feeding. When larvae have burrowed into their food, as occurs with potato seed pieces, they leave the food source before pupation. Duration of the pupal period is 7–14 days at 18–24°C.

Adult. The adults are grayish brown, with few distinctive markings. The male tends to bear stripes on the thorax and a mid-dorsal stripe on the abdomen, but stripes are usually lacking in females. Flies measure 4–5 mm long. The legs are black and the wings are unmarked but bear dark veins. Adults seek food and moisture soon after emerging, and feed on the flowers of numerous of plants. Fish meal, cottonseed meal, and other materials commonly used as fertilizer are quite attractive to ovipositing adults. The duration

Seedcorn maggot puparium.

of the pre-oviposition period in seedcorn maggot is 7–14 days. Mated adults live 30–40 days, with longevity increased during cool weather and shortened during the heat of summer. There are some reports that temperatures of 28–30°C are fatal to adults. Egg production is estimated at 100 per female, but this may not reflect field conditions.

The biology of seedcorn maggot was given by Reid (1940) and Miles (1948,1952). Methods of rearing were provided by Harris *et al.* (1966). Techniques to distinguish larvae, pupae, and adults of seedcorn maggot from several related *Delia* spp. were given by Brooks (1951). An interesting account of *Delia* ecology, and implications for management from a British perspective, were found in Finch (1989).

Damage

Damage to seeds and developing embryos may occur before the seedling breaks through the soil. Larvae are able to penetrate seeds as the seed coat splits. Rot often develops as they burrow in the cotyledon. Attack by several maggots may completely inhibit seed germination, but seedlings may suffer only deformity, or recover completely from attack by a single insect. Funderburk and Pedigo (1983), for example, documented that soybean plants were able to compensate for seedcorn maggot damage, and reduction in pod numbers per plant, by increased number of beans per pod. Despite the fact that flies are able to locate emerging seedling readily, and oviposit freely on seedlings, larval performance is superior on freshly planted seeds (Weston and Miller, 1989). Severe damage characteristically occurs during periods of cool and wet weather, when insect activity is favored but plant growth is inhibited.

There is a strong association between seedcorn maggot and plant disease, and it is believed that some

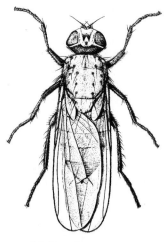

Adult seedcorn maggot.

microorganisms secrete attractive volatiles. Brooks (1951) classified this insect as one which normally was present only where more aggressive insects had occurred previously, or where mechanical damage or disease had damaged the plant. Beirne (1971), in reviewing several Canadian studies, also argued that seedcorn maggot was primarily saprophagous. In support of this, Everts *et al.* (1985) observed that in Colorado, seedcorn maggots were associated with Fusarium-infected onions, and that in choice tests, 78% of eggs were deposited on Fusarium-infected bulbs. Not all microbial organisms are attractive to seedcorn maggots, however, as Harman *et al.* (1978) found that *Chaetomium* fungi, which are antagonists of plant pathogenic fungi, inhibit oviposition. Also, Miles (1948) reported that cabbage, cauliflower, and leek grown in England were readily damaged by seedcorn maggot in the absence of disease. Similarly, Nair and McEwen (1973), working with radish in Ontario, concluded that attack by cabbage maggot was not a prerequisite for attack by seedcorn maggot.

Seedcorn maggot may also be important in transmission of plant disease. Harrison *et al.* (1980b) summarize data indicating that this insect, and many other species, assist in the overwintering and transmission of bacterial soft rot, *Erwinia carotovora*.

Management

Sampling. Seedcorn maggot flies are attracted to volatiles given off by microbial organisms associated with decay, and to other volatiles. Cone traps for adult monitoring can be baited with alcohol, odors of fermentation from honey-yeast or molasses solution, or oviposition stimulants such as meat and bonemeal (Reid, 1940; Strong and Apple, 1958; Funderburk *et al.*, 1984a). Enzymatic yeast hydrolyzate also is an attractive bait (Miller and Haarer, 1981). If sticky traps are used, the most attractive color is gray, followed by yellow, white, blue, and other colors (Vernon and Borden, 1983). Finch (1991) reported success with yellow water-pan traps.

Insecticides. The use of insecticides for control of seedcorn maggot is similar to onion and cabbage maggot, except that usually it is only the period of plant germination and early growth that is of concern. Therefore, there is less need for long-term protection, and for highly residual insecticides. Nevertheless, residual materials have been used, and insecticide resistance has resulted. Granular in-furrow and over-the-row applications are common. Seed treatment is often adequate. Eckenrode *et al.* (1973) present a good discussion of chemical treatments.

Host-Plant Resistance. Beans with a dark seed coat are less susceptible to injury than white varieties, possibly owing to the hardness of the seed coat and speed of germination (Vea and Eckenrode, 1976; Hagel et al., 1981). Seedling response to injury varies greatly among plants, with cantaloupe and watermelon being severely injured, snap bean and lima bean intermediate, and corn most resistant (Hough–Goldstein and Hess, 1984).

Cultural Practices. Cultural practices that normally are considered favorable for crop growth are, unfortunately, also sometimes suitable for seedcorn maggot. There are numerous observations suggesting that freshly tilled soil is attractive to ovipositing seedcorn maggots. Buried crop residues and rotting manure also are attractive, and reduced or no-tillage systems exacerbate fly problems due to the abundance of organic material (Hammond and Stinner, 1987). Egg and larval survival are enhanced when plants are frequently watered. Planting in warm soil, however, favors seedling growth more than it does maggot feeding, so delayed planting is advantageous if it can be timed to avoid the large (usually June) flight. If cover crops or other sources of organic matter are present, it should be disked at least four weeks before planting to allow adequate time for decomposition to occur (Hammond and Cooper, 1993). Anything that promotes rapid seed germination and seedling growth is considered to be beneficial in minimizing injury by seedcorn maggot.

Growers can sometime manipulate planting dates and escape the principal flights of seedcorn maggot because there is often a mid-season depression in maggot abundance. Such is the case in Washington, for example, where there are two flights of adults between May 1 and June 15, and lima beans planted after June 1 largely escape attack (Hagel et al., 1981).

Screens and row covers are effective protectants against a wide range of insect pests, but they can be ineffective against seedcorn maggot (Hough–Goldstein, 1987; Adams et al., 1990). Because pupae overwinter in the soil, even early spring placement of a covering will not protect against attack; rather, the insects emerge beneath the covering. This is especially true with seedcorn maggot, which feeds freely on manure and rotting plant material as well as a large number of crops.

Turnip Root Maggot
Delia floralis (Fallén)
(Diptera: Anthomyiidae)

Natural History

Distribution. Turnip root maggot is found in northern regions of North America, Europe, and Asia. It is common in Canada, including northern areas, but is infrequent in the United States. Turnip root maggot greatly resembles seedcorn maggot, *Delia platura* (Meigen), and in northern locations many records probably represent mixed populations.

Host Plants. The host range of turnip root maggot is quite similar to that of cabbage maggot, *Delia radicum* (Linneaus), but turnip maggot is more abundant in northern latitudes and where soils are light. Turnip maggot is known principally for its damage to rutabaga and turnip, which explains its common name. However, it damages most crucifers, including weed species. Brooks (1951) considered it to be a truly phytophagous insect, like cabbage maggot, rather than being attracted to stressed or plant disease-infected plants, like seedcorn maggot.

Natural Enemies. The biology of this insect is not well known, and the natural enemies are particularly poorly studied. It appears that predators and fungi associated with other *Delia* spp. also affect turnip root maggot. In Norway, Andersen and Sharman (1983) reported that over 40% of turnip root maggot eggs were consumed by predators in only 2–3 days. Plots with numerous carabids and staphylinids (both Coleoptera) had greater egg mortality, and insecticides interfered with predation.

Life Cycle and Description. Most information on turnip root maggot biology was gathered in Europe, but it seems consistent with the limited data available from North America. There is only one generation per year. Turnip root maggots overwinter as pupae, and flies from the overwintering population emerge throughout the summer months. Late July and early August are peak emergence times.

Egg. The eggs of turnip root maggot are white and elongate. One side is concave while the opposite side is convex. They are marked with longitudinal ridges, and measure about 1.1–1.2 mm long and 0.28–0.38 mm wide. They are laid in July and August, and are usually deposited in clumps in the soil around the base of plants. Females normally deposit about 200 eggs during their adult life of about 40 days (Havukkala and Virtanen, 1984). Eggs hatch 8–9 days after being deposited.

Turnip maggot larva, posterior view showing caudal spiracles.

Larva. Larvae also are whitish, undergo three instars, and attain a length of 9–10 mm before pupating. Duration of the larval stage is 5–8 weeks. The anterior larval spiracles have 14–16 lobes—a character useful for distinguishing this maggot from most related species. Larvae are found feeding on the host plant from August to October. At maturity they move in the soil a short distance from the feeding site and pupate.

Pupa. The puparium is oval, bluntly rounded at each end, and brown in color. The average length is about 6.5–7.5 mm. Pupae are the overwintering stage, and therefore are present from August until the following spring, a period of 8–10 months.

Adult. Flies generally resemble cabbage maggot adults, but are slightly larger, measuring 7–9 mm long. They also lack the tuft of bristles at the base of the hind femora that is distinctive in cabbage flies, and their color tends to be a lighter gray. Adults live 1–2 months, and usually begin to deposit eggs when they are about 9–10 days old.

The biology of turnip root maggot was given by Brooks (1951) and Miles (1952, 1955a). Brooks also provided characters to distinguish the adult, larval, and egg stages of this insect from related species. Rearing procedures for cabbage maggot (Finch and Coaker, 1969) also are suitable for turnip root maggot (Finch and Collier, 1989). An interesting account of *Delia* ecology, and implications for management from a British perspective, was found in Finch (1989).

Damage

Turnip root maggot can be quite damaging. Beirne (1971) reported that in Canada's Prairie Provinces up to 40% of commercial turnips and 80–90% of rutabagas had been made unsuitable for consumption owing to feeding damage by this insect. Larvae generally confine their feeding to roots, but sometimes work their way up into petioles of the lower leaves. They normally scarify the surface of the root, but rarely penetrating very deeply. For crops where the root is not harvested, such as cauliflower, the damage is much less, owing to both the indirect nature of the injury and the fact that larvae tend to occur so late in the season that the plants are fairly mature.

Management

Sampling. Yellow-sticky traps can be used to monitor abundance of adults. Flies are more likely to land on horizontal surfaces than vertical surfaces, however, so water-pan traps might be more suitable

(Finch and Collier, 1989). Allyl isothiocyanate is an important element in host acceptance, functioning as an oviposition stimulant (Havukkala and Virtanen, 1985), and has potential use in monitoring.

Insecticides. Management of turnip root maggot often depends on use of insecticides, usually applied as a granular formulation at planting to protect against larval injury.

Host-Plant Resistance. Variation in susceptibility between and among crops was studied by Alborn *et al.* (1985). In general, fast-maturing varieties seem especially prone to damage. Chinese cabbage, mustard, and some cauliflower cultivars were especially susceptible to injury, whereas kale and radish were quite resistant, and broccoli was intermediate. In an evaluation of 22 cauliflower selections, significant differences in susceptibility to injury were identified. There was a positive correlation between plant damage and number of eggs deposited by flies. Shaw *et al.* (1993) determined that turnip cultivars with relatively higher percent dry matter were more resistant to damage, but this conclusion is disputed (Finch and Thompson, 1992).

Cultural Practices. Weather and soil conditions affect damage potential. As is the case with seedcorn maggot, damage is worse in cool and wet years. Also, Beirne (1971) reported that in the Prairie Provinces, damage by turnip root maggot occurred principally on farms that were irrigated. In Norway, turnip root maggot was the principal crucifer pest on light soils, whereas cabbage maggot was the major pest on heavy soils (Alborn *et al.*, 1985).

FAMILY DROSOPHILIDAE—POMACE FLIES

Small Fruit Flies
Drosophila spp.
(Diptera: Drosophilidae)

Natural History

Distribution. Several *Drosophila* species affect vegetable crops in North America. Origin of the economically important pest species is uncertain, as they are nearly cosmopolitan in distribution. However, most are thought to be immigrants from the tropics. Among the important species are *Drosophila affinis* Sturtevant, *Drosophila busckii* Coquillett, *Drosophila melanogaster* Meigen, *Drosophila repleta* Wollaston, *Drosophila simulans* Sturtevant, and *Drosophila tripunctata* Loew. The importance of these species varies among crops, as well as geographically and temporally. However, they are quite similar in biology.

Host Plants. Although small fruit flies are normally found in association with "true" fruits, such as apple, banana, citrus, pineapple, and tropical species, the fruits and tubers of vegetable crops sometimes support *Drosophila* spp. Tomato, squash, cantaloupe, watermelon, and to a lesser extent cabbage and potato become infested, but almost any crop that bears fermenting tissue is attractive to ovipositing flies. Some species, such as *D. repleta*, also are attracted to animal excrement, though most fruit and vegetable-breeding *Drosophila* spp. do not display this behavior (Harrison *et al.*, 1954). The larvae of the aforementioned species are primarily yeast feeders, and when decay progresses to the state that bacteria and fungi become dominant, the host is no longer suitable for larval development.

Natural Enemies. Parasitoids seem to be an important natural enemies of *Drosophila* spp., though fungi and mites sometimes are present. Among the parasitoids known to attack *Drosophila* spp. are *Asobara tabida* (Nees), *Phaenocarpa drosophilae* Fischer (both Hymenoptera: Braconidae), *Spalangia drosophilae* Ashmead, *Cyrtogaster typherus* (Walker), and *Pachycrepoideus vindemiae* (Rondani) (all Hymenoptera: Pteromalidae). Probably the most important mortality agents, however, are beetles drawn to decaying fruit and which also ingest fly larvae, notably rove beetles (Coleoptera: Staphylinidae) and sap beetles (Coleoptera: Nitidulidae).

Life Cycle and Description. These flies breed very rapidly, so many generations can be completed annually if they are not interrupted by temperature extremes. A complete generation can be completed in about eight days when cultured at 29°C, and 11–13 days at 24°C. These *Drosophila* spp. reproduce best at cool to warm rather than hot temperatures, often thriving under autumn conditions rather than summer. The important *Drosophila* spp. apparently do not overwinter successfully in cold areas, surviving indoors on produce or reintroduced to temperate areas in the spring. Thus, in temperate areas the population density is very low in the spring, only moderately abundant in the summer due to high and unsuitable temperatures, and then building to very high densities in the autumn.

Egg. The female seeks cracks and wounds of fruit for oviposition. The eggs of *Drosophila* spp. are white, elliptical in form, and bear 2–4 filaments at one end, depending on the species. Similarly, filament length varies among species, but it is less than the length of the egg. The eggs measure 0.6–0.7 mm long and about 0.2 mm wide. The eggs are deposited at the surface of moist tissue; the filaments apparently keep the eggs from sinking. Fecundity is favored by cool temperatures. Females produce eggs at a rate of about 15 per day, with a mean total fecundity of about 430 eggs when reared at 25–30°C. However, when held at lower temperature of 19–25°C, egg production averages 26 per day and total fecundity increases to 940 eggs per female. Duration of the egg stage is 1.0–1.5 days.

Larva. There are three instars. Larvae increase in size from about 0.9 mm at hatching to 6.5 mm at maturity. The larva is elongate and tubular in form, tapering at the anterior end to a point. Larvae are white, except for black mouth hooks at the anterior end. The number of teeth on the mouth hooks increases from 1, to 2–3, and 9–12 during succeeding instars. The mouthparts (cephalopharyngeal skeleton) measure about 0.22, 0.45, and 1.0 mm long in instars 1–3, respectively. Large spiracles are evident on the posterior end of the larva. Duration of the larval stage is about four days at 29°C and 5.5 days at 24°C.

Pupa. The larva seeks a dry location for pupation, usually leaving the fruit and dropping to the soil. The puparium formed by the larva is whitish initially, but turns brown in 4–5 hours. It is basically oval, but bears two elongate-tubular structures which function as spiracles at the anterior end. The puparium measures 4.0–4.3 mm long and 1.4–1.5 mm wide. Duration of the pupal stage is about six days at 24°C and 5.5 days at 29°C.

Adult. The adult is yellowish or grayish with dark bands on the abdomen. The abdomen sometimes appears entirely dark. Eyes are large, and usually red in color. Wings are colorless. The fly measures

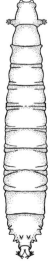

Small fruit fly larva.

about 2.5 mm long, though *D. melanogaster* is smaller, averaging 2.0 mm. Oviposition commences when the adult is about two day old, and continues for 20–30 days. Adult longevity is inversely related to temperature, and flies are most active early and late in the day. Males survive for about 40 days and females for 70 days when reared at 18°C. However, when cultured at 28°C, males survive for only 22 days and females for 28 days.

A very good introduction to *Drosophila*, including keys and species descriptions, was given by Sturtevant (1921). Wheeler (1987) provided a recent key to the genera. Wheeler (1981a,b) presented interesting biogeographic data on Drosophilidae. The publication by Demerec (1950) also contained a great deal of information, and though emphasizing anatomy and development, monitoring and rearing procedures also were provided. Ditman *et al.* (1936) provided a concise summary of *Drosophila* biology as it relates to crop damage.

Damage

The eggs of *Drosophila* spp. are deposited on fruit that is cracked or otherwise damaged. They do not attack undamaged produce. Moderately or completely ripe tomatoes are preferred over the green and pink stages (Collins, 1956). Attack usually occurs at harvest time because much produce is injured by the picking operation. The larvae develop very quickly, and if there is any delay in processing or consumption, larvae may attain a visible size and cause degradation or rejection of the crop. *Drosophila melanogaster* is normally the most abundant and damaging of the small fruit flies in vegetable crops. The injury is usually limi-

ted to tomato, and perhaps melon crops, but *Drosophila* spp. are not generally considered as serious pests.

Management

Sampling. Flies are attracted to light traps, yellow-sticky traps, and juice-baited jar or vial traps. Red containers are somewhat more attractive than other colors when jar or vial traps are used (Wave, 1964). However, fruit flies are detected easily by visual means because, though small in size, they allow close approach and are readily recognized.

Insecticides. Insecticides can be applied to crops to reduce the number of flies present and ovipositing on fruits at harvest time. This is most effective before fruit matures, however, because the number of effective insecticides which can be applied up to the day of harvest is limited, and insecticide residues must be minimized.

Cultural Practices. The presence of small fruit flies in a crop is favored by several factors, including the presence of weeds and grasses, because these provide shade for the flies, which do not prefer hot and sunny conditions; the proximity of a favorable breeding site, often a melon or tomato crop that has already matured, or piles of culled fruit; and injury to fruit, often from equipment or insect or hail damage, which provides oviposition sites. Small fruit fly problems are alleviated by early and frequent picking, which helps to eliminate some of the cracking associated with fruit maturity; the uniform watering of crops, which similarly reduces the frequency of cracking; and careful handling of fruit, which reduces breakage and limits access to oviposition sites by flies.

FAMILY OTIDIDAE—PICTUREWING FLIES

Cornsilk Fly
Euxesta stigmatias Loew
(Diptera: Otitidae)

Natural History

Distribution. Cornsilk fly is found in tropical areas of the western hemisphere. It occurs throughout the Caribbean Islands, Mexico, Central and South America, south to Bolivia and Paraguay, and in Florida. Historically, it has been a pest in the United States only in southernmost Florida, Puerto Rico, and the Virgin Islands. However, in recent years cornsilk fly has become a pest in south-central Florida, and there are sporadic reports of its occurrence from some other states.

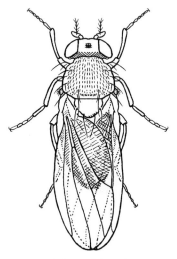

Adult small fruit fly.

Host Plants. Both larvae and adults feed on a wide variety of plants, including such vegetables as sweet corn, potato, and tomato; such field crops as field corn, sorghum, and sugarcane; and such fruit crops as atemoya, banana, guava, and orange. However, sweet corn and field corn are highly preferred, and are the only crops to be seriously damaged by cornsilk fly.

Natural Enemies. Little is known about natural enemies. The eggs are consumed by earwigs (Dermaptera), mites (Acarina), and minute pirate bugs (Hemiptera: Anthocoridae).

Life Cycle and Description. Phenology of these flies is inadequately documented, but they seem to be present throughout the year in southern Florida. The life cycle requires only about 30 days so apparently several generations occur annually. Overwintering potential is unknown.

Egg. The eggs are deposited mainly at the tip of the ear, on or near the silk at the point of emergence from the ear. Young ears are most preferred; decaying ears are avoided. In the absence of ears, the eggs may be deposited at the base of leaves or elsewhere, but survival rates are much lower than when they are deposited in ears. The eggs measure about 0.85 mm long and 0.16 mm wide. They are white and cylindrical, with ends that taper to a broadly rounded point. There are reports of eggs being laid singly, in a row and fastened end to end, or in clusters of up to 40 eggs. Females may cluster on corn ears and deposit hundreds of eggs in individual ears. Duration of the egg stage is 2–4 days.

Larva. Larvae are elongate and cylindrical, with a blunt, broadly rounded posterior that tapers to a pointed head which is equipped with a pair of mouth hooks. Mature larvae measure about 5.4 mm long. Larvae are whitish, and the ventral surface bears ridges and coarse spines. Duration of the larval stage averages about 7.5 days.

Pupa. Pupation commonly occurs on the corn silks inside the corn ears, but sometimes in the soil. The puparium is elongate and cylindrical, with the anterior end tapered to a blunt point and the body slightly flattened. It measures about 3.9 mm long and 1.4 mm wide. Initially, puparia are yellowish, become red-

dish-brown after a few hours and then dark-brown at maturity. Mean duration of the pupal stage is about 7.5 days.

Adult. The adult flies are metallic green or black, with reddish eyes, and wings banded with black. Males measure about 3.8 mm long, females about 4.2 mm. The adults feed on nectar, plant sap and glandular exudates, and also drink from dew and rain droplets. They tend to move their wings continuously, even while alighted. Adults frequent corn plant tassels and ears more than other plant parts. Mating occurs mostly at dusk and dawn.

The biology was described by App (1938), Seal and Jansson (1989, 1993), and Seal *et al.* (1995, 1996).

Damage

Damage is caused by the larval stage. Larvae feed on the silks, but the principal injury is caused to the developing kernels on the ear, where they often hollow out the kernels. Larvae may be found feeding along the entire length of the ear, or mostly at the tip. Larvae also bore into the cob, resulting in ear deformity, and introduce fungi beneath the ear sheath. Yield reductions of 95% are reported at peak levels of damage early in the season (Seal and Jansson, 1989; Seal *et al.*, 1996). Significant damage occurs even when insecticides are applied.

Management

Sampling and management tactics are poorly developed. Sampling should commence before tasselling is

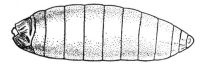

Cornsilk fly puparium.

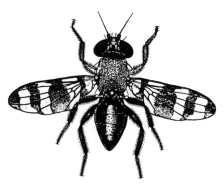

Adult cornsilk fly.

Cornsilk fly larva.

initiated. Adults can be detected in the late afternoon and early evening when they rest on the plants and mate on the tassel. During silking, silks should be checked for the presence of eggs. Growers in affected areas rely on frequent insecticide applications to maintain a toxic residue on the rapidly developing corn silk. Alternatives to insecticides are very few. Covering the tassel and ear with a bag can decrease the number of larvae found in ears, but as the silk must remain uncovered until the ear is pollinated, some oviposition is likely to occur on the young silks.

Sugarbeet Root Maggot
Tetanops myopaeformis (Röder)
(Diptera: Otitidae)

Natural History

Distribution. This native insect is found over a wide area of western North America. In the United States, it is known from Washington to Minnesota in the north, and from California to Colorado in the south. In Canada it occurs in the Prairie provinces.

Host Plants. Sugarbeet root maggot affects crops in the family Chenopodiaceae—beet, spinach, Swiss chard, and sugarbeet. Spinach appears the most suitable host, though it usually is not heavily infested because it is not normally grown in the summer months when flies are active. Not surprisingly, the herb garden orach, *Atriplex hortensis*, also support larval development, as it is also a member of this plant family. However, though flies oviposit on other chenopods such as lambsquarters, *Chenopodium album*, and Russian thistle, *Salsola kali*, larvae apparently fail to complete development. Flies also oviposit on other plants such as redroot pigweed, *Amaranthus retroflexus*; curly dock, *Rumex crispus*; and black nightshade, *Solanum nigrum*; but also with unsatisfactory larval development. As all the known host plants of sugarbeet root maggot have been introduced to North America, the natural host remains curiously unknown (Mahrt and Blickenstaff, 1979).

Natural Enemies. The natural enemies of sugarbeet root maggot are poorly documented.

Life Cycle and Description. There is one generation per year, but adult emergence in the spring is protracted, often with a small secondary emergence peak that may give the impression of a second generation. Mature larvae overwinter. In Alberta, Colorado, and North Dakota, maggots pupate in the early spring, adults emerge in May–June, eggs are deposited beginning about 10 days after adult emergence, and large larvae burrow deep into the soil in preparation for overwintering beginning in July–August.

Egg. The eggs are white, rounded at one end and tapering to a blunt point at the other end. They measure about 1.0 mm long and 0.25 mm wide, with one side convex and the opposite side concave. Eggs are deposited in clusters of about 10–25 in the soil at the base of plants at depths of up to 1 cm. Soil cracks are preferred oviposition sites. Mean fecundity is about 110 eggs, but some females produce up to 200 eggs. Duration of the egg stage is about 5–7 days.

Larva. Larvae feed below-ground on plant roots, with the specific depth influenced by soil moisture conditions, but the majority at depths of 12–18 cm. The larva is elongate and cylindrical, with the anterior end tapering to a point and the posterior end blunt. There are three instars. The first instar is whitish and its length ranges from 0.75 mm at hatching to about 2.1 mm just before molting. The second instar is white to ivory and measures 2.0–3.0 mm long. The third instar is creamy-white to light-yellow, and ranges 33–11.0 mm long. The mouthparts (cephalopharyngeal skeleton) darken as the maggot matures, and are useful for distinguishing instars. The mouthparts measure about 0.24, 0.41, and 0.72 mm long, respectively. The posterior end of the maggot terminates in a pair of elongate, spine-like spiracles which enlarge and darken as the maggot matures. Larvae overwinter as third instars, usually at depths of 10–25 cm, moving toward the surface in the spring. They require a cold period to terminate diapause and tolerate temperature of −5°C for brief period, but they survive better at warmer temperature (Whitfield and Grace, 1985). Post-diapause larvae develop at a temperature of about 8°C or greater (Whitfield, 1984).

Pupa. Pupation normally occurs at depths of 5–10 cm, with pupation occurring early in April even at northern latitudes. The brown puparium is elongate-elliptical and measures about 8 mm long. The pointed spiracles of the maggot are visible on the puparium. Duration of the pupal stage is 10–14 days.

Adult. Sugarbeet root maggot adults measure about 6.1 mm long for males and 8.2 mm for females. They are shiny black with yellowish-brown legs and face. The wings are light-gray with brown or smoky spots basally and near the mid-point of the wing. Wing spots originate at the leading edge and cross about one-half the width of the wing. Perhaps the most distinctive feature is the elongate, pointed tip of the abdomen in the female. The abdominal tip of the male, in strong contrast to the female, terminates abruptly, and gives a squared-off appearance. Peak adult emer-

Adult female of sugarbeet root maggot.

gence occurs at about 200 day-degrees when it is based on a developmental threshold of 9°C (Whitfield, 1984). Adults fly readily and disperse from field to field, but the air temperature of about 27°C is a threshold for their flight (Bechinski *et al.*, 1990). Mean duration of the adult stage at temperatures of 20–25°C is 5–10 days (range 4–14 days).

A description of sugarbeet root maggot was given by Gojmerac (1956). Hawley (1922b) and Harper (1962) described the life history. The larva was described by Bjerke *et al.* (1992). A key to otidid genera based on adults was provided by Steyskal (1987).

Damage

The injury is caused by the larval stage. Maggots feed on the roots, burrowing completely through young roots, but feeding the surface of large roots. Black and necrotic lesions are developed on the surface of roots where larvae are feeding. A single larva may destroy a young plant, and about 3–4 maggots can cause severe damage even to mature plants. Aboveground symptoms of feeding include wilting of plants and eventually skips or bare spots in the fields.

Management

Sampling. There is approximately a 10-day period between adult emergence and the initiation of oviposition, so it is possible to forecast damage by maggots from adult population density (Bechinski *et al.*, 1989). Adults densities are often determined with orange-sticky traps, though emergence, baited, waterpan, and blacklight traps operated during daylight hours also capture flies (Harper and Story, 1962; Swenson and Peay, 1969; Blickenstaff and Peckenpaugh, 1976). The timing of adult flights can also be predicted. In Idaho, forecasting of adult emergence can be accomplished by accumulating day-degrees above a threshold of 8.6°C starting on April 1. Because flies are

inactive at temperatures below about 25–27°C, however, this statistic must also be considered (Bechinski *et al.*, 1990).

Insecticides. Insecticides are directed at either the adult or larval stages for crop protection (Peay *et al.*, 1969). For adult suppression, foliar applications of insecticide are timed for peak-fly emergence. This application is preferred to the other common practice, which is to apply granular insecticides to the soil at planting. Soil-applied insecticides prevent damage effectively, but may be unnecessary in many fields.

Cultural Practices. The most important cultural practice is crop rotation, because crops grown in fields previously infested with root maggots are most likely to be damaged. Flies will disperse a kilometer or more, however, so distance and isolation are important factors. Crops grown in light, sandy soil are often reported to be most heavily infested, so soil characteristics are also a consideration in choosing fields for susceptible crops. For small plantings, screen and row covers can be used to deny flies access to young plants.

Biological Control. Little work has been directed toward finding biological control agents for this pest. However, Wozniak *et al.* (1993) reported that several species and strains of entomopathogenic nematodes (Nematoda: Steinernematidae) could infect sugarbeet root maggot larvae. Rates of infection were low, but many survivors were deformed and became incapable of reproduction.

FAMILY PSILIDAE—RUST FLIES

Carrot Rust Fly
Psila rosae (Fabricius)
(Diptera: Psilidae)

Natural History

Distribution. Carrot rust fly apparently originated in Europe, and was first discovered in North America in 1885 at Ottawa, Ontario. It reached the east coast (Maine) in 1893, and western North America in the 1920s. It is now widely distributed in the northeastern United States and eastern Canada, and in the northwestern United States and western Canada. It is absent from the arid central areas of Canada and the United States, as well as northern Canada and the southern United States. Carrot rust fly has also successfully invaded New Zealand. Generally, cool and moist habitats are favorable for this insect.

Host Plants. Carrot rust fly larvae feed on umbelliferous vegetable crops, including carrot, celery, celeriac, chervil, parsnip, and parsley, but this insect is considered principally a carrot pest. It also is recorded from umbelliferous herbs such as caraway, dill, and fennel, and from weeds such as wild carrot, *Daucus carota*, and water hemlock, *Conium maculatum*. The adults feed on the flowers of numerous plants.

The host range has been studied intensively in England, where about 100 umbelliferous weeds have been identified as suitable hosts. Only a few plants of this family have been determined to be unsuitable (Hardman and Ellis, 1982; Hardman *et al.*, 1990). Annual, biennial, and perennial species can support carrot rust fly, but fly larvae perform poorly on plant species that are not actively growing. Annual species tend to diminish in suitability as the season progresses. A small number of nonumbelliferous plants favor the growth of carrot rust fly larvae, particularly chicory, endive, and lettuce, but flies do not deposit eggs on such plants. Records indicating these Compositae as host plants apparently are based on unusual situations where ovipositing flies perceived chemical stimuli derived from the earlier presence of umbelliferous plants.

Natural Enemies. Much of our knowledge relating to natural population regulation has been derived from British studies. Life-table studies conducted by Burn (1984) demonstrate the importance of egg and early larval mortality in determining population trends. The egg loss was due to both desiccation and predation. The larval parasitoid *Chorebus gracilis* (Nees) (Hymenoptera: Braconidae) and the pupal parasitoids, *Eutrias tritoma* (Thomson) (Hymenoptera: Eucoilidae), and *Aleochara sparsa* Heer (Coleoptera: Staphylinidae), are the most abundant parasitoids, but none is considered a key element in population regulation. The wasps were introduced into Canada in the early 1950s, but failed to establish. The biology of these insects was given by Wright *et al.* (1947). Burn (1982) also reported egg predation by beetles (Coleoptera: Carabidae and Staphylinidae), and estimated that they often account for 20–40% egg loss.

Fungi, particularly *Entomophthora muscae*, cause epizootics in carrot rust fly. Also reported infecting adults are *Conidiobolus apiculatus* and *Erynia* sp. None of the fungi are known to regulate fly populations (Eilenberg and Philipsen, 1988). However, Eilenberg (1987) observed modified oviposition behavior by flies infected with *E. muscae* early in the adult stage, suggesting they contribute few eggs to the next generation.

Life Cycle and Distribution. The number of generations varies from one to three depending on climate. All generations may not be damaging, because one may occur before the crop is present in the field or after it is harvested. In Massachusetts, adults developing from overwintering stages usually become active during the spring in May–June, with first generation flies appearing in August (Whitcomb, 1938). Total generation time in the field is usually about 9–11 weeks. Stevenson (1981) measured egg-to-adult development periods of 59, 70, and 81 days at 20°, 17.5°, and 15°C, respectively. Summer flies initiate another generation by depositing eggs between August and September. Two periods of adult activity are also known from Quebec (Boivin, 1987). Getzin (1982) reported a third flight in October during studies in western Washington; this was also the case in nearby British Columbia (Judd and Vernon, 1985). Although the diapausing pupae are normally considered to be the overwintering stage, larvae also survive cold weather, and proceed in development without a period of diapause. This can result in a protracted or bimodal emergence of flies in the spring. In some cases, a higher proportion of larvae than pupae overwinter (Collier *et al.*, 1994). Developmental thresholds are quite low for this cold-adapted species; egg, larval, and pupal developmental thresholds are estimated at 4.5°, 2.0°, and 1.5°C, respectively (McLeod *et al.*, 1985).

Egg. The female deposits small clusters of eggs, normally 1–3 eggs per oviposition, in the soil around the base of food plants. A female may deposit up to 150 eggs during her life span, but 40 are considered average under field conditions and 100 eggs per female are commonly produced in the laboratory. The egg is white and elongate, measuring 0.6–0.9 mm long, but only 0.15–0.20 mm wide. It is marked with longitudinal ridges, and one end bears a short constricted section. Eggs hatch in 3–17 days, with the average usually 6–10 days. Temperatures of 15–20°C or less are considered optimal for both oviposition and egg hatch.

Larva. The larvae initially are colorless, become milky-white and then yellowish as they mature, and eventually attain a length of 6–9 mm. There are three instars. In typical fly fashion, the body is cylindrical, and the head tapers to a point and bears dark mouth hooks. Larvae often feed on lateral roots initially, then burrow into the main root as they grow larger. Duration of the larval period is variable, ranging from four to six weeks in the summer to over three months in winter. Mature larvae leave the root to pupate, usually pupating 4–5 days after reaching maturity.

Pupa. The puparium initially is yellow-brown but changes to brown at maturity, and measures 4.5–5.0 mm long. Puparia are found in the soil near the

Carrot rust fly larva.

Rust fly puparium.

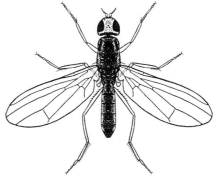

Adult carrot rust fly.

food plant, though some disperse distances of 10 cm before pupation, and most pupate at depths of 10–15 cm. Duration of the pupal stage is about 25 days in the summer, but extends for several months during the winter. The early portion of the pupal stage is sensitive to environmental conditions, and temperature exposure at this stage determines diapause induction. If young pupae are exposed to temperature below 10°C, they can enter diapause (Stevenson and Barszcz, 1991).

Adult.   Adults measure 4.5–5.0 mm long. The head and legs are yellowish-brown, and the thorax and abdomen are shiny black. A fine layer of yellowish hairs covers the thorax and abdomen. The wings are slightly iridescent. The tip of the abdomen tapers to a point in females, but is bluntly rounded in males. Cool weather and moist soil favors adult emergence.

Thorough treatment of carrot rust fly biology was given by Whitcomb (1938). A world review of biology and management was published by Dufault and Coaker (1987). Additional useful information and rearing procedures were provided by McLeod *et al.* (1985). Beirne (1971) gave a valuable Canadian perspective on this pest.

Damage

This is often considered as the most destructive pest of carrot. The larva is the only damaging stage. The larva mines the surface of the root, leaving trails or blotchy areas. The smaller roots or rootlets may be completely mined, but the principal economic effect is associated with the damage to the main root in crops such as carrot and parsnip. The feeding site often acquires a rust color owing to accumulation of rusty plant exudates; this is purported to be the basis of the insect's name. However, the foliage of affected plants also may develop a red or yellow color. Other signs of attack include wilting and forked or distorted

roots. The larval mines may be widely distributed, but tend to be concentrated in the lower or distal portions in carrot, and in the upper portion of the parsnip root. In celery, larvae will also burrow up into the crown and stalks, and in parsley they will mine the surface of the tap root and feed on the lateral roots.

Management

Sampling.   Various techniques have been developed to assess damage potential. Getzin (1982) used yellow-orange sticky traps for adult population sampling. Finch and Collier (1989) noted that a greater numbers of carrot rust flies were captured on the lower surface of yellow-sticky traps inclined at a 45° angle. Traps placed closer to the soil capture more flies than those that are elevated (Collier and Finch, 1990). Boivin (1987) used yellow-sticky traps for survey and quantitative estimates of fly abundance. Judd *et al.* (1985) reported successful use of sticky trap catches to monitor flies, resulting in appreciable reductions in insecticide use with nominal thresholds of about 0.4 flies per trap per 4–5 days.

Several authors have calculated day-degree accumulations necessary for carrot rust fly development, including Stevenson (1983), Judd and Vernon (1985), McLeod *et al.* (1985), and Boivin (1987). In England, developmental models were used to forecast the need for insecticide applications, and eliminated the need for several mid-summer insecticide applications on carrot (Finch, 1993).

Insecticides.   Insecticides, usually in a granular formulation, are often incorporated into the furrow at planting time to protect roots from attack by larvae. This may be followed by several foliar applications to suppress adults. Insecticide resistance has been a problem, with chlorinated hydrocarbons losing their effectiveness in the 1950s (Howitt and Cole, 1959). Organophosphate and carbamate products vary in efficacy, and there is evidence that resistance has developed to some of these materials (Stevenson, 1976b; Harris *et al.*, 1985).

Cultural Practices. Timing of planting and harvesting is commonly suggested as an effective means to avoid damage. This is more feasible for home garden production than commercial production. Carrot, for example, is a relatively short-season crop, so timing of planting can be adjusted to miss most of the flight of the flies from the overwintering insects, and harvesting planned to avoid the late summer flight. Cultivars vary somewhat in their maturity date, of course, and varieties that require long periods for growth are more susceptible to attack. Early season planting is particularly risky, as flies apparently select the largest carrots for oviposition. However, with long-season varieties, if planting is delayed to avoid the first flight of flies, then the carrots are susceptible to late season attack. To maximize production efficiency, commercial producers of carrots and other umbelliferous crops may stagger their planting dates, or at least have both an early season and a late season crop. Thus, it may be impossible for commercial carrot producers to avoid having at least some of the crop in a susceptible stage.

Because carrot rust fly is limited to umbelliferous crops, crop rotation is recommended. Although flies may move short distances, considerable benefit accrues from rotating fields. Intercropping of non-host crops such as onion, with a susceptible crop such as carrot, reduces the damage by carrot rust fly, but not all non-host plants are effective at disrupting fly injury (Uvah and Coaker, 1984).

Sanitation is also important. Crops left in the field may support overwintering insects, or be attractive oviposition sites in the spring.

Row covers can be used to prevent attack of crops by carrot rust fly in the spring. Finch (1993) indicated that though this is not economical under normal conditions, it is done for 10% of the British carrot crop to assure availability of insecticide-free carrots for young children.

Host-Plant Resistance. Some carrot varieties confer partial resistance to carrot rust fly. Finch (1993) indicated that partial resistance allows reduction in the intensity of insecticide application, and for this reason such partially resistant varieties now constitute two-thirds of the British carrot acreage. Partial resistance, combined with careful timing of harvest, is especially effective at reducing damage (Ellis *et al.*, 1987). Phenolic acid levels in roots are correlated with plant resistance to some insects, and Cole *et al.* (1988) devised a rapid technique to screen carrot cultivars for resistance to carrot rust fly.

FAMILY SYRPHIDAE—FLOWER AND BULB FLIES

Onion Bulb Fly
Eumerus strigatus (Fallén)

Lesser Bulb Fly
Eumerus tuberculatus Rondani
(Diptera: Syrphidae)

Natural History

Distribution. Onion bulb fly and lesser bulb fly are native to Europe. Apparently they were introduced to North America in the late 1800s—onion bulb fly was first recognized at Ottawa, Canada in 1904. Now, onion bulb fly is widespread in the United States and southern Canada, whereas lesser bulb fly is limited to the northern United States and southern Canada. These similar flies have been confused in the literature, and some damage by *E. tuberculatus* has been incorrectly attributed to *E. strigatus* (Latta and Cole, 1933). A third introduced species, *Eumerus narcissi* Smith, also is easily confused with the aforementioned species, but as it is not found in vegetables it is not to discussed further. Generally, onion bulb fly is more damaging in western North America, with lesser bulb fly more damaging in the east.

Host Plants. Vegetable crops attacked are onion and shallot, but more common hosts are bulbs of flowers such as amaryllis, hyacinth, iris, and narcissus. Reports of these insects attacking carrot and parsnip are almost certainly incorrect.

Natural Enemies. Natural enemies of these flies are not known.

Life Cycle and Description. Apparently there are two generations of these flies annually, although in Oregon a proportion of the second generation continues its development, producing a small third generation. Larvae are the overwintering stage, with adults from the overwintering flies present in April–May, and then again in June–August. There is no evidence that these two species differ in their biology.

Egg. The eggs are usually deposited in the soil near bulbs, generally at a depth of less than 1 cm, but sometimes on bulbs or foliage. The eggs measure about 0.7 mm long and 0.2 mm wide. The egg is somewhat elliptical, but one end tapers to a sharper point. The color is white. Eggs may be deposited singly or in clusters of up to 30. Duration of the egg stage is 4–5 days.

Larva. Larvae are wrinkled and flattened in appearance. Their general color is variable, ranging from white to greenish or reddish, depending on food substrate. Reddish respiratory processes project from the posterior end. Larvae attain a length of about 7–10 mm at maturity. Larvae disperse to bulbs, where they typically enter at the juncture of the foliage and bulb if areas of deterioration caused by plant disease are present. Healthy and wounded tissue lacking decay are not suitable for larval survival and growth. Larvae can attempt to feed on healthy tissue, and perhaps create openings that allow fungi or bacteria to gain entry, which in turn makes the damaged bulb suitable for the larvae. Larval development time is dependent not only on temperature, but on suitability of the food substrate. Martin (1934) reported that lesser bulb fly larvae initiating their feeding on diseased tissue completed their development in 22 days, whereas larvae that initially fed on healthy tissue (that later decayed) required an additional five days to complete the larval stage. Apparently not much development occurred until the bulb tissue began to decay.

Pupa. Pupation occurs in the soil, near the surface, in close association with the bulb. The puparium resembles the larva, except that it is relatively shorter and wider. It typically measures 6–8 mm long, and is gray or brown. Duration of the pupal stage is 9–13 days.

Adult. The adult is a robust fly, and is rather short (5.5–7.5 mm) and wide. Even the legs are thickened. The body is metallic bronze or green, and the abdomen bears three white bands that have a gap dorsally. Legs are blackish-green but bear some red and yellow on the base of the tibia. Wings are transparent. Differentiation of the flies is based on genitalia, and generally requires the services of an authority. Copulation normally commences within three days of adult emergence, and egg deposition commences in another

Adult onion bulb fly.

three days. Adults are observed to feed on flowers of Umbelliferae and Compositae. Adults often live 13–15 days. Total egg production is not well documented, but it is known to be at least 60 eggs per female. In all likelihood, egg production probably numbers in the 100s.

The biology of *E. strigatus* and *E. tuberculatus* were presented by Hodson (1927), based on research conducted in England. The biology of lesser bulb fly, based on work conducted in New York, was given by Martin (1934). Wilcox (1926) published a very nice study of bulb fly based on research in Oregon; the species involved is uncertain, but likely is onion bulb fly. Characters for differentiation of the species can be found in Latta and Cole (1933).

Damage

Larvae of both species are associated with bulb decay. Adults clearly prefer to oviposit on bulbs that are injured or infected with plant disease, and larval performance is enhanced by the presence of some disease. The exact role of the insect is uncertain, though there have been several attempts to elucidate the relationship of the flies with bulbs. It appears certain that the larvae require diseased tissue (Creager and Spruijt, 1935), but it is likely that larvae enlarge wounds and retard the normal healing process of the plant. Plants grown from seed do not incur injury, whereas those grown from bulbs are occasionally damaged. Commercial production most often involves growing plants from seed, so these insects tend to be home-garden pests.

Management

Recommendations for management usually involve sanitation. Bulbs displaying signs of decay should be destroyed, not just discarded. A water bath treatment consisting of 3 h exposure at 43°C is reported to rid bulbs of both bulb fly and nematodes. Bulbs infected with nematodes are attractive to flies, and nematode injury allows invasion by larvae. The observation by Doane (1983) that onion bulb flies are attracted to, and oviposit on, oatmeal that has been moistened

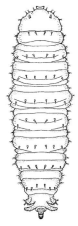

Onion bulb fly larva.

and buried suggests that this technique could be used for population monitoring, or even removal trapping.

FAMILY TEPHRITIDAE—FRUIT FLIES

Mediterranean Fruit Fly
Ceratitis capitata (Wiedemann)
(Diptera: Tephritidae)

Natural History

Distribution. Mediterranean fruit fly, commonly known as "medfly," probably originated in tropical Africa, but as its common name suggests, it first attracted attention in the Mediterranean region. It now is widespread in South America and Africa, and also occurs in southern Europe, Central America, and Australia. Southeast Asia is not infested, and most of North America is without permanent populations. In California, Texas and Florida, populations are occasionally detected and apparently eradicated. This pest is well-established in Hawaii, where it was first observed in 1910. Its potential range in North America likely includes all of the southern United States from Georgia to northern California, and all areas further south.

Host Plants. This insect has been known to develop successfully in the fruit of over 400 plants from numerous plant families, including several wild and ornamental plants that produce small berries. The most common North American host plants, however, are such stone, pome, and citrus fruits as almond, apple, apricot, nectarine, peach, pear, plum, grapefruit, orange, and tangerine. Other fruits at risk are fig, guava, grape, kumquat, loquat, lychee, passion fruit, quince, and persimmon. In general, thin-skinned, ripe fruits are preferred by ovipositing females. If the skin is cracked, other fruits such as avocado, banana, and papaya are attacked. Mediterranean fruit fly is not usually considered as a serious vegetable pest, but occasionally vegetables are attacked, and among those known to be hosts are eggplant, pepper, and tomato. Strawberry also is reported as a host.

Natural Enemies. Several general predators such as ants (Hymenoptera: Formicidae) and rove beetles (Coleoptera: Staphylinidae) have been reported to attack Mediterranean fruit flies, but only parasitoids are considered to be significant natural enemies. Many wasps were imported to Hawaii from western Africa, but most of them did not survive the trip or failed to establish when released. Additional research on parasitoids followed the introduction of oriental fruit fly, *Bactrocera dorsalis* (Hendel), into Hawaii in 1945, and many parasitoids introduced for oriental fruit fly

control also attack Mediterranean fruit fly. *Biosteres arisanus* (Sonan), *B. longicaudatus* Ashmead, and *Diaschasmimorpha tryoni* (Cameron) (all Hymenoptera: Braconidae) are the dominant parasitoid species in Hawaii, with the relative importance of individual species changing through the year (Wong and Ramadan, 1987). Together, they cause about 30–50% mortality in fruit fly larvae, but its effect varies considerably depending on the host fruit. In most countries where parasitoids have been introduced there has not been an appreciable reduction in damage by this fly. Host-specific parasitoids have not yet been identified (Headrick and Goeden, 1996). The taxonomy and biology of the parasitoids was discussed by Wharton and Gilstrap (1983).

Competition with other fruit flies seems to limit abundance of Mediterranean fruit fly. Its abundance and damage in Hawaii, especially at lower elevations, were diminished by the introduction of oriental fruit fly. This phenomenon has been observed elsewhere in the world with other fruit fly competitors.

Life Cycle and Description. The life cycle varies considerably in duration. Under ideal conditions a complete generation may be completed in about 18 days, but 30–40 days is common. The longevity of the life cycle under inclement conditions may be 3–4 months. Under favorable conditions in Hawaii 15–16 generations are estimated annually. Because the females are long-lived and continue oviposition over a considerable length of time, overlapping generations are common and virtually all stages are found throughout the year. The developmental threshold for egg, larval and pupal stages is about 9.7°C.

Egg. The eggs are deposited in small cavities under the epidermis of thin-skinned fruit, in clusters of about 3–9 eggs. These flies have the unusual habit of depositing eggs repeatedly into the same cavities, which may result in large aggregations, sometimes numbering up to 300 eggs of various ages in the same cavity. A small elevated crater may develop at the site of oviposition. Females deposit, on average, about 300 eggs during their life span, but some deposit more than 600 eggs. Duration of the egg stage is normally 2–6 days, but may be extended to 25 days under cool conditions. The white eggs are elongate-elliptical, measuring about 0.95 mm long. The end bearing the micropyle is marked by a small extension or tubercle. The egg is not completely symmetrical, as one side is strongly convex, whereas the opposite side is not convex or is slightly concave.

Larva. The shape of the Mediterranean fruit fly larva is similar to many other higher flies, bearing a

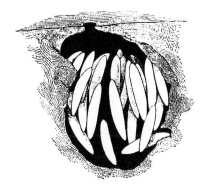

Mediterranean fruit fly eggs.

pointed head and a legless, cylindrical body that expands in diameter at the posterior region. The body color is usually creamy-white, but it may be tinted with other colors depending on the food source. There are three instars. The first instar measures only about 1 mm long, and bears spiracles only at the posterior end. The mouth hooks are very small, and bear two distinct teeth. The second instar is usually 3–5 mm long, though it may overlap the third instar in size. The second instar bears anterior spiracles on the second segment, in addition to the posterior spiracles. The mouth hooks in the second instar also have two teeth. The third instar measures 7–8 mm long, bears both anterior and posterior spiracles, and the mouth hooks consist principally of one large, sharply pointed tooth. The anterior spiracles possess 9–10 lobes. Larval development time is normally 6–10 days, but as for eggs, it may be extended to about a month in cool weather. The first two instars usually complete their development in 1–2 days, with the third instar requiring 2–8 days.

Mediterranean fruit fly larva.

Mediterranean fruit fly puparium.

Pupa. Pupation occurs in soil or among leaf debris. The puparium is elliptical, and measures about 4.0–4.5 mm long and 2 mm wide. The puparium consists of 11 visible segments. The color varies, but it usually is brown. Duration of the pupal stage ranges from 10 to 12 days during warm weather and from 25 to 30 days during cool weather.

Adult. Adults tend to emerge in the early morning hours. The adult is a fairly small fly, measuring 3.5–5.0 mm long. In general, its color is yellowish-brown. The dorsal surface of the thorax is creamy-white or yellow with black spots. The wings are mostly transparent, but marked with brownish-yellow blotches and black spots. The numerous narrow black stripes near the base of the wings are unusual, and this pattern assists with recognition. The yellowish-brown abdomen is marked with two silvery-white transverse bands and black hairs. The eyes are reddish purple. The males bear a pair of bristles, on setae originating near the top of the eyes, that are expanded and flattened at the tips. This character separates medfly males from all other members of the family. The adult stage is active, but most commonly it moves about by running. When disturbed, it makes short flights. However, the fly is commonly moved by wind to distances of 2 km or more. Adults must feed regularly, and perish within 4–5 days if they lack access to such food as fruit juice, plant sap, aphid honeydew, bird feces, yeasts, and leaf bacteria. Males and females may mate repeatedly, but it is not certain how frequently this occurs naturally. Males may form leks, territories where males aggregate for mating. Citrus trees seem to be especially favored cites for lek formation. Males produce a complex blend of pheromones that attract other males, which probably aids in lek formation. Females apparently are attracted to some components of the pheromone blend, but this is not conclusive. Females commence egg laying about 7–10 days after emergence. Oviposition averages about 5–6 eggs per day.

Complete treatment of Mediterranean fruit fly biology and management was given by Back and Pemberton (1918b), but many aspects were updated in Morse *et al.* (1995). Hendrichs and Hendrichs (1990) provided many important behavioral observations. Rearing procedures were given by Tanaka *et al.* (1969).

Damage

Females puncture the skin of fruit with their ovipositor and deposit eggs just beneath the surface at a depth of about 1 mm. The oviposition process often results in access to the fruit by plant pathogenic microorganisms. The larvae burrow deeply into the fruit, enhancing the movement of microorganisms. Infested

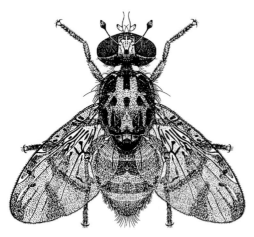

Adult Mediterranean fruit fly.

fruit usually drop from the plant and degrade into a rotten mass.

In addition to direct damage caused by flies and plant disease, Mediterranean fruit fly inflicts indirect economic injury by denying growers the ability to ship produce to unifested areas. Quarantines against movement of potentially infested crops are widespread throughout the world.

Management

Sampling. Traps are used to monitor the occurrence and abundance of Mediterranean fruit fly. Jackson traps, small cardboard sticky traps baited with trimedlure, are often used, but are neither very efficient nor reliable at low insect densities. The biological significance of trimedlure, which attracts only male flies, is obscure. Female populations are monitored using traps baited with food volatiles. Liquid protein hydrolysate baits deployed in McPhail traps, bell-shaped containers with a liquid reservoir (McPhail, 1937; 1939), are the standard technique for monitoring females. Unfortunately, McPhail traps are fragile, expensive, and labor-intensive to maintain. Alternative trap designs that overcome many of the disadvantages of McPhail traps have been developed (Heath et al., 1996).

Insecticides. In North America, where Mediterranean fruit fly is not firmly established, insecticides are used as a primary element of eradication programs. Insecticide is mixed with proteinaceous bait and sprayed by aircraft over large areas. This aerial treatment is supplemented with ground application of insecticide-containing bait applied to trees in the vicinity of fly infestations. The soil beneath trees known to be infested with flies receives additional insecticide. Eradication programs usually use release of sterile insects as a follow-up to insecticide treatment.

Cultural Practices. Few cultural practices are effective for Mediterranean fruit fly suppression (Aluja,

1996), and the most important tactic for North America is exclusion. In Hawaii, fruit is sometimes covered with bags to protect it from ovipositing females. Destruction of fallen fruit reduces reinfestation potential, but fruit must be collected before larvae complete their development and move to the soil. Mass trapping has been attempted, but most traps are not sufficiently attractive to be effective at reasonable cost. Rectangular yellow-sticky traps capture numerous flies, but these traps must be used at high densities and as they are not selective so they rapidly become covered with all types of insects. Placement of traps around the perimeter of orchards could eliminate flies that are invading from wild hosts.

Biological Control. The biological control organisms are generally considered to be incompatible with eradication programs unless they can be mass produced and applied repeatedly. Preliminary research has demonstrated that mass release of the parasitoid, *Diachasmimorpha tryoni* is compatible with sterile fly release, and may result in greater suppression of Mediterranean fruit fly than release of sterile flies alone (Wong et al., 1992). Also, application of the entomopathogenic nematode *Steinernema carpocapsae* to the soil surface at a rate of 5000 nematodes per square centimeter greatly reduced the emergence of Mediterranean fruit flies (Lindegren et al., 1990).

Sterile Insect Release. Mass release of laboratory-cultured, sterilized insects is used under the premise that wild females will mate with sterile males, and fail to produce offspring. This can be accomplished if sterile males are competitive with wild males, and the ratio of sterile to wild males is sufficiently high that the probability of wild females encountering a wild male is low. However, it is imperative that sterile flies originate from a vigorous colony, and that large number be available on a continuing basis. A ratio of 100-sterile flies for each wild fly is considered to be minimal for success. It has proven impossible, thus far, to eliminate well-established wild fly populations through release of sterile flies without the initial use of insecticide to reduce pest abundance. The release of sterile flies has few negative impacts, though the sterile females do sting the fruit in an attempt to deposit their eggs.

Melon Fly
Bactrocera cucurbitae (Coquillett)
(Diptera: Tephritidae)

Natural History

Distribution. Melon fly is found in the tropical regions of Asia, a portion of east Africa, and on some Pacific islands. In the United States, its distribution currently is limited to Hawaii. Melon fly has been recovered on several occasions in California, and though it has not become established on the mainland, California and Florida would likely be suitable environments.

Host Plants. Melon fly prefers such cucurbits as watermelon, cantaloupe, pumpkin, squash, and cucumber, but infests other vegetables including tomato, pepper, green bean, and cowpea. Wild hosts, particularly bitter melon, *Mormordica charantia*, can be important. Passion fruit, papaya, grape, and citrus also are suitable larval hosts, but it is the availability of vegetable crops that generally determines melon fly abundance (Vargas *et al.*, 1989; 1990).

The adult often is found associated with plants that are not larval food sources, but rather provide sustenance for the adults. Honeydew produced by homopterous insects and secretions of extrafloral nectaries attract flies to corn; castor bean, *Ricinus communis*; spiny amaranth, *Amaranthus spinosus*; rattlepod, *Crotolaria incana* and *C. mucronata*; and other cultivated and wild plants (Nishida and Bess, 1957).

Natural Enemies. Extensive research has been conducted around the world to identify effective natural enemies of this serious pest. The wasp *Psytallia fletcheri* (Silvestri) (Hymenoptera: Braconidae), a larval parasitoid, is the most effective biological control agent located. It was introduced into Hawaii from India in 1916 to provide biological suppression of melon fly. Studies conducted in Hawaii found that though *P. fletcheri* was effective at parasitizing melon fly larvae infesting wild cucurbits, it was relatively ineffective in crops (Liquido, 1991). Therefore, its use is limited to reduction of background populations only. Predators and parasitoids attack all stages of melon fly in Hawaii. A good review was provided by Nishida (1955).

Life Cycle and Description. A life cycle can be completed in about five weeks in warm climates but may require 3–4 months in cool climates. The adults can survive for months, and will continue to reproduce if fruit is available. In tropical climates, such as Hawaii, they are present throughout the year. Their abundance is determined mostly by availability of suitable host plant material, but they tend to be most common in summer and autumn (Vargas *et al.*, 1989; 1990).

Egg. Females deposit eggs in small batches, usually 5–20 eggs each. Females may produce 800–900 eggs over their life span. The white eggs are about 1.3 mm long and 0.25 mm wide, and are deposited in the fruit or vegetative parts of plants. The lower- and upper-developmental thresholds for egg development are

about 10° and 40°C, respectively (Meats, 1989). Eggs hatch, on average, in 1.3 days.

Larva. The larvae are white, and when they hatch are about the length of the egg. The larvae immediately begin to grow, and attain a length of about 2.5, 5.5, and 11.0 mm long during instars 1–3, respectively. The mouth hooks are pale in the first instar, but thereafter they are darkened and easily observed. The larvae complete their development in 5–8 days, with instars 1–3 requiring about 1, 1–2, and 2–3 days, respectively, in soft hosts such as papaya. Back and Pemberton (1914) reported that larvae developing in thick-skinned fruit such as cantaloupe and watermelon may remain in the fruit as full-grown larvae for several days before emerging. The lower- and upper-developmental thresholds for larvae are about 12° and 34°C, respectively (Meats, 1989).

Pupa. When mature, larvae leave the host, burrow into the soil up to a depth of 10 cm, and pupate. The adults emerge from the tan or yellow-brown puparium after about 10 days (range 7–14 days), and dig to the soil surface.

Adult. Newly emerged adults crawl to a sheltered location where they rest for 2–3 h before taking flight. Adults are yellow or yellow-brown, with yellow wings marked with brown bands. They measure about 6–9 mm long. Dark markings resembling a "T" occur on the abdomen, and two lateral and a medial yellow stripe are found on the thorax. The presence of wing bands and the dark abdominal markings can be used to differentiate melon fly from oriental fruit fly and Mediterranean fruit fly, respectively. The pre-oviposition period of adults is about 7.4 days, and they are capable of depositing eggs for about three months.

The adults spend most of their time associated with the adult host plants, seeking vegetable crops or other prospective larval hosts intermittently for oviposition. Fly activity in crop fields peaks in the afternoon hours. At dusk, males form aggregations called leks, and by making wing vibrations and releasing sex pheromone, they attract and copulate with females (Kuba and Koyama, 1985; Kuba and Sokei, 1988). Components of the sex pheromone were described by R. Baker *et al.* (1982).

Melon fly larva.

Adult melon fly.

Life history information was given by Vargas *et al.* (1984) and Nishida and Bess (1957). Descriptions of all stages, and keys for the differentiation of melon fly from oriental fruit fly, *Bactrocera dorsalis* (Hendel), and Mediterranean fruit fly, *Ceratitis capitata* (Wiedemann), were provided by Hardy (1949), White and Elson-Harris (1992), and Foote *et al.* (1993).

Damage

In the absence of biological or chemical control, melon fly is extremely damaging. For a period of time after the introduction of melon fly to Hawaii in the late 1800s, farmers stopped growing cucumbers, melons, and tomatoes because of the great damage. Melon fly larvae develop in blossoms, fruits, and some vegetative portions of plants. Among vegetative plant material, newly emerged seedlings and terminal shoots are preferred. Similarly, among fruits, immature fruit is usually selected. Larval feeding also opens the plant tissue to secondary invaders, both microbial and insect. Damage can also occur from egg laying even when larvae cannot survive because oviposition allows entry of microorganisms or causes deformities in the growing fruit. Some differences in damage among cucurbits exists. Seedling and stem damage is more common in watermelon and cantaloupe than in squash, cucumbers, and pumpkin. Blossom damage is serious among all cucurbits except cucumber. Both male and female blossoms of squash and pumpkin are affected, but in watermelon and cantaloupe the male blossom generally escapes attack (Back and Pemberton, 1918a).

Management

Sampling. Fruit may be sampled for eggs and larvae, and soil for pupae, but this is not done routinely because of high labor requirements and expense. A natural chemical constituent of some plants, raspberry ketone, is highly attractive to male melon flies, and a synthetic analog called cue-lure is equally or more attractive (Cunningham, 1989); the latter is useful with traps for sampling. Fluorescent yellow-sticky traps can also be used to sample adult densities (Katsoyannos, 1989; Cunningham, 1989), and odor and visual stimuli are sometimes combined in a single trap design (Economopoulos, 1989). However, the most widely used trap is the McPhail trap, a liquid-baited trap (McPhail, 1937).

Insecticides. Melon fly adults, unlike many insect pests, do not remain in contact with larval host plants for extended periods of time. Instead, they spend a great deal of time in weedy vegetation surrounding crop fields. A program of treating the surrounding vegetation, rather than the crops, was therefore developed. Treatment of border vegetation was much more effective than treatment of crops (Nishida, 1954). Bait sprays containing insecticide also can be used effectively for adult suppression. Sugar or protein hydrolysate is mixed with various insecticides to produce bait sprays (Nishida and Bess, 1957; Roessler, 1989). Insecticide and bait treatments have been combined with mass release of sterile male insects to eradicate melon fly from some Pacific islands (Steiner *et al.*, 1965a). Cue-lure can also be mixed with insecticide to attract and kill flies, but it is only effective against males (Cunningham *et al.*, 1970; Cunningham and Steiner, 1972).

Cultural Practices. Field sanitation is the most important element of cultural management. High melon fly populations result from continuous availability of larval food, which may be due either to continuous cropping or failure to dispose of crop residues. Destruction of infested or unmarketable vegetables in a timely manner is essential.

Protective coverings have long been used to deter oviposition by melon fly. Paper bags and newspapers are used to wrap individual fruit of large produce such as cantaloupe and watermelon. This is effective but tedious, and not useful for small-fruited vegetables. Also, it does not protect vines and flower buds. Row covers provide complete protection, but pollination may be interrupted.

Trap crops are not effective for protecting vegetables, probably due to the relatively wide host range and vagility of the insects.

Biological Control. Melon fly larvae are moderately susceptible to steinernematid nematode infection (Lindegren and Vail, 1986). There may be some use for this approach in suburban rather than agricultural environments because financial cost would be less of a consideration.

Oriental Fruit Fly
Bactrocera dorsalis (Hendel)
(Diptera: Tephritidae)

Natural History

Distribution. Oriental fruit fly occurs in tropical regions of Asia, including some Pacific islands. It became established in Hawaii about 1946. On occasion, oriental fruit fly appeared in California and Florida, but always was successfully eradicated. The environment in southern California, Texas, and Florida probably would be suitable for this pest, so it remains a threat. The taxonomy of this insect is uncertain; there likely are several species within the oriental fruit fly "complex." Thus, many elements of the life history, including host records, are suspect until confirmed.

Host Plants. Over 120 plants have been reported to serve as hosts of oriental fruit fly larvae, though many are attacked only during population outbreak conditions. Principal hosts are fruits such as avocado, apple, mango, peach, pear, citrus, coffee, and especially guava. Oriental fruit fly is so highly attracted to guava, and so effective at utilizing this host, that it has displaced Mediterranean fruit fly, *Ceratitis capitata* (Weidemann) as the principal pest of guava, and significantly lowered the overall density of Mediterranean fruit fly in Hawaii. Among vegetables, pepper, tomato, and watermelon are reportedly attacked.

The adult flies feed on secretions of extrafloral nectaries, honeydew, rotting fruit, bird dung, and other liquefied items. The adults survive only three days without water, and six days with water, but no sources of carbohydrate. The ability of flies to disperse long distances to obtain food is present in this species. However, unlike the clearly defined dispersal of melon fly, *Bactrocera curcurbitae* (Coquillett), between adult feeding and oviposition sites, it is uncertain how often oriental fruit flies relocate.

Natural Enemies. In the absence of natural enemies, which was the situation in Hawaii immediately after introduction of the fly, very high pest densities are attained. Searches were conducted in many countries for beneficial insects, and a predatory rove beetle (Coleoptera: Staphylinidae) and several hymenopterous parasitoids attacking the egg and larval stages were introduced. Apparently some of the wasps contributed materially to reducing population densities of this damaging fly, because fruit infestation levels declined by 1951 (Bess and Haramoto, 1961). The principal beneficial parasitoid is *Biosteres arisanus* (Sonan), but *Diachasmimorpha longicaudata* (Ashmead) and

Biosteres vandenboschi (Fullaway) (all Hymenoptera: Braconidae) also are considered important. Over a seven-year period on Oahu, Bess and Haramoto (1961) found that 68–79% of fly larvae in guava fruit were parasitized, though Vargas *et al.* (1993) reported only about 40–50% parasitism. Destruction of eggs by *B. arisanus* was sometimes considerable, up to 80%, probably owing to infection of eggs by bacteria and fungi following damage to the chorion by the parasitoid (Bess *et al.*, 1963). Rearing methods for *B. arisanus* have been developed (Harris and Okamoto, 1991). Parasitoid monitoring technology has also been studied (Stark *et al.*, 1991; Vargas *et al.*, 1991). Predation, especially by ants (Hymenoptera: Formicidae), was noted but not considered appreciable. Overall fruit fly densities were also determined significantly by availability of fruit and warm weather.

Life Cycle and Description. Oriental fruit fly can complete a generation in about 30 days. In tropical climates, many overlapping generations per year are reported. Fruit fly abundance typically coincides with availability of ripening fruit, though they tend to be most common in summer and autumn (Vargas *et al.*, 1989; 1990).

Egg. Oriental fruit fly eggs average about 1.17 mm long and 0.21 mm wide, which is slightly smaller than melon fly. The female may puncture fruit and deposit her eggs, or she may take advantage of cracks or other wounds, including the oviposition punctures of other flies. The principal reason that Oriental fruit fly is believed to have successfully displaced Mediterranean fruit fly in Hawaii is that Oriental fruit fly larvae are more competitive when both species of larvae inhabit the same fruit. Eggs may be deposited at a depth of 5–6 mm in soft fruit, whereas they may be very near the surface in hard fruit. The upper- and lower-developmental thresholds for eggs are estimated at 38° and 12°C, respectively (Meats, 1989). The average time for egg hatching is 1.6 days (Vargas *et al.*, 1984), but hatching may be extended to 20 days in cold weather.

Larva. Oriental fruit fly larvae are typical in form for tephritid fruit flies: cylindrical and broad posteriorly and tapering to point at the anterior end. There are three instars; all are whitish in color. The first instar ranges in size from about 1.2 mm to 2.3 mm, whereas the second ranges from 2.5–5.7 mm and third ranges from 7.0–11.0 mm. The upper- and lower-developmental thresholds for larvae are estimated at 34° and 11°C, respectively (Meats, 1989). Larval development generally requires about 7.8 days, though its development time can range from 6 to 35 days.

Pupa. Mature larvae leave infested fruit and enter the soil, usually at the base of affected trees, to pupate. The puparia are 3.8–5.2 mm long and vary in color from tan to brownish-yellow. Pupal development requires about 10.3 days.

Adult. The adult fruit fly has a yellow to orange abdomen marked with a black "T". The thorax is predominantly black but bears two-yellow stripes laterally. Oriental fruit fly lacks cross bands on its wings, and therefore is easily differentiated from melon fly. The dark abdominal markings serve to distinguish oriental fruit fly (and melon fly) from Mediterranean fruit fly. After adults emerge, a period of 6–12 days normally elapses before oviposition can occur. Copulation persists for 2–12 h. Males expel pheromone in a visible form resembling smoke (Ohinata *et al.*, 1982), similar to pheromone production by melon fly. Mating occurs at dusk in aggregations called "leks". Mating normally occurs at 4–5 day intervals. The adults continue to produce eggs for about two months. The female oriental fruit fly is more fecund than the related tephritids melon fly and Mediterranean fruit fly, and she produces an average of over 1400 eggs per female during a life span of about 80 days (Vargas *et al.*, 1984). The oviposition rate is reported to be about 130 eggs per day.

Keys for distinguishing all life stages of these species were provided by Hardy (1949), White and Elson–Harris (1992), and Foote *et al.* (1993).

Damage

The adult flies sting fruit during the process of oviposition. The presence of larvae is, of course, highly objectionable to consumers. Even if the larva fails to develop, fruit deformities may occur. Also, the oviposition wound is frequently a site for invasion by bacterial and fungal diseases. Other insects such as fruit flies (Hymenoptera: Drosophilidae) and sap beetles (Coleoptera: Nitidulidae) may attack fruit infested by larvae. As many locations lack Oriental fruit fly, quar-

Adult oriental fruit fly.

antine restrictions are frequently imposed that restrict the sale and transport of valuable produce.

Management

Sampling. Liquid bait traps such as the McPhail trap (McPhail, 1937; 1939) have long been used in fruit fly detection efforts, but they require considerable maintenance so there is a continuing effort to develop effective but inexpensive trap technology. Methyl eugenol is highly attractive to oriental fruit fly. Only males are attracted (Cunningham, 1989). Yellow and white sticky spheres may be useful for population monitoring (Vargas *et al.*, 1991). Growers are advised to spray crops when flies become abundant in traps.

Insecticides. Male suppression has been used to eradicate oriental fruit fly on some islands. Males are lured to substrate that is impregnated with the attractant methyl eugenol, where flies also come into contact with a lethal dose of insecticide. Baitinsecticide releasers are air-dropped or suspended from trees (Steiner *et al.*, 1965b; Koyama *et al.*, 1984). Release of sterile male insects also can be used for eradication (Steiner *et al.*, 1970), alone or in combination with other means of suppression. Insecticides are also mixed with protein hydrolyzate baits and sprayed on foliage to affect fly suppression (Roessler, 1989).

Cultural Practices. Fruit can be protected from flies by being wrapped in netting or bags, as is sometimes done for melon fly. Obviously this approach is limited to small-scale fruit or vegetable production.

Parsnip Leafminer
Euleia fratria (Loew)
(Diptera: Tephritidae)

Natural History

Distribution. Parsnip leafminer is found throughout the United States and eastern Canada. It is a native of North America. There is some confusion in the literature surrounding this insect. A very similar insect, *Euleia heraclei* (Linnaeus), occurs in Europe, where it is known as celery leafminer. *Euleia heraclei* does not occur in North America.

Host Plants. As its name suggests, this insect feeds on parsnip. Several wild umbelliferous hosts have been recorded, including honewort, *Cryptotaenia canadensis*; cow parsnip, *Heracleum lanatum*; water hemlock, *Conium maculatum*; *Oenanthe sarmentosa*; as well as a composite, rattlesnake root, *Prenanthes canadense* (Tauber and Toschi, 1965).

Natural Enemies. Several parasitoids were reared from parsnip leafminer larvae or pupae in California. Most abundant were *Diglyphus begini* (Ashmead) and

Chrysocharis sp. (both Hymenoptera: Eulophidae), but also recovered were *Achrysocharella* sp. (Hymenoptera: Eulophidae) and *Halticoptera* sp. (Pteromalidae) (Tauber and Toschi, 1964).

Life Cycle and Description. Apparently there is only one generation per year—the eggs are deposited in early spring and the generation is completed by mid- to late-summer. Overwintering is uncertain. Chittenden (1912c) reported that adults emerge in August, but it is likely that some diapause as pupae until the next year.

Egg. The elliptical egg is white and averages 0.72 mm long (range 0.64–0.74 mm) and 0.22 mm wide (range 0.18–0.23 mm). One end bears a protruding knob, which functions as a micropyle. Development time for the egg is 6–8 days.

Larva. Upon hatching, the larva burrows into leaf tissue and excavates a narrow tunnel. Within a few days, however, the larva begins to widen the tunnel, forming a blotch-like mine. Often, the mines coalesce to form a single blister containing several larvae. The larva is colorless or pale-yellow. Larvae are quite mobile, moving freely not only within the mine, but also exiting and re-entering mines. Larvae normally leave their mine due to deterioration of the food source, usually moving to a new leaf. Duration of the three instars is reported to be 4–5, 3–5, and 4–5 days, respectively. Larvae attain a length of about 7–8 mm at maturity.

Pupa. Mature larvae pupate on the foliage, or more commonly they drop to the soil and burrow to a depth of about 1 cm for pupation. The puparium measures 3.9 mm long (range 3.6–4.4 mm) and 1.7 mm wide (range 1.6–1.8 mm). Initially, the puparium is greenish, but turns yellowish with age. Adults emerge about 14–17 days after the puparia are formed.

Adult. The adults are brightly colored. The head, thorax, and legs are yellow, and the abdomen is pale green. The wings are marked with alternating bands of yellow and white. The wings also bear a black spot centrally near the anterior margin. The adult measures about 5 mm long. Adults mate within 12 h of emergence, and oviposition follows within three days. The female cuts a slit in leaf tissue with her ovipositor and deposits eggs within the parenchyma. Often 4–5 eggs are deposited in a row, usually near the leaf margin. Flies are relatively long lived, at least in the laboratory, and most of them survive longer than four months. It is reported that up to 210 eggs have been produced by a single female fly.

The most complete account of parsnip leafminer biology was given by Tauber and Toschi (1965). Chittenden (1912c), Phillips (1946), and Foote (1959) provided useful descriptive material.

Damage

The larvae form large blotch-like mines in the foliage of parsnip. The mines usually are rounded and blackish owing to the accumulation of frass within the mine. Chittenden (1912c) suggested a preference for mature plants, specifically those that have begun to produce seed. Damage is rarely severe.

Management

Parsnip leafminer should rarely require control, but if this is desired, application of insecticide should be made to the foliage in March or April, as soon as the plant begins active growth.

Pepper Maggot
Zonosemata electa (Say)
(Diptera: Tephritidae)

Natural History

Distribution. This native pest occurs throughout the eastern United States west to Texas and Kansas. However, it is most abundant along the east coast, from Massachusetts to South Carolina. In Canada its distribution apparently is limited to southern Ontario.

Host Plants. Pepper maggot develops successfully only in plants in the family Solanaceae. The vegetable crops injured are pepper and eggplant. Tomato has been reported to be a host, but this is rare. Solanaceous weeds that sometimes support pepper maggot are horsenettle, *Solanum carolinense*; silverleaf nightshade, *S. elaeagnifolium*; groundcherry, *Physalis* sp.; and possibly others. Insects developing in horsenettle are considerably smaller than those developing in pepper, indicating that pepper is a superior host (Foott, 1968a; Judd *et al.*, 1991).

Natural Enemies. Few natural enemies have been documented. The wasp *Opius sanguineus* (Ashmead) (Hymenoptera: Braconidae) has been reared from pepper maggot infesting horsenettle in Ontario, but not from maggots in crop plants.

Life Cycle and Description. So far as is known, there is a single generation throughout the range of this insect. The puparia overwinter, with adults emerging in June and July. The egg and larval stages are relatively brief, which require a total of only a month, and new puparia are formed and enter diapause during July–September.

Egg. Females insert eggs into fruit. The egg is unusual in shape, primarily elongate-oval but with one end narrowed, tapered, and curved. Altogether, the egg resembles crooked-neck squash. The stalk of the egg may be visible by close examination in the oviposition slit. The whitish egg measures 2.0–2.2 mm long. The width is about 0.30–0.35 mm at the bulbous end and only about 0.05 mm at the narrow end. Duration of the egg stage is about 8–10 days. The total fecundity of females is reported to average about 50 eggs, but some females produce up to 200 eggs.

Larva. The larva is typical in form for flies—a cylindrical body with the posterior end wide and blunt, and tapering gradually to a narrow, pointed head. Body length increases from 1.5 mm at hatching to 10–12 mm at maturity. The color changes from translucent white, when the larva is young, to opaque white and then yellowish at maturity. The larva feeds on the central fleshy region of the fruit, occasionally eating young seeds. Mean development time for larvae is about 18 days under field conditions, (range 12–22 days).

Pupa. The larva exits the fruit at maturity and drops to the soil, where it forms a puparium. The puparia are usually found within the upper 5–10 cm of soil, and persist from late summer or autumn until the next summer. The puparia are oval, measuring about 6–8 mm long and 2.8–4.2 mm wide, and flattened dorsoventrally. They are yellowish-brown to medium-brown.

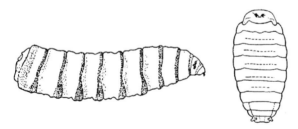

Pepper maggot larva (*left*) and puparium (*right*).

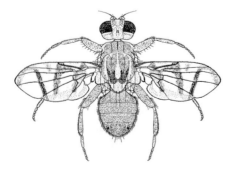

Adult pepper maggot.

Adult. The head, abdomen, and legs of the adult are pale yellow. The thorax is bright yellow and bears brownish stripes. The last abdominal segment bears a pair of small black dots. Dark bands cross the transparent wings, and the distal bands forms a "V" shape. The eyes are green. Males measure about 6.5 mm long, females about 7.5 mm. The adults often mate within the first day of emergence, but mate repeatedly. The pre-oviposition period is 6–7 days. Adults live about 20–40 days.

The biology of pepper maggot was given by Peterson (1923), Burdette (1935), and Foott (1963). Foott (1968b) described rearing procedures. Keys to adults were presented by Bush (1965) and Foote *et al.* (1993). An excellent summary of biology and management in Canada was presented by Howard *et al.* (1994).

Damage

Oviposition punctures occur on small fruit, usually with a diameter of 1–3 cm. As the fruit expands, the area around the puncture sinks, forming a shallow dimple. The larva normally burrows in the soft, central tissue or core, which becomes brown in color. Sometimes mining of the epidermal area is evident. Usually only a single maggot develops in each pepper, but several may develop within each eggplant. In the absence of controls, infestation levels of 90% were recorded, but this pest occurs only sporadically.

Damage by pepper maggot to peppers is similar to injury inflicted by pepper weevil, *Anthonomus eugenii* Cano. However, larvae of pepper weevil have distinct heads, unlike pepper maggot. Also, pepper weevil is found only to southern states whereas pepper maggot occurs in northern areas.

Management

Sampling. The adult populations can be monitored with yellow-sticky traps. Traps placed along the margin of fields indicate presence of flies, though they appear not to be reliable indicators of density. It is difficult to detect larvae in fruit until they produce an exit hole at maturity.

Insecticides. Insecticide applications are often recommended for pepper fields if adults are present. Foliar applications of insecticides usually are made at weekly intervals.

Cultural Practices. Sanitation is important in pepper maggot management. Removal and destruction of rotten and infested fruit is recommended to prevent future infestations. Alternate host such as horsenettle should be removed. Relocation of pepper fields to distant sites also is a helpful precaution.

Host-Plant Resistance. Some pepper varieties, particularly the thick-walled bell and cherry types, are most susceptible to injury, whereas the thin-walled pepper varieties are not preferred for oviposition and somewhat resistant to larval feeding.

FAMILY TIPULIDAE—CRANE FLIES

European Crane Fly
Tipula paludosa Meigen
(Diptera: Tipulidae)

Natural History

Distribution. The origin of European crane fly or "leatherjacket" is northern Europe, where it occurs from Finland to Great Britain. It was first found in North America in Labrador during 1952, followed by Nova Scotia in 1954, and British Columbia in 1965. The eastern populations have not expanded greatly, but on the west coast the population quickly spread south and east from Vancouver into western Washington and northwestern Oregon.

Host Plants. Larvae feed on numerous plants, though they are generally considered to be pests of pasture and turf grasses. Among vegetables attacked are beet, cabbage, cauliflower, corn, lettuce, pea, potato, rutabaga, turnip and likely others. Other crops such as barley, buckwheat, clover, flax, oat, rape, strawberry, tobacco, and wheat are also reported to be damaged occasionally, and various weeds are eaten. They can persist, but do not thrive, on decaying organic matter.

Natural Enemies. Probably the most important natural enemy of European crane fly in North America is the larval parasitoid *Siphona geniculata* (De Geer) (Diptera: Tachinidae), which was introduced from Europe. Several pathogens were identified in Europe or North America, including viruses, fungi, and protozoans, but they occur infrequently (Jackson and Campbell, 1975). Moles and birds are frequently cited as predators of leatherjackets, though there is no evidence that they can suppress populations. Weather seems to be more important than natural enemies, with survival or damage favored by cool, wet conditions in the spring and moist conditions in the summer, but negatively affected by cold winters.

Life Cycle and Description. There is a single generation annually, with this insect overwintering in the third instar. Larvae typically become fourth instars in April, pupate in August and then in rapid succession the adults emerge, eggs are deposited, and larvae begin to grow in September. The first two instars are completed by November, when larval growth slows for the winter.

Egg. The eggs of European crane fly are shiny purplish black, and elongate-oval. One side is slightly flattened. Eggs measure 1.1 mm long and 0.4 mm wide. They are deposited near the soil surface, usually within the upper 1 cm, and must be surrounded by moist soil or the eggs can desiccate. Duration of the egg stage is 11–15 days.

Larva. Larvae of European crane fly are called "leatherjackets," and like the larvae of most flies, are nearly cylindrical in form though they taper gradually at each end and particularly toward the anterior end. They are grayish or greenish-brown, and often bear spots. There are four instars. The larva grows from about 3 mm long at hatching to 40 mm at maturity. The duration of first and second instars is about 14 days each. The third instar is several months in duration because the winter is passed at this stage, and the final instar is also about 3–4 month long. Interestingly, during the final 6–8 weeks of larval life the leatherjacket does not feed, and remains relatively inactive. During the final few days of the larval stage, however, the larva descends deeper into the soil and prepares for pupation.

Pupa. Pupation occurs within the skin of the last instar larva, and so it resembles the mature larva in shape and size. Duration of the pupal stage is 2–3 weeks. When the adult is ready to emerge, the pupa wriggles to the soil surface where the empty puparium remains protruding about 2–3 cm from the soil.

Adult. The adults typically emerge at sunset, copulate, and produce eggs all in the same evening.

European crane fly larva.

The adults are fairly small, the males measuring about 14–19 mm long, the females about 19–25 mm. They are narrow-bodied, brownish, with very long legs. Emerging females are incapable of flight, but after depositing most of her eggs they can disperse by flying. Females deposit about 200–300 eggs, nearly all on the night of emergence as an adult. However, adult females can persist for 4–5 days, and males survive even longer, often 7 days.

The biology of European crane fly in North America was described by Jackson and Campbell (1975) and Wilkinson and MacCarthy (1967). Useful publications describing this insect in England include those by Rennie (1917) and Coulson (1962). Barnes (1937b) described crane fly culture. Keys to crane fly genera, including both larval and adult forms, were presented by Alexander and Byers (1981).

Damage

Surprisingly, the larvae feed not only below-ground on the roots and plant crowns, but also climb the foliage at night and attack the leaves and stems. When larvae are numerous, plants may be killed.

European crane fly larva, posterior view showing caudal spiracles.

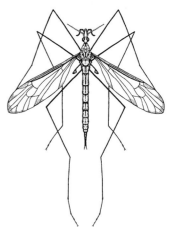

Adult European crane fly.

Management

Sampling. Larvae can be sampled by washing soil samples through sieves, and they can be flushed from the soil with hot water or salt solutions. Soil samples should be taken to a depth of about 10 cm. The adults are usually sampled by sweeping vegetation with a net, but they also can be captured with light traps.

Insecticides. Insecticides applied to the soil surface in either liquid or granular form is effective to suppress leatherjackets. Larvae also are reported to consume toxic baits (Wilkinson and MacCarthy, 1967).

Cultural Practices. Tillage can be very disruptive to overwintering larvae, though it rarely eliminates all leatherjackets. Flooding is more detrimental to larvae, but more difficult to accomplish. Dry soil is not attractive to ovipositing females.

Biological Control. The nematodes *Steinernema bibionis* and *S. carpocapsae* (both Nematoda: Steinernematidae), have been shown experimentally to infect larvae of European crane fly (Poinar, 1979), but apparently these potential biological control agents have not been evaluated under field conditions.

Order Hemiptera—Bugs

FAMILY COREIDAE—SQUASH AND LEAFFOOTED BUGS

Leaffooted Bugs
Leptoglossus spp.
(Hemiptera: Coreidae)

Natural History

Distribution. Several species of *Leptoglossus* affect crops; *L. phyllopus* (L.), *L. oppositus* (Say), and *L. clypealis* Heidemann are most common as vegetable pests. *L. phyllopus* is common in the southeastern states, extending north to New York and west to Missouri. *L. oppositus* occurs throughout the eastern United States, extending as far west as Minnesota and Arizona. *L. clypealis* is found from the Midwest to the Pacific Coast states (Slater and Baranowski, 1978). A related bug, *Phythia picta* (Drury), occasionally is found feeding on cucurbits and some other vegetables in California, Florida, and Texas (Baranowski and Slater, 1986).

Host Plants. Although commonly associated with squash and cantaloupe, *Leptoglossus* spp. feed on a variety of crops, including artichoke, bean, cucumber, pea, potato, southern pea, tomato, and watermelon. Field and fruit crops, particularly soybean and orange, also are damaged. These bugs are sometimes found feeding on weeds, and *L. phyllopus* was reported by several authors to have a strong preference for thistles. A list of occasional weed and crop hosts was given by Baranowski and Slater (1986).

Natural Enemies. Arnaud (1978) reported that *Leptoglossus* spp. were parasitized by the tachinids *Trichopoda pennipes* (Fabricius), *T. plumipes* (Fabricius), and *Trichopoda* sp. (all Diptera: Tachinidae).

Life Cycle and Description. The biologies of these insects have not been thoroughly studied. Only one generation per year is known. Apparently they overwinter as adults, emerging to feed on weed hosts in late spring.

Leaffooted bugs deposit rows of gold, bronze, or brown eggs, laid end to end on foliage or stem tissue, usually along a vein or stem. The length of an individual egg is about 1.4 mm, and the width about 1.0–1.15 mm. The eggs hatch in 5–7 days. The orange, red, or reddish-brown nymphs have 5 nymphal instars; 25–30 days is usually required for maturity.

The leaffooted bugs get their name from the peculiar flattening of the hind tibia, which resembles a leaf. They are usually about 20 mm long, and brown in color. *L. phyllopus* bears a fairly uniform, broad white band across the wing covers. *L. oppositus*, in contrast, lacks a white band, though small white spots may be present on some individuals. *L. clypealis* has a white band on the wing covers, but its pattern is jagged. The latter species is also distinguished by having a pronounced spine projecting forward from the head.

The biological information on leaf-footed bugs was provided by Chittenden (1899a, 1902), and a good description of *L. oppositus* nymphs by Chittenden (1902).

Damage

Leptoglossus spp. feed on foliar tissue and fruit. They may cause wilting and death of leaves, and deformities of fruit. Apparently only adult *L. phyllopus* affect

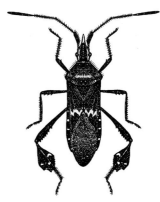

Adult leaffooted bug, *Leptoglossus clypealis*.

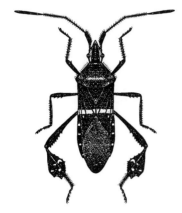

Adult leaffooted bug, *Leptoglossus phyllopus.*

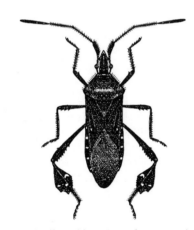

Adult leaffooted bug, *Leptoglossus oppositus.*

vegetables, with the immatures developing on thistle. With *L. oppositus*, however, nymphs are found on vegetable plants and this insect behaves much like squash bug, *Anasa tristis*; groups of insects of all stages aggregate on sheltered areas of the host plant. Chittenden (1902) reported that a toxic feeding secretion was produced by *L. oppositus.* A study in South Carolina compared the damage to southern peas by *L. phyllopus* and southern green stink bug, *Nezara viridula*, and found that three *L. phyllopus* per plant during early bloom stage could cause 54% total yield loss (Schalk and Fery, 1982). Late-bloom infestations, in contrast, caused only a 22% yield reduction. Stink bug feeding was more damaging, even with equivalent insect population densities, causing 100% and 74% yield loss at these two plant growth stages, respectively. (See color figures 136 and 137).

Management

Insecticides. *Leptoglossus* spp. are not difficult to control with insecticides. They most often cause injury when adults disperse into a susceptible crop, usually one with small fruit, where insecticides were not applied.

Cultural Practices. Because *L. phyllopus* breeds on thistle and other weeds, weed management has been suggested as an important factor in minimizing damage potential. The young of *L. oppositus*, however, are often found attacking tree fruit early in the season (Chittenden, 1902).

Squash Bug
Anasa tristis (De Geer)

Horned Squash Bug
Anasa armigera (Say)
(Hemiptera: Coreidae)

Natural History

Distribution. The squash bug, *Anasa tristis*, is found throughout the United States, and is known as a pest wherever it occurs. In Canada, it frequently is a pest in southern Ontario and Quebec and in southwestern British Columbia. Squash bug also occurs in most of Central America. The horned squash bug, *A. armigera*, has a more restricted distribution, occurring throughout the eastern United States west to Iowa and Texas, and is also known from Trinidad. Two other crop-feeding species of *Anasa*, *A. andresii* (Guerin) and *A. scorbutica* (Fabricius), are found in the southern states from California to Florida (Baranowski and Slater, 1986), but only rarely cause damage.

Host Plants. Squash bug has been reported to attack nearly all cucurbits, but squash and pumpkin are preferred for oviposition and support a high reproductive rate and survival (Bonjour *et al.*, 1993). There is considerable variation among species and cultivars of squash with respect to susceptibility to damage and ability to support growth of squash bugs. Beard (1935) noted that New World varieties were preferred. Bonjour and Fargo (1989) reported survival of squash bug to be 70, 49, 14, 0.3, and 0% when nymphs were reared to the adult stage on pumpkin, squash, watermelon, cucumber, and muskmelon (cantaloupe), respectively. Cook and Neal (1999) compared the development of nymphs on a preferred host (pumpkin) with a non-preferred host (cucumber) and found that the cucumber-fed nymphs perished, and that cucumber feeding prolonged nymphal survival only slightly long than ingesting water only. Vogt and Nechols (1993) reported that nymphal survival on green striped cushaw cultivar of *Cucurbita maxima* (winter squash), Waltham butternut cultivar of *C. moschata* (winter squash), and early prolific straightneck cultivar of *C. pepo* (summer squash) was 22, 17, and 49%, respectively. Field observations by Novero *et al.* (1962) indicated that butternut and sweet cheese cultivars of *C. moschata* were resistant to attack. Royal acorn and black zucchini cultivars of *C. pepo* were resistant and susceptible, respectively. Green striped cushaw and pink banana cultivars of *C. maxima* were intermediate in susceptibility to injury. In California, squash bug also caused injury to watermelon (Michelbacher *et al.*, 1955). Horned squash bug is reported to prefer wild cucurbits, but sometimes feeds on cultivated varieties (Slater and Baranowski, 1978); its host preferences among cultivated cucurbits is nearly identical to *A. tristis* (Chittenden, 1899). (See color figure 11.)

Natural Enemies. Several natural enemies of squash bug have been recorded, principally wasp egg parasitoids (Hymenoptera: Encyrtidae and Scelionidae). Up to 30% parasitism among eggs collected in Florida has been reported (Chittenden, 1899, 1908a; Tracy and Nechols, 1987, 1988; Nechols *et al.*, 1989; Vogt and Nechols, 1991, 1993). Chittenden (1908a) also mentioned a bacterial disease, but it must not be an important mortality factor because it was not mentioned by investigators in subsequent life-history studies. Cannibalism among nymphs is common, but this mortality factor has never been quantified.

The best known natural enemy is a common parasitoid of several hemipterans, *Trichopoda pennipes* (Fabricius) (Diptera: Tachinidae). Beard (1940) provided a detailed study of this parasitic fly, which also attacks the other *Anasa* spp. and some other coreids and pentatomids. The brightly colored adult fly is easy to recognize, having a gold and black thorax and an orange abdomen, with a prominent fringe of featherlike hairs on the outer side of the hind tibia. Flies develop principally in the adult bug, initially castrating the female, and then killing her when the fly emerges. In Connecticut, about 20% on the squash bugs were parasitized in late summer. In recent years, considerable research has been done on the relationship of *T. pennipes* and southern green stink bug, *Nezara viridula* (Linnaeus) (Hemiptera: Pentatomidae). (See the section on southern green stink bug for additional literature.)

Life Cycle and Description. A complete squash bug life cycle commonly requires from six to eight weeks. Squash bugs have one generation per year in the north, and two-to-three generations per year in southern locations such as Oklahoma (Fargo *et al.*, 1988; Palumbo *et al.*, 1991). In intermediate latitudes the early-emerging adults from the first generation produce a second generation, whereas the late-emerging adults go into diapause. In Kansas, the threshold for diapause induction is late July; adults emerging after this date exhibit progressively higher levels of diapause (Nechols, 1987). Both sexes overwinter as adults. The preferred overwintering site seems to be in cucurbit fields under crop debris, clods of soil, or stones, but occasionally adults also are found in adjacent wood piles or buildings.

Egg. Most squash bug eggs are deposited on the lower surface of leaves, though occasionally they occur on the upper surface or on leaf petioles. The elliptical egg is somewhat flattened on three sides, and initially is white or yellowish-brown, but its color soon changes to bronze. The average egg length is about 1.5 mm and the width about 1.1 mm (Chittenden, 1899). Females deposit about 20 eggs in each cluster, and the size of the cluster is not affected by type of host plant selected for oviposition (Bonjour *et al.*, 1990). Placement of single eggs or masses of up to 40 eggs is known. The eggs may be tightly clustered or spread to a considerable distance apart, but an equidistant spacing arrangement is commonly observed. The number of eggs produced by squash bug is greatly influenced by the quality of the host plant on which the female feeds. Bonjour *et al.* (1993) reported total fecundity per female as 889, 537, 240, and 72 for insects fed pumpkin, squash, watermelon, and muskmelon, respectively. Egg development time is about 7–9 days. (See color figure 262.)

Nymph. The five instars require about 33 days collectively for complete development. Bonjour and Fargo (1989) recorded average development times of about 3, 8, 6, 6, and 10 days, respectively, in Oklahoma. Ingestion of watermelon, a relatively unsuitable host plant, caused slightly longer development periods than pumpkin and squash. The nymph is about 2.5 mm long when it hatches, and light green in color. The second instar is initially about 3 mm long, and its color is light gray. The third, fourth, and fifth instars initially are about 4, 6–7, and 9–10 mm long, respectively, and darker gray. The youngest nymphs are rather hairy, but this decreases with each subsequent molt. In contrast, the thorax and wing pads are barely noticeable at hatching, but get more pronounced with each molt. Young nymphs are strongly gregarious—a behavior that dissipates slightly as the nymphs mature. (See color figure 135.)

Adult. The adult measures 1.4–1.6 cm long and is dark grayish brown. In many cases the edge of the abdomen is marked with alternate gold and brown patches. The adults are long-lived, and survive on an average of about 75–130 days, depending on availability and quality of food. A considerable portion of the adult period, about 10–20 days, occurs before oviposition (Bonjour *et al.*, 1993). Males live longer than females.

The biology of horned squash bug is inadequately known, but the observations that have been made (Chittenden, 1899) suggest a life cycle virtually identical to squash bug. Horned squash bug differs considerably in appearance, however. The nymphs of *A. armigera* are white until the fifth instar, when their body becomes variegated brown and yellow. Nymphs of *A. tristis*, in contrast, tend toward gray or brown. In the adult stage, *A. armigera* has a much broader prothorax, with sharp angles. The dorsal surface of the abdomen of *A. armigera* is orange, whereas in *A. tristis*, it is usually black (Chittenden, 1899). *A. armigera* also lacks the median-yellow stipe that is found on the head of *A. tristis* (Slater and Baranowski, 1978).

A good summary of squash bug biology was given by Beard (1940). Rearing techniques were provided by Woodson and Fargo (1991). A description and life history of *A. andresii* was provided by Jones (1916a).

Damage

This squash bug causes severe damage to cucurbits, because it secrets highly toxic saliva into the plant. The foliage is the primary site of feeding but the fruit is also fed upon. The foliage wilts, becomes blackened, and dies following feeding; this malady is sometimes called "anasa wilt." Often an entire plant or section of plant perishes while nearby plants remain healthy. The localized injury probably results from the gregarious behavior of the bugs. The amount of damage occurring on a plant is directly proportional to the density of squash bugs (Woodson and Fargo, 1991).

Management

Insecticides. Squash bug adults are unusually difficult to kill with insecticides. Although adult control can be accomplished if the correct material is selected, and it is advisable to target the more susceptible

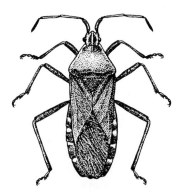

Adult squash bug.

nymphs. Squash bugs are not often considered a severe pest of large-scale cucurbit production, probably owing to the absence of suitable overwintering sites in well-managed crop fields and because the bug's effects are diluted by the vast acreage. Small fields and home gardens are most commonly damaged. The exception seems to be in Oklahoma, where two-to-three generations occur each year, and an additional generation allows attainment of higher bug densities and subsequent damage to commercial plantings. In Oklahoma, suppression of squash bug early in the year and maintenance at low densities are recommended (Fargo *et al.*, 1988).

Insecticides applied to the base of plants are especially effective, apparently because bugs tend to cluster there (W. Cranshaw, personal communication). Such applications may also minimize interference with pollinators. Pollinators, particularly honeybees, are very important in cucurbit production, and insecticide application can interfere with pollination by killing honeybees. If insecticides are to be applied when blossoms are present, it is advisable to use those with little residual activity, and to apply late in the day, when honeybee activity is minimal.

Cultural Practices. Adult squash bugs preferentially colonize larger, more mature plants. Thus, early-planted crops may be especially prone to attack. Numbers are also highest on plants during bloom and fruit set (Palumbo *et al.*, 1991). Use of early-planted crops as a trap crop has been proposed, but due to the high value of early-season fruit most growers try to get their main crop to mature as early as possible. The use of squash or pumpkin as a trap crop to protect less preferred host plants such as melons and cucumbers is reported to be effective (Chittenden, 1899). Trap crops have some benefit for squash bug control. Pair (1997) reported that small plantings of an attractive squash crop that were treated with systemic insecticide could suppress early-season populations on nearby cantaloupe, squash, and watermelon.

Plastic mulches provide good harborage for squash bugs, and densities are reported to be higher in the presence of white, black, and aluminized mulch (Cartwright *et al.*, 1990a). The tendency of squash bugs to aggregate in sheltered locations can be used to advantage by home gardeners. Smith (1910) recommended the placement of boards, large cabbage leaves, or other shelter for squash bugs, because they tend to congregate there during the day and are easily found and crushed. He also suggested hand picking, but the bugs are secretive and shy; they quickly hide if an observer approaches.

Row covers delayed colonization of squash, but bugs quickly invaded protected plantings in Oklahoma when covers were removed to allow pollination (Cartwright *et al.*, 1990a).

Removal of crop debris in a timely manner is very important. Squash bugs often are found feeding on old fruit or in abandoned plantings, so clean cultivation is essential to reduce the overwintering population.

FAMILY CYDNIDAE—BURROWER BUGS

Burrowing Bug
Pangaeus bilineatus (Say)
(Hemiptera: Cydnidae)

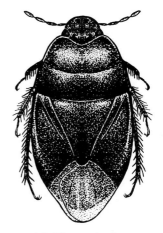

Adult burrowing bug.

Natural History

Distribution. This native species is found from New England and Quebec west to South Dakota and southern California, and south to Florida. It also occurs south through Mexico to Guatemala, and is known from Bermuda. It is the most common member of the family.

Host Plants. Burrowing bug is known as an occasional pest of pepper, spinach, strawberries, peanut, cotton, sugarcane and wheat. Crops are rarely reported to be damaged, though there are several records of injury to peanut in Texas; other cydnids are also implicated, with *P. bilineatus* the most numerous pest.

Natural Enemies. Few natural enemies are known from this insect. The strepsipteran *Triozocerca mexicana* Pierce parasitizes both nymphs and adults, and southern fire ant, *Solenopsis xyloni* McCook, feeds on eggs and young nymphs (Smith and Pitts, 1974).

Life Cycle and Description. This species is poorly known, undoubtedly due to its infrequent importance as a pest. In Texas, burrowing bugs overwinter as adults. They emerge from diapause and commence mating in March (Cole, 1988). Bugs enter crop fields in April, and deposit eggs singly in the soil. The bugs develop below-ground, with feeding injury caused by both adults and nymphs. Nymphs are present until autumn. Apparently there is a single generation per year. The bugs burrow into the soil to overwinter, and usually dig to a depth of at least 15 cm.

Pangaeus bilineatus is normally blackish, though sometimes reddish-brown. The adult is oval, and measures 5.2–7.8 mm long and about 3 mm wide. The legs are blackish-brown and the tibiae bear numerous stout spines. The front tibiae are modified for digging. The scutellum is large and triangular, extending over about one-half the length of the abdomen. Adults and nymphs release noxious chemical secretions when disturbed, a behavior that undoubtedly is a defensive response (Scheffrahn *et al.*, 1987). Although closely related to stink bugs (Pentatomidae), burrowing bug

is most likely confused with little negro bug, *Corimelaena pulicaria* (Germar) (Thyreocoridae), owing to the oval shape and blackish color.

Elements of biology were given by Gould (1931), Smith and Pitts (1974), and Cole (1988). Taxonomy was discussed, and bibliographic citations provided by Sailer (1954).

Damage

Burrowing bug damage is usually described as injury to young plants as the seedling emerges from seeds. The adults and nymphs feed below-ground, mostly on seeds and plant roots, resulting in plant damage or death. However, they also cause brown spots or "pitting" on developing peanut seeds (Smith and Pitts, 1974).

Management

Burrowing bug populations can be monitored with light traps that employ white light (Highland and Lummus, 1986), though there may not be a strong relationship between light trap catches and crop infestation levels. They also can be collected with pitfall traps (Huffman and Harding, 1980). In areas with a history of burrowing bug damage, insecticides can be applied to the soil to prevent injury to seedlings.

FAMILY LYGAEIDAE—SEED BUGS

Chinch Bug
Blissus leucopterus (Say)
(Hemiptera: Lygaeidae)

Natural History

Distribution. The chinch bug, *Blissus leucopterus*, is found through much of the eastern United States and southern Canada, west to about the Rocky Mountains. However, it is absent from the Gulf Coast region, where it is replaced by a closely related species, the southern chinch bug, *Blissus insularis* Barber. The two species have overlapping ranges in portions of the southern states from North and South Carolina through central Georgia and west to Texas. Southern chinch bug feeds only on lawn and forage grasses, particularly St Augustine grass, and is not a food-crop pest. Other species of *Blissus* occur in both eastern and western states but they are of little consequence. The *Blissus* spp. may have dispersed northwards from South America, but if it is so they apparently dispersed in pre-colonial times, as there is no record of their introduction.

The range of *B. leucopterus* can be subdivided because two discrete subspecies exist; *B. leucopterus hirtus* Montandon in the northeast, and *B. leucopterus leucopterus* (Say) in the central region of the eastern states. In eastern Canada, the New England states, and south to about northern Virginia and eastern Ohio, the northeastern form, *B. leucopterus hirtus*, is a pest of lawn grasses, but not of food crops. This subspecies is also called "hairy chinch bug" to distinguish it from the food crop-attacking subspecies, *B. leucopterus leucopterus*, which is known simply as "chinch bug." Chinch bug occurs from Virginia to Georgia in the east, extending to South Dakota and Texas in the west. Chinch bug generally is not damaging throughout its entire range, and is considered to be a pest mostly in the midwestern and southwestern states from Ohio in the east to South Dakota and Texas in the west.

Host Plants. Hosts of chinch bug consist solely of plants in the family Gramineae, but include both wild and cultivated grasses. It is known principally as a pest of such grain crops as barley, corn, millet, oat, rye, sorghum, and wheat, but oat is only marginally suitable. Among vegetables, only corn is damaged. However, it also damages forage grasses including sudangrass and timothy, and feeds on wild grasses such as foxtail, *Setaria* spp.; crabgrass, *Digitaria* spp.; and goosegrass, *Elusine indica* (Ahmad *et al.*, 1984).

Females select sorghum for oviposition over wheat and corn; barley is intermediate in preference. Crop suitability, as measured by development time, is similar to oviposition preference (M.T. Smith *et al.*, 1981).

Natural Enemies. Numerous natural enemies have been observed. Among avian predators feeding on chinch bugs are common birds as barn swallow, *Hirundo erythrogastra* Boddaert; horned lark, *Otocoris alpestris* (Linnaeus); meadowlark, *Sturnella magna* (Linnaeus); redwinged blackbird, *Agelaius phoeniceus* (Linnaeus); and kingbird, *Tyrannus tyrannus* (Linnaeus).

Insect predators of special importance are insidious flower bug, *Orius insidious* (Say) (Hemiptera: Anthocoridae), an assassin bug, *Pselliopus cinctus* (Fabricius) (Hemiptera:Reuviidae), and various ants (Formicidae). Lady beetles (Coleoptera: Coccinellidae) and lacewings (Neuroptera: Chrysopidae) frequently have been observed on plants infested with chinch bugs, but their effects are uncertain.

Parasitoids are sometimes found in these small insects, but rarely are they considered to be significant mortality factors. An egg parasite, *Eumicrosoma benefica* Gahan (Hymenoptera: Scelionidae), the nymphal or adult parasite *Phorocera occidentalis* (Walker) (Diptera: Tachinidae), and an unspecified, naturally occurring nematode have been reported. The egg parasitoid, which is found throughout most of the range of the chinch bug and is active during much of the season when chinch bug occurs, was reported to parasitize up to 46% of the eggs in Nebraska, so it may be of considerable value in biological control (Wright and Danielson, 1992). Also, chinch bugs are susceptible to infection by the entomopathogenic nematodes *Steinernema* spp. (Nematoda: Steinernematidae), but the dry environment inhabited by chinch bug is not very suitable for movement and infection by nematodes.

The most important natural mortality factor is fungal disease, particularly *Beauveria bassiana*. Interestingly, this fungus was intensively redistributed, particularly in Kansas, during the late 1880s in an effort to increase suppression. However, it eventually became apparent that the disease spread naturally, and that the effectiveness of the fungus was related more to weather than to the efforts of agriculturalists and entomologists to foster epizootics. The fungus is invasive and pathogenic at relative humidities of 30–100%, but fungal replication and conidia production require humidities of at least 75% (Ramoska, 1984). Clumps of such bunch grasses as little bluestem, *Andropogon scoparius*, serve to harbor not only overwintering bugs, but *Beauveria* as well (Krueger *et al.*, 1992), and may be important in initiating fungal epizootics. The food plant of the chinch bug affects susceptibility to *B. bassiana*, with a diet of corn and sorghum suppressing fungus development and bug mortality (Ramoska and Todd, 1985).

Weather. Weather has significant impact on chinch bugs. The overwintering period is moderately critical. The adults seek shelter in stubble and debris, but one of the most favorable locations is among the stems of bunch grasses. Bunch grasses provide food in the autumn before the onset of winter temperatures, and again in the spring before it is consistently warm and the bugs disperse. Bunch grasses also serve to break the wind by reducing desiccation and the severity of the wind, and by keeping excessive rainfall from the insects. Thus, in the absence of bunch grasses or similar shelter, survival can be poor. Heavy snow cover is favorable, keeping the bugs warmer and sheltered from the drying wind.

Summer weather is perhaps even more critical. Chinch bugs thrive in warm, dry conditions, at least in the midwestern states. Heavy rainfall can kill many bugs, and wet, humid weather fosters epizootics of fungal disease. In the southwest the situation is different because dry weather is usually assured, but the absence of summer rain causes premature senescence of plants, depriving bugs of green food late in the summer.

Life Cycle and Description. There are at least two generations per year throughout the range of *B. leucopterus leucopterus*. The first generation commences in the spring, with oviposition by overwintering adults, usually in April or May. The second generation begins in June–August. Second generation adults overwinter, often in the shelter of clump-forming wild grasses, dispersing in the spring to early-season crops, and then in early summer to later-developing crops where the second generation develops. Generations overlap considerably due to prolonged oviposition, and in the southwestern states there is some evidence of a third generation. A complete life cycle can occur in about 30–60 days.

Egg. The elongate-oval eggs are rounded at one end, truncate at the other, and measure about 0.85 mm long and 0.31 mm wide. The truncate end bears 3–5 minute tubercles, 0.1 mm long. The eggs are whitish initially, turning yellowish-brown after a few days and reddish before hatching. The eggs are deposited in short rows at the base of the plant on roots, on the lower leaf sheaths and stems, and on the soil near the plant. Females deposit eggs at a rate of 15–20 per day over a two- to three-week period, producing up to 500 eggs. Duration of the egg stage is about 16 days at 27°C and eight days at 31°C.

Nymph. There are five instars. Duration of the instars is about 5, 6, 5, 4, and 6 days for instars 1–5, respectively, when reared at 29°C. Under field conditions the development time may be extended, with a period of about 30–40 days considered normal, and 60 days not unusual. During the early instars the head and thorax are brown, and the legs are pale. These structures become darker as the nymphs mature, so the mature nymph is blackish. The first two segments of the abdomen are yellowish or whitish, the remainder is red except for the tip of the abdomen, which is black. The reddish abdomen becomes progressively

Chinch bug nymph.

darker, however, appearing almost black at nymphal maturity. The wing pads become visible in the third instar, but are difficult to discern. In the fourth instar the wing pads extend about half the width of the first abdominal segment, whereas in the fifth instar they extend to the third abdominal segment. The nymphal body lengths are about 0.9, 1.3, 1.6, 2.1, and 2.9 mm for instars 1–5, respectively. Nymphs prefer to feed in sheltered locations such as curled leaves and on roots, but are often found aggregated on the stem near the base of the plants. When not feeding they may hide under clods of soil and rubbish, or in loose soil. (See color figure 142.)

Adult. The body and legs of the adult are blackish. The wings of *B. leucopterus leucopterus* nearly attain the tip of the abdomen, and are white in color with a pronounced blackish spot found near the center and outer margin of the forewings. The adult measures 3.5–4.5 mm long. The black-spotted white wings serve to distinguish the chinch bug from similar crop-damaging species such as false chinch bug, *Nysius* spp. In the related form of hairy chinch bug, the wings generally are abbreviated, usually not extending beyond the middle of the abdomen.

There are numerous publications on chinch bugs, though many of the early works treated *B. leucopterus leucopterus*, *B. leucopterus hirtus*, and *B. insularis* as synonymous. A bibliography was published by Spike *et al.* (1994). Perhaps the most useful publications on biology were Webster (1907) and Luginbill (1922b), though Swenk (1925) presented a very good synopsis. Leonard (1966, 1968) revised the eastern *Blissus*, and provided descriptions and keys. Culture techniques were described by Parker and Randolph (1972), and Ramoska and Todd (1985).

Damage

Chinch bug is a plant sap-feeding insect, causing a reddish discoloration at the site of feeding and death

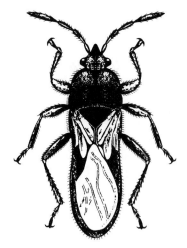

Adult chinch bug.

of that portion of the plant. Plant growth can be stunted, or plants killed when fed upon by numerous bugs. Their destructiveness is attributable, in part, to their gregarious nature. Not only do many bugs aggregate on certain plants, but they also disperse in tremendous numbers from field to field. Bugs at densities of 10–20 per young plant reduce growth rate of corn (Negron and Riley, 1985). When plants are infested while young they suffer more damage than if infested later in growth (Negron and Riley, 1990). It rarely is reported as a sweet corn pest, however.

Management

Insecticides. Granular and liquid insecticides are used to protect plants, particularly fields that are invaded by nymphs or adults dispersing from senescent early-season crops. Systemic insecticides can be applied at planting time, and either contact or systemic materials are used after the crop has emerged from the soil (Peters, 1983; Wilde *et al.*, 1986). Liquid insecticides should be directed to the base of the plants—a location favored by the insects.

Cultural Practices. Historically, most damage occurred when early-season small-grain crops were matured, and large numbers of first generation nymphs dispersed by walking to adjacent crops, usually corn. This was combated by erecting barriers, usually in the form of a ditch, between crops. Also, some destruction of overwintering bugs in wild grasses was accomplished by burning, though it was only 50% effective. These practices are largely obsolete, and insecticides are now used effectively. However, it is advisable to rotate among susceptible and nonsusceptible crops, and to grow susceptible crops in isolation from alternate hosts. Because the combination of small grains and corn leads to damage by chinch bugs, it may be desirable to eliminate one of these crops and thereby

eliminate an important foods from the chinch bug life cycle. In southern states the crop sequence is different, with both wheat and corn invaded by overwintering bugs early in the season (Negron and Riley, 1991).

Cultural practices that promote dense growth and shade will increase humidity and decrease chinch bug numbers. Thus, fertilization and irrigation can be detrimental to chinch bug survival.

Host-Plant Resistance. There is considerable difference among crops in susceptibility to injury, and within crops the level of resistance also is variable. For example, Stuart *et al.* (1985) examined resistance among wheat cultivars, though not because the bugs are a great problem in wheat but because they disperse from wheat to sorghum, where damage is more likely. They reported measurable levels of difference in longevity, development, and reproduction of bugs fed different varieties, but concluded that there was not a practical level of resistance. In contrast, Wilde and Morgan (1978) identified a resistant variety of sorghum, and Flint *et al.* (1935) reported considerable resistance in grain corn.

False Chinch Bug
Nysius niger Baker
Nysius raphanus Howard
(Hemiptera: Lygaeidae)

Natural History

Distribution. The taxonomy of these native insects is not completely resolved, and apparently more than one species is known as false chinch bug. The principal species known as false chinch bug, *Nysius niger* Baker, occurs throughout most of the United States and southern Canada. Another species, *Nysius raphanus* Howard, sometimes is common. Most reports in the literature attribute crop damage to *Nysius ericae* (Schilling), a species found only in Europe, or *N. raphanus*, with which *N. niger* is easily confused; *N. niger* is a major pest, despite the confusing reports (Slater and Baranowski, 1978; personal communication with R. M. B). Most reports of damage occur in the southern regions of the United States, particularly the arid southwest and Rocky Mountain regions. However, on occasion false chinch bug can reach high levels of abundance almost anywhere in its range.

Host Plants. False chinch bug has been reported to damage such vegetables as cabbage, carrot, celery, lettuce, mustard, radish, and turnip. Other crops injured include alfalfa, cotton, flax, oats, raspberry, strawberry, sugarbeet, sunflower, tobacco, and wheat. Despite its relatively wide host range, grain crops are not preferred, and damage is infrequent on any crop.

False chinch bug feeds principally on weeds, particularly mustards, and moves to crops when drought or herbivory destroy the preferred hosts. Numerous common weeds are suitable for nymphal development, including Virginia pepperweed, *Lepidium virginicum*; shepherdspurse, *Capsella bursa-pastoris*; Russian thistle, *Salsola kali*; spurge, *Euphorbia* spp.; and sagebrush, *Artemisia tridentata*. *Sisymbrium irio*, a mustard known as London rocket, is especially important in California. Some plants such as carpetweed, *Mollugo verticillata*; lovegrass, *Eragrostis* sp.; and blanketflower, *Gaillardia pulchella*, are reported to be selected as oviposition sites only during periods of extreme drought (Milliken, 1916).

Natural Enemies. This insect seems to be relatively free from natural enemies, though heavy rain is an important mortality factor.

Life Cycle and Description. The number of generations per year varies with its location and weather. One generation can be completed in about 29 days under favorable weather conditions. In Kansas, five generations are reported annually (Milliken, 1916), whereas 3–5 are observed in Canada (Beirne, 1972). The overwintering stage is variously reported to be the egg, nymph, and adult (Byers, 1973). Burgess and Weegar (1986) reported that eggs could be stored under refrigeration for up to five months, so it is possible that all stages overwinter successfully.

Egg. The elongate egg is about 1.5 mm long and 0.4 mm wide. It tapers toward both ends. One side is convex, and the opposite side nearly straight. Its color is pinkish white. They may be deposited in loose soil, among bits of organic debris, or among plant parts such as blossoms. The eggs may be laid singly or in small clusters of up to eight. Females produce, on average, one egg per day. Duration of the egg stage is reported to average eight days (range 6–12 days).

Nymph. There generally are five instars, though instar number has been observed to vary from four to six. The nymphal stages initially are pinkish white, but as they grow they darken to light-brown. Thin, dark longitudinal lines run the length of the body. The body is rather pear-shaped, with the abdomen considerably wider than the thorax. Nymphal development requires about 4, 4, 4, 4, and 5 days, respectively, for instars 1–5 (Milliken, 1918). Thus, total duration of the nymphal stage is about 21 days (range 16–27 days) (Burgess and Weegar, 1986).

Adult. The adult is about 3–4 mm long. Because of its general appearance and gregarious habits, it is confused with chinch bug, *Blissus leucopterus* (Hemiptera:

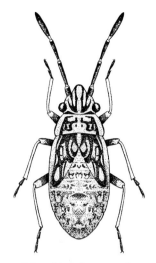

False chinch bug nymph.

Lygaeidae). However, in general color the false chinch bug is grayish-brown to blackish brown, bearing transparent wings marked with rows of small spots. In contrast, chinch bug is generally black, and marked with a large dark spot on each of the front wings. In food habits they also differs markedly. Chinch bug prefers grain and grass crops, but false chinch bug prefers broadleafed crops. The adult false chinch bugs commonly aggregate, both for feeding (Milliken, 1916) and mating (Byers, 1973). They begin egg production about 16 days after reaching maturity, and may survive for up to seven weeks. Mass dispersal is often observed, probably in search of food.

A description of false chinch bug was given by Riley (1873). The major observations on biology were reported by Milliken (1916, 1918). Rearing techniques were developed by Burgess and Weegar (1986). False chinch bug was included in the key by Slater and Baranowski (1978).

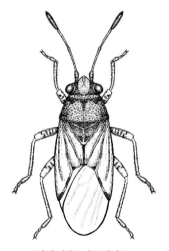

Adult false chinch bug.

Damage

These weed-feeding insects cause injury when they disperse, often in tremendous numbers, to nearby crop plants. Wene (1958), for example, described population densities of 5 per sq cm on vegetables in Texas. Because of their small size they often escape notice, and it is not until plants wilt that the insects are observed. They also move on and off the plant in response to weather, often seeking shelter beneath clods of soil during the heat of the day and at night (Leigh, 1961). Margins of fields are especially affected. The rapid wilting and death of plant tissue, especially following feeding by nymphs, caused Barnes (1970) to suggest that a toxin might be present in the saliva of this bug; this has not been investigated. Tappan (1970) described injury on several plants as minute necrotic lesions surrounded by chlorotic tissue. False chinch bug feeding on seed disrupts seed formation and inhibits seed germination (Wood and Starks, 1972). This insect also is capable of transmitting *Nematospora coryli*, a yeast pathogen of mustard (Burgess *et al.*, 1983).

Management

Sampling. Careful visual inspection of weeds, particularly cruciferous species, surrounding fields is suggested during periods of drought. During the warmest portion of the day it is advisable to examine sheltered areas, such as lower surfaces of foliage and beneath soil clods, for aggregations of bugs. False chinch bug tends to be highly aggregated in distribution, so systematic inspection is required. A sampling technology has been developed in Canada for adult *Nysius niger* that probably works for *N. raphanus*, also. Pivnick *et al.* (1991) successfully tested a yellow trap that used mustard oils (isothiocyanates) as a lure.

Insecticides. False chinch bug is controlled with foliar applications of insecticides. False chinch bug is not generally considered to be difficult to control, but not all insecticides work equally well. Owing to the irregular appearance of the insect, preventative applications should be made only if high densities be found on nearby weeds, and an ideal scenario would be to control the insects on the weeds before they disperse. Dispersal into crops is more likely to occur during periods of drought. Invasion of crops can sometimes be stopped with border treatments only, as the bugs usually disperse by walking rather than flying.

Cultural Practices. Weed management is especially critical for this insect. For example, in California large populations build up on London rocket during the spring. Nymphs and adults disperse to crops when this mustard is depleted by feeding, or the weeds are tilled. To avoid crop damage, weeds could be destroyed before insect populations develop to high levels, or perhaps the weeds could be killed at a time when crops susceptible to injury were not available (Barnes, 1970).

FAMILY MIRIDAE—PLANT BUGS

Alfalfa Plant Bug
Adelphocoris lineolatus (Goeze)
(Hemiptera: Miridae)

Natural History

Distribution. Alfalfa plant bug is native to Europe, and was first observed in North America in 1917 at Cape Breton Island, Nova Scotia, but it failed to spread. It then was accidentally imported to the United States with alfalfa, and introduced near Ames, Iowa in 1929. Thereafter, alfalfa plant bug spread across the United States and Canada, and was observed Minnesota in 1934; Illinois, Missouri and South Dakota in 1935; Nebraska and Wisconsin in 1936; Kansas and Manitoba in 1939; and Saskatchewan in 1947. It now is found in the northern United States and southern Canada from the Rocky Mountains to eastern Canada and the northeastern states. The southern limits to its range currently are Kansas in the midwest, and Virginia in the eastern states.

Host Plants. As suggested by its common name, this insect feeds principally on such forage legumes as alfalfa, birdsfoot trefoil, red clover, sainfoin, and sweet clover. However, it is an occasional pest of vegetables, including asparagus and potato. Often the occurrence of alfalfa plant bug in vegetable crops can be traced to nearby forage legumes or weeds which were harvested or otherwise became unattractive to the bugs. Hughes (1943) reported that alfalfa plant bugs normally deposited eggs only in alfalfa and sweet clover. However, in the absence of such plants, successful oviposition occurred in red clover; timothy; soybean; bindweed, Convolvulus sp.; and redroot pigweed, *Amaranthus retroflexus*.

Natural Enemies. A common parasitoid of mirids, *Peristenus pallipes* (Curtis) (Hymenoptera: Braconidae), is recorded from alfalfa plant bug. Day (1987) reported an undetermined species of mermithid nematode. Other parasitoids are known in Europe, and a program to import beneficial insects has been implemented. One imported species, *Peristenus digoneutis* Loan (Hymenoptera: Braconidae), has been succcessfully established in New Jersey (Day 1996). A key for identification of parasitoids affecting *Adelphocoris* was provided by Loan and Shaw (1987).

Life Cycle and Description. Two generations occur annually in Minnesota and, apparently, over most of this insect's range in North America. Total development time from the egg to the adult stage is estimated to require 33.5 days. Another two weeks are usually required before eggs are produced, so total generation time is about six weeks. The eggs are the overwintering stage. Hatching first occurs in May, and by mid-June first generation adults are present. Nymphs of the second generation can be observed in July, and adults in August. By mid-August overwintering eggs are produced. The two generations may overlap considerably, and in cool summers the development of the second generation is delayed so that second generation adults are not produced until September. All stages except the eggs are killed with the onset of cold weather.

In contrast, there is principally one generation per year in northern Saskatchewan. Most of the eggs laid by the spring generation enter diapause; only a few, probably less than 5%, develop into nymphs. A few second generation adults are produced, but there is not adequate time before the onset of cold weather for egg production to occur (Craig, 1963).

Egg. The eggs typically are deposited in alfalfa stems. During the first generation, they are usually deposited 25 cm or so above the soil surface, but the overwintering eggs are deposited in the less succulent alfalfa stems close to the soil surface. The eggs of alfalfa plant bug are about 1.38 mm long and 0.33 mm wide. They are slightly curved in shape, with one side concave and the opposite side convex, and are yellow in color. They are laid closely together in small groups, and about 10–30 eggs are deposited daily. The tip of the eggs protrudes slightly from the oviposition site. The incubation period of the eggs is 11–17 days, averaging 15 days.

Nymph. The nymphs are light green, and all except the first instar bear black spots on the femora. There are five instars. Wing pads are not evident during the first two instars, but in the third instar they become visible. The wing pads increase in length at each of the next molts, attaining the second abdominal segment in instar four, and the fourth abdominal segment in instar five. Mean development time (range) was reported as 4.9 (4–7), 2.9 (1–8), 2.5 (2–5), 3.2 (2–4), 4.9 (4–6) days for instars 1–5, respectively. These values can be prolonged significantly, perhaps doubled, during periods of inclement weather. Total nymphal development time is usually 18–20 days, but sometimes prolonged to 30 days. (see color figure 141).

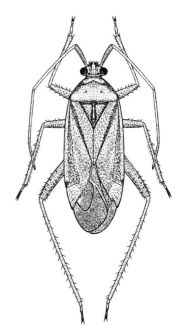

Adult alfalfa plant bug.

Adult. The adults are yellow-green, and marked with brown. This color pattern serves to distinguish them from the other common *Adelphocoris* species: rapid plant bug, *A. rapidus* (Say); and superb plant bug, *A. superbus* (Uhler). These latter species are principally dark-brown. The general body form (viewed from above) of alfalfa plant bug and other *Adelphocoris* spp. is fairly elongate, with the overall length nearly three times the maximum width. The *Lygus* spp., the other common plant bugs, tend to be much broader in general body form, only about twice as long as wide. Alfalfa plant bug is also quite large in size, usually measuring 7.4–10.0 mm long, with a mean length of about 8.0 mm. The antennae are yellowish-brown basally and reddish-brown distally. The pronotum bears two small dark spots. The edges of the scutellum are brown. The legs are yellowish-brown and the femora marked with black spots.

The biology of alfalfa plant bug was given by Hughes (1943) and Craig (1963).

Damage

Alfalfa plant bug removes the sap from plants, causing blossom and seed abortion, local lesions, tip die-back, and plant growth distortions. On alfalfa, the symptoms are similar to those induced by tarnished plant bug, *Lygus lineolaris* (Palisot de Beauvois), but not as severe (Grafius and Morrow, 1982), and also resemble the yellowing induced by potato leafhopper, *Empoasca fabae* (Harris) (Radcliffe and Barnes, 1970). It is principally a seed feeder, however, and is most damaging to alfalfa grown for seed.

Management

Sampling. Dispersion patterns and sampling protocols for alfalfa plant bug were studied in birdsfoot trefoil by Wipfli *et al.* (1992). The population of both nymphs and adults tends to be aggregated. The number of sweep net samples necessary to accurately determine population density varies from about 5 sets of 10 sweeps, to 20 sets of 10 sweeps, depending on bug density. Males of alfalfa plant bug have been shown to be attracted to traps baited with virgin females of tarnished plant bug, suggesting that a sex pheromone is involved in mate location (Slaymaker and Tugwell, 1984).

Insecticides. Foliar applications of chemical insecticides are used to suppress alfalfa plant bug. Conventional and ultra low-volume applications are effective, but populations can recover quickly. Recovery is due both to reinvasion of treated fields, and hatching of eggs developing in plant stems, a relatively protected location (Pruess, 1974). Thus, careful timing of application in advance of egg deposition, or use of residual insecticides, may be desirable.

Cultural Practices. Harvesting of alfalfa hay can induce relocation of plant bugs, especially if the population consists mostly of adults. Harvesting before development of adults greatly diminishes survival of plant bugs (Harper *et al.*, 1990). Vegetable crops located adjacent to alfalfa are susceptible to invasion by dispersing bugs, but the immigration rate is diminished by such barriers as roads and irrigation ditches (Schraber *et al.*, 1990).

Carrot Plant Bug
Orthops scutellatus Uhler
(Hemiptera: Miridae)

Natural History

Distribution. Carrot plant bug is native to Europe, where it is widely distributed. In North America, it was first discovered in New York in 1917, and since then it has spread throughout southern Canada and the northernmost states of the United States. The exception is the Rocky Mountain region, where this insect has spread southward, at high elevations, as far south as Arizona. This is a fairly common pattern of distribution for cold-adapted insects.

Host Plants. Carrot plant bug is restricted to the plant family Umbelliferae. Vegetable crops attacked are carrot, celery, parsnip, and parsley. It also feeds on dill, and on many wild umbelliferous plants such as poison hemlock, *Conium maculatum*; wild carrot, *Daucus carota*; and wild parsnip, *Pastinaca sativa*. In Massachusetts, this plant bug was reported to prefer celery over other plants. Unlike tarnished plant bug, *Lygus lineolaris* (Palisot de Beauvois), with which it is often associated, it prefers crops over weeds.

Natural Enemies. Natural enemies are not well-known. General predators such as lady beetles (Coleoptera: Coccinellidae), and likely many other insects, prey on the nymphs, but this has not been established.

Life Cycle and Description. There are three generations annually in Massachusetts. Duration of the life cycle, from egg to adult, is 6–7 weeks. This insect overwinters in the adult stage, hidden among weeds and tall grass. Adults emerge from overwintering in May, and egg laying occurs in early June, mid-July, and early September.

Egg. The eggs are deposited in the buds and other young tissue. They typically hatch in 4–10 days. The number of eggs produced per female is not known to exceed 48, but this seems to be inordinately low, and likely represents the consequences of feeding on an unsatisfactory diet.

Nymph. Young nymphs are pale green. There are five instars. Duration of the nymphal stage requires about 32–34 days, with about 6.7, 3.5, 5.0, 5.3, and 12.3 days required for instars 1–5, respectively.

Adult. The adult bug is quite distinctive in color, and readily separated from tarnished plant bug, *Lygus lineolaris*, with which it commonly co-occurs. Carrot plant bug is light-brown, but marked with dark-brown bands crossing the front wings (hemelytra). The scutellum is yellow, and surrounded by dark-brown. The legs are yellow basally, but become brownish to black distally.

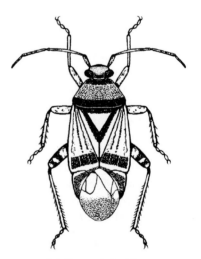

Adult carrot plant bug.

The only biological research conducted on this insect in North America was given by Whitcomb (1953). Henry and Froeschner (1988) gave a good account of geographic distribution.

Damage

This is a common plant bug on umbellifers, but damage is often trivial, or limited to seed crops. It is not considered to be a pest of commercial crops in North America, though in the past it did cause appreciable damage to celery in Massachusetts. Adults and nymphs feed on sap from leaves, buds, and seeds. In Massachusetts, carrot plant bug was shown to damage celery by feeding on the stalks; a darkening and destruction of the tissue, followed by invasion of plant pathogens, was associated with feeding by these plant bugs. Stunting of the plants also was reported.

Management

There generally is little need for control of this insect. Only when it first gained entry to North America was it observed to attain pest status. However, Whitcomb (1953) considered that population densities as low as one bug per 10 celery plants could produce damage. Carrot plant bug is readily suppressed with foliar insecticides, including botanicals.

Garden Fleahopper
Halticus Bractatus (Say)
(Hemiptera: Miridae)

Natural History

Distribution. Garden fleahopper is a native insect occurring widely in the eastern United States and Canada. It is known to occur as far west as the Rocky Mountain region, but it is relatively uncommon in the Great Plains. Its distribution extends southward through Central and South America to Argentina.

Host Plants. This fleahopper has a known host range that exceeds 40 different plants, but leguminous crops are most frequently damaged. Vegetable crops that host garden fleahopper include bean, beet, cabbage, celery, cowpea, cucumber, eggplant, lettuce, pea, pepper, potato, pumpkin, squash, sweet potato, and tomato. Field crops injured include alfalfa, clover, and sweet clover. Some of the common weeds that support garden fleahopper are beggartick, *Bidens* sp.; bindweed, *Convolvulus* spp.; burdock, *Arctium minus*; mallow, *Malva* spp.; pigweed, *Amaranthus* spp.; plantain, *Plantago* spp.; ragweed, *Ambrosia* spp.; smart-weed, *Polygonum* spp.; prickly lettuce, *Lactuca serriola*; thistle, *Carduus* spp.; and wood sorrel, *Oxalis stricta*.

Mirids often are facultative predators, occasionally feeding on small insects or eggs. However, garden fleahopper has only infrequently been observed in this role (Day and Saunders, 1990). (See color figure 16.)

Natural Enemies. Natural control has been poorly studied, but parasitic wasps can inflict high rates of mortality. In the United States, 50% mortality caused by *Leiophron uniformis* (Gahan) (Hymenoptera: Braconidae) has been observed, and this parasitoid was believed to be regulating bug densities in New Jersey. *Leiophron uniformis* attacks principally the nymphal stages. In Canada, *Peristenus clematidis* Loan (Hymenoptera: Braconidae) is known to parasitize garden fleahopper. An unidentified nematode and predatory mite also have been detected in the United States (Beyer, 1921; Day and Saunders, 1990). Beyer also reported several egg parasitoids, including *Anaphes perdubius* Girault and *Anagrus* sp. (both Hymenoptera: Mymaridae), and *Oligosita americana* Girault and *Paracentrobia subflava* Howard (both Hymenoptera: Trichogrammatidae), but gave no data on their effectiveness.

Life Cycle and Description. There appear to be five generations annually in Virginia, though there is considerable overlap among the generations, and all stages can be found through the warmer months. A life cycle can be completed in about 30 days. Overwintering occurs in the egg stage, with hatch of overwintering eggs occurring in April. In Florida, fleahoppers are present earlier in spring, and it has not been determined whether eggs hatch earlier or whether adults are the predominant overwintering form. However, adults have been captured over all months of the year except December, so overwintering of eggs is not essential under Florida's warm winter conditions. Temperatures above 32°C are reported to be unsuitable for fleahopper survival.

Egg. The eggs are normally inserted into the stems of vegetation. They are white to yellow, and measure about 0.7 mm long and 0.2 mm wide. The egg is curved in shape, with one side convex and the opposite side concave. The female deposits the eggs in feeding punctures. The end inserted into the plant tissue is rounded, whereas the end that is flush with the plant surface is truncate and level with the surface of the plant tissue. The female commences egg production about four days after mating, and deposits most of them during the evening hours. She produces 80–100 eggs during her life span, which averages about 30–50 days. Duration of the egg stage is about 14 days (range about 10–30 days).

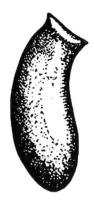

Garden fleahopper egg.

Nymph. The nymphs are green. Initially they are pale-green, but by the fifth and final instar they become dark-green. First and second instars are difficult to distinguish, differing principally in size. In the third instar, however, a black dot is present on the sides of the prothoracic segment; the black spots persist through the remaining nymphal stages. In the fourth instar, wing pads are apparent, and extend back over the first abdominal segment. During the last instar the wing pads extend over about half of the abdominal segments. In females which are brachypterous (short-winged) as adults, the wings pads in the fifth instar tend to be slightly less than one-half the length of the abdomen. In males, or females which are macropterous (long-winged) as adults, the wing pads in the fifth instar tend to be slightly longer than one-half the length of the abdomen. The mean duration (range) of each instar is about 8.3 (3–16), 9.9 (5–15), 7.8 (4–14), 5.8 (3–9), and 7.2 (5–12) days, respectively. Body length of each instar is about 0.7, 0.8, 1.0, 1.2, and 2.0 mm, respectively.

Adult. The adults are shiny black, with some yellow on the antennae and legs. Garden fleahopper adults occur in three forms: brachypterous (short-winged) females, macropterous (long-winged) females, and macropterous males. The males are thin, measuring about 1.9–2.1 mm long and 0.7 mm wide. The females are more robust, the brachypterous form measuring about 1.6 mm long and the macropterous about 2.2 mm long, with both forms about 1.0 mm wide. Females generally are brachypterous. The adults of all forms have greatly expanded hind femora and hop when disturbed. Thus, in both behavior and form (especially the brachypterous females) the fleahoppers resemble flea beetles. The long antennae of fleahoppers, which exceed the length of the body, help to distinguish fleahoppers from flea beetles, which have antennae less than half the length of the body. Fleahoppers and flea beetles also differ in their mouthpart configuration, having piercing-sucking and chewing mouthparts, respectively. Females normally mate within a day of attaining the adult state. (See color figures 138 and 139.)

The biology of garden fleahopper was given by Beyer (1921) and Cagle and Jackson (1947). Keys to plant bug genera can be found in Slater and Baranowski (1978) and keys to the species of *Halticus* in Henry (1983).

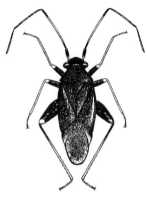

Male garden fleahopper.

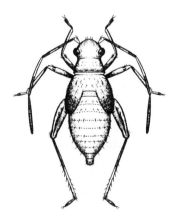

Garden fleahopper nymph.

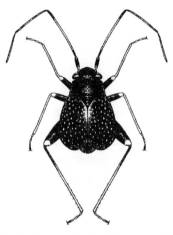

Garden fleahopper female, short-winged form.

Damage

Nymphs and adults frequent the stems and both the lower and upper surfaces of plant leaves, sucking the sap from individual cells and and causing their death. The result is a whitish or yellowish speckling on the foliage. Extensive feeding may cause stunting of plant growth and death of seedlings. Deposition of fecal material on the plant by both nymphs and adults also detracts from the appearance and marketability of vegetables.

This insect is rarely a pest of commercial vegetable crops because it is easily controlled with insecticides. However, it is commonly an early season nuisance in home gardens, especially those grown near wooded or in shaded areas of gardens.

Management

Suppression, when necessary, is easily accomplished with insecticides. Because fleahoppers commonly frequent weeds, where they may attain great abundance, such host plants should be monitored and sprayed or destroyed, if necessary. Also, garden fleahoppers sometimes build to high number in legume crops such as alfalfa and clover, so nearby vegetables may be at risk when the legume forage crops are harvested.

Rapid Plant Bug
Adelphocoris rapidus (Say)

Superb Plant Bug
Adelphocoris superbus (Uhler)
(Hemiptera: Miridae)

Natural History

Distribution.   Rapid plant bug occurs in the United States and Canada principally east of the Great Plains. In western North America it is replaced by superb plant bug, but the two species overlap in the Rocky Mountain region and Great Plains. Both species are native, and there is some thought that they are variants of the same species (Slater and Baranowski, 1978).

Host Plants.   These insects are known principally as pests of alfalfa and sweet clover, where they can attain high levels of abundance. Sometimes they move to vegetables, including bean, carrot, celery, and potato. They are also known from field crops such as cotton and sugarbeet, and such weeds as Canada thistle, *Cirsium arvense*; dock, *Rumex* sp.; and yarrow, *Achillea* sp.

Natural Enemies.   The little is known about the natural enemies of these insects. In Ontario, the parasitoid *Peristenus pallipes* (Curtis) (Hymenoptera: Braconidae) was recovered from up to 60% of first generation rapid plant bug nymphs but none from the second generation (Loan, 1965). Some general predators such as the assassin bug *Sinea diadema* (Fabricius) and the ambush bug, *Phymata fasciata* (Grey) were thought to affect rapid plant bug in Alberta (Lilly and Hobbs, 1956).

Description and Life Cycle.   There are two generations per year in northern areas of the United States. In Canada, there are two generations in southern regions, but only one further north. Overwintering occurs in the egg stage. In Minnesota, eggs of rapid plant bug hatch in May, and by mid-June first generation adults are numerous. Nymphs of the second generation are present in early July, and adults by early August. The second generation adults were observed to deposit eggs in August and September which overwintered. In Utah, superb plant bug exhibits a very similar phenology (Sorenson and Cutler, 1954). In contrast, in a single generation per year population in Alberta, first instars were most abundant in late May, instars 2–3 in mid-June, instars 4–5 in July, and adults from late July–September. Any surviving nymphs and adults die with the onset of winter. The time required for a complete generation is estimated about 55 days under warm conditions.

Egg.   The eggs are deposited singly in stems of vegetation. The eggs are greenish white or yellow, tend to become reddish near maturity, and measure about 1.2–1.4 mm long, and about 0.3 mm wide. The cylindrical, slightly curved eggs greatly resemble eggs of *Lygus* spp. (see section on tarnished plant bugs and pale legume bug), but in rapid plant bug the portion of the egg that reaches the surface of the plant, or cap, bears a distinct extension or spine. In superb plant bug, authors who have described eggs (Sorenson and Cutler, 1954; Lilly and Hobbs, 1956) failed to note or illustrate the egg spine or extension; perhaps this indicates that these insects were indeed separate species. Tarnished plant bug and pale legume bug also lack the spine. Incubation of the eggs stage requires about 15–20 days during the summer. Sorenson and Cutler (1954) reported fecundity averaging about 35 and sometimes attaining a level of 100 eggs per female. The larger number is likely more typical in nature.

Nymph.   There are five instars, the mean duration (range) of which is 3.8 (2–6), 2.4 (1–5), 1.8 (1–3), 3.0 (2–4), and 4.3 (3–5) days, for a total nymphal development time of about 15.4 days. The nymphs are brightly colored. They are principally yellow during instar one, but in later instars they are mostly red, with green on the ventral surfaces. In mature nymphs the head is dark. The abdomen bears a broad reddish-brown band near the tip. The antennae bear alternating light and dark bands, usually of red and yellow. The thorax and abdomen lack the dark spots found dorsally on the *Lygus* spp.

Adult. The adult measures 6.8–7.4 mm long. The head, legs, and prothorax of the adult plant bugs are yellow to red. In rapid plant bug the pronotum is yellow or yellowish-brown, though usually there are two large black spots near the hind margin of the pronotum. In superb plant bug, the pronotum is red, and this structure is seldom marked with dark spots. The hemelytra are brown except for the lateral edges, which are bordered narrowly with yellow in rapid plant bug, and red in superb plant bug. The antennae are black, and broadly ringed with yellowish-white. The adults are active and fly readily when disturbed. However, they rarely fly far or high, so dispersal is not rapid.

The biology of rapid plant bug was described by Hughes (1943), and nymphal morphology was provided by Webster and Stoner (1914). The biology of superb plant bug was given by Sorenson and Cutler (1954) and Lilly and Hobbs (1956).

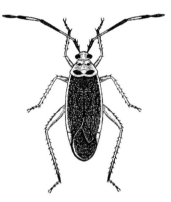

Adult rapid plant bug.

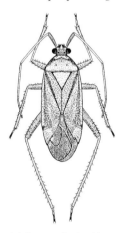

Adult superb plant bug.

Damage

These plant bugs are rarely considered to be significant pests, as they do not normally colonize vegetable crop plants. Damage usually occurs when the bugs disperse from a favored crop or weeds, often due to senescence or harvest. *Adelphocoris* plant bugs inflict damage on seeds and young tissue in much the same manner as *Lygus* spp. (see section on tarnished plant bugs and pale legume bug).

Management

Because these plant bugs are not often a pest, insecticidal control is rarely warranted. Foliar sprays are effective, if necessary. Effective management of alfalfa can do much to alleviate problems with these plant bugs. As they overwinter in alfalfa stems, close cropping of alfalfa in the autumn and feeding the hay to livestock reduces the overwintering population. Burning of stubble in the spring is sometimes recommended, but this practice also destroys predator populations (Lilly and Hobbs, 1962).

Tarnished Plant Bug
Lygus lineolaris (Palisot de Beauvois)

Western Tarnished Plant Bug
Lygus hesperus Knight

Pale Legume Bug
Lygus elisus Van Duzee
(Hemiptera: Miridae)

Natural History

Distribution. These insects are native to North America. Both in appearance and biology, these species are quite similar, and have long been plagued with taxonomic problems. Therefore, they are best considered as a group.

Tarnished plant bug, *Lygus lineolaris* (Palisot de Beauvois) is widely distributed and is considered damaging wherever it occurs. It is found throughout the United States, including Alaska. In Canada, it is found from Newfoundland to British Columbia, including the Yukon.

Western tarnished plant bug, *Lygus hesperus* Knight, and pale legume bug, *Lygus elisus* Van Duzee, are more common in western North America. Western tarnished plant bug is found only as far east as Alberta in Canada, and Wyoming and western Texas in the United States. Pale legume bug occurs as far east as Manitoba in Canada, and Illinois in the United States. The western tarnished plant bug and pale legume bug are considered to be damaging principally west of the Rocky Mountains.

Host Plants. These *Lygus* species have a very wide range of hosts. The host range of tarnished plant bug is greatest because it occurs over a wider geographic range and is well-studied; Young (1986) listed over 300 species. Over half of the cultivated-plant species grown in the United States are listed as host plants for tarnished plant bug. However, all three species are similar in their damage potential. The favored habitat seems to be crops, weeds associated with crops, and other early successional plant communities. Barlow *et al.* (1999) reported that adults of *L. hesperus* survived better on dicot weeds than on monocots.

Among vegetable crops injured are artichoke, asparagus, broccoli, cabbage, carrot, celery, corn, chard, Chinese cabbage, coriander, cowpea, cucumber, eggplant, endive, escarole, faba bean, fennel, horseradish, lettuce, lima bean, mustard, onion, parsnip, parsley, pea, pepper, potato, radish, salsify, snap bean, spinach, squash, sweet potato, Swiss chard, tomato, turnip, and watermelon. Field crops such as alfalfa, cotton, and safflower, as well as various fruit and flower crops frequently are damaged by *Lygus* plant bugs. Numerous weeds serve as hosts; among those frequently cited are aster, *Aster* spp.; curly dock, *Rumex crispus*; fleabane, *Erigeron* spp.; goldenrod, *Solidago altissima*; lambsquarters, *Chenopodium album*; pigweed, *Amaranthus* spp., ragweed, *Ambrosia* spp.; and shepherdspurse, *Capsella bursa-pastoris*. Although *Lygus* plant bugs generally are considered to be herbivores, they sometimes engage in predation of other insects, particularly eggs and young larvae (Cleveland, 1987), or even in cannibalism.

Natural Enemies. Several parasitoids of *Lygus* spp. are known, but the egg parasite *Anaphes iole* Girault (Hymenoptera: Mymaridae), and the nymphal parasites *Leiophron uniformis* (Gahan), *Peristenus pallipes* (Curtis), and *P. pseudopallipes* (Loan) (all Hymenoptera: Braconidae) are thought to be relatively important (Clancy and Pierce, 1966; Scales, 1973; Lim and Stewart, 1976; Jackson and Graham, 1983; Graham *et al.*, 1986; Day, 1987; Sohati *et al.*, 1989, 1992; Snodgrass and Fayad, 1991; Al-Ghamdi *et al.*, 1995). Other parasitoids (Hymenoptera: Ichneumonidae, Mymaridae, Scelionidae; Diptera: Tachinidae; Nematoda: Mermithidae) occur occasionally. The native parasitoids seem to be more effective at parasitizing *Lygus* on weeds than on crops. Several *Peristenus* species have been imported from Europe and released in the United States (Day *et al.*, 1990). The imported parasitoid *Peristenus digoneutis* Loan (Hymenoptera: Braconidae) is reported to have decreased tarnished plant bug abundance by 75% in the New Jersey. In addition,

it seems not to have affected native parasitoids (Day, 1996). A key for identification of parasitoids affecting these plant bugs was provided by Loan and Shaw (1987).

The predators of *Lygus* spp. are reported to include *Geocoris* spp. (Hemiptera: Lygaeidae), *Orius* spp. (Hemiptera: Anthocoridae), *Nabis* spp. (Hemiptera: Nabidae), *Podisus maculiventris* (Say) (Hemiptera: Pentatomidae), *Sinea diadema* (Fabricius) (Hemiptera: Reduviidae), and lady beetles (Coleoptera: Coccinellidae) (Leigh and Gonzalez, 1976; Whalon and Parker, 1978; Fleischer and Gaylor, 1987; Arnoldi *et al.*, 1991). Probably many other predators feed on *Lygus* bugs, particularly the nymphs.

Life Cycle and Description. These insects overwinter as adults. In cold climates they may remain on crop residues or weeds. Khattat and Stewart (1980) speculated that when the food plant became frozen or dry the insects resorted to predation and cannibalism. Emergence from overwintering is protracted, but adults can be found feeding and ovipositing on weeds as soon as the plants are available in the spring. Even in Canada, two generations are common, with the first generation adults produced in July and the second in August and September. In Quebec, a third generation sometimes is observed, with adults produced in late September and early October (Stewart and Khoury, 1976). There seems to be no accurate determination of generation number in warmer climates, but five generations have been suggested for North Carolina.

Egg. The egg is inserted into the plant tissue. The whitish egg is cylindrical, and slightly curved. The top of the egg, where it meets the surface of the plant tissue, is flattened and bears a small elliptical opening through which the hatching nymph escapes. The egg length is about 1.7 mm, and width is about 0.5 mm. They often are deposited in leaf petioles or at the base of the leaf blade, but the preferred location varies with the crop attacked. They usually are deposited singly, but sometimes more than one egg will be found in a single oviposition slit. Duration of the egg stage of *L. lineolaris* is reported to be 14.6, 7.6, and 6.6 days when incubated at 20°, 25°, and 30°C, respectively. In *L. hesperus* and *L. elisus*, incubation duration at uncontrolled temperature is reported to be 9.6 and 9.4 days, respectively. Khattat and Stewart (1977) estimated egg production at four to five eggs per day, and total production at nearly 100 eggs, based on laboratory rearing. However, Gerber (1995) reported mean (range) progeny production of 239 (25–532) for overwintered females, and 303 (6–704) for summer females, based on field collections of insects. The latter data, though higher than reported by most authors, is likely to be more reflective of normal fecundity.

Nymph. There are five instars. The young nymphs are greenish, with red on the antennae. In both *L. lineolaris* and *L. hesperus*, the nymphal color tends to be yellow-green, while in *L. elisus* the color is bluish-green. The early stages, especially instars 3–5, bear a pair of circular black spots dorsally on the first and second thoracic segments, and a single spot near the base of the abdomen. When reared at 25°C, mean duration (range) of the five instars in *L. lineolaris* was 4.8 (4–6), 3.1 (2–5), 3.3 (2–5), 3.3 (2–5), and 5.2 (4–6) days, respectively. Total nymphal development time was 31.5, 19.7, and 14.9 days at 20°, 25°, and 30°C, respectively (Ridgway and Gyrisco, 1960). In contrast, mean nymphal development times at uncontrolled temperature was 3.9, 3.2, 2.5, 3.2, and 4.9 days for *L. hesperus*, and 4.2, 2.9, 2.7, 3.7, and 5.1 days for *L. elisus* (Shull, 1933).

Adult. The adults measure about 4.0–6.0 mm long. Summer adults tend to be pale, mostly pale-green or yellow with brown or black markings. The level of dark pigmentation present in an adult is positively correlated with rearing temperature, and it increases with physiological age (Wilborn and Ellington, 1984). Overwintered adults are much darker than summer adults. The pronotum and scutellum are quite variable in color, varying from mostly yellow or light-green to black, but usually there is at least some light pigmentation at the tip (posterior) of the scutellum, often resembling a heart in shape. The front wings (hemelytra) may be light or dark, but they usually are lighter apically. Legs are yellowish, often marked with reddish-brown. Antennae are yellowish-brown to brownish-black. Females commence oviposition 7–10 days after molting to the adult stage. The adult longevity is estimated at about 30–40 days for males and about 40–60 days for females.

The other common species of plant bugs, *Adelphocoris* spp., tend to be fairly elongate in general body form (see sections on alfalfa plant bug and rapid plant bug). When viewed from above, *Adelphocoris* spp. are about three times as long as wide. In contrast, the *Lygus* spp. tend to be fairly oval, about twice as long as wide. (See color figure 140.)

Differentiation of the *Lygus* species is very difficult, and probably is better left to a specialist; Kelton (1975) and Schwartz and Foottit (1998) provided keys. Knight (1941) provided some readily observable morphological characteristics to distinguish among the species, but they seemed not to be definitive. Briefly, Knight suggested that the body of western tarnished plant bug and pale legume bug were chiefly pale or green, sometimes with dark markings. In contrast, the body of tarnished plant bug is said to be yellowish-brown to reddish or black. The western tarnished plant bug and pale legume bug are further distinguished by the color of the abdomen; it is uniformly green in *L. elisus* and marked with black in *L. hesperus*.

A comprehensive treatment of *L. lineolaris* biology was given by Crosby and Leonard (1914), and of *L. hesperus* and *L. elisus* by Shull (1933). Developmental biology for *L. lineolaris* was provided by Ridgway and Gyrisco (1960), *L. hesperus* by Champlain and Butler (1967), and *L. elisus* by Butler (1970). Examples of rearing techniques based on plant parts were presented by Khattat and Stewart (1977) and Slaymaker and Tugwell (1982); an artificial diet was given by Debolt (1982).

Damage

Various types of injury are inflicted by *Lygus* plant bugs. Examples include sunken or darkened lesions, tip dieback, gall formation, shedding of blossoms,

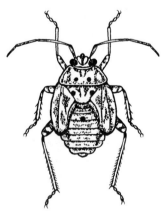

Tarnished plant bug nymph.

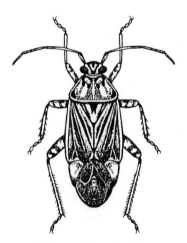

Adult tarnished plant bug.

inhibition of seed germination, and leaf distortion. On celery, the foliar discoloration and tissue destruction is sometimes called "blackheart." The *Lygus* plant bugs tend to prefer feeding on meristematic tissue, floral buds and immature blossoms, and immature fruit relative to leaves, stems, and mature fruit. The term "blasting" is sometimes applied to the destruction by *Lygus* bugs of floral tissue, which turns brown, shrivels, or drops from the plant. Wukasch and Sears (1981) working in Ontario, and Grafius and Morrow (1982) in Michigan, both studied the effects of plant bugs on asparagus. Bug feeding caused collapse of spears and distortion of ferns, and the amount of injury was proportional to bug density. The plants were not permanently injured, tending to recover unless plant pathogens invaded the feeding sites. In Idaho, seed production by carrots was affected by both western tarnished plant bug and pale legume bug (Scott, 1970). Production of seed by cabbage plants in Washington was found to be affected severely; a single pale legume bug could destroy 1.7 fruiting structures per day (Getzin, 1983). Research conducted in Quebec by Khattat and Stewart (1975) showed that flower buds, blossoms and setting pods of beans could be damaged by *L. lineolaris*, but that it was principally late-instar nymphs and adults that inflicted damage.

Management

Sampling. Plant bug populations in crops tend to be aggregated or clumped. This necessitates large plant samples if population assessments are to be accurate. In celery, for example, Boivin *et al.* (1991) used visual examination and fixed samples of about 75–125 plants per field in decision-making. However, they were able to reduce sampling intensity by developing sequential sampling protocols. They estimated treatment thresholds at 0.5–1.0 insect per plant depending on the stage of plant growth.

Sampling efficiency on snap beans was studied by Stewart and Khattat (1980), who compared sweep net and suction sampling. They found that the two methods were highly correlated, but that sweep net samples of nymphs tended to underestimate the population. Multiplication by a correction factor of 1.8 was suggested. Their data suggest that sweep net catches of 25 insects per 50 sweeps is equivalent to vacuum samples of one insect per 10 plants. These authors suggested that economic injury occurred when bug densities reached about two insects per 10 plants in the blossom or pod set stage, and four insects per 10 plants in the flower bud stage. In California, treatment thresholds for beans tend to be about 0.5–1.0 bugs

per sweep for blossom and other early fruiting stages, increasing to 1–2 bugs per sweep later in plant phenology.

Similar sampling studies have been conducted in lentil. Sweepnet, vaccum, and fumigation sampling produced similar estimates, though sweepnet sampling tended to underestimate nymphal populations. Afternoon sampling produced more accurate population estimates than morning sampling (Schotzko and O'Keeffe 1986). To obtain accurate estimates in this crop, 10 single-sweep samples are required for nymphs and five 25-sweep samples are required for adults (Schotzko and O'Keeffe 1989b).

Only limited work on trapping has been conducted. Plant bugs can be captured in light traps, but this is not considered to be an efficient monitoring technique. Slaymaker and Tugwell (1984) reported the capture of plant bugs, principally males, in virgin female-baited sticky traps. White sticky traps placed at a height of about 0.5 m are useful for detecting tarnished plant bugs in apple orchards (Boivin *et al.*, 1982), and reportedly work well in vegetable crops also (Boivin pers. comm.).

Insecticides. Chemical insecticides are commonly applied to the foliage of plants to protect them from plant bugs. The plant growth stages that especially require protection are the blossoms and young fruit, but owing to asynchronous blossoming and the desire to prevent damage to mature tissue, season-long protection is not uncommon. Also, the adults are highly mobile, often changing plants after each meal. Thus, there is continuous risk of invasion from weeds or untreated crops. Although numerous insecticides are effective for plant bug control, *Lygus* spp. are moderately difficult to kill with some insecticides (Martel *et al.*, 1986).

Cultural Practices. Vegetation management is very important in reducing damage potential by these highly mobile insects. Plant bug sources, both weeds and alternate crops, should be monitored and perhaps be managed. *Lygus* spp. may attain high numbers in crops such as alfalfa and cotton, and at maturity or harvest the insects disperse and invade susceptible vegetable crops. Thus, it may be desirable to avoid planting vegetables near vegetation that harbors plant bugs unless the source crop remains attractive to the plant bugs, or the source crop can be treated with insecticide.

Tractor-mounted vacuum devices reportedly have been used to reduce the number of plant bugs in some crops. Both adults and large nymphs are easily dislodged from vegetation, and therefore are good targets for such technology. However, population reductions seem to be incomplete and transient, so such equipment is generally viewed as unsatisfactory (Vincent and LaChance, 1993).

Biological Control. Mass releases of the egg parasitoid *Anaphes iole* have been evaluated for suppression of *Lygus hesperus*. Parasitoid releases can result in increase in egg parasitism, reduction in nymph abundance, and less fruit damage. The benefits are temporary, however, and frequent parasitoid releases at high rates are necessary to achieve effective damage suppression (Norton and Welter, 1996).

FAMILY PENTATOMIDAE—STINK BUGS

Brown Stink Bug
Euschistus servus (Say)
(Hemiptera: Pentatomidae)

Natural History

Distribution. This native insect is found throughout the United States and southern Canada, and extends south into Mexico. Two subspecies are recognized by some authors: *E. servus servus* (Say) occurs in the southern states, including Virginia, southern Illinois, Texas, Arizona, and southern California; *E. servus euschistoides* (Vollenhoven) occurs in the northern portions of the continent. They meet in a broad band of intergradation from Maryland to Kansas. *E. servus* is the most abundant member of the genus in southern states.

Host Plants. This species has a wide host range. Among the vegetables attacked are bean, cabbage, corn, cowpea, okra, pea, pepper, squash, and tomato. Other crops that serve as hosts include such field crops as alfalfa, clover, cotton, lespedeza, oat, soybean, sweet clover, and timothy, and such fruit crops as apple, citrus, peach, pear, raspberry, and tobacco. Some of the weeds fed upon by brown stink bug are cocklebur, *Xanthium* sp.; curly dock, *Rumex* sp.; fleabane, *Erigeron annuus*; goldenrod, *Solidago* sp.; horseweed, *Erigeron canadensis*; ragweed, *Ambrosia* sp.; mullein, *Verbascum thapsus*; pigweed, *Amaranthus* sp.; prickly lettuce, *Lactuca scariola*; yellow thistle, *Cirsium horridulum*; and Canada thistle, *Cirsium arvense*. Mullein is reported to be especially important as it is present early in the season before many other hosts are available.

Natural Enemies. Brown stink bug is parasitized by *Telenomus* and *Trissolcus* spp. (Hymenoptera: Scelionidae), *Hexacladia smithi* Ashmead (Hymenoptera: Encyrtidae) and several flies, including *Gymnosoma fuliginosum* Robineau-Desvoidy, *Trichopoda pennipes* (Fabricius), *Cylindromyia binotata* (Bigot), *C. euchenor* (Walker), *C. fumipennis* (Bigot), *Gymnoclytia unicolor*

(Brooks), *G. immaculata* (Macquart), *Euthera tentatrix* Loew (all Diptera: Tachinidae), and *Sarcodexia sternodontis* Townsend (Diptera: Sarcophagidae) (McPherson, 1982). Yeargan (1979) studied mortality of brown stink bug eggs in Kentucky, and reported that when egg masses were distributed in crops about 50% were parasitized, 12% were destroyed by predators, and 20% failed to hatch probably due to undetectable natural enemy feeding. When stink bugs were allowed to oviposit naturally on plants 60% were parasitized, 37% were destroyed by predators, and 2% failed to hatch. Obviously, natural enemies have the potential to destroy a high proportion of stink bug eggs.

Life Cycle and Description. There are two generations annually throughout the range of this species, though in northern locations many late-developing nymphs of the second generation perish with the onset of winter. The adult is the overwintering stage. Overwintered adults usually commence egg laying in early May and continue through mid-June. The adults from the first generation produce eggs beginning in August, and the nymphs from this generation mature, but do not produce eggs, before the onset of winter.

Egg. Females deposit eggs in clusters of about 20, but counts of 14 and 28 eggs per mass are especially common, probably reflecting ovariole number. Overwintered adults produce, on average, about 120 eggs per female, with over 500 being produced by an occasional individual. The egg production by first generation adults is lower, usually 40–80 per female. The eggs are yellowish-white, and slightly greenish initially. They are somewhat barrel-shaped, attached one end to a leaf, and with a row of 30–35 small processes ringing the upper end of the egg. During warm weather the incubation period is typically 5–6 days (range about 3–14 days).

Nymph. There are five instars, the mean durations of which were 3.7, 5.1, 4.9, 6.9, and 6.7 days, respectively, when reared on beans in Arkansas (Rolston and Kendrick, 1961). In contrast, Woodside (1946) reported mean instar durations of 5.3, 9.3, 10.3, 13.0, and 13.6 days when reared on peach in Virginia. The difference in total nymphal development time, 33 vs 51 days, is not likely due solely to food plant effects, because the long egg incubation time of about 9 days in the Virginia study suggests cooler rearing conditions. Nymphal lengths are reported to average 1.5, 2.4, 4.2, 8.5, and 10.4 mm, respectively. The young nymphs are light colored dorsally, with the thorax yellowish-brown and the abdomen white or yellow. The abdomen also bears red markings. The antennae are reddish-brown. The older nymphs have substantially the same appearance, though the abdomen is always yellow. The wing pads become evident in instar four, and the pads overlap abdominal segments in instar five.

Adult. Adults from the first generation do not all go on to produce eggs. Over 90% of insects reaching adulthood after August 1 enter reproductive diapause. Even adults produced in June may enter diapause, but the incidence is much lower. Adults successfully overwinter under plant debris and amongst weeds. In Arkansas, survival of adults sheltering in mullein was especially high. Open habitats, rather than wooded environments, seem to be favored (Jones and Sullivan, 1981). Longevity of overwintered adults in the spring usually is about two months, but some persist for the entire summer before perishing. The adult measures about 11–15 mm long, and is brown or grayish-yellow. The abdominal segments, when viewed from below, bear black spots at the lateral angles. The lateral edges of the pronotum, or "shoulders," are rounded. A key to distinguish stink bugs commonly affecting vegetables is found in Appendix A. (See color figure 144.)

The most complete life-history study was conducted by Rolston and Kendrick (1961), but McPherson (1982) provided an excellent summary of biology. Munyaneza and McPherson (1994) provided developmental data, description of the nymphal stages, and information on rearing. Aldrich *et al.* (1991b) provided information on the chemistry of a volatile that attracts nymphs and adults of brown stink bug and other *Euschistus* spp., as well as stink bug parasitoids.

Damage

Nymphs and adults feed on tender shoot tissues, buds, and fruit. Most damage to vegetables takes the form of deformed fruit, aborted blossoms, or death of young tissue. Sedlacek and Townsend (1988) and Apriyanto *et al.* (1989a,b) found that young corn plants

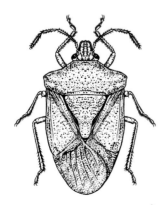

Adult brown stink bug.

were damaged by stink bug feeding. Symptoms included chlorotic lesions, tightly rolled leaves, wilting, stunting, increased tillering, delayed silk production, and smaller grain weight. The first instars, however, feed little or not at all—a common condition among stink bugs.

Management

Phillips and Howell (1980) stressed the importance of weeds in stink bug biology, and noted that damage was higher in weedy areas, especially following senescence of the weeds. Reduced tillage practices and presence of cover crops seem to contribute on increases in stink bug abundance, probably because weeds are more abundant. Thus, if stink bugs are prone to be a problem, it is important to monitor weed populations. Pheromones specific to *Eustichus* spp. can be used to capture and monitor abundance of this stink bug. Foliar insecticides are effective, with special care warranted to protect the blossoms and fruit.

Consperse Stink Bug
Euschistus conspersus Uhler
(Hemiptera: Pentatomidae)

Natural History

Distribution. Consperse stink bug occurs in western North America from British Columbia and Idaho south to California and Nevada. It is a native species.

Host Plants. Consperse stink bug damages vegetable and fruit crops. Known principally as pest of tomato, it also feeds on apple, apricot, blackberry, fig, loganberry, peach, pear, plum, and raspberry. It also is found occasionally in alfalfa, barley, cotton, sorghum, and sugarbeet. Among the numerous weeds known to support consperse stink bug are bracken fern, *Pteridium aquilinum*; dock, *Rumex* sp.; horehound, *Marrubium* sp.; mallow, *Malva* sp.; mullein, *Verbascum* sp.; plantain, *Plantago* sp.; mustard, *Brassica* sp.; and the thistles *Cirsium occidentale* and *C. mohavense*.

Natural Enemies. Stink bug eggs are parasitized by *Trissolcus* and *Telenomus* (Hymenoptera: Scelionidae). The level of parasitism varies greatly, and the first generation is typically lightly parasitized. Digger wasps, *Dryudella* sp. (Hymenoptera: Sphecidae), prey on nymphs, and spiders are important predators of adults.

Life Cycle and Description. In California, there are two generations annually, but only one generation is present in cooler areas. The adults overwinter in reproductive diapause and commence production of eggs in March or April. First generation nymphs are found from April to June, and adults from June to

September. Second generation eggs are present from July to August, nymphs from July to October, and adults until the subsequent spring.

Egg. Eggs are deposited in clusters of 7–28 eggs, with most clusters consisting of 13–15 eggs. They normally are arranged in four short rows of 3–4 eggs. Adults are not very discerning in their oviposition habits, selecting dry leaves and plants unsuitable for nymphal growth as well as suitable host plants. Initially pearly white in color, the barrel-shaped eggs bear a fringe of spines around the top. The eggs turn pink as they mature. Duration of the egg stage is about 5–30 days under field conditions, and averaged 6.2 days when reared at 27°C. Females may produce up to 640 eggs during their life span, though mean fecundity is about 225 eggs.

Nymph. There are five instars. The early instars are black and reddish, but later instars are various shades of yellow and brown. Duration of the nymphal stage ranges from 23–79 days under field conditions. When cultured at 27°C, mean development time was 25.6 days. Mean duration of the instars was 3.1, 5.6, 4.3, 4.9 and 7.7 days, respectively, for instars 1–5. Temperatures of about 27°C seems optimal for development and oviposition. The first instar does not feed, and young nymphs remain clustered around the egg cluster during this instar.

Adult. The adult measures 11–12 mm long and 6–6.5 mm wide. It is grayish-brown or greenish-brown dorsally, but sprinkled with numerous small black spots. The ventral surface is greenish, yellowish, or brownish. The legs are yellowish, though they also bear some black spots. The antennae are yellow or reddish, and darker at the tip. Adults overwinter under weeds and trash in the field. In the spring, overwintered adults feed on weeds and commence oviposition on both weeds and crop plants. Second generation adults commence egg laying after a pre-oviposition period of about 10–32 days.

The biology of consperse stink bug was given by Borden *et al.* (1952) and Hunter and Leigh (1965). Development was described by Toscano and Stern (1976). Toscano and Stern (1980) described reproductive biology. Alcock (1971) described aggregation and mating behaviors. Aldrich *et al.* (1991b) provided information on the chemistry of a volatile that attracted nymphs and adults of consperse stink bug and other *Euschistus* spp., as well as stink bug parasitoids. A key to distinguish stink bugs commonly affecting vegetables is also found in Appendix A.

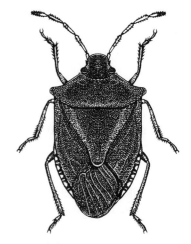

Adult consperse stink bug.

Damage

Stink bugs feed on developing fruit, usually green fruit, causing white corky tissue to form beneath the skin at the site of feeding (Michelbacher *et al.*, 1952). However, pink and red fruit also is readily attacked (Zalom *et al.*, 1997). Dimpling or distortion of the fruit is common. Stink bugs seem to favor the stem end of tomato fruit, but they also feed elsewhere. Some fruit drop also occurs and stink bugs are implicated in transmission of pathogenic yeast, though plant disease transmission has not been demonstrated specifically for consperse stink bug. Damage is often localized along the margin of a field because the bugs encounter these plants first as they disperse into a crop from overwintering shelter.

Management

Stink bugs can be sampled with a sweep net or by visual observation, often in association with a beeting sheet or tray. Insecticides are applied to foliage and young fruit to protect susceptible crops from injury, though stink bugs are quite difficult to kill. It is important to obtain good penetration of canopy foliage down to the soil level, because these stink bugs often are found on lower portions of the plant. Border treatments are especially important, because this is the portion of the field injured most severely.

Green Stink Bug
Acrosternum hilare (Say)
(Hemiptera: Pentatomidae)

Natural History

Distribution. This species, which is native to North America, occurs throughout the United States and southern Canada. It is less damaging, and therefore less well-known than southern green stink bug, *Nezara viridula* (Linnaeus). Although green stink bug is readily confused with southern green stink bug, the distribution of this latter species is limited principally to the southeastern states.

Host Plants.   Green stink bug has a wide host range, though it is known principally as a pest of soybean and tree fruit. Among vegetables it has been observed to feed upon are asparagus, cabbage, corn, cowpea, cucumber, eggplant, lima bean, okra, mustard, pea, pepper, squash, snap bean, tomato, and turnip. Beans, particularly lima bean, are often damaged. Other common hosts are fruit such as apple, apricot, blackberry, cherry, elderberry, grape, mulberry, orange, pear, and strawberry; trees such as ash, basswood, black cherry, black locust, dogwood, hackberry, honey locust, holly, maple, and redbud; and field crops such as alfalfa, cotton, and soybean. Green stink bug may also be found on such weeds as common elder, *Sambucus canadensis*; goldenrod, *Solidago* spp.; jimsonweed, *Datura stramonium*; mallow, *Malva* spp.; mullein, *Verbascum thapsus*; and rattlebox, *Crotalaria* sp. Schoene and Underhill (1933) suggested that suitability of host plants changed through the season, and that stink bugs relocated as necessary. Availability of appropriate wild hosts early in the season is a prerequisite for high densities of green stink bugs in crops. McPherson (1982) provided a comprehensive list of hosts. Schoene and Underhill (1933) and Jones and Sullivan (1982) provided less complete lists but they indicated relative preference in Virginia and South Carolina, respectively.

Natural Enemies.   Green stink bug is subject to attack by several natural enemies; McPherson (1982) and Jones *et al.*, (1996) provided lists of the insect enemies. Eggs are parasitized by several wasps, but the most important seem to be *Telenomus podisi* Ashmead, *Trissolcus euschitis* (Ashmead), and *T. edessae* (Fouts) (all Hymenoptera: Scelionidae), and *Anastatus reduvii* (Ashmead) (Hymenoptera: Eupelmidae). The late instar and adult bugs are attacked by *Trichopoda pennipes* (Fabricius) (Diptera: Tachinidae), a common parasitoid of hemipterans. Predators known to affect green stink bugs include predatory stink bugs (Hemiptera: Pentatomidae), green lacewings (Neuroptera: Chrysopidae) and various birds, particularly quail, *Colinus virginianus*. Yeargan (1979) studied the fate of stink bug eggs in two cropping systems and noted that green stink bug suffered less natural mortality than some other species of stink bugs. Also, mortality from unspecified chewing predators was more common than from parasitism, and mortality due to sucking predators was negligible.

Life Cycle and Description.   The number of generations appears to vary. In Virginia (Underhill, 1934) and Ontario and Quebec (Javahery, 1990) it is reported to be one, but in Arkansas (Miner, 1966) and Kansas (Wilde, 1969) apparently there are two generations annually. In Arkansas, the first generation often occurs on dogwood, with the second generation attacking principally soybean. It would be easy to overlook the first generation, which may explain some of the disagreement about generation number. Development time is long, especially the adult pre-oviposition period, so generation number is likely limited to one in the north, but the climate of southern states should allow a second generation. Photoperiod, rather than temperature, may determine generation number. Wilde (1969) reported that increasing day length stimulated egg production, whereas decreasing day length inhibited egg production. The life cycle requires about 100 days when stink bugs are cultured at 22°C. The adult is the overwintering stage. Adults become active in the spring when temperatures exceed about 21°C, and feed on young leaves and stems of trees.

Egg.   Oviposition in both Virginia and southern Canada does not occur until about mid-June, though in the midwestern states oviposition in early June has been observed. Eggs are normally deposited in clusters of about 30 eggs, though clusters of up to 69 eggs have been observed. The first egg cluster produced by a female is the largest, with subsequent clusters diminishing in size. The female is not very selective in her choice of oviposition site, with eggs deposited on both foliage and fruit. Mean egg production per female was estimated by Miner (1966) to average 60, but egg production of up to 170 per female was observed. The rearing conditions of Miner's study were suboptimal, and egg production likely suppressed. Javahery (1990) obtained oviposition of 130–150 eggs per overwintered female, and this is probably a better estimate of fecundity. The mean incubation period at 22°C is 12.7 days (range 9–19 days). Incubation length decreases as temperature increases, with eggs hatching in only six days at 30–33°C. The eggs are deposited on end and in fairly definite rows, but the rows are not as tightly interlocked as in most stink-bug species. They are light green when first produced, but turn yellow and then pink as they mature. The cap of the egg becomes red immediately preceding hatch. The eggs measure 1.3–1.5 mm long, and 1.1–1.3 mm in diameter. The egg is somewhat cup-shaped, the top being slightly broader than the base. The top is ringed with a row of 45–65 small flattened processes (Esselbaugh, 1946).

Nymph. The eggs from the overwintering adults begin to hatch in July, though overwintered females continue to produce eggs throughout the summer. Nymphs tend to remain aggregated during the first instar, but disperse thereafter. Mean duration (range) of the five instars is 7.0 (5–10), 8.9 (6–14), 7.9 (6–11), 8.9 (6–11), and 12.8 (8–20) days, respectively, when reared at 22°C on soybean (Miner, 1966). Development times are shorter when reared on apple, pear, and snap bean at 23°C, averaging 6, 6–7, 6–8, and 8–9 days, respectively, for instars 1–5 (Javahery, 1990). Simmons and Yeargan (1988a) studied development over a range of temperatures and observed fastest development at 27°C, with the egg to adult stage completed in about 40 days, whereas nearly 50 days were required at 24° and 30°C, and 70 days at 21°C. The nymphs measure about 1.6–2.0, 2.3–3.2, 4.0–5.2, 6.8–8.2, and 10.0–12.7 mm long for instars 1–5, respectively. In color, the dorsal surface of instars 1–4 tends to be brownish-black on the head and thorax, with a yellow spot centrally and yellow at the lateral margins of the thorax. The abdomen is marked with transverse black and light-blue stripes, and large black spots centrally. The lateral margins of the abdomen bear a row of semi-elliptical spots, blackish-bronze in color. The fifth instar differs from the earlier instars, as the abdomen lacks stripes, and is rather uniform yellow-green in color.

Adult. Nymphs from the first generation begin to attain the adult stage in August. Adults begin feeding within a few hours of attaining the adult stage, but mean days to copulation by adults is 22.3 (range 11–30 days). Mating requires only a few minutes, but sometimes persists for several hours. Females usually mate after each egg cluster is deposited. Another 21.7 (19–33) days are required, following copulation, until egg production commences. Duration of the adult stage is about two months during the summer, but several months during the winter. With the onset of cold weather, adults disperse to wooded areas. Overwintering occurs in leaf litter, and under bark of trees, in wooded areas. The adult bugs measure 13–19 mm long. They are largely uniform green in color, though the dorsal surface is darker than the ventral, and the edges of the head, pronotum, and abdomen are marked with yellow. The three distal segments of the antennae are marked with brownish-black.

Southern green stink bug, *N. viridula* (Linnaeus), is easily confused with green stink bug, *A. hilare*. However, the two species differ ecologically and morphologically. *Acrosternum hilare* is usually associated with trees and shrubs, rather than the herbaceous vegetation fed upon by *N. viridula*. Also, *A. hilare* is found widely in North America, though it is most abundant in the north. The two species can be differentiated by the shape of the abdominal spine. When viewed from below, *A. hilare* has a pointed spine protruding forward between the base of the hind legs whereas in *N. viridula*, the spine is rounded.

Male green stink bugs produce an aggregation pheromone that attracts both males and females, other stink bug species, and the tachinid parasitoid *Trichopoda pennipes* (Aldrich *et al.*, 1989). The pheromone consists of the same chemicals found in the pheromone of southern green stink bug, but the ratios are different.

The biology of green stink bug was given by Whitmarsh (1917), Underhill (1934), Miner (1966), and Javahery (1990). Developmental biology was described by Simmons and Yeargan (1988a). Nymphs were described by Whitmarsh (1917) and Decoursey and Esselbaugh (1962). Culture on plant tissue was provided by many authors, such as Wilde (1968); culture on artificial diet was presented by Brewer and Jones (1985). A key to the common stink bugs found in vegetables is included in Appendix A.

Nymph of green stink bug, dark form.

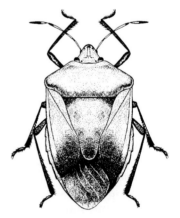

Adult green stink bug.

Damage

The green stink bug inserts its beak into fruit or foliar tissue and removes liquids. During feeding, the bug injects enzymes that liquefy plant tissue and cause the tissue at the feeding site to collapse. Subsequently, tissue adjacent to the feeding site continues to grow, but the tissues at the feeding site fail to grow, leading to dimples or similar deformities. Feeding sites also may become discolored, usually turning black and hardened, but sometimes white and spongy. Adults and nymphs inflict the same type of injury.

Damage potential varies considerably among insect stages. The first instar does not feed. Feeding frequency increases as the nymph matures, with the fifth instar feeding most frequently, an average of 3.4 times per day. Duration of each feeding session does not vary with bug age, however, averaging 1.5 h per session. The adults and fifth instars, and both sexes, displayed similar feeding behaviors (Simmons and Yeargan, 1988b). Plant responses to green stink bug feeding have not been well studied, but some states recommend treatment of beans if stink bug abundance attains 1 bug per 5 m of row.

Green stink bug harbors *Nematospora coryli*, a fungus that causes yeast-spot disease in beans and soybeans. Both adults and nymphs acquire the pathogen by feeding, retain it in their bodies, and excrete it in their saliva and feces (Foster and Daugherty, 1969; Clarke and Wilde, 1970). The pathogen greatly exacerbates the damage caused by green stink bug. Underhill (1934) reported the loss of entire lima bean crops due to this disease and its insect vector. Other fungi and bacteria can be transmitted by stink bugs (Russin *et al.*, 1988), so the vector potential of green stink bug may not be limited to transmission of *N. corylii*.

Management

Sampling. Stink bug populations tend to be highest at field margins, probably reflecting the tendency of adults to overwinter in wooded areas surrounding fields, and the early season feeding preference for trees. Sampling in soybean is usually accomplished by shaking the plant over soil or a drop cloth, and counting the insects dislodged, but visual examination may be appropriate for crops with less vegetation. In all crops, sampling for eggs is done by visual examination.

Insecticides. Stink bugs often are difficult to kill, but proper selection and timing of foliar insecticides can protect crops from injury. Residual materials are usually desirable because the stink bugs often develop outside the crop and enter at various times. It is most important to protect the crop during the blossom and early fruiting stage. Because bugs are entering from the crop margins, border treatments may be adequate.

Cultural Practices. Stink bug populations increase during the season, with maximum densities and damage occurring late in the growing season. Stink bugs commonly enter crop fields from adjacent vegetation, either woodlands or weedy areas. Thus, a common recommendation is to plant as early as possible to avoid peak insect abundance. However, in studies conducted with soybean, early plantings (early June) were as damaged as late (early July) (McPherson *et al.*, 1988). Jones and Sullivan (1982) suggested that destruction of black cherry and elderberry would greatly reduce abundance of green stink bug. Trap crops also are sometimes recomended for stink bugs. For information on this subject, see the discussion on trap crops in the section on southern green stink bug. Pole beans are considered more susceptible than bush varieties (Underhill, 1934).

Harlequin Bug
Murgantia histrionica (Hahn)
(Hemiptera: Pentatomidae)

Natural History

Distribution. A native of Mexico and Central America, harlequin bug was first observed in the United States in Texas, in 1864. Its appearance, coinciding with the occurrence of Union troops during the American Civil War, earned it the name "Sherman-bug" and "Lincolnite" in parts of the south. It rapidly spread throughout the southern states, and eventually reached northern locales such as Colorado, Iowa, southern Michigan, Pennsylvania, and Massachusetts. It is considered to be a serious pest only in southern states, however, and is not regarded as a problem in California. Harlequin bug is also a pest in Hawaii.

Host Plants. Harlequin bug is principally a pest of crucifers, attacking asparagus, broccoli, Brussels sprouts, cabbage, cauliflower, Chinese cabbage, collards, kale, kohlrabi, mustard, radish, rutabaga, turnip, and watercress. This bug is reported to be especially fond of horseradish. In the southernmost states, crucifers do not thrive during the summer months, and the bugs are forced onto other plants. Thus, they are sometimes found feeding on beans, okra, squash, tomato and many other vegetables, but this is usually due to lack of normal food. Harlequin bug feeds readily on cruciferous weeds such as wild mustard, *Brassica* spp.; shepherds purse, *Capsella bursa-pastoris*; and pepperweed, *Lepidium* spp.; and related mustard oil-containing plants such as members of the family Capparaceae. Other weeds common in crops, such as pigweed, *Amaranthus* spp.; and lambsquarter, *Chenopodium album*; are also fed upon, and reproduction occurs on these plants.

Natural Enemies. Harlequin bug appears to be relatively free of natural enemies, other than for egg parasites and general predators. The egg parasitoids are *Oencyryus johnsoni* (Howard) (Hymenoptera: Encyrtidae), *Trissolcus murgantiae* Ashmead, and *T. podisi* Ashmead (both Hymenoptera: Scelionidae). The best known species is *O. johnsoni*, which is reported frequently from harlequin bug eggs, and caused up to 50% mortality during a harlequin bug outbreak in Virginia (White and Brannon, 1939). This parasite is widely distributed, and apparently has other hosts. It attacks eggs in all stages of embryonic development, and prevents them from hatching (Maple, 1937). However, *O. johnsoni* is not the only effective parasite, as *T. murgantiae* was observed to parasitize 45% of harlequin bug eggs in North Carolina, at locations where *O. johnsoni* parasitized only 30% of eggs (Huffaker, 1941). Because of its effectiveness, *T. murgantiae* was introduced into California (DeBach, 1942).

Life Cycle and Description. Harlequin bug breeds continuously in the southern portions of its range. During mild winters all stages have been observed as far north as Virginia. In colder climates, only the adults survive the winter in sheltered locations. They seek shelter in and near fields, among overwintering crop plants, and in other organic debris such as dead leaves and bunches of grass. Two or three generations per year seems normal, but Paddock (1915a) indicated four generations in south Texas.

Egg. The adults begin depositing eggs about two weeks after becoming active in the spring. Eggs are deposited beneath leaves, usually in clusters of 12 arranged in two rows of six, at intervals of 5–6 days. As the female nears the end of her life, the egg batches get slightly smaller and the arrangement less regular. The eggs are barrel-shaped, and measure about 1.30–1.38 mm long and 0.90–0.92 mm in diameter. They are light gray or pale yellow, and generally are circled by two black bands. They may also bear small black dots or spots, and the top has a semicircular black marking. The average number of eggs is reported to be 115 per female. Egg deposition may occur over a period of about 40–80 days. Eggs hatch in about 4–5 days during warm weather, while 15–20 days may be required during cool weather.

Nymph. Upon hatching, young nymphs stay clustered near the old eggs for 1–2 days. The newly hatched nymphs are pale green with black markings, but soon become brightly colored—black or blue, with

Harlequin bug eggs on foliage.

red and yellow or orange markings. Paddock (1918) indicated that there were six instars in Texas, and that nymphal development could be completed in just 30 days. He gave average development times as 3.4, 3.2, 4.7, 4.7, 7.0, and 4.3 days, respectively, for the six instars developing under summer conditions. Under spring conditions, development times were increased by about 30%. In contrast, White and Brannon (1939) reported five instars and a development time requirement of about 40–60 days during the summer, and slightly longer, perhaps 70 days, during cool weather.

Adult. The adults usually live about 60 days, but may live considerably longer during the winter. They measure about 8.0–11.5 mm long. The adults are brightly colored, similar to the large nymphs, principally black and yellow or black and red. The color pattern varies, with the spring and summer bugs more brightly colored than the overwintering insects. As with many stink bugs, harlequin bugs produce a disagreeable odor if disturbed, and birds avoid eating them. (See color figure 143.)

The biology of harlequin bug was provided by Chittenden (1908b), Paddock (1918), and White and Brannon (1939). A key to distinguish stink bugs commonly affecting vegetables is found in Appendix A.

Damage

The piercing-sucking feeding behavior of this insect results in white blotches at the site of feeding. Wilting, deformity, and plant death may occur if insects are abundant. Mild winters are said to favor survival and subsequent damage (Walker and Anderson, 1933). Once considered the most serious crucifer pest in the south, this insect has been relegated to minor status in commercial production and persists mostly as a home garden pest.

Management

Chemical Control. Insecticides are applied to the foliage for suppression of this bug. Harlequin bug can be difficult to control with insecticides; targeting the young bugs and thorough coverage are recommended (Rogers and Howell, 1973). Soap applied alone or in

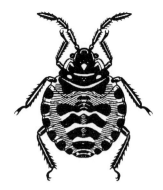

Harlequin bug nymph.

Adult harlequin bug.

combination with rotenone provided good control (Walker and Anderson, 1933; 1934).

Cultural Practices. Trap crops, usually consisting of early-planted mustard, rape, or kale are sometimes recommended to divert the overwintering bugs from the principal crop. Such trap crops must be sprayed or destroyed, however, or the adults will soon move to the main crop. Destruction of crop residues, on which the insect may overwinter in the north or oversummer in the south, is an important cultural practice to alleviate harlequin bug damage.

Host-Plant Resistance. Sullivan and Brett (1974) studied the relative susceptibility of different crucifer crops to harlequin bug in North Carolina. They reported that mustard and Chinese cabbage were the most susceptible; turnip, kale, rutabaga, and some radishes were intermediate; and cauliflower, cabbage, broccoli, collards, Brussels sprouts, kohlrabi, and most radish varieties were fairly resistant. Cabbages were the most resistant crop, but considerable variation among cultivars was evident.

Onespotted Stink Bug
Euschistus variolarius (Palisot de Beauvois)
(Hemiptera: Pentatomidae)

Natural History

Distribution. Onespotted stink bug is native to North America. It is distributed widely; in the United

States it is absent only from the southwestern states. In the midwest and other northern areas it is considered to be the most common stinkbug species (Esselbaugh, 1948). In Canada it is known from Quebec and Ontario west to British Columbia.

Host Plants. This species is reported to feed on numerous plants. Among vegetable crops attacked are asparagus, bean, cantaloupe, corn, cowpea, mustard, onion, pea, potato, squash, and tomato. Field crops that serve as hosts are alfalfa, clover, cotton, oats, rye, sugarbeet, timothy, tobacco, and wheat. Such fruit crops as gooseberry, grape, peach, pear, and raspberry were damaged. Some of the numerous weeds known to support this species are burdock, *Arctium* sp.; curly dock, *Rumex* sp.; goldenrod, *Solidago* sp.; horseweed, *Erigeron canadensis*; milkweed, *Asclepias* sp.; mullein, *Verbascum thapsus*; pigweed, *Amaranthus* sp.; ragweed, *Ambrosia* sp.; and thistle, *Cirsium* sp. Trees are commonly reported as hosts; among them are black walnut, elm, pine, poplar, sassafras, tulip tree, and willow.

Natural Enemies. Onespotted stink bug is parasitized by *Telenomus* and *Trissolcus* spp. (Hymenoptera: Scelionidae) and several flies, including *Gymnosoma fuliginosum* Robineau-Desvoidy, *Trichopoda pennipes* (Fabricius), *Cylindromyia binotata* (Bigot), *C. fumipennis* (Bigot), *Gymnoclytia occidua* (Walker), *Cistogaster immaculata* Macquart, and *Euthera tentatrix* Loew (all Diptera: Tachinidae) (McPherson, 1982).

Yeargan (1979) studied mortality of brown stink bug eggs in Kentucky, and reported that when egg masses were distributed in crops, about 50% were parasitized,13% were destroyed by predators, and 25% failed to hatch probably due to undetectable natural enemy feeding. When stink bugs were allowed to oviposit naturally on plants 71% were parasitized, 26% were destroyed by predators, and 1% failed to hatch. Natural enemies clearly have the potential to destroy a high proportion of stink bug eggs.

Life Cycle and Description. There is only a single generation per year throughout the range of this species. The eggs are deposited in late April to June, with adults produced by August. The adults overwinter, and both new and overwintering adults can be found late in the summer.

Egg. The eggs are barrel-shaped, and attached on end to the underside of a leaf. The upper end of the egg is ringed with a single row of 27–33 small spines. Initially, the eggs are light green, but soon turn creamy white. They measure about 1.1 mm long and 0.9 mm wide. Females deposit the eggs in clusters; the average number per cluster is about 14. The incubation period of eggs is about six days (range 3–10 days).

Nymph. There are five instars. The mean length (range) was reported by Munyaneza and McPherson (1994) to be 6 (6), 11 (9–14), 10 (8–12), 12.6 (7–15), and 10.6 (8–17) days, respectively. Esselbaugh (1948) reported mean (range) development times of 3.5 (2–11), 7.5 (4–16), 7.9 (5–17), 7.7 (5–13), and 10.0 (8–16), respectively. The first instars are grayish-brown on the thorax, the abdomen similar but lighter and marked with red. First instars measure about 1.5 mm long. Second instars are whitish, and conspicuously marked with black spots on the head and thorax and red on the abdomen. They measure about 2.8 mm long. Third instars are similar to second instars in appearance, except that the abdomen is greenish and they measure about 4.1 mm long. Fourth instars are greenish, with a row of red spots along the median line of the abdomen. The wing pads are evident in this instar, which measures about 6.7 mm long. The fifth instar has a dark yellow head and thorax, and a green abdomen. The wing pads are well-developed, overlapping the first three abdominal segments. This instar measures about 10.3 mm long. The nymphal instars of *E. variolaris* are very similar to those of brown stink bug, *E. servus* (Say).

Adult. The adult bugs are grayish-brown. They measure 11.0–15.0 mm long. The lateral edges of the pronotum, or "shoulders," generally taper to a sharp point. In males of this species the ventral surface of the tip of the abdomen is marked with a single dark spot, which serves as the basis for the common name. Because it is sex-specific, it is not a particularly useful diagnostic character. The adults overwinter under dry leaves and other plant debris, under logs, and beneath mullein leaves. Winter mortality is high, with 80% or more of the population perishing. Bugs begin mating a few days after emerging from diapause in the spring. The average time from copulation to egg laying is 27 days.

The biology and description of the nymphs were given by Parish (1934) and Munyaneza and McPherson (1994). The latter authors also gave information on rearing, and characters to distinguish onespotted stink bug nymphs from brown stink bug nymphs. A key to distinguish stink bugs commonly affecting vegetables is also found in Appendix A.

Damage

Nymphs and adults feed on tender shoot tissue, buds, and fruit. Most damage to vegetables takes the form of deformed fruit, aborted blossoms, or death of young tissue. Annan and Bergman (1988), Sedlacek and Townsend (1988), and Apriyanto *et al.*, (1989a,b)

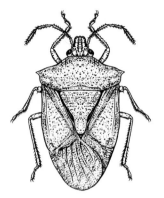

Adult onespotted stink bug.

found that young corn plants were damaged by stink bug feeding. Symptoms included are chlorotic lesions, tightly rolled leaves, wilting, stunting, increased tillering, delayed silk production, and smaller grain weight. Even a single day of feeding by stink bugs reduced corn growth and yield. The first instars, however, feed a little or not at all—a common condition among stink bugs.

Management

Phillips and Howell (1980) stressed the importance of weeds in stink bug biology, and noted that damage was higher in weedy areas, especially following senescence of the weeds. Reduced tillage practices and presence of cover crops, especially wheat, contributed to increase in stink bug abundance. Foliar insecticides are effective, with special care needed to protect the blossoms and fruit.

Say Stink Bug
Chlorochroa sayi (Stål)

Uhler Stink Bug
Chlorochroa uhleri (Stål)
(Hemiptera: Pentatomidae)

Natural History

Distribution. These stink bugs are native to western North America. Say stink bug is found throughout the west, from Montana and eastern Texas west to Caifornia and British Columbia. It has also been taken in Arkansas, which seems to be unusually far east for this species. Similarly, Uhler stink bug is found from the Saskatchewan and the Dakotas, Nebraska and New Mexico west to the Pacific Ocean.

Host Plants. Say and Uhler stink bugs feed on the fruit and seeds of many plants. They are known principally as a pest of grains, and prefer to attack the seed head of such crops as alfalfa, barley, oat, rye, and wheat. However, on occasion they damage such vegetables as asparagus, bean, cabbage, lettuce, pea, and tomato. They also feed extensively on such weeds as broom snakeweed, *Gutierrezia* spp.; lambsquarters, *Chenopodium album*; mallow, *Malva* spp.; musk thistle, *Carduus nutans*; pigweed, *Amaranthus* spp.; prickly pear cactus, *Opuntia* sp.; Russian thistle, *Salsola kali*; tansymustard, *Descurainia pinnata*; toadflax, *Linaria vulgaris*; tumblemustard, *Sisymbrium attissimum*; sage, *Artemisia* spp; saltbush, *Atriplex* spp.; and feathergrass, *Stipa* spp.

Natural Enemies. Several parasitoids are known. An egg parasitoid, *Telenomus utahensis* Ashmead (Hymenoptera: Scelionidae), is an important mortality factor of Say stink bug, sometimes causing 60% mortality or greater late in the season (Jubb and Watson, 1971a,b). When this wasp discovers an egg clutch, few if any eggs escape parasitism. Several generations of the parasitoid occur annually (Caffrey and Barber, 1919). The other known wasp parasitoids, *Telenomus podisi* Ashmead (Hymenoptera: Scelionidae) and *Ooencyrtus johnsoni* (Howard) (Hymenoptera: Encyrtidae), also attack eggs. Among fly species reared from Say stink bug are *Cylinromyia armata* Aldrich, *C. euchenor* (Walker), and *Gymnosoma fuliginosum* Robineau–Desvoidy (all Diptera: Tachinidae). Although the *Cylinromyia* spp. seem to be of little importance, the latter fly species attacks up to 25% of late instar nymphs and adults. Parasitoids of Uhler stink bug are less well-known, though *Telenomus utahensis* caused higher levels of parasitism in Uhler stink bug than in Say stink bug (Jubb and Watson, 1971b), and *G. fuliginosum* is also known from *C. uhleri*.

General predators, including the assassin bug *Sinea spinipes* (F.) (Hemiptera: Reduviidae), and the ambush bug, *Phymata erosa* Stål (Hemiptera: Phymatidae), feed on nymphs. The soft-winged flower beetle, *Collops bipunctatus* Say (Coleoptera: Melyridae), feeds on stink bug eggs. Songbirds and lizards sometimes consume both Say stink bug and Uhler stink bug (Knowlton, 1944; Knowlton *et al.*, 1946) despite the fact that some vertebrate predators learn to avoid these odor-producing and presumably distasteful insects.

Life Cycle and Description. Three or four generations of Say stink bug are known from New Mexico. The overwintering adults deposit eggs in late April or May, with first generation adults appearing in June, second in August, and third in September. A small fourth generation sometimes occurs, though many nymphs from this generation perish with the onset of cold weather. The adults from generations 2–4 enter diapause during late October or early November, and re-emerge the following spring. A complete generation requires about 80 days. Uhler stink bug's biology and morphology are largely undescribed, but presumably about the same as Say stink bug.

Egg. The eggs of Say stink bug are barrel-shaped, and deposited on end. The height is 1.1–1.2 mm, and diameter about 0.6–0.9 mm. When viewed from above, the egg is marked with three white circles which alternate with gray. On one side of each egg is an irregular patch of gray. Mean duration of the egg stage is about nine days (range 4–13 days), though during the warmest months egg hatch tends to occur in 5–7 days. Eggs darken in color immediately preceding hatch. Egg clusters are placed on the underside of some object. Overwintering adults may deposit eggs on the dead plants or rubbish comprising their overwintering quarters, but later generations tend to oviposit on green plants. They often are placed in parallel rows, usually two or four rows, with rows consisting of 7–20 eggs. This tendency is most pronounced when oviposition occurs on the narrow stems of grasses which constrain oviposition, whereas on broad surfaces the eggs may consist of several rows. The number of eggs per cluster averages about 26 (range 13–43 eggs).

Nymph. There are five instars in Say stink bug; the mean (range) duration is 5.0 (5), 8.2 (7–9), 6.4 (5–7), 8.2 (7–10), and 14.9 (13–17) days, respectively, during the summer months. Young nymphs usually remain on or near the egg mass for the duration of the first instar. The first instar measures 1.1–1.5 mm long. Its color is predominantly black, though a protruding area at the lateral margins of the thorax and abdomen bears some yellow pigment, and there are yellow-white spots on the abdomen. During the second instar, nymphs wander away from the egg mass to feed, but generally they remain fairly aggregated. The second instar measures 2.2–2.5 mm long and is colored like the first instar, except for also bearing yellow-white spots on the thorax. The later instars are solitary and move freely about the plant to feed. The nymphs feed during the morning and late afternoon, seeking shelter during the heat of the day and cool of the night. They also seek shelter during periods of high wind and rainfall. The third instar measures 3.1–3.2 mm long. The anterior region of the bug is black, but fading to dark-green toward the posterior. The lateral margins turn from white to orange during this instar, and some yellow spots remain on the abdomen. The fourth and fifth instars measure 5.2–6.6, and 8.9–10.6 mm, respectively. They are pale green, with the lateral margins white with an orange edge, and some yellow spots on the abdomen. The wing pads become evident during the final two instars, especially the latter.

Say stink bug nymph.

Adult Say stink bug.

Adult. The adults of Say stink bug are about 12–16 mm long. Their color is green, but the shade varies considerably. The lateral borders of the pronotum, three spots on the anterior border of the scutellum, and the apex of the scutellum are yellow, orange, or red. The apical portion of the front wings are marked with small purple flecks along the veins. The adult of Uhler stink bug greatly resembles Say stink bug, though it lacks purple flecks on the membraneous portion of the front wings and usually lacks the orange color common at the tip of the scutellum in Say stink bug. Newly emerged adults are unable to oviposit immediately. A period of 20–30 days normally is required before adults mate and begin to deposit eggs. Females deposit eggs over a period of about one month. Total fecundity is estimated to average 54 per female, with a maximum of 107 eggs. This seems to be a fairly small number of eggs, however, and may represent the adverse consequences of caging. Overwintering adults and first generation nymphs feed principally on early-developing native plants. The second generation adults feed extensively on grain, and cause considerable damage. Later generations also feed on native plants because grains have matured and have been harvested. Unlike the nymphs, which seek shelter during the heat of the day, adults are active throughout the daylight hours. The adults are long-lived, reportedly persisting for 3–4 months in the summer. During the winter, of course, adults must survive for several months. Adults are often found overwintering under plant debris, bales of hay, and dried livestock droppings. A key to distinguish stink bugs commonly affecting vegetables is found in Appendix A. (See color figure 146.)

The most complete research of Say stink bug biology, conducted in New Mexico, was published by Caffrey and Barber (1919). Patton and Mail (1935) contributed some observations from Montana. Buxton *et al.* (1983) described both species, and provided a key to close relatives.

Damage

Nymphs and adults prefer to suck liquids from seeds and fruits, but if these are not available they feed readily on young leaf and stem tissue. Once seeds, including grain, begins to harden, the bugs no longer are able to feed. Thus, rapidly growing seeds are preferred. Seed heads of grains that have been attacked acquire a dull yellowish-white color, in contrast to the green appearance of undamaged heads. Such seeds are hollow, or nearly so. When fruit is attacked, the tissue adjacent to the feeding puncture does not develop as the fruit grows, leaving a blemish in the form of a depression. Foliar tissue that has been fed upon becomes wilted or discolored, and often dies.

Management

These stink bugs are dependent upon weeds for development of the first generations of the season. Thus, suppression of weeds is a recommended management practice to suppress stink bugs. However, with the immense amount of rangeland supporting suitable food plants in the western states, it is improbable that weed management would be effective in many areas. Many growers, therefore, depend solely on application of foliar insecticides for stink bug control. It is useful, however, to monitor weedy areas and field margins for stink bugs, because they are infrequently abundant enough to damage vegetable crops.

Southern Green Stink Bug
Nezara viridula (Linnaeus)
(Hemiptera: Pentatomidae)

Natural History

Distribution. Southern green stink bug has a world-wide distribution. It is found on all continents where agriculture is practiced, but absent or rare from regions with cold winters. Its origin is probably eastern Africa, and was first observed in the western hemisphere in 1798. Distribution in North America is limited primarily to the southeastern United States— Virginia to Florida in the east, to Ohio and Arkansas in the midwest, and to Texas in the southwest. However, it has become established in Hawaii (in 1961) and California (in 1986), and occasional specimens have been found elsewhere outside the generally infested southeast. Southern green stink bug is a strong flier, and its range is expanding in many parts of the world.

Host Plants. The host range of this insect includes over 30 families of plants, though it shows a preference for legumes and crucifers. Preference among plants varies during the year, with this stink bug most attracted to plants that are producing pods or fruit. As plants senesce, the bugs move to more succulent hosts. Among vegetables attacked by southern green stink bug are artichoke, bean, Brussels sprouts, cabbage, cauliflower, collards, corn, cowpea, cucumber, eggplant, mustard, okra, pea, pepper, potato, radish, squash, sweet potato, tomato, and turnip. Outside of North America, this insect is most commonly known as "green vegetable bug," an indication of its most frequent host selection. However, it feeds readily on other crops, and is known to attack field crops such as corn, clover, cotton, peanut, soybean, sugarcane, rice, and tobacco as well as fruits such as blackberry, grapefruit, lime, mulberry, orange, and peach. In many tropical areas, this bug is considered a limiting factor in soybean production. Weeds commonly serve as hosts. Some of the wild plants fed upon by this bug are beggarweed, *Desmodium tortuosum*; castor bean, *Ricinus communis*; dock, *Rumex* sp.; nutgrass, *Cyperus esculentus*; lambsquarters, *Chenopodium album*; passion flower, *Passiflora incarinata*; pigweed, *Amaranthus* spp.; rattlebox, *Crotalaria usaramoensis*.; wild grape, *Vitus* sp.; and wild plum, *Prunus* sp.

Southern green stink bug, a long-lived and strong flier, moves readily among host plants. Some of the host associations are not true expressions of preference or suitability, but reflect availability. Also, host plants that are suitable for one stage may be unsuitable for another. Velasco and Walter (1992), for example, showed that castor bean was good for adult survival and egg production, but poor for nymphal survival. Corn also favors adult survival, but inhibits reproduction. Wild crucifers are not optimal hosts, but early in the season they are abundant and are the best hosts available, so they favor population increase. Soybean favors survival of both nymphs and adults, but is not especially attractive to these insects.

Natural Enemies. As might be expected of an insect with a world-wide distribution, numerous parasitoids and predators are known. Over 50 species of parasitoids are known, most as egg parasitoids, and most not specific to southern green stink bug (Jones, 1918a). The most common parasitoids in North America tend to be the egg parasitoid *Trissolcus basalis* (Wollaston) (Hymenoptera: Scelionidae) and the adult parasitoid *Trichopoda pennipes* (Fabricius) (Diptera: Tachinidae).

Trichopoda parasitizes high proportions of green stink bug populations on several crops, with the proportion parasitized increasing through the growing season (Todd and Lewis, 1976; Buschman and Whitcomb, 1980; McLain *et al.*, 1990). This insect is thought by some to be a complex of cryptic species, however, so the ecological relationship is uncertain. *Trichopoda* spp. were introduced into Hawaii and were credited with providing effective biological control on most crops.

Other North American parasitoids include *Anastatus* sp. (Hymenoptera: Eupelmidae), *Ooencyrtus* sp. (Hymenoptera: Encyrtidae), *Telenomus* spp. (Hymenoptera: Scelionidae), and others; Jones *et al.* (1996) provided a list of parasitoids. Hoffman *et al.* (1991b) reported on the successful introduction of *Trissolcus basalis* (Wollaston) (Hymenoptera: Scelionidae) to California to aid in the suppression of southern green stink bug.

Predators of southern green stink bug are numerous, and Stam *et al.* (1987) gave a detailed assessment of predation in Louisiana soybean. Common egg predators were red imported fire ant, *Solenopsis invicta* Buren (Hymenoptera: Formicidae), grasshoppers (Orthoptera: Acrididae and Tettigoniidae), and southern green stink bug nymphs. Common nymphal predators were big-eyed bug, *Geocoris* spp. (Hemiptera: Lygaeidae); damsel bug, *Reduviolus roseipennis* Reuter (Hemiptera: Nabidae); and other insects in the orders Hemiptera and Coleoptera. Spiders sometimes caused significant predation. Predators, including birds, were also discussed by Drake (1920).

The relative importance of egg predators and parsitoids was examined by Shepard *et al.*, (1994) in South Carolina. Stink bug egg clusters were introduced into several crops and their fate determined. The principal egg parasitoid was *Trissolcus basalis*, accouting for about 95% of parasitism. Predation (disappearance of eggs) and parasitism varied in relative importance among crops, seasons, and years; on average, predation and parasitism were about equally important. The action of natural enemies sometimes resulted in complete destruction of egg clusters.

Diseases of stink bug are poorly known. However, viruses affecting southern green stink bug were recently found in South Africa (Williamson and von Wechmar, 1995).

Life Cycle and Description. The southern green stink bug overwinters in the adult stage, and enters diapause when the length of photophase falls to 12–13 h. The number of generations annually is estimated at 4–5, with considerable overlapping late in the season. About 45 days are required for a complete life cycle during the summer months. Todd and Herzog (1980) indicated the following "typical" pattern of stink bug phenology in the southeastern United States: inactive adults overwinter until February or March, and when they become active they feed on crucifers and small grains; first generation nymphs and adults are present in March and April and feed on clover; second and third are present from May through July and feed on tobacco, corn, and vegetables, especially tomato; third generation adults disperse to soybean, where fourth generation nymphs and adults and fifth generation nymphs feed and develop until October or November; fifth generation adults disperse to crucifers or other late season hosts, and overwinter. With the onset of warm days in the spring, overwintered adults begin to copulate. The process is protracted, often lasting more than a day, and multiple matings are commonplace. Following 2–3 weeks of feeding the egg production commences.

Egg. The eggs are laid in clusters, generally on the lower surface of foliage, and with about 60 eggs (range 30–130 eggs) per cluster. Females normally produce 1–3 egg clusters, with total production ranging from 45 eggs when adults are provided only with relatively poor hosts, to 160 eggs when provided with good host plants. The eggs are deposited in regularly shaped, hexagonal clusters, with the individual eggs ordered in regular rows and glued together. They are affixed to the plant on end, with the top or visible end somewhat flattened, and the bottom end rounded. The eggs measure about 1.3 mm long and 0.9 mm wide. They are yellowish-white to pinkish-yellow and the top, or cap, is clearly indicated by a ring and 28–32 minute spines. The eggs darken in color as incubation proceeds, and hatching occurs after about 5 days. Hatching occurs synchronously in the egg cluster, so within 1.0–1.5 h all are hatched.

Nymph. There are five instars. At hatching the nymph is yellowish-orange, but soon it becomes brown. Each segment of the abdomen is marked, both dorsally and ventrally, with a pair of light spots. The appendages are yellow. The body length is about 1.6 mm. During most of this first instar the nymphs cluster tightly around the egg cluster and do not feed. Finally, just before the young insects are about to molt, the nymphs move from the cluster and commence feeding. In the second instar the head and thorax are black, with the abdomen reddish-black. Both the thorax and abdomen are marked with yellowish spots. The appendages are black. The body length is about 3.2 mm. Second instars remain near the egg cluster, and remain aggregated. The third instar is very similar to the second, differing principally in size, and measuring about 3.6 mm long. Third instars disperse away from the egg cluster, but remain aggregated. The fourth instar also may be relatively unchanged in appearance from the preceding instars, differing principally in body length of about 4.2 mm. However, the color of the fourth instar is highly variable, and may instead be greenish; the thorax is light green with black markings, and the abdomen is darker green with salmon shading and white spots. The appendages are brownish, with the tip of the antennae greenish. Unlike the preceding instars, fourth and fifth instars do not aggregate. The fifth instar is also highly variable, the head, thorax and wing pads ranging from light green to almost black. The abdomen tends to be colored light or dark, corresponding to the shade of the thorax, and marked with rose spots dorsally, and whitish spots laterally. The body length of fifth instars is about 10 mm. Mean duration (range) of instars 1–5 is about 4 (3–5), 5 (3–10), 5 (3–9), 8 (6–10), and 10 (7–13) days, respectively. Thus, total nymphal development time is about 32 days, and egg to adult development requires 35–37 days, depending on temperature and suitability of food. The optimal temperature for development is about 30°C.

Adult. The adult is generally uniform light-green, both dorsally and ventrally, though the ventral surface is paler. Adults measure about 13–17 mm long and 8 mm wide. The thorax is the widest part of the body,

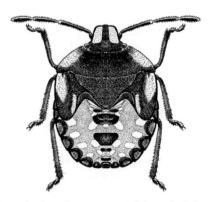

Nymph of southern green stink bug, dark form.

and is rounded laterally. Sometimes the adults differ in having yellow or brown on the thorax or scutellum, and the brown color may extend onto the forewings. During the summer months, females begin egg production about 14–20 days after attaining the adult stage. (See color figure 145.)

Southern green stink bug, *N. viridula*, is easily confused with green stink bug, *Acrosternum hilare* (Say). However, the two species differ ecologically and morphologically. *Acrosternum hilare* is usually associated with trees and shrubs, rather than the herbaceous vegetation fed upon by *N. viridula*. Also, though *A. hilare* is found widely in North America, it is most abundant in the north. The two species can be differentiated by the shape of the abdominal spine. When viewed from below, *A. hilare* has a pointed spine protruding forward between the base of the hind legs, whereas in *N. viridula* the spine is rounded. A key to distinguish stink bugs commonly affecting vegetables is found in Appendix A.

In coastal areas of the south, the adults of southern green stink bug are scarce during the winter months but not all are dormant, a few being found feeding on succulent plants. Even in these relatively warm areas, however, no nymphs are found. Overwintering occurs in sheltered locations, most commonly under bark of trees, but also beneath fallen leaves and in Spanish moss. Considerable winter mortality among southern green stink bug can occur at the northern limits of its range, and winter weather is considered to be an important variable in population abundance. High levels of mortality occur when bugs are exposed to -10 and $-5°C$ for 3 and 55 h, respectively (Elsey, 1993).

Southern green stink bug displays a complex chemical ecology. Nymphs produce a bifunctional pheromone that functions as an aggregation pheromone at low concentrations and a dispersant at high concentrations. Adult males produce an aggregation pheromone that attracts females, other males, and the parasitoid *Trichopoda pennipes* (Fabricius) (Diptera: Tachinidae). Presumably the principal function of this chemical is to enhance mate finding. However, there are visual, tactile, and acoustic stimuli that are a necessary prelude to mating (Harris *et al.*, 1982). Finally, stink bugs secrete a defensive chemical that apparently is repellent to predators.

Descriptions and summaries of life history were provided by Jones (1918a) and Drake (1920), the developmental biology was given by Harris and Todd (1980), and a useful review of behavior and ecology was presented by Todd (1989). Information on stink bug culture was provided by Harris and Todd (1981) and Brewer and Jones (1985).

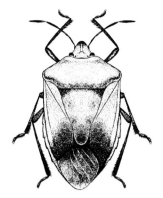
Adult southern green stink bug.

Damage

Southern green stink bug punctures plant tissue with its piercing-sucking mouthparts and removes sap, often feeding at night. Stem, leaf, blossom and fruit tissue may all be attacked, but fruiting structures are preferred and invariably the tissue selected is young and succulent. The feeding site, as it heals, becomes hard and darkened. Seeds may be shriveled, deformed, and shrunken, or may simply bear a dark mark and depression at the feeding site. Fruit may be deformed or dropped from the plant. Under high levels of feeding pressure, young plants and plant shoots may perish. Stink bugs are capable of introducing bacteria and yeasts into plants as they feed. Wounds also allow entry of secondary plant pathogens.

Vegetable crops are easily damaged by southern green stink bug. Studies of tomato fruit attack in Louisiana indicated that small fruit were preferred over large fruit, and green over red fruit (Lye and Story, 1988). There was an inverse linear relationship between bug density and fruit quality; fewer than two bugs on a tomato fruit for one day, or one bug for two days, resulted in decrease in fruit quality (Lye *et al.*, 1988a,b). In corn grown in Louisiana, as few as two bugs per ear of corn during early ear development could result in damage (Negron and Riley, 1987). Cowpea and lima bean were studied in Georgia, and two days of feeding by one bug on a pod significantly reduced pod size and mean seed weight (Nilakhe *et al.*, 1981a,b).

Management

Sampling. A comprehensive review of stink bug sampling was published by Todd and Herzog (1980), though soybean rather than vegetables is emphasized. Sweep nets and direct visual examination are techniques most useful for stinkbug sampling. Lye and Story (1989) studied southern green stink bug sampling in Louisiana tomatoes. They reported that bugs were moderately aggregated and that young fruit clusters were the most efficient sampling unit. They also developed a sequential sampling protocol.

Insecticides. Insecticides are commonly applied to foliage, especially at blossoming and pod set, to protect susceptible crops from stink bug damage. Systemic insecticides applied to the soil tend to be less effective (Chalfant, 1973a). Insecticide resistance has become a problem in some areas of the world, so steps to minimize development of resistant populations should be followed. Insecticides can interfere with naturally occurring biological control, but research has shown that insecticides with a short residual life have a minimal impact on egg parasitoids (Orr *et al.*, 1989).

Cultural Techniques. Consideration of early season feeding behavior is an important component of management. Cruciferous crops and weeds are potential sources of stink bugs for later legume crops, and should be monitored, treated, or destroyed. Legume cover crops or forage crops are another source of bugs that might invade vegetables, and should be treated similarly.

Farmers and entomologists have long attempted to use early-season crops, or late-season crops planted early, as trap crops to lure stink bugs from the principal crop. Early in the spring or late in the autumn cruciferous plants such as collards, mustard, radish, rape, and turnip are attractive to bugs, whereas late in the spring and during the summer legumes such as bean and cowpea are most attractive. Drake (1920) recommended radish and collards in the spring and rattlebox in the summer to protect tomato. McPherson and Newsom (1984) and Todd and Schumann (1988) reported that early-planted soybean or cowpea could serve as a trap crop for late-planted soybean. Stink bugs will disperse from trap crops, however, so usually they must be treated with insecticide before dispersal occurs in the adult stage.

Plant cultivars resistant to attack by stink bug generally are not available commercially, but genetic sources of resistance have been identified for cowpea (Schalk and Fery, 1986).

FAMILY THYRECORIDAE—NEGRO BUGS

Little Negro Bug
Corimelaena pulicaria (Germar)
(Hemiptera: Thyreocoridae)

Natural History

Distribution. This common native insect is found throughout the United States and southern Canada, except possibly the Maritime Provinces of Canada. Its distribution also includes Central America. In the western United States a related and similar-appearing species, *C. extensa* Uhler, occasionally feeds on crops.

Host Plants. This insect has been observed feeding on many plants. Among vegetables attacked are celery, potato, and sweet potato. It has also been associated with field crops such as chufa, clover, soybean, and sugarbeet. It is probably best known for damage to small fruits such as blackberry, grape, raspberry, and strawberry. Weeds are common hosts, incuding beggarstick, *Bidens* spp.; ragweed, *Ambrosia* spp.; redroot pigweed, *Amarnathus retroflexus*; wild carrot, *Daucus carota*; wild chervil, *Cryptotaenia canadensis*; plantain, *Plantago lanceolata*; common mullein, *Verbascum thapsus*, and others. Weeds appear to be preferred hosts.

Natural Enemies. Natural enemies seem to be unknown.

Life Cycle and Description. There is a single generation per year in the northern United States. The adult is the overwintering stage. Adults seek shelter under leaf litter and boards during the winter months. The adults are active early in the year, and eggs are deposited in May and June. Nymphs are found throughout the summer months.

Egg. The eggs are deposited singly. They are oval in shape and orange to red in color and measures about 0.6 mm long and 0.4 mm wide. Incubation is about 10–14 days.

Nymph. The young nymphs greatly resemble the adults, but the abdomen is colored red. There are five instars. Development time for the nymphs is about 30 days. Because the overwintered adults live through much of the summer and continue to oviposit, overlapping stages of development are found during the summer.

Adult. The adults are oval in overall form, and greatly resemble beetles. Examination of the mouthparts quickly separates the two groups, however, as the piercing-sucking mouthparts of the negro bug are easily observed. The scutellum of these insects, and other members of the family, are greatly enlarged and bluntly rounded, covering nearly the entire abdomen. This effectively hides all but the leading edge of the wings. Where the wings are exposed, along the perimeter of the abdomen, there is a whitish stripe. As its name suggests, the insect is almost entirely black, the principal exception being the aforementioned edges of the wings. This bug is quite small, and measures only 2.2–3.5 mm long. The antennae

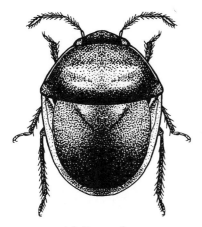

Adult negro bug.

are five segmented, and the mouthparts consist principally of a four-segmented labium.

The biology of this insect is poorly documented. Partial life histories were provided by Riley (1870a) and Davis (1893). McPherson (1982) gave a good summary, especially of host-plant records, and keys for identification. A key to distinguish stink bugs (including Negro bug) commonly affecting vegetables also is found in Appendix A.

Damage

This generally is a minor pest, but it does reach high numbers on occasion. On celery, the bugs are known to aggregate into small clusters and feed at the joint where the leaf petioles meet the stem. This causes the terminal leaves to wilt and die. The bugs then move progressively lower on the plant until even the youngest tissue is destroyed. In addition to its feeding injury, it is reputed to have a foul odor, and by walking on fruit, to impart a disagreeable taste (Riley, 1870a).

Management

These insect are infrequent pests, and are readily controlled with foliar applications of insecticides. Because little negro bugs feed predominantly on weeds, it is a good idea to keep weeds under control, or at least well-separated from vegetable and berry crops.

FAMILY TINGIDAE—LACE BUGS

Eggplant Lace Bug
Gargaphia solani Heidemann
(Hemiptera: Tingidae)

Natural history

Distribution. Eggplant lace bug occurs widely in the eastern United States north to about New Hamp-

shire and Iowa, and west to Kansas and Arizona, but it is not common in the Gulf Coast area. There are occasional reports from outside this range, including from Canada, but it is infrequently abundant in such areas. The range of eggplant lace bug also extends into Mexico. This appears to be a native insect.

Host Plants. As its common name suggests, this insect affects principally eggplant, though it sometimes affects potato and perhaps tomato. Not surprisingly, it is commonly found on weeds in the plant family Solanaceae such as horsenettle, *Solanum carolinense*, and silverleaf nightshade, *S. elaeagnifolium*. Eggplant lace bug occasionally is reported from hosts other than Solanaceae, including Compositae, Leguminosae, and Malvaceae.

Natural Enemies. Several common predators have been observed to attack eggplant lace bug, among them the lady beetles *Hippodamia convergens* Guerin–Meneville and *Coleomegilla maculata* De Geer (Coleoptera: Coccinellidae); the spined soldier bug, *Podisus maculiventris* (Hemiptera: Pentatomidae); the insidious flower bug, *Orius insidiosus* (Say) (Hemiptera: Anthocoridae); and various spiders.

Life Cycle and Description. In Virginia the complete life cycle takes about 20 days, and these lace bugs are active from May to November. Adults overwinter, often in clumps of grass but sometimes under bark. The number of generations is estimated at 6–8 per year, but overlap so completely that they are hard to distinguish.

Egg. The eggs are deposited on the lower surface of leaves in clusters containing over 100 eggs. The eggs are attached on end, and are not erect, but rather tilted at different angles. They are about 0.37 mm long and 0.18 mm wide. They are brown basally and greenish apically. The tips of the eggs are recessed and surrounded by a delicate, lace-like border. The female may require 4–5 days to complete the egg cluster, depositing 15–60 eggs at each oviposition. At completion of egg laying, a sticky secretion is spread over the cluster, and the female tends the eggs, leaving only to feed for brief intervals. The incubation period is 5–8 days.

Nymph. Nymphs remain clustered as they feed and grow. There are five instars, each with a duration of approximately two days. The first instar is light-yellow, with pink eyes, and antennae are about as long as the body. This instar measures 0.3–0.4 mm long. At the molt to the second instar, which measures about 0.8 mm long, the nymph acquires large spines laterally on the thorax and abdomen. The third instar is about the same length as the preceding stage, but is wider, bears spines dorsally, and displays evidence of wing pads. The fourth instar measures about 1.5 mm long, the thorax is greatly expanded laterally, the entire body is covered with spines, and the wing pads extend to about the second abdominal segment. The fifth instar is about 2.2 mm long, and like the preceding instars, yellowish in color. This stage also is adorned with spiny processes, most of which are brown in color.

Adult.   The adult eggplant lace bug is rather flat in general form, and mostly dark brown, though the legs are yellow. The front wings are distinctively lacy, with the basal cells more densely packed. The thorax is pronounced, flaring laterally but also extending posteriorly. The antennae are nearly as long as the body, and thickened apically. The adult measures about 4 mm long and 2 mm wide. Adults have a relatively brief period of oviposition, usually 4–5 days, after which they turn their attention to brood care.

Eggplant lace bug adults display complex social interactions, including maternal behavior. For example, the female remains with the eggs and nymphs as they develop. As nymphs deplete the food and become restless, the adult leads them to a new feeding site. Nymphs respond to an alarm pheromone that is released when the bugs are crushed by dispersing quickly, only to reassemble at another location, accompanied by the adult. The alarm pheromone was described by Aldrich *et al.* (1991a). The adults use wing fanning to deter predation, and also to communicate with nymphs. There is some evidence for an aggregation pheromone (Kearns and Yamamoto, 1981). Interestingly, females frequently deposit their eggs within egg clusters of other females. This relegates the recipient female to surrogate mother status, allowing the donor female to continue to produce more eggs without the responsibility for care of her offspring (Tallamy, 1985).

An excellent treatment of eggplant lace bug biology was given by Fink (1915), with supplementary information provided by Bailey (1951). Eggplant lace bug

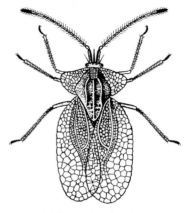

Adult eggplant lace bug.

was included in keys to the Hemiptera of Missouri (Froeschner, 1944) and the lace bugs of North Carolina (Horn *et al.*, 1979).

Damage

The adults and nymphs suck sap from the leaves of host plants. The first symptom of injury is large discolored blotches or patches, usually circular in shape. This reflects the feeding by the adult, and perhaps the very young nymphs, in the vicinity of the egg cluster. Later, as the nymphs disperse away from the egg cluster, the discoloration spreads until the entire leaf is involved, turning yellow and dry. Often every leaf of an infested plant is discolored as the insects move from leaf to leaf. The adults disperse from plant to plant.

Management

Although lace bugs are not often serious pests, they occasionally reach damaging levels. They are readily suppressed with insecticides applied to the foliage, but care must be taken to attain thorough coverage of the plant because these insects are found on the lower surfaces of the leaves.

Order Homoptera—Aphids, Leaf- and Planthoppers, Psyllids and Whiteflies

FAMILY ALEYRODIDAE—WHITEFLIES

Greenhouse Whitefly
Trialeurodes vaporariorum (Westwood)
(Homoptera: Aleyrodidae)

Natural History

Distribution. Greenhouse whitefly is found widely around the world, including most of the temperate and subtropical regions of North America, South America, Europe, Central Asia and India, northern and eastern Africa, New Zealand and southern Australia. It does not thrive in most tropical locations, and occurs in colder regions only by virtue of its ability to survive in winter in greenhouses. In the northern United States and Canada, it overwinters only in such protected locations, but in mild-winter areas of the southern states and in Hawaii and Puerto Rico, it survives outdoors throughout the year. The origin of this species is not clear, but is thought to be Mexico or the southwestern United States.

Host Plants. This species has a very wide host range, with over 300 species recorded as hosts. However, some hosts are more suitable. Vegetable plants often serving as good hosts are bean, cantaloupe, cucumber, lettuce, squash, tomato, eggplant, and sometime cabbage, sweet potato, pepper, and potato. Among greenhouse-grown vegetables, the most common hosts are tomato, eggplant, and cucumber. When adults land on favored host plants such as eggplant, they almost always remain to feed and oviposit; on a less preferred host such as pepper, they usually take flight after tasting the plant. Many ornamental plants serve as good hosts, including ageratum, aster, chrysanthemum, coleus, gardenia, gerbera, lantana, poinsettia, salvia, verbena, zinnia, and many others.

Natural Enemies. Natural enemies of greenhouse whitefly are numerous, but few are consistently effective, especially under greenhouse conditions (Gerling, 1990; Onillon, 1990). Greenhouse whitefly is attacked by the common predators of small insects, including minute pirate bugs (Hemiptera: Anthocoridae), some plant bugs (Hemiptera: Miridae), green lacewings (Neuroptera: Chrysopidae), brown lacewings (Neuroptera: Hemerobiidae), and lady beetles (Coleoptera: Coccinellidae). Parasitic wasps attacking greenhouse whitefly are largely confined to the family Aphelinidae, but many species are involved and they vary regionally. Those known from North America, including Hawaii, are *Encarsia formosa* Gahan, *Aleurodophilus pergandiella* (Howard), *Eretmocerus haldemani* Howard, *Prospaltella transvena* Timberlake, and *Aphidencyrtus aphidivorus* (Mayr). Although these agents exercise considerable control on whitefly populations in weedy areas or on crops where insecticide use is minimal or absent, they do not survive well in the presence of most insecticides. *Encarsia formosa* is used successfully under greenhouse conditions, and to a lesser extent field conditions, to affect biological suppression. (For more information, see the section on biological control.)

The pathogens of greenhouse whitefly are principally fungi, particularly *Aschersonia aleyrodis*, *Paecilomyces fumosoroseus*, and *Verticillium lecanii*. All occur naturally and can cause epizootics in greenhouses and fields, and also are promoted for use in greenhouses as bioinsecticides. *Aschersonia* is specific to whiteflies, *Verticillium* has a moderately wide host range, and *Paecilomyces* has a broad host range (Fransen, 1990; Osborne and Landa, 1992). For optimal development of disease, high humidity is required. *Aschersonia* is spread principally by rainfall, so often it fares poorly in greenhouse environments.

Life Cycle and Description. The development period from egg to adult requires about 25–30 days at 21°C, and 22–25 days at 24°C. Thus, because the pre-oviposition period of adults also is short, less than two days above 20°C, a complete life cycle is possible within a month. Greenhouse whitefly can live for months, and oviposition time can exceed the development time of immatures, resulting in overlapping generations. Optimal relative humidity is about 75–80%. The developmental threshold for all stages is about 8.5°C.

Egg. The eggs are oval, and suspended from the leaf by a short, narrow stalk. The eggs initially are green and dusted with white powdery wax, but turn brown or black as they mature. The eggs measure about 0.24 mm long and 0.07 mm wide. They are deposited on the youngest plant tissue, usually on the underside of leaves in an incomplete circular pattern. Up to 15 eggs may be deposited in a circle that measures about 1.5 mm in diameter. This pattern results from the female moving in a circle while she remains with her mouthparts inserted into the plant. This pattern is less likely on plants with a high density of trichomes, because plant hairs interfere with the oviposition behavior. Duration of the eggs stage is often 10–12 days, but eggs may persist for over 100 days under cool conditions. When cultured at 18°, 22.5°, and 27°C, egg development requires an average of 15, 9.8, and 7.6 days, respectively. Maximum fecundity varies according to temperature; optimal temperature is 20–25°C regardless of host plant. When feeding on eggplant, greenhouse whitefly produces over 500 eggs, and on cucumber and tomato about 175–200 eggs (Drost *et al.*, 1998).

Nymph. The newly hatched whitefly nymph is flattened, oval in outline, and bears functional legs and antennae. The perimeter is equipped with waxy filaments. The first instar measures about 0.3 mm long. It is translucent, usually appearing to be pale-green but with red eyes. After crawling one centimeter or so from the egg, it settles to feed and molt. Development of the first instar requires 6.5, 4.2, and 2.9 days, respectively, when cultured at 18°, 22.5°, and 27°C. The second and third nymphal stages are similar in form and larger in size, though the legs and antennae become reduced and nonfunctional. They measure about 0.38 and 0.52 mm long, respectively. Duration of the second instar requires about 4.3, 3.2, and 1.9 days, whereas third instars require 4.5, 3.2, and 2.5 days, respectively when cultured at 18°, 22.5°, and 27°C. The fourth nym-

Greenhouse whitefly egg.

phal stage, which is usually called the "pupa," differs in appearance from the preceding stages. The fourth instar measures about 0.75 mm long, is thicker and more opaque in appearance, and is equipped with long waxy filaments. The pupal stage actually consists of the fourth nymphal instar period, which is a period of feeding, plus the period of pupation, which is a time of transformation to the adult stage. Thus, pupation occurs within the cuticle of the fourth instar. Duration of the fourth instar and pupal periods are 8.7 and 5.9, 5.9 and 4.0, and 4.5 and 2.8 days, respectively, when cultured at 18°, 22.5°, and 27°C.

The form of the pupa is used to distinguish among whitefly species, and can be used to separate greenhouse whitefly from the similar-appearing silverleaf whitefly, *Bemisia argentifolii* Bellows and Perring, and sweetpotato whitefly, *B. tabaci* (Grennadius). Greenhouse whitefly is straight-sided when viewed laterally, ovoid, and lacks a distinct groove near the anal end of the body. In contrast, the *Bemisia* spp. are oblique-sided, irregularly oval, and possess a distinct groove in the anal region.

Individuals of greenhouse whitefly that develop on lightly or moderately pubescent leaves tend to be relatively large and to have four pairs of well-developed dorsal waxy filaments. In contrast, whiteflies developing on densely pubescent leaves tend to be smaller, and they bear more than four pairs of dorsal filaments. These morphological variations are not entirely consistent, and have led to considerable taxonomic confusion.

Adult. The adults are small, measuring 1.0–2.0 mm long. They are white, with the color derived from the presence of white waxy or mealy material, and have reddish eyes. They bear four wings, with the hind wings nearly as long as the forewings. The antennae are evident. In general form, viewed from above, this insect is triangular in shape, because the distal por-

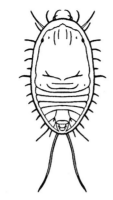

Greenhouse whitefly nymph.

Adult greenhouse whitefly.

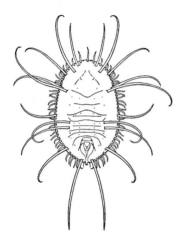

Greenhouse whitefly "pupa" (*top view*).

Greenhouse whitefly "pupa" (*side view*).

Damage

The adult and nymphal whiteflies use their piercing-sucking mouthparts to feed on the phloem of host plants. This results in direct damage, resulting in localized spotting, yellowing, or leaf drop. Under heavy feeding pressure wilting and severe growth reduction may occur. Whiteflies also secrete a large amount of sugary honeydew, which coats the plants with sticky material, and must be removed from fruit before it is marketed. The honeydew also provides a substrate for growth of sooty mold—a black fungus that interferes with the photosynthesis and transpiration of plants.

Greenhouse whitefly is, as its common name suggests, primarily a pest in greenhouses, and causes a serious limitation to the production of vegetables grown in such structures. However, it can also be a field pest, often in warmer climates but also in cool climates when seedlings contaminated with whiteflies are transplanted into the field. For example, in Hawaii field-grown tomatoes suffered a 5% reduction in fruit weight with just 0.7 whiteflies per square centimeter of leaf tissue, and a 5% reduction in grade-A fruit owing to contamination with honeydew at densities of about 8.3 whiteflies per square centimeter of leaf tissue (Johnson *et al.*, 1992).

Greenhouse whitefly is capable of transmitting viruses to plants, but is not considered to be a serious vector, particularly relative to the *Bemisia* spp. However, greenhouse whitefly transmits beet pseudoyellow virus to cucumber in greenhouse culture.

Management

Sampling. Although whitefly nymphs and adults can be detected readily by visual examination of foliage, most monitoring systems take advantage of the attraction of adults to yellow, and use yellow sticky traps to capture flying insects. Sticky cards or ribbons

tions of the wings are wider than the basal sections. The wings are held horizontally when at rest; this characteristic is useful for distinguishing this species from the similar-appearing silverleaf whitefly and sweetpotato whitefly, which hold their wings angled or roof-like when at rest. Mating may occur repeatedly, though females can also produce eggs without mating.

Detailed description of greenhouse whitefly was provided by Hargreaves (1914) and life history characteristics by van Roermund and van Lenteren (1992). Madueke and Coaker (1984) described temperature relations. A key to the common whitefly pests was published by Martin (1987).

are suspended at about the height of the crop for optimal monitoring. Traps must be placed close to plants or to the ground, or population densities will be underestimated (Gillespie and Quiring, 1992). An important aspect is to have traps dispersed widely, because whitefly distribution is not uniform within a crop. Whitefly flight peaks at about noon, but under greenhouse conditions it is independent of temperature if the basal flight temperature of 16–17°C is exceeded (Liu *et al.*, 1994). Heinz *et al.* (1992) suggested that sticky card traps could be subsampled by counting only the central vertical portion, thereby reducing labor and time for population estimation.

Insecticides. Applications of insecticides are often made to minimize the effects of whitefly feeding on crops in greenhouses. Greenhouse whitefly feeds on the lower surface of foliage and is sessile throughout most of its life, habits that minimize contact with insecticides, and resulting in frequent applications and effectiveness mostly against the adult stage. In greenhouse culture, application intervals of only 4–5 days are common, and sytemic insecticides are often used to increase the likelihood of insect contact with toxins. Thus, whitefly resistance to nearly all classes of insecticides is known, and rotation of insecticide classes is encouraged. Mixtures of insecticides are often used, which is indicative of high levels of resistance among whiteflies to insecticides. Field populations of greenhouse whitefly invariably are derived from greenhouse populations, and possess similar resistance to many insecticides. Applications of petroleum oils (Larew and Locke, 1990) and biological control agents (van Lenteren *et al.*, 1996) help to avoid difficulties with insecticide resistance.

Some insecticidal materials can be integrated into biologically based whitefly management systems. Selective materials that affect only adult and nymphal whiteflies, insect growth regulators, and insecticidal soaps are somewhat compatible with parasitoids and can be used when parasitoids are failing (Dowell, 1990).

Cultural Practices. Few cultural practices are available, but disruption of the whitefly population with host-free periods is important. Continuous culture of plants allows whiteflies to move from older to younger plants. Similarly, weeds may allow whiteflies to bridge crop-free periods, and should be eliminated. Culture of plants over white reflective mulch also decreases whitefly densities (Kelly *et al.*, 1989). Yellow sticky traps can be hung in greenhouses to capture adult whiteflies, thereby reducing whitefly density.

Biological Control. Seasonal inoculative release of the parasitoid *Encarsia formosa* Gahan into crops infested with greenhouse whitefly has been used extensively for suppression of whiteflies on greenhouse-grown vegetable crops. Excellent suppression of whiteflies is attainable, but on host plants such as cucumber and eggplant, which are very favorable for whitefly reproduction and have hairy leaves that interfere with parasitoid searching, frequent releases must be made. Alternatively, cucumber varieties with reduced trichome density have been developed which favor parasitism. Another critical factor is temperature, because low greenhouse temperatures are more suitable for whitefly activity than parasitoid activity. Daytime temperature of about 24°C seems to be optimal; temperature of 18°C or less suppresses parasitoid searching. A cold-tolerant *Encarsia* strain that is active at 13–17°C has also been used to overcome this temperature problem. Interference from pesticides can markedly affect parasitoid survival, so other pests such as mites must be managed biologically also. Lastly, release rates are important because if too many parasitoids are released the host whiteflies are driven nearly to extinction, leading to disappearance of the parasitoids; this is most likely to occur in small greenhouses. Alternatively, parasitoid releases can be made throughout the season, irrespective of whitefly presence. Use of *Encarsia formosa* for whitefly suppression was reviewed by van Lenteren *et al.* (1996). Although the protocols and technologies for whitefly management using *E. formosa* have been perfected for use in greenhouses, management under outdoor conditions awaits further research.

The fungus *Verticillium lecanii* is sometimes used commercially in Europe for whitefly and thrips suppression in greenhouses, though its success is strongly affected by humidity. Where humidity can be raised to a high level, epizootics can be induced in 1–2 weeks. Both young and adult stages are susceptible to infection.

Silverleaf Whitefly
Bemisia argentifolii Bellows and Perring

Sweetpotato Whitefly
Bemisia tabaci (Grennadius)
(Homoptera: Aleyrodidae)

Natural History

Distribution. Silverleaf whitefly and sweetpotato whitefly are closely related whitefly species, or possibly strains of the same species. They cannot be distinguished easily by appearance, though there are some biological differences. Much literature from tropical and subtropical environments around the world referring to sweetpotato whitefly probably pertains to silverleaf whitefly and other closely related species. Until the identities of these species are accurately determined, their distribution will remain unknown and some aspects of their biologies will remain confused. It appears, however, that sweetpotato whitefly has been in Florida since at least 1900, but it was not considered as a serious pest in the United States until about 1981 when it became very abundant in Arizona and southern California.

Silverleaf whitefly gained access to the southern United States in the 1980s; the first serious problem occurred in Florida in 1986 followed soon thereafter by similar problems in California, Arizona, and Texas. In the desert southwest, where both whiteflies occurred as serious vegetable and field crop pests, sweetpotato whitefly was displaced by silverleaf whitefly. Thus, sweetpotato whitefly appears to have been returned to insignificant pest status, but is often confused with the serious pest, silverleaf whitefly. The states of Arizona, California, Florida, and Texas are particularly affected by silverleaf whitefly because these are locations where silverleaf whitefly successfully overwinters and where crop and weed hosts persist throughout the year. Cooler climates do not experience major problems with silverleaf whitefly except when overwintering in greenhouses occurs or vegetable transplants originate from whitefly-infested areas.

Host Plants. Silverleaf whitefly has a very wide host range. In Florida alone, over 50 hosts have been reported, and there may be 500 worldwide. Sweet potato, cucumber, cantaloupe, watermelon, squash, eggplant, pepper, tomato, lettuce, broccoli and many other crops are hosts, but suitability varies. For example, sweet potato, cucumber, and squash are much more favorable for whitefly than broccoli and carrot (Coudriet *et al.*, 1985). Tsai and Wang (1996) rated suitability of five vegetable crops for whitefly population development as eggplant > tomato > sweet potato > cucumber > snap bean. Various weeds and field crops may favor survival of whiteflies during vegetable-free periods (Coudriet *et al.*, 1986). Wild lettuce, *Lactuca serriola*, and sowthistle, *Sonchus* spp., are examples of suitable weed hosts. Cotton, soybean, and to a lesser extent alfalfa and peanut are field crop hosts. (See color figure 10.)

Sweetpotato whitefly has a narrower host range than silverleaf whitefly. For example, silverleaf whitefly oviposits more readily on cabbage (Blua *et al.*, 1995), which is an important factor in overwintering.

As with most insects, host preference and suitability for whitefly growth and survival are highly, but not perfectly, correlated (Zalom *et al.*, 1995). Despite the lack of suitability among certain hosts for whitefly

growth and reproduction, even less suitable vegetable crops such as lettuce can sometimes be damaged when large numbers of dispersing whiteflies feed and oviposit. This is especially likely to happen when suitable crops supporting numerous whiteflies senesce or are harvested, forcing the whiteflies to disperse in search of food. Survival of whitefly may be enhanced by feeding on virus-infected plants, relative to healthy plants (Costa *et al.*, 1991). Deterioration of host-plant quality will often prompt dispersal of adults.

Weather. Silverleaf whitefly thrives under hot and dry conditions. Rainfall seems to decrease populations though the mechanism is not known. Silverleaf whitefly is not a strong flier, normally moving only short distances in search of young plant tissue. However, under proper weather conditions dramatic, long-distance flights involving millions or billions of insects are observed. Such flights normally occur in the morning as the sun heats the ground, and most insects move downwind.

Natural Enemies. Numerous predators, parasitoids, and fungal diseases of silverleaf whitefly are known (Gerling, 1990; Cock, 1993). The general predators usually associated with Homoptera such as minute pirate bugs (Hemiptera: Anthocoridae), green lacewings (Neuroptera: Chrysopidae), and lady beetles (Coleoptera: Coccinellidae) are important, as are many parasitic wasps, particularly in the genera *Encarsia* and *Eretmocerus* (both Hymenoptera: Aphelinidae) (Polaszek *et al.*, 1992). While these agents exert considerable control on whitefly populations in weedy areas or on crops where insecticide use is minimal or absent, they do not survive well in the presence of most insecticides.

Life Cycle and Description. Silverleaf whitefly can complete a generation in about 20–30 days under favorable weather conditions (Gerling *et al.*, 1986). In tropical countries up to 15 generations per year have been reported, and in the southwestern United States 12–13 generations occur annually.

Egg. The egg is about 0.2 mm long, elongate, and tapers distally; it is attached to the plant by a short stalk. The whitish eggs turn brown before hatching, which occurs in 4–7 days. The female deposits 90–95% of her eggs on the lower surfaces of young leaves (Simmons, 1994).

Nymph. All instars are translucent, and somewhat shiny. The flattened first instar is mobile, and is commonly called the "crawler" stage. It measures about 0.27 mm long and 0.15 mm wide. Its movement is usually limited to the first few hours after hatch, and to a distance of 1–2 mm (Price and Taborsky, 1992).

Duration of the first instar is usually 2–4 days. The feeding site is normally the lower surface of a leaf, but sometimes more than 50% of the nymphs are found on the upper surface, and feeding location seems not to affect survival (Simmons, 1999). The second and third instars are similarly flattened, but their leg segmentation becomes reduced and the legs nonfunctional. Duration of these instars is about 2–3 days for each. Body length and width are 0.36 and 0.22 mm, and 0.49 and 0.29 mm, for the second and third instar, respectively. Although the early portion of the fourth instar is similar to instars 2–3, the latter portion is sessile and called the "pupa." This term is not technically correct because some feeding occurs during this instar. The appearance of the fourth instar is variable, depending on the food plant; this stage tends to be spiny when develops on a hairy leaf but has fewer filaments or spines when feeding on smooth leaves. The fourth instar measures about 0.7 mm long and 0.4 mm wide. Duration of the fourth instar is about 4–7 days. Total pre-adult development time averages 15–18 days in the temperature range of 25–32°C, but increases markedly at lower temperatures. The lower- and upper-developmental thresholds are considered to be about 10° and 32°C (Natwick and Zalom, 1984). (See color figure 175.)

The form of the pupa is used to distinguish among whitefly species, and can be used to separate greenhouse whitefly from the similar-appearing greenhouse whitefly, *Trialeurodes vaporariorum* (Westwood), from the *Bemisia* spp. Greenhouse whitefly is straight-sided when viewed laterally, ovoid, and lacks a groove near the anal end of the body. In contrast, the *Bemesia* spp. are oblique-sided, irregularly oval, and possess a groove in the anal region.

Adult. The adult is white and measures about 1.0–1.3 mm long. The antennae are pronounced and the eyes red. Oviposition begins 2–5 days after emergence of the adult, often at about 5 eggs per day. Adults typically live 10–20 days and may produce about 50–150 eggs, though there are records of over 300 eggs per female. Females may produce male offspring without fertilization but males are common, so most females are probably fertilized. (See color figure 174.)

The biology and management of "sweetpotato" whitefly were comprehensively reviewed by Butler *et al.* (1986) and Cock (1986, 1993). A study of sweetpotato whitefly developmental biology on tomato was published by Salas and Mendoza (1995). Silverleaf whitefly was described by Bellows *et al.* (1994). The specific status of these whiteflies was discussed by Brown *et al.* (1995).

Another whitefly that easily can be confused with *Bemisia* spp. is greenhouse whitefly, *Trialeurodes vaporariorum* Westwood. They can be distinguished in the field by the manner in which they hold their wings. *Bemisia* spp. hold their wings roof-like over their body while at rest, whereas greenhouse whitefly holds its wings horizontally when at rest.

Damage

The adult and nymphal whiteflies use their piercing-sucking mouthparts to feed on the phloem of host plants. This results in direct damage, which is manifested in localized spotting, yellowing, or leaf drop. Under heavy feeding pressure wilting and severe growth reduction may occur.

Systemic effects also are common, with uninfested leaves and other tissue being severely damaged by whitefly feeding on other areas of the plant. A trans-

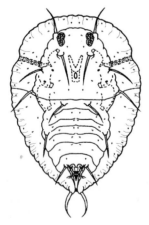

Silverleaf whitefly nymph.

Adult silverleaf whitefly.

located toxicogenic secretion by nymphs, but not by adults, is implicated (Yokomi *et al.*, 1990). The young and developing tissue is damaged by whiteflies while feeding on older tissue. Once the whiteflies are removed, new plant growth is normal if a disease is not transmitted. Damaged foliar tissue, however, does not recover once injured. Among leafy vegetables and crucifers, white streaking or discoloration, especially of veins, is common (Brown *et al.*, 1992; Costa *et al.*, 1993). In Texas, population densities of three adult whiteflies per leaf are estimated to inflict 10% yield reduction in cantaloupe (Riley and Sparks 1993). Other studies in Texas and Arizona (Riley and Palumbo, 1995a,b) demonstrated similar losses, and indicated that yields could be optimized if plants were treated with insecticide at whitefly densities of three adults per leaf or 0.5 large nymphs per 7.6 sq cm of leaf area.

A disorder called irregular ripening affects tomato fruit when whiteflies feed on tomato foliage (Schuster *et al.*, 1990). Although the tomato foliage is not damaged, the internal portions of the fruit do not ripen properly and the surface is blotched or streaked with yellow.

Squash silverleaf, a disorder responsible for the common name of *B. argentifolli*, has been known from Israel since 1963 but did not occur in the United States until about 1986, when silverleaf whitefly first became abundant in Florida. Silverleaf symptomology includes blanching of the veins and petioles, and eventually the interveinal areas of the leaf. The fruit of both yellow- and green-fruited varieties also may be blanched (Yokomi *et al.*, 1990; Schuster *et al.*, 1991).

In addition to direct damage, silverleaf whitefly also causes damage indirectly by transmitting plant viruses. Over 60 plant viruses, most belonging to a group called geminiviruses, are known to be transmitted to crops by silverleaf and sweetpotato whiteflies (Markham, 1994). Some viruses, such as tomato yellow leaf curl virus, cause more damage than the insect feeding alone, so the effects are devastating. Unfortunately, unlike the case with the phyotoxemia caused by the whitefly salivary secretions, once viruses are inoculated into the plant there is no recovery by the host even if the whiteflies are eliminated.

Lastly, whiteflies cause injury by excreting excess water and sugar in the form of honeydew. This sticky substance accumulates on the upper surface of leaves and fruit, and provides a substrate for growth of a fungus called sooty mold. The dark mold inhibits photosynthetic activity of the foliage, and may also render the fruit unmarketable unless it can be washed thoroughly and the residues are removed.

Management

Sampling. The distribution of whitefly life stages on cantaloupe was studied by Tonhasca *et al.* (1994) and Gould and Naranjo (1999). The eggs tend to be concentrated on young foliage and mature larvae on older foliage. Large nymphs are a considered good stage for population assessment, because they cannot move and are large enough to see without magnification. Adults sometimes are concentrated on lower leaves, but they move to young foliage during oviposition. Such distributions must be considered in population assessment before initiating management practices. Ohnesorge and Rapp (1986) recommended yellow sticky traps rather than direct visual counts for population estimates when insect densities were low. However, when Palumbo *et al.* (1995) evaluated several sampling methods in cantaloupe, they found that visual observation of the lower-leaf surface and vacuum sampling were less time consuming, and sometimes more precise, than yellow sticky traps. Sampling techniques were reviewed by Butler *et al.* (1986) and Naranjo (1996).

Insecticides. In southern states, where silverleaf whitefly can be the most important insect problem on some vegetable crops, frequent applications of insecticides are often made to minimize the direct and indirect effects of whitefly feeding. Whitefly resistance to nearly all classes of insecticides is known, and rotation of insecticide classes is encouraged. Mixtures of insecticides are often used, which is indicative of high levels of resistance. Most agriculturalists suggest that whitefly numbers be maintained at low levels because once they become abundant they are difficult to suppress; this, of course, exacerbates development of insecticide resistance. The phytotoxemia and disease transmission potential of this insect exaggerates its damage potential, further justifying frequent application of insecticides. Often the most effective approach to effective management involves regional or area-wide suppression based on a combination of insecticides, weed management, and crop management.

Silverleaf whitefly feeds on the lower surface of foliage and is sessile throughout most of its life—habits that minimize contact with insecticides. Frequent insecticide application also disrupts naturally occurring biological control agents. In an attempt to minimize the cost and disruptive effects of insecticides, and to reduce the evolution of insecticide resistance, soaps and oils have been extensively studied for whitefly control. The mechanism of control by surfactants such as soaps and oils is not clearly understood, but disruption of the insect cuticle, physical damage, and repellency are postulated. In any event, mineral and vegetable oils alone, or in combination with soaps and detergents, can provide some suppression of whiteflies. Combination of insecticide and oil often enhances whitefly control (Horowitz *et al.*, 1997). Suppression usually increases with concentration of the surfactants, but 0.5% detergent plus 0.5% vegetable oil, or 0.5% detergent alone, or 1% insecticidal soap alone, or 0.75–1.0% light-mineral oil are often recommended initially until the phytotoxicity potential is known. Oil has more residual activity than soaps or detergents; the former is also more repellent to adults (Liu and Stansly, 1995). Cucurbits and crucifers seem especially prone to foliage damage by surfactants, and damage occurs frequently under high-temperature conditions. High gallonage enhances coverage and pest population reduction, but increases cost of control (Butler and Henneberry, 1990a,b; Butler *et al.*, 1993). Neem products and other growth regulators affect immature insect survival only (Price and Schuster, 1991).

Biological Control. Although many predators, parasitoids, and fungal diseases are known to attack silverleaf and sweetpotato whitefly, no biotic agents are known to provide adequate suppression alone. Under greenhouse conditions, parasitoids can be released at high enough densities to provide some suppression, especially when insecticidal soap and other management techniques are also used (Parrella *et al.*, 1991). Under natural field conditions, parasitism does not usually build to high levels until late in the growing season (Cock, 1993). Insecticides often interfere with parasitoids and predators, of course, and effective use of biological control agents will probably be limited to cropping systems where broad-spectrum insecticide use is minimized, and other management techniques used which favor action of predators, parasitoids, and disease agents. Some new insecticides are quite selective, killing whiteflies and yet preserving most natural enemies. *Verticillium, Paecilomyces*, and other fungi similarly show some promise under greenhouse conditions, but are limited by low humidity under field conditions (Meade and Byrne, 1991; Cock, 1993).

Cultural Practices. Cultural controls can be vitally important in managing silverleaf whitefly. Incorrect crop management, in particular, can create or exacerbate whitefly problems. Whiteflies can move from crop to crop, and area-wide crop-free periods help diminish populations. Thus, prompt tillage of land and destruction of crop residues after crop maturity is recommended. Similarly, weeds can harbor whiteflies, whitefly-transmitted diseases, and whitefly parasitoids, so weed management is a consideration.

Row covers and other physical barriers can decrease infestation of crops, and infection with disease (Cohen and Berlinger, 1986). Screen hole sizes of about 0.19 sq mm or smaller are required to successfully exclude silverleaf whitefly (Bethke *et al.*, 1994). Colored and aluminum mulches provide only temporary reduction in whitefly abundance and disease transmission (Cohen and Berlinger, 1986) or none at all (Powell and Stoffella, 1993). Orozco-Santos *et al.* (1995) reported effective whitefly exclusion on cantaloupe by using row covers, and reduction in whitefly population levels with transparent mulch.

Host-Plant Resistance. Host-plant resistance offers considerable potential, but currently it is difficult to put into practice. Both very hairy and hairless cultivars are perhaps less suitable for parasitoid activity than plants with intermediate densities of plant hairs. Resistance to the toxic saliva and to viruses transmitted by whiteflies also are being sought among commercially acceptable hybrids. For example, McCreight and Kishaba (1991) evaluated numerous cucurbit species and cultivars for susceptibility to squash leaf curl virus. Susceptibility varied among and within cultivated species, but overall susceptibility among species was: *Cucurbita maxima* > *C. pepo* > *C. mixta* > *C. moschata*. All cultivated species and cultivars were at least moderately susceptible to squash leaf curl virus, and most were severely damaged in field tests. In contrast, many wild cucurbit species were unaffected. Among commercial tomato cultivars, there was less oviposition by whiteflies on plants with low trichome densities, but this relationship was not apparent for wild tomatoes (Heinz and Zalom, 1995). As the basis for resistance is understood, there is good possibility of incorporating at least partial resistance into commercial cultivars.

Disease Transmission. Growers generally rely on whitefly suppression to manage disease incidence. This is not entirely satisfactory, however, and roguing of virus-infected plants is often suggested to minimize within-field spread of viruses. As whiteflies may transmit disease from one crop to another, or from weeds to crops, vegetation management is important. As noted above, reflective mulches have not produced consistent economic benefits. Mineral and vegetable oils may inhibit virus transmission. Application of 10–15% commercial whitewash solution is reported to be as effective as mineral oil, and more effective than some pyrethroid insecticides, in reducing virus disease incidence; however, phytotoxicity is sometimes a problem (Marco, 1993). Row covers or other physical barriers can substantially prevent disease transmission (Costa *et al.*, 1994), but often they are not economical. Ultraviolet (UV) light-absorbing plastic has been suggested as a preferred medium for plastic greenhouses (Antignus *et al.*, 1996). Apparently, the elimination of UV light decreases the attraction of plants to whiteflies, or changes their feeding behavior, thereby suppressing feeding and disease transmission.

FAMILY APHIDIDAE—APHIDS

Artichoke Aphid
Capitophorus elaeagni (del Guercio)
(Homoptera: Aphididae)

Natural History

Distribution. This species is found in temperate regions throughout the world, but probably is of European origin. It is the most common and damaging aphid affecting artichoke in California.

Host Plants. The summer hosts of this aphid are artichoke and *Cirsium* and *Carduus* thistles. The winter hosts are *Elaeagnus* spp. trees and shrubs, principally Russian olive, *E. angustifolia*.

Natural Enemies. Natural enemies are not well known, but Lange (1941) reported that lady beetles (Coleoptera: Coccinellidae) and flower flies (Diptera: Syrphidae) commonly preyed on these aphids. A wasp parasitoid, *Ephedrus persicae* Froggott (Hymenoptera: Aphididae), is known. Also, an unspecified fungus that caused the aphid's body to turn orange has been observed.

Life Cycle and Description. In temperate regions this aphid spends the winter on *Elaeagnus* spp., dispersing to artichoke or thistles in the spring. This pattern is documented in Colorado, where artichoke and Canada thistle, *Cirsium arvense*, are suitable summer hosts. However, under the mild climatic conditions of California, where most North American artichoke is grown, the artichoke aphid remains on artichoke throughout the year.

This is a fairly small aphid, the females measure only about 1.6–2.0 mm long and the males 1.5–1.8 mm. In the wingless (apterous) viviparous form, the head and abdomen are greenish-yellow or frosty green with some darker patches dorsolaterally. The legs, antennae, and cornicles are pale though often tinged with brown or gray. The cornicles are quite long. On summer hosts the winged viviparous forms (alatae) are darker, bearing blackish antennae, head, and pronotum. The abdomen is yellowish-green or green, but with a large dark square patch dorsally, and three dusky patches laterally. The long cornicles are pale or greenish, but the tips are darker. The wing veins are dusky brown.

The sexual forms also are similar in color. In the wingless oviparous female, the yellowish body and head are suffused with reddish-brown or pink. A pair of large dark markings appears dorsally on the abdomen, accompanied by 5–6 dark spots laterally. The legs and cornicles are yellowish, but the cornicles are darker apically. In the winged male, the color of the head, thorax, and antennae is dark. The abdomen is yellowish, sometimes suffused with reddish-brown. The abdomen also bears a row of large dark spots dorsally and laterally. The cornicles are pale, but darker apically. The legs are brownish.

A good description of this poorly studied aphid can be found in Cottier (1953). A key to artichoke-infesting aphids was included in Blackman and Eastop (1984), and it was also included in the key of Palmer (1952).

Damage

Artichoke aphid is principally numerous during the summer months, when the populations completely cover the lower surface of foliage and cause wilting of the plant. Large quantities of honeydew are secreted, causing the plant to blacken. The effect of the aphid feeding and honeydew secretion is to delay maturation of the artichokes for up to several months (Lange, 1941). This species is capable of transmitting several plant viruses, including artichoke latent virus (Kennedy *et al.*, 1962).

Management

Insecticides applied for suppression of artichoke plume moth, *Platyptilia carduidactyla* (Riley), normally keep aphids at low levels. (For more information on aphid management, see the sections on melon aphid, *Aphis gossypii* Glover, or green peach aphid, *Myzus persicae* (Sulzer).)

Asparagus Aphid
Brachycorynella asparagi (Mordvilko)
(Homoptera: Aphididae)

Natural history

Distribution. This species was first observed in North America in 1969, in New York. Presumably it had been introduced accidentally from Eurasia, where it is found widely in eastern countries and along the Mediterranean. It dispersed (or was dispersed) quickly, attaining North Carolina in 1973, British Columbia in 1975, Missouri and Washington in 1979, Alabama, Georgia, Oklahoma and Idaho in 1981, and California in 1984. Although it is abundant across the northern portions of North America, and very abundant along the west coast including southern California, it is uncommon in the humid southeastern states. (See color figure 149.)

Host Plants. This aphid feeds only on species of *Asparagus*. In addition to garden asparagus, *Asparagus officinalis*, it is known to feed on ornamental *Asparagus* spp.

Natural Enemies. Many native predators, parasitoids, and insect diseases affected asparagus aphid. In New Jersey and Delaware, Angalet and Stevens (1977) documented 31 species of natural enemies—the most abundant are lady beetles (Coleoptera: Coccinellidae), green lacewings (Neuroptera: Chrysopidae), and a parasitoid of numerous aphid species, *Diaeretiella rapae* (M'Intosh) (Hymenoptera: Braconidae). Other species of some importance were brown lacewings (Neuroptera: Hemerobiidae), the predatory midge *Aphidoletes aphidimyza* Rondani (Diptera: Cecidomyiidae), flower flies (Diptera: Syrphidae), other wasps (Hymenoptera: Braconidae and Aphelinidae), and a fungus. From these studies the authors concluded that the natural enemies kept the population in check. In fact, this aphid has not proved to be a serious pest of asparagus in eastern states and provinces (Hayakawa *et al.*, 1990). Once attaining the arid western regions of North America, however, asparagus aphid developed into a severe pest, and seemed little influenced by natural enemies. This suggests that climate plays a significant role, perhaps in conjunction with weather-sensitive natural enemies such as fungi. Beginning in the mid-1980s, parasitoids were imported from southern Europe and introduced into California, where some establishment resulted (Daane *et al.*, 1995).

Life Cycle and Description. Oviparous aphids deposit overwintering eggs on the asparagus ferns in September or later. Beginning in about March, eggs hatch into aphids that develop into stem mothers (fundatrices). Fundatrices move to asparagus spears and give birth to about 18 nymphs. Subsequent generations may be apterous (wingless) or alate (winged). Sexual forms are produced in the autumn, they mate, and the females deposit eggs on the asparagus plant. Duration of a complete generation is about 15–19 days at 25°C.

Egg. The eggs initially are green, but turn shiny black within 1–2 days. Females produce, on average, 10.5 overwintering eggs during their life span. The elliptical eggs are deposited in the lower one-third of the asparagus canopy.

Nymph. The grayish-green nymphs exhibit four instars, the durations of which are about two days each, regardless of the morph or sex. Thus, nymphal development time averages about 8.0–9.5 days. Each female produces about 55 nymphs at 23°C, but only 27 and 9 at 14° and 32.5°C, respectively.

Adult. Both the alate and apterous adult viviparous aphids are present throughout the summer months, and reproduce parthenogenetically. The adults are relatively small, measuring 1.2–1.7 mm long, and have short antennae. They are elongate oval, and green or gray-green, often covered with a whitish waxy secretion. They blend in well with asparagus foliage, but impart a slightly bluish-gray tint when the infestation is heavy. The most important characters to distinguish viviparous asparagus aphid from other asparagus-infesting species are the inconspicuous cornicles and long cauda of *B. asparagi* (Stoetzel, 1990). Egg-producing females survive up to about 20 days.

The biology of asparagus aphid was discussed by Tamaki *et al.* (1983b), Wright and Cone (1988a,b), and Hayakawa *et al.* (1990). Keys to distinguish *B. asparagi* from other asparagus-infesting aphids were published by Blackman and Eastop (1984) and Stoetzel (1990).

Damage

Aphids feed on the new growth, causing shortening of the internodes, rosetting, dwarfing, and reduced root growth (Capinera, 1974a). Asparagus aphids deplete the sugars, particularly in the roots, and to a greater extent some other aphids (Leszczynski *et al.*, 1986). Heavily infested plants have a bushy or bonzai-like appearance. Aphid infestation can kill seedlings or small plants in a relatively short time. Older, well-established plantings may show damage and even death in the year following infestation by aphids, especially in a particularly cold winter. Freezing and aphid infestation are synergistic; together they decrease survival and vigor of dormant asparagus crowns greater than either aphid feeding or freezing alone (Valenzuela and Bienz, 1989). The threat of damage is much greater in western areas than in eastern North America.

Management

Sampling. Egg hatching can be predicted from temperature models. Eggs can be separated from foliage by washing with petroleum-cleaning solvent (Wright and Cone, 1988a). Nymphs can be extracted from plants by heat or methyl isobutyl ketone (Wright and Cone, 1983). Aphid distribution tends to be clumped, with most aphids in basal regions of the plants.

Wright and Cone (1986) studied the distribution of aphids and recommended a sample of about 140 branches per field for making management decisions.

Insecticides. Foliar insecticides can suppress aphids, but multiple applications may be necessary, especially under high-density conditions. Granular systemic insecticides may provide long-term control (Vernon and Houtman, 1983; Wildman and Cone, 1988). Some research has also been done to demonstrate the possibility of delivering insecticides to asparagus through the irrigation system (Wildman and Cone, 1986).

Cultural Practices. Several cultural practices for suppression of asparagus aphid were investigated in Washington (Halfhill *et al.*, 1984; Folwell *et al.*, 1990). Autumn and spring tillage reduce aphid overwintering but also reduce subsequent spear production. Mowing and herbiciding within asparagus fields can destroy early season aphids and delay the buildup of damaging populations. However, destruction of wild (volunteer) asparagus is the most important cultural practice available, because it eliminates overwintering sites and limits invasion of aphids into commercial, aphid-free fields. Herbiciding, burning, and removal of asparagus crowns by digging are all viable options to eliminate volunteer asparagus.

Bean Aphid
Aphis fabae Scopoli
(Homoptera: Aphididae)

Natural History

Distribution. The bean aphid, which is probably a complex of closely related species, apparently is native to Europe, but it has been spread to most temperate areas of the world except for Australia and New Zealand. Although it occurs in tropical areas of Africa, it is not a very serious pest in warm environments. Bean aphid occurs throughout the United States and southernmost Canada. (See color figures 19 and 151.)

Host Plants. This aphid feeds on a wide range of hosts, though it seems to favor plants in the family Chenopodiaceae as summer hosts. Vegetables attacked include asparagus, beet, carrot, celery, corn, fava bean, leek, lettuce, lima bean, onion, parsnip, pea, spinach, pea, rhubarb, and squash. It also attacks sugarbeet, and in Europe it is considered to be a very serious pest because it transfers viruses to this crop. Flowers such as nasturtium and dahlia commonly support this insect, as do many weeds, including curly dock, *Rumex crispus*; lambsquarters, *Chenopodium album*; and shepherdspurse, *Capsella bursa-pastoris*. The taxonomy of this insect is confused, and some host

records may prove to be due to other closely related species.

The winter, or primary, hosts of bean aphid are *Euonymus* spp. and *Viburnum* spp. In England, abundance of bean aphid is positively correlated with abundance of spindle tree, *Euonymus europaeus* (Way and Cammell, 1982). Similarly, in Colorado, bean aphid primarily affects crops grown adjacent to urban areas, where *Euonymous* and *Viburnum* spp. are cultivated as ornamental shrubs.

Natural Enemies. Fungi are an important mortality factor, often suddenly sweeping through high density populations. Several species of fungi are involved, but *Neozygites fresenii* apparently is most effective in suppressing bean aphid populations (Dedryver, 1978; Wilding and Perry, 1980; Rabasse *et al.*, 1982). Parasitism by native parasitoids, especially by *Lysiphlebus testaceipes* Cresson (Hymenoptera: Braconidae), is usually apparent when aphids are abundant. General predators such as green lacewings (Neuroptera: Chrysopidae), lady beetles (Coleoptera: Coccinellidae), and flower fly larvae (Diptera: Syrphidae) often are found feeding in bean aphid colonies. Ants commonly attend bean aphid, harvest honeydew, and apparently interfere with predators and enhance aphid survival (Capinera and Roltsch, 1981).

Natural enemies are believed to play an important role in the population cycles of bean aphid. When aphids are numerous on early season hosts they provide abundant food for natural enemies, which then reach high levels of abundance in late season populations, and greatly reduce the number of aphids that overwinter. Therefore, the aphid population in the following year is quite low. However, low early season populations fail to support natural enemies, so the aphids are free to reproduce and attain high late season and overwintering population densities. This, of course, means that the early season population will again be high. Because weather interacts with the aphids and their natural enemies, the cycling is not very predictable, so population-monitoring systems have been developed in England to aid in the prediction of aphid outbreaks.

Life Cycle and Description. The eggs hatch in the spring, often as early as February, and produce 1–2 generation of apterous (wingless) parthenogenetic females. This is followed by a generation of alate (winged) females that fly from the primary host to secondary (summer) host plants, where females reproduce parthenogenetically all summer. Alate and apterous forms are produced according to density and host-plant conditions, with high densities and depleted plants tending to produce alatae. In the autumn, in response to short-day conditions, alatae are produced that colonize the primary hosts. On the overwintering host, the females mate and lay eggs among cracks in bark and on bud axils.

Egg. In temperate areas, bean aphid overwinters as an egg on one of its primary hosts. Initially green, the eggs soon turn shiny black. In warmer areas, the aphids reproduce continuously without producing eggs.

Nymph. Nymphs are dark, but bear four pairs of transverse white bars on the dorsal surface of the abdomen. Most authors report four instars. Tsitsipis and Mittler (1976), for example, indicated durations of 2, 2, 1.5, and 2.5 days for instars 1-4, respectively, when reared at about 20°C. Ogenga–Latigo and Khaemba (1985), however, reported only three instars, with durations of about 2.3, 3.0, and 2.5 days. Total nymphal development time requires 5–10 days at 28–17°C, respectively.

Adult. Bean aphid is dark olive-green to dull-black in color. The body length is 1.8–2.4 mm in females, with males only slightly smaller. The appendages tend to be black, but the tibiae may be pale, in part. The wings are transparent. On some crops bean aphid may be confused with another black species, cowpea aphid, *Aphis craccivora* Koch; adults of this latter species are shiny black with whitish legs, and they average smaller in size.

Reproduction commences soon after attainment of the adult stage, usually a period of about 3–6 days. Both alate and apterous females reproduce. The adult produces about 85–90 nymphs during her reproductive period, which is estimated at 20–25 days. Most offspring are produced in the first 5–10 days of the reproductive period. Reproduction increases with temperatures, up to a threshold of about 24°C, and then decreases. The reproductive period is followed by a post-reproductive period of about seven days (Frazer, 1972). Apterous females give birth to more, and larger, nymphs than alatae (Dixon and Wratten, 1971).

The winged aphids disperse freely, but their eventual disposition depends largely on wind and wind-breaks, because they do not have strong powers of flight. Thus, leeward sides of hills and wind-breaks are the areas where aphids accumulate. They also are deposited more heavily on the edges of crops. Because small fields have proportionally more "edge," aphid mean density also tends to be higher in small plant-ings. Dispersal of up to 30 km sometimes occurs.

An excellent summary of bean aphid ecology was given by Cammell (1981). Tsitsipis and Mittler (1976) provided information on rearing aphids on both

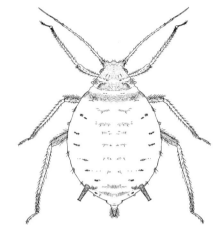

Adult female bean aphid, wingless form.

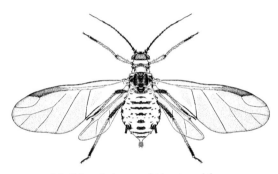

Adult female bean aphid, winged form.

plants and artificial media. Keys for identification of bean aphid, and most other common aphids, are found in Palmer (1952) and Blackman and Eastop (1984). Stoetzel *et al.* (1996) published a key for cotton aphids that is also useful to distinguish bean aphid from most other common vegetable-infesting aphids.

Damage

Aphids tend to reproduce rapidly and build to high numbers on plants that are actively growing. They concentrate their feeding on young tissue and may deplete nutrients needed for plant growth. In sugar-beet, for example, bean aphid infestation decreased foliage and root weights, and sucrose yield (Capinera, 1981). Symptoms of infestation include curling of leaves and stunted plants. Studies on bean suggest that aphid feeding during the preflowering stage is more damaging than later in development (Khaemba and Ogenga–Latigo, 1985). Research from England indicates that aphid suppression is warranted when 5% of faba bean plants are infested (Bardner *et al.*, 1978). In leafy vegetables such as celery, however, con-tamination of foliage with aphid bodies or honeydew is also a very important factor, as these conditions will cause crops to be rejected (Godfrey and Chaney, 1995).

Estimates of honeydew production indicate that aphids may produce up to 30–40 drops per day (Banks and Macaulay, 1964). Significant damage is, however, infrequent in North America.

Bean aphid is capable of transmitting numerous plant viruses, including many common viruses of vegetable crops. Kennedy *et al.* (1962) listed over 60 diseases, mostly stylet-borne viruses, transmitted by bean aphid. Although not usually considered as important a vector as many other species (Fereres *et al.*, 1993), some work indicated that this insect was quite effective at transmitting beet yellows virus (Kirk *et al.*, 1991).

Management

Sampling. This species often develops very high densities, perhaps 5000 aphids per plant. Because of the difficulties in dealing with such abundance, procedures for estimating aphid density tend to avoid counting actual numbers per plant, and focus instead on counts of subsamples, determination of number of leaves infested (Banks, 1954), or length of infestation (Capinera, 1981). In England, forecasts of aphid populations are made based on densities of overwintering eggs on spindle tree. Considerable savings in insecticide use are possible in years when aphid densities are low (Way *et al.*, 1977; Gould and Graham, 1977). Suction traps may also be used in forecasting, particularly in determining time of migration (Way *et al.*, 1981). There does not seem to be need for such a forecasting system in North America owing to the relatively minor pest status of this insect. To monitor invasion of celery, and most likely other vegetable crops, yellow water pan traps and yellow sticky traps can be used (Godfrey and Chaney, 1995), though sticky traps should be avoided because aphids are often damaged by the adhesive making identification difficult.

Insecticides. Insecticides are applied as foliar sprays or, for systemic insecticides, also as granules at planting time. However, use of granular materials in short-season crops such as beans is sometimes impossible, because the insecticide residue is too high at harvest time. In some crops, disruption of pollinators is a concern.

Bean aphid possesses both an alarm pheromone and a sex pheromone (Dawson *et al.*, 1990). The alarm pheromone causes the aphid to become disturbed, and may induce it to leave the host plant. This may prove useful in inhibiting virus transmission, increasing contact with insecticide residues, or by increasing aphid exposure to predators or unsuitable environmental conditions. The sex pheromone is active only at short distances, and its use is yet to be developed.

Biological Control. Research in England has demonstrated that introduction of entomopathogenic fungi could suppress aphid abundance on beans, and increase yield (Wilding, 1981). However, *Erynia neoaphidis* and *Neozygites fresenii* fungi are effective only at high aphid densities and under cool, moist weather conditions.

Cultural Practices. Faba beans have been examined for their resistance to bean aphid, and some cultivar differ in susceptibility. This species is more susceptible, however, than more primitive *Vicia* species (Holt and Birch, 1984).

Border plantings of scorpion weed, *Phacelia tanacetifolia* (Hydrophyllaceae), have been shown in England to enhance predation of bean aphid, and other aphid species, by flower flies (Diptera: Syrphidae). Adult flower flies require nectar for energy, and pollen for sexual maturation and egg production, with this flowering annual easily manipulated to provide these resources (Hickman and Wratten, 1996).

Bean Root Aphid
Smynthurodes betae Westwood
(Homoptera: Aphididae)

Natural History

Distribution. This aphid is found throughout the world, though it is not often recognized as a pest. Its origin seems to be the Mediterranean region.

Host Plants. The primary host of bean root aphid is pistachio and related trees in the genus *Pistacia*. Its secondary hosts are numerous dicotyledenous plants, particularly in the families Compositae, Leguminosae, and Solanaceae, but also including Chenopodiaceae, Cruciferae, Polygonaceae and others. Monocots (grasses) rarely serve as hosts. The principal crops affected are bean, parsnip, potato, and tomato.

Life Cycle and Description. Bean root aphid has a complex, two-year life cycle when *Pistacia* is involved, though it can also reproduce parthenogenetically on secondary hosts, bypassing the sexual stage. During the first year, nymphs developing from overwintered eggs initially form small red galls on the midvein of the *Pistacia* leaflet. At maturity, each of these aphids produces about 20 nymphs parthenogenetically which disperse to the margins of the leaflets and form new galls. Within the second set of galls, two additional generations are produced, resulting in about 35 aphids in each gall by autumn, when winged forms emerge from the galls and disperse to secondary hosts. Offspring of the dispersants penetrate the soil and feed on the roots. During the following spring winged forms are produced on the secondary plants which disperse back to the primary host and partheno-genetically produce male and female forms. The sexual forms mate and produce eggs.

Egg. Small black eggs are deposited in crevices on the trunk and major branches of *Pistacia*.

Adult. Wingless (apterous) adults are quite spherical in general body form. The antennae, head, legs, and abdomen are brownish or pinkish, and are covered with a moderately dense covering of short hairs. Cornicles are not present. This small aphid measures about 1.5–2.7 mm long. The antennae, head, and appendages of the winged (alate) form are dusky to black. The abdomen is green or olive, with a broad transverse black band across the dorsal surface of each segment. The short hairs also cover the body of this form and are white. Alatae are slightly larger, 2.2–2.7 mm long, and bear wings with dusky veins.

Bean root aphid was described by Cottier (1953). A summary of its life cycle was provided by Burstein and Wool (1993).

Damage

This species is rarely considered a pest, but when it is numerous it causes wilting and stunting of plants, and may cause deformities of roots and tubers.

Management

Management is rarely a consideration, but if it becomes necessary the approaches used for management of other root aphids such as lettuce root aphid, *Pemphigus bursarius* (Linnaeus) and the sugarbeet root aphids, *Pemphigus betae* Doane and *P. populivenae* Fitch, likely are applicable.

Bird Cherry-Oat Aphid
Rhopalosiphum padi (Linnaeus)
(Homoptera: Aphididae)

Natural History

Distribution. This species probably originated in North America, but is now found in Europe, Asia, and New Zealand as well as throughout the United States and southern Canada. (See color figure 150.)

Host Plants. Bird cherry-oat aphid may alternate between winter and summer hosts, or in mild-winter climates it can persist throughout the year on summer hosts. The winter, or primary, host is *Prunus* spp. The summer, or secondary, hosts are numerous species of grasses, including all the major cereals and pasture grasses. Sedges, iris, and a few other plants are occa-sionally reported as hosts, but these records are suspect. It is found principally on wheat, barley, oat, and rye, but in some locales it is a major pest of corn, including sweet corn.

Susceptibility of hosts to infestation and suitability for aphid reproduction have been assessed by many investigators, including Villanueva and Strong (1964), Dean (1973), Leather and Dixon (1982), and Weibull (1993). Among the best hosts are such wild or forage grasses as wheatgrass, *Agropyron* spp.; brome, *Bromus* spp.; ryegrass, *Lolium* and *Lilium* spp.; and wild oat, *Avena fatua*. Among crop plants barley seems to be optimal in most areas, but sweet corn is very suitable in the Pacific Northwest.

Natural Enemies. Bird cherry-oat aphid is susceptible to attack by many of the predators, parasitoids, and pathogens affecting other aphids. Lady beetles (Coleoptera: Coccinellidae), green lacewings (Neuroptera: Chrysopidae), brown lacewings (Neuroptera: Hemerobiidae), and flower flies (Diptera: Syrphidae) are the most common predators. In Europe, and presumably in North America, eggs suffer high mortality rates during the winter months owing to cold weather in most northern locations and predators in most mild locations. In small grain crops grown in Idaho, *Aphelinus varipes* (Forester) (Hymenoptera: Aphelinidae); *Aphidius ervi* Haliday, *Diaeretiella rapae* (M'Intosh), *Lysiphlebus testaceipes* (Cresson) and *Praon* sp. (all Hymenoptera: Aphidiidae) were recovered from bird cherry-oat aphid, with *A. varipes* most abundant (Feng *et al.*, 1992). Fungi have some significant effects on aphid populations, but this occurs mostly in irrigated fields. Precipitation, particularly heavy rainfall, also has direct negative effects on aphid survival during the summer months.

Life Cycle and Description. The biology of this aphid varies, depending on availability of primary hosts and weather conditions. A generation can be completed in about five days when reared at 26°C, but it requires about 22 days at 13°C. Longevity of the individual aphid is often 18–20 days. Aphids developing on primary hosts undergo 2–3 generations before overcrowding induces formation of winged (alate) forms which disperse to grasses. Aphids may reproduce continuously during the summer months, but reproduction is disrupted if temperature exceeds 30°C for even a few hours daily. Aphids feeding on wild grasses and grains may migrate back to *Prunus* and produce eggs in the autumn, or they may remain on Graminae through the winter months. In Manitoba, winged aphids are found dispersing in late May and June, and then again in October (Robinson and Hsu, 1963). In the northern Great Plains region of the United States, however, little overwintering occurs, this area instead becomes re-invaded annually by winged migrants from the south. Some individuals crawl down the plant stems and survive below-ground during cold weather (Kieckhefer and Gustin, 1967), and though this does not appear an important method of overwintering under the cold winter climate conditions of Idaho (Feng *et al.*, 1992), it likely is effective in warmer climates.

Egg. The egg is green initially, but turns black with age, and is deposited on the bark of *Prunus* trees. The egg measures about 0.57 mm long and 0.26 mm wide. In England, there is a steady but low rate of attrition in eggs during the winter months due primarily to predation, followed by a fairly sharp increase in egg loss in the spring, resulting in loss of 70–80% (Leather, 1981).

Nymph. There are four instars. The mean body length of nymphs is 0.6, 0.9, 1.2, and 1.4 mm for instars 1–4, respectively. Development time of nymphs reared at 26°C, which is about optimal for development, is 1.4, 1.1, 1.1, and 1.4 days for instars 1–4, respectively. Nymphs are green, but there is pronounced orange or rust color posteriorly near the base of the cornicles. First instar nymphs have only four antennal segments, instars two and three have five segments, and last instar and adults bear six segments. Aphids reared under cool conditions are greenish black and those reared under warmer conditions are light-green. Aphids feeding on *Prunus* live on bud and young leaf tissue; the latter may become deformed and form a pseudo-gall partly enclosing and protecting the aphid colony. Older leaf tissue also is attacked, and eventually some of the adults are winged. The shift to winged, dispersing forms coincides with a decrease in the nutritional quality of *Prunus* foliage (Dixon, 1971; Dixon and Glen, 1971).

Adult. The body of the wingless adults (apterae) is greenish or olive brown, with the head and prothorax yellowish-brown. The legs are green or darker, the cornicles green but with the tips dusky. A rusty colored patch is found around the base of the cornicles. The body length is 1–2 mm, averaging 1.7 mm. In the winged forms (alatae), the abdomen usually is light-green but darker medially, though sometimes it is entirely dark green. A rusty brown patch may be found around the base of the cornicles. The head and thorax are black, but the prothorax is green. The appendages are dusky to blackish. The body length is 1.8–2.0 mm. The prereproductive period of young adults is normally less than a day. Viviparous (parthenogenetic) adults produce 10–50 nymphs during their

life span, with host plant, plant age, and temperature affecting reproduction. However, normal fecundity is 40–50 at optimal temperature during a reproductive period of about 15 days. Young plants and cool weather enhance reproduction.

This species is easily confused with several grain-infesting species, but fortunately is not too readily confused with the other common corn-infesting aphid, corn leaf aphid, *Rhopalosiphum maidis* (Fitch). The body of bird cherry-oat aphid is ovate, whereas the body of corn leaf aphid is elongate. Also, the rust-orange color about the base of the spiracles distinguishes bird cherry-oat aphid from the solid blue-green color of corn leaf aphid.

This species was included in the aphid keys of Palmer (1952), Blackman and Eastop (1984), and Olsen *et al.* (1993). The biology of this insect in Europe was reviewed by Leather *et al.* (1989). Developmental biology was given by Villanueva and Strong (1964), Elliott and Kieckhefer (1989), and Michels and Behle (1989).

Damage

This species is known principally for its damage to small grains. It builds to high densities and damages grain plants directly, and it is also an important vector

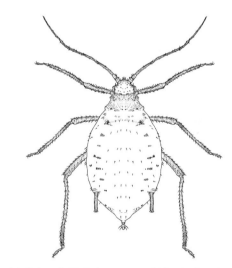

Adult female bird cherry-oat aphid, wingless form.

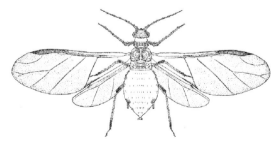

Adult female bird cherry-oat aphid, winged form.

of several plant viruses, most notably barley yellow dwarf. Both persistent and nonpersistent viruses are vectored (Kennedy *et al.*, 1962). However, in some areas it is the most abundant aphid on corn. Corn also serves as an important link or bridge for aphids and barley yellow dwarf virus between spring and autumn small grain crops (Blackmer and Bishop, 1991).

Management

This aphid is principally a pest of small grains, and sampling protocols have been developed for such crops, but not for corn. For example, Ekbom (1987) and Elliott *et al.* (1990) presented binomial or presence-absence protocols for *R. padi* sampling on small grains. Nevertheless, aphids may attain high numbers on sweet corn and stimulate application of foliar insecticides. There is evidence of resistance to certain insecticides in Europe.

Buckthorn Aphid
Aphis nasturtii Kaltenbach
(Homoptera: Aphididae)

Natural History

Distribution.   Buckthorn aphid is widespread in North America, but as a pest it is limited principally to eastern Canada and the northeastern United States. This species occurs throughout Asia and Europe. Its origin is uncertain, but likely is Europe.

Host Plants.   This insect feeds on numerous plants from many plant families. It is principally a pest of potato during the summer months. In Maine and New Brunswick it is often the most abundant aphid in potato fields. Other vegetables that may serve as summer hosts include bean, beet, cabbage, cucumber, parsnip, salsify, squash, and watercress. Interestingly, in Maine this little-known aphid has been reported to be the most abundant aphid on cucumber and squash. Other crops that support buckthorn aphid include buckwheat, strawberry, alsike and red clover, and tobacco. It also affects flowers such as dahlia, hollyhock, nasturtium, pansy, sunflower, and zinnia. Among the weeds supporting growth of these aphids are Canada thistle, *Cirsium arvense*; common plantain, *Plantago major*; dock, *Rumex* spp.; lambsquarters, *Chenopodium album*; pigweed, *Amaranthus retroflexus*; purslane, *Portulaca oleracea*; shepherdspurse, *Capsella bursa-pastoris*; sorrel, *Oxalis* spp.; smartweed, *Polygonum* spp.; and wild radish, *Raphanus raphanistrum*. The winter hosts are common buckthorn, *Rhamnus canthartica*, and alder-leaved buckthorn, *R. alnifolia*; both are shrubs, and the former is occasionally planted

as a hedge, but the latter is more important in overwintering.

Natural Enemies.   Buckthorn aphid has many natural enemies, though it is less heavily parasitized than potato aphid, *Macrosiphum euphorbiae* (Thomas). Among the parasitoids reared from buckthorn aphid are *Praon* spp., *Aphidius* spp., *Diaeretiella rapae* M'Intosh and *Ephedrus incompletus* (Provancher) (all Hymenoptera: Aphidiidae). The most important parasitoid, however, is *Aphidius nigripes* Ashmead (Hymenoptera: Aphidiidae) (Shands *et al.*, 1965). Most of the common lady beetles (Coleoptera: Coccinellidae), some lacewings (Neuroptera: Chrysopidae), flower flies (Diptera: Syrphidae), and the predatory midge *Aphidoletes aphidimyza* (Rondani) are reported to prey on buckthorn aphid. Ground beetles (Coleoptera: Carabidae) affect aphid populations, but not as greatly as canopy-level predators (Boiteau, 1986). Several fungi affect buckthorn aphid, sometimes causing population collapse (Shands *et al.*, 1965).

Life Cycle and Description.   In Maine, eggs overwinter on buckthorn and hatch in April. One or more generations consisting of apterous (wingless) parthenogenetic females are produced on the buckthorn leaves, but beginning with the second generation some alate (winged) females are also produced. In subsequent generations larger and larger proportions of alatae are produced, with the winged individuals dispersing to summer hosts. The aphids developing on summer hosts are mostly apterous until August when winged males and females are produced. The winged sexual forms disperse to buckthorn where the females deposit nymphs that, upon maturing, mate with the males and deposit eggs.

Egg.   Deposition of eggs on buckthorn commences in early September in Maine. They usually are deposited on the buds of alder-leaved buckthorn. The small oval eggs are green when first deposited, but soon turn black. They measure about 0.5 mm long and 0.25 mm wide. In a 24-year study conducted in Maine, egg density averaged 102 eggs per 100 buds (range 5–300 per 100 buds) in the autumn, and 72 eggs per 100 buds (range 2.5–300 per 100 buds) in the spring. Overwintering mortality of eggs ranged from about 30–90%.

Nymph.   Five instars were reported by MacGillivray and Anderson (1958). They reported mean development times of 1.9, 1.9, 1.4, 1.7, and 0.3 days, respectively, when reared at 20°C. Wang *et al.* (1997) reported only four instars, with mean development times of 2.4, 2.9, 2.7, and 2.0 days, respectively, when reared at 20°C, and 1.4, 1.4, 1.2, and 1.2 days when reared at 30°C. The lower developmental threshold was reported to be 6.5°C, and the upper threshold 35.6°C. Optimal range for development is 20–30°C.

Adult. The apterous (wingless) parthenogenetic female of buckthorn aphid is small, and measures 1.1–2.1 mm long. The body is flattened and yellow or green in color. The antennae are shorter than the length of the body, and the head lacks tubercles near the base of the antennae. The cornicles are short. The winged (alate) parthenogenetic form is similar, but tends to have a dark head, thorax, and abdominal tip. The body length is 1.5–2.4 mm. The sexual forms that are produced in autumn display sexual differences in coloration—the females are yellow but the males dark. Each adult female feeding on potato produces, on average, about 64 nymphs during a reproductive period averaging 20 days.

The biology of buckthorn aphid was not fully described, but good treatment was provided by Patch (1924) and Shands and Simpson (1971). MacGillivray and Anderson (1958) and K. Wang *et al.* (1997) gave developmental information. Buckthorn aphid was included in the keys by Palmer (1952) and Blackman and Eastop (1984).

Damage

This species can be quite numerous and damaging to potato and cucurbits in eastern Canada and the northeastern United States. It also causes deformity of buckthorn hedge foliage. It occurs on the lower surface of leaves, favoring the lower leaves of plants that are grown in full sunlight, but it disperses freely to terminal areas of plants growing in the shade. In this respect it differs considerably from potato aphid, which favors terminal growth as feeding sites. At high densities it causes stunting of potato, and even death. Buckthorn aphid also may transmit plant viruses, including potato leaf roll, beet yellows, and cucumber mosaic virus (Kennedy *et al.*, 1962).

Management

Sampling. The egg abundance on buckthorn was shown to be positively related to subsequent aphid abundance in potato, though the correlation is not strong (Shands and Simpson, 1971). Aphid populations normally are monitored by visual examination of crops or with yellow water pan traps.

Insecticides. Foliar insecticides are commonly applied for aphid suppression. Broad-spectrum insecticides are sometimes used because of the other pests associated with solanaceous crops. Chemical suppression is not usually recommended unless half of the leaves are infested. Soaps- and dish-washing detergents are also effective against aphids, but care should be taken not to burn the plants. It is important to apply the treatments carefully so that the lower leaf surfaces receive good coverage. Treatment of buckthorn hedges in the spring can destroy early season aphid populations and reduce infestation potential in crop plants.

Cultural Practices. Studies conducted in Maine demonstrated that planting practices could influence damage to potato. Delay of planting from early May to late May or early June resulted in up to 90% reduction in aphid infestation. Early hilling operations, wherein row ridges are heightened with soil from the row middles, similarly deprived dispersing aphids of young plant tissue, resulting in lower aphid densities (Shands *et al.*, 1972). Delayed planting was also shown to be beneficial in New Brunswick (Boiteau, 1984). Reflective mulches, particularly aluminum mulch, are sometimes recommended for disruption of aphid invasion of crops. Evaluation of aluminum mulch in potato, however, showed that the beneficial effect of the reflective mulch was slight or of short duration, and therefore impractical (Shands and Simpson, 1972). Destruction of buckthorn, or treatment with insecticide, can decrease the risk of crop infestation.

Cabbage Aphid
Brevicoryne brassicae (Linnaeus)
(Homoptera: Aphididae)

Natural History

Distribution. Cabbage aphid is found throughout the United States and Canada, and in most other countries with a temperate climate. This aphid was probably introduced accidentally to North America from Europe by early colonists, as records of its damage to crops extend back to the late 1700s. (See color figure 157.)

Host Plants. Virtually all crops in the family Cruciferae are suitable hosts for cabbage aphid, but feeding is restricted to this family. The vegetable crops most severely affected are broccoli, Brussels sprouts, cabbage, and cauliflower, but collards, kale, kolhrabi, mustard, rape, and turnip are occasionally attacked. Beirne (1972) reported that in western Canada the principal injury of cabbage aphid was caused to early cabbage and cauliflower, whereas in eastern Canada turnip was the principal crop injured. Seed heads of radish are suitable hosts, as are numerous cruciferous weeds, especially *Brassica* spp. and shepherdspurse, *Capsella bursa-pastoris*. The mustard-oil glycoside, sinigrin, functions as a feeding stimulant for cabbage aphid, as it does for other insects that specialize in feeding on crucifers (Moon, 1967). (See color figure 20.)

Natural Enemies. Aphids normally are attacked by numerous predators and parasites, but cabbage aphid seems less susceptible to natural control than most species. *Diaretiella rapae* (MacIntosh) (Hymenoptera: Braconidae) is invariably associated with this aphid, and often builds up to high levels late in the season, when aphids are most numerous. However, its ability to attain densities sufficient to suppress cabbage aphid populations is hampered by the presence of several species of hyperparasites (Pimentel, 1961; Oatman and Platner, 1973). Flower fly larvae (Diptera: Syrphidae), principally *Allograpta*, *Syrphus* and *Scaeva* spp., and predatory cecidomyiid *Aphidoletes aphidimyza* (Rondani) (Diptera: Cecidomyiidae) larvae are generally reported to be the most important natural control agents (Oatman and Platner, 1973; Raworth *et al.*, 1984). Lady beetles (Coleoptera: Coccinellidae), while sometimes found feeding on cabbage aphids, are unusually ineffective (George, 1957). Some authors have suggested that predators and parasitoids decreased cabbage aphid population size only after aphid population growth rate is reduced by host plant deterioration or other factors. Fungi sometimes affect cabbage aphids, but causes epizootics only at very high aphid densities.

Life Cycle and Description. Cabbage aphid may have numerous generations per year, depending on climate; 20 are reported from southern California. In the north, this species produces sexual forms and overwinters in the egg stage, whereas in the south sexual forms and eggs are not observed. Trumble *et al.* (1982b) reported that eggs were the overwintering stage in northern California, but that crop debris and weeds served as the overwintering site for females in the southern portion of the state. Unlike many aphids, cabbage aphid does not disperse to an alternate host to overwinter.

Egg. Egg deposition occurs just one day after mating, and the oviparous (egg-laying) aphids may live for a month or more. The egg production is not well-studied, but Herrick and Hungate (1911) estimated 5–7 eggs per female. The eggs are initially pale-yellow or yellow-green, but they become shiny black within a few days of deposition. They are generally found on the underside of leaves, and measure 0.65 mm long and 0.15 mm wide. Unlike many aphids, cabbage aphids do not alternate between summer and winter hosts; the eggs are found on the plants fed upon by the summer populations. Eggs typically hatch in April.

Nymph. There are four instars, each about 2–3 days in duration. Nymphal bodies initially lack the grayish-white, waxy exudate that is so typical of this species, but acquire this soon after they commence feeding. The waxy material is easily removed, revealing a grayish-green insect with two rows of black bars along the back.

Adult. During the spring and summer months, the ensuing adults are all female and they reproduce parthenogenetically. Adult wingless (apterous) females give birth to 30–50 nymphs in their life span, which is about 30–40 days. Their reproduction initially is high, but drops off markedly as they mature. Most aphids found during the summer months are wingless females; only at high densities or when the host plant deteriorates are winged forms produced. The wingless females are grayish green, with a dark head, and pale-brown legs. The cornicles are dark and measure about 0.16 mm long. There is a double row of dark bars on the back, and the body is covered with a white powder or mealy secretion. The body measures 1.6–2.6 mm long, averaging about 2.5 mm.

The winged (alate) parthenogenetic females are similarly dull green with a dark head, but the legs are dark-brown. The dorsal surface of the abdomen is marked with a single row of dark bars, and the veins of the transparent wings are dusky to black. The cornicles are dusky-green to black. The entire body of this insect is dusted with a fine white powder. The winged parthenogenetic female measures 1.6–2.8 mm long, averaging about 1.9 mm. The winged forms, though not strong fliers, quickly disperse over crop fields and to adjacent fields. The winged females have a shorter life span than the wingless form; the former persists only about 10 days. They also produce relatively few offspring, about 6–8.

In the autumn months, especially September and October, egg-laying forms may be produced. Apparently the oviparous forms can be produced both by winged and wingless summer forms. The oviparous female has a pale green or greenish-yellow body, with a row of dark bars located centrally on the back. This form is wingless and measures about 2.2 mm long. Males are also produced at this time, but unlike the wingless oviparous females, the males have transparent wings. Males are greenish-brown or yellowish, and have dark legs and a double row of dark bars on the back. The body measures about 1.3 mm long.

Cabbage aphid is easily confused with turnip aphid, *Lipaphis erysimi* (Kaltenback), though turnip aphid generally can be distinguished by the very sparse occurrence of waxy exudate on the body and longer cornicles. In humid climates, however, there tends to be greater accumulation of waxy secretions on the turnip aphid's body. Therefore, Blackman and

Eastop (1984) recommended using the shape of the cauda—a structure found at the tip of the abdomen—to differentiate these two species. When viewed from above, the cauda of cabbage aphid is triangular, about as wide as it is long. In contrast, the cauda of turnip aphid is slender, about twice as long as it is wide.

A comprehensive treatment of cabbage aphid was provided by Herrick and Hungate (1911). Brief description of these aphids and keys for their identification are found in Palmer (1952) and Blackman and Eastop (1984). Cottier (1953) also provided a description of cabbage aphids.

Damage

Cabbage aphid populations, if not controlled, often build to very high densities. Heavily infested plants acquire a grayish appearance due to the mass of aphid bodies on the foliage. Honeydew and sooty mold are often evident. High densities also cause the leaves to wrinkle and curl, usually cupping downward. Under dry conditions, aphids cause the plants to wilt, and leaf tissue that has been fed upon may turn yellow. Cabbage aphid prefers the youngest tissue and highest portions of the plant, but may occur on both the upper and lower surface of foliage. Flower heads of seed crops may be attacked, reducing the setting of seed. Contamination of the plants with honeydew and

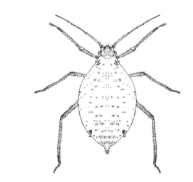

Adult female cabbage aphid, wingless form.

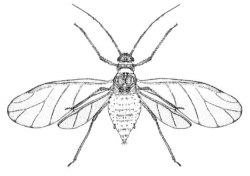

Adult female cabbage aphid, winged form.

aphids can cause considerable loss. In California, cabbage aphid is the most serious aphid pest of Brussels sprouts and broccoli because they are hard to remove from the head of the plant and thereby contaminate the produce (Trumble, 1982b; Pickel *et al.*, 1983).

Wolfenbarger (1967) estimated the impact of aphid densities on cabbage yield. He projected that when aphid densities averaged one per square inch (0.16 per sq cm) on the most heavily infested leaf of each plant, the proportion of plants in a field that was aphid-free would be 64%. Further, he estimated that when there was an average aphid density as low as one per plant, the proportion of unmarketable heads would be 12%.

In addition to the direct effects of feeding by aphids on plant growth, and the damage caused by aphid contamination of foliage, cabbage aphid can also be a vector of plant viruses. Over 30 viruses are known to be transmitted by *B. brassica*. Cauliflower mosaic and cabbage ring spot virus transmission were studied by Broadbent (1954); cabbage aphid transmits cauliflower mosaic more effectively because this virus concentrates in the young tissue of the plant, which is the preferred feeding site of the aphid. Host-plant resistance offers considerable potential, but currently it is difficult to put into practice. Both very hairy and hairless cultivars are likely to be less suitable for parasitoid activity than plants with intermediate densities of plant hairs.

Management

Sampling. Aphid distribution within crop fields is highly aggregated when overall densities are low, but become more uniform when aphid densities increase (Trumble, 1982a). Populations of winged aphids are more effectively sampled with cylindrical sticky traps than by water pan traps, but such populations are not consistently correlated with on-plant densities (Trumble *et al.*, 1982a) and aphids removed from sticky traps are difficult to identify. Sequential sampling protocols were presented by Wilson *et al.* (1983a).

Insecticides. Commercial producers usually depend on chemical insecticides to maintain an insect-free crop. Granular forms of systemic insecticides are sometimes used at planting, and may provide aphid suppression for several weeks. However, systemic and non-systemic insecticides are also applied to the foliage, and frequent applications may be necessary to maintain a clean crop. Many crucifer crops have curly leaves or form heads, and once aphids are established within such protected locations it becomes difficult to obtain control. Thus, most growers prefer to apply insecticides as a preventative measure, rather than waiting for infestation to occur and then seeking to cure the problem. It is often difficult to attain good control of cabbage aphid during cool weather (<21°C), especially with insecticides that volatilize or are translocated within the tissue (Vail *et al.*, 1967b).

Cultural Practices. As the eggs and other forms of aphids often overwinter on crop residue, destruction of crop remains after harvest is strongly recommended. This recommendation is often disregarded by home gardeners because the crucifers are quite cold tolerant, and it is sometimes possible to leave plants in the field and to harvest them throughout the winter months. If this is done, however, efforts should be made to bury plant debris deeply before new crops are put into the field.

The presence of weed hosts, or cruciferous oilseed crops such as rape, may exacerbate problems with cabbage aphid. In contrast, interplanting crucifers with nonhost crops may decrease aphid problems (Gliessman and Altieri, 1982). In studies conducted in New Zealand, aphid populations were decreased when cabbage was interplanted with pollen-rich flowering plants due to the increased predatory activities of flower flies (Diptera: Syrphidae) (White *et al.*, 1995).

Crop condition influences cabbage aphid abundance. Wearing (1972) reported that intermittent water stress was favorable and continuous water stress was detrimental to aphid reproduction and survival. van Emden (1966) indicated that high nitrogen and low potassium levels in foliage favored aphids.

Host-Plant Resistance. Several strains of crucifer crops have been observed to exhibit measurable levels of host-plant resistance. However, there are also numerous biotypes of the aphid, and no plant varieties have yet been developed that were permanently resistant to aphid attack. Also, varieties that are resistant to attack by cabbage aphid may be more suitable for other insects. For example, Way and Murdie (1965) reported that "glossy" Brussels sprouts, strains lacking the superficial waxy bloom commonly found on cabbages, supported fewer cabbage aphids than did the normal waxy strains. However, green peach aphids, *Myzus persicae* (Sulzer), were more abundant on glossy than waxy strains.

Carrot Root Aphids
Dysaphis crataegi (Kaltenbach)
Dysaphis foeniculus (Theobald)
(Homoptera: Aphididae)

Natural History

Distribution. These species are native to Europe and Asia, but are now widely distributed around the world, including North America. *Dysaphis crataegi* is found in the northern United States, including California, and perhaps in some parts of southern Canada. *Dysaphis foeniculus* seems to be limited to areas of west of the Rocky Mountains. In North America, economic problems seem to be limited to carrot production in California. Other *Dysaphis* species are occasionally reported to be associated with Umbelliferous plants, particularly *D. apiifolia* (Theobold), but evidence for association with carrot is weak.

Host Plants. The primary, or winter host, of *D. crataegi* is hawthorn, *Crataegus* spp. The secondary, or summer hosts, are species of Umbelliferae. *Dysaphis foeniculus* apparently remains on Umbelliferae throughout the year, and *D. crataegi* may do so in areas with mild climate. Among vegetables known to support these aphids are carrot, celery, fennel, and perhaps parsnip. They also may be found on wild Umbelliferae.

Natural Enemies. The natural enemies of these aphids seem to be poorly known. As the aphids are regularly attended by ants, which often protect aphids from predation, the level of predation may be relatively low.

Life Cycle and Description. These species are rather poorly known. They probably have a life cycle similar to bean aphid, *Aphis fabae*, differing principally in their choice of host plants. Also, *D. crataegi* forms a curling, thickening, and reddish discoloration of hawthorn leaf tissue, a structure termed a pseudo-gall. In appearance, both species of aphids tend to be yellowish-green to grayish-green, and often bear a light coating of wax. They range in size from 1.5–2.5 mm.

Information on biology and identification was provided by Stroyan (1963), Blackman and Eastop (1984), and McKinlay (1992).

Damage

These aphids form dense colonies at the top of the root, slightly below the soil surface, and extending upward to the base of the stems. The attending ants often construct earthen "shelters" over the aphid colonies. Aphid feeding weakens the plant, resulting in stunting. Both species are implicated in the transmission of plant viruses (Kennedy *et al.*, 1962). Although these aphids are potentially serious virus vectors in Europe, they are considered infrequent pests in North America.

Management

Crop rotation and sanitation are recommended to help prevent increase of these aphids.

Coriander Aphid
Hyadaphis coriandri (Das)

Honeysuckle Aphid
Hyadaphis foeniculi (Passerini)
(Homoptera: Aphididae)

Natural History

Distribution. Coriander aphid is found in the Mediterranean regions of Europe and Africa and east to central and southern Asia. It was first found in North America in 1997, in Florida. Honeysuckle aphid has a similar Eurasian origin, but is now cosmopolitan, having invaded North and South America, Australia, New Zealand, and southern Africa. It invaded North America in 1900s or earlier, as it was known from California and Connecticut before 1910. Honeysuckle aphid is suspected of being a complex of species.

Host Plants. These species cause damage by infesting Umbelliferae, including such vegetables as carrot, celery, fennel, parsley, and parsnip, but also spices such as caraway, *Carum carvi*; cumin, *Cuminum cyminum*; dill, *Anthum graveolens*; and sweet cumin, *Pimpenella anisum*. Coriander aphid, but not honeysuckle aphid, affects coriander. These food crops are summer, or secondary, hosts. Alternate secondary hosts include water parsnip, *Berula erecta*; water hemlock, *Cicuta* sp.; poison hemlock, *Conium maculatum*; hemlock parsley, *Conioselinum chinense*; wild chervil, *Cryptotaenia canadensis*; and cow parsnip, *Heracleum* sp. The primary host of coriander aphid is unknown, but for honeysuckle aphid it is, as its common name suggests, honeysuckle. Various honeysuckle species are suitable, including *Lonicera caprifolium*, *L. ciliosa*, *L. etrusca*, *L. japonica*, *L. periclymenum*, and *L. xylosteum*. A common alternate primary host is snowberry, *Symphoricarpos* sp.

Natural Enemies. The natural enemies of these aphids are not well-documented, but likely they are similar to those of other leaf-feeding aphids such as green peach aphid, *Myzus persicae* (Sulzer). The wasp parasitoids known to attack honeysuckle aphid are *Aphidius salicis* Haliday, *Diaeretiella rapae* (M'Intosh), and *Ephedrus persicae* Froggott (all Hymenoptera: Braconidae).

Life Cycle and Description. These *Hyadaphis* spp. seem quite similar in appearance and biology, though as noted earlier, the primary host of coriander aphid is unknown. Honeysuckle aphid occurs on both summer and winter hosts throughout the summer months in Colorado. Sexual forms are found on the summer hosts and winter hosts in the autumn. The following description is based on honeysuckle aphid.

Egg. The eggs initially are greenish, but soon turn black. The elliptical eggs measure about 0.51 mm long and 0.22 mm wide. They are deposited on the primary host.

Adult. The female developed from the overwintering egg is pale-yellow with greenish splotches on the abdomen and often orange around the base of the cornicles. The legs, antennae, and cornicles are darker distally. The body length is about 1.5 mm. Wingless (apterous) parthenogenetically reproducing summer adults are greenish or grayish-green, and dusted with white, waxy particles. The abdomen near the base of the cornicles is darker. The antennae, legs, and cornicles are black. These aphids measure about 1.5–2.0 mm long. In the winged (alate) parthenogenetic form, the head and thorax are dark green. The abdomen is green with mottled areas of darker green, and with dark patches laterally and at the base of each cornicle. The base of the cornicles and the distal portions of the legs are dark. The wings are transparent though the veins are dusky-brown. The alatae measure only about 1.3–1.5 mm long. The female is yellowish with a brownish head, the appendages dusky. Her body length is 0.84–1.15 mm. The male is similar to the winged alatae in appearance, though perhaps a bit darker and slightly smaller is size.

Honeysuckle aphid was described by Essig (1938), Palmer (1952), Cottier (1953), Bodenheimer and Swirski (1957), and Blackman and Eastop (1984). Coriander aphid was treated by Bodenheimer and Swirski (1957) and Blackman and Eastop (1984); the latter publication included a key to differentiate these species.

Damage

These aphids contaminate the foliage, rendering it unmarketable. They may also build to particularly high densities on flower heads. Although honeysuckle aphid attains high densities on honeysuckle, it is not the species responsible for the witches broom deformities that plague honeysuckle in northern areas; this deformity is generally caused by *Hyadaphis tataricae* (Aizenberg) (Voegtlin, 1984). Honeysuckle aphid is capable of transmitting over 20 plant viruses, including celery crinkle leaf mosaic, celery mosaic, and celery leaf spot. Although coriander aphid is not a known vector of plant viruses, as yet it is poorly studied.

Management

These are generally minor pests which are suppressed, if necessary, by foliar application of insecticides. Care should be taken to treat the underside of the foliage.

Corn Leaf Aphid
Rhopalosiphum maidis (Fitch)
(Homoptera: Aphididae)

Natural history

Distribution. This species may be of Asiatic origin, but now has a world-wide distribution. Corn leaf aphid does not persist in areas of severe winters, so it is absent from most of Canada, and northern Europe and Asia, and re-invades large areas of temperate regions annually. In North America, it is normally found on corn through most of the United States, but is relatively uncommon on this host in the northwestern states, preferring instead to feed on barley. In Canada, it is found principally in southern areas of the eastern and prairie provinces. (See color figure 154.)

Host Plants. Corn leaf aphid feeds on numerous grasses in addition to corn, and is considered to be a serious pest of cereal grains due to its ability to transmit virus diseases. Although corn is the only vegetable crop affected, several field crops are fed upon, including barley, chufa, oat, rye, millet, sorghum, Sudan grass, sugarcane, and wheat. Barley is the most important early season host. While the leaf blades are rolled, aphid colonies develop within the furled-barley leaves. However, once the corn is about 30-days old it also becomes very suitable for aphids; the tassels forming within the furled leaves are particularly suitable for aphid population growth. As the tassel extends, aphids disperse over the entire plant to feed.

Among weeds and prairie grasses known to serve as hosts are barnyardgrass, *Echinochloa crusgalli*; buffalo grass, *Buchloe dactyloides*; crabgrass, *Digitaria sanguinalis*; foxtail, *Setaria* spp.; gramagrass, *Bouteloua* spp.; and Johnsongrass, *Sorghum halepense* (Robinson and Hsu, 1963; Kieckhefer, 1984).

Natural Enemies. These aphids feed on exposed areas of the plant and often are subject to significant levels of predation and parasitism. Numerous species of lady beetles attack corn leaf aphid. In Ontario, for example, Foott (1973) reported that *Hippodamia convergens* Guerin–Meneville, *H. tredecimpunctata* (Say), and *Coleomegilla maculata* Timberlake were the dominant lady beetle predators, but *Adalia bipunctata* (Linnaeus), *Cycloneda sanguinea* Linnaeus, *Hippodamia parenthesis* (Say), and *Coccinella transversoguttata* Brown (all Coleoptera: Coccinellidae) also were present. Despite their abundance, however, the lady beetles were numerous and effective predators only after the aphid populations attained high and damaging densities.

Other common predators include various flower flies (Diptera: Syrphidae), predatory midges (Diptera: Cecidomyiidae), minute pirate bugs (Hemiptera: Anthocoridae), and lacewings (Neuroptera: Chrysopidae).

Several wasps parasitize corn leaf aphid. *Lysiphlebus testaceipes* (Cresson) (Hymenoptera: Braconidae) is reportedly the most widespread and common parasitoid, but *Aphelinus varipes* (Foerster) (Hymenoptera: Aphelinidae) was an important parasitoid in Texas (Gilstrap *et al.*, 1984). Other parasitoids include *Diaeretiella rapae* (M'Intosh) and *Ephedrus persicae* (Froggatt) (both Homoptera: Aphidiidae) and *Aphelinus asychis* Walker (Hymenoptera: Aphelinidae).

Life Cycle and Description. This species routinely overwinters as viviparous females (displaying parthenogenetic reproduction) in the southern states, north to about northern Texas. During mild winters, there is evidence that overwintering may occur at least as far north as southern Illinois. Overwintering does not occur in northern states and in Canada. Oviparous (egg producing) forms and males are rare, though in Pakistan eggs overwinter on *Prunus* sp. The northern areas of North America are assumed to be invaded annually (Rose *et al.*, 1975), with the timing of invasion and the number of subsequent generations in an area a function of weather. The short life cycle of this aphid, normally 6–12 days, allows production of 20–40 generations per year in such southern locations as Texas and about nine generations in Illinois, but fewer further north.

Nymph. There are four instars. The young nymph initially is pea green, with red eyes and colorless antennae and legs, and measures about 0.5 mm long. Body length of this species increases to about 0.9, 1.1, and 1.3 mm, the green body becomes darker, and the appendages gain some dark pigmentation as the nymph progresses through instars 2–4. Mean development times are 4.5, 4.5, 4.5, and 4.7 days, respectively, for instars 1–4 among nymphs destined to grow into apterae, when cultured at 11°C. The corresponding development times for 19° and 29°C are 1.7, 1.8, 1.7, and 1.9 days, and 1.2, 1.2, 1.5, and 1.7 days, respectively. Development times for alatae are quite similar, except that instar four tends to require an additional day (Elliott *et al.*, 1988).

Adult. The viviparous females generally are bluish-green, though they become darker with time, some becoming almost black. Fine white powder covers the entire body. The winged form (alatae) have a black thorax and head, whereas in the wingless form (apterae) only the head is black. The winged form also tends to bear three black spots laterally on the abdomen, and the base of the cornicles is enveloped in a small area of purple or black. In both female forms the appendages are black. The alatae measure about 1.7 mm long, the apterae about 2.4 mm. The adult male aphid measures about 1.5 mm long. The head, thorax, and legs are black. The abdomen is dark bluish-green, with a dark spot laterally on each segment.

Elements of corn leaf aphid biology were found in Wildermuth and Walter (1932), Foott (1977), and Steiner *et al.* (1985), and descriptive material in Davis (1909), Wildermuth and Walter (1932), and Cottier (1953). Kring (1985) described the instars. Corn leaf aphid was included in keys to small grain aphids published by Pike *et al.* (1990) and Olsen *et al.* (1993).

Damage

Corn leaf aphid is commonly found feeding on the tassel and silk of the corn plant in addition to the leaves. It interferes with pollen production and fertilization, resulting in poor kernel fill of the ears. Infestation also can cause a delay in plant maturity and reduced-plant size (Bing *et al.*, 1991). Honeydew secreted by the aphids supports growth of sooty mold fungus, causing an unsightly appearance of the ears.

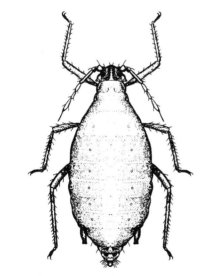

Adult female corn leaf aphid, wingless form.

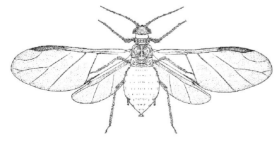

Adult female corn leaf aphid, winged form.

In general, corn plants are very tolerant of aphids, and only extremely dense infestations cause injury. If soil moisture conditions are inadequate, however, then the damage by the aphids is increased, and yield reductions are more likely.

The ability to transmit plant viruses greatly exacerbates the damage potential of this aphid. Several diseases, including barley yellow dwarf, beet yellows, cucumber mosaic, lettuce mosaic, maize streak, and maize stripe are transmitted. Of particular concern is transmission of maize dwarf mosaic—a serious disease of sweet corn (Straub, 1984). Although failing to colonize most vegetable crops, corn leaf aphid is one of the most abundant aphids collected in crops, and is implicated in the transmission of numerous stylet-borne viruses (Hander *et al.*, 1993).

Management

Sampling. Sampling for corn leaf aphids in young corn is usually done by visual examination of the whorls. As the aphids disperse from the whorls, they are readily apparent on other plant structures. As populations increase and honeydew accumulates, remote detection of aphid populations is possible by examination of photographs taken by aircraft using infrared-sensitive film. Sooty mold growing on honeydew impedes the reflectance of infrared radiation (Wallen *et al.*, 1976). It is advisable to monitor populations in barley early in the season, as incidence in this favored crop may reflect potential incidence in corn later in the season. Populations of alate aphids may be monitored with yellow water-pan traps or sticky traps (Straub, 1982; Hander *et al.*, 1993), though aphids collected on sticky traps are often damaged and difficult to identify.

Insecticides. Insecticides applied to corn for corn leaf aphid control prior to complete tassel development may increase corn yields, though in many areas corn leaf aphid is not viewed as a serious threat so insecticides are not applied. Systemic insecticides applied to the young plants or to the soil soon after plant emergence from the soil are particularly effective (Foott, 1974, 1975).

Cultural Practices. Early plantings can escape injury, especially in northern areas where aphids do not overwinter. In New York, for example, corn planted before June 10 escaped injury by corn leaf aphid and maize dwarf virus, but incidence of aphids and disease increased thereafter (Straub and Bothroyd, 1980). Little work on the management of corn leaf aphid as a virus vector has been reported, other than the assessment of corn varieties for resistance, but approaches discussed in the sections on green peach aphid, *Myzus persicae* (Sulzer), and melon aphid, *Aphis gossypii* (Glover), are applicable.

Corn Root Aphid
Aphis maidiradicis Forbes
(Homoptera: Aphididae)

Natural history

Distribution. Corn root aphid occurs throughout the United States east of the Rocky Mountains. It is most abundant and troublesome in the mid-western states. Corn root aphid likely occurs in southern Canada also, but is not reported to be a pest there. Though likely a complex of closely related species, for the purpose of this discussion it is treated as a single species.

Host Plants. Corn and cotton are the only crops injured by this species. However, corn root aphid is reported to have a wide host range, existing principally on weeds. Although some of the host records may be erroneous because it is difficult to distinguish this species from some other *Aphis* species, it is reported to feed on aster, *Aster* spp.; crabgrass, *Digitaria sanguinalis*; dock, *Rumex crispus*; foxtail, *Setaria* spp.; pigweed, *Amaranthus* spp.; plantain, *Plantago* spp.; portulaca, *Portulaca oleracea*; prairie sagewort, *Artemisia frigida*; ragweed, *Ambrosia* sp.; smartweed, *Polygonum* sp.; sunflower, *Helianthus* spp.; tansymustard, *Descurainia* sp.; and many others.

Natural Enemies. There seems to be little known about natural enemies of this insect, though fungi are known to attack colonies.

Life History and Description. In Illinois, hatching occurs during April and May. The aphids reproduce parthenogenetically during the spring and summer months. There are perhaps 16–22 generations per year, with a generation completed in just 6–8 days. Sexual reproduction occurs only in the autumn, followed by overwintering in the egg stage.

Egg. The eggs are collected by ants, commonly the cornfield ant, *Lasius alienus* (Förster) (Hymenoptera: Formicidae), and tended by the ants in their burrows throughout the winter. The ants pile them in their chambers and move them in accordance with the moisture and temperature conditions of the soil. The eggs are black or greenish black, elliptical in shape, and measure about 0.8 mm long and 0.4 mm wide.

Nymph. Upon hatching the young aphid nymphs, which are all wingless viviparous (parthenogenetic) females, are transferred by ants to the roots of weeds, where they normally complete 2–3 generations. The

nymphs initially are pale green, but as they mature they become grayish, eventually attaining a length of about 2 mm. The nymphs usually complete their growth in about 7–14 days, and almost immediately they commence reproduction. Davis (1909), working during the cooler weather of October, reported development times of 2–4, 5–6, 9–14, and 13–21 days, respectively, for instars 1–4.

Adult. The various forms of the adult aphids differ in appearance. The wingless viviparous females are predominantly bluish-green, but bear a white flocculence on their body and have a black head and transverse bars on the thorax. In winged viviparous females the head and thorax are dark-brown or black, and the abdomen is light-green with three distinct black spots on each side. The wings are dusky. The oviparous females are wingless, and measure about 2.2 mm long. The head, and at least part of the thorax are dark, but the abdomen is bluish-gray and tinged with pink. Males also may be either winged or wingless. Wingless males are about 1.75 mm long, whereas the winged male aphid measures about 1.5 mm long. In both forms of the male, the head, thorax, and appendages are black and the abdomen is greenish but marked with 2–3 dark transverse bars on the apical (and sometimes basal) segments.

Reproduction during the summer is parthenogenetic. Viviparous, or parthenogenetically reproducing females give birth to 40–50 nymphs, averaging about five nymphs per day. Later, as corn plants develop, the ants transfer the aphids to the roots of corn seedlings. The aphids are redistributed and protected from predators and disease throughout the summer by the attending ants. The incentive for the ants to tend the aphids is, of course, the production of honeydew by the aphids. During the summer months a mixture of winged and wingless females develop and feed on the corn roots. The winged forms crawl through the ant tunnels to the surface, and disperse to new locations. If they alight in the vicinity of ants they are seized and carried below-ground by the ants and placed on a root, giving rise to a new infestation. Sometimes the aphids are transported up to 50 m to infest new locations. Males and egg-producing (oviparous) females are produced in the autumn, when sexual reproduction occurs and overwintering eggs are produced and deposited on the roots of corn or weeds.

Description and biology of corn root aphid were given by Davis (1909, 1917). Ecology of the cornfield ant, including its interrelationship with corn root aphid, was described by Forbes (1908).

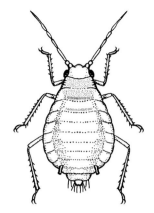

270a Adult female corn root aphid, wingless form.

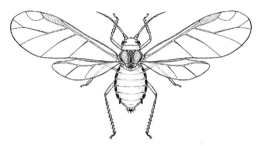

270b Adult female corn root aphid, winged form.

Damage

Aphids cluster on the roots of corn and the underside of the crown at the base of the roots, where their feeding inhibits growth of young corn plants. Heavily infested plants are yellow or brown and stunted. Plants infested later in the season are more tolerant of aphid feeding and are not damaged unless the plant is under additional stress, such as lack of adequate water.

Management

Sampling. The distribution of corn root aphids is governed by their attending ants. Ant colonies vary in size and frequency in corn fields, depending largely on whether insecticide application was made to earlier-grown crops of corn. In the absence of insecticide and presence of earlier crops of corn, individual ant colonies may be large, each encompassing areas of 50 sq m, and supporting high densities of aphids.

Insecticides. Insecticides applied to the soil for other root-feeding insects such as corn rootworms species typically kills ants and therefore aphids are not a problem. In decreased insecticide production systems, especially if accompanied by reduced tillage, the aphids become more of a problem.

Cultural Practices. Tillage disrupts ant nests, thereby decreasing the abundance of aphids. Deep cul-

tivation is necessary because ant nests may occur at a depth of 18 cm or greater in the soil. Plowing and disking to a depth of 15 cm destroys most ant nests. Crop rotation can be beneficial because few other crops are suitable hosts. However, it is advisable not to plant corn adjacent to fields previously infested because the ants can relocate aphids to nearby fields.

Cowpea Aphid
Aphis craccivora Koch
(Homoptera: Aphididae)

Natural History

Distribution. This species is known from all continents where agriculture occurs. It can be a serious pest in both temperate and tropical areas, especially the latter in Africa and Asia. Cowpea aphid appears to have been introduced from Europe to the western hemisphere, though the date is unknown. It now occurs throughout the United States and southern Canada. (See color figure 152.)

Host Plants. Cowpea aphid feeds on numerous plants, but tends to be most abundant on plants in the family Leguminosae. Summer hosts include such vegetables as asparagus, carrot, cowpea, kidney bean, lettuce, and lima bean. Other crops affected include alfalfa, apple, citrus, hairy indigo, hairy vetch, pinto bean, red clover, rye, sesbania, sweet clover, and wheat. Common weeds serving as hosts include dandelion, *Taraxacum officinale*; lambquarters, *Chenopodium album*; pigweed, *Amaranthus* spp.; shepherdspurse, *Capsella bursa-pastoris*; dock, *Rumex* spp.; goldenrod, *Solidago* spp.; kochia, *Kochia scoparia*; pepperweed, *Lepidium* spp.; and Russian thistle, *Salsola kali*. Palmer (1952) reported the overwintering host in Colorado as locust trees, *Robinia* sp., but overwintering aphids usually persist on succulent plants in warmer areas.

Natural Enemies. The natural enemies of cowpea aphid include many of the common aphid predators such as lady beetles (Coleoptera: Coccinellidae), lacewings (Neuroptera: Chrysomelidae and Hemerobiidae), and flower flies (Diptera: Syrphidae). (See the section on green peach aphid, *Myzus persicae* (Sulzer) for a more complete discussion on aphid predators.) Parasitoids are also known, including *Ephedrus persicae* Froggatt, *Praon abjectum* (Haliday), and *Diaeretiella rapae* (M'Intosh). The latter, a cosmopolitan species attacking numerous aphids, is well known. Some biological information on this parasitoid is available in Wheeler (1923) and Schlinger and Hall (1960).

Life Cycle and Description. Sexual reproduction is absent or unproven from cowpea aphid populations in North America; overwintering normally occurs in greenhouses and in warm regions. As is the case with most aphids, their development time is very short. Kaakeh and Dutcher (1993) reported mean generation times of as few as 8.4 days on favored hosts such as cowpea, to 12.2 days on hairy vetch.

Nymph. These aphids are black, but are lightly dusted with whitish wax, resulting in a gray appearance. The appendages are whitish. Cowpea aphid has four instars. The duration of development is largely a function of temperature, and secondarily of diet. The effects of photoperiod and humidity are slight. Radke *et al.* (1973) indicated that mean duration of the four instars was 3.3, 2.1, 1.4, and 1.6 days, respectively, at 24°C and 12 h photophase. Nymphal development time of aphids fed with several orchard cover crops was studied by Kaakeh and Dutcher (1993), who reported that though both host plant and temperature would influence development rate, most temperature and host combinations resulted in similar nymphal development times, about 5–8 days. The lower threshold for aphid development is estimated at 4.0–5.5°C. The aphids typically form colonies on the rapidly growing plant tissues. These aphids are often attended by ants. In response to disturbances such as predation, cowpea aphid produces and responds to alarm pheromone (Yang and Zettler, 1975).

Adult. These are dark-colored aphids, but the appendages are whitish with black tips. The cornicles are black. The adults are shiny black, sometimes with a fine dusting of white powder. Their shine distinguishes them from the similar bean aphid, *Aphis fabae* Scopoli. Winged (alate) and wingless (apterous) forms are similar in appearance, with a body length of about 1.9–2.5 mm and 1.5–2.2 mm for the two forms, respectively. The wing veins of the alatae are brownish. The adult pre-reproductive period is usually only about one day in duration. The reproductive period is more variable, ranging from 7–9 days at 30–24°C, to about 40 days at 13°C. Aphids developing at warmer temperatures produce fewer offspring. Fecundity was estimated by Radke *et al.* (1973) at 12–17 nymphs per apterous female at 30°C, to 30–45 nymphs at 13°C. However, Elliott and McDonald (1976) observed higher fecundities: 84 nymphs per alate female and 96 nymphs per apterous female.

Developmental biology was given by Radke *et al.* (1973), Elliott and McDonald (1976), and Abdel–Malek *et al.* (1982). A brief description of this aphid, and keys for its identification, are available in Palmer (1952)

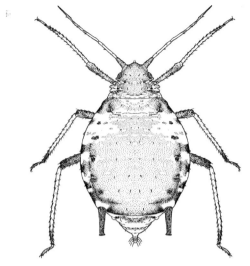

Adult female cowpea aphid, wingless form.

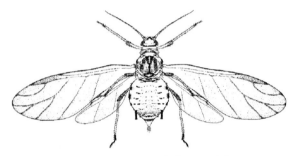

Adult female cowpea aphid, winged form.

and Blackman and Eastop (1984). Cottier (1953) also provided a good description. Stoetzel *et al.* (1996) published a key for cotton aphids that is also useful to distinguish cowpea aphid from most other common vegetable-infesting aphids.

Damage

Cowpea aphid colonizes various plant tissues, including stems, pods, young leaves and old leaves, with the stem tissue most preferred (Srikanth and Lakkundi, 1988). The aphid removes plant sap, disrupting the normal plant growth pattern, including reduction in root growth and nodulation. At high aphid densities, plants are deformed and stunted, and seed set is reduced. Cowpea aphid does not produce toxins or other persistent effects, so that once the aphids are removed, normal plant growth rates are resumed. In fact, a degree of compensatory growth occurs, with plants exposed to short-term aphid infestations displaying an increased rate of growth once aphids are removed. This allows plants that support aphids for short periods of time to recover and produce crop yields equivalent to uninfested plants (Hawkins *et al.*, 1986).

The principal damage by this species is disease transmission. Over 30 stylet-borne and persistent plant viruses, including diseases of bean, beet, cowpea, onion, pea, cucurbits, and crucifers are transmitted (Kennedy et al., 1962). For example, cowpea aphid was reported to be one of the most common potyvirus vectors in squash cultivated in Arkansas, despite failing to colonize and establish large colonies on the crop (Hander et al., 1993). Exposure of this aphid to alarm pheromone does not result in decreased virus transmission (Yang and Zettler, 1975).

Cowpea aphid can also be beneficial in some agricultural systems. In the southeastern United States, leguminous cover crops such as sesbania, hairy indigo, hairy vetch, and red clover are planted in pecan orchards to support high densities of cowpea aphid. In pecan production systems cowpea aphids do no harm, but serve as valuable sources of food for general insect predators such as lady beetles (Coleoptera: Coccinellidae), which then also prey on the pecan aphid complex (Kaakeh and Dutcher, 1993).

Management

Sampling. Cowpea aphid can be sampled in a manner similar to other aphids. Winged adults typically are monitored with yellow water pan or sticky traps, and wingless adults and nymphs with sweep nets or by direct visual observation. Due to the difficulty in counting many tightly clustered aphids, methods of population estimation have been developed that are based on degree of substrate inhabited by aphids (Srikanth and Lakkundi, 1988). Studies in Africa have demonstrated significant reductions in insecticide use on cowpea when decisions on whether or not to spray insecticide were based on population sampling rather than calendar-based plans (Afun et al., 1990). Aphid monitoring is considered in greater detail in the sections on green peach aphid, *Myzus persicae* (Sulzer), and melon aphid, *Aphis gossypii* Glover.

Insecticides. Insecticides are commonly used to protect susceptible crops, particularly cowpea, from this aphid. Both systemic and contact insecticides are applied, and systemic materials are often applied at planting. In North America, cowpea is grown extensively only in the southeastern United States, where it is susceptible to infection with plant viruses. Insecticides are used to limit disease transmission, but there is some doubt about the effectiveness of this approach because many of the virus diseases are stylet-borne, and rapidly transmitted.

Host-Plant Resistance. Some cultivars of cowpea are resistant to cowpea aphid (Pathak, 1988). The basis

for resistance seems to be related to interference with feeding, and the inability of aphids to colonize and to produce progeny (Givovich et al., 1988). In greenhouse studies comparing resistant and susceptible varieties, aphid survivorship and reproduction were decreased when fed on foliage of a resistant cultivar (Annan et al., 1997).

Foxglove Aphid
Aulacorthum solani (Kaltenbach)
(Homoptera: Aphididae)

Natural History

Distribution. Foxglove aphid occurs throughout the United States and southern Canada. As a pest, its effect is greatest in the eastern regions of the continent. Foxglove aphid is found almost world-wide, but is probably of European origin.

Host Plants. Although having a very wide host range, foxglove aphid is a pest principally of potato. In the northernmost potato-growing areas in eastern North America it is sometimes the dominant aphid species on this crop, but in other areas it is displaced by green peach aphid, *Myzus persicae* (Sulzer); potato aphid, *Macrosiphum euphorbiae* (Thomas); and buckthorn aphid, *Aphis nasturtii* Kaltenbach. Foxglove aphid also occurs on a few other vegetables including celery, chervil, lettuce, pea, and tomato, and also on alsike, red, and white clover. In Asia, but not North America, it frequently colonizes soybean. Fruit crops that on occasion support foxglove aphid include apple, raspberry, and strawberry. It may be found on such flowers as calla lily, cineraria, Easter lily, foxglove, gladiolus, pansy, salvia, tulip, and violet. Among the numerous weeds known to support foxglove aphid are bittersweet, *Solanum dulcamara*; buttercup, *Ranunculus* spp.; cinquefoil, *Potentilla* spp.; common chickweed, *Stellaria media*; common plantain, *Plantago major*; dock, *Rumex* spp.; fall dandelion, *Leontodon autumnalis*; orange hawkweed, *Hieracium aurantiacum*; king devil, *Hieracium floribundum*; oxeye daisy, *Chrysanthemum leucanthemum*; pigweed, *Amaranthus* spp.; purslane, *Portulaca oleracea*; shepherdspurse, *Capsella bursa-pastoris*; and smartweed, *Polygonum* spp. The overwintering hosts of foxglove aphid in cold-winter climates are foxglove, which is the basis of its common name, and hawkweed species, which apparently are more important hosts. In Europe, additional overwintering hosts are known, but in North America, woody plant hosts are not required for overwintering by either the egg or adult stages.

Natural Enemies. The natural enemies of foxglove aphid are not well documented in North

America. Among the predators of foxglove aphid are the common lady beetles (Coleoptera: Coccinellidae), some lacewings (Neuroptera: Chrysopidae), flower flies (Diptera: Syrphidae), and the predatory midge *Aphidoletes aphidimyza* (Rondani). In Maine, several parasitoid species have been collected, including *Aphidius nigripes* Ashmead, *Praon* spp., *Monoctonus* sp., and *Aphidius* spp. (all Hymenoptera: Aphidiidae), and *Aphelinus semiflavus* Howard (Hymenoptera: Eulophidae) (Shands *et al.*, 1965). In Europe, parasitoids reportedly exert good control of foxglove aphid (Robert and Rabasse, 1977), and this is likely the case in North America also. In addition, fungi affect these aphids.

Life Cycle and Description. Foxglove aphid can remain on its primary hosts, foxglove and hawkweed species, throughout the year. Alternatively, it can disperse during the summer to crops and weeds. In either case, the population peaks in August or September. In cold-winter climates such as Maine, overwintering occurs in the egg stage. In climates with warmer winters, such as New Jersey, aphids may overwinter as adults.

Egg. The eggs are normally shiny, black, and elliptical in form. They measure about 0.6 mm long and 0.3 mm wide. They are found principally on foxglove and hawkweed. They are deposited by oviparous adults in October and November, and hatch in March.

Nymph. There are five instars. Mean nymphal development time is 2.4, 1.8, 2.4, 2.7, and 3.2 days for instars 1–5, respectively, at 21°C. After the 12–13 day developmental period, apterous adults commence parthenogenetic reproduction almost immediately.

Adult. The general body form of the adults is best described as pear-shaped, with the posterior portion greatly expanded relative to the anterior portion. The tubercles at the base of the antennae are pronounced but straight-sided, facing neither inward as in green peach aphid, nor outward, as in potato aphid. The apterous females measure 1.5–3.0 mm long. They are principally yellow, yellowish-green, or green, with the head and thorax also pale. In some specimens the dorsal surface is darker. The abdominal cornicles are moderately long, thin, and tapered, measuring 0.4–0.7 mm long; they vary in color from almost colorless to dark, and with dark tips. The reproductive period lasts about 15–30 days, with females producing an average of 60 offspring per female (range 25–81). Adults perish a few days after completing reproduction. The alate females are similar in size, measuring 2.0–2.9 mm long. The abdomen is pale green, olive green, or green, but the head and thorax vary from dusky-yellow to almost black with a brown tinge. The alatae usually bear dark bars or patches dorsally on the abdomen, and there are dark patches at the base of the cornicles. As in the apterous form, the cornicles are long and thin, measuring 0.5–0.7 mm long and varying in color. The veins of the wings are dark.

A brief treatment of foxglove aphid biology was found in Patch (1928). A morphological description was presented by Cottier (1953), and phenology and host plant relations by Wave *et al.* (1965). Developmental biology was given by MacGillivray and Anderson (1958). Foxglove aphid was included in the keys of Palmer (1952) and Blackman and Eastop (1984).

Damage

Aphid populations cause potato leaves to curl, especially if the leaves are young when attacked. Foxglove aphid apparently produces toxic saliva, as it causes plant deformity and stunting at lower densities than other potato-infesting aphids. This aphid also transmits numerous plant viruses.

Management

Management of foxglove aphid has not been well-studied, probably because this aphid is usually only

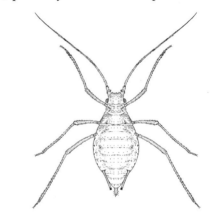

Adult female foxglove aphid, wingless form.

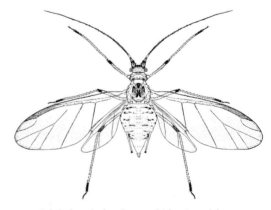

Adult female foxglove aphid, winged form.

one component of the aphid complex, and often a minor component. The sampling and management techniques discussed in the sections on green peach aphid, potato aphid, and buckthorn aphid are likely applicable to foxglove aphid.

Green Peach Aphid
Myzus persicae (Sulzer)
(Homoptera: Aphididae)

Natural history

Distribution. Green peach aphid is found all over the world, including the tropics and temperate latitudes. In the United States, it is considered to be a pest in all states. In Canada, it occurs from Nova Scotia to British Columbia, where it is viewed as a pest principally due to its ability to transmit plant viruses (Beirne, 1972). In addition to attacking plants in the field, green peach aphid readily infests vegetables and ornamental plants grown in greenhouses. This allows high levels of survival in areas with inclement weather, and favors ready transport on plant material. When young plants are infested in the greenhouse and then transplanted into the field, fields are not only inoculated with aphids, but insecticide resistance may be introduced. The aphids are also reported to be transported long distances by wind and storms. The origin is Asia. (See color figure 153.)

Host Plants. Green peach aphid feeds on hundreds of host plants in over 40 plant families. However, it is only the viviparous summer stages that feed so widely; the oviparous winter stages are much more restrictive in their diet choice. In temperate latitudes the primary or overwintering hosts are trees of the genus *Prunus*, particularly peach and peach hybrids, but also apricot and plum. During the summer months the aphids abandon their woody hosts for secondary or herbaceous hosts, including vegetable crops in the families Solanaceae, Chenopodiaceae, Compositae, Cruciferae, and Cucurbitaceae. Vegetables that are reported to support green peach aphid include artichoke, asparagus, bean, beets, broccoli, Brussels sprouts, cabbage, carrot, cauliflower, cantaloupe, celery, corn, cucumber, fennel, kale, kohlrabi, turnip, eggplant, lettuce, mustard, okra, parsley, parsnip, pea, pepper, potato, radish, spinach, squash, tomato, turnip, watercress, and watermelon. Crops differ in their susceptibility to green peach aphid, but it is actively growing plants, or the youngest plant tissue, that most often harbors large aphid populations (Heathcote, 1962). In warmer climates the aphids do not seek out overwintering hosts, but they persist as active nymphs and adults on hardy crops and weeds.

Broadleaf weeds can be very suitable host plants for green peach aphid, thereby creating pest problems in nearby crops. Tamaki (1975), for example, estimated that about 3–16 million aphids per acre were produced on weeds growing on the floor of peach orchards. Up to one-third of the aphids feeding on weed species carried beet western yellows virus (BWYV) (Tamaki and Fox, 1982). Peach trees are not a host of BWYV, so weeds are obviously good reservoirs for plant virus. Common and widespread weeds such as field bindweed, *Convolvulus arvensis*; lambsquarters, *Chenopodium album*; and redroot pigweed, *Amaranthus retroflexus*, are often cited as important aphid hosts (Annis *et al.*, 1981).

Natural Enemies. Hundreds of natural enemies have been recorded, principally lady beetles (Coleoptera: Coccinellidae), flower flies (Diptera: Syrphidae), lacewings (Neuroptera: mainly Chrysopidae), parasitic wasps (Hymenoptera: Braconidae), and entomopathogenic fungi (mainly Entomophthorales). van Emden *et al.* (1969) provided a long list of beneficial organisms. Most are general predators, and move freely among green peach aphid, other aphids, and even other insects. In some cases the natural enemies are influenced by the host plant, crop cultural practices, and environmental conditions (Tamaki *et al.*, 1981). Green peach aphid, and many other aphids, may occasionally be cannibalistic or predatory (Banks *et al.*, 1968); however, this behavior occurs only under stressful condition and is not known to be an important factor in population regulation. Quantitative data generally are lacking for the influence of most natural enemies. Weather also reportedly contributes a significant change in aphid numbers, including direct mortality (Beirne, 1972), but this also is poorly documented.

The ephemeral nature of aphid infestation in many crops is believed to prevent the beneficial organisms from consistently locating the aphids and reproducing in a timely manner. Nevertheless, anyone who has frequently observed green peach aphid at high densities probably has observed sudden population decreases following the appearance of lady beetles, wasp parasitoids, or entomopathogenic fungi. For example, green peach aphid infesting spring-harvested spinach crops in Arkansas and Oklahoma is suppressed late in the growing season by *Erynia neoaphidis* fungus. Unfortunately, the disease epizootic often occurs too late to keep aphids from attaining high numbers, and fungus-infected aphids remain attached to foliage, providing a serious contaminant of spinach foliage (McLeod *et al.*, 1998). Various studies that selectively excluded or killed beneficial organisms have demonstrated the explosive reproductive potential of these aphids in the absence of biological control agents, thus demonstrating their value in reducing damage potential. In greenhouse crops, where environmental conditions and predator, parasitoid, and pathogen densities can be manipulated, biological suppression is more effective and consistent.

Integration of chemicals with natural enemies offers promise for enhanced protection from aphid damage. Shean and Cranshaw (1991) demonstrated that *Aphelinus semiflavus* Howard (Hymenoptera: Encyrtidae) and *Diaeretiella rapae* (McIntosh) (Hymenoptera: Braconidae) differed significantly in their relative susceptibility to insecticides, depending on the chemical evaluated. Also, these parasitoids, while in the mummy stage, were less susceptible to insecticide toxicity than green peach aphid. In general, however, insecticide use in crops is more disruptive to parasitoids than to aphids, leading to larger aphid populations. Sublethal doses of some insecticides also increase aphid reproduction (Lowery and Sears, 1986).

Despite the beneficial nature of these biotic agents, virus diseases can be effectively transmitted by very low aphid densities. In crops susceptible to aphid-borne virus disease, natural enemies alone are probably destined to be relatively ineffective in preventing damage.

Life Cycle and Description. The life cycle varies considerably, depending on the presence of cold winters. van Emden *et al.* (1969) provided a useful review of the life cycle. Development can be rapid, often 10–12 days for a complete generation, and with over 20 annual generations reported in mild climates. Where suitable host plants cannot persist, the aphid overwinters in the egg stage on *Prunus* spp. In the spring, soon after the plant breaks dormancy and begins to grow, the eggs hatch and the nymphs feed on flowers, young foliage, and stems. After several generation on *Prunus* spp., dispersants from overwintering hosts deposit nymphs on summer hosts. In cold climates, adults return to *Prunus* spp. in the autumn, where mating occurs and eggs are deposited. All generations except the autumn generation culminating in egg production are parthenogenetic.

Egg. The eggs are deposited on *Prunus* spp. trees. The eggs measure about 0.6 mm long and 0.3 mm wide, and are elliptical. They initially are yellow or green, but soon turn black. Mortality in the egg stage sometimes is quite high.

Nymph. Nymphs initially are greenish, but soon turn yellowish, greatly resembling viviparous adults. Horsfall (1924) studied the developmental biology of viviparous aphids on radish in Pennsylvania. He reported four instars in this aphid, with the duration of each averaging 2.0, 2.1, 2.3, and 2.0 days, respectively. Parthenogenetic females gave birth to offspring 6–17 days after birth, with an average age of 10.8 days at first birth. The length of reproduction varied considerably, but averaged 14.8 days. The average period of life was about 23 days, but this was under caged conditions where predators were excluded. The daily rate of reproduction averaged 1.6 nymphs per female. The maximum number of generations observed annually during these studies was determined to be 20–21, depending on the year. In contrast, MacGillivray and Anderson (1958) reported five instars with a mean development time of 2.4, 1.8, 2.0, 2.1, and 0.7 days, respectively. Further, they reported a mean reproductive period of 20 days, mean total longevity of 41 days, and mean fecundity of 75 offspring.

Adult. Up to eight generations may occur on *Prunus*, but as aphid densities increase winged forms are produced, which then disperse to summer hosts. Winged (alate) aphids have a black head and thorax, and a yellowish-green abdomen with a large dark patch dorsally. They measure 1.8–2.1 mm long. Winged green peach aphids seemingly attempt to colonize nearly all plants available. They often deposit a few young and then again take flight. This highly dispersive nature contributes significantly to their effectiveness as vectors of plant viruses.

The offspring of the dispersants from the overwintering hosts are wingless, and each produces 30–80 young. The wingless (apterous) aphids are yellowish or greenish. They measure about 1.7–2.0 mm long. A medial and lateral green stripes may be present. The cornicles are moderately long, unevenly swollen along their length, and match the body in color. The appendages are pale. The rate of reproduction is positively correlated with temperature, with the developmental threshold estimated to be about 4.3°C. As aphid densities increase or plant condition deteriorates, winged forms are again produced to aid dispersal. The nymphs that give rise to winged females may be pinkish. The dispersants typically produce about 20 offspring, which are always wingless. This cycle is repeated throughout the period of favorable weather.

In the autumn, in response to change in day length or temperature, winged male and female aphids are produced which disperse in search of *Prunus*. Timing is an important factor, as foliage on the *Prunus* hosts is physiologically optimal as leaves begin to senesce. Females arrive first and give birth to wingless (apterous) egg-laying forms (oviparae). Males are attracted to oviparae by a pheromone, capable of mating with several females, and the eggs are produced. The oviparous female deposits 4–13 eggs, usually in crevices in and near buds of *Prunus* spp. The oviparous female is 1.5–2.0 mm long, and pinkish in color.

Parthenogenic reproduction is favored in the many parts of the world, where continuous production of crops provides suitable host plants throughout the year, or where weather allows survival on natural (noncrop) hosts. The average temperature necessary for survival of active forms of green peach aphid is estimated at 4–10°C. Plants that readily support aphids during the winter months include beet, Brussels sprout, cabbage, kale, potato, and many winter weeds.

In the Pacific Northwest, both yellow and green strains coexist (Tamaki *et al.*, 1982). The yellow strain is holocyclic—a sexual generation in the autumn produces overwintering eggs. The green strain is anholocyclic—no sexual generation is produced. The egg-producing populations are less tolerant of cold weather, and deposit eggs on *Prunus*. The anholocyclic populations remain active throughout the year by feeding during the winter on weeds growing adjacent to warm springs, drainage ditches, and slopes exposed to solar radiation.

There are several synthetic diets suitable for aphid culture, such as the one described by Mittler *et al.* (1970). Cottier (1953) provided a good description of green peach aphid. Keys for identification of green peach aphid, and many other common aphids, were found in Palmer (1952), and Blackman and Eastop (1984). Stoetzel *et al.* (1996) published a key for cotton aphids that is also useful to distinguish green peach aphid from most other common vegetable-infesting aphids.

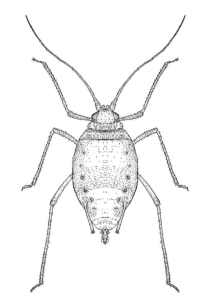

Adult female green peach aphid, wingless form.

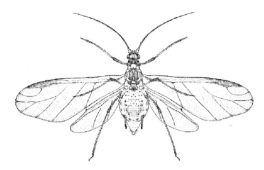

Adult female green peach aphid, winged form.

Damage

Green peach aphids can attain very high densities on young plant tissue, causing water stress, wilting, and reduced-growth rate of the plant. Prolonged aphid infestation can cause appreciable reduction in yield of root crops and foliage crops. Early season infestation is particularly damaging to potato, even if the aphids are subsequently removed (Petitt and Smilowitz, 1982). Contamination of harvestable plant material with aphids, or with aphid honeydew, also causes loss. In Arkansas, mild winters allow good overwintering survival of green peach aphid on spinach, thereby leading to contamination problems (McLeod, 1987). However, green peach aphid does not seem to produce the high volume of honeydew observed with some other species of aphids. Blemishes to the plant tissue, usually in the form of yellow spots, may result from aphid feeding. Leaf distortions are not common except on the primary host. Contamination of vegetables by aphids sometimes presents quarantine problems (Stewart *et al.*, 1980), and fumigation techniques have been developed that kill the insects without causing harm to the vegetables.

The major damage caused by green peach aphid is through transmission of plant viruses. Indeed, this aphid is considered by many to be the most important vector of plant viruses all over the world. Nymphs and adults are equally capable of virus transmission (Namba and Sylvester, 1981), but adults, by virtue of being so mobile, probably have greater opportunity for transmission. Both persistent viruses, which move through the feeding secretions of the aphid, and nonpersistent viruses, which are only temporary contaminants of aphid mouthparts, are effectively transmitted. Kennedy *et al.* (1962) listed over 100 viruses transmitted by this species. Some of the particularly damaging diseases include potato leafroll virus and potato virus Y to Solanaceae, beet western yellows and beet yellows viruses to Chenopodiaceae, lettuce mosaic virus to Compositae, cauliflower mosaic and turnip mosaic viruses to Cruciferae, and cucumber mosaic and watermelon mosaic viruses to Cucurbitaceae. A discoloration in potato tubers, called net necrosis, occurs in some potato varieties following transmission of potato leafroll.

Management

Management options for green peach aphid are very similar to procedures discussed for melon aphid, *Aphis gossypii* Glover.

Sampling. Degree-day models using a developmental threshold of 4°C can be used to predict various phenological events such as egg hatch and immigration of alate aphids (Ho *et al.*, 1998). Yellow traps, particularly water pan traps, are commonly used for population monitoring. Sequential sampling plans for green peach aphid on potato were developed by Hollingsworth and Gatsonis (1990). Sampling of aphid natural enemies in this crop were studied by Mack and Smilowitz (1980).

Insecticides. Despite the several options potentially available, many vegetable crop producers are dependent on insecticides for suppression of green peach aphid abundance. Systemic insecticide applications are especially popular at planting time. Most of them provide long-lasting protection against aphid population buildup during the critical and susceptible early stages of plant growth (Powell, 1980) and some provide protection for three months (Palumbo and Kerns, 1994).

Green peach aphid is often a pest of cold-weather crops like spinach. Aphids are inherently difficult to kill with contact insecticides because they are often found under the leaves or on new, sheltered growth. Cold weather (less than about 20°C) exacerbates the problem because of less volatilization (fumigation) by the insecticide (Wolfenbarger, 1972). Even systemic insecticides, which kill aphids feeding under the leaf when applied to the upper surface, are much less effective at cool temperatures (McLeod, 1991).

Biological Control. Mackauer (1968) noted that because green peach aphid was able to develop at lower temperatures than its parasitoids, the wasps were very beneficial only in benign climates or where temperature could be controlled, as in some greenhouses. Indeed, there has been considerable success using parasitoids, the entomopathogenic fungus *Verticillium lecanii*, and the predatory midge *Aphidoletes aphidimyza* (Diptera: Cecidomyiidae) for greenhouse-grown vegetables, especially in Europe (Gilkeson and Hill, 1987; Milner and Lutton, 1986).

In the field, biological control agents may be differentially affected by the cropping system. For example, Tamaki *et al.* (1981) found that the wasp *Diaeretiella rapae* (Hymenoptera: Braconidae) was more effective in broccoli, whereas lady beetles (Coleoptera: Coccinellidae) and bigeyed bug (Hemiptera: Lygaeidae) predators were more effective on radish.

Cultural manipulations may benefit predators and parasitoids. In Washington, bands placed around the trunks of peach trees provided good harborage for predators that may suppress the aphids in the spring, thereby reducing the number dispersing to vegetables (Tamaki and Halfhill, 1968). In California asparagus, a brown lacewing (Neuroptera: Hemerobiidae) consistently reduces green peach aphid populations, but benefits from application of supplemental food sprays (Neuenschwander and Hagen, 1980). In New Zealand, pollen levels were supplemented by interplanting flowering plants with cabbage, increasing predation of aphids by flower flies (Diptera: Syrphidae) (White *et al.*, 1995).

Cultural Practices. The overwintering behavior of green peach aphid, which in many areas is restricted to *Prunus* or other relatively restricted sites, has fostered research on techniques to reduce aphid abundance and disease transmission to vegetables by either removing the overwintering site or by eliminating the aphids before they disperse to vegetables. Destruction of peach and apricot trees (often found in association with houses), and treatment of trees with dormant oil and insecticide, have been used in western states to disrupt aphid population increase and disease transmission (Powell and Mondor, 1976). Similarly, vegetable and flower plants grown in greenhouses during the winter months have been shown to be an excellent source of infestation during the following spring (Bishop and Guthrie, 1964), and incidence of leafroll in potatoes grown in Idaho is directly related to the abundance of aphids in home gardens. In Colorado, inspection of garden centers and treatment of seedlings found infested with aphids are important elements of the overall potato leafroll reduction effort. As is usually the case with aphids, green peach aphid populations tend to be higher when plants are fertilized liberally with nitrogen fertilizers (Jansson and Smilowitz, 1986).

Small arthropods such as green peach aphid are susceptible to injury by secretions of glandular leaf hairs found on some plants, including wild potato (Tingey and Laubengayer, 1981). Younger aphids are particularly susceptible. Although this is a promising source of plant resistance, this valuable trait has yet to be incorporated into popular commercial cultivars.

Disease Transmission. Because some of the virus diseases transmitted by green peach aphid are persistent viruses, which typically require considerable time for acquisition and transmission, insecticides can be effective in preventing disease spread in some crops. Research in Minnesota (Flanders *et al.*, 1991) showed that potato leafroll virus was transmitted within the potato crop principally by wingless aphids moving from plant to plant. Infected seed potatoes are the principal source of leafroll in most potato crops, so planting disease-free seed is obviously an important step in minimizing the incidence of the disease. Growers commonly inspect fields for signs of disease, and remove and destroy infected and nearby plants— a process called "roguing." This procedure reduces the ability of aphids to spread disease from plant to plant. Insecticides may not keep winged aphids from alighting in a crop and quickly transmitting nonpersistent virus, but they can certainly prevent the secondary spread of virus within a crop by colonizing aphids. As is the case with melon aphid, however, insecticide resistance is a severe problem in many areas. Application of mineral oil (Ferro *et al.*, 1980; Lowery *et al.*, 1990) and use of aluminum or white plastic mulch (Wyman *et al.*, 1979) decrease virus transmission. (See melon aphid for further discussion of disease transmission.) Aphids that are not effectively repelled by reflective mulch seem to thrive on mulched crops (Zalom, 1981a) and exhibit high rates of reproduction. Therefore, even in mulched crops some aphid control is necessary.

Green peach aphid is quite responsive to alarm pheromone, which is normally produced when aphids are disturbed (Phelan *et al.*, 1976). Application of alarm pheromone has shown the potential to disrupt virus transmission (Gibson *et al.*, 1984), but this has yet to become an operational technology. A sex pheromone is also known from this aphid, but it functions only at short distances, and has not yet proved useful in aphid management (Dawson *et al.*, 1990).

Lettuce Aphid
Nasonovia ribisnigri (Mosley)
(Homoptera: Aphididae)

Natural History

Distribution.   This species is of European origin. It was detected in the eastern United States and eastern Canada during the early 1970s, and in the northwestern United States and British Columbia shortly thereafter. It has become a problem in California but remains absent from the Great Plains region. Lettuce aphid also is established in South America. Like many aphids, it is favored by cool weather, and seems likely to remain a pest mostly in the northern areas of North America.

Host Plants.   Lettuce aphid alternates between its primary, or winter, hosts and secondary, or summer, hosts. The winter hosts are *Ribes* spp., principally gooseberry but also red and black currant. Summer hosts are more varied, consisting of numerous plants in the family Compositae, including hawksbeard, *Crepis* spp.; hawkweed, *Hieracium* spp.; sowthistle, *Sonchus* sp.; and nipplewort, *Lampsana communis*. Some members of the family Scrophulariaceae (*Euphrasia* spp., *Veronica* spp.) and Solanaceae (*Nicotiana* spp., *Petunia* spp.) are also colonized. Among the vegetable crops injured are chicory, raddicio, and lettuce.

Natural Enemies.   Few natural enemies of lettuce aphid are documented, though the parasitoid *Monoctonus crepidus* (Haliday) (Hymenoptera: Aphidiidae) is known to attack lettuce aphid in Canada. It seems likely that common aphid predators such as lacewings (Neuroptera: Chrysopidae and Hemerobiidae), flower flies (Diptera: Syrphidae), and lady beetles (Coleoptera: Coccinellidae) are important natural mortality agents.

Life Cycle and Description.   Following egg hatch, parthenogenetically reproducing females and offspring are found on currant and gooseberry plants in British Columbia from March or April–May. Migration to secondary hosts occurs during late May and June, and many generations develop during the summer months. Migration back to primary hosts occurs in September and October, followed by mating, oviposition and overwintering in the egg stage. The eggs are deposited on buds, twigs, and foliage of overwintering hosts. Under mild-winter conditions, some aphids can remain on lettuce through the winter months.

These aphids are pale-green to apple-green when fed on *Ribes*. Winged dispersants (alatae) bear extensive black or dark pigmentation, often including dark head, thoracic regions and cornicles, dark patches or bars on the abdomen, and dark bands on the legs. The wings lack pigmentation. The wingless forms (apterae) developing on lettuce are pale-yellow to green, and sometimes reddish; they also are marked with darker patches, though the head and thorax are not uniformly dark. The antennae and cornicles are long in this species. Alatae measure 1.3–3.2 mm long and apterae 1.5–3.1 mm. The sexual forms are similar, averaging slightly smaller.

Lettuce aphid and its damage were described by Forbes and Mackenzie (1982). Description and keys were found in Heie (1979) and Blackman and Eastop (1984).

Damage

Lettuce aphid can be a serious pest because, unlike other lettuce-infesting aphid species, it can colonize

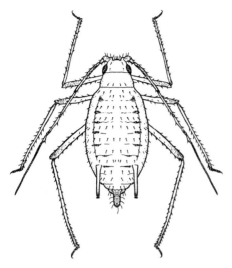

Adult female lettuce aphid, wingless form.

the interior portions of developing lettuce heads, making detection and removal difficult. Both field- and greenhouse-grown lettuce may be affected, and though the interior leaves are infested, exterior leaves are more heavily populated. In addition to contamination of lettuce heads, lettuce aphids cause injury by transmitting such lettuce-infecting viruses as cucumber mosaic. The early season aphid colonies occurring on *Ribes*, though small, cause leaf curl and retardation of shoot growth. On gooseberry, lettuce aphids transmit gooseberry veinbanding virus. Kennedy *et al.* (1962) reported that several other viruses also were transmitted by lettuce aphid.

Management

Sampling. Plants should be inspected starting with the seedling stage, and the interior as well as exterior leaves should be examined. Most of the aphids are found on the outer layer of the head (wrapper leaves), relative to the outer and the inner leaves. Infestations normally begin with field margins, so this area should be sampled most intensively (MacKenzie and Vernon, 1988).

Insecticides. When lettuce crops are grown from transplants rather than directly seeded, the young plants are often treated with an insecticide before transplanting as a precautionary measure to guard against field infestation. Once aphids are detected, insecticide applications are usually made on a regular basis until harvest. Systemic insecticides, including neem products, are often recommended (MacKenzie *et al.*, 1988; Lowery *et al.*, 1993).

Cultural Practices. If head interiors become infested it is important to destroy the crop by tillage rather than to simply abandon the field, as this reduces

the potential to spread to other fields. Weeds can also be important sources of aphids.

Host-Plant Resistance. Host-plant resistance has been examined extensively in Europe. Although partial resistance to lettuce aphid is not uncommon, varieties resistant to lettuce aphid often are not equally resistant to other aphids infesting lettuce, such as potato aphid, *Macrosiphum euphorbiae* (Thomas), and lettuce root aphid, *Pemphigus bursarius* (Linnaeus). The exception seems to be iceburg lettuce, which displays considerable resistance to all three aphid species (Dunn and Kempton, 1980b). Some resistance among *Ribes* cultivars also has been observed (Keep and Briggs, 1971).

Lettuce Root Aphid
Pemphigus bursarius (Linnaeus)
(Homoptera: Aphididae)

Natural History

Distribution. This species is likely of European origin, but is now found widely around the world, including the Middle East, Central Asia, Siberia, North and South Africa, and Australia in addition to North America. It thrives in temperate environments, and is found in southern Canada and in northern regions of the United States, and in California.

Host Plants. The primary, or winter, host of lettuce root aphid is poplar, *Populus* spp. The Lombardy poplar, *Populus nigra* var. *italica* is the most common host, but other varieties of *P. nigra* and of *P. deltoides* also are recorded as hosts. The secondary, or summer hosts of lettuce root aphid are normally plants in the family Compositae such as wild lettuce, *Lactuca seriola*; sow thistle, *Sonchus arvensis*; and dandelion; *Taraxacum officinale*. Plants from other families such as redroot pigweed, *Amaranthus retroflexus* (Amaranthacea); shepherdspurse, *Capsella bursa-pastoris* (Cruciferae); lambsquarter, *Chenopodium album* (Chenopodiacea); and white campion, *Lychnis alba* (Caryophilaceae) have been reported to support small populations of these aphids (Alleyne and Morrison, 1977b), but these records are suspect. Vegetable crops damaged by lettuce root aphid include chicory, endive, and lettuce.

Natural Enemies. Several natural enemies are known from Quebec (Alleyne and Morrisson, 1977c) and from Europe (Dunn, 1960). The natural enemy complex in Quebec varies considerably depending upon the host plant and location of the aphids on the host. It is difficult for large predators to gain access to aphids feeding within intact galls, but galls are sometimes damaged by birds, and small predators can gain entry through the gall exit hole. The pirate bugs *Anthocorus* spp. (Hemiptera: Anthocoridae), the flower flies *Syrphus ribesii* (L.) and *Metasyrphus musculus* (Say) (both Diptera: Syrphidae), the chaemaemyiid *Leucopis pemphigae* Malloch (Diptera: Chamaemyiidae), and the lacewings *Hemerobius* and *Chrysopa* spp. (Neuroptera: Hemerobiidae and Chrysopidae) feed on aphids within galls. Aphids on the surfaces of poplar and lettuce leaves also are subject to predation by such insects as *Chrysopa* sp. lacewings; spined assassin bug, *Sinea diadema* (Fabricius) (Hemiptera: Reduviidae); the damsel bug *Nabis ferus* (Linnaeus) (Hemiptera: Nabidae); and such lady beetles as *Adalia* sp., *Coccinella* spp., *Coleomegilla maculata* (De Geer), and *Hippodamia* spp. (all Coleoptera: Coccinellidae). Aphids feeding on roots are less exposed to predation, but *Thaumatomyia glabra* (Meigen) and *T. bistriata* (Meigen) (both Diptera: Chloropidae), and *Sphaerophoria menthastri* (Linnaeus) (Diptera: Syrphidae) were observed in Quebec. An aphid fungal disease readily infects aphids belowground, and protozoa and a nematode were occasionally observed. In England, similar natural enemies were observed, though the species are different. In addition, however, rove beetles (Coleoptera: Staphylinidae) and ground beetles (Coleoptera: Carabidae) are aphid predators in England, and further study in North America likely would demonstrate the importance of these predators.

Life Cycle and Description. Lettuce root aphid normally overwinters in the egg stage on poplar trees, hatching in March–April. The females hatching from these eggs are large, rotund, and highly fecund. They reproduce parthenogenetically, and after two generations on poplar, winged forms (alatae) disperse to summer hosts. Dispersal of alatae usually occurs in June–July, and on summer hosts the aphids produce about three generations of parthenogenetic wingless aphids (apterae). Beginning in about August, winged aphids are again produced, with dispersants migrating to the overwintering hosts. On poplar, small nonfeeding male and egg-producing female aphids are produced, mating occurs, and eggs are produced. However, some aphids can also remain on lettuce or weed roots throughout the winter, colonizing nearby plants in the spring.

Egg. The egg is elongate oval, initially greenish-white but eventually orange, and enveloped in slender strands of waxy material. The egg measures about 0.48 mm long and 0.23 mm wide. They are deposited in cracks and crevices of poplar bark at various heights up to at least 9 m on the tree.

Nymph. Nymphs infesting foliage are green, and initially are found within poplar leaf petiole galls caused by feeding of the first generation adult. The nymphs have five instars, and may require about 50–60 days in the spring to complete their development. Nymphs also are present on summer hosts, where their development time is much shorter, and where they respond to increasing density and declining host quality by producing winged progeny that disperse to poplar trees (Judge, 1968; Dunn, 1974). Nymphs on roots of summer hosts are not green; rather, they are light-gray and covered with bluish wax. Nymphs are not capable of burrowing through soil, but they move freely through cracks and over the surface of the soil, sometimes moving from plant to plant in this manner. Nymphs of all instars can overwinter below-ground on summer hosts, often occurring at depths of 10–20 cm (Alleyne and Morrison, 1978a).

Adult. The adults of this species lack elongate cornicles; they are pore-like or absent. Adults possess relatively short antennae and legs. The adult female resulting from overwintering eggs is enveloped in bluish-white wax. It is greenish, with dark appendages, and measures 2.0–2.2 mm long. The feeding of this adult induces formation of a gall on the leaf petiole of poplar. The gall measures 6–18 mm in diameter and can eventually contain about 100 aphids, or more. The adults emerging from the gall are winged (alate), disperse to summer hosts, and immediately begin reproduction. The new females produce 15–25 nymphs within the first hour, which give rise to wingless (apterous) adults. Wingless adult females produced during the remainder of the year are oval and yellowish-white in color, and measure 1.6–2.5 mm long. The short antennae are grayish to dark with black head. The dorsal surface of the body bears six longitudinal rows of circular wax glands, which produce waxy threads over the body. The entire aphid also is dusted with a fine white powder. The population of aphids peaks in August, at which time some nymphs move closer to the surface and molt into winged adults. Winged adult females (alatae) are gray, green, or brownish, with darker head, antennae, thorax and markings on the abdomen. The wing veins are dusky brown. The alatae also produce a waxy secretion, and measure 1.7–2.2 mm long. The reproductive forms produced late in the season on poplar are orange and lack functional mouthparts. The sexual forms are quite small, the males measuring 0.65 and the females 0.75 mm long. The abdominal cavity of the female is filled with a single egg that is deposited in a bark crevice.

The biology of this aphid was described by Dunn (1959a,b) and Alleyne and Morrison (1977b). Cottier (1953) provided a good morphological description. A

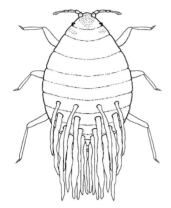

Adult female lettuce root aphid, wingless form.

method of rearing was presented by McLean and Kinsey (1961). Lettuce root aphid was included in the keys of Palmer (1952) and Blackman and Eastop (1984).

Damage

Feeding by lettuce root aphid results in wilting, yellowing and stunting of leaves. Rootlets may turn brown and die, and roots are covered with masses of bluish white, woolly material if aphids are present. Heavy infestations can cause softening and collapse of the lettuce head, and the death of the plant.

Management

Insecticides. Insecticides may be applied to the soil for protection of lettuce varieties that are susceptible to root aphid attack. Contact and systemic materials, liquid and granular formulations, and planting-time as well as side-dressed applications can be made, but effectiveness varies among products, and most insecticides work better as a preventative (Gentile and Vaughan, 1974; Alleyne and Morrisson, 1977a).

Cultural Practices. Destruction of old lettuce plants, particularly root stumps, and alternate hosts is often recommended to decrease the likelihood that aphids will be present when future crops are planted. Although the practice of sanitation always has considerable merit, it is most effective if soil temperatures are warm. This is because at cool temperatures lettuce root aphids can survive for months without food (Dunn, 1959a; Alleyne and Morrison, 1978a). Also, the presence of Lombardy poplars near lettuce fields is detrimental to aphid management, and planting of Lombardy poplars in lettuce production areas should be avoided, but few advocate destruction of these short-lived trees if they already exist.

Soil water management can be important in minimizing aphid damage. Adequate and even levels of soil moisture promote rapid growth of lettuce, enabling the plants to mature quickly and to escape the eventual population increase of the aphids. Plants also are more tolerant of injury if they be provided with adequate moisture. Constant soil moisture also limits cracking of the soil, limiting access to the roots by aphids and interplant movement.

Host-Plant Resistance. Alleyne and Morrison (1978b) evaluated lettuce varieties for susceptibility to aphid attack in Quebec. The varieties varied considerably in their susceptibility to damage, but some commercial romaine (cos) and crisphead varieties displayed some resistance. Dunn and Kempton (1980b) identified additional resistant varieties.

Melon Aphid
Aphis gossypii Glover
(Homoptera: Aphididae)

Natural History

Distribution. Melon aphid is widely distributed, and is known from tropical and temperate regions throughout the world except northern Canada and northern Asia. In the United States, it is a regular a pest in the southeast and southwest, but is occasionally damaging everywhere. In Canada, it is widespread in the eastern portion of the country, extending at least as far west as Manitoba; frequent injury is limited to southern Ontario and Quebec. As melon aphid occasionally overwinters in greenhouses, and may be introduced into the field with transplants in the spring, it has potential to be damaging almost anywhere. Its origin is in Europe. (See color figure 158.)

Host Plants. Melon aphid has a very wide-host range. Patch (1925) reported it occurring on 45 species from 25 plant families in Maine. Goff and Tissot (1932) provided a list of 60 host plants in Florida. Blackman and Eastop (1984) reported 700 host plants worldwide. However, the taxonomy of this species is uncertain, so some records may be incorrect. Among cucurbit vegetables, it can be a serious pest on watermelons, cucumbers, and cantaloupes, and to a lesser extent squash and pumpkin. Other vegetable crops seriously affected are asparagus, pepper, eggplant, and okra. Some other important crops injured regularly are citrus, cotton, and hibiscus. The overwintering, or primary, host in cold climates was long thought to be limited to live forever, *Sedum purpureum* (Patch, 1925). However, Kring (1955, 1959) demonstrated that *Sedum* was not a host of melon aphid, but a closely related species. Kring showed that catalpa, *Catalpa bignonioides*, and rose of sharon, *Hibiscus syriacus*, were the overwintering hosts in Connecticut. In the south, overwintering eggs are not commonly produced and overwintering hosts are more numerous, including dock,

Rumex crispus; *Lamium amphlexicaule*; boneset, *Eupatorium petaloiduem*; and citrus, *Citrus* spp. (Goff and Tissot, 1932). In California and North Carolina, Michelbacher *et al.* (1955) and Sorensen and Baker (1983) indicated overwintering on cold-tolerant plants, but did not mention specific hosts other than spinach. Isely (1946) noted the existence of host races; aphids reared on cotton could be transferred successfully to okra but not to cucurbits. This inability to transfer to other hosts has subsequently been shown for several other combinations.

Natural Enemies. Many of the natural enemies known to be effective against other aphids also attack melon aphid: lady beetles (Coleoptera: Coccinellidae), syrphid flies (Diptera: Syrpidae), and braconid wasps (Hymenoptera: Braconidae). Ants are commonly found associated with melon aphid but they are there to collect honeydew, and may even hinder predation by other insects. The wasp, *Lysiphlebus testaceipes* (Cress) (Hymenoptera: Braconidae), was noted by Goff and Tissot (1932) to be especially effective, somewhat causing up to 99% parasitism. Fungi also are sometimes observed to affect melon aphid.

Life Cycle and Description. The life cycle differs greatly between north and south. In the north, female nymphs hatch from eggs in the spring on the primary hosts. They may feed, mature, and reproduce parthenogenetically (viviparously) on this host throughout the summer, or they may produce winged females that disperse to secondary hosts and form new colonies. The dispersants typically select new growth to feed upon, and may produce both winged (alate) and wingless (apterous) female offspring. Under high density conditions, deterioration of the host plant, or upon arrival of autumn, production of winged forms predominates. During periods stressful to the host plant, small yellow or white forms of the aphid are also produced; Kring (1959) considered this the aestivating form. Late in the season, winged females apparently seek primary hosts, and eventually both males and egg-laying (oviparous) females are produced. They mate and females deposit yellow eggs—eggs are the only overwintering form under cold conditions. Under warm conditions, a generation can be completed parthenogenetically in about seven days.

In the south, and at least as far north as Arkansas, sexual forms are not important. Females continue to produce offspring without mating so long as weather allows feeding and growth. Unlike many aphid species, melon aphid is not adversely affected by hot weather. Melon aphid can complete its development and reproduces just in a week. So numerous generations are possible under suitable environmental conditions.

Egg. When first deposited, the eggs are yellow, but they soon become shiny black. As noted earlier, the eggs normally are deposited on catalpa and rose of sharon.

Nymph. The nymphs vary in color from tan to gray or green, and often are marked with dark head, thorax and wing pads, and with the distal portion of the abdomen dark green. The body is dull in color because it is dusted with wax secretions. The nymphal period averages about seven days.

Adult. The wingless (apterous) parthenogenetic females are 1–2 mm long. The head is greenish, but the body is quite variable in color—light green mottled with dark green is most common, but also occurring are whitish, yellow, pale green, and dark green forms. The legs are pale with the tips of the tibiae and tarsi black. The cornicles also are black. Small yellow forms apparently are produced in response to crowding (Tamaki and Allen, 1969) or plant stress (Kring, 1959). Winged (alate) parthenogenetic females measure 1.1–1.7 mm long. The head and thorax are black, and the abdomen is yellowish green except the tip of its abdomen, which is darker. The wing veins are brown. The egg-laying (oviparous) female, as well as the male, are dark purplish green.

The duration of the adult's reproductive period is about 15 days, and the post-reproductive period five days. These values vary considerably, mostly as a function of temperature. The optimal temperature for reproduction was considered about 21°C by Isely (1946) and 27°C by Goff and Tissot (1932). Viviparous females produce a total of about 70–80 offspring at a rate of 4.3 per day. Goff and Tissot (1932) reported that cool weather reduced the rate of reproduction, but not total fecundity, because longevity was extended. However, Isely (1946) indicated that total offspring numbers were twice as high at 18–20°C than 26–28°C.

Goff and Tissot (1932) and Isely (1946) gave detailed information on melon aphid biology, and a review was provided by Slosser *et al.* (1989). This aphid was described by Cottier (1953), and keys for identification of melon aphid, and many other common aphids, can be found in Palmer (1952) and Blackman and Eastop (1984). Stoetzel *et al.* (1996) published a key for cotton aphids that was also useful to distinguish melon aphid from most other common vegetable-infesting aphids.

Damage

Melon aphids feed on the underside of leaves, or on growing tip of vines, sucking nutrients from the plant.

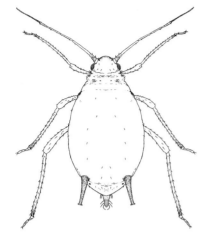

Adult female melon aphid, wingless form.

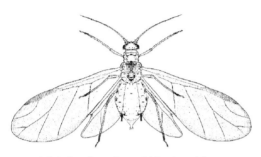

Adult female melon aphid, winged form.

The foliage may become chlorotic and die prematurely. Their feeding also causes great distortion and leaf curling, hindering photosynthetic capacity of the plant. In addition, they secrete a great amount of honeydew, which provides a substrate for growth of sooty mold, so the quality of fruit may be impaired and the photosynthetic capacity of foliage further hindered. Watermelon may be severely injured when average infestation levels exceed 10 aphids per leaf (Riley, pers. comm.). In California, melon aphid is often a late-season (September–October) problem, because the cool weather hampers the activities of natural enemies more than it does the aphids (Michelbacher *et al.*, 1955).

Melon aphid effectively transmits potyviruses, though it is only one of dozen of species implicated in the spread of plant viruses in cucurbits. Relative to green peach aphid, *Myzus persicae* (Sulzer), another common virus vector in California cucurbits, melon aphid is little effective at cool temperatures and low humidities (Fereres *et al.*, 1992). Cucumber mosaic, watermelon mosaic two, and zucchini yellow mosaic viruses are transmitted in California despite applications of insecticide and oil sprays, probably because the viruses can be transmitted within 15 seconds (Perring *et al.*, 1992).

Management

Insecticides. If insecticides are used to suppress melon aphid, care should be taken to obtain thorough cover of foliage. Leaf distortions caused by aphid feeding provide excellent shelter for the insects, so systemic insecticides are useful. Electrostatic sprayers have sometimes been shown to increase control of melon aphid and other insects inhabiting the lower surface of leaves. Electrostatic sprayers use small droplets of highly concentrated insecticide that receive an electric charge as they leave the sprayer. These small droplets better penetrate foliage and adhere to the plant. Some studies have shown adequate or improved aphid control with less insecticide through the use of electrostatic sprayers (Kabashima *et al.*, 1995).

Excessive and unnecessary use of insecticides should be avoided. Early in the season, aphid infestations are often spotty, and if such plants or areas are treated timely, great damage can be prevented later in the season. In California and elsewhere, use of insecticides for other, more damaging insects sometimes leads to outbreaks of melon aphid. Inadvertent destruction of beneficial insects is purported to explain this phenomenon (Michelbacher *et al.*, 1955). Slosser *et al.* (1989), however, suggested that photoperiod, daylight length, and host-plant quality may interact with insecticides to stimulate aphid population change. Bartlett (1968) reported direct stimulation of aphid reproduction by a sublethal dose of insecticide. Resistance by melon aphid to chlorinated hydrocarbon, organophosphate, and pyrethroid insecticides is widespread, and population outbreaks are often related to excessive insecticide use (Hollingsworth *et al.*, 1994).

Melon aphids transmit viruses to crops that they do not colonize. Insecticides have little effect on virus transmission by non-colonizing, transient aphids, though insecticides can prevent secondary transmission within crops where colonization occurs.

Biological Control. Smith (1897) noted that natural enemy populations could completely eliminate melon aphid populations in New Jersey by mid-season. Conradi (1906) suggested planting of a winter nurse crop, preferably rape, in areas where cucurbits will be cultivated in the subsequent summer. He claimed that cabbage aphid and its attendant predators and parasites could overwinter successfully on the rape, and the beneficial organisms would reliably suppress melon aphid following dispersal to the new cucurbit crop. Chambers (1986) obtained good suppression of melon aphid in greenhouses with releases of a syrphid, *Metasyrphus corollae* (F.) (Diptera: Syrphidae). As this insect is not available commercially, a more suitable insect predator for greenhouse releases might be the predatory midge *Aphidoletes aphidimyza* (Rondani) (Diptera: Cecidomyiidae).

Cultural Practices. The wide host range of melon aphid makes crop rotation a difficult tactic to implement successfully. Also, crops grown down-wind from infested fields are especially susceptible because aphids are weak fliers and tend to be blown about. Infested crops should be destroyed immediately after harvest to prevent excessive dispersal, and it may be possible to destroy overwintering hosts if they are weeds. If continuous cropping is implicated in retention of aphid populations, a crop-free period is needed. Row covers can be used to inhibit development of aphid populations (Costa *et al.*, 1994). Time of planting may influence potential aphid population increase (Slosser, 1989).

Disease Transmission. As noted above, it is difficult to disrupt transmission of nonpersistent viruses with insecticides, so total dependence on insecticides is not advised. Row covers, whitewash sprays, and reflective mulches or coarse net covers are helpful in delaying or reducing disease transmission (Adlerz and Everett, 1968; Cohen, 1981; Marco, 1986; Webb and Linda, 1992), but this is expensive to implement on a large scale. Growing crops under ultraviolet light-absorbing plastic is reported to reduce the abundance of melon aphids, and alleviate disease transmission, apparently because aphid alighting or feeding behavior is disrupted (Antignus *et al.*, 1996). Basky (1984) demonstrated reduction in aphid number and incidence of cucumber and watermelon mosaic viruses among seedling cucumbers when transparent or blue plastic mulch was compared with black plastic or unmulched plots. Chalfant *et al.* (1977) also demonstrated the benefits of aluminum and plastic mulch for suppression of watermelon mosaic virus.

Elimination of alternate hosts of both aphids and the virus diseases is often key to disease management; both weeds and crop plants can harbor the disease and vectors (Perring *et al.*, 1992). Romanow *et al.* (1986) evaluated transmission of watermelon mosaic virus to cantaloupe cultivars by melon aphid. They reported that a cultivar resistant to population increase of melon aphid was also less susceptible to virus transmission, despite the fact that there was more interplant movement by aphids on resistant cultivars.

Transmission of nonpersistent viruses such as cucumber mosaic virus can sometimes be decreased by coating the foliage with vegetable or mineral oil. Oil is postulated to inhibit virus acquisition and transmission by preventing virus attachment to the aphid's mouthparts, or to reduce probing behavior (Loebenstein and Raccah, 1980). Oil seems to be most effective

when the amount of disease in an area that is available to be transmited to a crop is at a low level. When disease inoculum or aphid densities are at high levels, oils may be inadequate protection (Umesh *et al.*, 1995). Also, some plants may be damaged by oil applications, especially during hot weather (Marco, 1993).

Pea Aphid
Acyrthosiphon pisum (Harris)

Blue Alfalfa Aphid
Acyrthosiphon kondoi Shinji
(Homoptera: Aphididae)

Natural history

Distribution. The origin of pea aphid is likely Europe or Asia, though it is now found throughout the world in regions with temperate climates. In North America, it was first noted about 1878, and became first serious pest problem about 1900 when it caused extensive damage in the mid-Atlantic states from New Jersey to Virginia, and in eastern Canada from Nova Scotia to Quebec. By the 1950s it had spread throughout the United States and Canada. (See color figure 148.)

Blue alfalfa aphid is native to Asia, but it has spread to Australia, New Zealand, and South America in addition to North America. First observed in California in 1974, it now is widely distributed—blue alfalfa aphid reached Nebraska in 1979, Kentucky and Georgia in 1983, and Maryland in 1992. This species is not yet known as a pest in Canada.

Host Plants. Pea aphid and blue alfalfa aphid are known principally as pests of Leguminosae. As they are efficient vectors of plant viruses, however, they also cause loss in crops on which they normally do not feed, such as cucurbits.

Pea aphid is prone to develop races or subspecies with slightly different host ranges, so populations may differ somewhat in their damage potential to specific legumes (Auclair and Srivastava, 1977). Pea is the most suitable vegetable host for this species, and faba bean and lentil are sometimes damaged. Other hosts that are important in pea aphid biology are alfalfa, sweet pea, vetch, and such clovers as alsike clover, red clover, white clover, and sweet clover.

Blue alfalfa aphid is not widely recognized as a vegetable pest, but Ellsbury and Nielson (1981) demonstrated that pea, lentil, and cowpea were suitable hosts. For the pea aphid, many legume forage crops are suitable hosts.

Natural Enemies. Many natural enemies of pea aphid have been documented (Fluke, 1929), and most also will attack blue alfalfa aphid. Flower fly (Diptera: Syrphidae) and lady beetle larvae (Coleoptera: Coccinellidae) species are most numerous among the natural enemies, and sometimes important in regulating aphid population densities. However, fungi (Wilding, 1975; Pickering and Gutierrez, 1991) and parasitic wasps (Hymenoptera: principally Braconidae) also are very important, with introduced species of *Aphidius* normally dominant. The braconid *Aphidius ervi* Haliday is often reported to be the most important parasitoid, and was observed in North America soon after the introduction of pea aphid, probably being introduced simultaneously. Initially it did not provide effective suppression, and other introductions of beneficial species were made, apparently including additional subspecies or races of *A. ervi*. Among the other species introduced were *A. smithi* Sharma and Subba Rao; *A. eadyi* Stary, Gonzalez, and Hall; *A. urticae* Haliday; and *A. staryi* Chen, Gonzalez, and Luhman. Apparently there has been displacement among species and strains of parasitoids, with the native *A. pisivorus* Smith displaced by imported *A. smithi*, and then the latter species displaced by *A. ervi* (Gonzalez *et al.*, 1995b). Blue alfalfa aphid has proved especially susceptible to parasitism by *A. ervi*, but damage by pea aphid also has been reduced in the western states. Good sources of information on pea aphid parasites are Mackauer and Finlayson (1967), Angalet and Fuester (1977), and Marsh (1977).

A rather predictable cycle of events is observed in many pea aphid populations. After building to high densities on either forage legumes or peas in the spring months, predators, parasites, and fungi take a toll. Both the natural enemy densities and weather are unfavorable at mid-summer, and aphid populations decline. Warm, humid weather, in particular, favors development of fungal epizootics among pea aphids. With the subsequent collapse of natural enemy populations and the return of cooler, more favorable weather for aphids in the autumn, pea aphid populations typically increase again. In Canada and other northern latitudes the aphid populations may not attain high densities until late summer, so there is only one population peak.

Blue alfalfa aphid displays similar cycles, but tends to be favored by cooler weather; thus, it is most common in the early spring and is displaced by pea aphid as the season progresses. Because it is present early in the season, predator and parasitoid populations often are low, and have little suppressive effect. The same biological control agents usually affect both species of aphids, and the presence of blue alfalfa aphid early in the year seems to favor the increase of some agents such as the fungus *Pandora neoaphidis*, which then has

its greatest impact on the later-developing pea aphid population (Pickering and Gutierrez, 1991).

Weather. In addition to the effect of weather on disease incidence, there also can be direct effects of weather on pea aphids. In cold climates, which are normally characterized by abundant snowfall, egg survival rates are high and overwintering success is not highly variable. In mild-winter areas, such as Oregon and Washington, overwintering success is more variable. Mild winters favor the survival of overwintering females and eggs. During cold winters, however, when temperatures fall below freezing, neither adults nor eggs survive well. In addition to the cold temperatures, the lack of snow cover is thought to reduce the overwintering success rate. Also, aphids are especially susceptible to destruction by such adverse weather owing to heavy rainfall in the early part of the season, soon after egg hatch (Cooke, 1963). The lower developmental threshold for pea aphid is about 5.5°C, whereas its principal parasitoids is 6–7°C (Campbell and Mackauer, 1975). The higher developmental threshold of the parasitoids allows the aphid populations to increase to higher densities during cool spring weather.

Life Cycle and Description. *Acyrthosiphon* aphids complete their life cycle quickly. These aphids can reach maturity and begin reproduction 10–12 days after birth. The number of generations completed annually by pea aphid is estimated at about 13 in Indiana, 14–15 in Wisconsin, 15 or more in Washington and Oregon, and 20–22 in Virginia. The overwintering stage of pea aphid varies with climate; in cold regions the eggs overwinter, in warm areas females persist, and in temperate climates both eggs and females can be found during winter months. In blue alfalfa aphid, overwintering is similar. Unlike many species of aphids, these species do not migrate to a woody host for overwintering. However, they do commonly disperse from annual legumes in the summer to perennial legumes such as alfalfa and clover in the autumn, so the difference in behavior is not great (Evans and Gyrisco, 1956).

Egg. The egg is elliptical, and measures about 0.75–0.80 mm long and 0.35–0.40 mm wide. Initially pale to bluish-green, the eggs soon turn shiny black. At the latitude of Indiana, eggs commonly hatch in March, but under the mild winter conditions of Washington egg hatch may occur throughout the winter months. In the spring, after developing on alfalfa or clover for several generations, overwintered populations produce winged females which disperse to uninfested plants. New hosts include both peas and forage legumes. Egg biology is similar for blue alfalfa aphid.

Nymph. Both species display four instars. Instar duration of pea aphid is about 2.7, 2.6, 2.7, and 3.6 days, respectively, at 15°C (Sharma *et al.*, 1976). Duration of blue alfalfa aphid instars is similar, about 2.7, 2.6, 2.7, and 4.3 days, respectively, at 15°C (Kodet *et al.*, 1982). Optimal temperatures are about 10–15°C for survival and fecundity, but 20–25°C for rapid development. Throughout development the nymphs are pale-green. The body length increases from about 0.9, to 1.2, 1.7, and 2.0 mm in instars 1–4, respectively. The tip of the cornicles, distal portion of the tibia, and the tarsi are blackish throughout development. The instars are best separated by cornicle length, with lengths of <0.21, 0.21–0.36, 0.37–0.58, and >0.58 mm in instars 1–4, respectively, of pea aphid (Hutchison and Hogg, 1983).

Adult. Pea aphids are relatively large, the apterae measuring about 2.3–2.7 mm and the alatae about 3.0–3.5 mm long. In color, both forms are pale-green with black at the tips of the rather long cornicles, though in the alate form the head and thorax tend toward yellow-brown. Both forms also have long legs, principally green but with the tip of the tibia, and the entire tarsus, blackish in color. The wings of the alatae are colorless. Pea aphid sometimes exhibits red body color rather than green. The basis for the color forms is genetic, and in crosses the red form is dominant. Red color forms rarely are found on pea, but are sometimes found on red clover and alfalfa (Sandstrom, 1994). For most of the year, pea aphid reproduces parthenogenetically (viviparously), females giving birth without mating. Reproducing females may be either winged (alate) or wingless (apterous). Females typically begin production of offspring 10–12 days after birth, or about 2.5 days after attainment of the adult stage. Mean number of nymphs produced per female is estimated at about 70–100, though some individuals produce up to 150 offspring. The rate of reproduction is highest early in life, with as many as 14 nymphs being produced per day by each female. Near the end of her life (mean of 40 days), an aphid may be producing less than one nymph per day. The average rate of reproduction is estimated at 7 per day, but apterous females produce 8.5 per day whereas alatae produce only 3.8 per day (Cooke, 1963). The aphids seem to form only small colonies on pea, with frequent dispersal of alatae as densities increase. Maturation of pea plants also results in development of alatae, and their dispersal.

Blue alfalfa aphid is similar in appearance to pea aphid. However, the blue alfalfa aphid averages slightly smaller in size, apterae measuring about 2.5–2.7 mm long and alatae measuring about 2.5 mm.

Although this species tends to have a slightly blue-green appearance, this is not a useful diagnostic character. An important diagnostic character is antennal color. In pea aphid the distal portion of each antennal segment is dark, giving the appearance of black bands spaced widely on the antennae. In contrast, the antennae of blue alfalfa aphid are uniformly dark.

Sexual forms are produced by pea aphid in the autumn in response to shorter day length and cooler temperature, and males and females may be alate or apterous. The females greatly resemble the apterous viviparous females, differing principally in size. Oviparous females measure only about 2.0–2.2 mm long. The greenish males also bear brown coloration on the thorax, and usually brown or yellowish-brown on the head and abdomen. The males measure less than 2.0 mm long. Sexual forms are rare at the latitude of northern California, Tennessee, Virginia, and southward, but in Indiana and Oregon both eggs and viviparous forms are found during the winter months. Production of sexual forms is less predictable than in many other species of aphids. For example, oviparous females may be produced by either alate or apterous females, and the same female may produce both viviparous and oviparous forms alternately. Fecundity of oviparous forms seems unknown, but based on examination of ovaries, at least 25 eggs are produced by each female. Cooke (1963) suggested that this value may be the maximum, due to inclement weather late in the year. Blue alfalfa aphid seems to have a similar biology.

The aphids produce and respond to alarm pheromone. Pheromone production normally results when aphids are disturbed, and the standard response by aphids is to walk away from the source of the pheromone or to drop from the plant. Responsiveness to pheromone is increased when the substrate is vibrated (Clegg and Barlow, 1982). Young instars are less responsive (Roitberg and Myers, 1978).

Description of pea aphid development and biology were given by many authors; among the most complete were Davis (1915) and Cooke (1963). Developmental biology of pea aphid was provided by Hutchison and Hogg (1984), and of blue alfalfa aphid by Kodet *et al.* (1982). Culture of pea aphids on plants was described by Halfhill (1967), and on artificial diet by Akey and Beck (1971), and Cloutier and Mackauer (1975). A description of blue alfalfa aphid can be found in Takahashi (1965); life table studies of blue alfalfa aphid were given by Poswal *et al.* (1990).

Damage

Pea and blue alfalfa aphids cause direct injury by removal of plant sap from leaves, stems, blossoms,

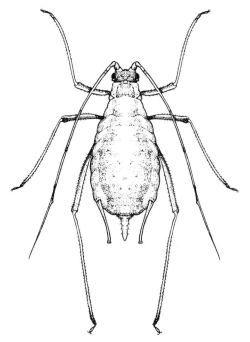

Adult female pea aphid, wingless form.

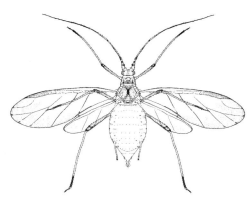

Adult female pea aphid, winged form.

injury. Plants infested before the reproductive stage recover from damage if the aphids are removed, resulting in no reduction in yield. The treatment threshold was estimated to be about 2–3 aphids per stem tip, or 9–12 aphids per sweep (Maiteki and Lamb, 1985b).

The *Acyrthosiphon* species commonly vector plant viruses. Over 30 viruses are known to be transmitted by pea aphid. Blue alfalfa aphid is less well-known, but also it is a moderately efficient vector. As these species develop large populations on many legume hosts, particularly extensively planted forages like alfalfa, they provide numerous potential vectors to cropping areas. In some instances they are important vectors in crops on which they do not normally feed. For example, watermelon and zucchini yellow mosaic viruses were transmitted to cucurbits by both pea aphid and blue alfalfa aphid (Castle *et al.*, 1992). They are very important in vectoring legume viruses also, including pea enation mosaic—one of the most prevalent virus diseases of peas (McEwen *et al.*, 1957).

Management

Sampling. Aphids may be sampled by various methods, including direct visual examination of plants, sweeping, vacuum sampling, and trapping. Sequential sampling procedures for aphids in peas were developed by Maiteki and Lamb (1987), who showed that population counts from 20 to 35 plant tips could be substituted for fixed samples of 100 plants. Sweep net sampling was recommended by Schotzko and O'Keeffe (1989a) in lentil fields.

Insecticides. Both soil-applied and foliar insecticides are commonly used to protect plants from injury by pea aphid. Systemic and non-systemic products are used. The susceptibility of aphids to foliar insecticide is increased slightly when alarm pheromone is released near aphid-infested plants, presumably because the elevated activity level associated with pheromone release increases contact with insecticide (El-Agamy and Haynes, 1992).

Cultural Practices. Crop management is a key factor in preventing damage by pea aphid and blue alfalfa aphid. As alfalfa and clover are the common overwintering sites, they serve as the source of inoculum for surrounding areas. Such crops can be sprayed with insecticide or harvested before aphid populations produce alate forms, resulting in the destruction of most potential dispersants. Destruction of crop residue by burning also disrupts aphid populations by destroying overwintering aphids.

Disease Transmission. Forage legumes may serve as a source of plant viruses for both legume and non-legume vegetable crops, and the *Acyrthosiphon*

and pods. Aphids feeding on foliage disrupt the flow of plant photosynthates to the plant roots. Nitrogen-fixing bacteria inhabiting the nodules on roots of legumes are affected by availability of phytosynthates. Thus, aphid feeding decreases nitrogen fixation and plant growth. Both young and mature plants are affected (Sirur and Barlow, 1984). Damage may not be apparent if aphid populations are low, but at high densities foliage may turn brown, pods are only partially filled, or fewer pods are produced. Under extreme conditions young plants are killed.

Pea plants may support aphids throughout the growing season, but not all stages of plant growth are equally susceptible to damage. Maiteki and Lamb (1985a), working in Manitoba, showed that the flowering and pod-filling stages were most susceptible to species serve as effective vectors. For example, the incidence of bean yellow mosaic virus is related to the proximity of clover (Hagel and Hampton, 1970). Also, emigration of aphids from alfalfa to bean increases following cutting of alfalfa (Stoltz and McNeal, 1982).

Pea leaf roll, pea enation mosaic, and some other *Acyrthosiphon*-transmitted viruses are circulative viruses, persisting for the duration of the vector's life. Such viruses usually require extensive feeding for successful transmission; thus, insecticides can be helpful in protecting against disease. Research in Idaho showed that incidence of pea leaf roll virus could be decreased by application of systemic insecticides (Stoltz and Förster, 1984). Insecticides are less useful for stylet-borne viruses, which are transmitted quickly. Interestingly, stylet-borne virus transmission is often decreased by application of oils to plants, but oil does not greatly decrease transmission of circulative viruses such as pea enation virus (Peters and Lebbink, 1973). The differential effectiveness of oil is likely due to the tendency of oil to inhibit probing by transient alatae, but not feeding by colonizing aphids.

Host-Plant Resistance. Peas differ in their susceptibility to pea aphid (Harrington, 1941; Newman and Pimentel, 1974; Bintcliffe and Wratten, 1980; Soroka and Mackay, 1991). Resistance is reflected in differences in aphid weight, longevity, time to maturity, and progeny and honeydew production on many plant cultivars, and in plant tolerance to aphids in some varieties. Although aphid performance differs among pea cultivars, none are immune to attack. Faba bean cultivars, though differing with aphid performance, similarly display little practical resistance to pea aphid (Frazer *et al.*, 1976). Pea plant morphology affects aphid success by indirectly influencing search effectiveness of aphid predators (Kareiva and Sahakian, 1990; Eigenbrode *et al.*, 1998), but this form of resistance has not been exploited by pea breeders.

Potato Aphid
Macrosiphum euphorbiae (Thomas)
(Homoptera: Aphididae)

Natural History

Distribution. Potato aphid occurs throughout the United States and southern Canada. As a serious pest, however, its range is mostly the northeastern and northcentral United States, and eastern Canada west to Ontario. Potato aphid is a highly variable species, and may eventually be shown to be a species complex. The origin of potato aphid is thought to be North America, though it is now found widely around the world. (See color figure 147.)

Host Plants. Potato aphid is polyphagous, though its predominant hosts are potato, tomato, and sometimes corn in the summer, and wild or cultivated rose in the winter and spring. Sometimes spinach and lettuce are heavily infested during the autumn months. Potato aphid also are found to feed on such diverse vegetable crops as asparagus, beet, celery, chicory, corn, cucumber, eggplant, horseradish, kale, lettuce, mustard, pea, parsnip, pepper, potato, pumpkin, rhubarb, spinach, sweet potato, tomato, turnip, and watercress, but many of these plants are suitable for brief period of times, usually during the seedling stage. Potato aphid may be found in association with other crops such as clover, field corn, hops, peach, pawpaw, soybean, strawberry, sugarbeet, sunflower, and tobacco. It also infests such flowers as canna, geranium, gladiolus, hollyhock, iris, lily, poppy, rose, rudbeckia, and tulip. Among the weed hosts are black nightshade, *Solanum nigrum*; groundcherry, *Physalis* spp.; hairy nightshade, *Solanum villosum*; hoary cress, *Lepidium draba*; jimsonweed, *Datura stramonium*; lambsquarters, *Chenopodium album*; matrimony vine, *Lycium* sp.; morningglory, *Ipomoea purpurea*; pepper vine, *Solanum jasminoides*; pigweed, *Amaranthus* spp.; pepperweed, *Lepidium* spp.; plantain, *Plantago* spp.; ragweed, *Ambrosia* spp.; round-leaved mallow, *Malva rotundifolia*; shepherdspurse, *Capsella bursa-pastoris*; sow thistle, *Sonchus oleraceus*; smartweed, *Polygonum* spp.; wild lettuce, *Lactuca* sp.; and winter cress, *Barbarea vulgaris*. These aphids commonly move from host to host as the quality of the plants deteriorates due to seasonal changes. They are capable, however, of feeding indefinitely on the same hosts if the plants remain nutritionally suitable. Host associations were given by Patch (1915), Smith (1919), and Landis *et al.* (1972). Although rose is normally reported to be the overwintering host in cold climates, there are occasional reports that either apterae (wingless forms) or eggs are found during the winter months in cold climates on such diverse plants as asparagus, raspberry, and various weeds.

Natural Enemies. Potato aphid has many natural enemies, as is commonly the case with aphids. Most of the common lady beetles (Coleoptera: Coccinellidae), some lacewings (Neuroptera: Chrysopidae), flower flies (Diptera: Syrphidae), and the predatory midge *Aphidoletes aphidimyza* (Rondani) are reported to prey on potato aphid (Walker *et al.*, 1984b; Perring *et al.*, 1988; Dean and Schuster, 1995). Ground beetles (Coleoptera: Carabidae) affect aphid populations, but not as greatly as canopy-level predators (Boiteau, 1986).

Among the parasitoids of potato aphid are *Aphidius*, *Diaeretiella*, *Ephedrus*, and *Praon* spp. (all Hymenoptera: Aphidiidae) and *Aphelinus* and *Dahlbominus* spp. (both Hymenoptera: Eulophidae). Only *Aphidius nigripes* Ashmead is regularly abundant, accounting for over 90% of the parasitism of potato aphid in Maine over an 11-year period (Shands *et al.*, 1965). In studies of potato aphid attacking tomato in North Carolina, Walgenbach (1994) reported that parasitism rates were inversely density dependent, with parasitism higher when aphid densities were low. *Aphidius nigripes* also was the dominant parasitoid in North Carolina, California (Sullivan and van den Bosch, 1971), Ohio (Walker *et al.*, 1984b), and Quebec (Brodeur and McNeil, 1994).

Several closely related species of fungi affect potato aphid, and are considered suppressive (Shands *et al.*, 1962). Frequency of fungal epizootics is positively correlated with aphid density (Boiteau, 1986). High levels of rainfall, though perhaps aiding in development of epizootics, are not a necessary prerequisite (Shands, 1962). Soper (1981) discussed the significance of fungi for aphid control, emphasizing the importance of not interfering with natural disease outbreaks by applying fungicides or other suppressive chemicals.

Life Cycle and Description. In northern areas, potato aphid has a sexual component to its life cycle and overwinters in the egg stage. In the spring, potato aphid feeds on rose, where 2–6 generations are completed. The aphids abandon the rose in the summer months and fly or walk to other suitable hosts, where several additional generations occur and high densities are attained. In the autumn, the winged forms disperse, usually back to rose. On rose, autumn migrant females, oviparous (egg-laying) females and males are found, mating occurs, and overwintering eggs are produced by oviparous females. In mild-winter, southern regions at least as far north as Virginia, the sexual forms are not produced, or the egg is not the only stage of overwintering, as the aphid can reproduce parthenogenetically (viviparously) throughout the year in such climates.

Nymph. Smith (1919) reported four instars. In all instars the body is yellowish-green or yellowish-pink, the cornicles are long and with dark tips. Body length is 0.77–0.85, 0.96–1.11, 1.56–1.75, and 1.74–2.05 mm for instars 1–4, respectively. However, MacGillivray and Anderson (1958) indicated that there were five nymphal instars, the mean duration of which was 1,7, 1.9, 2.1, 2.4, and 1.5 days, respectively, for apterous forms.

Adult. Potato aphids differ considerably in appearance, not only because of the different sexual forms but because they typically produce two discrete color forms—one green and the other pink. It is not surprising, therefore, that one of the common names formerly applied to this species is the "pink and green aphid." The most common form is the adult wingless (apterous) parthenogenetic form, which predominates during the summer months. The body is green or pink, and free of dark markings. The cornicles are quite long, and with dark at the tips. The tubercles at the base of the antennae diverge or point outward, unlike the other common potato-infesting aphids. This aphid, in its apterous form, measures about 3.0–4.0 mm long, making it the largest of the common potato-infesting aphid species. The adult winged (alate) parthenogenetic form also is abundant in the summer, especially when aphid densities are high or the nutritional quality of the host plant declines. This form has the same pink or green body with cornicles that are darker distally, but bears transparent wings with dusky veins and is slightly smaller in size than apterae, about 2.1–3.4 mm long.

In Virginia, winged viviparous females disperse from winter hosts such as kale and spinach during March and April to young warm-season plants such as weeds and potato. They often remain on these hosts until the plants deteriorate, often from overcrowding by aphids. The hot weather does not favor the aphids, and their numbers greatly diminish, but begin to increase again in the autumn. They are abundant in the winter on cool-season plants such as spinach, crucifers, and weeds. Viviparous females live for about 30 days, with 10 days in the nymphal stage and the remainder as adults. Occasionally they will survive up to 50 days. They produce, on average, about 50 young aphids, but sometimes up to 80 offspring, at about 2.5 per day.

The sexual forms also differ in appearance, both between sexes and in comparison with the parthenogenetic forms. In the oviparous female, wings are absent. The head and thorax are whitish, the abdomen pinkish or greenish, and the tibiae are dark. This form measures only about 2.15 mm long. The male has wings. The male also has a dark head and thorax, with a brownish green abdomen and dark appendages. The male measures less than 1.6 mm long.

The biology of potato aphid was described by Patch (1915), Houser *et al.* (1917), and Smith (1919). A good description appeared in Cottier (1953). Keys that included this aphid were found in Palmer (1952) and Blackman and Eastop (1984).

Damage

Young tissue, usually the growing tip of the plant, is attacked first by potato aphids. As the aphids multiply they spread over the entire plant, removing plant sap

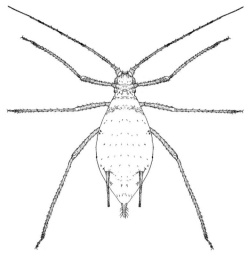

Adult female potato aphid, wingless form.

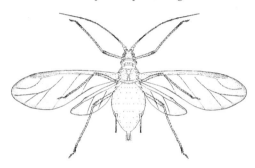

Adult female potato aphid, winged form.

throughout. The leaves take on a distorted appearance, with the leaf edges turned downward. This form of injury is typical of many aphids on a variety of plants, but is especially evident on potato infested with potato aphid. Potato plants may die back from the tip downward, and heavy infestation can kill potato plants. Leaves may be covered with honeydew, which fosters growth of sooty mold. At times, potato aphid can be the most important pest of potato.

On tomato, pepper, and eggplant the leaf deformities are less apparent though stunting of the growing tips is evident. Another difference is that the blossoms are especially preferred in the latter hosts, and blossom drop or fruit deformities may occur. Walgenbach (1997) reported that potato aphid infestation resulted in higher sunscald damage to tomato fruit, presumably owing to decreased shading by foliage. Also, damage by stink bugs (Hemiptera: Pentatomidae) and leaffooted bugs (Hemiptera: Coreidae) was increased in the presence of potato aphids, perhaps because these hemipterous insects also feed on aphids. Despite the many ways in which potato aphids damage tomato, Walgenbach (1997) determined that greater than 50% of the leaves had to be infested before tomato yields were significantly depressed.

In the northwestern United States, potato aphid is very abundant on corn (Landis *et al.*, 1972), often covering the lower surface of leaves. Despite its great abundance, it does not cause plant deformities and is not considered to be a serious pest on this crop.

In lettuce, potato aphid may be a contaminant that reduces marketability, particularly the ability to ship the crop to other nations (Hinsch *et al.*, 1991).

Potato aphid can transmit numerous plant viruses (Kennedy *et al.*, 1962), though it is not considered to be an especially effective vector. Among the viruses transmitted to potato and tomato are cucumber mosaic virus and potato virus Y.

Management

Sampling. There is an aggregated distribution pattern of aphids in tomato. Walgenbach (1994) developed a binomial sampling plan, and suggested sampling of the third most recently expanded tomato leaf, with 50 leaves examined from each of five locations within a field. Sampling tomato leaflets for the presence or absence of aphids was recommended by Wittenborn and Olkowski (2000). They found that sampling the leaflet rather than the entire leaf, which may consist of 5–13 leaflets, increased accuracy. A sequential sampling plan was proposed by Walker *et al.* (1984a). Suction traps have been used to capture winged forms of potato aphid, and though useful for determining presence of the aphids, they are not very predictive of crop infestation levels (Elliott, 1981). Periodic placement of plants in pots in fields, or "plant traps" has also been used in monitoring, and data obtained from plant traps correlates with suction trap estimates (Elliott, 1980). Population estimates are usually attained by visual examination, though yellow water pan traps are useful for estimating invasion by alate aphids.

Insecticides. Foliar insecticides are often applied for aphid suppression. Broad-spectrum insecticides are usually used because of the other, often more serious, pests associated with solanaceous crops. As they are not important virus vectors, however, high numbers of aphids can be tolerated. Chemical suppression is not usually recommended unless half of the leaves are infested. Soaps, detergents, and oils can be used against aphids, but care should be taken not to burn the plants.

Cultural Practices. Studies conducted in Maine demonstrated that planting practices could influence damage to potato. Delay of planting from early May to late May or early June, resulted in up to 90% reduction in aphid infestation. Early hilling operations, wherein row ridges are heightened with soil from the row middles, similarly deprived dispersing aphids of young plant tissue, resulting in lower aphid densities (Shands *et al.*, 1972). Delayed planting was also shown to be beneficial in New Brunswick (Boiteau, 1984).

Cultural practices can also interfere with aphid host selection behavior. Reflective mulches, particularly aluminum mulch, are sometimes recommended for disruption of aphid invasion of crops. Evaluation of aluminum mulch in potato, however, showed that the beneficial effect of the reflective mulch was slight or of short duration, and therefore impractical (Shands and Simpson, 1972). The undersowing of potato with rye grass also affects aphids, resulting in decreased aphid densities in fields with grass, and no loss in crop yield (McKinlay, 1985).

In northern climates, aphids deposit overwintering eggs upon wild and cultivated rose, and develop on these plants in the spring. Therefore, it is desirable to destroy, or to treat with an insecticide, such overwintering sites before the aphids disperse to crops. Increasingly, aphids are shipped northward from southern climates on young plants, or overwinter in northern areas in greenhouses. Care should be taken to keep from introducing aphids into fields on transplants.

Host-Plant Resistance. There are significant differences in susceptibility to aphid infestation among tomato (Gentile and Stoner, 1968c), lettuce (Reinink and Dieleman, 1989), and to a lesser extent potato, varieties. Obrycki *et al.* (1983) suggested that though high densities of glandular trichomes are detrimental to aphids, plant varieties with moderately high trichome densities might be preferable because such varieties are more compatible with the action of predators and parasitoids. The basis for resistance in lettuce is uncertain, but butterhead varieties are less susceptible to infestation.

Rice Root Aphid
Rhopalosiphum rufiabdominalis (Sasaki)
(Homoptera: Aphididae)

Natural History

Distribution. Rice root aphid may be of Asian origin, though it now has nearly world-wide distribution in tropical and subtropical areas, and to a lesser extent in temperate countries. It occurs widely in the United States, but is rarely a pest, and is injurious only in warmer climates such as California, Texas, and the southeastern states. It is not known to occur naturally in Canada, though it has been imported on plant products.

Host Plants. The primary, or overwintering, host is *Prunus* spp. Secondary, or summer, hosts are numerous monocots in the families Gramineae and Cyperaceae, but also some dicots including Solanaceae and Cucurbitaceae. It is a serious pest only of upland rice, and prospers on flooded rice only during the dry period between floodings. It has been found in association with the roots of cotton, but no serious injury has been reported. Among vegetables, it has been recorded from the roots of potato in California, from squash and pepper in Florida, and elsewhere in the world on bean, okra, and watercress. Its status as a pest in Florida is potentially serious, but limited thus far to plants grown in greenhouses under aeroponic culture, wherein the roots are suspended in air and moistened with a nutrient mist system (Etzel and Petitt, 1992), and on bromeliads.

Natural Enemies. The natural enemies of rice root aphid are poorly documented. The enemies of other root-infesting aphids such as lettuce root aphid, *Pemphigius bursarius* (Linnaeus), are probably similar to those attacking rice root aphid. The aphid-infecting fungus *Verticillium lecanii* also affects rice root aphid (Etzel and Petitt, 1992).

Life Cycle and Description. Rice root aphid can alternate between *Prunus* and summer hosts, but in most parts of the world it persists continuously on roots of plants in the family Gramineae. In North America it persists on roots of grains at least as far north as Illinois (Jedlinski, 1981), and overwintering in the egg stage has not been observed. Elsewhere, populations overwintering on *Prunus* disperse during May and June to summer hosts, where up to 20 generations occur before winged forms, including males, are produced that migrate back to winter hosts. The alate (winged) males mate with apterous (wingless) oviparae (egg-laying females) on *Prunus*, and eggs are deposited. Generation time varies from about 20 days at 15°C to only 8 days at 25–30°C.

Nymph. Nymphs feeding on rice roots undergo four instars, with mean total nymphal development time reported to be 11.3, 6.4, and 4.8 days when aphids are reared at 15°, 20°, and 25°C, respectively. Mean instar development time is 0.9, 1.1, 1.2, and 1.5 days for instars 1–4, respectively, when cultured at 25°C.

Adult. Parthenogenetically reproducing apterous adults are redish-green, olive-green or dark-green, usually bearing reddish color around and between the cornicles. The appendages, including the cornicles, are dusky to black. The alate form is similarly colored, except that the head and thorax are black. Both forms measure 1.2–2.2 mm long. This species is distinguished from the other *Rhopalosiphum* spp. by the presence of long and fine hairs on the antennae and body.

High temperatures adversely affect longevity and reproduction of wingless (apterous) adults. Adults survive an average of 24 days at 15°C, but only 9.4 days at 30°C. Similarly, adults cultured at 15–25°C typically produce 50–60 nymphs, but at 30°C they produce only about 30 nymphs. Winged adults (alatae) produce only 5–6 offspring.

Rice root aphid was discussed by Doncaster (1956), and it was included in a brief review of rice-infesting aphids published by Yano *et al.* (1983). Developmental biology was described by Tsai and Liu (1998). Rice root aphid was included in keys by Palmer (1952; as *Rhopalosiphum splendens* (Theobald)), Richards (1960), and Blackman and Eastop (1984).

Damage

Rice root aphid infestations result in discoloration and rotting of plant roots, wilting and stunting of the above-ground portions of the plants, and sometimes in death of the host. Rice root aphid also is a minor vector of barley yellow dwarf virus (Jedlinski, 1981). As new technologies such as aeroponics are applied to greenhouse vegetable production rice root aphid may become a more important vegetable pest.

Management

Thus far, rice root aphid is not a species that often warrants control on vegetable crops. Strategies for chemical suppression are the preventative approaches used for other root-feeding aphids such as lettuce root aphid, *Pemphigius bursarius* (Linnaeus), and the sugarbeet root aphids, *P. betae* Doane and *P. populivenae* Fitch. Under greenhouse conditions, especially where aeroponic culture is used, new suppression techniques must be developed.

Sugarbeet Root Aphids
Pemphigus betae Doane
Pemphigus populivenae Fitch
(Homoptera: Aphididae)

Natural History

Distribution. There has been considerable disagreement over the status of the *Pemphigus* species associated with crops in the plant family Chenopodiaceae. *Pemphigus betae* Doane and *P. populivenae* Fitch are the names used most commonly in conjunction with this pest, though only one species with different forms or biotypes may be involved. Sugarbeet root aphid is native, and found principally in the range occupied by narrow-leaved poplar, *Populus angustifolia*—the principal overwintering host in the sexual life cycle. As a pest, it ranges principally the western states and provinces, including the Great Plains area. (See color figure 155.)

Host Plants. During the summer moths sugarbeet root aphid usually develops on roots of plants in the family Chenopodiaceae. Thus, the principal crops affected are beet, chard, quinoa, spinach, and sugarbeet. Weeds, particularly lambsquarters, *Chenopodium album*, and redroot pigweed, *Amaranthus retroflexus*, but also dock, *Rumex* spp., and smartweed, *Polygonum* spp., also support sugarbeet root aphid. Many crops and weeds have been reported to be summer hosts, but most old records are suspect owing to the difficulty in distinguishing among the several similar-appearing *Pemphigus* spp. Of 31 species of plants tested under greenhouse conditions, Harper (1963) cultured aphids readily on beet and sugarbeet, whereas chard and spinach were less suitable, and alfalfa was a poor host; the other species were unsuitable.

The overwintering, or primary, hosts of sugarbeet root aphid are poplar trees, primarily narrow-leaved cottonwood, *Populus angustifolia*, and balsam poplar, *P. balsamifera*, though in California *P. trichocarpa* is a suitable host. The *Pemphigus* form designated *betae* is reported to create leaf galls on the underside of the leaf blade and has an opening from the gall on the upper surface of the leaf. In contrast, the form designated *populivenae* is reported to create leaf galls on the upper surface of the leaf and has the opening on the lower leaf surface.

Natural Enemies. Many predators feed on these aphids, both within the gall and below-ground. Among the predators found in galls are the flower fly *Syrphus bigelowi* Curran (Diptera: Syrphidae); *Leucopis pemphigae* Malloch (Diptera: Chamaemiidae); and the lady beetle *Scymnus* sp. (Coleoptera: Coccinellidae). Found below-ground feeding on aphids are *Scymnus* spp. and *Hippodamia convergens* Guerin–Meneville (both Coleoptera: Coccinellidae); *Thaumatomyia glabra* (Meigen) (Diptera: Chloropidae); and several genera and species of flower flies (Diptera: Syrphidae). In the autumn, aphid dispersants to poplar trees fall prey to numerous predators including various ants (Hymenoptera: Formicidae); the pirate bug *Anthocoris antevolens* White (Hemiptera: Anthocoridae); and such lady beetles as *Coccinella* spp. and *Adalia bipunctata* Say (all Coleoptera: Coccinellidae). Of all the natural enemies, the fly *T. glabra* is perhaps most frequently associated with root aphid colonies, but there is no evidence that it regulates aphid populations. A fungal disease, *Erynia neoaphidis*, sometimes causes epizootics, and seemingly causes aphids to ascend to the soil surface (Summers and Newton, 1989).

Life Cycle and Description. Sugarbeet root aphid passes the winter in two forms, either as eggs on poplar trees or as wingless parthenogenetic females on the roots of plants. Those overwintering in the egg stage hatch in early spring, aphids feed on the overwintering host, and their offspring   disperse. Several parthenogenetic generations occur annually on the summer or secondary host. In the autumn the aphids disperse back to poplars, produce sexual forms, and after mating the female produces an overwintering egg. In Alberta, hatching occurs in April, offspring are produced in June and July, dispersal to beets occurs in late June and July, dispersal back to poplars and production of sexual forms begins in September, and eggs are produced soon thereafter.

As noted earlier, however, some individuals fail to disperse to poplars, and remain below-ground on roots throughout the winter. Maxson (1948) reported that the aphids developing from overwintering females produced seven generations in Colorado. He found that there were three wingless generations on beet during the spring, followed by three additional generations consisting of both winged and wingless forms. The wingless aphids remained on the beets, forming the basis of the adult-overwintering population, whereas the winged aphids dispersed to poplar trees. The seventh generation is comprised of males and females; the females deposit overwintering eggs in cracks and crevices of poplar bark.

Egg.   The eggs overwinter on and beneath the bark of poplar trees, often in the lower regions of the trees. The egg is white initially, becoming yellowish-white with time, and then black. The elliptical egg measures about 0.54 mm long and 0.26 mm wide. The egg is usually deposited in a mass of filamentous waxy material. Females produce only a single egg, which must be exposed to cold weather before it will hatch.

Nymph.   The nymphs, which display four instars, are pale-yellow to dark-greenish.

Adult.   The adult developing from an overwintering egg is wingless (apterous) and parthenogenetic, producing nymphs without mating. She is yellowish-green, with dark appendages. The body measures 2.3–2.5 mm long, and some wax secretion is found on the abdomen. This form feeds at the base of the leaf blade on poplar trees, and induces formation of a swelling or gall. The gall occurs along the midrib and is hollow. This structure provides protection and nourishment for both the adult and her offspring. The gall can only be induced to form on young, growing leaves. This female remains within the gall and continues to reproduce for 4–6 weeks. The fecundity of these females is estimated to average 163 nymphs, with some individuals producing over 200 offspring.

The offspring of the first females are winged (alate) and parthenogenetic. These aphids have a black head and thorax, and a yellowish or greenish abdomen. The appendages are dark. The body measures about 1.5 mm long. The wings are unmarked except for a dark area along the leading edge of the forewing. After dispersing from the poplar gall, these females each produce an average of 13.3 nymphs (range 11–16 nymphs), which are deposited on summer-host plants.

Additional generations of parthenogenetically reproducing apterous females occur on the summer hosts. This is the form that is damaging to crops. It is whitish to yellowish-green, with dark appendages and a dark head. A great amount of white, waxy, hair-like secretion is present on the abdomen. This form measures 1.5–2.5 mm long. The ability of this form to create colonies on beet and other susceptible plants is apparently related to their ability to gain access to plant roots. Plants growing in dry soil, especially in heavy soil, are most susceptible to infestation because cracks forming in the soil provide the aphids with ready access to the roots. The reproductive ability of this form has not been determined under field conditions, but Harper (1963) showed that a single aphid feeding below-ground for six weeks could result in a population of several hundred to over 1000 aphids. Under laboratory conditions mean fecundity was about 50–65 progeny per female, depending on temperature (Campbell and Hutchison, 1995). The optimal temperature for reproduction is about 25–27°C. This form of the aphid, unlike most aphid species, seems unable to produce winged aphids that disperse to other summer hosts.

Late in the season the aphids on the summer hosts produce a winged parthenogenetic form that disperses to poplar and produces 5–6 offspring. This form greatly resembles the winged forms produced by the gall-forming females, but may be even darker in color, and bears tufts of waxy secretion on the thorax. The progeny of these aphids are sexual forms: males and oviparous females. This form can be produced both by summer populations derived from eggs overwintering on poplar and derived from adults overwintering below-ground on roots.

The sexual forms lack wings and functional mouthparts. The female is yellowish-green with a dusky head and thorax. The appendages are whitish. The body measures only about 0.8 mm long. The males are even smaller, measuring about 0.6 mm long. They are yellowish-brown, with darker appendages. They mate and the female deposits her egg in September or October in bark cracks or beneath loose bark. Each female deposits only one egg. The females are larger than the males. The females and males may live for 8–12 and 3–5 days, respectively.

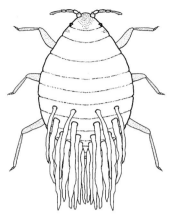

Adult female sugarbeet root aphid, wingless form.

The biology of sugarbeet root aphid was described by Grigarick and Lange (1962) and Harper (1963). Rearing was described by Campbell and Hutchison (1995). Sugarbeet root aphid was included in the keys by Palmer (1952), Smith (1981), and Blackman and Eastop (1984).

Damage

Aphids infest the roots of summer-host plants. Infested roots are unable to take up adequate water and nutrients, disrupting normal plant growth. The principal signs of infestation are yellowing and wilting of the above-ground portions of the plant. Root size and sugar content are decreased. These aphids also transmit beet curly top, beet mosaic, and beet yellows virus.

Management

Infested beets and spinach are easily detected by the presence of the bluish-white, filamentous masses of aphids on the roots. Such occurrences are rare except where host plants are grown continuously. Complete removal of summer-host plants greatly disrupts the population cycle, alleviating the need for aphid suppression. Summers and Newton (1989) recommended a three-year period between susceptible crops. Irrigation can be used to prevent soil from cracking, thus depriving aphids of easy access to the roots. Insecticides applied to the soil at planting or shortly afterward will eliminate damage.

Turnip Aphid
Lipaphis erysimi (Kaltenbach)
(Homoptera: Aphididae)

Natural History

Distribution. Turnip aphid is found in both tropical and temperate areas throughout the world. It is often confused with cabbage aphid, however, so distributional records and economic impact of both species are sometimes incorrectly reported. Turnip aphid is found throughout the United States, where it sometimes is more important than cabbage aphid, *Brevicoryne brassicae* (Linnaeus), as a vegetable pest. In Canada, it is found in the southernmost regions of the country from the east coast to the west coast. Its origin is Europe. (See color figure 156.)

Host Plants. This aphid has been found associated with all *Brassica* crops, including such vegetables as broccoli, Brussels sprouts, cabbage, cauliflower, collards, kale, kohlrabi, mustard, and turnip. Other cruciferous plants such as the vegetables radish and watercress, and the weed shepherdspurse, *Capsella bursa-pastoris*, also are known as suitable hosts. Records from lettuce, bean, and onion are likely due to incorrect identifications.

Natural Enemies. Turnip aphid is host to virtually the same parasitoids and predators found attacking cabbage aphid. *Diaeretiella rapae* (MacIntosh) (Hymenoptera: Braconidae) is frequently reported to be an important biological suppressant, but in rapidly maturing crops such as radish apparently there is insufficient time for the parasitoids to attain high levels of abundance. Various lady beetles (Coleoptera: Coccinellidae), flower flies (Diptera: Syrphidae), lacewings (Neuroptera: Chrysopidae), and the fly *Aphidoletes aphidimyza* Rondani (Diptera: Cecidomyiidae) prey on the aphids (Davis and Satterthwait, 1916a; Paddock, 1915b; Pimentel, 1961). Fungus epizootics sometime occur when aphid population densities are high, especially during the autumn months.

Life Cycle and Description. The turnip aphid is very prolific. Under the temperate conditions of Indiana, Davis and Satterthwait (1916a) reported that 11–25 generations could be produced annually. Under the warmer conditions of Texas, Paddock (1915b) observed a maximum of 35 generations. Aphid longevity is usually 20–40 days, and reproduction begins about six days after aphids reach maturity. Wingless parthenogenetic females typically produce 80–100 young, often at 4–6 per day. Winged parthenogenetic females produce fewer offspring.

Egg. An egg-producing (oviparous) female form is known and was briefly described by Cottier (1953). The egg deposition rarely occurs, however, with nymph-producing (viviparous) forms occurring on crops and weeds throughout the year, even in cold climates. Both Paddock (1915b) and Davis and Satterthwait (1916a), who conducted detailed studies of this insect, were unable to observe the egg stage. Oviparous females have been observed on several continents, however (Kawada and Murai, 1979). Therefore eggs, though rarely observed, may occur on crops.

Nymph. There are four instars. Nymphs are pale greenish-yellow, and average about 0.6, 0.9, 1.0, and 3.3 mm long during the four instars, respectively.

Adult. The wingless (apterous) adult females are whitish-green or green, and measure about 1.4–2.4 mm long. The dorsal surface of the thorax and abdomen of wingless adult females are marked with two rows of dark bands, which coalesce into a single band on the distal abdominal segments. The legs are pale with dusky joints. The antennae are dark and the cornicles pale with dusky tips. The entire body is lightly covered with a white secretion.

The winged (alate) adult females are pale green, with a black head and thorax. The last three abdominal segments are marked dorsally by narrow black bands. The sides of the abdomen also bear black patches. The legs are brownish to blackish, and the antennae are black. The cornicles are dusky yellowish, with a small black area around the base. The wings are transparent, but marked with conspicuous black veins. Winged females measure about 1.4–2.2 mm long.

Males have also been observed occasionally, but they are too infrequent for their biology to be known. The wingless male is quite small, measuring 1.2–1.3 mm long. It is olive-green to brown.

Turnip aphid generally can be differentiated from cabbage aphid by the very sparse occurrence of waxy exudate on the body, and by the slightly longer cornicles, 0.23 mm, of the former. In humid climates, however, there tends to be greater accumulation of waxy secretions on the aphid body. Therefore, Blackman and Eastop (1984) recommended using the shape of the cauda—a structure found at the tip of the abdomen, to differentiate these two species. When viewed from above, the cauda of cabbage aphid is triangular, about as wide as long. In contrast, the cauda of turnip aphid is slender, about twice as long as wide.

The biology of turnip aphid was described by Paddock (1915b) and Davis and Satterthwait (1916a). Keys for identification of turnip aphid were provided by

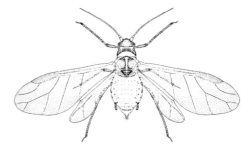
Adult female turnip aphid, winged form.

Palmer (1952) and Blackman and Eastop (1984). A good description was provided by Cottier (1953).

Damage

Turnip aphids feed mainly on the undersides of leaves, but under high aphid densities or during the winter months the aphids may move to the center of the plant and feed on both the upper and lower surfaces of tender young foliage. They also feed readily on the stem tissue of the flowers. Heavy infestations can kill plants, and even light infestations can cause the leaves to cup, with yellowing of the foliage occurring where aphids are concentrated. Stunting of the plants is common, but contamination of foliage with aphid bodies, cast skins, and honeydew is the principal type of direct damage.

Turnip aphid may also serve as a vector of stylet-borne plant viruses. Kennedy *et al.* (1962) listed over 10 viruses transmitted, including cabbage black ring spot, cauliflower mosaic, and radish mosaic. Turnip mosaic virus also may be transmitted.

Management

Sampling. Comparison of sampling methods was reported by Trumble *et al.* (1982a), but there was little correlation between aphid water pan or sticky traps and overall aphid populations on plants. However, there was a significant correlation between water pan trap catches of winged aphids and populations of winged aphids on plants, suggesting that such traps could be used to indicate when winged dispersants were entering a field. This information is critical for the prevention of population establishment and the transmission of viruses. Populations within fields are aggregated when densities are low, becoming more uniform as densities increase.

Insecticides. Sampling and chemical management of turnip aphid are virtually the same as discussed for cabbage aphid. Although sanitation is perhaps the most important factor in turnip aphid management, insecticides are used extensively. Systemic and nonsystemic insecticides are commonly

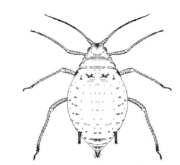
Adult female turnip aphid, wingless form.

applied in liquid or granular formulations, often at short intervals.

Host-Plant Resistance. Varietal differences occur among turnip cultivars in ability to support turnip aphid (and cabbage aphid) populations (Kennedy and Abou–Ghadir, 1979). As the aphid-resistant variety is only moderately resistant to aphids, and is more susceptible to attack by caterpillars, variety selection many not be a satisfactory management tactic.

Disease Transmission. The techniques used to prevent transmission of plant viruses by aphids has been discussed under melon aphid. Turnip aphid produces an alarm pheromone when disturbed (Phelan *et al.*, 1976). Alarm pheromones have been shown to inhibit virus transmission experimentally (Gibson *et al.*, 1984), but thus far there is no practical application of this technology.

Willow-Carrot Aphid
Cavariella aegopodii Scopoli
(Homoptera: Aphididae)

Natural History

Distribution. Willow-carrot aphid is widespread in the temperate areas of the world. In North America it occurs mostly in the northern United States and probably in southern Canada. It is a pest principally along the west coast, from Washington to California. Its origin seems to be Asia.

Host Plants. Willow-carrot aphid alternates between primary hosts, which are fed upon during the autumn and spring months, and secondary hosts, which are fed upon during the summer. The primary host is willow, *Salix* spp. (Dunn, 1965; Kundu and Dixon, 1994). Willow-carrot aphid feeds on Umbelliferae during the summer months. Vegetables attacked include carrot, celeriac, celery, chervil, coriander, fennel, parsnip, parsley, and parsnip. Also infested are the herbs anise, dill, and caraway, and numerous umbelliferous weeds such as wild carrot, *Daucus carota*; cow parsnip *Heracleum lanatum*; angelica, *Angelica sylvestris*; lovage, *Ligusticum* spp.; and water hemlock, *Circuta* spp. (Palmer, 1952; Dunn and Kirkley, 1966).

Natural Enemies. The enemies of willow-carrot aphid are similar to those affecting most other aphids. Lady beetles (Coleoptera: Coccinellidae), flower flies (Diptera: Syrphidae), and anthocorid bugs (Hemiptera: Anthocoridae) were the dominant predators observed in England. An unspecified fungal disease and parasitoids were also observed.

Weather. Probably the most important factor affecting aphid populations in England is weather. Precipitation, either at normal levels or above normal, suppresses aphid abundance, especially when it occurs during the period of dispersal by winged forms (alatae). Severe winter weather has the potential to kill aphids overwintering on secondary hosts, both cultivated and wild Umbelliferae; this greatly reduces the population size in the subsequent crop season (Dunn, 1965).

Life Cycle and Description. The life cycle of this insect has been thoroughly studied only in England, but it is likely very similar in the temperate areas of North America. Overwintering occurs on willow, with egg hatch as early as February. In the spring, several generations may occur on willow, followed by dispersal to summer hosts in the plant family Umbelliferae. During autumn, aphids migrate back to willows and oviposit. In England, some individuals seem to overwinter successfully in the adult and immature stages on umbellifers that remain in the soil and do not die back completely during the winter months (Gratwick, 1992).

Egg. Females of willow-carrot aphid apparently do not discriminate among the species of willows when they select oviposition sites in the autumn. The overwintering eggs are shiny and black, and are found around the bud axils.

Nymph. Young aphids feed through the thin bark of new shoots if the buds are not yet emerged from willow, later moving to the leaves and catkins. The form developing in the spring is rusty brown or red, and measures 1.8–2.0 mm long. Nymphs of summer populations developing on umbellifer foliage are green and elongate, and are very difficult to discern.

Adult. Parthenogenetic reproduction occurs on willow, and starting in about May alatae (winged forms) are produced. Apterae (wingless forms) are green or yellowish-green, and measure 1.8–2.2 mm long. They are also greenish, but have black or dark coloration on the head, thorax, and abdomen, with the abdominal markings in the form of irregular bars. The aphids measure 1.2–2.0 mm long. The legs and cornicles of both apterous and alate forms are pale or dusky, and darker distally. Most alatae migrate to, and colonize, Umbelliferae. However, some disperse to, and colonize, other willows. Concurrently, some aphids apparently overwinter on umbellifers, probably carrot, and migrate to willow in the spring. The number of aphids on willow, regardless of the source, dissipates through the summer months, and essentially is eliminated by autumn.

Winged aphids arrive on, and infest, secondary hosts over a period of several weeks, with peak invasion occurring about early June. Peak numbers occur about three weeks after the June influx owing to reproduction, and then are decreased due to emigration and the activity of predators and parasitoids. Aphid numbers increase on secondary hosts again in the autumn, immediately preceding migration to primary hosts. Males are produced before migration. Not all aphids depart, of course, some remain on the secondary hosts for the winter. In the autumn, after dispersal back to willow, oviparous sexual forms are produced and mating occurs. Females then deposit overwintering eggs on willows. The egg-producing form is unlike the summer forms, but resembles the early spring aphids and is rusty brown and black.

The work by Dunn (1965) and Dunn and Kirkley (1966) presented the most complete information on willow-carrot aphid biology. Cottier (1953) provided a good description. Keys for identification of willow-carrot aphid, and many other common aphids, were provided by Palmer (1952) and Blackman and Eastop (1984).

Damage

There are two forms of damage caused by willow-carrot aphid—direct feeding injury and disease transmission. Direct feeding causes development of reddish or yellow foliage, and distortion of foliar tissue. Honeydew also accumulates on foliage. For parsley, contamination of the foliage with aphids makes the crop unmarketable. These are considered to be relatively minor problems.

The importance of willow-carrot aphid is related to its ability to transmit a circulative virus complex, known as carrot motley dwarf, to carrots. Virus infection causes carrot foliage to turn red or yellow and to be stunted. Reservoirs of this virus, in addition to carrot, cilantro, and chervil, are the many other secondary umbelliferous hosts fed upon by willow-carrot aphid during the summer months. Of lesser importance is the ability of *C. aegopodii* to transmit parsnip yellow fleck virus, anthriscus yellow virus (Elnagar and Murant, 1976), and many other stylet-borne viruses (Kennedy *et al.*, 1962).

Management

Insecticides. Insecticides may be applied in the spring as aphids move from primary to secondary hosts and 2–3 applications are adequate. Either systemic or contact insecticides may be used. In the absence of high insect and disease pressure, which is the case in North America if crops are managed properly, such treatments should not be necessary.

Cultural Practices. The most critical element of cultural control is sanitation (see section on disease transmission, below). Overwintering sources of aphids and disease must be eliminated.

Other aspects of cultural control are helpful, but less important. Attempts to identify carrot cultivars with resistance to willow-carrot aphid have not been very successful (Dunn, 1970). However, susceptibility of carrots to the carrot motley dwarf virus complex varies considerably among commercially available cultivars (Watson and Falk, 1994). Chemical attractant and repellent research (Chapman *et al.*, 1981) has demonstrated that willow-carrot aphid is sensitive to odors, but this information has not yet been put to practical use. In research conducted in England, intercropping of carrots with onions did not affect invasion of carrots by willow-carrot aphid (Uvah and Coaker, 1984), but plant density was important. As the distance between carrot rows decreased, there was a corresponding diminution of aphid numbers.

Disease Transmission. Although carrot motley dwarf virus can be very damaging, in some areas it can be managed effectively. Carrot cultivars vary widely in their susceptibility to infection by carrot motley dwarf virus, so cultivar selection is important. In California, the proximity of overwintering carrots is directly related to the severity of the disease. Weeds are not important sources of aphids or disease in this production area. Overwintering carrots should be eliminated, isolated from new production sites, or treated with insecticide to reduce the number of vectors available to transmit the virus (Watson and Falk, 1994).

In Washington, secondary spread of the virus within carrot fields is especially important (Howell and Mink, 1977). Virus was introduced into carrot fields in the spring by *C. aegopodii*, as was the case in California. However, there can be high levels of secondary transmission within the carrot fields during the summer months. Howell and Mink (1977) suggested that green peach aphid, *Myzus persicae* (Sulzer), was implicated in secondary spread though this seems unlikely because it is not a known vector of this disease.

FAMILY CICADELLIDAE—LEAFHOPPERS

Aster Leafhopper
Macrosteles quadrilineatus Forbes
(Homoptera: Cicadellidae)

Natural History

Distribution. Aster leafhopper is a native to North America, where it is found in nearly all states and provinces. It is most common, however, in the central states and provinces. Also, it overwinters poorly in cold areas. Most areas with aster leafhopper problems are invaded annually by leafhoppers originating in the southern Great Plains. In the mild-climate northwest, however, leafhoppers are able to overwinter successfully, and long-distance dispersal is not an important factor.

Host Plants. Aster leafhopper has a wide host range, but the plants suitable for maintenance of adults are not always suitable for reproduction and the development of nymphs. It tends to overwinter on grains such as wheat and barley, and on grasses, clover, and weeds, then dispersing to vegetables in the summer months. The vegetable crops damaged by aster leafhopper include carrot, celeriac, celery, corn, lettuce, parsley, potato, and radish, but among vegetables only lettuce is consistently suitable for leafhopper reproduction. Other crops fed upon are barley, clover, dill, field corn, flax, oat, rice, rye, sugarbeet, and wheat.

In Washington, Hagel *et al.* (1973) indicated that the most important breeding areas were mixtures of clover and pasture grasses, followed by clover, sweet corn, oats, carrots, lawn grasses, rye, field corn, and various weeds. Among the weeds favored were fleabane, *Erigeron* spp.; ragweed, *Ambrosia* spp.; dandelion, *Taraxacum officinale*; wild lettuce, *Lactuca canadensis*; tumble mustard, *Sisymbrium altissimum*; and lambsquarters, *Chenopodium album*. Low, sparse, and young vegetation provided the best habitat. Meade and Peterson (1967) indicated that large crabgrass, *Digitaria sanguinalis*; horseweed, *Conyza canadensis*; barnyardgrass, *Echinochloa crusgalli*; fowlmeadow grass, *Poa palustris*; and barley, wheat, and oats were especially suitable for reproduction in Minnesota. Carrot, dill, potato, and radish were important adult food plants, but were not good breeding hosts. Based on laboratory choice tests, McClanahan (1963) ranked flax and wheat among the most preferred host plants.

Natural Enemies. Natural enemies seem not to be well studied, nor very important in the population ecology of aster leafhopper. The most important enemies are the parasitoids *Pachygonatopus minimus* Fenton, *Neogonatopus ombrodes* Perkins, and *Epigonatopus plesius* Fenton (all Hymenoptera: Dryinidae). The best known among these is *P. minimus*, which caused up to 37% parasitism. The wingless female of this dryinid is also predatory on leafhoppers (Miller and DeLyzer, 1960, Barrett *et al.*, 1965).

Life History and Description. This insect overwinters in the egg stage in northern locations, and in the adult stage in warmer climates. In Manitoba and Maine there are three generations per year, whereas three to four are reported from Washington, and up to five generations may occur in more favorable midwestern locations (Westdal *et al.*, 1961). As the generations overlap and are initiated by both overwintering eggs and migrating leafhoppers, it is difficult to discern the generations. Total generation time requires about 27–34 days.

Egg. The eggs are deposited in leaf, petiole, or stem tissue, often near the juncture of the leaf blade and stem. They are deposited singly, but often in short rows of up to five eggs. They require a moist environment and perish if the foliar tissue desiccates. The eggs are translucent when first produced, but soon turn white. They average 0.80 mm long (range 0.73–0.87 mm) and 0.23 mm wide. They are slightly curved, with one side concave and the opposite side convex, and taper to a blunt point at each end. The incubation period of the egg is about 7–8 days.

Nymph. Newly hatched nymphs are nearly white, but soon become yellow and gain brownish markings, including dark markings on the head. There are five instars, the duration of which are about 3–4, 4–5, 3–4, 4–6, and 5–7 days, respectively, when reared at 21–25°C. The body length measures about 0.6–1.0, 1.2, 2.0, 2.5, and 3.0 mm in instars 1–5, respectively. As the nymphs mature they gain spines on the hind tibiae, the number increasing from about 6–7 to 8–9. The tip of the abdomen also bears spines. The wing pads are indistinct through instar three, but are apparent in instar four and overlap the abdominal segments in instar five.

Adult. The adults are small, the male measuring 3.2–3.4 mm long, and the female 3.5–3.8 mm. These insects are light green, with the front wings tending toward grayish-green and the abdomen yellowish-green. There are six pairs of black spots, including some that are elongated almost into bands, starting at the top of the head and extending along the front of the head almost to the base of the mouthparts. The six pairs of spots are the basis for the other common name of this insect, "six-spotted leafhopper." Strong winds moving north in the spring transport adults into mid-western and northern crop production areas annually. Adults usually arrive in advance of egg hatch by overwintering populations, and populations of long-distance dispersants greatly exceed resident leafhoppers. The arrival time in the north varies, but arrival dates in May are common.

Good documentation of the northward dispersal by aster leafhopper was provided by Wallis (1962), who worked in the western Great Plains, the western edge of the migration path. He reported overwintering of the adult leafhoppers only in Texas, though north of Texas eggs may overwinter. By May, northward movement was evident, with adults present in Kansas, Nebraska, and South Dakota where no nymphs were found previously. By June, the leafhoppers progressed northward into Montana and North Dakota, and westward into Colorado and Wyoming.

The biology of aster leafhopper is available from many sources, but some of the more important contributions were provided by Hagel and Landis (1967), Nielson (1968), and Hoy et al. (1992). Osborn (1916) described the nymphal stages. Hagel and Landis (1967) and Hou and Brooks (1975) gave rearing procedures. A key for *Macrosteles* spp. was given by Beirne (1952).

Damage

Leafhoppers pierce leaf tissue of plants and remove the sap. The feeding punctures cause death and discoloration of individual plant cells, resulting in a yellow, speckled appearance in affected plants. This feeding damage, while unsightly, is minor in comparison to the damage caused to numerous vegetable crops by transmission of aster yellows by leafhoppers.

Aster yellows is a plant disease caused by a phytoplasma, and is transmitted almost exclusively by aster leafhopper. Such crops as carrot, celery, cucumber, lettuce, potato, pumpkin, and squash are affected. Losses of 50–100% are reported due to this disease. Phytoplasma-infected plants are discolored, stunted, and deformed. On carrots, for example, the symptoms are red or yellow foliage and excessively hairy, bitter-tasting roots. On lettuce, symptoms are chlorosis, stunting, and lack of head formation.

Management

Sampling. Leafhoppers are easily collected with sweep nets, especially from grasses and grain fields. Sequential sampling protocols have been developed for sweep net sampling in carrot (O'Rourke et al., 1998). Yellow-sticky traps are also useful and easy to use. Light traps equipped with fans for suction also have been used effectively to capture leafhoppers. Cool and wet weather limit leafhopper activity and decrease the ability to sample effectively. Wind also interferes with population assessment (Durant, 1973).

In addition to sampling for leafhopper abundance, it also is desirable to determine the proportion of leafhoppers that harbor the phytoplasma. Formulas based on both insect number and disease incidence, the aster yellows index, have been developed to trigger control measures before the pathogen is widely transmitted to susceptible crops. Leafhoppers are collected before they enter an area, fed on aster plants, and the plants read for disease. This works effectively to alert large areas, such as entire states, but is not useful for local prediction. Disease incidence varies both spatially and temporally (Mahr et al., 1993). Further information on the aster yellows index, and action thresholds for several crops, were given in Foster and Flood (1995).

Insecticides. Insecticides commonly are used to kill leafhoppers, and thereby to minimize disease transmission. As there are protracted acquisition and incubation times associated with this disease, chemical-based disease suppression is feasible (Eckenrode, 1973; Koinzan and Pruess, 1975). Insecticides are especially effective in the absence of long-distance dispersal by leafhoppers. Systemic insecticides are often favored due to their persistence, but contact insecticides can be effective also (Thompson, 1967; Henne, 1970). Insecticides are often applied at 5–7 day intervals. As it takes 10–15 days for infected plants to show signs of infection, it is not necessary to treat plants just before harvest.

Cultural Practices. Crop varieties differ in their susceptibility to infection with aster yellows; this is well-studied both for carrots and lettuce, two of the more susceptible crops. Cultural manipulations can also enhance resistance. In studies conducted in Minnesota, Zalom (1981b) demonstrated very significant reductions in disease incidence where aluminum foil mulch was used. Straw mulch was equally effective (Setiawan and Ragsdale, 1987). Row covers, where economically feasible, should provide good protection against leafhoppers and disease transmission (Lee and

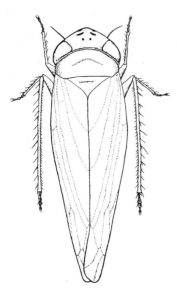

Adult aster leafhopper.

Robinson, 1958). Destruction of weed species known to harbor aster yellows is desirable.

Disease Transmission. Aster leafhopper acquires the phytoplasma by feeding on infected perennial and biennial weeds, or crop plants. Acquisition requires a prolonged period of feeding, usually at least two hours, before the leafhopper is infected. Usually less than 2% of dispersant leafhoppers become infected. There is evidence that the phytoplasma multiplies in the body of the leafhopper, and there is an incubation period of about two weeks in nymphs and 6–10 days in adults before the insects are capable of transmitting aster yellows. Leafhoppers remain infective for the duration of their life, but the phytoplasma is not transmitted between generations through the egg stage. Nielsen (1968) provided a good summary of aster yellows from an entomological perspective.

Beet Leafhopper
Circulifer tenellus (Baker)
(Homoptera: Cicadellidae)

Natural History

Distribution. Beet leafhopper originated in the Mediterranean region and apparently was introduced to the western hemisphere by Spanish explorers. It spread throughout Central and South America, the Caribbean area, and to Hawaii and Australia. In North America the beet leafhopper is common and is damaging throughout the western United States from southwest Texas to Washington. It also occurs in low numbers in the eastern states where it is not considered to be a pest. Beet leafhopper can remain fairly local, feeding throughout the year on crops and weeds in a small geographic area. However, it commonly is quite dispersive, moving north to British Columbia and east to the Great Plains area. Even within the generally infested area west of the Rocky Mountains, there is considerable annual movement from Arizona to Utah and Colorado.

This species is well-adapted for life in the desert. It accepts many plant species, develops rapidly, and disperses readily to find a new food source. Although more than one generation may develop on desert annuals, adults of the first generation display a strong tendency to disperse even in the presence of succulent hosts. The major breeding areas are the San Joaquin Valley of California; the lower Colorado River area of southern California, southwestern Arizona, southern Nevada, and southern Utah; the Rio Grand River area of New Mexico and Texas; the lower Snake River plains of Idaho and Oregon; the Columbia River area

of Oregon and Washington; and some small, scattered areas in western Colorado, northern Utah, and northern Nevada. In most cases, heat and drought eventually force the leafhoppers to disperse to cooler and moister areas, where they tend to populate long-lived summer hosts. However, in some areas where irrigation is practiced, particularly in southern California, a succession of crops and weeds allows good survival without the need for dispersal.

Host Plants. Beet leafhopper often has complex host-plant requirements, and they vary regionally. During the spring, rapid growth of annual weeds that serve as overwintering host plants allows the population of leafhoppers to increase rapidly. They then disperse to summer hosts, which may support them until autumn when they return to their winter-spring annual hosts. In some cases perennial "hold-over" plants are fed upon after the summer hosts until winter annuals become available; they persist on the latter until spring. Many publications have addressed host associations and damage, but an excellent overview of host relations was given by Cook (1967).

The overwintering and spring plants are mostly plants in the family Cruciferae, and include bittercress, *Erysimum repandum*; pepperweed, *Lepidium* spp.; tansymustard, *Descurainia pinnata*; flixweed, *D. sophia*; hedgemustard, *Sisymbrium irio*; and tumblemustard, *S. altissimum*; but also plants from other families including filaree, *Erodium* spp.; plantain, *Plantago* spp., and others. Summer hosts are often members of the family Chenopodiaceae such as Russian thistle, *Salsola kali*; halogeton, *Halogeton gomeratus*; lambsquarters, *Chenopodium album*; nettleleaf goosefoot, *Chenopodium murale*; and sagebrush, *Atriplex* spp.; but include others such as perennial pepperweed, *Lepidium alyssoides*. Hold-over hosts include perennial saltbush, *Atriplex* spp.; sagebrush, *Artemisia* spp.; rabbitbrush, *Chrysothamnus* spp.; snakeweed, *Gutierrezia* spp.; and perennial pepperweed, *Lepidium alyssoides*.

Crop plants damaged by beet leafhopper are bean, beet, cantaloupe, cucumber, pepper, spinach, sugarbeet, Swiss chard, squash, tomato, and watermelon. Sugarbeet, beet, spinach, and Swiss chard are most favorable for beet leafhopper growth; the other crops are fed upon but are poor hosts. All of the aforementioned crops can be injured, but the cucurbit crops are least susceptible.

The effects of host plants on beet leafhopper biology are variable. Harries and Douglass (1948) compared nymphal development time on several common host plants including tumblemustard, filaree, Russian thistle, flixweed, bittercress, pepperweed, and sugarbeet, and reported similar development times, 14.5–17.1 days when reared at 32°C. In contrast, however, sugarbeet was a much superior plant with respect to oviposition.

Natural Enemies. The principal natural enemies are little-known natural parasitoids including big-headed flies (Diptera: Pipunculidae), wasp egg parasitoids (Hymenoptera: Mymaridae and Trichogrammatidae), wasp nymphal parasitoids (Hymenoptera: Dryinidae), and twisted-wing parasites (Strepsiptera: Halictophagidae). Of these, the strepsipterans are least important, because they do not induce much mortality and infect a small percentage of the leafhopper population. The dryinids (*Gonatopus* spp.) are not often reared from leafhoppers, suggesting low levels of parasitism, but as the adults are predatory, the impact of these wasps is difficult to gauge. In California, the egg parasitoid *Polynema eutettixi* Girault (Mymaridae), and the nymphal and adult parasitoid *Pipunculus industrius* Knab (Pipunculidae), were reported to be most important by Stahl (1920). However, Meyerdirk and Hessein (1985), in a study of the egg parasitoids in California, indicated that *Anagrus giraulti* Crawford (Hymenoptera: Mymaridae) and *Aphelinoidea* sp. (Hymenoptera: Trichogrammatidae) were the most important of the five egg-attacking species collected. Although egg parasitoids sometimes cause up to 90% parasitism of eggs during the summer, and overwintering populations of beet leafhopper may experience up to 25% parasitism, parasitoids are often not reliable for suppressing beet leafhopper. *Pipunculus* and other parasitoids are often left behind when leafhoppers disperse, so dispersing leafhoppers initially escape from many of their important natural enemies. Several parasitoids have been introduced to the western United States from the Mediterranean area, particularly northern Africa; however, they have failed to establish (Clausen, 1978).

Birds, spiders, and other general predators also affect leafhopper populations. The predator most commonly observed to feed on these leafhoppers is western big eyed bug, *Geocoris pallens* Stål (Hemiptera: Lygaeidae), but green lacewings (Neuroptera: Chrysopidae) also prey on these leafhoppers. Unspecified fungal diseases affect overwintering populations, and horsehair worms (Nematomorpha) have been observed occasionally (Severin, 1933).

Weather. As leafhoppers inhabit harsh environments during the winter, they are susceptible to high levels of mortality during this period. In northern areas, the overwintering adults cannot feed during cold periods, so many die due to starvation and dehydration. In southern areas, there is less cold stress, but adults also suffer during foggy weather. Also, desiccation of host plants is a major cause of winter mortality in southern desert areas.

Life Cycle and Description. A complete generation requires 1–2 months. Several generations develop annually. In northern areas such as Washington, Oregon, Idaho, and Utah, three generations is the norm. In warmer areas such as California and Arizona, five generations is usual. Normally, 1–2 generations occur in the spring months on the early season host plants, with an additional 2–3 generations on summer hosts. Fertilized females overwinter, become inactive when it is cold, and active again when weather is favorable. Males perish during the winter months.

Egg. Egg deposition commences about the time the winter host plants begin their spring growth. The eggs are deposited singly within a slit in the tissue of the leaves and stems. The petiole and leaf midrib are preferred sites, but leaf margins are sometimes selected. The eggs are elongate and slightly curved, with the posterior end tapering almost to a point. They measure 0.06–0.07 mm long and 0.018 mm wide, and are white to yellow. Each female may deposit 300–400 eggs, though Meyerdirk and Moratorio (1987) reported only about 100 eggs per female when reared at 20–25°C and up to about 200 eggs at 32°C. Eggs hatch in 5 to 40 days, depending on temperature. Harries and Douglass (1948) reported mean egg durations of 26.3, 11.9, 7.4, and 5.6 days at 18°, 24°, 30°, and 35°C.

Nymph. At hatching, the nymphs are transparent white. They begin to feed almost immediately, and within a few hours acquire a greenish color. There are 5–6 instars, and the larger instars usually are spotted with black, red, and brown on the thorax and abdomen. Mean head widths are 0.33, 0.43, 0.54, 0.37, and 0.84 mm for instars 1–5, respectively. Maximum body lengths average 1.13, 1.52, 1.88, 2.38, and 3.20 mm, respectively, for instars 1–5. Development time normally ranges from 2–6 weeks, but at 32°C the mean duration of instars 1–5 is 3.4, 2.3, 2.4, 3.1, and 3.7 days, respectively.

Adult. The adult measures 3.4–3.7 mm long and about 1 mm long. Adults are variable in color. The color is lighter at high temperature. During the summer they are usually uniform whitish or greenish, in the autumn they acquire some dark spots dorsally, particularly on the forewings. During the winter they become mostly dark. During the summer months, mating occurs within a few days of adult emergence, whereas during the autumn this period is prolonged. Oviposition may commence 5–10 days after mating, or may be protracted during cool weather. Mean oviposition rates exceed 10 eggs per day at temperature above

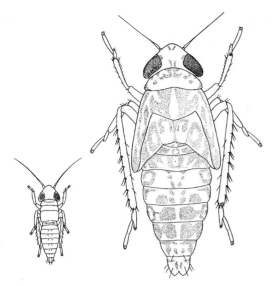

Beet leafhopper nymphs.

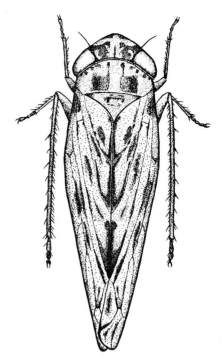

Adult beet leafhopper, dark form.

32°C, but the rate decreases at lower temperature. Duration of the adult stage is reported to be up to 4–5 months during the winter, though Harries and Douglass (1948) reported maximum mean longevity of about 82 days when overwintering females were held at about 15.5°C. Similarly, Meyerdirk and Moratorio (1987) reported mean longevity of about 40–90 days at various temperatures, but some individuals lived more than four months. Overwintering adults benefit from availability of water, but atmospheric humidity is not a significant factor.

The biology and physical description were given by Severin (1930, 1933) and Cook (1967). Phenology and host relations were presented by Douglass and Cook (1954) and Cook (1967). Physical ecology was treated by Harries and Douglass (1948).

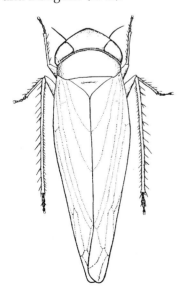

Adult beet leafhopper, light form.

Damage

Beet leafhopper is known as a serious pest in the western states principally because it transmits curly top virus. It is the only known vector of this plant disease. Signs of infection vary greatly among crops, but vein clearing, rolling or curling of leaves, stunting of the plant, and discoloration are common. Young plants are most injured, and often fail to produce fruit, or produce deformed fruit. Commercial vegetable production is infrequent in some southwestern areas owing to a high incidence of curly top.

Beet leafhopper also transmits *Spiroplasma citri*, causing ailments known as stubborn disease in citrus and brittle root in horseradish (Liu *et al.*, 1983; O'Hayer *et al.*, 1984). A mycoplasma-like (mollicute) organism is also transmitted to vegetable crops (Golino *et al.*, 1987, 1989; Shaw *et al.*, 1993).

Management

Sampling. Visual examination and sweeping of plants are usually used to detect beet leafhopper. The mere presence of leafhoppers in susceptible crops usually stimulates insecticide applications. In some locales, sampling also is directed to breeding areas, so advance warning of leafhopper densities and damage potential can be obtained. The presence of high densities in rangeland breeding areas sometimes results in insecticide treatment to prevent dispersal of leafhoppers to crop production regions.

Insecticides. Breeding is often confined to relatively small areas, making it feasible to apply insecticides and to eliminate a high proportion of the leafhoppers before they disperse to crops. Applications of insecticide to such areas can be made in the spring, before overwintered females have deposited eggs, or after the eggs have hatched but before the leafhoppers have matured into adults. More commonly, however, insecticides are applied directly to susceptible crops in areas that are prone to attack by leafhoppers. A common practice is to apply systemic insecticides at planting to protect seedlings.

Cultural Practices. Broadleaf weeds are common hosts of beet leafhopper, and it is desirable to replace them on rangeland with nonhost perennial grasses. Similarly, it is important to prevent rangeland from being damaged by overgrazing and fire, as this fosters the development of annual broadleaf weeds. As annual broadleaf weeds are replaced by late successional plants such as sagebrush, *Artemisia tridentata*, the abundance of leafhoppers drops markedly (Fox, 1938). Similarly, weed suppression in areas with crops deprive leafhoppers of good alternate hosts. Russian thistle is particularly hazardous because it is a good host.

Host-Plant Resistance. Varieties that are resistant to curly top are available for most crops. This approach to manage the pest is widely used, but sometimes the resistant varieties bear horticultural characteristics that make them less suitable for commercial production, or resistance level varies due to horticultural practices such as date of planting (Martin and Thomas, 1986). Resistance, at least in tomatoes, appears to be due to lower frequency of feeding on the resistant varieties (Thomas and Martin, 1971; Thomas and Boll, 1977). Resistance in beans is temperature sensitive, with high temperature reducing resistance (Silbernagel and Jafri, 1974).

Disease Transmission. Curly top survives the winter both within the leafhopper and within some host plants. The leafhopper acquires the disease through feeding. About three hours of feeding are usually necessary for acquisition, and an incubation period of 4–7 h is necessary before successful transmission. Both nymphs and adults are vectors. Once infected, the leafhopper remains infected for the duration of its life. It cannot transmit the disease to its progeny through the egg, however. The proportion of the leafhopper population that is capable of transmitting the virus varies from 4–80%, thus greatly affecting damage potential. Different strains of the virus also display varying pathogenicity.

Leafhoppers feed on, and infect, numerous hosts. Apparently they reject, or select plants for prolonged feeding, only after an extended period of feeding during which curly top virus can be transmitted (Thomas, 1972). Insecticides often fail to kill leafhoppers before disease transmission occurs, but within-field transmission can be greatly reduced due to mortality of the vectors.

Corn Leafhopper
Dalbulus maidis (DeLong and Wolcott) (Homoptera: Cicadellidae)

Natural History

Distribution. The corn leafhopper occurs throughout the western hemisphere from the United States to Argentina. It is most destructive in tropical areas. In the United States it is damaging in the southern states from North Carolina to California. Apparently corn leafhopper overwinters only where host plants are available continuously, such as the Gulf Coast states and southern California, and disperses northward annually. It has been recovered as far north as Ohio. Its origin is likely Mexico.

Host Plants. The only crop affected by this insect is corn, and the life cycle is completed principally on this plant or other *Zea* spp. However, in the southeastern states gamagrass, *Tripsacum dactyloides*, can also support populations of corn leafhopper (Pitre, 1970) and in Mexico several *Tripsacum* spp. are hosts. Although corn leafhopper feeds on other crops such as barley, oat, rye, sorghum, sudangrass, sugarcane, and wheat, and even upon nongrass crops such as carrot and beet, these plants are not considered to be suitable for complete development of the life cycle.

Natural Enemies. The natural enemies of corn leafhopper are poorly known. Some ant species apparently prey upon them, because when ant populations are decreased leafhopper numbers increase (Perfecto, 1991). In Nicaragua, *Anagrus* sp. (Hymenoptera: Mymaridae) and *Paracentrobia* sp. (Hymenoptera: Trichogrammatidae) parasitize corn leafhopper eggs (Gladstone *et al.*, 1994). In Mexico, *Gonatropus bartletti* Olmi (Hymenoptera: Dryinidae) and *Eudorus* sp. (Diptera: Pipunculidae) attack nymphs, and the fungi *Beauveria bassiana* and *Metarhizium anisopliae* infect nymphs and adults (Vega and Barbosa, 1990; Vega *et al.*, 1991).

Life Cycle and Description. The life cycle can be completed in about 21 days, so several generations occur annually in warm climates. Two complete generations can be completed during the normal growth period of corn (Todd *et al.*, 1991). Adults overwinter under cold conditions, and persist for up to seven months without breeding. They may feed on various grasses during this period, but Larsen *et al.* (1992) suggested that corn leafhopper could survive during the winter in Mexico without host plants if free water is available. These leafhoppers are gregarious and secrete honeydew, so they often are attended by ants.

Egg. The eggs are deposited singly, but often in rows of up to eight, in the veins on the upper surface of young foliage in the whorl. Duration of the egg stage is five days at 24°C.

Nymph. Nymphs feed on the base of the leaf, or on the stem at the juncture of leaf and the stem. There usually are five instars, but occasionally a sixth is observed. Average nymphal body length is 0.8, 1.1, 1.7, 2.1, and 2.8 mm for instars 1–5, respectively. Mean nymphal development time is 2.5, 2.6, 2.5, 3.1, and 3.8 days for instars 1–5, respectively, when reared at 27°C. Nymphal develpment at variable temperatures usually requires about 15–18 days (range 11–19 days).

Adult. The general body color of this insect is yellowish-white, with the pronotum yellowish. However, the top of the head is marked with two large black spots, and sometimes with additional dark pigmentation. The front wings are translucent and extend beyond the tip of the abdomen. Nielson (1968) reported the length of the adult male to measure 3.50–4.00 mm, the female 4.00–4.10 mm. Tsai (1988), however, gave values of 2.52–3.36 mm for males and 2.73–3.50 mm for females. Females oviposit at 21°C, but not at 18°C or lower. According to Davis (1966) females produce, on average, about 150 eggs and live for 25–50 days. Tsai (1988), however, reported fecundities of 400–600 eggs per female depending on temperature, and obtained mean adult longevity of 77 days when leafhoppers were reared at 27°C.

The biology was given by Davis (1966) and Tsai (1988). Triplehorn and Nault (1985) provided a description and a key to the members of the genus.

Damage

The adults and nymphs remove sap, causing wilting and yellowing. Honeydew production by leafhoppers results in extensive growth of sooty mold (Bushing and Burton, 1974). Feeding by 4–5 adult leafhoppers causes severe wilt symptoms within five days, though the plants recover if the insects are removed (Hildebrand, 1954). However, this leafhopper is important mostly because it is a disease vector. It is the most important leafhopper pest of corn, because it transmits corn stunt spiroplasma, maize bushy stunt mycoplasma, and maize rayado fino

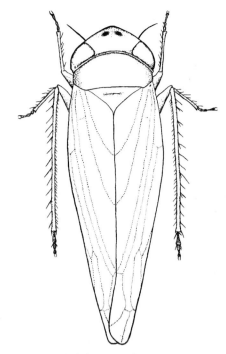

Adult corn leafhopper.

virus. Symptoms of disease include discoloration, striping, and stunting. Severity of the disease depends on how early in the growth of the plant infection occurs, with greatest damage resulting from early infection (Hao and Pitre, 1970).

Management

Sampling. Nymphs and adults can be detected by visual examination of corn plants, concentrating attention on the whorl and base of the leaves. Adults can also be sampled with yellow water-pan traps (Vega *et al.*, 1990). Although there is a positive relationship between insect abundance and disease incidence, yellow-sticky trap catches were shown not to be a good predictor of disease potential (Power *et al.*, 1992).

Insecticides. Systemic insecticides applied to the soil at planting or post-planting can provide protection against population establishment and increase during the early portion of the growing season. Planting time treatment, but not mid-season treatment, also reduces the incidence of corn stunt (Bhirud and Pitre, 1972). Infestation later may require foliar application of insecticides, with particular attention directed to application into the whorl.

Cultural Techniques. Cultural manipulations have frequently been reported to affect the incidence of corn leafhopper or disease, but the findings are not always consistent. Intercroppings of bean and corn were reported to have higher abundance of leafhoppers than corn monocultures by Perfecto and Sediles

(1992) and corn plantings with low plant density to have higher incidence of corn stunt (Power, 1989). In contrast, however, Castro *et al.* (1992) observed no effect of interplanting and density. Late plantings of corn are more subject to attack, and larger fields experience a lower incidence of attack. In California, crops of corn separated by a crop-free period experience little damage, but when cropped continuously leafhopper numbers attain high levels and considerable damage results (Bushing and Burton, 1974).

Disease Transmission. Diseases transmitted by corn leafhopper can cause severe economic losses. Significant damage to corn occurred in the late 1970s in Florida (Tsai, 1988), and in Nicaragua in the mid-1980s (Hruska *et al.*, 1996). The basis for the disease outbreaks is not apparent, but in Nicaragua a possible explanation is the introduction of irrigated agriculture, which allows the disease to survive at high levels through the dry season. Corn stunt spiroplasma also survives in leafhoppers during the winter months, and its presence in the leafhopper increases survival rates at low temperature (Ebbert and Nault, 1994). Only a few hours of feeding are required for the insect to acquire corn stunt, but leafhoppers are not able to acquire the virus for 15–18 days after the disease is inoculated into a plant (Granados *et al.*, 1968).

Potato Leafhopper
Empoasca fabae (Harris)
(Homoptera: Cicadellidae)

Natural History

Distribution. Potato leafhopper is found throughout the humid, low-altitude regions of eastern United States, occurring as far west as eastern Colorado (DeLong, 1931). It occurs in eastern Canada, including the Prairie Provinces, but is most damaging in southern Ontario. Potato leafhopper successfully overwinters in Gulf Coast States from Louisiana to Florida and disperses northward annually. Leafhoppers typically arrive with warm fronts in midwestern states during April to mid-May, and in northern states and the Canadian provinces during June (Medler, 1957). In late summer and autumn months they are carried southward again by cold fronts (Taylor and Reling, 1986). Apparently it is a native species.

Host Plants. Potato leafhopper feeds on over 200 wild and cultivated plants, though fewer species are suitable for nymphs than adults, and males have a wider host range than females (Lamp *et al.*, 1994). Vegetable hosts include bean, broad bean, cowpea, cucumber, eggplant, Jerusalem artichoke, lima bean, potato, pumpkin, rhubarb, squash, and sweet potato.

Also attacked are field crops such as alfalfa, clover, soybean, sugarbeet, sunflower and tobacco, as well as woody plants such as apple, cherry, hickory, maple, oak, redbud, rose, and walnut. The most suitable crop hosts are alfalfa, bean, cowpea, and potato. In much of the economic entomology literature this species is referred to as "bean leafhopper," which is indicative of a major host preference and damage potential. Indeed, 61% of the plant species fed upon by potato leafhopper are legumes. Poos and Smith (1931) evaluated the oviposition behavior of potato leafhopper in several crop plants and reported the following preference (from most preferred to least): potato, cowpea, dahlia, non-pubescent soybean, alfalfa, bean, pubescent soybean, and red clover. Lamp *et al.* (1984) studied leafhopper survival on various weed; nymphs developed successfully on smartweed, *Polygonum pennsylvanicum*; pigweed, *Amaranthus retroflexus*; shepherdspurse, *Capsella bursa-pastoris*; dandelion, *Taraxacum officinale*; and carpetweed, *Mollugo verticillata*. Castor bean, *Ricinus communis*; and pokeweed, *Phytolacca* sp. have also been observed to support nymphal development (Beyer, 1922), and many other weed hosts are known (Lamp *et al.*, 1994).

Many woody hosts are of significance only for first-generation leafhoppers; apparently their suitability declines as plant tissue matures (Poos and Wheeler, 1943). However, feeding and overwintering apparently can occur on loblolly pine (Taylor *et al.*, 1993); during early spring, as adult leafhoppers terminate reproductive diapause, they shift their feeding to deciduous trees and legumes (Taylor and Shields, 1995). (See color figure 12.)

Natural Enemies. Biological agents affecting potato leafhopper are either few in number, or not well studied. Egg parasitoids (Hymenoptera: Mymaridae and Dryinidae) sometimes are abundant. Generalist predators such as spiders, lacewings, nabids, lady beetles, and ants destroy these leafhoppers, and the fungus *Erynia radicans* is effective during warm, moist weather (Beyer, 1922). Natural control of leafhoppers was reviewed by DeLong (1971).

Life Cycle and Description. Where potato leafhopper overwinters, the adult is the overwintering stage. During the winter months the adult leafhopper is in reproductive diapause. After leafhoppers invade an area in the spring, several overlapping generations may occur. Four generations are known from Ohio, but the last is small. Six generations are possible in Virginia (Poos, 1932) and are documented in Florida (Beyer, 1922). Even as far north as Ontario, 2–3 generations may occur. Invading leafhoppers sometimes produce a generation on deciduous hosts before moving to annual crops, because the latter have not yet emerged from the soil (Flanders and Radcliffe, 1989).

Egg. The eggs are transparent to pale yellow, and measure about 1 mm long. They are inserted into the veins and petioles of leaves, usually at about 2–6 per day. Total egg production is about 200–300 per female. They hatch, on average, in 10 days, but hatching occurs over a range of 7–20 days.

Nymph. Nymphal development requires 8–25 days, depending largely on temperature. The lower temperature threshold for development is estimated to be 8.4°C, and the upper threshold to be 29°C. The average development time is usually 15 days, with 2.6, 2.3, 2.3, 2.5, and 4.7 days required for development of instars 1–5, respectively. Beyer (1922) provided description of the instars. Briefly, nymphal instars one and two are difficult to distinguish, and are separated principally by the larger size of the latter stage, 1.0 versus 1.2 mm, respectively. In instars three to five, the wing pads develop and are used for differentiation. The wing pads extend over the first, second, and fourth abdominal segment in instars three, four, and five, respectively.

Adult. The adults are pale green, and marked with a row of white spots on the anterior margin of the pronotum. They average 3.5 mm long. Examination of the male genitalia is required to distinguish this species

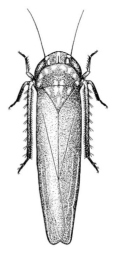

Adult potato leafhopper.

from closely related *Empoasca* species. Adults emit sounds that apparently represent intraspecific communication, but the sounds are barely audible to the human ear (DeLong, 1971). Adult longevity is typically 30–60 days. Adults normally mate within 48 hours after emergence. The post-mating pre-oviposition period is 3–8 days.

Potato leafhopper biology was given by many authors, including Fenton and Hartzell (1923), DeLong (1938), Simonet and Pienkowski (1980), and Hogg (1985). A bibliography was published by Gyrisco *et al.* (1978).

Damage

Potato leafhoppers feed on phloem or mesophyll tissue (Backus and Hunter, 1989) and secrete a toxin into the plant. Plant respiration is increased and photosynthesis is decreased by leafhopper feeding (Ladd and Rawlins, 1965). Feeding results in curling, stunting, and yellowing of potato foliage. The chlorotic tissue eventually becomes necrotic, initially at the leaf margins. The damage is called "hopperburn," because the plant appears to have been singed by fire. The toxin is not systemic, and the level of damage is directly proportional to the number of leafhoppers feeding. Reduction in crop yield is often significant. In potatoes, though the normal number of tubers may be produced, they are very small. Beans similarly stop production of pods once attacked, and existing pods usually develop incompletely. Damage is exacerbated by drought. Potato leafhopper is not known to transmit plant pathogens.

Management

Insecticides. Standard practice for many vegetable growers is to apply insecticides to the foliage on

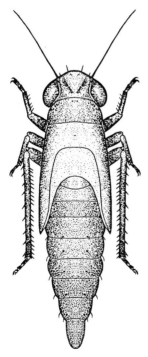

Potato leafhopper nymph.

a regular schedule to prevent injury. Systemic insecticides applied at planting also can provide excellent long-term control of leafhoppers and prevent hopperburn. However, as few as two well-timed foliar applications aimed to control peak nymphal populations can be adequate to prevent injury by leafhoppers (Radcliffe, 1982). Subsequent reinvasion, of course, may warrant additional treatments. Cancelado and Radcliffe (1979) estimated that insecticide treatment would be economic at densities of one leafhopper per potato leaf. Cucurbits are susceptible to hopperburn, but are usually not severely damaged unless the plants are also water stressed.

Cultural Practices. The wide host range of potato leafhopper and highly dispersive nature of the insects eliminates consideration of crop rotation and many other cultural practices. Consideration should be given to proximity of infested crops, however, especially if they are reaching maturity or are about to be harvested. Harvesting alfalfa, for example, could drive both adults and nymphs from the alfalfa in search of food. Leafhopper nymphs do not survive long without vegetation, so clean cultivation between fields is helpful, but adult movement will not be deterred by this tactic. Row covers protect small plantings from dispersing leafhoppers.

Host-Plant Resistance. Plant resistance offers potential for leafhopper management. Leafhoppers are affected by the hairiness of foliage and petioles. In general, pubescent soybean varieties are less suitable for potato leafhopper, but the presence of hooked leaf hairs (trichomes) that impale nymphs is more important than trichome density (Poos and Smith, 1931). Leaf hairiness is also a major, but not exclusive, component of resistance in eggplant (Poos and Haenseler, 1931). Glycoalkaloids also have been implicated in resistance, and glandular trichomes associated with wild *Solanum* impede leafhopper mobility and feeding, resulting in death (Radcliffe, 1982). Nymphs are especially subject to death from glandular exudates (Tingey and Laubengayer, 1981). Some potato, bean, and alfalfa varieties display considerable resistance but none are immune to damage (DeLong, 1971).

Western Potato Leafhopper
Empoasca abrupta DeLong
(Homoptera: Cicadellidae)

Natural History

Distribution. This leafhopper is found principally in the southwestern states, but occurs as far north as Oregon and as far east as Missouri.

Host Plants. In addition to potato, this insect attacks various vegetables including artichoke, bean, beet, celery, corn, cucumber, eggplant, lettuce, melon, okra, parsnip, pepper, spinach, sweet potato, tomato, and turnip (Poos and Wheeler, 1943). Several field crops such as alfalfa, red clover, sugarbeet, and sunflower also are hosts.

Life Cycle and Description. Western potato leafhopper overwinters in the adult stage. Generation time is estimated to be about 43 days, including an adult pre-oviposition time of 10 days, egg incubation time of 14 days, and a nymphal development period of 19 days (Michelbacher *et al.*, 1955). The kidney-shaped eggs are deposited singly into host tissue just below the surface. Females produce up to 17 eggs per day, and a total of 217 eggs per female (Poos, 1932). There are five nymphal instars. The first instar is pale and almost colorless, but its color intensifies with each molt until the adult stage is reached. The adult is green marked with white, and is about 3.5 mm long. It closely resembles the more well-known potato leafhopper, *Empoasca fabae* Harris; the male genitalia must be examined to separate the species accurately (DeLong, 1931).

Damage

Nymphs and adults feed on the lower surface of foliage. They prefer shady areas; thus, the older, lower leaves are often most heavily damaged. The contents of individual mesophyll cells are destroyed, producing a stippled appearance on the upper surface of the leaves. The loss in photosynthetic tissue, while appearing slight, may decrease quality and yield of plants. Damage also accrues when fecal material accumulates on produce. The dramatic "hopperburn" symptoms produced by *E. fabae* are not produced by *E. abrupta*.

Management

Insecticides. Western potato leafhopper is not considered to be difficult to control with insecticides. Foliar applications are standard practice. Unlike some other species that transmit plant viruses or have toxic saliva, moderate numbers of these leafhoppers can be tolerated in many crops.

Cultural Practices. Crop management has long been considered critical to prevent development of large leafhopper populations. Crop residues should be destroyed promptly to prevent continuous breeding. The leafhoppers are quite dispersive, so new crops planted adjacent to infested fields are prone to problems.

FAMILY DELPHACIDAE—PLANTHOPPERS

Corn Delphacid
Peregrinus maidis (Ashmead)
(Homoptera: Delphacidae)

Natural History

Distribution. Although now found throughout the world in tropical areas, this species is likely a native to Africa. In the United States it is reported to occur as far north as Washington D.C. and Ohio, and south to Florida and Texas. As a pest, however, it is best known from subtropical areas such as southern Florida and Texas. It also is common in Hawaii and the Caribbean, including Puerto Rico.

It has been suggested that corn delphacid was introduced to Mesoamerica in pre-Columbian times, and was responsible for crop destruction and collapse of the Mayan civilization. However, post-Columbian introduction of this species also has been suggested (Nault, 1983).

Host Plants. Among cultivated crops, corn delphacid feeds principally on corn and sorghum. Sugarcane is sometimes suggested as a host, but there is no evidence that it is a good host plant. Forage grasses such as pangolagrass, napiergrass, and vaseygrass are poor hosts (Namba and Higa, 1971), but nymphs are able to complete their development on some weed grasses such as barnyardgrass, *Echinochloa crusgalli*; itchgrass, *Rottboellia exaltata*; and goosegrass, *Eleusine indica* (Tsai, 1996).

Natural Enemies. Several natural enemies have been imported into Hawaii, where corn delphacid was especially damaging. Some attack principally sugarcane leafhopper, *Perkinsiella saccharicida* Kirkaldy, and only slightly affect corn delphacid. Egg parasites, *Anagrus* spp. (Hymenoptera: Mymaridae) were established in Hawaii for suppression of corn delphacid, but their effectiveness is uncertain. A predator, *Cyrtorhinus lividipennis* Reuter (Hemiptera: Miridae), is reported to be an important suppressive agent in Hawaii (Liquido and Nishida, 1985).

This delphacid is attended by ant species, though the attending ants differ in the various regions of the world. The ants collect honeydew produced by the delphacids, sometimes building shelters for the delphacids, and fending off wasp parasitoids (Dejean *et al.*, 1996).

Life Cycle and Description. Several overlapping generations occur during the summer or growing period. A complete generation requires about 30 days.

Egg. The eggs are most often found along the midrib of leaves, especially on the basal and on the lowest portions of leaves on the plant. The eggs are inserted into a slit cut by the female, and 2–4 (usually two) eggs per oviposition site. The female deposits a drop of glue-like material after oviposition, sealing in the eggs. However, egg deposition also occurs below-ground on corn plants measuring less than about 30 cm in height. On such young plants, females oviposit in roots and in the below-ground portion of the stem. Access to these locations occurs through ant burrows and cracks in the soil. The eggs are reported to measure 0.75–1.1 mm long and 0.18–0.28 mm wide. They are curved in shape, tapering slightly at both ends, and resemble a banana in form. The eggs initially are transparent white, but as they approach hatch the dark embryo is visible. Duration of the eggs stage is normally 7–10 days. The mean oviposition rate is 19.6 eggs per day, with a lifetime fecundity of about 605 eggs.

Nymph. There normally are five instars, but the number may range from 4–6 depending on temperature and perhaps nutrition. The general body color is whitish, becomes yellowish by the third instar, but slightly darker dorsally during the latter instars. The eyes of the nymphs are reddish. The wing pads are first apparent during the fourth instar, extending back to about the middle of the first abdominal segment. In the fifth instar the wing pads extend to the third abdominal segment. The nymphs were reported to measure about 0.9, 1.3, 2.0, 2.8, and 3.2 mm long, respectively, in instars 1–5 by Quaintance (1898a). However, Tsai and Wilson (1986) indicated larger lengths, 1.4, 1.6, 2.3, 2.9, and 3.9 mm for instars 1–5, respectively. Keys to the nymphal instars were also provided by Tsai and Wilson (1986). Mean instar-specific development times for insects cultured on corn at 27°C is 3.8, 3.0, 3.1, 3.1, and 4.2 days, respectively, for instars 1–5. Total nymphal development time when fed on corn was reported to require 65, 27, and 18 days at 15.6°, 21.1°, and 26.7°C. Although development time was rapid when nymphs were fed on corn; survivorship was higher when nymphs were provided with sorghum and barnyardgrass. Curiously, adult longevity was short when fed on barnyardgrass (Tsai, 1996). Young nymphs frequent leaf sheaths and other sheltered locations, presumably in response to the higher moisture levels found there. Older nymphs move freely about the plant. Both nymphs and adults leap freely when disturbed, but also they tend to move about the plant to avoid the source of disturbance. Nymphs are highly gregarious in habit, and often are found in association with adults.

Adult. The adults occur in short-winged (brachypterous) and long-winged (macropterous) forms. Both

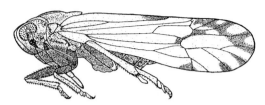

Adult corn delphacid.

males and females occur in each form. The short-winged forms measure about 3 mm long, whereas the body of the long-winged forms measure about 2–2.6 mm. The body color is greenish or brownish-yellow. The front wings (when present) are greenish or brownish, darker distally. The abdomen is smoky, with lateral yellow stripes. The legs are pale or brownish. Duration of the adult stage is about 30 days for both males and females at 21–27°C.

The general biology and appearance were described by Quaintance (1898a) and Verma (1955), but Tsai (1996) and Tsai and Wilson (1996) gave more detailed descriptions of morphology and development.

Damage

Corn delphacid affects corn both directly and indirectly. Direct injury includes removal of plant sap, tissue damage resulting from oviposition, and production of honeydew which supports the growth of sooty mold. Excessive growth of sooty mold impedes photosynthesis. The principal damage by these insects, however, results from the transmission of plant disease. Corn delphacid is the vector of two important diseases of corn—maize mosaic virus and maize stripe virus. Both nymphs and adults are capable of virus transmission, which results in yellowing and stunting of the corn. For maize strip virus, nymphs acquire the virus after feeding for about four hours and can transmit the disease after 4–5 days, and for the period of their life. They also can transmit the virus transovarially (Tsai and Zitter, 1982). For this insect to transmit maize mosaic, 10 days is required, and the insect also remains capable of disease transmission (Carter, 1941).

Management

Sampling. Fields are normally invaded by winged forms of corn delphacid, with highest insect densities found along the margins of fields. New corn fields are most commonly invaded by insects from older corn fields, so densities are highest at the juncture of new and old crops. When higher densities occur, the insects are more likely to be found in the center of fields. Therefore, field margins should be the principal focus of sampling programs (Takara and Nishida, 1983).

Insecticides. Insecticides are used to prevent invading insects from establishing and reproducing in corn fields. Contact and systemic foliar insecticides, and systemic insecticides applied to the soil, are effective (Nishida, 1978; Tsai *et al.*, 1990).

Cultural Practices. The strong association of corn delphacid with corn suggests that crop rotation will limit damage, especially if new crops are relocated at a considerable distance from previous crops. However, some grass weeds also support corn delphacid. If these grasses cannot be eliminated, they should be monitored for delphacid populations and sprayed, if necessary, to prevent the insects from infesting the crop.

FAMILY PSEUDOCOCCIDAE—MEALYBUGS

Pink Hibiscus Mealybug
Maconellicoccus hirsutus (Green)
(Homoptera: Pseudococcidae)

Natural History

Distribution. This insect is common in the tropical areas of the world, including Africa, Southeast Asia, and northern Australia. It was noted in Hawaii in 1983 and California in 1999. Pink hibiscus mealybug was first reported from the Caribbean region in 1994, where it now infests many islands including Puerto Rico and the Virgin Islands. It seems inevitable that pink hibiscus mealybug will infest Florida, and potentially much of the southern United States north to Virginia and northern California.

Host Plants. As its common name suggests, hibiscus is a favorite host of this mealybug. In fact, many woody ornamental, fruit, and forest trees support high population densities of hibiscus mealybug. Among the important economic hosts are such tropical fruits as avocado, banana, carambola, citrus, custard apple, grape, guava, mango, mulberry, passion fruit, and soursop; and ornamentals such as croton, heliconia, and hibiscus. It is also a minor pest of cotton. Vegetables are infested mostly when the mealybug population density is high, with incidence of infestation declining as overall densities decrease. Among the vegetable hosts most susceptible to infestation are bean, beet, carrot, cowpea, cucumber, okra, pepper, pigeon pea, squash, and tomato, but among the other vegetables occasionally infested are asparagus, cabbage, lettuce, onion, potato, pumpkin, sweet potato, and yam.

Natural Enemies. In Asia and Africa, many predators and parasitoids are reported to attack pink hibiscus mealybug, and some have been introduced to other countries to implement biological suppression

(Mani, 1989). Nevertheless, it remain a pest in Egypt and India (Mani, 1989). The parasitoid *Anagyrus kamali* Moursi (Hymenoptera: Encyrtidae) was apparently accidentally introduced to Hawaii simultaneously with the mealybug about 1982, and has held this pest in check. When pink hibiscus mealybug attained the Caribbean area in the mid-1990s, considerable damage was caused in Grenada, Trinidad, and elsewhere until *Anagyrus kamali* could be successfully established. Many parasitoids and predators attacking other mealybugs can attack pink hibiscus mealybug as well.

Life Cycle and Description. In subtropical climates a generation is completed in about 30–40 days and about 10 generations occur annually. Complete development from egg to the adult stage normally occurs in 25–26 days, but it is temperature dependent. In Egypt, reproduction is usually parthenogenetic, but males are sometimes produced, and both sexual and parthenogenetic reproduction occurs in many populations. In the Caribbean region, biparental reproduction apparently occurs exclusively (Williams, 1996). Hibiscus mealybug survives during the cold weather in all stages, but the egg stage is particularly hardy.

This insect is characterized by the presence of waxy white cotton-like secretions, so infested plants have a white fuzzy appearance. If the cottony material is removed, however, the eggs, nymphs, and adults are revealed to be pink.

Egg. The female produces a cottony egg sac, which is attached to the host plant. The oval egg sac is about twice as long as wide and consists of loose fibers and eggs internally and matted fibers externally. Each egg sac contains 80–650 eggs, which turn pink before hatching. The oval eggs measure about 0.35 mm long and 0.20 mm wide. Eggs hatch in 3–9 days.

Nymph. The nymphs are elongate-oval. Initially they are orange but then turn pink. Newly hatched nymphs (crawlers) are mobile, but soon settle and begin feeding. There are three instars in females. Mean duration of the female instars is about 6.7, 6.5, and 7.9 days for instars 1–3, respectively. In males, there are two nymphal stages followed by two "pupal" stages. Mean duration of the male instars is 6.6, 6.5, 1.0, and 5.6 days, respectively. Three pairs of long legs and moderately long six-segmented antennae are evident in nymphs, and the anal region bears a pair of stout hairs. The piercing-sucking mouthparts are narrow and difficult to observe. Occasionally, waxy secretion is found in the posterior region.

Pupa. In males, the third and fourth instars are nonfeeding stages in which the nymph transforms into a winged adult. The third instar (puparium) is somewhat elongate and resides in a loose mass of fine white filaments. It measures 1.1–1.5 mm long and 0.35–0.45 mm wide. The fourth instar (pupa) is brownish and shows evidence of wing formation. The antennae are directed posteriorly, and closely resembles the adult, though lacking the terminal abdominal filaments of the adult male. It measures about 1.25 mm long and 0.4 mm wide.

Adult. The adult female is elongate oval, bears three pairs of relatively small legs, and short but apparent nine-segmented antennae. Occasionally, waxy secretion is found in the posterior region. The female is pink and measures 2.0–3.0 mm long and 0.9–2.0 mm wide. It bears stout hairs at the posterior end of the body, and is wingless. The adult female may disperse away from the terminal growth of the plant if it is withering and unsuitable for feeding. The pre-oviposition period varies from 0.5–6 days, followed by an ovipositional period of 4–8 days. Oviposition may occur at the terminal portion of the plant, but during cool weather more sheltered locations are sought for oviposition.

The adult male is slightly smaller than the female in size. The male is pinkish, elongate and narrow in body form. It bears one pair of wings and three pairs of moderately long legs. The antennae are prominent and ten-segmented. The tip of the abdomen bears a pair of long, stout filaments that are white.

A detailed synopsis of pink hibiscus mealybug biology was given by Mani (1989) and Williams (1986, 1996). Ezzat (1958) and Williams (1996) provided a technical description of the species, and the latter author also presented a key to the species of *Maconellicoccus*.

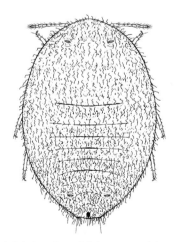

Adult female pink hibiscus mealybug.

Damage

These mealybugs may infest any portion of plants, including the below-ground portions. However, the stems and terminal shoot tissues are favored. They secrete a toxic saliva that causes various symptoms in the host plant. Typical symptoms are severe malformation of shoots and leaves, including twisting and crinkling of the foliage. Growth is stunted and tip growth may be bushy rather than elongate. Infested flowers drop and fruit is not produced, or they are small and malformed. In hibiscus, one of the most preferred hosts, galls are produced on terminal growth.

Management

As hibiscus is the preferred host, these plants should be monitored for the presence of pink hibiscus mealybug in an area. If natural enemies are established in an area, particularly *Anagyrus kamali*, the mealybug should not be a severe pest. In the absence of natural enemies, or if natural enemy activity is disrupted by pesticides, this mealybug can be damaging. Insecticides can provide some control, particularly if systemic insecticides are applied. Application of granular formulations may be necessary for protection of crops with susceptible below-ground product, such as potato. The waxy secretions produced by the mealybugs greatly reduce the effectiveness of contact insecticides. Ants attend pink hibiscus mealybug, so elimination of ants can favor suppression of mealybug by natural enemies. Some natural enemies of pink hibiscus mealybug, including the lady beetle *Cryptolaemus montrouzieri* Mulsant (Coleoptera: Coccinellidae), are available from commercial insectaries and can be released into affected areas to supplement naturally occurring biological control.

FAMILY PSYLLIDAE—PSYLLIDS

Potato Psyllid
Paratrioza cockerelli (Sulc)
(Homoptera: Psyllidae)

Natural History

Distribution. This native insect pest occurs principally in the Rocky Mountain region of North America, from Arizona and New Mexico north to Alberta and Saskatchewan. However, it occasionally has been reported as far west as California and British Columbia and as far east as Minnesota and Quebec. Potato psyllid also occurs in Mexico. It overwinters only in the southernmost United States and in Mexico, so occurrence in northern regions is largely a function of wind-borne dispersal.

Host Plants. Potato psyllid is also sometimes known as "tomato psyllid," and these common names accurately indicate the most preferred hosts. However, this psyllid can develop on other solanaceous vegetables, particularly pepper and eggplant. Solanaceous weeds such as groundcherry, *Physalis* spp.; black nightshade, *Solanum nigrum*; buffalo bur, *Solanum rostratum*; and matrimony vine, *Lycium halimifolium*; also support reproduction. Matrimony vine is believed to be an important overwintering host in southern Arizona. Adults have been collected from numerous plants, including conifer trees, but this is not an indication of the true host range. With the possible exception of field bindweed, *Convolvulus arvensis*, non-solanaceous plants seem to be relatively unimportant, and even solanaceous weeds are not as important as crop plants. (See color figure 18.)

Natural Enemies. Several predators and parasites of potato psyllid are known, though there is little documentation on their effectiveness. Under rather artificial conditions the larvae and adults of lady beetles (Coleoptera: Coccinellidae), lacewings (Neuroptera: Chrysopidae), flower flies (Diptera: Syrphidae), big-eyed bugs (Hemiptera: Lygaeidae), minute pirate bugs (Hemiptera: Anthocoridae), and damsel bugs (Hemiptera: Nabidae) consume psyllid nymphs and adults (Pletsch, 1947). The wasps *Metaphycus psyllidus* Compere (Hymenoptera: Encyrtidae) and *Tetrastichus triozae* Burks (Hymenoptera: Eulophidae) parasitize potato psyllid. The latter species feeds externally on nymphs and attacks other psyllid species as well, and though sometimes abundant, is often absent from psyllid populations.

Weather. Weather is an important element of biology and damage potential. These psyllids seem to be adapted for warm, but not hot weather. List (1939a), for example, demonstrated how prolonged exposure to 32°C depressed egg production and egg hatching. Temperatures of about 21–27°C are considered optimal. Dispersal northward occurs when higher temperature is attained; temperatures of about 32°C and higher are deleterious to psyllids. Heavy precipitation may be detrimental, but this is less certain. Because psyllids disperse passively over long distances, wind patterns also are critical in determining their occurrence, and account for much of the variability in damage. Once invading psyllids have established, they multiply and cause greatest damage at moderate temperature, and populations fail to develop to damaging levels if the temperature is high.

Life Cycle and Description. A generation can be completed in 20–30 days, depending on temperature.

The number of generations varies considerably among regions. In Montana, only three generations can be completed before frost kills the host plants. In northern Utah, 3–4 generations are known, and in Colorado there are 4–7 generations. Once psyllids invade an area, however, prolonged oviposition by adults causes the generations to overlap, so it is difficult to distinguish generations.

Potato psyllid overwinters in southern Arizona, southern Texas, and northern Mexico, where populations build to high densities in early spring. By early summer the psyllids disperse from the spring breeding areas northward to Utah and Colorado and shortly thereafter to Nebraska and eastern Wyoming. In July, as the temperature becomes hot in these mid-summer breeding areas, psyllids disperse northward to the northern states and Canadian provinces. Psyllids reappear in the overwintering areas between October and November, presumably dispersing southward from northern locations.

Egg. The eggs are deposited principally on the lower surface of leaves, usually near the leaf edge, but some eggs can be found everywhere on suitable host plants. The eggs are oval and borne on thin stalks which connect one end of the egg to the leaf. They initially are light-yellow, and become dark yellow or orange with time. The egg measures about 0.32–0.34 mm long, 0.13–0.15 mm wide, and with a stalk of 0.48–0.51 mm. Eggs hatch 3–6 days after deposition.

Nymph. At hatching, the young nymph quickly escapes from the egg, then crawls down the egg stalk and searches for a place to feed. Nymphs are found mostly on the lower surface of leaves, sometimes on the upper surface, and almost never on the stems. Nymphs, and also adults, produce large quantities of whitish particulate excrement which may adhere to the foliage and fruit. Nymphs are elliptical when viewed from above, but very flattened in profile, appearing almost scale-like. There are five nymphal instars. Nymphal body widths were given by Pletsch (1947) as 0.23, 0.35–0.41, 0.53–0.58, 0.70–0.82, and 1.27–1.32 mm for instars 1–5, respectively, but body lengths were not provided. The body width measurements given by Rowe and Knowlton (1935) tend toward the lower end of the aforementioned size range; they also give nymphal lengths as 0.36, 0.50, 0.70, 1.10, and 1.60, respectively. Initially the nymphs are orange, but become yellowish-green and then green as they mature. The compound eyes are reddish and quite prominent. During the third instar the wings become evident in the form of pads which remain light in color and become more pronounced with each molt.

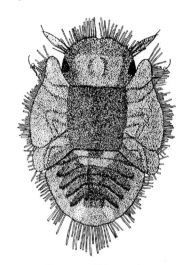

Potato psyllid nymph.

A short fringe of wax filaments is present along the lateral margins of the body. Nymphs tend to remain essentially sedentary during their development, and prefer sheltered, shaded locations. Mean duration (range) of the instars was reported by Knowlton and Janes (1931) to be 2.8 (1–5), 2.4 (1–5), 2.5 (1–4), 2.7 (1–5), and 4.9 (3–9) days, respectively, for instars 1–5. Thus, nymphal development time required 12–21 days, though the average was about 15 days. (See color figure 183.)

Adult. The adults are quite small, and measure only about 2.5–2.75 mm long. In general body form they resemble miniature cicadas, largely because they hold their wings angled and roof-like over their body. They have two pairs of transparent wings. The front wings are considerably larger than the hind wings. The antennae are moderately long, about the length of the thorax. Body color ranges from pale green at emergence, to dark green or brown within 2–3 days, and gray or black thereafter. White or yellow lines are found on the head and thorax, and whitish bands on the first and terminal abdominal segments. Adults are active and easily disturbed, a distinct contrast to the nymphal stage. The pre-oviposition period of adults is normally about 10 days, with oviposition continuing for about 20 days. Total adult longevity is normally 25–35 days. Pletsch (1947) reported mean longevity of females to be 20, 23, 29, and 14 days when reared on tomato, potato, eggplant, and pepper, respectively. He also observed mean fecundity of 67, 258, 187, and 53 eggs on these same hosts. Knowlton and Janes (1931), working with potato foliage, reported mean fecundity in various trials of 300–400 eggs, with some individuals producing 1100–1350 eggs.

The biology of potato psyllid was described by many authors including Knowlton and Janes (1931),

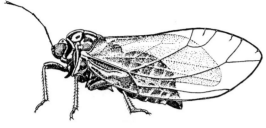

Adult potato psyllid.

Adult potato psyllid.

List (1939a,b) and Wallis (1946, 1955). However, Pletsch (1947) provided the most complete summary and Cranshaw (1993) published an annotated bibliography. Keys to the North American Psyllidae were given by Tuthill (1943).

Damage

Feeding by nymphs of potato psyllid causes a disease known as "psyllid yellows" (Eyer, 1937; Arslan *et al.*, 1985). Psyllids apparently produce a toxin or growth regulator that disrupts plant growth; a plant pathogen is not involved. Even a few psyllids can affect growth, but large populations are necessary for extensive disruption of plant growth. Symptoms of the disease include upward rolling of the young leaves, yellowing along the midrib and leaf margins, and sometimes induction of a purple color. Stem elongation is suppressed, plants are dwarfed, exceptionally small tubers and fruits are formed, and tubers sprout prematurely. In some areas, yield losses of 20–50% are not uncommon in the absence of treatment, and complete losses have been reported in potato. This insect is considered to be the principal pest of tomatoes in some Rocky Mountain areas. Greenhouse-grown tomatoes

are also affected. Elimination of psyllids from infested plants results in partial recovery of plants. Other psyllid species are occasionally reported from potatoes, but only *P. cockerelli* occurs in large numbers and causes damage. Beet leafhopper, *Circulifer tenellus* (Baker), induces similar symptoms in tomato, but this disease (called "curly top") is caused by a plant virus.

Management

Sampling. The adult populations are commonly sampled with the assistance of a sweep net, but egg and nymphal numbers are assessed by visual examination of foliage. The adults also can be sampled with yellow water-pan traps. As for many insects that are not strong fliers, psyllid populations typically are highest at field edges initially, but if not controlled they eventually spread throughout the crop.

Insecticides. Insecticides are usually applied, once psyllids are found in a field, with applications continuing at least until mid-season. Good coverage is important because psyllids are commonly found on the underside of foliage. Once plants are mature, and bear abundant foliage, they can withstand some infestation with no significant yield loss. For some other small insects such as aphids and whiteflies, application of insecticidal soap or dilute dishwashing detergent can suppress populations, but adequate protection from psyllids is infrequent. Even with conventional insecticides, this insect tends to be difficult to manage.

Cultural Practices. Early-planted crops are more susceptible to injury than crops planted at mid-season or late, principally because only early crops are present when psyllids disperse into areas in the spring. However, part of the reason early season crops are more injured is that psyllids do not thrive under the hot weather conditions found later in the summer, and large plants from early-season plantings protect the psyllids from the sun and heat (Wallis, 1955). Sanitation also is important because if discarded potatoes are allowed to sprout they can serve as temporary hosts.

Order Hymenoptera—Ants and Sawflies

FAMILY ARGIDAE—SAWFLIES

Sweetpotato Sawfly
Sterictiphora cellularis (Say)
(Hymenoptera: Argidae)

Natural History

Distribution. This native species occurs throughout the eastern United States, occurring as far west as eastern Nebraska and Texas. It has been known as a serious pest only in the Chesapeake Bay region, specifically Maryland, Virginia and North Carolina, and near the mouth of the Mississippi River, in Louisiana and Mississippi.

Host Plants. Sweetpotato sawfly is known only from *Ipomoea* spp., and rarely has been collected from anything other than sweet potato.

Natural Enemies. A larval parasitoid, *Schizocerophaga leibyi* Townsend (Diptera: Tachinidae) is reported to be an effective biological control agent, having been reared from 60–70% of sweetpotato sawfly cocoons. Chapman and Gould (1929) attributed the small size of the autumn generation principally to the actions of this fly. A wasp parasitoid, *Boethus schizoceri* (Riley and Howard) (Hymenoptera: Ichneumonidae) also is known from sweetpotato sawfly.

Life Cycle and Description. Apparently there are three generations annually in Virginia, with larvae present in July, August, and September. A generation can be completed in about 28 days. This insect is presumed to overwinter in the pupal stage.

Egg. The eggs are inserted within the leaf tissue in rows along the principal veins on the lower surface of the leaf. They cause the leaf tissue to bubble or blister, and each blister contains a single egg. Rows of 6–12 eggs are common, and a single leaf may contain over 200 eggs. The egg is whitish, and has a somewhat flattened oval shape. The egg measures about 1.6 mm long at hatching, but because sawfly eggs enlarge before hatching they must be smaller when first deposited. Eggs hatch in 6–7 days.

Larva. The larva is yellowish-green and adorned with numerous black, blunt spines. The larva bears seven pairs of abdominal prolegs, but they are short and obscure, the terminal prolegs especially cryptic. Apparently there are five instars. Duration of the larval stage is about 10 days. At maturity the larva measures about 12 mm long.

Pupa. When the larva completes its growth and is ready to pupate it drops to the soil and constructs a brownish silken cocoon at, or near, the soil surface. The cocoon is oval in shape and measures about 14 mm long and 5 mm wide. The cocoon frequently has soil and leaf debris attached to the outside. The pupal stage apparently has not been described. The adult emerges from the cocoon about 9–12 days after cocoon formation.

Adult. The adults are reported to be short-lived and weak fliers. The body of adult males is shiny black, and the wings dusky. The female is similar in color except that her abdomen is orange-red. In both sexes the legs are dark basally but lighter or colorless distally. Females measure about 7 mm long, with a wingspan of about 12 mm. The male is slightly smaller

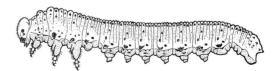

Sweetpotato sawfly larva.

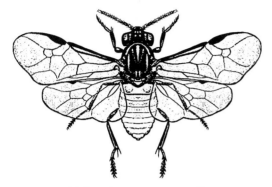

Adult male sweetpotato sawfly.

in size. The antenna of the male is branched whereas that of the female is unbranched.

The biology of sweetpotato sawfly was described by Chapman and Gould (1929).

Damage

The larvae are defoliators of sweet potato, consuming leaf tissue but leaving the stems and basal portion of the principal veins. As adults are not very dispersive, they tend to deposit many eggs in small areas. This leads to severe defoliation in very localized sections of fields, and sometimes death of numerous larvae as they exhaust the food supply. Repeated defoliation by sawfly may reduce yield by 50%.

Management

This species is infrequent, and apparently has caused no large-scale damage since the early 1900s. Observations from early in the 1900s suggested that damage was limited to the sweet potato cultivar "Big Stem Jersey," and changes in cultivar selection by sweet potato growers might have accounted for the demise of this insect as a pest. Foliar insecticides are effective, and crop rotation is recommended due to the poor dispersive ability of the adults.

FAMILY FORMICIDAE—ANTS

Red Imported Fire Ant
Solenopsis invicta Buren
(Hymenoptera: Formicidae)

Natural History

Distribution.   Red imported fire ant was accidentally introduced to the United States at Mobile, Alabama between 1933 and 1941. It gradually spread over most of the southeast, occupying much of Georgia, Alabama, Mississippi, and Louisiana by 1957, and then adjacent states from North Carolina to Texas by 1985. Isolated infestations occur elsewhere periodically, including in California, and this species seems destined to occupy all of the southern half of the United States except for very arid areas. Red imported fire ant also has successfully invaded Puerto Rico. Activity decreases during cold weather, and this species seemed ill-adapted for cold-weather conditions, which was limiting northward spread. The northern limits now are southern Tennessee and southern Oklahoma. The origin of Solenopsis invicta is South America, where it is most abundant in southwestern Brazil and in Paraguay.

Red imported fire ant is not the only fire ant in North America, but it is the only species that commonly affects vegetable crops. Two native species, Solenopsis geminata (Fabricius) or tropical fire ant, and S. xyloni McCook or southern fire ant, also occur in the southeastern states, with S. xyloni also extending to California. Another immigrant species, S. richteri Forel or black imported fire ant, was also introduced from South America, but likely from areas south of the homeland of red imported fire ant. Black imported fire ant is quite similar to red imported fire ant, though darker in color, and so initially the two species were confused. Black imported fire ant has a very restricted distribution, and is limited to northern Alabama and Mississippi, and western Tennessee. Red imported fire ant tends to replace most other species of ants when it invades a new area, including the aforementioned Solenopsis spp. (See color figure 179.)

Host Plants.   These ants are omnivorous, and feed on a mixture of plant and animal matter. They are effective predators of insects, spiders, earthworms, and other small invertebrates. Plant feeding is limited, and often occurs when ants are deprived of other food. Nevertheless, fire ants are known to feed on such vegetables as bean, cabbage, corn, cucumber, eggplant, okra, potato, and sweet potato, and on other crops such as young citrus trees, peanut, sorghum, soybean, and sunflower. Okra fruit is particularly at risk. Sweet plant exudates and honeydew from homopterous insects are readily consumed. (See color figure 22.)

Natural Enemies.   Many natural enemies of red imported fire ant are known, but few seem to be effective under either South American or North American conditions. Competition from other ants is believed to be one of the most important factors limiting abundance and dispersal in their native land, but North American ants seem to be ineffective competitors of fire ants, at least within disturbed habitats. Colony survival, particularly of incipient colonies, is also affected by competition among fire ant nests. Minim workers from new colonies often raid brood from nearby nests, boosting size of receiving colonies and allowing new colonies to achieve maturity sooner. Although raided colonies disappear, workers and queens from the losing colonies may be amalgamated into the receiving colonies (Tschinkel, 1992).

Among the promising pathogens of fire ant are the microsporidian *Thelohania solenopsae* and the fungus *Beauveria bassiana*. Neither has demonstrated consistent effectiveness in North America, though there is some evidence that *Thelohania* may be important in South America.

A South American parasitoid, *Pseudacteon* sp. (Diptera: Phoridae), which decapitate ants during the final stages of development has been introduced to the southern states, and related species are being introduced. An unusual South American parasitic ant, *Solenopsis daguerri* (Santschi) (Hymenoptera: Formicidae), invades fire ant colonies and takes over control of the fire ant colony, possibly through food diversion. This and other parasites are being investigated for potential to achieve fire ant suppression. Natural enemies have been discussed by Jouvenaz (1983, 1986), Banks *et al.* (1985), and Wojcik (1986a).

Life Cycle and Description. Ants are colonial insects, with the number of ants per colony depending on colony age and availability of food. Activities of the colony members are highly structured, with some members (castes) specialized for reproduction, tending of the young, and obtaining food. The seasonal reproductive cycle usually begins in March, and eggs from mature colonies give rise to both worker and reproductive castes. Workers predominate, with the maximum proportion of sexual forms (about 10%) produced in June. A life cycle can be completed in about 30 days. Longevity of colonies is uncertain, though individual queens can live 6–7 years.

Egg. Colonies may be initiated by a single mated queen or several queens. Mating flights may occur at any time, but peak abundance is during May–August. Flights are preceded by rain, normally within 1–2 days, and at least 80% relative humidity. After the mating flight the queen usually breaks off her wings and begins excavation of a burrow within four hours of her mating flight. Initially the burrow consists of a vertical tunnel 6–12 cm deep into soil, and small cells. The queen lays 15–20 eggs within 2–3 days and produces 20–125 eggs by the time the first larvae hatch. The newly deposited egg is whitish, oval in shape, and measures about 0.2 mm wide and 0.3 mm long. As the egg approaches maturity it acquires the larval form. Only about half of the eggs are fertile, the remaining serve as food for the young larvae. Mean duration of the egg is 8.4, 5.2, 5.0, and 5.8 days when cultured at 25°, 30°, 32°, and 35°C; optimal develop-

ment of eggs and other stages occurred at about 32°C (O'Neal and Markin, 1975b). Once the colony is firmly established, the queen can produce up to 2000 eggs per day.

Larva. There are four instars, and all stages are whitish, C-shaped, and lack distinct appendages. As described by Petralia and Vinson (1979), first instars are hairless, and measure 0.27–0.42 mm long. Head capsule width is 0.14–0.16 mm. Second instars have only a few simple hairs, measure 0.42–0.57 mm long, and have a head capsule width of 0.16–0.19 mm. Third instars have moderately numerous short hairs of various types, measure 0.59–0.91 mm long, and have a head width of 0.20–0.25 mm. Fourth instars are similar to third instars in general appearance, though the body hairs are slightly longer. The body length of fourth instars is 0.8–1.8 mm, and the head width is 0.26–0.32 mm. Head capsule widths increase slightly during both the third and fourth instars. Duration of instars 1–4 was 1, 1, 2, and 3 days, respectively, when cultured at 32°C (O'Neal and Markin, 1975b). Porter (1988) reported that duration of the larval stage decreased from 28 to 11 days as temperature increased from 24° to 35°C. The castes are quite similar in appearance during the larval stage; the principle difference is size. Mean size is greatest among larvae destined to be sexual forms, followed by major, minor and minim workers. Development time is proportional to size, with the larger reproductives requiring nearly twice as much time as the minim workers, and the other castes intermediate.

Pupa. Pupal development is often reported to require about 6–8 days, though Porter (1988) indicated that duration of the pupal stage decreased from 28 days when cultured at 21°C to less than 7 days at 35°C. The pupa greatly resembles the adult in form, though it is whitish. Size varies considerably, depending on the caste of the adult form. The pupae, along with the eggs and larvae, comprise the ant brood, and are moved about within the colony by workers according to environmental conditions.

Adult. A total of 20–30 days is normally required between egg deposition and emergence of the first mature worker ants, called minim workers. Within 90 days of colony founding, the colony may consist of 200 minims and a few minor workers. Porter and Tschinkel (1986) described the importance of minims in colony founding and success. Within five months there may be 1000 minor and a few major workers. Within seven months the colony may consist of 6000–14,000 workers, and may contain about 3% major workers. The number of ants per colony is estimated at 11,000, 30,000, and 60,000 workers at 1, 1.5 and 3 years. After three years the colony is considered to be mature, and workers number about 230,000 per colony. Abundance typically decreases each winter, however, and reproduction ceases at northern latitudes and decreases in more southern locations. Colonies may begin to produce reproductives at 5–7 months, but normally it is older colonies that produce most reproductives.

Adult worker ants vary considerably in size, ranging from about 1.6–6 mm long. They lack wings, and vary in color from light reddish-brown to dark-brown. The gaster is sometimes marked with an orange spot. Assignment of workers to castes is based on arbitrary head width categories, not on distinct morphological or behavioral differences. Adult reproductives bear two pairs of wings, with the front wings longer than the hind wings. The females resemble the workers in color, but the males are black except for pale antennae.

The physical structure of the nest changes markedly over time. As noted earlier, initially it is a small burrow with chambers. As workers are produced, however, the tunnels are enlarged and additional chambers are produced. Within 90 days of colony founding, a soil surface mound is produced, typically 5–7 cm in height and 3–7 cm in diameter. At this time there are 5–15 vertical tunnels extending a meter or more into the soil down to the water table. Mounds in areas with sandy soil tend to be relatively flat, whereas in clay soil they can be 0.5–1.0 m in height. Large colonies may construct several mounds. They may continue to use the same mound for several years, or may construct new ones. The mound enables the colony to optimize environmental conditions for the brood. During cool weather the brood is moved to the sunny side of the mound near the surface, while during hot weather the brood is found deep within the underground tunnels.

Food for the colony is collected by foraging workers, which often forage for considerable distances from the colony, with the distance dependent on colony food requirements. Maximum foraging occurs between the soil temperature (5 cm depth) of 21–35°C, but some foraging can occur between 10–37°C. Foraging is aided by the construction of tunnels leading back to the nest; these often extend 15–25 m from the mound. Foraging ants search randomly for food, but once food is located, ants returning to the tunnels deposit trail pheromones that serve to recruit other ants. Soon there is a stream of ants leading to the food. Food obtained by the foraging workers is quickly distributed throughout the colony. Food typically passes from foragers to nurses, and then to larvae and queens. Colony organization and coordination is maintained by secretion of pheromones. Several pheromones associated with fire ants have been identified, including the brood, trail, queen tending, and nestmate recognition pheromones.

There are two types of colonies: single-queen (monogyne), and multiple-queen (polygyne). The monogyne colonies contain about 100,000–240,000 workers, fight with other colonies, and tend to be situated farther apart, with mound densities of 100–350 per hectare. In contrast, polygyne colonies contain 20–60 queens and 100,000–500,000 workers, do not fight with other polygyne colonies, and have densities of 500–2000 mounds per hectare. Polygyne colonies tend to erect lower mounds than monogyne colonies, and certain smaller workers which are lighter in color. Polygyne colonies are becoming more common in the southern United States.

The fire ants were described by Buren (1972). Hung et al. (1977) provided keys to the common species of fire ants. Creighton (1950) presented a comprehensive treatment of North American ants, though red imported fire ant was not distinguished in this treatment. Wheeler and Wheeler (1990) provided a good key to North American genera. Biology of fire ant was reviewed by Lofgren et al. (1975), Vinson and Greenberg (1986), and Vinson (1997). Vander Meer (1988) discusses caste structure. A laboratory rearing method was described by Williams et al. (1980). Bibliographies were published by Banks et al. (1978), Wojcik and Lofgren (1982), and Wojcik (1986b). Zoogeography was summarized by Buren et al. (1974).

Damage

Red imported fire ant thrives in disturbed habitats such as cropland and pastureland. Thus, agricultural environments are particularly likely to be infested, but similar sites such as roadsides, irrigation ditches, and athletic fields also support high densities. Red imported fire ant does not survive well in climax plant communities, particularly if dense shade is provided by trees and shrubs.

An interesting feature of these ants is that they are unable to ingest solid food. They place solid food near

Adult worker red imported fire ant.

the mouth of fourth instar larvae, which then secret digestive enzymes. The liquefied food is then passed around the colony. Nevertheless, fire ants readily collect solid materials, including plant tissue. Young seedlings, in particular, can be killed by fire ants. For example, Adams (1983) reported destruction of over 50% of eggplant seedlings in Florida. Damage was in the form of stem girdling and destruction of the growing point of the young plants. Seedling injury is the most frequent form of damage, but ants also feed below-ground on seeds, roots and tubers; they remove bark from young citrus trees; and cause abortion of okra flowers by feeding at the base. Plant injury occurs most commonly when ants are deprived of food, as when land is first cultivated. Fire ants are not normally considered to be primary plant pests except for okra, where damage can be severe and frequent. However, they affect crop production indirectly because their mounds interfere with equipment operations, and laborers are reluctant to harvest heavily infested crops due to the toxic sting of fire ants.

Fire ants are well-known for their venomous sting which induces a burning sensation in victims—in fact, this is the basis of their common name. Although the individual sting is much less toxic than that of bees and many other stinging insects, the abundance of the ants and the tendency of victims to receive multiple stings represents a real threat to human health. Humans are not the only animals at risk. Reptiles, ground-nesting birds, young mammals, and even fish have been reported to be killed by fire ants. Animals reproducing during the warmest periods of the year are at great risk because ant foraging is reduced during cool weather. Rates of reproduction of several forms of wildlife reportedly increase following area-wide suppression of fire ants. This is also of interest to livestock producers, because newborn livestock are at risk; blinding of young calves is the most common form of livestock injury. The impact of fire ants on plant, animal and human health was reviewed by Adams (1986), Lofgren (1986), Jemal and Hugh–Jones (1993), and Allen *et al.* (1994).

Damage is somewhat offset by the beneficial predatory behavior of red imported fire ant. Predation of fire ants on sugarcane borer, *Diatraea saccharalis* (Fabricius) (Lepidoptera: Pyralidae), boll weevil, *Anthonomus grandis* Boheman (Coleoptera: Curculionidae), and lone star tick, *Amblyomma americanum* (Linneaus) (Acari: Ixodidae), is particularly well-documented. However, predation occurs on many species of insects, particularly caterpillar larvae on plants and fly larvae breeding in animal manure, and red imported fire ant is usually considered to be a beneficial insect within the context of cotton and sugarcane production sys-

tems. Reagan (1986) reviewed the beneficial aspects of fire ants.

Management

Sampling. Fire ants and their mounds are easily detected, and visual examination of fields is adequate for most purposes. Density is usually expressed as number of mounds per unit of land area, though colony size is an important variable.

Insecticides. Insecticides are effective for ant suppression, but reinvasion of treated areas from untreated areas can occur, sometimes giving the impression that insecticides were not effective. Ants are often treated by application of toxin-treated bait. The attractive component of the bait is normally soybean oil, and the granular carrier can be any of a relatively inert substances, often corncob grit. The toxin is usually a slow-acting insecticide or an insect growth regulator. Slow induction of mortality is advantageous, because it allows the toxin to be transported through the colony to the brood and perhaps even to the queens. Unfortunately, few bait-based insecticides can be legally applied to vegetable crops owing to failure by manufacturers to register their products for this use. Treatment of field perimeters is allowable, however, and foraging ants may collect bait deposited in perimeter areas and return it to colonies located within crop fields.

An alternative method of suppression is application of granular, dust, or liquid insecticide to mounds, or to tilled soil, but mortality is often less complete than with bait formulations. Ants from mounds treated with insecticide may relocate following treatment, necessitating one or more reapplications of insecticide before the colony is eliminated.

Cultural Practices. There are few effective non-chemical approaches to managing fire ants. Suppression of other insects is helpful because it deprives ants of their principal food supply. Individual mounds can be treated with boiling water, but at least 10 liters of water are necessary to penetrate most mounds, and its effectiveness is limited. Boiling water also kills nearby plants. Physical destruction of mounds by digging or tilling, and shoveling two mounds together in hopes of stimulating a fatal fight, are of little value. Large individual plants can be protected from foraging ants by placing a wide barrier of adhesive around the base of the plant. This approach is often best preceded by wrapping the trunk or stem with tape or foil to prevent the adhesive from disfiguring the plant or damaging the plant tissue.

Biological Control. The diseases and parasites discussed under "natural enemies" are being inoculated for permanent establishment and are not generally available from commercial sources, but other organisms have been suggested for biological suppression if applied regularly. In particular, entomopathogenic nematodes (Nematoda: Steinernematidae and Heterorhabditidae) and the straw itch mite, *Pyemotes tritici* (Lagrexe–Fossat and Montane) have been suggested for biological suppression. However, performance of these biotic agents have been disappointing under field conditions.

Order Lepidoptera—Caterpillars, Moths and Butterflies

FAMILY ARCTIIDAE—WOOLLYBEAR CATERPILLARS AND TIGER MOTHS

Banded Woollybear
Pyrrharctia isabella (J. E. Smith)
(Lepidoptera: Arctiidae)

Natural History

Distribution. A native species, banded woollybear is found throughout the United States and southern Canada. Banded woollybear is the best known of the woollybears, because in American folklore its color pattern is said to foretell the severity of forthcoming winter weather. Banded woollybear, though common, is more of a curiosity than a pest.

Host Plants. Banded woollybear consumes a wide breadth of flora, though damage to economic plants is infrequent, and therefore poorly documented. It is recorded from beet, corn and pea, and probably nibbles on nearly any garden vegetable, but its normal hosts are weeds and wild native plants. The normal foods are asters, *Aster* sp.; dandelion, *Taraxacum officinale*; dock, *Rumex* spp.; goldenrod, *Solidago* spp.; plantain, *Plantago* spp.; sweetclover, *Melilotus*; and some grasses. In the only broad study of banded woollybear feeding, a no-choice test, over 90 plant species from over 50 plant families were consumed to some degree (Shapiro, 1968).

Natural Enemies. Several natural enemies are known, but they have not been the subject of much study, so their relative importance are to be determined. Fly parasitoids associated with banded woollybear include *Euexorista futilis* (Osten Sacken), *Exorista mella* (Walker), *Hubneria estigmenensis* (Sellers), *Mystacella* spp., *Parachaeta fusca* Townsend, *Thelaria* spp., and *Winthemia datanae* (Townsend) (all Diptera: Tachini-

dae). Among wasps known to attack this woollybear are *Apanteles flavovariatus* Muesebeck (Hymenoptera: Braconidae); *Coccygomius pedalis* (Cresson), *Dusona crassicornis* (Provancher), and *Enicospilus glabratus* (Say) (all Hymenoptera: Ichneumonidae). A cytoplasmic polyhedrosis virus has been isolated from banded woollybear larvae; infected larvae displayed significantly slower rates of development (Boucias and Nordin, 1978).

Life Cycle and Description. The phenology of banded woollybear is not well known, but there appears to be at least two generation in the southern United States, and only a single generation in the northern states and southern Canada (Goettel and Philogene, 1978a). Thus, there are reports of a single generation in Ontario, Quebec, the New England states, and New York, but two generations in Illinois.

Egg. Banded woollybear deposits spherical eggs in clusters of 50–100 eggs. Duration of the egg stage is about six days at 27°C.

Larva. Larval development is rather variable in this species. When reared on artificial diets, the larvae display 7–10 instars, with development times of about 3, 3, 3.2, 3.9, 4.9, 6.9, 14, 21 (and if present) 21 days, respectively for instars 1–9 (Goettel and Philogene, 1978b). These same authors (Goettel and Philogene, 1979) also provided head capsule measurements for larvae with 7–10 instars. Head capsule widths for larvae with eight instars, a common number, are 0.4, 0.5, 0.8, 1.2, 1.7, 2.3, 2.9, and 3.7 mm, respectively. Instar number is influenced by photoperiod but not by temperature; exposure of larvae to greater than 14 h photophase results in an increased number of instars (Goettel and Philogene, 1978a).

The banded woollybear is thickly clothed with stout bristles. The head and body are black, with black or reddish-brown spines covering the body. Typically the caterpillar's bristles are brown at the middle of the body, and black at both the anterior and posterior ends of the caterpillar. Young larvae tend to be about two-thirds black, with the amount of black dissipating as larvae mature. Sometimes only the head end remains black, and in California caterpillars are sometimes uniformly brown. The larva overwinters and is often observed in the autumn as it disperses in search of a suitable overwintering shelter. If cold weather commences early in the year, the larvae are relatively immature, and thus possesses a disproportionately large amount of brown coloration as compared to years with long summers, when the larvae are more mature. Thus, there is some meteorological basis to the American belief that banded woollybear foretells the weather. Banded woollybear larvae tend to have mostly one type of hair, of equal length, and have a rather short-cropped look as compared to some other long-haired species such as yellow woollybear, *Spilosoma virginica* (Fabricius). There may be a few longer hairs at both the anterior and posterior ends of the caterpillar, but this does not detract from the overall impression of uniform hair length. The mature larva attains a length of about 30 mm, and has a tendency to roll into a ball if disturbed.

Pupa. Banded woollybear overwinters as a larva in leaf debris, under loose bark and similar shelter, and in the spring it feeds briefly before pupating. The pupal case is constructed principally from hairs from the caterpillars body, spun loosely together with silk. The coccoon is dark because it consists mainly of black and brownish body hairs. The pupa is light to dark brown. Pupation usually requires 14–21 days.

Adult. The moth is yellowish-orange to light brown, with one to several dusky spots on the front wings and two on each hind wing, and has a wingspan of about 45 mm. In males, the hind wing tends to be paler. The abdomen is marked with three rows of black spots. Banded woollybear courtship is unusual in that it involves acoustical signals produced by

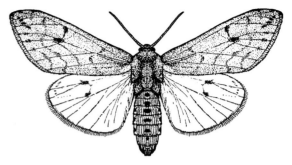

Adult banded woollybear.

females in response to male production of pheromone (Krasnoff *et al.*, 1987; Krasnoff and Yager, 1988).

The biology of this species is poorly documented. Banded woollybear biology was discussed by Saunders (1873) and a method of culture was given by Goettel and Philogene (1978b).

Damage

Larvae are defoliators, but they feed principally on weeds and other noncrop plants. Young larvae are gregarious, feeding together on the underside of foliage and skeletonizing the plant tissue. Larger larvae disperse and feed sporadically, creating irregular holes in foliage.

Management

It is highly unusual to experience banded woollybear in enough abundance to warrant concern. However, they are easily killed with foliar insecticides if it becomes necessary.

Saltmarsh Caterpillar
Estigmene acrea (Drury)
(Lepidoptera: Arctiidae)

Natural History

Distribution. This native insect is found widely in North and Central America. It is abundant enough to be damaging everywhere in the United States, and in Canada it damaged crops in Ontario and Quebec. It is most serious as a pest in the southern United States, particularly the southwest.

Host Plants. Saltmarsh caterpillar's peculiar common name is derived from initial description as a pest of salt-grass hay grown in the vicinity of Boston. This is an anomaly, and despite the wide host range of this insect, grasses are not particularly preferred. Broadleaf weeds are the normal host plants, but larvae commonly disperse from these late in the growing season to damage vegetable and field crops. Vegetables

Banded woollybear larva.

injured include asparagus, bean, beet, cabbage, carrot, celery, corn, lettuce, onion, pea, tomato, turnip, and probably others. Field crops damaged are alfalfa, clover, cotton, soybean, sugarbeet, and tobacco. The favored weed host seems to be pigweed, *Amaranthus* spp., but many others may be consumed, including anglepod, *Gonolobus* sp.; sicklepod, *Cassia tora*; dog fennel, *Eupatorium capillifolium*; ground cherry, *Physalis* spp.; and mallow, *Anoda* sp.

Natural Enemies. Saltmarsh caterpillar larvae frequently are parasitized, particularly by tachinids (Diptera: Tachinidae). In Arizona, the most common parasitoids were *Exorista mellea* (Walker) and *Leschenaultia adusta* (Loew), but *Gymnocarcelia ricinorum* Townsend and *Lespesia archippivora* (Riley) were also observed (Taylor, 1954). Jackson *et al.* (1970) documented the biology and importance of *L. adusta*. Arnaud (1978) reported additional species of tachinids associated with saltmarsh caterpillar. Hymenopteran parasitoids are known from both the larval and egg stages (Taylor, 1954; Taylor and Stern, 1971), and include *Apanteles diacrisiae* Gahan (Braconidae); *Therion fuscipenne* (Norton), *T. morio* (Fabricius), *Casinaria genuina* (Norton), *Hyposoter rivalis* (Cresson) (all Ichneumonidae); *Psychophagus omnivorus* (Walker), *Tritneptis hemerocampae* Vierick (both Pteromalidae); *Anastatus reduvii* (Howard) (Eupelmidae); and *Trichogramma semifumatum* (Perkins) (Trichogrammatidae). A cytoplasmic polyhedrosis virus is known (Langridge, 1983), but there are little data on importance. General predators such as lady beetles (Coleoptera: Coccinellidae), softwinged flower beetles (Coleoptera: Melydridae), and assassin bugs (Hemiptera: Reduviidae) prey on these caterpillars, but are not thought to be very important in population regulation (Young and Sifuentes, 1959).

Life Cycle and Description. Total generation time requires 35–40 days under ideal conditions, but most reports from the field suggest about six weeks between generations. The number of generations per year is estimated at one in the northern states to 3–4 in the south. Overwintering reportedly occurs in the mature larval stage, with pupation early in the spring. Saltmarsh caterpillars usually are infrequent early in the season, but may attain high numbers by autumn.

Egg. The eggs are nearly spherical in shape, and measure about 0.6 mm in diameter. Initially they are yellow, but soon become grayish. Females commonly produce 400–1000 eggs in one or more clusters. It is not unusual to find a single egg mass containing 1200 eggs. Eggs hatch in 4–5 days.

Larva. There are 5–7 instars. This description is based on Hinds (1904), who observed 5 instars in Texas. Upon hatching the larvae are about 2 mm long, brown in color, and bear numerous long hairs over the entire length of the body. During this stage, and the subsequent instar, larvae feed gregariously on the lower leaf surface, usually failing to eat entirely through the leaf. Larvae attain a length of about 10 mm during the first instar. Second instars display longitudinal stripes, usually brown, yellowish, and white, and the body hairs become darker. Larvae attain a length of about 15 mm. During the third instar, larvae become darker, but a consistent color pattern is not apparent. Larvae attain a length of about 30 mm. In the fourth and fifth instars, larvae maintain the same general appearance as earlier stages, but grow to a length of about 45 and 55 mm, respectively. Larvae usually are dark, but sometimes are yellowish-brown or straw colored. The larvae are marked by long body hairs, and these too range from cream or grayish to yellowish-brown to dark-brown. Although they are decidedly hairy, the hairs are not as dense or stiff as those found in woollybear larvae. Duration of larval development was 24–37 days. In contrast, Young and Sifuentes (1959) and Capinera (1978b) reported six instars in Mexico and Colorado, respectively. Development time of the six instars was about three, two, two, two, three, and eight days, respectively, for a larval period of 20–22 days, depending on diet. However, some studies have reported longer larval periods, up to about 45 days.

Larvae are active dispersers, a habit that is relatively uncommon among caterpillars. Most commonly, late instar larvae are found individually or in large numbers ambling over the soil, searching for suitable food. Damage to margins of crop fields often occurs as such larvae desert drying weeds for irrigated crops. Stracener (1931) reported that young larvae drop readily from plants when disturbed, spin a strand of silk, and are blown considerable distances by wind. Frequency of distribution by wind is unknown. (See color figure 87.)

Pupa. Pupation occurs on the soil among leaf debris, in a thin cocoon formed from silken hairs interwoven with caterpillar body hairs. The dark brown

Saltmarsh caterpillar larva.

pupa measures about 30 mm long. Duration of the pupal stage is about 12–14 days.

Adult.   Adults are fairly large moths, measuring 3.5–4.5 cm in wingspan, and are distinctive in appearance. They are predominantly white, though generally the wings bear numerous, small, irregular black spots. The hind wings of the male are yellow; those of the female are white. The underside of the male's front wings may also be tinted yellowish. Most of the abdominal segments are yellow, and bear a series of large black spots dorsally. Mating occurs the evening following emergence, and egg deposition the next evening. Females usually live only 4–5 days, but may produce more than one cluster of eggs. (See color figure 216.)

Accounts of saltmarsh caterpillar biology were provided by Hinds (1904), Stracener (1931), and Young and Sifuentes (1959). Rearing methods were given by Dunn *et al.* (1964) and Vail *et al.* (1967a).

Damage

Larvae are defoliators. Young larvae feed gregariously and skeletonize foliage. Older larvae are solitary and eat large holes in leaf tissue. On celery, most feeding is restricted to leaf tissue, unlike many other lepidopterous species, which may also feed readily on petioles (stalks) (Jones and Granett, 1982). Older larvae may disperse long distances in search of food, sometimes moving in large numbers. Commonly this is associated with maturation of cotton or weeds in the autumn. Thus, these caterpillars tend to be damaging to fall-planted vegetable crops. Foliage consumption at least doubles with each succeeding instar, and mature larvae can consume over 13 sq cm of thick-leaved foliage, such as sugarbeet, daily (Capinera, 1978b). Capinera *et al.* (1987) measured bean foliage consumption by each instar, and recorded over 400 sq cm of foliage consumed during the life of a caterpillar. Further, they estimated that 1.0–1.5 mature

caterpillars per plant could inflict 20% defoliation, a level adequate to cause yield loss.

Management

Insecticides are commonly used to suppress saltmarsh caterpillars if they become abundant in vegetable crops. Baits are not effective. Most damage occurs at field margins as larvae disperse into crops from nearby senescent vegetation. Both chemical insecticides and *Bacillus thuringiensis* are recommended. Physical barriers, including ditches or trenches with steep sides, can be used to interrupt invasion of crops by caterpillars.

Yellow Woollybear
Spilosoma virginica (Fabricius)
(Lepidoptera: Arctiidae)

Natural History

Distribution.   Yellow woollybear is a native insect, and is found throughout the United States and southern Canada. Yellow woollybear's range as a pest is generally restricted to the Great Plains region to the west coast. Even within this area, however, it infrequently is numerous enough to be damaging.

Host Plants.   Yellow woollybear is a very general feeder, and reported from over 100 different plants. Yellow woollybear has been observed to damage such vegetable crops as asparagus, bean, beet, cabbage, cantaloupe, carrot, cauliflower, celery, corn, eggplant, lima bean, parsnip, pea, potato, pumpkin, radish, rhubarb, squash, sweet potato, Swiss chard, turnip, and watermelon. Other economic plants damaged include field crops such as alfalfa, peanut, and sugarbeet; fruits such as blackberry, cherry, currant, gooseberry, grape, and raspberry; and flowers such as canna, dahlia, geranium, hollyhock, hyacinth, and verbena. Among weeds fed upon are dandelion, *Taraxacum officinale*; dock, *Rumex* sp.; pigweed, *Amaranthus* spp.; lambsquarters, *Chenopodium album*; plantain, *Plantago major*; Russian thistle, *Salsola kali*; Spanish needle, *Bidens bipinnata*; and sunflower, *Helianthus* spp.

Natural Enemies.   The number of parasitoids found in association with this insects is quite large, though the importance of these natural enemies has not been well-studied. Arnaud (1978) listed several tachinids reared from yellow woollybear, including *Aplomya caesar* (Aldrich), *Blondelia hyphantriae* (Tothill), *Bombyliopsis abrupta* (Wiedemann), *Carcelia diacrisiae* Sellers, *C. reclinata* (Aldrich and Webber), *Compsilura concinnata* (Meigen), *Exorista mella* (Walker), *Gymnocarcelia ricinorum* Townsend, *Hubneria estigmenensis*

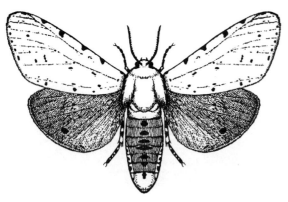

Adult male saltmarsh caterpillar.

(Sellers), *Lespesia aletiae* (Riley), *L. frenchii* (Williston), *Mericia ampelus* (Walker), *Thelaira americana* Brooks, and *Winthemia datanae* (Townsend) (all Diptera: Tachinidae). Wasps reared from yellow woollybear include *Apanteles diacrisiae* Gahan, *A. scitulus* Riley, (both Hymenoptera: Braconidae); *Coccygomius sanguinipes* (Vierick), *Cratichneumon unifasciatorius* (Say), *Vulgichneumon subcyaneus* (Cresson), *Therion morio* (Fabricius), *Hyposoter rivalis* (Cresson), *Enicospilus glabratus* (Say) (all Hymenoptera: Ichneumonidae); *Psychophagus omnivorous* (Walker), *Tritneptis hemerocampae* Vierick (both Hymenoptera: Pteromalidae); *Elachertus marylandicus* Girault, *E. spilosomatis* Howard (both Hymenoptera: Eulophidae); *Telenomus nigriscapus* Ashmead, and *T. spilosomatis* Ashmead (both Hymenoptera: Scelionidae). The fungus *Beauveria bassiana* has been reported to cause low levels of mortality, and a granulosis virus has been observed (Boucias and Nordin, 1977).

Life Cycle and Description. There likely are three generations of yellow woollybear annually, despite the numerous reports of only two generations. The discrepancy is due to the overlapping flights of the moths from overwintering pupae with those of the spring generation. In the most complete study of yellow woollybear population dynamics, conducted in Iowa, the apparent spring flight of moths was shown to consist of two reproductive populations, each represented by separate peaks in abundance within the overall spring flight period. There also was a late summer flight which produced overwintering pupae (Peterson *et al.*, 1993). Elsewhere, moths also are abundant in spring (April–June) and autumn (July–October), but there is considerable geographic variation in the timing of flights. For example, the early-season flight activity occurs in mid-April in Arkansas and North Carolina, but not until late May in Maine. Late-season flight occurs in August in Maine, September in North Carolina, and October in Arkansas. The pupal stage reportedly overwinters throughout this insect's range.

Egg. The spherical eggs of yellow woollybear are yellow, 0.6 mm in diameter, and deposited in clusters of 50–200 both on plant foliage and on inedible substrata. Duration of the egg stage is about seven days. (See color figure 256.)

Larva. Upon hatching, the larvae are hairy and bluish white. During the first two instars the larvae feed gregariously, and then disperse. As they mature, they retain their hairy characteristic, and develop both long, fine, soft hairs and shorter, stout bristles. The hairs are not so thick as to hide the larval body, which is quite variable in coloration. The most common color for larvae is yellow, which is the basis for the common

name, but they also may be cream, light brown or dark brown. Also, there is a dark line along each side of the caterpillar, and the membrane between each body segment tends to be marked by dark pigment. The head is principally yellow, and this character is useful to distinguish it from saltmarsh caterpillar, *Estigmene acrea* (Drury), which tends to have a black head. Larvae attain a length of up to 5 cm at maturity. Larvae reared on bean foliage at 25°C required about 39 days to progress through nine instars (Capinera *et al.*, 1987). Instar duration was 3.1, 2.4, 3.9, 4.9, 3.6, 3.6, 3.6, 4.6, and 10.0 days for instars 1–9, respectively. Although Peterson *et al.* (1993) suggested that this long development time or large instar numbers might be indicative of a suboptimal host, Dethier (1987) similarly reported larval development times of 35 days on suitable host plants. Larvae often move to another plant after completing a meal, even though the plant is relatively suitable for growth and development. They spend less than 1% of their time in eating, 2.5% in wandering, and the remaining of their existence in resting (Dethier, 1987). (See color figure 88.)

Pupa. Pupation occurs in plant debris, under bark of trees, and in other sheltered locations. The pupal case is constructed from the larval hairs, which is held together loosely with silk. Duration of the pupal stage is 7–14 days, and the reddish-brown pupa measures about 15–16 mm long.

Adult. Adults are medium-size moths measuring about 38–50 mm in wingspan. The wings are white, but the front wings bear a small black spot near the center, and the hind wings usually are with three black spots. The head and thorax are covered with white scales. The abdomen is yellow-orange with three rows of black spots, one row dorsally and one on each side.

The biology of yellow woollybear is not well documented. Brief treatment of yellow woollybear was provided by Riley (1871), Marsh (1912b), and Maxson (1948).

Damage

Larvae are defoliators. Young larvae are gregarious, and tend to feed together on the underside of foliage

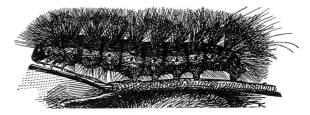

Yellow woollybear larva.

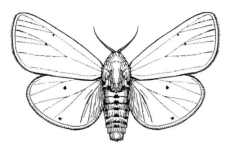

Adult yellow woollybear.

and skeletonize the plant tissue. Larger larvae disperse and feed sporadically, creating irregular holes in foliage. When larvae are particularly numerous, and succulent vegetation scarce, many of the large larvae remain on crops and inflict injury. This most often occurs in irrigated cropland when adjacent weedy vegetation senesces, or dries up due to drought. Typically it is only the late summer generation that attains densities adequate to inflict injury. Capinera *et al.* (1987) measured bean foliage consumption by each instar, and recorded over 300 sq cm of foliage consumed during the life of a caterpillar. Further, they estimated that 1.2–2.2 mature caterpillars per plant could inflict 20% defoliation, a level adequate to cause yield loss.

Management

Yellow woollybear is common among weeds growing along roadsides, fence rows, and irrigation ditches. Larvae disperse into crops only when native or weedy vegetation is depleted or otherwise unsuitable. Such sources of infestation should be monitored. Larvae are easily killed with foliar insecticides, though this is rarely warranted. Treatment of the source of infestation or the borders of crops is generally adequate to prevent damage. Burning of crop residues in the autumn is sometimes recommended to destroy overwintering larvae and pupae because they are located aboveground and very susceptible to fire. However, it is normally better to leave such organic matter on the soil surface, or tilled into the soil; the exception might be ditch banks or other small areas that are not tilled.

FAMILY GELECHIIDAE—LEAFMINER MOTHS

Eggplant Leafminer
Tildenia inconspicuella (Murtfeldt)
(Lepidoptera: Gelechiidae)

Natural History

Distribution. Eggplant leafminer is present in the southeastern and midwestern United States north to about New Jersey and Iowa, and west to Nebraska and Texas. This species has a rather confused history, but records from most western states appear to be due to related species. Eggplant leafminer is a native insect.

Host Plants. Eggplant leafminer apparently limits its attacks to eggplant and horsenettle, *Solanum carolinense*. Horsenettle is the natural host.

Natural Enemies. Several parasitoids are known, and apparently effective in keeping this insect from becoming very numerous. Among the common species are *Apanteles epinotiae* Viereck, *Bracon gelechiae* Ashmead, *Cardiochiles* sp., *Macrocentrus delicatus* Cresson, *Orgilus mellipes* Say, *Agathis gibbosa* (Say) (all Hymenoptera: Braconidae); and *Chrysonotomyia* sp. and *Miotropis* sp. (both Hymenoptera: Eulophidae). *Cirrospiloides bicoloriceps* (Girault) and *Campoplex phthorimaeae* (Cushman) also parasitize this leafminer. A eumenid, *Parancistrocercus fulvipes* (Saussure) (Hymenoptera: Eumenidae), has been observed to prey upon larvae by digging them from their tunnels. Gross and Price (1988) observed mean parasitism rates of about 33% in Illinois, and equivalent levels of parasitism on larvae developing in both eggplant and horsenettle.

Life Cycle and Description. Eggplant leafminer can complete its development, from the egg to the adult stage, in about 25 days when cultured at 27°C. In Illinois, eggplant leafminer was active from June–September, and underwent three generations.

Egg. The eggs are deposited singly on the leaf surface, with deposition occurring on both the upper and lower surfaces, but the lower surface is heavily favored. They are somewhat cylindrical, but with rounded ends. Mean egg length (range) is 0.34 mm (0.30–0.38 mm); mean width (range) is 0.19 mm (0.16–0.21 mm). Egg color is yellow. Duration of the egg stage is about seven days.

Larva. Larvae burrow within the leaf along the edge of the leaf blade. If the larvae do not hatch at the leaf edge, they may construct a small linear mine near the egg, but soon relocate to the leaf margin. Other than to move to the leaf edge, larvae do not leave the mine and do not web together foliage. Larvae form a blotch-shaped mine and deposit feces and silk within the mine. There are five instars, with mean head capsule widths of 0.17, 0.27, 0.38, 0.57, and 0.76 mm for instars 1–5, respectively. The larva initially is white or pale yellow except for the head and thoracic shield, which are brown. By the third instar the larva acquires a brownish or greenish color, and in the fourth or fifth instar becomes dark green, turquoise, or dark blue. The thoracic legs are light in

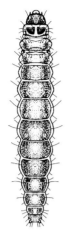

Eggplant leafminer larva.

Adult eggplant leafminer.

color. Mature larvae attain a length of 7–8 mm and are slightly flattened in form.

Pupa. When ready to pupate, mature larvae spin down to the soil on a strand of silk. Pupation normally occurs within a silken cocoon in the soil, and usually quite close to the soil surface. The pupa is dark blue when first formed, but becomes dark brown with maturity. Eggplant leafminer pupae have no distinctive features, resembling most moth pupae. They measure 3.4–5.2 mm long and 1.0–1.7 mm wide.

Adult. The adult is a small grayish brown moth that is marked with yellowish brown. The forewings, but especially the hind wings, bear a long fringe of hairs. The wingspan measures 10–14 mm. Adults are nocturnal.

Eggplant leafminer is quite similar in appearance to potato tuberworm, *Phthorimaea operculella* (Zeller), and they share common hosts. However, eggplant leafminer tends to mine leaf margins, whereas potato tuberworm attacks the central areas, particularly the area of the main veins. Also, eggplant leafminer does not leave the mine to web leaves. Poos and Peters (1927) gave the morphological characters useful in distinguishing the species, but accurate determination is best accomplished by an authority.

The biology of eggplant leafminer was given by Jones (1923) and Gross (1986). A key that included eggplant leafminer larvae (as *Keferia glochinella* Zeller) is provided by Capps (1946).

Damage

Though commonly found mining eggplant in southern states, eggplant leafminer is not considered to be a serious pest. The leaf edge is preferentially mined, acquiring a dry, sometimes swollen, blotch.

Management

This insect is not known to be a serious pest and its presence should not normally be cause for concern. However, insecticides applied to the foliage should be effective if suppression is warranted.

Potato Tuberworm
Phthorimaea operculella (Zeller)
(Lepidoptera: Gelechiidae)

Natural History

Distribution. Potato tuberworm occurs widely in the United States, but normally is absent the northernmost states and Canada. It is most common in southern areas, particularly California and the southeastern states north to Maryland. Potato tuberworm appears to be native to North America, but is now spread throughout the world. Areas with warm, dry climates such as southern Europe, northern and southern Africa, India, Australia, and Central and South America all experience problems with potato tuberworm. The transport of tubers infested with insects causes extensive dissemination of this pest, and also results in occurrence records where this insect does not exist permanently.

Host Plants. This insect feeds almost entirely on members of the plant family Solanaceae. Vegetable crops supporting potato tuberworm include eggplant, pepper, potato, and tomato, though potato is the only frequent host. Tobacco is occasionally affected, and potato tuberworm is sometimes called "tobacco splitworm" when it is associated with this host. Solanaceous weeds such as bittersweet, *Solanum dulcamara*; black nightshade, *S. nigrum*; groundcherry, *Physalis* spp.; henbane, *Hyoscyamus* sp.; horsenettle, *S. carolinense*; jimson weed, *Datura stramonium*; and matrimony vine, *Lycium europaeum*; also serve as hosts. A world-wide host list was provided by Das and Raman (1994).

Natural Enemies. Natural enemies affect the egg, larval, and pupal stages of potato tuberworm, though they are much more effective when the tuberworms are feeding on the aerial portions of the plant rather than within tubers. Among the parasitoids known to affect potato tuberworm are numerous species of Braconidae, Encyrtidae, Eulophidae, Ichneumonidae, Mymaridae, Pteromalidae, Scelionidae, and Trichogrammatidae (all Hymenoptera). Reported to be the most abundant in Virginia were *Bracon gelichiae* Ashmead (Braconidae) and *Campoplex* sp. (Ichneumonidae) (Hofmaster, 1949). In California, *Apanteles dignus* Muesebeck (Braconidae) was the dominant parasitoid (Oatman and Platner, 1989), though in an earlier report by the same authors (Oatman and Platner, 1974), *Agathis gibbosa* (Say), *A. scutellaris* Muesebeck, and *Campoplex phthorimaeae* (Cushman) (all Hymenoptera: Braconidae) were most common. Many of the parasitoids of tuberworm were discussed and pictured by Graf (1917). There have been several attempted introductions of parasitoids to North America, but with few successes (Clausen, 1978).

Other natural enemies are less important. Several general predators have been noted to feed on tuberworm, including the ants *Pheidole* and *Lasius* spp. (Hymenoptera: Formicidae), pirate bugs (Hemiptera: Anthocoridae), shield bugs (Hemiptera: Pentatomidae), and rove beetles (Coleoptera: Staphylinidae). Diseases have been noted (Briese and Mende, 1981; Trivedi and Rajagopal, 1992), but seem to be of little natural significance.

Weather. Weather is thought to affect the abundance of potato tuberworm. Summers that are unusually warm and dry favor increase in tuberworm populations (Langford and Cory, 1932; Hofmaster, 1949).

Life Cycle and Description. A life cycle may be completed in 15–90 days, resulting in about five generations annually in both California and Virginia; in other locations around the world the number of generations is reported to range from 2 to 13 annually. In warm climates, the generations overlap and cannot be distinguished easily. Potato tuberworm normally cannot withstand freezing, so in cold climates overwintering survival by larvae is poor except within potatoes in storage or in cull piles.

Egg. The eggs are deposited singly or in poorly defined clusters, usually on the underside of leaves. If deposited on potatoes in storage, however, the egg clusters tend to be larger, up to 30 eggs. Tubers tend to be heavily infested if they are exposed, that is, not covered with soil. There also are reports of oviposition on soil adjacent to plants (Traynier, 1975). The egg is elliptical, and measures about 0.48 mm long and 0.36 mm wide. Initially white in color, they turn yellow and acquire a distinct iridescence with age. Duration of the egg stage is only about five days during the summer but may reach 30 days during cool weather.

Larva. At hatching, larvae normally begin to burrow almost immediately. Larvae normally mine the leaves, but occasionally the petioles and stems, and sometimes burrow into tubers. The older or lower leaves are preferred. Larvae often plug the entrance to their burrow with excrement, but extrude the cast skins and head capsules. Sometimes considerable amounts of silk are produced by larvae, usually when they are forced to traverse the leaf surface, but also to plug larval burrows and to web together leaves. There are four larval instars. Mean head capsule widths are about 0.20, 0.36, 0.60, and 1.13 mm for instars 1–4, respectively. Body lengths are about 1.1, 2.0, 4.5, and 7.0 mm, respectively. Mean duration (range) of the instars is about 3.5 (2–6), 2.5 (2–3), 3.1 (2–4), and 7.3 (5–12) days, respectively. Initially white in color with a black head and thoracic plate, the larva acquires additional color as it grows. In the mature larva, the head, thoracic plate and thoracic legs are black. The body is principally white, with pink or greenish-pink dorsally. There are five pairs of prolegs. The anal plate is yellow. Duration of the larval period may require only 14 days during the summer, but up to 70 days during the winter months.

Pupa. Pupation occurs in the soil, or just beneath the epidermis of the leaf or tuber. Before pupation, the larva spins a silk cocoon that is usually covered with leaf trash, fecal material, soil particles, and other debris. Initially white or yellow, the pupa eventually becomes dark mahogany. The form of the pupa is typical of Lepidoptera, wider at the anterior end, tapering

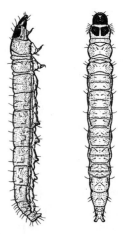

Potato tuberworm larvae.

Potato tuberworm pupa.

Adult potato tuberworm.

Adult potato tuberworm.

to a point at the posterior end, and with the partially developed wings twisted ventrally. The tip of the abdomen bears a hook and a circle of spines. It measures about 6 mm long. Mean duration (range) of the pupal period is 11.6 days (range 8–14 days).

Adult. The adult stage is a small grayish-brown moth with a wingspan of 12–16 mm. The wings, especially the hind wings, are fringed. The front wings are marked with dark spots, which usually coalesce to form a dark longitudinal streak or a row of dark spots. The wings, abdomen, and legs also are tinged with yellow scales. The moths are nocturnal, hiding during the day beneath debris and clods of soil. Mating occurs within 2 days of moth emergence. Oviposition is usually completed in 6–17 days, with females each producing about 150–250 eggs. Longevity rarely extends beyond 21 days. A sex pheromone has been identified and can be used for trapping under field conditions (Persoons *et al.*, 1976).

Potato tuberworm is easily confused with eggplant leafminer, *Tildenia inconspicuella* (Murtfeldt). These species are similar in appearance and have overlapping host range. Potato tuberworm moths are usually larger, with yellow scaling more distinct and forming longitudinal streaks, but accurate differentiation is best accomplished by an authority. Poos and Peters (1927) presented differences in the genitalia of adults, and discussed procedures to distinguish the other stages. Eggplant leafminer does not attack potato or tobacco though it shares eggplant and horsenettle with potato tuberworm. On eggplant and horsenettle, mines of potato tuberworm begin at the midrib or one of the principal veins, whereas mines of eggplant leafminer begin near the leaf margin.

Excellent treatment of potato tuberworm biology was given by Graf (1917) and Poos and Peters (1927). The reports of Clarke (1901) and Hofmaster (1949) also

were useful. A brief world-wide review of this insect was published by Trivedi and Rajagopal (1992). Culture of potato tuberworm was described by Platner and Oatman (1968).

Damage

Leaf mining is the most common habit of potato tuberworm, but mining of the tuber is the most damaging. Mining normally is restricted to the foliage so long as it is green and succulent. Larvae may also mine the stems, usually working downward. If the tuber is attacked, the mining may occur near the epidermis, or "skin" of the tuber, or larvae may burrow deeply. The tunnels in potato tubers normally fill with fungus (Graf, 1917). Tunneling not only destroys the food quality of the tubers, but also the sprouting potential of tubers that are used for propagation. In tomato, foliage is initially attacked, but larvae can mine through the fruit stem into the fruit (Gilboa and Podoler, 1995).

Management

Sampling. Pheromone traps are effective for monitoring potato tuberworm populations, and usually there is a good correlation between trap catches and damage levels (Shelton and Wyman, 1979b; Yathom *et al.*, 1979). Pheromone-baited water pan traps are more effective than pheromone-baited sticky traps (Bacon *et al.*, 1976), though funnel traps seem to be as effective as water traps (Raman, 1988). Sticky traps are prone to be covered with dust, thereby reducing catch (Kennedy, 1975). Larval sampling was discussed by Horne (1993), and a binomial sequential sampling plan for tuberworm in tomato was developed by Gilboa and Podoler (1995).

Insecticides. Tuberworm often is controlled by application of insecticide to foliage (Bacon, 1960), though in some parts of the world resistance to insecticides is a problem (Collantes *et al.*, 1986). Also, insecticides interfere with predators and parasitoids of tuberworm, which can be quite effective, so it is prudent to determine that tuberworm is present in potentially damaging numbers before implementing an insecticide-based management effort (Shelton *et al.*, 1981). Integration of chemical insecticides with cultural practices is effective (Fuglie *et al.*, 1993). Biological insecticides, particularly the bacterium *Bacillus thuringiensis*, are recommended for protection of potato tubers in storage, but not usually in the field. Suppression in the field is possible, but several applications may be required (Broza and Sneh, 1994). A granulosis virus has been used experimentally as a suppressive bioinsecticide under field conditions (Kroschel *et al.*, 1996).

Cultural Practices. Cultural practices can greatly affect susceptibility of potato to potato tuberworm. Overwintering population tend to be low, with tuberworm populations increasing through the year. Thus, areas where more than one potato crop are cultivated tend to experience greater loss by tuberworm, and greatest damage occurs late in the season. In some regions, potato production is limited to the spring months to eliminate the nearly year-long availability of potatoes for tuberworm breeding.

Sanitation is extremely important in potato tuberworm management. Potatoes held in storage or in cull piles are potential sources of infestation. Similarly, potatoes left in the field, volunteer plants, and solanaceous weeds can support tuberworms. Harvested potatoes should not be left in the field overnight as this is when oviposition occurs.

If vines are killed before senescence and tubers harvested soon thereafter, the level of tuber infestation is low. Delayed harvest increases the exposure of tubers to ovipositing moths. Infestation of tubers is especially likely if there are cracks in the soil, allowing access by tuberworm. Soil depths of 5 cm or more protect tubers from infestation. Sandy soil can also be a problem if rainfall washes away soil, exposing tubers. Irrigation practices greatly affect soil condition, with furrow irrigation producing more cracks than overhead irrigation. Frequent irrigation helps to prevent soil cracking. Hilling of the soil, wherein soil is scraped from between the rows and deposited at the base of the plants, helps to deny access by tuberworm to tubers (Langford, 1933; Shelton and Wyman, 1979a,b; Von Arx *et al.*, 1990). Deep planting of potato seed, and

culture of varieties that do not produce shallow tubers, also reduce incidence of tuberworm damage.

Some differences in tuber susceptibility or suitability exist among cultivars (Fenemore, 1980). Oviposition preference, percent pupation, and moth fecundity are affected, but the significance of preference diminishes when moths are confronted with a no-choice situation.

The sex pheromone can be used to manipulate populations. Mass trapping can be used to reduce damage in the field. Trapping, and disruption of mating by saturation of the atmosphere and confusion of the moths, works best for potatoes in storage (Raman, 1988).

Tomato Pinworm
Keiferia lycopersicella (Walsingham)
(Lepidoptera: Gelichiidae)

Natural History

Distribution. Tomato pinworm was first found in the United States in southern California in 1923. It now occurs regularly as a field pest in warm areas such as California, Arizona, Texas, and Florida. It also is known from Hawaii, the Caribbean, and Central and South America. Tomato pinworm commonly overwinters in greenhouses, and may be shipped northward in the spring on seedlings cultured in warm areas, so the potential area of infestation is quite large. The origin of tomato pinworm is thought to be Central America.

Host Plants. Tomato pinworm develops only on plants in the family Solanaceae. Tomato is undoubtedly the preferred host, but eggplant and potato also serve as suitable hosts (Schuster, 1989). The role of weeds in tomato pinworm biology is less certain, but among the weeds reported to support pinworm are bitter nightshade, *Solanum dulcamara*; black nightshade, *S. nigrum*; horsenettle, *S. carolinense*; and silverleaf nightshade, *S. elaeagnifolium* (Elmore and Howland, 1943; Batiste and Olson, 1973).

Natural Enemies. Over 20 species of wasp parasitoids of tomato pinworm are known, including species in the families Braconidae, Ichneumonidae, Eulophidae, Pteromalidae, Benthylidae, and Trichogrammatidae (Oatman and Platner, 1989). In southern California, six species are common, but the dominant species there, and elsewhere, usually is *Apanteles dignus* Muesebeck (Hymenoptera: Braconidae). Other important parasitoids include *Apanteles scutellaris* Muesebeck, *Chelonus phthorimaeae* Gahan, *Parahormius pallidipes* (Ashmead) (all Hymenoptera: Braconidae), *Sympiesis stigmatipennis* Girault (Hymenoptera: Eulophidae); and *Campoplex phthorimaeae* (Cushman) (Hymenoptera: Ichneumonidae) (Oatman *et al.*, 1979). Oatman (1970), working in California, observed that parasitism of pinworm increased from practically undetectable in April to about 70% in June; six species contributed to pinworm suppression. Many of the parasitoids affecting tomato pinworm also are associated with potato tuberworm, *Prthorimaea operculella* (Zeller). In Florida, Pena and Waddill (1983) similarly observed an increase in parasitism rates as the season progressed, and also observed the egg parasitoid, *Trichogramma pretiosum* Riley (Hymenoptera: Trichogrammatidae) to be important, particularly in fields with low pinworm densities.

Life Cycle and Description. A generation can be completed in just 30 days under summer conditions. The number of generations is estimated at 7–8 annually in California, with 4–5 occurring during the summer months. The pinworm does not enter diapause, but development slows greatly during cool weather, with the pupal stage most important for survival during cool weather or the absence of hosts.

Egg. The eggs of tomato pinworm are deposited on both the upper and lower surface of leaves in small clusters of 3–7. The small eggs are elliptical, though somewhat flattened where they attach to the substrate. They measure 0.30–0.45 mm long and 0.20–0.25 mm wide, and are very difficult to detect under field conditions. When first deposited they are light yellow, but gradually turn light orange. The lower developmental threshold for eggs is estimated at 11.4°C. Duration of the egg stage is 4–7 days.

Larva. Upon hatching, young larvae spin a small web of silk on the leaf surface, then dig beneath the web and enter the leaf. There are four larval instars. The lower developmental threshold for larvae is about at 11°C. Mean head capsule width (range) is 0.15 (0.14–0.16), 0.25 (0.23–0.28), 0.37 (0.36–0.39), and 0.56 (0.52–0.61) mm, respectively, for instars 1–4. Young larvae are yellowish-gray with a brown head capsule, and measure about 0.85 mm long. As the larva matures it develops dorsally on the abdominal segments a darker pigmentation which is initially orangish or brownish, but eventually purplish. The dark region is irregular in shape, but contains two small circular light spots, and two elongate light spots that connect to the light-colored background. The mature larva measures 5.8–7.9 mm long. The first two instars are leaf miners, but they become too large for mining and third instars construct a leaf fold in which the last two instars dwell. Leaf mining pinworms deposit nearly all their fecal material in a single mass at the entrance to the mine.

This characteristic is useful to distinguish pinworm mining from feeding by dipterous leaf miners, *Liriomyza* spp., because the latter species deposit their feces throughout the mines. (See color figure 90.)

Pupa. Mature larvae drop to the soil on a strand of silk, spin a loosely woven pupal cell intermingled with soil particles, and pupate. Pupae are found at or near the soil surface, about 90% within the upper 1 cm of soil. Cool weather is often passed in the pupal stage. The pupa is green when first formed, but soon turns brown. The pupa measures 4–5 mm long. Duration of the pupal stage is normally 8–20 days.

Adult. The moth is grayish in general appearance, with a wingspan of 9–12 mm. The oval forewings, though mostly gray, are marked with diffuse orangish or brownish longitudinal spots or streaks. The hind wing is a more uniform yellowish-brown, narrow in form, and pointed apically. The forewings, but especially the hind wings, are heavily fringed. Adults live only about 7–9 days and are nocturnal. Mating commences within 24 h of emergence, and most mating occurs shortly after sunset (McLaughlin *et al.*, 1979). Most of the eggs are deposited within 3–4 days of adult emergence.

The biology of tomato pinworm was described by Thomas (1936) and Elmore and Howland (1943). Thermal relations were given by Weinberg and Lange (1980) and Lin and Trumble (1985). Rearing was described by Schuster and Burton (1982) and Burton and Schuster (1986).

Damage

Young larvae mine the leaves of host plants, forming narrow serpentine or straight mines. Later, as larvae mature they leave the mines, fold the leaves, and dwell within the folds, causing large blotch-shaped mines adjacent to each leaf fold. Most important, however, is the tendency of larvae to enter fruit from the area of the stem. In the fruit they may feed shallowly beneath the skin of the fruit, causing blotches, or feed within the fruit, usually in the core. In general, there is a positive relationship between numbers of pinworm larvae and fruit damage, though it is much stronger in the fruit of the lower canopy of the plant than the upper region (Pena *et al.*, 1986).

In warm weather tomato-producing areas such as southern California and Florida, fruit can be severely injured, with infestation levels of 70–100% recorded in the absence of cultural and chemical management of pinworm. In northern areas, such injury can also occur in greenhouse tomatoes.

Management

Sampling. Wellik *et al.* (1979) suggested that the lower canopy stratum, particularly the large fruit, be sampled for pinworm. Pena *et al.* (1986) also suggested sampling the lower canopy because of the strong relationship between foliar injury and yield loss in this stratum. They also indicated that population densities of less than one larva per plant could cause economic loss in Florida due to the high market value of winter tomatoes.

Pheromone traps can also be used to monitor populations (Wyman, 1979) and to time insecticide applications. Van Steenwyk *et al.* (1983), for example, recommended an insecticide application threshold of 10 moths per trap per night. Toscano *et al.* (1987), however, reported a relatively poor relationship between trap catches and larval populations. Sex pheromone can also be applied to crops to confuse the males and disrupt mating (Van Steenwyk and Oatman, 1983).

Insecticides. Applications of insecticide are often made to foliage to prevent increase in pinworm number and damage to fruit (Batiste *et al.*, 1970b; Schuster, 1982). Weisenborn *et al.* (1990) reported that weekly applications of insecticide made after a threshold of about 0.5 larvae per plant was attained provided protection of tomato in California. Application of granular systemic insecticides often is ineffective (Schuster, 1978), and use of the bacterium *Bacillus thuringiensis* is not recommended. Insecticide resistance has developed in many areas (Brewer *et al.*, 1993).

Cultural Practices. Sanitation is a very helpful practice. Tomato plants and cull tomatoes should be destroyed immediately after harvest or pinworm will continue to breed. If crops are grown sequentially, it is highly desirable to grow new crops some distance from old sites. Infested tomato transplants moved from southern to northern areas also serve to transport pinworm into new areas or fields (Batiste *et al.*, 1970a).

FAMILY HESPERIIDAE—SKIPPERS

Bean Leafroller
Urbanus proteus (Linnaeus)
(Lepidoptera: Hesperiidae)

Natural History

Distribution. Bean leafroller is a tropical species, but apparently is a native to the southeastern United States. It is found throughout Florida and in coastal areas from South Carolina west to eastern Texas. It also invades most of the southeastern states, even attaining New England during warm years, and regularly invades the southernmost areas of the Southwest. It cannot tolerate prolonged freezing temperatures, however, and in the United States it persists only in the southern coastal plain, perhaps only southern Florida. The range of bean leafroller also includes Mexico, Central America and the Caribbean, and south to Argentina.

A closely related species, *Urbanus dorantes* Stoll, has expanded its range from the tropical Americas to include the permanent range of *U. proteus* in the United States. Found throughout Florida, and southern Texas and Arizona, it is readily confused with bean leafroller, particularly because it also feeds on legumes, including bean (Heppner, 1975). Thus far, however, it is not particularly abundant. (See color figure 21.)

Host Plants. Bean leafroller feeds exclusively on legumes, and normally is found inhabiting open, disturbed habitats. Vegetable hosts include cowpea, lima bean, pea, and snap bean. Other hosts include soybean; wisteria, *Wisteria* sp.; trefoil, *Desmodium* spp.; butterfly pea, *Clitoria* spp.; and hog peanut, *Amphicarpa bracteata*.

Natural Enemies. Natural enemies are poorly documented. Wasp and fly parasitoids were observed in Colombia (van Dam and Wilde, 1977). Two tachinids with a very wide-host range, *Lespesia aletiae* (Riley) and *Nemorilla pyste* (Walker), have been reported from bean leafroller (Arnaud, 1978). In Florida, *Chrysotachina alcedo* (Loew) (Diptera: Tachinidae) was reared from larvae, and predation was observed by a *Polistes* sp. wasp (Hymenoptera: Vespidae) and *Euthyrhynchus floridanus* (Linnaeus) stink bugs (Hemiptera: Pentatomidae). Also, a nuclear polyhedrosis virus was found to infect and to kill up to 40–50% of larvae late in the season when larvae were numerous (Temerak *et al.*, 1984).

Life Cycle and Description. The bean leafroller can complete its life cycle in about 30 days during warm weather. They breed in southern Florida throughout the year, but are relatively infrequent in northern Florida until June, and only become abundant late in the season, usually September–October. The number of generations is not known, but has been estimated at 3–4 in Florida. In the northern areas invaded during the summer months there usually is only a single generation. In Florida, large numbers of adults are frequently observed migrating southward in the autumn (Balciunas and Knopf, 1977); they make a similar northward migration in the spring and summer, but it is less apparent.

Egg. The eggs of bean leafroller may be deposited singly on the lower epidermis of foliage, but often are found in small clusters of 2–6 eggs. Also, clusters of up to 20 eggs have been observed, and on occasion they may be stacked upon one another to form a column. Initially the eggs are white, but soon turn yellow. The egg is a slightly flattened sphere, and marked with about 12 vertical ridges. It measures about 1 mm in diameter and about 0.8 mm long. Eggs hatch in about 4.5 days when held at 24°C and 2.8 days at 29°C.

Larva. The larva of bean leafroller increases in size from about 2 mm at hatching to 30 mm at maturity. Initially the larva cuts a small triangular patch at the edge of the leaf, folds over the flap, and takes up residence within this shelter. The larva leaves the shelter to feed, and lines the shelter with silk. These flaps are used until the third or fourth instar, when the larva constructs a larger shelter formed by folding over a large section of the leaf or webbing together two separate leaves. Larvae feed nocturnally. Eventually, the larva pupates within the leafy shelter. There are five instars. Mean head capsule widths (range) for the five instars are 0.6 (0.5–0.7), 1.1 (1.0–1.2), 1.8 (1.5–2.1), 3.2 (2.9–3.5), and 4.7 (4.2–5.2) mm, respectively. When maintained at 24°C the mean duration is about 2.8, 3.1, 3.5, 5.6, and 15 days, respectively, for instars 1–5, whereas instar duration at 29°C is about 2.0, 1.7, 2.2, 2.8, and 5.9 days.

Initially the larva is yellowish with a brownish-black head and prothoracic shield, and this general color pattern is maintained though the markings become more distinct as the larva matures, and the larva may also acquire considerable green color. With the molt to the second instar the dorsal surface of the insect is marked with numerous small black spots. Beginning with the third instar, lateral yellow lines become quite distinct. The last two instars are similar to the preceding: brownish-black head, black prothoracic shield, yellowish body sprinkled with black spots but lighter below, and yellow lateral lines. Also evident are orange spots on the head near the base of the mandibles, and red on the ventral portion of the thoracic segment. The body tapers sharply toward both the anterior and posterior ends. Perhaps the most striking attribute of this insect, but a character shared with other members of the family Hesperiidae, is the greatly enlarged head, which is connected to the body by a narrow "neck." (See color figure 96.)

Bean leafroller larvae.

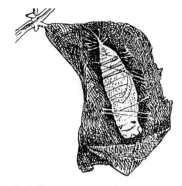

Bean leafroller pupa, partially encased in leaf.

Pupa. The larva pupates on the plant within the shelter formed by the larva from leaf material. The pupa measures about 20 mm long and about 6 mm wide. The pupa is yellow to brown, and is covered by a bluish-white pubescence. Duration of the pupal stage is about 17.1 day at 24°C and only 8.7 days at 29°C.

Adult. The bean leafroller moth is fairly large for a hesperiid, measuring about 50 mm in wingspan. The most pronounced feature is the prolonged extensions or "tails" of the hind wings. Not surprisingly, the butterfly is commonly known as "long-tailed skipper." The front and hind wings are chocolate brown dorsally and pale brown ventrally. The front wings also are marked with 5–7 squares or rectangular white spots. Green iridescent scales are found dorsally on the wings and body. The presence of green scales serves to differentiate bean leafroller from the other bean-feeding skipper, *Urbanus dorantes*. Skippers are active principally at dawn and dusk, often seen darting from flower to flower seeking nectar. (See color figure 202.)

This insect has not been well studied. The most complete description was provided by Quaintance (1898a), but Young (1985) provided valuable observations. Phenology and developmental biology were given byGreene (1970a, 1971a). Greene (1970b) made reference to culture on standard bean-based artificial diet, but the relative success of this approach is not clear.

Damage

Larvae are defoliators, feeding only on leaf tissue of legumes. Greene (1971b) determined that larvae consumed about 0.5, 1.3, 5.1, 26.1, and 162.4 sq cm of foliage during instars 1–5, respectively. Beans can tolerate up to about 30% leaf loss without reduction in yield, so Greene estimated that about 4.4 larvae must complete their development on a "typical" bean plant with 2175 sq cm of foliage to inflict a damaging level of defoliation. As about one-half of the individuals perish in each life stage, Greene estimated that densities of

Adult bean leafroller.

140 eggs or 70 first instar larvae per plant must occur to cause damage.

Management

Sampling. The distribution of eggs and larvae is clumped (Shepard, 1972). Populations are normally sampled by visual observation because the larvae are sheltered within leaf folds and are difficult to dislodge by sweeping, and as the leaf damage caused by shelter-building activity is readily apparent.

Insecticides. Insecticides applied to the foliage at the first sign of damage are very effective for leafroller suppression. This should only be necessary for late-season bean crops. The microbial insecticide *Bacillus thuringiensis* currently is not recommended.

FAMILY LYCAENIDAE—HAIRSTREAK BUTTERFLIES

Gray Hairstreak
Strymon melinus Hübner
(Lepidoptera: Lycaenidae)

Natural History

Distribution. This insect is found throughout the United States and southern Canada. Its range also extends south through Central America to South America. Apparently it is absent from the Caribbean.

Host Plants. The caterpillar stage of gray hairstreak feeds on the widest array of plants of any butterfly, but few are crop plants. Serious damage is limited to bean and cotton. Because of its injury to the latter plant, this species is also known as "cotton square borer." Vegetable crops eaten include cowpea, lima bean, okra, pea, and snap bean. This insect is sometimes associated with such field crops as alfalfa, hops, lespedeza, and sweet clover. Numerous other

plants are consumed by this insect, including such Leguminocae as *Astragalus*, *Casia*, *Desmodium*, *Lupinus*, and *Trifolium* spp.; such Malvaceae as *Hibiscus*, *Malva*, and *Sphaeralcea* spp.; such Polygonaceae as *Eriogonum*, *Polygonum* and *Rumex* spp; and many others. Although the preferred habitat is open areas inhabited by early successional plants and shrubs, most agricultural weeds are exempt from attack by this insect. Scott (1986) provided a list of larval hosts. Adults collect nectar from goldenrod, *Solidago* spp.; dogbane, *Apocynum* spp.; milkweed, *Asclepias* spp.; sweet clover, *Melilotus* spp.; and other flowering plants (Opler and Krizek, 1984). (See color figure 32.)

Natural Enemies. Natural enemies generally are effective at suppression of gray hairstreak numbers, so little crop damage occurs. In Texas, for example, the first generation is abundant, but in subsequent generations caterpillars tend to be less and less abundant as parasitoids take a toll. The most important parasitoid is a gregarious species, *Apanteles thecloe* Riley (Hymenoptera: Braconidae). This species attacks the caterpillar, with larvae emerging from the host to pupate in white silken cocoons on the back of the caterpillar. Other, less important parasitoids are *Octosmicra* sp. (Hymenoptera: Chalcididae), *Aplomya theclarum* (Scudder), and *Lespesia* sp. (both Diptera: Tachinidae). Predators are poorly known, but adults are taken by robber flies (Diptera: Asilidae).

Life Cycle and Description. There are two generations annually in the northern portion of gray hairstreak's range, and 3–4 generations in the central and southern states. About 40 days are required for a complete generation. In Texas, butterflies are observed in all months except December and January. The winter is passed in the pupal stage. Females begin oviposition in February or March in the south, but not until May in the North.

Egg. The eggs are spherical in shape, but slightly flattened. They are pale-green to yellowish-white. They measure 0.6–0.7 mm in diameter, and 0.2–0.3 mm long. They are deposited singly, usually on the underside of leaves. Caged adults have been observed to produce 40–70 eggs each, but this is believed to be an underestimate of true fecundity because this species fares poorly in captivity. Oviposition occurs most frequently in the morning and early evening. Mean duration of the egg stage is 5.5 days (range 4–6 days).

Larva. There are five instars. The mean (range) duration of each instar is 3.6 (3–5), 3.4 (3–5), 3.5 (3–6), 4.8 (3–6), and 7.7 (5–11) days, respectively. Thus, total larval development requires about 23 days. At hatch, larvae measure only about 1 mm long, but even-

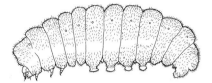

Gray hairstreak larva.

tually grow to measure 12–16 mm long and 4.5–6.5 mm wide. Initially the larvae are slender, but with each succeeding molt the larva becomes relatively broader, so at maturity it is rather stout. The caterpillar in all instars is covered with stout hairs or bristles. The body color is generally green, but sometimes brownish. Larvae normally feed on the leaf surface during the first three instars, and thereafter display a tendency to bore into pods or other reproductive tissues. (See color figure 97.)

Pupa. Near the completion of larval development the larva attaches itself with silk to the plant, usually adjacent to the feeding site. After a quiescent period of 2–3 days the larva molts into the pupa. The pupa is yellow to brown, and bear blackish spots that vary in size, frequency, and intensity. It measures 7.5–10.5 mm long. The sides of the pupa bear spines arranged regularly in rows. Duration of the pupal stage is about 9.6 days (range 8–11 days).

Adult. The adults usually emerge in the morning, and 2–3 days are required before oviposition commences. The butterflies are small, measuring only 25–32 mm in wingspan. In general appearance they are blue-gray. The dorsal surface of the wing is steel-gray, with the lower surface pale-gray. The hind wing is marked dorsally with a reddish-orange spot. The wings are fringed with long scales, and each hind wing bears two thin extensions or tails—one long and the other short. Adults live at least 10 days in confinement, but likely live considerably longer in nature. In southern latitudes, butterflies apparently are dormant during the brief cold periods of winter, and become active during warm weather. Adult males perch on bushes and trees and patrol their territory to watch for intruders. They are aggressive about defending their territory not only from males of the same species, but other butterflies, wasps, and even hummingbirds (Alcock and O'Neill, 1986). (See color figure 203.)

A comprehensive treatment of gray hairstreak biology was given by Reinhard (1929).

Damage

Foliage consumption by early instar larvae is insignificant, but damage may occur as larvae burrow into

Adult gray hairstreak.

bean or okra pods. On beans, larvae often burrow only partially into the pod, then move to another location, which may be on the same or another pod. Only in home gardens is this damage likely to be significant.

Management

These insect cause only incidental damage and infrequently require suppression. Foliar insecticides are generally very effective.

FAMILY LYONETIIDAE—LYONETID MOTHS

Morningglory Leafminer
Bedellia somnulentella (Zeller)

Sweetpotato Leafminer
Bedellia orchilella Walsingham
  (Lepidoptera: Lyonetiidae)

Natural History

Distribution. Morningglory leafminer apparently is a native to Eurasia, but now it is widely distributed in the United States. Morningglory leafminer likely occurs in southern Canada but it is not recorded as a pest there. It is found in most areas of the world, including both tropical and temperate climates. Sweetpotato leafminer is associated with sweet potato in Hawaii, but its origin is uncertain, and it does not occur elsewhere in the United States or Canada.

Host Plants. Larvae of both species feed on plants in the family Convolvulaceae. Among vegetables, only sweet potato is attacked. Wild hosts include bindweed, *Convolvulus* spp., and morningglory, *Ipomoea* spp.

Natural Enemies. Morningglory leafminer is frequently parasitized by *Apanteles bedelliae* Viereck (Hymenoptera: Braconidae). Also reared from morningglory leafminer is *Spilochalcis albifrons* Walsh (Hymenoptera: Chalcididae), but there is some disagreement whether this wasp is a primary parasite or hyperparasite.

A parasitoid identified as *Omphale metallicus* Ashmead (Hymenoptera: Eulophidae) attacks sweetpotato leafminer. In Hawaii this parasitoid is thought to be a significant mortality factor; nevertheless, leafminer can be a serious problem in the absence of insecticides. *Apanteles bedelliae* was introduced to Hawaii to aid in suppression of sweetpotato leafminer, and is reported to be effective (Zimmerman, 1978).

Life Cycle and Description. These species are similar, though because there are some differences in their biology they are treated separately.

In morningglory leafminer, total development time, from the egg to adult stage, requires about 30 days at 18°C, but only 16–17 days at 29°C. Two generations are reported annually in New York.

Egg. The eggs of morningglory leafminer are described as translucent white and a flattened sphere in shape. They measure about 0.3 mm long and 0.2 mm wide. Eggs generally are deposited on the lower surface of leaves, and usually adjacent to leaf veins. They are placed singly or in small clusters of 2–3 eggs. Embryonic development requires about 4.5 days at 27°C.

Larva. Upon hatching, larvae of morningglory leafminer burrow directly into the leaf tissue. They are yellowish-gray though they turn greenish with age. Larvae develop dorsal pink spots during the third instar which disappear in the fourth instar, and are replaced by reddish tubercles in the fourth and fifth instars. White tubercles are added during the fifth instar. First instar larvae appear to be legless, but both thoracic legs and prolegs are evident in subsequent instars. Head capsule widths are 0.09, 0.15, 0.22, 0.31, and 0.45 mm in instars 1–5, respectively. Larvae are somewhat flattened in appearance and the anal prolegs project posteriorly, forming a fork. The first two instars form serpentine mines, but young third instars leave the mine, construct a lose webbing of silk on the lower surface of the leaf, and re-enter the leaf tissue to form a blotch-like mine. The remaining instars also function as blotch miners. An interesting feature of the feeding behavior is that except while molting the larvae use the silk as a point of attachment while feeding, insert the anterior portion of their body into the mine but leave the posterior portion protruding from the mine. At this point the larva voids its fecal material externally. The larva usually forms several blotch mines during its development. Fecal material is voided from the blotch and usually hangs down in a continuously webbed chain. Larval development time is about 11 days at 27°C. (See color figure 91.)

Pupa. Newly formed pupae are greenish with a mottled-red pattern but the red gradually fades and the pupae become greenish or brownish. The posterior end of the pupa is anchored to silk webbing. The anterior end is also supported by silk threads, however, so the pupa is positioned parallel to the leaf rather than hanging from the substrate by its anal end. Duration of the pupal stage is about 4.5 days at 27°C.

Adult. Adults of morningglory leafminer are grayish brown and with fringed wings. The wingspan is about 11.5 mm. Females begin ovipositing soon after mating and deposit about 170 eggs (range 80–324 eggs). Most eggs are laid about 1–8 days after oviposition commences, at about 20 eggs per day. Longevity of adults is about 25 days.

Sweetpotato leafminer is a relatively unimportant pest on a minor crop. Thus, it is perhaps not surprising that little about this insect is known, and that phenology is undocumented. As is the case with morningglory leaf miner, a complete generation requires about 30 days. In Hawaii, it likely occurs whenever its host is available.

Egg. The eggs of sweetpotato leafminer are laid singly in a crevice along a vein, usually on the lower surface of a leaf. The eggs are oval and flattened, and measure 0.3 by 0.2 mm in size. In color the eggs are whitish to reddish and highly iridescent. Duration of the egg stage is about eight days.

Larva. The larval stage of sweetpotato leafminer is small, growing from about 0.4 mm to 7.0 mm long as it matures. The larva tunnels between the upper and lower leaf surfaces as it feeds. The tunnels are quite long and narrow, unlike morningglory leafminer where the mine becomes blotch-like as the larva reaches maturity. As the food supply is exhausted the larva leave its tunnel for a new feeding position and creates a new tunnel. When not in the tunnel the larva moves with a looping motion. The caterpillar is greenish throughout its development, and unlike many Lepidoptera the head, prothoracic and anal shield are not darkly pigmented, though they may have a brownish tint. Moderately sized hairs are sparsely distributed on the body. Duration of the larval stage is about 10 days.

Sweetpotato leafminer larva.

Adult sweetpotato leafminer.

Sweetpotato leafminer pupa.

Pupa. When the sweetpotato leafminer larva is mature it leaves the mine and spins a few silken strands which are used to anchor the pupa to the outside of the leaf. The pupa is dark-green when first formed, assuming a brown color with black spots as it matures. The head bears a dark angular projection, and the eyes are black. The pupa measures about 3.5 mm long. Duration of the pupal stage is about six days.

Adult. The moth of sweetpotato leafminer is grayish-brown, and lacks distinctive markings. The wings, however, are unusual in form. The forewing tapers to a point distally, and the trailing edge bears a long fringe. The hind wing is very narrow, but bears a fringe of long hairs on all sides. The wingspan of the moth is about 7.5 mm.

Morningglory leafminer was described by Shorey and Anderson (1960) and Parrella and Kok (1977). Biology of this species in Egypt seems to be nearly identical (Tawfik *et al.*, 1976). Clemens (1862) gave a good description of the larvae. Biology of sweetpotato leafminer was described by Fullaway (1911); Zimmerman (1978) added useful observations.

Damage

The larvae of both species mine the foliage of sweet potato, leaving long-winding tunnels containing fecal material. As noted earlier, the larva of morningglory leafminer changes its behavior and forms blotch mines later in life. If infestations are heavy the leaves acquire a withered or seared appearance. Sweet potato is quite tolerant of foliar injury, so leafminers is not usually considered to be a serious pest. Parrella and Kok (1977) suggested that hot weather was unfavorable for survival of morningglory leafminer, which might explain why it rarely was reported to be damaging.

Management

Foliar insecticides, particularly systemic materials, can be applied for leaf miner suppression. Except perhaps in Hawaii, these species are rarely abundant, and insecticide applications are used sparingly so as not to disrupt naturally occurring parasitoids.

FAMILY NOCTUIDAE—ARMYWORMS, CUTWORMS, LOOPERS, STALK BORERS AND NOCTUID MOTHS

Alfalfa Looper
Autographa californica (Speyer)
(Lepidoptera: Noctuidae)

Natural History

Distribution. Alfalfa looper is native to North America, and is western in distribution. In the United States it occurs occasionally in Nebraska and Kansas, and frequently in all states further west. It is most damaging along the West Coast. In Canada its distribution is similar, and is considered a pest only in British Columbia and Alberta. Alfalfa looper is also known from Mexico.

Host Plants. This insect feeds on numerous plants. Vegetable crops reportedly injured include beet, bean, cabbage, carrot, cantaloupe, celery, cucumber, lettuce, onion, pea, potato, radish, rhubarb, spinach, squash, tomato, turnip, watermelon, and probably others. The most frequently damaged vegetables are lettuce, bean, and the crucifer crops. Field crops damaged include alfalfa, cotton, red clover, sweet clover, flax, white clover, sugarbeet, and sunflower; as its common name suggests, alfalfa is the most common host. Grasses and grains are not eaten, with the exception of rare feeding on corn. Fruit crops such as apple, currant, gooseberry, raspberry may be damaged. Many of the hosts recorded in the literature are based on outbreak conditions, when hordes of larvae, after totally consuming their preferred hosts, disperse in search of food. Under such conditions, relatively unpreferred plants such as the aforementioned fruit crops may be eaten, but this is not the normal situation. Numerous common weeds support larval development, including dock, *Rumex crispus*.; lambsquarters, *Chenopodium album*; wild lettuce, *Lactuca canadensis*; and many others.

Natural Enemies. Mortality due to natural enemies was estimated at 70–80% in California (Puttarudriah, 1953), of which parasitoids accounted for 30–35% and the remaining from undetermined diseases. The principal parasitoid was *Voria ruralis* Fallén (Diptera: Tachinidae). Also present were *Microplitis* spp. and *Apanteles yakutatensis* Ashmead (Hymenoptera: Braconidae), and *Campoletes sonorensis* (Cameron) and *Patroclides montanus* (Cresson) (Hymenoptera: Ichneumonidae). Other parasitoids, of uncertain importance, were noted by Hyslop (1912a). A nuclear polyhedrosis virus is very common in alfalfa looper populations, usually killing the larvae as they reach maturity.

Life Cycle and Description. The number of generations per year is reported to be two in Alberta and 2–3 in Washington. A complete life cycle is estimated to require 30–40 days. Overwintering occurs in the pupal stage.

Egg. The eggs are deposited singly, usually on the underside of the foliage. They are pale yellow, hemispherical in shape, and marked with narrow vertical ridges. The eggs normally hatch in 3–5 days.

Larva. Larvae have only three pairs of abdominal prolegs, and greatly resemble cabbage looper, *Trichoplusia ni* (Hübner). The body is broader at the posterior end and narrower at the anterior end. As is the case with cabbage looper, the mature larva is predominantly green, but is usually marked with a distinct white stripe on each side. Dorsally, the larva bears several narrow, faint white stripes clustered into two broad white bands. When larvae hatch they measure less than 2 mm long, but at completion of the five

Alfalfa looper larva.

instars they attain a length of about 25–35 mm. Unlike the cabbage looper, the alfalfa looper tends to have dark thoracic legs, and a dark bar on the side of the head, or even an entirely black head capsule. Also, alfalfa looper lacks the small, nipple-like structures located ventrally on the third and fourth abdominal segments of cabbage looper. Alfalfa looper can be distinguished from bilobed looper, *Megalographa biloba* (Stephens), and celery looper, *Anagrapha falcifera* (Kirby) by the absence of numerous dark microspines on the body of alfalfa looper. A key to common vegetable-feeding loopers can be found in Appendix A. Duration of the larval stage is about 14 days. (See color figure 37.)

Pupa. The mature larva spins a loose, whitish silk cocoon, often incorporating leaves into the structure, and pupates within. The pupa is 18–20 mm long, and its color is blackish brown. Duration of the pupal stage is about 10–14 days.

Adult. The moth measures 35–45 mm in wingspan. The front wings are gray with irregular light-brown and dark-brown patches. Near the center of each forewing there is a prominent silvery white or yellow-colored mark, which in general shape is said to resemble the hind leg of a dog. The hind wing is light brown or gray basally, and dark brown distally. The edge of both front and hind wings is marked with a series of dark spots. Unfortunately, the color pattern of the moths is not completely diagnostic, as there are other species with similar markings. LaFontaine and Poole (1991) should be consulted for keys to moths.

The biology of alfalfa looper is not well-cataloged. Hyslop (1912a) and Parker (1915) provided accounts of its ecology. Alfalfa looper is easily reared on bean-based artificial diet (Shorey and Hale, 1965). The larvae are included in the keys of Crumb (1956), Okumura (1962), and Capinera (1986). Adults are included in the keys of Eichlin and Cunningham (1978) and Capinera and Schaefer (1983). (See color figures 222 and 223.)

Damage

Larvae of alfalfa looper are defoliators. Initially, young larvae may be gregarious and skeletonize

Adult alfalfa looper.

leaves. They soon disperse, however, and chew irregular holes in foliage. They feed on the underside of leaves. A sign of their presence is large quantities of wet fecal matter adhering to foliage. When larvae are abundant they may disperse in large numbers from favored to less-favored plants. Alfalfa is often the source of such infestations.

Management

Alfalfa looper infrequently is a vegetable pest, though in the southwest it sometimes damages lettuce. Foliar application of *Bacillus thuringiensis* or chemical insecticides is effective. Careful monitoring is recommended, especially if late-season lettuce is grown in proximity to alfalfa or cotton. Lettuce is sensitive to injury, and an average of 0.25–0.5 larvae per plant likely warrants suppression. Components of the sex pheromone are known, and can be used to bait traps for population monitoring (Steck *et al.*, 1979). Moths also can be captured in blacklight traps.

Army Cutworm
Euxoa auxiliaris (Grote)
(Lepidoptera: Noctuidae)

Natural History

Distribution. This native insect is abundant in the Great Plains and Rocky Mountain regions in the United States and Canada. It has been recorded from all states west of the Mississippi River, and as far east in Canada as Ontario, but it attains high densities only in semiarid areas.

Host Plants. Army cutworm has been reported to feed on numerous plants. It is known principally as a pest of small grains, perhaps because these crops dominate the landscape where army cutworm occurs. Among vegetable crops, it has been reported to damage beet, cabbage, celery, corn, onion, pea, potato, radish, rhubarb, tomato, and turnip. Other crops injured include such fruit crops as apple, apricot, blackberry, cherry, currant, gooseberry, peach, plum, prune, raspberry, and strawberry, and such field crops as alfalfa, barley, clover, flax, rye, sanfoin, sunflower, sweet clover, timothy, vetch, and wheat. Army cutworm also feeds on noncultivated plants such as bluegrass, *Poa* spp.; bromegrass, *Bromus* spp.; buffalograss, *Buchloe dactyloides*; gramagrasses, *Bouteloua* spp.; field pennycress, *Thlaspi arvense*; dandelion, *Taraxacum officinale*; lambsquarters, *Chenopodium album*; and lupine, *Lupinus* spp.

Natural Enemies. Many natural enemies have been found associated with army cutworm, and both hymenopterous parasitoids and disease have been documented to cause considerable mortality. Walkden (1950), working in the central Great Plains, reported mortality trends over a 20-year period and observed parasitism levels of up to 33% and disease incidence of up to 57%. Not surprisingly, incidence of disease was greatest at high armyworm population densities. Snow (1925) reported 30% parasitism in Utah. In a three-year study in Oklahoma, researchers found that less than 12% of larvae were parasitized, with most parasitism due to two species, *Meteorus leviventris* (Wesmael) and *Apanteles griffini* Viereck (both Hymenoptera: Braconidae) (Soteres *et al.*, 1984). A polyembryonic wasp, *Copidosoma bakeri* (Howard) (Hymenoptera: Encyrtidae), causes larvae to consume more food, to become larger, and live longer; this can result in the appearance of artificially high rates of parasitism, which sometimes exceeds 50% (Byers *et al.*, 1993).

Among the other parasitoids known from army cutworm are such wasps as *Cotesia marginiventris* (Cresson), *A. militaris* Walsh, *Chelonus insularis* Cresson, *Macrocentrus incompletus* Muesebeck, *Microplitis feltiae* Muesebeck, *M. melianae* Viereck, *Rogas* sp., *Zele melea* (Cresson) (all Hymenoptera: Braconidae); *Campoletis flavicincta* (Ashmead), *C. sonorensis* (Cameron), *Diphyus nuncius* (Cresson), *Exetastes lasius* Cushman, and *Spilichneumon superbus* (Provancher) (all Hymenoptera: Ichneummonidae). Flies known to parasitize this species include *Bonnetia comta* (Fallén), *Euphorocera claripennis* (Macquart), *Mericia* spp., *Peleteria* sp., *Periscepsia cinerosa* (Coquillett), *P. helymus* (Walker), and *P. laevigata* (Wulp) (all Diptera: Tachinidae).

Several viruses are known to infect army cutworm, including entomopox, granulosis, and nonoccluded viruses (Jackson and Sutter, 1985; McCarthy *et al.*, 1975; Sutter, 1972, 1973). The relative importance of each is uncertain, but the granulosis virus is unusually pathogenic.

Life Cycle and Description. There is a single generation per year throughout the range of this insect. The eggs are deposited on soil in August–October.

They hatch in autumn or early winter, and larvae over-winter, feeding actively in the spring. Pupation occurs about a month before adults appear. Adults first become active in April–May in southern locations such as Kansas and Texas, whereas in more northern locations such as Alberta and Montana they may not appear until June–July. The moths migrate from the plains, where the larvae develop, to higher elevations in the Rocky Mountains, where the adults feed on nectar from flowers. The adults return to the plains in September–October.

Egg. The eggs are deposited singly or in small clusters just beneath the soil surface on a solid substrate (Pruess, 1961). Soil particles adhere to the eggs so they are difficult to detect in the field. In shape, the eggs are a slightly flattened sphere, measuring about 0.6 mm in diameter and 0.5 mm in height. The egg is white to yellow initially, becomes gray to brown as the embryo matures. The egg is marked with about 18 very narrow ridges that radiate from the apex. Survival of eggs apparently is affected by moisture, and Seamans (1928) suggested that above-average rainfall in late summer and autumn assured good insect survival and damaging populations in the subsequent year; this concept appears not to have been independently confirmed, however. Field-collected females were reported by Pruess (1963) to produce 200–300 eggs, with the potential to produce about 500 eggs. However, Jacobson and Blakeley (1959) suggested that 1000–2500 eggs could be produced by a female based on laboratory studies in which larvae were fed dandelion, a highly suitable host.

Larva. The eggs hatch in the autumn or early winter but the larvae are usually not noticed until spring when they increase in size and begin to consume considerable foliage. There are 6–7 instars, with head capsule widths of 0.26–0.30, 0.40–0.45, 0.65–0.72, 1.04–1.21, 1.70–2.10, and 2.90–3.40 mm, respectively, for instars 1–6 among larvae with only six instars (Jacobsen and Blakeley, 1959). In comparison, head capsule widths of 0.25–0.30, 0.36–0.43, 0.55–0.70, 0.88–1.28, 1.40–1.90, 1.95–2.50, and 2.95–3.55 mm were reported for instars 1–7, respectively, in larvae with seven instars (Sutter and Miller, 1972). Additional instars apparently occur when larvae feed on less suitable host plants. Duration of the instars is estimated at 16–48, 13–73, 4–70, 3–42, 6–11, 4–18, and 9–25 days for instars 1–7, respectively (Burton *et al.*, 1980). The body color of the larvae is grayish-brown, but bears numerous white and dark brown spots. There usually is evidence of three weak light-colored dorsal stripes. Laterally, it tends to be a broad dark band, and the area beneath the spiracles

Army cutworm larva.

is whitish. The head is light brown with dark spots. Larvae attain a length of about 40 mm. They usually are found beneath the surface of the soil, emerging in late afternoon or early evening to feed. On cloudy days, however, they may be active during the daylight hours. Larvae assume a migratory habit when faced with food shortage, and numerous larvae proceed in the same direction, consuming virtually all vegetation in their path. It is this dispersive behavior that is the basis for their common name, and larvae are observed to disperse over 4 km. (See color figure 38.)

Pupa. Pupation occurs in the soil, in a cell prepared by the larva. The walls of the cell are formed with salivary secretion, which hardens and provide a degree of rigidity. The depth of pupation varies according to soil and moisture conditions, but it may be any depth up to 7.5 cm. The larva spends about 10 days in the cell before pupation. Duration of pupation is 25–60 days. The pupa is dark brown, and measures about 17–22 mm long and 6 mm wide.

Adult. The adults measure 35–50 mm in wingspan. They are quite variable in appearance, with five named subspecies (Pruess, 1967), but moths generally assume two basic forms. One common form has the leading edge of the forewing marked with a broad yellowish stripe, and the remainder of the wing blackish but marked with white-rimmed bean-shaped and round spots, and a light transverse line. In another common color form the forewing is mottled brown, bearing bean-shaped and round spots but lacking bands and stripes. In all cases the hind wings are brownish with dark veins, and darker distally. The brown body of the moth is quite hairy. (See color figures 220 and 221.)

As earlier noted, the adults are migratory, dispersing from the plains to the mountains annually (Pruess, 1967; Pruess and Pruess, 1971). In transit and in the mountains they feed on nectar from flowering plants (Kendall *et al.*, 1981). They are nocturnal, and seek shelter during the daylight hours. They have the habit of aggregating in houses, automobiles, and other sheltered locations where they become a nuisance, soil walls, and induce allergic reactions among some individuals (Storms *et al.*, 1981). They also may aggregate in natural shelters in mountainous regions, where they become prey for bears (Chapman *et al.*, 1955; Mattson *et al.*, 1991). In the Rocky Mountain region they are commonly called "miller moths."

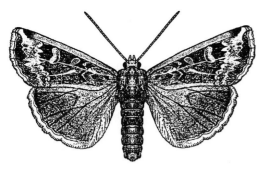

Adult army cutworm, light form.

Adult army cutworm, dark form.

An excellent summary of army cutworm biology was given by Burton *et al.* (1980). Rearing procedures using vegetation were provided by Blakeley *et al.* (1958) and using artificial diet by Sutter and Miller (1972). Sex pheromones have been identified (Struble and Swailes, 1977b; Struble, 1981b). Larvae are included in keys by Whelan (1935), Walkden (1950), and Capinera (1986), and are included in a key to armyworms and cutworms in Appendix A. Moths are included in pictorial keys of Okumura (1962), Rings (1977a), and Capinera and Schaefer (1983).

Damage

These insects principally are pests of small grain crops grown in arid regions, though many irrigated crops also are at risk. Larvae readily climb plants to consume foliage, eating holes in vegetation initially, and eventually destroying the entire plant. Although they burrow into the soil during the daylight hours, they do not normally feed below-ground. However, when succulent food is in short supply they follow the plant stem down into the soil. When food supplies are exhausted numerous larvae may disperse in search of additional food.

Management

Sampling. Adults can be captured in light and pheromone traps. However, males are attracted to the sex pheromone only during the autumn flight. Pheromone traps positioned at a height of 1 m or lower are more effective than those placed higher (Swailes and Struble, 1979). Larvae can be recovered from soil by raking through the top 5–7 cm.

Insecticides. Persistent insecticides can be applied to vegetation to kill army cutworm larvae when they emerge from the soil to feed; *Bacillus thuringiensis* is not effective (McDonald, 1979; Bauernfeind and Wilde, 1993). Larvae also accept bran bait containing insecticide.

Cultural Practices. Cultural manipulations are not generally effective to prevent oviposition because moths deposit eggs on barren soil. Delayed planting of crops can be effective, however, as larvae complete their development on weeds or starve before crops are planted. If larvae are dispersing, creation of deep ditches with steep sides, or filled with running irrigation water, may prevent invasion of fields.

To protect plants grown in the home garden, barriers are occasionally used to decrease access by cutworms to seedlings. Metal- or waxed-paper containers with both the top and bottom removed can be placed around the plant stem to deter consumption. Aluminum foil can be wrapped around the stem to achieve a similar effect. Because larvae burrow and feed below the soil line, the barrier should be extended below the soil surface.

Armyworm
Pseudaletia unipunctata (Haworth)
(Lepidoptera: Noctuidae)

Natural History

Distribution. Armyworm is a native species occurring throughout North America. It is most common, however, in the eastern United States and Canada west to the Rocky Mountains. It does not overwinter at northern latitudes such as Canada and the northern states. Rather, armyworm disperses northward each spring, principally along the Mississippi River Valley, and then disperses southward during the autumn, principally along the east coast (McNeil, 1987). It is often called "true armyworm" to distinguish it from other armyworms such as fall armyworm, *Spodoptera frugiperda* (J.E. Smith), and yellowstriped armyworm, *Spodoptera ornithogalli* (Guenée). Armyworm also occurs in Central and South America, southern Europe, central Africa, and western Asia.

Host Plants. Armyworm generally prefers to oviposit and feed upon plants in the family Gramineae, including weedy grasses. Thus, such grain and grass crops as barley, corn, millet, oats, rice, rye, sorghum, sugarcane, timothy, and wheat may be consumed, as well as wild or weed grasses. During periods of abundance larvae feed more generally, damaging such vegetables as artichoke, bean, cabbage, carrot, corn, celery, cucumber, lettuce, onion, parsley, parsnip, pea, pepper, radish, sweet potato, watermelon, and others. Field crops damaged during such population outbreaks include alfalfa, dry bean, and sugarbeet in addition to the aforementioned grain crops. Adults feed on nectar of various flowers and sometimes feed on other sweet foods such as ripe and decaying fruit.

Natural Enemies. The importance of natural enemies, especially parasitoids, has been studied, though nearly all data are derived from the periods of high armyworm density, which is not typical for this insect. Krombein *et al.* (1979) listed 35 species of Hymenoptera reared from armyworm, of which 19 species are braconids and 12 species are ichneumonids. Similarly, Arnaud (1978) listed 35 species of tachinids from armyworm. Breeland (1958) studied armyworm in Tennessee, and gave a long list of natural enemies known to affect armyworm around the world. He reported rates of parasitism to be 30–40% in Tennessee studies. The most important parasitoids found by Breeland during 1956 were *Glyptapanteles militaris* (Walsh) and *Rogas terminalis* (Cresson) (both Hymenoptera: Braconidae), which accounted for 27% and 5% of total parasitism, respectively; *Winthemia rufopicta* (Bigot) (Diptera: Tachinidae), which accounted for 12% parasitism; and *Eniscospilus merdarius* Gravenhorst (Hymenoptera: Ichneumonidae), which accounted for 9% parasitism. In contrast, during 1957, *W. rufopicta* accounted for only 1.5% parasitism; parasitism by *A. militaris* and *R. terminalis* increased to 36% and 22%, respectively; and parasitism by *E. merdarius* dropped to 4.5%. Interestingly, *Hyposoter* sp. (Hymenoptera: Ichneumonidae) was not observed in 1956 but accounted for nearly 20% of total parasitism in 1957. Guppy and Miller (1970) provided keys to the immature stages of armyworm parasitoids.

Predators readily consume armyworm larvae. Ground beetles (Coleoptera: Carabidae) are especially effective because larvae spend most of their time in association with soil, but various predatory bugs (Hemiptera: various families), ants (Hymenoptera: Formicidae), and spiders (Araneae: Lycosidae and Phalangiidae) also feed on armyworm (Clark *et al.*, 1994). Avian predators are often credited with destruction of armyworms. The bobolink, *Dolichonyx oryzivorus* (Linnaeus), prospers during outbreak years and has sometimes been called the "armyworm bird." Other birds of note include the crow, *Corvus brachyrhynchos* Brehm, and starling, *Sturnus vulgaris* Linnaeus.

Diseases commonly infect armyworms, especially during periods of high density. Bacteria and fungi, particularly the fungus *Metarhizium anisopliae*, are reported in the literature. In Arkansas, Steinkraus *et al.* (1993a) reported an epizootic caused by the fungus *Furia virescens*, and a significant incidence of mermithid nematodes, with mortality by the nematodes estimated at 13.5%. However, undoubtedly the most important diseases are viruses; several granulosis, cytoplasmic polyhedrosis, and nuclear polyhedrosis viruses often kill virtually all armyworms during periods of outbreak, especially when larvae are also stressed by lack of food or inclement weather (Tanada, 1959, 1961).

Weather. Armyworm attains high densities irregularly, often at 5–20 year intervals. The exact cause is unknown, but outbreaks often occur during unusually wet years and are preceded by unusually dry years. Armyworm is not well-adapted for hot temperature; survival decreases markedly when temperatures exceed about 30°C. Consequently, at southern latitudes populations are higher early and late in the year, but at northern latitudes it is a mid-season pest.

Life Cycle and Description. Larvae apparently overwinter at least as far north as Tennessee, though they are unsuccessful in the northernmost states and in Canada, which are invaded annually by moths dispersing northward. In the south, all stages may be found during the winter months (Moran and Lyle, 1940). The number of generations varies among locations, but two generations occur in Ontario, 2–3 occur in Minnesota and New York, 4–5 are reported in Tennessee, and 5–6 in southern states (Knutson, 1944; Frost, 1955; Breeland, 1958; Guppy, 1961; Chapman and Lienk, 1981). In Tennessee, moth flights are observed in March–May, June, July–mid-August, September, and November. In New York, moths are common from March-September, but sometimes as late as November. A complete generation requires 30–50 days.

Egg. Females deposit eggs in clusters consisting of 2–5 rows, in sheltered places on foliage, often between the leaf sheath and blade, especially on dry grass. Often females seem to deposit large numbers in the same vicinity, resulting in very high densities of larvae in relatively small areas of a field. Nevertheless, the eggs are very difficult to locate in the field. The eggs are white or yellowish, but turn gray immediately before hatching. They are spherical, and measure about 0.54 mm (range 0.4–0.7 mm) in diameter. The egg surface appears to be shiny and smooth, but under high magnification fine ridges can be observed. Their clutches are covered with an adhesive secretion that is opaque when wet but transparent when dry. As the adhesive material dries it tends to draw together the foliage, almost completely hiding the eggs. Mean duration of the egg stage is about 3.5 days at 23°C, and 6.5 days at 18°C, but the range is 3–24 days over the course of a season. Hatching rates are affected by temperature, with cool weather more favorable for embryonic survival. In Tennessee, about 98% egg hatch occurs in early spring and autumn, with hatching rates dropping to less than 30% during the summer; this probably accounts for the evolution of the dispersal behavior in this species.

Larva. Larvae normally display six instars, though up to nine instars have been observed. Mean head capsule widths (range) are 0.34 (0.30–0.37), 0.55 (0.49–0.63), 0.94 (0.83–1.12), 1.5 (1.29–1.70), 2.3 (2.08–2.56), and 3.3 (3.04–3.68) mm, respectively, for instars 1–6. Head capsule widths increase slightly with increased temperature up to about 30°C (Guppy, 1969). Larvae attain a body length 4, 6, 10, 15, 20, and 35 mm, respectively, during instars 1–6. Except for the first instar, which is pale with a dark head, the larvae of armyworm are marked with longitudinal stripes throughout their development. The head capsule is yellowish or yellow-brown with dark net-like markings. The body color is normally grayish-green, but a broad dark stripe occurs dorsally and along each side. A light subspiracular stripe often is found laterally beneath the dark stripe. Development time varies with temperature. During summer larvae complete their development in about 20 days, but this is extended to about 30 days during the spring and autumn, and greatly prolonged during winter. Instar-specific development times recorded during early summer in Tennessee are 2–3, 2–3, 2–4, 2–3, 4–5, and 7–10 days for instars 1–6, respectively. The larvae tend to disperse upward following hatching, where they feed on tender leaf tissue. If disturbed, they readily extrude silk and spin down to the soil. Larvae in instars 3–6 are active at night, seeking shelter during the day on the soil beneath debris or clods of soil. (See color figure 39.)

Pupa. Larvae pupate in the soil, often under debris, at depths of 2–5 cm. Pupation occurs in an oval cell that contain a thin silken case. The pupa is moderate in size and robust, measuring 13–17 mm long and 5-6 mm wide. The pupa is yellowish-brown initially, but soon assumes a mahogany brown color. The tip of the abdomen bears a pair of hooks. Duration of

Armyworm larva.

Head capsule of armyworm larva.

Armyworm pupa.

the pupal stage is 7–14 days during summer but longer early and late in the season, sometimes lasting 40 days.

Adult. The adult is a light reddish-brown moth, with a wingspan measuring about 4 cm. The forewing is fairly pointed, appearing more so because a transverse line of small black spots terminates in a black line at the anterior wing tip. The forewing is also marked with a diffuse dark area centrally containing 1–2 small white spots. The hind wings are grayish, and lighter basally. Adults are nocturnal. Mating commences 1–3 days after moths emerge from the soil, and usually 4–7 hours after sunset. Eggs are normally deposited within a 4–5 day period (range 1–10 days). Females produces an average of 4.9 egg masses (range 1–16 masses). Breeland (1958) reported that about 450 eggs are produced by each female (range 15–1350 eggs) in Tennessee, but Guppy (1961) found a mean fecundity of 1450 (range 250–1900). Feeding is necessary for normal oviposition. Mean longevity at warm temperature is about 9 days in males and 10 days in females (range 3–25 days) whereas at cool temperature mean longevity of males is 19 days and females 17 days. (See color figure 224.)

An excellent treatment of biology was given by Breeland (1958). Other informative publications include Davis and Satterthwait (1916b), Walton and Packard (1947), Pond (1960), and Guppy (1961). Devel-

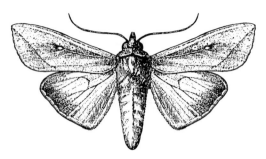

Adult armyworm.

opmental biology was presented by Guppy (1969). A sex pheromone is described by Steck *et al.* (1982) and female response by Turgeon *et al.* (1983). Armyworm is included in the larval keys of Walkden (1950), Crumb (1956), Rings and Musick (1976), Oliver and Chapin (1981), and Capinera (1986). Larvae also are included in a key to armyworms and cutworms in Appendix A. Moths are included in the keys by Rings (1977a) and Capinera and Schaefer (1983), and pictured in Rockburne and Lafontaine (1976) and Chapman and Lienck (1981).

Damage

Larvae initially skeletonize foliage, but by the third instar they eat holes in leaves, and soon afterwards consume entire leaves. Larvae of armyworm are notorious for appearing out of nowhere to inflict a high level of defoliation. This occurs for several reasons: a highly clumped distribution of young larvae, with most of the crop uninfested until larvae are nearly mature and highly mobile; a tendency by larvae to feed on grass weeds preferentially, only moving to crops after the grass is exhausted; occurrence of a preponderance of feeding, about 80%, in the last instar; the nocturnal behavior of larvae, which makes them difficult to observe during the day; and the gregarious and mobile behavior of mature larvae, which form large aggregations or bands (hence the common name "army" worm). As earlier noted, grasses and grains are preferred, but as these plants are consumed larvae disperse, often in large groups, to other plants. During outbreaks, few plants escape damage.

Management

Sampling. Adults can be captured with blacklight traps. A sex pheromone has been identified and can be used for population monitoring (Kamm *et al.*, 1982; Lopez *et al.*, 1990). Light-colored pheromone traps capture more moths than dark-colored traps (Hendrix and Showers, 1990). It is advisable to examine crop fields for larvae, especially if moths have been captured in light or pheromone traps. Fields should be examined at dawn or dusk, because larvae are active at this time. If it is necessary to check fields during the day, it is important to sift through the upper surface of the soil and under debris for resting larvae.

Insecticides. Larvae consume wheat bran or apple pomace baits treated with insecticide, but foliar and soil-applied insecticides are also effective, and used frequently (Musick and Suttle, 1973; Harris *et al.*, 1975a; Harrison *et al.*, 1980a).

Cultural Techniques. Cultural practices have limited effect on armyworm abundance due to their highly dispersive behavior. However, grass weeds are a focal point of infestation, and should be eliminated, if possible. Not surprisingly, no-till and minimum tillage fields experience greater problems with armyworm than conventional tillage fields (Harrison *et al.*, 1980a; Willson and Eisley, 1992). Proximity to small grain crops is considered to be a hazard owing to the preference of moths for such crops, and the suitability of grains for larval development. In Virginia, destruction of winter cover crops by herbicide application is more favorable to armyworm survival than mowing of cover crops, apparently because predators are more disrupted by herbicide treatment (Laub and Luna, 1992). Before availability of effective insecticides, deep furrows with steep sides were sometimes plowed around fields to prevent invasion by dispersing armyworm larvae. Although this approach remains somewhat useful, but it is rarely practiced.

Biological Control. Some suppression of armyworm can be achieved with *Bacillus thuringiensis*, though it is not as effective as with some other caterpillars. Larvae also are susceptible to infection by the entomopathogenic nematode *Steinernema carpocapsae*, though neonate larvae are fairly resistant (Kaya, 1985).

Bean Leafskeletonizer
Autoplusia egena (Guenée)
(Lepidoptera: Noctuidae)

Natural History

Distribution. This insect is a native to tropical areas of the western hemisphere. The northern limits of its range are the Gulf Coast states and California. Its distribution extends southward through Central and South America to Brazil, and the Caribbean Islands.

Host Plants. As suggested by its common name, the principal host is bean. However, other vegetables sometimes consumed include cabbage, carrot, and celery. Soybean, tobacco, mint, spearmint, comfrey, chrysanthemum, marigold, hollyhock, verbena, and larkspur are other economic plants eaten.

Natural Enemies. Several parasitoids were found to affect bean leafskeletonizer in California; *Copidosoma truncatellum* (Dalman) (Hymenoptera: Encyrtidae) was the only parasitoid affecting a significant proportion of the leafskeletonizers (Lange, 1945). An undetermined fungal disease was also noted among the larvae. There has not been adequate study to determine the relative importance of the natural enemies.

Life Cycle and Description. The complete life cycle requires about 45 days. The number of generations appears to be undetermined, but Lange (1945) reported "several" in California.

Egg and Larva. The duration of the egg is about five days. Larvae have five instars, the duration of which is 4.4, 3.4, 4.0, 4.1, and 7.8 days, respectively. Mean (range) of head capsule widths are 0.25 (0.25–0.25), 0.45 (0.42–0.46), 0.93 (0.66–1.10), 1.40 (1.21–1.57), and 2.16 (2.03–2.32) mm for instars 1–5, respectively. The body color is green, and pale white stripes usually are evident. The head is green, but marked with black spots at the base of setae. Numerous black tubercles are found dorsally on the body. The body bears microspines, but they are quite small and difficult to detect. The larva attains a length of about 22–33 mm at maturity. The lack of nipple-like structures on abdominal segments three and four separates bean leafskeletonizer from cabbage looper, *Trichoplusia ni* (Hübner), and soybean looper, *Pseudoplusia includens* (Walker). The lack of dark lateral bands on the head helps to distinguish bean leafskeletonizer from most of the other common loopers. A key to common vegetable-feeding loopers can be found in Appendix A.

Pupa. Larvae roll the foliage, form a silken cocoon within the rolled leaves, and pupate within. The pupa measures about 18 mm long. Duration of the pupal stage averages about 13.7 days.

Adult. The front wings of the adult are distinctively rusty brown, blending into gray, cream, or yellowish-brown. The front wings lack the distinctive silver or white spot found centrally in most loopers. The hind wings are light grayish-brown basally and darker brown distally. Under laboratory conditions the adults survived only about 5.3 days.

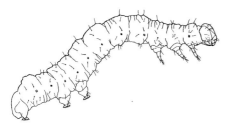

Bean leafskeletonizer larva.

The only detailed report on biology was provided by Casanova (1977). Useful information and diagnostic characters were given by Crumb (1956), Eichlin and Cunningham (1978), and LaFontaine and Poole (1991).

Damage

Both the leaf tissue and pods of beans are eaten. Although generally regarded as only a minor pest, this insect reportedly was very damaging at times to several varieties of beans in California. In Puerto Rico it is most damaging to tobacco.

Management

Foliar insecticides are effective if needed.

Beet Armyworm
Spodoptera exigua (Hübner)
(Lepidoptera: Noctuidae)

Natural History

Distribution. Beet armyworm is a tropical insect, and is native to Southeast Asia. It is now found around the world, except South America. It was first discovered in North America about 1876, when it was found in Oregon. It rapidly spread to Hawaii (1880), California (1882), Colorado (1899), Texas (1904), Arizona (1916), Mississippi (1920), and Florida (1924). It has since spread to Mexico and the Caribbean (Mitchell, 1979). As it is a tropical insect and lacks a diapause mechanism, it can overwinter successfully only in warm areas or in greenhouses. Daytime temperatures below 10°C are deleterious, and it rarely overwinters in areas where frost kills its host plants. Thus, overwintering is generally limited to Arizona, Florida, and Texas. Despite its inability to overwinter in most of the United States, beet armyworm nevertheless invades the southern half of the United States (Maryland to Colorado to northern California, and south) annually. Sometimes it is found as far north as New York and Ontario in the east, and British Columbia in the west. It is rarely viewed as a serious pest anywhere but the southern states, however, except sometimes in greenhouses.

Host Plants. This insect has a wide host range, occurring as a serious pest of vegetable, field, and flower crops; even trees are sometimes attacked. Among susceptible vegetable crops are asparagus, bean, beet, broccoli, cabbage, cauliflower, celery, chickpea, corn, cowpea, eggplant, lettuce, onion, pea, pepper, potato, radish, spinach, sweet potato, tomato, and turnip. Field crops damaged include alfalfa, corn, cotton, peanut, safflower, sorghum, soybean, sugarbeet, and tobacco.

Weeds also are suitable for larval development, including such common plants as lambsquarters, *Chenopodium album*.; mullein, *Verbascum* sp.; pigweed, *Amaranthus* spp.; purslane, *Portulaca* spp.; Russian thistle, *Salsola kali*; parthenium, *Parthenium* sp.; and tidestromia, *Tidestromia* sp. Although its host range is wide, there are significant differences in suitability even among hosts considered to be suitable. For example, in a comparison among diets consisting of sugarbeet, pigweed, or lambsquarters, sugarbeet-fed larvae had the shortest development time and the highest fecundity, lambsquarter had the longest development time and the lowest fecundity, and pigweed was intermediate (Al–Zubaidi and Capinera, 1986).

Natural Enemies.   Numerous native natural enemies have adapted to this pest. Among the most common parasitoids are *Chelonus insularis* Cresson, *Cotesia marginiventris* (Cresson), and *Meteorus autographae* (Muesbeck) (all Hymenoptera: Braconidae), and the tachinid *Lespsia archippivora* (Riley) (Diptera: Tachinidae) (Oatman and Platner, 1972; Tingle *et al.*, 1978; Ruberson *et al.*, 1994). Predators frequently attack the eggs and small larvae; among the most important are minute pirate bugs, *Orius* spp. (Hemiptera: Anthocoridae); big-eyed bugs, *Geocoris* spp. (Hemiptera: Lygaeidae); damsel bugs, *Nabis* spp. (Hemiptera: Nabidae); and a predatory shield bug, *Podisus maculiventris* (Say). Pupae are subject to attack, especially by the red imported fire ant, *Solenopsis invicta* Buren. Fungal diseases, *Erynia* sp. and *Nomurea rileyi*, and a nuclear polyhedrosis virus also inflict some mortality (Wilson, 1933, 1934; Harding, 1976b; Ruberson *et al.*, 1994). The important mortality factors vary among crops and geographic regions. None except the nuclear polyhedrosis virus are highly specific to beet armyworm, which may explain why they are not especially effective. Virus is considered to be the most important mortality factor in Mexico (Alvarado–Rodriguez, 1987).

Life Cycle and Description.   Seasonal activity varies considerably according to climate. In warm locations such as Florida, all stages can be found throughout the year, though development rate and overall abundance are decreased during the winter months (Tingle and Mitchell, 1977). The life cycle can be completed in just 24 days, and six generations have been reared during five months of summer weather in Florida (Wilson, 1934). However, generation times of 50–126 days have been observed, with a total of five generations annually, in southern California (Campbell and Duran, 1929).

Egg.   Eggs are laid in clusters of about 50–150 per mass. Females may deposit over 1200 eggs during their life time, but normal egg production is about 300–600.

Beet armyworm egg.

They usually are deposited on the lower surface of the leaf, and often near blossoms and the tip of the branch. The individual egg is circular when viewed from above, but when examined from the side it is slightly peaked, tapering to a point. The eggs are greenish to white, and are covered with a layer of whitish scales that gives the egg mass a fuzzy or cottony appearance. They hatch in 2–3 days during warm, but the incubation period is extended to about four days in cool weather. The developmental threshold for eggs is estimated at 12.4°C.

Larva.   Normally there are five instars, though additional instars are sometimes found. Duration of the instars under warm (summer) conditions is reported to be 2.3, 2.2, 1.8, 1.0, and 3.1 days, respectively (Wilson, 1932), and at constant 30°C instar development time was reported by Fye and McAda (1972) to be 2.5, 1.5, 1.2, 1.5, and 3.0 days, respectively. Total larval development time is also influenced by diet quality (Al–Zubaidi and Capinera, 1984). The developmental threshold for larvae is estimated at 13.6°C. Only 1 mm long at hatching, the larvae attain a mean length of 2.5, 5.8, 8.9, 13.8, and 22.3 mm during instars 1–5, respectively (Wilson, 1932). Head capsule widths average 0.25, 0.45, 0.70, 1.12, and 1.80 mm, respectively.

The larvae are pale green or yellow during instars 1–2, but acquire pale stripes during instar three. During instar four, larvae are darker dorsally and possess a dark lateral stripe. Larvae during instar five are variable in appearance, and tend to be green dorsally, with pink or yellow color ventrally, and a white stripe laterally. A series of dark spots or dashes is often present dorsally and dorsolaterally. Sometimes larvae are very dark. The spiracles are white with a narrow black border. The body is practically devoid of hairs and spines. In the western states, the larva of beet armyworm is easily confused with clover cutworm, *Discestra trifolii* (Hufnagel), but beet armyworm lacks the black pigment adjacent to the spiracles that is so evident in clover cutworm. In the southern states, the larva of beet armyworm is easily confused with southern armyworm, *Spodoptera eridania* (Cramer), but southern armyworm can be distinguished by the presence of a

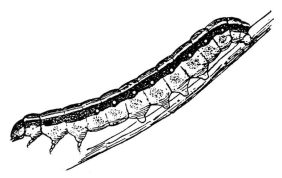

Beet armyworm larva.

Adult beet armyworm.

large dark spot laterally on the first abdominal segment that disrupts the lateral stripe. Beet armyworm occasionally bears a spot laterally, but if present it occurs on the mesothorax, not on the first abdominal segment.

Initially, the larvae of beet armyworm are gregarious, feeding as a group and skeletonizing plant foliage. As they mature, larvae become solitary and quite mobile, often traveling from plant to plant. Cannibalism may occur when larvae are at high densities or feeding on food low in nitrogen (Al–Zubaidi and Capinera, 1983). (See color figures 40 and 41.)

Pupa. Pupation occurs in the soil. The larva generally constructs a pupal chamber near the soil surface, digging only about 1 cm beneath the surface. The chamber is constructed from sand and soil particles held together with an oral secretion that hardens when it dries. The pupa is light brown and measures about 15-20 mm long. Duration of the pupal stage is 6–7 days during warm weather. Fye and McAda (1972) reported a pupal duration of 5.1 days at 30°C.

Adult. The moths are moderately sized, the wingspan measuring 25–30 mm. The front wings are mottled gray and brown, and normally with an irregular banding pattern and a light colored bean-shaped spot. The hind wings are a more uniform gray or white color, and trimmed with a dark line at the margin. Mating occurs soon after emergence of the moths, and oviposition begins within 2–3 days. Oviposition extends over a 3–7 day period, and the moths usually perish within 9–10 days of emergence. (See color figure 225.)

An overview of biology was given by Wilson (1932), and developmental biology by Wafa *et al.* (1969), Fye and McAda (1972), and Ali and Gaylor (1992). Brown and Dewhurst (1975) provided detailed description of all stages, and a comprehensive list of host plants. Rearing technology was discussed by many authors, including Cobb and Bass (1975) and Hartley (1990). A sex pheromone has been identified (Persoons *et al.*, 1981; Mitchell *et al.*, 1983). Effects of irradiation and

potential for release of sterile insects has been investigated (Debolt and Wright, 1976). Larvae are included in keys by Okumura (1962), Oliver and Chapin (1981), and Capinera (1986), and are included in a key to armyworms and cutworms in Appendix A. Adults are included in a key by Capinera and Schaefer (1983). Heppner (1998) provided keys to the adults and larvae of North American *Spodoptera*.

Damage

Larvae feed on both foliage and fruit. Young larvae feed gregariously and skeletonize foliage. As they mature, larvae become solitary and eat large irregular holes in foliage. They also burrow into the crown or center of the head on lettuce, or on the buds of crucifers. As a leaf feeder, beet armyworm consumes much more cabbage tissue than diamondback moth, *Plutella xylostella* (Linnaeus), but it is less damaging than cabbage looper, *Trichoplusia ni* (Hübner) (East *et al.*, 1989). This insect also is regarded as a serious pest of celery in California, and damage is directly correlated with abundance of late-instar larvae late in the season. However, damage to foliage and petioles (stalks) during the first half of the growing season is of little consequence because these plant parts are removed at harvest (van Steenwyk and Toscano, 1981). Tomato fruit is most susceptible to injury, especially near fruit maturity, but beet armyworm is not considered to be as threatening to tomato as is corn earworm, *Helicoverpa zea* (Boddie) (Zalom *et al.*, 1986a). Larvae not only damage tomato fruit, but may appear as contaminants in processed tomato (Zalom and Jones, 1994).

Management

Sampling. Pheromone traps can be used to detect the presence of adult beet armyworm (Trumble and Baker, 1984; Mitchell and Tumlinson, 1994). Visual sampling for damage and larvae, combined with an action threshold of 0.3 larvae per plant, was used successfully on cabbage in South Texas to determine the need for crop treatment with insecticides (Cartwright *et al.*, 1987). A binomial sequential sampling program for armyworm-damaged tomato fruit was developed by Wilson *et al.* (1983b). Egg distribution on tomato was studied by Zalom *et al.* (1983). Regular monitoring of crops, probably about twice per week, is recommended because adults frequently invade from surrounding crops or weeds (Edelson *et al.*, 1988).

Insecticides. In the southeast and southwest, the relatively high abundance of beet armyworm has stimulated frequent application of insecticides. Chemical insecticides and *Bacillus thuringiensis* are commonly applied to foliage to protect against defoliation. Insecticide resistance is a major problem in management of this insect, possibly because it attacks crops such as flowers, cotton, and vegetables crops that are treated frequently with insecticides (Brewer *et al.*, 1990; Brewer and Trumble, 1994). Beet armyworm abundance is favored by frequent insecticide use, and it is considered to be a secondary or induced pest in some crops (Eveleens *et al.*, 1973). Also, intensive use of insecticides for beet armyworm control in vegetables such as celery has stimulated outbreaks of other pests, principally American serpentine leafminer, *Liriomyza trifolii* (Burgess).

Beet armyworm larvae are susceptible to management with other products such as neem formulations (Prabhaker *et al.*, 1986). Eggs can be killed with petroleum oil (Wolfenbarger *et al.*, 1970), and both eggs and young larvae can be controlled with foliar applications of 5% cottonseed oil, though this concentration is damaging to some plants (Butler and Henneberry, 1990).

Pheromones can also be used to disrupt mating and to inhibit or eliminate reproduction. Saturation of the atmosphere around beet armyworm-susceptible crops has been estimated to decrease mating by 97% (Wakamura and Takai, 1992) and production of eggs and larvae by 57% and 95%, respectively (Mitchell *et al.*, 1997).

Cultural Practices. Host-plant resistance in several crops has been studied for its contribution to beet armyworm pest management. In tomato, for example, resistance is correlated with total glycoalkaloid concentration in the fruit tissue. However, leaf tissue does not have any effective antibiotic chemistry, so larvae are able to develop on plants even if they have unsuitable tomato fruit (Eigenbrode and Trumble, 1994). The future for beet armyworm-resistant celery is most promising (Meade and Hare, 1991; Diawara *et al.*, 1996).

Biological Control. Several insect pathogens may prove to be useful for suppression of beet armyworm. A nuclear polyhedrosis virus isolated from beet armyworm is fairly effective as a bioinsecticide under greenhouse conditions, where inactivation by ultraviolet light in sunlight is not a severe problem (Smits *et al.*,

1987). It is as effective as commonly used insecticides (Gelernter *et al.*, 1986), but it is not commercially available. The fungus *Beauveria bassiana* has the same attributes and limitations (Barbercheck and Kaya, 1991). Entomopathogenic nematodes (Rhabditida: Steinernematidae and Heterorhabditidae) successfully infect both larvae and adults of beet armyworm, and infected adults can fly short distances, helping to spread the pathogens (Timper *et al.*, 1988). Use of nematodes is similarly constrained by environmental conditions, but these biological control agents are available commercially.

Bertha Armyworm
Mamestra configurata Walker
(Lepidoptera: Noctuidae)

Natural History

Distribution. Bertha armyworm occurs widely in western North America, but is known principally as a pest in Canada's Prairie Provinces and British Columbia. This native species occurs from Ontario west to British Columbia, and south through the Rocky Mountain states to Mexico. It is usually associated with dry grassland areas.

Host Plants. Although considered to be a general feeder, this species prefers plants in the families Cruciferae and Chenopodiaceae. The importance of bertha armyworm as a crop pest has grown as the popularity of canola, an important oilseed crop, has increased. However, it has long been known as a pest of potatoes in southern British Columbia. It is reported from several vegetables including bean, beet, cabbage, cauliflower, corn, lettuce, pea, potato, rhubarb, Swiss chard, tomato, and turnip. Other crops injured include alfalfa, canola, flax, red clover, sugarbeet, sweet clover, sunflower, tobacco, and wheat. Several flower crops are reported injured, including geranium, gladiolus, hollyhock, larkspur, poppy, petunia, sunflower, and zinnia. Among common weeds consumed are lambsquarters, *Chenopodium album*; mustard, *Brassica* spp.; and Russian thistle, *Salsola kali*.

Natural Enemies. Parasitoids and diseases are known to affect bertha armyworm larvae. Of the several parasitoids known from this insect, *Banchus flavescens* Cresson (Hymenoptera: Ichneumonidae) and *Athrycia cinerea* (Coquillett) (Diptera: Tachinidae) are most common, but it has proved to be difficult to accurately assign causes to much of the mortality observed among larvae (Wylie and Bucher, 1977; Turnock, 1988). Among other parasitoids found in association with bertha armyworm are *Apanteles xylinus* (Say) (Hymenoptera: Braconidae), *Ichneumon canadensis* Cresson

(Hymenoptera: Ichneumonidae), *Eulophus* sp. and *Euplectris bicolor* (Swederus) (both Hymenoptera: Eulophidae); and *Panzeria ampelus* (Walker), *Exorista mella* (Walker), and *Phryxe pecosensis* (Townsend) (all Diptera: Tachinidae).

Diseases known from bertha armyworm include a nuclear polyhedrosis virus, a fungus, and a microsporidian. The nuclear polyhedrosis virus occurs infrequently at low armyworm densities, but its incidence increases at higher host densities. For example, Erlandson (1990) reported occurrence of over 95% infection of late-instar larvae in Saskatchewan. Fairly high doses of virus are needed to infect late-instar larvae, so higher levels of infection occur if the virus is distributed early, among young larvae. The virus probably survives in the soil and is distributed naturally by blowing soil (Bucher and Turnock, 1983).

Life Cycle and Description.   There is a single generation annually in Canada, and apparently elsewhere. This species overwinters in the pupal stage, with moths emerging principally in June and July to mate and oviposit. Larvae mature in late summer and autumn, and pupate in the soil.

Egg.   The eggs are deposited in clusters of 50–500 on the lower surface of leaves. They are white initially, but turn brown with age, and black just before hatching. In shape, the eggs resemble a slightly flattened sphere, and bear about 38 ridges radiating from the apex. Duration of the egg stage is about 10, 5.5, 4, and 3 days at 15°, 20°, 25°, and 30°C (Jones and Heming, 1979). Fecundity is reported to be 700–900 eggs per female when fed favorable host such as canola, but only about 280 eggs when fed less suitable hosts such as potato.

Larva.   The larval stages are variable in appearance. There are six instars. During the first four instars the larva is green with narrow, white or yellow dorsal and subdorsal stripes, and has the appearance and habits of loopers. They are solitary, and nocturnal in their feeding. If disturbed, young larvae quickly spin down from the plant on a strand of silk. The later instars may be greenish with conspicuous stripes, or brownish to blackish with less apparent stripes. The darker forms tend to occur where larvae feed in exposed habitats, with the green form often found feeding in sheltered locations. Diet also influences color. Mature larvae usually have narrow dorsal and subdorsal stripes that are yellow. An irregular orangish or brownish band may be present below the spiracles. The mature larva measures about 32 mm long, and the head capsule width is 3.1–3.2 mm. Bailey (1976a) studied larval development on several natural diets at 20°C and reported development times of about

3.1, 2.2, 2.3, 2.4, 2.5, and 8.5 days for instars 1–6, respectively, and total larval development time of about 20.5 days. The head of the mature larva is yellowish to orange with darker spots.

Pupa.   Pupation occurs in the soil, usually at a depth of 5–15 cm. The pupa is reddish brown, and measures about 18–20 mm long and 5–6 mm wide. The posterior end bears a pair of unusually long spines with curved tips. Although nearly all pupae remain in diapause, invariably a few continue their development and emerge late in the season, only to perish due to cold weather. Pupae usually perish during the winter in Manitoba in the absence of snow cover on the soil, but about half of the pupae survive if 5–10 cm of snow is present (Lamb *et al.*, 1985). The pupal stage is intolerant of warm conditions; exposure to temperatures of 25°C or higher results in sterilization of adults (Bucher and Bracken, 1977). Duration of the pupal stage is commonly 8–9 months.

Adult.   The moths measure about 35–40 mm in wingspan. The forewing is predominantly gray, bearing patches of brown, black, olive, and white scales. The bean-shaped spot near the center of the forewing is mostly white, as is an irregular transverse line near the wing tip. The hind wing is grayish white basally, and darker distally. Moths are nocturnal, and feed avidly at flowers. The calling behavior of females, and hence mating, occur during the latter two-thirds of the scotophase. This is different from most noctuids, which often mate early in the dark period (Howlader and Gerber, 1986). (See color figure 226.)

A comprehensive description and summary of biology is lacking; the report by King (1928b) is the most complete. A sex pheromone has been described (Underhill *et al.*, 1977; Struble *et al.*, 1984). Beirne (1971) provided a good historical treatment in Canada. Thermal biology was described by Bailey (1976a,b) and Bodnaryk (1978). Bucher and Bracken (1976) described culture techniques. The larva is included in keys to noctuid larvae by Crumb (1956) and Godfrey (1972). Capinera and Schaefer (1983) included the adult in a pictorial key of moths.

Damage

This is one of the species considered to be a "climbing cutworm" because it spends little time in association with the soil. Larvae initially feed on foliage, creating irregularly shaped holes in the leaves. As they mature, however, they will feed on several other plant parts, including the flowers of clover; pods of canola, pea, and bean; bolls of flax; silk and ears of corn; fruit of tomato; and the head of cabbage.

Management

Sampling. Moths can be captured in blacklight and pheromone traps, but light traps are not considered to be efficient (Bucher and Bracken, 1979). Pheromone traps are useful, though usually not completely specific for bertha armyworm (Steck *et al.*, 1979; Struble *et al.*, 1984; Landolt, 2000). Although pheromone traps give good indications of broad trends an armyworm abundance, larval populations in adjacent fields may differ considerably, depending on the relative attractiveness of the crops when the eggs are being deposited (Turnock, 1987). Thus, it is advisable to sample plants for larvae, using caution to avoid having the larvae drop to the soil. Turnock and Bilodeau (1985) described sampling techniques for larvae.

Insecticides. Application of chemical insecticides to foliage provides good suppression of larvae. Larvae are only weakly susceptible to the common formulations of *Bacillus thuringiensis* (Morris, 1986; Trottier *et al.*, 1988). Neem products act as growth regulators and feeding deterrents (Isman, 1993).

Cultural Methods. Weed control is an important component of bertha armyworm management because weeds such as lambsquarters and wild mustard attract ovipositing females and serve as a source of larvae, which damage crops only after the preferred weeds are consumed. Therefore, it is advisable to keep crops free of weeds during the flight period of adults, so they do not oviposit on favored weeds within otherwise nonpreferred crops.

Tillage in the autumn can be destructive to overwintering pupae. Mortality can result from direct damage to pupae from tillage equipment, with increased exposure of pupae to cold temperatures, and to differential mortality to parasitoids. Parasitoids seem to be less affected than their armyworm hosts by tillage of the soil (Turnock and Bilodeau, 1984).

Biological Control. Entomopathogenic nematodes (Nematoda: Heterorhabditidae and Steinernematidae) have been evaluated for suppression of bertha armyworm larvae under laboratory conditions (Morris and Converse, 1991). Although bertha armyworm is more susceptible than some other soil-dwelling insects under laboratory conditions, field evaluation is needed before nematodes can be recommended for armyworm suppression.

Bilobed Looper
Megalographa biloba (Stephens)
(Lepidoptera: Noctuidae)

Natural History

Distribution. This native insect is found throughout North and South America, including Hawaii. It is relatively infrequent, however, in western Canada and the northwestern United States. Apparently it cannot overwinter in northern latitudes and reinvades the northern United States and southern Canada each summer.

Host Plants. Bilobed looper is reported to feed on several families of plants, but only a few crops are affected. Among vegetables crops eaten are bean, cabbage, and lettuce. Other crops accepted include alfalfa, clover, and tobacco, as well as some ornamental plants such as geranium, gladiolus, ivy, and salvia. Weeds consumed include hedge nettle, *Stachys* sp.; sunflower, *Helianthus* sp.; vervain, *Verbena* sp.; and yellow thistle, *Cirsium horridulum*.

Natural Enemies. The natural enemies of bilobed looper are unknown.

Life Cycle and Description. Adults, eggs, and larvae have been found during the period of January–June in Florida, with most larvae observed in late spring (Martin *et al.*, 1981a). Presumably they disperse northward during the spring and summer months.

Egg. The eggs of bilobed looper are hemispherical in shape, white in color, and bear ridges that radiate vertically. They hatch in 3–5 days, usually during the morning hours.

Larva. There are five instars, with development times of about three, two, three, three, and four days, respectively, when reared at 25°C. Total larval development time is about 13, 16, and 25 days at 30°, 25°, and 20°C, respectively. Head capsule widths for the five instars are about 0.2, 0.4, 0.7, 1.2, and 1.9 mm, respectively. The larvae attain a length of about 30 mm at maturity. In form, larvae resemble most other related loopers; the body is distinctly broader at the posterior end and tapers toward the head. The general color of the larva is green, but there is a dark green dorsal stripe, 3–4 weak white lines running parallel to the dorsal green line, and a thin white line on each side just above the lateral spiracles. The head is green, and has a strong black band on each side of the head. The thoracic legs are normally black, and three pairs of thoracic legs are present. The larva of bilobed looper is easily confused with other loopers, but because it bears black thoracic legs and black bars on the side of the head it is most easily confused with alfalfa looper, *Autographa californica* (Speyer). In contrast to alfalfa looper, however, bilobed looper is weakly marked with stripes, and bears microspines on the abdomen. A key to common vegetable-feeding loopers can be found in Appendix A. (See color figure 42.)

Pupa. Pupation occurs in a thin, nearly transparent silk cocoon that is attached to the host plant or

Bilobed looper larva.

nearby vegetation. Pupae of bilobed looper are variable in color, usually mottled black with irregular tan or light green areas. Pupation requires about 6, 9, and 15 days at 30°, 25°, and 20°C, respectively. (See color figure 266.)

Adult. The wingspan of the bilobed looper moth measures about 4 cm. The forewing is irregularly marked with pale-brown to medium-brown, and a silver bilobed spot is located near the center of the wing. The hind wing is gray to tan basally and darker brown distally. The prereproductive period of the adult is estimated at 2–4 days, and the reproductive period at 5–6 days. Adults often mate more than once. Total egg production is about 500 eggs per female. Adult longevity is estimated at 8.5 days for males and 10.0 days for females. (See color figure 227.)

Developmental biology and rearing information were provided by Beach and Todd (1988). Adult and larval descriptions, and keys to differentiate bilobed looper from related species, were provided by Eichlin and Cunningham (1978), Crumb (1956), Capinera and Schaefer (1983), and Capinera (1986). Bilobed looper was also included in the key to vegetable-atttacking loopers in Appendix A.

Damage

Bilobed looper is a defoliator, and mature larvae can consume considerable quantities of foliage during the final days of larval life. However, it rarely is sufficiently abundant to be cause for concern.

Management

Moth populations can be monitored with blacklight traps. A sex pheromone component that attracts numerous noctuids in the subfamily Plusiinae, *cis*-7-dodecenyl acetate, also is attractive to bilobed looper

Adult bilobed looper.

(Roelofs and Comeau, 1970). Larvae are readily controlled with foliar applications of insecticides, including *Bacillus thuringiensis*.

Black Cutworm
Agrotis ipsilon (Hufnagel)
(Lepidoptera: Noctuidae)

Natural History

Distribution. Black cutworm is found widely around the globe, though it is absent from some tropical regions and cold areas. Its origin is uncertain. It is more widespread, and damaging, in the northern than the southern hemispheres. It is found annually throughout the United States and southern Canada, but apparently does not overwinter in northern states and Canada. There is strong evidence that black cutworm disperses northward from the Gulf Coast region each spring.

Long distance dispersal of adults has long been suspected in Europe, China, and North America. The basic pattern is to move north in the spring, and south in the autumn. Studies in the United States demonstrated northward displacement of moths during the spring in the range of 1000 km in 2–4 days when assisted by northward flowing air (Kaster and Showers, 1982; Showers *et al.*, 1989; Smelser *et al.*, 1991). Similar displacement to the south and southwest has been documented in the autumn (Showers *et al.*, 1993).

Host Plants. Black cutworm has a wide host range. Among vegetables injured are artichoke, asparagus, bean, beet, broccoli, cabbage, cantaloupe, carrot, cauliflower, celery, Chinese cabbage, corn, cowpea, cucumber, eggplant, garbanzo, garlic, kale, kohlrabi, lettuce, mustard, okra, onion, pea, pepper, potato, sweet potato, radish, spinach, squash, tomato, turnip, and watermelon. This species also feeds on alfalfa, clover, cotton, rice, sorghum, strawberry, sugarbeet, tobacco, and sometimes grains and grasses. Suitability of several grasses and weeds for larval development was studied by Busching and Turpin (1977); among the relatively suitable plants were bluegrass, *Poa pratensis*; curled dock, *Rumex crispus*; lambsquarters, *Chenopodium album*; yellow rocket, *Barbarea vulgaris*; and redroot pigweed, *Amaranthus retroflexus*. The preference by black cutworm for weeds is sometimes quite pronounced. Genung (1959), for example, demonstrated how black cutworm, the dominant cutworm species in southern Florida during the winter, would avoid feeding on beans if spiny amaranth, *Amaranthus spinosus*, was present. Adults feed on nectar from flowers. Deciduous trees and shrub such as linden, wild plum, crabapple, and lilac are especially attractive to moths (Wynne *et al.*, 1991).

Natural Enemies. Numerous species of natural enemies have been associated with black cutworm, but data on their relative importance are scarce. However, Puttler and Thewke (1970) collected black cutworm larvae in Missouri and reported 69% parasitism, so natural enemies probably exact a significant toll on cutworm populations. Among the wasps known to attack this cutworm are *Cotesia marginiventris* (Cresson), *Microplitis feltiae* Muesebeck, *Microplitis kewleyi* Muesebeck, *Meteorus autographae* Muesebeck, *Meterorus leviventris* (Wesmael) (all Hymenoptera: Braconidae); *Campoletis argentifrons* (Cresson), *Campoletis flavicincta* (Ashmead), *Hyposoter annulipes* (Cresson), and *Ophion flavidus* Brulle (all Hymenoptera: Ichneumonidae). Larvae parasitized by *Meteorus leviventris* (Wesmael) consume about 24% less foliage and cut about 36% fewer seedlings (Schoenbohm and Turpin, 1977), so considerable benefit is derived from parasitism in addition to the eventual death of the host larva. Other parasitoids known from black cutworm include flies often associated with other ground-dwelling noctuids, including *Archytas cirphis* Curran, *Bonnetia comta* (Fallén), *Carcelia formosa* (Aldrich and Webber), *Chaetogaedia monticola* (Bigot), *Eucelatoria armigera* (Coquillett), *Euphorocera claripennis* (Macquart), *Gonia longipulvilli* Tothill, *G. sequax* Williston, *Lespesia archippivora* (Riley), *Madremyia saundersii* (Williston), *Sisyropa eudryae* (Townsend), and *Tachinomyia panaetius* (Walker) (all Diptera: Tachinidae). Predatory ground-dwelling insects such as ground beetles (Coleoptera: Carabidae) apparently consume numerous larvae (Best and Beegle, 1977; Lund and Turpin, 1977). Although Genung (1959) indicated that 75–80% of cutworms could be killed by a granulosis virus, there is surprisingly little information on epidemiology and of natural pathogens. Rather, such pathogens as viruses, fungi, bacteria, and protozoa from other insects have been evaluated for black cutworm susceptibility; in most cases only relatively weak pathogens have been identified (Ignoffo and Garcia, 1979; Grundler *et al.*, 1987; Johnson and Lewis, 1982a,b). An entomopathogenic nematode, *Hexamermis arvalis* (Nematoda: Mermithidae), is known to parasitize up to 60% of larvae in the midwestern states (Puttler and Thewke, 1971; Puttler *et al.*, 1973).

Life Cycle and Description. The number of generations occurring annually varies from 1–2 in Canada to 2–4 in the United States. In Tennessee, moths are present in March–May, June–July, July–August, and September–December. Based on light trap collections, moths are reported to be abundant in Arkansas during May–June and September–October (Selman and Barton, 1972), and in New York they occur mostly in June–July (Chapman and Lienk, 1981). However, light traps are not very effective during the spring flight, and underestimate early season densities (Willson *et al.*, 1981; Levine *et al.*, 1982). Thus, the phenology of black cutworm remains uncertain, or perhaps is inherently variable owing to the vagaries associated with long-range dispersal. Overwintering has been reported to occur in the pupal stage in most areas where overwintering occurs, but larvae persist throughout the winter in Florida. Pupae have been known to overwinter as far north as Tennessee, but apparently are incapable of surviving farther north (Story and Keaster, 1982a). Thus, moths collected in the midwestern states in March and April are principally dispersing individuals that are past their peak egg production period (Clement *et al.*, 1985). Nonetheless, they inoculate the area and allow production of additional generations, including moths which disperse north into Canada. Duration of the life cycle is normally 35–60 days.

Egg. The egg is white initially, but turns brown with age. It measures 0.43–0.50 mm long and 0.51–0.58 mm wide and is nearly spherical in shape, with a slightly flattened base. The egg bears 35–40 ribs that radiate from the apex; the ribs are alternately long and short. The eggs normally are deposited in clusters on foliage. Females may deposit 1200–1900 eggs. Duration of the egg stage is 3–6 days.

Larva. There are 5–9 instars, with a total of 6–7 instars most common. Head capsule widths are about 0.26–0.35, 0.45–0.53, 0.61–0.72, 0.90–1.60, 2.1–2.8, 3.2–3.5, 3.6–4.3, and 3.7–4.1 mm for instars 1–8, respectively. Head capsule widths are very similar for instars 1–4, but thereafter those individuals that display 8–9 instars show only small increments in width at each molt and eventually attain head capsule sizes no larger than those displaying only 6–7 instars. Larval body length is reported to be 3.5, 5.3–6.2, 7, 10, 20–30, 30–45, 50, and 50 mm for instars 1–8, respectively. Duration of the larval stage is normally 20–40 days. Mean duration of instars 1–6 was reported to be 6.0, 5.0, 4.6, 4.3, 5.6, 4.0 days, respectively, at 22°C. Larval development is strongly influenced by temperature, with the optimal temperature about 27°C. Humidity is less important, but instars 1–5 thrive best at higher humidities.

In appearance, the larva is rather uniformly colored on the dorsal and lateral surfaces, ranging from light gray or gray-brown to nearly black. On some individuals, the dorsal region is slightly lighter or brownish, but the larva lacks a distinct dorsal band. Ventrally, the larva tends to be lighter in color. Close examination of the larval epidermis reveals that this species

Black cutworm larva.

Adult black cutworm.

bears numerous dark, coarse granules over most of its body. The head is brownish with numerous dark spots. Larvae usually remain on the plant until the fourth instar, when they become photonegative and hide in the soil during the daylight hours. In these latter instars they also tend to sever plants at the soil surface, pulling the plant tissue below-ground. Larvae tend to be cannibalistic. (See color figure 43.)

Pupa. Pupation occurs below-ground at a depth of 3–12 cm. The pupa is 17–22 mm long and 5–6 mm wide, with dark brown. Duration of the pupal stage is normally 12–20 days.

Adult. The adult is fairly large in size, with a wingspan of 40–55 mm. The forewing, especially the proximal two-thirds, is uniformly dark brown. The distal area is marked with a lighter irregular band, and a small but distinct black dash extends distally from the bean-shaped spot. The hind wings are whitish to gray, and the veins marked with darker scales. The adult pre-oviposition period is about 7–10 days. Moths select low-growing broadleaf plants preferentially for oviposition, but lacking these, they deposit eggs on dead plant material. Soil is an unsuitable oviposition site (Busching and Turpin, 1976). (See color figure 228.)

The life cycle of black cutworm was described by Harris *et al.* (1962b) and Abdel-Gawaad and El-Shazli (1971). Developmental data were provided by Satterthwait (1933), Luckmann *et al.* (1976), Archer *et al.* (1980), and Beck (1988). Laboratory culture on artificial media has been described (Reese *et al.*, 1972; Blenk *et al.*, 1985). Rings *et al.* (1974b) published a bibliography. The larva was described, and included in a key, by Crumb (1929, 1956). Larvae are also included in keys published by Okumura (1962), Rings (1977b), Oliver and Chapman (1981), and Capinera (1986), and is included in a key to armyworms and cutworms in Appendix A. Moths are included in keys of Rings (1977a) and Capinera and Schaefer (1983).

Damage

This species occurs frequently in many crops, and is one of the best-known cutworms. Despite the frequency of occurrence, however, it tends not to appear in great abundance, as is known in some other cutworms and armyworms. Black cutworm is not consid-

ered to be a climbing cutworm, most of the feeding occurring at soil level. However, larvae will feed above-ground until about the fourth instar. Larvae can consume over 400 sq cm of foliage during their development, but over 80% occurs during the terminal instar, and about 10% in the instar immediately preceding the last (Satterthwait, 1933). Thus, little foliage loss is possible during the early stages of development. Once the fourth instar is attained, larvae can do considerable damage by severing young plants, and a larva may cut several plants in a single night. Plants tend to outgrow their susceptibility to injury. Showers *et al.* (1983) demonstrated that corn at the one-leaf stage was very susceptible to damage, but that by the four- or five-leaf stage plant yield was not decreased by larval feeding. Levine *et al.* (1983) showed that leaf feeding and cutting above the soil line were less damaging to corn than cutting at the soil surface, and that subterranean damage was very injurious.

Management

Sampling. Adult populations can be monitored with both blacklight and sex pheromone traps. However, several authors have noted the inefficiency of light traps. Light traps are relatively effective in the summer and autumn, but the late-season generations generally pose little threat to crops. Pheromone traps are more effective during the spring flight, when larvae present the greatest threat to young plants (Willson *et al.*, 1981). Trap color affects moth capture rate, with white and yellow traps capturing more than green traps (Hendrix and Showers, 1990).

Large larvae burrow in the soil, and are difficult to observe. However, larvae can be sampled with bait traps, and this is most effective before emergence or planting of seedlings. Various trap designs have been studied, but many employ a container sunk into the soil with the upper lip at the soil surface. The container is baited with fresh plant material and/or bran, and with vermiculite so that the larvae can attain shelter. Larvae are effectively captured in baited containers if the vermiculite is not very near the surface (Story and

Keaster, 1983), and catches are enhanced if a screen cylinder, which provides a visual stimulus to the cutworms, is suspended above the baited container (Whitford and Showers, 1984). If plants are present in the field they compete with the bait in the traps, and trap efficiency declines markedly. The distribution of larvae in the spring is random (Story and Keaster, 1982b).

Insecticides. Persistent insecticides are commonly applied to plants and soil for black cutworm suppression, but surface rather than subsurface soil applications are desirable (Foster *et al.*, 1990). Larvae readily accept insecticide-treated bran and other baits (Sechriest and Sherrod, 1977; Gholson and Showers, 1979). Application of systemic insecticides to seeds also provides some protection against larval injury (Levine and Felsot, 1985; Berry and Knake, 1987). *Bacillus thuringiensis* is not usually recommended for cutworm control.

Cultural Practices. Black cutworm larvae feed readily on weeds, and destruction of weeds can force larvae to feed exclusively on crop plants, exacerbating damage. Thus, it is often recommended that weeds not be tilled or treated with herbicide until larvae are matured. Timing is important, however, because prolonged competition between crop and weed plants can reduce crop yield (Engelken *et al.*, 1990). Presence of flowering weeds also can be beneficial by supporting prolonged survival of parasitoids (Foster and Ruesink, 1984). In contrast, reduced tillage cropping practices, which often produce higher weed populations, seem to result in increased abundance of black cutworm and higher levels of cutting in corn (Johnson *et al.*, 1984; Tonhasca and Stinner, 1991; Willson and Eisley, 1992). This may be due, in part, to the tendency of moths to oviposit on weeds; weedy fields tend to have higher cutworm populations (Sherrod *et al.*, 1979).

Black cutworm populations also tend to be higher in wet areas of fields, and in fields that are flooded. Black cutworm has been known, at times, as "overflow worm," due to its tendency to become abundant and damaging in fields that are flooded by overflowing rivers (Rockwood, 1925).

In the home garden, barriers are sometimes useful to prevent damage to seedlings by cutworms. Metal or waxed-paper containers with both the top and bottom removed can be placed around the plant stem to deter consumption. Aluminum foil can be wrapped around the stem to achieve a similar effect. As the larvae burrow and feed below the soil line it is necessary to extend to barrier below the soil surface. Because black cutworm moths, which easily circumvent such barriers, are active during the growing season; this procedure alone may have little value. Use of netting or row covers, in addition to larval barriers, can prove more effective.

Biological Control. Entomopathogenic nematodes (Nematoda: Steinernematidae and Heterorhabditidae) will infect and kill black cutworm larvae, but their populations normally need to be supplemented to realize high levels of parasitism (Capinera *et al.*, 1988; Levine and Oloumi-Sadeghi, 1992). Their effectiveness is related to soil moisture conditions (Baur *et al.*, 1997).

Bronzed Cutworm
Nephelodes minians Guenée
(Lepidoptera: Noctuidae)

Life History

Distribution. Bronzed cutworm is widespread in distribution, occurring throughout the United States except for the southernmost tier of states. It also is found in southern Canada, from the Maritime Provinces to British Columbia. Despite the broad distribution of this species, its economic impact is limited to the eastern portion of its range, as far west as the Rocky Mountains. Also, it rarely is known to be damaging south of Kansas, Missouri, and Virginia. It is a native species.

Host Plants. Bronzed cutworm larvae feed on grasses and such grain crops as barley and wheat. It is most frequently considered a pest of pasture and lawn grasses, especially *Poa* spp., and occasionally it damages field crops such as clover and sugarbeet. It commonly damages corn in the midwestern states, and when preferred plants are exhausted it may feed on other vegetables. On occasion, larvae also are observed to climb fruit trees and feed on the buds and leaves.

Natural Enemies. Parasitoids and predators, though observed, seem less significance as mortality factors than viral diseases. Wasps known to attack bronzed cutworm include *Rogas terminalis* (Cresson), *Apanteles rufocoxalis* Riley (both Hymenoptera: Braconidae), and *Campoletis oxylus* (Cresson) (Hymenoptera: Ichneumonidae). Among parasitic flies reared from bronzed cutworm are *Aplomya trisetosa* (Coquillett), *Euexorista futilis* (Osten Sacken), *Phryxe pecosensis* (Townsend), and *Tachinomyia variata* Curran (all Diptera: Tachinidae). Western yellowjacket, *Vespula pensylvanica* (Saussure), is reported to prey on bronzed cutworm moths (Warren, 1990). A polyhedrosis virus has long been considered to be an important mortality factor (Walkden, 1937), but a granulosis virus also has been reported (Steinhaus, 1957).

Life Cycle and Description. There is a single generation per year over the entire range of this insect.

Adults are present in the autumn. In New York, moth flights consistently occur in September (Chapman and Lienck, 1981), in Minnesota they occur in late August and early September (Knutson, 1944), and in the central Great Plains their flights occur in September and October (Walkden, 1950). The eggs overwinter; however, egg hatch occurs early in the year, often in January and February. Larvae complete their development in April or May, and become quiescent until July or August, when pupation occurs.

Egg. Moths are reported to scatter eggs singly at the surface of the soil. In shape, the egg has a slightly compressed sphere. It measures about 0.93 mm wide and 0.77 mm high. The egg is marked with about 250 minute ribs that radiate outward from the center. In color, the egg is initially grayish, but soon acquires a pinkish tint and then a hint of purple. Duration of the egg stage, which is the overwintering form, is quite variable, but Walkden (1937) reported a mean of 127 days (range 98–145 days).

Larva. There are 6–7 instars. Head capsule widths are about 0.5, 0.8, 1.2, 2.1, 3.0, and 4.3 mm for instars 1–6, respectively. Mean development time for larvae with six instars is 30.0, 10.5, 9.0, 8.2, 13.5, and 158.3 days, respectively. For larvae exhibiting seven instars, the duration of the first five instars is the same as with six-instar larvae, but duration of instar six is 15.6 days, and instar seven is 147.3 days. Total larval development time is estimated to be 230 days. Body length increases from 3–5 mm in the first instar to about 35–45 mm at maturity. The mature instar is very distinctive in appearance, with a shiny bronze body and five sharply defined, broad stripes running to the length of the body. The stripes are whitish to yellowish. The head is orangish-brown. The first four instars differ in background color, in that they are green instead of bronze, but also are marked with longitudinal stripes as found in the latter instars. (See color figure 44.)

Pupa. Mature larvae form a small cell in the soil for pupation. The pupa is brown and measures 23–33 mm long and 8–11 mm wide. Duration of the pupal stage is reported to average 27.3 days (range 24–34 days).

Adult. The moth is reddish brown, the front wings marked with an irregular dark-brown band crossing the wing centrally. Both the front and hind wings may display a reddish or violet tint. The moth measures about 35–50 mm in wingspan. Adults live for about 14 days, and commence oviposition when about two-day old. Based on dissection of eggs from adults, oviposition potential of about 1000 eggs is estimated. (See color figure 229.)

Detailed description and biology of bronzed cutworm was given by Crumb (1926) and Walkden (1937). A bibliography was published by Rings *et al.* (1974a). A comprehensive key to larvae of the Noctuidae, including this species, was presented by Crumb (1956). It is also included in less inclusive keys for caterpillar pests in Nebraska (Whelan 1935) and Colorado (Capinera, 1986), and in a key to armyworms and cutworms in Appendix A. Moths are included in pictorial keys by Rings (1977a) and Capinera and Schaefer (1983).

Damage

Larvae are defoliators, and consume the leaves and stems of young plants. As they are present early in the year they normally damage only early-season plants.

Management

Larvae have been controlled successfully with applications of residual insecticides to the soil and foliage. *Bacillus thuringiensis* is not often recommended for cutworms. Although there seems to be no report of experimentation with baits, treated bran would likely prove effective.

Cabbage Looper
Trichoplusia ni (Hübner)
Lepidoptera: Noctuidae

Natural History

Distribution. The origin of cabbage looper is uncertain, but it is now found in Africa, Asia, Europe, and in the Americas. In North America, it is found

Bronzed cutworm larva.

throughout Canada, Mexico, and the United States wherever crucifers are cultivated. Overwintering in the United States apparently occurs only in the southernmost states, however. It is somewhat erratic in occurrence—typically very abundant one year, and then scarce for 2–3 years; this is likely due to nuclear polyhedrosis virus and differing success in dispersal from southern to northern areas. Cabbage looper is highly dispersive, and is sometimes found at high altitudes and far from shore. Flight ranges of approximately 200 km have been estimated.

Host Plants. Cabbage looper feeds on a wide variety of cultivated plants and weeds. As its common name implies, it feeds readily on crucifers, and has been reported damaging broccoli, cabbage, cauliflower, Chinese cabbage, collards, kale, mustard, radish, rutabaga, turnip, and watercress. Other vegetable crops injured include beet, cantaloupe, celery, cucumber, lima bean, lettuce, parsnip, pea, pepper, potato, snap bean, spinach, squash, sweet potato, tomato, and watermelon. Additional hosts are flower crops such as chrysanthemum, hollyhock, snapdragon, and sweetpea, and field crops such as cotton and tobacco. With such a wide range of crops serving as suitable hosts, it is not surprising that an extremely wide range of broadleaf weeds also serve as hosts (Eichlin and Cunningham, 1978; Soo Hoo *et al.* 1984), though the weeds vary somewhat with geographic locale, depending on availability. Surprisingly few common agricultural weeds are frequent hosts; among those that are suitable are lambsquarters, *Chenopodium album*; wild lettuce, *Lactuca* spp.; dandelion, *Taraxacum officinale*; and curly dock, *Rumex crispus*.

Despite the wide host range of cabbage looper, not all hosts are equivalent. For example, Elsey and Rabb (1967) compared the suitability of collards and tobacco and reported differing suitabilities. Adults preferred to oviposit on collards when given a choice, and early instar survival was higher on collards. However, once the third instar was attained, survival was equivalent on both hosts, and pupal weights were only slightly diminished when larvae were reared on tobacco. Sutherland (1966) studied the growth rate of cabbage looper fed on a wide variety of crops and weeds. He found relatively small differences among the different crucifer crops, and many other vegetable crops and some weeds were equally suitable for larval growth. Soo Hoo *et al.* (1984) conducted one of the most complete studies of relative suitability, and reported that only about one-third of the plants tested were suitable for complete development of larvae.

A survey of looper pests infesting crops in Alabama revealed that though cabbage looper could be recovered from numerous hosts (clover, cotton, crucifers, peanut, soybean, sweet potato, tomato), most were found on cotton and crucifer crops. Soybean looper, *Pseudoplusia includens* (Walker), an insect easily confused with cabbage looper and having a similar broad host range, occurred predominately on soybean (Canerday and Arant, 1966).

The adults feed on nectar from a wide range of flowering plants, including clover, *Trifolium* spp.; goldenrod, *Solidago canadensis*; dogbane, *Apocynum* spp.; sunflower, *Helianthus* spp.; and others.

Natural Enemies. Cabbage looper is attacked by numerous natural enemies, and the effectiveness of each seems to vary spatially, temporally, and with crop environment. Most studies noted the effectiveness of wasp and tachinid parasitoids, and a nuclear polyhedrosis virus (NPV). Predation has not been well-studied except in cotton.

In studies conducted on collards in North Carolina, Elsey and Rabb (1970a) observed considerable variation in the impact of natural enemies between years. They determined that *Trichogramma* (Hymenoptera: Trichogrammatidae) egg parasitoids were not very important; they were reared from less than 5% of the eggs. They identified no major mortality factors until the fifth instar, despite the presence of considerable 'disappearance' during this period. Either predators or weather could account for these larval deaths. During the latter instars, *Voria ruralis* (Fallén) (Diptera: Tachinidae), an endoparasite attacking the medium- or large-size larvae, was the dominant cause of death, accounting for an average of about 53% mortality. Elsey and Rabb (1970b) presented the biology of this important parasitoid. *Trichoplusia ni* NPV caused about 12% mortality, and undetermined fungi about 10%. *Copidosoma truncatellum* (Dalman) (Hymenoptera: Encyrtidae) was the other significant mortality factor, but accounted for only 6–7% mortality. *Copidosoma truncatellum* oviposits in cabbage looper eggs, emerging from and killing the mature larvae or prepupae.

In studies conducted in California involving cabbage, Oatman and Platner (1969) reported that egg parasitism of cabbage looper by *Trichogramma*, while variable, could reach about 35%. Larval parasitism averaged 38.9%, and tended to increase toward the end of the year. The tachinid *V. ruralis* was the dominant parasitoid, and was especially abundant in the autumn and winter months. The other principal parasitoids, especially during summer and autumn, were *C. truncatellum* and *Hyposoter exiguae* (Viereck) (Hymenoptera: Ichneumonidae). The latter species is a solitary endoparasite that attacks small larvae. A total of 24 species of parasitoids were observed: 14 wasps and 10 flies. Despite the abundance of parasitoids, however, the authors concluded that *T. ni* NPV was the key factor affecting populations.

In Ontario, Harcourt (1963a) indicated that *C. truncatellum* was the most important parasitoid. No data were provided, but cabbage looper populations were said to be "frequently destroyed" by NPV.

One of the most complete studies of cabbage looper natural enemies was conducted in California by Ehler (1977a,b), on cotton. He determined that the egg and early larval stages experience most of the generational mortality, and that predators and *C. truncatellum* were the most important elements contributing to this mortality. During the early larval instars, the minute pirate bug, *Orius tristicolor* (White) (Hemiptera: Anthocoridae), the big-eyed bug, *Geocoris palens* Stål (Hemiptera: Lygaeidae), and the damsel bug, *Nabis americoferis* Carayon (Hemiptera: Lygaeidae) were the predators responsible for most of the mortality. Ehler documented several mortality factors during the middle larval instars. The parasitoid *Microplitis brassicae* Muesebeck (Hymenoptera: Braconidae), a solitary endoparasite attacking small larvae, was the dominant mortality factor, but rarely exceeded 20% mortality. Other factors included *H. exiguae* and *T. ni* NPV, but both at levels of less than 10% mortality. During the late larval instars *C. truncatellum* inflicted 40–50% mortality, and *V. ruralis* and *T. ni* NPV each caused less than 10% mortality. Pupal mortality was insignificant.

Predation is rarely studied as it is very difficult to measure accurately. Barry *et al.* (1974) attempted to assess the potential of selected predators of cabbage looper by using caged populations in Missouri. They reported that the damsel bug *Nabis alternatus* Parshley (Hemiptera: Nabidae) was most effective, the big-eyed bug *Geocoris punctipes* (Say) (Hemiptera: Lygaeidae) was intermediate, and the green lacewing *Chrysoperla carnea* (Stephens) was relatively ineffective in predation of cabbage looper on soybean. Sutherland (1966) suggested that general predators, including yellowjackets (Hymenoptera: Vespidae) and birds, could be important mortality factors.

The *T. ni* NPV is well-studied. Larvae normally die within 5–7 days of consuming virus inclusion bodies. Early signs of larval infection are a faint mottling of the abdomen in the area of the third to the sixth abdominal segments. This is followed by a more generalized blotchy appearance, and the caterpillar eventually becomes creamy white swollen, and limp. Death usually follows within hours following the limp condition, and caterpillars are often found hanging by their prolegs. Dark blotches appear after death, and the integument becomes very fragile and eventually ruptures. The body contents, heavily contaminated with new inclusion bodies, drip onto foliage where they can be consumed by other larvae (Semel, 1956; Drake and McEwen, 1959). Hofmaster (1961) reported that looper populations in Virginia were highest during dry weather because rainfall assisted the spread of NPV, and that this virus greatly suppressed loopers. In New York, Sutherland (1966) indicated that though *T. ni* NPV was an important mortality factor, natural incidence did not appear adequate to protect crops from damage.

Life Cycle and Description. The number of generations completed per year varies from two to three in Canada, five in North Carolina, from five to seven in California. The generations overlap considerably, and therefore are indistinct. Development time (egg to adult) requires 18–25 days when insects are held at 32–21°C, respectively (Toba *et al.*, 1973), so at least one generation per month could be completed successfully under favorable weather conditions. There is no diapause present in this insect, and though it is capable of spending considerable time as a pupa, it does not tolerate prolonged cold weather. It reinvades most of the United States and all of Canada annually after overwintering in southern latitudes. The lower limit for development is about 10–12°C, and 40°C is fatal to some stages. Cabbage looper is considered to be a warm-weather insect; even in areas where it successfully overwinters it rarely occurs in high numbers until there has been adequate time for 2–3 spring generations. Sutherland (1966) conducted a survey of entomologists along the Atlantic coast, and reported that looper populations were present year-round as far north as coastal South Carolina, and that looper infestations commenced in North Carolina and Maryland in May, in New Jersey in June, and in New York in July. Pennsylvania was not infested until August. Subsequent research by Chalfant *et al.* (1974) clarified the winter activity patterns of cabbage looper in the southeastern United States: continuous activity and reproduction occur only in the part of Florida south of Orlando; the part of Georgia south of Byron as well as southeast South Carolina have intermittent adult activity during the winter months, depending on weather; all points north of this have no winter activity.

Egg. Cabbage looper eggs are hemispherical in shape, with the flat side affixed to foliage. They are deposited singly on either the upper or lower surface of the leaf, though clusters of 6–7 eggs are not uncommon. The eggs are yellowish-white or greenish, bear longitudinal ridges, and measure about 0.6 mm wide and 0.4 mm high. They hatch in about two, three, and five days at 32°, 27°, and 20°C, respectively, but require nearly 10 days at 15°C (Jackson *et al.*, 1969).

Cabbage looper egg.

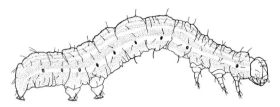

Cabbage looper larva.

Larva. Young larvae initially are dusky white, but become pale green as they commence feeding on foliage. They are somewhat hairy initially, but the number of hairs decreases rapidly as larvae mature. Larvae have three pairs of prolegs, and crawl by arching their back to form a loop and then projecting the front section of the body forward. The mature larva is predominantly green, but is usually marked with a distinct white stripe on each side. The thoracic legs and head capsule are usually pale green or brown. Dorsally, the larva bears several narrow, faint white stripes clustered into two broad white bands. In some cases the mature larva is entirely green. The body is narrower at the anterior end, and broadens toward the posterior. It measures 3–4 cm long at maturity. Cabbage looper is easily confused with other loopers, but can be distinguished from most by the presence of small, nipple-like structures (vestigial prolegs) located ventrally on abdominal segments three and four. Soybean looper, *Pseudoplusia includens* (Walker), also bears these structures, but usually has dark thoracic legs. Also, under high magnification it is possible to observe microspines on the body of soybean looper—a feature lacking from cabbage looper.

The number of instars was given as 4–7 by Shorey *et al.* (1962), but many authors indicated only five. McEwen and Hervey (1960) gave mean head capsule width measurements as 0.29, 0.47, 0.74, 1.15, and 1.79 mm, respectively, for instars 1–5. Larval development required 17.8 and 19.9 days when reared on bean and held at 23° and 32°C, respectively. When reared on cabbage at the same two temperatures, larval development required 19.9 and 20.8 days, respectively (Shorey *et al.*, 1962). Development was also studied by Toba *et al.* (1973), who determined that the number of larval instars could be increased from five to six by exposing the larvae to cooler temperature. Cool temperature also resulted in lowered egg production by ensuing adults. (See color figure 45.)

Pupa. At pupation, a white, thin, fragile cocoon is formed on the underside of foliage, in plant debris, or among clods of soil. The pupa contained within is initially green, but soon turns dark brown or black.

The pupa measures about 2 cm long. Duration of the pupal stage is about 4, 6, and 13 days at 32°, 27°, and 20°C, respectively.

Adult. The front wings of the cabbage looper moth are mottled gray-brown; the hind wings are light brown at the base, with the distal portions dark-brown. The forewing bears silvery white spots centrally: a "U"-shaped mark and a circle or dot that are often connected. The forewing spots, though slightly variable, serve to distinguish cabbage looper from the most other crop-feeding noctuid moths. The moths have a wingspan of 33–38 mm.

After a pre-oviposition period of about 1–2 days, females begin depositing their eggs, initially at about 80 per night. During the adult stage, which averages 10–12 days, 300–600 eggs are produced by females having access to food, and less than 100 when only water is provided (Shorey, 1963). Moths are considered seminocturnal because feeding and oviposition sometimes occurs about dusk. They may become active on cloudy days or during cool weather, but are even more active during the night-time hours. They oviposit readily at temperatures as low as 15.6°C (Henneberry and Kishaba, 1967), but flight activity is higher on warmer evenings (Sutherland, 1966). (See color figures 230 and 231.)

Rearing procedures was given by McEwen and Hervey (1960) for plant-based culture, and Shorey and Hale (1965) for artificial diet. Cabbage looper larvae are included in the keys by Okumura (1962) and Capinera (1986), and are included in a key to common loopers in Appendix A. Adults are included in the keys of Rings (1977a) and Capinera and Schaefer (1983).

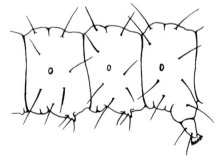

Abdominal segments of cabbage looper larva showing vestigial prolegs on segments three and four.

Adult male cabbage looper.

Damage

Cabbage looper replaced imported cabbageworm, *Pieris rapae* (Linnaeus), as the dominant cabbage caterpillar in the 1950s, apparently due to greater susceptibility of the latter to most insecticides. In recent years, diamondback moth has emerged as a more important caterpillar pest than cabbage looper; nevertheless, *T. ni* can be a serious problem. In studies conducted in South Carolina, diamondback moth was the major caterpillar pest in the spring crucifer crop, whereas cabbage looper predominated in the autumn crop (Reid and Bare, 1952). In Florida and Texas, however, spring populations of cabbage loopers can be damaging.

Cabbage loopers are leaf feeders, and in the first three instars they confine their feeding to the lower leaf surface, leaving the upper surface intact. The fourth and fifth instars chew large holes, and usually do not feed at the leaf margin. For cabbage, however, they feed not only on the wrapper leaves, but also may bore into the developing head. Larvae consume three times their weight in plant material daily (McEwen and Hervey, 1960). Feeding sites are marked by large accumulations of sticky, wet fecal material.

Despite their voracious appetite, larvae are not always as destructive as presumed. In California studies, feeding on celery during the first one-half of the growing season did not constitute loss because these petioles were routinely stripped from the plant at harvest (van Steenwyk and Toscano, 1981). With cabbage, moderate defoliation before head formation is similarly irrelevant. In Texas, average population densities of 0.3 larvae per plant justify control (Kirby and Slosser, 1984). In New York, Ohio, and Ontario, a density of 0.5 larvae per plant has been used as a treatment threshold (Shelton *et al.*, 1983b). In Florida and Georgia, one new feeding site per head is considered the damage threshold (Workman *et al.*, 1980). Recent work

in Canada suggested that an appropriate action threshold was 40% of plants infested (Dornan *et al.*, 1995). Cabbage looper can be a serious contaminant of fresh market broccoli and processed peas.

Management

Sampling. Various sampling strategies have been developed for cabbage looper, and many approaches include consideration of the other crucifer-feeding caterpillars. Fixed sample units of at least 40 plants are sometimes recommended. However, sequential sampling (Shepard, 1973a) and variable intensity sampling (Hoy *et al.*, 1983) protocols have been developed to minimize the amount of sampling required to make appropriate management decisions. Dornan *et al.* (1995) recommended a binomial (presence-absence) approach because it eliminated counting and insect identification.

Blacklight traps and pheromone traps have been used in an attempt to predict looper population densities. Moth catches are monitored effectively by light traps (Hofmaster, 1961), but NPV, spread by rain, affects larval abundance and damage, thereby reducing predictability. The cabbage looper sex pheromone has at least seven chemical components, but not all are required to elicit attraction (Linn *et al.*, 1984). Pheromone releasers and blacklight traps can be combined to increase moth catches, an approach that has been studied for area-wide suppression of cabbage loopers (Gentry *et al.*, 1971). Although numerous moths have been trapped by such techniques, and insects significantly decreased, suppression has not proven to be adequate to protect lettuce from damage (Debolt *et al.*, 1979).

Insecticides. Insecticide resistance has become a problem in cabbage looper control, but susceptibility varies widely among locations (Shelton and Soderlund, 1983). Botanical insecticides such as rotenone are less effective against cabbage looper than other cabbage-feeding Lepidoptera (Dills and Odland, 1948), but neem functions as both a feeding deterrent and growth regulator (Isman, 1993).

Biological Control. Microbial insecticides currently play a role in cabbage looper management, and their potential role has yet to be fully realized. *Bacillus thuringiensis* has long been used for effective suppression of cabbage looper (Kennedy and Oatman, 1976; Gharib and Wyman, 1991; Leibee and Savage, 1992), and has the advantage of not disrupting populations of beneficial insects. *T. ni* NPV is effective (Hall, 1957), but it has not been commercialized because of the narrow host range. Home gardeners sometimes collect loopers dying of *T. ni* NPV, grind up the larval cadavers, and concoct their own effective microbial control agent. A nuclear polyhedrosis virus from alfalfa looper, *Autographa californica* (Speyer), has a wide host range, including cabbage looper (Jaques, 1977; Vail *et al.*, 1980; Tompkins *et al.*, 1986); it likely will become a useful tool for cabbage looper management.

Mass release of *Trichogramma* spp. has been investigated for cabbage looper suppression. Looper egg parasitism can be increased several-fold by careful timing of parasitoid release (Oatman and Platner, 1971). Effectiveness varies among crops, however. This approach was most suitable in tomato, but also effective in crucifers and pepper (Martin *et al.*, 1976b).

Cultural Practices.   Some differences in crucifer susceptibility have been observed. In New York, Dickson and Eckenrode (1975) found few significant differences, but red cabbages tended to be more resistant than kale or Chinese cabbage. In Wisconsin, Chinese cabbage, mustard, rutabaga, and turnip were less preferred for oviposition, whereas cabbage, Brussels sprouts, and collards were highly preferred. Unfortunately, there was no correlation between crops and varieties resistant to cabbage looper, and resistance to imported cabbageworm (Radcliffe and Chapman, 1966). Among cabbage cultivars studied in North Carolina, mammoth red rock and savoy perfection drumhead cultivars are considered to be relatively resistant, but this resistance dissipated under heavy insect feeding pressure. Interestingly, in this case the resistant varieties received high numbers of cabbage looper eggs, but larval survival was poor (Chalfant and Brett, 1967). In studies of broccoli susceptibility in Virginia, Vail *et al.* (1991) found that early maturing varieties were less subject to attack than late maturing varieties. Row covers, where economically practical, are effective at preventing cabbage looper moths from depositing eggs on crops.

Celery Looper
Anagrapha falcifera (Kirby)
(Lepidoptera: Noctuidae)

Natural History

Distribution.   Celery looper is found throughout the United States and southern Canada. There is some question whether this native species overwinters in the northern United States and Canada. Research conducted in Iowa suggested that celery looper did not overwinter successfully, but was carried into the area in the spring when the appropriate weather patterns developed (Peterson *et al.*, 1988). This is highly plausible, as many other noctuids similarly overwinter in the south and disperse northward annually. However,

there also are reports of this insect overwintering in the north in the larval stage, and adult activity was reported in New York from April to November (Chapman and Lienk, 1981).

Host Plants.   The host range of this insect is poorly known, but it appears that celery looper feeds on a wide variety of plants. Among vegetables damaged are celery, beet, cabbage, carrot, lettuce, and pea. Corn has been reported to be a host plant, but this is questionable. It is an occasional pest of sugarbeet and has been reported to feed on cranberry and hollyhock. Weeds fed upon include dandelion, *Taraxacum officinale*; plantain, *Plantago* sp.; and a burdock, *Arctium lappa*. Coquillett (1881) reported oviposition on grass, so larvae also may develop on unknown grasses. Adults have been observed taking nectar from clover and lilac blossoms. Moths have also been found to be contaminated with pollen from several plant taxa, including *Quercus*, Rosaceae, and *Pyrus*, indicating an extensive range of adult host plants (Lingren *et al.*, 1993).

Natural Enemies.   Little is known concerning the insect enemies of celery looper. A nuclear polyhedrosis virus is widespread and highly pathogenic to celery looper, and may be the key factor that limits abundance of this insect. This virus is unusual in that it is also pathogenic to numerous other species of Lepidoptera, affecting over 30 species in 10 families (Hostetter and Puttler, 1991).

Life Cycle and Description.   Most reports suggest 2–3 generations annually in northern states (Knutson, 1944). However, the generations overlap and it is difficult to discern population dynamics solely from capture rates of adults. For example, Chapman and Lienk (1981) presented data from New York showing continuous occurrence of moths in all but the coldest months. Peterson *et al.* (1988) conducted a study in Iowa that included determination of ovary development, which aids in assessing the age of insects. Based on these studies, there were four generations annually.

Egg.   The eggs are milky white, and measure about 0.5 mm in diameter. They are somewhat flattened, and bear vertical ridges. They may be deposited singly or in small groups on either the upper and lower surfaces of vegetation. Duration of the egg stage is 4–6 days.

Larva.   There likely are five instars, with each of about 3–5 days duration. Total larval development time is usually about 21 days, and larvae attain a length of about 35 mm. The larvae are green, and tend to have a weak dark longitudinal line dorsally, accompanied by three narrow whitish lines on each side of the dorsal line. The thoracic legs are pale colored, and the abdominal tubercles located above the lateral spiracles are not black. The most distinctive marking is a narrow white line on each side, running through the lateral spiracles. The skin bears numerous minute spines, called microspines, over most of its surface. The spiracles are ringed with black pigment. If there is a dark bar on the head, it is indistinct. The caterpillar is more robust posteriorly, and bears three pairs of prolegs. It is easily confused with other common loopers. Celery looper can be distinguished from bilobed looper, *Megalographa biloba* (Stephens), and alfalfa looper, *Autographa californica* (Speyer) by its pale thoracic legs and absence of pronounced dark, lateral bars on the head; the latter species generally have dark bars and dark thoracic legs. Celery looper can be distinguished from cabbage looper, *Trichoplusia ni* (Hübner) by the presence of small, nipple-like structures located ventrally on the third and fourth abdominal segments, and the absence of dark microspines on the abdomen of the latter. (See color figure 46.)

Pupa. The mature larva spins a thin, white silken cocoon on the underside of a leaf or amongst debris, and pupates within. The pupa is blackish brown, and measures 13–15 mm long. Duration of the pupal stage is 10–20 days.

Adult. The moth measures 3.5–5 cm in wingspan. The front wings are dark, usually purplish brown and reddish brown, but the wings are marked distally with a silvery band. Many members of the noctuid subfamily Plusinae bear a spot near the center of the forewing, and this species is no exception. For celery looper, however, the spot is silver, and is drawn out into a curving line that terminates at the posterior margin of the wing. Although not completely unique to celery looper, this pattern can serve to differentiate this moth from other common vegetable pest such as cabbage looper, alfalfa looper, and bilobed looper. The front wings are also abnormally widened distally, due to a conical projection on the trailing edge of the forewing. The hindwings are yellowish brown, with darker brown bands and a whitish border. (See color figures 232 and 233.)

The biology of celery looper is poorly documented. Some elements of the biology were given by Coquillett (1881), Chittenden (1902), and Peterson *et al.* (1988). Information on rearing and artificial diet were found in Treat and Halfhill (1973). Keys for the differentiation of celery looper moths from related species were given by Rings (1977a), Eichlin and Cunningham (1978), and Capinera and Schaefer (1983). Larvae are included in the keys of Crumb (1956), Capinera (1986), and in a key to common vegetable-feeding loopers in Appendix A.

Adult celery looper.

Damage

The larva is a defoliator, eating holes in the leaves of lettuce, celery, and other crops. It has been known to be destructive in Florida (Ball *et al.*, 1932), but generally this insect is considered to be a minor pest. It sometimes is a serious contaminant of peas harvested for canning and freezing.

Management

Moths of this species can be monitored with a black-light trap. Butler *et al.* (1977) provided information on trapping celery looper with a sex pheromone. Insecticides applied to commercial or home garden crops for other insect pests are adequate to keep celery looper at very low levels of abundance. The microbial insecticide *Bacillus thuringiensis* is effective.

Clover Cutworm
Discestra trifolii (Hufnagel)
(Lepidoptera: Noctuidae)

Natural History

Distribution. Clover cutworm occurs throughout the United States except for the southeastern states. It also occurs throughout southern Canada and in Alaska. Apparently it is a native species, though it is also reported to occur in Europe and Asia.

Host Plants. This species is an occasional pest of several crops, preferring crops and weeds in the plant family Chenopodiaceae. Among vegetables attacked are beet, cabbage, lettuce, onion, pea, spinach, tomato, and turnip. Field crops consumed include alfalfa, cotton clover, flax, canola (rape), sugarbeet, and sunflower. The incidence of clover cutworm on sugarbeet explains the alternative common name, "striped beet caterpillar." However, weeds are the usual host. Weeds known to be suitable larval food plants include kochia, *Kochia scoparia*; lambsquarters, *Chenopodium album*; purslane, *Portulaca* sp.; and Russian thistle, *Salsola kali*.

Natural Enemies.   Walkden (1950) reported that of field collected larvae, 12% died from parasitic Hymenoptera, 16% from parasitic Diptera, and 22% from pathogens. Santiago–Alvarez and Federici (1978) reported *Euplectrus* sp. (Hymenoptera: Eulophidae) and *Euphorocera tachinomoides* Townsend (Diptera: Tachinidae) attacking larvae in southern California. Other wasps parasitizing clover cutworm are *Apanteles plathypenae* Muesebeck, *Meteorus leviventris* (Wesmael) (both Hymenoptera: Braconidae), and *Enicospilus merdarius* (Gravenhorst) (Hymenoptera: Ichneumonidae). Other parasitic flies include *Euphorocera claripennis* (Macquart) and *Lespesia archippivora* (Riley) (both Diptera: Tachinidae) (Arnaud, 1978). Both granulosis and a nuclear polyhedrosis viruses are known from clover cutworm larvae, and Federici (1978) reported the granulosis virus to be especially important in regulating insect density. Federici (1982) also described an unusual rickettsia-like organism in clover cutworm larvae, but there is no indication that this pathogen occurs frequently.

Life Cycle and Description.   There are three generations annually in Colorado and Kansas. Moth flights occur in late May, early July, and late August–early September. Pupae from the third generation overwinter. Knutson (1944) and Ayre *et al.* (1982a) suggested only two flights of adults in Minnesota and Manitoba, respectively, though the "second" flight period was protracted and may represent two overlapping periods of adult activity. Ayre *et al.* (1982a) suggested that the level of diapause induction in pupae was a critical determinant in overwintering survival in Manitoba; they speculated that if the last generation develops early in the season, diapause is not induced and the insects proceed with a generation that is not completed before the onset of winter.

Egg.   The eggs are deposited singly or in small clusters on the underside of leaves. They are white to pale yellow. They resemble a slightly flattened sphere and are equipped with ribs that radiate out from the top of the egg. The number of eggs produced by females is not well documented but seems to be in excess of 500 eggs, and is likely much greater. When reared on artificial diet, they generally produced 650–1700 egg per female (Santiago–Alvarez *et al.*, 1979). Duration of the egg stage is 4–5 days.

Larva.   The larvae are dull green. There may be a weak white line dorsally along the length of the body. The most distinctive character, however, is the combination of a broad lateral yellowish or pinkish band below the spiracles, and black pigmentation surrounding the spiracles. The black pigmentation forms a

Clover cutworm larva.

series of black spots immediately above the lateral band, and serves to distinguish this caterpillar from beet armyworm, *Spodoptera exigua* (Hübner), a species that is superficially quite similar (Capinera, 1986). Larvae attain a length of 35–40 mm at maturity. Duration of the larval stage is 16–22 days under warm conditions, but it may be extended to nearly 50 days by cool weather. (See color figure 47.)

Pupa.   Mature larvae burrow into the soil to a depth of about 2–3 cm to pupate. The pupa is reddish brown and measures 13–14 mm long. Duration of the pupal stage is 10–20 days during the spring and summer generations, but about 150 days for the overwintering population.

Adult.   The adults have a wingspan of 31–35 mm. The front wings are yellowish brown, and heavily marked with darker and lighter spots. The hind wings are grayish basally with a diffuse darker brown band distally. The compound eyes of this moth bear hairs—a feature that is usefull for distinguishing it from some similar species. Females produce a sex pheromone that has been identified and synthesized (Struble and Swailes, 1977a). (See color figure 234.)

This insect is not well-studied. Marsh (1913) gave a brief account of its biology. Walkden (1950) gave additional notes and provided a key to noctuids in the central Great Plains. Crumb (1956) described the mature larva and provided a key to larvae. The larva was also included in a key by Capinera (1986), and in a key to armyworms and cutworms in Appendix A. The moth was included in pictorial keys developed by Rings (1977a) and Capinera and Schaefer (1983). Santiago–Alvarez *et al.* (1979) reported a suitable artificial diet.

Damage

The larval stage defoliates plants, though it appears to favor weeds over crops. It is not a ground-dwelling species and does not sever plants at the soil surface, but climbs plants to feed on leaves. During periods of great abundance it has caused significant damage, and has been reported to assume a gregarious, dispersive "armyworm" habit.

Management

Natural enemies, especially pathogens, generally serve to keep the population in check. Adult populations can be monitored with pheromone traps (Struble and Swailes, 1977a; Swailes and Struble, 1979; Ayre *et al.*, 1982a). This species is reported to be easy to control with foliar applications of insecticides, though *Bacillus thuringiensis* is not often recommended. This species is a plant-inhabiting, climbing cutworm; therefore, the mechanical barriers recommended for protection of seedlings against ground-dwelling species are of little value. It is possible to deny access by ovipositing moths through use of netting and row covers, however.

Corn Earworm
Helicoverpa zea (Boddie)
(Lepidoptera: Noctuidae)

Natural History

Distribution. Corn earworm is found throughout North America except for northern Canada and Alaska. It tends to be less abundant west of the Rocky Mountains, and is infrequently a pest in Canada's Prairie Provinces. It also occurs in Hawaii and the Caribbean islands. Corn earworm is common in South America, persisting to a southern latitude of about 40°. Its origin is uncertain, but likely is native to North America.

In the eastern United States, corn earworm does not normally overwinter successfully in the northern states. It is known to survive as far north as about 40° north latitude, or about Kansas, Ohio, Virginia, and southern New Jersey, depending on the severity of winter weather (Blanchard, 1942). However, it is highly dispersive, and routinely spreads from southern states into northern states and Canada (Hardwick, 1965b; Fitt, 1989; Westbrook *et al.*, 1997). Thus, areas have overwintering, both overwintering and immigrant, or immigrant populations, depending on location and weather. In the relatively mild Pacific Northwest, corn earworm can overwinter at least as far north as southern Washington.

Host Plants. Corn earworm has a wide host range; hence, it is also known as "tomato fruitworm," "sorghum headworm," "vetchworm," and "cotton bollworm." In addition to corn and tomato, perhaps its most favored vegetable hosts, corn earworm also attacks artichoke, asparagus, cabbage, cantaloupe, collards, cowpea, cucumber, eggplant, lettuce, lima bean, melon, okra, pea, pepper, potato, pumpkin, snap bean, spinach, squash, sweet potato, and watermelon. Not all are good hosts, however. Harding (1976a), for example, studied relative suitability of crops and weeds in Texas, and reported that though corn and lettuce were excellent larval hosts, tomato was merely a good host, and broccoli and cantaloupe were poor. Other crops injured by corn earworm include alfalfa, clover, cotton, flax, oat, millet, rice, sorghum, soybean, sugarcane, sunflower, tobacco, vetch, and wheat. Among field crops, sorghum is particularly favored. Cotton is frequently reported to be injured, but this generally occurs only after more preferred crops have senesced. Fruit and ornamental plants may be attacked, including ripening avocado, grape, peaches, pear, plum, raspberry, strawberry, carnation, geranium, gladiolus, nasturtium, rose, snapdragon, and zinnia. In studies conducted in Florida, Martin *et al.* (1976a) found corn earworm larvae on all 17 vegetable and field crops studied, but corn and sorghum were most favored. In cage tests earworm moths preferred to oviposit on tomato over a selection of several other vegetables that did not include corn.

Such weeds as common mallow, *Malva neglecta*; crown vetch, *Coronilla varia*; fall panicum, *Panicum dichotomiflorum*; hemp, *Cannabis sativa*; horsenettle, *Solanum* spp.; lambsquarters, *Chenopodium album*; lupine, *Lupinus* spp.; morningglory, *Ipomoea* spp.; pigweed, *Amaranthus* sp.; prickly sida, *Sida spinosa*; purslane, *Portulaca oleracea*; ragweed, *Ambrosia artemisiifolia*; Spanish needles, *Bidens bipinnata*; sunflower, *Helianthus* spp.; toadflax, *Linaria canadensis*; and velvetleaf, *Abutilon theophrasti* have been reported to serve as larval hosts (Ditman and Cory, 1931; Roach, 1975; Sudbrink and Grant, 1995). However, Harding (1976a) rated only sunflower as a good weed host relative to 10 other species in a study conducted in Texas. Stadelbacher (1981) indicated that crimson clover and winter vetch, which may be both crops and weeds, were important early season hosts in Mississippi. He also indicated that cranesbill, *Geranium dissectum* and *G. carolinianum*, were particularly important weed hosts in this area. In North Carolina, especially important wild hosts were toadflax and deergrass, *Rhexia* spp. (Neunzig, 1963). Gross and Young (1977) documented some of the differences in suitability among various natural hosts relative to development time, weight gain, and fecundity.

Adults collect nectar or other plant exudates from numerous plants. Lingren *et al.* (1993) studied adult host associations in Texas and Oklahoma, and reported that trees and shrub species were especially frequented. Among the hosts identified were *Citrus, Salix, Pithecellobium, Quercus, Betula, Prunus, Pyrus* and other Rosaceae, and Asteraceae. Callahan (1958) also presented a long list of adult hosts. The quality or quantity of nectar affects potential fecundity of moths, with such plants as alfalfa; red and white clover; milkweed, *Asclepias syriaca*; and Joe–Pye weed, *Eupatorium purpureum* proving especially suitable in Virginia (Nuttycombe, 1930).

Natural Enemies. Although numerous natural enemies have been identified, they usually are not effective at causing high levels of earworm mortality or preventing crop injury. For example, in a study conducted in Texas, Archer and Bynum (1994) reported less than 1% of the larvae were parasitized or infected with disease. However, eggs may be heavily parasitized (Oatman 1966a, Campbell *et al.*, 1991). *Trichogramma* spp. (Hymenoptera: Trichogrammatidae), and to a lesser extent *Telenomus* spp. (Hymenoptera: Scelionidae), are common egg parasitoids. Also, natural control agents can affect populations late in the season (Roach, 1975). Exotic parasitoids and predators have been introduced to North America in hopes of gaining better natural control of *H. zea*, but thus far the imported beneficials have failed to establish successfully (King and Coleman, 1989).

Common larval parasitoids include *Cotesia* spp., and *Microplitis croceipes* (Cresson) (all Hymenoptera: Braconidae); *Campoletis* spp. (Hymenoptera: Ichneumonidae); *Eucelatoria armigera* (Coquillett) and *Archytas marmoratus* (Townsend) (Diptera: Tachinidae). However, additional wasp and fly species have, on occasion, been reported from corn earworm (Arnaud, 1978; Krombein *et al.*, 1979). In Mississippi, Lewis and Brazzel (1968) observed only *M. croceipes* to be abundant regularly.

General predators often feed on eggs and larvae of corn earworm; over 100 insect species have been observed to feed on *H. zea*. Among the common predators are lady beetles such as convergent lady beetle, *Hippodamia convergens* Guerin–Meneville, and *Coleomegilla maculata* De Geer (both Coleoptera: Coccinellidae); softwinged flower beetles, *Collops* spp. (Coleoptera: Melyridae); green lacewings, *Chrysopa* and *Chrysoperla* spp. (Neuroptera: Chrysopidae); minute pirate bug, *Orius tristicolor* (White) (Hemiptera: Anthocoridae); and big-eyed bugs, *Geocoris* spp. (Hemiptera: Lygaeidae) (King and Coleman, 1989). Birds can also feed on earworms, but rarely are adequately abundant to be effective (Barber, 1942).

Within season mortality during the pupal stage seems to be slight (Kring *et al.*, 1993), and though overwintering mortality is often very high the mortality is due to adverse weather and collapse of emergence tunnels rather than to natural enemies. A nematode, *Chroniodiplogaster aerivora* (Nematoda: Diplogasteridae), occurs naturally in midwestern states, but it is a weak pathogen (Steinkraus *et al.*, 1993b). In Texas, *Steinernema riobrave* (Nematoda: Steinernematidae) has been found to be an important mortality factor of prepupae and pupae, but this parasitoid is not yet generally distributed. Similarly, Khan *et al.* (1976) found *Heterorhabditis heliothidis* (Nematoda: Heterorhabditidae) parasitizing corn earwom in North Carolina, but it has not been found widely. Both of the latter species are being redistributed, and can be produced commercially; so in the future they may assume greater importance in natural regulation of earworm populations.

Epizootics caused by pathogens may erupt when larval densities are high. The fungal pathogen *Nomuraea rileyi*, and the *Helicoverpa zea* nuclear polyhedrosis virus are commonly involved in outbreaks of disease, but the protozoan *Nosema heliothidis* and other fungi and viruses also have been observed.

Life Cycle and Description. This species is active throughout the year in tropical and subtropical climates, but becomes progressively more restricted to the summer months with increasing latitude. In northeastern states dispersing adults may arrive as early as May or as late as August due to the vagaries associated with weather; thus, their population biology is variable. The number of generations is usually reported to be one in northern areas such as most of Canada, Minnesota, and western New York (Knutson, 1944; Beirne, 1971; Chapman and Lienk, 1981), two in northeastern states (Prostak, 1995), two to three in Maryland (Ditman and Cory, 1931), three in the central Great Plains (Walkden, 1950) and northern California (Okumura, 1962), four to five in Louisiana (Oliver and Chapin, 1981) and southern California (Okumura, 1962), and perhaps seven in southern Florida and southern Texas. The life cycle can be completed in about 30 days.

Egg. The eggs are deposited singly, usually on leaf hairs and corn silk. The egg is pale green when first deposited, becomes yellowish and then gray with time. The shape varies from slightly dome-shaped to a flattened sphere, and measures about 0.5–0.6 mm in diameter and 0.5 mm high. They bear 21–31 ridges radiating from the center. Fecundity ranges from 500–3000 eggs per female. The eggs hatch in about 3–4 days.

Larva. Upon hatching, larvae wander about the plant until they encounter a suitable feeding site, normally the reproductive structure of the plant. Young larvae are not cannibalistic, so several larvae may feed together initially. However, as larvae mature they become very aggressive, killing and cannibalizing other larvae. Consequently, only a small number of larvae are found in each ear of corn. Normally, corn earworm displays six instars, but five is not uncommon and seven to eight have been reported. Mean head capsule widths are 0.29, 0.47, 0.77, 1.30, 2.12, and 3.10 mm, respectively, for instars 1–6. Larval lengths are estimated at 1.5, 3.4, 7.0, 11.4, 17.9, and 24.8 mm, respectively. Development time averaged 3.7, 2.8, 2.2, 2.2, 2.4, and 2.9 days, respectively, for instars 1–6 when reared at 25°C. Butler (1976) reported that the cultured earworm on corn at several temperatures, reporting total larval development times of 31.8, 28.9, 22.4, 15.3, 13.6, and 12.6 days at 20°, 22.5°, 25°, 30°, 32°, and 34°C, respectively.

The larva is variable. Overall, the head tends to be orange or light brown with a white net-like pattern, the thoracic plates black, and the body brown, green, pink, or sometimes yellow or mostly black. The larva usually bears a broad dark band laterally above the spiracles, and a light yellow to white band below the spiracles. A pair of narrow dark stripes often occurs along the center of the back. The close examination reveals that the body bears numerous black thorn-like microspines. These spines give the body a rough feel when touched. The presence of spines and the light-colored head serve to distinguish corn earworm from fall armyworm, *Spodoptera frugiperda* (J.E. Smith), and European corn borer, *Ostrinia nubilalis* (Hübner). These other common corn-infesting species lack the spines and bear dark heads. Tobacco budworm, *Heliothis virescens* (Fabricius), is a closely related species in which the late instar larvae also bear microspines. Although it is easily confused with corn earworm, it rarely is a vegetable pest and never feeds on corn. Close examination reveals that in tobacco budworm larvae the spines on the tubercles of the first, second, and eighth abdominal segments are about half the height of the tubercles, but in corn earworm the spines are absent or up to one-fourth the height of the tubercle. Younger larvae of these two species are difficult to distinguish, but Neunzig (1964) gave a key to aid in separation. (See color figure 48.)

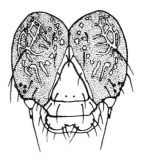

Head capsule of corn earworm larva.

Pupa. Mature larvae leave the feeding site and drop to the ground, where they burrow into the soil and pupate. The larva prepares a pupal chamber 5–10 cm below the surface of the soil. The pupa is mahogany-brown, and measures 17–22 mm long and 5.5 mm wide. Duration of the pupal stage is about 13 days (range 10–25 days) during the summer, but 250 days or more for the overwintering pupae. (See color figure 265.)

Adult. Because pupation occurs rather deep in the soil, the moths have difficulty digging to the surface unless the tunnel created by the larvae as they dig into the soil remains intact. As with the larval stage, adults are quite variable. The front wings of the moths usually are yellowish-brown, and often bear a small dark spot centrally. The small dark spot is especially distinct when viewed from below. The forewing also may bear a broad dark transverse band distally, but the margin of the wing is not darkened. The hind wings are creamy white basally and blackish distally, and usually bear a small dark spot centrally. The moth

Corn earworm larva.

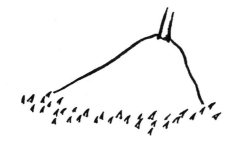

Spinules associated with tubercles on abdomen of corn earworm larva. The minute spines are less than one-fourth the height of the tubercle.

Corn earworm pupa.

measures 32–45 mm in wingspan. Adults are reported to live for 5–15 days, but may survive for over 30 days under optimal conditions. The moths are principally nocturnal, and remain active throughout the dark period. During the daylight hours they usually hide in vegetation, but sometimes they can be seen feeding on nectar. Oviposition commences about three days after emergence, and continues until death. Fresh-silking corn is highly attractive for oviposition but even ears with dry silk can receive eggs. Fecundity varies from about 500–3000 eggs, though feeding is a prerequisite for high levels of egg production. Females may deposit up to 35 eggs per day. (See color figure 235).

The biology of corn earworm was presented by several authors; among the most complete were Quaintance and Brues (1905), Ditman and Cory (1931), Brazzel *et al.* (1953), Hardwick (1965b), and Neunzig (1969). An extensive bibliography was published by Kogan *et al.* (1978). Keys to *Helicoverpa* adults and larvae were provided by Hardwick (1965b). Keys to differentiate earworm from similar crop-infesting larvae were given by many authors, including Whelan (1935), Walkden (1950), Frost (1955), Okumura (1962), Oliver and Chapin (1981), and Capinera (1986). Corn earworm is also included in a key to armyworms and cutworms in Appendix A. Keys to moths can be found in Rings (1977a) and Capinera and Schaefer (1983). Artificial diet and rearing procedures have been developed (Burton, 1970; Singh and Moore, 1985).

Damage

Some consider corn earworm is the most costly crop pest in North America. It is more damaging in areas where it successfully overwinters, however, because in northern areas it may arrive too late to inflict extensive damage. It often attacks harvested portions of valuable crops. Thus, larvae often are found associated with such plant structures as blossoms, buds, and fruits. When feeding on lettuce, larvae may burrow

Adult corn earworm.

into the head. On corn, its most common host, young larvae tend to feed on silks initially, and interfere with pollination, but eventually they usually gain access to the kernels. They may feed only at the tip, or injury may extend half the length of the ear before larval development is completed. Such feeding also enhances development of plant pathogenic fungi, and is attractive to sap beetles (Coleoptera: Nitidulidae). If the ears have not yet produced silk, larvae may burrow directly into the ear. They usually remain feeding within a single ear of corn, but occasionally they abandon the feeding site and begin search for another. Larvae also can damage whorl-stage corn by feeding on the young, developing leaf tissue. Survival is better on more advanced stages of development, however (Gross *et al.*, 1976). Young fields adjacent to favored, more mature plants are likely to experience low rates of egg deposition (Weisenborn and Trumble, 1988). On tomato, larvae may feed on foliage and burrow in the stem, but most feeding occurs on the tomato fruit. Larvae commonly begin to burrow into a fruit, feed only for a short time, and then move on to attack another fruit. Tomato is more susceptible to injury when corn is not silking; in the presence of corn, moths will preferentially oviposit on fresh corn silk. Other crops such as bean, cantaloupe, cucumber, squash, and pumpkin may be injured in a manner similar to tomato, and also are less likely to be injured if silking corn is nearby.

Management

Sampling. Eggs and larvae often are not sampled on corn because eggs are very difficult to detect, and larvae burrow down into the silks, out of the reach of insecticides, soon after hatching. Sampling protocols have been developed for larvae on corn, however (Hoffmann *et al.*, 1996b). On tomato, eggs tend to be placed on the leaves immediately below the highest flower cluster (Zalom *et al.*, 1983). Thus, sampling protocols, including both fixed and sequential sampling procedures, have been developed for this crop (Hoffman *et al.*, 1991a). A common procedure is to examine the leaves beneath all flower clusters on 20–30 plants per field, but the sampling effort can be reduced significantly using sequential sampling. Similarly, sampling protocols for fruit damage have been developed (Wilson *et al.*, 1983b).

Moths can be monitored with blacklight and pheromone traps. Both sexes are captured in light traps whereas only males are attracted to the sex pheromone. Both trap types give an estimate of when moths invade or emerge, and relative densities, but pheromone traps are easier to use because they are selective. The pheromone is usually used in conjunction with an inverted cone-type trap. Trap designs and commercial source of pheromone lures affect trap catches (Gauthier et al. 1991, Lopez *et al*,. 1994). Generally, the presence of 5–10 moths per night is sufficient to stimulate pest control practices (Foster and Flood, 1995). Light traps have also been investigated for removal of moths from cropping areas. Some protection of small plantings can be attained, but this approach is ineffective for large areas or when moth densities are high (Barrett *et al.*, 1971). Pheromone components also can be released in an area to confuse the moths and disrupt mating; this has been demonstrated experimentally (Mitchell *et al.*, 1975) but has not yet come into commercial practice.

Insecticides. Corn fields with more than 5% of the plants bearing new silk are susceptible to injury if moths are active. Insecticides are usually applied to foliage in a liquid formulation, with particular attention to the ear zone, because it is important to apply insecticide to the silk. Systemic insecticides are not effective because the plant does not translocate insecticide effectively to reproductive tissues (Russell *et al.*, 1993). Considerable economic benefit has been documented for chemical suppression of earworm on tomato in North Carolina (Walgenbach and Estes, 1992). Insecticide applications are often done at 2–6 day intervals, sometimes as frequently as daily in Florida. Because it is treated frequently and over a wide geographic area, corn earworm has become resistant to many insecticides (Fitt, 1989; Kanga *et al.*, 1996). Susceptibility to *Bacillus thuringiensis* also varies, but the basis for this variation in susceptibility is uncertain (Stone and Sims, 1993). Mineral oil, applied to the corn silk soon after pollination, has insecticidal effects. Application of about 0.75–1.0 ml of oil 5–7 days after silking can provide good control in the home garden (Carruth, 1942; Barber, 1942). Corn earworm moths, like many moths, feed readily on baits containing sweet material such as sucrose. Some work has been conducted to demonstrate that baits containing insecticide can attract and kill moths (Ditman, 1937; Creighton and McFadden, 1976), but this has yet to be developed into a practical technique.

Cultural Practices. Cultural procedures have some application for corn earworm management. This dispersive species moves readily from weeds to crops, and among crops, as their host plants become more or less suitable. Thus, effective management is best considered on an area-wide basis (Graham *et al.*, 1972). Trap cropping is often suggested for this insect; the high degree of preference by ovipositing moths for corn in the green silk stage can be used to lure moths from less preferred crops. Lima beans also are relatively attractive to moths, at least as compared to tomato (Pepper, 1943). However, it is difficult to maintain attractant crops in an attractive stage for protracted periods. In southern areas where populations develop first on weed hosts and then disperse to crops, treatment of the weeds through mowing, herbicides, or application of insecticides can greatly ameliorate damage on nearby crops (Snodgrass and Stadelbacher, 1994). In northern areas, it is sometimes possible to plant or harvest early enough to escape injury. Throughout the range of this insect, population densities are highest, and most damaging, late in the growing season. Tillage, especially in the autumn, can significantly reduce overwintering success of pupae in southern locations (Barber and Dicke, 1937).

Biological Control. Several insect pathogens have been evaluated for suppression of corn earworm. A nuclear polyhedrosis virus isolated from corn earworm is efficacious and has been sold commercially for larval suppression on non-food crops (Young and McNew, 1994). Application of virus to weed hosts over an area-wide basis early in the season has been shown to reduce earworm population increase (Bell and Hayes, 1994; Hayes and Bell, 1994). Several other viruses, isolated from other caterpillars, are possible biological control agents if they attain commercial development (Young and McNew, 1994). In addition to the nuclear polyhedrosis virus, the fungus *Nomuraea rileyi*, the bacterium *Bacillus thuringiensis*, and steinernematid nematodes all provide some suppression (Oatman *et al.*, 1970; Ignoffo *et al.*, 1978; Mohamed *et al.*, 1978; Bartels and Hutchinson, 1995). Entomopathogenic nematodes, which are available commercially, provide good suppression of developing larvae if they are applied to corn silk; this has application for home garden production of corn but not commercial production (Purcell *et al.*, 1992). Soil surface and subsurface applications of nematodes also can affect earworm populations because larvae drop to the soil to pupate (Cabanillas and Raulston, 1994, 1995, 1996). This approach may have application for commercial crop protection, but larvae must complete their development before they are killed, so some crop damage ensues.

Trichogramma spp. (Hymenoptera: Trichogrammatidae) egg parasitoids have been reared and released for suppression of *H. zea* in several crops. Levels of parasitism averaging 40–80% have been attained by such releases in California and Florida, resulting in fruit damage levels of about 3% (Oatman and Platner, 1971). The host crop seems to affect parasitism rates, with tomato being an especially suitable crop for parasitoid releases (Martin *et al.*, 1976b).

Host-Plant Resistance. Numerous varieties of corn have been evaluated for resistance to earworm, and some resistance has been identified in commercially available corn varieties (McMillian *et al.*, 1975; Story *et al.*, 1983; Archer *et al.*, 1994). Resistance is derived from physical characteristics such as husk tightness and ear length, which impede access by larvae to the ear kernels, or chemical factors such as maysin, which inhibit larval growth (Douglas, 1947). Host-plant resistance thus far is not completely adequate to protect corn from earworm injury, but it may prove to be a valuable component of multi-faceted pest management programs. Varieties of some crops are now available that incorporate *Bacillus thuringiensis* toxin, which reduces damage by *H. zea*, and this likely will assume importance as a plant resistance mechanism (Benedict *et al.*, 1996).

Darksided Cutworm
Euxoa messoria (Harris)
(Lepidoptera: Noctuidae)

Natural History

Distribution. Darksided cutworm, a native species, is found throughout most of Canada and the northern United States. Although it occurs as far south as South Carolina, Oklahoma, and southern California, it is a pest only in northern climates.

Host Plants. This species has been reported to feed on numerous plants including such vegetables as bean, broccoli, Brussels sprouts, cabbage, cauliflower, Chinese cabbage, corn, cucumber, onion, pea, pepper, potato, radish, rutabaga, sweet potato, tomato, and turnip. In Canada, it is sometimes known as a field crop pest, injuring barley, flax, oat, rye, sugarbeet, tobacco, and wheat. The larvae also climb readily, and damage the blossom and leaf buds of apple, currant, grape, peach, and other trees and shrubs.

Natural Enemies. Bucher and Cheng (1971) studied mortality factors associated with darksided cutworm. They reported that bacterial, microsporidian, and other diseases collectively accounted for about 30% mortality, whereas nuclear polyhedrosis virus and fungus pathogens occurred only at low levels. The other major mortality factor was parasitism, which caused about 15% mortality. Over 35% of cutworm larvae successfully pupated in these studies, which were conducted in Ontario. In another study, Cheng (1977) noted 10 species of parasitoids, including *Meteorus leviventris* (Wesmael) (Hymenoptera: Braconidae); *Eutanyacra suturalis* (Say), *Arenetra rufipes vernalis* Walley, *Campoletis flavicinctus* (Ashmead), *Enicospilus* sp. (all Hymenoptera: Ichneumonidae); *Copidosoma bakeri* (Howard) (Hymenoptera: Encyrtidae); *Muscina stabulans* (Fallén) (Diptera: Muscidae); *Win-*themia rufopicta* (Bigot), and *W. deilephilae* (Osten Sacken) (both Diptera: Tachinidae). However, most of the parasitism was due to a single species, *C. bakeri*. Other parasitoids known to affect darksided cutworm include *Apanteles laeviceps* Ashmead, *A. militaris* Walsh, *Meteorus communis* (Cresson) (all Hymenoptera: Braconidae); *Spilichneumon superbus* (Provancher), *Diphyus euxoae* Heinrich, *Campoletis* sp. (all Hymenoptera: Ichneumonidae); *Sarcophaga cimbicis* Town (Diptera: Sarcophagidae); *Bonnetia comta* (Fallén), and *Aphria ocypterata* Townsend (both Diptera: Tachinidae) (Cheng, 1977; 1981). Cheng (1973a) observed predation of larvae by ground beetles (Coleoptera: Carabidae) and by birds, and suggested that rodents were important predators.

Life Cycle and Description. There is a single generation throughout the geographic range of this species. Overwintering occurs in the egg stage. In Tennessee the eggs hatch in late January–late March, larvae complete their development in spring and early summer, and pupation occurs in June–August. Moths emerge beginning in September until about mid-October, and deposit overwintering eggs. In Ontario, the life cycle is similar but compressed owing to the shorter period of favorable weather. Thus, eggs hatch in late March–early May, pupation occurs in July–September, and adults are present in August–October.

Egg. The eggs are deposited singly or in small clusters of up to 30 in the soil at a depth of 6–12 mm. They are difficult to locate in the soil because soil particles adhere to the chorion. They initially are iridescent white, become yellow and then brown as the embryo develops. They are elliptical in shape, and measure about 0.55–0.63 mm in diameter and 0.38–0.45 mm in height, and bear slight ridges radiating from the center. Duration of the egg stage is 5–7 months.

Larva. There are 6–8 instars. Crumb (1929) reported only six instars, but Cheng (1973a) reported mostly seven instars. Mean head capsule widths are 0.28, 0.39, 0.64, 0.95, 1.46, 2.18, and 3.10 mm for instars 1–7, respectively. The range of body lengths during the larval instars is about 2.2–3.0, 3.5–4.8, 5.5–7.8, 9.0–14.1, 16.2–23.0, 21.2–33.2, and 37.2–43.7 mm, respectively. Duration of the instars is about 17, 11, 10, 10, 10, 11, and 47 days, respectively, for a total larval development time of about 117 days (the last instar includes a lengthy nonfeeding prepupal period of about 30 days). Larvae are grayish, with irregular brownish longitudinal lines subdorsally and laterally. The common name "dark sided" is not particularly appropriate, and refers to the fairly indistinct, wavy dark band above the spiracles. Laterally, below the spiracles, is found a whitish band. The head is orange-brown with darker spots. During the early instars larvae tend to feed on the upper surface of the leaves, leaving the lower epidermis intact. By the fourth instar, however, larvae consume leaves entirely. Large larvae, instars 6–7, may cut plants off at the soil surface. Throughout development, larvae tend to feed at night. When not feeding they may rest on the lower parts of plants or under leaves, but in the final instar they tend to hide in the soil when not eating. When larvae seek shelter in the soil the depth at which they bury themselves varies, and is related to soil moisture. If the soil is moist they are buried just beneath the surface; under dry conditions they may dig to depths of 7–10 cm. (See color figure 49.)

Pupa. After larvae complete the feeding period they dig into the soil and create a pupal cell at a depth of 7.5–10 cm. As noted above, the larvae remain quiescent for about 30 days, then pupate. The pupa initially is yellowish brown, but becomes dark brown toward maturity. The pupa measures about 16 mm long and 5 mm wide. Duration of the pupal stage is about 22 (range 18–28 days) days.

Adult. The adult, like the larva, is not distinctly marked. The front wings are grayish brown, and sometimes marked with dark transverse lines. A bean-shaped spot is evident on the forewing. The hind wings are grayish white, but darker distally and along the veins. The moth measures 30–40 mm in wingspan. Moths are most active 1–2 h after sunset, but mating apparently occurs shortly after midnight and oviposition occurs in the early morning. The female inserts her ovipositor into the soil to deposit eggs. In the laboratory, females produce about 1300 eggs. Adult longevity is usually 13–14 days. (See color figure 236.)

The biology of darksided cutworm was most completely described by Cheng (1973a), though Crumb (1929) gave a detailed description and some biological observations. Hinks and Byers (1976) and Belloncik et al. (1985) gave information on rearing. Sex pheromone components were identified by Struble et al., 1977; and Struble and Byers, 1987. A bibliography was published by Rings et al. (1975b). Larvae are included in keys by Crumb (1929), Whelan (1935), Rings (1977b), and Capinera (1986), and are included in a key to armyworms and cutworms in Appendix A. Adults are included in keys by Rings (1977a) and Capinera and Schaefer (1983).

Damage

Larvae feed on the leaves and stems of young plants, sometimes causing complete defoliation and death of the plant. Although widespread, this insect historically has been principally a pest of tobacco in southern Ontario. Elsewhere it is only an occasional pest.

Management

Sampling. Larvae are difficult to sample, especially when they are young. Bucher and Cheng (1970) recommended transplanting a few small attractive plants such as tobacco into a field to serve as bait because larvae quickly accumulate in the area of such favored food plants. Both blacklight and pheromone traps can be used to monitor population density of adults. However, though both types of traps show similar population trends, pheromone traps are much more attractive (Cheng and Struble, 1982). There is some indication that the sex pheromone can also be used to disrupt mating, but this has not been fully evaluated (Palaniswamy et al., 1984).

Insecticides. Insecticides are often used when this insect becomes troublesome. Insecticides may be applied to young plants or directly to the soil; in the latter case it may be applied only to the surface or may be incorporated (Cheng, 1971; 1973b; 1980; 1984). Also, it is sometimes possible to apply the insecticide to the cover crop grown before vegetables, or to apply the insecticide postplanting as a rescue treatment following unexpected invasion by cutworms (Harris et al., 1975b). Darksided cutworm is tolerant of some soil insecticides (Harris et al., 1962a). The microbial insecticide *Bacillus thuringiensis* was evaluated as a control agent for darksided cutworm by Cheng (1973c). Although young larvae are susceptible, by the time they attain the fourth instar susceptibility drops. *Bacillus thuringiensis* is not usually recommended for suppression of cutworms.

Cultural Practices. Barriers are sometimes used to reduce access by cutworms to seedlings grown in the home garden. Metal or waxed-paper containers, with both the top and bottom removed, can be placed around the plant stem to deter consumption. Aluminum foil can be wrapped around the stem to achieve a similar effect. The barrier should be extended below the soil surface because larvae burrow and feed below the soil line.

Dingy Cutworm
Feltia jaculifera (Guenée)
Feltia subgothica (Haworth)
(Lepidoptera: Noctuidae)

Natural History

Distribution. The name "dingy cutworm" is applied to at least two, and possibly four, species in

North America. Possibly the most important is *F. jaculifera*, also known as *F. ducens* Walker. Second in importance usually is *F. subgothica*, but *F. tricosa* Lintner and *F. herilis* (Grote) are very similar in appearance and biology. The literature on these species is terribly confused, and they usually are treated as a complex of co-occurring species (Chapman and Lienk, 1981). The distribution of this cutworm complex is most of the United States and southern Canada. They are absent only from southern Florida and from California and adjacent desert areas (Rings *et al.*, 1975). They are most damaging in the midwestern and eastern states, and eastern Canada.

Host Plants. Dingy cutworm is commonly known as a corn pest, but also feeds on such vegetables as bean, cabbage, celery, cucumber, lettuce, onion, pea, squash, sweet potato, and tomato. Among other crops injured are alfalfa, blue grass, clover, flax, horseradish, raspberry, sweetclover, tobacco, and wheat. A wide variety of weeds are suitable food for dingy cutworm, including aster, *Aster ericoides*; chickweed, *Stellaria* sp.; goldenrod, *Solidago* sp.; mullein, *Verbascum* sp.; plantain, *Plantago* sp.; and yellow dock, *Rumex crispus*.

Natural Enemies. Collection of larvae in spring from the central Great Plains has shown that natural enemies may account for 28% mortality, with most attributable to parasitoids (Walkden, 1950). Among the wasps parasitizing dingy cutworm are *Aleiodes aciculatus* Cresson, *Chelonus sericeus* (Say), *Apanteles griffini* Viereck, *Microplitis feltiae* Muesbeck, *Meteorus leviventris* (Wesmael), *Spilichneumon superbus* (Provancher) (all Hymenoptera: Braconidae), and *Copidosoma bakeri* (Howard) (Hymenoptera: Encyrtidae). Flies known to parasitize dingy cutworms include *Euphorocera claripennis* (Macquart), *Gonia frontosa* Say, *G. fuscicollis* Tothill, *Triachora omissa* (Aldrich), and *Winthemia quadripustulata* (Fabricius). A nematode, *Hexamermis arvalis* (Nematoda: Mermithidae) infects young dingy cutworms in the autumn and emerges in the spring, killing the larvae (Puttler and Thewke 1971). Diseases known to affect dingy cutworm include the fungi *Beauveria* sp. and *Metarhizium anisopliae* and an unspecified virus (Crumb, 1929).

Life Cycle and Description. There is a single generation of dingy cutworm annually. Moth flight occurs in July–September, followed immediately by oviposition. Larvae hatch and become partly grown before the onset of winter. Dingy cutworm overwinters in the larval stage, with larval development normally completed in March–May. The larvae remain quiescent in the soil until August, when pupation occurs, followed immediately by emergence of the adults. *Feltia jaculifera* usually lags behind the other species in appearing as an adult. In New York, *F. jaculifera* normally first occurs in early August, whereas the other species comprising the dingy cutworm complex first appear in mid-July (Chapman and Lienk, 1981).

Egg. The eggs reportedly are deposited on vegetation and on the soil surface. However, Balduf (1931) described deposition of eggs into flower heads of sunflowers. The eggs are oval, whitish to light brown, and the surface is marked by about 36 (*F. jaculifera*) or 56 (*F. subgothica*) narrow ridges. They measure about 0.60–0.65 mm long, 0.5 mm wide, and 0.36–0.38 mm in height. Duration of the egg stage is 5–21 days, depending on weather conditions, but normally 6–11 days. Based on dissections, females appear to be capable of producing about 800 eggs (range 500–1220 eggs). However, when Stanley (1936) captured moths feeding at flowers and confined them, he observed egg production of only about 100 eggs per female.

Larva. Larvae display 6–7 instars, with 6 being normal. Head capsule widths are 0.30, 0.4–0.5, 0.55–0.80, 0.80–1.25, 1.30–1.90, 1.90–2.50, and 2.30–2.80 mm for instars 1–7, respectively. Duration of the instars is normally 3–12, 4–12, 4–12, 5–17, 8–119, 15–160, and 18–222 days, respectively. Total larval development time is 250–350 days. Body size increases from about 3 mm to 22–32 mm over the course of development. The body color is grayish brown. An interrupted blackish stripe is found dorsolaterally, bounding a broad lighter dorsal band. The head is brownish gray, and marked with darker regions. The larvae of dingy cutworm are extremely hardy, able to tolerate long periods without food. They rarely display a tendency to climb, preferring to feed along the soil surface. Larvae

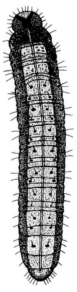

Dingy cutworm larva.

are normally considered to be foliage feeders, and in the early spring this is certainly the case. However, Balduf (1931) described pollen consumption by first instar larvae, and Duffus *et al.* (1983) indicated that larvae may feed until the fourth instar on sunflower heads. (See color figure 50.)

Pupa. Larvae prepare a cell in the upper 1–2 cm of the soil, and pupate within. The pupa is brown, and measures about 18 mm long and 6 mm wide. Duration of the pupal stage is 12–35 days.

Adult. The moths are brownish and measure 35–42 mm in wingspan. The front wings are grayish-brown marked with darker brown and a hint of purple. The hind wings may be whitish basally and brown distally, or uniformly brown. The thorax and abdomen are gray to brown. Adults live for 5–10 days. Moths reportedly are affected by weather, with significant effects on reproductive success. Specifically, *F. jaculifera* is reported to be favored by dry weather during autumn, especially in October, because moths are most likely to fly and oviposit. In the case of *F. subgothica*, dry weather during September is favorable (Stanley, 1936). (See color figure 237.)

Sex pheromones have been described for *F. jaculifera* (Byers and Struble 1990). It appears, based on pheromone studies, that there is a genetic substructuring of the nominal species known as *F. jaculifera*. The 'pheromonal strains' respond to at least four slightly different pheromones (Byers *et al.*, 1990), but the genetic differences among strains are, as yet, too small to be considered different species (Gooding *et al.*, 1992).

A detailed description of dingy cutworm is provided by Crumb (1929), and it was included in keys by Whelan (1935), Crumb (1956), Rings (1977b), Capinera (1986), and in a key to armyworms and cutworms in Appendix A. The moths were included in keys by Rings (1977a), Oliver and Chapin (1981), and Capinera and Schaefer (1983). Developmental data were provided by Walkden (1950). Chapman and Lienk (1981) made valuable observations on the species complex comprising dingy cutworm. A bibliography on dingy cutworm was published by Rings *et al.*, 1975a.

Damage

Larvae damage young plants in the spring, usually by cutting the seedlings off at the soil surface. In a survey of midwestern corn fields conducted between 1979 and 1981, dingy cutworm was the second most abundant cutworm encountered, following only black cutworm, *Agrotis ipsilon* (Hufnagel) in abundance (Story *et al.*, 1984). On occasion, larvae have been observed

Adult dingy cutworm.

to ascend plants, including trees, to feed on buds and young foliage.

Management

Sampling. The adult populations may be monitored with light or pheromone, but because plant damage does not occur for several months after adult activity, larval monitoring is needed. Duffus *et al.* (1983) compared five types of larval sampling protocols and recommended sack trapping—the collection of larvae from beneath squares of plastic or burlap.

Insecticides. Cutworms can be controlled by application of persistent insecticides to soil or plants, or by application of baits such as bran that have been treated with insecticide. *Bacillus thuringiensis* is not usually recommended for control of this insect.

Cultural Practices. If seedlings are to be transplanted into the home garden, larger plants are preferred, because they are less likely to be irreparably damaged by cutworms. Transplanted plants can be protected if surrounded by a barrier such as a can or waxed-paper container with the bottom removed. Aluminum foil wrapped around the base of the seedling also deters cutting by larvae.

Fall Armyworm
Spodoptera frugiperda (J.E. Smith)
(Lepidoptera: Noctuidae)

Natural History

Distribution. Fall armyworm is a native to the tropical regions of the western hemisphere from the United States to Argentina. It normally overwinters successfully in the United States only in southern Florida and southern Texas, but during warm winters it may survive along the Gulf Coast and in southern Arizona. It is commonly found in the Caribbean, including Puerto Rico. Fall armyworm is a strong flier, and disperses to long distances annually during the summer months. It is recorded from virtually all states east of the Rocky Mountains, from Arizona and California, and from southern Ontario. As a regular and serious pest, its range tends to be mostly the southeastern states, though it is feared by sweet corn growers as far north as the New England states.

Host Plants. This species seemingly displays a very wide host range, with over 80 plants recorded, but clearly prefers grasses. The most frequently consumed plants are field corn and sweet corn, sorghum, Bermudagrass, and grass weeds such as crabgrass, *Digitaria* spp. When the larvae are very numerous they defoliate the preferred plants, acquire an "armyworm" habit and disperse in large numbers, consuming nearly all vegetation in their path. Many host records reflect such periods of abundance, and are not truly indicative of oviposition and feeding behavior under normal conditions. Vegetables, other than sweet corn which is frequently at risk, are only occasionally damaged, but include a wide range of crops such as asparagus, bean, beet, cabbage, chickpea, cowpea, corn, cucumber, kale, onion, pea, pepper, potato, rutabaga, spinach, sweet potato, tomato, turnip, and watermelon. Field crops are frequently injured, including alfalfa, barley, Bermudagrass, buckwheat, cotton, clover, corn, oat, millet, peanut, rice, ryegrass, sorghum, sugarbeet, sudangrass, soybean, sugarcane, timothy, tobacco, and wheat. Other crops occasionally injured are apple, grape, orange, papaya, peach, strawberry and many flowers. Among the weeds known to serve as hosts are bentgrass, *Agrostis* sp.; crabgrass, *Digitaria* spp.; Johnsongrass, *Sorghum halepense*; morningglory, *Ipomoea* spp.; nutsedge, *Cyperus* spp.; pigweed, *Amaranthus* spp.; and sandspur, *Cenchrus tribuloides*. In studies conducted in Honduras, fall armyworm larvae preferred *Amaranthus* foliage over both corn and sorghum (Portillo *et al.*, 1996b). Pencoe and Martin (1981) measured development on several grasses found in Georgia, and determined that large crabgrass, *Digitaria sanguinalis*; goosegrass, *Eleusine indica*; vaseygrass, *Paspalum urvillei*; and coastal Bermudagrass, *Cynodon dactylon* were very suitable hosts whereas yellow nutsedge, *Cyperus esculentus*; the sedge, *Cyperus globulosus*, and Texas panicum, *Panicum texanum*, were relatively poor hosts.

There is some evidence that fall armyworm strains exist, based primarily on their host plant preference. One strain feeds principally on corn, but also on sorghum, cotton and a few other hosts if they are found growing near the primary hosts. The other strain feeds principally on rice, Bermudagrass, and Johnsongrass. Some reproductive isolation exists between the strains, even when both occur in the same area (Pashley, 1988).

Natural Enemies.   Cool, wet springs followed by warm, humid weather in the overwintering areas favor survival and reproduction of fall armyworm,

allowing it to escape suppression by natural enemies. Once dispersal northward begins, the natural enemies are left behind. Therefore, though fall armyworm has many natural enemies, a few act effectively enough to prevent crop injury.

At least 53 species of parasitoids are known from throughout the range of fall armyworm, mostly in the families Braconidae and Ichneumonidae (both Hymenoptera), and Tachinidae (Diptera), but not all occur in North America. The known parasitoids, nearly all of which attack the larval stage, were listed by Ashley (1979). The parasitoids most frequently reared from larvae in the United States are *Cotesia marginiventris* (Cresson) and *Chelonus texanus* (Cresson) (both Braconidae), species that are also associated with other noctuid species. In a study conducted in Georgia, for example, *C. marginiventris* was the most abundant parasitoid collected in 1990, attaining up to 34% parasitism early in the season, but *Chelonus insularis* Cresson (Hymenoptera: Braconidae) assumed dominance late in the season (Riggin *et al.*, 1992). During the second year of the study, however, the most abundant parasitoid was *Archytas marmoratus* (Townsend) (Diptera: Tachinidae), followed closely by *Ophion flavidus* Brulle and *Aleiodes laphygmae* (Gahan) (both Hymenoptera: Ichneumonidae). Also, in a subset of the study, the same authors found *Aleiodes laphygmae* and *Ophion flavidus* to be the dominant parasitoids (Riggin *et al.*, 1993). Luginbill (1928) and Vickery (1929) described and pictured many of the fall armyworm parasitoids.

The predators of fall armyworm are general predators that attack many other caterpillars. Among the predators noted as important are various ground beetles (Coleoptera: Carabidae); the striped earwig, *Labidura riparia* (Pallas) (Dermaptera: Labiduridae); the spined soldier bug, *Podisus maculiventris* (Say) (Hemiptera: Pentatomidae); and the insidious flower bug, *Orius insidiosus* (Say) (Hemiptera: Anthocoridae). Vertebrates such as birds, skunks, and rodents also consume larvae and pupae readily. Predation may be quite important, as Pair and Gross (1984) demonstrated loss of pupae to predators at 60–90% in Georgia.

Numerous entomopathogens, including viruses, fungi, protozoa, nematodes, and a bacterium are associated with fall armyworm (Gardner *et al.*, 1984), but only a few cause epizootics. Among the most important are the *S. frugiperda* nuclear polyhedrosis virus (NPV), and the fungi *Entomophaga aulicae*, *Nomuraea rileyi*, and *Erynia radicans*. Incidence of NPV reached 50–60% in Louisiana, but disease typically appears too late to alleviate high levels of defoliation. A most interesting pathogen is the ectoparasitic nematode *Noctuidonema guyanense* (Nematoda: Aphelenchoididae). This nematode is a weak pathogen, having a debilitating effect on its host. Although fall armyworm is the principal host, it is associated with many other Lepidoptera (Rogers *et al.*, 1991; Simmons and Rogers 1996).

Life Cycle and Description. The life cycle is completed in about 30 days during the summer, but 60 days in the spring and autumn, and 80–90 days during the winter. The number of generations occurring in an area varies with the appearance of the dispersing adults. The ability to diapause is not present in this species. The population often spreads northward at about 480 km (300 miles) per generation. In Minnesota and New York, where fall armyworm moths do not appear until August (Knutson, 1944; Chapman and Lienk, 1981), there may be but a single generation. The number of generations is reported to be one to two in Kansas (Walkden, 1950), three in South Carolina (Luginbill, 1928), and four in Louisiana (Oliver and Chapin, 1981). In coastal areas of north Florida, moths are abundant from April to December, but some are found even during the winter months (Tingle and Mitchell, 1977). With the aid of certain weather patterns, dispersal northward may be much faster than just described. For example, during 1973 a redistribution of moths from Mississippi to Ontario occurred in about two days, a distance of 1600 km (1000 mile), with the aid of strong surface winds (Rose *et al.*, 1975).

Egg. The egg of fall armyworm is dome shaped; the base is flattened and the egg curves upward to a broadly rounded point at the apex. It is well-marked with 47–50 ridges that radiate outward from the apex. The egg measures about 0.4 mm in diameter and 0.3 mm in height. They are deposited on hosts and non-hosts; in the latter case the larvae disperse, often with the help of a strand of silk, which allows them to be blown a considerable distance by wind. The female typically produces several egg masses during her oviposition period, with deposition occurring at night, and on larger plants if provided a choice between large and small. The number of eggs per mass varies considerably but it is often 100–200, and total egg production per female averages about 1500, with a maximum of over 2000. They are sometimes deposited in layers, but most are spread over a single layer, and are attached to foliage. The female also deposits a layer of grayish scales between the eggs and over the egg mass, imparting a furry or moldy appearance. Initially the eggs are grayish green, but they soon turn brown. Egg masses are deposited beneath leaves when the moth density is low, but oviposition becomes indiscriminate at high densities. The period of incubation is only 2–3 days during the summer months.

Fall armyworm egg.

Larva. There usually are six instars in fall armyworm. Head capsule widths are about 0.35, 0.45, 0.75, 1.3, 2.0, and 2.6 mm, respectively, for instars 1–6. Larvae attain lengths of about 1.7, 3.5, 6.4, 10.0, 17.2, and 34.2 mm, respectively, during these instars. Young larvae are greenish with a black head, the head turning orangish in the second instar. In the second, but particularly the third instar, the dorsal surface of the body becomes brownish, and lateral white lines begin to form. In the fourth to the sixth instars the head is reddish-brown, mottled with white, and the brownish body bears white subdorsal and lateral lines. Elevated spots occur dorsally on the body; they are usually dark, and bear spines. The face of the mature larva also is marked with a white inverted "Y" and the epidermis of the larva is rough or granular in texture when examined closely. However, this larva does not feel rough touch, as does corn earworm, *Helicoverpa zea* (Boddie), because it lacks the microspines found in the similar-appearing corn earworm. In addition to the typical brownish form, its brown dorsal coloration may be replaced with green. In the green form, the dorsal elevated spots are pale rather than dark. Larvae are most active in the morning, late afternoon, and evening, and tend to conceal themselves during the brightest time of the day. Duration of the larval stage tends to be about 14 days during the summer and 30 days during cool weather. Mean development time was determined to be 3.3, 1.7, 1.5, 1.5, 2.0, and 3.7 days for instars 1–6, respectively, when larvae were reared on corn at 25°C (Pitre and Hogg, 1983). However, total larval development time was extended when larvae were fed less suitable hosts: from 13.5 days on corn, to 18.9 days on soybean, and 22.3 days on cotton. As development time increased, pupal weights and survival rates decreased. (See color figure 51.)

Fall armyworm larva.

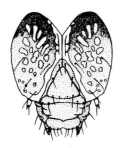

Head capsule of fall armyworm.

Pupa. Pupation normally takes place in the soil at a depth of 2–8 cm. The larva constructs a loose cocoon, oval in shape and 20–30 mm long, tying together particles of soil with silk. If the soil is too hard, larvae may web together leaf debris and other material to form a cocoon on the soil surface. The pupa is reddish-brown, and measures 14–18 mm long and about 4.5 mm wide. Duration of the pupal stage is about 8–9 days during the summer, but reaches 20–30 days during the winter in Florida and Texas. Unlike many noctuids, the pupal stage of fall armyworm cannot withstand protracted periods of cold weather. For example, Wood *et al.,* (1979) studied winter survival of the pupal stage in Florida, and found 51% survival in southern Florida, but only 27.5% in central Florida, and 11.6% in northern Florida.

Adult. The moths, with a wingspan of 32–40 mm, are quite variable in appearance. In the male moth, the forewing is shaded gray and brown, with a triangular white spots at the tip and near the center of the wing. The front wings of females are less distinctly marked, ranging from a uniform grayish-brown to a fine mottling of gray and brown. The hind wing is iridescent silver-white with a narrow dark border in both sexes. Adults are nocturnal, and are most active during warm, humid evenings. On the first night of emergence they may feed, but do not mate. They feed on nectar from many plants, usually during early eve-

ning. Females may mate repeatedly, but only once per night. After a pre-oviposition period of 3–4 days, the female normally deposits most of her eggs during the first 4–5 days of life, but some oviposition occurs for up to three weeks. The oviposition period tends to be shorter under warm conditions, sometimes as short as one day, and longer under cool conditions. Duration of adult life is estimated to average about 10 days (range about 7–21 days). (See color figures 238, 239, and 240.)

A comprehensive account of the biology of fall armyworm was published by Luginbill (1928), and an informative synopsis by Sparks (1979). Ashley *et al.* (1989) presented an annotated bibliography. Fall armyworm was included in many larval identification guides, such as the keys by Whelan (1935), Walkden (1950), and Crumb (1956), and the pictorial keys by Okumura (1962), Rings (1976), Oliver and Chapman (1981), and Capinera (1986). It also was included in a key to armyworms and cutworms in Appendix A. Fall armyworm was included in the keys to moths by Rings (1977a) and Capinera and Schaefer (1983), and pictured by Chapman and Lienk (1981). Heppner (1998) provided a key to the adults of North American *Spodoptera*. A sex pheromone has been described (Tumlinson *et al.*, 1986). Culture of this insect was easily accomplished with a bean-based diet (Perkins, 1979).

Damage

A highly visible form of damage by larvae is consumption of foliage. Young larvae initially consume leaf tissue from one side, leaving the opposite epidermal layer intact. By the second or third instar, larvae begin to make holes in leaves, and eat from the edge of the leaves inward. Feeding in the whorl of corn often produces a characteristic row of perforations in the leaves, though the larvae quickly produce a ragged appearance as they grow and feed. Larval densities are usually reduced to 1–2 per plant when larvae feed in close proximity to one another, due to cannibalistic behavior. Older larvae cause extensive defoliation,

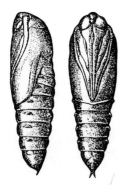

Fall armyworm pupae.

Adult fall armyworm.

often leaving only the ribs and stalks of corn plants, or a ragged, torn appearance. Total leaf consumption by larvae exceeds 100 cm. The proportion of defoliation is estimated at 0.1%, 0.6%, 1.1%, 4.7%, 16.3%, and 77.2% during instars 1–6, respectively. As over three-fourths of the defoliation occurs during the last instar, the presence of larvae is easily overlooked and damage occurs suddenly. Marenco *et al.* (1992) studied the effects of fall armyworm injury to early vegetative growth of sweet corn in Florida. They reported that the early whorl stage was least sensitive to injury, the mid-whorl stage intermediate, and the late whorl stage was most sensitive to injury. Further, they noted that mean densities of 0.2–0.8 larvae per plant during the late whorl stage could reduce yield by 5–20%.

Larvae also burrow into the growing point (bud, whorl, etc.), destroying the growth potential of plants, or clipping the leaves. In corn, they sometimes burrow into the ear, feeding on kernels in the same manner as corn earworm, *Helicoverpa zea.* Unlike corn earworm, which tends to feed down through the silk before attacking the kernels at the tip of the ear, fall armyworm feed by burrowing through the husk on the side of the ear. Ear damage is of greater concern to sweet corn growers than foliage feeding.

Management

Sampling. Moth populations can be sampled with blacklight and pheromone traps; the latter are more efficient (Starratt and McLeod, 1982). Pheromone traps should be suspended at canopy height, preferably in corn during the whorl stage. The type of trap selected for population monitoring can have significant effect on moth catches; plastic canister styles are most desirable based on both the number of moths captured and the ease of trap servicing (Adams *et al.,* 1989). Such catches are not necessarily good indicators of density, but indicate the presence of moths in the area. Once moths are detected, it is advisable to search for eggs and larvae. A search of 20 plants in five locations, or 10 plants in ten locations, is generally considered to be adequate to assess the proportion of plants infested. Hoffman *et al.* (1996b) compared fixed sample size with sequential sampling protocols for caterpillar pests of corn in New York, and reported substantial savings in time by using this technique for classification of infestation, relative to fixed samples of 100 plants. Sampling to determine larval density often requires large sample sizes, especially when larval densities are low or larvae are young (Mitchell and Fuxa, 1987), so it is not often used.

Insecticides. Insecticides are usually applied to sweet corn in the southeastern states to protect against damage by fall armyworm, often as frequently as daily during the silking stage. In Florida, fall armyworm is the most important pest of corn. It is often necessary to protect both the early vegetative stages and reproductive stage of corn. As larvae feed deep in the whorl of young corn plants, a high volume of liquid insecticide may be required to obtain adequate penetration. Insecticides may be applied in the irrigation water if it is applied from overhead sprinklers (Sumner *et al.,* 1991). Granular insecticides are also applied over the young plants because the particles fall deep into the whorl. Baits and ultra low volume techniques are less frequently used. Some resistance to insecticides has been noted, with resistance varying regionally (Harrell *et al.,* 1977; Young, 1979; All *et al.,* 1986). Foster (1989) reported that keeping the plants free of larvae during the vegetative period reduced the number of sprays needed during the silking period. The grower practice of concentrating the sprays at the beginning of the silking period instead of spacing the sprays evenly provided little benefit.

Cultural Techniques. The most important cultural practice, employed widely in southern states, is early planting and/or early maturing varieties. Early harvest allows many corn ears to escape the higher armyworm densities that develop later in the season (Mitchell, 1978). Reduced tillage seems to have little effect on fall armyworm populations (All, 1988), though delayed invasion by moths of fields with extensive crop residue has been observed, thus delaying and reducing the need for chemical suppression (Roberts and All, 1993).

Host-Plant Resistance. Partial resistance is present in some sweet corn varieties, but it is inadequate for complete protection. Resistance is largely due to non-preference by larvae, but some antibiosis is present (Wiseman *et al.,* 1981, Wiseman and Widstrom, 1986).

Biological Control. Although several pathogens have been shown experimentally to reduce the abundance of fall armyworm larvae in corn, only *Bacillus thuringiensis* now is feasible, and success depends on having the product on the foliage when the larvae first appear. Natural strains of *Bacillus thuringiensis* tend not to be very potent, and genetically modified strains improve performance (All *et al.,* 1996). An interesting and unusual approach to biological control involves the application of mass-produced parasitoid larvae, *Archytas marmoratus* (Diptera: Tachinidae). The fly larvae are mechanically extracted from the female flies, suspended in aqueous solution, and sprayed onto plants (Gross and Johnson, 1985; Gross *et al.,* 1985).

Green Cloverworm
Hypena scabra (Fabricius)
(Lepidoptera: Noctuidae)

Natural History

Distribution. This native species is found widely in eastern North America. It is recorded from all states east of the Great Plains, and there are occasional records from the Rocky Mountain region. As a pest, however, it is best known from the soybean-growing areas of the midwest and southeast where it has an abundance of suitable host material. In Canada, green cloverworm is known from southern Ontario, but it rarely causes serious damage.

Host Plants. Larvae of green cloverworm develop successfully only on plants in the family Leguminosae. They have been observed to feed on weeds and crops from other plant families, but this occurs only after legumes have been consumed. Vegetable crops eaten include bean, cowpea, faba bean, lima bean, and pea. Field crops suitable for development include alfalfa, alsike clover, crimson clover, red clover, white clover, lespedeza, birdsfoot trefoil, velvet bean, and soybean. Pedigo *et al.* (1973) indicated that the most common food plants are soybean, alfalfa, clovers, field bean, lima bean, and pea, in that order. Adults feed on the nectar from blossoms.

Because of the preference for soybean, most of this insect's biology and management recommendations have been derived from soybean-based research, but in large measure the findings should be applicable to related crops.

Natural Enemies. Many natural enemies are known, with their significance varying according to cloverworm population density. In Iowa, research has shown that during low (endemic) densities parasitoids, and to a lesser extent predators, are relatively important. During outbreak (epidemic) densities, resulting from invasion by many migrating moths early in the year, the entomopathogenic fungus *Nomuraea rileyi* becomes a key factor. The effectiveness of the fungus is principally dependent on presence of high densities of larvae in the second generation. The fungal disease, but not the other mortality factors, is capable of controlling the cloverworm population (Pedigo *et al.*, 1983; Thorvilson and Pedigo, 1984).

Among the parasitoids commonly attacking green cloverworm are several wasps (Hymenoptera: mostly Braconidae and Ichneumonidae) and flies (Diptera: Tachinidae) (Lentz and Pedigo, 1975; Mueller and Kunnalaca, 1979; Bechinski and Pedigo, 1983; Bechinski *et al.*, 1983a; Daigle *et al.*, 1988). The most abundant larval parasitoids are *Cotesia marginiventris* (Cresson) and *Rogas nolophanae* Ashmead (both Hyme-

noptera: Braconidae). Egg parasitism is infrequent, but predation of eggs and young larvae by *Nabis americoferus* Carayon and *N. roseipennis* Reuter (both Hemiptera: Nabidae) is documented (Sloderbeck and Yeargan, 1983b). Predation assumes greater importance in the pupal stage, when such predators as ground beetles (Coleoptera: Carabidae), field crickets (Orthoptera: Gryllidae), and rodents inflict heavy mortality. In addition to the aforementioned entomopathogenic fungus, a granulosis virus sometimes occurs (Carner and Barnett, 1975; Daigle *et al.*, 1988)

Life Cycle and Description. There normally are three generations annually in Iowa, with four flights of moths present in May, June–July, August, and September. The fourth flight may not be evident in some years. Green cloverworm fails to overwinter successfully in cold climates such as Iowa, and reinvades the northern states each spring. The green cloverworm is reported to overwinter in the pupal and adult stages in the south, and as far north as southern Ohio. The overwintering of this species has not been intensively studied in southern states, but it remains reproductively active throughout the year along the Gulf Coast. It is thought to overwinter in the south as far north as southern Virginia, Kentucky, southern Missouri, and most of Texas. In the spring, when sustained winds blow from the south-central states northward, green cloverworm moths are carried into northern areas (Wolf *et al.*, 1987). The length of the life cycle is about 40 days during the summer months.

Egg. Females normally deposit 200–230 eggs, but up to 670 eggs have been recorded from a single female. They are deposited singly. The egg initially is greenish, becomes speckled with orange or red, and then purplish gray just before hatching. The egg is a slightly flattened sphere in shape; the base, in particular, is flattened. The egg measures about 0.51 mm in diameter and 0.35 mm in height, and bears 14–19 readily discernible ridges. Hatching occurs 2–5 days after oviposition.

Larva. There are 6–7 instars, the larvae growing from about 1.5 mm to over 30 mm long as they mature. The larvae are green throughout their development. Larvae bear a pair of longitudinal white stripes along each side, with a less distinct along the back, but they are fairly indistinct until about the third instar. During the terminal instar the white stripes fade, the insect appearing almost entirely green. One of the most distinctive features is the presence of only four pairs of prolegs. The larva walks with a looping motion. Mean head capsule widths (range) for the larvae are 0.23 (0.13–0.28), 0.35 (0.32–0.43), 0.57 (0.48–0.70), 0.89

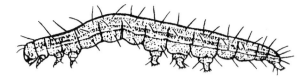

Green cloverworm larva.

Adult green cloverworm.

(0.66–1.00), 1.24 (0.70–1.50), 1.69 (1.35–2.00), and 1.88 (1.73–2.30) mm for instars 1–7, respectively. Duration of the instars was estimated by Stone and Pedigo (1972) to be about 3.1, 1.4, 1.9, 2.1, 2.5, 3.7, and 5.5 days, respectively, for a total larval duration of about 19 days when reared on soybean. Hill (1925) reported an average larval development period of 22.8 days when fed alfalfa. Larvae are solitary in their feeding behavior.

Pupa. As the larvae are near completion of their development they spin a loose web in preparation for pupation. Pupation may occur in the plant canopy, whereby larvae usually web together a leaflet and pupate within a leaf fold. Most larvae, however, drop to the soil and pupate at the surface or just beneath the surface. The pupa is brown to brownish black, and measures 11–15 mm long. Duration of the pupal stage is 9–12 days.

Adult. The adults are mottled grayish-brown with black and silver markings. The male and female differ somewhat in appearance, however. The male has less distinctive markings, bearing about three irregular, transverse black lines across the forewing. The female also has transverse black markings but with greater contrast, and silver and reddish brown areas distally. The hind wings are blackish brown. Wingspan measures 27–34 mm. The mouthparts of both sexes protrude, forming a distinctive snout. Moths hide in vegetation during the daylight hours. They become active at dusk, and reportedly feign death if disturbed, suddenly dropping to the soil with their wings folded. Once they take flight they are strong fliers, and their flight may continue until the early morning hours. Oviposition commences about 4–5 days after adult emergence, and continues for 10 days or longer. Oviposition may occur on both the upper and lower leaf surfaces. Apparently moths prefer to oviposit on leaf surfaces that contain leaf hairs, and deposit eggs preferentially on the lower leaf surfaces of alfalfa and clover because of the greater pubescence. On soybean, which is hairy on both surfaces, the females do not discriminate between locations. Moths preferentially deposit eggs on pubescent varieties of soybean relative to glabrous varieties (Pedigo, 1971). (See color figure 244.)

A good summary of green cloverworm biology was given by Pedigo *et al.* (1973), but a more detailed description, particularly of insect morphology, was found in Hill (1925). The larva was included in keys by Crumb (1956), Oliver and Chapin (1982), Capinera (1986), and in a key to loopers in Appendix A. The adult occurred in a key by Capinera and Schaefer (1983).

Damage

Larvae feed principally on the leaf tissue between the main veins of leaves. Most authors indicate that pods, blossoms, or stem tissue are rarely consumed. Larvae usually feed from the lower leaf surface, and instars 1–2 or 1–3 do not eat completely through the leaf tissue, but leave the upper epidermis intact. Each larva eventually consumes over 100 sq cm of bean leaf tissue, with about 90% occurring in the last two instars. As the beans are very tolerant of defoliation, withstanding about 30% leaf tissue loss before yields are depressed, at least 5–6 mature larvae likely are necessary to inflict damage.

In Delaware, green cloverworm larvae were frequently observed to feed on small pods of lima bean (Burbutis and Kelsey, 1970). However, the beans were very tolerant of injury, and larval densities of up to 8 per plant did not suppress yield.

Management

Sampling. The adult populations can be monitored with blacklight traps, though more males than females are captured. Eggs are deposited principally on the upper surface of leaves. Egg dispersion is random. A sequential sampling plan for eggs was developed by Buntin and Pedigo (1981). Sweep nets are usually used to sample larvae. Larval dispersion is aggregated, and a sequential sampling protocol was presented by Bechinski *et al.* (1983b).

Insecticides. Green cloverworm rarely attains pest status in vegetable crops, but can be controlled easily with foliar insecticides. Also, insecticide-containing baits are effective (Morgan and Todd, 1975). *Bacillus thuringiensis* products are not usually recommended.

Cultural Practices. Cloverworm is most abundant late in the season, in late-maturing cultivars, and in narrow-row plantings (Buschman *et al.*, 1981). Planting date apparently has little influence on damage (McPherson *et al.*, 1988). Tillage practices similarly have few consistent effects on green cloverworm populations (Sloderbeck and Yeargan, 1983a; Thorvilson *et al.*, 1985a).

Green cloverworm oviposits readily in alfalfa, and the first generation often occurs in this crop or clover before soybean or bean are available. Alfalfa is harvested frequently, however, and harvesting results in mortality of most larvae. Thus, alfalfa acts as a "sink" for the cloverworm population, causing a decline in abundance. It is the presence of soybean that generally leads to the great abundance of green cloverworm late in the season (Buntin and Pedigo 1983). Alfalfa also acts as an early season source for parasitoids and disease (Thorvilson *et al.*, 1985b).

Glassy Cutworm
Apamea devastator (Brace)
(Lepidoptera: Noctuidae)

Natural History

Distribution. This native species occurs throughout the United States and southern Canada except for the southeastern states. Its range also includes South America.

Host Plants. Glassy cutworm is principally a grass-feeding species, and crop damage is most likely done when crops follow sod or are planted into fields heavily infested with grassy weeds. In Oregon, Kamm (1990) reported that bentgrass, *Agrostis tenus*; ryegrass, *Lolium perenne*; and wild oats, *Avena fatua*; were attractive to ovipositing females. Vegetables reported injured by glassy cutworm include beet, bean, cabbage, corn, lettuce, and radish. Among other crops injured are alfalfa, barley, bluegrass, fescue, oat, strawberry, timothy, tobacco, and wheat.

Natural Enemies. Several parasitoids of glassy cutworm larvae are known, though in general there is little information about natural population regulation of this species. Kamm (1990) reported that *Lissonota montana* (Cresson) (Hymenoptera: Ichneumonidae) and *Nowickia latianulum* (Tothill) (Diptera: Tachinidae) collectively caused 30–48% parasitism in Oregon. Other wasp parasitoids include *Macrocentrus crassipes* Muesebeck (Hymenoptera: Braconidae), *Pter-*

ocormus ambulatorius (Fabricius), and *Spilichneumon inconstans* (Cresson) (both Hymenoptera: Ichneumonidae). Among other fly parasitoids known from glassy cutworm are *Gonia aldrichi* Tothill and *G. frontosa* Say (both Diptera: Tachinidae).

Life Cycle and Description. There is one generation annually. The winter is passed in the larval stage, with pupation beginning in May. Moth emergence begins in June, but peak abundance usually is during late July or August. Moths may be present until October, and produce eggs that hatch into the overwintering larval stage.

Egg. The egg stage of this poorly known species seems to be undescribed. In Minnesota, Knutson (1944) reported that egg laying was completed by the end of August, but Kamm (1990) inferred from adult and larval data that oviposition occurred over several months in Oregon. Duration of the egg stage is about 12–21 days.

Larva. The larva feeds entirely below-ground, or at least below the plant litter on the soil surface. The body of this cutworm is largely unpigmented, and many authors noted that this grayish larva resembles a white grub (Coleoptera: Scarabaeidae) in general appearance. The mature larva measures about 35–40 mm long. The head is reddish brown, and measures 4.5 mm wide. A large prothoracic plate is also present, and is reddish-brown but with a darker margin. Duration of the larval stage is several months, depending on weather. (See color figure 52.)

Pupa. The larva prepares a pupal cell several centimeter below the surface of the soil. The reddish-brown pupa is about 18–20 mm long and 5 mm wide. Duration of pupation is not well-documented but there are reports of 15–60 days in the literature, with the lower value more typical.

Adult. The adult is light gray to brownish gray in general, with extensive amounts of dark brown mottling. A narrow white transverse line is usually present distally on the forewing, with a series of dark triangles located along the inner margin of the transverse line. The hind wing are brownish, and darker distally. The wingspan is 35–45 mm. A sex pheromone produced by females has been identified (Steck *et al.*, 1977, 1980b). (See color figure 241.)

Glassy cutworm larva.

The most complete description of this insect is found in Crumb (1929). Knutson (1944) and Kamm (1990) also provided useful observations. A bibliography was published by Rings and Arnold (1974). Keys including the larva of this species were given by Crumb (1929, 1956), Whelan (1935), Rings (1977b), and Capinera (1986). It is also included in a key to armyworms and cutworms in Appendix A. The moth was included in pictorial keys by Rings (1977a) and Capinera and Schaefer (1983).

Damage

The larva lives below-ground, feeding on roots and the base of plant stems. Plants are readily killed by this type of injury, and the first sign of injury usually is wilting plants.

Management

Glassy cutworm is not a common pest unless crops are planted into fields that previously had been pasture or grass sod. The problem normally dissipates within 2–3 years after destruction of the grass. Population monitoring is most easily accomplished with pheromone traps because the other stages are associated with the soil and difficult to detect.

Chemical insecticides are useful for prevention of injury, but are most effective when placed in the furrow at planting. Baits are not very effective because larvae remain below-ground and have little contact with bait. Mechanical barriers such as metal cans with the top and bottom removed are often recommended for prevention of cutworm damage in home gardens. For glassy cutworm, a burrowing species, the lower edge of the barrier must be sunk well below the soil surface to become an effective deterrent to feeding.

Granulate Cutworm
Agrotis subterranea (Fabricius)
(Lepidoptera: Noctuidae)

Natural History

Distribution. This species is native to the western hemisphere, and principally tropical in distribution. Although occasionally found as far north as Nova Scotia and Minnesota, it appears not to breed at these latitudes, and is not known from the northwestern states. It is common south of the Ohio River and regularly damaging in the southernmost states from Georgia to California. It also occurs in Central and South America and the Caribbean.

Host Plants. Granulate cutworm feeds on a wide range of plants. Among vegetables attacked are bean, beet, broccoli, Brussels sprouts, cabbage, carrot, cauliflower, celery, corn, cowpea, eggplant, kale, lettuce, onion, pea, pepper, potato, radish, spinach, sweet potato, tomato, turnip, and watermelon. Other crops reported injured include alfalfa, clover, cotton, lespedeza, peach, peanut, sorghum, soybean, strawberry, tobacco, vetch, and wheat. Some of the weeds observed to support larvae include thorny amaranth, *Amaranthus spinosus*; cocklebur, *Xanthium* sp.; dandelion, *Taraxacum* sp.; passion vine, *Passiflora incarnata*; plantain, *Plantago* sp.; and shepherdspurse, *Capsella bursa-pastoris*.

Natural Enemies. Considering the importance of this cutworm in southern states, surprisingly little is known about natural enemies. Among the wasps known to parasitize granulate cutworm are *Apanteles griffini* Viereck, *Chelonus insularis* Cresson, *Meteorus laeventris* (Wesmael), *M. laphygmae* Viereck, *Microgaster feltiae* Meusebeck, *Zele mellea* (Cresson) (all Hymenoptera: Braconidae), *Campoletis flavicincta* (Ashmead) and *Simphion merdarius* (Gravenhorst) (both Hymenoptera: Ichneumonidae). Fly parasitoids known from this cutworm include *Bonnetia comta* (Fallén), *Gonia crassicornis* (Fabricius), *G. longipulvilli* Tothill, *Lespesia archippovora* (Riley), and *Spallanzania hebes* (Fallén) (all Diptera: Tachinidae). A microsporidian disease was reported from Florida (Adlerz, 1975) and a granulosis virus is known (Hamm and Lynch, 1982), but the importance of natural pathogens is uncertain.

Life Cycle and Description. Granulate cutworm is active continuously in the south; adults, eggs and larvae have been collected during all months in Louisiana. Nevertheless, there seems to be seasonality to reproduction, as unmated females are found mostly from May–November. Total abundance similarly is greatest in June–November. In Tennessee, three complete generations are reported, with overwintering insects emerging in March. In addition to egg production about March, peaks in egg production occur in May, July and September. The pupae from the September generation overwinter. The complete life cycle requires 50–70 days.

Egg. The eggs are deposited singly or in small clusters on the upper surface of foliage. Females produce about 800–1600 eggs. They are hemispherical, with 36–40 narrow ridges radiating from the apex, and measure 0.60–0.71 mm in diameter and about 0.50 mm in height. Initially they are white, but darken with age. Normally they hatch in 3–5 days.

Larva. Young larvae initially remain on the foliage during both day and night, but after a few days they begin to hide beneath plant debris or soil during the daylight hours, feeding only at night. The larva buries itself very shallowly, even remaining partially exposed during the day. The number of instars varies from 5–7, but six instars is most common. Mean (range) head capsule widths are 0.31 (0.30–0.32), 0.48 (0.45–0.50), 0.81 (0.73–0.86), 1.31 (1.17–1.57), 1.98 (1.62–2.26), and 2.93 (2.70–3.19) mm for instars 1–6, respectively. Mean (range) duration of the instars is 3.4 (3–5), 2.5 (2–4), 3.1 (3–4), 3.1 (2–5), 4.8 (3–7), and 7.7 (5–10) days, respectively. Total larval development time is about 25 days for larvae with six instars, but 22 days for five-instar larvae and 32 days for seven-instar larvae. The body length measures about 2.0–3.5, 5–6, 12, 17, 22, and 30–37 mm long during instars 1–6, respectively. Larvae with six instars consume about 150 sq cm of foliage but the longer-lived larvae that undergo seven instars consume considerably more, up to 240 sq cm. This cutworm is grayish to reddish brown, with each abdominal segment bearing dull yellowish oblique marks subdorsally. A weak gray line occurs laterally below the spiracles, accompanied by spots of white or yellow. The head is yellowish to brownish. (See color figure 53.)

Pupa. Pupation occurs in the soil, usually at a depth of 3–12 cm. The pupae are dark brown or mahogany, and measure 15–21 mm long and 5–6 mm wide. Duration of the pupal stage is 10–20 days.

Adult. Moths begin mating about one day after emergence, and peak oviposition occurs 2–3 nights after mating (Cline and Habeck, 1977). Longevity of adults is 10–20 days, averaging about 14 days. The moth is medium in size, with a wingspan of 31–43 mm. The color of the forewing varies considerably in its shades of brown and gray, but it is often yellowish-brown and distinctly lighter distally. The forewing bears distinct bean-shaped and round spots centrally, and these spots are linked by a small but sharply defined black bar. The hind wings are white, but dusky marginally and along the veins. (See color figures 242 and 243.)

Biology was described by Jones (1918b), and Snow and Callahan (1968), with the most complete morphological description by Crumb (1929). Culture techniques were described by Lee and Bass (1969). Keys including the larva of this species were given by Crumb (1929, 1956), Whelan (1935), Okumura (1962), Oliver and Chapin (1981), and in a key to armyworms and cutworms in Appendix A. The moth was included in a pictorial key by Capinera and Schaefer (1983).

Damage

This is the most important cutworm pest of vegetables in the Gulf Coast region, and also quite important in California. It damages seedlings by cutting off the stem at the soil surface, older plants by climbing and feeding on foliage, and injures such plants as tomato, watermelon, and eggplant by feeding on, or burrowing into, the fruit. Due to its surface-feeding behavior, granulate cutworm is sometimes a major component of the "rindworm" complex affecting cucurbit fruit. This type of damage usually occurs when fruit are in contact with soil, a common habitat of the larva. Young larvae, through about the second instar, feed on the lower leaf surface and skeletonize leaves. Thereafter, they consume entire leaves.

Management

Moth populations can be monitored with blacklight traps. Larvae can be controlled by application of insec-

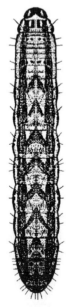

Granulate cutworm larva.

Adult granulate cutworm.

ticides delivered as liquid, granule, or bait formulation. Baits, particularly bran-based, seem particularly effective (Morgan and French, 1971). *Bacillus thuringiensis* is not usually recommended for cutworms. Many cutworm species have a great affinity for weedy fields, but granulate cutworm seems to lack this association. Mechanical barriers can provide some protection from dispersing larvae to seedlings in the home garden. However, moths often are active during the growing season and easily circumvent such barriers. Therefore, it may also be necessary to use netting or row cover material to deny access to plants by ovipositing moths.

Okra Caterpillar
Anomis erosa Höbner
(Lepidoptera: Noctuidae)

Natural History

Distribution. This species is found in South America, Africa, southern Asia, and Australia in addition to North America. It is generally considered to be a southeastern species, and indeed it is most common there. However, it has been reported from as far north as Massachusetts and Montreal, Canada, and as far west as Kansas and Texas. Its distribution also extends southward through Mexico and the Caribbean region.

Host Plants. Okra caterpillar feeds primarily on plants in the family Malvaceae. Okra is the only vegetable crop affected, but its caterpillar also feeds on such ornamental or weedy plants as rose-of-sharon, *Hibiscus syriacus*; swamp rose, *Hibiscus moscheutos*; cotton rose, *Hibiscus mutilabilis*; chinese mallow, *Hibiscus sinensis*; roselle, *Hibiscus sabdariffa*; flour-of-an-hour, *Hibiscus trionum*; velvet leaf, *Abutilon theophrasti*; flowering maple, *Abutilon striatum*; hollyhock, *Althaea rosea*; and round-leaved mallow, *Malva rotundifolia*. It also can be found on cotton, and though it is not generally thought to be a common pest of this crop, it very closely resembles cotton leafworm, *Alabama argillacea* (Hübner), so it may be misidentified and its abundance underestimated (Creighton, 1936). Okra caterpillar is also reported to feed on *Peperomia* sp., family Piperaceae.

Natural Enemies. The natural enemies of okra caterpillar are mostly generalists that attack other caterpillars. For example, paper wasps, *Polistes* spp. (Hymenoptera: Vespidae) are commonly feed on larvae, as do ground beetles (Coleoptera: Carabidae), stink bugs (Hemiptera: Pentatomidae), and assassin bugs (Hemiptera: Reduviidae). Parasitoids of okra caterpillar include *Trichogramma* sp. (Hymenoptera: Trichogrammatidae), *Apanteles bedelliae* Viereck

(Hymenoptera: Braconidae), *Itoplectis conquisitor* (Say) (Hymenoptera: Ichneumonidae), *Copidosoma truncatellum* (Dalman) (Hymenoptera: Encyrtidae), *Syntomosphyrum esurus* (Riley) (Hymenoptera: Eulophidae), and *Eusisyropa blanda* (Osten Sacken) Diptera: Tachinidae). Creighton (1936) indicated that *S. esurus* was the most important parasitoid in his studies conducted in Florida.

Life Cycle and Description. The life cycle of this insect is poorly documented but has been observed in the field from March–October in northern Florida and throughout the year in southern Florida. As a life cycle can be completed in about 35 days, several generations are possible annually. It is thought to overwinter in the pupal stage, but this is not satisfactorily proven.

Egg. The egg is basically spherical, but flattened at the point of attachment. It measures about 0.8 mm in diameter, and bears 31–38 ribs, but most ribs fade before reaching the apex of the egg. Initially whitish, the egg soon turns greenish. They are deposited singly on foliage. Duration of the egg stage is about four days (range 3–6 days).

Larva. The larva is reported to display 5–7 instars. Mean duration of the larval stage is about 16 days (range 13–22 days). It grows from about 2 mm long at hatching to about 35 mm long at maturity. The larva feeds during daylight hours, moves with a typical looper-like gait, and frequents the underside of the leaves. The larva is yellowish-green or green, with a dark stripe dorsally, though it acquires white dorsolateral stripes along the length of its body. As the larva matures, an irregular and broad yellow stripe develops laterally above the spiracles. There are four pairs of prolegs, three abdominal plus the anal prolegs, present on the large larvae. The first pair of abdominal prolegs is reduced in size, relative to the other pairs. About two days before termination of the larval stage the larva folds a section of leaf and anchors it with silk; this is the site of pupation. (See color figure 54.)

Pupa. Initially, the pupa is bright green, but soon turns brown and eventually almost black. It measures about 15 mm long. The tip of the abdomen possesses

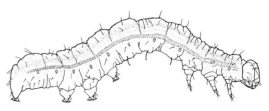

Okra caterpillar larva.

Adult okra caterpillar.

short bristles and hooks which anchor the pupal case to silk webbing. Duration of the pupal stage is about 12 days (range 8–15 days).

Adult.   The body of the adult is yellowish or rust. The forewing is marked with irregular yellowish, rust, and gray areas, though the hind wings are brownish, and darker distally. Adults commence oviposition about five days after emergence, and may continue egg production for 25 days. Adults are nocturnal.

The biology of okra caterpillar was described by Chittenden (1913b), Dozier (1917), and Creighton (1936). The larva was included in the key by Crumb (1956). (See color figure 245.)

Damage

The larvae eat large irregular hole in the leaves of okra, sometimes defoliating entire plants. It is not a common pest, even in the deep south, where it is most abundant.

Management

This caterpillar rarely is abundant enough to warrant suppression, but insecticides applied to the foliage are effective, especially if applied when the larvae are young.

Pale Western Cutworm
Agrotis orthogonia Morrison
(Lepidoptera: Noctuidae)

Natural History

Distribution.   This native species is found in semi-arid regions of the western United States and Canada. Its distribution is largely restricted to the western edge of the Great Plains and eastern portions of the Rocky Mountains, including southern Alberta and Saskatchewan, most of Montana, Wyoming, Colorado and Utah, and northern New Mexico. It also occurs in portions of adjoining states, and has been particularly troublesome in western Kansas.

Host Plants.   Larvae generally feed on grasses and grain crops, and also on some weeds. However, they occasionally have been known to damage vegetables, including bean, beet, carrot, corn, onion, potato, and tomato. Most commonly damaged are the small grain crops such as wheat, barley, rye, oats, and millet, but other field crops such as alfalfa, flax, Sudan grass, sugarbeet, and sweet clover may be fed upon. Among weeds known to be consumed are sunflower, *Helianthus annuus*; tumble mustard, *Sisymbrium altissimum*; Russian thistle, *Salsola kali*; wild lettuce, *Lactuca scariola*; mallow, *Malva* sp.; and dandelion, *Taraxacum* spp.

The adults feed on nectar from flowers, preferring goldenrod, *Solidago* spp., sunflower, *Helianthus* spp., and rabbit brush, *Chrysothamnus* spp. In the absence of the preferred nectar sources other flowers are used, including snakeweed, *Gutierresia* sp.; Canada thistle, *Cirsium arvense*; fleabane, *Erigeron* spp.; and Russian thistle, *Salsola kali*.

Natural Enemies.   Natural enemies are believed to play an important role in the occurrence of this species, but it is the interaction of weather and natural enemies that is critical. Larvae of this cutworm normally spend most of their time below-ground. However, wet weather in the spring months causes larvae to move to the soil surface, where they can be attacked by parasitoids and predators. The proportion of the population lost to natural enemies varies from about 20–70%. Among the wasps known to parasitize pale western cutworm are *Meteorus leviventris* Wesmael, *Chelonus* sp., *Zele* sp. (all Hymenoptera: Braconidae); *Apanteles griffini* Viereck, *Paniscus* sp. (both Hymenoptera: Ichneumonidae); and *Copidosoma bakeri* (Howard) (Hymenoptera: Chalcididae). Other parasitoids include *Bonnetia comta* (Fallén), *Mericia* sp., *Gonia aldrichi* Tothill, *G. longiforceps* Tothill, *G. longipulvilli* Tothill, *Peleteria texensis* Curran, and *Periscepsia rohweri* (Townsend) (all Diptera: Tachinidae); *Anthrax molitor* Lowen, *Villa alternata* (Say), *V. willistoni* (Coquillett), and *Poecilanthrax sackenii* (Coquillett) (all Diptera Bombyliidae). Numerous avian and insect predators have been observed to feed on cutworm larvae; among the insects are leaffooted bugs (Hemiptera: Coreidae), assassin bugs (Hemiptera: Reduviidae), and ambush bugs (Hemiptera: Phymatidae), ground beetles (Coleoptera: Carabidae), and predatory wasps (Hymenoptera: Sphecidae and others). The role of pathogens is uncertain; fungi and viruses seem to be unimportant, but bacterial diseases are sometimes suggested to be a significant mortality factor.

Weather.   The abundance of pale western cutworm is directly related to precipitation patterns. Outbreaks of pale western cutworm rarely occur in areas with more than 12 cm of precipitation during the period of May–July, when pale western cutworm is in the larval stage. Within the semi-arid area of western North America inhabited by this cutworm, spring periods with fewer than 10 precipitation events exceeding 6.4 mm are followed by increases in cutworm number during the following year. Similarly, spring periods with more than 15 such precipitation events are followed by population decreases. The effect of the precipitation is to drive the larvae to the surface of the soil, where they are susceptible to attack by predators and parasitoids. Higher moisture levels may also favor spread of disease among the insects, but this is less certain.

Life Cycle and Description.   There is only a single generation per year throughout the range of this insect. Eggs are laid in the autumn and hatch in the winter or early spring. Larvae feed until early June and then enter a quiescent prepupal period that may last for 40–50 days if the weather is warm. Pupation occurs in July or August, with moths common in late August–October. Not surprisingly, the active period of this species is shorter in the north, approximately April–September in Alberta, whereas in New Mexico the insects are active from February–October.

Egg.   Eggs are deposited in the soil at a depth of 6–12 mm, apparently singly or in small clusters. They are white, turn yellowish-gray and then slightly bluish as the embryo matures. In shape the egg is a slightly flattened sphere, measuring about 1 mm in diameter and 0.8 mm in height. The egg bears 27–32 ridges radiating from the apex. Incubation requires 30–50 days in the field, and embryos require a cold period before hatching. Under laboratory conditions, embryo development requires 11, 14, 21, and 33 days at 30°, 25°, 20°, and 15°C, respectively. They must have contact with moisture or high humidity in order to hatch. As noted above, hatching occurs early in the spring.

Larva.   The larvae feed below-ground for their entire life. They normally display 6–8 instars. Jacobson (1971) gave mean development times of 8, 6, 7, 7, 8, and 14 days (excluding the prepupal part of the terminal instar of another 13 days) for instars 1–6, respectively, when reared at 20°C. In contrast, Parker *et al.* (1921) gave mean development times of 11.2, 8.0, 9.4, 9.8, 11.3, 14.4, 22.6, and 29.6 days (excluding the prepupal period) for instars 1–8, respectively, under unspecified insectary conditions. Thus, larval development time in the latter study, which averaged 118 days, was more than twice in the former study, where development required only 50 days. The difference is even greater if development times at warmer tempera-

Pale western cutworm larva.

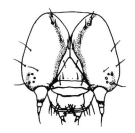

Head capsule of pale western cutworm.

ture are compared. Jacobson (1971) gave mean larval development times of only 29 and 24 days at 25° and 30°C, respectively. Head capsule widths are 0.27–0.31, 0.35–0.45, 0.48–0.67, 0.70–0.98, 1.00–1.50, 1.60–2.20, 2.30–2.80, 2.90–4.00 mm, respectively, for instars 1–8 (Sutter *et al.*, 1972). Mean body lengths are reported to be 1.4, 4.0, 5.9, 9.3, 15.9, 24.6, 31.1, and 36.6 mm, respectively. The larva is gray or bluish gray, and is largely free from distinctive markings. The head capsule, which is yellowish-brown, bears two dark vertical bars. The larva attains a length of about 35–40 mm at maturity. At maturity the larva digs deeper into the soil, usually 5–15 cm, and forms an earthen cell with the aid of salivary secretions. While in this cell, the prepupa shrivels to a length of about 20–25 mm and its color changes to yellowish white. A period of inactivity follows, 30–75 days in duration, that is quite long compared to similar species, and likely is an adaptation allowing the insect to escape the heat and dryness of the summer in its hostile environment. (See color figure 55.)

Pupa.   Pupation occurs within the cell formed by the larva. The pupa is yellowish initially, turning brown with time. The pupa measures about 13–19 mm long. Duration of the pupal period is 20–40 days.

Adult.   The adult is an attractive moth, gray with yellowish and brownish spots on the forewing, and white on some of the veins. The hind wings are whitish, but darker distally. The body is robust and clothed with long scales. The wingspan of the moth is 25–40 mm. Although the moths are considered to be principally nocturnal, females fly and oviposit in late afternoon, and both sexes begin feeding at flowers about sunset. Females select loose soil for oviposition, avoiding hard or crusted surfaces. Fecundity is not well documented, but apparently females can produce 300–400 eggs. Adult activity usually ceases by mid-

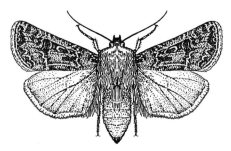

Adult pale western cutworm.

night, principally because it becomes cool, but there are reports of mating occurring at this time. Length of adult life in the field is uncertain, but appears to be 7–14 days. (See color figure 246.)

The biology of pale western cutworm was described by Parker *et al.* (1921), Cook (1930), Sorenson and Thornley (1941), and Jacobson (1971). Rearing was described by Sutter *et al.* (1972). A sex pheromone was described by Struble and Swailes (1978). Keys to the larval stage were found in Whelan (1935), Walkden (1950), Capinera (1986), and in a key to armyworms and cutworms in Appendix A. Adults were included in a key by Capinera and Schaefer (1983).

Damage

The larvae feed below-ground just above the growing point of plants. Young larvae are not large enough to completely sever the plant, but this is accomplished routinely once the third instar is attained. This feeding usually results in death of the plant. In grain crops, entire fields may be killed. The larvae are reported to move underground from plant to plant, often feeding on just the small below-ground section of each seedling. They rarely move above the soil, and are not known to disperse long distances in an "armyworm" fashion in search of food.

Management

Sampling. The eggs and small larvae are very difficult to detect. Large larvae can be recovered from soil, but their densities are usually assessed indirectly from plant damage. Adult densities can be monitored easily with light traps.

Damaging populations are predicted on the basis of pattern of precipitation during the larval stage. The days of the larval stage when at least 6.4 mm of precipitation occur—termed "wet days"—are tabulated, and if they are less than 10, cutworm densities are expected to increase during the next year. In contrast, if the number of "wet days" exceeds 15, then cutworm populations are expected to decrease (Seamans, 1935). Two successive years of dry weather are required to cause high and damaging population densities.

Insecticides. Persistent insecticides applied to the foliage or soil in a liquid formulation provide good control of pale western cutworm. Larval mortality in fields treated with insecticide is not rapid, often requiring several days. The use of persistent insecticides is very important because larvae are inactive and remain below-ground for 3–5 days during molting cycles. Thus, if insecticides are to be effective, they must be persistent enough to remain until the larvae resume activity and come into contact with insecticide (Byers *et al.*, 1992; Hill *et al.*, 1992). Bait formulations generally provide only partial control because larvae generally feed below the soil surface and therefore have little contact with baits (DePew, 1980; McDonald, 1981a). *Bacillus thuringiensis* is not usually recommended for control of cutworms.

Cultural Practices. Cultural methods of management have been developed for grain production systems, but they may have applicability for other cropping systems. Moths infrequently deposit eggs in soil that has a crust, apparently because females cannot penetrate hard soil with the tip of their abdomen. Therefore, a widely practiced technique is to allow a crust to form on the soil in the autumn, before oviposition. Another common practice is to till the soil 10–14 days before planting. This destroys weeds and other alternate hosts on which larvae may be feeding, and causes them to starve before the new crop germinates. A practice that is often suitable for vegetable production is to increase irrigation frequency during periods of cutworm abundance; frequent irrigation in the spring is deleterious to larval survival.

Barriers are sometimes used to reduce access by cutworms to plants grown in the home garden. Metal or waxed-paper containers with both the top and bottom removed can be placed around the plant stem to deter consumption. The barrier should be extended below the soil surface because larvae may burrow below the soil line. For pale western cutworm, the requirement that the lower edge of the barrier be deeply recessed in the soil is especially important, because larvae burrow actively in the soil, rarely coming to the soil surface except during wet weather.

Plantain Looper
Autographa precationis (Guenée)
(Lepidoptera: Noctuidae)

Natural History

Distribution. This native species is eastern in distribution. It is found widely in the United States east of the Great Plains, though it is infrequent in the southernmost states. Occasionally, plantain looper is found as far west as Kansas, Nebraska, and Wyoming. Similarly, in Canada it is known from Nova Scotia to Manitoba.

Host Plants. This insect sometime feeds on such vegetables as bean, cabbage, and parsnip, but is more commonly associated with such weeds as burdock, *Arctium lappa*; common morningglory, *Ipomoea purpurea*; dandelion, *Taraxacum officinale*; lambsquarters, *Chenopodium album*; plantain, *Plantago* spp.; thistle, *Carduus* sp. and *Cirsium* sp.; wild lettuce, *Lactuca* sp.; and wild sunflower, *Helianthus* sp. It has also been found feeding on hollyhock.

Natural Enemies. Natural enemies of this insect are unknown.

Life Cycle and Description. Larvae are the overwintering stage, and apparently there are 2–3 generations annually (Knutson, 1944; Chapman and Lienk, 1981). Moths are present in New York from May until November with a reduction in abundance near the end of June, which probably signifies the completion of the first generation (Chapman and Lienk, 1981). A complete life cycle requires about 30–37 days for completion (Khalsa *et al.*, 1979).

Egg. The egg stage of this little-known insect seems to be undescribed.

Larva. The number of instars usually is six, but occasionally seven are observed. Head capsule widths are about 0.3, 0.5, 0.7, 1.0, 1.3, 2.0, and 2.7 mm for instars 1–7, respectively. Total duration of the larval stage is 17–20 days, with the length of individual instars being about 3.0, 3.0, 2.0, 3.5, 2.8, and 2.3 days, respectively, for instars 1–6. The mature larva is green and, like most of the other loopers, bears three white lines on each side of the back and a white lateral line slightly above the lateral spiracles. The white lines are pale in overwintering larvae but distinct in summer larvae. There tends to be dark shading above the lateral line. The body appears to lack microspines, but there is at least a subdorsal strip of these minute structures. The thoracic legs generally are black. The head bears a broad black line on each side; sometimes the lines are broad and cover the entire head. The larva measures about 30 mm long at maturity. This insect closely resembles alfalfa looper, *Autographa californica*, and bilobed looper, *Megalographa biloba*, in appearance, and is reliably distinguished by examination of the larval mandibles (see Crumb, 1956, or Eichlin and Cunningham, 1978 for a key). However, the geographic range is generally adequate for differentiation from alfalfa looper, and if microspines are readily apparent there is great likelihood that it is bilobed looper.

Pupa. Duration of the pupal stage is about 6–7 days.

Adult. The moth is similar to alfalfa looper in general appearance, with the forewing bearing a silvery white central spot shaped roughly like a "dog leg." However, the "foot" is weakly connected to the "leg," or even disconnected, in plantain looper. The background color of the forewing varies from gray to dark-brown. The hind wing is light brown basally and light to dark-brown distally. The wingspan of this moth is about 35 mm. Mating typically occurs about two days after emergence from the pupa. Females oviposit over a period of about 14 days and produce over 2000 eggs per female. Total adult longevity is estimated at about 19 days.

Key elements of the biology of this insect can be found in Khalsa *et al.* (1979), with additional information and keys in Crumb (1956), Eichlin and Cunningham (1978), and Chapman and Lienk (1981). A key to some common vegetable-feeding loopers, including plantain looper, can be found in Appendix A.

Damage

This insect feeds on the underside of leaves during the first three instars, causing a skeletonizing effect. Thereafter, larvae eat large, irregular holes in leaves. Khalsa *et al.* (1979) demonstrated that this insect consumed almost as much as the more damaging soybean looper, *Pseudoplusia includens*. However, few plantain loopers are usually found attacking crops, so ovipositional preference keeps this insect from becoming a serious pest.

Management

Moths of this species are attracted to light traps. They also can be captured in traps baited with phenylacetaldehyde (Cantelo *et al.*, 1982). This insect normally does not warrant suppression. Foliar insecticides are effective, if needed.

Adult plantain looper.

Potato Stem Borer
Hydraecia micacea (Esper)

Hop Vine Borer
Hydraecia immanis (Guenée)
(Lepidoptera: Noctuidae)

Natural History

Distribution. Potato stem borer is native to Europe, northern Asia, and Japan. It was first found in North America in 1905 in Nova Scotia, but has since spread through eastern Canada west to Manitoba, and is occasionally damaging throughout this geographic range. In the United States potato stem borer is known from the northeastern and midwestern states. Hop vine borer, in contrast, is a native species, found across southern Canada and northern United States from coast to coast. Whereas potato stem borer tends to be more common in Canada, hop vine borer occurs frequently in the United States. Both species assumed greater importance as crop pests starting in the 1970s and 1980s, though the cause is unknown.

Host Plants. Potato stem borer is polyphagous, but it is known principally as a pest of potato, corn, and rhubarb. Crops attacked include barley, corn, hops, onion, potato, raspberry, rhubarb, strawberry, sugarbeet, tomato, and wheat. Several grasses can serve as hosts, including bromegrass, *Bromus sp.*; reed canary grass, *Phalaris arundinacea*; orchardgrass, *Dactylis glomerata*; and quackgrass, *Agropyron repens*. Hemp nettle, *Galeopsis tetrahit*; curly dock, *Rumex crispus*; and possibly other swamp or marsh dwelling plants are suitable broadleaf hosts (Giebink *et al.*, 1992).

The major hosts of hop vine borer are hops and native perennial grasses, and this insect is a serious pest principally in areas where cultivated or wild hops grow. Increasingly, however, it has become a pest of corn. Among weed hosts preferred by ovipositing females are foxtail, *Setaria* spp.; quackgrass, *A. repens*; and to a lesser extent large crabgrass, *Digitaria sanguinalis*; barnyardgrass, *Echinochloa crusgalli*; and fall panicum, *Panicum dichotomiflorum*. Larvae can survive on curly dock, but growth is poor.

Natural Enemies. Several native parasitoids of potato stem borer are known. Egg parasitoids include *Telenomus* sp. (Hymenoptera: Scelionidae), *Trichogramma retorridum* (Girault) (Hymenoptera: Tichogrammatidae), and *Centrodora* sp. (Hymenoptera: Mymaridae). Parasitoids reared from larvae include *Lydella radicus* Townsend (Diptera: Tachinidae), *Diadegma* sp., *Campoletis* sp., *Ectopimorpha luperinae* Cushman, and *Glypta* sp. (all Hymenoptera: Ichneu-

monidae). Reared from pupae are *Therion* sp. and *Pterocormus* sp. (both Hymenoptera: Ichneumonidae), but these species likely attack the larval stage. The most effective parasitoid in Ontario is *Lydella radicus*, and parasitism levels of 25–60% have been reported (West *et al.*, 1983), but the other species seem to contribute little to the overall level of parasitism. Additional parasitoids have been imported from Europe and released in Canada, including *Macrocentrus blandus* Eady and Clark (Hymenoptera: Braconidae) and *Lydella stabulans* Fallén (Diptera: Tachinidae).

Natural enemies of hop vine borer are less well known, but several predators and parasitoids were identified in New York. Among the ground beetle predators are *Calosoma calidum* Fabricius, *Harpalus pensylvanicus* De Geer, *Pterostichus lucublandus* Say, *Pterostichus stygicus* Say, and *Amara impuncticollis* Say (all Coleoptera: Carabidae). Hawley (1918) suggested that the ground beetles consumed the egg, larval, and pupal stages of hop vine borer. Parasitoids identified from New York included *Microplitis gortynae* Riley, *Aenoplex* sp., and *Synaldis* sp. (all Hymenoptera: Braconidae), and *Lespesia frenchii* Williston (Diptera: Tachinidae).

Life Cycle and Description. Potato stem borer and hop vine borer are similar in biology and appearance. They display one generation per year, with the egg serving as the overwintering stage. Eggs hatch in April–May, pupation typically occurs in July, and adults are found from late July–September.

Egg. The eggs are laid in 2–3 parallel rows between the stem and leaf sheath of grasses with a split leaf sheath. The number of eggs ranges from about 30–300 per clutch (Levine, 1986a). In shape, the eggs are a flattened sphere. They measure 0.64–0.82 mm in diameter and 0.31–0.51 mm in height. The edges of the eggs are marked with about 100 narrow, branching ridges, but ridges are absent from the center of the egg. They are white when first deposited, but turn reddish-brown, and then black just before hatching. The egg cluster is covered with a transparent film. They are often deposited in August–September and hatch in April–May—a duration of about eight months.

Larva. There are six instars. Mean (range) of head capsule widths is about 0.33 (0.30–0.38), 0.60 (0.41–0.76), 0.97 (0.89–1.11), 1.43 (1.25–1.75), 2.24 (2.00–2.50), and 3.50 (2.75–4.06) mm, respectively, for instars 1–6. Larval development time is about 30–70 days when reared at 24–27°C. Deedat *et al.* (1983), for example, gave mean instar-specific development times of potato stem borer as 4.9, 3.8, 3.8, 3.2, 4.4, and 11.5 for instars 1–6, respectively. In the field, however, larval

Hop vine borer eggs.

Adult potato stem borer.

Adult hop vine borer.

periods are reported to be longer, about 6–9 weeks in Wisconsin (Giebink *et al.*, 1984) and 9–12 weeks in New York (Hawley, 1918). The developmental threshold of hop vine borer is about 5°C, slightly lower than that of potato stem borer, which is about 7°C (Giebink *et al.*, 1985). The larva is whitish, but the early instars bear rose to purplish bands on the thoracic and abdominal segments. Thereafter the bands fade in hop vine borer, whereas potato stem borer larvae bands tend the remain evident. The head capsule is yellow in potato stem borer but brown in hop vine borer. The larvae measure about 2–3 mm at hatch, but eventually attain a length of 30–50 mm.

Pupa. Pupation occurs in the soil at a depth of 2–4 cm. Pupation normally occurs in July, with a duration of 4–6 weeks in the field in New York, but only 20–30 days under laboratory conditions. The pupa is dark brown, and measures 15–28 mm long. The tip of the abdomen bears two short spines.

Adult. The adult is light brown, with pinkish or greenish tints. In hop vine borer the front wings bear broad light-colored transverse lines bordered with brown, and some darker gray to olive brown shaded areas. In potato stem borer the overall color and pattern are similar, but the transverse lines on the front wings are narrow and dark. The hind wings are grayish with dark veins. The wingspan is 40–50 mm. The moths were pictured by Rings and Metzler (1982).

Hop vine borer larva.

Moths emerge in August–September, mate and begin egg production within a few days of emergence. Fecundity is estimated at 400–1200 eggs.

Life history of potato stem borer was given by Deedat *et al.* (1983). Life history of hop vine borer was given by Hawley (1918) and Giebink *et al.* (1984). Developmental biology of both species was described by Giebink *et al.* (1985). Culture of *Hydraecia* spp. on artificial diet was described by West *et al.* (1985) and Giebink *et al.* (1985). A sex pheromone has been described for potato stem borer (Teal *et al.*, 1983; Burns and Teal, 1989).

Damage

Larvae feed initially on grasses growing as weeds among or near crop plants, then usually switch to larger grasses such as corn, or broadleaf plants such as hops, potato, or curly dock. On perennial grasses the feeding occurs above-ground, but after the feeding switch at about the fourth instar larvae tunnel below-ground into the base of the stem and roots. Some of the perennial grasses and other plants may have sizable underground rhizomes, roots, or stems that allow complete larval development, but this aspect of larval biology is poorly known.

Damage to corn by potato stem borer was described by Deedat and Ellis (1983), who reported that over 90% of seedling corn plants were infested in some fields. Small plants, such as two-leaf stage seedlings, may be completely severed by the entry of potato stem borer, and this damage resembles cutworm injury. With larger seedlings, however, larvae may burrow within the stem, feeding just above the roots, until larvae attain the fifth or sixth instar. Such mature larvae tend to remain below-ground, outside the stem, entering only to feed. Early signs of larval feeding are leaf or plant wilting; later signs are death and disintegration of the plant. Young plants perish within a few days of larval attack. Plants that have attained the eight-leaf stage are slow to wilt and die, but eventually perish. Larvae often destroy 3–4 plants during the course of their development.

Management

Sampling. Adults can be sampled with blacklight traps, and the sex pheromone of these species may eventually prove useful. Levine (1989) used temperature summation, about 1700 degree-days above a threshold of 5.3°C, to estimate peak moth flight by hop vine borer. Larvae can be sampled by dissecting seedlings, but the below-ground portion should also be included in the sample. Wilting plants are a good indication of infestation by larvae.

Insecticides. Persistent insecticides applied to crop plants and soil can provide some suppression of larvae, including those that disperse into crop fields from nearby weedy vegetation. However, better crop protection can be attained by applying insecticide directly to the source of many larvae, weedy fence rows (Deedat *et al.*, 1982). Insecticidal control alone is often inadequate if the borers are abundant in the proximity of a susceptible crop.

Cultural Practices. Cropping practices can help alleviate injury by *Hydraecia* spp. Of foremost importance is weed management. Larvae often invade crop fields from weedy fence rows, resulting in considerable damage along field margins. Thus, insecticidal treatment of the crop periphery, or destruction of grasses and weeds by burning or application of herbicides, can reduce injury. The critical period for weed management is early in the season, typically April or May before susceptible annual crops are available. The presence of wild hops is often related to the occurrence of hop vine borer, whereas potato stem borer is positively affected by the presence of marshy areas where several alternate host plants may occur. Potato stem borer is most likely to be widespread in fields that are heavily infested with grasses (Deedat *et al.*, 1982).

Redbacked Cutworm
Euxoa ochrogaster (Guenée)
(Lepidoptera: Noctuidae)

Natural History

Distribution. Redbacked cutworm is widely distributed in northern climates, occurring in Asia as well as Canada and northern parts of the United States. It is found south to New Mexico in the Rocky Mountain region, but elsewhere is restricted to more northern latitudes, extending only as far south as Missouri. Although redbacked cutworm has damaged crops throughout Canada, it is most abundant, and damaging, in the northern Great Plains and westward, including British Columbia, Washington, and Oregon.

Host Plants. Redbacked cutworm feeds on numerous crops. Among vegetables injured are asparagus, bean, beet, broccoli, cabbage, cantaloupe, cauliflower, cucumber, lettuce, onion, pea, radish, squash, tomato, and turnip. Many other vegetables probably are unreported. Other crops injured include alfalfa, alsike, barley, canola, flax, mustard, oat, sugarbeet, sunflower, sweetclover, and wheat. As might be expected from an insect with such a varied diet, there are occasional reports of injury to flowers and fruit trees.

Natural Enemies. Natural enemies, particularly diseases, seem to play a significant role in redbacked cutworm biology. This cutworm displays periods of abundance lasting 2–4 years, followed by periods of scarcity persisting for at least two years. Duration of abundance typically are preceded by warm, dry weather in late summer and followed by similar warm weather in spring. Warm and dry weather apparently allows the moths to feed freely in the summer, and minimizes the impact of fungal disease in the spring. Diseases and parasitoids apparently contribute to the cutworm-free period that typically follows population increase. King and Atkinson (1928) suggested that diseases were the most effective factor in reducing outbreaks, but diseases of redbacked cutworm apparently have escaped serious study.

Among the parasitoids known to affect redbacked cutworm are *Gonia aldrichi* Tothill, *G. fuscicollis* Tothill, *Bonnetia comta* (Fallén), *Periscepsia helymus* (Walker) (all Diptera: Tachinidae); *Villa alternata* (Say), *V. fulviana* (Say), *V. lateralis* (Say), *Poecilanthrax alcyon* (Say), *P. willistonii* (Coquillet) (all Diptera: Bombyliidae); *Eutanyacra suturalis* (Say), *Diphyus euxoa* Heinrich, *D. apiculatus* (Walkley), *Exetastes obscurus* Cresson, *Spilichneumon superbus* (Provancher), *Campoletis atkinsoni* (Viereck), *C. australis* (Viereck), *Netelia* sp., *Gravenhorstia propinqua* (Cresson) (all Hymenoptera: Ichneumonidae); *Apanteles lalticola* (Ashmead), *A. aeviceps* (Ashmead), *A. griffini* Viereck, *Microplitus kewleyi* Muesebeck, *Meteorus vulgaris* Cresson, *M. laeviventris* (Wesmael) (Hymenoptera: Braconidae); *Copidosoma bakeri* (Howard) (Hymenoptera: Encyrtidae); and *Agamermis* sp. (Nematoda: Mermithidae). Not all species are abundant and important in population regulation, and their relative importance varies among locations and years. Nevertheless, Canadian studies indicate that the larval parasitoids *Gonia aldrichi*, *Meteorus vulgaris*, and *Campoletis atkinsoni*, and the egg parasitoid *Copidosoma bakeri*, are among the more important species (Schaaf *et al.*, 1972).

Predators also affect redbacked cutworm populations. In Alberta, Frank (1971) reported 21 species of ground beetles (Coleoptera: Carabidae), and several species of rove beetles (Coleoptera: Staphylinidae) feeding on eggs, larvae, or pupae. Spiders also are thought to feed on this cutworm.

Life Cycle and Description. There is one generation per year. Adults are active in late July and throughout August. They deposit eggs just beneath the soil surface, and the eggs overwinter. They hatch in April, larvae develop over a 6–8 week period, and pupation normally occurs in June.

Egg. The eggs are deposited in August and September in loose soil. Heavy, crusted, and wet soil tends to be avoided. They are whitish initially, become bluish as the embryo completes its development. Embryonic development is completed in 8, 10, 14, and 20 days at 30°, 25°, 20°, and 15°C, respectively. Eggs do not hatch upon completion of embryonic development, the embryos remaining in diapause until spring. When held at 5°C for 3–4 months the eggs can hatch. They are quite resistant to desiccation (Jacobson, 1970).

Larva. Larvae tend to live below-ground during the daylight hours, but come to the surface, and even climb plants, at night. The grayish larvae are distinguished by the presence of brick-red stripes along the back, separated by a narrow pale stripe centrally. Laterally, a black stripe borders the red stripes. The head and prothoracic plate are yellowish-brown, though the head bears dark brown submedial arcs. Normally there are six instars, but when reared at 30°C only five instars develop. Mean duration (range) of the instars is estimated at 5.6 (5–7), 4.0 (3–7), 3.7 (3–8), 5.8 (3–8), 6.1 (4–11), 14.0 (6–22) days, respectively, at 20°C. Total larval development time (including the prepupal portion of the last instar) is reported to be about 72, 39, 32, and 22 days at 15°, 20°, 25°, and 30°C, respectively. Larvae attain a length of about 38 mm at maturity.

Pupa. Larvae pupate in the soil at depths of 2.5–5.0 cm. The pupa is reddish-brown and measures about 2 cm long. Pupal development times are about 36.8 (33–42), 21.5 (18–23), 14.3 (13–16), and 12.6 (12–14) days at 15°, 20°, 25°, and 30°C, respectively.

Adult. The moths measure about 35–40 mm in wingspan. The front wings are variable and tend toward four basic types. One common wing pattern is uniformly dark reddish-brown with bean-shaped and round spots on the front wings bearing a light border. Another other common wing pattern is much lighter in color, often bearing a grayish bar along the leading edge of the wing. This latter color form also tends to have a black bar connecting the two spots; the spots similarly have a light margin. The hind wing in both forms is grayish basally, with brown distally and along the major veins. The major color forms of the adult were pictured by Hardwick (1965a). Females produce a sex pheromone (Struble 1981a, Palaniswamy *et al.*, 1983) beginning about seven days after adult emergence. Males respond to the pheromone soon after sunset (Struble and Jacobson, 1970). The pre-oviposition period is 6–13 days. Adults normally survive for about 20 days. (See color figure 247.)

A brief summary of redbacked cutworm, including management options, was given by King (1926). The importance of redbacked cutworm in Canada was described by Beirne (1971). Developmental biology was given by Jacobson (1970). A pictorial key that included redbacked cutworm adults was published by Capinera and Schaefer (1983). It was also included in a comprehensive treatment of North American noctuids (LaFontaine, 1987).

Damage

Larvae generally feed at the soil surface or slightly below the surface, though on fruit crops they may climb the plant to feed on leaf and blossom buds. On herbaceous plants, young larvae eat holes and notch the leaf tissues, whereas older larvae sever the plant stem at the soil surface. The spotty distribution of damage is easily confused with poor seed germination, but the presence of severed, shriveled plants indicates cutworms. Larvae are most damaging during hot, dry weather because the plants tend to be stressed. As their distribution in the field tends to be aggregated, larvae usually damage most plants in an area, leaving few individual plants available to compensate for damage. Thus, damage tends to be almost directly proportional to abundance (Ayre, 1990). This species is considered by some to be the most destructive cutworm in Canada.

Management

Sampling. Both blacklight and pheromone traps give good estimates of moth activity, though blacklight traps capture moths earlier, and pheromone traps capture moths longer (Steck *et al.*, 1980a; Ayre *et al.*, 1982b; Gerber and Walkof, 1992). The sex pheromone, when applied in large quantities to an area, can also be used to disrupt mating. The distribution of larvae in peppermint fields is generally clumped (Danielson and Berry, 1978), and this is likely the condition in other crops. In peppermint, 20–40 soil samples were necessary to determine whether or not a crop required treatment.

Insecticides. Persistent insecticides are used for redbacked cutworm suppression. Most products are effective when applied to foliage, and some when applied to soil (McDonald, 1981b). Some populations of redbacked cutworm have been known to develop measurable resistance to insecticides (Harris, 1976). As redbacked cutworm comes to the soil surface at night to feed, baits containing insecticide are very effective.

Cultural Practices. Tamaki *et al.* (1975) documented the abundance of cutworm larvae in weedy fields and weedy sections of fields in Washington. Perennial weeds such as field bindweed, *Convolvulus arvensis*, and Canada thistle, *Cirsium arvense*, either favor larval survival or attract ovipositing moths, leading to increased larval abundance.

Tillage of fallow fields during the oviposition period deters egg laying because moths seem to select weedy patches. Also, clean cultivation early in the season, before planting, can cause young larvae to starve.

Mechanical barriers are sometimes recommended for protection of plants in the home garden. Metal or waxed-paper containers with the top and bottom removed may be placed over susceptible seedlings to deter feeding by larvae. It is helpful to extend the barrier well below the soil surface to minimize access by burrowing larvae.

Biological Control. Steinernematid and heterorhabditid nematodes can infect larvae in the soil, but redbacked cutworm larvae appear to be less susceptible than some other species (Morris and Converse, 1991). *Bacillus thuringiensis* is not usually recommended for suppression of cutworms.

Southern Armyworm
Spodoptera eridania (Cramer)
(Ledidoptera: Noctuidae)

Natural History

Distribution. This insect is native to the Americas, occurring widely in North, Central and South America and the Caribbean. In the United States, southern armyworm is found principally in the southeastern states. Although its range extends west to Kansas, New Mexico, and California, this species is of little consequence in western states.

Host Plants. Southern armyworm has a very broad host range. Some of the vegetables injured are beet, cabbage, carrot, collards, cowpea, eggplant, okra, pepper, potato, squash, sweet potato, tomato, and watermelon. Other crops injured include avocado, citrus, peanut, sunflower, velvet bean, and tobacco. Numerous weeds are consumed, but pigweed, *Amaranthus* spp.; and pokeweed, *Phytolacca americana*; are especially favored, and grasses are rarely eaten. There are several reports of armyworm infestations beginning with these two weeds, and adjacent crops experience damage only after the more favored weeds are consumed. (See color figure 1.)

Natural Enemies. Several wasp parasitoids commonly associated with caterpillars of other species, including *Cotesia marginiventris* (Cresson), *Chelonus insularis* Cresson, *Meteorus autographae* Muesebeck, and *M. laphygmae* Viereck (all Hymenoptera: Braconidae) also attack southern armyworm. *Meteorus autographae* was the dominant parasitoid in a Florida study (Tingle *et al.*, 1978). Also reared from southern armyworm are *Campoletis flavicincta* (Ashmead) and *Ophion flavidus* Brulle (both Hymenoptera: Ichneumonidae); *Euplectrus platyhypenae* Howard (Hymenoptera: Eulophidae); *Choeteprosopa hedemanni* Braeur and Bergenstamm, *Euphorocera claripennis* (Macquart), *Gonia crassicornis* (Fabricius), *Winthemia quadripustulata* (Fabricius), and *W. rufopicta* (Bigot) (all Diptera: Tachinidae). Predators certainly must be an important factor in southern armyworm biology, but this aspect seems undocumented. Larvae are susceptible to infection by the fungus *Beauveria bassiana* (Gardner and Noblet, 1978).

Life Cycle and Description. The number of generations is estimated at four annually in Florida. Insects may be present throughout the insect's range from March until October, but they are most commonly observed in late summer and autumn. In northern Florida, moths can be found throughout the year, withstanding several days of freezing weather (Mitchell and Tumlinson, 1994). About 30–40 days are required for a complete generation.

Egg. The shape of the egg is a flattened sphere. Eggs measure about 0.45 mm in diameter and 0.35 mm in height. They bear about 50 slender ribs which radiate outward from the center. The eggs are greenish initially, turning tan as they age. They are laid in clusters, and covered with scales from the body of the moth. Duration of the egg stage is 4–6 days.

Larva. The larvae display six instars as they grow to attain a length of about 35 mm. The head capsule widths are about 0.25–0.30, 0.40–0.50, 0.60–0.80, 0.95–

Southern armyworm larva.

1.15, 1.35–1.85, and 2.35–2.85 mm, respectively (Redfern, 1967). Larvae are green or blackish-green with a uniform light brown or reddish-brown head throughout their development. Larger larvae bear a narrow white line dorsally, and additional stripes laterally. Each side normally bears a broad yellowish or whitish stripe that is interrupted by a dark spot on the first abdominal segment, though in some cases this spot is weak. A series of dark triangles is usually present dorsolaterally along the length of the body. Larvae usually are found on the lower surface of leaves, and are most active at night. Duration of the larval stage is normally 14–20 days. (See color figure 56.)

Pupa. The larvae pupate in the soil, usually burrowing to a depth of 5–10 cm. The pupa is mahogany brown and measures about 16–18 mm long and 5–6 mm wide. Duration of the pupal period is 11–13 days.

Adult. The moth measures 33–38 mm in wingspan. The front wings are gray and brown, with irregular dark brown and black markings. The wing pattern is highly variable. Some individuals bear a pronounced bean-shaped spot near the center of the wing, whereas others lack the spot or instead bear a broad black band extending from the center of the wing to the margin. The hind wings are opalescent white.

The biology of southern armyworm is poorly documented, but Chittenden and Russell (1910) gave key features. Additional description was given by Crumb (1929). The larval keys developed by Levy and Habeck (1976), Passoa (1991), and Heppner (1998) are useful to distinguish southern armyworm from related species. It was also included in a key by Oliver and Chapin (1981). Heppner (1998) provided a key to the adults of North American *Spodoptera*. Rearing techniques were provided by Redfern and Raulston (1970). A sex pheromone has been identified and evaluated in the field (Mitchell and Tumlinson, 1994).

Damage

This is one of the most common pests of southeastern vegetable gardens. Larvae are defoliators and feed gregariously while young, often skeletonizing leaves. As they mature they become solitary, and also bore readily into fruit, often damaging tomato. When

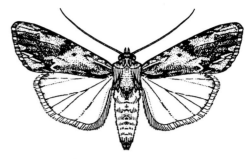

Adult southern armyworm.

stressed by lack of food they eat the apical portions of branches, bore into stem tissue, and attack tubers near the surface of the soil. High densities and lack of food sometimes prompt larvae to move in great number, the basis of the "armyworm" designation, feeding on all vegetation in their path. Southern armyworm is more damaging to cassava than black cutworm, *Agrotis ipsilon* (Hufnagel), given equal numbers of insects (Pena and Waddill, 1981).

Management

Insecticides. Southern armyworm is best controlled with foliar insecticides when larvae are small. Insecticides vary considerably in their toxicity to larvae (Aziz, 1973). This insect is difficult to control with botanical insecticides (Valles and Capinera, 1993). Berger (1920) reported some success at southern armyworm suppression by application of bran bait containing insecticide. However, this is useful principally for large, mobile larvae that have left the plant and are on the soil surface.

Host-Plant Resistance. There is limited information that suggested differences in susceptibility among sweet potato cultivars to armyworm damage (Habeck, 1976). Conventional sources of insect resistance in corn have little effect on southern armyworm (Manuwoto and Scriber, 1982).

Soybean Looper
Pseudoplusia includens (Walker)
(Lepidoptera: Noctuidae)

Natural History

Distribution. Soybean looper is native to the western hemisphere, where it is found from the United States to Argentina. It is a tropical or subtropical insect, and overwinters successfully only in warm climates. In the United States, it survives the winter in southern Florida and southern Texas. Not surprisingly, damaging populations are largely restricted to the southeastern and south-central states. The moths are highly dispersive, however, and this insect has been captured throughout the United States, except for the Pacific Northwest. It is rare in Canada, but has been recovered from southern Ontario and Quebec, and from Nova Scotia.

Host Plants.   This insect is known principally as a defoliator of soybean, but it reportedly has a fairly wide host range. Vegetables reported to be consumed are asparagus, bean, broccoli, celery, collards, corn, cowpea, garlic, lettuce, mustard, okra, pea, pepper, potato, sweet potato, tomato, watercress, and watermelon. Other host plants include field crops such as alfalfa, cotton, peanut, sunflower, and tobacco, and flower crops such as aster, begonia, calendula, carnation, chrysanthemum, geranium, and poinsettia. Weeds also are suitable hosts, including such common species as pigweed, *Amaranthus* sp.; cocklebur, *Xanthium pennsylvanicum*; horseweed, *Egeron canadensis*; wild sunflower, *Helianthus* spp.; pepperweed, *Lepidium virginicum*; and dock, *Rumex* spp. Herzog (1980) provided a long list of host plants.

Soybean looper has long been confused with cabbage looper, *Trichoplusia ni*; the larvae are difficult to separate accurately. Some of the aforementioned records may reflect such misidentifications, though this insect certainly does occur on vegetables, particularly tomato. It is invariably the dominant looper on soybean in the southeast, but cabbage looper is the dominant looper species on vegetables. Martin *et al.* (1976a) presented interesting data from Florida, collected both from field cages and crop fields, demonstrating the preference of soybean looper for field crops.

Natural Enemies.   Soybean looper is host to numerous parasitoids, pathogens, and predators. In Georgia, it is common for 40% of larvae to succumb to a parasitoid or disease (Beach and Todd, 1985). In Louisiana, such natural mortality seems to be even higher, but the effectiveness of these biotic control agents varies considerably with time, location, and cropping system (Burleigh, 1972). Nearly all research has emphasized soybean, so results of some studies may not be applicable to vegetables. It is of interest to note that Martin *et al.* (1981b) reported higher levels of parasitism on soybean looper in vegetables than in soybean during studies conducted in northern Florida.

The most important parasitoids tend to be *Copidosoma truncatellum* (Dalman) (Hymenoptera: Encyrtidae), *Meteorus autographae* Muesebeck, *Apanteles scitulus* Riley, and *Cotesia marginiventris* (Cresson) (all Hymenoptera: Braconidae). Several other wasp and fly (Diptera: Tachinidae) parasitoids exert only low levels of mortality (Burleigh, 1971 and 1972; Beach and Todd, 1985; Daigle *et al.*, 1990).

Fungi are important mortality factors for soybean looper. *Entomophthora gammae*, *Massospora* sp., and *Nomuraea rileyi* are especially common. A nuclear polyhedrosis virus also is known. Often the pathogens are not effective until late in the season, or until looper densities are high (Burleigh, 1972; Harper and Carner, 1973).

Many general predators such as the lady beetle *Coleomegilla maculata* (Coleoptera: Coccinellidae); big-eyed bug, *Geocoris* spp. (Hemiptera: Lygaeidae); the damsel bug *Nabis roseipennis* Reuter (Hemiptera: Nabidae); and the ground beetle *Calosoma* spp. (Coleoptera: Carabidae) feed on soybean looper. All stages of development are subject to predation (Richman *et al.*, 1980; Brown and Goyer, 1984).

Life Cycle and Description.   Total generation time for soybean looper is estimated at about 27–34 days during the summer months. In most of Florida, moths remain active throughout the year, but in north the period of adult activity is restricted (Mitchell *et al.*, 1975). Moths disperse northward annually, attaining Georgia in June and July, and South and North Carolina in August and September. There likely are 3–5 generations annually in the southern areas of soybean looper's range, and two per year in most northern latitudes such as North Carolina.

Egg.   The eggs are slightly flattened spheres, laid singly with a flattened side affixed to the foliage. They measure about 0.6 mm in diameter and 0.4 mm in height, and their close examination reveals numerous ridges radiating from the top of the egg. There are about 34 ridges per egg, but unlike many other noctuid eggs, the ridges are not sharply defined and easy to count. Initially, the egg is white, but turns light brown with maturity. Duration of the egg stage averages 4.1 days at 27°C. Females deposit about 300–600 eggs. Egg deposition is affected by adult food sources, cotton nectar being especially favorable (Jensen *et al.*, 1974). Maximum egg production occurs at about 29°C (Mason and Mack, 1984).

Larva.   The larvae are green, and eventually attain a length of about 30 mm. The body is marked with thin white stripes dorsally and along the sides. In most respects soybean looper closely resembles cabbage looper, *Trichoplusia ni*, including the presence of only three pairs of prolegs and the presence of nipple-like vestigial prolegs on abdominal segments three and four. Soybean looper differs in having minute microspines, but this character is of marginal diagnostic value because the microspines are difficult to detect. The thoracic legs of soybean looper are often dark, and sometimes can be used to differentiate a specimen from cabbage looper. A reliable character for differentiation of these species is structure of the mandible (see Eichlin and Cunningham (1978) for an illustration of this character). The head is grayish-green, and may be marked with dark bands laterally. Generally there are six larval instars. According to Mitchell (1967), mean duration (range) of the instars is 3.7 (3–5), 2.1 (2–3), 2.2 (1–5), 2.6 (1–5), 2.9 (1–5), and 6.2 (4–11) days, respectively, for instars 1–6 when reared at 27°C. Head capsule widths for the six instars are about 0.25, 0.39, 0.60, 0.96, 1.45, and 2.18 mm, respectively. However, Shour and Sparks (1981) noted that the total number of instars could vary, and that total larval duration was 15.4, 16.8, or 20.7 days depending on whether there were 5, 6, or 7 instars, respectively. Strand (1990) described the basis for instar variation in soybean looper. (See color figure 57.)

Pupa. Larvae spin a loose silk cocoon on the foliage and pupate within. Duration of the pupal stage is about seven days (range 3–11 days).

Adult. The moth's front wings are marked with various shades of brown and gray, but the overall effect is dark. The silvery white marking at the center of the forewing, which is sometimes said to resemble a "dog leg," usually has a detached "foot." The hind wings are light brown basally and dark brown distally. The moth is readily confused with plantain looper, *Autographa precationis*, but the forewing of plantain looper is rusty red, whereas that of soybean looper has a brassy reflection. Wingspan of soybean looper moths is 30–39 mm. Moths begin oviposition 3–4 days after emergence, and adult longevity is typically 5–10 days. Moth activity is highest between 10 p.m. and 2 a.m. (Mitchell 1973).

A review of soybean looper biology was given by Herzog (1980). Rearing procedures were given by Hartley (1990). A multi-component sex pheromone was described by Linn *et al.* (1987). Keys to adults and larvae can be found in LaFontaine and Poole (1991). Larvae also were included in the key of Crumb (1956), and adults in Eichlin and Cunningham (1978). Soybean looper is included in the key to vegetable-feeding loopers found in Appendix A.

Soybean looper larva.

Damage

Larvae feed principally on leaf tissue. Foliage consumption of soybean, which probably is similar for green bean, is 0.2, 0.8, 1.8, 7.9, 15.5, and 55.7 sq cm for the six instars, respectively (Reid and Greene, 1973). Larvae also sometimes feed on the pods of legumes, and the silk and kernels of corn (Janes and Greene, 1970).

Management

Sampling. Soybean looper moths can be captured with light and pheromone traps (Tumlinson *et al.*, 1972), or combination of light and pheromone traps (Mitchell *et al.*, 1975). Larval sampling has been studied extensively in soybean, but methodology for vegetables has not been developed. Herzog (1980) reviewed soybean looper sampling.

Insecticides. Damage is usually prevented by application of insecticide to foliage. However, insecticide resistance has been reported to be a serious problem (Leonard *et al.*, 1990). Population increases of soybean looper have been reported following foliar application of certain insecticides; it seems likely that destruction of beneficial insects by insecticides allows the loopers to attain elevated levels of abundance (Shepard *et al.*, 1977). Soil-applied systemic insecticides have little effect on parasitism (McCutcheon *et al.*, 1990).

Biological Control. Natural enemies are often adequate to maintain soybean looper at non-damaging densities on vegetable crops. If this level of control is not adequate, *Bacillus thuringiensis* can be applied. A nuclear polyhedrosis virus has been demonstrated to be effective under field conditions, but it is not available commercially (McLeod *et al.*, 1982).

Spotted Cutworm
Xestia adela Franclemont
Xestia dolosa Franclemont
(Lepidoptera: Noctuidae)

Natural History

Distribution. These species have long been known as *Xestia* (*Amathes*) *c-nigrum* (Linnaeus), a Eurasian species. However, they now have been differentiated into two other species: *Xestia adela* and *Xestia dolosa* (Franclemont, 1980). *Xestia adela* occurs throughout Canada and the United States, including Alaska, whereas *X. dolosa* occurs only in the eastern United States and Canada west to about North Dakota. They remain difficult to differentiate and are best treated together. Hudson (1982) suggested that these species are derived from separate introductions of *X. c-nigrum*.

Host Plants. Spotted cutworm is a general feeder, consuming various flowers, vegetables, fruit trees, and other plants. Among vegetables attacked are asparagus, beet, cabbage, carrot, cauliflower, celery, corn, lettuce, onion, pea, potato, rhubarb, tomato, and turnip. Fruits consumed include apple, cranberry, currant, gooseberry, raspberry, and pear. Occasional damage is incurred by barley, clover, flax, oat, rye, sugarbeet, tobacco, and wheat. Various weeds are consumed, including Canada thistle, *Cirsium arvense*; chickweed, *Stellaria* sp.; goldenrod, *Solidago* sp.; lambsquarters, *Chenopodium album*; morningglory, *Ipomoea* sp.; redroot pigweed, *Amaranthus retroflexus*; smartweed, *Persicaria* sp.; sunflower, *Helianthus* sp.; ferns and grasses.

Natural Enemies. Flies known to parasitize spotted cutworm include *Euphorocera claripennis* (Macquart), *Winthemia quadripustulata* (Fabricius), and *W. rufopicta* (Bigot) (all Diptera: Tachinidae). Among the wasp parasitoids of spotted cutworm are *Apanteles yakutatonsis* Ashmead (Hymenoptera: Braconidae), *Eutanyacra succincta* (Brulle) (Hymenoptera: Ichneumonidae), *Dibrachys carus* (Walker) (Hymenoptera: Pteromalidae), and *Euplectrus frontalis* Howard (Hymenoptera: Eucharitidae).

Life Cycle and Description. The number of generations is variable. There reportedly are three generations annually in Tennessee, with moths present in April–May, July, and September–October (Crumb, 1929), though evidence for the middle generation is weak. In Washington, two flight generally are observed, but a small third flight late in the year is possible (Howell, 1979). In New York, Minnesota, and elsewhere, flights of moths occur in June–July and in August–October (Chapman and Lienk, 1981; Knutson, 1944); this seems to be the most common pattern of occurrence. Overwintering occurs in the larval stage, with pupation in early spring. Most damage to crops results from feeding by larvae of the summer generation, which normally pupate in August.

Egg. The eggs are normally deposited on the underside of foliage. They are pinkish, and in shape resemble a slightly flattened sphere. They measure 0.60–0.65 mm in diameter and 0.50–0.61 mm in height. Eggs bear about 30 ridges, of which about 13–14 extend from the base to the center; the remainder terminate before they reach the top. They reportedly are deposited singly, in rows, or in clusters of up to 200 eggs. Duration of the egg stage is 6–12 days.

Larva. Spotted cutworm displays seven instars. Head capsule widths are about 0.35, 0.55, 0.8, 1.3, 1.7, 2.1, and 3.0 mm for instars 1–7, respectively. Body length is about 2–4, 6, 0.9–1.5, 12–16, 14–20, 28, and 30–36 mm, respectively. The general body color is brown or gray, and it lacks distinct stripes. The most diagnostic character is a series of subdorsal triangles or wedges along the length of the body. The triangles are distinct posteriorly, but decrease in size toward the head. The head is light brown with dark markings. Duration of the larval stage in the winter period is 5–6 months, whereas it is only about one month during the summer. The larvae of *X. adela* and *X. dolosa* are structurally indistinguishable. However, *X. dolosa* has a longer developmental period and attains a greater size at larval maturity. When *X. dolosa* is reared at 24°C, the larval, pupal, and egg-adult periods were 27.6, 15.8, and 43.3 days, respectively. In contrast, the corresponding periods were 21.0, 12.1, and 32.4 days for *X. adela*. Maximum larval weights when reared at 24°C were 0.99 g for *X. dolosa* but only 0.56 g for *X. adela*. Not surprisingly, head capsule widths of *X. dolosa* tend to be slightly larger than *X. adela*. Although head size is about the same for both species through instar four, the widths are about 1.8 and 1.7 mm in instar five, and 3.0 and 2.4 in instar six for *X. dolosa* and *X. adela*, respectively (Hudson, 1983). Larvae of all ages are prone to climb, but older larvae climb and feed only at night, dropping to the ground and seeking shelter during the daylight hours (Olson and Rings, 1969). (See color figure 58.)

Pupa. Pupation occurs in the soil. The pupa is about 18 mm long and 6 mm wide. It is brown. Duration of the pupal stage is 17–41 days.

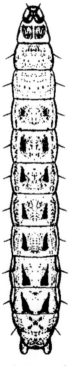

Spotted cutworm larva.

Adult. The adult is a medium-sized moth, measuring 29–43 mm in wingspan. The moth of *X. dolosa*, which has a wingspan of 37–43 mm, is slightly larger than *X. adela*, with a wingspan of 29–38 mm. The front wings are purplish or reddish brown, with dark brown to black basally. At about the mid-point of the leading edge of the forewing is a light brown or tan-colored triangle. On average, *X. adela* tends to be slightly darker than *X. dolosa*. The hind wings are white to light gray, often darker along the outer margin. Moths of *X. dolosa* and *X. adela* fly at about the same periods each year. However, as might be expected from the longer developmental period of *X. dolosa*, the flight period of this species begins at a slightly later date than *X. adela*. Adults live 2–3 weeks in the field. (See color figure 248.)

These species are poorly known. Crumb (1929) gave the most complete account, but Howell (1979) provided valuable data from the west coast. Hudson (1982) supplied developmental data. Franclemont (1980) and Hudson (1982) clarified the status of the North American species comprising dingy cutworm. Rings and Johnson (1977) published a bibliography. Spotted cutworm larvae were included in keys by Whelan (1935), Crumb (1956), Rings (1977b), Capinera (1986), and are included in a key to armyworms and cutworms in Appendix A. Moths were included in keys by Rings (1977a) and Capinera and Schaefer (1983).

Damage

Damage to crops results from feeding by both the overwintering larvae in the spring, and from the summer generation later in the growing season. The overwintering larvae are perhaps best known for their damage to fruit, because larvae readily climb vines and trees to feed on buds during the spring months, when vegetation is scarce. Although the larvae can sever young plants at ground level, they also feed on foliage and burrow into tomato fruits. At high densities larvae may assume a gregarious, dispersive "armyworm" habit. Spotted cutworm has been reported to be an early season rangeland pest, destroying nearly

Adult spotted cutworm.

all forbs and grasses in some areas (Launchbaugh and Owensby, 1982).

Management

Sampling. Light traps can be used to monitor populations of adults. A pheromone-based lure has also been identified, though it is not completely specific (Landolt, 2000). Although these are considered to be climbing cutworms, and larvae spend a good deal of time searching or feeding on foliar tissue, they also hide beneath plant debris and beneath the soil surface. Thus, it is important to rake the surface of the soil to search for larvae if cutworm injury is suspected.

Insecticides. Insecticides are commonly used to control cutworms. Application of persistent insecticides to the base of the plant can be effective, as can foliar treatment of leafy vegetables. Bran and other organic baits treated with insecticide are effective methods of cutworm suppression. *Bacillus thuringiensis* is not usually recommended for this pest.

Cultural Practices. Barriers have been used to prevent access by cutworms to buds and foliar tissues (Wright and Cone, 1983b), but in commercial crop production this has more applicability to perennial crops such as fruit than to annual crops such as vegetables. In the home garden, however, transplanted plants can be protected by a barrier such as a can or waxed paper container with the bottom removed. Aluminum foil wrapped around the base of the seedling also deters cutting by larvae. Ditches with steep sides and metal barriers can be effective impediments to prevent dispersal of cutworm larvae into crops.

Cutworm problems often develop in weedy fields or portions of fields infested with weeds. It is advisable to till, or otherwise destroy weeds, 10–14 days in advance of planting, as this should cause small larvae to starve. If seedlings are to be transplanted into a field or garden, larger plants are preferred because they are less likely to be irreparably damaged by cutworms.

Stalk Borer
Papaipema nebris (Guenée)
(Lepidoptera: Noctuidae)

Natural History

Distribution. This native insect occurs throughout the eastern United States and southern Canada east of the Rocky Mountains. It rarely is abundant in the southern states and along the western margin of its distribution, occurring as a pest principally in the midwestern states.

There are nearly 50 species of *Papaipema* found in northeast and northcentral United States and adjacent

Canada. Although, *P. nebris* is the dominant pest, on occasion related species such as *P. cataphracta* (Grote), the burdock borer, have been reported to damage crop plants. Accurate identification of these insects is difficult in the larval stage owing to their similarity in appearances and habits. Based on light trap collections in Iowa, Peterson *et al.* (1990) concluded that *P. nebris* was by far the most abundant species in this group.

Host Plants. Stalk borer has a very wide-host range, with almost 200 plant species recorded as hosts. In the spring the larvae burrow into grass stems, but as they grow the larvae move to nearby plants with thicker stems. Among the vegetable crops injured are asparagus, bean, cantaloupe, cauliflower, celery, corn, eggplant, parsnip, pea, pepper, potato, rhubarb, spinach, and tomato. Other crops sometimes injured include alfalfa, barley, cotton, oat, red clover, rye, sugarbeet, sweet clover, timothy, and wheat. Fruit damaged by stalk borer include apple, blackberry, currant, gooseberry, strawberry, peach, and plum. Shade trees can also be injured, including catalpa, elm, maple, poplar, and willow (Solomon, 1988). Among the numerous flower crops that have been reported to be damaged are anemone, canna, carnation, cosmos, daisy, gladiolus, hollyhock, iris, larkspur, lily, peony, phlox, purple coneflower, rose, and rose mallow. Some of the common weeds supporting stalk borer larvae, are cattail, *Typha* spp.; dock, *Rumex* spp.; Kentucky bluegrass, *Poa pratensis*; dogbane, *Apocynum androsaemifolium*; groundcherry, *Physalis* spp.; goldenrod, *Solidago* spp.; lambsquarters, *Chenopodium album*; quackgrass, *Agropyron repens*; ragweed, *Ambrosia* spp.; smartweed, *Polygonum* spp.; sunflower, *Helianthus* spp.; thistle, *Cirsium* spp.; wildrye, *Elymus canadensis*, and many others. Ragweed is often suggested as the favored host. Some common weeds such as milkweed, *Asclepias syriaca*; and velvetleaf, *Abutilon theophrasti*; are not suitable. Small-stemmed grasses often induce larval wandering (Alvarado *et al.*, 1989).

Natural Enemies. Stalk borer is attacked by many of the general predators that are found attacking other caterpillars. Because stalk borers often move among host plants, they likely are more susceptible to predation than some borers. Among the known predators are ground beetles (Coleoptera: Carabidae), lady beetles (Coleoptera: Coccinellidae), minute pirate bugs (Hemiptera: Anthocoridae), stink bugs (Hemiptera: Pentatomidae), and damsel bugs (Hemiptera: Nabidae).

Parasitoids may be more important natural enemies than predators, but the significance of individual parasitoids varies among localities and habitats. Of the more than 20 parasitoids known to attack stalk borer, *Lydella radicis* (Townsend) (Diptera Tachinidae) was reported to be the most important in Iowa (Decker, 1931), parasitizing up to 70% of larvae. The most important wasp parasitoid in Iowa, and probably the second most important parasitoid, was *Apanteles papaipemae* Muesebeck (Hymenoptera: Braconidae); this parasitoid attained levels of parasitism up to 38%, but averaged about 10%. In contrast, the important parasitoids in Ohio were *Lixophaga thoracica* (Curran) (Diptera: Tachinidae) in corn, *Sympiesis viridula* (Thompson) (Hymenoptera: Eulophidae) in potato and ragweed, and *Lissonata brunnea* (Cresson) (Hymenoptera: Ichneumonidae) and *Gymnochaeta ruficornis* Williston (Diptera: Tachinidae) in ragweed (Felland, 1990). Lasack *et al.* (1987) reported only low levels of parasitism in Iowa.

Weather. Weather also influences abundance of stalk borer. Both excessive rainfall and hot, dry weather during the spring when larvae are hatching and moving from host to host are reported to reduce larval survival markedly (Decker, 1931; Lasack *et al.*, 1987).

Life Cycle and Description. There is a single generation annually. The egg is the overwintering stage. Hatching occurs in April–June, followed by a larval development period of 60–90 days. Pupation occurs in late summer and autumn, with moths present and oviposition occurring during August–October. Population monitoring in Iowa demonstrated that nearly all moths were found during September (Bailey *et al.*, 1985).

Egg. The eggs are deposited singly or in clusters of up to 100 on the stems and leaves of dead grasses and weeds. The preferred site of oviposition is within rolled leaves, leaf sheaths, or cracks and crevices, particularly of narrow-leaved perennial grasses (Levine, 1985; Highland and Roberts, 1989). In shape, the egg is a slightly flattened sphere. It measures about 0.6 mm in diameter and 0.45 mm in height. It is white when first deposited, but turns grayish or amber with age. Narrow ridges, about 50 in number, radiate outward from the center of the egg. They normally hatch after a period of 7–9 months, usually in April or May. Under controlled conditions, post-diapause eggs require a mean of 12.4, 14.6, 18.8, 27.5, and 41.7 days for development when held at 24°, 21°, 18°, 16°, and 13°C, respectively (Levine, 1983).

Larva. Larvae initially mine leaves, or if feeding on grass then the stem may be attacked. However, larvae relocate to plants with large-diameter stems such as ragweed as these hosts become available. The shift among host plants occurs principally during instars 4–6. Depending on food availability, larvae may be forced to move repeatedly before completing their development. The larva is cylindrical in form, but tapering toward both the anterior and posterior ends. The head and thoracic shield are dark brown or black during the first two instars, and yellowish thereafter, though marked with a dark narrow band laterally. The larva bears pairs of prolegs on the third-sixth abdominal segments in addition to anal prolegs. The body color is brown with a broad white stripe dorsally and on each side. The lateral stripes are interrupted by a large brown spot in the region behind the thoracic legs. The stripes fade at maturity, the larva assuming a whitish or purplish color. At maturity, the larva attains a length of about 27 mm. Larval development entails 6–17 instars, with larvae usually displaying 7–9 instars. Head capsule widths for larvae with seven instars are 0.25, 0.38, 0.57, 0.87, 1.3, 2.0, and 2.9 mm, respectively, for instars 1–7. In contrast, head capsule widths for larvae with nine instars are 0.25, 0.34, 0.46, 0.63, 0.87, 1.2, 1.6, 2.1, and 2.9 mm, respectively, for instars 1–9. Development time for larvae with seven instars when reared at 27°C is 4, 3.6, 3.7, 4.5, 10, 15, and 28 days for instars 1–7, respectively, producing a mean larval development time of 68 days. Development time for larvae with nine instars when reared at 27°C is 4, 3.4, 3.9, 4.2, 5.3, 6.2, 9.0, 14.1, and 26 days for instars 1–9, respectively, producing a mean larval development time of 76 days. Lasack and Pedigo (1986) indicated a developmental threshold of about 5.1°C. Cannibalism is common among stalk borer larvae, the larger individuals usually killing and consuming the smaller.

Pupa. When larval development has been completed, larvae usually move to the soil and prepare a small cell just beneath the surface, though in some hosts such as corn pupation often occurs within the stem of the host. After a prepupal period of 1–6 days, pupation occurs. The pupa is typical in form—elongate, broadly rounded anteriorly, tapering posteriorly, and terminating in a pair of small curved spines. It is reddish brown or brown and measures 16–22 mm long and 5–7 mm wide. Duration of the pupal stage is normally 22–29 days.

Adult. The adult is medium in size, with a wingspan of 25–40 mm. The general color is grayish-brown, but close examination shows that the wings are dark-brown and covered with scales bearing white tips. The basal two–thirds of the front wings tend to be darker, and separated from the distal portion by a thin white transverse line. The hind wings are paler, similar to the distal portion of the forewings. Some moths also bear spots on the forewings—three small whitish spots about one–third the distance to the wing tip and a bean-shaped yellowish or whitish spot just beyond

Stalk borer larva.

the mid-point of the wing. Moths are nocturnal, and begin copulation and oviposition within three nights of emergence. The period of oviposition averages about 10 days (range 4–23 days), and is followed by death within a few days (mean 2.4 days, maximum 9 days). Females may produce 200–500 eggs daily, with average fecundity reported to be about 900 and maximum egg production just over 2000. Adults seem to be weak fliers, making only short flights.

An excellent treatment of stalk borer biology was given by Decker (1931). Temperature relations were given by Levine (1983). Egg diapause was described by Levine (1986b). A key for identification of larvae was provided by Crumb (1956) and Capinera (1986). A guide to common stalk boring caterpillars also is included in Appendix A.

Damage

Historically, damage has been sporadic and limited to border rows of crops. In recent years, however, as reduced tillage practices have become more widespread in corn production areas, population densities have increased and damage has become more frequent. Young larvae enter a variety of hosts in the spring, but often choose grass plants, because these tend to predominate early in the season. Larvae usually enter the plant by burrowing into the stem, but they may also mine leaves. Leaf mining in young corn plants causes no significant loss, but when larvae burrow into the whorl, causing its death (called "dead heart"), significant damage occurs (Bailey and Pedigo, 1986). The youngest corn seedlings are most susceptible to injury, and little damage is observed once corn attains the six-leaf stage (Levine *et al.*, 1984; Davis and Pedigo, 1990b, 1991). If a food source of a single stem is exhausted, larvae move to other stems or plants. The entrance to the plant may be anywhere along the stem, and is usually made obvious by the large entrance hole. The stem is often completely hollowed out, causing the distal portions of the plant to perish. Stalk borer larvae sometimes feed on tissue of woody plants such as trees, but only the soft terminal tissue is damaged.

Management

Sampling.   Moths can be attracted with blacklight traps, though mostly males are captured (Bailey *et al.*,

1985). The distribution of eggs and young larvae is highly aggregated, but larval distribution is altered and it becomes more uniform as larvae disperse. Many samples are required to estimate density, especially in wild grasses. However, it is not always necessary to dissect grass stems to locate larvae, because infested stems wilt and discolor (Davis and Pedigo, 1989).

Insecticides. Persistent insecticides, including systemic materials, can be applied to rows of crop plants at the margin of crop fields to reduce damage by invading larvae. However, if crop fields support grass, particularly in the autumn when eggs are deposited, the entire field may require treatment. Davis and Pedigo (1990a) demonstrated that treatment of weedy areas adjacent to crop fields, especially if timed to coincide with larval hatch, could provide good crop protection.

Cultural Practices. The abundance of stalk borer is directly related to the availability of preferred weedy host plants in or near crop fields. Thus, field edges and small fields are more likely to experience damage. Stinner *et al.* (1984) noted the preference by ovipositing moths for grasses within crop fields, especially fields that were grown under reduced tillage practices. However, the presence of broad-leaf weeds also has been shown to be correlated with increased abundance of stalk borer (Pavuk and Stinner, 1991). Reduced tillage practices often result in higher weed densities within crop fields, and greater damage by stalk borer (Willson and Eisley, 1992; Levine, 1993). Thus, destruction of weeds and grasses at field margins is recommended to reduce the invasion potential by larvae dispersing from weeds, but weeds within fields must also be suppressed.

Striped Grass Looper
Mocis latipes (Guenée)
(Lepidoptera: Noctuidae)

Natural History

Distribution. Striped grass looper is native to the western hemisphere, where it occurs commonly from the southern United States south to Brazil, including the Caribbean. On occasion, it is reported from as far north as Labrador and as far south as Argentina, but its occurrence always is limited to east of the Rocky Mountains and Andes Mountains. In the United states it is damaging only in the Gulf Coast region, where warm weather favors the survival of this tropical insect. Other species of the genus occur in the same area, but are not thought to be important pests.

Host Plants. As implied by its common name, this species feeds only on grasses. Several crops support striped grass looper, including bahiagrass, Bermuda-

grass, corn, guineagrass, millet, oat, pangolagrass, paragrass, rice, ryegrass, St. Augustine grass, Sudan grass, sugarcane, sorghum, and wheat. Among vegetable crops, only sweet corn is injured. Wild grasses reported as hosts include barnyardgrass, *Echinochloa crusgalli*; bluestem, *Andropogon* spp.; crabgrass, *Digitaria* spp.; goosegrass, *Elusine indica*; Johnsongrass, *Sorghum halepense*; panicum, *Panicum* spp.; and others.

Natural Enemies. Although there is extensive information on natural enemies affecting striped grass looper in Central and South America (e.g., Cave, 1992), data from North America are limited. Hall (1985) reported larval and pupal parasitism rates of 7.0–44.6% with a mean value of 29% in southern Florida sugarcane fields. Sarcophagid and tachinid flies seem to be most important, and include *Sarcodexia sternodontis* Townsend and *Sarcophaga* sp. (both Diptera: Sarcophagidae); *Archytas marmoratus* (Townsend), *Attacta brasiliensis* Schiner, *Belvosia bicincta* Robineau-Desvoidy, *Chetogena* sp., *Eucelatoria armigera* (Coquillett), *Euphorocera claripennis* (Macquart), and *Lespesia aletiae* (Riley) (all Diptera: Tachinidae). Among the wasp parasitoids are *Brachymeria ovata* (Say), *B. robusta* (Cresson), *Spilochalcis sanguineiventris* (Cresson), *Spilochalcis* n. sp. (Hymenoptera: Chalcididae); *Copidosoma truncatellum* (Dalman) (Hymenoptera: Encyrtidae); *Apanteles scitulus* Riley, *Meteorus autographe* Muesebeck, *Microplitis maturus* Weed (Hymenoptera: Braconidae); *Coccygomimus aequalis* (Provancher), *Enicospilus purgatus* (Say), *Gambrus ultimus* (Cresson) (Hymenoptera: Ichneumonidae); and *Trichogramma* sp. (Hymenoptera: Trichogrammatidae). The ectoparasitic nematode, *Noctuidonema guyanense* (Nematoda: Aphelenchoidea) is found on the bodies of moths (Rogers and Marti, 1993), but causes few detectable effects on the insects.

Predation may be an important factor, if overlooked element in striped grass looper population dynamics. In addition to predation by such usual predators as lady beetles (Coleoptera: Coccinellidae), there are frequent reports of predation by *Anolis* spp. lizards and songbirds such as eastern meadowlark, *Sturnella magna*, and redwing blackbird, *Agelaius phoeniceus*.

Pathogens associated with striped grass looper include the fungi *Beauveria bassiana* and *Nomuraea rileyi* and a virus, but the importance of these microorganisms has not been determined.

Life History and Description. Striped grass looper is found throughout the year in southern Florida and Puerto Rico; elsewhere in the United States it is abundant principally during late summer and autumn. There is no period of diapause in this species. Duration of the complete life cycle has been given by most authors as 40–60 days, though it varies with temperature. The developmental threshold for this insect is reported to be 13.7°C, and it tolerates temperature of 20–30°C readily. Thus, several generations occur annually in the south.

Egg. The eggs are deposited singly or in small groups of up to five or six eggs on either the upper or lower leaf surface, usually just before midnight. The egg is pale green initially, darkening with age. It is hemispherical, with the base flattened. The egg is marked by the presence of 29–33 ridges, which radiate from the center. It measures about 0.65 mm in diameter and 0.55 mm in height. Duration of the egg stage is 3–5 days at 25–28°C.

Larva. The larvae are slender and elongate throughout development. Initially, the larvae measure only about 3 mm long, eventually attaining a length of 55–70 mm. The body is yellowish or tan, but marked with numerous, narrow, brown and black stripes. The stripes extend from the body onto the head. In many specimens a dark brown or black band encircles the body between the first and second, and second and third abdominal segments. Viewed from the side, these bands appear to be dorsal spots. The larva bears three pairs of prolegs, and moves with a looping motion. Duration of the larval stage is 20–40 days. Alvarez and Sanchez (1981) reported mean (range) development times of 2.0 (2–2), 2.5 (2–3), 2.7 (2–5), 3.5 (3–4), 1.8 (1–3), and 4.9 (4–5) days, respectively, for instars 1–6 when cultured at 30°C. They also reported a mean development time for larvae of 17.4 days, but did not include the prepupal period in this calculation. There are 6–7 instars, with mean (range) head capsule widths of 0.38 (0.33–0.41), 0.56 (0.49–0.63), 0.90 (0.79–1.01), 1.40 (1.13–1.54), 1.70 (1.64–2.00), 2.3 (2.20–2.50), and 3.01 (2.95–3.26) mm for instars 1–7, respectively (Ogunwolu and Habeck, 1975). Larvae tend to be found singly on blades of grass, and drop to the soil surface if disturbed. They are active principally at night. (See color figure 431.)

Pupa. Larvae fold the leaves and pupate within. The pupa is covered with a soft, flimsy cocoon. The pupa measures 16–21 mm long and bears a waxy bloom which imparts a whitish or bluish color. Duration of the pupal stage is 6–12 days. Alvarez and

Sanchez (1981) reported a mean pupal period of 6.7 days at 30°C.

Adult. The moths are grayish tan or gray in general color, with dark lines and circular markings. Like other *Mocis* spp., striped grass looper males have a dark spot on the lower margin of the forewing about one-third the distance to the outer margin. However, most specimens of *M. latipes* bear a dark area or spot at about the mid-point along the outer margin of the forewing, a character that helps distinguish *M. latipes* from the other *Mocis* spp. Striped grass looper moths superficially resemble adults of velvetbean caterpillar, *Anticarsia gemmatalis* Hübner. However, the transverse stripe across the forewing of velvet-bean caterpillar moths terminates at the apex, whereas in striped grass looper it runs parallel to the outer wing margin (Gregory *et al.*, 1988). The wingspan of striped grass looper moths is 35–40 mm. Adults fly most actively during the early evening hours, though mating frequency is highest near midnight. Adults survive for 10–20 days when provided with food. They typically produce 200–300 eggs.

The biology of striped grass looper was summarized by Genung and Allen (1974), Reinert (1975), and Dean (1985). Larval and pupal descriptions can be found in Ogunwolu and Habeck (1979). Striped grass looper is included in the key to looper pests of vegetables found in Appendix A.

Damage

Over 40 species of grasses are fed upon, but other plants seem to be relatively immune to attack. The first and second instars feed on epidermal leaf tissue only, but later instars notch the leaf margins. When abundant, larvae completely defoliate grass plants, leaving only the midribs and stems. Larvae feed at night, and remain curled on the soil or in clumps of grass during the day. Although not regularly abundant, during some seasons they can be devastating pests, causing high levels of defoliation.

Management

Moths are attracted to light (Gregory *et al.*, 1988) and to sugar-based baits (Landolt, 1995). It is also possible to attract males to pheromone formulations, though velvetbean caterpillar may be attracted to the same chemicals (McLaughlin and Heath, 1989; Landolt and Heath, 1989). Larvae often are difficult to detect because they hide during the day, and their presence is observed only following damage. Grass weeds are often the site of initial infestation in crop fields, so grass weeds should be destroyed to dis-

Striped grass looper larva.

courage moths from ovipositing within crops. If larval numbers are high or damage is imminent, application of insecticides to crop foliage is recommended.

Sweetpotato Armyworm
Spodoptera dolichos (Fabricius)

Velvet Armyworm
Spodoptera latifascia (Walker)
(Lepidoptera: Noctuidae)

Natural History

Distribution. These native armyworms are found in eastern states, primarily along the Gulf Coast. They also are found in the Caribbean and in Central and South America. Adults and sometimes larvae of sweetpotato armyworm, *Spodoptera dolichos* (Fabricius), may occur as far north as Kentucky and Maryland. Velvet armyworm, *Spodoptera latifascia* (Walker), rarely is numerous outside the Gulf Coast area and is more damaging than sweetpotato armyworm in Florida and Central America.

Host Plants. These insects are general feeders, and only occasionally damage vegetables in the United States. Among vegetable crops damaged are asparagus, bean, corn, cowpea, pepper, potato, sweet potato, turnip, and probably others. Sweetpotato armyworm is known to damage cotton, and has also been called the "larger cotton cutworm." Velvet armyworm also is a common pest of ornamental plants in Florida.

In studies conducted in Honduras, velvet armyworm oviposited preferentially on *Amaranthus* spp. and *Ixophorus unisetus* weeds relative to corn and sorghum; however, only amaranths, *Amaranthus* spp., were good hosts for larvae (Portillo *et al.*, 1996b). Other weeds known to sustain larvae are sea purslane, *Thrianthema portulacastrum*; sena, *Cassia leiophila*; morningglory, *Ipomoea* sp.; *Melampodium divaricatum*; and purslane, *Portulaca oleracea* (Portillo *et al.*, 1991, 1996a).

Natural Enemies. Sweetpotato armyworm is parasitized by *Winthemia quadripustulata* (Fabricius) (Diptera: Tachinidae). Velvet armyworm is parasitized by *Archytas marmoratus* (Townsend) and *Winthemia* sp. (Diptera: Tachinidae), *Chelonus* sp. (Hymenoptera: Braconidae), *Euplectrus plathypenae* Howard (Hymenoptera: Eulophidae), and *Trichogramma* sp. (Hymenoptera: Trichogrammatidae). Velvet armyworm is also known to be infected with a *Vairimorpha* sp. (Microsporidia: Nosematidae) and a nuclear polyhedrosis virus.

Life Cycle and Description. The biology of these insects is poorly known. Larvae have been collected from July–November, and "several" broods of *S. doli-*

chos are reported to occur annually in Tennessee (Crumb, 1956). Development from the egg to adult stage requires 35–50 days.

Egg. The egg closely resembles that of yellowstriped armyworm, *Spodoptera ornithogalli*. It is a slightly flattened sphere, and measures about 0.46 mm in diameter and 0.36 mm in height. The egg bears about 50 narrow ridges that diverge from the center. They are laid in clusters of 200–500 eggs on the undersides of leaves, and bear scales from the female's abdomen. Apparently females produce 500–3000 eggs. Duration of the eggs stage is 4–8 days.

Larva. Larvae are gregarious during the early instars but disperse thereafter. There likely are 5–7 instars in both species. Crumb (1929) gave head capsule widths of 0.27, 0.41, 1.0–1.1, 1.5–1.6, 2.2–2.5, and 3.1–3.3 mm for instars 1–6 of *S. dolichos*. Santos *et al.* (1960) indicated head capsule widths of 0.3, 0.45, 0.75, 0.95, 1.4, 2.20, and 3.25 mm for instars 1–7 in *S. latifascia*. Duration of the larval stage is reported to be 23–30 days in *S. dolichos* and 15–27 days in *S. latifascia*. Instar-specific development time is reported to be about three, four, four, four, three, four, and five days for instars 1–7, respectively, in *S. latifascia*. Larvae of *S. latifascia* completed their larval period in about 15 days when fed cotton, and 18 days when soybean; no larvae completed development on lettuce. Larvae are variable, ranging from light gray or green to blackish, and well marked with spots and stripes. Larvae bear prominent dark triangular spots subdorsally along the abdomen, consistent with many other *Spodoptera* spp., and lateral yellowish lines are usually present both above and, to a lesser extent, below the spiracles. The subspiracular yellowish line is not interrupted by a spot on the first abdominal segment, as is usually the case with southern armyworm, *S. eridania*, but the supraspiracular line does bear a dark spot. Dark subdorsal markings found on the mesothorax are small and semicircular in *S. latifascia* but large and trapezoidal in *S. dolichos*. In contrast, the mesothoracic markings of yellowstriped armyworm, with which these species are easily confused, are triangular. Larvae attain body lengths of 43 and 48 mm in *S. dolichos* and *S. latifascia*, respectively, which makes them quite large for the genus. (See color figures 60 and 62.)

Sweetpotato armyworm larva.

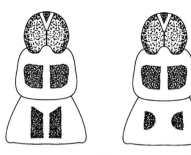

Anterior region of sweetpotato (*left*) and velvet (*right*) armyworms.

Velvet armyworm larva.

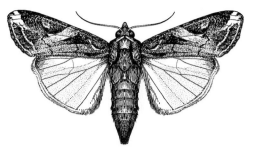

Adult sweetpotato armyworm.

Adult velvet armyworm.

Pupa. Larvae pupate in the soil. The pupal stage is dark brown and measures 20–30 mm long. Duration of the pupal stage is 9–20 days.

Adult. The adults are grayish-brown moths with few distinguishing characters, but heavily mottled forewings. The forewing of sweetpotato armyworm usually bears an irregular orangish streak distally, and the thoracic region is marked with two broad-dark bands running from the head to the abdomen; the moths of velvet armyworm lack these characters. In both species, the hind wing is white, with a narrow dark band along the distal edge. The wingspan of these moths is 40–50 mm. (See color figure 249.)

Descriptions of sweetpotato armyworm are found in Crumb (1929, 1956). Velvet armyworm was described by Levy and Habeck (1976), developmental biology was given by Habib *et al.*, 1983, and the sex pheromone discussed by Monti *et al.* (1995). Larvae of both species were described briefly by Passoa (1991), and keys for many *Spodoptera* were presented by Levy and Habeck (1976), Passoa (1991), and Heppner (1998). These species also occur in a key to Louisiana noctuids (Oliver and Chapin, 1981) and in a key to armyworms and cutworms in Appendix A. Keys to adults were presented by Todd and Poole (1980) and Heppner (1998). King and Saunders (1984) discussed biology in Central America.

Damage

Larvae are defoliators, and because of their large size inflict considerable damage late in the larval stage. They can function as cutworms, severing young plants at the soil surface, and also may burrow into tomato and other soft fruits.

Management

Larvae are most often controlled by application of foliar insecticides. These insects are not frequent pests, however. In Central America, velvet armyworm sometimes feeds on corn and sorghum, but if weeds are present the larvae of this insect feed preferentially on them. Corn and sorghum, because they are not suitable for larval development, serve as sink habitat, whereas certain broadleaf weeds serve as a source habitat (Portillio *et al.*, 1991).

Tobacco Budworm
Heliothis virescens (Fabricius)
(Lepidoptera: Noctuidae)

Natural History

Distribution. Tobacco budworm is a native species and is found throughout the eastern and southwestern United States. Like corn earworm, *Helicoverpa zea* (Boddie), it generally overwinters successfully only in southern states. However, it occasionally survives in cold climates in greenhouses and other sheltered locations. Tobacco budworm disperses northward annually, and can be found in New England, New York, and southern Canada during the late summer. It also occurs widely in the Caribbean, and sporadically in Central and South America.

Host Plants. Tobacco budworm is principally a field-crop pest, attacking such crops as alfalfa, clover, cotton, flax, soybean, and tobacco. However, it sometimes attacks such vegetables as cabbage, cantaloupe, lettuce, pea, pepper, pigeon pea, squash, and tomato, especially when cotton or other favored crops are abundant. Tobacco budworm is a common pest of geranium and other flower crops such as ageratum, bird of paradise, chrysanthemum, gardenia, geranium, petunia, mallow, marigold, petunia, snapdragon, strawflower, verbena, and zinnia. Weeds serving as a host for larvae include beardtongue, *Penstemon laevigatus*; beggarweed, *Desmodium* spp.; bicolor lespedeza, *Lespedeza bicolor*; black medic, *Medicago lupulina*; cranesbill, *Geranium dissectum*; deergrass, *Rhexia* spp.; dock, *Rumex* spp., groundcherry, *Physalis* spp.; Japanese honeysuckle, *Lonicera japonica*; lupine, *Lupinus* spp.; morningglory, *Ipomoea* spp.; a morningglory, *Jacquemontia tamnifolia*; passionflower, *Passiflora* sp.; prickly sida, *Sida spinosa*; sunflower, *Helianthus* spp.; toadflax, *Linaria canadensis;* and velvetleaf, *Abutilon theophrasti* (Brazzel *et al.*, 1953; Neunzig, 1963; Graham and Robertson, 1970; Roach, 1975; Harding, 1976a; Stadelbacher, 1981; Pair, 1994; Sudbrink and Grant, 1995). In Georgia, Barber (1937) determined that tobacco budworm developed principally on toadflax during April–May for 1–2 generations, followed by one generation on deergrass during June–July and 2–3 generations on beggarweed during July–October. In Mississippi, cranesbill was identified as the key early season host plant (Stadelbacher, 1981). In southern Texas, cotton is the principal host, but such weeds as wild tobacco, *Nicotania repanda*; vervain, *Verbena neomexicana*; ruellia, *Ruellia runyonii*; and mallow, *Aubitilon trisulcatum*, are important hosts early or late in the year (Graham *et al.*, 1972). In cage tests and field studies conducted in Florida and which did not include cotton, tobacco was more highly preferred than other field crops and vegetables, but cabbage, collards, okra, and tomato were attacked (Martin *et al.*, 1976a).

Natural Enemies. Numerous general predators have been observed to feed upon tobacco budworm. Among the most common are *Polistes* spp. wasps (Hymenoptera: Vespidae); big-eyed bug, *Geocoris punctipes* (Say) (Hemiptera: Lygaeidae); damsel bugs, *Nabis* spp. (Hemiptera: Nabidae); minute pirate bugs, *Orius* spp. (Hemiptera: Anthocoridae), and spiders.

Several parasitoids also have been observed, and high levels of parasitism have been reported (Lewis and Brazzel, 1968; Tingle *et al.*, 1994). The egg parasitoid *Trichogramma pretiosum* Riley (Hymenoptera: Trichogrammatidae) can be effective in vegetable crops. Other important parasitoids are *Cardiochiles nigriceps* Viereck in vegetables and *Cotesia marginiventris* (Cresson) in other crops (both Hymenoptera: Braconidae). Effectiveness of the parasitoids varies among crops (Martin *et al.*, 1981). Other species known from tobacco

budworm include *Archytas marmoratus* (Townsend) (Diptera: Tachinidae); *Meteorus autographae* Muesebeck (Hymenoptera: Braconidae); *Campoletis flavicincta* (Ashmead), *C. perdistinctus* (Viereck), *C. sonorensis* (Cameron), *Netelia sayi* (Cushman) and *Pristomerus spinator* (Fabricius) (all Hymenoptera: Ichneumonidae).

Pathogens are also known to inflict mortality. Among the known pathogens are microsporidia, *Nosema* spp., fungi such as *Spicaria rileyi*, and nuclear polyhedrosis viruses. In a study conducted in South Carolina, *Spicaria* was a more important mortality agent than natural incidence of virus, and was considered to be one of the most important natural mortality agents (Roach, 1975).

Life Cycle and Description. Moths emerge during March–May in southern states, followed by 4–5 generations through the summer, with overwintering commencing in September–November. Four generations have been reported from northern Florida (Chamberlin and Tenhet, 1926) and North Carolina (Neunzig, 1969), and at least five from Louisiana (Brazzel *et al.*, 1953). Moths have been collected in New York in July–September, but at such northern latitudes it is not considered to be a pest (Chapman and Lienk, 1981). This species overwinters in the pupal stage.

Egg. The eggs are deposited on blossoms, fruit, and terminal growth. They are spherical, with a flattened base. They measure 0.51–0.60 mm wide and 0.50–0.61 mm long. They initially are whitish to yellowish white, turning gray as they age. Narrow ridges radiate from the apex of the egg, and number 18–25. Eggs of tobacco budworm are nearly indistinguishable from those of corn earworm. At high magnification, however, the primary ribs of tobacco budworm eggs can be observed to terminate before they reach the rosette of cells surrounding the micropyle; in corn earworm at least some primary ribs extend to the rosette (Bernhardt and Phillips, 1985). Females normally produce from 300–500 eggs, but fecundity of 1000–1500 eggs per female have been reported from larvae cultured on artificial diet at cool temperature (Fye and McAda, 1972).

Larva. Tobacco budworm larvae have 5–7 instars, with five or six most common. Head capsule widths for larvae that develop through five instars measure 0.26–0.31, 0.46–0.54, 0.92–0.99, 1.55–1.72, 2.38–2.87 mm for instars 1–5, respectively. Larval lengths are 1.1–4.0, 4.2–8.0, 8.7–14.7, 18.5–25.6, and 23.3–35.6 mm for these same instars. Head capsule widths for larvae that develop through six instars measure 0.26–0.31, 0.36–0.53, 0.72–0.85, 1.12–1.25, 1.60–1.72, and 2.40–2.82 mm for instars 1–6, respectively. Larval lengths are 1.4–4.1, 3.0–7.0, 7.5–9.2, 12.0–15.8, 19.5–24.3, and 25.5–36.0 mm for these same instars. Development time was studied by Fye and McAda (1972) at various temperatures. When cultured at 20°C, development required about 4.6, 2.6, 3.1, 3.7, 10.1, and 9.8 days for instars 1–6, respectively. At 25°C, larval development times were 3.1, 2.0, 1.9, 2.1, 5.7, and 2.5 days, respectively. Young larvae are yellowish or yellowish-green with a yellowish-brown head capsule. Later instars are greenish with dorsal and lateral whitish bands, and with a brown head capsule. Many of the bands may be narrow or incomplete, but a broad, lateral subspiracular band is usually pronounced. Body color is variable, and pale green or pinkishforms, or dark reddish or maroon forms are sometimes found. Larvae are very similar to corn earworm. As in corn earworm, its body bears numerous black thorn-like microspines. These spines give the body a rough feel when touched. Early instars are difficult to separate from corn earworm; Neunzig (1964) gave distinguishing characteristics. Starting with the third instar, close examination reveals tubercles with small thorn-like microspines on the first, second, and eighth abdominal segments that are about half the height of the tubercles. In corn earworm the microspines on the tubercles are absent or up to one-fourth the height of the tubercle. Larvae exhibit cannibalistic behavior starting with the third or fourth instar, but are not as aggressive as corn earworm. (See color figure 61.)

Pupa.   Pupation occurs in the soil. Pupae are shiny reddish brown, become dark brown before emergence of the adult. The pupa averages 18.2 mm long and 4.7 mm wide. Duration of the pupal stage is reported to be about 22 days at 20°C, 13.0 days at 25°C, and 11.2 days at 30°C. Diapause is initiated by either low temperature or short day length (Henneberry *et al.*, 1993; Henneberry, 1994).

Adult.   The moths are brownish, and lightly tinged with green. The front wings are crossed transversely by three dark bands, each of which is often accompanied by a whitish- or cream-colored border. Females tend to be darker. The hind wings are whitish, with the distal margin bearing a dark band. The moths measure 28–35 mm in wingspan. The pre-oviposition per-

Tobacco budworm pupa.

iod of females is about two-days long. Longevity of moths is reported to range from 25 days when held at 20°C, to 15 days at 30°C. A sex pheromone has been identified (Tumlinson *et al.*, 1975).

The biology of tobacco budworm was given by Neunzig (1969) and Brazzel *et al.* (1953). The larva was included in keys by Okumura (1962) and Oliver and Chapin (1981); the latter publication also pictured the adult stage. Tobacco budworm also is included in a key to armyworms and cutworms in Appendix A. Larvae are readily cultured on bean-based rearing media or other diets (King and Hartley, 1985).

Damage

Larvae bore into buds and blossoms (the basis for the common name of this insect), and sometimes the tender terminal foliar growth, leaf petioles, and stalks. In the absence of reproductive tissue, larvae feed read-

Adult tobacco budworm.

Tobacco budworm larva.

Spinules associated with tubercles on abdomen of tobacco budworm larva. The minute spines are about one-half the height of the tubercle.

ily on foliar tissue. Neunzig (1969) infested tobacco with both tobacco budworm and corn earworm, and observed very similar patterns and levels of injury by these closely related species. In California, both budworm and earworm burrow into the heads of developing lettuce. Entry of larvae into fruit increases frequency of plant disease. Research in southern Arkansas tomato fields indicated that though tobacco budworm was present from May–July, they were not nearly as abundant or damaging as corn earworm (Roltsch and Mayse, 1984).

Management

Sampling. Large cone-shaped wire traps baited with sex pheromone lures are commonly used to capture tobacco budworm moths (Hartstack et al., 1979). Smaller bucket traps can be used to capture these moths, but they are not very efficient (Lopez et al., 1994).

Insecticides. Foliar insecticides are commonly used in crops where tobacco budworm damage is likely to occur. However, destruction of beneficial organisms often results, and this is thought to exacerbate budworm damage. Also, resistance to insecticides is widespread, particularly in crops where pyrethroid use is frequent (Kanga et al., 1995; Greenstone, 1995). Larvae also consume bait formulated from cornmeal and insecticide (Creighton et al., 1961).

Cultural Techniques. Early season destruction of weeds with herbicide or mowing, or destruction of larvae on the weeds by treatment with insecticides, can reduce tobacco budworm population size later in the year (Bell and Hayes, 1994; Snodgrass and Stadelbacher, 1994).

Biological Control. The microbial insecticide *Bacillus thuringiensis* is effective against budworm (Johnson, 1974; Stone and Sims, 1993). *Heliothis* nuclear polyhedrosis virus has been used effectively to suppress tobacco budworm on field crops (Andrews et al., 1975) and on early season weed hosts (Hayes and Bell, 1994). Tobacco budworm also is susceptible to nuclear polyhedrosis virus from alfalfa looper, *Autographa californica* (Speyer) (Vail et al., 1978; Bell and Romine, 1980). Release of *Trichogramma* egg parasitoids has been shown to be beneficial in some vegetable crops (Martin et al., 1976b).

Host-Plant Resistance. Although there is little evidence for natural resistance to tobacco budworm among many crops, cotton is being genetically engineered to express resistance (Benedict et al., 1996). Enhanced resistance to larval survival by cotton should result in lower insect pressure on nearby vegetable crops.

Variegated Cutworm
Peridroma saucia (Hübner)
(Lepidoptera: Noctuidae)

Natural History

Distribution. Variegated cutworm is found in many areas of the world. It occurs throughout the western hemisphere, and in Hawaii, and also portions of Europe, Asia, and North Africa. The origin of this insect is uncertain, but is thought to be Europe, where it was described in 1790. First observed in North America in 1841, it is now abundant in southern Canada and the northern United States, where it is often considered to be the most damaging cutworm pest of vegetables.

Host Plants. This cutworm has an extremely wide host range. Unlike some cutworms that expand their dietary range only when confronted with overpopulation and starvation, variegated cutworm feeds readily on numerous plants. Among vegetables attacked are asparagus, bean, beet, Brussels sprouts, cabbage, cantaloupe, carrot, cauliflower, celery, Chinese cabbage, collards, corn, cowpea, cucumber, garbanzo, globe artichoke, kale, lettuce, lima bean, mustard, onion, pea, pepper, potato, sweet potato, Swiss chard, radish, rhubarb, rutabaga, spinach, squash, tomato, turnip, and watermelon (Rings et al., 1976b). Based on frequency of reports, variegated cutworm is most likely to be found damaging beet, cabbage, lettuce, potato, and tomato. Variegated cutworm also is known to damage fruit trees, including apple, apricot, avocado, cherry, currant, gooseberry, grape, lemon, mulberry, orange, plum, raspberry, and strawberry. Other crops injured include alfalfa, barley, clover, corn, cotton, flax, hops, mint, sunflower, sugarbeet, sweet clover, tobacco, wheat, and many flower crops. Weeds are occasionally consumed, but seem not to be preferred. Some of the weeds eaten are jimsonweed, *Datura* sp.; dock, *Rumex* sp.; dogfennel, *Eupatorium capillifolium*; plantain, *Plantago* sp.; ragweed, *Ambrosia* sp.; and shepherdspurse, *Capsella bursa-pastoris*.

Natural Enemies. Numerous natural enemies are known for variegated cutworm. Walkden (1950) reported mortality of variegated cutworm over a 20-year period in the central Great Plains, and frequently observed 20–75% mortality, with wasp and fly parasitoids accounting for most of the deaths among larvae. In a study of cutworms in Oklahoma, Soteres et al. (1984) reported six species of Braconidae, three species of Ichneumonidae, and one species of Eulophidae, but among the Hymenoptera only *Ophion* sp. (Ichneumonidae) accounted for more than 5% mortality. In the same study, 12 species of Tachinidae were observed, but among the Diptera only *Archytas apicifer* (Walker) and *Peleteria texensis* Curran (both Tachinidae) accounted for more that 5% mortality. In Oregon, Coop and Berry (1986) reported eight species of Hymenoptera from variegated cutworm larvae. The parasitoid, *Meteorus communis* (Cresson) (Braconidae), was recovered from about 35% of the intermediate age larvae. Not only did the parasitism affect abundance of cutworm in subsequent generations, but caused a 93% reduction in foliage consumption by parasitized larvae. A study of variegated cutworm in Hawaii demonstrated 33–80% parasitism, with most of the parasitism due to *Hyposoter exiguae* (Viereck) (Ichneumonidae), *Cotesia marginiventris* (Cresson) and *Meteorus laphygmae* Viereck (both Braconidae) (all Hymenoptera) (Hara and Matayoshi, 1990).

Other Hymenoptera known to parasitize variegated cutworm include *Apanteles xylinus* (Say), *Chelonus insularis* Cresson, *C. militaris* (Walsh), *Meteorus autographae* Muesebeck, *M. leviventris* (Wesmael), *Microplitis feltiae* Muesebeck, *Rogas perplexus* Gahan, and *R. rufocoxalis* Gahan (all Braconidae); *Campoletis sonorensis* (Cameron), *Enicospilus merdarius* (Gravenhorst), *Nepiera fuscifemora* Graf, and *Ophion flavidus* Brulle (all Ichneumonidae); and *Dibrachys canus* (Walker) (Pteromalidae). Other Diptera parasitizing this cutworm are *Archytas aterrimus* (Robineau–Desvoidy), *A. cirphis* Curran, *Bonnetia comta* (Fallén), *Carcelia* spp., *Chaetogaedia monticola* (Bigot), *Clausicela opaca* (Coquillett), *Eucelatoria armigera* (Coquillett), *Euphorocera claripennis* (Macquart), *E. omissa* (Reinhard), *Gonia longipulvilli* Tothill, *G. porca* Williston, *G. sequax* Williston, *Lespesia archippivora* (Riley), *Madremyia saundersii* (Williston), *Peleteria texensis* (Curran), *Periscepsia helymus* (Walker), *P. laevigata* (Wulp), *Voria ruralis* (Fallén), *Winthemia leucanae* (Kirkpatrick), *W. quadripustulata* (Fabricius), and *W. rufopicta* (Bigot) (all Tachinidae).

The fungus *Metarhizium anisopliae* and viruses can inflict mortality, but incidence is often low. Both a granulosis (Steinhaus and Dineen, 1960) and a nuclear polyhedrosis virus (Harper, 1971) are known to affect this cutworm.

Life Cycle and Description. There are 2–4 generations annually, with two generations common in northern states and southern Canada, and 3–4 in southern areas. Flights of moths are protracted, and generations are difficult to discern based on adult populations. Overwintering may occur in the pupal stage, or perhaps the larval stage, but there also is evidence that moths migrate into northern areas from southern latitudes each spring, and return to the south in the autumn. In Iowa, 3–9% of moths collected during the spring flight were considered to have emerged from overwintering pupae, with the balance migrating from southern latitudes (Buntin *et al.*, 1990). The April–May spring flight in Iowa tends to be followed by generations and adult flights in June–July, August–September, and October–November. Total duration of the life stages is usually 35–70 days. Overwintering in Canada is unlikely (Ayre, 1985) except in relatively warm areas such as coastal British Columbia.

Egg. The egg of variegated cutworm is hemispherical; the egg is flattened at the point of attachment to a leaf or plant stem. The surface of the egg is marked with ridges, about 42 in number, radiating from the center. The egg measures about 0.55–0.58 mm in diameter and 0.40–0.45 mm in height. Initially, they are white, but soon turn brownish. The developmental threshold for the egg stage is estimated at 3.0–6.0°C by various authors. Duration of the egg stage is 4–6 days in warm weather (20–30°C), but 10 days when held at 15°C. Eggs are deposited in clusters, often numbering several hundred per egg mass. Females may deposit 1200–1400 eggs during their lifespan.

Larva. There normally are six instars. The developmental threshold for the larval stage was estimated at 2.6–6.7°C by various authors. Mean duration of the instars is reported to be 6.5, 4.6, 4.8, 4.7, 6.7, and 16.8 days for instars 1–6, respectively, at 15°C. When reared at 25°C, instar durations are reduced to 3.1, 1.9, 2.2, 2.2, 2.9, and 8.4 days, respectively (Shields, 1983). Head capsule widths are 0.30–0.35, 0.46–0.62, 0.80–1.00, 1.20–1.65, 1.9–2.6, and about 3.0–3.2 mm, respectively. Body lengths are estimated at 2.0–3.0, 3.6–6.5, 5.3–9.0, 12–16, 25–28, and 35–46 mm for instars 1–6, respectively. Body color is brownish gray to grayish black. The most distinctive character is a dorsal yellow or whitish spot, which is present on each of the first four abdominal segments, often on the first six segments, though this character may be absent in early instars. Less distinctive is the black "W"-shaped mark on the eighth abdominal segment of the last instar, and light brown or tan on the posterior end of the body. An inconspicuous black line is often present laterally above the spiracles. An orangish-brown line may connect the spiracles, and below the spiracles there usually is some irregular yellowish or orangish coloration. The head is orangish-brown and marked with darker spots. (See color figure 63.)

Pupa. The larva forms a cell in the soil and pupates near the soil surface. The pupa is mahogany brown,

Variegated cutworm larva.

and measures 15–23 mm long and 5–6 mm wide. The developmental threshold of the pupal stage was estimated at 4.3–8.5°C by various authors. Duration of the pupal stage is about 33 and 13 days at 15° and 25°C, respectively.

Adult. The adult is fairly large in size, measuring 43–50 mm wingspan. The front wings normally are grayish brown, tinged with reddish and shaded centrally and distally with darker brown. The background color varies, however, from dark brown to yellowish-brown. A bean-shaped spot and a smaller round spot are usually evident centrally. The hind wings are iridescent or pearly white with brown veins and brown shading marginally. The head and thorax are dark brown, whereas the abdomen is lighter brown. Females produce a sex pheromone to attract males (Struble *et al.*, 1976). Oviposition commences 7–14 days after emergence of the adults. (See color figure 250.)

Description and biology of variegated cutworm was given by Chittenden (1901), Crumb (1929), and Walkden (1950). Developmental biology was given by Simonet *et al.* (1981) and Shields (1983). Rearing methods were described by Finney (1964) and Harper (1970). A bibliography was published by Rings *et al.* (1976a). The larva was included in keys by Whelan (1935), Crumb (1956), Okumura (1962), Rings (1977b), Oliver and Chapin (1981), and Capinera (1986). It is also included in a key to armyworms and cutworms in Appendix A. The moth was included in pictorial keys by Rings (1977a) and Capinera and Schaefer (1983).

Damage

The larvae cause considerable mortality to seedlings by cutting off the plant at the soil surface. Larvae also are defoliators, and though they commonly frequent low-growing herbage, they readily climb trees to feed on buds and foliage. Young larvae may remain on the foliage during the daylight hours, but feed principally during the evening hours. Large larvae often hide in the soil or other sheltered locations during the daytime hours, moving to exposed areas of foliage in the evening to feed. Larvae may burrow into tomato fruit and the heads of cabbage and cauliflower. At high

densities larvae may assume a gregarious, dispersive "armyworm" habit, but this is uncommon. It also invades greenhouses frequently.

The larvae consume about 125 sq cm of sugarbeet or 160 sq cm of potato foliage during their larval development (Capinera, 1978c; Shields *et al.*, 1985). Potato and many other plants can tolerate some defoliation without significant yield decrease. Late in the season, after tuber formation is nearly complete, potato can withstand up to 75% leaf loss without yield decrease. At this time up to 40 variegated cutworms per plant can be tolerated. However, at an earlier period in the season such as at full bloom, plants may be able to tolerate only three cutworms per plant (Shields *et al.*, 1985).

Management

Sampling. The adult populations can be monitored with blacklight and pheromone traps. Captures by the two types of traps are correlated, but pheromone traps capture larger numbers from the spring generation, whereas blacklight traps capture larger numbers at other times of the year (Willson *et al.*, 1981). Pheromone traps, though not completely species-specific (Ayre *et al.*, 1983), provide considerable selectivity relative to blacklight traps, thereby reducing labor requirements associated with population monitoring.

Larval populations are difficult to sample. Young larvae may be found clustered on foliage, but older larvae tend to hide in sheltered locations or burrow beneath the surface of the soil during the daylight hours. If plants are severed at the soil surface, or have disappeared, it is important to rake the soil surface and search for cutworm larvae.

Insecticides. Insecticides are commonly recommended to protect young plants from cutting damage, and older plants from defoliation and fruit injury. Insecticide applications are directed at the foliage or soil, the latter because the larvae often seek shelter there. Insecticides differ greatly in their effectiveness, and larger larvae are considerably more difficult to kill (Harris *et al.*, 1977). Insecticide-treated bran baits are effective against variegated cutworm. *Bacillus thuringiensis* is not recommended. Variegated cutworm larvae are sensitive to neem products, which act as feeding deterrents and disrupt larval growth and survival (Koul and Isman, 1991; Isman, 1993).

Cultural Practices. Cutworm problems often develop in weedy fields or portions of fields infested with weeds. It is advisable to till, or otherwise destroy weeds, 10–14 days in advance of planting, as this should cause small larvae to starve. If seedlings are to be transplanted into a field or garden, larger plants are preferred, because they are less likely to be irrepar-

Adult variegated cutworm.

ably damaged by cutworms. Transplanted plants derive considerable protection if surrounded by a barrier such as a can or waxed paper container with the bottom removed. Aluminum foil wrapped around the base of the seedling also deters cutting by larvae.

Biological Control. Variegated cutworm is susceptible to infection by entomopathogenic nematodes (Nematoda: Steinernematidae and Heterorhabditidae) (Morris and Converse, 1991), but demonstration of practical use under field conditions is lacking.

Western Bean Cutworm
Loxagrotis albicosta (Smith)
(Lepidoptera: Noctuidae)

Natural History

Distribution. This native cutworm occurs in the Rocky Mountain and western Great Plains regions of the United States and Canada, extending east to Nebraska, Oklahoma, and Texas. Its range has expanded, apparently following adoption of corn as a food plant in the 1950s, and likely will continue to spread in the Great Plains. Western bean cutworm appears to be absent from the Pacific Coast States, but is known from Mexico.

Host Plants. Originally known as a minor pest of bean, and perhaps of tomato, the host range of western bean cutworm apparently expanded during the 1950s to include corn. Acreage in the western states devoted to bean and corn also expanded during this period, so it is not clear whether host range expansion occurred, or whether damage became more noticeable as the availability of suitable hosts increased. It is now considered to be a locally important pest of grain corn and dry beans in western states, but also injures sweet corn, snap beans, and rarely tomato. Other legume crops may receive eggs and support partial larval development, but are considered to be poor hosts. Weed hosts include fruit of groundcherry, *Physalis* spp., and black nightshade, *Solanum nigrum*, though these are mostly suitable after larvae are partly grown (Blickenstaff and Jolley, 1982). (See color figure 31.)

Natural Enemies. Few natural enemies have been reported, though this is more likely due to lack of study than absence of predation and parasitism. The wasp *Apanteles laeviceps* Ashmead (Hymenoptera: Braconidae) has been reared from western bean cutworm. Common predators such as lady beetles (Coleoptera: Coccinellidae), minute pirate bugs (Hemiptera: Anthocoridae), damsel bugs (Hemiptera: Nabidae); big-eyed bugs (Hemiptera: Lygaeidae), green lacewings (Neuroptera: Chrysopidae), and spiders consume western bean cutworm eggs and larvae under laboratory conditions (Blickenstaff, 1979).

Life Cycle and Description. There is a single generation per year. Adults emerge in July and early August, with eggs found throughout this period. Larvae develop during late summer, and mature larvae overwinter. Pupation occurs in May or June of the following year.

Egg. The eggs are deposited on the upper surface of leaves and within the whorl of corn plants. They are nearly spherical, but flattened at the point of attachment to the foliage. They are white with a thin red line around the top initially, and turn pinkish or purplish gray at maturity. The eggs are about 0.7 mm in diameter, and well-marked with ridges that radiate from the center. The mean number of eggs per cluster is reported to be 52 (range 21–195 eggs). The incubation period is 4–7 days. (See color figure 255.)

Larva. Upon hatching, larvae remain near the egg mass for several hours and then disperse, often to the tassel of corn where they feed on pollen. Larvae normally feed at night, but they can be observed feeding during the daylight hours if it is cloudy. On small plants, particularly on beans, larvae may move to the soil during the daylight hours to seek shelter. The larvae initially are brownish and marked with weak stripes. As they mature they lose the stripes and become pinkish-brown dorsally, and grayish laterally. The thoracic plate bears broad dark brown or black stripes. The number of instars varies from 5–8, but six is normal. Mean development time requires 31 days (range 21–45 days). The mature cutworm measures about 35 mm long. (See color figure 64.)

Pupa. The larvae burrow into the soil in the autumn, mostly to a depth of 7–15 cm, and create a small pupal cell from saliva and soil particles. However, they remain in the larval stage until late May or June, when pupation occurs. The pupa is dark brown and measures about 20 mm long.

Adult. The moth is brownish, with a wingspan of 35–40 mm. The forewing is distinctively marked with a broad whitish or tan stripe along the anterior margin. The center of the forewing generally is well-marked with a small round spot basally and a bean-shaped spot distally. The spots are white or margined with

Western bean cutworm larva.

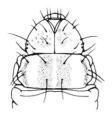

Head and prothorax of western bean cutworm.

Adult western bean cutworm.

white, and connected by a short black bar. The hind wing is white, but with delicate brown lines at the margin and along the veins. Adults live for a relatively brief period, averaging 7.2 days for males and 9.2 days for females. The pre-oviposition period is about four days, with females depositing about 400 eggs. (See color figure 251.)

The biology was described by Hoerner (1948), Hagen (1962, 1976), and Blickenstaff (1979). A sex pheromone has been described (Klun *et al.*, 1983). Western bean cutworm was included in the larval keys by Crumb (1956) and Capinera (1986), and is included in a key to armyworms and cutworms in Appendix A. The adult was included in the key by Capinera and Schaefer (1983).

Damage

On corn, larvae feed on developing pollen. As the tassel matures, larvae disperse and feed on pollen that has collected on the foliage and on the leaves. If pollen is not available, larvae feed on corn silk. As ears develop, larvae feed beneath the leaf sheath on both silk and kernels. Unlike corn earworm, *Helicoverpa zea* (Boddie), which tends to attack the ear tip, larvae of western bean cutworm are likely to burrow randomly into the ear, attacking all areas. Also, western bean cutworm larvae, unlike corn earworm, are not cannibalistic, so several larvae may be found feeding on a single ear. Leaf and stalk feeding are trivial, but silk feeding can inhibit pollination.

When feeding on bean, larvae initially remain near the top of the plant, feeding on buds and young leaves. Larger larvae feed on pods, but usually do not burrow into the pod. Instead they wander and feed on devel-

oping beans through the seed pod at several locations. When not feeding, large larvae may seek shelter in the soil.

Management

Sampling. Moths can be captured with both blacklight and pheromone traps. Blacklight traps capture moths earlier in the season than pheromone traps, but both trap types correlate well with damage potential. Although trap captures and overall damage potential are highly correlated, there is high among-field variation in damage (Mahrt *et al.*, 1987).

Insecticides. Application of insecticides to foliage is effective, though it is important to time the application early in the larval period, so applications are usually timed shortly after peak moth flight. If moth populations are not being monitored, insecticide should be applied before all the corn tassels have emerged from the whorl. Applications of insecticide-treated bait and soil-applied systemic insecticides are not very effective (Hoerner, 1948; Blickenstaff and Peckenpaugh, 1981).

Cultural Practices. Timing of planting affects susceptibility of corn to western bean cutworm infestation. Moths oviposit preferentially on plants that are silking, avoiding early and late planted corn. Emergence of moths from the soil is enhanced by high moisture conditions and high sand content; the rate of emergence from heavy and dry soil is low.

Yellowstriped Armyworm
Spodoptera ornithogalli (Guenée)

Western Yellowstriped Armyworm
Spodoptera praefica (Grote)
(Lepidoptera: Noctuidae)

Natural History

Distribution. Yellowstriped armyworm, *Spodoptera ornithogalli* (Guenée), is common in the eastern United States as far west as the Rocky Mountains, and occurs in southern Canada. However, it also is reported from southwestern states, including California. The distribution of this native insect includes Mexico, Central and South America, and many Caribbean islands. As a pest, however, its occurrence is limited principally to the southeastern states. Western yellowstriped armyworm, *Spodoptera praefica* (Grote), is known only from the western states, principally California and Oregon. It also is a native insect. In California, *S. praefica* is much more important than *S. ornithogalli*.

Host Plants. These species are very general feeders, reportedly damaging many crops. Among vegetable crops injured are asparagus, bean, beet, cabbage, cantaloupe, carrot, corn, cucumber, lettuce, onion, pea, potato, rhubarb, rutabaga, salsify, sweet potato, tomato, turnip, and watermelon. Other crops damaged include alfalfa, blackberry, cotton, clover, grape, lentil, peach, rape, raspberry, sorghum, soybean, sugarbeet, sweetclover, sunflower, tobacco, wheat, and several flower crops. Some of the weed species known to be suitable hosts are castorbean, *Ricinus communis*; dock, *Rumex* sp.; gumweed, *Grindelia* sp.; horse nettle, *Solanum carolinense*; horseweed, *Erigeron canadensis*; jimsonweed, *Datura* sp.; lambsquarters, *Chenopodium album*; morningglory, *Ipomoea* sp.; plantain, *Plantago lanceolata*; prickly lettuce, *Lactuca scariola*; and redroot pigweed, *Amaranthus retroflexus*.

In California, western yellowstriped armyworm develops for 1–2 generations on rangeland plants, preferring storksbill, *Erodium* sp.; and foxtail, *Setaria* sp. early in the year. However, as these plants senesce the armyworms move to irrigated areas, where they prefer alfalfa and morningglory. Alfalfa is a very suitable host, and such fields become heavily infested and serve as important sources of armyworms for other crops later in the season. A similar phenomenon was reported in Washington, where larvae developed on thistle, *Cirsium* spp.; wild lettuce, *Lactuca* spp.; wild mustard, *Brassica* spp.; and goosefoot, *Chenopodium* spp.; and then moved to crops as the weeds senesced (Halfhill, 1982).

Natural Enemies. Several wasp parasitoids affect *S. ornithogalli* including *Rogas laphygmae* Viereck, *R. terminalis* (Cresson), *Zele mellea* (Cresson), *Chelonus insularis* Cresson and *Apanteles griffini* Viereck (all Hymenoptera: Braconidae). Also, *Euplectrus plathypenae* Howard (Hymenoptera: Eulophidae) attacks larvae and causes a cessation of feeding within two days (Parkman and Shepard, 1981). Thus, this parasitoid is particularly valuable at minimizing damage. Numerous flies have been found to parasitize these armyworms including *Archytas* spp., *Choeteprosopa hedemanni* Brauer and Bergenstamm, *Euphorocera omissa* (Reinhard), *E. tachinomoides* Townsend, *Lespesia aletiae* (Riley), *L. archippivora* (Riley), *Omotoma fumiferanae* (Tothill), *Winthemia quadripustulata* (Fabricius), and *W. rufopicta* (Bigot) (all Diptera: Tachinidae). A nuclear polyhedrosis virus is highly pathogenic to larvae, and survivors that do not succumb exhibit reduced fecundity (Hostetter *et al.*, 1990; Young, 1990). Undoubtedly, predators are important, but unlike the situation with western yellowstriped armyworm, their effect has not been quantified.

Not surprisingly, western yellowstriped armyworm also is host to many parasitic insects. Among wasps associated with this caterpillar are *Chelonus insularis*

Cresson, *Cotesia marginiventris* (Cresson) (both Hymenoptera: Braconidae), *Hyposoter exiguae* (Viereck), and *Pristomeres spinator* (Fabricius) (both Hymenoptera Ichneumonidae). Flies parasitizing western yellowstriped armyworm include *Archytas californiae* (Walker), *Eucelatoria armigera* (Coquillett), *Euphorocera claripennis* (Macquart), and *Lespesia archippovora* (Riley) (all Diptera: Tachinidae). Despite the numerous parasitoids, the first three species listed above, all wasps, account for 95% of recorded parasitism in alfalfa. The relative importance of the individual parasitoid species varies geographically (Miller, 1977). Of great interest, however, is the impact of predation. Bisabri–Ershadi and Ehler (1981) reported that over 96% of total mortality occurred in the egg and early larval stages, and most was attributed to predation. The most important predators were minute pirate bug, *Orius tristicolor* (White) (Hemiptera: Anthocoridae); big-eyed bugs, *Geocoris* spp. (Hemiptera: Lygaeidae); and damsel bugs, *Nabis* spp. (Hemiptera: Nabidae). The plant bug *Lygus hesperus* Knight (Hemiptera: Miridae) was a facultative predator, often feeding on armyworm eggs. A nuclear polyhedrosis virus is found in western yellowstriped armyworm, and when larvae occur at high densities it may be a significant mortality factor.

Life Cycle and Description. There apparently are 3–4 generations annually, with broods of adults present in March–May, May–June, July–August, and August–November. Some of the latter brood of yellowstriped armyworm and all members of the latter brood of western yellowstriped armyworm overwinter as pupae rather than emerging as adults. Although eggs, larvae and adults of yellowstriped armyworm may be present in autumn or early winter they cannot withstand cold weather, and perish. Development time, from egg to adult, is about 40 days.

Egg. The eggs are greenish to pinkish brown and bear 45–58 small ridges. In shape, the egg is a slightly flattened sphere, measuring 0.46–0.52 mm in diameter and 0.38–0.40 mm in height. Females typically deposit clusters of 200–500 eggs, usually on the underside of leaves. Total fecundity was determined to be over 3000 eggs under laboratory conditions. They are covered with scales from the body of the adults. Duration of the egg stage is 3–5 days at warm temperature.

Larva. The larvae initially are gregarious in behavior, but as they mature they disperse, sometimes spinning strands of silk upon which they are blown by the wind. There usually are six instars, though seven instars have been reported. Head capsule widths are about 0.28, 0.45, 0.8–1.0, 1.4–1.6, 2.0–2.2, and 2.8–3.0 mm, respectively, for instars 1–6. The larva grows from about 2.0 to 35 mm long over the course of its development. In California, mean development times for instars 1–7 of *S. praefica* were 2.5, 2.8, 3.0, 3.4, 4.5, 5.2, 6.9 days, respectively, during the second generation. There was a total larval development time of 26.8 days during the second generation, but it was shortened to 21.3 days during the third. Coloration is variable, but mature larvae tend to bear a broad brownish band dorsally, with a faint white line at the center. More pronounced are black triangular markings along each side, with a distinct yellow or white line below. A dark line runs laterally through the area of the spiracles, and below this is a pink or orange band. Dark subdorsal spots are found on the mesothorax of yellowstriped armyworm, and the triangular shape of these spots aids in distinguishing this insect from sweetpotato armyworm, *Spodoptera dolichos*, and velvet armyworm, *S. latifascia*, in eastern states. The head is brown but has extensive blackish markings. In western yellowstriped armyworm the blackish markings on the head form an arc on each side that extend from the mouth to the back of the head; in yellowstriped armyworm the band is less definite dorsally. Duration of the larval stage is 14–20 days, with the first three instars requiring about two days each and the last three instars requiring about three days each. (See color figure 65.)

Pupa. Larvae pupate in the soil within a cell containing a thin lining of silk. The reddish brown pupa measures about 18 mm long. Duration of the pupal stage is 9–22 days, normally averaging 12–18 days.

Adult. The moths measure 34–41 mm in wingspan. The front wings are brownish-gray with a complicated pattern of light and dark markings. Irregular whitish bands normally occur diagonally near the center of the wings, with additional white coloration distally near the margin. The hind wings of yellowstriped armyworm are opalescent white, with a narrow brown

margin. In western yellowstriped armyworm the hind wings are similar but tend to be tinted with gray, and the underside bears a dark spot centrally. In *S. praefica*, both sexes are similar in appearance, whereas in *S. ornithogalli* the sexes are dimorphic. Under laboratory conditions average longevity of adults is 17 days, with most egg production completed by the tenth day (Adler *et al.*, 1991). (See color figure 252.)

The most complete description of *S. ornithogalli* and its biology was given by Crumb (1929), with additional comments by Crumb (1956). Blanchard (1932) decribes *S. praefica*. van den Bosch and Smith (1955) and Okumura (1962) described both species, and their biology in California. Keys for identification are also found in these references. Keys for separation of *Spodoptera* adults can be found in Todd and Poole (1980) and Heppner (1998). Larvae can be distinguished using the keys of Passoa (1991) and Heppner (1998). Rearing on artificial diet was described by Adler and Adler (1988).

Damage

Larvae damage plants principally by consumption of foliage. The small, gregarious larvae tend to skeletonize foliage but as the larvae grow and disperse they consume irregular patches of foliage or entire leaves. However, they also feed on the fruits of tomato, cotton, and other plants. Larval consumption of soybean was estimated by King (1981) to total 115 sq cm; this is an intermediate value relative to some other lepidopterous defoliators.

Management

Insecticides. Insecticides are applied to foliage to prevent injury by larvae. The microbial insecticide *Bacillus thuringiensis* can be applied to kill armyworms, but should be applied when the larvae are young, as they become difficult to control as they mature. Larvae consume bran bait containing insecticide.

Cultural Methods Proximity of crops to rangeland-containing weed hosts, or to alfalfa, may be important factors predisposing vegetable crops to injury.

Yellowstriped armyworm larva.

Adult yellowstriped armyworm.

At high densities, especially if alfalfa hay is mowed, larvae sometimes disperse simultaneously and invade nearby vegetable fields. Physical barriers such as trenches can be used to deter such dispersal.

Zebra Caterpillar
Melanchra picta (Harris)
Lepidoptera: Noctuidae

Natural History

Distribution. This native insect is found in southern Canada and the northern United States from the Atlantic to the Pacific Coast. Nowhere is it considered to be a major pest, yet it can be a fairly regular nuisance, and a significant component of the defoliator pest complex of several crops.

Host Plants. Zebra caterpillar feeds on several vegetable plants, and has been recorded as a pest of asparagus, bean, beet, broccoli, cabbage, celery, corn, lettuce, parsnip, pea, potato, rutabaga, spinach, tomato, and turnip; cabbage seems preferred. It also attacks flowers such as aster, hydrangea, and sweetpea, field crops such as alfalfa, clover, rape, sugarbeet, and tobacco, and trees such as apple, plum, and willow. As might be expected from an insect with such a broad host range, zebra caterpillar feeds on many weeds. Much of the economic entomology literature reports this insect as a pest of sugarbeet.

Natural Enemies. A nuclear polyhedrosis virus (NPV) of zebra caterpillar causes marked decreases in abundance, especially when larval densities are high (Adams *et al.*, 1968). Signs and symptoms of infection include loss of appetite, sluggish behavior, and reduced larval growth rate. Mortality occurs within 14 days of ingesting the polyhedral inclusion bodies, and dead larvae can be found hanging from foliage, suspended by their hind prolegs. Soon after death, larval color darkens, and the cadavers turn black and rupture. Body contents, including virus polyhedra, contaminate the foliage, enhancing spread of the disease. There is some evidence that other armyworms and cutworms may be infected by *M. picta* NPV.

Parasitoids known to attack zebra caterpillar include *Limneria annulipes* Harris (Hymenoptera: Ichneumonidae), *Microplites mamestrae* Weed (Hymenoptera: Braconidae), and *Winthemia quadripustulata* (Fabricius) (Diptera: Tachinidae) (Li *et al.*, 1993).

Life Cycle and Description. There are two generations per year over most of the range of this insect, generally occurring in June–July and August–October. A complete life cycle requires about 60 days during the summer. Zebra caterpillar overwinters in the pupal stage.

Egg. The moths are highly fecund, often depositing 1200 white, ovoid, slightly flattened eggs in clusters of 100–200, usually on the underside of leaves. As might be expected from an insect with a northern distribution, eggs do not hatch at high temperature, 32°C and higher. Excellent survival occurs at about 27°C, and hatching time is about five days. At 21°C, excellent hatching also occurs, but development is delayed to about six days. Hatching is inhibited at 15°C, and development time extended to about 12 days. At 10°C, eggs fail to develop.

Larva. The mature larva is boldly colored black, yellow, and white, and is immediately recognizable. The sides are yellow and white, with a vertical row of short-black stripes running the length of the body. There is a black band dorsally, separating the yellow and zebra-stripe lateral bands. The underside of the caterpillar is orange or red. Young larvae are very difficult to identify; they are principally black and green. However, they feed gregariously until they are about one-half grown, and this behavior helps to distinguish them. There are six instars. Head capsule widths are about 0.3, 0.5, 0.9, 1.5, 2.1, and 3.0 mm for the six instars, respectively. The average larval length is about 3.5, 6, 10, 17, 27, and 37 mm, respectively. Development time averages 3.5, 2.9, 3.3, 3.0, 3.4, and 7.2 days for larvae reared at 27°C. The last 3–4 days of the terminal instar is usually spent in the soil, where the larva prepares to transform into a pupa. (See color figure 66.)

Pupa. The pupa is dark brown, and measures about 2 cm long. The males weigh about 400 mg and the females are slightly heavier, about 450 mg. Pupal development time is about 30 days. This is the normal overwintering stage.

Adult. The adult wingspan measures about 3.5–4.5 cm. The front wings are chocolate brown, with a weak gray spot centrally. The hind wind is white, but bears a narrow brown band at the wing margin. Moths begin to emerge as early as 45 days after eggs are deposited, but usually about two months is required for a complete generation to occur. The pre-oviposition period of moths is about two days. Moths may continue to deposit eggs for a period of up to two weeks, although most oviposition occurs within one week. Adults perish after 10–12 days. (See color figure 253.)

Zebra caterpillar larva.

Zebra caterpillar pupa.

Adult zebra caterpillar.

The biology of zebra caterpillar was given by Tamaki *et al.* (1972) and Capinera (1979a).

Damage

The larvae are leaf-feeders, and initially they feed gregariously. They may make small holes or skeletonize foliage early in their development, but they soon become voracious, eating large holes in foliage. Larvae often completely consume individual leaves, leaving only the stem or petiole before moving to new food. Foliage consumption and damage potential on sugarbeet were given by Capinera (1979).

Management

These insects are not difficult to kill with foliar applications of chemical insecticides or *Bacillus thuringiensis*. They also are susceptible to the botanical insecticide neem, which functions as a feeding deterrent and growth regulator (Isman, 1993). Because of their seasonal biology, they sometimes develop to damaging levels late in the season, after the threat of insect damage is generally past. Thus, continued vigilance is suggested if the insects have been observed earlier in the season. Zebra caterpillar is susceptible to infection

by *Autographa californica* NPV (Capinera and Kanost, 1979).

FAMILY OECOPHORIDAE—OECOPHORID MOTHS

Parsnip Webworm
Depressaria pastinacella (Duponchel)
(Lepidoptera: Oecophoridae)

Natural History

Distribution. Parsnip webworm was originally known from Europe, and apparently was introduced to North America some time before 1869. Its distribution now includes southern Canada from Nova Scotia to British Columbia, and the northern United States south to Maryland and Arizona.

Host Plants. In addition to feeding on parsnip, this insect feeds on several umbelliferous weeds, including cow parsnip, *Heracleum lanatum*; and angelica, *Angelica* spp. Records of wild carrot, *Daucus carota*, are doubtful.

Natural Enemies. Parasitism varied from 0–100% in Iowa (Gorder and Mertens 1984), but was limited to the larval stage. *Apanteles depressariae* Muesebeck (Hymenoptera: Braconidae) is the principal parasitoid, though others have been collected on occasion.

Life Cycle and Description. There seems to be only a single generation per year, though in Iowa a few eggs are deposited in late summer, suggesting the possibility of a partial second generation. Egg to adult development times are reported to be 38 days in Iowa and 62 days in Nova Scotia.

Egg. The eggs are deposited principally on foliage, and to a lesser extent on flower stalks. They are elliptical, but slightly rectangular. They are white, and ribbed longitudinally. Egg length averages about 0.56 mm (range 0.36–1.14 mm). Duration of the egg stage is about four days.

Larva. Upon hatching the young larvae bore into the blossom to feed. They prefer unexpanded flowers, and generally distribute themselves so that there is only one insect per flower head (Thompson and Price, 1977). The larval color is generally greenish-yellow dorsally with yellow laterally and ventrally, but sometimes tends toward blue-gray. The head is black, and the length of the body is well-marked with rows of raised black spots. There are six larval instars. Head capsule widths are about 0.1, 0.3, 0.5, 0.7, 1.0, and 1.6 mm, respectively for instars 1–6. Larval development requires about 21 days, with duration of the instars about three, two, three, four, three, and eight

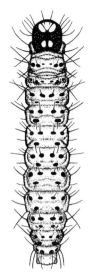

Parsnip webworm larva.

Adult parsnip webworm.

days, respectively. The larva, at maturity, attains a length of 16–18 mm. Larvae rest within the flower or other sheltered locations on the plant, surrounded by a tunnel webbed from silk. When disturbed, the larvae retreat within the tunnel, and if pursued wriggle violently and drop to the soil. Larvae feed preferentially on plants low in furanocoumarins, chemicals that function as toxicants or deterrents for many insects (Zangerl and Berenbaum, 1993).

Pupa. Pupation occurs within a silken sheath, generally within the stalk of the food plant. The pupa is brown. Duration of the pupal stage is about 13 days.

Adult. The adults fly at night, but are not attracted to lights. The moths are fairly large for this group of moths, with wing lengths of 9.5–13.0 mm. The front wings are yellowish-brown, marked with gray, whereas the hind wings are grayish. Overwintering occurs in the adult stage, with the adults in reproductive diapause. Egg production under laboratory conditions averages 470 eggs per female (range 170–830 eggs). Females deposit more eggs on larger plants (Zangerl and Berenbaum, 1992).

The most complete account of parsnip webworm biology was given by Gorder and Mertins (1984). Methods for rearing were developed by Nitao and Berenbaum (1988), who also provided data on developmental biology. Brittain and Gooderham (1916) gave a good morphological description of this insect, except that some of the measurements are incorrect.

Damage

Parsnip webworm feeds primarily on the flowers and seeds of parsnip, and is of concern only where seed production is desired. The larvae web together the flower heads and feed within, leaving the flower structure a mass of webbing and feces.

Management

This insect should rarely warrant control efforts, but foliar insecticides applied before opening of blossoms can prevent larval feeding.

FAMILY PAPILIONIDAE—CELERYWORMS AND SWALLOWTAIL BUTTERFLIES

Black Swallowtail
Papilio polyxenes Fabricius

Anise Swallowtail
Papilio zelicaon Lucas
(Lepidoptera: Papilionidae)

Natural History

Distribution. Black swallowtail, also known as "American swallowtail," is found in southern Canada and the eastern United States as far west as the Rocky Mountains. Its range also extends south into Mexico and northern South America. Anise swallowtail, also known as "western swallowtail," inhabits southwestern Canada and the western United States from the Rocky Mountains to the Pacific coast. The ranges of these native species overlap in the Rocky Mountain region.

Host Plants. Larvae of these species are commonly known as "parsleyworm" or "celeryworm" because they feed on plants in the family Umbelliferae. Vegetable crops fed upon by larvae include carrot, celery, fennel, parsley, and parsnip. The herbs anise, caraway, and dill also are eaten. Numerous umbelliferous weeds serve as suitable hosts, including angelica, *Angelica* spp.; cow parsnip, *Heracleum* spp.; lovage, *Ligusticum* spp.; water hemlock, *Cicuta maculata*; and wild carrot, *Daucus carota*. There are some chemical similarities between Umbelliferae and Rutaceae, the plant family containing citrus. Thus, occasionally these swallowtails are observed feeding on citrus, particularly orange trees, or other Rutaceae. This has occurred in California with greater frequency since fennel has been planted widely at lower elevations. Nowhere are these swallowtails considered to be a serious pest on citrus, however. Adults visit a variety of flowers to obtain nectar, they seem particularly fond of milkweed, *Asclepias* spp.; thistle, *Cirsium* spp.; and red clover, *Trifolium pratense*. Relative suitability of some plant hosts for black swallowtail larvae was provided by Finke and Scriber (1988).

Natural Enemies. Avian predators are an important source of mortality for adults. Resting butterflies are especially susceptible to predation, and during inclement weather they spend more time roosting, thus incurring higher levels of predation (Lederhouse *et al.*, 1987). Larvae are attacked by several predatory insects such as shield bugs (Hemiptera: Pentatomidae), damsel bugs (Hemiptera: Nabidae), and assassin bugs (Hemiptera: Reduviidae). A wasp parasitoid, *Trogus pennator* (Fabricius) (Hymenoptera: Ichneumonidae) attacks the larvae of both swallowtail species; black swallowtail, and likely anise swallowtail, are attacked by the fly parasitoids, *Buguetia obscura* (Coquillett), *Compsilura concinnata* (Meigen), and *Lesesia frenchii* (Williston) (all Diptera: Tachinidae).

Life History and Description. These two species are very similar in life history, differing principally in the appearance of adults, as described below. There are two generations annually in northern regions of New York, but three generations in warmer climates. Overwintering occurs in the pupal stage, and in warm climates some pupae from the second generation, as well as those from the third generation, enter diapause and emerge the following spring.

Egg. The spherical eggs measure about 1 mm in diameter, and are pale green or cream initially, developing a reddish-brown cap with age. Eggs are deposited singly on leaves and flowers of host plants. Egg deposition begins about 2–4 days after adults emergence at 40–50 per day. Total number of eggs produced is estimated at about 200–400 per female, with oviposition occurring over about a 13-day period (Blau, 1981). Eggs hatch in 4–9 days.

Larva. The color pattern among individuals and among the five instars is variable, but the general pattern follows. The first three instars are bird-dropping mimics, principally black with a median, dorsal spot resembling a saddle on the third and fourth abdominal segments. The third instar also bears some lateral orange and white markings. In the fourth instar, the caterpillar is principally black, but the anterior and

Black swallowtail larva.

posterior of each body segment is edged in green, and orange or yellow spots are located near the center of each body segment. The fifth instar is predominantly green, with black restricted to the center of each segment; this stage also has the orange spots found on the earlier instar. Larval development time is 10–30 days, depending on temperature. Larvae have an orange, eversible gland that resembles horns or antennae when extended. Located dorsally behind the head, the gland is exposed only when the caterpillar is disturbed. The gland, called an osmeterium, releases volatile chemicals that deter predation by some, but not all, insect predators (Berenbaum *et al.*, 1992). (See color figure 98.)

Pupa. The pupa is attached with silk to a plant stem. The posterior end is attached closely to the plant, but the anterior end hangs away from the stem, at about a 30 degree angle, suspended by silk strands. The pupa is green or light brown, but also bears irregular black marks that help camouflage the pupa. The pupa measures 2.5–3.0 cm long. Except when the pupal stage is overwintering, its duration is 9–18 days.

Adult. The adults of black swallowtail are principally black, but a row of yellow spots is found near the margin of the wings. In males, there is also a row of larger yellow spots parallel to the marginal spots, but is located more centrally. In females, the interior row is greatly reduced. Both sexes bear, at the posterior margin of the hind wing, an elongation of the wing that forms a "tail," a row of diffuse bluish spots, and a multi-colored "eyespot." (See color figure 204.)

The adults of anise swallowtail may greatly resemble black swallowtail, but more commonly they have the interior row of yellow spots greatly expanded, filling the central region of the wings with yellow. The net result is that the western species, anise swallowtail, is predominantly yellow, whereas the eastern species, black swallowtail, is predominantly black.

The adults of both species are quite large, the wingspan measuring over 7 cm. Adults are active during most of the day, and males are territorial, defending areas against other males. They often perch on elevated objects to maintain a good view of their territory, or they may patrol an open area, looking for females with which to mate. Males often frequent hilltops, and virgin females fly to hilltops to seek mates.

Adult black swallowtail.

The biology of these species was given by Chittenden (1912d), Fisher (1980), Scott (1986) and Opler and Krizek (1984). Culture techniques were described by Carter and Feeny (1985).

Damage

Larvae are large and can consume considerable quantities of foliage toward the end of their development. This should be of concern only in the home garden, because the butterflies are not sufficiently abundant ever to threaten commercial cultivation of an umbelliferous crop.

Management

Owing to their infrequent occurrence, control of larvae should not be necessary. For at least the past 100 years, hand picking has been recommended in the home garden, and this recommendation remains valid. In instances where plot size is too big to make this practical, larvae can be killed easily with foliar insecticides, either chemical or *Bacillus thuringiensis*.

FAMILY PIERIDAE—CABBAGEWORMS, WHITE AND SULFUR BUTTERFLIES

Alfalfa Caterpillar
Colias eurytheme Boisduval
(Lepidoptera: Pieridae)

Natural History

Distribution. This native species can be found throughout the United States, and often disperses to southern Canada during the summer. Formerly limited to central and western North America east of the Appalachian Mountains, the eastern dispersal of alfalfa caterpillar in the early 1900s was helped by the clearing of forests and widespread culture of alfalfa. It remains most common, and damaging, in the southwest.

Host Plants. Alfalfa caterpillar feeds on several forage legumes, particularly alfalfa, sweet clover, white clover, and hairy vetch. Red clover is not suitable for development, and adults rarely oviposit on this plant. Wild hosts include many species of milkvetch, *Astragulus* spp.; trefoil, *Lotus* spp.; clover, *Trifolium*; and vetch, *Vicia* spp. Among vegetable crops sometimes consumed are bean and pea, but not cowpea.

The adults feed on a succession of nectar-producing flowers, though the sequence varies between localities. In northern Virginia, the preferred flowers are dandelion, *Taraxacum* spp., and winter cress, *Barbarea* spp., during the first generation; followed by dogbane, *Apocynum* sp., and clover, *Trifolium* spp., during the second generation; then by milkweeds, *Asclepias* spp., in the third generation; and goldenrods, *Solidago* spp., aster, *Aster* spp., and tickseed sunflower, *Bidens coronata*, during the fourth to fifth generations (Opler and Krizek, 1984).

Natural Enemies. Natural enemies often are effective to keep this insect from becoming very abundant, and consequently, it generally is not a serious pest. An egg parasitoid, *Trichogramma* sp., occasionally destroys 50% of the eggs. As it develops rapidly, completes two generations in the very time it takes the host to complete a single generation, this egg parasitoid builds to high levels by the end of the season when caterpillar populations are high (Wildermuth, 1914). Young caterpillar larvae are attacked frequently by *Apanteles flaviconchae* Riley (Hymenoptera: Braconidae), with over 50% of them destroyed during some years. Michelbacher and Smith (1943) reported that when this parasitoid was abundant during the first generation of its host, the alfalfa caterpillar did not reach damaging levels. Other braconid parasitoids of lesser importance include *Apanteles cassianus* Riley, *A. medicaginis* Muesebeck, *Meteorus autographae* Muesebeck, *M. laphygmae* Viereck, and *M. leviventris* (Wesmael). Other wasps known from alfalfa caterpillar are *Itoplectis vidulata* (Gravenhorst), *Thyrateles instabilis* (Cresson), *Nepiera benevola* Gahan, *Hyposoter exiguae* (Vierick), and *Pristomerus spinator* (Fabricius) (all Hymenoptera: Ichneumonidae). A pupal parasite, *Pteromalus puparum* Linnaeus) (Hymenoptera: Pteromalidae), can be very important, sometimes parasitizing up to 60% of the pupae (Wildermuth, 1920). Among other parasitoids known from alfalfa caterpillar are *Euphorocera claripennis* (Macquart), *E. omissa* (Reinhard), and *Lespesia archippivora* (Riley) (all Diptera: Tachinidae).

The importance of predators has not been determined, but several are known. Ants (Hymenoptera: Formicidae) and lady beetles (Coleoptera: Coccinellidae) kill the caterpillar larvae, and robber flies (Diptera:

Asilidae) capture adults. Pupae of alfalfa caterpillar are sedentary, and therefore easy prey, and are consumed by the softwinged flower beetle, *Collops vittatus* Say (Coleoptera: Melyridae), larvae of corn earworm, *Helicoverpa zea* Boddie (Lepidoptera: Noctuidae), and likely many others.

A disease caused by a cytoplasmic polyhedrosis virus sometimes causes spectacular alfalfa caterpillar population collapse late in the season. However, as happens with many insect diseases, these become apparent only during seasons when other natural enemies are ineffective and caterpillar densities are high.

Life Cycle and Description. The adults are observed from March–May until late November over most of the ranges of this species. The number of generations varies with latitude and elevation. In northern regions and in the Rocky Mountains, only two generations are reported annually. However, three to five generations occur over most of the United States, five to six are reported from southern California and Arizona, and seven generations occur in Louisiana. During mid-summer a complete life cycle can be completed in 30 days. In northern regions this species survives periods of cold weather in the pupal stage, though survival in northern latitudes is poor. In southern regions, alfalfa caterpillar tends to pass the cold periods in the larval stage.

Egg. The egg of alfalfa caterpillar measures about 1.2–1.5 mm in long and 0.4–0.6 mm wide. The egg is spindle-shaped, tapering to a point at each end, and marked with 18–20 raised longitudinal ridges which are connected with smaller cross-ridges. They are deposited on end to the upper surface of foliage, and normally singly. They are greenish-white initially, but turn reddish after the second day. Duration of the egg stage is 3–7 days, with mean development time of 2.5–3.0 days at 26–32°C. Mean fecundity in the field is about 350 eggs during the adult lifespan of about two weeks. However, under laboratory conditions and exceptionally in the field, the female can live for over a month and deposit 700 eggs.

Larva. There are five instars. Body length is reported to be 1.5–3.0, 3.0–5.5, 4.5–8.0, 8.0–15, and 15–

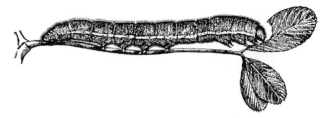

Alfalfa caterpillar larva.

29.0 mm in instars one to five respectively (Michelbacher and Smith, 1943). Larvae are primarily uniform green. A white lateral line normally runs the length of the body; a thin yellow, pink, or red line often occurs within the white line. Dorsally, a dark dorsal line and pair of fine white subdorsal lines may also be found. In some specimens the lines are absent, and the body is completely green. Mean larval development time during instars 1–5 is 2.6, 2.2, 1.8, 2.3, and 4.3 days when cultured at 27°C, and 2.1, 1.8, 1.7, 2.1, and 2.9 days when cultured at 32°C, respectively. (See color figure 92.)

Pupa. The pupa is angular, bearing keel-like projections on the thorax and tapering to a point at both the anterior and posterior ends. It is light green or pink, measures about 17.5 mm long, and bears a yellow line and three black dots on each side of the abdomen. The pupa is anchored loosely to a plant stem, anterior end pointed upward, by means of a thread and a posterior anchor. Duration of the pupal stage is often 7–10 days, but averages 4–5 days at 27–32°C.

Alfalfa caterpillar egg.

Alfalfa caterpillar pupa.

Adult. The adults usually are orange-yellow butterflies with a broad black band distally on the upper surface of each wing. A large black spot is also found near the mid-point of each forewing. In females, but not males, the black marginal band contains light spots. The undersurface of each wing, in both sexes, is uniformly pale yellow. The wingspan measures about 50–58 mm in males, and about 62–64 mm in females. The upper surface of the male's wings reflects ultraviolet light. This light reflection is visible to butterflies and serves to separate alfalfa butterfly from a co-occurring species, *Colias philodice* Godart, and to minimize hybridization (Silberglied and Taylor, 1978), though pheromones are also involved in mating (Sappington and Taylor, 1990). The adults of alfalfa butterfly are quite variable in appearance, and white variants are sometimes observed. Also, larvae exposed to cold temperature produce, after a lag of one generation, adults with wings that are mostly yellow, but containing some orange. Larvae exposed to warm temperature, in contrast, produce adults with orange wings (Tuskes and Atkins, 1973). Common sulfur butterfly, *C. philodice*, is most easily confused with alfalfa butterfly, but it is yellow rather than yellow-orange. Hybrids between these two species are not uncommon.

The biology of alfalfa caterpillar was described by Wildermuth (1914, 1920), though Floyd (1940) and Michelbacher and Smith (1943) added important observations. A useful synopsis was included in Opler and Krizek (1984). Methods of rearing were given by Taylor *et al.* (1981). (See color figure 199.)

Damage

Alfalfa caterpillar larvae are defoliators. Young larvae initially feed on the leaf surface or make small holes in leaves, but soon consume large quantities of leaf material. Although they are recorded from several types of bean and pea, damage to vegetable crops is usually minor. The pupal stage is an imortant contaminant of peas processed for freezing and canning.

Adult male alfalfa caterpillar.

Management

Vegetables generally are not at risk unless the caterpillars are very abundant on other crops or weeds. Thus, it is useful to monitor nearby alfalfa crops for the presence of caterpillars. The larvae and pupae are difficult to detect visually on foliage, and a sweep net is often used for sampling. If larvae are abundant in alfalfa, especially early in the season, it is often beneficial to harvest the alfalfa early because such harvesting can result in caterpillar mortality, and subsequent crops may avoid infestation. Infested crops can also be treated with insecticide or *Bacillus thuringiensis* applied to the foliage.

Imported Cabbageworm
Pieris rapae (Linnaeus)
Lepidoptera: Pieridae

Natural History

Distribution. Imported cabbageworm is found principally in temperate regions of Europe, Africa, Asia, and North America, but is now firmly established in Australia and New Zealand, southern Mexico, and Hawaii. Imported cabbageworm was first observed in North America in 1860 at Quebec City, Canada. It spread to Massachusetts and New Jersey by 1869, reached Ohio in 1875, and by 1886 it was found in the Gulf Coast and Rocky Mountain states. *P. rapae* attained British Columbia in 1898 and the Pacific Coast by 1901. It is now widespread in North America. It is a strong flier and also is spread long distances annually by strong winds. Few cabbageworms reportedly survive the winter in most of Canada, but much of the country is invaded annually by dispersants from the United States or from southern Canada (Beirne, 1971).

Pieris rapae is easily confused with other common cabbage white butterflies: *Pontia protodice*, southern cabbageworm; *Pieris napi*, mustard white; band *Ascia monuste* (Linnaeus), southern white. Before introduction of imported cabbageworm, *P. napi* (Linnaeus) was the dominant cabbage butterfly in the north, and *Pontia protodice* (Boisduval and LeConte) was the principal cabbage-feeding butterfly in the south. Both have been largely replaced by *P. rapae*, though they sometimes co-occur on cultivated crucifers or on weeds. A key for the differentiation of these species is included in Appendix A.

Host Plants. Larvae of this insect feed widely on plants in the family Cruciferae, but occasionally on a few other plant families that contain mustard oils. Vegetable crops attacked include broccoli, Brussels sprouts, cabbage, cauliflower, collards, horseradish, kale, kohlrabi, mustard, radish, turnip, and watercress. Also attacked are flowers such as nasturtium and sweet alyssum, and weeds such as field pennycress, *Thlaspi arvense*; pepperweed, *Lepidium* spp.; wild mustard, *Brassica kaber*; horsehair mustard, *Conringia orientalis*; and yellow rocket, *Barbarea vulgaris*. Adults sip nectar from flowers, and are commonly seen at mustards, *Brassica* spp.; dandelion, *Taraxacum officinale*; aster, *Aster* spp.; purple heliotrope, *Heliotropum* spp.; thistle, *Cirsium* spp.; and both weedy and cultivated crucifers (Harcourt, 1963a). Sea kale is reported to be attractive for oviposition, but larvae fail to complete their development on this plant (Richards, 1940).

Natural Enemies. Imported cabbageworm is subject to numerous predators, parasitoids, and diseases. General predators such as shield bugs (Hemiptera: Pentatomidae), ambush bugs (Hemiptera: Phymatidae), and vespid wasps (Hymenoptera: Vespidae) attack them, as do many insectivorous birds. Chittenden (1916a), for example, noted 90% predation of overwintering pupae by birds. However, parasitoids are considered to be much more important mortality factors. Harcourt (1963a) identified three important species in Ontario. *Cotesia glomeratus* (L.) (Hymenoptera: Braconidae) attacks the early instars, and emerges from the mature larva as it prepares to pupate. *Phryxe vulgare* (Fallon) (Diptera: Tachinidae) attacks mature larvae and emerges from the host pupa. *Pteromalus puparum* (L.) (Hymenoptera: Pteromalidae) attacks and kills cabbageworm pupae. *Cotesia glomeratus* has long been considered to be the most important parasitoid in Canada and in the northern United States. *Cotesia glomeratus* is readily observed in the field, searching diligently on foliage for larvae. Dead cabbageworm larvae are often found with clusters of 20–30, *C. glomeratus* cocoons attached. However, in Europe, *Cotesia rubecula* (Marsall) (Hymenoptera: Braconidae) is the most important parasitoid, and in recent years it has become established in North America where it is now assuming a dominant role (Godin and Boivin, 1998b,c).

In contrast, tachinids were more important in California, particularly *Madremyia saundersii* (Williston) (Diptera: Tachinidae) (Oatman, 1966b). As observed in Canada, however, *C. glomeratus* was also a significant larval mortality factor in California. A low level of egg parasitism by *Trichogramma pretiosum* Riley (Hymenoptera: Trichogrammatidae) occurred, but failure of eggs to hatch, which was attributed to infertility caused by cool weather, resulted in more deaths than did egg parasitoids. The pupal parasitoid *P. puparum* was also effective in California, but this is not always the case. In some locations, relatively low levels of parasitism have been observed (Oatman

and Platner, 1969; Ru and Workman, 1979; Lasota and Kok, 1986); high rates tend to occur late in the season.

Virus and fungal diseases of imported cabbageworm have been reported, but the predominant natural disease in a granulosis virus (GV). *Pieris rapae* GV occurs most commonly under high density conditions, and often among late instar larvae after they have consumed the exterior foliage of plants and are forced into close contact. Harcourt (1963a) observed over 90% mortality of larvae due to natural occurrence of this disease. In the early stages of infection, larvae are inactive and paler in color. As the disease progresses, the caterpillar body turns yellow, and tends to appear bloated. After death, the body blackens, the integument ruptures, and the liquefied body contents ooze on the plant foliage. Rainfall has a major roll in assisting the spread of the virus on the plant, and from the soil to the plant. Beirne (1971) suggested that outbreak of granulosis virus in the second generation of cabbageworm often prevented the third annual generation from causing extensive damage.

Life Cycle and Description. The complete life cycle of this insect requires 3–6 weeks, depending on weather. Godin and Boivin (1998a) reported that about 320 degree-days above a threshold of 10°C was required to complete a generation in Quebec. The number of generations reported annually is two-to-three in Canada, three in the New England states, three-to-five in California, and six-to-eight in the south. Imported cabbageworm can be found throughout the year in the south.

Egg. The eggs are laid singly, usually on the outer leaves of plants. About 70–85% of the eggs are deposited on the lower surface of the leaves, where the larvae also tend to feed. The egg measures 0.5 mm wide and 1.0 mm long, and initially it is pale white, but eventually turns yellowish. The egg is laid on end, with the point of attachment flattened and the distal end tapering to a blunt point. The shape is sometimes described to resemble a bullet. The egg is strongly ribbed, with 12 longitudinal ridges, and hatches in about five days (range 2–8 days) during August.

Larva. The larva is green, velvety in appearance, and bears five pairs of prolegs. There are five instars. Head capsule widths are about 0.4, 0.6, 0.97, 1.5, and 2.2 mm, respectively. Body lengths at maturity of each instar averages 3.2, 8.8, 14.0, 20.2, and 30.1 mm, respectively. The larva requires about 15 days (range 11–33 days) to complete its development during August. Average (and range) development time for each instar at 19°C was observed to be 4.5 (2.5–6), 3.0 (1.5–5), 3.3

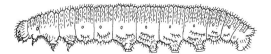

Imported cabbageworm larva.

(2–5), 4.1 (3–6.5), and 7.8 (5–18) days, respectively. All larval stages except the first instar bear a narrow yellow line running along the center of the back; this stripe is sometimes incomplete on the early instars. A broken yellow line, or series of yellow spots, also occurs on each side. The mature lava typically wanders before pupation, spending an average of 1.8 days (range 1–4 days) without feeding, and finally spins a silk pad or platform, where pupation occurs. (See color figure 93.)

Pupa. Pupation often occurs on the food plant, but cabbageworm may leave the plant to pupate in nearby debris, especially when larval densities are high. The chrysalis is about 18–20 mm long, and varies in color, usually matching the background; yellow, gray, green and speckled brown are common. A sharply angled, keel-like projection is evident dorsally on the thorax, and dorsolaterally on each side of the abdomen. At pupation, the chrysalis is anchored by the tip of the abdomen to the silk pad, and a strand of silk is loosely spun around the thorax. The silk line serves to anchor the anterior portion of the chrysalis, and also keeps the head from hanging down. Pupation during the summer generations lasts, on average, 11 days (range 8–20 days). The chrysalis is the overwintering stage, however, so its duration may be prolonged for months if the pupa diapauses. The proportion of pupae that diapause increases as autumn progresses, so that at the time of the final generation all pupae are in diapause. (See color figure 267.)

Adult. Upon emergence from the chrysalis the butterfly has a wingspan of about 4.5–6.5 cm. It is white above with black at the tips of the forewings. The front wings are also marked with black dots—two in the central area of each forewing in the female, and one in case of males. When viewed from below, the wings generally are yellowish, and the black spots usually show through the wings. The hind wing of each sex also bears a black spot on the anterior edge. The black spots and the yellow coloration may be reduced or absent from both sexes, especially in the spring generation. The body of the butterfly is covered with dense hair, which is colored white in females, but darker in males. The adult typically lives about three weeks, and may be active very early in the spring. The female produces 300–400 eggs. The adult is very active during the daylight hours, often moving from

Adult male imported cabbageworm.

Adult female imported cabbageworm.

the crop to flowering weeds to feed. This movement tends to result in a preponderance of the eggs being deposited on the edges of crucifer fields unless there are flowering weeds contained within the crop. (See color figure 200.)

The biology of imported cabbageworm was given by Chittenden (1916a), Wilson *et al.* (1919), Richards (1940), and Harcourt (1962, 1963a). Culture was described by Webb and Shelton (1988).

Damage

The larvae defoliate crucifer crops, sometimes killing young plants. Severe damage to young plants often prevents head formation even when the caterpillars are later removed, so early season protection is important (Wilson *et al.*, 1919). If left unchecked, cabbageworm often can reduce mature plants to stems and large veins. Although they prefer leafy foliage, larvae may burrow into the heads of broccoli and cabbage, especially as they mature. Larvae are often immobile and difficult to dislodge, and they may be overlooked when cleaning produce. Larvae produce copious quantities of fecal material which also contaminate and stain produce.

Imported cabbageworm larvae are present, and potentially damaging, throughout most of the period that crucifer crops are under culture. In northern latitudes such as Wisconsin (Radcliffe and Chapman, 1966) and Ontario (Harcourt, 1963a) damage typically occurs during summer months, in southern California they are most abundant during the autumn (Oatman and Platner, 1969), and in southern latitudes such as South Carolina (Reid and Bare, 1952) and Louisiana (Smith and Brubaker, 1938), damage occurs during the autumn, early winter- and spring-month periods that coincide with peak crucifer production.

Management

Sampling. Harcourt (1962) studied the distribution of imported cabbageworm on crops. He suggested that one-half of each plant be examined visually for various stages. Recommended sample sizes were 20 plants for eggs, 30 for young larvae, 40 for mid-age larvae, 50 for large larvae, and 70 for pupae. Larvae often rest along the principal leaf vein, and are very difficult to see because their body color closely matches the background. Damage and fecal material is often the most visible indication of infestation. The presence of highly visible butterflies suggests future problems. Shelton *et al.* (1994) compared the benefits of sequential and variable-intensity sampling for cabbageworm management, and recommended the latter as being more reliable and requiring fewer samples.

Insecticides. Imported cabbageworm are readily killed by foliar application of insecticides, including the bacterial insecticide *Bacillus thuringiensis* and some botanical insecticides (Hamilton and Gemmell, 1934; Huckett, 1934, 1946; Dills and Odland, 1948).

Biological Control. Several microbes have been investigated for control of imported cabbageworm, and have the potential to be developed as microbial insecticides. The imported cabbageworm granulosis virus (*Pieris rapae* GV) suppressed cabbageworm larvae in the laboratory (Payne *et al.*, 1981) and in a field test, but required 4–10 days to inflict mortality, and was not superior to control provided by *Bacillus thuringiensis* (Jaques, 1973). The nuclear polyhedrosis virus (NPV) from alfalfa looper (*Autographa californica* NPV), a granulosis virus from cabbage butterfly (*Pieris brassicae* GV), and the microsporidian *Vairimorpha necatrix* were shown by Jaques (1977) and Tompkins *et al.* (1986) to suppress imported cabbageworm, but the population reduction was not superior to that achieved using *Bacillus thuringiensis*. These pathogens have commercial potential because they are not very host specific, and also suppress another important crucifer pest, cabbage looper, *Trichoplusia ni* (Hübner). Home gardeners sometimes collect dead or dying caterpillars, macerate them in water, and spray the resulting suspension onto foliage as a home-made biological insecticide.

Flower flies (Diptera: Syrphidae) consume the eggs and small larvae of imported cabbageworm and numerous other insects. Populations of flower flies have been manipulated to increase predation of cabbageworm by interplanting cabbage with pollen-rich flowering plants. Although aphid populations on cabbage were suppressed, imported cabbageworm populations were not decreased significantly (White *et al.*, 1995).

Host-Plant Resistance. Crucifer crops differ in their susceptibility to attack by imported cabbageworm. Chinese cabbage, turnip, mustard, rutabaga, and kale are less preferred than cabbage, collards, Brussels sprouts, broccoli, and cauliflower (Harrison and Brubaker, 1943; Radcliffe and Chapman, 1966). Some cultivars of certain crops also have moderate levels of resistance to infestation by imported cabbageworm. One resistance character is due to, or correlated with, dark green, glossy leaves. This character imparts resistance to imported cabbageworm and other caterpillars, but increases susceptibility to flea beetle injury (Dickson and Eckenrode, 1980). The red color found in many crucifer varieties also affects imported cabbageworm. Cabbage butterflies avoid ovipositing on red cabbage varieties (Radcliffe and Chapman, 1966). However, larval survival is favored by red cabbage. Thus, while important genetic material has been identified, in most cases existing varieties are not a practical solution to caterpillar problems. Research conducted in Virginia on susceptibility of broccoli cultivars, and where pest abundance was relatively low, indicated that the principal factor in susceptibility to attack was the date of plant maturity; early maturing varieties were less infested (Vail *et al.*, 1991).

Cultural Practices. Herbs are sometimes recommended as companionate or repellent plants for vegetable cultivation. Herbs are hypothesized to give off odors that repel ovipositing vegetable pests, or prevent them from locating the vegetables. In most cases this has not been investigated critically. However, for imported cabbageworm, there is ample evidence that herbs impart no benefit, and some herbs are associated with increased cabbageworm infestation (Latheef and Ortiz, 1983a,b). Paper caps early in the season, and row covers later, are effective in preventing oviposition by imported cabbageworm butterflies.

Mustard White
Pieris napi (Linnaeus)
(Lepidoptera: Pieridae)

Natural History

Distribution. This insect is found both in North America and Eurasia, where it has a decidedly northern range. In the eastern United States, it occurs principally in the northernmost states, but in the West its range extends as far south as New Mexico and California. It is widespread in Canada and Alaska, extending north to the edge of the Arctic tundra. Mustard white has declined in abundance in the northeastern United States, a phenomenon commonly attributed to competition with imported cabbageworm. However, adults frequent shaded areas, whereas imported cabbageworm butterflies prefers sunny fields, so competition may not completely explain the decline in mustard white abundance. It has also been suggested that change in habitat availability could account for this population decline.

Host Plants. Larvae feed on wild and cultivated Cruciferae. Weeds such as Virginia pepperweed, *Lepidium virginicum*; toothworts, *Dentaria* spp.; and the *Brassica* and *Descurainia* mustards are suitable larval hosts. Females oviposit on yellow rocket, *Barbarea vulgaris*, and field pennycress, *Thlaspi arvense*, but larvae cannot complete their development on these plants. Vegetables attacked include Brussels sprouts, cabbage, radish, turnip, watercress, and probably others. Adults take nectar from flowers, preferring *Geranium* spp. if they are available.

Life Cycle and Description. One to three generations are reported; two is most common. In areas with three generations, many members of the second generation enter diapause.

Egg. The eggs are pale yellow or white, tinged with green, and bear 13 ridges. The eggs are elongate, tapering to a blunt point distally. They measure about 1.2 mm long and 0.45 mm wide. They are deposited singly, on end, to the underside of leaves and on stems of hosts. The female apparently produces about 300 eggs over the course of her life, which may be of several weeks' duration.

Larva. Larvae are covered with short, fine hairs, imparting a velvety appearance. They are green with small black spots, with a green stripe dorsally and yellow spots along each side. Mustard white larvae are easily confused with other pierid butterfly larvae; a key to separate these species is provided in Appendix A.

Pupa. The chrysalis is green, gray, or tan flecked with black. It bears frontal and lateral projections similar to imported cabbageworm. Light-colored dorsal and lateral stripes may occur on the chrysalis. The mustard white undergoes diapause in the pupal stage.

Adult. The wingspan is about 4–5 cm. This insect is white dorsally, and sometimes bears one or two black spots on the front wings. The lower side of the

Adult male mustard white.

hind wing is usually yellow, as is the tip of the forewing. The most distinctive feature is the darkening along the wing veins; this is especially evident on the hind wings, and most pronounced on the underside of the hind wings. This insect has two color forms. The summer generation tends toward nearly white, while the spring generation has pronounced bands of gray along the veins. A key for separation of this species from other common cabbage white butterflies is included in Appendix A.

The biology of mustard white was given by Opler and Krizek (1984), and Scott (1986).

Damage

Larvae feed on the foliage, and produce small to large holes depending on larval size. They rarely are numerous enough to warrant cancern.

Management

Mustard white is rarely a serious pest, and should respond to management practices developed for imported cabbageworm.

Southern Cabbageworm
Pontia protodice (Boisduval and LeConte)
(Lepidoptera: Pieridae)

Natural History

Distribution. Southern cabbageworm, also known as the checkered white, is found throughout the United States and south into Mexico. It is native to North America. As its common name suggests, it is principally a southern insect, and occurs in the North only sporadically. It is rare in the New England states and southern Canada.

Host Plants. Southern cabbageworm attacks a variety of cultivated crucifers, but is much less abundant than imported cabbageworm, *Pieris rapae* (Linnaeus). Apparently southern cabbageworm, which formerly was quite abundant in some localities, has been largely displaced by *Pieris rapae*. Vegetable crops known to be attacked are cabbage, cauliflower, Chinese cabbage, radish, and turnip. The host range is probably equivalent to imported cabbageworm, but because southern cabbageworm is not a serious pest it has not been well studied. Wild crucifers that support the growth of southern cabbageworm include pepperweed, *Lepidium* spp.; yellow rocket, *Barbarea vulgaris*; shepherdspurse, *Capsella bursa-pastoris*; tansy-mustard, *Descurainia pinnata*; field pennycress, *Thlaspi arvense*; hoary cress, *Cardaria draba*; black mustard, *Brassica nigra*; and others.

Life Cycle and Description. Three or four generations occur annually. Southern cabbageworm overwinters in the pupal stage.

Egg. The eggs are pale yellow initially, but turn orange as they mature. The are deposited singly on flowers or leaves of the larval host plants.

Larva. Larvae are pale blue-green or bluish-gray, with numerous small black spots. They bear a pair of yellow stripes running the length of the back, and a stripe on each side. The head is yellow with reddish spots. A key to distinguish this caterpillar from similar pierid larvae feeding on crucifers is provided in Appendix A. (See color figure 94.)

Pupa. The chrysalis is light blue-grey and speckled with black. In general shape, the chrysalis resembles that of imported cabbageworm, bearing sharp keel-like structures on the back. This is the overwintering form.

Adult. The adults have a wingspan of about 3.5–5 cm. They are similar in appearance to imported cabbageworm, but the southern cabbageworm butterflies are more heavily marked with scattered grayish-brown or black spots, both dorsally and ventrally, on their white background. The female, in particular, may be dark, with up to 50% of the wings darkened. The males, in contrast, bear only a few dark spots, but are still more heavily marked than *Pieris rapae*. The first generation butterflies, which emerge in the spring following diapause, may be smaller and very light in color. Adults can be seen feeding on nectar from a wide variety of flowering plants, and are highly dispersive. A key for differentiating this species from other cabbage white butterflies is included in Appendix A.

The biology of southern cabbageworm was given by Opler and Krizek (1984), and Scott (1986).

Damage

Larvae prefer to feed on flowers and flower buds, but they also eat leaves. On cabbage, they restrict their

Adult male southern cabbageworm.

Adult female southern cabbageworm.

feeding to the outer leaves. This makes them much less damaging than imported cabbageworm, which bore into the head.

Management

Southern cabbageworm butterflies were observed to be highly attracted to fluorescent orange, and to be captured with sticky traps, especially when the traps were positioned close to the soil surface (Capinera, 1980). The methods discussed for management of imported cabbageworm are also appropriate for southern cabbageworm.

Southern White
Ascia monuste (Linnaeus)
(Lepidoptera: Pieridae)

Natural History

Distribution. Southern white resides in Florida, the Gulf Coast areas of the southern states, southern Texas, Mexico, and most of Latin America. This butterfly exhibits strong migratory tendencies, moving northward in the summer and southward in the winter, but in North America it rarely becomes numerous anywhere other than subtropical, coastal areas. They are more dispersive under high-density conditions.

Host Plants. Larvae feed on plants in the families Cruciferae, Bataceae, Capparidaceae, and Tropaeolaceae. Principal hosts are saltwort, *Batis maritima*, in coastal regions and pepperweed, *Lepidium virginicum*, elsewhere. Vegetables fed upon are cabbage, cauliflower, collards, kale, radish, and turnip. Other food plants include beach cabbage, *Cakile maritima*; spider flower, *Cleome spinosa*; clammy weed, *Polanisia* sp. nasturtium, *Tropaeolum majus*; and others. Adults frequent a variety of flowering plants to obtain nectar.

Life Cycle and Description. The number of generations is poorly documented, but at least three are known annually, and adults are active year-round in southern Florida and Texas. Reproductive diapause likely occurs during the winter months.

Egg. The eggs are pale-yellow, elongate, and bear 11 longitudinal ridges. They may be laid singly, in small groups, or in cluster of up to 50, depending on the host. On average, eggs are laid in clusters of 16 on young leaves. The female may deposit 800–1000 eggs over the course of her life, which is typically 8–10 days in duration.

Larva. Young are often gregarious, and such larvae grow faster than solitary individuals. Larvae are mottled gray or brownish green, often with a purplish hue, and bear five longitudinal orange or yellow-green bands running the length of the body. Rows of black spots also occur laterally. The head is yellow and orange, and marked with black spots. The larva attains a length of about 3–4 cm. A key for differentiating this caterpillar from other crucifer-feeding pierids is provided in Appendix A. (See color figure 95.)

Pupa. The chrysalis is ivory white with numerous black markings, and bears one orange-yellow band dorsally, and another on each side. The general shape is similar to that of imported cabbage butterfly, but the projections are greatly blunted in the southern white, and appear to be nothing more than large bumps. The length is about 2.5 cm.

Adult. The adult has a wingspan of about 6–8 cm. The male butterfly is almost pure white dorsally, but has a black, deeply indented band along the distal edge of the front wing. On the underside of the male, the hind wings are tan, as are the tips of the front wings. The female has two rather distinct color forms. The long-lived winter form is principally white, but with wider black bands than the male, and a single black spot on the forewings. During the summer months, the female is much darker, brown to dark-gray; such butterflies tend to have a broader dark band on the forewings, but less scalloping of the band. Caterpillars raised under short-day conditions produce the dark adult form. A key for differentiating this

Adult male southern white.

Adult female southern white.

species from other cabbage white butterflies is included in Appendix A. (See color figure 201.)

The biology of southern white was given by Opler and Krizek (1984) and Scott (1986).

Damage

Larvae feed on foliage, and eat small and large holes in leaves. In coastal areas they sometimes defoliate small plantings.

Management

Hayslip *et al.* (1953) considered southern white to be a minor pest in Florida. Management techniques appropriate for imported cabbageworm should be suitable for this insect also.

FAMILY PTEROPHORIDAE—PLUME MOTHS

Artichoke Plume Moth
Platyptilia carduidactyla (Riley)
(Lepidoptera: Pterophoridae)

Natural History

Distribution. This native insect is found widely in North America. Artichoke plume moth is damaging throughout the west, from British Columbia to southern California, and east to Montana and New Mexico. It is also known from eastern North America, but in Canada it is known only as far east as Ontario, and in the United States it occurs as far east as New York and North Carolina but is absent from most southeastern states.

Host Plants. Artichoke plume moth attacks thistles in the family Compositae. Several weedy *Cirsium* species are suitable hosts, and some are preferred over globe artichoke. Other weeds also known to be hosts are milk thistle, *Silybum marianum*; and at least on one occasion Napa thistle, *Centaurea melitensis*. Despite the reported preference for weed species, artichoke is regularly attacked in California, where most artichoke is cultivated. Artichoke can serve as a host during the entire year, but a sequence of weedy thistle species is usually required in non-agricultural habitats. Lange (1950) and Turner *et al.* (1987) provided lists of known-host plants.

Natural Enemies. Several natural enemies are known, including a variety of Hymenoptera (Braconidae, Ichneumonidae), Diptera (Tachinidae), lacewing (Neuroptera: Hemerobiidae), rove beetles (Coleoptera: Staphylinidae), the whirligig mite, *Anystis agilis* (Banks) (Acarina), as well as spiders and birds. In California, *Diadegma acuta* (Viereck) (Hymenoptera: Ichneumonidae) is the most widespread and effective parasitoid, attacking larvae in their burrows. Other parasitoids were documented by Lange (1941) and Bragg (1994). Life table analyses conducted by Goh and Lange (1980b) suggested that much of the natural mortality occured soon after egg hatch. Predation and drowning were important factors at this time; once larvae began to burrow they were much less affected by such mortality agents.

Life Cycle and Description. There are three generations annually in California, where considerable overlapping of generations occurs, and all stages of development can be found throughout the year. The number of generation in New York was reported to be one, whereas two were reported from Minnesota. Development time is quite long, requiring about 110–140 days for a complete life cycle, so it seems improbable that there was more than one generation in most of the North.

Egg. The oval eggs are glossy and yellow, and measure 0.52–0.66 mm long and 0.26–0.35 mm wide. They are deposited singly and externally on various plant tissues, especially the undersides of leaves, and they may be attached either on the side or on end. Females prefer to attach their eggs among leaf hairs, and avoid smooth substrates. Mean production was

estimated at 170 eggs (range 70–300 eggs) by Lange (1941) but at about 250 eggs by Bari and Lange (1980). Duration of the egg stage varies with temperature, but averages 15 days (range 8–24 days) during the spring in California.

Larva. Upon hatching, the larva measures about 1 mm long, and bears 4 pairs of prolegs. Larvae often feed on young foliage during the first two instars, and then burrow into the stalks or flower buds. Lacking suitable young foliage, however, they immediately begin to burrow. There are four instars. Mean duration (range) of the instars is 8 (8), 5.2 (5–6), 7.1 (5–12), and 19.4 (16–26) days, respectively. Thus, larval development times of 40–60 days are common, depending on weather. The larva is yellowish for the first three instars, and whitish during the fourth instar. During all instars the head, true legs, cervical shield, and anal plate are black. The head capsule widths are 0.24, 0.43, 0.67, and 1.06 mm for instars 1–4, respectively. Body length increases to about 12 mm at maturity, and the larva is rather stout.

Pupa. The pupa varies from yellow to brown, and measures about 10–12 mm long. The abdomen bears tooth-like spines that project posteriorly. Pupation does not always occur within a cocoon, but a thin silken cocoon tends to be present if the larva pupates in an exposed location such as among old senescing leaves and leaf litter. Often the naked pupa is found within the burrow of the larva. Duration of the pupa is about 24 days (range 22–28 days).

Adult. The adult is a small yellowish-brown moth with a wingspan of 18–30 mm. The forewing is lobed, with a cleft extending inward about one-third the length of the wing. The forewing also bears a dark triangular mark adjacent to the cleft, and extending to the leading edge of the forewing. The hind wing is even more divided, consisting of three large lobes;

Artichoke plume moth larva.

Artichoke plume moth pupa.

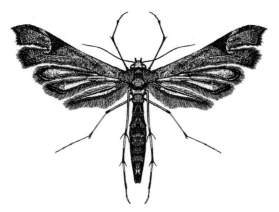

Adult artichoke plume moth.

the color is uniform brown. The legs are light in color except that they bear dark bands at the joints. Adults normally are not active during the day, and tend to rest on the underside of foliage, but they make short flights if disturbed. The temperature threshold for flight is about 8–10°C. Mating usually occurs within three days of emergence, and adults display a pre-ovipositional period of about 3–8 days. Adults live for up to 30 days. Females produce a sex pheromone (Klun *et al.*, 1981) with maximal pheromone release about 4 h after sunset (Haynes *et al.*, 1983).

The most complete accounts of artichoke plume moth biology were supplied by Lange (1941, 1950). Bari and Lange (1980) provided information on developmental biology.

Damage

The larvae may feed on any portion of the plant, but they are usually found in the developing flower head, or bud, of the artichoke. Small larvae may burrow through the tissue of the outer bracts. As the larvae mature, they tunnel into the inner portions of the fruit. Larvae sometimes feed on the leaf tissue, especially the newly formed foliar tissue at the center of the plant, and sometimes burrow into the stalks and crown, including short distances below-ground. Artichoke plume moth is often considered to be a serious limiting factor in commercial artichoke production.

Management

Sampling. Distribution of eggs and larvae was studied by Goh and Lange (1980a). During the vegetative or prebloom stage, most eggs and young larvae were found on the young leaf tissue at the center of the plant. As blossoms are produced, however, eggs are deposited on the leaves just below the flower heads, and virtually all the larvae are found within the buds. Moths can be captured with light traps. Sex pheromone can be used to bait traps for population monitoring, but

can also be used to permeate the atmosphere and disrupt mating (Haynes *et al.*, 1981).

Insecticides. Foliar insecticides, including *Bacillus thuringiensis*, are often applied to protect artichoke from artichoke plume moth. The preferred materials affect both the adult and larval stages, and this often limits use of *Bacillus thuringiensis* to tank mixes with chemical insecticides. Insecticide use sometimes induces outbreaks of twospotted spider mite, *Tetranychus urticae* Koch.

Cultural Practices. Sanitation is an important element in artichoke plume moth management. Infested artichokes should be shredded, deeply buried, fed to livestock, or otherwise destroyed and not discarded in the field, because moths emerge successfully from infested buds. Similarly, destruction of plants after harvest can be beneficial if they are infested. Weedy thistles should be eliminated if found growing near cultivated artichoke as these weeds can provide inoculum for crops.

Biological Control. The bacterium *Bacillus thuringiensis* and the entomopathogenic nematode *Steinernema carpocapsae* have been shown to provide effective suppression of artichoke plume moth larvae. The dense vegetative growth and tunneling behavior of larvae provide suitable microhabitat for nematode survival during the prebloom period, but not during the fruiting period (Bari and Kaya, 1984). At planting time, artichoke cuttings can be soaked in a suspension of entomopathogenic nematodes to reduce the likelihood of planting insects into a field along with the new crop.

FAMILY PYRALIDAE—BORERS, BUDWORMS, LEAFTIERS, WEBWORMS AND SNOUT MOTHS

Alfalfa Webworm
Loxostege cereralis (Zeller)
(Lepidoptera: Pyralidae)

Natural History

Distribution. This native insect is primarily western in distribution, occurring from British Columbia to Quebec in the north and from California to western Texas in the south. It is highly dispersive, however, and sometimes attains the northeastern states. As a pest, its range is limited to the western Great Plains and Rocky Mountain region.

Host Plants. Alfalfa webworm has a broad host range, very similar to that of the better-known beet webworm, *Loxostege sticticalis* (Linnaeus), with which it is sometimes confused. Its common name is misleading, because though females oviposit in alfalfa and larvae consume alfalfa foliage, it is not routinely associated with this crop. In general, plants in the family Chenopodiaceae are preferred. However, when this insect is abundant and the favored plants are no longer available, a wide range of plants can be injured. Among vegetables susceptible to injury are bean, beet, cabbage, cantaloupe, carrot, eggplant, lettuce, parsnip, pea, pepper, spinach and likely many others. Grasses normally are avoided, but corn can be fed upon if nothing else is available. Alfalfa and sugarbeet are the field crops most often damaged during periods of great abundance. Among the weeds fed upon by alfalfa webworm, at least during periods of abundance, are dock, *Rumex* sp.; lambsquarters, *Chenopodium album*; common mallow, *Malva neglecta*; ragweed, *Ambrosia* sp.; redroot pigweed, *Amaranthus retroflexus*; saltbush, *Atriplex* sp.; Russian thistle, *Salsola kali*; sunflower, *Helianthus annuus*; and sweetclover, *Melilotus* sp. (Hoerner, 1933). In moth oviposition and larval food preference tests, Capinera *et al.* (1981b) reported that kochia, *Kochia scoparia*, and Russian thistle were favored for oviposition, though some eggs were deposited on all weeds tested, whereas only lambsquarters, redroot pigweed, and shepherdspurse, *Capsella bursa-pastoris*, were fed upon. Maxson (1948) reported finding eggs on dandelion, *Taraxacum officinale*; field bindweed, *Convolvulus arvensis*; kochia; lambsquarters; plantain, *Plantago major*; purslane, *Portulaca oleracea*; nightshade, *Solanum* sp.; ragweed, *Ambrosia* sp.; redroot pigweed; and saltbush. Adults feed on nectar from dandelion and field bindweed. (See color figure 9.)

Natural Enemies. Several natural enemies are described, though none are known to be particularly effective. In Colorado, for example, Hoerner (1933) reported that only 3% of the larvae were parasitized. Among the parasitoids are *Cremnops vulgaris* (Cresson), *Meteorus campestris* Viereck (both Hymenoptera: Braconidae), *Aplomya caesar* (Aldrich), *A. trisetosa* (Coquillett), *Lespesia archippivora* (Riley), and *Phryxe vulgaris* (Fallén) (all Diptera: Tachinidae).

Life Cycle and Description. In Canada, 2–3 generations are reported annually, but in the United States three generations seem to be normal. A generation can be completed in about 40 days under optimal conditions. As with beet webworm, mature larvae overwinter and the proportion of each generation that diapauses increases as the summer progresses. In Colorado, the first flight of moths occurs in May, followed by additional flights in June–July and September. However, Capinera *et al.* (1981a) suggested that the first two flights might both result from protracted emergence of overwintering larvae.

Egg. Moths begin oviposition within a few days of emergence, and continue to oviposit for about two weeks. The flattened, oval eggs are about 1 mm long and 0.7 mm wide. They are white initially, but eventually become yellow. They usually are deposited in clusters of 2–20 overlapping eggs on the underside of foliage. Although individual eggs greatly resemble those of beet webworm, the clustering arrangement of alfalfa webworm serves to distinguish them from the single-row oviposition pattern of beet webworm. Eggs of alfalfa webworm hatch in 4–6 days during warm weather. (See color figure 258.)

Larva. The larval stage has six instars. Larvae increase in length from about 3 mm at hatching to about 25 mm at maturity. Mean head capsule widths are 0.24, 0.41, 0.64, 1.01, 1.36, and 1.75 mm during instars 1–6, respectively. Development time of larvae fed sugarbeet foliage and reared at 27°C was 2.7, 1.8, 1.9, 2.3, 2.4, and 5.5 days, respectively, for instars 1–6. The optimal temperature for larval development is about 30°C, and larvae fail to develop at 15°C. Young larvae are pale yellowish-green, but they become darker as they mature. The head and thoracic plate are dark during instars 1–3, becoming irregularly colored with light and dark blotches thereafter. During the final instar the larva bears a broad whitish dorsal line along its body, bordered by broad black dorsolateral stripes, a distinct contrast with the dark dorsal line of beet webworm. Most abdominal segments bear six dark spots with light centers dorsally, with a single dark hair arising from each. Larvae sometimes spin a mat of silk on foliage, where they rest during the daylight hours. They also produce strands of silk which are used to web together foliage and to construct a long thin tube into which they may retreat if disturbed. The silk tube often connects to clods of soil or other protected retreats. Duration of the larval stage is about 17 days at 30°C, increasing to about 25 days at 25° and 35°C. Duration of the larval stage is extended to several months for larvae that enter diapause. During diapause they remain below-ground within the silken pupation cell and then pupate in the spring. (See color figure 67.)

Pupa. Larvae enter the soil to a depth of up to 2.5 cm when they are ready to pupate, where they construct a silk-lined cell. The pupa is yellowish initially, turns dark-brown with age, and measures about 25–30 mm long. The posterior, pointed end of the pupa bears eight small spoon-shaped appendages. This character serves to distinguish alfalfa webworm from beet webworm, which bears eight spines instead.

Duration of the pupal stage is often 14–21 days under field conditions, but requires only 8–9 days when reared at constant temperature of 25–35°C.

Adult. The moth is fairly small, measuring only about 2.5–3.0 cm in wingspan. It is grayish brown in general color, though containing some irregular black marks and a transverse cream-colored band on the distal margin of the forewings. Adults are attracted to light, and during periods of abundance tremendous numbers aggregate around light sources. Many observers have described the appearance of the moth flights as being equivalent to a blinding snowstorm.

Alfalfa webworm moths are often confused with adults of beet webworm, *Loxostege sticticalis* (Linnaeus). However, they are easily differentiated by viewing the underside of the wings. Both species have a narrow dark line along the distal edge of the wings, but whereas the line is complete in beet webworm is it broken in alfalfa webworm. (See color figure 205.)

The biology of alfalfa webworm was given by Hoerner (1933), Maxson (1948), and Capinera *et al.* (1981a,b). Keys to the moths were included by Munroe (1976) and Capinera and Schaefer (1983). The larva was included in the key to the genus, *Losostege*, by Allyson (1976), and in the field key by Capinera (1986); Capinera *et al.* (1981a) pictured the instars.

Damage

Larvae injure plants by consuming foliage. Young larvae skeletonize leaves, but older larvae consume leaf tissue completely, often leaving only the stems and large veins. They often web together leaves and feed within the clustered foliage. Damage potential varies greatly among crops. For example, larvae consume over 50 sq cm of sugarbeet foliage during their development, but only about 13 sq cm of alfalfa foliage (Capinera *et al.*, 1981b).

Adult alfalfa webworm.

Management

Sampling. The eggs and larvae can be found by visual examination of plants, but both of these life stages are difficult to detect. Moths are attracted to light and can be captured in light traps. Adults also can be flushed during the day, especially toward evening.

Insecticides. Alfalfa webworm populations are usually suppressed by application of insecticide to foliage. *Bacillus thuringiensis* affords some control. Populations are infrequently damaging, so insecticides should not be applied unless high densities of adults or larvae are observed.

Cultural Practices. Several practices can alleviate webworm damage. Tillage can disrupt and destroy overwintering larvae within their silken tubes in the soil. Destruction of preferred weeds before adult oviposition flights can minimize the deposition of eggs within crops. Destruction of weeds after eggs hatching, however, tends to drive larvae to nearby crop plants. Construction of deep furrows with steep or slippery sides has sometimes been recommended to stop the advance of dispersing larvae.

Beet Webworm
Loxostege sticticalis (Linnaeus)
(Lepidoptera: Pyralidae)

Natural History

Distribution. Beet webworm is found in the northern regions of Europe and Asia, and apparently is an immigrant from Europe. In the United States and Canada, beet webworm is present from coast to coast, but is a pest principally in western sugarbeet-growing areas from Alberta and Manitoba in the north to Utah and Kansas in the south (Pepper, 1938).

Host Plants. Known principally as a pest of sugarbeet, this insect also feeds readily on table beet and chard. During periods of abundance over 80 species of plants are damaged, but infestation is normally limited to *Beta vulgaris* and to certain weeds. Among the vegetables occasionally injured are cabbage, cantaloupe, carrot, cucumber, garlic, lettuce, mustard, onion, pea, potato, pumpkin, rhubarb, spinach, and turnip. Grasses, including corn, are rarely eaten. Weeds readily consumed include lambsquarters, *Chenopodium album*; redroot pigweed, *Amaranthus retroflexus*; and Russian thistle, *Salsola kali*. These weeds commonly serve as the preferred oviposition site of moths, with larvae dispersing to other, less preferred plants, when the weeds perish or are consumed. Larvae have been reared successfully on such diverse flora as alfalfa;

onion; lambsquarters; sagebrush, *Artemesia* sp.; sunflower, *Helianthus annuus*; and Canada thistle, *Cirsium arvense* (Pepper and Hastings, 1941).

Natural Enemies. Several parasitoids are known from beet webworm in North America. Numerous other species have been identified in Europe and Asia, but none have been imported. Wasps seem to be the most important mortality agent, though this has been little studied. Among the fly parasitoids are *Aplomya caesar* (Aldrich), *Euphorcera omissa* (Reinhard), *Lespesia archippivora* (Riley), *L. ciliata* (Macquart), and *Stomatomyia parvipalpis* (Wulp) (all Diptera: Tachinidae).

Predators also are detrimental to webworm survival. Among the insect predators are potter wasps (Hymenoptera: Vespidae), digger wasps (Hymenoptera: Sphecidae), robber flies (Diptera: Asilidae), and damsel bugs (Hemiptera: Nabidae). Numerous species of birds, particularly blackbirds (Icteridae), have been cited as contributing to webworm mortality, but no assessments of impact are available for North America.

Weather. Weather has been implicated repeatedly in the development of outbreak populations of beet webworm, and subsequent extensive damage. Damaging populations are generally limited to the Great Plains and Rocky Mountain region by too much precipitation to the east and too little precipitation to the west. Populations survive well in areas with 2.5–6.5 cm of precipitation monthly during the growing season. Also, mean temperature greater than 13°C is deleterious, which limits the southern occurrence of webworms (Pepper, 1938). The condition of pre-pupae as they enter the winter is considered critical. High temperature during autumn hastens larval development, shortens the feeding period, and reduces larval weight. Thus, cool weather during autumn promotes development of large larvae that have high fecundity during the following spring (Bykova, 1984).

Life Cycle and Description. One generation requires 30–40 days, and 3–4 generations occur annually. Moths first appear in June; thereafter, all stages of development are present until cold weather. Mature larvae overwinter in the soil, and though few larvae from the first generation diapause, an increasing proportion of larvae from each generation enters diapause as the season progresses.

Egg. Beet webworm eggs are flattened, oval in shape and measure about 1 mm long and 0.7 mm wide. They occasionally are deposited singly, but more often in small clusters, usually in a single row, with individual eggs overlapping slightly. They are usually found on the underside of leaves, though females sometimes deposit eggs near succulent plants on dry twigs and clods of soil. Females each deposit 200–300 eggs, and

oviposit over a broad temperature range of about 20–32°C. Duration of the egg stage under field conditions is normally 3–5 days.

Larva. Larvae are mostly green or yellowish green, but sometimes darker. A pronounced dark stripe is located dorsally, and a broken dark stripe on each side; each dark stripe is bordered on each side by a white stripe. Despite the presence of the stripes, perhaps the most striking feature is the numerous white circular spots on the body segments. Each circular marking consists of a dark spot from which protrudes a hair, and is surrounded by a white ring. Larvae display five instars and grow about 4 mm long at hatching to 20 mm at maturity. Total larval development time is 12, 18, and 29 days when reared at a constant temperature of 32°, 26°, and 22°C, respectively. Under normal field conditions, duration of the larval stage is about 17–20 days. Larval feeding occurs over a temperature range of 15–44°C, but the optimal temperature is about 30°C. Larvae often construct a silken tube leading from a protected area on the host plant or in the soil, to the feeding site, and sometimes web together leaves. This webbing behavior is the basis for the common name. (See color figure 68.)

Pupa. At larval maturity insects enter the soil and construct a silk-lined cell that varies from 2.5–5.0 cm long. The cell is oriented vertically, with the uppermost end within 1 cm of the surface. Pupation occurs within the cell, with pupae changing from yellow to brown as they mature. They measure about 12 mm long. Duration of the pupal period is related to temperature, with development periods of 35, 16, 9, and 6 days when reared at 18°, 22°, 27°, and 32°C. Duration of the pupal stage under normal field conditions is about 11 days. Not all larvae that enter the soil proceed to immediate pupation, as many enter diapause in the prepupal stage. The proportion of each generation entering diapause varies; in Montana the proportion of first generation larvae that enters diapause varies from 0.5–60%. Induction of diapause is related to day length; short day length (less than 13 h of photoperiod)

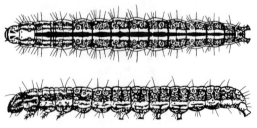

Beet webworm larvae.

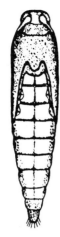

Beet webworm pupa.

Adult beet webworm.

induces diapause (Khomyakova *et al.*, 1986). Pupation of overwintering larvae (prepupae) occurs in the spring. The posterior, pointed end of the pupa bears eight small spines. This character serves to distinguish beet webworm from alfalfa webworm, *Loxostege cerealis* (Zeller), which bears eight small spoon-shaped appendages instead.

Adult. Emergence of moths from the overwintering population occurs in May–July in Montana. In some years, large synchronous emergences follow protracted periods of warm weather. Other years, brief periods of favorable weather interspersed with unfavorable weather result in protracted emergence. The moths are grayish brown in general color, with irregular dark and light markings crossing the forewings. Most prominent of the markings are a dark border distally on the front wing bordered by a cream-colored band. When at rest, the wings are folded back to give the triangular form typically found in the family Pyralidae. The wingspan of beet webworm moths is about 21–22 mm. The moths may disperse in great aggregations, and are attracted to lights. Heavy flights do not necessarily precede high larval populations because infertility is common among females. Adults are commonly seen collecting nectar. A female-produced sex pheromone was identified by Struble and Lilly (1977). (See color figure 206.)

Beet webworm moths are often confused with adults of alfalfa webworm. However, they are easily differentiated by viewing the underside of the wings. Both species have a narrow dark line along the distal edge of the wings, but whereas the line is complete in beet webworm is it broken in alfalfa webworm.

Beet webworm was described by Gillette (1905), Marsh (1912c), Paddock (1912), Pepper and Hastings (1941), and Maxson (1948). Adults were included in

keys by Munroe (1976) and Capinera and Schaefer (1983). The larva was included in the field key by Capinera (1986).

Damage

The first two instars feed on the underside of foliage, skeletonizing the leaves. Large larvae consume holes in foliage, eventually eating all except the principal veins and stems. At high densities, fourth instar or older larvae may disperse long distances in dense aggregations, a behavior typical of "armyworms." It is under these high density, and dispersing conditions that so many plants are destroyed. Damage to crop plants also results when preferred weed species are exhausted and larvae are forced to seek alternate food sources.

Management

Sampling. The eggs and larvae can be found by visual examination of plants, but both of these life stages are difficult to detect. Moths are attracted to light and can be captured in light traps. Adults also can be flushed during the day, especially toward evening. Thus, adult population census is an important element of population monitoring. However, sterility is common in adults, and their presence does not necessarily indicate impending damage; rather, it should serve as a stimulus to initiate careful monitoring of eggs and larvae.

Insecticides. Beet webworm populations are usually suppressed by application of insecticide to foliage. *Bacillus thuringiensis* provides some control. Populations are infrequently damaging, so insecticides should not be applied unless high population densities of larvae are observed.

Cultural Practices. Several practices can alleviate webworm damage. Tillage can disrupt and destroy overwintering larvae within their silken tubes in the soil. Destruction of preferred weeds before adult oviposition flights can minimize the deposition of eggs within crops. Destruction of weeds after egg hatching, however, tends to drive larvae to nearby crop plants. Crops planted into land immediately after alfalfa are at greater risk because not only it is a suitable host but in the later stages of its growth cycle it is often interspersed with numerous weeds.

Cabbage Budworm
Hellula phidilealis (Walker)
(Lepidoptera: Pyralidae)

Natural History

Distribution. Cabbage budworm is a tropical insect. It is a common pest of cabbages in the Caribbean and in Central and South America, and also occurs in West Africa. In the United States, it occurs in Florida, Texas, Arizona, and Hawaii.

Host Plants. Budworm attacks several crucifer crops, including cabbage, cauliflower, Chinese cabbage, radish, and turnip.

Natural Enemies. This insect has not been well studied, and few natural enemies are known. The parasitic fly *Nemorilla maculosa* Meigen (Diptera: Tachinidae) is reported to parasitize budworm.

Life Cycle and Description. Total generation time of cabbage budworm is about 30 days, and it breeds continuously in tropical areas, including southern Florida.

Egg. The eggs are deposited singly or in small clusters, often along the leaf midrib. The oval eggs are white when first deposited, but soon turn brown, and hatch in about three days. They measure about 0.5 mm in diameter, and bear a longitudinal ridge. The female produces 3–4 egg clusters, producing eggs at about 16 per day over a period of about five days. Total egg production averages about 65 per female, but may reach 160 per female.

Larva. Newly emerged larvae feed initially on the lower surface of the leaf, and then tunnel into a petiole or leaf vein. Frass is ejected from the entrance hole as the larvae feed. Feeding tunnels are lined with silk threads. The larval developmental period averages about 16 days. There usually are six instars, with average development times of 4.2, 2.2, 2.9, 2.1, 2.3, and 5.6 days, respectively. Larvae are creamy white with three reddish-brown longitudinal stripes dorsally. At maturity, they measure about 10–14 mm long. The head capsule is black. About two days before pupation the larvae cease feeding and spin a whitish silken cocoon, and pupate within.

Pupa. Pupation may occur on the plant, near the exit hole, or nearby in the soil. The pupa is about 9 mm long, yellow-brown in color, and covered with a gray waxy secretion. The pupal period is about 10 days.

Adult. Adults are dimorphic. The males are dark brown with an undulating wavy fringe on the forewings, whereas the females are light brown and lack the fringe. The hind wings are whitish, becoming darker at the margins. The moths are quite similar in appearance to cabbage webworm, *Hellula rogatalis* (Hulst), but they lack the yellowish tint on the forewings. Adult longevity averages 10 days, and the moths are nocturnal.

The biology of cabbage budworm was given by Cadogan (1983) and King and Saunders (1984).

Damage

Larvae bore into leaf stalks, stems, and the growing points of plants. Destruction of the growing point causes the plant to produce several small heads rather than one large one; these small heads have no commercial value. Damage to stems results in stunted plants. It is considered to be a serious pest in parts of the Caribbean, but elsewhere it is only occasionally important. It is a pest of small plots and gardens, particularly at low altitude.

Management

Effective insecticidal control, including use of *Bacillus thuringiensis*, requires good penetration into the foliage canopy. As insects quickly bore into plant tissue, it is important to treat before the infestation becomes severe. Sanitation also is very important. Many problems stem from crop residues or volunteer plants, especially Chinese cabbage.

Cabbage Webworm
Hellula rogatalis (Hulst)

Oriental Cabbage Webworm
Hellula undalis (Fabricius)
(Lepidoptera: Pyralidae)

Natural History

Distribution. Cabbage webworm, *Hellula rogatalis*, apparently is an American insect, and is found in the southern states from Virginia and Florida to California. Occasionally it is reported from a more northern location, such as Nova Scotia, but is not a pest in northern climates. When first studied in Georgia in the late 1800s, it was confused with the closely related *H. undalis*. This latter species, sometimes called the oriental cabbage webworm, is found in tropical and subtropical areas of Europe, Africa, and Asia. It also is found in Hawaii.

Host Plants.   These insects feeds on several crucifer crops, including broccoli, cabbage, collards, kale, kohlrabi, mustard, radish, rutabaga, and turnip. Studies by Latheef and Irwin (1983) failed to demonstrate significant preference among collards, kale, mustard, and turnip. Weed hosts include shepherdspurse, *Capsella bursa-pastoris*; and purslane, *Portulaca oleracea*.

Natural Enemies.   Despite a report by Chittenden and Marsh (1912) of several fly (Diptera: Tachinidae) and wasp (Hymenoptera: Braconidae and Ichneumonidae) parasitoids, more recent studies have failed to identify natural enemies of cabbage webworm (Ru and Workman, 1979; Kok and McAvoy, 1989).

Life Cycle and Description.   Cabbage webworm generation time is 43, 34, and 23 days at 26°, 30°, and 35°C, respectively. They breed continuously in southern Florida and Hawaii, but in more temperate areas such as Virginia they become numerous enough to cause damage only during the autumn months.

Egg.   The eggs are gray or yellowish-green when first deposited, but turn pink as embryonic development proceeds. The flattened eggs, which are often marked with a distinct nipple at one end, measure about 0.3 mm wide and 0.5 mm long. Oviposition does not occur at 15°C, though eggs incubated at this temperature can hatch. Temperatures of 20–30°C are suitable both for oviposition and egg hatching. Eggs hatch after about three days. They are normally deposited singly or in small masses on the terminal leaves.

Larva.   The first and second instars feed singly as a leaf miner between the upper and lower epidermis. Initially, the larva is yellowish gray, but soon takes on an appearance resembling the mature larva. At about the third instar, larvae begin to web and fold the foliage. Much of the damage occurs on the lower surface of the leaf, but large larvae also feed on the leaf midrib. There are five instars. The head capsule widths are 0.2, 0.3, 0.5, 0.8, and 1.2 mm, respectively. Average body length for the five instars is 0.9, 2.4, 4.8, 6.4, and 9.9 mm, respectively. Development time of the larval instars at 30°C is 2.8, 2.4, 1.9, 2.4, and 4.3 days, respectively. The mature larva is yellowish-gray, with five brownish purple longitudinal bands running the length of the insect, and attains a maximum length of 15 mm. The head capsule is black. The body is sparsely covered with moderately long yellow or light brown hairs, and tapers at both the anterior and posterior ends. (See color figure 69.)

Cabbage webworm larva.

Pupa.   Pupation occurs in the soil, in a webbed cocoon comprising grains of soil. The duration is 5.0–5.5 days at 30°C. The pupa is yellowish-brown, and measures about 7–9 mm long.

Adult.   The moth consists of yellowish-brown front wings which are marked with white bands and a dark kidney-shaped spot. The hind wings are grayish white, though the margin is darker. The wingspan of the moth is about 18–21 mm. Oviposition begins about 3–5 days after moths emerge. About 150–300 eggs are produced by each female. Adult longevity is normally 7–14 days.

The biology of cabbage webworm was presented by Chittenden and Marsh (1912) and McAvoy and Kok (1992).

The oriental cabbage webworm, *H. undalis*, is very similar in appearance to *H. rogatalis* (Munroe, 1972). The genitalia are used to distinguish these species. The biology of *H. undalis* was given by Youssef *et al.* (1973) and Sivapragasam and Aziz (1992), and is virtually identical to that of *H. rogatalis*.

Damage

Larvae initially mine the leaves, eventually webbing and rolling the foliage. They may cause enough damage to destroy the growing tip of plants. Thus, the problem is most severe with young plants (Smith and Brubaker, 1938). Webworms also sometimes burrows into veins, causing death of the leaf beyond the point of feeding.

Cabbage webworm is normally a minor component of the lepidopterous defoliator complex of crucifers. However, Latheef and Irwin (1983) suggested that it had the potential to become one of the most serious pests in Virginia, especially of autumn-grown crops. Kok and McAvoy (1989) reported that webworm com-

Adult cabbage webworm.

prising 43% of the larvae affected broccoli during a 1987 study in Virginia, making it the most abundant defoliator.

Management

Insecticidal control can be difficult owing to the cryptic feeding behavior of the larvae, and the tendency to feed on the rapidly expanding terminal growth. To keep the terminal tissue protected with chemical insecticides or *Bacillus thuringiensis*, frequent applications are required, at least once in a week. Evaluation of trap crops in India (Srinivasan and Moorthy, 1992), using early planted mustard to attract insects and thereby to reduce damage to cabbage, was effective for both diamondback moth, *Plutella xylostella* (Linnaeus) and oriental cabbage webworm, *H. undalis*; this is likely effective for *H. rogatalis,* also. In Virginia, late-maturing cultivars tend to be more heavily infested than early maturing varieties (Vail *et al.,* 1991).

Celery Leaftier
Udea rubigalis (Guenée)

False Celery Leaftier
Udea profundalis (Packard)
(Lepidoptera: Pyralidae)

Natural History

Distribution. Celery leaftier, also known as "greenhouse leaftier," is found throughout the United States. Its distribution is favored by its adaptability to both indoor- and outdoor-plant cultivation. As a celery pest, it has proved to be numerous, and destructive, in all major celery-growing regions including California, Florida, Michigan, and New York. It is known to cause damage in the field in southern Ontario, but in most of Canada it is known only as a greenhouse pest. This insect also occurs in Central and South America.

False celery leaftier, which occurs only on the Pacific Coast from British Columbia to California, is a much less important pest, and is poorly known. It closely resembles celery leaftier.

Host Plants. In the field, celery leaftier is principally a pest of celery, but also has damaged sugarbeet and lettuce, and feeds on bean, beet, cabbage, cauliflower, kale, parsley, and probably other crops. In the greenhouse, its host range is quite large, and includes ageratum, anemone, calendula, carnation, cineraria, cucumber, dahlia, daisy, geranium, lettuce, sweetpea, snapdragon, rose and violet, but chrysanthemum is most important. Plants in the family Compositae seem preferred. A long list of greenhouse

hosts was given by Weigel *et al.* (1924). However, neither the flowers nor the vegetables are usually attacked under field conditions, and Ball *et al.* (1935) suggested that the luxuriance of the forced, greenhouse-grown plants was the most important factor in allowing larvae to develop on these plants.

Weeds are important hosts. In Florida, key elements in celery leaftier biology are redroot pigweed, *Amaranthus retroflexus*, and spiny amaranth, *A. spinosus*. These species and to a lesser extent several other weed hosts are important in maintaining the leaftiers through the celery-free summer period. Other known hosts are plantain, *Plantago* spp.; wild lettuce, *Lactuca* spp.; cowslip, *Caltha palustris*; tickweed, *Verbesina virginica*; and water hemp, *Acnida* sp. In California, sugarbeet similarly provides a suitable host during the celery-free summer months. Essig (1934) reported that false celery leaftier fed extensively on hawksbeard, *Crepis* sp.

Natural Enemies. Studies in Florida documented the importance of several insect parasitoids of celery leaftier (Ball *et al.,* 1935), though the most important was an egg parasite, *Trichogramma* sp. (Hymenoptera: Trichogrammatidae). It could account for 70–90% mortality, but was only effective during warm weather. During the summer months the egg parasites also use *Amaranthus*-infesting webworms as hosts. Of lesser importance were the larval parasitoids, *Casanaria infesta* Cresson (Hymenoptera: Ichneumonidae) and *Cotesia marginiventris* (Cresson) (Hymenoptera: Braconidae). These normally accounted for less than 5% mortality.

Ball *et al.* (1935) provided unusually good documentation on the importance of birds as predators of celery leaftier in Florida. They contended that during the spring northward migration, birds faced difficulty in finding adequate food supplies. Thus, celery leaftier and celery looper, *Anagrapha falcifera*, which can be abundant in celery during the spring months, are a prime food source. Small fields were often maintained free of caterpillars by birds, principally palm warbler, *Dendroica palmarum*, and tree swallow, *Tachycineta bicolor*.

Life Cycle and Description. Under ideal conditions the celery leaftier may complete its life cycle in a month, and this occurs under greenhouse conditions. Under field conditions, development is delayed, depending on weather. In California, a generation requires one month in the summer, two months in the spring and autumn, and over three months during the winter. Thus, five to six generations occur annually, with three to four during the period of June–December and two during the remainder of the year (Campbell, 1927). In Illinois, four generations occur under field conditions.

Egg. Celery leaftier eggs are about 0.8 mm long and 0.6 mm wide, spherical, and slightly flattened. The eggs are shiny, and whitish initially, growing darker as they mature. The eggs are deposited on the underside of leaves singly or in small groups of up to 12. As is common with pyralids, the eggs overlap one another. The incubation period is usually about six days (range 4–10 days). Most eggs are deposited just below the outer layer of plant canopy.

Larva. The larvae are pale yellowish-white and measure only about one mm long initially, but soon become pale green, with a single dark green line dorsally and bordered on each side by a broad whitish band. This striped pattern persists throughout the larval period. The color of the ventral surface of the larva tends toward yellow. The head, thoracic shield, and thoracic legs are pale, though the shield bears a dark, oval spot on each side. The larvae are sparsely covered with long hairs. Larvae feed on the lower surface of leaves, often rolling leaves or webbing them together. In most respects, larvae behave like other webworms, including the tendency to retreat into their webbed shelter or to wriggle violently and spin down from the plant on a strand of silk if disturbed. At maturity, the larvae measure only about 17–19 mm long. There are five larval instars, each requiring 2–5 days for completion. The total development time of the larval stage averages about 21 days (range 15–30 days). Head capsule widths for the instars are about 0.20, 0.27, 0.40, 0.67, 1.05 mm, respectively.

Pupa. Pupation normally takes place in a thin, whitish silk cocoon amongst folded leaves. The pupa is smoky brown, and measures 8.5–9.0 mm long. Dura-

tion of the pupal stage averages 10 days (range 6–16 days).

Adult. The moth of celery leaftier is small, the wingspan measuring only 15–21 mm. The front wings are light-brown, suffused with reddish-brown, and are irregularly marked with black lines. The hind wings are grayish, becoming brown distally. The outer margin of the fore and hind wings are marked with a row of small, dark spots. Adults live for several days; males tend to survive about 4–5 days and females 9–10 days. They are nocturnal, and remain hidden during the daylight hours. Mating occurs almost immediately upon adult emergence, and oviposition begins within 24 h of emergence. Eggs, which tend to be deposited principally during the first few days of adulthood, are deposited at night.

The eggs, larvae, pupae, and moths of false celery leaftier are not easily differentiated from celery leaftier, and in nearly all respects the biology is the same. Munroe (1966) gave characters to distinguish the moths. False celery leaftier averages slightly larger than celery leaftier, the wingspan ranging from 19–25 mm. The larvae also are slightly larger. Mean head capsule widths for the five instars are 0.23, 0.34, 0.54, 0.82, and 1.23 mm, respectively.

The field ecology of celery leaftier was best described by Ball *et al.* (1935); developmental biology was provided by Fletcher and Gibson (1901) and Weigel *et al.* (1924). False celery leaftier is poorly studied; Tamaki and Butt (1977) gave the only substantive biological information on this insect.

Damage

Larvae feed mostly on the underside of foliage. Depending on the thickness of the plant tissue they may eat the entire leaf, only the lower surface, or leave the foliage skeletonized. They also cover the leaves with a thin layer of silk. The silk webbing may be used to draw portions of the leaf together. Celery leaftier larvae may work their own way down to the youngest tissue, the "heart" of celery, and feed on both leaves

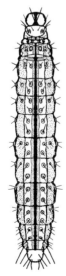

Celery leaftier larva.

Adult celery leaftier.

and petioles (Stone *et al.*, 1932). Jones and Granett (1982) indicated that false celery leaftier also damaged the heart of celery in California.

Management

Moths of celery leaftier are attracted to phenyl-acetaldehyde, and this chemical may be useful with some types of traps for population monitoring (Cantelo *et al.*, 1982). They also come to light traps. Foliar insecticides are applied for leaftier suppression, and growers are encouraged to target the young larvae for control because they are easier to kill. This insect is only occasionally threatening, however.

Cross-striped Cabbageworm
Evergestis rimosalis (Guenée)
(Lepidoptera: Pyralidae)

Natural History

Distribution. Cross-striped cabbageworm is recorded principally from North America east of the Rocky Mountains, but is infrequent in the northernmost states and eastern Canada. This insect is known as a common pest only in the southeastern states and in Central America, and even in the southeastern states it usually is a minor component of the defoliator complex. It is likely a native insect, though its close relatives are native to Europe. Cross-striped cabbageworm was accidentally introduced to Australia and Jamaica.

Host Plants. Broccoli, Brussels sprouts, cabbage, collards, kale, rutabaga, and turnip are among the vegetable crops known as hosts. Brussels sprouts and collards tend to support relatively large numbers of cabbageworms, and cabbage and kale relatively few. Other crucifers such as rape, and many weeds, likely support this insect.

Natural Enemies. Several parasitoids are known, including *Cotesia congregata* Say, *C. xylina* (Say), and *C. orobenae* Forbes (all Hymenoptera: Braconidae). Except for *C. orobenae*, the importance of the parasitoids has not been thoroughly assessed, and predators and diseases are undocumented. Mays and Kok (1997) reported on *C. orobenae*, and showed that the occurrence of this parasitoid closely tracked the abundance of its host.

Life Cycle and Description. Development time, from egg to adult stages, ranges from 61 days at 20°C to 18 days at 35°C. The number of annual generations is three in Illinois, and four in Maryland. Cross-striped cabbageworm tends to be most abundant during the autumn generation in the north. In the south,

however, it can be quite abundant during the winter- and spring-cropping period.

Egg. The eggs are laid on the underside of leaves in small masses, usually from 3–5 to about 25. They are yellow, flattened, and overlap slightly. They are oval, measure about 1.2 mm long and 0.9 mm wide. They hatch in about 12.4 days when reared at 20°C and 1.8 days at 35°C, hatching most readily at 20–30°C.

Larva. Four instars are reported for this species, with mean head capsule widths (range) of 0.34 (0.32–0.38), 0.56 (0.52–0.60), 1.04 (0.98–1.10), and 1.66 (1.58–1.71) mm, respectively, for instars 1–4. Mean duration of the instars, when cultured at 25°C, is 3.0, 2.0, 2.0, and 5.8 days, respectively. Small larvae are gray, with black tubercles bearing stout hairs. Larger larvae are bluish-gray dorsally, and with numerous transverse black bands. Transverse lines are relatively rare among caterpillars, and serve as a key diagnostic feature for this insect. The prominent tubercles are gray, marked with black. A yellow line occurs along each side of the caterpillar. The underside of the caterpillar is green, mottled with yellow. The mature larva measures about 15–17 mm long. Larval development requires 2–3 weeks. (See color figure 70.)

Pupa. Transformation into a pupa occurs in the soil, near the surface, in a small cocoon covered with particles of sand. The yellowish-brown pupa measures about 10–12 mm long. Duration of this stage is 9–11 days when reared at 30–20°C.

Adult. The adult has a wingspan of about 25 mm. The front wings are straw-colored, but also marked with olive or purplish brown, and crossed by narrow transverse lines. The hind wings are transparent and whitish, with a darker band at the margin. The pre-oviposition period is 3–6 days, and the oviposition period 6–14 days, when reared at 20–30°C. Females oviposit readily at 20–30°C, but oviposition frequency decreases at higher and lower temperatures. Adults live for over 20 days if held under cool conditions, but survive for only 5–11 days at 30°C or above. (See color figure 207.)

Cross-striped cabbageworm larva.

Adult cross-striped cabbageworm.

The biology of cross-striped cabbageworm was given by Chittenden (1902) and Mays and Kok (1997). A key including the adult stage of cross-striped cabbageworm was contained in Munroe (1973). Cross-striped cabbageworm is included in the key to "cabbageworms" in Appendix A.

Damage

Larvae feed on foliage, creating small holes. They prefer the terminal buds. If disturbed, they have a tendency to drop from the foliage on a silken thread. Larvae also may burrow into the center of developing heads.

Management

Cross-striped cabbageworm, along with several other lepidopterous defoliators, exhibits a positive response to nitrogen fertilization of host plants. However, indications are that unlike the case with piercing-sucking insects, it is not the increased nitrogen levels *per se*, but the increased foliar biomass that favors the abundance of the cabbageworms (Jansson *et al.*, 1991b). In general, methods used for management of imported cabbageworm, *Pieris rapae* (Linnaeus), are appropriate for this insect.

Diamondback Moth
Plutella xylostella (Linnaeus)
(Lepidoptera: Plutellidae)

Natural History

Distribution. Diamondback moth is probably of European origin but has become rather cosmopolitan, and is now found throughout the Americas and in Europe, Southeast Asia, Australia, and New Zealand. It was first observed in North America in 1854, in Illinois, but spread quickly. It was found in Massachusetts and Maryland by 1870, had spread to Florida and the Rocky Mountains by 1883, and was reported from British Columbia by 1905. In North America, dia-

mondback moth is now recorded everywhere that cabbage is grown, even as far north as Canada's Northwest Territories. By virtue of its ability to feed on cruciferous weeds, diamondback moth is sometimes abundant even in some areas where cruciferous crops do not occur. However, it is highly dispersive, and is often found in areas where it cannot successfully overwinter, including most of Canada (Beirne, 1971).

Host Plants. Diamondback moth attacks only plants in the family Cruciferae. Virtually all cruciferous vegetable crops are eaten, including broccoli, Brussels sprouts, cabbage, Chinese cabbage, cauliflower, collards, kale, kohlrabi, mustard, radish, turnip, and watercress. Not all are equally preferred, however, and collards usually chosen by ovipositing moths relative to cabbage. Several weeds are important hosts, especially early in the season before cultivated crops are available. Yellow rocket, *Barbarea vulgaris*; shepherdspurse, *Capsella bursa-pastoris*; pepperweed, *Lepidium* spp.; and wild mustards, *Brassica* spp. are commonly cited as important weed hosts.

Natural Enemies. A comprehensive analysis of diamondback moth mortality factors was conducted in Ontario by Harcourt (1960, 1963b). Large larvae, pre-pupae, and pupae often were killed by the parasitoids *Microplitis plutellae* (Muesbeck) (Hymenoptera: Braconidae), *Diadegma insulare* (Cresson) (Hymenoptera: Ichneumonidae), and *Diadromus subtilicornis* (Gravenhorst) (Hymenoptera: Ichneumonidae). All are specific on *P. xylostella*. In Ontario, *D. insulare* was considered most important except during diamondback moth population outbreaks when the other species assumed greater importance. More recent studies conducted in Quebec are consistent with the Ontario studies (Godin and Boivin, 1998b). *Diadegma insulare* was also important in California (Oatman and Platner, 1969; Kennedy and Oatman, 1976). Nectar produced by wildflowers is important in determining parasitism rates by *D. insulare* (Idris and Grafius, 1995). Egg parasites are poorly known. However, a *Trichogramma* egg parasitoid is reported in Japan (Wakisaka *et al.*, 1992) and also from Florida (Leibee, pers. comm.). Fungi, granulosis virus, and nuclear polyhedrosis virus sometimes occur in high density diamondback moth larval populations.

Weather. Harcourt (1960, 1963b) and Wakisaka *et al.* (1992) found that a large proportion of young larvae were often killed by rainfall. However, the most important factor determining population trends was reported to be adult mortality. Adult survival was thought to be principally a function of weather, though this hypothesis has not been examined rigorously.

Life Cycle and Description. Total development time from the egg to pupal stage averages 25–30 days, depending on weather (range about 17–51 days). In Ontario, diamondback moth is present from May to October, but is most abundant in July and September. There are 4–6 generations annually, but discrete broods are not apparent (Harcourt, 1957). In Quebec, Godin and Boivin (1998a) reported adults and eggs beginning in early June, and estimated 3–4 generations annually. In Colorado, the number of annual generations is estimated to be seven, and overwintering survival is positively correlated with the abundance of snowfall (Marsh, 1917). There is continuous breeding in the southern states, so the number of generations is likely 12–15 per year.

Egg. Diamondback moth eggs are oval and flattened, and measure 0.44 mm long and 0.26 mm wide. They are yellow or pale-green, and are deposited singly or in small groups of 2–8 eggs in depressions on the surface of foliage, or sometime on other plant parts. Females may deposit 250–300 eggs early in the year, but the number decreases in later generations by 90%; average total production is probably 150 eggs. Development time averages 5.6 days (range 4–8 days).

Larva. Diamondback moth has four instars. Average and range of development time is about 4.5 (3–7), 4 (2–7), 4 (2–8), and 5 (2–10) days, respectively. Throughout their development, larvae remain quite small and active. If disturbed, they often wriggle violently, move backward, and spin down from the plant on a strand of silk. Overall length of each instar rarely exceeds 1.7, 3.5, 7.0, and 11.2 mm, respectively, for instars 1–4. Mean head capsule widths for these instars are about 0.16, 0.25, 0.37, and 0.61 mm. Larval body form tapers at both ends, and a pair of prolegs protrudes from the posterior end, forming a distinctive "V". The larvae are colorless in the first instar, but thereafter they become green. The body bears relatively few hairs, which are short in length, and most are marked by the presence of small white patches. There are five pairs of prolegs. Initially, the feeding habit of first instar larvae is leaf mining, though they are so small that the mines are difficult to notice. The larvae emerge from their mines at the conclusion of the first instar, molt beneath the leaf, and thereafter feed on the lower surface of the leaf. Their chewing results in irregular

patches of damage, and the upper leaf epidermis is often left intact. (See color figure 71.)

Pupa. Pupation occurs in a loose silk cocoon, usually formed on the lower or outer leaves. In cauliflower and broccoli, pupation may occur in the florets. The yellowish pupa is 7–9 mm long. The duration of the cocoon averages about 8.5 days (range 5–15 days), but during this time the insect is in the pre-pupal rather than the pupal stage.

Adult. The adult is a small, slender, grayish-brown moth with pronounced antennae. It is about 6 mm long, and marked with a broad cream or light-brown band along the back. The band is sometimes constricted to form one or more light-colored diamonds on the back, which is the basis for the common name of this insect. When viewed from the side, the tips of the wings can be seen to turn upward slightly. Moths usually mate at dusk, immediately after emergence from the cocoon. Flight and oviposition take place from dusk to midnight, and moths can be found feeding at blossoms on nectar. Adult males and females live about 12 and 16 days, respectively, and females deposit eggs for about 10 days. The moths are weak fliers, usually flying within 2 m of the ground, and not flying long distances. However, they are readily carried by the wind. The adult is the overwintering stage in temperate areas, but moths do not survive cold winters, as is found in most of Canada. They routinely re-invade these areas each spring, evidently aided by southerly winds (Smith and Sears, 1982).

Detailed biology of diamondback moth can be found in Marsh (1917) and Harcourt (1955, 1957, 1963b). A survey of the world literature was published by Talekar *et al.* (1985). Rearing techniques were provided by Biever and Boldt (1971) and Liu and Sun (1984). Chow *et al.* (1974) identified two major components of the sex pheromone, and Chisholm *et al.* (1983) evaluated a four-component mixture, but the exact blend remains undetermined.

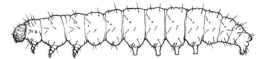

Diamondback moth larva.

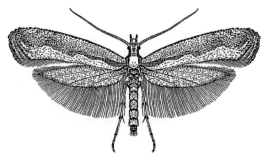

Adult diamondback moth.

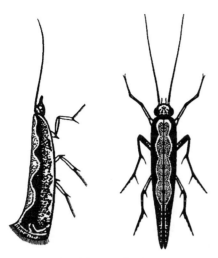

Adults of diamondback moth.

Damage

Damage is caused by larval feeding. Although the larvae are very small, they can be quite numerous, resulting in complete removal of foliar tissue except for the leaf veins. They are particularly damaging to seedlings, and may disrupt head formation in cabbage, broccoli, and cauliflower. The presence of larvae in florets can result in complete rejection of produce, even if the level of plant tissue removal is insignificant.

Diamondback moth was long considered a relatively insignificant pest. Its impact was overshadowed by such serious defoliators as imported cabbageworm, *Pieris rapae* (Linnaeus), and cabbage looper, *Trichoplusia ni* (Hübner). However, in the 1950s the general level of abundance began to increase, and by the 1970s it became troublesome to crucifers in some areas. Although this shift in abundance has been attributed to increased availability of alternate weed hosts or destruction of parasitoids, insecticide resistance was long suspected to be a component of the problem. This was confirmed in the 1980s as pyrethroid insecticides began to fail, and soon thereafter virtually all insecticides were ineffective (Leibee and Capinera, 1995). Relaxation of insecticide use, which can be implemented by use of thresholds to trigger applications, rotation with *Bacillus thuringiensis* and particularly elimination of pyrethroid use, can return diamondback moth to minor pest status by favoring survival of parasitoids.

Management

Sampling. Populations are usually monitored by making counts of larvae, or by the level of damage. In Texas, average population densities of up to 0.3 larvae per plant are considered to be below the treatment threshold (Kirby and Slosser, 1984; Cartwright *et al.*, 1987). In Florida and Georgia, treatment is recommended only when damage equals or exceeds one hole per plant (Workman *et al.*, 1980). When growers monitor fields and subscribe to these treatment thresholds rather than trying to prevent any insects or damage from occurring in their fields, considerably fewer insecticide applications are needed to produce a satisfactory crop. Harcourt (1961) studied the distribution of various life stages on cabbage, and recommended a minimum plant sample size of 40–50 except for the egg stage, where 150 plants should be examined for accurate population estimates. Shelton *et al.* (1994) compared the benefits of sequential and variable-intensity sampling for diamondback moth management, and recommended the latter as being more reliable and requiring fewer samples.

Pheromone traps can be used to monitor adult populations, and may predict larval populations 11–21 days later. However, because of variation among locations, each crop field requires independent evaluation (P. B. Baker *et al.*, 1982).

Insecticides. Protection of crucifer crops from damage often requires application of insecticide to plant foliage, sometimes as frequently as twice per week. However, resistance to insecticides is widespread, and includes most classes of insecticides including some *Bacillus thuringiensis* products. Rotation of insecticide classes is recommended, and the use of *B. thuringiensis* is considered especially important because it favors survival of parasitoids. Even *B. thuringiensis* products should be rotated, and current recommendations generally suggest alternating the *kurstaki* and *aizawa* strains because resistance to these microbial insecticides occurs in some locations (Tabashnik *et al.*, 1990). Mixtures of chemical insecticides, or chemicals and microbials, are often recommended for diamondback moth control (Leibee and Savage, 1992). This is due partly to the widespread occurrence of resistance, but also because pest complexes often plague crucifer crops, and the insects vary in susceptibility to individual insecticides.

Sex pheromone, though not usually considered to be an insecticide, can also be used as a chemical crop protectant. Continuous release of pheromone has been investigated as a technique to suppress diamondback moth mating activity. McLaughlin *et al.* (1994) reported that the number of insecticide treatments could be reduced from 13–15 to only three when crops were grown in the continuous presence of diamondback moth pheromone.

Cultural Practices. Rainfall has been identified as a major mortality factor for young larvae, so it is not surprising that crucifer crops with overhead sprinkle irrigation tend to have fewer diamondback moth larvae than drip or furrow-irrigated crops. Irrigation also tends to disrupt oviposition. Best results were obtained with daily evening applications (McHugh and Foster, 1995).

Crop diversity can influence abundance of diamondback moth. Larvae generally are fewer in number, and more heavily parasitized, when crucifer crops are interplanted with another crop or when weeds are present. This does not necessarily lead to reduction in damage, however (Bach and Tabashnik, 1990). Surrounding cabbage crops with two or more rows of preferred hosts such as collards and mustard can delay or prevent the dispersal of diamondback moth into cabbage crops (Srinavasan and Moorthy, 1992).

Crucifer transplants are often shipped long distances before planting, and diamondback moth may be included with the transplants. In the United States, many transplants are produced in the southern states, and then moved north as weather allows. Cryptic insects such as young diamondback moth larvae are sometimes transported, and inoculated in this manner. The transport of insecticide-resistant populations also may occur (Wyman, 1992). Every effort should be made to assure that transplants are free of insects before to planting.

Host-Plant Resistance. Crucifer crops differ somewhat in their susceptibility to attack by diamondback moth. Mustard, turnip, and kohlrabi are among the most resistant crucifers, but resistance is not as pronounced as it is for imported cabbageworm and cabbage looper (Radcliffe and Chapman, 1966). Varieties also differ in susceptibility to damage by diamondback moth, and a major component of this resistance is the presence of leaf wax. Glossy varieties, lacking the normal waxy bloom and therefore green rather than grayish green, are somewhat resistant to larval feeding (Dickson *et al.*, 1990; Stoner, 1990; Eigenbrode *et al.*, 1991). Larvae apparently spend more time searching, and less time feeding, on glossy varieties. Glossy varieties also tend to have fewer imported cabbageworm larvae and cabbage aphids, but more cabbage flea beetles.

European Corn Borer
Ostrinia nubilalis (Hübner)
(Lepidoptera: Pyralidae)

Natural History

Distribution. First found in North America near Boston, Massachusetts in 1917, European corn borer quickly spread to the Great Lakes region. By 1948 it

was established throughout the midwestern corn-growing region and eastern Canada. It now has spread as far west as the Rocky Mountains in both Canada and the United States, and south to the Gulf Coast states. European corn borer is thought to have originated in Europe, where it is widespread. It also occurs in northern Africa. The North American European corn borer population is thought to have resulted from multiple introductions from more than one area of Europe. Thus, there are at least two, and possibly more, strains present. The presence of an eastern or New York strain, and a midwestern or Iowa strain, is evident because different pheromone blends are required to capture moths from each population. Both strains sometimes occur in the same area (Eckenrode *et al.*, 1983).

Host Plants. European corn borer has a very wide host range, attacking practically all robust herbaceous plants with a stem large enough for the larvae to enter. However, the eastern strain accounts for most of the wide host range, the western strain feeding primarily on corn. Among vegetable crops injured are beet, broccoli, celery, corn, cowpea, eggplant, lima bean, pepper, potato, rhubarb, snap bean, spinach, Swiss chard, and tomato. Vegetables other than corn tend to be infested if they are abundant before corn is available, or late in the season when senescent corn becomes unattractive for oviposition; snap and lima beans, pepper, and potato are especially damaged. In North Carolina, for example, potato is more attractive than corn at peak emergence of the first moth flight, and more heavily damaged (Anderson *et al.*, 1984). Other crops sometimes attacked include buckwheat, grain corn, hop, oat, millet, and soybean, and such flowers as aster, cosmos, dahlia, gladiolus, hollyhock, and zinnia. Corn is the most preferred host, but many thick-stemmed weeds and grasses also support European corn borer, especially if they are growing amongst, or adjacent to, corn. Some of the common weeds infested include barnyardgrass, *Echinochoa crusgalli*; beggarticks, *Bidens* spp.; cocklebur, *Xanthium* spp.; dock, *Rumex* spp.; jimsonweed, *Datura* spp.; panic grass, *Panicum* spp.; pigweed, *Amaranthus* spp.; smartweed, *Polygonum* spp.; and others. A good list of host plants was given by Caffrey and Worthley (1927).

Natural Enemies. Native predators and parasites exert some effect on European corn borer populations, but imported parasitoids seem to be more important. Among the native predators that affect the eggs and young larvae are the insidious flower bug, *Orius insidious* (Say) (Hemiptera: Anthocoridae); green lacewings, *Chrysoperla* spp. (Neuroptera: Chrysopidae); and several lady beetles (Coleoptera: Coccinellidae) (Jarvis and Guthrie, 1987; Andow, 1990). Insect predators often eliminate 10–20% of corn borer eggs. Avian predators such as downy woodpecker, *Dendrocopos pubescent* (Linnaeus); hairy woodpecker, *D. villosus* (Linnaeus); and yellow shafted flicker, *Colaptes auratus* (Linnaeus) have been known to eliminate 20–30% of overwintering larvae.

Native parasitoids include *Bracon caulicola* (Gahan), *B. gelechiae* Ashmead, *B. mellitor* Say, *Chelonus annulipes* Wesmael, *Macrocentrus delicatus* Cresson, and *Meteorus campestris* Viereck (all Hymenoptera: Braconidae); *Gambrus ultimus* (Cresson), *G. bituminosus* (Cushman), *Itoplectis conquisitor* (Say), *Campoletis flavicincta* (Ashmead), *Nepiera oblonga* (Viereck), *Rubicundiella perturbatrix* Heindrich, *Vulgichneumon brevicinctor* (Say) (all Hymenoptera: Ichneumonidae); *Dibrachys carus* (Walker) and *Eupteromalus tachinae* Gahan (both Hymenoptera: Pteromalidae); *Syntomosphyrum clisiocampe* (Ashmead) (Hymenoptera: Eulophidae); *Scambus pterophori* (Ashmead) (Hymenoptera: Hybrizontidae); *Trichogramma nubilale* Ertle and Davis and *T. minutum* Riley (both Hymenoptera: Trichogrammatidae); and *Archytas marmoratus* (Townsend) and *Lixophaga* sp. (both Diptera: Tachinidae). Although many species of native parasitoids are known, native parasitoids rarely cause high levels of corn borer mortality.

Exotic parasitoids numbering about 24 species have been imported and released to augment native parasitoids. About six species have successfully established. Among the potentially important species is *Lydella thompsoni* Herting (Diptera: Tachinidae), which may kill up to 30% of second generation borers in some areas, but has disappeared or gone into periods of low abundance in other areas. Other exotic parasitoids that sometimes account for more than trivial levels of parasitism are *Eriborus terebrans* Gravenhorst (Hymenoptera: Ichneumonidae), *Simpiesis viridula* (Hymenoptera: Eulophidae), and *Macrocentris grandii* Goidanich (Hymenoptera: Braconidae) (Burbutis *et al.*, 1981; Andreadis, 1982a; Losey *et al.*, 1992). A comprehensive review of biological control agents imported in the first half of the 1900s was published by Baker *et al.* (1949).

Several microbial disease agents are known from corn borer populations. The common fungi *Beauveria bassiana* and *Metarhizium anisopliae* are sometimes observed, especially in overwintering larvae. The most important pathogen seems to be the microsporidian, *Nosema pyrausta*, which often attains 30% infection of larvae and sometimes 80–95%. It creates chronic, debilitating infections that decrease longevity and fecundity of adults, and decreases survival of larvae that are under environmental stress (Hill and Gary, 1979; Andreadis, 1984). Unfortunately, *N. pyrausta* also infects the parasitoid *M. grandii* (Andreadis, 1982b).

Life table studies conducted on corn borer populations in Quebec with a single annual generation perhaps provide insight into the relative importance of mortality factors (Hudon and LeRoux, 1986c). These workers demonstrated that egg mortality (about 15%) was low, stable and due mostly to predators and parasites. Similarly, mortality of young larvae, due principally to dispersal, dislodgement, and plant resistance to feeding was fairly low (about 15%) but more variable. Mortality of large larvae during the autumn (about 22%) and following spring (about 42%) was due to a number of factors including frost, disease and parasitoids, but parasitism levels were low. Pupal mortality (about 10%) was low and stable among generations. The factor that best accounted for population trends was considered survival of adults. Dispersal and disruption of moth emergence by heavy rainfall are thought to account for high and variable mortality (68–98%, with a mean of 95%), which largely determines population size of the subsequent generation. Overall generation mortality levels were high, averaging 98.7%.

Weather. There are many reports that weather influences European corn borer survival. Heavy precipitation during egg hatch, for example, is sometimes given as an important mortality factor (Jarvis and Guthrie, 1987). Low humidity, low nighttime temperature, heavy rain, and wind are detrimental to moth survival and oviposition. However, during a 10-year, three-state study, Sparks *et al.* (1967) reported no consistent relationship between weather and survival.

Life Cycle and Description. The number of generations varies from one to four, with only one generation occurring in northern New England and Minnesota and in northern areas of Canada, three to four generations in Virginia and other southern locations, and usually two generations in the northern United States and southern Canada. In many areas generation number varies depending on weather, and there is considerable adaptation for local climate conditions even within strains. For example, though the developmental rates of single-generation strains are lower than multiple-generation strains, at northern locations such as Prince Edward Island the single-generation strain develops quickly (Dornan and Stewart, 1995). European corn borer overwinters in the larval stage, with pupation and emergence of adults in early spring. Diapause apparently is induced by exposure of last instar larvae to long days, but there also is a genetic component. Moth flights and oviposition usually occur during June–July and August–September in areas with one to two generations. In southern locations with three generations, moth flights and oviposition typically occur in May, late June, and

August. In locations with four generations, adults are active in April, June, July, and August–September.

Egg. The eggs are deposited in irregular clusters of about 15–20 (range 5–50 eggs). They are oval, flattened, and creamy white, usually with an iridescent appearance. They darken to a beige or orangish tan color with age. They normally are deposited on the underside of leaves, and overlap like shingles on a roof or fish scales. The eggs measure about 1.0 mm long and 0.75 mm wide. The developmental threshold for eggs is about 15°C. Eggs hatch in 4–9 days.

Larva. Larvae tend to be light brown or pinkish-gray dorsally, with a brown to black head capsule and a yellowish-brown thoracic plate. The body is marked with round dark spots on each body segment. The developmental threshold for larvae is about 11°C. Larvae normally display six instars, but 4–7 instars have been observed. Head capsule widths are about 0.30, 0.46, 0.68, 1.03, 1.66, and 2.19 mm in instars 1–6, respectively. Mean body lengths during the six instars are about 1.6, 2.6, 4.7, 12.5, 14.5, and 19.9 mm, respectively. For populations with only five instars, mean head capsule widths are 0.29, 0.44, 0.80, 1.27, and 2.00 mm, respectively. Young larvae tend to feed initially within the whorl, especially on the tassel. When the tassel emerges from the whorl, larvae disperse downward where they burrow into the stalk and the ear. Mortality tends to be high during the first few days of life, but once larvae establish a feeding site within the plant survival rates improve. Larvae in the final instar overwinter within a tunnel in the stalk of corn, or in the stem of another suitable host. Duration of the instars varies with temperature. Under field conditions in New York, development time was estimated at 9.0, 7.8, 6.0, 8.8, 8.5, and 12.3 days for instars 1–6, respectively, for a mean total development period of about 50 days. In contrast, during the next year development time at the same site was 4.4, 4.3, 4.6, 5.8, 8.5 and 9.0 days for the six instars, for a mean total larval development period of about 35 days (Caffrey and Worthley, 1927). (See color figure 77.)

Pupa. Pupae usually occur in April or May, and then later in the year if more than one generation occurs. The pupa is normally yellowish-brown. The pupa measures 13–14 mm long and 2–2.5 mm wide in males and 16–17 mm long and 3.5–4 mm wide in females. The tip of the abdomen bears 5–8 recurved spines that are used to anchor the pupa to its cocoon. The pupa is ordinarily, but not always, enveloped in a thin cocoon formed within the larval tunnel. Duration of the pupal stage under field conditions is usually

European corn borer larva.

European corn borer pupa.

about 12 days. The developmental threshold for pupae is about 13°C.

Adult. The moths are fairly small, with males measuring 20–26 mm in wingspan, and females 25–34 mm. Female moths are pale yellow to light brown, with both the forewing and hind wing crossed by dark zigzag lines and bearing pale, often yellowish, patches. The male is darker, usually pale brown or grayish brown, but also with dark zigzag lines and yellowish patches. Secondary host plants and adjacent grassy areas play a significant role in the mating behavior of adults, as adults rest and mating takes place in such areas of dense vegetation, called "action sites." Retention of droplets from rainfall and dew in this dense vegetation stimulates the sexual activity of females.

Moths are most active during the first 3–5 h of darkness. The sex pheromone has been identified as 11-tetradecenyl acetate, but eastern and western strains differ in production of Z and E isomers. The western strain produces a blend that approximates 97:3 Z:E, whereas the eastern strain uses a blend of 3:97 Z:E. The pre-oviposition period averages about 3.5 days. Duration of oviposition is about 14 days, with oviposition averaging 20–50 eggs per day. The female often deposits 400–600 eggs during her life span, though there are also estimates of mean fecundity of about 150 eggs in some locations. Total adult longevity is normally 18–24 days. (See color figures 208 and 209.)

Brindley and Dicke (1963), Brindley *et al.* (1975), Hudon and LeRoux (1986a,b) and Hudon *et al.* (1989) published reviews of the biology and management of European corn borer. Detailed biology was presented in Vinal and Caffrey (1919), and Caffrey and Worthley (1927). Sex pheromone blends were identified (Klun and Robinson, 1971; Kochansky *et al.*, 1975; Showers *et al.*, 1974). Beck (1987) reviewed corn borer seasonal biology, and offered interesting insight into the pheromonal races or strains. Rearing procedures were given by Reed *et al.* (1972). European corn borer was included in the larval key by Capinera (1986) and the moth key by Capinera and Schaefer (1983). A key to stalk borers associated with corn in southern states was presented by Dekle (1976); this publication also included pictures of the adults. A key to pyralid borers

Adult male European corn borer.

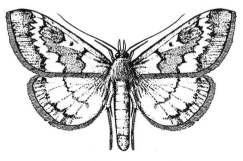

Adult female European corn borer.

was also included by Stehr (1987). A key to common stalk boring caterpillars is included in Appendix A.

Damage

This is a very serious pest of both sweet and grain corn, and before the availability of modern insecticides this insect caused very marked reductions in corn production. Young larvae feed on tassels, whorl, and leaf sheath tissue; they also mine midribs and eat pollen that collects behind the leaf sheath. Sometimes they feed on silk, kernels, and cobs, or enter the stalk. Older larvae tend to burrow into the stalk and sometimes the base of the corn ear, or into the ear cob or kernels. Feeding by older larvae is usually considered to be most damaging, but tunneling by even young larvae can result in broken tassels. The presence of 1–2 larvae within a corn stalk is tolerable, but the presence of any larvae within the ear of sweet corn is considered intolerable by commercial growers, and is their major concern. European corn borer is considered to be the most important sweet corn pest in northern production areas, and second-generation borers are the principal source of ear damage. Heavily tunneled stalks of grain corn suffer from lodging, reducing the capacity for machine harvesting. Lodging is not a serious threat to sweet corn. Boring by corn borers also allows several fungi to affect corn plants.

In crops other than corn, the pattern of damage is variable. European corn borer larvae damage both the stem and fruit of beans, pepper, and cowpea. The temporal occurrence of fruit affects susceptibility to injury, of course; in Wisconsin, snap beans 14–30 days from harvest were susceptible to damage by larvae, but young plants and fruit near harvest suffered little damage (Sanborn *et al.*, 1982b). In celery, potato, rhubarb, Swiss chard, and tomato, it is usually the stem tissue that is damaged. In beet, spinach, and rhubarb, leaf tissue may be injured. Entry of borers into plant tissue facilitates entry of plant pathogens. The incidence of potato blackleg caused by the bacterium *Erwinia carotovora atroseptica*, for example, is higher in potato fields with stems heavily infested by corn borers. Direct damage by corn borers to potato vines, however, results in negligible yield loss (Nault and Kennedy, 1996c).

Management

Sampling. Moths can be sampled with blacklight and pheromone traps, and catches by these traps are correlated (Legg and Chiang, 1984; Welty, 1995). Pheromones attract only males, whereas both sexes are captured in traps with a blacklight. Blacklight traps tend to be more reliable, but light traps can capture many other insects, necessitating a great amount of sorting. Pheromone-baited water pan traps seem to be the most efficient method of adult monitoring (Thompson *et al.*, 1987; Stewart, 1994). Trap catches are usually used to initiate intensive in-field scouting for egg masses, as moth catches are only roughly correlated with density. Plant phenology can be used to predict corn borer development. In New York, for example, peak flight of the first brood of moths corresponds to bloom of elderberry, *Sambucus canadensis*, and peak second brood flight corresponds to peak bloom of hydrangea, *Hydrangea paniulata grandiflora* (Straub and Huth, 1976). Thermal summations are also highly predictive (Jarvis and Brindley, 1965). Moths seek shelter during the daylight hours in dense grass and weeds near corn fields. Flushing moths from such habitats gives an estimate of population densities (Sappington and Showers, 1983). Eggs can be sampled by visual examination, but this is a very time-consuming effort. Similarly, larval populations can be estimated from visual examinations, particularly of whorls during the first generation. A sequential sampling protocol for larvae was developed for potato (Nault and Kennedy, 1996b), and Hoffmann *et al.* (1996b) described a sequential sampling plan based on infested plants.

Insecticides. Liquid formulations of insecticide are commonly applied to protect against damage to sweet corn, particularly from the period of early tassel formation until the corn silks are dry. Recommendations vary from a single application before silking, to weekly applications (Ferro and Fletcher–Howell, 1985). Liquid applications are usually made to coincide with egg hatching in an effort to prevent infestation. If corn borers are present in a field, however, the critical treatment time is just before the tassels emerge, or at tassel emergence from the whorl. This plant growth period is significant because the larvae are active at this time and more likely to contact insecticide. A popular alternative to liquid insecticides is the use of granular formulations, which can be dropped into the whorl for effective control of first generation larvae because this is where young larvae tend to congregate. Insecticide is more persistent when applied in a granular formulation (Straub, 1983). Botanical insecticides such as rotenone and ryania are moderately effective against young corn borers, but must be applied frequently (Turner, 1945). In grain corn, insecticide applications for suppression of second generation corn borers can be made outside the corn fields in areas of thick grass, or action sites, where adults tend to aggregate (Showers *et al.*, 1980). This approach has not been assessed for sweet corn. For borer suppression on potato, a single application of insecticide timed to coincide with the presence of first instar larvae provides optimal yield (Nault and Kennedy, 1996a).

Cultural Practices. Destruction of stalks, the overwintering site of larvae, has long been recognized as an important element of corn borer management. Disking is not adequate; plowing to a depth of 20 cm is necessary for destruction of larvae. Mowing of stalks close to the soil surface eliminates more than 75% of larvae, and is especially effective when combined with plowing (Schaafsma *et al.*, 1996). Minimum tillage procedures, which leave considerable crop residue on the surface, enhance borer survival.

Diversified cropping is detrimental to corn borer population survival. Intercropping with red clover, for example, resulted in lower borer density (Lambert *et al.*, 1987).

Early planted corn is taller and attractive to ovipositing female moths, so late planting has been recommended, but this is useful mostly in areas of only a single generation per year. If a second generation occurs, such late-planted corn is heavily damaged. Planting border rows of a highly attractive variety of corn to surround a less attractive variety has been investigated in France (Derridj *et al.*, 1988). The attractive variety, especially if it is an early flowering cultivar, receives most of the eggs of moths dispersing into the field. If treated with insecticide or destroyed, this border row trap could provide protection for the main corn crop.

Soil conditions can affect corn borer oviposition patterns. Research conducted in Ohio demonstrated that corn grown in rich organic soils were not as attractive to moths as low-protein plants grown in conventionally fertilized soil (Phelan *et al.*, 1996).

Host-Plant Resistance. Extensive breeding research has been conducted, and resistance has been incorporated into grain corn, especially against first generation borers. A principal factor in seedling resistance to young larvae is a chemical known as DIMBOA, which functions as a repellent and feeding deterrent (Klun *et al.*, 1967). It has proven difficult to incorporate the known resistance factors into sweet corn without degradation of quality. However, some progress has been made in producing commercially acceptable resistant cultivars, especially when host-plant resistance is complemented by use of other suppressive tactics such as application of *Bacillus thuringiensis* (Bolin *et al.*, 1996).

Pepper cultivars differ in their susceptibility to corn borer. Hot pepper cultivars are most resistant, and most green bell peppers are susceptible.

Biological Control. Biological control has been attempted repeatedly in sweet corn and other vegetables susceptible to European corn borer attack. *Bacillus thuringiensis* products can be as effective as many chemical insecticides, but often prove to be less effective than some (Bartels and Hutchison, 1995). Most single-factor approaches, with the exception of newer formulations of *Bacillus thuringiensis*, have proven to be erratic. Release of native*Trichogramma* spp. (Hymenoptera: Trichogrammatidae), for example, provides variable and moderate levels of suppression (Andow *et al.*, 1995). In Massachusetts, an egg parasitoid normally associated with a related *Ostrinia* species in China was released. This new parasitoid, *Trichogramma ostriniae* (Hymenoptera: Trichogrammatidae), may prove useful for augmentative biological control programs, but seems susceptible to disruption by adverse weather (Wang *et al.*, 1997). The effect of egg parasitoids is enhanced by application of *Bacillus thuringiensis* (Losey *et al.*, 1995; Mertz *et al.*, 1995). Application of pathogens such as *Nosema pyrausta* and *Vairimorpha necatrix* (Microsporida: Nosematidae) has been proven to have benefit under experimental conditions (Lewis *et al.*, 1982), but a commercial product has not been developed.

Garden Webworm
Achyra rantalis (Guenée)
(Lepidoptera: Pyralidae)

Natural History

Distribution. This webworm is found throughout the eastern United States west to the Rocky Mountains, and also in California. Despite its wide distribution, it rarely is damaging except in the southern Great Plains region. Garden webworm also occurs in eastern Canada, in most of Mexico, and throughout the Caribbean. It is native to North America.

Host Plants. Garden webworm is quite similar to alfalfa webworm, *Loxostege cereralis* (Zeller), and beet webworm, *L. sticticalis* (Linnaeus), in most aspects of its biology, including host range. Though its biology is not well documented, it is known to attack such vegetables as bean, beet, cabbage, cantaloupe, cucumber, chard, corn, cowpea, eggplant, lettuce, onion, potato, pumpkin, squash, sweet potato, and tomato. Field crops including alfalfa, clover, pea, soybean, and sugarbeet are injured. If allowed to become abundant in alfalfa, larvae may disperse when the alfalfa is harvested and may damage nearby crops, including such crops as corn and cotton, which are not normally eaten. Grain crops generally are avoided. Weed hosts include dock, *Rumex* spp.; lambsquarters, *Chenopodium album*; *Parthenium* sp.; pigweed, *Amaranthus* spp.; ragweed, *Ambrosia* spp.; saltbush, *Atriplex patula*;

smartweed, *Polygonum* spp.; and sunflower, *Helianthus* spp. Pigweed and lambsquarters are regarded as favorite weed hosts.

Natural Enemies. Several parasitoids of garden webworm are known, but there is little information on their relative importance. Among the wasps are *Cremnops vulgaris* (Cresson), *C. haematodes* (Brulle), *Apanteles conanchetorum* Viereck, *A. pyraloides* Muesebeck, *Cardiochiles explorator* (Say) (all Hymenoptera: Braconidae), and *Phytodietus rufipes* (Cresson) and *Diadegma pattoni* (Ashmead) (both Hymenoptera: Ichneumonidae). Among the fly parasitoids are *Eusisyropa blanda* (Osten Sacken), *E. boarmiae* (Coquillet), *Hyphantrophaga hyphantriae* (Townsend), *Lespesia archippivora* (Riley), *Lixophaga variablis* (Coquillet), *Nemorilla pyste* (Walker), *Patelloa leucaniae* (Coquillet), *Pseudoperichaeta erecta* (Coquillet), *Stomatomyia parvipalpis* (Wulp) and *Winthemia quadripustulata* (Fabricius) (all Diptera: Tachinidae).

Life Cycle and Description. A complete life cycle requires about 40 days. The number of annual generations is thought to be 3–4 throughout its range. In Texas there are four flights of adults: they occur in May, late June–early July, early August, and mid-September. However, because of overlapping generations it is difficult to discern separate flights.

Egg. The oval eggs initially are nearly transparent, becoming cream or yellowish in color. They measure about 0.64 mm wide and 1.1 mm long. The flattened, overlapping eggs are deposited in clusters of 8–20, usually on the underside of foliage. Total fecundity is estimated at 300–400 eggs. Duration of the egg stage is 2–5 days, but averages 2.8 days.

Garden webworm eggs on foliage.

Larva. The larva generally is pale green dorsally and yellowish-green ventrally. It measures 21–24 mm long at maturity. The head and thoracic plate are yellowish but marked with dark spots. The thoracic plate markings are in the form of two dark bars on each side. The body is well-marked with raised, dark spots enclosing a paler area from which emerges a stout black hair. This is especially evident in the lateral spots below the spiracles, which appear almost ring-like. There are six such spots on each abdominal segment. A light mid-dorsal stripe occurs dorsally along the center of the abdomen. Mean head capsule widths (range) are 0.25 (0.20–0.26), 0.37 (0.32–0.40), 0.58 (0.55–0.70), 0.92 (0.80–1.00), 1.10 (1.05–1.20), and 1.34 (1.25–1.50) mm, respectively for instars 1–6. Mean overall body lengths (range) are 2.1 (1.6–2.5), 3.7 (2.2–4.3), 6.1 (4.0–9.3), 9.5 (6.6–13), 11.0 (9–14), and 13.4 (9–20) mm, respectively. During the first instar, larvae normally are gregarious, but disperse afterwards. If disturbed, larvae wriggle violently, drop from the plant on a strand of silk, or rapidly retreat within a silken tube, if one is present. Duration of the larval stage requires 14–28 days, averaging 16 days, and is followed by a prepupal period of 1–3 days.

Pupa. The mature larva pupates in the soil within a silk-lined cell, usually under debris or near the soil surface. The pupal case measures about 12.5 mm long and 3.5 mm wide, and is encrusted with soil. It is closed at the bottom but open at the tip, allowing easy escape of the moth. The pupa is light to dark brown and measures about 8.0–9.5 mm long. The tip of the abdomen bears three stout spines. Duration of the pupal stage requires 4–13 days, averaging 8.7 days.

Adult. This small moth has a wingspan of about 17–43 mm, males averaging smaller than females. It is yellowish-brown or reddish-brown, but bears lighter and darker markings. The forewing and hind wing usually are crossed by a light irregular band, and sometimes a dark band. The hind wing is yellow-

Adult garden webworm.

ish. The moth is quite variable in coloration, which may account for the many times it has been described as a new species. Adults feed on nectar from various flowers and commence oviposition about 3–6 days after emergence.

Elements of garden webworm biology were given by Sanderson (1906), Sanborn (1916), Kelly and Wilson (1918), Poos (1951), and Smith and Franklin (1954). Capps (1967) provided a description of several stages. A key to some webworm larvae, including garden webworm, was published by Allyson (1976). A key to the adults was provided by Munroe (1976).

Damage

Larvae feed only on the leaf epidermis during the first two instars, skeletonizing the tissue. Thereafter they consume the entire leaf. Eventually, they defoliate plants, consuming all except the stems and major veins. Webworms usually wrap young leaves in a loose web and feed within the protection of the web. During periods of abundance, entire plants are shrouded in webbing. This webbing is not diagnostic, however, because beet webworm and alfalfa webworm also display this behavior.

Management

Infestations often result from the presence of weeds in crop fields. Destruction of weeds can deter oviposition. The other major source of webworms is alfalfa fields, as adults disperse at maturity and larvae disperse when the alfalfa is cut. Early harvesting of alfalfa, especially if the larvae are young and incapable of long distance dispersal in the absence of food, can reduce webworm numbers. Alfalfa can also be treated with insecticide before harvest if it is heavily infested. Insecticidal suppression of webworms is normally accomplished by foliar applications, though this is seldom warranted. The bacterium *Bacillus thuringiensis* provides some control.

Garden webworm larva.

Garden webworm pupa.

Hawaiian Beet Webworm
Spoladea recurvalis (Fabricius)

Spotted Beet Webworm
Hymenia perspectalis (Hübner)

Southern Beet Webworm
Herpetogramma bipunctalis (Fabricius)
(Lepidoptera: Pyralidae)

Natural History

Distribution. Hawaiian beet webworm, *Spoladea recurvalis* (Fabricius), is found throughout the world in tropical and subtropical regions. In North America, it is found in the southern states from Florida west to California and Hawaii, and also in Puerto Rico. It sometimes causes damage as far north as Virginia, but it cannot overwinter there and must reinvade annually; thus it inflicts injury only late in the season. Its origin is uncertain, but it is not native to North America.

Spotted beet webworm, *Hymenia perspectalis* (Hübner), and southern beet webworm, *Herpetogramma bipunctalis* (Fabricius), are closely related to Hawaiian beet webworm. They are not well known, but also are tropical pests with a wide geographic range. In the United States, they similarly are southern insects but can occasionally cause injury as far north as Illinois and Virginia.

Host Plants. Hawaiian beet webworm is largely restricted to plants in the family Chenopodiaceae. Among vegetable crops, beet, chard, spinach, and New Zealand spinach are normally injured. Sugarbeet is readily attacked, but this crop is rarely grown in the warm environments favored by Hawaiian beet webworm. When they face starvation, this insect may feed on other crops, but it is much less damaging than the better-known webworms—beet webworm, *Loxostege sticticalis* (Linnaeus); alfalfa webworm, *L. cerealis* (Zeller); and garden webworm, *Achyra rantalis* (Guenée). Weed hosts are pigweed, *Amaranthus* spp., lambsquarters, *Chenopodium album*, and purslane, *Portulaca oleracea*. Pigweed is preferred even over cultivated hosts. Spotted beet webworm and southern beet webworm display the same dietary preferences as does Hawaiian beet webworm. Tingle *et al.* (1978) indicated that southern beet webworm was the dominant caterpillar on *Amaranthus hybridus* in Florida corn fields during the late summer months.

Natural Enemies. Hawaiian beet webworm is known to be parasitized by *Cotesia marginiventris* (Cresson) (Hymenoptera: Braconidae), *Venturia infesta* (Cresson) (Hymenoptera: Ichneumonidae), *Argyrophylax albincisa* (Weidemann), *Chaetogaedia monticola*

(Bigot), *Eucelatoria armigera* (Coquillett), and *Nemorilla pyste* (Walker) (all Diptera: Tachinidae).

Not surprisingly, the parasitoid complex attacking spotted and southern beet webworms is similar to Hawaiian beet webworm. *Venturia infesta* (Cresson) (Hymenoptera: Ichneumonidae) has been reared from all three webworms. Also, *Apanteles mimoristae* (Muesebeck) (Hymenoptera: Braconidae) is associated with spotted beet webworm, whereas *Gambrus ultimus* (Cresson), *Temelucha* sp. (both Hymenoptera: Ichneumonidae) and *C. marginiventris* (Hymenoptera: Braconidae) attack southern beet webworm. *Argyrophylax albincisa* (Weidemann) is associated with spotted beet webworm and *Nemorilla pyste* (Walker) with southern beet webworm (both Diptera: Tachinidae).

Life Cycle and Description. In Hawaii, Hawaiian beet webworm is active throughout the year and about 10 generations occur annually. During warm weather, one generation is completed in about 30 days. The biology of spotted and southern beet webworms is less certain, but seems to be basically the same as Hawaiian beet webworm except as noted below.

Egg. The egg is elliptical, iridescent white, and flattened, and may be deposited singly or in rows of up to several eggs. The egg measures 0.6 mm long, 0.4 mm wide, and 0.25 mm in height. They most often are laid on the lower surface of leaves adjacent to large veins. Duration of the egg stage is about four days.

Larva. Young larvae of Hawaiian beet webworm feed beneath the leaves, and occasionally spin light webs in which they rest. The larvae initially are pale green, though a few dark spots are found on the head and thoracic plate. As larvae attain maturity they develop broad, irregular, whitish lateral stripes that contrast strongly with a dark stripe dorsally. The larvae also become uniformly pinkish or rust-colored as they prepare to pupate. The head capsule increases in width from 0.25 to 1.2 mm as the larva grows. The body bears numerous stout hairs over the length of its body, but lacks the dark spots or rings found with such hairs on many webworms. The larval development period is normally 9–13 days. (See color figure 73.)

Larvae of spotted beet webworm also are green, but as its common name suggests, bear numerous spots. The thorax and abdomen are equipped with raised-dark spots from which arise dark hairs. The head bears purplish dots, though the center is unmarked. The thoracic plate has black borders.

Larvae of southern beet webworm are dark-green with a dark, nearly black, head. The thoracic plate is similarly dark except for the central area. The body bears numerous large, dark, raised spots from which

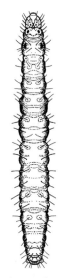

Hawaiian beet webworm larva.

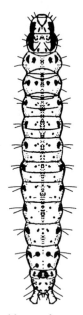

Spotted beet webworm larva.

arise black hairs. The spots are light in color centrally, however, resulting in a ring-like appearance and causing this insect to resemble garden webworm, *Achyra rantalis* (Guenée). (See color figure 72.)

Pupa. Mature larvae drop to the soil to pupate. They burrow slightly beneath the soil surface and form firm, compact, elliptical cocoons of silk covered with grains of soil. The pupa is light brown and the posterior end is equipped with terminal spines bearing hooked tips. The pupa measures about 10 mm long

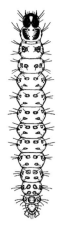

Southern beet webworm larva.

and 2.5 mm wide. Duration of the pupal stage is 7–14 days.

Adult. The moth of Hawaiian beet webworm is dark brown, and often tinted purple. The front and hind wings are marked with a white transverse band that nearly crosses the wings. The forewing also bears one elongate and two small white spots distally. The margin of the front wings is alternating dark and light. Narrow light bands are found on the abdomen. The wingspan measures 17–23 mm. (See color figure 210.)

Spotted beet webworm is similar to Hawaiian beet webworm in general appearance. The spotted species is lighter brown, however, with a reddish tint, and the white wing markings are less discrete and pronounced. The wingspan is about 20 mm.

Southern beet webworm moths are light yellowish-gray, sometimes with an iridescent purplish cast. The front wings bear three dark spots along the leading edge, but the markings are not very distinct. The hind wings are not distinctly marked. The abdomen is

Hawaiian beet webworm pupa.

darker. The wingspan is 22–26 mm. (See color figure 214.)

The biology of Hawaiian beet webworm was given by Marsh and Chittenden (1911b) and Walker and Anderson (1939). Spotted beet webworm was described by Davis (1912) and Chittenden (1913a). Southern beet webworm was described by Chittenden (1911a).

Damage

Young larvae remain on the lower surface of foliage, not eating entirely through the leaves. When nearly mature, however, they consume the entire leaf. Unlike some other webworm species, webbing is not very pronounced.

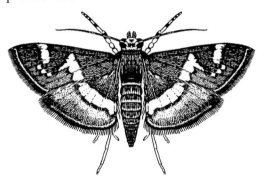

Adult Hawaiian beet webworm.

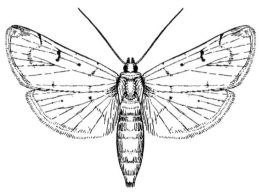

Adult southern beet webworm.

Adult spotted beet webworm.

Management

Infestations often result from the presence of weeds in crop fields. Destruction of weeds can deter oviposition. Insecticidal suppression of webworms is normally accomplished by foliar applications, though this is seldom warranted except, perhaps, for Hawaiian beet webworm in Hawaii. The bacterium *Bacillus thuringiensis* provides some control.

Lesser Cornstalk Borer
Elasmopalpus lignosellus (Zeller)
(Lepidoptera: Pyralidae)

Natural History

Distribution. This species occurs widely in the western hemisphere, but not elsewhere. It is known from much of the southern United States, though in the eastern regions it occurs much farther north than in western states. The northernmost limits seem to be Massachusetts to southern Iowa in the east, and Oklahoma to southern California in the west. It is found throughout Central and South America. Despite its wide range, damage is limited principally to sandy soil, so it tends to cause injury in the coastal plain of the southeastern states from South Carolina to Texas.

Host Plants. Lesser cornstalk borer damages several crops grown in the southeast, though it is mostly a pest of peanut, sorghum, and soybean. Among vegetable crops injured are bean, beet, cabbage, cantaloupe, corn, cowpea, lima bean, pea, pepper, sweet potato, tomato, and turnip. Legume and grass crops are most often damaged. Field crops injured are corn, chufa, millet, oat, rice, rye, sorghum, peanut, soybean, sudan grass, sugarcane, and wheat. It also infests crabgrass, *Digitaria sanguinalis*, wiregrass, *Elusine indica*; and Johnsongrass, *Sorghum halepense*.

Natural Enemies. Several natural enemies of lesser cornstalk borer are known, though they are not thought to be major determinants of population trends. Smith and Johnson (1989) constructed life tables for populations in Texas, and identified survival of large larvae as the key element in generation survival, but the causative factor remained unidentified. The predominant parasitoids are *Orgilus elasmopalpi* Muesebeck and *Chelonus elasmopalpi* McComb (both Hymenoptera: Braconidae), *Pristomerus spinator* (Fabricius) (Hymenoptera: Ichneumonidae), and *Stomatomyia floridensis* Townsend (Diptera: Tachinidae) through most of the range of lesser cornstalk borer. Other parasitoids sometimes are present include *Bracon gelechiae* Ashmead (Hymenoptera: Braconidae), *Geron aridus* Painter (Diptera: Bombyliidae), and *Invreia* spp. (Hymenoptera: Chalcididae) (Johnson and Smith, 1981; Funderburk *et al.*, 1984b; Smith and Johnson, 1989). Parasitoids rarely cause more than 10% mortality.

Among the predators thought to be important mortality factors are a ground beetle, *Plilophuga viridicolis* LeConte (Coleoptera: Carabidae); *Geocoris* spp. bugs (Hemiptera: Lygaeidae); and larval stiletto flies (Diptera: Therevidae).

Pathogens are commonly present in lesser cornstalk borer populations. The most important pathogen appears to be a granulosis virus, but a *Beauveria* sp. fungus, microsporidia, and mermithid nematodes also have been found (Funderburk *et al.*, 1984b).

Weather. Lesser cornstalk borer seems to be adapted for hot, xeric conditions, and therefore tends to be more abundant and damaging following unusually warm, dry weather. Mack *et al.* (1993) used data from Alabama and Georgia to develop a predictive equation that forecasts the potential for crop injury and the need to monitor crops. It is based on the concept of "borer-days." Borer-days is calculated as the sum of days during the growing season in which the temperature equals or exceeds 35°C *and* the precipitation is less than 2.5 mm, less the number of days in which the temperature is less than 35°C *and* the precipitation equals or exceeds 2.5 mm. Thus, it is the sum of the number of hot, dry days less the number of cooler, wetter days. If the number of borer-days equals or exceeds 10, damage is likely. If borer-days equals 5–9, damage is possible and fields should be scouted. The relationship between borer-days and larval abundance is nonlinear, and small increases in borer-days beyond 10 results in large increases in larval abundance. Reduction in larval feeding at high soil moisture levels was documented by Viana and da Costa (1995).

Life Cycle and Distribution. There are three complete generations annually in Georgia, and some members go on to form a fourth generation. Other southeastern states also experience 3–4 generations, but in the southwest there are only three generations annually. Activity extends from June to November, with the generations overlapping considerably and little evidence of breaks found between generations. Overwintering apparently occurs in the larval and pupal stage; diapause is not present. A complete life cycle usually requires 30–60 days.

Egg. The egg is oval, measuring about 0.6 mm long and 0.4 mm wide. When first deposited, the egg is greenish, soon turns pinkish, and eventually reddish. The female deposits nearly all her eggs below the soil surface, adjacent to plants. A few, however, are placed on the surface or on leaves and stems

(Smith *et al.*, 1981). Duration of the egg stage is 2–3 days.

Larva. Larvae live in the soil, constructing tunnels from soil and excrement tightly woven together with silk. They leave the tunnel to feed in the basal stalk area or just beneath the soil surface, returning and constructing new tunnels as they mature. Thus, tunnels often radiate out from the stem of the food source, just below the soil surface, to a depth of about 2.5 cm. Normally there are six instars, though 5–7 have been observed. During the early instars, larvae are yellowish-green, with reddish pigmentation dorsally, tending to form transverse bands. As the larva matures, whitish longitudinal stripes develop, so that by the fifth instar they are pronounced. The mature larva is bluish-green, but tends toward reddish-brown, with fairly distinct yellowish-white stripes dorsally. The head capsule is dark, and measures about 0.23, 0.30, 0.44, 0.63, 0.89, and 1.2 mm wide, respectively, for instars 1–6. Larval lengths are about 1.7, 2.7, 5.7, 6.9, 8.8, and 16.2 mm, respectively. Mean development time is estimated at 4.2, 2.9, 1.4, 3.1, 2.9, and 8.8 days respectively, for instars 1–6. Total larval development time varies widely, but normally averages about 20 days. (See color figure 78.)

Pupa. At maturity, the larva constructs a pupal cell of sand and silk at the end of one of the tunnels. The cocoon measures about 16 mm long and 6 mm wide. The pupa is yellowish initially, turning brown

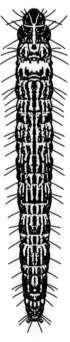

Lesser cornstalk borer larva.

and then almost black just before the adult emerges. It measures about 8 mm long and 2 mm wide. The tip of the abdomen is marked by a row of six hooked spines. Pupal development time averages about 9–10 days (range 7–13 days).

Adult. Moths are fairly small, measuring 17–22 mm in wingspan. Sexual dimorphism is pronounced. The forewing of the male moth is yellowish centrally, bordered by a broad dark band bearing purplish scales. In females, however, the entire forewing is dark, sometimes almost black, but also bears reddish or purplish scales. The thorax is light in males, but dark in females. The hind wings of both sexes are transparent with a silvery tint. Adults are most active at night when the temperature exceeds 27°C, relative humidity is high, and there is little air movement. Such conditions are optimal for mating and oviposition. Indeed, these activities cease if temperature falls below 18–20°C. Mack and Backman (1984) studied adult longevity and fecundity in relation to temperature. Longevity varied from 20 days if held at 17°C, to eight days if held at 32–35°C. Mean fecundity varied from only about 20 eggs per female when reared at 17°C, to a maximum of 110 eggs at 30°C, then decreased at higher temperatures. Mean fecundity per day was estimated at about 12 eggs. However, these values included an unknown proportion of individuals that did not reproduce. Other reports indicate that fecundity is about 100–450 eggs among females that successfully reproduce, with an average of about 200 eggs per female. Adult longevity under field conditions is estimated at about 10 days.

The biology was described by Luginbill and Ainslie (1917) and a review was published by Tippins (1982). Developmental data were given by Dupree (1965) and Leuck (1966). Rearing was described by Chalfant (1975). A sex pheromone blend was identified by Lynch *et al.* (1984b). A key to stalk borers associated with corn in southern states was presented by Dekle (1976); this publication also includes pictures of the adults. A guide to common stalk boring caterpillars also is found in Appendix A.

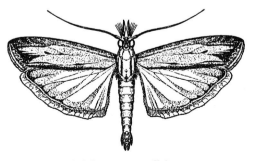
Adult lesser cornstalk borer.

Damage

Damage is caused by the larval stage which feeds upon, and tunnels within, the stems of plants. Normally the tunneling is restricted to the basal region of stalks, including the below-ground portion, and girdling may occur. In affected plants wilting is one of the first signs of attack, but buds may wither, and stunting and plant deformities are common. Plant death is not uncommon, and infested areas of fields often have a very thin stand.

Management

Sampling. The egg stage is difficult to sample because they are small and resemble sand grains. Eggs can be separated by flotation, however (Smith *et al.*, 1981). Larval populations are aggregated, and can be separated from soil by sieving or flotation (Mack *et al.*, 1991; Funderburk *et al.*, 1986). Adults are attracted to light traps, but are difficult to monitor with this technique because the moths of lesser cornstalk borer are difficult to distinguish from many other species. This is especially true of the females, which are less distinctive than the males. Pheromone traps have been used successfully to monitor adult populations, and adults can be flushed from fields by beating the vegetation. Adult pheromone trap catches and flush counts are correlated (Funderburk *et al.*, 1985). Adult and larval counts are often highly correlated, indicating that flush counts can be used to predict the abundance of larvae in subsequent weeks (Mack *et al.*, 1991). Loera and Lynch (1987) successfully used pheromone trapping to monitor moth populations in bean, and reported that trap heights of about 0.5 m were most effective.

Insecticides. Insecticides applied for suppression of lesser cornstalk borer are usually applied in a granular formulation in the seed furrow or in a band over the seed bed. Liquid formulations can also be applied, but it is important that they be directed to the root zone.

Cultural Practices. Lesser cornstalk borer damage is largely restricted to sites with sandy soil. Modified planting practices have long been used to minimize crop loss in such locations. Populations tend to increase over the course of a season, so some damage can be avoided by early planting. Tillage and destruction of weeds are recommended before planting because this helps to destroy larvae that may be present in the soil and damage seedlings, the stage most susceptible to destruction. However, crop culture that uses conservation tillage (i.e., retention of crop residue at the soil surface) experiences little injury from lesser cornstalk borer feeding because the larvae feed freely on crop residue and other organic matter, sparing the young crop plants (All and Gallaher, 1977; All *et al.*, 1979).

Limabean Pod Borer
Etiella zinckenella (Treitschke)
(Lepidoptera: Pyralidae)

Natural History

Distribution. Limabean pod borer is found throughout the world, but is most common in the tropics and subtropics. It occurs in the warmer regions of the temperate zone, but is absent from cold climates such as northern Europe. In North America, limabean pod borer is widespread in the western United States, from Washington to southern California and east to Colorado and Texas. Limabean pod borer is known from certain eastern states, such as Florida, North and South Carolina, and Maryland, but is not considered to be a serious pest in the East. It is a problem, however, in the Caribbean, including Puerto Rico. In Canada, though limabean pod borer has been detected, it is not considered to be a field pest. It was first observed in the United States in California, in 1885, but its origin is unknown.

Host Plants. This insect limits its attack to plants in the family Leguminosae. Vegetables damaged by lima bean pod borer include cowpea, faba bean, snap bean, lima bean, pea, and pigeon pea. Other legumes such as lupine, *Lupinus* spp.; rattlebox, *Crotalaria sagittalis* and *C. incana*; locoweed, *Astragalus antiselli*; and milkvetch, *Astragalus trichopodus*, also serve as hosts. In California, severity of limabean pod borer damage is inversely related to the distance between crops and wild hosts, particularly lupine (Stone, 1965). In Puerto Rico, *Crotalaria* spp. are important alternate hosts (Segarra-Carmona and Barbosa, 1988).

Natural Enemies. Limabean pod borer is not heavily parasitized in North America. Several native egg and larval parasitoids (Hymenoptera: various families) have been detected in California and Washington, but none are effective. Numerous parasitoid species are known from Europe (Parker, 1951; Stone, 1965), but attempts to introduce parasitoids into California and Puerto Rico during the 1930s were unsuccessful (Clausen, 1978).

Life Cycle and Description. The number of generations varies according to the weather. The complete life cycle requires at least 60 days, and often considerably longer during cool weather. In California, one generation usually occurs during March–June on lupines or other wild host plants, and then another 2–4 generations occur during June–December on crop

plants or perennial lupines. Starting in about July, some of the mature larvae diapause in their cocoons. The proportion of larvae entering diapause in July is small, only about 8%, but the proportion increases to 53% in August, 89% in September, and 100% in October and later. Pupation of overwintering larvae occurs from January to March. Emergence of adults from overwintering larvae begins in March, but it is protracted.

Egg. The egg is oval and measures about 0.6–0.7 mm long and 0.3–0.45 mm wide. The egg is white when first deposited, but turns pink and then gray as the embryo develops. Duration of the egg stage is about 15.4 days, but varies from 5 to 33 days depending on temperature. The eggs are deposited singly or in small groups of up to 12. They are deposited on the flower, stems, or pod petiole. Estimates of egg production vary widely. Laboratory-reared moths often produce only 50–90 eggs, whereas field-collected moths may produce about 140–260 eggs. The latter values are probably much better estimates of fecundity.

Larva. The larva is white or cream at hatching, and measures only 1 mm long. Young larvae immediately bore into a bean pod and develop internally. The entrance hole into the pod usually heals, leaving no indication that a larva was feeding within. There are five instars, with mean head capsule widths of about 0.15, 0.35, 0.65, 1.05, and 1.60 mm, respectively. Duration of the instars is about three, three, three, three, and four days, respectively. The mature larva measures

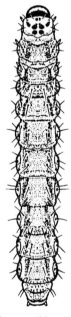

Limabean pod borer larva.

12–17 mm long and is pinkish or tan. The head and pronotum are yellow with black markings. The larva has five pairs of prolegs in addition to three pairs of thoracic legs. Duration of the larval stage is usually about 35 days, but it may vary from 13 to 65 days. At maturity, the larva eats through the wall of the pod, exits, and drops to the soil. (See color figure 79.)

Pupa. The mature larva burrows into the soil to a depth of 1–5 cm and spins a cocoon. The soil particles adhere to the cocoon, so if dug from the soil the cocoon is an elongate cylinder of soil measuring 15–20 mm long and 6–8 mm wide. The duration of the pre-pupal period is variable, but during the summer months it is typically 8–24 days. During the winter, of course, it is greatly prolonged, because this is the overwintering stage. The pupa, which eventually develops in the cocoon, is 8–10 mm long and 2.5–3 mm wide. Usually, it is amber or light brown. Duration of the pupal period is about 36 days (range 16–101 days).

Adult. The adult is a small brownish-gray moth, with a wingspan of 24–27 mm. The most distinctive features are the forward-protruding mouthparts, a characteristic feature of pyralid moths, and a broad white band along the leading edge of the forewings. There is also a transverse yellowish band slightly anterior of the mid-point of the wing. Copulation commences about 24 h after emergence. The pre-oviposition period is usually 4–6 days, and adult longevity is about 10–12 days.

The biology of limabean pod borer was described by Hyslop (1912b), Abdul-Nasr and Awadalla (1957), and Stone (1965). Rearing procedures were outlined by Hattori and Sato (1983).

Damage

The larvae feed on buds and blossoms, and burrow into the pods of legumes to feed on the developing seeds. They typically feed on only a portion of a seed and then move on to attack adjacent seeds. Silk and fecal material accumulate in the pods. In the case of

small pods, larval feeding usually causes the pod to drop from the plant, but large pods remain attached. Larvae sometimes leave one pod and move to another to continue feeding, especially following pod drop. Market standards often cause loss in excess of the direct feeding injury by larvae, because even low levels of damaged pods or beans are considered to be undesirable. Formerly, this insect was considered to be an important pest in the western United States, but with the introduction of modern insecticides it has assumed minor status. In tropical climates, however, it remains a serious pest of beans and soybeans.

Management

Sampling. Light traps can be used to sample populations of moths. Also, a sex pheromone has been identified and used successfully to trap moths in the field (Toth *et al.*, 1989).

Insecticides. Residual insecticides provide good control of borers. As the larvae feed internally, it is essential that the insecticides be on the vegetation at the time moths are ovipositing and eggs are hatching. Protracted emergence of adults and multiple overlapping generations often necessitate numerous applications of insecticides. *Bacillus thuringiensis* is not usually recommended for this insect.

Cultural Practices. Because larvae overwinter in the soil beneath legumes, tillage can reduce emergence in the spring. Autumn plowing to a depth of at least 20 cm is recommended. Early season planting is also helpful because the crop can reach maturity before pod borers attain high densities.

Host-Plant Resistance. Host-plant resistance is a viable option for some crops. Several pigeon pea and soybean cultivars have been shown to exhibit considerable resistance to attack (Cruz, 1975; Armstrong, 1991; Talekar and Lin, 1994).

Melonworm
Diaphania hyalinata Linnaeus
(Lepidoptera: Pyralidae)

Natural History

Distribution. Melonworm is most abundant in tropical climates, where cucurbits can grow during the winter months. It occurs throughout most of Central and South America and the Caribbean. The United States is the northern limit of its permanent range, and its wintertime occurrence generally is limited to south Florida and perhaps south Texas. Melonworm disperses northward annually. Its distribution during the summer months is principally the southeastern

Adult limabean pod borer.

states, extending west to Texas and, sometimes, north to New England and the Great Lakes region. Melonworm has been reported in North Carolina as early as July (Smith, 1911).

Host Plants. Melonworm is restricted to feeding on cucurbits. The host preferences are nearly identical to that of pickleworm, *Diaphania nitidalis* (Stoll); both wild and cultivated cucurbits may be attacked. Summer and the winter squash species are good hosts. Pumpkin is of variable quality as a host, probably because pumpkins are bred from several *Cucurbita* species. The *Cucumis* species—cucumber, gerkin, and cantaloupe—are attacked but not preferred. Watermelon is almost never eaten. (See color figure 23.)

Natural Enemies. Natural enemies of melonworm are nearly the same as those of pickleworm. All of the parasitoids found by Pena *et al.* (1987b) to attack pickleworm also attacked melonworm: *Apanteles* sp., *Hypomicrogaster diaphaniae* (Muesebeck), *Pristomerus spinator* (Fabricius) (all Hymenoptera: Braconidae), *Casinaria infesta* (Cresson), *Temelucha* sp. (both Hymenoptera: Ichneumonidae), and undetermined trichogrammatids (Hymenoptera: Trichogrammatidae) (Pena *et al.*, 1987b; Capinera, 1994). However, additional species parasitized melonworm, including *Gambrus ultimus* (Cresson), *Agathis texana* (Cresson) (both Hymenoptera: Ichneumonidae) and an undetermined fly (Hymenoptera: Tachinidae). The tachinids known from melonworm are *Nemorilla pyste* (Walker) and *Stomatodexia cothurnata* (Wiedemann). Studies conducted in Puerto Rico (Medina–Gaud *et al.*, 1989) reported levels of parasitism reaching 24%. Generalist predators such as *Calosoma* spp. and *Harpalus* (both Coleoptera: Carabidae), the soldier beetle *Chauliognathus pennsylvanicus* De Geer (Coleoptera: Cantharidae), and the red imported fire ant *Solenopsis invicta* Buren (Hymenoptera: Formicidae) have also been reported to be mortality factors.

Life Cycle and Description. The melonworm can complete its life cycle in about 30 days. It is present throughout the year in southern Florida, where it is limited mostly by availability of host plants. It disperses northward annually, usually arriving in northern Florida and other southeastern states in June or July, where not more than three generations normally occur before cold weather kills the host plants.

Egg. Melonworm moths deposit oval, flattened eggs in small clusters, averaging 2–6 eggs per mass. Apparently they are deposited at night on buds, stems, and the underside of leaves. Initially, they are white, but soon become yellow. They measure about 0.7 mm long and 0.6 mm wide. Hatching occurs after 3–4 days. (See color figure 257.)

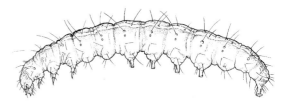
Melonworm larva.

Larva. There are five instars. Total larval development time is about 14 days, with mean (range) duration the instars about 2.2 (2–3), 2.2 (2–3), 2.0 (1–3), 2.0 (1–3), and 5.0 (3–8) days, respectively. Head capsule widths are about 0.22, 0.37, 0.62, 1.04, and 1.64 mm, respectively (Smith *et al.*, 1994). Larvae attain lengths of about 1.5, 2.6, 4.5, 10, and 16 mm in instars 1–5, respectively. Neonate larvae are colorless, but by the second instar larvae assume a pale yellow-green color. They construct a loose silken structure under leaves which serves to shelter them during the daylight hours. In the fifth instar, larvae have two subdorsal white stripes extending the length of the body. The stripes fade or disappear just before pupation, but they are the most distinctive characteristic of the larvae. Smith (1911) provided a complete description of larvae. (See color figure 76.)

Pupa. Before pupation, larvae spin a loose cocoon on the host plant, often folding a section of the leaf for added shelter. The melonworm cocoon is much better formed than the cocoon of pickleworm, and the melonworm's preference for green foliage as a pupation site also serves to differentiate the insects. The pupa is 12–15 mm long, about 3–4 mm wide, and fairly pointed at each end. It is light to dark brown. The pupal stage persists for 9–10 days.

Adult. The moth's wingspan is about 2.5 cm. The wings are pearly white centrally, and slightly iridescent, but are edged with a broad band of dark brown. Moths frequently display brushy hairpencils at the tip of the abdomen when at rest. Melonworm moths differ from pickleworm as they remain in the crop during the daylight hours. While they are generally inactive during the day, they fly short distances when disturbed. Smith (1911) provided a detailed account of melonworm biology. Rearing techniques were given by Elsey *et al.* (1984) and Valles *et al.* (1991). (See color figure 211.)

Damage

Melonworm feeds principally on foliage, especially if foliage of a favored host plant such as summer or winter squash is available. Usually the leaf veins are left intact, resulting in lace-like plant remains. However, if the available foliage is exhausted, or the plant

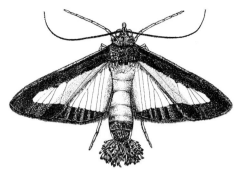

Adult melonworm.

is a less preferred species such as cantaloupe, the larva may feed on the surface of the fruit, or even burrow into the fruit. As happens with pickleworm, growers sometimes refer to these insects as "rindworms," because they cause scars on the surface of melons. In a study of melonworm damage potential to summer squash conducted in south Florida, melonworm caused a 23% yield loss due to foliage damage (indirect loss) and a 9–10% yield reduction owing to fruit damage (direct loss) (McSorley and Waddill, 1982). Kelsheimer (1949) considered this insect to be the most important pest of cucurbits in Florida.

Management

Sampling. Pheromone production by female moths peaks at about sunset (Valles and Capinera, 1992). The sex pheromone has been identified (Raina *et al.*, 1986), but is not available commercially. Moths are not attracted to light traps.

Insecticides. Historically, melonworm was considered to be a very damaging pest, but because it feeds preferentially on foliage it is easy to control with a variety of insecticides. In tropical areas it often is considered more damaging than pickleworm. In temperate areas, and especially in commercial vegetable production areas, it is treated as only a minor pest. In insecticide-free cucurbit production and in home gardens, melonworm can cause serious damage.

Pollinators, particularly honeybees, are very important in cucurbit production, and insecticide application can interfere with pollination by killing honeybees. If insecticides are to be applied when blossoms are present, it is advisable to use insecticides with little residual activity, and to apply insecticides late in the day, when honeybee activity is minimal.

Biological Control. In addition to chemical insecticides, *Bacillus thuringiensis* is commonly recommended for suppression. The entomopathogenic nematode *Steinernema carpocapsae* provides only moderate suppression because the nematodes do not survive long on the foliage, where larvae are found resting and feeding (Shannag and Capinera, 1995).

Cultural Practices. Row covers can be used effectively to exclude melonworm adults (Webb and Linda, 1992). Intercropping of corn and beans with squash was shown to reduce damage by melonworm (Letourneau, 1986). Because melonworm prefers squash to most other cucurbits, trap cropping has been suggested, and of course destruction of crop residue which may contain melonworm pupae is recommended (Smith, 1911). Early plantings, except in tropical areas where melonworm overwinters, often escape serious damage.

Pickleworm
Diaphania nitidalis (Stoll)
(Lepidoptera: Pyralidae)

Natural History

Distribution. Pickleworm is a tropical insect that occurs widely in Central and South America and the Caribbean. In the United States, it routinely survives the winter only in south Florida and perhaps south Texas. Pena *et al.* (1987a) documented the overwintering biology in south Florida, but overwintering has been observed as far north as Sanford, in central Florida, during mild winters. Pickleworm is highly dispersive, and invades much of the southeast each summer. North Carolina and South Carolina regularly experience crop damage by pickleworm, but often this does not occur until August or September. In contrast, northern Florida and southern Georgia are flooded with moths each year in early June as warm, humid tropical summer weather conditions become firmly established. Although it regularly takes one or two months for the dispersing pickleworms to move north from Florida to the Carolinas, in some years they reach locations as far north as Michigan and Connecticut. Presumably they are assisted in their northward dispersal by favorable wind patterns. However, there is some evidence that picklworms also can overwinter in particularly mild coastal areas of southeastern states (M. Jackson, pers. comm.). In Canada, pickleworm has occasionally been found in southern Ontario. In Puerto Rico, it is more common in the mountains than at low elevations, and is not found at all in dry areas of the island (Wolcott, 1948).

Host Plants. Pickleworm feeds only on cucurbits, but both wild and cultivated species are suitable hosts. Creeping cucumber, *Melothria pendula*, is considered to be an important wild host. Wild balsam apple, *Mormordica chorantia*, which has also been reported to be a host, is of questionable significance (Elsey *et al.*, 1985). Summer and the winter squash species are good hosts. Pumpkin is considered of variable quality as a host, probably because pumpkins are bred from several *Cucurbita* species. The *Cucumis* species—cucumber, gerkin, and cantaloupe—are attacked but not preferred. Among all cucurbits, summer squash is most preferred, and most heavily damaged. Cultivars vary widely in susceptibility to attack, but truly resistant cultivars are unknown (Dilbeck *et al.*, 1974). Cucurbits are intolerant of cold weather. Although diapause is unknown in pickleworm, it is the lack of host plants during the winter months that functionally limits the distribution of pickleworm. (See color figures 2 and 26.)

Natural Enemies. Pickleworm has several natural enemies, but none reliably suppress damage. Generalist predators such as *Calosoma* spp. and *Harpalus* (both Coleoptera: Carabidae); the soldier beetle *Chauliognathus pennsylvanicus* De Geer (Coleoptera: Cantharidae); and the red imported fire ant, *Solenopsis invicta* Buren (Hymenoptera: Formicidae); have been reported to be important mortality factors. Also, several parasitoids are known, including *Apanteles* sp., *Hypomicrogaster diaphaniae* (Muesebeck), *Pristomerus spinator* (Fabricius) (all Hymenoptera: Braconidae); *Casinaria infesta* (Cresson), *Temelucha* sp. (both Ichneumonidae); and undetermined trichogrammatids (Pena *et al.*, 1987b; Capinera, 1994). The braconid *Cardiochiles diaphaniae* Marsh (Hymenoptera: Braconidae) has been imported from Colombia and released into Florida and Puerto Rico in an attempt to obtain higher levels of parasitism (Smith *et al.*, 1994).

Life Cycle and Description. The pickleworm can complete its life cycle in about 30 days. Over much of its range, multiple and overlapping generations may occur. The number of generations was estimated to be four in Georgia (Dupree *et al.*, 1955) and two or three in North Carolina (Fulton, 1947).

Egg. The eggs are minute, measuring only about 0.4–0.6 mm wide and 0.8 mm long. The shape varies from spherical to flattened. Their color is white initially, but changes to yellow after about 24 hours. The eggs are distributed in small clusters, usually 2–7 per cluster. They are deposited principally on the buds, flowers, and other actively growing portions of the plant. Hatching occurs in about four days (Smith, 1911). Elsey (1980) estimated egg production to be 300–400 per female.

Larva. There are five instars. Total larval development time averages 14 days. Mean duration (range) of each instar is about 2.5 (2–3), 2 (1–3), 2 (1–3), 2.5 (2–3), and 5 (4–7) days, respectively. Head capsule widths for the five instars are about 0.25, 0.42, 0.75, 1.12, and 1.65 mm, respectively (Smith *et al.*, 1994). Body lengths average 1.6, 2.5, 4.0, 10, and 15 mm during instars 1–5, respectively. Young larvae are nearly white with numerous dark gray or black spots. The dark spots are

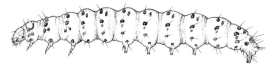

Pickleworm young larva.

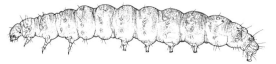

Pickleworm mature larva.

lost at the molt to the fifth instar. Larval color during the last instar is somewhat variable, depending largely on the insect's food source. For example, they tend to be orange when feeding on blossoms, green when feeding on stem tissue, and white when feeding on fruit. Before pupation larvae tend to turn a dark copper color. When mature, larvae often attain a length of 2.5 cm. Smith (1911) provided a good description of the larval instars. (See color figures 25, 28, 74, and 75.)

Pupa. Pupation usually occurs in a leaf fold; often dead, dry material is used. There is only poor evidence of a cocoon, usually just a few strands of silk. The pupa is elongate, measuring about 13 mm long and 4 mm wide. It is light brown to dark brown, and tapers to a point at both ends. Pupation usually lasts about 8–9 days.

Adult. Emerging moths fly during much of the evening hours, but most flight occurs 3–5 h after sundown, with peak flight at approximately midnight (Valles *et al.*, 1991). The female moth produces a pheromone that attracts males, with peak production occurring at 5–7 h after sunset (Klun *et al.*, 1986; Elsey *et al.*, 1989; Valles *et al.*, 1992). Moths are fairly distinctive in appearance. The central portion of both the front and hind wings is a semi-transparent yellow color, with an iridescent purplish reflection. The wings are bordered in dark brown. The wing expanse is about 3 cm. Both sexes often display brushy hairpencils at the tip of the abdomen. Moths are not found in the field during the daylight hours, and probably disperse to adjacent wooded or weedy areas during the heat of the day. Moths do not produce eggs until they are several days old. (See color figure 212.)

Good sources of information on pickleworm biology were supplied by Dupree *et al.* (1955), Fulton (1947), Quaintance (1901), and Smith (1911). Rearing techniques were provided by Elsey *et al.* (1984), Robinson *et al.* (1979), and Valles *et al.* (1991).

Damage

Pickleworm may damage summer and winter squash, cucumber, cantaloupe, and pumpkin. Water-

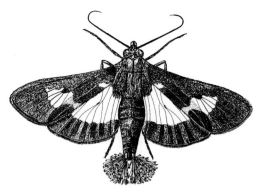

Adult pickleworm.

melon generally is not a host. The blossom is a favored feeding site, especially for young larvae. In plants with large blossoms, such as summer squash, larvae may complete their development without entering fruit. They may also move from blossom to blossom, feeding and destroying the plant's capacity to produce fruit. Very often, however, the larva burrows into the fruit. The larva's entrance is marked by a small hole, through which frass is extruded. The presence of the insect makes fruit unmarketable, and fungal or bacterial diseases often develop once entry has occurred. If larvae burrow into fruit just before harvest, their presence is difficult to detect, yet a considerable amount of larval growth and feeding damage may occur. When all blossoms and fruit have been destroyed, larvae attack the vines, especially the apical meristem. Cantaloupe is not a preferred host, and larvae often seem reluctant to burrow into the fruit. Rather, they feed on the surface or "rind," causing scars. Thus, pickleworm is sometimes referred to as "rindworm."

Management

Sampling. It is very difficult to scout for this insect and predict its appearance. Moths are not attracted to light traps, and pheromone traps have had limited success (Elsey *et al.*, 1991; Valles *et al.*, 1991). Pheromone lures are not currently available commercially. Brewer and Story (1987) developed sampling plans for pickleworm larvae in squash. They suggested that the most reliable sampling unit was the large green staminate flower bud. However, the small eggs, night-flying behavior, and inability to trap the insect reliably has led most growers to depend on preventative applications of insecticides.

Insecticides. Cucurbit producers in areas where pickleworm damage is likely to occur usually apply chemical insecticides from the onset of fruiting through harvest. The internal feeding behavior of larvae, which is so difficult to detect at harvest, causes particular emphasis on prevention of damage. In areas

that are on the fringe of the normal range there are many seasons when damage do not occur, but producers apply insecticides as a preventative measure because prediction of occurrence is so difficult. Botanical insecticides can be effective (Arant, 1942).

Pollinators, particularly honeybees, are very important in cucurbit production, and insecticide application can interfere with pollination by killing honeybees. If insecticides are to be applied when blossoms are present, it is advisable to use insecticides with little residual activity, and to apply late in the day, when honeybee activity is minimal.

Biological Control. The entomopathogenic nematode *Steinernema carpocapsae* has been shown to effectively suppress pickleworm injury in squash (Shannag *et al.*, 1994). Nematode survival is quite good in large-blossomed squash, where the nematodes can kill the young pickleworm before it burrows into the fruit. This approach is probably ineffective for species with small, open blossoms such as cucumber, however, because the nematodes die quickly when exposed to sunlight. *Bacillus thuringiensis* can kill pickleworm, but is usually not recommended because the internal feeding behavior puts the larvae beyond the reach of a stomach-active toxin.

Cultural Practices. It is possible to cover plants with screen or row covers to prevent moths from depositing eggs on the foliage (Webb and Linda, 1992). However, as the plants must be pollinated, usually by honey bees, some allowance must be made to leave the plants uncovered. Given the night-flying behavior of the moths and the daytime activities of honeybees, this is not a difficult task on a small planting but prohibitive on large acreage.

Some growers are able to prevent plant injury through careful timing of their cropping cycle. By planting early, it is often possible to harvest part of the crop before pickleworms appear. Usually the crop is eventually infested, so some yield is lost. Plowing under of the crop residue is recommended to destroy pupae in the leaf debris (Smith, 1911).

The presence of aluminum or polyethylene mulch of various colors was shown by Dupree (1973) not to influence pickleworm damage to squash. Wolfenbarger and Moore (1968) reported that white, black, and aluminum mulches did not reduce pickleworm infestation of squash as compared to unmulched crop, but that white mulch produced lower levels of fruit injury than aluminum mulch.

In a comparison of monocultural and polycultural crop production systems conducted by Letourneau (1986), no difference in abundance of pickleworm on squash was observed. In the same study, distribution of melonworm, *Diaphania hyalinata* (L.) was significantly lower in polycultures, a common response for an insect with a restricted host range. The differential response between the two species is likely due to the more active, dispersive nature of pickleworm.

Smith (1911) reported that squash could be used as a trap crop to keep pickleworm from attacking cantaloupe, a less preferred host. He recommended that destruction of squash blossoms, or even the entire plant, be done periodically to keep pickleworms from exhausting the food supply and then moving onto adjacent cantaloupes. In contrast, Dupree *et al.* (1955) reported unsatisfactory results with trap cropping.

Purplebacked Cabbageworm
Evergestis pallidata (Hufnagel)
(Lepidoptera: Pyralidae)

Natural History

Distribution. This is a decidedly northern species, and is found throughout most of Canada except for British Columbia, and as far south as Virginia and Kentucky in the eastern United States, and northern Arizona in the West. It is likely of European origin, but has been in North America at least since 1869. It is recorded as a pest principally in Canada's Maritime Provinces.

Host Plants. Purplebacked cabbageworm feeds on a variety of cruciferous plants, including broccoli, Brussels sprouts, cabbage, cauliflower, Chinese cabbage, horseradish, kale, kohlrabi, radish, rutabaga, and turnip. Horseradish and turnip seem to be most favored by this insect. Although moths deposit eggs on shepherdspurse, *Capsella bursa-pastoris*; and sheep sorrel, *Rumex acetosella*; larvae do not develop successfully on these plants.

Natural Enemies. Few natural enemies are known. The wasps *Bracon montrealis* Morrison and *Meteorus autographae* Muesebeck (both Hymenoptera: Braconidae) have been reared from this caterpillar.

Life Cycle and Description. There is only a single generation annually in Newfoundland, but two generations per year in Virginia. Overwintering occurs as a mature larva (prepupa) in the cocoon.

Egg. The female deposits small batches of about 3–12 eggs on the underside of host plant foliage. The bright yellow, oval eggs are about 1.1 mm long and 0.8 mm wide, flattened, and overlap like fish scales. The eggs darken markedly just before hatching. Larvae hatch in 4–8 days.

Larva. Newly hatched larvae are whitish green, and measure 1.5–2.0 mm long. The body bears numerous dark tubercles, each bearing one or more hairs. Head capsule widths are about 0.3, 0.5, 0.9, and 1.6 mm for the four instars, respectively. Duration of the instars is reported to be about 9.1, 8.9, 10.3, and 25.7 days, followed by a protracted prepupal period. Larvae are nocturnal, and are usually found hiding between leaves during the day. Mature larvae are robust, bristly, and darker in color, normally olive-green to purple-brown, and measure about 20–22 mm long. There is a conspicuous yellow band on each side, with a narrow white band beneath, and the larva is colored ash-gray or greenish below. The body tapers at both the anterior and posterior ends. The larvae feed on the underside of leaves, then drop to the soil to prepare a cocoon.

Pupa. The cocoon is oval, about 12–15 mm long, and covered with soil particles. The winter is passed as a prepupa in the cocoon, with pupation occurring in the spring. The pupa is light to dark brown.

Adult. The moth escapes through a loosely constructed end of the cocoon. The moth has a wingspan of about 22–28 mm. The front wings are straw yellow, with irregular, narrow dark lines crossing the wing. The hind wings are whitish or pale yellow with darker margins.

Morris (1958), Munroe (1973), and Howard *et al.* (1994) provided the biology of purplebacked cutworm.

Damage

Larvae generally eat holes in the leaves, webbing them together, but also attack the crown and even the roots of such crops as rutabaga. Morris (1958) reported that it was a serious pest in Newfoundland, but its abundance varied widely from year to year.

Management

Moths can be attracted to traps baited with phenylacetaldehyde, which offers the potential for population monitoring (Cantelo *et al.*, 1982). Chemical insecticides and *Bacillus thuringiensis* can be applied against the larvae, but this is usually a minor pest as compared to other crucifer-feeding caterpillars. Spring tillage can destroy the cocoon, and deep tillage can prevent the moths from emerging. Early planted turnip can be used as a trap crop to help protect cabbage and rutabaga.

Sod and Root Webworms
Crambus and others
(Lepidoptera: Pyralidae)

Natural History

Distribution. There are several native pasture-dwelling webworms that occasionally damage crops.

Commonly they are called sod webworms, because they usually are associated with pasture and lawn grasses. They occur throughout the United States and southern Canada. Among the species known to cause damage are corn root webworm, *Crambus caliginosellus* Clemens; silverstriped sod webworm, *Crambus praefectellus* (Zincken); larger sod webworm, *Pediasia trisectus* (Walker); and striped sod webworm, *Fissicrambus mutabilis* (Clemens).

Food Plants. Sod and root webworms feed principally on pasture and sod grasses in the family Gramineae. However, if pasture or sod is tilled and the ground is planted to non-grass crops, they too may be injured by the residual webworm population. In addition to grasses such as bluegrass, corn, orchardgrass, rye, timothy, and wheat, some webworms have been known to attack alfalfa, cabbage, clover, mint and tobacco. Consumption of the latter hosts is unusual. Weed grasses such as crabgass, *Digitaria sanguinalis*, and even some broad-leaf weeds such as sheep sorrel, *Rumex acetosélla*, and aster, *Aster ericoides*, are consumed by larvae. Corn root webworm displays a particular preference for plantain, *Plantago lanceolata*, and oxeye daisy, *Chrysanthemum leucanthemum*.

Natural Enemies. The natural enemies are not well known, but their impact is thought to be significant. Cockfield and Potter (1984) estimated 75% reduction in eggs within 48 h due to predation. Among those thought to be important are the mite egg predators *Hypoaspis* sp., *Cosmolaelaps* sp. (both Acari: Laelapidae), and *Parasitus* sp. (Acari: Parasitidae); the ground beetles *Anisodactylus rusticus* Say, *Amara cupreolata* Putzeys, *A. familiaris* Duftschmidt, *Calathus opaculus* LeConte, and *Stenolophus rotundata* LeConte (all Coleoptera Carabidae); the rove beetles *Meroneura venustula* (Erichson), *Neohypnus* sp., *Philonus* sp., and *Tachyporus jocosus* Say (all Coleoptera: Staphylinidae) and ants, especially *Phedole tysoni* Forel. Birds also are common predators, and where webworms are abundant the sod or soil often is heavily disturbed by birds probing for larvae. Flies are not uncommon parasitoids, including *Aplomya caesar* (Aldrich), *A. confusionis* (Sellars), and *Stomatomyia floridensis* (Townsend) (all Diptera: Tachinidae). Wasp parasitoids known from larger sod webworm include *Macrocentrus crambi* (Ashmead), *M. crambivorus* Viereck, *Apanteles crambi* Weed, *Orgilus detectiformis* Viereck (all Hymenoptera: Braconidae), and *Diadegma obscurum* (Cresson) (Hymenoptera: Ichneumonidae). *Macrocentrus crambi* and *M. crambivorus* have also been reared from corn root webworm. Species reared from striped sod webworm include *Apanteles terminalis* Gahan, *A. ensiger* (Say), and *M. crambi* (all Hymenoptera: Braco-

nidae) and *Campoletis argentifrons* (Cresson) (Hymenoptera: Ichneumonidae).

Life Cycle and Description. Following is a description of larger sod webworm, but the other sod webworm species are similar except for the larger size of *P. trisectus*. Overwintering occurs in the larval stage. In Iowa and Tennessee, these larvae give rise to flights of moths in June, followed by an additional generation-producing flights of moths in August. Light trap catches from Ontario indicate two flights of moths for numerous sod webworm species (Arnott, 1934). Three generations are thought to occur for some sod webworms in the midwestern states, however.

Egg. The eggs are dropped individually and randomly by females, while either at rest or while flying. The eggs are quite small, though large relative to other sod webworms. They measure 0.45–0.56 mm long and 0.31–0.36 mm wide. The eggs are white initially, and turn yellow with age. They are elongate-oval, with one end more broadly rounded than the other. Duration of the egg stage is 5–7 days. Dry air is lethal to eggs (Morrison *et al.*, 1972), but they tolerate a wide range of temperatures (Matheny and Heinrichs, 1971).

Larva. The larvae move downward upon hatching, usually hiding between the blades of grass. They produce silk readily, and if disturbed spin down on a silken thread. As the larvae increase in size they produce a silken tube or tubes beneath the surface of the soil, to which sand and soil particles adhere, as shelter. The larvae may leave the tube to feed, or if food is convenient it remain at least partially within the tube while feeding. The larvae may undergo 7–10 instars, but normally display eight. Duration is about 3.0, 3.1, 3.0, 4.0, 3.6, 4.1, 5.9, and 9.9 days respectively, for instars 1–8. Head capsule widths average 0.21, 0.31, 0.45, 0.67, 0.99, 1.32, 1.55, and 2.20 mm and body lengths 1–2, 2.5–3.5, 3.7–5.5, 6.0–8.5, 10–12, 12–18, 18–24, and 21–28 mm, respectively, for instars 1–8. The young larvae are reddish-brown with a blackish head capsule, but at instar four and thereafter the head becomes yellowish-brown. The larvae bear reddish raised plates from which setae arise. The thoracic and caudal plates tend to be brownish or black until the final instar when they become lighter. Larvae that overwinter can do so in nearly any instar, but this normally is accomplished by instars 2–5.

Larger sod webworm larva.

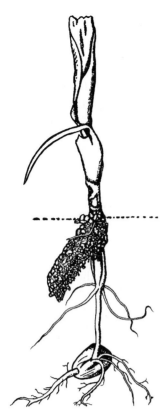

Web enclosing webworm larva attached to seedling beneath the soil surface.

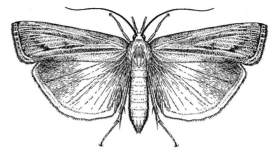

Adult larger sod webworm.

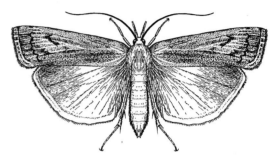

Adult striped sod webworm.

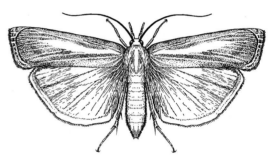

Adult silverstriped sod webworm.

Pupa. The larvae abandon their feeding tubes at maturity and pupate in the soil nearby. The pupal cell is oval and lined with silk. It measures about 14 mm long and 6 mm wide. The pupa measures about 11 mm long and 3 mm wide. The cell is placed very near the soil surface so the moth has no trouble in escaping. Pupation may require 5–15 days, but 7–8 days is normal.

Adult. The moth's front wings and body are yellowish-gray or yellowish-brown, the hind wings lighter and silvery at the base. A line of dark scales usually extends along the mid-line of the wing almost the entire length, but then turns up to the anterior tip. This moth measures about 21–35 mm in wingspan. Moths are nocturnal, but begin activity at dusk. They hide during the day, usually on the underside of broad-leaf weeds. They apparently require water or dew, but have not been observed to feed on flowers. Longevity of adults is normally 7–10 days, though some individuals have survived for nearly a month. Females are believed to produce 200–250 eggs. (See color figure 213.)

The biology of corn root webworm was described by Runner (1914), striped sod webworm by Ainslie (1923a), silverstriped webworm by Ainslie (1923b),

and larger sod webworm by Ainslie (1927). Forbes (1904) and Ainslie (1922) provided a brief description of several sod webworms. Artificial diets have been developed by Ward and Pass (1969) and Dupnik and Kamm (1970).

Damage

Damage often occurs when larvae feed on the leaves of grasses, but feeding can also occur at the soil line or even on the roots. Under high density conditions or if there is a shortage of leaf material, the plant stems and growing point of grasses may be eaten and the plants killed. Larvae commonly chew pits into the side of underground stems or leave the foliage ragged. Plant mortality is most common during periods of drought, and consumption of leaf material by webworms in the pasture environment often goes unnoticed during periods of adequate rainfall.

Management

Sampling. Moths are highly attracted to light and can be captured in blacklight traps. Larval populations are assessed by careful examination of the soil surface.

Insecticides. Insecticides are needed only when crops immediately follow sod or pasture infested with webworms. Webworms normally redistribute themselves within a year, dispersing from crop plants to grass-dominated areas. Liquid and granular insecticides can be applied at, or shortly after, planting to protect the seedlings.

Cultural Practices. Rotation from sod or pasture to crops, especially corn, is risky if webworms have been abundant. Disking and tilling can destroy overwintering larvae, but intense soil disturbance is necessary. Both autumn and spring tillage are suggested for effective suppression.

Southern Cornstalk Borer
Diatraea crambidoides (Grote)
(Lepidoptera: Pyralidae)

Natural History

Distribution. Southern cornstalk borer is, as its common name suggests, found predominantly in the southeastern states. It may be found as far north as Maryland and southern Ohio, and as far west as Kansas and New Mexico. However, these are the geographic extremes, and as a pest its range is mostly limited to the southeastern states from Alabama and Florida to Virginia. It frequently has been confused with related species, particularly sugarcane borer, *Diatraea saccharalis* (Fabricius), so some records of occurrence are suspect. Southern cornstalk borer apparently is native to the southeastern United States, but also occurs in Mexico and South America.

Host Plants. This insect is known principally from corn, but occasionally damages sorghum. It may also be found in some of the wild grasses with thick stems such as Johnsongrass, *Sorghum halepense*; *Paspalum* spp.; and panic grass, *Panicum* spp.

Natural Enemies. Mortality factors of southern cornstalk borer are not well documented. However, an undetermined fungus seems to be quite important in affecting survival of overwintering larvae. *Trichogramma* sp. (Hymenoptera: Trichogrammatidae) egg parasitoids as well as wasp parasitoids of the larval and pupal stages have been noted on occasion, particularly *Syntomosphyrum clisiocampae* (Ashmead) (Hymenoptera: Eulophidae) and *Temelucha ferrunginea* (Davis) (Hymenoptera: Ichneumonidae). The parasitic flies *Lixophaga diatraeae* (Townsend) and *L. sphenophori* (Villeneuve) (both Diptera: Tachinidae) also have been reared from this stalk borer. Probably several general predators consume larvae, and the goldenrod soldier beetle *Chauliognathus pennsylvanicus* De Geer (Coleoptera: Cantharidae) is among those insects known to attack larvae and pupae within the tunnels of the corn stalk.

Life Cycle and Description. There are two generations per year, with the larval stage overwintering in the base of the corn stalk. In Virginia and North Carolina, pupation of overwintering larvae commences in late April and May. Adults and eggs of the first generation are found in May and early June, larvae from May to July, and pupae in July. Second generation adults and eggs occur in July and early August, followed by the larval stage which persists until the following spring.

Egg. The egg is flattened and oval. It measures about 1.6 mm long and 1.0 mm wide. They are deposited singly or in an overlapping, shingle-like fashion in small clusters. Initially, there may be as many as 20 eggs in a single cluster, but over time the female deposits smaller and smaller egg masses until she is depositing only single eggs. Deposition usually occurs on the upper surface of leaf blades. The eggs are whitish when first deposited, but gradually assume a yellow or orange-yellow color. Duration of the egg stage averages about nine days in the spring and eight days during the summer (range about 7–13 days).

Larva. Larvae increase in size from about 1.5 mm long to about 25 mm, and display 5–6 instars during development. Newly hatched larvae are brownish, with a black head. Each body segment is darker toward the posterior, producing a transverse banded pattern. The appearance of the mature larvae differs slightly between seasons. Summer larvae bear a yellowish-brown to brown head capsule, with each body segment bearing several large dark spots dorsally and laterally on a whitish background. Overwintering larvae have a thicker, more robust appearance with the spots markedly lighter, barely darker than the whitish body color. Larvae of southern cornstalk borer are difficult to differentiate from other *Diatraea* spp. A key to the caterpillars boring in corn stalks is found in Appendix A. Mean development time of the first generation, or summer brood, is about 30 days (range 24–38 days). Second brood larvae, of course, persist for about nine months. Larvae tunnel freely within the stalk, with spring generation larvae usually feeding upward from the point of entrance. Overwintering larvae, however, eventually move down to the taproot to pass the winter in a more sheltered location. As they near the end of the larval stage the larvae eat to the

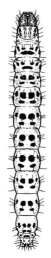

Southern cornstalk borer larva, summer form.

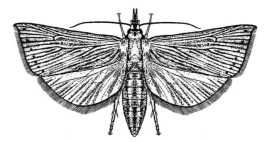

Adult southern cornstalk borer.

outer edge of the stalk, but leave a thin covering of the outermost layer of the stem in place or a few strands of silk across the exit hole. This presumably provides the mature larva and pupa protection from predators, but still allows the adult to escape. Second generation larvae apparently do not feed in the spring, but clear out the tunnel near the exit hole in preparation for pupation.

Pupa. Pupation takes place within the tunnel produced by the larva. The pupa is mahogany brown and measures about 13–17 mm long and 3.5 mm wide. Duration of the pupal stage averages about 20 days in the spring and 12 days during the summer.

Adult. The moths are yellowish-brown except for the hind wings, which are white. Also, there is a small black spot near the center of the forewing, and the wing veins are darker brown. The tips of the front wings of the male moth are sharply angled rather than broadly rounded, approaching 90°, whereas the females are more rounded. The moths measure 25–35 mm in wingspan. As is characteristic of many pyralids, the palpi are enlarged and project forward, imparting the appearance of a pointed head. Oviposition usually commences two days after emergence, and extends for 3–4 days. The moths are nocturnal, and are not strong fliers. The females, in particular, fly only short distances. Females usually deposit about 200 eggs during their life span of 4–6 days, with nearly all eggs produced during the first two nights of oviposition.

A good summary of southern cornstalk borer biology was given by Leiby (1920). The reports by Phillips *et al.* (1921) and Cartwright (1934) are basically abbreviated versions of the same information. A key to the

Diatraea larvae can be found in Peterson (1948), and Stehr (1987), and to the adults in Dyar and Heinrich (1927). A key to stalk borers associated with corn in southern states is presented by Dekle (1976); this publication also includes pictures of the adults. A key to common stalk boring caterpillars also is found in Appendix A.

Damage

The corn plant may be attacked at various stages of growth by southern cornstalk borer larvae, resulting in different types of injury. Early damage occurs when larvae feed on the unfolded leaves at the tip of the plant. If fewer than four larvae are present the result may be only ragged foliage bearing irregular holes. If several larvae are present, however, the bud or growing point of the plant may be killed, stunting the plant and preventing production of a functional tassel. Larvae may also burrow through the veins of leaves and into the stalk. Tunneling larvae are found most frequently at the base of the stalk near the soil line and brace roots. Stalks may support up to 15 larvae, but usually no more than 2–3 attain the pupal stage. Plants that have been tunneled by larvae are susceptible to wind damage.

Management

Insecticides. Insecticides are not usually required for management of southern cornstalk borer. If necessary, insecticides can be applied in granular form to the whorl, where many larvae contact the insecticide. Insecticides can also be applied to the foliage, though because older larvae burrow within the plant, it is difficult to achieve high levels of suppression unless the insecticide is applied when the larvae are young. Systemic insecticides, particularly granular formulations applied to the soil, also are sometimes recommended.

Cultural Practices. Several cultural practices can alleviate damage by southern cornstalk borer. Destruction of stubble or lifting stubble from the soil can eliminate overwintering larvae. Burying stubble deeply can also prevent moths from escaping in the spring.

Modified planting dates are sometimes recommended because late-planted corn can escape infestation. In Virginia and North Carolina, corn planted in June generally escapes damage by first generation insects, but is susceptible to infestation by the second generation. In general, the earlier the planting the more heavily it will be infested. Because moths, particularly females, do not fly far after emergence, crop rotation has considerable benefit.

Southwestern Corn Borer
Diatraea grandiosella Dyar
(Lepidoptera: Pyralidae)

Natural History

Distribution.   Southwestern corn borer is a native to Mexico, and appears to have been first found in the United States in New Mexico in 1891. By 1931 it had spread to the nearby states of Arizona, Texas, and Oklahoma. By 1970 the northern limits of its distribution were Kansas, Missouri, and Kentucky and the eastern limits were Tennessee and Alabama. Thus, it is found widely in the southern United States east of California.

Host Plants.   This species attacks only grasses, with corn as the principal host. Sorghum and sorghum hybrids, broomcorn, sugarcane, and pearl millet are other crops that sometimes serve as hosts. Among weeds, Johnsongrass, *Sorghum halepense*, may also serve as an alternate host, but development time is much longer and fecundity greatly reduced on this plant (Aslam and Whitworth, 1988).

Natural Enemies.   Several natural enemies are known, though their value is open to question. In Texas, egg or early instar mortality is critical to survival of the spring population, whereas in the summer generation survival is most affected by mortality of large larvae and diapausing larvae (Knutson and Gilstrap, 1989a,b, 1990). Among the egg parasitoids are *Trichogramma* spp. (Hymenoptera: Trichogrammatidae), and a larval parasitoid of *Diatraea* spp., *Apanteles diatraeae* Muesebeck (Hymenoptera: Braconidae) (Knutson and Gilstrap, 1989a,b). Parasitism by *Trichogramma* spp. usually occurs late in the oviposition period at low levels, but sometimes 20% or more of the eggs are parasitized (Moulton *et al.*, 1992). General predators such as insidious flower bug, *Orius insidiosus* (Say) (Hemiptera: Anthocoridae), and lady beetles such as *Hippodamia* spp. (Coleoptera: Coccinellidae) are common in corn fields but have little impact because they often seem to be out of synchrony with susceptible stages of southwestern corn borer.

Pathogens such as the fungus *Beauveria bassiana*, and to a fewer extent *Bacillus* sp. bacteria, also affect larvae, particularly the overwintering population. Knutson and Gilstrap (1989a) reported that up to 45% of large larvae were infected with *Beauveria*. Rolston (1955) indicated that unspecified nematodes were sometimes observed, and Knutson and Gilstrap (1990) reported minor incidence of *Heterorhabditis* sp. (Nematoda: Heterorhabditidae).

Mortality also occurs during the winter months when the larval overwintering cell is penetrated by stalk rot fungi and termites, apparently because such penetration allows seepage of water into the cell. Thus, heavy rainfall, and heavy and wet soils are detrimental to overwintering survival, and southwestern corn borer typically is more destructive in areas with sandy soil. Considerable overwintering mortality also results from bird predation, often 50–80% of the total mortality. The yellowshafted flicker, *Colaptes auratus* (L.), is the most important avian predator (Black *et al.*, 1970). Rodents also sometimes consume overwintering larvae, but this occurs erratically.

Life Cycle and Description.   In most locations, southwestern corn borer exhibits 2–3 generations per year. In Mississippi, moths are present in late April–early May, late June–early July, and in August. In northern Arkansas, corresponding moth flights appear in June, July, and August–September. In Texas, moths are found in late May–June, mid-July–early August, and sometimes in late August–early September. A generation normally requires 40–50 days. In most areas, larvae of the first or spring generation do not display diapause, but larvae from the subsequent generation(s) may enter diapause depending on photoperiod and temperature present when they are developing. Photoperiod of less than about 15 h often induces diapause. Larvae developing after early August usually enter diapause in Missouri.

Egg.   The eggs of southwestern corn borer are oval, measuring about 0.8 mm wide and 1.3 mm long. They are flattened, and deposited in an overlapping manner, resembling scales of a fish. The eggs are creamy white, but bear three parallel, transverse, orange-red lines. They are deposited principally on the upper surface of foliage, but also on the lower surface, and occasionally on the stem (Poston *et al.*, 1979). The number of eggs per mass varies from one to several, but averages 3–5 per cluster. Initially, egg masses tend to consist of several eggs, but the number diminishes as the females age, so by the fourth day of oviposition most females are depositing but single egg. The incubation period of the eggs is 4–7 days.

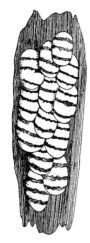

Southwestern corn borer eggs.

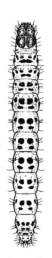

Southwestern corn borer larva, summer form.

Larva. Most larvae pass through six instars, but 5–8 instars have been observed. Mean head capsule widths for instars 1–6 are 0.34, 0.51, 0.89, 1.34, 1.76, and 2.24 mm, respectively (Jacob and Chippendale, 1971). Mean duration of instars fed corn was determined to be 3.1, 2.9, 3.0, 4.2, 4.8, and 7.0 days for instars 1–6, respectively, when reared under variable insectary conditions. Larvae fed corn required about 22 days for development, whereas those fed sorghum, millet, and Johnsongrass required about 29, 49, and 45 days, respectively. Larval survival rates are much higher on young corn than old plant material due to the absence of succulent tissue on mature corn. Larvae in the first two instars are whitish but have a reddish prothorax and dorsal reddish stripe along the abdomen. The third and succeeding instars of nondiapausing larvae are distinctly marked with black spots on a yellowish white background. Larvae that overwinter do not acquire the black spots, rather appearing mostly yellowish-white, or marked only with faint brownish spots. Larvae are difficult to distinguish from other *Diatraea* stalk borers. Larvae generally feed externally on leaf or husk tissue during the first three instars, then bore into the stalk for the remainder of their larval development period. Once inside the stalk, larvae tunnel for about 7–30 cm. Following tunneling, nondiapausing larvae prepare an exit hole for the adult, and pupate within the tunnel. Larvae about to enter diapause feed downward, or leave the feeding site and crawl downward externally, moving to the base of the stalk and entering the taproot where they construct a crude cell in preparation for overwintering. As part of the pre-diapause feeding behavior, larvae may girdle the stalk a few centimeters above the soil line. Only southwestern corn borer, among the several *Diatraea* spp. affecting corn, girdles the stalk. Girdling apparently is an acquired trait associated with dispersal into colder climates. Girdling is infrequent in Mexico, but common in the United States, where increased overwintering survival is associated with this behavior. Larvae tend to be cannibalistic, especially as they prepare for diapause, and though several larvae may develop in a single stalk usually only one successfully overwinters. In the spring, larvae clean the escape tunnel in preparation for emergence of the adult stage. They close off the tunnel exit with strands of silk, probably to deter entry by predators that might attack the pupal stage. (See color figure 80.)

Pupa. Pupation usually occurs in the overwintering cell in the taproot, but occasionally above-ground within the tunneled area of the stalk. The pupa is yellowish-brown with diffuse dark bands, especially dorsally. The tip of the abdomen is broadly rounded and bears thick spines. The pupa measures 13–25 mm long and 3.5 mm wide. Duration of the pupal stage averages about 14.8 days (range 11–20 days).

Adult. The moths are buff or tan in color, with seven faint narrow lines on each forewing that terminate in a minute dark spot. The hind wings are white with buff-colored veins. The male moth measures 15–30 mm in wingspan, whereas the female measures 30–38 mm. The palpi are prominent, projecting forward from the head in a manner common among pyralid moths. The adult stage is short lived, persisting for about 4–5 days, and does not feed. Adult females are ready to oviposit within 24 h of emergence. They are active at night, particularly 1–2 h before and after midnight. Females produce and release a sex pheromone during the first three days after emergence. Females produce 300–400 eggs. (See color figure 215.)

The biology of southwestern corn borer was reviewed by Davis *et al.* (1933), Rolston (1955), Henderson and Davis (1969), and Chippendale (1979). Phenology was also presented by Walton and Bieberdorf (1948) and Knutson et al. (1982). Developmental biology was given by Whitworth and Poston (1979), Knutson et al. (1989), and Ng et al. (1993). An artificial diet for borer culture was presented by Whittle and Burton (1980). A bibliography was published by Morrison et al. (1977). A key to the *Diatraea* larvae can be found in Peterson (1948) and Stehr (1987), and to the moths in Dyar and Heinrich (1927). Southwestern corn borer was included in the larval key by Capinera (1986) and the moth key by Capinera and Schaefer (1983). A key to stalk borers associated with corn in southern states was presented by Dekle (1976); this publication also includes pictures of the adults. A guide to the common stalk boring caterpillars is found in Appendix A.

Damage

All stages of the plant may be injured by feeding of southwestern corn borer larvae. Early instars of the first generation feed on leaf tissue, especially new tissue within the whorl of young corn plants. This can result in destruction of the terminal bud (a condition called "dead heart"), loss of apical dominance, and development of lateral buds. Such plants are stunted and bushy. Early instars of the second generation feed mostly on leaf sheaths, the husk, shank, kernels, and cobs of ears. In sweet corn it is difficult to detect the presence of the larvae in the ear until the husk is removed. Late instars of all generations bore within the stalk. Stalk damage may result in stunting if it occurs early in the growth of the plant; in more mature plants tunneling may disrupt translocation of nutrients to the ears, causing decrease in kernel size. Larvae about to enter diapause also girdle the stalk internally just above soil level. Girdling increases the likelihood that plants break. Stalk breakage is not often a problem in sweet corn due to early harvesting, but is a severe threat to grain corn. Damage potential to grain corn was presented by Whitworth et al. (1984).

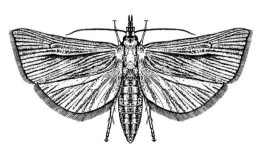

Adult female southwestern corn borer.

Management

Sampling. Moths can be taken at blacklight traps, but they are not strongly attracted to light. Several types of traps baited with sex pheromone can be used to monitor populations (Knutson et al., 1987), but they differ in their ability to capture moths. The nonsticky *Heliothis* trap (Davis et al., 1986) and bucket trap (Goodenough et al., 1989) are most effective. The distribution of eggs and small larvae are highly aggregated, but older stages are progressively less clumped. Sequential sampling protocols have been developed (Poston et al., 1983; Overholt et al., 1990).

Insecticides. Several protocols are available for insecticide-based suppression of larvae (Daniels, 1978; Buschman et al., 1985). Liquid formulations of insecticides are applied to the foliage to kill young larvae before they burrow into the stalk. Fairly precise timing or multiple applications may be necessary to produce good larval suppression, and application of insecticide in overhead irrigation systems is effective. Granules may be applied into the whorls, because larvae tend to aggregate on this succulent tissue. Systemic insecticides applied to the soil at planting are effective for first generation borers, and may also be applied to foliage for second generation infestations.

Cultural Practices. Several cultural practices are used to minimize the effects of southwestern corn borer on corn production. Early planting is often recommended, because damage to the growing point of the corn plant is minimized or prevented. Cultivation, especially if done early in the winter, prevents adults from emerging successfully (Archer et al., 1983). Lifting stubble from the soil exposes larvae to more severe overwintering conditions, and also reduces adult emergence in the spring. Corn planted at high densities is more likely to suffer girdling, than corn at low densities, despite a tendency for equal incidence of infestation; the basis for this disparity is unknown (Zepp and Keaster, 1977).

Host-Plant Resistance. Corn cultivars possessing considerable resistance to corn borer feeding have been located and incorporated into commercial varieties, with mixed results (Davis et al., 1991; Ng et al., 1990; Thome et al., 1992). Resistance is attributable to both limited damage by larvae and non-preference by adults (Ng et al., 1990).

Biological Control. The wasp *Pediobius furvus* (Gahan) (Hymenoptera: Eulophidae) was imported from Africa to attack nocutid and pyralid borers, and evaluated against southwestern corn borer in Texas. Low levels of borer parasitism were obtained under field conditions, and the wasp failed to overwinter, thus limiting its usefulness to augmentative releases

(Overholt and Smith, 1989). Southwestern corn borer is also susceptible to infection by the bacterium *Bacillus thuringiensis* and the alfalfa looper baculovirus (Davis and Sikorowski, 1978; Nolting and Poston, 1982), but these materials have not come into general use for southwestern corn borer.

Sugarcane Borer
Diatraea saccharalis (Fabricius)
(Lepidoptera: Pyralidae)

Natural History

Distribution. This species is a native to the western hemisphere, but not to the United States. It apparently was introduced into Louisiana about 1855, and has since spread to the other Gulf Coast states. It inhabits only the warmer parts of these states, however. Sugarcane borer also occurs throughout the Caribbean, Central America, and the warmer parts of South America south to northern Argentina.

Host Plants. Sugarcane borer attacks plants in the family Gramineae. Though principally a pest of sugarcane, this insect also feed on other crops such as corn, rice, sorghum, and sudangrass. Many wild or weed grasses are suitable hosts, including Johnsongrass, *Sorghum halepense*; *Paspalum* sp.; *Panicum* spp.; *Holcus* sp.; and *Adropogon* sp.

Natural Enemies. The importance of natural enemies in corn-cropping systems is not known, because most studies involve only sugarcane. Ants are reported to be important predators of sugarcane borer in sugarcane fields, capable of reducing damage by over 90% (Bessin and Reagan, 1993). Although much of the attention has been focused on red imported fire ant, *Solenopsis invicta* Buren, other species such as *Pheidole dentata* Mayr and *P. floridana* Emery (all Hymenoptera: Formicidae) also are important (Adams *et al.*, 1981).

Effective parasitoids are not established in the United States. Egg parasitoids, *Trichogramma* sp. (Hymenoptera: Trichogrammatidae), are possibly the most important naturally occurring parasitic insects. Although they are not very abundant early in the season, by autumn they may inflict almost complete destruction of borer eggs. The most important imported parasitoid is *Agathis stigmaterus* (Cresson) (Hymenoptera: Braconidae), which was reported by King *et al.* (1981) to affect, on average, less than 12% of borers. *Lixophaga diatraeae* (Diptera: Tachinidae) has the potential to cause high levels of parasitism, but does not persist well (see biological control, below). A wasp introduced from India, *Cotesia flavipes* Cameron (Hymenoptera: Braconidae), is an important

late-season parasitoid late in the summer within Florida. Other parasitoids include *Orgilus elasmopalpi* Muesebeck, *Apanteles diatraeae* Musebeck, *Apanteles impunctatus* Musebeck (all Hymenoptera: Braconidae), *Euplectrus plathypenae* Howard, and *Syntomosphyrum clisiocampe* (Ashmead) (both Hymenoptera: Eulophidae).

The comparative assessment of natural enemies in sugarcane and sorghum conducted by Fuller and Reagan (1988) probably offers some insight into the role of natural enemies in corn, because cultural practices in sorghum and corn are similar. Predator densities were higher in sugarcane owing to the greater abundance of red imported fire ant. However, *Orius* spp. pirate bugs (Hemiptera: Anthocoridae), lacewings (Neuroptera: Chrysopidae), tiger beetles (Coleoptera: Cicindelidae), spiders, and foliage-dwelling ground beetle larvae (Coleoptera: Carabidae) were more abundant in sorghum fields. Suppression of predators with soil-applied insecticide affected predation in both agroecosystems, with borer populations 40–60% higher where predator abundance was reduced.

Weather. An inverse relationship between rainfall and borer abundance has been reported from both Louisiana and Puerto Rico. Heavy rainfall, and particularly winter rainfall resulting in flooding, depresses borer survival (Holloway *et al.*, 1928). This is thought to result from prolonged emersion of stalks containing overwintering larvae in flood water. Also, young larvae living in the whorl of corn or sugarcane are quite tolerant of short-term emersion, but heavy rainfall while they are dispersing could lead to death because they are washed from the plants. In addition to rainfall, cold winter temperature is reported to depress larval survival rates in Louisiana.

Life Cycle and Description. Overwintering occurs in the larval stage, with pupation in the spring. In Louisiana and Texas, adults become active by April or May, and the borer population continues to cycle until autumn. Development time is highly variable, so the generations overlap considerably, obscuring population trends. There is potential for 4–5 generations to occur annually, but moths are abundant only in spring and autumn (Fuchs and Harding, 1979), so perhaps there are fewer generations. During the summer a complete generation may require only 25 days, whereas during the winter over 200 days are needed.

Egg. The eggs are flattened and oval, measuring about 1.16 mm long and 0.75 mm wide. They are deposited in clusters, and overlap like the scales of a fish. An egg cluster may contain from 2 to 50 eggs, with eggs deposited on both the upper and lower surface of leaves. The eggs are white initially, but turn orange with age and then acquire a blackish hue just before hatching. Duration of the egg stage is 4–6 days. Mean fecundity is about 700 eggs when borers are reared on corn and sugarcane, but only about 425 eggs when fed Johnsongrass (Bessin and Reagan, 1990).

Larva. The eggs within a cluster hatch about the same time, or at least within a few hours of one another. Larvae tend to congregate in the whorl of corn plants and begin feeding almost immediately. They may feed through the leaf tissue or tunnel through the midrib. After the first or second molt they burrow into the stalk. The larvae display both summer and winter forms. The larvae are whitish with a brown head, but the summer form also bears large brown spots on each body segment whereas the winter form lacks spots. A stout hair originates in each of the spots, or for the winter form, from the location where the spot might appear. Larvae during the winter are rarely found in corn; sugarcane and stalks of large grasses are more suitable and preferred. Instar number is quite variable. There are reports of about 3–10 instars, but 5–6 is normal. Holloway *et al.* (1928) reported instar duration of about 3–6, 4–8, 6–9, 4–6, and 4–9 days for instars 1–5, respectively, for larvae fed sugarcane. When reared on artificial diets, most larvae tend to display six instars. Roe *et al.* (1982) reported mean head capsule widths of about 0.29, 0.40, 0.62, 0.93, and 1.32 mm for instars 1–5 in larvae that had six instars; head capsule measurements were not reported for the final instar but probably were about 1.75 mm. Larval development time usually requires 25–30 days during warm weather and 30–35 days during cool weather except, of course, during the winter when development is arrested. Larvae attain a length of about 2–4, 6–9, 10–15, 15–20, 20–30 mm during instars 1–5, respectively. Larvae of sugarcane borer are easily confused with southern cornstalk borer, and definitive separation involves microscopic examination of the mouthparts. Sugarcane borer, however, is much less likely than southern cornstalk borer, *Diatraea crambidoides* (Grote), to be found infesting corn. (See color figure 81.)

Pupa. Pupation occurs within the plant, in a tunnel created by the larva. The larva cleans and expands the tunnel before pupation, leaving only a thin layer of plant tissue for the moth to break through at emergence. The pupa is elongate and slender, and yellowish-brown to mahogany brown. It measures 16–20 mm long and bears prominent pointed tubercles on the distal segments. Duration of the pupal stage is usually 8–9 days, but under cool conditions may extend for up to 22 days.

Adult. The adult is a yellowish or yellowish-brown moth with a wingspan that measures 18–28 mm in males and 27–39 mm in females. The forewing also bears numerous narrow brown lines extending to the length of the wing. The hind wing of females is white, but in males it is darker. The adults are nocturnal, and remain hidden during the daylight hours. Oviposition commences at dusk and continues throughout the evening. Females may deposit eggs for up to four days, but often less. Duration of the adult stage is 3–8 days.

The biology of sugarcane borer was described by Holloway *et al.* (1928) and a bibliography was authored by Roe (1981). Several wheat germ-based diets are suitable for rearing (Roe *et al.*, 1982). A key to the *Diatraea* larvae can be found in Peterson (1948), and Stehr (1987), and to the adults in Dyar and Heinrich (1927). A key to stalk borers associated with corn in southern states was presented by Dekle (1976); this publication also includes pictures of the adults. A key to common stalk boring caterpillars also is included in Appendix A.

Damage

Although generally regarded as a potentially serious pest of sugarcane, other crops are rarely at risk.

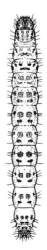

Sugarcane borer larva, summer form.

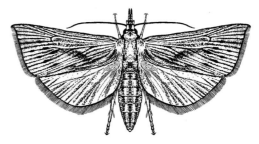

Adult female sugarcane borer.

Sugarcane borer is a minor pest of sweet corn even in Florida, where the weather favors its survival and sugarcane is found abundant (Kelsheimer *et al.*, 1950). Damage by sugarcane borer to grain corn was described by Flynn and Reagan (1984) and Flynn *et al.* (1984). Larvae injure corn in two ways. Early in the season they attack the whorl, feeding on the young developing tissue. If such damage is light, the result may be only series of holes across the leaf blade. If such damage is extensive however, the growing point of the plant may be killed and plant growth stunted. This condition is called "dead heart." Later in the season the larvae descend to the stalk and burrow into it. Large larvae tunnel through the stalk, causing the plant to be prone to breakage. On occasion, especially during the second generation, larvae may burrow into corn ears (Rodriguez-del-Bosque *et al.*, 1990).

Management

Sampling. Sampling protocols have not been devised for sweet corn because this is a relatively minor pest. Monitoring of adult populations, egg density, and foliar feeding by young larvae are advisable if sweet corn is cultured in the vicinity of sugarcane. Larval distribution in sugarcane was described by Hall (1986).

Insecticides. Insecticides can be applied to the foliage of sugarcane, providing significant yield increases even in the presence of predation and resistant varieties (Bessin *et al.*, 1990). Insecticides should be applied while the larvae are young, before they burrow into the stalk. However, some control is possible even later, possibly because larvae leave their tunnel during the process of pushing out excrement.

Cultural Practices. Sugarcane is the principal host of sugarcane borer, and proximity of corn to sugarcane is an important determinant of borer abundance in corn. Moths deposit more eggs on sugarcane than corn when these hosts are in close proximity, and avoid pubescent cultivars (Sosa, 1990). It is advisable to destroy cane trash in the winter as it reduces overwintering by larvae, but the practice of burning does not always kill borers deep within the stalks. Borers overwinter within corn stalks, but usually only late-planted corn is suitable. Some sugarcane cultivars display considerable resistance to sugarcane borer (Bessin *et al.*, 1990; Bessin and Reagan, 1993), which presumably can reduce overall abundance of borers and infestation potential in corn. Grain corn varieties with resistance to sugarcane borers also have been identified (Maredia and Mihm, 1991).

Biological Control. The Caribbean region and tropical areas of South America have been surveyed extensively for natural enemies. Many species were introduced into the United States, but few could be established (Clausen, 1978). *Agathis stigmatera* (Cresson) (Hymenoptera: Braconidae) was successfully imported from Argentina and Peru, and though it is well-established in both Florida and Louisiana, its effect on sugarcane borer is minimal. The fly, *Lixophaga diatraeae* (Townsend) (Diptera: Tachinidae) was imported and released repeatedly, but tends to disappear or dissipate after a few years. In some countries, augmentative releases are used to attain high levels of parasitism in sugarcane borer, and this has been attempted in Louisiana (King *et al.*, 1981). Some authors have claimed success with augmentative releases of *Trichogramma* spp. (Hymenoptera: Trichogrammatidae), but this has proven difficult to implement in the United States (Long and Hensley, 1972).

Sweetpotato Vine Borer
Omphisa anastomasalis (Guenée)
(Lepidoptera: Pyralidae)

Natural History

Distribution. Sweetpotato vine borer is widespread in Asia where it is destructive in such countries as China, India, Indonesia, Japan, Philippines, and Vietnam. In the United States its distribution is limited to Hawaii, where it was first observed in 1900.

Host Plants. This species is associated with plants in the family Convolvulaceae. It is destructive only on sweet potato, but other *Ipomoea* spp. are common hosts. In Hawaii it is also reported from *Stictocardia campanulata*.

Natural Enemies. Larval parasitoids known from Hawaii include *Chelonus blackburni* Cameron (Hymenoptera: Braconidae), *Enytus chilonis* Cushman and *Pristomeris hawaiiensis* Perkins (both Hymenoptera: Ichneumonidae). Other parasitoids are known from Asia.

Life Cycle and Description. Phenology of sweetpotato vine borer is not documented. Considering the importance of the insect and crop, relatively little work has been done on this species.

Egg. The eggs of sweetpotato vine borer are elliptical with a flat base, measuring about 0.6 mm long, 0.5 mm wide, and 0.35 mm in height. They are greenish and laid singly or in small clusters of 2–3 in crevices on leaves, petioles, and stems. The incubation period is 5–7 days.

Larva. After hatching, larvae bore into the vine and move toward the base of the plant as they feed.

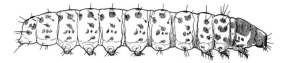

Sweetpotato vine borer larva.

Larvae are only about 1 mm long at hatching and are whitish with a black head and prothoracic shield. Mature larvae attain a length of about 30 mm and may be yellowish-white or light purple. Large larvae are marked with brownish tubercles, which appear as spots over most of the body. The intersegmental membranes tend to be yellowish-brown, especially in the anterior region of the body. The head and prothorax are brownish and the body bears scattered stiff hairs. There are six instars. The duration of the larval stage is usually 30–35 days, but values of 21–92 days have been reported.

Pupa. Pupation normally occurs within the base of the vine, but occasionally larvae pupate in the tubers if they are close to the soil surface. The larva spins a thin web-like cocoon and cuts an exit hole for the moth before pupation. The pupa is light to medium brown and measures about 16 mm long and 3 mm wide. Duration of the pupal stage is 14–18 days.

Adult. The moth is white, but is heavily marked with yellowish brown. The base of the front wings bear a large, irregular dark spot. The abdomen is brown dorsally. The distal area of the front wings and the entire hind wings are marked with irregular dark lines. The wingspan is 30–40 mm. Moths are nocturnal and females produce sex pheromone. Adults survive for about 10 days, during which females deposit about 300 eggs.

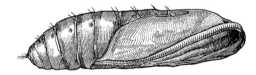

Sweetpotato vine borer pupa.

Adult sweetpotato vine borer.

Fullaway (1911) and Talekar and Pollard (1991) provided the biology of sweetpotato vine borer. Yoshiyasu (1975) described the adult stage.

Damage

Larvae mine the vines of sweet potato, disrupting the flow of water and photosynthates. Infested vines show weak growth, poor foliage development, and poor tuber development. Larvae also may bore into the upper portions of tubers. Yield reductions are directly related to infestation levels, and yield loss of 30% or more are common in Asia. In Hawaii, heavy infestations have been reported to kill plants (Talekar and Cheng, 1987; Talekar and Pollard, 1991). When infestation levels are low the feeding damage is much less pronounced. Sometimes, the only outward evidence of larval feeding is the accumulation of fecal material near the opening of the larval tunnel, which usually is at the crown of the plant.

Management

Insecticides. Insecticides can be applied for larval suppression, but the mining behavior of this insect requires that insecticides be in place at hatching, before larvae burrow into the tissue, or be systemic and translocated to the tissues where larvae feed. In areas of the world where sweetpotato vine borer is common, insecticide use often results in very sizable yield increases. Farmers are often discouraged from using insecticides, however, because of the low value of the crop.

Host-Plant Resistance. Considerable effort has been directed to screening varieties for resistance to sweetpotato vine borer. Although some cultivars display resistance, the yield potential of these selections is low, and additional work is needed by breeders to combine the insect resistance with high yield characteristics.

FAMILY SESIIDAE—VINE BORERS AND CLEARWING MOTHS

Squash Vine Borer
Melittia cucurbitae (Harris)

Southwestern Squash Vine Borer
Melittia calabaza Duckworth and Eichlin
(Lepidoptera: Sesiidae)

Natural History

Distribution. Squash vine borer, *M. cucurbitae*, is found throughout the United States east of the Rocky

Mountains, and is also known from Central America. In Canada it occurs in southern Ontario. In the southwest, *M. cucurbitae* is replaced by *M. calabaza* (Eichlin and Duckworth, 1988). This latter species is poorly studied, but seems to have a biology similar to *M. cucurbitae*; the only known exception is host range, as noted below.

Host Plants.   Squash vine borer feeds on wild and cultivated species of *Cucurbita*. Summer squash, *Cucurbita pepo*, and some winter squash, *C. maxima*, are most preferred for oviposition by adults, and most suitable for larval development. Some pumpkin, *C. mixta*, are intermediate in suitability. Winter squash derived from *C. moschata* does not support complete larval development. The weeds *C. texana* and *C. andreana*, are very suitable hosts, while *C. okeechobeensis* is intermediate and several cucurbit weed species are unsuitable (Howe and Rhodes, 1973). Other cucurbit crops such as cucumber and melon may be attacked, but this occurs mostly when the more favored hosts are not present (Friend, 1931).

Southwestern squash vine borer, *M. calabaza*, while generally having a host range similar to *M. cucurbitae*, was able to complete larval development in *C. moschata* and was not found in *C. mixta* (Eichlin and Duckworth, 1988).

Natural Enemies.   The natural enemies of squash vine borer are not well-studied. General predators such as robber flies (Diptera: Asilidae) are reported to prey upon adults, and a wasp egg parasitoid, *Telenomus* sp. (Hymenoptera: Scelionidae) is known. Larvae seem to be relatively free of natural enemies (Friend, 1931).

Life Cycle and Description.   The squash vine borer completes its life cycle in about 60 days. In the southern states it has two generations per year. In New York, New Jersey and other cold climates only a single generation occurs regularly. In Ohio, most borers diapause after the first generation, but some go on to complete a second generation (Smith, 1893). There has been considerable confusion concerning the number of generations in northern states, because in the field it is difficult to discern a single protracted generation from two overlapping generations. In North Carolina, the first generation is reported to occur from April to June and the second from July to September (Smith, 1910). In Connecticut, squash vine borer may not be observed until June or July and persist until September (Friend, 1931). Thus, each generation may remain active for two or three months.

Egg.   Oval eggs, reddish brown in color, are glued to the plant tissue and are slightly flattened at the point of attachment. They measure about 1.1 mm long and 0.85 mm wide. Each female is capable of producing 150–200 eggs. They are distributed singly on all parts of the plant except the upper surface of the leaves; most appear to be attached to the basal region of the vine, probably because this tissue is oldest and therefore exposed longer. The basal part of leaf petioles receives a moderate number of eggs. Some eggs are deposited in cracks in the soil near the base of the plant. Eggs hatch in 10–15 days and larvae soon bore into the plant. Those that hatch on the vine usually remain in this location to feed, whereas those hatching on leaf petioles eventually work their way to the base and enter the vine.

Larva.   Larvae are white with a dark-brown or black head. Initially, the larval body is markedly tapered posteriorly and equipped with numerous large hairs. As the larva matures it loses its tapered shape and hairy appearance and acquires a dark thoracic shield. There are four instars, with head capsule widths of about 0.4, 0.7, 1.3, and 2.0 mm, respectively. The larva requires 4–6 weeks to complete its development and eventually attains a length of about 2.5 cm. The mature larva exits the plant tissue and burrows into the soil to a depth of 4–5 cm where pupation and formation of a black cocoon occurs. (See color figures 27 and 89.)

Pupa.   The pupa is mahogany brown and the anterior end is equipped with a short but sharply pointed cocoon breaker, a structure used to cut a hole in the cocoon when the moth is ready to emerge. There is some disagreement whether squash vine borer always pupates soon after producing the cocoon. Friend (1931), working in Connecticut, reported that the fully grown larva overwintered in the cocoon, and pupated in the spring before producing the single northern generation. In contrast, R. I. Smith (1910) maintained that among two-generation populations in North Carolina the pupa overwintered. In any event, a distinctive, heavy-bodied moth eventually emerges from the cocoon.

Adult.   The wingspan of the moth is about 3 cm. The front wings are blackish and tinged with olive-green while the hind wings are colorless. The abdomen is usually dull orange and marked with black

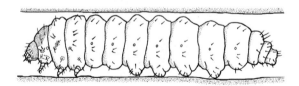

Squash vine borer larva.

Squash vine borer pupa and cocoon.

dots dorsally, though on some specimens the abdomen is entirely black. The hind legs are prominently tufted and are orange and black. Adults are capable of oviposition about three days after emergence. Moths are active during the daylight hours. (See color figure 254.)

Friend (1931) gave a detailed summary of life history. Good rearing techniques for this insect are unknown.

Damage

The damage caused by these insects is accurately described by the common name—squash vine borer. The larva spends almost its entire life feeding within the plant stem. Because larvae feed within plant tissues they are hidden from view and easily overlooked. However, upon close examination frass can be found accumulating beneath small entrance holes. The presence of one, or even several, larvae is not always deadly to the plant. However, up to 142 larvae have been removed from a single plant, and obviously such large numbers can disrupt the physiology of the plant beyond its ability to compensate successfully for the feeding injury. Often the first sign of infestation is wilting of the plant in the heat of the day while other plants remain turgid. If the vine is thin or heavily

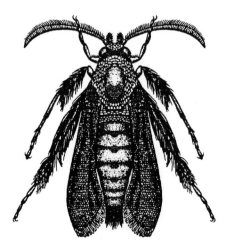

Adult squash vine borer.

infested the portion beyond the feeding site of the larva(e) may be killed. Infestation of fruit can occur. Commercial cucurbit production rarely suffers significant damage by squash vine borer. These borers seem to plague small plots especially severely; almost all home gardeners have had some unpleasant experience with this pest.

Management

Sampling. Shallow pans painted yellow and filled with water can attract and capture squash vine borer moths. Although water pan traps are recommended by investigators in Ontario for population monitoring, it is not known whether such traps are attractive enough to be used as a method of population reduction (McLeod and Gualtieri, 1992). A sex pheromone produced by females has been identified (Klun *et al.*, 1990). Male moths of *M. cucurbitae* may be attracted to traps baited with the insect's sex pheromone, but thus far pheromone technology has not been demonstrated to improve management practices. In fact, one study reported increased injury of squash plantings when pheromone was released; the biological basis for this is uncertain (Pearson, 1995). *M. calabaza* also is attracted to some components of the *M. cucurbitae* pheromone.

Insecticides. Squash vine borer can be difficult to control with chemical insecticides, because it is protected from contact with most insecticides by its burrowing activity. Therefore, it is imperative that insecticide residues be in place during oviposition so that when the larva hatches from the egg it has an opportunity to make contact with a lethal dose.

Pollinators, particularly honeybees, are very important in cucurbit production, and insecticide application can interfere with pollination by killing honeybees. If insecticides are to be applied when blossoms are present, it is advisable to apply insecticides late in the day, when honeybee activity is minimal.

Cultural Practices. It has long been known that in the north, early planted crops experience minimal injury, so early planting is recommended. Also, the moths preferentially attack certain species such as summer squash; this can be used as a trap crop if the plants are then destroyed or sprayed, thereby protecting less preferred cucurbit species.

Cucurbit plants are very resilient, recovering well from injury. If accumulated frass can be located adjacent to vine tissue, indicating the presence of active feeding, larvae can be killed or removed with a knife without severely affecting growth of the remaining tissue. Vines also develop additional roots if covered with soil, which can compensate for damage by borers.

Covering the vines near the base of the plant discourages oviposition near critically important tissue. Dying vines should be removed and destroyed (Britton, 1919).

Biological Control. Using a medicine dropper or similar device, the bacterium *Bacillus thuringiensis*, and the entomopathogenic nematode *Steinernema carpocapsae*, can be injected into squash vines to kill existing larvae or to prevent their establishment. The hollow, moist vines are especially conducive to the spread and survival of nematodes.

FAMILY SPHINGIDAE—HORNWORMS AND SPHINX MOTHS

Sweetpotato Hornworm
Agrius cingulatus (Fabricius)
(Lepidoptera: Sphingidae)

Natural History

Distribution. Sweetpotato hornworm is native to the western hemisphere, where it is found throughout the tropical and subtropical areas. In the United States, it occurs throughout the southern states, and is found throughout the year in Florida and southern Texas. Each summer the moths are found as far north as Arkansas, and occasionally a stray is found in such northern locations as Michigan and Nova Scotia. Sweetpotato hornworm also occurs in Hawaii and Puerto Rico.

Host Plants. Larvae feed on various species of *Ipomoea* in the plant family Convolvulaceae. Among vegetable crops, only sweet potato is injured. The adults feed on nectar from various deep-throated flowers.

Natural Enemies. The eggs of this species are heavily parasitized by *Trichogramma semifumatum* (Perkins) (Hymenoptera: Trichogrammatidae). Several fly parasitoids are known, including *Agryophylax* spp., *Belvosia bifasciata* (Fabricius), *Chaetogaedia monticola* (Bigot), *Drino inca* (Townsend), *D. incompta* (Wulp), and *D. rhoeo* (Walker) (all Diptera: Tachinidae) (Arnaud, 1978). The aforementioned tachinids attack the larval stage and have a wide host range; they also have been reared from many other species of Lepidoptera.

Life Cycle and Description. The biology of this insect is not well documented. The number of generations is reported to be at least two in North Carolina, and probably more in Gulf Coast states.

Egg. The eggs are deposited singly on the underside of the foliage. They are nearly spherical in shape, whitish or greenish, and measure about 1.35 mm in diameter. The female is reported to produce about 30–40 eggs. Duration of the egg stage is 6–10 days.

Larva. Young larvae measure only about 3 mm long when they hatch, and are greenish-white with white granulations. The head, however, is greenish-yellow. There are five instars and the larvae attain a length of about 115 mm at maturity. Head capsule widths are about 0.5, 1.0, 1.8, 3.2, and 6.0 mm, respectively. Duration of the larval stage is about 30–60 days, depending on temperature. The appearance of the larva is variable, the body color ranging from light green or brown to almost black. Typically, in the early instars the body is green and marked by oblique black stripes along each side on a whitish or light background, or with a narrow white band along the lower edge of the oblique stripes. The oblique lines may also form "V"-shaped lateral markings. The spiracles and anal appendage, or "horn" are blackish. In the final instar, the larva may remain green or becomes mostly brown. The green form has dark bands on the head capsule, and the oblique bands are found in the earlier instars; the posterior oblique band terminates in the horn. In the brown form, the body is gray with brown spots. The lateral oblique lines are weak, lighter beneath, and less likely to be "V"-shaped markings. The black spiracles are surrounded by a dark ring, and give the appearance of very large spiracles. The prolegs are dark brown or purplish. As in the green form, the head is marked with dark bands and the posterior oblique lines terminate in the dark horn. (See color figures 85 and 86.)

Pupa. At maturity the larva pupates in the soil. The pupa is dark brown or mahogany red and measures 55–65 mm long. Duration of this stage is 20–30 days. (See color figure 269.)

Adult. The moth is sometimes known as the "pink-spotted hawk moth." The adult is fairly typical in form for sphingids, heavy bodied with long pointed front wings, and short hindwings. The front wings are dark gray mottled with brown and black. The hind wings are pinkish basally with 2–3 dark bands that run parallel to the wing border. The abdomen bears a dark band dorsally and transverse pink or rose transverse bars on each side. The moth measures about 45 mm long and 90–100 mm in wingspan. Moths are often observed feeding from flowers at dusk.

Sweetpotato hornworm larva.

Adult sweetpotato hornworm.

The biology of the sweetpotato hornworm was given by Fullaway (1911), with additional notes provided by Hodges (1971). Larvae were described by Dyar (1895).

Damage

The larvae of this species feed on foliage. They rarely are abundant enough to be considered a serious pest, though sometimes they have been so abundant as to defoliate entire fields and to acquire the "armyworm"-like habit of dispersing in groups (Watson, 1944).

Management

Larvae are easily suppressed with foliar insecticides, including the microbial insecticide *Bacillus thuringiensis*.

Tobacco Hornworm
Manduca sexta (Linnaeus)

Tomato Hornworm
Manduca quinquemaculata (Haworth)
(Lepidoptera: Sphingidae)

Natural History

Distribution. The tobacco and tomato hornworms are very similar in appearance, biology, and distribution. However, the tobacco hornworm, *Manduca sexta* (Linnaeus), is more common in the southern United States, especially the Gulf Coast states. Its range extends northward to New York, and even sometimes into southern Ontario. The tomato hornworm, *Manduca quinquemaculata* (Haworth), in contrast, is uncommon along the Gulf Coast, but relative to tobacco hornworm it is more likely to be encountered in northern states and southern Canada. These are native species, occurring from the Atlantic to the Pacific Oceans, though for tobacco hornworm the range extends south into Central and South America, and the Caribbean.

Host Plants. These insects feed only on solanaceous plants, particularly tomato and tobacco. They are recorded from other vegetables such as eggplant, pepper, and potato, but such feeding is unusual. Several solanaceous weeds are reported to serve as hosts, including groundcherry, *Physalis* spp.; horsenettle, *Solanum carolinense*; jimsonweed, *Datura stramonium*; and nightshade, *Solanum* spp.; but wild hosts are unimportant larval food sources relative to crops. Adults imbibe nectar from flowers of several plants such as catalpa, *Catalpa speciosa*; daylily, *Hemerocallis* sp.; hollyhock, *Althaea rosea*; jimsonweed; four o'clock, *Mirabilis jalapa*; mallow, *Hibiscus lasiocarpus*; mimosa, *Albissia julibrissin*; and tobacco.

Natural Enemies. Many natural enemies are known from tobacco and tomato hornworm. Lawson (1959) provided a good summary of their importance in North Carolina. Natural enemies of the egg stage include *Trichogramma* sp. (Hymenoptera: Trichogrammatidae) and *Telenomus* sp. (Hymenoptera: Scelionidae), but these are thought to be of little importance. Of greater importance is the stilt bug *Jalysus spinosus* (Say) (Hemiptera: Berytidae) and other general predators such as big-eyed bugs (Hemiptera: Lygaeidae) and lacewings (Neuroptera: Chrysopidae).

The larvae often are parasitized by *Cotesia congregatus* (Say) (Hymenoptera: Braconidae), and they sometimes are seen bearing clusters of white pupal cases formed by this wasp, attached to their bodies. In North Carolina, 30–40% of hornworms may be parasitized by this wasp. In contrast to this one species of wasp accounting for most of the parasitism by Hymenoptera, numerous insects in the order Diptera, particularly the Tachinidae, attack tobacco and tomato hornworms. Records of tachinid parasitism from tomato hornworm include *Compsilura concinnata* (Meigen), *Drino incompta* (Wulp), *D. rhoeo* (Walker), *Lespesia frenchii* (Williston), *Winthemia leucanae* (Kirkpatrick), and *W. quadripustulata* (Fabricius). Records of tachinids from tobacco hornworm include *Carcelia* spp., *Drino incompta*, *D. rhoeo*, *Lespesia* spp., *Metavoria orientalis* Townsend, and *Winthemia quadripustulata*. In North Carolina, the proportion of hornworm larvae parasitized by tachinids increases during the season, sometimes attaining 100%. The number of tachinid eggs per hornworm larva also increases seasonally, indicating a greater abundance of flies. Also of great significance as a mortality agent of larvae is *Polistes* spp. (Hymenoptera: Vespidae). These wasps kill and consume, or sometimes carry away to provision their nest, a very high proportion of larvae. Apparently, *Polistes* spp. is the most important larval mortality factor in North Carolina.

Pathogens sometimes affect hornworms. A bacterial disease which causes the larva to blacken and to droop is sometimes reported to kill larvae, but this does not occur regularly, and is observed mostly at very high larval densities. Larvae also are susceptible to infection by the fungus *Entomophaga aulicae*, but this disease also does not occur widely (Morrow *et al.*, 1987).

Vertebrates may prey on the pupae. Skunks and moles are particularly important. Tachinids emerge from the prepupal and pupal stages, but attack only the larvae.

Life Cycle and Description. Although these insects are common, they are not serious pests except in home gardens. Thus, there is surprisingly little information about these insects, and particularly about tomato hornworm. The following information applies specifically to tobacco hornworm except where noted, but apparently is equally applicable to tomato hornworm.

The number of annual generations ranges from 1 to 2 in Canada to 3–4 in northern Florida, but two generations per year is common over most of the ranges of these species. The proportion of insect that enter diapause increases from about 5% in June to 95% in mid-August, as day length decreases. In northern Florida the insects are active from April to November, but they are abundant only for the first two generations because many pupae enter diapause. In North Carolina they occur from mid-May to October, and evidence of a third generation is poor. Throughout their range, hornworms overwinter in the pupal stage, and in the northernmost parts of their range they may not overwinter successfully, occurring only after dispersal northward during the summer. The life cycle can be completed in 30–50 days, but often is considerably protracted.

Egg. The eggs are spherical to oval, and measure 1.25–1.50 mm in diameter. They are smooth and vary in color from light green or yellow when they are early in development, to white at maturity. They are deposited principally on the lower surface of foliage, but also on the upper surface. Mean fecundity was reported by Madden and Chamberlain (1945) to be about 250–350 eggs with seasonal fluctuations and maximum egg production at mid-season. However, Yamamoto (1968) showed that with adequate adult nutrition, fecundity of nearly 1400 eggs per female could be attained. Duration of the egg stage is 2–8 days, but averages five days.

Larva. The larva is cylindrical and bears five pairs of prolegs in addition to three pairs of thoracic legs. The most striking feature of the larva is a thick-pointed structure or "horn" located dorsally on the terminal abdominal segment. The "horn" is as long as the body in newly hatched larvae, but becomes relatively smaller as the larva matures. Young larvae are yellowish-

white but during the second instar become green with white lines laterally. The tobacco hornworm develops seven straight oblique whitish lines laterally. The white lines are edged with black on the upper borders, and the "horn" is usually red. The tomato hornworm is superficially similar, but instead of the seven oblique lateral bands it bears eight whitish or yellowish "V"-shaped marks laterally, and pointing anteriorly. The "V"-shaped marks are not edged in black. Also, in tomato hornworm the "horn" tends to be black. The body of both species is usually light-green, but occasionally dark-brown or blackish forms occur. There normally are five instars, but occasionally six are observed. The mean head capsule width is 0.8, 1.2, 2.0, 3.0, and 5.0 mm for instars 1–5, respectively. Corresponding mean larval body lengths are 6.7, 11.2, 23.4, 49.0, and 81.3 mm, respectively. Larval development time averages about 20 days, but ranges from 13–44 days depending on temperature. Mean development time for larvae reared under insectary conditions in northern Florida is 3.4, 2.9, 3.0, 3.9, and 6.6 days for instars 1–5, respectively. (See color figures 82 and 83.)

Pupa. Mature larvae drop to the soil at maturity and burrow to a depth of 10–15 cm. There they form a pupal cell measuring about 7 cm long and 4 cm wide, and pupate within. The interval between entering the soil and pupation is usually 4–8 days. The pupa is large and elongate-oval, but pointed at the posterior end. It measures 45–60 mm long and 13–14 mm wide. The pupa bears a pronounced maxillary loop, a structure which encases the mouthparts. The maxillary loop in tobacco hornworm extends back about one-fourth the length of the body, whereas in tomato hornworm it is longer, usually extending for about one-third the length of the body. The color of the pupa is brown or reddish-brown. Duration of the pupal stage is protracted and variable. Pupal duration often exceeds 100 days, even in the summer generations, but may be as short as 15 days, and 21 days is about average. Overwintering pupae do not emerge synchronously, with emergence spread from May to early

Tobacco hornworm larva.

Tomato hornworm larva.

Tobacco hornworm pupa.

Tomato hornworm pupa.

August; a few even diapause through two winters. (See color figure 268.)

Adult. The adults of both species are large moths with stout, narrow wings, and a wingspan of 80–130 mm. The front wings are much longer than the hind wings. The females are larger than the males and can be differentiated by the narrower antennae. Both species are dull-grayish or grayish-brown, though the sides of the abdomen usually are marked with six orange-yellow spots in tobacco hornworm and five spots in tomato hornworm. The hind wings

of both species bear alternating light and dark bands. Adults become active at sunset, when they can be observed feeding at flowers. Although some adults remain active throughout the night, most activity occurs early in the evening and just before dawn. Both hornworm species likely produce sex pheromone, but only that of the tobacco hornworm has been identified. The pre-oviposition period of moths is about two days, and eggs are deposited for about 4–8 days. (See color figures 217 and 218.)

A detailed description of tobacco hornworm biology was given by Madden and Chamberlain (1945), and a more abbreviated but useful description of both species was established by Gilmore (1938). Diapause development was reported by Rabb (1966). Rearing was described by several authors, including Yamamoto (1968, 1969), Stewart and Baker (1970), and Bell and Joachim (1976). Keys to the adults were included in Hodges (1971).

Damage

Larvae are defoliators, and usually attack the upper portion of plants initially. Rather than chewing holes in the leaf, they usually consume the entire leaf. Sometimes they attack green fruit. As the larvae of hornworms attain such a large size, they are capable of high levels of defoliation. Madden and Chamberlain (1945) calculated the consumption of tobacco foliage by larvae and reported that 5.2, 9.7, 33.5, 175.4, and 1941.4 sq cm of foliage were eaten by instars 1–5, respectively, for a total of 2165 sq cm. Notice that about 90% of the foliage consumption occurred during the

Adult tobacco hornworm.

Adult tomato hornworm.

final instar, and though the exact value of foliage consumption would differ on tomato, the pattern would be the same. Larvae blend in with the foliage and are not easy to detect. Thus, it is not surprising that they often were not observed until they caused considerable damage at the end of the larval period.

Management

Sampling. Moths are attracted to light and can be captured in light traps. Light traps also were used to attempt suppression of hornworm populations, and though some reduction was noted, this approach did not prove practical (Gentry *et al.*, 1967). Isoamyl salicylate is attractive to both species of hornworms (Scott and Milam, 1943). Visual examination of foliage usually is recommended for monitoring of larval populations. Young larvae of tobacco hornworm tend to be found in the upper regions of the plant, whereas larvae of tomato hornworm tend to be lower; differential flight and oviposition behavior between the species is implicated (Yamamoto, 1972).

Insecticides. Chemical insecticides or *Bacillus thuringiensis* are applied to the foliage for larval suppression (Creighton *et al.*, 1961).

Cultural Practices. The pupae are large and are not buried very deeply in the soil, so greater than 90% mortality is caused by normal soil tillage practices. Hand picking and destruction of larvae is often practical in the home garden.

To take advantage of the preference of *Polistes* wasps for hornworm larvae, wasp shelters or nesting boxes were placed in tobacco fields to encourage the wasps, and wasp colonies were relocated into tobacco (Lawson *et al.*, 1961). Although wasp predation was inadequate to prevent damage to tobacco, this approach might be satisfactory for tomato.

Tobacco and tomato hornworms thrive on tobacco plants that are allowed to revegetate after harvest of the leaves, leading to high populations during the next year. Destruction of tobacco stalks, or inhibition of sprouting by application of plant growth regulators, greatly reduces hornworm populations in subsequent seasons (Rabb, 1969). Although not documented, timely destruction of tomato crop residue likely would have similar beneficial effects.

Whitelined Sphinx
Hyles lineata (Fabricius)
(Lepidoptera: Sphingidae)

Natural History

Distribution. This native insect is probably the most widely distributed sphingid moth in North America. It is found throughout the United States and southern Canada, and its range extends south into Central America and the Caribbean.

Host Plants. This insect reportedly has a wide host range, but feeds principally on weeds. Its presence among garden plants often results in the assumption that it is developing at the expense of crops when it is actually grazing on an understory of competing plants, especially portulaca, *Portulaca* spp. If preferred host are eliminated, of course, larvae attempt to feed on nearly any nearby plant. Among vegetables reportedly injured are beet, cantaloupe, lettuce, tomato, turnip, and watermelon. Fruits such as apple, currant, gooseberry, grape, pear, and plum are also listed among hosts. Other plants consumed include bitter dock, *Rumex obtusifolius*; evening primrose, *Oenothera* spp.; fuchsia, *Fuchsia* spp.; four o'clock, *Mirablis* spp.; and willow herb, *Epilobium* spp. The adults take nectar from several of flowering plants.

Natural Enemies. Several flies are known to parasitize the larvae, including *Compsilura concinnata* (Meigen), *Drino incompta* (Wulp), *Winthemia deilephilae* (Osten Sacken), and *W. quadripustulata* (Fabricius) (all Diptera: Tachinidae). Although there is no quantitative data on incidence of parasitism, it is evident from the literature that larvae frequently are parasitized by tachinids. Many mature larvae, when collected, are observed to have tachinid eggs adhering to their bodies, usually located dorsally behind the head. Also attacking whitelined sphinx is a pupal parasitoid, *Brachymeria robusta* (Cresson) (Hymenoptera: Chalcididae).

Life Cycle and Description. There are two generations annually, with the pupa being the overwintering stage. Adults are usually observed in June and September.

Egg. The eggs are oval, and yellow-green initially, but become bluish later in development. They are deposited on host foliage. Eggs normally hatch in 6–7 days.

Larva. There are six instars. At hatching the young larva measures about 4 mm long, and measures 6–7, 12–13, 22–23, 37–40, and 55–60 mm long at the start of the subsequent instars, respectively. At maturity, the larva attains a length of 75–90 mm. Development time for the instars is about 8, 4, 5, 5, 4, and 11 days, respectively. During the initial five instars, larvae are green or black, with an yellowish to orange-brown head. The posterior end bears a "horn" that is usually black or yellow and black. A horizontal yellowish subdorsal line is present along each side. Although larvae are fairly variable in color during the early instars,

Whitelined sphinx larva, dark form.

Whitelined sphinx pupa.

variation is exceptionally marked in the terminal instar. In the sixth instar the green form has a yellowish-green body with a subdorsal row of pale spots bordered above and below with a black line, and brightly colored spots around the spiracles. The corresponding black form has a blackish body with three narrow yellow lines dorsally. (See color figure 84.)

Pupa. Pupation often occurs at the soil surface in a loosely constructed cocoon of brown color, but some larvae apparently pupate without forming a cocoon, and some enter the soil to pupate. The pupa is light brown, and measures about 44–48 mm long. Duration of the pupal stage is 30–40 days.

Adult. The adults are more active at dusk, but can also be observed feeding during the day, hovering at flowers while sipping nectar. The moth has a wingspan of 60–90 mm. The body is dull brown with white lines, running the length of the head and thorax. The abdomen bears white and dark brown spots dorsally. The olive front wings are marked with white-lined veins and a whitish stripe, extending from the base to the tip of the wing. The hind wings are dark brown, with a rosy band across the middle. (See color figure 219.)

Life history information on white-lined sphinx is quite limited. Soule (1896), Cooley (1905), and Eliot and Soule (1921) provided notes on this species.

Adult whitelined sphinx.

Damage

The mature larvae of white-lined sphinx are quite robust, and can consume large quantity of foliage. However, they normally limit their feeding to portulaca, moving onto crop plants only when their favored food is completely consumed, and they are faced with starvation. Much of the larva's reputation for damaging crops seems to stem from the habit of resting or nibbling on crop plants. There are no reports of this species completing its larval development on vegetable crops.

Management

Moths are attracted to light, and can be captured in light traps. The presence of adults does not necessarily indicate future problems with crops, however, unless portulaca or another favored weed is also present. Larval infestation of vegetable crops can be avoided by preventing portulaca from growing amongst crop plants. Should portulaca become extensively established, it should not be killed or removed if infested with larvae, as this action will force the larvae to feed on the crop.

FAMILY TORTRICIDAE— LEAFROLLER MOTHS

Pea Moth
Cydia nigricana (Fabricius)
(Lepidoptera: Tortricidae)

Natural History

Distribution. Pea moth is native to Europe, and was first observed in North America in 1893 at Toronto, Canada. Probably, the insect had been in residence for a number of years before attracting attention. It quickly spread across the northern United States, causing great damage in the Great Lakes region in the early 1920s, and reached Washington in 1926. It inflicted considerable damage in both Washington and British Columbia in 1933. Now found throughout the northern United States and southern Canada, it is rare in the Prairie Provinces, and infrequently is a serious pest anywhere in North America. It remains a serious problem in Europe, however, and is known from Japan.

Host Plants. This insect feeds only on plants in the family Leguminosae. The only crop injured is pea, but several other legumes are suitable hosts, including Canada pea, *Vicia cracca*; common vetch, *Vicia angustifolia*; hairy vetch, *Vicia villosa*; lupines, *Lupinus* spp.; Scotch broom, *Cytisus scoparius*; spring vetch, *Vicia sativa*; sweet pea, *Lathyrus odoratus*; yellow vetchling, *Lathyrus pratensis*; and possibly other members of these plant genera. Hanson and Webster (1936) reported that the most suitable host was pea, followed by sweet pea, and then the vetches. Larvae feeding on lupines and Scotch broom experience high mortality.

Natural Enemies. Several natural enemies have been imported from Europe and introduced into Canada and the United States. Although some species failed to establish, the parasitoids *Ascogaster quadridentata* Wesmael, *Phanerotoma fasciata* Provancher (Hymenoptera: Braconidae), and *Glypta haesitator* Gravenhorst (Hymenoptera: Ichneumonidae) became established. *Ascogaster quadridentata* proved to be very important, causing over 70% parasitism in British Columbia. Although the parasitoids undoubtedly are important in checking the abundance of pea moth, changes in cropping practices also are important. Specifically, the dramatic decline in the production of dried peas is of major significance, because the major host plant is now much less abundant. Also, harvesting of peas while they are green results in the destruction of larvae before most are able to mature. Clausen (1978) summarized the parasitoid introduction program. Pea moth seems to be free of important diseases. However, it has been shown to be susceptible to infection by a granulosis virus isolated from a related insect, codling moth, *Cydia pomonella* (Linnaeus) (Payne, 1981).

Life Cycle and Description. Generally there is a single generation per year, though in Washington a few insects apparently go on to form a second generation. In England, a small proportion of the insects have been observed to have a two-year life cycle, larvae spending two winters and a summer in the soil; this likely occurs in North America as well. Oviposition occurs in June and July, followed by development of larvae until September. Larvae overwinter, with pupation occurring in the spring.

Egg. Moths become active in June, and after a preoviposition period of 8–9 days (range 5–13 days), eggs are deposited. Oviposition occurs principally on the upper half of the plant, and eggs are scattered on or near the flowers and pods. When deposited on the leaf surface, the lower surface is preferred. The egg is flattened, oval, and measures about 0.7 mm long and 0.5 mm wide. The lower theshold for development is about 8.5–9.5°C, with the upper threshold about 30°C. Initially, the eggs are whitish and somewhat translucent, but two pink streaks develop after 2–3 days. Duration of the egg stage is about nine, six, and three days at 15°, 20°, and 25°C, respectively.

Larva. The larva is whitish with a black head and thoracic plate during the first two instars. Afterwards it assumes a yellowish color, but retains the dark markings. The body bears short and sparse hairs. It grows from a length of about 1.5 mm to 12–15 mm. Duration of the larval stage, which includes five instars, is 18–35 days. The first instar is very active, searching extensively on the plant for suitable food. When the larva locates a suitable pod it burrows in, the entry hole developing into a small brown blister. All stages of pods are attacked, except for the very old, dry pods. Usually 2–4 seeds are destroyed in each pod, though occasionally more than one larvae enter a pod, resulting in greater damage. Mature larvae drop from the plant and enter the soil, usually to a depth of 5–7 cm. They spin a thick silken cocoon measuring about 10 mm long and 4.5 mm wide, and enter diapause. The larva remains in the cocoon until spring, when they escape the old cocoon, and move closer to the soil surface. They then spin a second, much thinner cocoon, and pupate.

Pupa. The pupa is dark brown and measures 7–8 mm long and about 1.7 mm wide. The dorsal surface of the abdominal segments is armed with rows of spines, structures which aid the escape of the pupa from the cocoon when the moth is about to emerge. Duration of the pupal stage is 12–14 days.

Adult. The adult is a small brownish-gray moth with a wingspan of about 12 mm. The forewing is also marked with black, and spots of white, particularly along the leading edge of the wing. The hind wing is

Pea moth larva.

Adult pea moth.

nearly uniform brownish gray, but is bordered with a light-colored fringe with a thin, dark inner line. The moths tend to be active only in the afternoon, and a threshold of about 18°C must be exceeded before flight occur. Flight also is reported to be favored by high humidity. Female moths emerge with nearly mature ovaries, mate upon emergence, and commence egg laying almost immediately. Females disperse freely in search of suitable oviposition sites, but males are less dispersive. Males are attracted to a female-produced sex pheromone that has been identified, and synthesized, for use in population monitoring.

Wright and Geering (1948) provided detailed biology of pea moth, though the observations were made in England. The work of Hanson and Webster (1936) in Washington, though less complete, likely reflected American and Canadian conditions. Lewis and Sturgeon (1978) gave important infomation on egg ecology.

Damage

The larvae damage the pea crop by tunneling into the pod and feeding on one or more seeds in each pod. Often the larva does not consume the entire seed before attacking another. Silken webbing and fecal material also are found in the pea pod. In addition to yield loss caused by destruction of pea seed, and quality loss caused by contamination of undamaged seed with insect-damaged seed, additional loss may occur when the larvae contaminate peas harvested for processing (freezing or canning). The latter problem is minimal with this insect because it is not hidden within the pea seed.

Damage potential varies with intended use of the peas. Pea moth damage is of little consequence for peas grown as livestock food because the quality standards are low. For processing peas, the damage potential is significant, but commercial processors have established procedures to eliminate damaged peas and insects. Pea grown for seed is another instance where damage potential is great, because even a small level of damage, 1.0–1.5%, is sufficient to inflict loss to farmers due to downgrading of the seed quality.

Management

Sampling. In England, a great effort has gone into refinement of a pheromone-based monitoring system to farmers of pea moth problems and to time insecticide applications carefully. At least two pheromone traps are placed in each field, and when 10 or more moths are collected in either trap for two consecutive nights the farmer can be assured that moths are present in sufficient number to warrant control measures (Gratwick, 1992). After consulting weather data and estimating the time of egg hatch, farmers can apply insecticide to kill first instar larvae at hatch. Usually a delay of 10–15 days is recommended between initiation of sustained moth catches and application of insecticide to allow for oviposition and egg hatch. This system has found good acceptance in England (Wall et al., 1987; Wall, 1988).

Release of large quantities of the sex pheromone can also be used to disrupt chemical communication in pea moth and disrupt mating (Bengtsson et al., 1994). This potential crop protection system is not yet operational.

Insecticides. Insecticides are applied to foliage to kill first instar larvae before they enter the pea pod, so timing is critical. This is accomplished most easily with determinant pea varieties, because the synchrony of fruiting allows growers to protect pods with fewer applications of insecticides. With indeterminant varieties, or when planting of determinant varieties is staggered, more applications are required to protect the crop from damage. If crops remain in the field in June, they may require two insecticide treatments at about 10 day intervals to protect them adequately, though early crops may escape injury. Application of granular formulations of systemic insecticides is reported to be less satisfactory than foliar application (Thompson and Sanderson, 1977).

Cultural Practices. Several cultural practices help reduce damage from pea moth (Wright and Geering, 1948). Tillage can destroy the larvae overwintering in the soil. Disking the soil twice is commonly recommended, especially if the field, or a nearby field, is to be planted to peas during the following growing season. Destruction of wild vetches is desirable to decrease the potential of pea moth to develop on these alternate hosts. Thick vegetation, either in the form of leafy pea varieties or weedy vegetation, exacerbates pea moth problems, probably by providing shelter for the moths. Thus, weeds should be destroyed and pea varieties with minimal foliage should be favored.

Peas that mature early may also escape attack, so early planting or selection of early maturing varieties is desirable. Crops that are harvested before mid-June may escape attack. It is also beneficial to either pick all peas or to quickly destroy remaining pods after harvest, as this can minimize successful development of pea moth in the field. Peas that are grown for livestock food seldom receive insecticide applications, so they may serve as a source of infestation. Cultivation of fresh market and processing peas at a considerable distance away from livestock peas is suggested.

Order Orthoptera—Grasshoppers and Crickets

FAMILY ACRIDIDAE—GRASSHOPPERS

American Grasshopper
Schistocerca americana (Drury)
(Orthoptera: Acrididae)

Natural History

Distribution. This grasshopper is found widely in the eastern United States, west to Iowa and Texas. Occasionally it is reported from southern Canada and the New England states, but is likely a migrant there. Even in the midwestern states, where it is common, the resident population receives a regular infusion of dispersants from southern locations. In the southeast it is quite common, and is one of the few grasshopper species to reach epidemic densities. American grasshopper seems to be native to North America, but its distribution also extends south through Mexico to northern South America.

Host Plants. The adults of American grasshopper tend to be arboreal in habit, and considerable feeding by adults occurs on forest, shade, and fruit trees. The nymphs, however, feed on numerous grasses and broadleaf plants, both wild and cultivated. During periods of abundance, almost no plants are immune to attack, and vegetables, grain crops, and ornamental plants are injured. American grasshopper consumes bean, corn, okra, and yellow squash over some other vegetables when provided with choices (Capinera, 1993b), but free-flying adults normally avoid low-growing crops such as vegetables—corn being a notable exception.

Natural Enemies. The natural enemies of *S. americana* are not well known. Birds such as mockingbirds, *Mimus polyglottos polyglottos* (Linnaeus), and crows, *Corvus brachyrhynchos brachyrhynchos* Brehm have been observed to feed on these grasshoppers. Fly larvae, *Sarcophaga* sp. (Diptera: Sarcophagidae) are sometimes parasitic on overwintering adults.

Life Cycle and Description. American grasshopper has two generations per year and overwinters in the adult stage. In Florida, eggs produced by overwintered adults begin to hatch in April–May, producing spring generation adults by May–June. This spring generation produces eggs that hatch in August–September. The adults from this autumn generation survive the winter.

Egg. The eggs of *S. americana* initially are light orange, turning tan with maturity. They are elongate-spherical, widest near the middle, and measure about 7.5 mm long and 2.0 mm wide. The eggs are clustered together in a whorled arrangement, and number 75–100 eggs per pod, averaging 85. They are inserted into the soil to a depth of about 4 cm and the upper portion of the oviposition hole is filled by the female with a frothy plug. Duration of the egg stage is about 14 days. The nymphs, upon hatching, dig through the froth to attain the soil surface.

Nymph. Normally there are six instars in this grasshopper (Kuitert and Connin, 1952; Capinera, 1993a), though Howard (1894) reported only five instars. The young grasshoppers are light green. They are extremely gregarious during the early instars. At low densities the nymphs remain green throughout their development, but normally gain increasing amounts of black, yellow, and orange coloration commencing with the third instar. Instars can be distinguished by their antennal, pronotal, and wing development. The first and second instars display little wing development but have 13 and 17 antennal segments, respectively. In the third instar, the number of antennal segments increases to 20–22, the wings begin to display weak evidence of veins, and the dorsal length of the ventral lobe of the pronotum is about 1.5 times the length of the ventral surface. Instar four is quite similar to instar three, with 22–25 antennal segments, though the ratio of the length of the dorsal to ventral surfaces of the pronotal lateral lobe is 2:1. In instar five there are 24–25 antennal segments, and the wing tips assume a dorsal rather than ventral orientation, but the wing tip does not exceed the first abdominal segment. In the sixth instar there are 24–26 antennal segments and the wing tips extend beyond the second abdominal segment. The overall body length is about 6–7, 12–13, 16–18, 22–25, 27–30, and 35–45 mm for instars 1–6, respectively. Development time is about 4–6, 4–6, 4–6, 4–8, 6–8, and 9–13 days for the corresponding instars when reared at about 32°C. (See color figure 168.)

Adult. The adult is rather large but slender bodied, and measures 39–52 and 48–68 mm long in the male and female, respectively. A creamy white stripe normally occurs dorsally from the front of the head to the tips of the forewings. The front wings bear dark-brown spots, the pronotum dark stripes. The hind wings are nearly colorless. The hind tibiae normally are reddish. Overall, the body color is yellowish-brown or brownish with irregular lighter and darker areas, though for a week or so after assuming the adult stage a pinkish or reddish tint is evident.

Adults are active, flying freely and sometimes in swarms. They normally are found in sunny areas, but during the warmest periods of the day move to shade. Adults are long-lived, persisting for months in the laboratory and apparently in the field as well. This can lead to early season situations where overwintered adults, all instars of nymphs, and new adults are present simultaneously. Mild winters favor survival of overwintering adults and apparently lead to population increase. (See color figure 167.)

The biology of American grasshopper was described by Howard (1894) and Kuitert and Connin (1952).

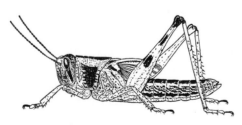

American grasshopper nymph.

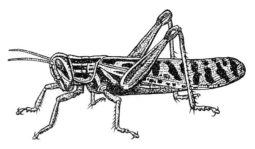

Adult American grasshopper.

Damage

American grasshopper is a defoliator, and eats irregular holes in leaf tissue. Under high density conditions it can strip vegetation of leaves, but more commonly leaves plants with a ragged appearance. Adults display a tendency to swarm, and high densities can cause severe defoliation.

As American grasshopper is a strong flier, it also sometimes becomes a contaminant of crops. When the late-season crop of collards in the southeast is harvested mechanically, American grasshopper may be incorporated into the processed vegetables. Although most grasshoppers can be kept from dispersing into crops near harvest by treating the periphery of the crop field, it is much more difficult to prevent invasion by American grasshopper because it may fly over any such barrier treatments.

Management

Sampling. Sampling methods are not well defined. Populations normally originate in weedy areas such as fence rows and abandoned fields. Thus, margins of fields are first affected and where monitoring should be concentrated. It is highly advisable to survey weedy areas in addition to crop margins if grasshoppers are found, as this gives an estimate of the potential impact if the grasshoppers disperse into the crop. Also, it is important to recognize that this species is highly dispersive in the adult stage, and can fly hundreds of meters or more to feed.

Insecticides. Foliar application of insecticides can suppress grasshoppers, but they are difficult to kill, particularly as they mature. Bait formulations are not usually recommended because these grasshoppers spend little time on the soil surface, preferring to climb high in vegetation.

Cultural Practices. Land management is an important element of *S. americana* population regulation. Grasshopper densities tend to increase in large patches of weedy vegetation that follow the cessation of agriculture or the initiation of pine tree plantations. In both cases, the mixture of annual and perennial forbs and grasses growing in fields that are untilled seems to favor grasshopper survival, with the grasshoppers then dispersing to adjacent fields as the most suitable plants are depleted. However, as abandoned fields convert to dense woods or the canopy of pine plantations shades the ground and suppresses weeds, the suitability of the habitat declines for these grasshoppers.

Disturbance or maturation of crops may cause American grasshopper to disperse, sometimes over long distances, into crop fields. Therefore, care should be taken not to cut vegetation or till the soil of fields harboring grasshoppers if a susceptible crop is nearby. Planting crops in large blocks decreases the relative amount of crop edge, and the probability that a crop plant within the field will be attacked.

Differential Grasshopper
Melanoplus differentialis (Thomas)
(Orthoptera: Acrididae)

Natural History

Distribution. This native grasshopper occurs widely in the central and western regions of the United States, and in northern Mexico; in Canada it occurs in southern Saskatchewan and British Columbia. Within the United States it is absent from the Atlantic and Gulf Coast region, except that it occurs in the Pennsylvania, New Jersey, and Maryland area. It is infrequent in the Pacific Northwest area. Also, within the large geographic area generally inhabited by differential grasshopper, it is rare in arid environments.

Host Plants. The host plants preferred by differential grasshopper are tall broadleaf plants such as those typically associated with fence rows, irrigation ditches, and fallow fields. It prefers plants in the family Compositae such as ragweed, *Ambrosia* spp.; sowthistle, *Sonchus asper*; sunflower, *Helianthus annuus*; and prickly lettuce, *Lactuca scariola*; though it will feed on other broadleaf plants such as kochia, *Kochia scoparia*; and smartweed, *Polygonum* sp.; and on such grasses as bermudagrass, *Cynodon dactylon*; slender oat, *Avena barbata*; barley, *Hordeum* sp.; and Johnsongrass, *Sorghum halepense*. In North Dakota alfalfa fields, differential grasshopper reportedly ate kochia, *Kochia scoparia*; quackgrass, *Agropyron repens*; squirreltail grass, *Hordeum jubatum*; bristly foxtail, *Setaria* spp.; and field bindweed, *Convolvulus repens*, in addition to alfalfa (Mulkern *et al.*, 1962). On prairie, they ate mostly stickseed, *Lappula echinata*; wavyleaf thistle, *Cirsium undulatum*; quackgrass, *Agropyron repens*; and pepperweed, *Lepidium densiflorum* (Mulkern *et al.*, 1964).

Crops sometimes injured include alfalfa, clover, corn, cotton, soybean, sugarbeet, timothy, and small grains such as barley and wheat. Differential grasshopper is not normally an important vegetable pest. It occurs among vegetables if weeds are present within, or adjacent to, crops. However, during periods of great abundance all vegetables are at risk, because under such conditions virtually all green vegetation may be consumed. (See color figure 29.)

Natural Enemies. The natural enemies of the crop-feeding *Melanoplus* spp. are quite similar. For information on natural enemies of differential grasshopper, see the section on natural enemies of migratory grasshopper, *Melanoplus sanguinipes* (Fabricius).

Weather. Weather affects the distribution of differential grasshopper, but in a manner somewhat different from some other grasshopper species. Differential grasshopper is associated with dense vegetation, so it follows that it would thrive in areas with adequate moisture to support lush growth of plants. It is a common, and damaging, species on the eastern edge of the Great Plains, where rainfall is plentiful, and relatively infrequent along the drier western edge of this region. Wakeland (1961) documented the expansion of differential grasshopper populations into areas of the northern Great Plains dominated by migratory grasshopper, *Melanoplus sanguinipes* (Fabricius), a species that is better adapted to dry conditions. Migratory grasshopper was supplanted by differential grasshopper, as long-term mean precipitation levels increased and temperatures decreased in this area. When weather returned to normal, however, migratory grasshopper resumed its status as the dominant species.

High levels of precipitation are not entirely advantageous for differential grasshopper. Precipitation during the warm months leads to outbreak of disease in differential grasshopper populations. This is a short-term response, and disease outbreaks occur only when grasshoppers are abundant. Differential grasshopper seems to be more susceptible to disease than some other species, including migratory grasshopper (Wakeland, 1961). Precipitation accompanied by cool weather during the hatching period is also detrimental to differential grasshopper, as with all grasshoppers, largely because it disrupts feeding during the critical early life of the grasshopper. The late onset of winter can favor grasshopper population increase, because it allows adults additional time to produce eggs.

Life Cycle and Description. A single generation occurs annually, with the egg stage overwintering. This is a late-season species, with eggs hatching about three weeks after those of twostriped grasshopper, *Melanoplus bivittatus* (Say), and two weeks after *Melanoplus sanguinipes* (Fabricius). In Colorado, eggs hatch in June, usually within a two-week period. Nymphs complete their development in July–August; adults are present from August to October.

Egg. The eggs are creamy white, yellowish, or light brown. They are elongate-cylindrical, and taper to a blunt point at each end. The eggs measure about 4–5 mm long and 0.85 mm in diameter. They are clustered within an elongate, cylindrical pod consisting of about 40–200 eggs arranged in four columns and held together by frothy material. They are deposited in the soil, and the upper portion of the pod is plugged with additional froth. They are normally deposited among the roots of grasses and weeds, especially along edges of fields, and in moist soil. Embryonic development occurs in the autumn after eggs are deposited, but the embryo enters diapause at about the point of 50% development, and must endure a period of cold before commencing growth.

Nymph. The nymphs normally develop during the warmest period of the summer, and complete their development in about 30 days, though development times are extended by cool weather. There are six instars, but the nymphs are not distinctive in appearance. First instars are greenish, yellowish or brownish with an indication of a black stripe on the outer face of the hind femora. Instars 2–6 are similar but possess a curved-dark stripe extending from the back of the eye across the pronotum, and bordered below by a narrower white stripe. The black stripes on the femora are pronounced. The hind tibiae are light green or gray. The nymphs increase in body length from 5.3–6 mm in the first instar, to 5.2–6.8, 9.4–12.6, 12–14, 18–21.5, and 22–32 mm in instars 2–6, respectively. The number of antennal segments is 12, 14–17, 19–20, 21–22, 25–26, and 26 in instars 1–6, respectively.

Adult. The adults are large grasshoppers, the males measuring 28–34 mm long, the females 32–44 mm. They display more color variability than most grasshoppers. They may be principally brownish-green, or yellow, or almost entirely black, though the yellow form is most abundant. The most distinctive feature of these grasshoppers is the row of black marks, arranged in herring-bone fashion on the outer face of the hind femora. These markings are not so evident in the infrequent black form, which instead bears four white blotches on its otherwise black hind femora. The front wings in all except the black form are uniform grayish or brownish; the hind wings in all forms are colorless. The males bear large and boot-shaped cerci. (See color figure 164.)

Like many other grasshoppers, differential grasshopper tends to roost on elevated locations at night. This allows them to bask in the morning sun, and to assume activity early in the day. Bushes and other tall vegetation are favorite perches.

Aspects of the biology of differential grasshopper were treated by many authors, including Parker and Shotwell (1932), Parker (1939), and Shotwell (1941). Kaufmann (1968) gave some interesting biological information, but due to low rearing temperature the relevance of this study to field biology is questionable. An excellent synopsis was presented by Pfadt (1994e), who also pictured all stages of development. *Melanoplus differentialis* is included in many grasshopper keys, including those by Blatchley (1920), Dakin and Hays (1970), Helfer (1972), Capinera and Sechrist (1982), and Richman *et al.* (1993). This species was also included in a key to grasshopper eggs by Onsager and Mulkern (1963). Rearing of *Melanoplus* species was described by Henry (1985).

Damage

This species seemingly has benefited from agricultural practices more than most grasshoppers, with grasshopper survival increased by the abundance of weeds associated with crops, and the irrigation practices of western farms. It also readily exploits disturbed sites in cities and towns. Unlike some of the arid environment-loving species, its numbers and damage may increase following long-term increases in precipitation. The damage caused by differential grasshopper principally takes the form of leaf removal. Plants may be completely defoliated, or left ragged. As this grasshopper tends to roost in elevated locations at night, where they may nibble while rest-

Adult male differential grasshopper.

Male cercus from differential grasshopper.

ing, trees and shrubs outside the normal dietary range are sometimes severely injured.

Management

Management of the various *Melanoplus* spp. grasshoppers is substantially the same. For information on differential grasshopper management, see the section on management under migratory grasshopper, *Melanoplus sanguinipes* (Fabricius).

Eastern Lubber Grasshopper
Romalea microptera (Beauvois)
(Orthoptera: Acrididae)

Natural History

Distribution. This native grasshopper is common in the southeastern states from North Carolina to eastern Texas, including the entire peninsula of Florida.

Host Plants. Eastern lubber grasshopper has a broad host range. Jones *et al.* (1987) indicated that at least 26 species from 15 plant families containing shrubs, herbs, broadleaf weeds, and grasses are eaten. Watson (1941) indicated preference for pokeweed, *Phytolaca americana*; tread-softly, *Cnidoscolus stimmulosus*; pickerel weed, *Pontederia cordata*; lizard's tail, *Saururus* sp.; sedge, *Cyperus*; and arrowhead, *Sagittaria* spp. Although its preferred habitat seems to be low, wet areas in pastures and woods and along ditches, lubbers disperse long distances during the nymphal period. Lubbers are gregarious and flightless, their migrations sometimes bring large numbers into contact with crops where they damage vegetables, fruit trees, and ornamental plants. Lubbers seemingly display little preference among vegetable crops, feeding widely on whatever is available. In choice tests they favor broccoli, Brussels sprout, carrot, pea, and squash relative to other common vegetables. Watson and Bratley (1940) indicated preference for corn, cowpea, and peanut under field conditions. Also, they seek out and defoliate amaryllis, Amazon lily, crinum, narcissus, and related plants in flower gardens. In Florida, lubbers sometimes damage young citrus trees.

Natural Enemies. The natural enemies of lubber grasshoppers are poorly documented. Vertebrate predators such as birds and lizards learn to avoid these insects due to the production of toxic secretions. Naive vertebrates gag, regurgitate, and sometimes die following consumption of lubbers. However, loggerhead shrikes, *Lanius ludovicianus* Linnaeus, capture and cache lubbers by impaling them on thorns and the barbs of barbed-wire fence. After 1–2 days, the toxins degrade and the dead lubbers becomes edible to the shrikes (Yosef and Whitman, 1992). Undetermined

flies and nematodes have been reported from lubbers, and it is possible to infect lubbers experimentally with the grasshopper-infecting nematode *Mermis nigrescens* (Poinar, 1979).

Life Cycle and Description. There is a single generation annually, with the egg stage overwintering. These grasshoppers are long-lived, and either nymphs or adults are present throughout most of the year in the southern portions of Florida. In northern Florida and along the Gulf Coast they may be found from March–April to about October–November.

Egg. The eggs of lubber grasshoppers are yellowish or brown. They are elongate elliptical and measure about 9.5 mm long and 2.5 mm wide. The are laid in neatly arranged clusters or pods which consist of rows of eggs positioned parallel to one another, and held together by a secretion. Normally there are 30–50 eggs in each pod. Watson (1941) reported that ovipositing females preferred mixed broadleaf tree-pine habitats with intermediate soil moisture levels. He indicated that they avoided both lowland, moist, compact soil and upland dry, sandy soil. The female deposits the pod in the soil at a depth of 3–5 cm and closes the oviposition hole with a frothy secretion or plug. The plug allows the young grasshoppers easy access to the soil surface when they hatch. Duration of the egg stage is 6–8 months.

Nymph. Young nymphs are highly gregarious, and remain gregarious through most of the nymphal period, though the intensity dissipates with time. Normally there are five instars, though occasionally six instars occurs. The nymphs are mostly black with a narrow median yellow stripe along the pronotum and abdomen, the edges of the pronotum, and on the lower side of the abdomen. The legs are well-marked with red. Their color pattern is distinctly different from the adult stage, and so the nymphs commonly are mistaken for a different species than the adult form. The early instars can be distinguished by a combination of body size, the number of antennal segments, and the form of the developing wings. The nymphs measure about 10–12, 16–20, 22–25, 30–40, and 35–45 mm long during instars 1–5, respectively. Antennal segments, which can be difficult to distinguish even with magnification, number 12, 14–16, 16–18, 20, and 20 per antenna during instars 1–5, respectively. The shape of the wing pads immediately behind the pronotum changes slightly with each molt. During the first instar the ventral surface is broadly rounded; during the second instar the ventral edges begin to narrow and point slightly posteriorly, and also acquire slight indication of wing veins; during the third instar the ventral edges of the wing pads

Eastern lubber grasshopper nymph.

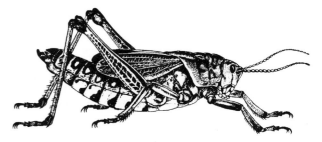

Adult eastern lubber grasshopper.

are markedly elongate, point strongly posteriorly, and the veins are pronounced. At the molt to the fourth instar the orientation of the small, developing wings shifts from pointing downward to upward, and posteriorly. In instar four the small front and hind wings are discrete and do not overlap, though the front wings may be completely or partly hidden beneath the pronotum. In instar five, the slightly larger wings overlap, appearing as a single pair of wings. Nymphs can complete instars 1–4 in about seven days each, with the terminal instar requiring 10 days. However, under cool conditions 60 days are required for nymphal development. (See color figure 170.)

Adult. Adults are colorful, but their color pattern varies. In strong contrast to the nymph, the adult eastern lubber normally is mostly yellow or tawny, with black on the distal portion of the antennae, on the pronotum, and on the abdominal segments. The front wings extend two-thirds to three-fourths the length of the abdomen. The hind wings are short and are incapable of providing lift for flight. The front wings tend to be pink or rose centrally, whereas the hind wings are entirely rose. Darker forms of this species also exist, wherein the yellow color becomes the minor rather than the major color component, and in northern Florida a predominantly black form is sometimes found. Adults attain a large size, males measuring 43–55 mm long and females often measuring 50–70 mm, sometimes 90 mm. Both sexes stridulate by rubbing the forewing against the hind wing. When alarmed, lubbers spread their wings, hiss, and secrete foul-smelling froth from their spiracles. They can expel a fine spray of toxic chemicals to a distance of 15 cm. The chemical discharge from the tracheal system is believed to be an anti-predator defense, consisting of chemicals both synthesized and sequestered from the diet. Vertebrate, but not invertebrate, predators are affected (Whitman *et al.*, 1991; 1992). (See color figure 169.)

The life history and ecology of eastern lubber grasshopper are poorly documented. Rehn and Grant (1961) provided important descriptive notes.

Damage

Lubber grasshoppers are defoliators, consuming the leaf tissue of numerous plants. They climb readily, and as they are gregarious they can completely strip foliage from plants. More commonly, however, they eat irregular holes in vegetation and then move on to another leaf or plant.

Management

Management practices are not well developed. Insecticides applied to the foliage or directly to the grasshopper can be lethal. However, due to their large size they often prove difficult to kill. Insecticide treatment is more effective for young grasshoppers. As they are dispersive, and may continue to invade an area even after it is treated with insecticide, it is difficult to afford protection to plants. Bait formulations have not been evaluated.

Migratory Grasshopper
Melanoplus sanguinipes (Fabricius)
(Orthoptera: Acrididae)

Natural History

Distribution. This native grasshopper is extremely adaptable. It is found in every state in the continental United States, and in every province in Canada. It is absent from only the northernmost, coldest regions of Canada, and from southern Florida and Texas.

Host Plants. This species feeds on a wide range of food plants, and occurs in numerous habitats. Relative to the other common crop-feeding *Melanoplus* spp., migratory grasshopper is more tolerant of arid, shortgrass environments. It tends to prefer annual broadleaf plants, but eat grasses. Dry plant material is an important element of the diet in addition to succulent leaf tissue (McKinlay, 1981). Many authors have noted that the population abundance of migratory grasshopper is correlated with availability of annual broadleaf plants. Among the preferred plants are dandelion, *Taraxacum officinale*; stinkweed, *Thlaspi arvense*; Johnsongrass, *Sorghum halepense*; Kentucky bluegrass, *Poa pratensis*; shepherdspurse, *Capsella bursapastoris*; pepperweed, *Lepidium* spp.; tansymustard, *Descurainia sophia*; western wheatgrass, *Agropyron smithii*; winter mustard, *Sisymbrium irio*; young Russian thistle, *Salsola kali*; and young rabbitbrush, *Chrysothamnus* spp. (Pfadt, 1949; Scharff, 1954). Among the preferred weeds eaten in North Dakota alfalfa fields were awnless brome-grass, *Bromus inermis*; kochia, *Kochia scoparia*; field sowthistle, *Sonchus arvensis*; field bindweed, *Convolvulus arvensis*; and Russian thistle, *Salsola kali* (Mulkern, 1962). On prairie, however, the preferred host plants were Kentucky bluegrass, *Poa pratensis*; leadplant, *Amorpha canescens*; white sage, *Artemisia ludoviciana*; and western ragweed, *Ambrosia psilostachya* (Mulkern et al., 1964).

Migratory grasshopper does not normally infest vegetable crops, but prefers to inhabit weedy areas along fences, irrigation ditches, roadsides, and in pastures. However, as favored food plants become over-mature, desiccated, or depleted, grasshoppers move into vegetable crops. This is especially likely during periods when grasshoppers are extremely abundant. Among the vegetable crops reported to be injured by migratory grasshopper are asparagus, bean, cabbage, carrot, cauliflower, celery, cucumber, lettuce, melon, onion, pea, radish, squash, tomato and watermelon. Field crops are more often injured, particularly alfalfa, barley, corn, oat, and wheat. However, as happens with vegetables, when grasshoppers are numerous buckwheat, clover, flax, millet, rye, young sorghum, soybean, sugarbeet, timothy, and tobacco may be damaged. Even fruits such as apple, cherry, currant, grape, peach, plum and strawberry, as well as numerous flowers and shrubs are attacked during periods of abundance.

Pfadt (1949) studied host preferences and nymphal survival on many rangeland grasses, weeds and some field crops. Not surprisingly, there was a strong positive relationship between preference and survival. Among the plants most suitable for nymphal survival were wheat, sunflower, alfalfa, corn and barley, accounting for the reputation of this species as a severe pest in central and western North America. (Alfalfa is an unusual host, however, because though it is quite suitable for large nymphs, it is but a poor source of food for the youngest of the species.) Several broadleaf weeds including dandelion; downy chess, *Bromus tectorum*; tumblemustard, *Sisymbrium altissimum*; slimleaf scurfpea, *Psoralea tenuiflora*; and prickly lettuce, *Lactuca scariola* were quite satisfactory for survival, though not as suitable as the crop plants. Among the least suitable plants were common prairie grasses. Diet also affected fecundity, with favored food such as dandelion resulting in production of a mean value of 3.5 egg pods per female during a three-week period, whereas grasshoppers fed a mixture of prairie grasses produced only 0.3 pods per female.

Natural Enemies. Many insects parasitize or prey on *Melanoplus* grasshoppers. The most comprehensive listing of arthropod natural enemies was published by Rees (1973); this publication also contains keys to many of the important species.

The most important parasitoids are nymph- and adult-attacking flies (Diptera) in the families Anthomyiidae, Nemestrinidae, Sarcophagidae, and Tachinidae, though egg parasitoids (Hymenoptera: Scelionidae) also cause mortality in grasshopper populations. In Oregon, over 70% of migratory grasshoppers were parasitized by the nemestrinid, *Neorhynchocephalus sackenii* (Williston), resulting in decreased longevity and reproduction. Other *Melanoplus* spp. also are affected by this fly, though grasshopper populations on rangeland, not cropland, are usually affected (Prescott, 1960). In a study of migratory grasshopper parasitism conducted in Ontario, the incidence of parasitism reached about 7% by the end of September, with *Blaesoxipha hunteri* (Hough) and *B. atlantis* (Aldrich) (both Diptera: Sarcophagidae) the most effective parasitoids (Smith, 1965). Sanchez and Onsager (1994), using more accurate methods to estimate parasitism of migratory grasshopper, reported generation parasitism levels of 15–41% in Montana, with anthomyids and sarcophagids accounting for 50% and 35% of the parasitism, respectively.

Among the most important predators are sphecid wasps (Hymenoptera: Sphecidae). Adult sphecids capture and paralyze nymphal and adult grasshoppers, bury them within cells in the soil, and deposit an egg on the surface of the grasshopper. Upon hatching, the larva devours the paralyzed grasshopper. Predatory beetles (Coleoptera) attack the egg, nymphal and adult stages of grasshoppers, and include ground beetles (Carabidae), tiger beetles (Cicindelidae), soldier beetles (Cantharidae), and blister beetles (Meloidae). Blister beetles are most important, though because the grasshopper egg pod is the stage destroyed, and the predatory activities are hidden below-ground, their effect is often not appreciated. Parker and Wakeland (1957) summarized the results of several studies on egg pod predation in western states; during the period 1938–1940, for example, an average of 8.8% of egg pods were destroyed by blister beetles. Flies also are important predators, particularly robber (Asilidae) and bee flies (Bombyliidae). Robber fly larvae and *Gryllus* spp. field crickets (Orthoptera: Gryllidae) occasionally attack egg pods, and robber fly adults routinely attack nymphs and adults of grasshoppers, though other insects also are taken. Robber flies undoubtedly are important predators under rangeland conditions, but predation rates of grasshoppers in cropping systems has not been determined. Also, the propensity of robber flies to capture other predators such as sarcophagids significantly decreases their value (Rees and Onsager, 1982). Bee fly larvae are predatory on grasshopper eggs and on other insects. In western studies, an average of 6.2% of egg pods were destroyed by bee fly larvae (Parker and Wakeland, 1957).

Birds are known to be important predators of grasshoppers. They are among the most important sources of food for many avian species due to their large size and abundance. The great abundance of grasshoppers in the spring coincides with the period when most birds are nesting. Birds forage freely on them in open areas such as grasslands, with some species consuming 65–150 grasshoppers per day (McEwen, 1987). Although avian predators significantly decrease the abundance of grasshoppers in grasslands, it is less certain that they forage freely in crops.

Microbial pathogens can be quite important mortality agents, especially when weather conditions are suboptimal for grasshoppers, or when grasshoppers are very abundant. The principal microbial pathogens of grasshoppers are fungi, viruses, protozoans, and nematodes and nematomorphs.

The fungus *Entomophthora grylli* causes "summit disease," a behavior wherein grasshoppers ascend vegetation, cling to the uppermost point, and perish. In some areas, particularly near bodies of water, many dead grasshoppers can be found attached to plants. *Melanoplus* spp. are susceptible to one pathotype of the fungus, and significant grasshopper population decreases have been linked to the incidence of this fungal disease. Infection normally occurs when nymphs contact spores that are sheltered in the soil. Spores produced in grasshoppers dying due to this disease remain in cadavers or soil for protracted periods of time. Attempts to manipulate this pathogen have met with mixed results (Carruthers *et al.*, 1997). This is the only common naturally-occurring fungus of grasshoppers.

Several viruses called entomopoxviruses affect grasshoppers. One such virus, *Melanoplus sanguinipes* entomopoxvirus, affects the crop-feeding *Melanoplus* spp. and American grasshopper, *Schistocerca americana* (Drury). The virus disease spreads naturally by cadaver feeding. Infected grasshoppers are pale colored and lethargic, have prolonged developmental periods, and often perish. These diseases are quite rare in the field (Streett *et al.*, 1997).

Several types of protozoa are associated with grasshoppers, including amoebae, eugregarines, and neogregarines, but the most important are microsporida. Species of *Nosema* are most common, and *Nosema locustae* has been developed as a microbial insecticide.

Nosema spp. affect feeding, development, reproduction and survival, and are transmitted by ingestion (Johnson, 1997). However, they infrequently appear at high levels in natural populations.

Nematodes are important mortality factors of grasshoppers in South America, New Zealand and Australia, but in North America only *Mermis nigrescens*, *Agamermis decaudata*, and *Hexamermis* spp. affect them. *Mermis nigrescens* is most important, and is unique in that nematodes crawl from the soil onto vegetation to deposit desiccation-resistant eggs. The eggs hatch when consumed by grasshoppers, and the resulting larvae kill the grasshoppers, return to the soil, and continue the life cycle. These nematodes apparently do not thrive in arid areas, as the adults only emerge during wet periods, and are more common in irrigated croplands than dry rangeland. Sometimes they parasitize up to 70% of grasshoppers in an area. Nematomorphs, commonly called horsehair worms, resemble nematodes but tend to be much longer. They are rare, possibly because part of their life cycle must occur in water, but they attract considerable attention because of their large size (Baker and Capinera, 1997).

Weather. Migratory grasshopper is greatly influenced by weather. Through most of its range longevity and reproduction are limited by shortage of warm weather. Thus, abnormally warm and dry periods of about three years stimulate increase in their numbers. Warm weather during spring and autumn is particularly important. Cool and cloudy weather in the spring inhibits feeding by young nymphs, and results in high mortality. Also, adults have the potential to be long-lived and highly fecund, but their reproductive effort is normally terminated prematurely by the onset of cold weather. When summers are hot or prolonged, development proceeds faster or longer, resulting in greater egg production. In southern areas grasshoppers are less limited by shortage of warm weather, but are more affected by shortages of food. Therefore, occurrence of precipitation early in the season to provide luxurious foliage, especially broadleaf weed vegetation, is an important prerequisite for population increase (Capinera and Horton, 1989).

Life Cycle and Description. In most areas of its range, migratory grasshopper produces a single generation, and overwinters in the egg stage. However, in southern portions of its range two generations may occur annually. Eggs hatch relatively early, usually beginning in early June but about a week after twostriped grasshopper, *Melanoplus bivittatus*, begins to hatch. Hatching is protracted and may require up to six weeks in an area, resulting in asynchronous development of the population. Early hatching individuals mature early in the summer and have adequate time for reproduction whereas late-hatching individuals are handicapped by the onset of cold weather.

Egg. Eggs are yellowish, elongate-eliptical and slightly curved. One side convex and the opposite side concave, causing the egg to resemble a banana. They measure about 4.5 mm long and 1.2 mm wide. They are arranged in two columns within a frothy egg pod. Egg pods contain 18–20 eggs per pod, and are buried to a depth of about 35 mm. The pods are curved, about 25 mm long, and 3–4 mm in diameter. The upper portion contains only froth, allowing ready escape of the nymphs when they hatch. Females can produce 7–10 pods if fed high-quality diets; up to 20 per female is recorded, usually at 2–3 day intervals. Thus, fecundity of about 200 eggs per female is possible. Natural habitats vary greatly in suitability, however, and reproductive potential is sometimes affected. In Montana, for example, Sanchez and Onsager (1988) estimated that females were able to produce only three to four egg pods annually during the two years of their study. Egg pods are inserted into the soil among the roots of plants. Migratory grasshopper is more likely to deposit pods within crop fields, particularly among wheat stubble, than other common crop-feeding *Melanoplus* spp. Under normal weather conditions about 80% of embryonic development occurs in the autumn before the onset of diapause, with the remainder of development in the spring after a period of cold weather terminates egg diapause.

Nymph. Nymphs can develop over a range of about 22–37°C, but early instars suffer high mortality at both extremes in temperature. Nymphal development requires about 42, 27, and 23 days when cultured at a constant temperature of 27°, 32°, and 37°C. Diet also affects development rate. When reared at 30°C, favorable plants like tansymustard allow complete nymphal development in about 29 days, whereas nymphs developing on less suitable plants like western wheatgrass require about 42 days. Most field observations suggest that nymphal development requires 35–45 days. Normally there are five instars,

though if migratory grasshoppers are cultured at low temperature six instars is most common.

The nymphs are tan or gray, occasionally greenish, throughout nymphal development. They bear a curved black stripe that extends from behind each eye across the pronotum. The lower edge of the stripe is bordered in white. The outer face of the hind femur is marked with an interrupted black stripe. Body length in instars 1–5 is 4–6, 6–8, 8–11, 11–16, and 16–23 mm, respectively. The number of antennal segments is 12–13, 15–17, 18–20, 21–22, and 22–24 in the corresponding instars. A detailed description of the nymphal stages is found in Shotwell (1930).

Adult. The adult is a medium-sized species, measuring 20–26 mm long in males and 20–29 mm in females. They are grayish-brown, and often tinged with reddish-brown. A broad black stripe extends back from the eye and about two-thirds of the length of the pronotum. The front wings are grayish-brown or brown, usually with a row of brown spots centrally. The hind wings are colorless. The hind femora are not distinctly marked. The hind tibiae are greenish-blue or red. The cerci of males are broad and flat, and turn dorsally at the tip. The subgenital plate is elongate, and bears a notch and grove apically. Females have a pre-oviposition period of 2–4 weeks. Adults normally live 60–90 days, though with good food and weather, and living under low density conditions, longevity may be extended considerably. They can mate repeatedly. (See color figure 165.)

Nymphs and adults are affected by daily change in temperature. Activity levels at the soil surface, including feeding, are at their maximum when the air temperature is 18–25°C. This often results in a peak in feeding in late morning, followed by cessation of feeding at mid-day due to excessively hot temperature, and then perhaps a secondary peak in feeding in the afternoon as temperatures cool. Mass flights by adults take place only if air temperature is high, often about 29°C, but high densities and light wind also are required. When it is hot, grasshoppers tend to climb upward to escape the soil, which is usually considerably warmer than the air temperature. However, they also tend to roost on tall vegetation at night, as this allows them to be warmed by sunlight early in the morning, thus extending their period of activity.

Because of its importance as a field crop pest over wide areas of central North America, there is an extensive literature on *M. sanguinipes*. Important aspects of biology were given by Parker (1930,1939), Shotwell (1941), Pfadt (1949), Parker *et al.* (1955), Pfadt and Smith (1972), Onsager and Hewitt (1982), and many others. A very good synopsis was presented by Pfadt

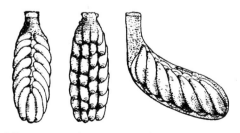
Migratory grasshopper egg pods, various aspects.

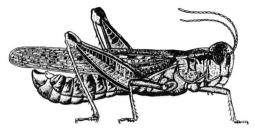

Adult male migratory grasshopper.

Male migratory grasshopper, tip of abdomen.

(1994b), who also pictured all stages of development. *Melanoplus sanguinipes* was included in many grasshopper keys, including those by Blatchley (1920), Dakin and Hays (1970), Helfer (1972), Capinera and Sechrist (1982), and Richman *et al.* (1993). *Melanoplus sanguinipes* was also included in a key to grasshopper eggs by Onsager and Mulkern (1963). A synopsis of migratory grasshopper, including keys to related Canadian Orthoptera, was given by Vickery and Kevan (1985). Rearing of *Melanoplus* species was described by Henry (1985).

Damage

Migratory grasshopper is a defoliator, often completely removing leafy vegetation and leaving only stem tissue. Sometimes other tissue is eaten; heads of wheat may be clipped, for example. Migratory grasshopper thrives on rangeland that has a high density of broadleaf weeds, so it often moves from grazing land to nearby irrigated crops. In this behavior it differs from some other species, particularly differential grasshopper, *Melanoplus differentialis* (Thomas), and twostriped grasshopper, *M. bivittatus* (Say), which favor the taller undisturbed vegetation usually associated with fences and irrigation ditches, and usually do not develop high numbers on grazing land. Migratory grasshopper is often quite dispersive, and of course this behavior is the basis for its common name. When they are developing at high densities, the weather is abnormally warm, and a light wind is present, swarms of grasshoppers may disperse tens or even hundreds of kilometers and descend without warning to cause immense damage. With the availability of modern insecticides and aircraft for application, such potential disasters can be dealt with quickly

and efficiently. When such insecticide and effective application technologies are not available, or where environmentally sensitive land or crops are concerned, grasshopper swarms can be disastrous. This species is the most important grasshopper pest in western North America.

Management

Sampling. Grasshopper populations are usually assessed by visual observation. A sweepnet is a useful tool to aid in collection, and its use is a prerequisite to identify the species complex. It is important to determine if grasshoppers collected from non-crop areas are crop-feeding species, because there are many non-pest grasshoppers that restrict their feeding to grasses or weeds. It is advisable to monitor nearby uncultivated land, particularly weedy areas, in addition to crop plants, due to the tendency of the pest species, to invade crops later in the season.

Insecticides. Liquid formulations of insecticides are commonly applied to foliage to protect against damage. As grasshoppers rarely develop in crops, but instead invade from weedy areas, it is often the edges of crop fields that are most injured. Therefore, application of insecticide to the borders of crop fields is often adequate to protect an entire field. It is even better to apply insecticides to the developing grasshopper populations in weedy areas before they move to crops. This not only minimizes damage to crop plants, but often results in younger grasshoppers being targeted for elimination. Younger grasshoppers are more susceptible to insecticides, with large nymphs and adults sometimes difficult to kill.

Application of insecticide-treated bait is an effective alternative to foliar treatments for *Melanoplus* spp, because these grasshoppers spend considerable time on the soil where they come into contact with baits. Bait formulations are bulky and more difficult to apply than liquid products, so they are not often used, but have the advantage of limiting exposure of crops to insecticide residue and of minimizing mortality of beneficial insects such as predators and parasitoids due to insecticide exposure. Also, the total amount of insecticide active ingredient that is necessary to obtain control is usually considerably less when applied by bait, because the grasshoppers actively seek out and ingest the bait. Finally, for relatively expensive products that must be ingested to be effective, such as microbial insecticides, baits are the most effective delivery system.

The attractant used most commonly for grasshopper bait is flaky wheat bran, though other products such as rolled oats are sometimes suggested. No additives, other than insecticide (usually 5% active ingredient), are necessary because the wheat bran is quite attractive to *Melanoplus* grasshoppers. Other additives such as sawdust, water, vegetable or mineral oil, molasses, amyl acetate, salt, or sugar have been suggested, but provide little or no additional benefit over dry bran. The bait should be broadcast widely to maximize the likelihood of grasshopper contact, and be applied while they are in the late instars as the adults ingest less bait. Shotwell (1942) and Cowan and Shipman (1940) provided excellent information on formulation of grasshopper baits; Mukerji *et al.* (1981) gave an interesting perspective of bait use on rangeland.

Cultural Practices. Elimination of weeds within, and adjacent to, crops is the most important cultural practice, and can have material benefit in preventing damage to crop borders. However, during periods of weather when grasshoppers become numerous they may move long distances and invade crops.

Tillage is an effective practice for destruction of eggs, particularly in migratory grasshopper, which is especially likely to deposit eggs among crop plants. Deep tillage and burial are required, as shallow tillage has little effect. All the crop-feeding *Melanoplus* species deposit some eggs in crop fields, especially during periods of abundance, but it is fence row, irrigation ditch, field edge, and roadside areas that tend to be the favorite oviposition sites, so tillage is not entirely satisfactory unless other steps are taken to eliminate grasshopper egg pods from these areas that cannot be tilled. Although providing suppressive effects, deep tillage is not consistent with the soil and water management practices in many areas, so it may not be a good option.

Row covers, netting, and similar physical barriers can provide protection against grasshoppers. This approach obviously is limited to small plantings, and can interfere with pollination. Also, grasshoppers are capable of chewing through all except metal screening, so this approach can not guarantee complete protection.

Biological Control. The opportunities for biological control are limited. Historically, poultry were found to consume many grasshoppers and could provide considerable relief if the grasshopper-infested garden was small or moderate in size and the birds were plentiful. This remains a viable option for some people, and turkeys are usually considered most suitable among poultry. The birds may also inflict some direct damage to plants, however, so introduction of poultry is probably most viable when grasshoppers are plentiful and threatening.

The microsporidian pathogen *Nosema locustae* is well-studied as a microbial control agent of *Melanoplus*

spp. (Ewen and Mukerji, 1980; Johnson and Pavlikova, 1986; Johnson and Henry, 1987; Bomar *et al.*, 1993), and is available commercially. It is fairly stable, and easily disseminated to grasshoppers on bait. However, its usefulness is severely limited by the long period of time that is required to induce mortality and reduction in feeding and fecundity. Also, the level of mortality induced by consumption of *Nosema* is quite low (Johnson and Dolinski, 1997), often imperceptible. It is best used over very large areas, not just on individual farms, and should be applied at least one year in advance of the development of potentially damaging populations. More rapid suppression of grasshoppers is attainable by applying very high levels of *Nosema* (Capinera and Hibbard, 1987), but this is not usually considered to be an economic approach, and commercial products are not prepared in this manner.

Fungi have also been investigated for grasshopper suppression, and a grasshopper strain of *Beauveria bassiana* has been effective in some trials (Moore and Erlandson, 1988; Johnson and Goettel, 1993; Jaronski and Goettel, 1997). Behavioral thermoregulation by grasshoppers, wherein they bask in the sun and raise their body temperatures, is potentially a limiting factor for use of fungi. Basking grasshoppers easily attain temperature in excess of 35°C; such high temperature decreases or even prevents disease development in infected grasshoppers (Inglis *et al.*, 1996). Inconsistent quality control in production of fungi also limits use of these organisms for grasshopper control.

Redlegged Grasshopper
Melanoplus femurrubrum (De Geer)

Southern Redlegged Grasshopper
Melanoplus propinquus Scudder
(Orthoptera: Acrididae)

Natural History

Distribution. The redlegged grasshopper, *Melanoplus femurrubrum* (De Geer), is known from nearly all of the United States and southern Canada, and its distribution also extends south through most of Mexico. In parts of the southeast, on the coastal plain from eastern North Carolina to southern Mississippi and Louisiana, and including all of Florida, it is replaced by a very similar form, *M. propinquus* Scudder, which appears to be a separate species (Dakin, 1985). These are native insects.

Host Plants. These species are polyphagous, feeding on a broad range of plants and apparently preferring a dietary mixture over a single food plant, and broadleaf plants over grasses. The preferred habitat is tall vegetation in pastures, fence rows, along irrigation ditches and roadways, and in fallow agricultural fields which have become weedy. The redlegged grasshoppers are known throughout North America for damage to crops, attacking alfalfa, barley, birdsfoot trefoil, clover, corn, lespedeza, oat, orchardgrass, soybean, timothy, tobacco, and vetch in addition to vegetables. Among vegetable crops, bean, beet, cabbage, and potato seem to be most frequently injured, but nearly any crop may be fed upon. Among the uncultivated plants eaten are aster, *Aster* spp.; Kentucky bluegrass, *Poa pratensis*; brown knapweed, *Centaurea jacea*; cinquefoil, *Potentilla argentea*; dandelion, *Taraxacum officinale*; fleabane, *Erigeron divergens*; goldenrod, *Solidago canadensis*; kochia, *Kochia scoparia*; Russian thistle, *Salsola kali*; smooth brome, *Bromus inermis*; sweet clover, *Melilotus officinalis*; wavyleaf thistle, *Cirsium undulatum*; western ragweed, *Ambrosia psilostachya*; and likely many others. The weeds fed upon most frequently in North Dakota alfalfa fields were reported to be kochia, *Kochia scoparia*, field bindweed, *Convolvulus arvensis*; awnless bromegrass, *Bromis inermis*; and foxtail, *Setaria* spp. (Mulkern *et al.*, 1962). On prairie, redlegged grasshopper ate primarily Kentucky bluegrass, *Poa pratensis*; western ragweed, *Ambrosia psilostachya*; golden aster, *Chrysopsis villosa*; flixweed, *Descurainia sophia*; and leadplant, *Amorpha canesens* (Mulkern *et al.*, 1964). Bailey and Mukerji (1976) were able to culture redlegged grasshopper on nearly all plants tested, though they completed development more rapidly when provided with plants on which they preferred to feed.

Natural Enemies. The natural enemies of the crop-feeding *Melanoplus* spp. are quite similar. For information on natural enemies of redlegged grasshopper, see the section on natural enemies of migratory grasshopper, *Melanoplus sanguinipes* (Fabricius).

Weather. Like most grasshoppers, redlegged grasshopper is favored by hot weather. Long-term periods of drought and hot weather favor population increase, especially in northern areas that normally are cooler. A certain amount of precipitation is necessary to provide adequate food for the grasshoppers, of course, but prolonged cool, wet weather, especially during the period of eggs hatch, is detrimental for survival. The late onset of winter can favor grasshopper population increase because it allows adults additional time to produce eggs.

Life Cycle and Description. There is only a single generation per year throughout the range of these species, with the egg stage overwintering. Eggs hatch in late spring and adults are present from July until they are killed by heavy frost, though their numbers decrease steadily throughout the season. The following information has been derived from studies of *M. femurrubrum*, but likely applies equally well to *M. propinquus*.

Egg. The eggs are elongate-cylindrical, and widest at the middle. They measure 4.1–4.6 mm long and 0.9–1.5 mm in diameter. Their color is yellowish-brown or creamy-white. They are deposited in structures called pods, which consist of two columns of eggs arranged in parallel rows. The pod, which is secreted by the female during oviposition, consists of frothy material secreted between, and covering, the eggs. The pod is a curved cylinder in form, and measures about 20–25 mm long and 3–5 mm in diameter. It is buried in the soil, and normally contains 20–26 eggs per pod. The upper portion of the pod consists solely of froth, and the young grasshoppers chew their way through this material to escape from the soil. Eggs are often deposited among the roots of grasses and weeds, particularly along the edges of crop fields. Females can produce 300 eggs during their life time. These are not early season grasshoppers. Hatching occurs about three weeks after hatching of twostriped grasshopper, *Melanoplus bivittatus* (Say), two weeks after migratory grasshopper, *Melanoplus sanguinipes* (Fabricius), and about the same time as differential grasshopper, *Melanoplus differentialis* (Thomas), with which redlegged grasshopper may co-occur. The period of hatching is extended, however, so nymphs can be found during most of the summer.

Nymph. Development of the nymphal stage normally requires about 40 days, during which there usually are 5–6 instars. Throughout their development they are yellowish, but marked with a broad black stripe that extends across the face, eyes and prothorax, and onto the abdomen. They also bear a second, less discrete black stripe that is below, but parallel to the aforementioned stripe, and that arches across the lateral lobe of the prothorax. The outer face of the femora is marked with a broad black stripe. The underside is yellowish. The hind tibiae are yellow or gray, and bear black spines. The overall body length of nymphs is 4.0–5.6, 6.2–7.2, 7.4–9.7, 10.0–15.5, and 16.5–22.5 mm for instars 1–5, respectively. Antennal segment numbers increase from 12–14 to 15–16, 17–19, 22–24, and 24–26 in the corresponding instars. Bellinger and Pienkowski (1989) reported mostly 6–7 instars in Virginia. They reported total nymphal development times for grasshoppers displaying a total of six instars to average 67, 42, 31, and 29 days when cultured at 26.5°, 30°, 35°, and 38°C, respectively. Nymphs developing through seven instars required 5–10 additional days, depending on the rearing temperature. Mean instar development time was reported to be 3.6, 4.1, 4.7, 5.0, 5.8, and 7.9 days for instars 1–6 when cultured at 35°C, an optimal temperature. Nymphs change their behavior in response to temperature, with feeding commencing at 20–24°C, and nymphs moving to elevated perches to escape the heat when the air or soil temperatures get high. Like most grasshoppers, they tend to remain inactive, even at high temperature, in the absence of sunlight.

Adult. The adults are medium-sized grasshoppers, the males measuring 17–23 mm long, and the females 18–27 mm. They are reddish-brown or grayish-brown dorsally and yellow or yellowish-green ventrally. The front wings lack distinct markings; the hind wings are colorless. The lateral lobe of the pronotum is usually marked with a distinct black area. The outer face of the hind femora are yellowish but bear an indistinct dark stripe. The hind tibiae are almost always deep red with black spines, and are the basis for the common name of these grasshoppers. In the male, the tip of the abdomen is rather bulbous, with the subgenital plate bearing a broad "U"-shaped depression apically. The cerci narrow rapidly from the base, with the distal third narrow and the tip either angled (*M. femurrubrum*) or rounded (*M. propinquus*).

The males, but not the females, of these two species are easily separated by the genitalia. In *M. femurrubrum*, the cerci are pointed at the tip, formed by an acute angle. The cerci at the mid-point are relatively broad, about one-half the maximum width at the base. The furcula in this species diverge from the base and then converge distally, the space between the arms of the furcula forms a "U" shape. In contrast, in *M. propinquus* the tips of the cerci are rounded, and the cerci are relatively narrow at the mid-point, about one-third the maximum width at the base. The furcula diverge more strongly, the space between the arms forms a "V" shape (Dakin, 1985).

Adults normally roost at night on the tops of tall grasses and weeds. Early in the morning they crawl down the plant and resume feeding when the air temperature warms, often moving along the soil in search of food. As happens with nymphs, the adults ascend vegetation to escape high temperature. In the evening, they perch again on elevated roosts, and remain there until they are warmed by sunlight in the morning. Females have a pre-oviposition period of 9–15 days before they commence egg laying. Redlegged grasshopper is a fairly strong flier and can fly 10 m if disturbed. (See color figure 163.)

Despite its importance, comprehensive treatment of *M. femurrubrum* biology is lacking. Some important aspects were given by Parker (1939), Shotwell (1941)

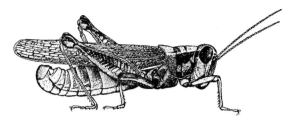

Adult male redlegged grasshopper.

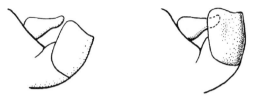

Male redlegged grasshopper, tip of abdomen.

and Bellinger and Pienkowski (1989). A very good synopsis was presented by Pfadt (1994d), who also pictured all stages of development. A summary of redlegged grasshopper biology, including keys to related Canadian Orthoptera, was given by Vickery and Kevan (1985). *Melanoplus femurrubrum* was included in many grasshopper keys, including those by Blatchley (1920), Dakin and Hays (1970), Helfer (1972), Capinera and Sechrist (1982), and Richman *et al.* (1993). Blatchley (1920) includes *M. propinquus*; and Dakin and Hays (1970) treated the two redlegged grasshoppers as subspecies. *Melanoplus femurrubrum* was also included in a key to grasshopper eggs by Onsager and Mulkern (1963). Rearing of *Melanoplus* spp. was described by Henry (1985).

Damage

Redlegged grasshopper is a defoliator, often removing all leaf tissue and leaving only plant stems. Lower densities leave plant ragged or tattered. Redlegged grasshopper is a common component of the grasshopper complex that affects plants growing along the margins of fields, though it causes extensive damage only during periods of very high density. Redlegged grasshopper is capable of developing high densities and migratory tendencies during periods of drought, and may be found mixed into swarms of migratory grasshopper, *Melanoplus sanguinipes* (Fabricius).

Management

Management of the various *Melanoplus* spp. grasshoppers is substantially the same. For information on redlegged grasshopper management, see the section on management under migratory grasshopper, *Melanoplus sanguinipes* (Fabricius).

Twostriped Grasshopper
Melanoplus bivittatus (Say)
(Orthoptera: Acrididae)

Natural History

Distribution. This native grasshopper is widely distributed in northern North America. In the United States it extends from the Atlantic to the Pacific Oceans, and is absent only from the Gulf Coast region. In Canada it occurs from Nova Scotia to British Columbia.

Host Plants. This species is adaptable, and is found in a variety of habitats. However, it is most abundant in moist meadows, dense vegetation along water courses, and in disturbed, weedy areas. In the Great Plains region it is abundant in moist tallgrass regions, but uncommon in the drier shortgrass prairie. Twostriped grasshopper feeds on both grasses and broadleaf plants, but prefers the latter and fares poorly in habitats lacking broadleaf plants. Plants in the families Compositae and Cruciferae seem to be preferred. Among the uncultivated vegetation consumed is arrowleaved colt's foot, *Petasites sagittatus*; burdock, *Arctium lappa*; dandelion, *Taraxacum officinale*; dog mustard, *Eruscastrum gallicum*; flixweed, *Descurainia sophia*; needleleaf sedge, *Carex eleocharis*; leadplant, *Amorpha canescens*; oxeye daisy, *Chrysanthemum leucanthemum*; pepperweed, *Lepidium densiflorum*; plantain, *Plantago major*; redtop, *Agrostis alba*; sand dropseed, *Sporobolus cryptandrus*; Canada thistle, *Cirsium arvense*; sunflower, *Helianthus* spp.; wavyleaf thistle, *Cirsium undulatum*; mustard, *Brassica* spp; and others. Mulkern *et al.* (1962) determined the plants consumed by twostriped grasshoppers in North Dakota alfalfa fields, and reported that the plants most often consumed, after alfalfa, were kochia, *Kochia scoparia*; wild oat, *Avena fatua*; awnless bromegrass, *Bromus inermis*; flixweed, *Descurainia sophia*; marsh elder, *Iva xanthifolia*; and quackgrass, *Agropyron repens*. On prairie, however, the plants most often consumed were Kentucky bluegrass, *Poa pratensis*; leadplant, *Amorpha canescens*; and western ragweed, *Ambrosia psilostachya* (Mulkern *et al.*, 1964). Survival rates and body weights of twostriped grasshopper are higher, and development times shorter, on mixed diets than on single hosts (MacFarlane and Thorsteinson, 1980). Bailey and Mukerji (1976) were able to culture twostriped grasshopper on nearly all plants tested, though they completed development more rapidly when provided with plants on which they preferred to feed.

Twostriped grasshopper commonly infests vegetable and field crops, though most injury is limited to field margins. Among vegetable crops injured are beet, cabbage, chicory, corn, lettuce, onion, potato and likely others. Field crops such as alfalfa, birdsfoot trefoil, clover, young barley and oat, timothy, vetch and the immature seedheads of wheat also are fed upon. Flowers and ornamental plants likewise are attacked. (See color figure 29.)

Natural Enemies. The natural enemies of the crop-feeding *Melanoplus* spp. are quite similar. For information on natural enemies of twostriped grasshopper, see the section on natural enemies of migratory grasshopper, *Melanoplus sanguinipes* (Fabricius).

Weather. Survival and population increase in grasshoppers are favored by hot weather. High numbers tend to occur after a period of years with abnormally hot and dry weather during the spring and summer months. This is especially true in northern areas, where it tends to be cooler. Enough precipitation is required to provide adequate food for grasshoppers, of course, but protracted periods of rainfall during egg hatch, especially if accompanied by cool weather, disrupt feeding by young grasshoppers and induce high mortality. Late onset of winter can favor grasshopper population increase because it allows adults additional time to produce eggs.

Life Cycle and Description. Over most of its range, twostriped grasshopper displays one generation annually, with the egg stage overwintering. In Colorado, eggs begin to hatch in June, though hatching can occur over a four-to-six week period. Nymphs may be present until September, but adults appear beginning in July. Oviposition commences in August and continues until adults are killed by cold weather. At higher elevations in British Columbia, a two-year life cycle is reported (Beirne, 1972).

Egg. The egg is elongate-cylindrical, with the ends tapering to blunt points. The eggs are olive in color and measure about 5.0 mm long and 1.2 mm in diameter. The reported values of number of eggs per pod and total fecundity vary considerably among studies. Drake *et al.* (1945) reported 69.7 per pod and a mean total of 129 eggs per female, whereas Smith (1966) reported 43.3 eggs per pod and a total of 355. The number of pods produced per female ranges from 4–15, with a mean interval between oviposition of four days (Smith 1966), but this assumes good weather. Eggs are arranged in columns of four within a frothy secretion; the egg structure is called an egg pod. The pods are curved, measure 30–38 mm long and 6–7 mm in diameter. They are inserted into the soil at a depth of 2–5 cm, and topped with a frothy plug. Favorite oviposition sites are along fence rows, ditch banks, and pastures with compact, undisturbed soil. Pods are often inserted among the roots of plants, and a soil moisture content of 10–20% is preferred. The act of oviposition requires about two hours. The egg is the overwintering stage, and typically persists in the soil for 7–8 months. Embryonic development begins in the summer and autumn after oviposition, and is 60–80% complete before embryos enter diapause for the winter. However, they can be induced to hatch if exposed to about 5°C for 90 days.

Nymph.   Nymphal development normally requires 30–50 days. Most nymphs display 5–6 instars, but seven instars occurs occasionally. Nymphal development time when fed lettuce or alfalfa and cultured at 21°C is about 10, 8.5, 10, 11, and 14 days, respectively, for instars 1–5 (Langford, 1930). The young nymphs initially are dark-brown or greenish, but gain a distinct dark stripe along the pronotum behind each eye at the third instar. The wing development is poor until instar three, and the developing wings is pointed downward. At instar four the wing orientation is reversed, with the wings oriented upward, but also pointing posteriorly. In the fifth instar the wings are quite evident and extend out to at least the second abdominal segment. The number of antennal segments is 12–13, 17–18, 19–22, 23–25, and 24–26 for instars 1–5, respectively. Corresponding body lengths are 5.0–6.6, 7.4–10.4, 9–14, 15–21, and 20–17 mm. Nymphs are found on the soil and seek food each morning, but may ascend plants to escape the heat of the soil by noon. Like the adult stage, nymphs can perch on elevated roosts at night, and can sun themselves in the morning and in cool weather to attain optimal body temperature.

Adult.   This is a fairly large and robust species. Males measure 23–29 mm long, females 29–40 mm. The general body coloration is olive or brownish-green dorsally and yellowish or yellowish-green ventrally. The head and pronotum tend to be darker, usually olive green. A narrow but distinct yellow stripe passes from behind each eye along the pronotum and forewings, extending nearly to the wing tips. The stripes are often bordered below with black, especially on the anterior portions of the body. The stripes come together posteriorly in the forewings, forming a "V" shape. It is this pair of yellow stripes that is the basis for the common name of this grasshopper. The front wings are usually uniform in color except for the stripes, and the hind wings colorless. The hind femora are yellow with a dark stripe along the outer face. The hind tibiae are variable, usually reddish but also greenish, yellowish, and purplish, and equipped with black spines. The male cerci are short, broad, and boot-shaped. Adults seek crop borders and roadsides for oviposition. Mated females have a 7–14 day pre-

Adult female twostriped grasshopper.

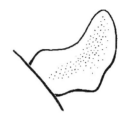

Twostriped grasshopper male cercus.

oviposition period, after which they oviposit within the roots of grasses and weeds. Duration of the oviposition period is about 30 days (range 15–55 days). (See color figure 166.)

Aspects of the biology of twostriped grasshopper were treated by many authors, including Parker (1939), Shotwell (1941), Church and Salt (1952), and Smith (1966). An excellent synopsis was presented by Pfadt (1994e), who also pictured all stages of development. A summary of twostriped grasshopper biology, including keys to related Canadian Orthoptera, was given by Vickery and Kevan (1985). *Melanoplus bivittatus* was included in many grasshopper keys, including those by Blatchley (1920), Dakin and Hays (1970), Helfer (1972), Capinera and Sechrist (1982), and Richman *et al.* (1993). This species was also included in a key to grasshopper eggs by Onsager and Mulkern (1963). Rearing of *Melanoplus* spp. was described by Henry (1985).

Damage

Twostriped grasshopper consumes the leaves of numerous plants. Damage is greatest in areas adjacent to weeds, and along fence rows, irrigation ditches, roadsides, and fallow fields. Damage is exacerbated by drought, which apparently increases nymphal survival rates and decreases the amount of weed vegetation available to the grasshoppers. Although the grasshoppers feed at night if it is sufficiently warm, where nights are cool these grasshoppers tend to perch on elevated objects. This behavior allows them to be warmed by the light from the setting and rising sun, and maximizes their period of activity. This also results in nibbling on the resting substrate by the grasshoppers; these grasshoppers feed on the bark of bushes and young trees, and even damage shingles on buildings and eat holes in vinyl window screens while perching. The nymphs and adults are fairly dispersive, and walk tens or even hundreds of meters in the search for food. At high densities they show propensity to swarm, which is expressed by band formation in the nymphal stage and flight by adults. The temperature threshold for flight is 30–32°C. Ascending to heights of 200–500 m, and flying with the wind, swarming adults can disperse long distances.

Management

Management of the various *Melanoplus* spp. grasshoppers is substantially the same. For information on twostriped grasshopper management, see the section on management under migratory grasshopper.

FAMILY GRYLLIDAE—FIELD CRICKETS

Fall Field Cricket
Gryllus pennsylvanicus **Burmeister**

Spring Field Cricket
Gryllus veletis **(Alexander and Bigelow)**

Southeastern Field Cricket
Gryllus rubens **Scudder**
  (Orthoptera: Gryllidae)

Natural History

Distribution. Field crickets are extremely difficult to distinguish based on appearance, so for many years they have been grouped into one species, *Gryllus assimilis* (Fabricius). As the significance of the calling behavior (chirping) became known, some of the species have been distinguished, but much work remains. The biology and damage potential of the various field crickets is confused by the problems with identification. The most economically significant North American species are fall field cricket, *Gryllus pennsylvanicus* Burmeister; spring field cricket, *Gryllus veletis* (Alexander and Bigelow); and southeastern field cricket, *Gryllus rubens* Scudder; though other species may be locally important. The fall and spring field crickets are most abundant in the northern states and southern Canada, whereas the southeastern field cricket is known from the southeastern states. These are native insects.

Host Plants. These field crickets are found widely in grassy fields, pastures, weedy areas, roadsides and lawns, and occasionally feed on foliage, flowers and fruit of crop plants. Among the vegetable crops damaged are bean, beet, carrot, cabbage, cantaloupe, cucumber, lettuce, parsnip, pea, potato, pumpkin, squash, sweet potato, tomato, watermelon, and likely others. Other crops known to be injured by field crickets are alfalfa, barley, corn, cotton, flax, rye, strawberry, sweetclover, and wheat. Flowers and seeds of weeds also are suitable food for crickets, and among those known to be consumed are foxtail, *Setaria* spp.; lambsquarters, *Chenopodium album*; pigweed, *Amaranthus* spp.; ragweed, *Ambrosia* spp.; Russian thistle, *Salsola kali*; sunflower, *Helianthus* spp.; and wheatgrass, *Agropyron* spp. Russian thistle is reported to be particularly suitable, and foliage as well as flowers and seeds are eaten. The foliage of broadleaf plants is preferred over grasses.

Natural Enemies. Many natural enemies of field crickets are known, though their importance is poorly documented. Eggs are parasitized by *Ceratotelia marlattii* Ashmead and *Paridris brevipennis* Fouts (both Hymenoptera: Scelionidae), with 20–50% of the eggs in South Dakota reportedly parasitized by *C. marlatti* and 1–5% by *P. brevipennis*. Nymphs and adults are killed by *Exoristoides johnsoni* Coquillett and *Euphasiopteryx ochracea* (Bigot) (both Diptera: Tachinidae), *Sarcophaga kelleyi* Aldrich (Diptera: Sarcophagidae), mermithid nematodes (Nematoda: Mermithidae), and horsehair worms (Nematomorpha). Gregarine parasites (Sporozoa) infest the guts of field crickets (Zuk, 1987), but have little effect on their host other than to extend the cricket developmental period. Among the predators that take advantage of cricket abundance are many birds, but particularly crows, *Corvus bachyrhynchos brachyrhynchos* Grehm; and ring-necked pheasant, *Phasianthus torquatus* Gmelin; as well as snakes, toads, and gophers.

Life Cycle and Description. Field crickets differ in their life history, but the aforementioned species exemplify the common life cycles. *Gryllus pennsylvanicus* and *G. veletis* have one generation per year, whereas *G. rubens* has two generations annually. In South Dakota and most of its range, *G. pennsylvanicus* eggs overwinter, hatching occurs in the spring, nymphs develop in early summer, adults appear beginning in July, and eggs are deposited in August and September. The presence of adults in the late summer and autumn is the basis for its common name of "fall" field cricket. In Michigan, adults chirp from early August until mid-November (Alexander and Meral, 1967).

In the same area, *G. veletis* nymphs overwinter in the later instars, the adult stage is attained in May–June, eggs are produced in May–June, hatching is completed by the end of August, and nymphs complete part of their development before the onset of winter.

The presence of adults during spring and early summer is the basis for its common name of "spring" field cricket. In Michigan, adults chirp from mid-May until early August. Thus, though these two field cricket species occupy the same habitat, there is seasonal separation of life stages.

In Louisiana and most of its range, *G. rubens* overwinters as nymphs, the adult stage is attained by April when eggs are produced, and nymphs are mature by July. Second generation adults begin to appear and produce eggs in late July–August, and nymphs develop until winter, failing to molt to the adult stage until the subsequent spring.

Egg. Eggs are deposited in firm, damp soil, usually within the upper 2 cm of soil. Under dry conditions crickets may deposit their eggs within the cracks formed as moist soil hardens. The eggs are elongate-cylindrical, with bluntly rounded ends. They also are slightly curved, with one side convex and the opposite side concave. Eggs normally measure about 3 mm long and 0.6 mm wide. They are light yellow or cream. As the embryo develops it may change shape and color, and increase slightly in size. Such eggs become barrel-shaped, and measure about 4 mm long and 0.85 mm wide. They are deposited singly, but several are usually placed in close proximity. The number of eggs produced by each female ranges from about 150–400. Except for overwintering eggs, most hatch in about two weeks.

Nymph. The young cricket hatching from an egg is faced with the difficult task of burrowing through soil to reach the surface. The form of the cricket that escapes the egg is called the "vermiform larva," and differs from the following nymphal stages by being encased in a transparent membrane that immobilizes the appendages. Once the vermiform larva wriggles to the surface the membrane is shed, the legs are freed, and the young first instar cricket is able to walk and jump. The cricket undergoes several molts, growing at each stage. In *G. pennsylvanicus* and *G. veletis* normally there are 8–9 instars, though there may be more under adverse conditions. Mean duration of instars under field temperature in South Dakota was reported to average about 8.5, 8.2, 8.9, 8.8, 9.3, 10.7, 11.9, 12.0, and 14.6 days for instars 1–9, respectively. In *G. rubens* the number of instars averages 10, but the same pattern of latter instars requiring more time for development is apparent. Total nymphal development time usually requires 80–90 days in the cooler northern environments inhabited by *G. pennsylvanicus* and *G. veletis*, and 70–80 days in the warmer environments inhabited by *G. rubens*.

The nymphs initially are brownish, but marked with black and sometimes yellow in the thoracic area. They resemble the adults, and bear long antennae and cerci, but lack wings. With each succeeding molt the nymphs become darker. The ovipositor begins to appear in instars 3–4, the wing pads in instar six. By instar eight both the ovipositor and wing pads are apparent. Body length is about 3, 3.3–3.9, 4.2–5.0, 5–6, 7–8, 7.3–8.5, 9–12, 12–18, and 13.5–20 mm for instars 1–9, respectively.

Adult. The adult cricket is mostly shiny black, though the front wings may be brownish-black. The front wings usually cover the abdomen or extend slightly beyond the tip. The hind wings, however, are variable long, and wing length determination has both environmental and genetic components (Walker, 1987). In short-winged forms at least half of the abdomen is covered, but such insects are incapable of flight. Long-winged forms tend to be a small component of the population (Veazey *et al.*, 1976; Harrison, 1979), and though they are capable of flight, they do not often fly (Walker and Sivinski, 1986; Walker, 1987). Body length of adults is 15–26 mm. Females bear a long ovipositor, which often equals the length of her body. The width of the head is greater than the pronotum in *G. pennsylvanicus* and *G. veletis*, whereas in *G. rubens* the head is narrower than the pronotum. Males of these cricket ordinarily space themselves in a field and remain rather sedentary for all of their adult lives. They produce a song that attracts roving females. The song of *G. pennsylvanicus* and *G. veletis* is intermittent, which consists of about 150–240 discrete chirps per minute at 29°C, with chirps comprised of four pulses at about 25 pulses per second. In contrast, *G. rubens* produces a nearly continuous song, consisting of about 60 pulses per second at 29°C, with a considerably louder song than the other species. Females usually mate within 1–4 days of attaining the adult stage, and normally they mate repeatedly during the oviposition period. Most females commence oviposition from 7 to 14 days after attaining the adult stage, though some require up to 30 days. Egg production continues for the life of the cricket, with 50–60 days considered to be about average adult longevity.

Crickets are usually active at night, though they may venture forth during daylight during cloudy weather, and in late afternoon. Males normally call at night, but in cold weather may call during warmer daylight hours. Typically, crickets hide beneath debris during the day, but sometimes seek shelter in soil cracks or excavate small chambers in the soil. They sometimes appear to be gregarious, because they are found clustered, but this is simply a reflection of preference for a habitat with limited availability.

Alexander (1957) discussed the taxonomy of eastern field crickets and included keys to separate most species based on morphology; however, the fall and spring field crickets were treated as spring and fall broods of "*pennsylvanicus*." Nickle and Walker (1974) provided keys to the southeastern species. An excellent treatment of fall and spring field cricket biology (as *G. assimilis*) was provided by Severin (1935), with similar treatment of (probably) *G. rubens* (as *G. assimilis*) by Folsom and Woke (1939). Calling behavior of *G. pennsylvanicus* and *G. veletis* was described by Alexander and Meral (1967), and of *G. rubens* by Doherty and Callos (1991). Synopses of fall and spring field crickets, including keys to related Canadian Orthoptera, were given by Vickery and Kevan (1985). Culture of field crickets was described by Winewriter and Walker (1988).

Damage

Field crickets are omnivorous, and feed on a variety of plant and animal matter. They may consume the roots, stems, leaves, flowers, fruits and seeds of plants, but the flower, fruit and developing seeds are preferred. Once mature, seeds are no longer suitable. They also eat nearly any dead insect they encounter, sometimes are reported to be cannibalistic, and gnaw on animal and plant products such as fur, wool, linen, and cotton. They are rarely considered to be serious plant pests, and only when exceedingly abundant. Spring field cricket, in particular, is questionable as a plant pest, because it is not very gregarious.

Field crickets also display predatory behavior which offsets their occasional tendency to feed on valuable plants. For example, grasshopper eggs, flea

Adult field cricket, long-winged form.

beetle adults, fly puparia, caterpillar pupae, and insects from spider webs are among the documented animal material eaten by nymphs and adults (Burgess and Hinks, 1987).

Management

Suppression of crickets is rarely necessary, but application of insecticide-treated bran bait is effective if needed. Often its application can be limited to the edge of fields, where crickets dispersing into crops would encounter and ingest the poison bait. A formula for preparing baits was given by Severin (1935), but modern bait formulations usually eliminate molasses and other additives, employing only coarse bran and insecticide. Insecticides can also be applied to sawdust and distributed broadcast or in a band around the margin of a field; in such cases crickets perish from contact rather than ingestion of the bait, and this approach is most appropriate when cricket densities are very high (Blank *et al.*, 1985). Foliar applications of insecticides can also be effective if they are residual, but are less selective. Soil tillage can destroy eggs, and clean cultivation deprives overwintering insects of shelter.

FAMILY GRYLLOTALPIDAE—MOLE CRICKETS

Shortwinged Mole Cricket
Scapteriscus abbreviatus Scudder

Southern Mole Cricket
Scapteriscus borellii Giglio–Tos

Tawny Mole Cricket
Scapteriscus vicinus Scudder
(Orthoptera: Gryllotalpidae)

Natural History

Distribution. These mole crickets were inadvertently introduced to the southeastern United States in about 1900. Shortwinged mole cricket, *Scapteriscus abbreviatus* Scudder, was first observed at Tampa, Florida in 1899, but separate introductions were discovered near Miami in 1902 and Brunswick, Georgia in 1904. Southern mole cricket, *Scapteriscus borellii* Giglio–Tos (known until recently as *S. acletus* Rehn and Hebard), was similarly introduced to major seaports, beginning with Brunswick in 1904, and followed by Charleston, South Carolina in 1915, then Mobile, Alabama in 1919, and finally Port Arthur, Texas in 1925. Tawny mole cricket, *Scapteriscus vicinus* Scudder, was first observed at Brunswick, Georgia in 1899. The origin of these crickets is uncertain, but Argentina and Uruguay are likely sources, because they occur in these areas of southern South America.

In the years since introduction to the United States, the *Scapteriscus* spp. have expanded their ranges, but they differ considerably in their current distribution. Shortwinged mole cricket, which is flightless, remains fairly confined to the southern Florida and southern Georgia–northeast Florida introduction sites, though it also occurs in Puerto Rico and the Virgin Islands. It has been redistributed in southern Florida, but is largely found in coastal areas. In contrast, southern mole cricket is now found from North Carolina to eastern Texas, including the northern regions of Georgia and Alabama and the entire peninsula of Florida, and recently was detected in Yuma, Arizona. Tawny mole cricket is somewhat intermediate in its spread; it occurs from North Carolina to Louisiana, and throughout Florida, but thus far remains restricted to the southern coastal plain.

These are not the only mole crickets found in North America, but they are most damaging. For example, a native species, the northern mole cricket, *Neocurtilla hexadactyla* (Perty), is widely distributed in the eastern states west to about South Dakota and Texas, and including southern Ontario, but is not a pest. European mole cricket, *Gyllotalpa gryllotalpa* (Linnaeus), has been introduced from Europe into the northeastern states, but is of minor significance. Changa, *Scapteriscus didactylus* (Latreille), invaded Puerto Rico from South America before 1800, and has caused considerable damage to crops on this island.

Host Plants. Mole crickets are omnivorous, feeding on animal as well as plant material. Several studies have indicated that when provided with grass or collected from grass-dominated habitats, southern mole cricket is less damaging than tawny mole cricket. Southern mole cricket feeds mostly on other insects, whereas tawny mole cricket is principally herbivorous (Matheny, 1981; Matheny *et al.*, 1981; Walker and Ngo, 1982). Both species are associated with tomato and strawberry fields (Schuster and Price, 1992). Among vegetable crops reported to be injured are beet, cabbage, cantaloupe, carrot, cauliflower, collards, eggplant, kale, lettuce, onion, pepper, potato, spinach, sweet potato, tomato, and turnip. Other plants injured include chufa, turf, and pasture grasses, peanut, strawberries, sugar cane, tobacco, and such flowers as coleus, chrysanthemum, and gypsophila. Among the turf grasses, bahiagrass and Bermudagrass are commonly injured by tawny mole cricket, whereas St. Augustine grass and Bermudagrass are favored by shortwinged mole cricket. Mole crickets also feed on weeds such as pigweed, *Amaranthus* spp.

Natural Enemies. Few natural enemies of *Scapteriscus* mole crickets exist naturally in North America. Among the natural enemies are amphibians such as toads, *Bufo* spp.; birds such as sandhill cranes, *Grus canadensis*; and mammals such as armadillos, *Dasypus novemcinctus*. They, and the few predatory insects that attack crickets such as tiger beetles (Coleoptera: Cicindelidae), are not effective. Therefore, several natural enemies have been introduced from South America (Parkman *et al.*, 1996). The most effective introduced beneficial insect is the parasitoid *Ormia depleta* (Wiedemann) (Diptera: Tachinidae), which was imported from Brazil. This fly is attracted to the calls of male mole crickets. Its release has resulted in reduced mole cricket injury in southern Florida (Frank *et al.*, 1996). A less effective parasitoid is *Larra bicolor* Fabricius (Hymenoptera: Sphecidae), which was imported from Bolivia but seems to be constrained by availability of suitable adult food sources in Florida (Frank *et al.*, 1995). An entomopathogenic nematode, *Steinernema scapterisci*, was introduced from Uruguay (Nguyen and Smart, 1992). It is fairly specific to mole crickets, persists readily under Florida's environmental conditions, and is dispersed by crickets. Field collections consistently show infection levels of 10% or greater (Parkman *et al.*, 1993a,b; Parkman and Smart, 1996), and infected crickets die within 10–12 days.

Life Cycle and Description. Southern and tawny mole cricket are quite similar in appearance and biology. Shortwinged mole cricket differs in appearance owing to the short wings but also in behavior, because it has no calling song and the short wings render it incapable of flight. Typically, the eggs of these three species are deposited in April–June, and nymphs predominate through August. Beginning in August or September some adults are found, but overwintering occurs in both the nymphal and adult stages. Maturity is attained by the overwintering nymphs in April, and eggs are produced at about this time. A single generation per year is normal, though in southern Florida there are two generations in southern mole crickets and an extra peak of adult flight activity in the summer, resulting in spring, summer, and autumn flights from the two generations (Walker *et al.*, 1983).

Egg. The eggs are deposited in a chamber in the soil adjacent to tunnels. The chamber is usually constructed at a depth of 5–30 cm below the soil surface. It typically measures 3–4 cm in length, width, and height. The eggs are oval to bean-shaped, and initially measure about 3 mm long and 1.7 mm wide. The eggs increase in size as they absorb water, eventually attaining a length of about 3.9 mm and a width of 2.8 mm. The color varies from grey or brownish. They are deposited in a loose cluster, often numbering about 25–60 eggs. Duration of the egg stage is about 10–40 days. Total fecundity is not certain, but over 100 eggs have been obtained from a single female, and the mean number of egg clutches produced per female was reported to be 4.8 (Hayslip, 1943).

Nymph. Hatchlings are whitish initially but turn dark within 24 hours. They may consume the egg shell or cannibalize siblings, but soon dig to the soil surface. The juvenile stages resemble the adults, but nymphs have poorly developed wings. The number of instars is variable, probably 8–10 (Hudson, 1987). Nymphs and adults create extensive below-ground tunnel systems, usually within the upper 20–25 cm of soil. When the soil is moist and warm they tunnel just beneath the surface, but crickets tunnel deeper if the weather becomes cooler or the soil dries. They come to the surface to forage during the evening, usually appearing shortly after dusk if the weather is favorable.

Adult. Mole crickets have peculiar enlarged forelegs that are used for digging in the soil. The foretibiae have large blade-like projections, called dactyls, and the number and arrangement of dactyls are diagnostic. These crickets also bear antennae which are shorter than the body. Females lack a distinct ovipositor. Both sexes have elongate cerci at the tip of the abdomen. The male produces a courtship song that is attractive to females; they normally call during the night. Except for the shortwinged mole cricket, the male enlarges the entrance to his burrow, forming a horn-shaped opening, in preparation for calling. This increases the volume of the call, and allows flying females to locate males. Mating occurs within the male's burrow in the soil, and apparently the female may usurp the burrow after mating.

Shortwinged mole cricket bears front wings that are shorter than the pronotum. The front wings cover the hind wings, which are minute. The body is mostly whitish or tan in color, though the pronotum is brown mottled with darker spots. Also, the abdomen is marked with a row of large spots dorsally, and smaller spots dorsolaterally. These crickets measure 22–29 mm long. The two dactyls on the foretibiae are slightly divergent, and separated at the base by a space equal to at least half the basal width of a dactyl. Shortwinged mole cricket makes no calling song, producing only a weak 1–5 pulse chirp during courtship. (See color figures 159 and 171.)

Southern mole cricket has long hind wings that extend beyond the tip of the abdomen. The front wings are longer than the pronotum, about two–thirds the length of the abdomen. They are broad and rounded at the tips. This cricket is brown, with the dorsal surface of the pronotum often quite dark. As with short-winged mole cricket, in southern mole cricket the two dactyls on the foretibiae are separated at the base by a space equal to at least half the basal width of a dactyl. Thus, these two species can be distinguished by the wing length. Southern mole cricket produces a calling song that consists of a low-pitched ringing trill at about 50 pulses per second. It usually is emitted during the first two hours after sunset. (See color figure 161.)

Tawny mole cricket is quite similar to southern mole cricket in general appearance, with moderately long front wings and long hind wings, a yellowish brown body, and a dark pronotum. It can be distinguished from southern mole cricket by dactyl form. The tibial dactyls are nearly touching at the base, separated by less than half the basal width of a dactyl. Tawny mole cricket produces a loud, nasal trill at about 130 pulses per second during the first 90 minutes after sunset. (See color figure 160.)

Summaries of mole cricket life history were given by Worsham and Reed (1912), Thomas (1928), Hayslip (1943), and Walker (1984), though biology of short-winged mole cricket is poorly documented. Keys to North American and Caribbean area mole crickets were provided by Nickle and Castner (1984).

Damage

The crickets usually damage seedlings, feeding above-ground on foliage or stem tissue, and below-ground on roots and tubers. Girdling of the stems of seedling plants at the soil surface is a common form of injury, though young plants are sometimes severed and pulled below-ground to be consumed. Additional

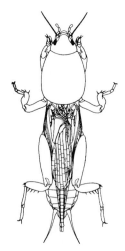

Adult tawny mole cricket.

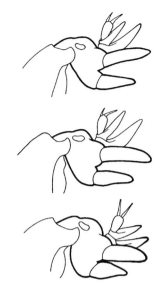

Dactyls of mole crickets: shortwinged (*top*); southern (*middle*), and tawny (*lower*).

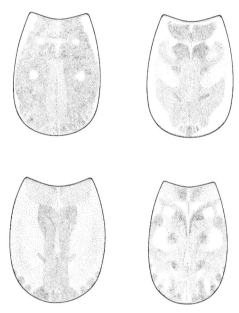

Pronotal patterns of mole crickets: southern mole cricket (*top left and right*), tawny mole cricket (*lower left*), short-winged mole cricket (*lower right*).

injury to small plants is caused by soil surface tunneling, which may dislodge seedlings. Southern mole cricket does much more tunneling injury than tawny mole cricket.

Management

Sampling. Various approaches to population estimation have been developed. A commonly used, but less reliable technique, is the assessment of population density by the frequency of soil surface tunneling. Tunneling is affected by soil moisture levels, and is most appropriate for assessing nymphs. A more consistent, but labor intensive, approach for estimation of nymph and adult abundance is flushing with about a 0.5% aqueous solution of dishwashing soap. Soil flushing is affected by soil moisture conditions, with greater cricket extraction efficiency as the soil approaches field capacity (Hudson, 1989). Flushing with synergized pyrethrin insecticide solution is equally effective (Hudson, 1988). A soil washing apparatus also has been developed to separate crickets from soil (Fritz, 1983). Adults can be captured with sound traps that use electronic sound synthesizers to lure crickets to a catching device, usually a large funnel.

Insecticides. Liquid and granular formulations of insecticides are commonly applied to the soil to suppress mole crickets. Insecticide application should be followed by irrigation, because the insecticide needs to enter the root zone of the plants to be most effective. Bait formulations are also useful. Various baits have proven effective, but most contain wheat bran, cottonseed meal, or some other grain product plus 2–5% toxicant. Also, addition of 5–15% water and 2–5% molasses to the grain-toxicant mixture are sometimes recommended (Thomas, 1928; Walker, 1984).

Cultural Practices. Most injury to vegetable transplants occurs on small plants, so placement of larger plants is suggested as a strategy to avoid injury (Schuster and Price, 1992). Crickets can quickly invade crop land that has been fumigated or otherwise cleared of crickets, so isolation from sources of crickets, or planting in large blocks of land with proportionally little edge, is desirable (Poe, 1976).

Host-Plant Resistance. Efforts have been made to find turf and pasture grass varieties that are resistant to attack by mole crickets. If grass varieties contain antibiotic properties, or otherwise limit the reproductive abilities of mole crickets, this can translate into fewer crickets seeking food within vegetable crops, because grass-containing fields are a principal source of mole crickets. Thus far, strains have been identified which are fairly tolerant of feeding or which are not preferred by mole crickets, primarily the finer textured grass selections, but considerable improvement in these grasses is needed before they can affect cricket population biology.

Biological Control. Biological control of mole crickets can be enhanced by the application of the entomopathogenic nematode *Steinernema carpocapsae*. This nematode can be cultured and applied in the same manner as an insecticide, and is fairly persistent. It is more effective when applied to adults than applied to nymphs.

FAMILY TETTIGONIIDAE—SHIELD-BACKED CRICKETS

Mormon Cricket
Anabrus simplex Haldeman

Coulee Cricket
Peranabrus scabricollis (Thomas)
(Orthoptera: Tettigoniidae)

Natural History

Distribution. These similar insects are native to western North America. Mormon cricket occurs widely, with a range that includes southern British Columbia to Manitoba in the north, and south to northern California and northern New Mexico. As a persistent pest, however, its range is limited to the Rocky Mountain and Great Basin regions. Coulee cricket is more limited in occurrence, and is known from Montana, eastern Washington, northeast Oregon, and southern British Columbia.

Host Plants. These crickets are often considered to be omnivorous, but despite their wide host range they display some specific preferences unless confronted by starvation. Among the vegetables damaged by crickets are bean, beet, cabbage, cantaloupe, carrot, cauliflower, Chinese cabbage, corn, lettuce, onion, potato, pumpkin, radish, rutabaga, salsify, spinach, tomato, turnip, and likely others. Not readily eaten are pea and mustard. Other crops susceptible to injury are alfalfa, barley, clover, flax, millet, oat, sugarbeet, sweetclover, timothy, and wheat. Due to the nature of the cropping systems in the areas inhabited by these crickets, alfalfa and wheat are most often injured.

Over 400 species of grasses, forbs, trees and shrubs were reported by Swain (1944) to be eaten by Mormon cricket. Most of these records occurred during the arid "dust-bowl" era of the 1930s when cricket densities were extremely high. Thus, they are not typical of cricket feeding behavior. Crickets often feed preferentially on the flowers and seed-heads of plants, ignoring the leaf material. Seed-head consumption is especially pronounced in grasses, though grasses are a minor component of cricket diet. Forbs such as bitterroot, *Lewisia rediviva*; wild onion, *Allium* spp.; arrowleaf balsamroot, *Balsamorhiza sagittata*; wild mustard, *Brassica* spp.; tumblemustard, *Sisymbrium altissimum*; and lupine, *Lupinus* spp.; are preferred by nymphs. In the adult stage, crickets eat mostly big sagebrush, *Artemisia tridentata* (MacVean, 1987; Redak *et al.*, 1992). In an analysis of Mormon cricket diet in Colorado, Ueckert and Hansen (1970) reported that the diet consisted of forbs 50%, arthropods 21%, fungi 16%, grasses 6%, clubmoss 5%, and grasslike plants 2%.

Crickets actively prey on other insects, including cicadas, ants, aphids, and beetles if they have the opportunity to catch them. They are quick to consume injured or dead crickets as well, and one of the distinctive characteristics about these crickets is their tendency to stop and feed on comrades that have been crushed on roadways by vehicles. As the healthy crickets remain on the highways to feed on fallen crickets, they often become crushed also, resulting in long, dark, greasy road slicks consisting of pulverized crickets.

Natural Enemies. Predators are perhaps the best-known mortality factor associated with crickets. Destruction of crickets in 1848 by California gulls, *Larus californicus*, saved the early Mormon settler's grain crops; a fact commemorated by a large statue of the gulls in Salt Lake City. Gulls are not the only vertebrates attracted to these insects when they become numerous, and among the avian predators most commonly observed feeding on crickets are crows, *Corvus brachyrhynchos*; hawks, *Falco* spp. and possibly others; meadowlarks, *Sturnella magna*; and blackbirds, various species (Wakeland, 1959). Mammals such as coyotes, *Canis latrans*; ground squirrels, *Citellus* spp.; and kangaroo rat, *Dipodomys* spp. also feast on crickets when they are abundant. Also, the wasps *Palmodes laeviventris* (Cresson) and *Tachysphex semirufus* (Cresson) (both Hymenoptera: Sphecidae) capture crickets and feed them to their young. Despite the frequency at which predation is observed, there is little evidence that predators are normally effective at maintaining crickets at low densities, or capable of suppressing crickets during periods of population outbreak.

Parasitism is surprisingly uncommon in cricket populations. Only the egg stage is parasitized with any degree of frequency, and though levels of up to 50% parasitism have been reported, it is usually quite low. The parasitoids responsible for attacking eggs are *Sparaison pilosum* Ashmead (Hymenoptera: Scelionidae) and *Oencyrtus anabrivorus* (Hymenoptera: Encyrtidae). A fly, *Sarcophaga harpax* Pandelle (Diptera: Sarcophagidae), has been reared from adult crickets, but occurs infrequently.

Pathogens vary greatly in their effect on crickets. The microsporidian *Heterovesicula* (*Vairimorpha*) *cowani* can naturally infect substantial proportions of Mormon cricket populations, and causes rapid mortality when young crickets ingest spores (MacVean and Capinera, 1991, 1992; Lange *et al.*, 1995). *Heterovesicula* appears to be the most important pathogen of crickets. The report that the grasshopper-infesting microsporidian *Nosema locustae* can infect Mormon cricket (Henry and Onsager, 1982) seems to be premature. The nematode *Agamaspirura anabri* (Nematoda) (Christie, 1930) and the horsehair worm *Gordius robustus* (Nematomorpha) (Thorne, 1940) have also been observed in crickets. *Gordius* was reported to be quite common in crickets near standing water, because part of the horsehair worm's life cycle takes place in water; unfortunately, water is not plentiful in the habitat of these crickets.

Life Cycle and Description. The life cycle and description of Mormon cricket and coulee cricket are nearly identical. Normally there is one generation per year, though there are reports of eggs at high altitudes remaining in diapause for an entire year, resulting in a two-year life cycle. They overwinter, with egg hatch occurring in March–May, often while snow remains on the ground. The nymphs are present until June when adults begin to emerge and start egg production. Adults usually perish by late August, often earlier. About 100 days is required for the nymphal and adult stages to be completed. Mormon cricket and coulee cricket are differentiated by the texture of the dorsal surface of the pronotum; it is punctate or rough in coulee cricket but smooth in Mormon cricket.

Egg. The egg is elliptical and measures about 7–8 mm long and 2.0–2.5 mm wide. Initially, brown in color, it soon turns whitish and then gray. They are deposited in the soil singly or in small clusters at a depth of 6–25 mm during the summer, where they remain until spring. One end of the egg, where the head of the embryo is located, swells slightly before egg hatching. Sometimes they are deposited around the base of plants, but more often bare soil is favored, including the mounds of ants. Females deposit, on average, about 85 eggs, but up to 160 per female has been observed. They complete their embryonic development in the summer and autumn, before entering diapause. Thus, they are ready to emerge early in the spring, and begin to hatch when soil temperature attain about 5°C, a much lower temperature than the threshold of development for hatch of grasshoppers. (See color figure 264.)

Nymph. Upon hatching from the soil, the nymphs are dark, resemble the adults, and measure about 6 mm long. There are seven instars, and by the time they attain the last instar they are about 30 mm long. The initial instars are black with white along the lateral edge of the posterior end of the pronotum, and the ovipositor of the female is not apparent. As they attain the fourth instar, however, they acquire green, red, purple, or brown color and in the female the ovipositor becomes increasingly obvious. Mean duration (range) of development period of crickets cultured at 21–26°C is 9.5 (7–12), 7.4 (4–14), 5.1 (4–7), 6.5 (4–10), 5.6 (3–9),

5.6 (5–7), and 10.4 (5–15) days, respectively, for instars 1–7. Thus, the mean total development time of nymphs is estimated at 50 days (range 43–58 days) but weather, and probably density, can significantly affect development rate. Mature nymphs have small wings but the wings do not protrude from beneath the pronotum.

The crickets tend to aggregate, seeking shelter together beneath bushes and debris during inclement weather and at night. Once they reach the third or fourth instar crickets the aggregations begin to move long distances, with numerous crickets coalescing into groups, which move in bands. The density of crickets in bands may be 10–30 per sq m, but sometimes much lower. The width of a band is often 300 m or more, but only 10 m deep, with crickets moving in the same direction along the entire width. The crickets all seem to move independently and consistently in the same direction, often at 1 km per day. There is no indication that they follow one another, and the basis of orientation is unknown. Bands moving in different directions sometimes converge and then emerge without loss of individual band integrity.

Adult. The adult is very similar to the mature nymph in form and color but larger, measuring 35–45 mm long. Also, the sword-shaped ovipositor of the adult female is longer, and the short wings of the adult male protrude from beneath the pronotum and are used as a stridulatory organ. As happens with nymphs, the adults may cluster under shelter both during the evening and inclement weather. They can also climb into bushes to escape the hot soil during excessively warm weather. Adults continue to move in bands in the same manner as nymphs, stopping only to eat and oviposit. (See color figure 162.)

Reproduction commences 10–14 days after attaining the adult stage. Males call from perches on vegetation during the morning hours. Females compete for the attention of males, mount the males, and are inseminated. Males are selective in their choice of partners, often choosing the largest female with which to mate. During insemination the male provides the female, attached to the sac containing sperm, a large proteinaceous mass that protrudes from her genital opening. While the sperm is draining into the female's reproductive system the proteinaceous mass provides a meal for the female. This is a significant investment on the part of the male, as the sperm and its accompanying protein meal represents up to 25% of his body weight (Gwynne, 1984). Breeding usually occurs in hilly areas where vegetation is sparse.

Mormon cricket (and presumably coulee cricket) exist in solitary and gregarious forms. The aforemen-

Adult female Mormon cricket.

tioned description applies mostly to the gregarious form. Only this damaging form has been thoroughly studied. The solitary form occurs at low density between periods of population outbreak and in areas where crickets do not become numerous. In contrast to the gregarious form, solitary crickets are green, and do not aggregate or form bands. Sexual behavior is also reversed, with females choosing among males. This change in mating behavior seems to be related to better nutrition of crickets when they are not at high densities (Gwynne, 1993).

Biology of Mormon cricket was described by Cowan (1929), and of coulee cricket by Melander and Yothers (1917). The impact of Mormon cricket was summarized by Wakeland (1959), and a modern review was given by MacVean (1987). A key to Mormon and coulee crickets, and their near relatives, was published by Rentz and Birchim (1968). Pfadt (1994c) provided a concise summary of Mormon cricket biology and pictures all stages of development. Synopses of Mormon cricket and coulee cricket, including keys to related Canadian Orthoptera, were given by Vickery and Kevan (1985).

Damage

Mormon and coulee crickets occur in arid sagebrush rangeland, and generally cause little injury unless they move into irrigated cropland. In earlier times, when settlers had to be nearly self-sufficient,

vegetable gardens were critically important to ranchers, and crop losses caused by Mormon and coulee crickets were a significant threat to the existence of western communities. Now, however, vegetable production is much less significant in these arid lands, and control technologies have improved markedly, so cricket importance has declined. Crickets remain a threat, however, and when bands of crickets invade lush crop vegetation, they can cause serious defoliation.

Management

Sampling.   Cricket bands are easily detected when they cross roads, and their presence in an area rarely is a surprise. However, they move rapidly and their course of travel is unpredictable, so when crickets are discovered control efforts are usually directed at the bands before they enter crop-growing areas. Some areas, usually mesas, seem to support continuous breeding populations, and serve as a source of crickets for nearby regions.

Insecticides.   Persistent insecticides are sometimes applied by aircraft to foliage in areas supporting nymphal populations or migrating bands. An alternative is to apply insecticide-treated bait. The preferred bait is flaky wheat bran, and it may be applied dry or with 10% water, but other additives such as molasses do not increase effectiveness (Cowan and Shipman, 1940).

Cultural Practices.   In earlier times, a common practice to prevent invasion of crop fields by cricket bands was to surround the crop with ditches possessing steep sides; crickets falling into such ditches had great difficulty regaining the soil surface. Similarly, vertical barriers of metal topped by a deflector served to prevent crickets from entering areas surrounded by such "cricket fences."

Order Thysanoptera—Thrips

Bean Thrips
Caliothrips fasciatus (Pergande)
(Thysanoptera: Thripidae)

Natural History

Distribution. Bean thrips apparently is a native species that has spread to many other parts of the world, including Asia, Europe, and South America. In North America it is known principally from the western United States, including Oregon, Idaho, and Wyoming south to California and Arizona. It is most damaging in the dry interior valleys of California, and arid regions of nearby states. However, its distribution also extends eastward throughout the southern states to South Carolina and Florida, and south into Mexico.

Host Plants. Bean thrips is reported from numerous plants, though some are incidental hosts, on which the adult can feed but not reproduce successfully. Vegetable crops that support bean thrips include asparagus, bean, beet, cabbage, cantaloupe, carrot, cauliflower, corn, fennel, garlic, kale, leek, lettuce, melon, onion, pea, pepper, potato, radish, Swiss chard, tomato, and turnip. Severe damage is generally limited to beans, cantaloupe, lettuce, and pea. Other hosts include such field crops as alfalfa, clover, cotton, and hops; fruit such as apple, avocado, fig, grape, orange, peach, pear, persimmon, plum, prune, and tangerine; and ornamentals such as canna, California poppy, geranium, gladiolus, hollyhock, iris, nasturtium, and sunflower. Many wild hosts are known, including trees, grasses, and weeds. Among common weeds known to support bean thrips are field bindweed, *Convolvulus arvensis*; milkweed, *Asclepias* spp.; mallow, *Malva parviflora*; mullein, *Verbascum virgatum*; redroot pigweed, *Amaranthus retroflexus*; common sowthistle, *Sonchus oleraceus*; and prickly lettuce, *Lactuca scariola*. In California, the abundance of prickly lettuce is one of the most important elements in the ecology of bean thrips. Bailey (1937) provided a more complete list of host plants.

Natural Enemies. Field studies suggest that only about 40% of thrips attain the adult stage. Much of the mortality results from the feeding habits of general predators such as minute pirate bug, *Orius tristicolor* (White) (Hemiptera: Anthocoridae), the larvae of the predatory thrips *Aelothrips fasciatus* (Linnaeus) and *A. kuwanae* Moulton (Thysanoptera: Thripidae), the lacewing *Chrysopa californica* Coquillette (Neuroptera: Chrysopidae), and the convergent lady, *Hippodamia convergens* Guerin–Meneville (Coleoptera: Coccinellidae). However, the most important natural enemy seems to be the internal parasitoid *Ceranisus russelli* (Crawford) (Hymenoptera: Eulophidae). Levels of parasitism by *C. russelli* observed in the field have been quite variable rates, but up to 70% have been noted. An unspecified nematode was found in bean thrips nymphs, but the importance of this observation is uncertain.

Weather. Bean thrips suffer direct mortality due to weather conditions. They are largely restricted to areas with mean winter temperature of about 20°C or greater. However, during the winter the adults can survive temperature as low as −9°C for short periods of time. Rainfall is also important. Heavy rains dislodge larvae from the plant, washing them to the soil where they perish. Summer rainfall in excess of about 10 cm is detrimental, but winter rainfall also is detrimental to overwintering adults. Thus, rainfall is likely the principal factor restricting bean thrips damage to arid regions of western states.

Weather also indirectly affects thrips populations. Bean thrips require an abundance of early-season and late-season weeds if they are to attain high densities. In the absence of spring or autumn rains, the thrips populations are checked by the shortage of suitable host plants. Thus, a critical balance of rainfall is required; enough early and late in the season to assure adequate food, but not so much in the summer and winter to cause direct mortality from being crushed and drowned.

Life Cycle and Description. This species has several overlapping generations annually. Six generations are estimated from California. Bean thrips overwinters in the adult stage. In California, overwintered adults produce the first generation, generally in March, on weed hosts. Overwintered adults usually perish by late April or May. Adults from the first and second generations may find the weeds suitable, or may move to alfalfa to feed. By late June, however, thrips begin dispersal to numerous cultivated crops, and weeds contained within the crops, where additional generations develop. As crops mature, thrips disperse again to late-planted crops, where an additional generation develops. The first generation in the spring requires about six weeks for completion, but as the weather warms the generations are completed in about three weeks.

Egg. The egg is bean-shaped, bluntly rounded at both ends, with one side concave and the opposite side converse. The egg measures about 0.2 mm long and 0.1 mm wide. The egg is white and is inserted by the female into the leaf tissue. The fecundity of females has been poorly studied, but females commonly produce five or more eggs per day when held in captivity.

Larva. There are two larval stages in bean thrips. They are quite small in size, measuring about 0.3 and 0.9 mm long, respectively. Duration of the first instar is about 5.4 and 4.3 days at 21° and 32°C, respectively. Duration of the second instar is about 8.2 and 3.8 days at 21° and 32°C, respectively. Thus, larval development requires about 13.6 days at relatively cool temperature and only about 8.1 days at relatively warm temperature. The larvae of the bean thrips tend to be reddish yellow or pink. When they first hatch, however, they are nearly translucent white, and as they mature they gain more red coloration, becoming almost entirely crimson in color at the end of the larval stage. Larvae lack wings, and are found feeding on foliage, often in association with adults. The larvae typically carry a drop of liquid excrement at the tip of their slightly upturned abdomen. In most respects they resemble the adult stage. After completing the two larval instars the insect drops to the soil and burrow to depths up to 35 cm. Most are found at depths of 7–15 cm, with burrowing behavior affected by soil type, moisture level, and shading. Dry, fine, and sandy soil is not conducive to thrips burrowing and survival.

Prepupa and Pupa. Two additional immature stages of development, the prepupal and pupal stages, are passed in the soil. In general appearance, they greatly resemble the larval and adult stages. They differ from both, however, in having partly developed wings, or wing pads, and by not feeding. The prepupa and pupa are orange, with crimson markings on the thorax and abdomen. They measure about 1.0 mm long during the prepupal stage, but shrink slightly to about 0.8 mm during the pupal stage. Also, they can be differentiated from each other in that the wing pads are much larger in the pupal stage. Duration of the prepupal stage is only 1–2 days, with 1.6 days the normal development period at 21°C, and 0.9 days at 32°C. The pupal stage is slightly longer, requiring 9.3 and 2.4 days at 21° and 32°C, respectively.

Adult. The adult is minute, the female measuring only about 1.0 mm long and the male 0.9 mm. The body color is grayish black, and the front wings are banded with two alternating light and dark areas. The hind wings are entirely dark. Both sets of wings are fringed, as is normal for insects of this order. When the wings are folded against the body, which is the usual condition, the body appears to bear two whitish bands across the central region. The eight-segmented antennae and the legs also are banded with alternating light and dark areas. Adults begin to copulate and commence oviposition 2–4 days after emergence. A sex ratio of about two females per one male is normal for this species. The adult overwinters within curled leaves, on the underside of pubescent foliage, under the old coverings of scale insects, and in other sheltered locations. Plants that remain green throughout the winter are favorite overwintering locations. If disturbed, the adults slowly move, but little or no feeding occurs during this period.

The most complete treatment of bean thrips biology was supplied by Bailey (1933a). However, a later treatment by Bailey (1937) and an earlier report by Russell (1912) also contain useful information. This species is included in a key to common vegetable-infesting thrips in Appendix A.

Damage

Damage results from larvae and adults feeding on the underside of foliage. In the process of feeding, the thrips extrude their stylets, puncture cells, and drain the liquid contents. Apparently the larvae feed more extensively and inflict more damage than do adults. The insects feed gregariously, so there tends to be a great amount of tissue destruction localized on individual leaves, and these leaves turn whitish and often are shed by the plant. The underside of the

Adult female bean thrips.

foliage becomes covered with black spots of fecal material. Injury tends to be concentrated on leaves of intermediate age, though adults tend to move to newer tissue as it becomes available. As the foliage is depleted, thrips move to bean pods and stem material to feed. The scar resulting from insertion of the eggs into foliage seems to cause little injury.

Management

Sampling. Thrips populations are best assessed by close visual observation of susceptible crops and suitable alternate hosts, such as early season weeds.

Insecticides. Formerly a serious pest of several crops, bean thrips have been reduced to minor status by the introduction of modern insecticides. Foliar contact and systemic insecticides are generally effective, but because the prepupal and pupal stages are sheltered in the soil, persistent insecticides or multiple applications are necessary. Thorough coverage of the lower surface of the leaves is also important.

Cultural Practices. The most important cultural practice is weed management. Elimination of early season weeds in, or adjacent to, crops can reduce thrips populations. Irrigation ditches and other sites where weeds may survive during winter and periods of drought should be examined for thrips, and be treated with insecticides if necessary. Overhead irrigation can decrease bean thrips densities, and reduce the need for insecticides.

Grass Thrips
Anaphothrips obscurus (Müller)
(Thysanoptera: Thripidae)

Natural History

Distribution. This is the most common and widespread of the thrips species found on corn. It occurs throughout the United States and southern Canada.

Host Plants. Among vegetable crops, only corn is affected. Other crops infested are barley, oat, timothy,

and wheat. Numerous forage and weed grasses known to be suitable hosts include *Agrostis* spp., *Arrhenatherusm* sp., *Avena* sp., *Bromus* spp., *Elymus* spp., *Festuca* spp., *Lolium* sp., *Panicum* spp., and *Poa* spp.

Natural Enemies. The natural enemies of grass thrips are poorly known because this minor pest has been studied infrequently. See the sections on onion thrips, *Thrips tabaci* Lindeman, or western flower thrips, *Thrips occidentalis* Pergande, for discussion of natural mortality factors.

Life Cycle and Description. Adult females overwinter at the base of grass stems just above the soil. In the spring, as grass begins to grow, egg deposition begins. The number of annual generations is estimated at eight or nine in Massachusetts, with a generation requiring 2–4 weeks.

Egg. The eggs are deposited in the tender leaf tissue. Duration of the egg stage is 10–15 days during spring, but only about seven days during summer. Females are thought to produce, on average, 50–60 eggs during their life span, but individuals produce over 200 eggs. The egg laying continues throughout spring and summer until freezes occur, but they do not overwinter successfully.

Larva. The larvae are active, but tend to seek sheltered places to feed such as beneath leaf sheaths or within flowers. Larvae are long and slender in form and wingless. There are two larval instars. Larvae increase in size from about 0.25 mm at hatching to about 1.0 mm at maturity. Larvae resemble adults but have a smaller head, much narrower than the thorax. They also have shorter antennae. They are brownish or pinkish.

Pupa. There are two pupal instars, the prepupal and pupal stages. They occur in a sheltered location,

Grass thrips larva.

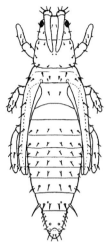

Grass thrips pupa.

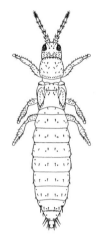

Adult female grass thrips, wingless form.

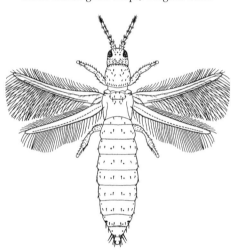

Adult female grass thrips, winged form.

often at the base of a leaf sheath. The pupal stages are sluggish and do not feed. The antennae are folded back over the head. The thorax bears wing cases which are long for pupae destined to produce winged adults, or short if giving rise to wingless individuals.

Adult. Adults feed more openly than do larvae, often feeding on the leaves rather than within the leaf sheaths. Adults have both winged and wingless forms. Most overwintering females are wingless, but the winged forms soon predominate, accounting for about 90% of the thrips during the spring months. The proportions shift over the course of the summer, with wingless forms common late in the season, accounting for about 98% of the thrips. The winged individuals are larger, about 1.5 mm long, bear two pairs of fringed wings, and are brown. The wingless form is shorter, measuring about 1.0 mm long, and may lack signs of wings or may bear short protuberances (wing buds). The wingless forms are pink. The antennae of adults protrude forward, as in the larvae, but the head is relatively wide, nearly as wide as the thorax. Males are infrequently found, the females generally reproducing parthenogenetically. The males are not as slender as the females, the eyes are located more dorsally, and the tip of the abdomen bears paired copulatory structures that are absent from the females. Females oviposit for a period of 4–6 weeks.

The biology of grass thrips was given by Fernald and Hinds (1900) and Cary (1902), but important observations were made by Kamm (1972). This species is included in a key to common vegetable-infesting thrips in Appendix A.

Damage

Larvae and adults puncture individual cells and remove the sap. This kills the cells, and imparts a silv-ery appearance to the tissue. The growing point or top of the plant is most often affected, usually while still immature, resulting in a condition called "silvertop." This thrips is said to be highly mobile and very destructive to grasslands in Oregon (Kamm, 1971, 1972). However, grass thrips typically do not persist in corn, and cause little damage. Usually thrips disperse into young corn in June and within a generation they again disperse. Thrips injury to corn usually occurs only if the corn is also under moisture stress.

Management

Grass thrips tend to be a transient problem, and rarely require action. Presence of grasses, including small grains, may predispose a corn crop to infestation. Destruction of grasses in the autumn or winter may eliminate the overwintering stage. If corn is rapidly growing, recovery from thrips feeding is likely. Foliar insecticides are effective for suppression.

*Color figure 17 is referred to on page 539.

Melon Thrips
Thrips palmi Karny
(Thysanoptera: Thripidae)

Natural History

Distribution. Until the mid-1970s the distribution of melon thrips was limited to Southeast Asia. In recent years it has spread throughout Asia, and to many Pacific Ocean islands, North Africa, Australia, Central and South America, and the Caribbean. In the United States it was first observed in Hawaii in 1982, Puerto Rico in 1986, and Florida in 1990. It has the potential to infest greenhouse crops widely, but under field conditions melon thrips likely will be limited to tropical areas.

Host Plants. Melon thrips is a polyphagous species, but is best known as a pest of Cucurbitaceae and Solanaceae. Among vegetables injured are bean, cabbage, cantaloupe, chili, Chinese cabbage, cowpea, cucumber, bean, eggplant, lettuce, melon, okra, onion, pea, pepper, potato, pumpkin, squash, and watermelon. Tomato is reported to be a host in the Caribbean, but not in the United States or Japan. Tsai *et al.* (1995) reported that cucurbits were more suitable than eggplant, whereas pepper was less suitable than eggplant. Other crops infested include avocado, carnation, chrysanthemum, citrus, cotton, hibiscus, mango, peach, plum, soybean, tobacco, and others. (See color figure 17.)

Natural Enemies. Natural enemies, particularly predators, are quite important in the ecology of melon thrips. There is strong indication that melon thrips abundance and damage are increased by application of some insecticides (Etienne *et al.*, 1990). Among the most important predators observed in Hawaii were the predatory thrips *Franklinothrips vespiformis* (Crawford) (Thysanoptera: Aeolothripidae) and especially the minute pirate bug, *Orius insidiosus* (Say) (Hemiptera: Anthocoridae). Other predators in Hawaii were *Curinus coeruleus* (Mulsant) (Coleoptera: Coccinellidae), *Rhinacoa forticornis* Reuter (Hemiptera: Miridae), and *Paratriphleps laevisculus* Champion (Hemiptera: Anthocoridae). Other predators and parasitoids are known in Asia (Hirose, 1991; Hirose *et al.*, 1993; Kajita, 1986). The parasitoid, *Ceranisus menes* Walker (Hymenoptera: Eulophidae), shows particular benefit in many asian studies, and this wasp has been introduced to Florida (Castineiras *et al.*, 1996a). Fungi known to affect melon thrips include *Beauveria bassiana*, *Neozygites parvispora*, *Verticillium lecanii*, and *Hirsutella* sp. (Castineiras *et al.*, 1996b).

Life Cycle and Description. A complete generation may be completed in about 20 days at 30°C, but it is lengthened to 80 days when the insects are cultured at 15°C. Melon thrips are able to multiply during any season that crops are cultivated but are favored by warm weather and suppressed by senescent crops. In southern Florida they were damaging on both autumn and spring vegetable crops (Seal and Baranowski, 1992; Frantz *et al.*, 1995). In Hawaii, they also became numerous on vegetables during the summer growing season (Johnson, 1986). (See color figure 180.)

Egg. The eggs are deposited in leaf tissue, in a slit cut by the female. One end of the egg protrudes slightly. The egg is colorless to pale white, and bean-shaped in form. Duration of the egg stage is about 16 days at 15°C, 7.5 days at 26°C, and 4.3 days at 32°C.

Larva. The larvae resemble the adults in general body form though they lack wings and have a smaller body size. There are two instars during the larval period. Larvae feed gregariously, particularly along the leaf midrib and veins, and usually on older leaves. Larval development time is determined principally by the suitability of temperature, but host plant quality also has an influence. Larvae require about 14, 5, and 4 days to complete their development at 15°, 26°, and 32°C, respectively. At the completion of the larval instars the insect usually descends to the soil or leaf litter, where it constructs a small earthen chamber for a pupation site.

Pupa. There are two instars during the "pupal" period. The prepupal instar is nearly inactive and pupal instar is inactive. Both instars are nonfeeding stages. The prepupae and pupae resemble the adults and larvae in form, except that they possess wing pads. The wing pads of the pupae are longer than that of the prepupae. The combined prepupal and pupal development time is about 12, 4, and 3 days at 15°, 26°, and 32°C, respectively.

Adult. Adults are pale yellow or whitish, and lack dark pigmentation but bear numerous dark setae on the body. A black line, resulting from the juncture of the wings, runs along the back of the body. The population is heavily weighted toward females. The slender fringed wings are pale. The hairs or fringe on the anterior edge of the wing are considerably shorter than those on the posterior edge. The thrips measure 0.8–1.0 mm in body length, with females averaging slightly larger than males. Unlike the larval stage, the adults tend to feed on young growth, and so are found on new leaves. Adult longevity is 10–30 days for females and 7–20 days for males. Development time varies with temperature, with mean values of about 20, 17, and 12 days at 15°, 26°, and 32°C, respectively. Females produce up to about 200 eggs, but average about 50 per female. Both mated and virgin females deposit eggs.

Careful examination is required to distinguish melon thrips from other common vegetable-infesting species. The *Frankliniella* species are easily separated, because their antennae consist of eight segments, whereas in *Thrips* species there are seven antennal segments. To distinguish melon thrips from onion thrips, *Thrips tabaci* Lindeman, it is helpful to examine the ocelli. There are three ocelli on the top of the head, in a triangular formation. A pair of setae are located near this triangular formation, but unlike the arrangement found in onion thrips, the setae do not originate within the triangle. Also, the ocelli bear red pigment in melon thrips, whereas they are grayish in onion thrips. In general, the basic body color of adult melon thrips is yellow, but in onion thrips it is yellowish gray to brown. The range of melon thrips in North America is quite restricted, so this also should aid in diagnosis.

The most complete summary of melon thrips biology and management was presented by Girling (1992). A detailed description of melon thrips is found in Bhatti (1980); Layland *et al.* (1994) also provided some diagnostic characters. Developmental biology was given by Tsai *et al.* (1995). Keys for identification of common thrips were presented by Palmer *et al.* (1989) and Oetting et al. (1993). Also, this species is included in a key to common vegetable-infesting thrips in Appendix A.

Damage

Melon thrips cause severe injury to infested plants. Leaves become yellow, white or brown, and then crinkle and die. Heavily infested fields sometimes acquire a bronze color. Damaged terminal growth may be discolored, stunted, and deformed. Densities from 1–10 per cucumber leaf have been considered to be the threshold for economic damage in some Japanese studies. However, studies in Hawaii suggested a damage threshold of 94 thrips per leaf early in the

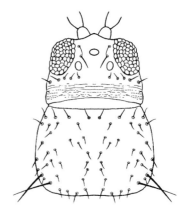

Anterior region of adult melon thrips.

growth of the plant (Welter *et al.*, 1990). Feeding usually occurs on foliage, but on pepper, a less suitable host, flowers are preferred to foliage. As the melon thrips prefer foliage, they are reported to be less damaging to cucumber fruit than western flower thrips, *Frankliniella occidentalis* (Pergande) (Rosenheim *et al.*, 1990). Nevertheless, fruits may also be damaged; scars, deformities, and abortion are reported. In Hawaii, thrips were observed to attain higher densities on cucumber plants infected with watermelon mosaic virus, but it was not determined whether the plants were more attractive to adults, or more suitable for survival and reproduction (Culliney, 1990).

In addition to direct injury, melon thrips are capable of inflicting indirect injury by transmitting some strains of tomato spotted wilt virus and bud necrosis virus.

Management

Sampling. Larvae and adults are collected from foliage. Adults tend to move toward young foliage, with nymphs tending to be clustered on foliage inhabited by adults several days earlier. Adults can also be sampled with sticky and water pan traps. Blue and white are attractive colors for thrips, and have been used to trap melon thrips (Layland *et al.*, 1994; Kawai and Kitamura, 1987). However, yellow has also been suggested to be an attractive color (Culliney, 1990).

Insecticides. Foliar insecticides are frequently applied for thrips suppression, but at times it has been difficult to attain effective suppression. Various foliar and drench treatments, alone or combined with oil, have achieved some success (Seal and Baranowski, 1992; Seal *et al.*, 1993; Seal, 1994) though it is usually inadvisable to apply insecticides if predators are present. The eggs, which occur in the foliar tissue, and the pupae, which reside in the soil, are relatively insensitive to insecticide application.

Adult female melon thrips.

Cultural Techniques. Several cultural practices apparently affect melon thrips abundance, but few have been evaluated in the context of North American agriculture. Physical barriers such as fine mesh and row cover material can be used to restrict entry by thrips into greenhouses, and to reduce the rate of thrips settling on plants in the field (Kawai and Kitamura, 1987).

Organic mulch is thought to interfere with the colonization of crops by winged thrips. Plastic mulch also is reported to limit population growth, but it is uncertain whether this is due to reduced rates of invasion or denial of suitable pupation sites. Crop stubble was not an effective deterrent (Litsinger and Ruhendi, 1984).

The effects of intercropping potato with onion on melon thrips populations was evaluated by Potts and Gunadi (1991). Although aphid and aphid-borne disease incidence were decreased in such potato plantings, the density of thrips on potatoes was increased. Thus, the benefits of such cropping practices are largely a function of which pests are likely to be most important in an area.

Heavy rainfall is thought to decrease thrips numbers (Etienne *et al.*, 1990). However, there seems to be no evidence that overhead irrigation is an important factor in survival.

Biological Control. The predatory mite *Neoseiulus cucumeris* (Oudemans) has been investigated for suppression of melon thrips (Castineiras *et al.*, 1997). The mite density is correlated with thrips density, but within-plant distribution differs among the two species, suggesting that though the mites may increase in numerical abundance they are unlikely to drive the thrips to extinction.

Host-Plant Resistance. Nuessly and Nagata (1995) reported that susceptibility to injury varied among pepper cultivars. They reported that though sweet and jalapeno types were sensitive to foliar injury, cubanelle and cayenne types produced acceptable size and quality fruit. This is the reverse of injury susceptibility to western flower thrips, so in areas with mixed thrips populations growers cannot rely solely on plant selection to avoid damage.

Onion Thrips
Thrips tabaci Lindeman
(Thysanoptera: Thripidae)

Natural History

Distribution. Onion thrips is believed to have originated near the eastern end of the Mediterranean Sea, or perhaps India. It was first observed in North America in 1872, and by the early 1900s had spread throughout the United States and southern Canada. It is easily transported on plant material, and redistribution of onion thrips occurs frequently with commercial shipment of bulbs and plants. It is now found throughout the world. (See color figure 181.)

Host Plants. Onion thrips has a wide host range, reportedly feeding on over 300 plants. In Hawaii, for example, 66 plants from 25 families were found to support onion thrips (Sakimura, 1932). It has been found to infest such vegetables as asparagus, bean, beet, cabbage, cantaloupe, carrot, cauliflower, celery, cowpea, cucumber, garlic, kale, leek, mustard, onion, parsley, pea, pepper, pigeon pea, potato, pumpkin, spinach, squash, sweet potato, tomato, and turnip. Under field conditions, the most serious problems occur on onion, followed by cosmetic injury to cabbage and edible-podded pea. In greenhouse cultivation of vegetables, onion thrips sometimes causes severe injury to tomato and cucumber. Field crops such as alfalfa, cotton, oat, soybean, sugarbeet, wheat, and tobacco also may support onion thrips. Ornamental crops such as rose and carnation may be injured, especially when grown under greenhouse conditions. Many common weeds support onion thrips, including amaranth, *Amaranthus palmeri*; dandelion, *Taraxacum officinale*; mullein, *Verbascum thapsus*; goldenrod, *Solidago canadensis*; ragweed, *Ambrosia* spp.; kochia, *Kochia scoparia*; sage, *Salvia* sp.; sunflower, *Helianthus annuus*; smartweed, *Polygonum* spp.; and yellow nutgrass, *Cyperus esculentus* (Chittenden, 1919; Doederlein and Sites, 1993). (See color figures 13, 14, and 15.)

Onion thrips is the most important insect pest of onion, and the dominant thrips species on onion. However, it is not the only species attacking onion, and in the southern states it is sometimes a relatively small component of the thrips fauna, being supplanted by western flower thrips, *Frankliniella occidentalis* (Pergande) (Bender and Morrison, 1989; Doederlein and Sites, 1993), and possibly by tobacco thrips, *Frankliniella fusca* (Hinds) (D.G. Riley, personal communication).

Natural Enemies. No natural enemies of significance are known. Numerous species of lady beetles (Coleoptera: Coccinellidae), lacewings (Neuroptera: Chrysopidae), and flower flies (Diptera: Syrphidae) have been observed attacking onion thrips, but none regularly are effective enough to provide suppression. Insidious plant bug, *Orius insidiosus* (Say), and minute pirate bug, *O. tristicolor* (White) (both Hemiptera: Anthocoridae), are among the most effective of the predators, because their small size allow them to pursue the thrips between the closely appressed leaves of the onion plant, but these predators are rarely abundant enough to suppress thrips populations. Parasitoids have been introduced from Southeast Asia, where they parasitize a high proportion of onion thrips and other thrips species. In Hawaii, *Ceranisus* *brui* (Vuillet) (Hymenoptera: Eulophidae) was successfully introduced, but introductions of other *Ceranisus* spp. failed (Clausen, 1978).

Weather. Weather is reported by many authors to be important in determining thrips abundance and damage. The combination of abnormally high temperature and low precipitation stimulates thrips reproduction and/or enhances survival. High temperature speeds up the life cycle, increasing the biotic potential of populations. Heavy rain is considered to be an important mortality factor, and is easy to observe large decreases in thrips abundance following significant rainfall events (Harding, 1961).

Life Cycle and Description. The life cycle of onion thrips is completed rapidly. Based on field work conducted in Iowa, development time from the egg to adult stages is estimated at only 10 days in July, and 20 days in September. Similar values were obtained by Watts (1934) in South Carolina. Sakimura (1932) studied this insect in Hawaii, and observed longer developmental periods, but these studies were conducted during the relatively cool winter months. In any event, the potential number of generations is great under ideal conditions. The actual number of generations is estimated at ten per year in southern climates but only two in the north. In temperate areas, overwintering occurs in the adult stage. Adults survive on winter wheat, alfalfa, clover, other crops and weeds, but not on soil in the absence of living plants (Chambers and Sites, 1989). Some of these crops, particularly wheat, are also favorable for oviposition and larval development in the spring, and provide inoculum for nearby vegetables (Shirck, 1951; North and Shelton, 1986b). In warmer climates nymphs may survive the winter, or reproduction may continue throughout the year.

Egg. The female cuts slits in the foliage and deposits eggs just beneath the epidermis, with one end of the egg protruding slightly. The egg is colorless or yellowish white. It measures about 0.26 mm long and 0.12 mm wide, a size that is quite large considering the small size of the female. The egg is bean-shaped, with one side being strongly concave whereas the opposite side is strongly convex. Duration of the egg stage is estimated at 5–7 days when held at 21°C, but only three days at 25°C.

Larva. The larvae of thrips (sometimes called nymphs) resemble the adults in appearance, though the larvae are colorless or yellowish-white. There are two instars. The first instar measures about 0.4 mm long, whereas the second instar measures about 0.9 mm long. The larvae can be differentiated by the tarsal arrangement—first instars bear tarsi with two claws, but these are absent from second instars. A less reliable, but more convenient, character for distinguishing the instars is their color; first instars tend to be colorless to white, whereas second instars tend to be yellowish. Both the first and second instars bear antennae with four segments. Larval duration varies from 3–11 days, with about 8–10 days required for development at 21°C and 5–6 days required at 25°C. Watts (1934) reported development times of 2.0 and 2.8 days, respectively, for the first and second instars. Larvae usually confine their feeding on onion to the youngest foliage at the center of the plant, but are more widely distributed on cabbage.

Pupa. After feeding for the first two instars, the larva almost always leaves the foliage and descends to the soil, where it constructs a small earthen chamber. Occasionally, it descends only to some vegetative structure such as a leaf axil. The active larval stages are followed by relatively inactive prepupal and pupal instars, which may last about 1.0 and 2.5 days, respectively, at warm temperature. This non-feeding period requires about 5–7 days at 21°C, and 4–6 days at 25°C. These stages are colorless to yellow, and measure about 0.7 mm long. Setae are much more abundant on the prepupal and pupal stages than in the larvae, and their wing pads are visible. The wing pads of the prepupae extend back to the second abdominal segment. The wing pads of the pupae are considerably larger, extending back to the eighth abdominal segment. Antennae of the prepupal stage project forward, whereas those of the pupal stage orient back over the thorax.

Adult. The adults do not occur in an equal sex ratio. Male thrips are relatively rare, and seem to be more plentiful in the autumn (Sakimura, 1932) and in the western hemisphere (Kendall and Capinera, 1990). Females generally reproduce parthenogenetically. Females are about 1.0–1.2 mm long, and yellowish or yellowish-brown. Patches of dark-brown occur on the thorax and abdomen. The overall body color is lighter in the summer and darker in the winter. The wings are colorless, and extend back to about the sixth abdominal segment. As is typical of thrips, there are two pairs of long, narrow wings, each bearing a fringe of hairs. Male onion thrips are slightly smaller than females, and have nine abdominal segments instead of the ten found in females. A pre-oviposition period in adults of about three days was noted by Watts (1934), but it was estimated at 5–7 days by Sakimura (1932). Adult females live about a month, and typically deposit an average of 30–40 eggs (maximum of about 100 eggs) at 1.5–2.5 per day.

Differentiation of onion thrips from other common vegetable-infesting thrips requires close examination. The adult female of onion thrips has seven antennal segments, a character that is useful to distinguish this species from co-occuring *Frankliniella* thrips; females of western flower thrips, *F. occidentalis*, and tobacco thrips, *F. fusca*, have eight antennal segments. To distinguish onion thrips from melon thrips, which also has seven antennal segments, it is helpful to examine the ocelli. There are three ocelli on the top of the head, in a triangular formation. In onion thrips a pair of setae originate from within this triangular formation, unlike the arrangement found in melon thrips, where the setae do not originate within the triangle. The ocelli also tend to be gray in onion thrips, but bear red pigment in melon thrips. Although the body color of onion thrips is quite variable, some gray or brown is usually present on the body of adults in addition to yellow, whereas in melon thrips the body is uniformly yellow. Melon thrips has a very resticted geographic range, and so should infrequently be confused with onion thrips.

Good accounts of onion thrips biology were given by Chittenden (1919) and Horsfall and Fenton (1922). Developmental parameters were given by Sakimura (1932), Watts (1934), and Edelson and Magaro (1988). Quaintance (1898b) provided a useful morphological description. A key for identification of common thrips was given by Palmer *et al.* (1989). This species also is included in a key to common vegetable-infesting thrips in Appendix A.

Damage

The principal form of damage caused by onion thrips results from the piercing of cells and removal of cell contents by larvae and adults. In onions, this leads to an irregular or blotchy whitening of the leaves, a condition sometimes termed "blast." Heavy levels of feeding injury disrupt the hormonal balance of the plant, causing the leaves to curl and twist, and the foliage to be stunted (Kendall and Bjostad, 1990).

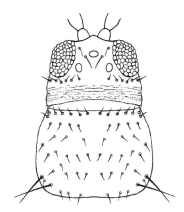

Anterior region of onion thrips.

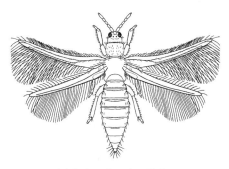

Adult female onion thrips.

Such damage decreases onion bulb size, and may even lead to death of the plant. Silvering or whitening of the pods on edible-podded peas also is attributed to onion thrips (Shelton and North, 1987). On cabbage, feeding by thrips causes a bronze discoloration and rough texture, and the cabbage heads may fail fresh market standards (North and Shelton, 1986a).

Crops such as cabbage also may fail processing standards due to contamination of products such as sauerkraut with thrips bodies (Shelton *et al.*, 1982). As happens with onion, the thrips feed in sheltered locations, here between the leaves that form the cabbage head. The tips of asparagus spears may be heavily infested with thrips following dispersal from nearby weeds or crops. Careful management of weeds and crop borders can alleviate such problems (Banham, 1968).

The relationship between thrips numbers and onion yield has been the subject of much study. Although Mayer *et al.* (1987) found no relationship between thrips abundance and yield of dry onions in Washington, studies in Colorado (Kendall and Capinera, 1987), Texas (Edelson *et al.*, 1986, 1989), and Quebec (Fournier *et al.*, 1995) demonstrated yield decreases associated with thrips feeding. For thrips injury to be significant, feeding must occur during the mid-season period of rapid bulb expansion; early and late season feeding has little or no effect on yield. Also, there is a threshold effect. Thrips densities of 1–2 per leaf or 30 per plant generally must be reached for injury to occur. To complicate matters further, however, onion responses are modified somewhat by weather, particularly moisture, and by onion variety. Sweet onions are particularly susceptible to bulb size reduction. Unlike the situation with storage onions, there is virtually no tolerance of thrips on scallions (green onions) because the tops as well as the developing bulb are marketed and consumed (Kawate and Couglin 1995).

Onion thrips may also affect plant disease incidence. The fungus *Alternaria porri* causes a foliar disease of onions called purple blotch. Although the fungus does not depend on thrips for transport or inoculation, feeding wounds can serve as a penetration site for the purple blotch fungus, and disease incidence is increased in the presence of thrips (McKenzie *et al.*, 1993). Onion thrips is also implicated in the transmission of tomato spotted wilt virus to several vegetable crops (Greenough and Black, 1990). Tomato spotted wilt virus is acquired by thrips during the larval stage, but the insects remain infected and capable of transmitting the virus for the duration of their life (Sakimura, 1963).

Management

Sampling. Edelson *et al.* (1986) studied the dispersion of thrips among onion plants and reported a clumped distribution. Distribution within plants also is non-uniform. Thrips densities are normally determined by visual examination of plants. Although there is variation among counts attributable to different observers (Theunissen and Legutowska, 1992) there is a strong correlation between visual estimates and actual population densities (Edelson, 1985a). Onion thrips is normally found at the basal area of onion leaves or in the leaf folds, and it is necessary to pull the leaves apart slightly to observe the entire population. This is especially true early in the season and early in thrips development, while later the thrips are more prone to move apically. Sunny weather, maturity, and the need for flight are reputed to account for the change in thrips distribution (Sites *et al.*, 1992). Sequential sampling (Shelton *et al.*, 1987) and binomial (presence-absence) sampling (Fournier *et al.*, 1994) plans have been developed to reduce the effort associated with thrips sampling on onion.

Thrips are active and are readily dispersed by wind when they fly. Because thrips readily disperse from crop to crop, it is useful to employ sticky traps to monitor thrips flights, or to monitor their densities in crops that provide potential inoculum. White- or yellow-sticky traps can be used for monitoring thrips in flight, and thrips on foliage can also be sampled by using heat to drive than from the foliage, as with a Berlese funnel (Shelton and North, 1986; Doederlein and Sites, 1993).

Insecticides. Insecticides are frequently used for thrips suppression, especially during the period of rapid bulb expansion when plants are most susceptible to injury, or late in the season when thrips have the potential to reach very high levels of abundance and feeding injury is obvious. Foliar applications are made as frequently as twice per week in commercial onion production. It is difficult to obtain excellent control because of the cryptic nature of thrips feeding. Many onion and cabbage producers, concerned that they will be unable to eliminate thrips from sheltered feeding locations, apply insecticides as a preventive measure, in advance of potential problem development. Especially for cabbage, once the leaves begin to cup and form a head, effective insect control is difficult. Insecticides that produce toxic fumes are desirable, because they penetrate into crevices where thrips hide, but these materials are limited in availability. On occasion, insecticides are applied to the soil or plastic mulch beneath plants because the thrips descend to the soil to pupate, where they contact the insecticide (Pickford, 1984). Some, but not all, systemic insecticides effectively suppress thrips for at least part of the season (Getzin, 1973; Sinha *et al.*, 1984). Insecticide resistance occurs in many locations.

Biological Control. Introduction of exotic eulophid parasitoids has had some success (see above, natural enemies), but this approach to biological control has not been fully exploited. Release of predators that are easily cultured, particularly lacewings (Neuroptera: Chrysopidae), has not been very successful in the field. Predatory mites (Acari: Phytoseiidae) have been found to provide suppression of onion thrips in greenhouse culture (Bakker and Sabelis, 1989), as have releases of *Orius* spp. (Hemiptera: Anthocoridae). It is important to release adequate numbers of predators at the first sign of thrips infestations.

Cultural Practices. Crop management can influence the nature of thrips injury in several ways. For example, the proximity of susceptible crops to thrips sources is important. Damage sometimes occurs when thrips disperse in large numbers into susceptible crops. This often results when an early season crop such as oats or wheat reaches maturity, or when a crop is cut at mid-season, as is the case with alfalfa and clover (Banham, 1968; Shelton and North, 1986).

Mulches can influence the abundance of thrips and the transmission of plant viruses. In studies conducted in Louisiana, aluminum-surfaced mulch reduced the incidence of tomato-spotted wilt transmission by thrips to tomato and pepper by about 60–80% (Greenough and Black, 1990). This approach toward disease management has been studied much more with aphid vectors (see section on Melon Aphid), but most of this technology is probably applicable to thrips-transmitted diseases.

Intercropping can have some benefit for onion thrips management. Despite its wide host range, there are clearly preferred hosts, principally onion. For example, Uvah and Coaker (1984) alternated rows of onion with various ratios of carrot rows, and found that the presence of carrots decreased abundance of thrips. This occurred despite the fact that carrot is a nominal host of thrips.

Sanitation is very important. Long ago, Horsfall and Fenton (1922) noted the ability of thrips to disperse from contaminated overwintering plants left in the field, or from transplanted onion bulbs taken from storage, to newly seeded onions. Now, with the availabilty of rapid transportation, thrips are often moved with plant material, and then inadvertently inoculated into fields. For example, Schwartz *et al.* (1988) found that nearly all batches of onion transplants shipped from Texas to Colorado were contaminated with thrips. Sporadic incidence of insecticide resistance among Colorado onion fields apparently was related to different sources of onions and thrips, and different pesticide exposure histories. Also, in some northern areas greenhouses are a source of thrips in the spring.

Host-Plant Resistance. The cryptic nature of thrips feeding on onion has long made chemical control difficult, and has stimulated the search for resistant varieties. Characteristics associated with resistance are round leaves and open or spreading plant architecture, attributes sometimes found in white onion varieties. It has been speculated that this plant architecture affords less opportunity for thrips to hide between leaves, hastening their predation by other insects (Jones *et al.*, 1935; Coudriet *et al.*, 1979). However, differences in plant chemistry have also been suggested (Saxena, 1975) to account for this difference. Indeed, differences in the ratio of adult and larval thrips among onion varieties (Coudriet *et al.*, 1979) could indicate physiological differences in suitability for thrips growth and survival. Some studies, however, report that the basis for resistance to onion thrips is feeding tolerance by some onion cultivars.

Resistance has also been identified among cabbage varieties (Shelton *et al.*, 1983a;1988; 1998; Hoy and Kretchman, 1991), but the basis for resistance is uncertain, and no clear patterns have emerged that would allow prediction of resistant types of cabbage. It is interesting to note, however, that some "resistant" cultivars support as many thrips as the susceptible cultivars, but on resistant plants the thrips feed principally on the outer leaves that are discarded at harvest, thus causing little damage (Stoner and Shelton, 1988). Whereas insecticide treatments alone sometimes fail to keep thrips from damaging thrips injury-susceptible cabbage in New York, when insecticides are used with varieties moderately susceptible to thrips injury, the combination is effective at preventing damage (Shelton *et al.*, 1998).

Tobacco Thrips
Frankliniella fusca (Hinds)
(Thysanoptera: Thripidae)

Natural History

Distribution. Tobacco thrips is widely distributed in eastern Canada and the United States, west to about the Rocky Mountains. However, it is most abundant, and most often recorded as a pest, in the southeastern states. It is a native species.

Host Plants. Vegetable hosts include bean, beet, cantaloupe, carrot, corn, cowpea, cucumber, onion, pea, pepper, potato, tomato, and watermelon. Because tobacco thrips can vector tomato spotted wilt virus, among vegetable crops it is known principally as a pest of tomato. Tobacco thrips is better known as a field crop-infesting insect, infesting alfalfa, barley several types of clover, cotton, lespedeza, peanut, rye, tobacco, vetch, wheat, and occasionally corn and oats. Winter grains such as rye and wheat, and volunteer peanut, apparently are suitable overwintering hosts. Several weeds have been reported to support tobacco thrips, such as Bermudagrass, *Cynodon dactylon*; blue toadflax, *Linaria canadensis*; broomsedge, *Andropogon virginicus*; buttercup, *Ranunculus* sp.; cocklebur, *Xanthium* sp; crabgrass, *Digitaria* sp.; cutleaf evening primrose, *Oenothera laciniata*; dandelion, *Taraxacum officinale*; dog fennel, *Eupatorium capillifolium*; false dandelion, *Pyrrhopappus carolinianus*; feathergrass, *Leptochloa filiformis*; Johnsongrass, *Sorghum halepense*; little barley, *Hordeum pusillum*; rabbit tobacco, *Gnaphalium obtusifolium*; sand blackberry, *Rubus cuneifolius*; shepherdspurse, *Capsella bursa-pastoris*; spiny sowthistle, *Sonchus asper*; wild lettuce, *Lactuca* sp.; wild radish, *Raphanus raphanistrum*; wood sorrel, *Oxalis* spp.; and a grass, *Brachiaria extensa*.

Natural Enemies. The natural enemies of tobacco thrips have not been well documented, but likely are the same as those associated with western flower thrips, *Frankliniella occidentalis*. A nematode, *Thripenema fuscum* (Tylenchida: Allantonematidae), was observed to parasitize up to 68% of thrips in Florida, suggesting that this may be an important mortality factor in some cropping systems (Tipping *et al.*, 1998). Insidious flower bugs, *Orius insidiosus* (Say) (Hemiptera: Anthocoridae), also have been observed to be important, and heavy rainfall is detrimental.

Life Cycle and Description. Tobacco thrips tend to be abundant in a crop during the spring and summer (McPherson *et al.*, 1992). In Florida, tomato blossoms are infested during April–June (Salguero Navas *et al.*, 1991a), but the thrips are abundant later further north. Several generations are present annually in Florida, including about three during the winter months (Toapanta *et al.*, 1996). Eddy and Livingstone (1931) reported five generations annually from South Carolina. Unlike the situation in Florida, where reproduction occurs during the winter months, in Georgia, South Carolina, and Louisiana overwintering occurs only in the adult form. The life cycle requires about 15–21 days for completion.

Egg. The egg is inserted into the foliage with an end protruding slightly. The egg measures about 0.25 mm long, is white, and bean-shaped. Mean duration of the egg stage is 6.7 days (range 3–10 days).

Larva. The immature stages consist of two larval and two pupal instars. Only the larval instars feed on foliage. The first instar measures about 0.23 mm long, the second instar 0.60–1.17 mm. Both instars are yellowish or whitish. Mean duration of the first larval instar is about 1.1 and 1.0 days at 25° and 35°C, respectively. Mean duration of the second larval instar is about 4.7 and 2.6 days at 25° and 35°C, respectively. Larvae tend to feed in cryptic habitats such as blossoms and terminal growth, and rarely in open, exposed areas.

Pupa. At the completion of the larval instars the mature larva drops to the soil to pupate. The prepupa is 0.5–0.6 mm long, the pupa 0.6–1.2 mm. The color of these latter instars is yellow, but unlike the larval instars they display wing pads. Mean duration of the prepupa is about 1.1 and 0.8 days at 25° and 35°C, respectively. Mean duration of the pupa is about 1.4 days at both 25° and 35°C.

Adult. The adult stage is variable but often has two pairs of fringed wings. The fringe on the anterior edge of the wings is markedly shorter than the fringe on the posterior edge. The adults measure 1.0–1.3 mm long. The head and thorax are light brown or yellowish-brown, but the abdomen is dark-brown. The antennae consist of eight segments. Long-winged (macropterous) and short-winged (brachypterous) forms occur, with brachypterous forms dominant during the winter months. The wings of the macropterous form usually do not attain the tip of the abdomen, but wing length is quite variable. Wings of the brachypterous forms barely extend to the first abdominal segment. Feeding behavior of the adults is similar to that of the larvae. The mean pre-oviposition period is about 2.6 days (range 2–6 days). Lowry *et al.* (1992) reported adult female longevity of about 6–10 days, and fecundity of 13–24 eggs per female, but these values seemed small and might reflect a poor host or suboptimal rearing conditions. When fed cotton, mean

longevity of mated females is 27–47 days and fecundity about 20–60 eggs per female (Eddy and Livingstone, 1931). Fertilized females produce both female and male offspring, though females are favored; unfertilized females produce only males.

Distinguishing tobacco thrips from other vegetable-infesting thrips requires careful examination. Antennal structure can be used to separate the *Thrips* spp. because their antennae consist of seven segments, whereas in *Frankliniella* there are eight segments. Separation of western flower thrips from tobacco thrips is accomplished by examining the eighth dorsal plate on the abdomen. In western flower thrips there is row of short hairs of approximately equal length along the posterior edge, whereas in tobacco thrips the hairs at the center of the posterior edge are shorter or absent.

The biology of tobacco thrips was presented by Hooker (1907) and Eddy and Livingstone (1931), but additional observations were made by Newsom *et al.* (1953), Chamberlin *et al.* (1992) and Chellemi *et al.* (1994). Developmental data were given by Watts (1934) and Lowry *et al.* (1992). Culture methods were described by Kinzer *et al.* (1972). Keys for identification were included by Palmer *et al.* (1989) and Oetting *et al.* (1993). Also, this species is included in a key to common vegetable-infesting thrips in Appendix A.

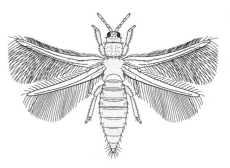

Adult female tobacco thrips.

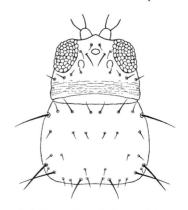

Anterior region of tobacco thrips.

Damage

Vegetable seedlings can be damaged by this thrips when they disperse to young annual crops from maturing perennial crops such as alfalfa or clover. Thrips feed and deposit eggs into the young tissue, causing young leaves to curl upward and older leaves to acquire a silvery or speckled, and crinkled, appearance (Webb, 1995). Buds and other young tissue may be killed, giving the seedling a scorched or burnt appearance. Destruction of terminal growth may disrupt apical dominance, producing an excessively bushy, branched growth form. Tobacco thrips may be found in blossoms, but unlike its co-occurring species western flower thrips, it is primarily a leaf feeder. Direct feeding injury has been studied best in peanuts, where insect suppression has been shown to increase yields slightly or not at all (Tappan and Gorbet 1981, Tappan 1986, Lynch *et al.*, 1984a). Similarly, direct injury to vegetables in infrequent, but because tobacco thrips now transmits tomato spotted wilt virus, its importance as a vegetable pest has escalated greatly. As happens with western flower thrips, virus acquisition occurs in the larval stage. After a latent period of 4–18 days, adults remain capable of transmitting tomato spotted wilt throughout their life (Sakimura, 1963). Weeds are important in the overwintering of both the thrips and virus (Hobbs *et al.*, 1993; Johnson *et al.*, 1995).

Management

Sampling. Distribution of tobacco thrips in tomato blossoms was studied by Salguero Navas *et al.* (1994). Populations were aggregated or randomly dispersed, varying with year of sampling. A binomial (presence or absence) sampling protocol was developed and shown to be useful for populations with less than an average of two thrips per blossom. These authors recommended collecting 16–18 flowers to assess the need for population suppression. Washing of plant material with a dilute sodium hypochlorite and soap solution produced higher thrips estimates than visual searches (Burris *et al.*, 1990)

Management. Management of this species is normally accomplished in conjunction with western flower thrips suppression, because *Frankliniella occidentalis* is generally more numerous and damaging (see the discussion under western flower thrips for more detail), they are similar in ecology, and because they occur together in mixed populations. Other thrips species, particularly flower thrips, *Frankliniella tritici* (Fitch), also may be present, but because they are not an important vectors of tomato spotted wilt virus they are considered to be of little importance.

Western Flower Thrips
Frankliniella occidentalis (Pergande)
(Thysanoptera: Thripidae)

Natural History

Distribution. Formerly restricted mostly to the western United States and Canada, by 1980 this native thrips had spread east to Georgia. Subsequently, it has spread throughout the United States and into southern Canada, and to other continents. It also has become a serious pest in Hawaii. Range expansion has undoubtedly been enhanced by movement of ornamental plants and vegetable seedlings from southern nurseries. It survives best in warm climates, and overwinters outdoors on growing plants along the west coast and throughout the southeastern states. Normally it is not thought to overwinter in very cold climates, but to re-invade these areas annually from greenhouses, or via introduction of seedlings from southern areas. However, the report of overwintering in Pennsylvania under leaf debris and in bare soil (Felland *et al.*, 1993) suggests a significant degree of cold hardiness. (See color figure 182.)

Host Plants. Western flower thrips apparently has an exceedingly wide host range. However, plant suitability varies seasonally and even geographically. Also, from an economic perspective the most important hosts are those that support both thrips reproduction and virus disease multiplication. Western flower thrips occurs on several vegetable crops, including cucumber, onion, pepper, potato, lettuce, and tomato. Tomato is most seriously injured directly by the thrips, through oviposition, but both lettuce and tomato are seriously damaged by tomato spotted wilt virus transmitted by thrips. Under greenhouse conditions, cucumber and pepper also are readily damaged. Field crops on which western flower thrips occurs include alfalfa, canola, crimson and white clover, millet, peanut, rye, vetch, and wheat. Several fruit crops have been reported to serve as hosts, such as apple, blackberry, blueberry, peach, pear, and plum. Among the weeds that serve as good hosts are such common species as black nightshade, *Solanum nigrum*; cheese weed, *Malva palviflora*; daisy fleabane, *Erigeron annuus*; dandelion, *Taraxacum officinale*; false dandelion, *Pyrrhopappus carolinianus*; jimson weed, *Datura stramonium*; galinsoga, *Galinsoga parciflora*; lambsquarters, *Chenopodium album*; lantana, *Lantana camara*; pigweed, *Amaranthus* spp.; prickly lettuce, *Lactuca serriola*; sorrel, *Oxalis* spp.; sowthistle, *Sonchus oleraceus*; and wild radish, *Raphanus raphanistrum* (Stewart *et al.*, 1989; Yudin *et al.*, 1986; Chamberlin *et al.*, 1992; Bautista and Mau, 1994; Chellemi *et al.*, 1994), but numerous other species also can serve as hosts. In Hawaii, the blossoms of woody legumes growing near cultivated fields serve as a major source of thrips.

Natural Enemies. Considering the abundance of western flower thrips and the severity of their injury to plants, surprisingly little is known about natural enemies. Minute pirate bugs, particularly *Orius tristicolor* (White), feed voraciously on western flower thrips, and there is good evidence that they suppress thrips populations in vegetable crops (Salas–Aguilar and Ehler, 1977; Letourneau and Altieri, 1983). *Ceranisus* spp. (Hymenoptera: Eulophidae) parasitize the immature stages, but are not generally abundant. Fungal epizootics caused by *Verticillium* and *Entomophthora* have been observed in moist climates. Nematodes in the genus, *Thripenema* (*Howardula*) (Nematoda: Allantonematidae), appear to be frequent associates of western flower thrips (Wilson and Cooley, 1972; Heinz *et al.*, 1996), and induce sterility in their hosts (Poinar, 1979). Impact of these nematodes has been inadequately studied, but an incidence of 88% has been reported from California (Heinz *et al.*, 1996).

Life Cycle and Description. In mild climates these thrips readily overwinter as adults and nymphs on many crops and weeds. However, in relatively cold climates such as northern Texas they overwinter on hardy crops such as alfalfa and winter wheat (Chambers and Sites, 1989). As noted above, they apparently can also survive in leaf debris and soil under the cold-weather conditions of Pennsylvania (Felland *et al.*, 1993). Toapanta *et al.* (1996) estimated 3–5 generations per year in north Florida, with populations highest in spring and a smaller peak in autumn. They can complete one generation in 15 days, so under ideal conditions many more generations are possible. A temperature of about 30°C seems to be optimal for population growth.

Egg. The eggs are deposited in young vegetative tissue, with one end protruding slightly. They are shaped like a bean, white in color, and measure about 0.25 mm long. Duration of the egg stage is reported to require 5–15 days in the field, but mean duration is only 2.6 days at 25°C. Females were reported by Gaum *et al.* (1994) to produce about 9–10 eggs during their life span when cultured on cucumber, and Lowry *et al.* (1992) reported that fecundity was 14–23 eggs per female when fed peanut. Because western flower thrips have preferred hosts, and usually include pollen in their diet, the aforementioned estimates of egg production may be artificially low. Thus, Lublinkhof and Foster (1977), recorded 43–95 eggs per female cultured on bean, and Trichilo and Leigh (1988) observed fecundity of about 130 eggs per female on cotton but 190 eggs on cotton supplemented with pollen.

Larva. Development of the immature thrips requires two larval instars (also sometimes called nymphs), which are feeding stages, followed by non-feeding prepupal and pupal stages. The temperature threshold for larval development is about 9.4°C. Larval development time may require 9–12 days in the field, extending to 60 days during the winter. However, when reared at a constant temperature of 25°C, the first and second larval instars require only 2.3 and 3.7 days, respectively. Larvae and adults are somewhat gregarious, often feeding together in small groups. At maturity, the larvae drop to the ground to pupate.

Pupa. The prepupal and pupal stages are reported to require 1–3 and about 3–10 days, respectively, under field conditions. At a constant temperature of 25°C, however, thrips complete their pupal development in an average of 1.1 and 2.7 days, respectively. The prepupa is distinguished by the presence of short-wing pads and erect antennae. The pupae have long wing pads that reach almost to the tip of the abdomen, and antennae that are bent backward along the head. Prepupae and pupae may be found on the surface, under debris, or in cracks and crevices to a depth of 7–10 cm.

Adult. The adults have fully formed, fringed wings and measure 1.2–1.9 mm long, averaging 1.5 mm. The fringe along the anterior edge is markedly shorter than the posterior edge. Body color varies from yellow to brown. The antennae bear eight segments. The adults commonly live about 20–30 days, but some persist providing 40–70 days. Females may mate, or reproduce parthenogenetically. Unmated females produce only males, whereas mated females produce both sexes. The offspring of mated females are female-biased, usually on the order of 2:1. Females mate immediately upon emergence, and repeatedly over the course of their life. Dispersal by adults usually occurs when their food plants become unsuitable, which commonly results from drought, maturity, or harvesting.

Distinguishing western flower thrips from other vegetable-infesting thrips requires careful examination. Antennal structure can be used to separate the *Thrips* spp., because their antennae consist of seven segments, whereas the antennae in *Frankliniella* bear eight segments. Separation of western flower thrips from tobacco thrips, *Frankliniella fusca* (Hinds), is accomplished by examining the eighth dorsal plate on the abdomen. In western flower thrips there is row of short hairs of approximately equal length along the posterior edge, whereas in tobacco thrips the hairs at the posterior edge of the plate are shorter or absent centrally.

Both nymphs and adults produce an alarm pheromone, and respond to it by moving away from the source of the pheromone, and usually by dropping from the plant. The pheromone is released in droplets of anal fluid (Teerling *et al.*, 1993).

Thrips biology was given by Bailey (1933b), Bryan and Smith (1956), Lublinkhof and Foster (1977), Gaum *et al.* (1994), and van Rijn *et al.* (1995). Rearing techniques were described by Teulon (1992) and Doane *et al.* (1995). Keys that included western flower thrips were presented by Palmer *et al.* (1989) and Oetting *et al.* (1993). Also, this species is included in a key to common vegetable-infesting thrips in Appendix A.

Damage

This species, as its common name suggests, prefers an interstitial habitat such as within flowers or in leaf clusters; only rarely is it found in exposed locations. It typically feeds on pollen grains and on the ovary of flowers, resulting in malformed, stunted, or discolored fruit. In cucumber, for example, western flower thrips feeding causes silvery, web, or streak-like scarring, which may be accompanied by fruit malformation (Rosenheim *et al.*, 1990). When they feed on foliage, they cause distortion of expanding leaves and mottling or speckling of mature leaves. On onion, their

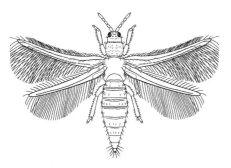

Adult female western flower thrips.

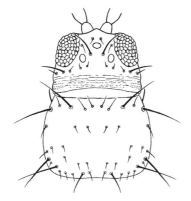

Anterior region of western flower thrips.

feeding injury is similar to the effects of feeding by onion thrips, *Thrips tabaci* Lindeman.

Thrips also deposit eggs in small fruits, inducing deformities. In the absence of flowering plants, however, they oviposit readily on such plants as non-flowering lettuce. Salguero Navas *et al.* (1991a) documented the damage to tomato fruit caused by oviposition, typically reflected by a dimple or indentation surrounded by a light-colored halo. Western flower thrips was much more damaging than some other tomato-infesting thrips at comparable densities. In southern Texas, western flower thrips has been reported to damage onions along with onion thrips (Bender and Morrison, 1989). These thrips also are found associated with onion blossoms, where they enhance pollination and seed set. Only at very high densities, approximately 9,000–10,000 thrips per onion seed head, is damage likely to occur (Carlson, 1964).

There is a strong association between the prevalence of western flower thrips and tomato spotted wilt virus. In this thrips species, oviposition preference is as important as feeding preference, because only nymphs are capable of virus acquisition. Thus, selection of oviposition sites by females determines the likelihood of the thrips developing into a virus vector. Once infected, the thrips remain capable of transmitting the virus for the remainder of their life. Tomato spotted wilt virus-infected weeds are the major host of virus in vegetable fields unless susceptible crops are cultivated continuously (Cho *et al.*, 1986, 1987).

Western flower thrips has become a particularly serious pest of vegetable and ornamental greenhouse crops. Its short development time, wide host range, cryptic feeding habits, and particularly its tendency to evolve insecticide resistance rapidly, make it well suited for inhabiting commercial greenhouses.

Despite the severity of the western flower thrips-virus disease problem on some crops, these thrips are not entirely detrimental. The thrips also feed on mites, and serve as important alternate hosts for some larger predators. Thus, in cropping systems that are not particularly susceptible to thrips or virus injury, the presence of low to moderate numbers of thrips can be beneficial (Gonzalez *et al.*, 1995a).

Management

Sampling. Thrips densities in blossom samples are often made in the field, by visual examination of the plant or by shaking the blossom or other vegetative material over a tray, but the precision of this type of population estimate is quite low. A better estimate is gained by submerging the plant sample in 70% ethanol, or in sodium hypochlorite and soap solution, and shaking it to dislodge the insect. Rummel and Arnold (1989), for example, found that thrips counts were 5–6 times higher when sampled by washing vegetation. A sample unit of 10 blossoms from each of five areas in a field is considered optimal (Cho *et al.*, 1995). Thrips densities can also be estimated with the use of sticky traps. Yellow, white, and blue traps are generally most attractive to western flower thrips (Yudin *et al.*, 1987; Vernon and Gillespie, 1990). Trap efficiency is increased by highly contrasting background color; yellow in front of a violet background, for example, is highly attractive (Vernon and Gillespie, 1995). Thrips can also be captured in water traps, and higher captures are made when certain volatile chemicals are added to the trap (Teulon *et al.*, 1993). Sampling has been reviewed by Shipp (1995).

In tomato, western flower thrips are most abundant in blossoms on the upper half of plants, and at field margins. Nymphs are more abundant in blossoms in the lower regions of plants (Salguero Navas *et al.*, 1991a). Thrips populations, especially nymphal populations, are aggregated. A binomial, or presence-absence, sampling program has been developed for tomato when thrips densities average less than 1.4 per blossom; 16–18 blossom samples are used to estimate abundance and suppression is initiated only when greater than 50% of the blossoms are infested (Salguero Navas *et al.*, 1994).

Insecticides. Insecticides are commonly applied to the foliage and blossoms of vegetables to minimize feeding and oviposition damage and to limit disease transmission. However, insecticide resistance is a widespread phenomenon (Zhao *et al.*, 1995). The severity of the resistance problem in the field is exacerbated by the ability of thrips to infest and escape from greenhouses, where insecticide use is frequent. Rotation of insecticide classes is frequently recommended to forestall development of resistance. If insecticides are used, a common practice is to apply two treatments about five days apart because the eggs are within the plant tissue and the prepupal and pupal stages are beneath the soil, and thus relatively immune.

Cultural Practices. Barriers are sometimes recommended for insect exclusion, including such small species as thrips. However, the small size of western flower thrips (width of males is about 184 microns; width of females about 245 microns) requires extremely fine screen if thrips are to be denied access to plants. This fact excludes most standard materials from

consideration as screens (Bethke and Paine, 1991). However, low barriers can be used under field conditions to limit thrips dispersal and invasion of crops (Yudin *et al.*, 1991). Also, walk-in tunnels or greenhouses covered with ultraviolet light-absorbing plastic are less infested by thrips than ultraviolet light-reflecting coverings, apparently due to reduced attraction of plants grown in flitered light or to modified thrips feeding behavior (Antignus *et al.*, 1996).

Sanitation is an important element in thrips management. Weeds can serve as important alternate hosts of both thrips and virus diseases, and their presence should be minimized. Also, if seedlings are used to initiate a crop, care should be taken to assure that they are free from thrips. The proximity of greenhouses is another consideration, as this may be a principal source of crop infestation by thrips, especially in cold climates where overwintering success by thrips is limited.

Biological Control. Considerable emphasis has been placed on development of biological control agents for thrips-infesting greenhouses, including western flower thrips. Such beneficial organisms as the parasitic wasp *Ceranisus menes* (Hymenoptera: Eulophidae) (Loomans *et al.*, 1995); the minute pirate bug, *Orius laevigatus* (Hemiptera: Anthocoridae) (Chambers *et al.*, 1993); the foliage-dwelling predatory mites *Amblyseius cucumeris* (Oudemans) and *A. degenerans* Berlese (Acari: Phytoseiidae) (van Houten and van Stratum, 1995); the soil-dwelling predatory mites *Geolaelaps* sp. (Acari: Laelapidae) (Gillespie and Quiring, 1990); and entomopathogenic nematodes (Nematoda: Heterorhabditidae and Steinernematidae) (Chyzik *et al.*, 1996) have been studied. Some beneficial organisms, particularly the mite *A. cucumeris*, are used in commercial greenhouse vegetable production. Factors such as temperature, photoperiod, and crop type sometimes limit success, even under greenhouse conditions. For example, mite predators are much more effective in suppressing western flower thrips on pepper than on cucumber, presumably due to interference with mite searching behavior by the numerous trichomes found on cucumber leaves (Shipp and Whitfield 1991). Several common entomopathogenic fungi such as *Beauveria bassiana*, *Metarhizium anisopliae*, *Paecilomyces fumosroseus*, and *Verticillium lecanii* also have been used, but the level of suppression is only moderate. The expense and dispersal tendencies of most beneficial organisms thus far have limited their use to greenhouses.

Other Invertebrate Pests

CLASS ACARI—MITES

Banks Grass Mite
Oligonychus pratensis (Banks)
(Acari: Tetranychidae)

Natural History

Distribution. This species occurs principally in the southern United States, particularly in the southern Great Plains. However, it also can occur in northern areas, and in the Great Plains region it commonly causes injury as far north as Nebraska. In the northwestern states, it is sometimes called the timothy mite. Banks grass mite also is known from Hawaii and Puerto Rico, Central America, and Africa. Apparently it is native to North America.

Host Plants. This mite feeds predominately on wild grasses and on grasses grown as field crops. The crops injured include Bermudagrass, bluegrass, corn, sorghum, and wheat. However, dates are reported to be injured in southern California. Among vegetables, only sweet corn is damaged, principally in the southern Great Plains and southern Rocky Mountain areas. Among the many wild grasses that serve as hosts for Banks grass mite are wheatgrass, *Agropyron* spp.; gramagrass, *Bouteloua* spp.; bromegrass, *Bromus* spp.; wildrye, *Elymus* spp.; panicum, *Panicum* spp.; and sorghum, *Sorghum* spp.

Natural Enemies. Predators and disease agents both can be important in population biology, though predators are considered most important. The most important predator is considered to be the predatory mite, *Neoseiulus fallacis* (Garman) (Acari: Phytoseiidae), which can consume an average of 15 mites per day. However, other phytoseiids may assume great importance at times, or in certain locations. Pickett and Gilstrap (1984) provided information on the phytoseiid complex attacking Banks grass mite in Texas,

including a key to the important species. Also important as predators of Banks grass mite are the small black lady beetles, *Stethorus* spp. (Coleoptera: Coccinellidae); the six-spotted thrips, *Scolothrips sexmaculatus* (Thysanoptera: Thripidae); the lacewings, *Chrysoperla* spp. (Neuroptera: Chrysopidae); and minute pirate bugs, *Orius* spp. (Hemiptera: Anthocoridae).

The fungi *Neozygites* spp. are common pathogens at times, especially when conditions are moist and cool. Dick and Buschman (1995), working in Kansas, reported that though *Neozygites* sp. appeared frequently, the fungus required an ambient humidity of at least 80% and often occurred too late in the season to prevent crop injury.

Life Cycle and Description. The number of life cycles varies with location. In the northern states, 6–7 generations is common, though more occur in the south. The life cycle requires about 8–25 days depending on temperature. Although adults may overwinter and become active during the winter in mild areas of the country, eggs are not produced until spring.

Egg. The egg is pearly white when first deposited, but eventually turns yellowish-brown. They normally are laid singly on the underside of the leaf. They may also be deposited on the webbing that covers the leaf. Just before hatching, the eyespots of the developing larvae become evident. Duration of the egg stage is 2–5 days under summer conditions, but may require up to 36 days in Washington.

Larva. The larva is whitish or pinkish-white initially. As it matures, it becomes green. The six-legged larval stage typically persists for 2–3 days under good conditions, but may require up to 17 days in northern locations.

Nymph. The protonymph and deutonymph stages each may require only 1–2 days for development, but under cool conditions development time for each instar may extend for 13–18 days. The eight-legged nymphal stages are pale green or bright green, depending on their food.

Adult. The eight-legged adult stage is greenish with darker pigmentation laterally. Banks grass mite is quite similar to twospotted spider mite, *Tetranychus urticae* Koch, in appearance, but *T. urticae* has dark pigmentation mostly in the anterior half of the body whereas in *O. pratensis* the pigmentation extends back to the posterior tip of the body. Banks grass mite is slightly smaller and its body is more flattened than twospotted spider mite. Banks grass mite measures about 0.4 mm long, with the females averaging larger, the males smaller. The abdomen tapers to a point in males whereas in the female it is bluntly rounded. Overwintering forms are orangish or pinkish. Adults live for about 7–23 days under warm conditions, and up to 48 days under cool conditions. The pre-oviposition period is only 1–2 days, with peak oviposition at about 6th day and a steady decrease thereafter. Females produce 7–14 eggs per day, with a total fecundity of 75 to 150 eggs. Fecundity is higher when mites feed on mature corn or moisture-stressed corn (Feese and Wilde, 1977), and Banks grass mite thrives under hot, dry conditions. Optimum temperature is about 36–37°C. Light and low humidity favor passive aerial dispersal by adults, which are aided in their dispersal by production of silk threads (Margolies, 1987).

The biology of Banks grass mite was given by Tan and Ward (1977). Temperature relations were described by Perring *et al.* (1984). A useful synopsis was provided by Jeppson *et al.* (1975). Banks grass mite is included in a key to vegetable-feeding mites in Appendix A.

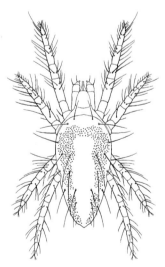

Adult male Banks grass mite.

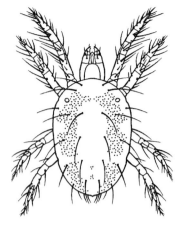

Adult female Banks grass mite.

Damage

Damage is typical of mite infestations. Feeding initially causes the appearance of minute yellow spots on the undersides of leaves, particularly the lower leaves on plants. With continued feeding, leaves turn yellow and then necrotic and brownish as damage increases. The mite infestation progresses upwards on the plant, with older leaves suffering premature death. Young leaves may be deformed, and leaf tissue can be contaminated with webbing. Serious infestation may occur relatively early in the season, once the plant is half grown, whereas twospotted spider mite tends to attain high densities on corn only after flowering has commenced. Damage also is found in dry areas, where wheat and range grasses serve as overwintering hosts. In contrast, twospotted spider mite infestations often occur in more moist locations such as along rivers, where broadleaf plants such as alfalfa may serve as overwintering hosts.

Management

Sampling. Mite infestations should be checked regularly, especially if mites are detected early in the season. Populations usually are monitored visually, though a hand lens with 10 × magnification is generally needed for accurate assessment, and even greater magnification is helpful for species confirmation. Populations of natural enemies should also be considered in determining the need to suppress mite populations.

Cultural Practices. Drought stress commonly favors mite population increase, so proper irrigation is an important preventative treatment. Late planting of corn alleviates infestation in some areas, and excessive fertilization with nitrogen should be avoided. Presence of mites in winter-grown wheat, an important overwintering host, is positively related to subsequent infestation levels in corn (Holtzer *et al.*, 1984).

Insecticides. Application of insecticides to foliage is sometimes necessary to prevent damage. In areas where both Banks grass mite and twospotted mite occur, it is important to determine the pest species present before application of insecticides because they differ in their susceptibility to various products. Banks grass mite is resistant to insecticides in some areas, but is generally considered easier to suppress than twospotted spider mite. Sometimes Banks grass mite populations increase following application of insecticides for other insect pests. If possible, it is desirable to suppress caterpillar pests with the microbial insecticide *Bacillus thuringiensis* because such applications do not cause destruction of mite predator populations and increase in mite abundance.

Broad Mite
Polyphagotarsonemus latus (Banks)
(Acari: Tarsonemidae)

Natural History

Distribution. Broad mite is found throughout the tropical regions of the world, and is a greenhouse pest in many temperate areas. In the United States, it is a minor citrus pest in Florida, California, and Hawaii, and occasionally injures greenhouse and field-grown vegetables in those states and in Georgia and Texas. Its origin is likely southern Asia.

Host Plants. The host range of broad mite is extensive with plants from over 60 families known to be injured (Gerson, 1992). The principal exception is the grasses, which are not fed upon. Among the important crop plants injured are cotton, citrus, mango, papaya, passion fruit, rubber, tea, and tobacco. Among the vegetables reported to be injured are beet, bean, chard, cucumber, eggplant, pepper, potato, and tomato. Damage is especially severe in bell pepper. Ornamental plants also are affected, including azalea, balsam, begonia, chrysanthemum, gerbera, dahlia, fuchsia, zinnia and likely many others. Weeds such as spiny pigweed, *Amaranthus spinosus*; beggartick, *Bidens* spp.; jimsonweed, *Datura* sp.; and galinsoga, *Galinsoga* sp.; serve as important reservoirs for mites during crop-free periods. Broad mites also can develop on plant pollen (Pena, 1992).

Natural Enemies. Predators have been observed to be an important elements of field ecology. The natural enemies are listed by Gerson (1992), but *Amblyseius* spp. and *Typhlodromus* spp. (both Acari: Phytoseiidae) are most important (Badii and McMurtry, 1984; Pena *et al.*, 1989). Sometimes these predators are able to maintain suppression of mite populations on crops, including pepper and citrus. Among dis-

eases, only the fungus *Hirsutella nodulosa* has been noted as a naturally occurring pathogen.

Life Cycle and Description. The complete life cycle requires only about seven days under favorable conditions. Optimal conditions for growth and reproduction are about 24°C and high humidity. Reproduction does not occur below about 13°C and above about 34°C. Under such favorable conditions 20–30 generations may occur annually. They are found reproducing throughout the year in southern Florida, but are most abundant during the summer months. Temperature of about 25°C and humid conditions are optimal. Hot, dry weather in the autumn and early winter is disruptive to broad mite populations in California. Diapause is unknown in this species.

Egg. The eggs are laid singly and are found on the lower surfaces of young apical leaves and in flowers. The eggs are elongate-oval, and translucent. The upper surface is covered with rows of hemispherical projections, whereas the ventral, flat surface is attached to the substrate. The projections are whitish, though the egg itself is nearly transparent. Hatching occurs after two days when maintained at 25°C and high humidity. Female fecundity averages 40–50 eggs (range 30–75 eggs).

Larva. Larvae are small, flattened, and oval. They are whitish and bear six legs. Larvae measure about 0.12–0.18 mm long. The general appearance is similar to that of the adult stage. The larval stage feeds for one day, then molts to a quiescent stage usually called the pupa.

Pupa. The pupal (also called nymphal) stage is similarly brief, lasting only about 1–2 days, and occurs within the old larval cuticle. The body form is similar to that of the larva and adult, though the legs are eight in number and are larger than in the larval stage.

Adult. Adults also are flattened and broadly oval, particularly the females. They possess four pairs of legs, the anterior two pairs being widely separated from the posterior pairs. The legs are well-equipped with spines. The females bear relatively short legs, usually measuring less than the width of the body, but males bear longer legs, with some about as long as the length of the body. The posterior end of the body bears a few hairs, but lacks long or stout spines. The color of the mites varies, but often is whitish, yellow or yellowish-green. These organisms are quite small; body length is 0.2–0.3 mm in females, and males only half as large. Females usually have an indistinct white stripe along the center of the back, though males lack this character. Adult males emerge before females. Males grasp female pupae with their genital organs and carry them upwards in the plant to younger tissue. Therefore, males effectively select the oviposition sites of females. Females normally copulate immediately upon emerging from the pupal stage, though they may mate later. Adult males persist for about seven days, whereas females survive for about 10 days. The ratio of males to females varies among populations, but usually is weighted toward females, often at a ratio of 4:1. Unmated females produce only males, which may then mate with their mother, assuring production of females. Both sexes are very active, though they do not use their hind legs for walking. Adults disperse short distances within and among plants by walking, but disperse long distances by being blown by wind, and by attaching to, and "hitch-hiking" on winged insects such as aphids and whiteflies.

A description of broad mite was given by Lavoipierre (1940). Gerson (1992) provided an excellent review of broad mite biology and management. Temperature and humidity responses were studied by Jones and Brown (1983). A simple key to common vegetable-feeding mites is found in Appendix A.

Damage

The damage caused by broad mite varies among host plants. In general, the mite seems to secrete a plant growth regulator or toxin, as it induces plant deformities as it feeds. Among the common symptoms are distortion of apical tissue, shortening of internodes, stem swelling, darkening, blistering, shriveling, and curling of leaves. Fruit and flower development may be inhibited, or fruit may be deformed, split, or russeted. Broad mite injury is commonly confused with herbicide toxicity, micronutrient deficiency, and virus diseases (Cross and Bassett, 1982; Pena and Bullock, 1994).

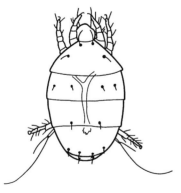

Adult broad mite.

The presence of only a few mites may induce injury. As few as 10 mites per plant may damage pepper, potato, and bean. Young pepper plants support higher rates of mite population increase, and are more easily damaged (de Coss–Romero and Pena, 1998). Damage may result at some distance from the site of feeding, and may persist for weeks after the mites are removed by application of insecticide. These characteristics suggest the presence of toxins secreted by the mites.

Management

Sampling. Mites are found on the lower surface of young foliage, and in flowers. Populations often are highly aggregated.

Insecticides. Various insecticides and acaricides are effective against broad mite (Schoonhoven *et al.*, 1978), but some classes of pesticides such as pyrethroids are not satisfactory and may cause population increases. Pesticides normally are applied to the foliage, and often at short intervals due to rapid rate of population growth. Sulfur is commonly recommended for suppression of broad mite infestations (Pena, 1988).

Cultural Practices. Broad mites cannot survive long in the absence of suitable plants. Thus, clean cultivation between crops decreases the incidence of infestation, as does the elimination of nearby weeds. The wide host range of broad mite, however, makes it difficult to eliminate mites through cultural practices.

Biological Control. The fungi *Beauveria bassiana*, *Hirsutella thompsonii*, and *Paecilomyces fumosoroseus* have been shown to infect and kill broad mite (Pena *et al.*, 1996). The predatory mites *Neoseiulus californicus* (McGregor) and *N. barkeri* (Hughes) (both Acari: Phytoseiidae) were studied as inoculative agents by Fan and Petitt (1994) and Pena and Osborne (1996), and found to have potential for broad mite population regulation, particularly under greenhouse conditions.

Bulb Mites
Rhizoglyphus echinopus (Fumouze and Robin)
Rhizoglyphus robini Claparede (Acari: Acaridae)

Natural History

Distribution. Several species of mites are associated with the bulb mite complex. Misidentification of mites is common, so the validity of many reports is uncertain. However, there is evidence for vegetable crop damage by *Rhizoglyphus echinopus* (Fumouze and Robin) and *R. robini* Claparede (Manson, 1972). These species are found throughout the world but do not appear to be native to North America.

Host Plants. Bulb mites are associated with bulbs and roots of numerous plants. They are most commonly reported to be associated with bulbs of flowering crops such as dahlia, hyacinth, iris, lily, narcissus, and tulip. Among vegetable crops, onion, garlic, and leek are occasionally damaged. Bulb mites also have been known to feed on beet, celery, and potato, and perhaps other vegetables with large roots or tubers, but they are not considered a pest of these crops except in storage.

Life History and Description. Mites of the genus *Rhizoglyphus* are relatively large, usually measuring 0.5–0.8 mm and sometimes up to 1.0 mm long, and about 0.3 mm wide. They are oval, smooth in general appearance, and with a whitish or colorless body. Long hairs are sometimes found on the body, especially at the posterior end of the mite. There normally are four pairs of thickened, short, light-brown legs. The legs bear hairs or spines, though they are less dense than in the spider mites. The first larval stage has only three pairs of legs. The eggs are oval, translucent and whitish, and are about one half the size of the adult female. There are four immature stages. The life cycle is completed in 17–27 days at 18–24°C, and 9–13 days at 20–27°C. Under cooler conditions, however, the life cycle may require over 100 days. Optimal growth and development occurs at 22–26°C. The developmental threshold is about 9.7°C. Longevity is about 40–70 days. The female deposits eggs singly, producing up to 700. They hatch in 8–15 days. The mites are abundant in soil, and often associated with decaying plant material. The hypopal stage sometimes appears in mite colonies; this small (0.3 mm long), dark-colored, and sluggish form will attach to insects and thereby be transported to new locations.

The ecology of *R. robini* was studied by Gerson *et al.* (1985) in Israel. They reported that both clay and loam soils were suitable for the mites, but that sandy soils were not. Summer soil temperatures were too high for the mites unless the soil was irrigated. Temperatures of about 35°C caused sterility. Thus, mite populations tended to increase during the autumn and winter, declining in the spring. Virtually any organic material was suitable for mite maintenance, but manure was stimulatory.

Description and biology of bulb mites can be found in Hodson (1928), Hughes (1961), and Gerson *et al.* (1983). Manson (1972) provided keys to some species and clarified some synonymies. Bulb mites are also included in a key to vegetable-feeding mites in Appendix A. Mite culture techniques were given by Gerson *et al.* (1991).

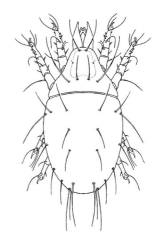

Adult bulb mite, *Rhizoglyphus echinopus*.

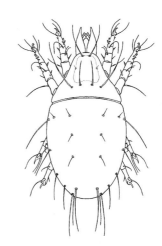

Adult bulb mite, *Rhizoglyphus robini*.

Damage

Most astigmatid mites feed on fungi or decaying-plant tissue, and are not considered to be primary plant pests. However, bulb mites may invade and enlarge bulb tissue previously invaded by fungi. Mechanical injury, which may result from tillage or injury by other pests, also may create an environment suitable for bulb mite population growth, but this form of damage is not as conducive for mite population growth as is plant disease (Okabe and Amano, 1991). Bulb mites apparently inhibit the recovery of bulbs from injury, and aid in the transport of fungi. Rawlins (1955), working in New York, reported that *Rhizoglyphus robini* (reported as *R. solani*) fed on roots, destroying field-grown onions at both the seedling and mature bulb stages. Latta (1939) also reported root damage, but by *R. echinopus* (reported as *R. hyacinthi*) on lily. *R. echinopus* usually is considered to be a pest of stored products, particularly those that have not been dried properly. They are often found in association with rot in stored potatoes. However, there are enough reports of plant damage associated with this species to be very concerned about its abundance.

Management

Insecticides. Bulb mites are quite tolerant to pesticides (Knowles *et al.*, 1988). Soil fumigants may be applied to reduce the incidence of fungi and mites (Jefferson *et al.*, 1956). In some studies, fungicide application reduces disease, but not mite abundance (Ascerno *et al.*, 1981). Suppression of disease alone often increases crop yield, supporting the belief that bulb mites are not primary pests. However, Ascerno *et al.* (1983) suggested that if mites attained a threshold of abundance, it was necessary to suppress them if fungicidal suppression of disease was to be effective.

Cultural Practices. Bulbs that are lightly infested can be purged of mites before planting; a hot water bath at 43°C for one hour is adequate (Weigel and Nelson, 1936). Mulching with transparent polyethylene raises the temperature of soil, and can be used to reduce the abundance of both fungi and mites (Gerson *et al.*, 1981).

In fields, where bulb mites have been observed, prolonged fallow periods are recommended to eliminate the food supply of the mites. It is important to eliminate organic residues under such circumstances, both crop remains and manure, and to avoid following with a susceptible crop.

Twospotted Spider Mite
Tetranychus urticae Koch

Strawberry Spider Mite
Tetranychus turkestani Ugarov and Nikolski

Tumid Spider Mite
Tetranychus tumidus Banks
(Acari: Tetranychidae)

Natural History

Distribution. The wide distribution of spider mite species is due largely to their small size and the ease with which they are inadvertently transported in commerce. Their wide host range also allows them to establish readily in new locations. Recently, their growing tolerance to many insecticides (acaricides) has exacerbated their tendency to survive quarantine treatments, further enhancing spread. The most widespread is twospotted spider mite, *Tetranychus urticae* Koch, which is found throughout the world in temperate and subtropical locations. Strawberry spider mite, *Tetranychus turkestani* Ugarov and Nikolski, is found in Europe and Asia in addition to North America. Tumid spider mite, *Tetranychus tumidus* Banks, is commonly found from the southern United States south through Central America and the Caribbean to Brazil. These mites are most damaging in the warmer, southern areas of North America, though twospotted spider mite survives temperate climates better than the other species. Within greenhouses, these species can be very abundant anywhere in North America, though twospotted mite is the most troublesome of the group. Other tetranychid mites which are known from vegetables, but are less common as pests, include desert spider mite, *Tetranychus desertorum* Banks in the southern United States and Pacific spider mite, *Tetranychus pacificus* McGregor, in the western states. The taxonomy of spider mites is uncertain, and name changes are frequent.

Host Plants. These mites affect numerous vegetable crops, including bean, beet, cantaloupe and other melons, carrot, celery, corn, cucumber, eggplant, parsley, pea, pepper, squash, sweet potato, tomato, and likely others. Tomato, bean, and cucurbit crops are affected most often, and twospotted spider mite is the most likely species to cause injury. Most of these mites are also known from cotton, soybean, strawberry, tree fruit crops, and ornamental plants. Of these four species, however, twospotted spider mite is the most likely to affect tree fruits and ornamentals, with the other species normally associated with low-growing crops.

Natural Enemies. The natural enemies of spider mites are numerous and diverse. Among the most important are predatory mites, particularly *Phytoseiulus persimilis* Athias-Henriot (Acari: Phytoseiidae); lady beetles, particularly *Stethorus* spp. (Coleoptera: Coccinellidae); dusty-wings (Neuroptera: Coniopterygidae); pirate bugs, particularly *Orius* spp. (Hemiptera: Anthocoridae); some plant bugs (Hemiptera: Miridae); and thrips, particularly *Scolothrips sexmaculatus* (Pergande) (Thysanoptera: Thripidae) (Huffaker *et al.*, 1969). Also, fungi sometimes cause epizootics (Dick and Buschman, 1995). The natural enemies of tetranychids were reviewed by McMurtry *et al.* (1970). van de Vrie *et al.* (1972) discussed population ecology, including population regulation.

Life Cycle and Description. With a life cycle of only 8–12 days at 30°C and about 17 days at 20°C, over 20 generations may develop annually, though conditions rarely allow this rate of population cycling. Overwintering may occur on many hosts in warm-winter climates, and in cold winter areas forage legumes and greenhouses often shelter these pests, but adult twospotted spider mite females also pass the winter under leaves or other organic debris in a state of diapause. The development time of the immature stages is 4–5 days at 30–32°C, but is extended to about 16–17 days when the temperature is 15°C at night and 28°C during the day. (See color figure 172.)

Egg. The eggs are whitish and spherical in form. They measure about 0.10–0.15 mm in diameter. They are often deposited singly on the lower surface of foliage, but sometimes the upper, and the leaf surfaces are covered with strands of silk. Females oviposit at 5–6 eggs per day, for a total of 60–120 eggs. Duration of the egg stage is about three days at 30°C and 6–7 days at 20°C.

Larva. The first instar is called the larva, and is colorless initially but yellowish or pinkish after feeding. The body is nearly spherical in shape, and bears three pairs of legs. The terminal portion of the larval stage is a non-feeding period called the nymphochrysalis or protochrysalis. Duration of the first instar is 1–2 days at 30°C and 2–3 days at 20°C.

Nymph. There are two nymphal instars—the protonymph and the deutonymph. These stages are easily separated from the larva because they bear four pairs of legs. They tend to be green or red. As in the larval stage, the terminal portion of each nymphal period is a non-feeding period called the deutochrysalis and teliochrysalis, respectively. Duration of each instar is 1–2 days at 30°C and about three days at 20°C.

Adult. Adults are 0.4–0.5 mm long, males averaging slightly less than females and are usually less abundant than females. Like the nymphs, adults bear four pairs of relatively long legs. They also have numerous long hairs on their legs, and long but sparse hairs on their body. Females tend to be oval in body shape, males elongate-oval or diamond-shaped. In twospotted spider mite and strawberry spider mite the actively feeding female is usually greenish, with dorsolateral dark spots. The two species sometimes may be distinguished by the number of dark spots, which is two in twospotted but four in strawberry spider mite. In tumid spider mite the female body color is reddish with dark lateral markings. Overwintering females of twospotted spider mite and strawberry spider mite become orangish-red. Color is not a very reliable character with tetranychid mites; accurate determination depends on examination of tarsal characteristics and genitalia of males.

Males are attracted to immature females by a sex pheromone, perform extensive mating rituals (Cone *et al.*, 1971a,b) and may mate repeatedly. The pre-oviposition period of females is 1–2 days. Fertilized females produce both male and female offspring; unfertilized females produce only males. Duration of the adult stage is normally about 30 days except when overwintering. The pre-oviposition period of adults is less than a day at 30°C, and 1–2 days at 20°C. Adults disperse by crawling and by wind-borne dispersal.

The biology of tumid spider mite was described by Saba (1974). Laing (1969) and Carey and Bradley (1982) gave development and temperature relations for *T. urticae, T. turkestani* and *T. pacificus.* Behavior and ecology of *Tetranychus* spp. was also described by Huffaker *et al.* (1969) and van de Vrie *et al.* (1972). Comprehensive accounts of twospotted spider mite were given by Ewing (1914) and of strawberry spider mite by Cagle (1956); biology of carmine spider mite was described by Hazan *et al.* (1974). Descriptions and keys are found in Jeppson *et al.* (1975) and Kono and Papp (1977), Baker and Tuttle (1994), and Bolland *et al.* (1998). A simple key to mites found on vegetables is included in Appendix A.

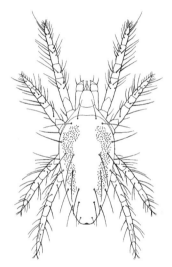

Adult male twospotted spider mite.

Adult female twospotted spider mite.

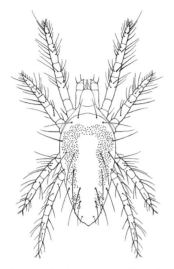

Adult male strawberry spider mite.

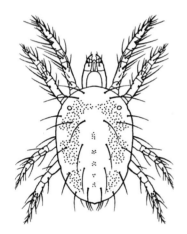

Adult female strawberry spider mite.

Damage

Spider mites generally feed on the lower leaf surface, though twospotted spider mite affects the upper surface of some host plants. They pierce individual cells with their stylets, withdrawing the cell contents. Twospotted spider mite can feed on 18–22 cells per minute, resulting in many dead cells, and often a speckled appearance. Leaf transpiration is accelerated, and affected leaves may dry and drop from the plant. Yellowing and speckling are the most common early plant responses to feeding, though reddening may also occur. Injection of plant growth regulators or interference with growth regulators during feeding is also reported. Wilting, tissue death, leaf deformity, and abcission are characteristics of prolonged and high-density infestations. Disruption of photsynthesis results in stunting of plant growth and reduced-fruit yields. Mite products such as webing, eggs, cast skins,

and fecal material also detracts from the cosmetic quality of plants.

Before the 1940s, spider mites were infrequently considered to be serious pests, but since then they have assumed major pest status in some crops. Apparently mite problems are induced by crop management practices, particularly the use of broad-spectrum insecticides (see section on "insecticides"). Also, the suitability of crops for mites is greatly enhanced when mites develop on plants which receive excessive nitrogen fertilization, grow in a dusty environment, or are stressed by inadequate moisture and high temperature. These environmental factors can convert plants which might be only poor hosts into very good hosts, resulting in mite population increase and crop damage.

Management

Sampling. Visual examination of foliage for leaf stippling on the upper surface, and mites and webbing on the lower surface, is the usual method of sampling. Older, or lower, leaves are usually examined. Infestations tend to be clumped initially, with clumping decreasing as the crop matures and females disperse to younger foliage (Perring *et al.*, 1987).

Insecticide. Chemical insecticides (acaricides) are commonly used in greenhouses to prevent injury by mites, though natural enemies usually are capable of maintaining spider mite densities at low levels on crops grown under field conditions. In the field, insecticides directed at other pests can induce mortality among natural enemies of mites, causing increase in spider mite populations. Therefore, considerable effort is now directed at managing pests without disrupting natural control of mites, often by use of selective insecticides. Avoidance of early season applications of insecticides is also encouraged. Certain classes of insecticides, particularly pyrethroids, are especially disruptive (e.g., Rock, 1979). If chemicals are applied for mite suppression, thorough coverage of the plants is essential (Hagel and Landis, 1972). Frequent application of insecticides has led to many cases of resistance among spider mites. Resistance has not developed to oils, but oils are effective mostly against eggs, and frequent application leads to phytotoxicity among vegetable crops. Insecticidal soaps are similarly useful, though eggs are not entirely susceptible (Osborne and Petitt, 1985). Neither oil nor soap should be used at temperatures above 32°C. Sulfur is applied to some crops, but not to cucurbits, and not if temperature exceed 32°C.

Cultural Practices. Cultural practices have considerable influence on mite damage. Water stress disrupts the physiology of the host plant, making it more suitable for mite survival and population increase. Dry, dusty conditions also favor mite survival, because blowing dust interferes with the predators of mites more than it does the mites; the latter are partially protected beneath their silk webbing. Thus, overhead sprinkler irrigations may alleviate mite problems. Excessive nitrogen fertilization of crops also favors mite population increase. Weeds and senescent crops can be important sources of mites, as can winter legume forage crops. Plant cultivars vary considerably in resistance to mites, and in some crops this offers excellent opportunity to manage mites, whereas in others there seems to be little inherent resistance (Childs *et al.*, 1976).

Biological Control. The biological suppression of mites has been well-developed for greenhouse crops, but is infrequently practiced for annual crops grown outdoors. The predatory mite *Phytoseilus persimilis* Athia–Henriot (Acari: Phytoseiidae), is commonly used to suppress twospotted spider mite in greenhouses. Effective use of *P. persimilis* involves maintenance of a low level of pest mites so that the predators do not starve, and the distribution of pests (prey) uniformly in the greenhouse so that the predatory mites also become widely distributed. Supplemental release of predatory mites may be needed to maintain a favorable ratio of predators to prey, often between 1:6 and 1:25. The maintenance of stable predator-prey-host plant relations is not a simple task, and even seemingly benign environmental changes like variation in light intensity within the greenhouse affect stability (Nihoul, 1993). The successful use of *P. persimilis* for twospotted spider mite suppression in strawberry fields (e.g., Decou, 1994) demonstrates the potential for use in field-grown vegetable crops.

Tomato Russet Mite
Aculops lycopersici (Massee)
(Acari: Eriophyidae)

Natural History

Distribution. Tomato russet mite was first observed in Florida in 1892, but was not viewed as a serious pest in North America until about 1940 when it caused considerable damage in California. Following its appearance in California, it quickly spread eastward across the United States, attaining New York and Georgia in 1953. It is also known from Hawaii. The origin of this mite is uncertain, though not native to North America, and it is now found throughout the world in both temperate and tropical latitudes. It apparently overwinters outdoors successfully at south-ern latitudes, but in cold areas it survives only in greenhouses or is reintroduced with seedling plants shipped from the south.

Host Plants. This species feeds principally on plants in the family Solanaceae. Among vegetables, it is not only primarily a pest of tomato, but also affects eggplant, pepper, potato, and tomatillo. Other solanaceous plants that can serve as hosts include petunia; black nightshade, *Solanum nigrum*; cape gooseberry, *Physalis peruviana*; Chinese thornapple, *D. ferox*; hairy nightshade, *S. villosum*; jimson weed, *Datura stramonium*; popolo, *S. nodifolium*; small-flowered nightshade, *S. nodiflorum*; and silverleaf nightshade, *S. elaeagnifolium*. In general, the *Lycopersicon* spp. are very suitable, whereas the *Nicotiana* spp. are unsuitable. The only non-solanaceous plants known to support this mite are such *Convolvulus* species as bindweed and morningglory spp.

Natural Enemies. The principal natural enemies of tomato russet mite are predatory mites, but predatory thrips and a cecidomyiid have been noted. Among specific examples of predators found in North America are *Typhlodromus* spp. (Acari: Phytoseiidae) (Anderson, 1954); *Seiulus* sp. (Acari: Phytoseiidae), black hunter thrips, *Leptothrips mali* (Fitch) (Thysanoptera: Phlaeothripidae), an unspecified predatory gall midge (Diptera: Cecidomyiidae) (Bailey and Keifer, 1943); *Euseius concordis* (Chant) (Acari: Phytoseiidae), *Pronematus ubiquitus* (McGregor), and *Homeopronematus anconai* (Baker) (both Tydeidae) (Royalty and Perring, 1987).

Life Cycle and Description. This mite reproduces whenever weather permits; in California this tends to be May–November. Their persistence is dependent on the availability of green plants, as there is not an over-wintering form. A generation may be completed in about seven days (range 6–13 days).

Egg. The eggs are deposited among the leaf hairs or crevices, and on the stem tissue. The eggs are spherical in shape and whitish or yellowish in color. They measure about 0.45–0.60 mm in diameter. The incubation time of eggs is just two days, but 3–4 days is more common.

Nymph. The nymphs normally do not move far from the point of hatching, but tend to aggregate along the edges of leaves. Nymphs are elongate, broad anteriorly and tapering to a blunt point posteriorly. They are fairly featureless in shape, but bear two pairs of cephalothoracic legs. Long hairs are found at the tip of the abdomen. Careful examination reveals that the abdomen bears numerous minute rings. There are two instars which measure 0.09–0.10 and 0.14–0.16 mm long, respectively. Duration of the instars is 1–2 days each.

Adult. The adult form is similarly tapered or wedge-shaped in general appearance. The adults measure 0.15–0.20 mm long and 0.05 mm wide. They are yellow-orange. The abdomen is covered dorsally by plate-like ridges, the anterior portion of the body with a cephalothoracic shield. The adults also bear two pairs of cephalothoracic legs and long hairs at the tip of the abdomen. Adult females have a prereproductive period of about two days. Reproductive capacity is about 40–50 eggs, and capacity to increase is great. Bailey and Keifer (1943) estimated that with females producing about two eggs per day, one mated female could result in a population of 350 mites within 21 days and the population would double every 3–4 days. Rice and Strong (1962) indicated that life processes were favored by temperature of about 27°C and relatively low humidity (about 30%). Adult longevity is about 16–22 days. Eggs from unfertilized females develop into males; those from fertilized females develop into females.

The biology of tomato russet mite was described by Bailey and Keifer (1943), Anderson (1954), Rice and Strong (1962), and Perring and Farrar (1986); the latter is a comprehensive review. An interesting general discussion of eriophyids was given by Keifer (1946).

Damage

Mites injure the plant by puncturing the surface cells with their needle-like chelicerae. Injury usually appears first on the tomato stalk at the soil surface and spreads upwards on both stalk and leaf tissue. The surface feeding produces a bronzing or russeted appearance on both the stems and leaves. Infested leaves first curl at the leaf edges, then become dry and drop from the plant (Royalty and Perring, 1988).

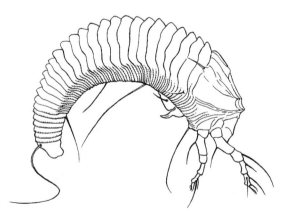

Adult tomato russet mite.

Unlike many mites, eriophyids feed readily on the upper surface of leaves and tolerate direct sunlight. Fruit and blossoms are rarely infested, but fruit may be scalded by the sun following loss of leaves. Injury is usually first observed in small patches of the field, but wind-borne dispersal of the mites soon results in widespread infestation.

Management

Sampling. These minute mites are barely visible without magnification, making field scouting difficult. Thus, most monitoring is done by watching for damage, with confirmation of mite presence accomplished microscopically. Mite infestations are usually detected when fruit are present but still small.

Insecticides. Sulfur in dust or wettable form has long been used for mite suppression, but other insecticides and acaricides also are effective (Perring and Farrar, 1986; Royalty and Perring, 1987).

Cultural Practices. Mites are readily dispersed by people and equipment, so care should be taken to minimize traffic within infested fields. Weeds and plants in greenhouses are important in overwintering. In California, early plantings are more likely to be infested than later plantings, but type of irrigation does not affect infestation levels (Zalom et al., 1986b). There is some indication of differences in susceptibility among commercially available tomato cultivars (Kamau et al., 1992).

CLASS COLLEMBOLA—SPRINGTAILS

Garden Springtail
Bourletiella hortensis Fitch
(Collembola: Sminthuridae)

Natural History

Distribution. The springtails of North America consist of both native and introduced species, though the species most commonly cited as causing injury also occur in Europe and likely are immigrant species. Probably less than 20 species are suspected of causing crop injury in North America. Among the most injurious is garden springtail, *Bourletiella hortensis* Fitch (Collembola: Sminthuridae), which accounts for most reports of injury in eastern North America. However, *Onychiurus armatus* Tullberg, *Onychiurus pseudarmatus* Folsom (both Collembola: Onychiuridae), and *Ceratophysella armata* Nicolet (Collembola: Hypogastruridae) are responsible for most reports of injury in western areas. The latter species are poorly studied.

Host Plants. Springtails feed mostly on dead plant tissue and fungal hyphae. Some also feed on other animals such as nematodes and other springtails, and a few attack growing plants. Many vegetable crops are reported to be injured, including bean, beet, broccoli, cabbage, cantaloupe, carrot, cauliflower, celery, cucumber, lettuce, lima bean, onion, parsnip, pea, potato, pumpkin, radish, spinach, squash, tomato, turnip, and watermelon. Other crops such as sugarbeet, rye, tobacco, vetch, and wheat, as well as turf grass and ornamental annual plants are occasionally attacked.

Natural Enemies. Many arthropods consume springtails, including spiders (Araneae), centipedes (Chilopoda), mites (Acari), and beetles (Coleoptera: Carabidae and Staphylinidae). Vertebrates such as toads, lizards, and birds also consume springtails. Parasitism by nematodes and disease agents including microsporidia and fungi is well-documented, but seems to be of little importance as mortality factors.

Life Cycle and Description. Springtails are primitive arthropods, considered by some to be insects and by others to be their own class of organisms. They are widespread and numerous. They may be the most abundant soil-dwelling animals, but can be outnumbered by soil-dwelling mites. Springtails bear a ventral tube (collophore) on the first abdominal segment, and many bear a jumping organ (furcula) on the third and fourth abdominal segment. It is the jumping organ and the leaping behavior that it imparts, of course, that is the basis of its common name "springtail."

Egg. The eggs may be laid singly or in clusters, and are often deposited in crevices or other sheltered locations. They are spherical, measuring about 0.2 mm in diameter. Duration of the egg stage is about 10 days.

Juvenile. Juvenile springtails closely resemble the adults, differing principally in size and the lack of sexual characters. They normally pass through 5–8 instars before attaining adulthood, with instar durations of about four days each.

Adult. There are two basic body types present among springtails. One body type is represented by *Bourletiella hortensis* Fitch and other members of the family Sminthuridae. In sminthurid springtails the first four abdominal segments are fused to form a globular mass, and overall the body is short and plump. The other basic body form is represented by the *Onychiurus* spp. and members of many common families. In these latter springtails the abdominal segments are clearly differentiated, and the body is relatively long and thin. The presence of antennae, six

abdominal segments, a collophore, a furcula, and three pairs of legs serve to distinguish springtails from the numerous other small animals, including immature insects, inhabiting soil and leaf litter. They differ greatly in color, with some species brightly colored or ornately decorated. *B. hortensis*, for example, is mottled blue to purple brown in body color, with purplish antennae, and sometimes with yellow areas. In contrast, *O. armatus* is entirely whitish. Springtails range in size from 0.5–8.0 mm long, though generally 1–3 mm, and are wingless. The period of activity is variable among springtails, with some active during the day but most active principally at night.

The adult stage continues to molt, sometimes up to 40 times, but they do not change in appearance. The duration of each instar is often about six days for young adults, and lengthen to about 10 days after instar 15. Many springtails live for more than a year, and mean fecundity is often about 400 eggs per female. Sexual dimorphism is rare, and though parthenogenesis occurs in some species, many springtails engage actively in mating. Males normally deposit their sperm externally, in small packets (spermatophores) suspended on stalks. Females either take these spermatophores up willingly, or are guided to them by males. In some species the males have grasping organs on the antennae, which are used to grasp the female's antennae and to back her into a spermatophore. (See color figure 173.)

Several pheromones are documented in springtails, including sex, alarm, and aggregation pheromones, though they are poorly known. The aggregation pheromone is probably quite important in springtail biology, as aggregation is a regular and widespread feature of their behavior. Interestingly, springtails can respond to the pheromone produced by other springtail species, but respond most strongly to pheromone produced by their own species.

Reviews of springtail biology were published by Christiansen (1964) and Hopkin (1997). A comprehensive key to American springtails was provided by Christiansen and Bellinger (1998). Other useful keys included a key to genera (Christiansen, 1990), a pictoral key to genera (Scott, 1961), and regional keys for Iowa (Mills, 1934) and New York (Maynard, 1951).

Damage

Damage by springtails is usually observed aboveground, often in the form of small holes in the young leaves, and injury that greatly resembles feeding by flea beetles (Coleoptera: Chrysomelidae). However, they also feed on stems and roots. Garden springtail tends to feed readily above-ground, whereas *Ony-*

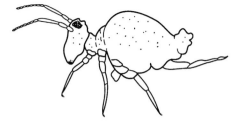

Adult garden springtail.

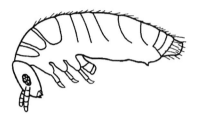

Adult *Onychiurus* springtail.

chiurus spp. feed mostly below-ground. Although injury is sometimes localized within regions of fields, sometimes entire crops are destroyed by these small but numerous arthropods (Folsom, 1933; Scott, 1964; Getzin, 1985).

Some of the damage caused by springtails is offset by their beneficial fungus-feeding activities. They feed on plant pathogenic fungi and some have suggested that they could be used to suppress fungi affecting plant roots of container-grown crops (Hopkin, 1997). In natural environments they are also quite important in decomposition of organic matter, increasing the rates at which nutrients are released from decomposing vegetation.

Management

Sampling. Springtail populations can be assessed by visual observation, pitfall traps, or by soil sampling. In the latter case, springtails are usually separated from the soil by driving them from the soil with heat.

Insecticides. Insecticide treatments normally are recommended only in fields that have a history of problems or where cultural practices increase the possibility of injury (cultural practices). Soil treatments, usually applied as granular or liquid formulation in the furrow or in a band over the row, are common practices. Seed treatments also can be effective, but only with certain insecticides (Getzin, 1985). Foliar applications can be effective for some species, but are not often used unless prolonged cool weather is hampering seedling growth.

Cultural Practices. Soil high in organic matter is favorable to springtail population growth, and such fields are more likely to experience injury. Injury also follows application of organic matter, minimum tillage practices favoring retention or crop residue, and conversion of uncultivated fields to crop fields. Also favoring injury are furrow irrigation and soils that tend to produce deep soil cracks in response to drying, because cracks afford good shelter during periods of drought (Scott, 1964).

CLASS DIPLOPODA—MILLIPEDES

Garden Millipede
Oxidus gracilis Koch
(Diplopoda: Paradoxosomatidae)

Natural History

Distribution. Several species of millipedes are reported to be injurious, or present in high enough numbers to be considered a nuisance. Probably, the most important is garden millipede, *Oxidus gracilis* Koch (Diplopoda: Paradoxosomatidae), which is abundant in southern and western states, and in greenhouses elsewhere. It apparently was accidentally introduced from the tropics, though it may have arrived circuitously via Europe, where it has also become a greenhouse pest.

Among other common millipedes that sometimes may be pests are *Diploiulus caeruleocinctus* (Wood) and *Diploiulus latistriatus* (Curtis) (both Diplopoda: Julidae), which occur in the eastern and midwestern United States and adjacent areas of Canada, though the distribution of *D. latistriatus* also extends west to Washington and British Columbia. Both are immigrants from Europe. Another European immigrant species is *Polydesmus inconstans* Latzel (Diplopoda: Polydesmidae), which is similarly found across the North American continent along the United States–Canada border. North America also has numerous native species that may become abundant locally, but they rarely are damaging. Millipedes are poorly known animals, with almost all of our information based on studies in northern Europe.

Host Plants. Millipedes normally eat dead plant material, usually in the form of leaf litter. However, they occasionally graze on roots and shoots of seedlings, algae, and dead arthropods and molluscs. They are selective in their consumption of leaf litter, preferring some leaves over others. They also tend to wait until leaves have aged, and are partially degraded by bacteria and fungi. Thus, they function principally as decomposers, hastening the break-up of leaf material into smaller pieces, and incorporating the organic matter into the soil. Whether they derive most of their nutritional requirements from the organic substrate or the microorganisms developing on the substrate is uncertain. Millipedes also tend to consume their own feces, and many species fare poorly if deprived of this food source.

Many vegetable crops have been reported to be injured by millipedes, including bean, cabbage, carrot, cauliflower, corn, cucumber, lettuce, parsnip, pea, potato, radish, tomato, and turnip. In addition, annual flower crops such as coleus, geranium, sweet pea, as well as other greenhouse-grown plants may be damaged.

Natural Enemies. Despite the formidable chemical defenses of millipedes, several natural enemies are known. Small vertebrate predators such as shrews, frogs, and lizards eat millipedes. Invertebrate predators such as scorpions (Arachnida), ground beetles (Coleoptera: Carabidae), and rove beetles (Coleoptera: Staphylinidae) also consume millipedes, though ants are usually deterred. Various disease-causing agents such as fungi, iridoviruses, rickettsia, and protozoa are documented from millipedes, though they seem to be only sporadically effective. Although flies have been reared from millipedes in Europe, they seem not to be important parasitoids in North America.

Life Cycle and Description. The name "millipede" suggests that these animals have 1000 legs, and though the presence of numerous legs is a characteristic of this group, none have more than 375 pairs and most have considerably fewer. In differentiating this group from the similar-appearing centipedes and symphylans, the presence of 2 pairs of legs per body segment is the key character used to identify millipedes. Symphylans also can be pests (see section on garden symphylan, *Scutigerella immaculata* (Newport)), but centipedes are carnivorous and therefore beneficial.

Millipedes are quite diverse morphologically, though they all consist of a long chain of rather uniform body segments, and lack wings. Some are rather short, and may be covered with feather or scale-like adornment. Others look greatly like woodlice, and even roll into a ball in the manner of pillbugs. Most, however, are elongate and thin in general body form. There are 3 basic body regions: the head, which bears a pair of moderately long antennae; the body, consisting of numerous leg-bearing segments, and which normally are rather cylindrical but sometimes bears prominent lateral projections; and the telson, or posterior body segments bearing the anus. The integument is very hard.

The life cycle of millipedes is often long. Many live for a year, but some persist for 2–4 years before attaining maturity. In a *Julus* sp. studied in a temperate environment, oviposition took place in April, with

instar 6–7 attained by winter and instar 8–9 by the second winter. They overwintered as instar 9–11, then mated and oviposited the following spring, their third year of life, before they perished. Following is a description that is based, as noted, mostly on garden millipede, a species with a one-year development time.

Egg. The creamy white, yellow or brownish eggs are deposited in the soil, usually in clusters. A glutinous material causes them to adhere to one another. They are nearly spherical in shape, measuring 0.35–0.4 mm wide and about 0.40 mm long. The female may create a chamber or cell for her eggs. Garden millipede deposits about 50–300 eggs in a cluster, and they can be found throughout the summer months. Duration of the egg stage is 9–10 days.

Juvenile. The first instar juvenile millipedes bear only a few segments and three pairs of legs, but body segments, legs, and ocelli are added with each molt. By counting the number of rows of ocelli and adding one, the instar can be estimated for many millipedes. Thus, fourth instar millipedes have three rows of ocelli, fifth instars have 4 rows, etc. (Hopkin and Read, 1992). Maturity is often attained after about 10 instars, but some species continue to molt as adults.

In garden millipede, most individuals develop through eight instars before they attain the adult stage. The number of pairs of legs present in the juveniles is about 3, 6, 11, 16, 22, 26, 28, and 30 for instars 1–8, respectively. Body length is about 0.5, 1.5, 3.6, 4.1, 4.8, 7.4, 12.4, and 20 mm during the corresponding instars. Development time is 1, 11–18, 13–18, 16–30, 20–38, 28–46, 42–60, during instars 1–7, with the final (8th) instar generally overwintering, though in some cases instar 7 overwinters.

Adult. The adult millipedes vary considerably in size, often measuring from 10–30 mm long, but in some species exceeding 100 mm. Their color ranges from whitish to brown and black. Their sexes are separate. The external genitalia of adult millipedes are located between the second and third pairs of legs. Some adult millipedes have the ability to molt from a sexually active adult to an intermediate stage which is not functional sexually. Parthenogenesis occurs in some species and some populations, but this is not usual. Millipedes lack a waxy cuticle and are susceptible to desiccation. They have glands, with openings usually located laterally, which secrete chemicals that are toxic and may immobilize predatory arthropods like spiders and ants.

Garden millipede is probably flattened dorsoventrally, and bears 30 or 31 pairs of legs in males

Adult garden millipede.

Millipede in coiled position.

and females, respectively. They generally measure 18.5–22.2 mm long and 2.0–2.5 mm wide. Initially, they are light brown but gradually attain a dark brown coloration, and sometimes are bordered with yellow. Adults seem to live for about 2 months in the spring or summer, and like all millipedes, they are intolerant to dry conditions. Garden millipede is nocturnal.

The biology on millipedes was reviewed by Hopkin and Read (1992) and Blower (1974), and a good synopsis including keys to North American families provided by Hoffman (1990). Chamberlin and Hoffman (1958) gave an annotated list of North American millipede species. Causey (1943) presented the life history of garden millipede.

Damage

Millipedes occasionally damage seedlings of vegetable crops, but generally they are not primary pests. Destruction of seedling carrot, lettuce, and tomato was observed in Pennsylvania, for example, in both spring and autumn crops grown under shadecloth (Horsfall and Eyer, 1921). Reports of injury to some crops, such as damage to potato tubers in New York (MacLean and Butcher, 1934) were subsequently shown to be related to the presence of potato scab fungus (Butcher, 1936), with millipedes feeding selectively on scab-infested tissue but not on healthy potato tubers. Once they commence feeding on the tuber, however, damage can be substantial. Indeed, in most cases of substantial injury, feeding by millipedes has followed damage by another insect or infection with a plant disease. Damage usually occurs within the context of cool weather and soil rich in organic matter.

Millipedes also burrow into fruits of such crops as cucumber, melon, squash, and tomato where the fruit comes in contact with the soil. Similarly, millipedes are sometime found feeding within the heads of cabbage, cauliflower, and lettuce.

Millipedes sometimes are viewed as a severe nuisance as a result of exceptional abundance in an inappropriate location such as in yards, homes, or commercial or food processing facilities. Millipedes can exist in tremendous quantities in the soil and become a problem only when they come to the surface and disperse as a group. This often occurs following abnormally large rainfalls (O'Neill and Reichle 1970), though hot and dry conditions are also sometimes suspected to be a stimulus for dispersal (Appel, 1988).

Management

Insecticides. Liquid, dust, and granule formulations of residual insecticides have been used with success to kill millipedes by application to soil, though not all products work equally well. On occasion, bait formulation consisting of wheat bran, molasses, and toxicant has been recommended, but this does not consistently meet with success. Barrier treatments of insecticides are sometimes recommended to disrupt the movement of millipedes from breeding areas (Rust and Reierson, 1977).

Cultural Practices. Damage by millipedes often occurs when seedlings are grown under cool, early spring conditions or within cold frames or greenhouses where growth is not optimal. Delay in planting until weather is more favorable for plant growth is often recommended as a means to minimize injury by this, and other, seedling pests. Heavy application of manure or other organic matter also predisposes seedlings to injury by millipedes and several other pests because they are attracted to organic matter, their principal food source. Conditions that favor high ambient humidity, including coarse mulch, vegetative groundcover, and thick weeds, favor survival and activity of millipedes, so these should be removed (Appel, 1988).

CLASS ISOPODA—PILLBUGS AND SOWBUGS

Common Pillbug
Armadillidium vulgare (Latreille)
(Isopoda: Armadillidiidae)

Dooryard Sowbug
Porcellio scaber Latreille
(Isopoda: Porcellionidae)

Natural History

Distribution. Common pillbug, *Armadillidium vulgare* (Latreille), and dooryard sowbug, *Porcellio scaber* Latreille, are found throughout North America. They are cosmopolitan species, apparently originating in Europe, but now found throughout the world. They have adapted well to humans and human habitations, and are often considered to be anthropophilic, but they also survive well in forests and grasslands, particularly if they can find shelter beneath logs and rocks.

The pillbugs and sowbugs are collectively known as woodlice in Europe. This term conveniently depicts their relatedness and preferred habitat, and deserves wider recognition and use in North America. There are numerous species of woodlice in North America, though many seem to be immigrants from Europe. For example, Hatchett (1947) reported 10 species of woodlice in Michigan, Drummond (1965) reported 28 species from Florida, and Hatch (1949) found 13 species dwelling within greenhouses in the northwest. In addition to common pillbug, *A. vulgare* (Latreille), and dooryard sowbug, *P. scaber* Latreille, other common and potentially important woodlice include *A. nasatum* Budde–Lund (Isopoda: Armadillidiidae); *Oniscus asellus* Linneaus (Isopoda: Oniscidae); *P. laevis* Latreille, *P. dilatatus* Brandt, *Metoponorthus pruinosus* (Brandt), and *Trachelipus rathkei* (Brandt) (all Isopoda: Porcellionidae). (See color figures 184 and 185.)

Host Plants. Pillbugs and sowbugs generally feed on dead plant material, though they also accept dead animal remains and dung, and occasionally ingest bacteria, fungi, and living plants. They are best viewed as decomposers, similar to earthworms, breaking down plant material and mixing it with mineral particles to produce soil. However, they are not the first organisms to attack leaf litter, waiting until microorganisms have begun the degradation process. Also, they sometimes have the unfortunate habit of grazing on plants, particularly seedlings. Pierce (1907) reported damage to seedlings of bean, cowpea, pea, radish, lettuce, mustard, tomato, and other crops in Texas, particularly by *A. vulgare*. In a study of *A. vulgare* in natural California grasslands, pillbugs consumed green leaves of milk thistle, *Silybum marianum*, and ox tongue, *Picris echioides*, though dead leaves of both plants were greatly preferred (Paris, 1963). In Britain, the leaves, stems, and fruit of cucumber were reportedly damaged by *A. nasatum* (Goats, 1985).

Natural Enemies. Woodlice in North America and elsewhere are parasitized by *Melanophora roralis* (Linnaeus) (Diptera: Tachinidae). *Porcellio scaber* is especially often attacked, with levels of up to 10% parasitism observed. Several other species of tachinids are known from Europe, with most being fairly specific in host range. Predation and cannibalism are known to occur, but it is uncertain whether these are important mortality factors in nature. Lizards, salamanders, shrews, spiders, centipedes, and ground beetles (Coleoptera: Carabidae) eat woodlice. An iridovirus has been found to occur in woodlice populations in California. In addition to causing a slight blue to purple discoloration in infected woodlice, the longevity of infected hosts is greatly reduced when woodlice are infected with iridovirus (Federici, 1980). Fungus, nematode, and protozoan parasites seem to be of little importance (Federici, 1984).

Life Cycle and Description. Woodlice belong to the class Crustacea, whereas insects are in the class Insecta. Though superficially similar to insects because they have a rigid exoskeleton and jointed appendages, there are some important differences. As in insects, the body of woodlice is divided into three major regions—the head, which bears the antennae and mouthparts; the thorax or pereion which bears the legs but never wings; and the abdomen or pleon. The head bears two pairs of antennae instead of one found in insects, but one pair of the antennae in woodlice is greatly reduced in size and therefore is not often observed. The pereion (thorax) consists of seven segments instead of three found in insects, with each segment bearing a pair of legs ventrally. The pleon (abdomen) consists of six segments, but invariably it is much smaller than the pereion. The ventral surface of the pleon bears plate-like structures, and is an important site for gas exchange. A terminal pair of tail-like appendages, called uropods, may be located at the tip of the pleon. Uropods are present in sowbugs but are absent in pillbugs. Sowbugs cannot completely roll into a ball, though pillbugs are capable of this behavior. As the pillbugs can roll into a ball, they are sometimes called "rolly-pollys."

Egg. The female woodlouse carries her eggs and young about with her in a special compartment, called the marsupium, on the underside of her body. Fertilized eggs are inserted into the marsupium where the embryos (and later the young) obtain water, oxygen and nutrients from a nutritive fluid called, appropriately, marsupial fluid. They may be up to 0.7 mm in diameter, and in some species over 100 eggs may be produced. The eggs persist for 3–4 weeks, then hatch, but the young remain in the marsupium for another 1–2 weeks before crawling out. They are only 2 mm long at this stage of development. Woodlice commonly produce offspring 1–3 times per year, with spring and autumn broods most common. Woodlice often survive for longer than a year, with longevity of 2–5 years not uncommon.

Young. Hatchett (1947), working in Michigan, found that *A. vulgare* and *P. scaber* normally produced 20–40 young, and 1–3 broods per season. Brood size was positively correlated with female size. The young are highly gregarious, and sometimes cannibalistic. Once they have left the female they molt, usually within 24 hours, acquiring a seventh pereion segment. After an additional 14 days a second molt occurs, and a seventh pair of legs is produced. Thereafter they do not change in morphology, other than to increase in size. The interval between molts is 1–2 weeks until the age of about 20 weeks, and molting continues irregularly for the remainder of their life, including adulthood. Molting occurs in two stages, with the posterior portion of the body shedding its old skin first, followed about three days later by the anterior portion.

Adult. A length of 8.5–18.0 mm is eventually attained as woodlice reach adulthood. The width of the body is about half of its length. Woodlice are somewhat flattened and elongate-oval, 7 pairs of legs and 13 body segments are apparent, and they have long, jointed antennae. Eyes are evident on the side of the head. They are brownish or grayish in general body color, though often marked with areas of black, red, orange or yellow. Hatchett (1947) reported that woodlice attained a larger size during summers when precipitation was above normal. Woodlice are nocturnal.

A key to the genera of North American woodlice was published by Muchmore (1990). The key to British woodlice published by Sutton (1972) generally applies to North American woodlice found in crop, yard, and greenhouse situations, because our damaging woodlice are of European origin. Good treatments of North American woodlice biology were given by Hatchett (1947) and Paris (1963). Sutton *et al.* (1984) provided important information on biology of common pillbug, with Nair (1984) treating dooryard sowbug. Edney (1954), Sutton (1972) and Warburg (1993) also provided good overviews of woodlice ecology and physiology. Van Name (1936) described and pictured

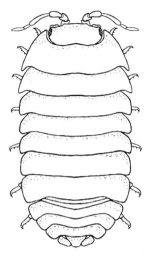

Adult common pillbug.

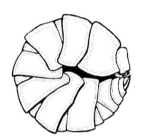

Pillbug in coiled position.

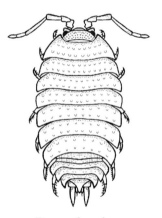

Dooryard sowbug.

each of the aforementioned species, and in conjuction with supplements (Van Name, 1940, 1942) described most other woodlice in North America. Sutton (1972) provided culture methods.

Damage

Woodlice occasionally attack seedlings aboveground, feeding especially on stems and young leaves, and below-ground, feeding on roots. Older plants also may be attacked, but this is generally of little consequence. Cool wet weather favors the activity of wood-lice but inhibits the growth of seedlings, resulting in greater likelihood of seedling damage.

Management

Woodlice can be sampled by objects that provide shelter, such as upturned-flower pots, or items of food, such as bread dough. Woodlice are most common in soils with neutral or alkaline pH, good crumb structure, high organic matter content, and where soil bacteria and other macro-decomposers such as earthworms and millipedes flourish. They tend to be absent from acid and waterlogged soil. Due either to the disturbance or lack of shelter, woodlice are virtually absent from thoroughly tilled land. On the other hand, straw or other coarse mulch provides good habitat for woodlice and can lead to crop damage. Though infrequently damaging, woodlice can be suppressed with liquid, granular, dust, and bait applications of insecticide applied to the soil around seedlings, or to protected habitats where woodlice tend to aggregate. Bait formulations developed for slugs and snails are sometimes recommended for woodlice.

CLASS GASTROPODA—SLUGS AND SNAILS

Slugs
Deroceras, Limax, Milax spp. and others
(Mollusca: Gastropoda: various families)

Natural History

Distribution. Although North America has numerous native slugs, none of the native species are serious crop pests. The slugs that damage crops have been introduced accidentally, mostly from Europe. The major seaports on the east coast of North America were known to be infested with European slugs by the mid-1800s. Slugs are easily and routinely transported long distances in commerce, then spread relatively slowly by their own means. Thus, their distribution is sometimes discontinuous, often being limited to seaports and metropolitan areas. Of the troublesome slugs, only the veronicellid slugs, which are of tropical origin, have not been introduced from Europe. Getz and Chichester (1971) gave an account of introduced species. (See color figures 186–189.)

The slugs in the family Agriolimacidae are the most widespread, and in some areas most damaging. Foremost is gray garden slug, *Deroceras reticulatum* (Müller). It is found throughout southern Canada and in the United States, though it is uncommon in the Gulf Coast region. Marsh slug, *Deroceras laeve* (Müller) is

also found widely, including the Gulf Coast area. It is thought to occur naturally in North America.

The family Limacidae also contains important species. Spotted garden slug, *Limax maximus* Linnaeus, occurs in the northern states south to Virginia and California, and in portions of southern Canada. Tawny garden slug, *Limax flavus* (Linnaeus), is widespread.

Greenhouse slug, *Milax gagates* (Draparnaud), is the only slug in the family Milacidae that is important in North America. It is quite destructive to crops in California, and elsewhere in greenhouses, but is absent from the southeast. However, the subterranean slug, *Tandonia budapestensis* (Hazay), is a relatively recent introduction to North America that may assume serious status at some time in the future.

The family Arionidae contains several slugs of importance. Black slug, *Arion ater rufus* (Linnaeus) is common in southern Canada, the northernmost areas of the United States, and in California. Banded slug, *Arion circumscriptus* Johnston, and garden slug, *Arion hortensis* Férussac, are pests in gardens and greenhouses, particularly in California. Among the Arionidae, *Arion subfuscus* (Draparnaud) is the most widespread in the United States, and can cause damage in many habitats.

Slugs in the family Veronicellidae usually are of little importance in North America, though quite damaging in more tropical locations. In Hawaii, *Laevicaulis alte* (Férussac), introduced from Australia or Southeast Asia but originating in Africa, is a pest of gardens. A native species, *Leidyula floridana* (Leidy), affects home gardens in Gulf Coast states and the Caribbean.

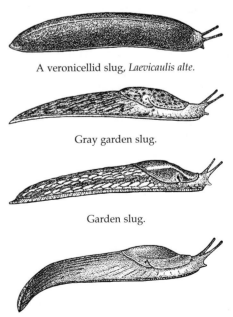

A veronicellid slug, *Laevicaulis alte.*

Gray garden slug.

Garden slug.

Greenhouse slug.

Host Plants. Slugs eat plant tissue, decaying organic matter, and occasionally animal tissue. Although they seemingly feed on a wide range of plants, their pattern of preferences is much like that of insects, and often is related to the allelochemicals found in the plants. Among the vegetable crops injured frequently are bean, beet, Brussels sprouts, cabbage, carrot, cauliflower, celery, corn, cucumber, eggplant, lettuce, lima bean, melon, pea, potato, radish, tomato, and turnip. Fruits such as strawberry, and to a lesser extent currant and gooseberry, may be injured. Slugs also consume barley, clover, oat, wheat and other field crops, but unlike the situation in Europe, they are not normally considered to be serious pests of these crops in North America.

Slugs are opportunistic feeders. In addition to crops, slugs eat flowers, weeds, grasses and other materials. Pallant (1972) studied food preferences of *Deroceras reticulatum* in a British grassland, and found it to feed principally on the dominant grass species. However, it was discriminatory, not feeding on plants solely on the basis of availability. Also, it fed on mites, collembolans, and earthworms. Fox and Landis (1973) reported predation by *D. laeve* on aphids and moth eggs.

Natural Enemies. Several invertebrates are potentially important natural enemies of slugs (Stephenson and Knutson, 1966; Godan, 1983). *Tetanocera* spp. marsh flies (Diptera: Sciomyzidae) paralyze and parasitize slugs. Several beetles, including ground beetles (Coleoptera: Carabidae), rove beetles (Coleoptera: Staphylinidae), and larval fireflies (Coleoptera: Lampyridae) are thought to be important.

Protozoan pathogens of slugs, particularly *Tetrahymena* spp., reduce longevity and fecundity, and may cause death. Field studies frequently demonstrate about 10–50% infection, and sometimes field colonies are eliminated by these pathogens. Fungi affect slugs, especially their eggs, but this is believed to be mostly an artifact of laboratory culture, and not very important under field conditions. Numerous helminths are associated with slugs, including trematodes, cestodes, and nematodes; some helminths are parasitic on slugs, but many use slugs as intermediate hosts and eventually infect livestock and humans.

Vertebrates also consume slugs, particularly birds and toads. Domestic poultry is sometimes suggested as a means of reducing slug abundance.

Life Cycle and Description. Slugs are molluscs, but unlike the closely related snails, their shell is vestigial and internal. Their eyes are rounded knobs borne on filamentous stalks that may be retracted into the head. Slugs secrete from pores in the body, and especially from the anterior ventral surface of the body, a slimy secretion called mucus. Located behind the head is a swelling, or mantle, which contains the thin shell and a respiratory opening or breathing pore.

Slugs have phases of growth that are marked by changes in their reproductive status, but often are not indicated outwardly other than by a gradual increase in size. The first phase, or infantile period, marks the time from egg hatch to differentiation of the sexual organs. In the second phase, or juvenile period, the male genital tract develops to sexual maturity. In the third phase, or adult period, sperm are produced and the slug is ready for copulation. Later in this period the female genital tract develops and eggs are produced. In the final phase, or senile period, the slug senesces and the genital tracts deteriorate.

Unlike insects, slugs are hermaphrodites, each individual possessing both male and female sexual organs. Nevertheless, self-fertilization is rare, and slugs normally pair for copulation. The mating process normally requires only a few minutes, followed by a period of several days to weeks before oviposition commences. Eggs are deposited in small cavities in the soil, usually in small clusters. They are spherical or elliptical. In the case of *Deroceras reticulatum*, they are translucent, bluish white, and slightly iridescent. They measure 2.0–3.0 mm long and 1.7–2.0 mm wide, and sometimes bear a nipple-like extension. The clusters of eggs are held together with a transparent secretion. Young slugs initially are transparent, but soon take on a pinkish color, and then become darker as they feed. Individuals slugs produce about 100–200 eggs during their life span.

Slugs attain reproductive maturity within a few months, but they continue to grow in size and continue to reproduce, sometimes for a year or more. Longevity varies greatly among slugs, with *D. reticulatum* living 9–13 months, *Arion* spp. living 7.5–12 months, and *Limax* spp. living 24–36 months.

Slugs have a protracted period of development, and development ceases in the absence of favorable environmental conditions. Thus, discrete generations are not present, with all stages present during much of the year. Up to three overlapping generations may be present simultaneously. In England, Hunter (1968) reported that *D. reticulatum* had two generations per year, *A. hortensis* had one generation annually, and *Tandonia budapestensis* required two years for a single generation. Others have observed different patterns of abundance. In Ontario, for example, Rollo and Ellis (1974) observed only one generation of *D. reticulatum* annually in corn fields, with overwintering in the egg stage, and egg production only in the autumn. During the summer months slug populations appear to be low, but adults emerge following heavy rains, suggesting that they disperse deep into the soil in response to dry conditions. In the same region but in uncultivated areas where vegetation holds an insulating layer of snow, all stages overwinter.

Slugs are not well-adapted to life on land, surviving only where water is fairly abundant. Gray garden slug, for example, has a body water content of over 80%, and its eggs of over 85%. Loss in body water content of 20% results in death. Water is replenished both by drinking and by direct absorption of water through the skin. Thus, cool and wet environments favor existence of slugs. Feeding and dispersal are limited principally to periods of darkness, though the period of peak activity varies among species, and some remain active into the morning hours. Slugs may also be active during daylight hours, if it is cloudy or foggy. Some slugs avoid the soil surface, and remain belowground. Although slugs can dig through the soil, their activity and damage potential are greatly favored by natural spaces in the soil, as results when heavy soil is plowed and left in large aggregates.

Comprehensive treatment of pest slugs was presented by Godan (1983) and Port and Port (1986). A key to common slugs is given in Appendix A. More complete keys were given by Kono and Papp (1977), Stange (1978), and Godan (1983). Burch and Pearce (1990) published a key to North American genera of slugs and snails. Barnes and Weil (1944) provided brief description of, and a key to, British slugs. Useful accounts of *D. reticulatum* and *L. maximus* were given by Hawley (1922a) and Barker and McGhie (1984), respectively. Lovett and Black (1920) described slugs in Oregon. Other good sources of information on slugs were provided by Henderson (1989, 1996). Godan (1983) discussed the elements of slug culture.

Damage

Slugs may feed on plant tissue, causing serious defoliation injury to leafy vegetables such as lettuce, and to seedlings of many vegetables. Seedlings are often cut off at the soil surface by slugs. They also feed below-ground on roots and tubers of beet, carrot, potato, radish, and turnip, with potato most frequently injured. Slugs usually feed on the surface of potato tubers and tomato and eggplant fruits, but sometimes dig deeply into the flesh of these vegetables. Contamination of vegetables with slugs, or with their slime trails, can be as damaging as defoliation in commercial vegetable production. Slugs can also serve as vectors of bacterial and fungal diseases of plants (Dawkins *et al.*, 1985; 1986). Although there is a direct relationship between slug abundance and crop damage, crops vary greatly in their susceptibility to damage (Barratt *et al.*, 1994).

Management

Sampling. Direct assessment of population density is difficult because slugs are nocturnal, and usually hide below-ground during daylight hours. Damage assessment is a useful approach for plants that have already emerged from the soil, but is useless for assessment of slug populations before plant emergence. Soil-washing techniques can be used to separate slugs from soil (Rollo and Ellis, 1974), but this approach is very labor intensive. Therefore, traps are usually used to census populations. The most common type of trap uses a board or inverted flower pot bottom as an attractive shelter under which slugs can aggregate during the day. Catches can be increased by baiting the trap with grain or such grain products as bran. The bait can also be poisoned to ensure that the slugs will not disperse. Beer is highly attractive to slugs, and they often enter containers with beer and drown (Smith and Boswell, 1970). The efficiency of traps suffers from dependency on slug mobility, which is influenced by weather. For maximum slug activity, soil temperatures at a depth of 30 cm should exceed 7°C, air temperature should attain at least 9°C, and soil moisture content should be at least 26% (Young *et al.*, 1993). The accuracy of traps for estimating slug density has been questioned (Young, 1990), but it remains the most convenient and widely used method.

Pesticides. Few pesticides cause consistent and high levels of slug mortality. Metaldehyde and methiocarb (mesurol), and to a lesser extent other carbamate insecticides, are usually recommended (Airey, 1986). Metaldehyde is the traditional favorite, but after temporary immobilization slugs may recover from low or moderate levels of exposure to metaldehyde, especially in the presence of moisture (Cragg and Vincent, 1952). Although these pesticides may be applied as sprays and dusts (Howitt and Cole, 1962), they are usually formulated as baits (Prystupa *et al.*, 1987; Hammond *et al.*, 1996). Dispersal of pesticide on bait is more effective than dispersal without bait (Barnes and Weil, 1942). Slug species vary in their propensity to accept bait (Airey, 1986), though this variable is usually ignored. Pesticides are not completely satisfactory because effectiveness varies with weather, and nontarget organisms may be affected. Slugs are not very mobile, so broadcast applications are most effective. Application of pesticide to seed has been evaluated and can be effective, but the hazard to birds is high.

Cultural Practices. Slugs are a greater problem on heavy soil, though in the presence of irrigation slug populations can increase on light soils also. Slug problems are greatest if soil remains in large aggregates, which provide slugs with shelter. If soil is finely tilled or compacted, it reduces the amount of shelter available to slugs, and plant seedling survival is higher (Hunter, 1967; Stephenson, 1975).

Sanitation also is important. In the home garden, it is important to eliminate boards, stones, debris, and weedy areas that may provide shelter for slugs. In commercial agriculture, residual organic matter contributes greatly to slug problems by providing both food and shelter. Thus, mulching and minimum tillage practices exacerbate slug problems (Tonhasca and Stinner, 1991; Barratt *et al.*, 1994; Hammond *et al.*, 1996, 1999). Planting crops into former pastureland can also be hazardous.

Other cultural practices can alleviate slug problems. The use of drip irrigation rather than overhead sprinkling systems lowers the ambient humidity and makes an area less suitable for slugs. For planting beds or particularly prized specimen plants, copper screen or metal flashing can be used as a barrier. Copper reacts with the mucus produced by slugs, producing an electric charge. The use of row covers to protect crops from insects have received considerable study, and in some cases may protect plants against slugs. However, slugs can burrow through soil and penetrate such barriers, perhaps explaining the failure of covers to protect vegetables in New Zealand (Evans *et al.*, 1997). Slugs are reluctant to cross barriers of diatomaceous earth, but will do so under wet conditions. Bordeaux mixture, a copper sulfate and lime mixture, and copper sulfate alone deters slugs, but plant toxicity may occur with copper sulfate.

Biological Control. Although there are numerous microbial, invertebrate, and vertebrate natural enemies of slugs, few have shown much promise for manipulation to achieve biological suppression. One exception seems to be the nematode *Phasmarhabditis hermaphrodita* (Nematoda: Rhabditidae). This nematode kills slugs within 1–3 weeks of application, kills a wide range of species, and can be cultured on artificial medium (Wilson *et al.*, 1993a,b; 1995). A ground beetle, *Abax parallelepidedus* Piller and Mitterpacher (Coleoptera: Carabidae), feeds readily on slugs and is under investigation in Europe for suppression of slugs in the field and in enclosures; it also seems amenable to mass culture (Asteraki, 1993; Symondson, 1994).

Snails
Cepaea, Helix, Rumina spp. and others
(Mollusca: Gastropoda: various families)

Natural History

Distribution. Native snails are widespread in North America, but usually only imported snails are

of consequence as plant pests. Many of the snails are of European origin, but some originated in tropical countries. Unlike the situation with the closely related slugs, which were accidental introductions, the snails were in many cases (at least the larger species) deliberately introduced to North America to serve as a source of food. Unfortunately, such introductions were made without legal and environmental considerations, and have proved to be costly. Within North America, the redistribution of snails results mostly from movement of nursery stock.

The snails in the family Helicidae have proven to be most adaptable to North America, and most damaging to plants. The brown garden snail, *Helix* (or *Cryptomphalus*) *aspersa* Müller was introduced to California in the 1850s to serve as a source of escargot. It has adapted well to California and is very troublesome as a pest of crops and ornamentals. It also occurs now along the west coast north to British Columbia, in most southeastern states and along the east coast north to New Jersey, but has not developed the serious pest status found in California. Because it also is edible, the singing snail, *H. aperta* Born, apparently was introduced deliberately into southern California and Louisiana, where it damages vegetable and flower crops. The Roman snail, *H. pomatia* Linnaeus, a very popular edible snail in Europe, was introduced to Michigan. The milk snail, *Otala lactea* (Müller), another popular food snail, is established in most of the southern states, but has been seriously damaging only in California. The brown-lipped snail, *Cepaea nemoralis* (Linnaeus) and the white-lipped snail, *C. hortensis* (Müller), though edible were apparently introduced to northeastern North America for their ornamental value. However, some have argued that *C. hortensis* is an endemic species. The *Cepaea sp.* have proven to be more of a nuisance than a pest. The white garden snail, *Theba pisana* (Müller), was introduced to southern California some time before 1918, and proved to be very damaging. Though apparently eradicated, it occasionally reappears. Many other species have been introduced but have not developed into serious pests, or have been successfully eradicated. A history of helicid introductions to North America was given by Mead (1971a).

Family Achatinidae contains at least one snail of note. The giant African snail, *Achatina fulica* Bowdich, which is a native of Africa and has been established throughout Southeast Asia, is well-established in Hawaii, where it was introduced in the 1930s. Also, it was introduced into south Florida in 1966, and successfully eradicated, but remains a threat to southern states.

The decollate snail, *Rumina decollata* (Linnaeus), originated in the Mediterranean region, but has been introduced to many areas of the southern United States, and now is common from California to Florida. Small colonies occur as far north as Pennsylvania. It is perhaps most interesting of the "pest" snails, because it is omnivorous, and quite effective at killing small brown garden snails. It is considered to be beneficial in areas suffering from brown garden snail infestations, but is a minor pest of plants in other areas. Presently, *Rumina* is placed in the family Subulinidae.

The family Bradybaenidae contains one species of increasing importance in North America, *Bradybaena similaris* (Rang). Originally found in Asia, it has spread to the Caribbean and South America, and now occurs in Hawaii and the Gulf Coast region. However, though it feed on vegetable and ornamental plants it also feeds on algae, and is considered beneficial because it cleans the algae from leaves, twigs and bark of shrubs and trees. (See color figures 190–198.)

Host Plants. Snails tend to be omnivorous, consuming decaying organic matter, foliage, and fruits. Not all vegetation is consumed equally, and preferences vary among snail species. Among the vegetables known to be severely injured by snails are bean, beet, cauliflower, chicory, cucumber, endive, pea, pumpkin, and spinach. Flowers and other ornamental plants are

Decollate snail.

Brown garden snail.

A bradybaenid snail, *Bradybaena similaris*.

frequent hosts, and lawn grasses are commonly fed upon by brown garden snail.

Natural Enemies. The natural enemies of snails are substantially the same as those affecting slugs (see section on slugs). However, predatory snails also can be important mortality factors affecting phytophagous snails. The rosy predator snail, *Euglandina rosea* (Férussac), occurs in the southeastern states, and has been redistributed to Hawaii and other countries where giant African snail occurs. The decollate snail, *R. decollata*, is more widespread and effective as a snail predator, but as noted earlier, also can feed on plants and native snail species.

Life Cycle and Description. Snails are typical molluscs, dwelling in a shell that consists principally of calcium carbonate. The shell grows in size as the snail matures by deposition of more shell at the mouth of the shell opening. Snails secrete an acidic substance from their foot which dissolves calciferous materials and allows them to extract calcium from the soil, though they also obtain calcium by direct ingestion. Lack of calcium in the environment limits the health and growth of snails. The fleshy body tissue may be withdrawn into the shell for protection, but part is extruded during normal bodily functions like eating and moving. The eyes are at the tips of filamentous tentacles which can be extruded or retracted back into the body.

Snails are better able to withstand long periods of dryness than of slugs because they can withdraw into their shell. They may withdraw and become dormant for months during periods of dryness. They may secrete several layers of dried mucus over the opening of the shell to aid water retention during periods of inactivity. The body also secretes slimy mucus which aids the snail in movement. Snails are active mostly at night, and may hide during daylight hours, often beneath rocks, refuse, or dense vegetation. These sites also are preferred for overwintering. Some species, however, remain in the open and attached to their food plant or another substrate even during the day, secure because they are protected by their hard shell. Most snails, like slugs, are hermaphroditic, containing the sexual organs of both sexes within a single body. However, they normally mate with another snail, and self-fertilization is rare. Snails deposit eggs in nest holes that they dig into loose soil.

The following account of brown garden snail by Basinger (1931) serves to illustrate the basics of snail biology. Mating requires 4–12 hours. Though initiated at night, snails are often observed copulating in the morning due to the duration of fertilization. Both individuals deposit eggs, beginning 3–6 days after mating. Damp, loose soil is preferred for oviposition. The snail uses its foot to shovel soil upwards, digging a hole about 2–3 cm deep and 1–2 cm in diameter. The eggs are deposited singly but adhere to one another, forming a loose mass. The number of eggs deposited at one time varies from about 30–120, averaging 86. The snail fills the nest hole with soil after oviposition and deposits excrement atop the hole. The entire oviposition process may require 24 h for completion. Eggs may be deposited as often as monthly under ideal environmental conditions, but no eggs are deposited during the winter months even in California. Thus, total fecundity is estimated at 400–450 eggs annually.

The eggs hatch in about two weeks. The young snails are miniatures of the adult form, but measure only about 5 mm in diameter. Two years may be required for the snail to reach maturity. Snails add calcareous material to the opening of the shell continuously; as the shell is enlarged, it spirals around the old shell tissue. Brown garden snails attain a diameter of 16–20 mm within one year, but 26–33 mm by the second year. In addition to becoming reproductively dormant during cold weather, they also are inactive during dry weather. The snails are a mixture of grayish yellow and brown, with the brown concentrated into spiral bands. The adult typically possesses 4–5 whorls, and may attain a maximum diameter of 38 mm.

Brown garden snails are nocturnal. They feed on organic matter in the soil, bark from trees, and especially on vegetation. Nearly anything growing in a vegetable or flower garden can be consumed. They normally feed only within the temperature range of 5–21°C.

Comprehensive treatment of pest snails, including their culture, was presented by Godan (1983). A key to common snails is found in Appendix A; more complete keys were provided by Burch (1960), Kono and Papp (1977), and Godan (1983). Burch and Pearce (1990) published a key to North American genera of slugs and snails. Useful discussion of helicid snails in California was given by Gammon (1943); the helicids *Achatina* and *Rumina* were treated by Mead (1971a,b). An interesting treatment of white garden snail and the eradication campaign developed for it were supplied by Basinger (1927). Common predatory snails in Florida were described by Auffenberg and Stange (1986). *Rumina decollata* was considered by Fisher and Orth (1985).

Damage

The damage caused by snails is very similar to that inflicted by slugs, though in most of North America they are much less damaging. Snails may feed on plant tissue, chewing irregular holes in leaves and causing serious injury to leafy vegetables such as lettuce, and to seedlings of many vegetables. Seedlings are often cut off at the soil surface by slugs. Contamination of vegetables with snails, or with their slime trails, is often considered to represent damage. Commercial vegetables normally are not subject to attack by snails.

Management

The approaches to management of snails is not significantly different from that used for slugs (see section on slugs).

CLASS SYMPHYLA—SYMPHYLANS

Garden Symphylan
Scutigerella immaculata (Newport)
(Symphyla: Scutigerellidae)

Natural History

Distribution. Garden symphylan has a wide distribution in North America, and can be a pest under both greenhouse and field conditions. It is most common and damaging along the west coast, from southern British Columbia to southern California. However, at times it has been damaging in other western states such as Idaho, Utah, and Colorado, as well as most of the midwestern states and New England. Garden symphylan is infrequent in arid regions such as the desert southwest and Great Plains. It also is rare in the southeast. This species is not the only symphylan to cause plant injury, but most symphylan damage in crops is justifiably attributed to this species (Waterhouse, 1970). *S. immaculata* also occurs in Europe and North Africa,

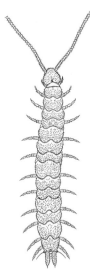

Garden symphylan.

and has been introduced into Argentina. It likely is a native to Europe.

Host Plants. Garden symphylan has a wide host range. Among vegetables fed upon are asparagus, beet, broccoli, carrot, celery, corn, cowpea, cucumber, eggplant, lettuce, lima bean, onion, parsley, pea, potato, pumpkin, radish, rhubarb, snap bean, spinach, and tomato. Not all vegetable crops are equally susceptible, with asparagus, snap bean, and lima bean commonly mentioned as damaged. Field crops injured include clover, dry beans, field corn, hop, and sugarbeet. Flowers and other ornamentals also are affected. Garden symphylan feed on yeast, algae, dead insects, and decaying vegetation.

Natural Enemies. Various centipedes (Chilopoda) have been shown to devour garden symphylan (Wymore, 1931; Waterhouse, 1969). Predatory beetles (Coleoptera: Carabidae) have been suggested to be mortality factors, but there are no data related to their effect on symphylans. Fungi have been reported to decimate symphylan populations under laboratory conditions (Getzin and Shanks, 1964), but have not been observed to cause mortality in the field.

Life Cycle and Description. Symphylans are not true insects, belonging instead to the class Symphyla in the phylum Arthropoda. They resemble centipedes, and the garden symphylan has often been called the "garden centipede." Nevertheless, they are not centipedes, which belong to the class Chilopoda, and are best thought of as insect relatives. In most respects their biology is insect-like.

Garden symphylan prefer cool soil temperature, about 18°C, but the temperature range of 12–20°C is suitable. They may be found near the soil surface, and to depths of over one meter. Symphylans retreat to greater depths in the soil as the soil warms in the summer, but migrate upwards again in the autumn. They are more likely to be found near the surface if growing plants are present, though if the environmental conditions are unfavorable, they may disperse for episodes of molting and egg laying, and return only to feed (Edwards, 1959; 1961). Symphylans apparently line the runways or channels through which they travel in the soil with a layer of fine silk. They seem unable to dig their own passageways through the soil. Rather, symphylans inhabit earthworm burrows, natural crevices, and openings in the soil left by the decay of roots. Packed soil is not favorable. Soil moisture levels affect symphylans, with optimal levels for both plants and symphylans about the same. Relative humidities of less than 95% are deleterious to survival (Waterhouse, 1968). Garden symphylans generally require about 90 days for a complete generation, but temperature may affect development rate considerably.

Egg. The eggs are deposited in tunnels in the soil. They are deposited in small clusters of 4–25, but usually 9–12 eggs. They are spherical and measure about 0.5 mm in diameter. They bear minute ridges on their surface. Initially, the eggs are white, but darken to brownish or greenish as they mature. Eggs can be found through most of the year in California, but are most abundant in the spring. At this time they hatch in 7–10 days. They were reported to hatch in 39.8, 24.9, and 12.8 days when held at 10°, 20°, and 25°C, respectively (Berry, 1972).

Immature. Young symphylans measure about 0.75 mm long, and have 10–11 dorsal body plates, six pairs of legs, and six antennal segments. The posterior segment also bears a pair of cerci, and the body is sparsely clothed with long hairs. Young symphylans remain near the site of hatching, and feed little or perhaps not at all, until the first molt. Additional pairs of legs, body segments, and antennal segments are added as the symphylans molt and grow. The immatures have seven instars. The number of antennal segments is typically about 6, 13, 15, 17, 19, 22, and 25 for instars 1–7, but the numbers are somewhat variable (Waterhouse, 1968). Duration of the first instar is only 2–3 days, but the other instars persist from 10 to 14 days. Development rates are highly variable, but Berry (1972) reported total instar development time are about 120.0, 65.8, and 40.4 days at 10°, 20°, and 25°C, respectively.

Adult. The adults vary in length from 5 mm to 8 mm. They have 12 pairs of legs and 14 body segments, and though they continue to molt after becoming adults they do not acquire additional body segments or appendages. They continue to add antennal segments up to at least 50–59 in instar 24. Young adults, however, usually bear about 25–40 bead-like antennal segments. Compound eyes are absent, and these insects depend heavily on their long antennae for orientation. The anal body segment bears a pair of pointed cerci, each producing silk. The color of garden symphylan is normally white, but is affected by the food plant on which it feeds. Thus, it acquires a pink tint when feeding on radish, red when feeding on garden beet, and brownish when feeding on decaying vegetation. Duration of the adult is probably 9–10 months but there are records of individuals living for more than four years. The sexes are difficult to distinguish; Filinger (1931) provided useful characters. Egg production is reported to cycle, with a two-month period of high egg production followed by a 3–4-month period of relatively low egg output, and this is repeated several times over the course of the symphylan life span (Berry, 1972).

Good summaries of garden symphylan biology were given by Filinger (1931) and Wymore (1931), but the most complete description was provided by Michelbacher (1938). Michelbacher (1942) provided a key to the *Scutigerella* spp. Culture methods were given by Shanks (1966) and Ramsey (1971).

Damage

Both immatures and adults feed on roots and root hairs. Symphylans chew into the large roots, creating small holes or pits, but completely consume fine roots. Warty and corky growths may appear on carrot, beet, and other vegetables that develop below-ground. Heavy infestations result in severe root pruning. Plants are stunted, grow slowly, and may wilt or even perish. When asparagus is blanched by throwing soil over the emerging shoots the asparagus stems may be damaged. Foliage is not normally injured, because symphylans do not leave the soil. However, when foliage comes in contact with soil it may be eaten. In greenhouse studies, Eltoum and Berry (1985) found that though five symphylans per young bean plant did not affect plant growth, densities of 10 or 20 symphylans disrupted the physiology and weight gain of the seedlings.

Management

Insecticides. This arthropod has proven difficult to suppress with many chemical insecticides. Fumigants applied before planting can be effective but they are costly and may kill only the individuals near the surface. Insecticides mixed into the soil at planting also can be effective (Howitt, 1959; Howitt *et al.*, 1959; Sechriest, 1972). Broadcast rather than banded treatments are preferred by some, but adequate suppression is usually attained with the banded treatments (Gesell and Hower, 1973). Dipping of transplants has been proposed as a method of insecticide delivery (Berry and Crowell, 1970).

Cultural Practices. Soil type affects symphylan abundance and damage to crops. Loam soils are more suitable for symphylans than sandy and clay soils. High organic matter, good moisture holding capacity, and the ability of the soil to form crevices favor the occurrence of symphylans (Edwards, 1958). As moist soil favors symphylans, cultivation of the soil to increase surface drying can drive the symphylans deeper into the soil and decrease feeding on plants.

Crop management also affects symphylan abundance. Historically, garden symphylans are serious greenhouse pests, but they are largely limited to greenhouses where plants were grown in the ground, or where pots were in direct contact with soil. If plants are grown in sterile soil, or not in direct contact with soil, few problems can develop. Flooding can destroy symphylans, and in California flooding for 1–2 weeks during the summer is recommended before planting the autumn crops. Asparagus is quite tolerant of flooding, and in California good control of symphylans has been attained by submerging fields for 3–4 weeks during the winter (Wymore, 1931).

There has been only limited work on identification of plant varieties resistant to garden symphylan. Simigrai and Berry (1974) observed significant differences among commercially available cultivars of broccoli, and suggested that this approach might hold some promise in other crops as well.

Beetle

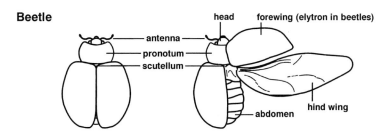

Grasshopper

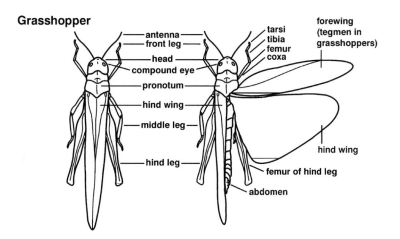

Bug

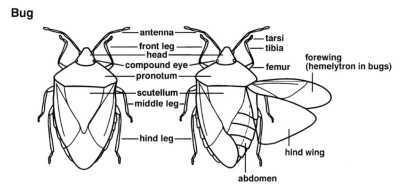

Caterpillar

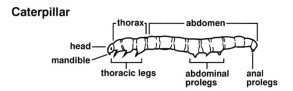

Slug

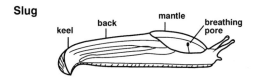

Aphid

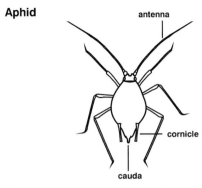

Principal features of some common types of vegetable pests.

Head and representative chewing mouthparts (grasshopper)

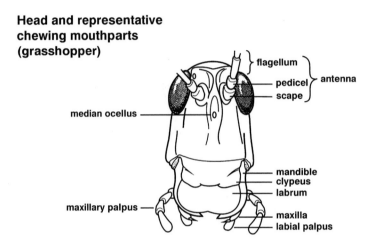

Head and representative piercing-sucking mouthparts (cicada)

Mouthparts exploded to show components

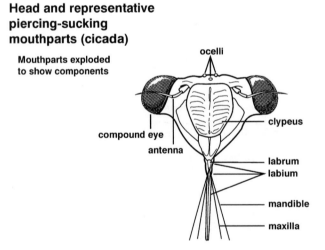

Mouthparts of representable chewing and piercing-sucking insects.

Appendix A

Other Aids for Pest Identification

KEY TO MAJOR ORDERS OF INSECTS AFFECTING VEGETABLE CROPS

1. Insect bearing wings 2
   Insect lacking wings 15
2. Front wings entirely or partly membranous, with veins evident 3

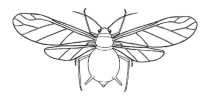

Example of insect with membranous wings and wing veins: aphid.

Front wings hard or thickened, veins often lacking 11

Example of insect with hardened front wings: scarab beetle.

3. Wings covered with scales, mouthparts long and tubular but coiled
   Butterflies and moths; order Lepidoptera
   Wings lacking scales 4
4. Only one pair of wings present, hind wings reduced to small knob-like structures
   Flies; order Diptera
   Two pairs of wings present, though hind wings may be smaller 5

5. Wings narrow and bearing long fringe around margin; body less than 5 mm long
   Thrips; order Thysanoptera

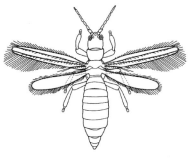

Insect with narrow wings and fringe: thrips.

Wings broad, without fringe; body greater than 5 mm long 6
6. Chewing mouthparts; tarsi with five segments
   Ants, bees, wasps, and sawflies; order Hymenoptera

Example of insect with chewing mouthparts: grasshopper

Piercing-sucking mouthparts; tarsi with 2–3 segments 7

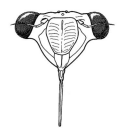

Example of insect with piercing-sucking mouthparts.

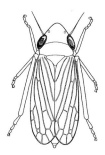

Thin-bodied insect with pointed head and wing veins: leafhopper.

7. Mouthparts arising at front of head
  Plant bugs, stink bugs, and others; order
  Hemiptera
  Mouthparts arising beneath head 8

Piercing-sucking mouth-
parts arising at front of head.

Piercing-sucking mouth-
parts arising beneath head.

8. Front wings opaque, white, with veins not
  apparent; body not exceeding 2 mm long
  Whiteflies; order Homoptera, family
  Aleyrodidae

Opaque, veinless wings: whitefly.

Front wings not opaque white; veins evident;
body size larger 9
9. Front wings appearing transparent; veins
  few or many; body robust 10

Front wings pigmented but veins evident;
body relatively long and thin; head often
tapering to blunt or acute point
  Leafhoppers and planthoppers; order
  Homoptera, families Cicadellidae and
  Delphacidae

10. Dorsal surface of posterior end of abdomen
  with short- or long-tubular projections (cornicles)
  Aphids; order Homoptera, family Aphididae

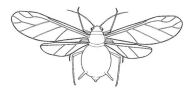

Abdomen with dorsal projections (cornicles): aphid.

Dorsal surface of abdomen lacking apical
projections; insect hops when disturbed
  Psyllids; order Homoptera, family Psyllidae
11. Tip of abdomen with forceps-like cerci
  Earwigs; order Dermaptera

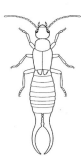

Abdomen with forceps-like cerci: earwig.

Tip of abdomen lacking cerci or cerci not
forceps-like 12
12. Mouthparts adapted for chewing 13
  Mouthparts adapted for piercing-
  sucking . . . . . . go back to 7

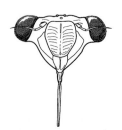

Chewing mouthparts.

Piercing-sucking mouthparts.

13. Front wings slightly thickened but bearing branched veins; hind legs appreciably larger than other legs or front legs adapted for digging 14

    Front wings thickened and often with ridges or grooves but lacking branched veins; hind legs about same size as other legs
    Beetles; order Coleoptera

Thickened front wings and legs about equal in size: beetle.

14. Antennae short, usually less than half length of body; wings often long
    Grasshoppers and mole crickets; order Orthoptera, families Acrididae and Gryllotalpidae
    Antennae long, usually extending beyond tip of abdomen; in vegetable-feeding species, front wings often are abbreviated
    Crickets; order Orthoptera, families Gryllidae and Tettigoniidae

15. Body constricted at juncture of thorax and abdomen to form narrow "waist", antennae elbowed
    Wingless ants; order Hymenoptera

Abdomen constricted and antennae elbowed: ant.

    Body not constricted at abdomen; antennae not elbowed 16

16. Antennae apparent; legs usually normal in length 17
    Antennae absent or not apparent; legs usually short or absent 24

17. Abdomen with a spring-like structure (furcula) beneath abdomen
    Springtails; class Collembola

Spring-like structure beneath abdomen: springtail.

    Abdomen lacking furcula 18

18. Body with about 10 segments and at least six pairs of legs; dwelling in soil
    Symphylans; class Symphyla

    Body not as described above; if similar, antennae not apparent 19

19. Body flattened; apparently legless; clinging to plant without moving 20
    Body not flattened; legs present or absent but insect mobile 21

20. Long waxy filaments protruding dorsally
    Immature whiteflies; order Homoptera, family Aleyrodidae

Flattened body with waxy filaments protruding: immature whitefly.

    Short waxy filaments confined to margin of body
    Immature psyllids; order Homoptera, family Psyllidae

Flattened body with waxy filaments at margin of body: immature psyllid.

21. Abdomen with a pair of tubular projections (cornicles) at posterior; body robust; often present in groups along with winged adults
    Wingless or immature aphids; order Homoptera, family Aphididae

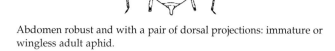

Abdomen robust and with a pair of dorsal projections: immature or wingless adult aphid.

    Abdomen without cornicles 22

22. Body narrow, minute in size, about 1 mm long; short piercing-type mouthparts; tarsi without claws
    Immature thrips; order Thysanoptera

Insect very small with narrow body, and tarsi lacking claws: immature thrips.

Mouthparts of thrips.

Body larger in size; mouthparts normally long piercing-sucking type or chewing type; tarsi with claws 23

23. Mouthparts piercing-sucking and located at front of head; antennae with 4–5 segments; cerci absent
    Immature stink bugs and plant bugs; order Hemiptera
    Mouthparts for chewing; antennae with seven or more segments; small cerci present
    Immature crickets and grasshoppers; order Orthoptera

24. Legs present 26
    Legs absent 25

25. Body lacking distinct head capsule; mouth hooks present; body usually tapers strongly toward head
    Immature flies (maggots); order Diptera

Tapered, legless body with small head: immature fly.

Body with distinct head capsule; chewing mouthparts present
    Immature weevils; order Coleoptera, families Bruchidae and Curculionidae

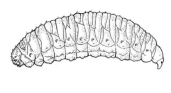

Body legless but with distinct head: immature seed beetle or weevil.

26. Body with fleshy prolegs on abdomen 27
    Body lacking prolegs, except perhaps anal appendages resembling prolegs
    Immature beetles except weevils; order Coleoptera

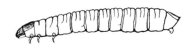

Body with legs on thorax but not prolegs on abdomen except fleshy protuberance on terminal segment: immature beetle.

27. Body with 2–5 pairs of prolegs
    Immature moths and butterflies (caterpillars); order Lepidoptera

Body with 2–5 pairs of prolegs on abdomen: immature moth.

Body with at least six pairs of prolegs
    Immature sawflies; order Hymenoptera

Body with at least six pairs of prolegs on abdomen: immature sawfly.

Note: This simple key was modified from Howard *et al.* (1994). More complete keys were found in Arnett (1985), Stehr (1987, 1991) and many introductory texts on entomology.

KEY TO COMMON STINK BUGS AFFECTING VEGETABLES

1. Base of mouthparts (beak) not closely pressed to head, and at least the basal segment of the beak thickened
   Predatory stink bugs

Base of mouthparts (beak) not closely pressed to head; beak thick: predatory stink bugs.

   Base of mouthparts pressed close to head; beak slender throughout
   Plant-feeding stinkbugs 2

Base of mouthparts (beak) pressed close to head; beak slender: plant-feeding stink bugs.

2. Scutellum large, covering most of abdomen; small bug measuring 2.2–3.5 mm long, and black in color
   Little negro bug, *Corymelaena pulicaria* (Germar)

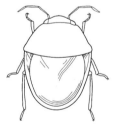

Scutellum large, covering abdomen: negro bug.

   Scutellum not covering most of abdomen; medium to large bugs measuring > 5 mm long 3
3. Color black or reddish-brown; medium-sized bug measuring 5–8 mm long; tibiae with numerous stout spines
   Burrowing bug, *Pangaeus bilineatus* (Say)
   Color not uniformly dark; generally larger than 8 mm long; lacking numerous spines on tibiae 4

4. Color black and red, or black and yellow; medium-sized bug measuring 8.0–11.5 mm long
   Harlequin bug, *Murgantia histrionica* (Hahn)
   Color not black and red, nor black and yellow; size large, usually > 12 mm 5
5. Body color predominantly green 6
   Body color predominantly brown 9
6. When viewed from below, second abdominal segment without spine extending forward between the base of the hind legs; body green with a yellow or red spot at the posterior tip of the scutellum and often additional spots along the anterior border of the scutellum; body length 13–15 mm; distribution is limited to western North America 7

Ventral surface of stink bugs. Spine absent from between base of hind legs (*left*); spine with sharp point extending forward between base of hind legs (*center*) and with dull point (*right*). Scent gland located near base of middle legs elongate (*middle*) and not elongate (*right*).

   When viewed from below, second abdominal segment with spine extending forward between the base of the hind legs; body color uniformly green; body size viable but often exceeding 15 mm; distribution is principally eastern North America 8
7. Distal region (membrane) of front wing with some purple flecks along vein; tip of scutellum and marginal areas often orange
   Say stink bug, *Chlorochroa sayi* (Stål)
   Distal region (membrane) of front wing without purple flecks along vein; tip of scutellum and marginal areas usually not orange
   Uhler stink bug, *Chlorochroa uhleri* (Stål)
8. Spine on second abdominal segment extending forward between base of hind legs with sharp point; scent gland opening located near the base of the middle pair of legs elongate, measuring at least three times as long as wide; body size 13–19 mm
   Green stink bug, *Acrosternum hilare* (Say)
   Spine on second abdominal segment with dull point; scent gland opening located near the base of the middle pair of legs not elongate, measuring about twice as long as wide; body size 14–17 mm

Southern green stink bug, *Nezara viridula* (Linnaeus)

9. Sides of pronotum generally acutely pointed 10
   Sides of pronotum rounded 11
10. Abdominal segments, when viewed from below, with slim black markings at lateral angles
    Brown stink bug, *Euschistus servus* (Say)

Ventral surface of abdomen with slim dark spots at lateral angles of segments.

Abdominal segments, when viewed from below, without black markings at lateral angles
   Onespotted stink bug, *Euschistus variolaris* (Palisot de Beauvois)

11. Dark spots at base of front wing and elsewhere discrete and relatively large, encompassing several small and darker punctures; body size moderate, measuring smaller, 11–12 mm long; distribution limited to west of Rocky Mountains
    Consperse stink bug, *Euschistus conspersus* Uhler
    Dark spots on front wing and elsewhere not large and discrete, usually limited to punctures; body size averaging larger, 11–15 mm long; widespread in North America
    Brown stink bug, *Euschistus servus* (Say)

Note: In this key "stink bug" is broadly defined to include not only Pentatomidae but the related Cydnidae and Thyreocoridae.

KEY TO COMMON ARMYWORMS AND CUTWORMS

1. Caterpillars that might be confused with armyworms and cutworms
   Caterpillar with a single large spine or "horn" at tip of abdomen
   Hornworms (Lepidoptera: Sphingidae)

Hornworm.

Caterpillar covered with dense covering of long hairs ("woollyworms")
   Woollybears and saltmarsh caterpillar (Lepidoptera: Arctiidae)

Long hairs of woollybear caterpillar.

Caterpillar producing significant amount of silk, or webbing together leaves with strands of silk
   Leaftiers and webworms (usually Lepidoptera: Pyralidae)
Caterpillar burrowing into plant stems and roots
   Borers (Lepidoptera: some Pyralidae, Noctuidae, and Sesiidae)
Caterpillar on crucifers (cabbage and related plants) usually pierid larvae (Lepidoptera: Pieridae), loopers, or diamondback moth
Caterpillar with 3–4 pairs of abdominal prolegs
   Loopers (some Lepidoptera: Noctuidae)

Looper.

Caterpillar with six or more pairs of abdominal prolegs
   Sawflies (Hymenoptera: Argidae)

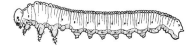

Sawfly.

Caterpillars with five pairs of prolegs on the abdomen, having a tendency to feed at or just below the soil surface (so-called subterranean cutworms) or on plant foliage (so-called climbing cutworms), and sometimes having a strong tendency to disperse in aggregations when densities are high (so-called armyworms)
   Armyworms and cutworms (Lepidoptera: Noctuidae) 2

Typical armyworm or cutworm.

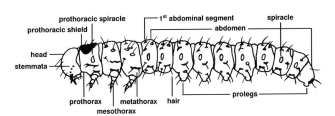

Diagram of typical caterpillar.

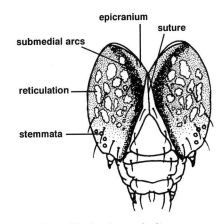

Caterpillar head capsule diagram.

2. Caterpillar with a uniform tan or pink abdomen, lacking stripes or bands 3
   Caterpillar with a dark abdomen, or with stripes or bands 6
3. Sutures on top of head bordered by distinct, dark brown bars forming submedial arcs 4
   Top of head without submedial arcs 5
4. Submedial arc in the form of a discrete narrow band; body lacking other dark markings

Pale western cutworm, *Agrotis orthogonia*
Morrison

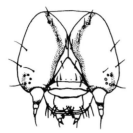

Pale western cutworm head capsule.

Submedial arc broad or dark band spreading
over most of head; body with dark flecks
forming weak stripes laterally and dorsally
   Granulate cutworm, *Agrotis subterranea*
   (Fabricius)
5. Prothoracic shield with irregular central
   dark bands
      Western bean cutworm, *Loxagrotis albicosta*
      (Smith)

Western bean cutworm prothorax.

Prothoracic shield without central dark bands
   Glassy cutworm, *Apamea devastator* (Brace)
6. Caterpillar with yellow and black transverse
   bands laterally; head reddish
      Zebra caterpillar, *Melanchra picta* (Harris)

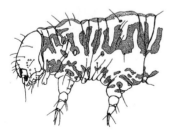

Zebra caterpillar transverse bands.

Caterpillar lacking transverse bands, though
longitudinal stripes may be present 7
7. Abdomen dark but with a row of four or
   more distinct whitish or yellowish spots
   mid-dorsally
      Variegated cutworm, *Peridroma saucia* (Hübner)

Dorsal spots on variegated caterpillar.

Abdomen lacking row of light-colored
   mid-dorsal spots 8
8. Abdomen marked dorsolaterally with paired
   black triangular spots 9
   Abdomen lacking paired triangular spots . . . 13
9. Largest triangular black spots located
   posteriorly on abdomen
      Spotted cutworm, *Amathes c-nigrum* (L.)

Dorsal spots on spotted cutworm.

Triangular black spots on abdomen fairly
   uniform in size 10
10. Abdominal segments with a prominent
    yellowish subspiracular line interrupted by
    a large dark spot on the first abdominal
    segment
       Southern armyworm, *Spodoptera eridania*
       (Cramer)

Lateral view of southern armyworm showing light-colored lateral
line interrupted by dark area.

Adominal segments without prominent yellowish
   subspiracular line or, if line is present, line extends
   without interruption through first abdominal
   segment 11
11. Dark markings found dorsolaterally on
    mesothorax are triangular in shape; triangular
    marks on abdomen are bisected by a thin white
    line
       Yellowstriped armyworm, *Spodoptera
       ornithogalli* (Guenée)

Dorsal spots of yellowstriped armyworm.

Mesothoracic markings are semicircular or trapezoidal in shape; abdominal dark marks without white line 12

12. Mesothoracic dark dorsolateral markings semicircular
   Velvet armyworm, *Spodoptera latifascia* (Walker)
   Mesothoracic dark dorsolateral markings trapezoidal
   Sweetpotato armyworm, *Spodoptera dolichos* (Fabricius)

Mesothoracic markings of sweetpotato armyworm (*left*) and velvet armyworm (*right*).

13. Head with white bands at margins of sutures, forming an inverted "V" or "Y" when viewed from the front 14
   Head lacking white bands along sutures 15

14. White bands form inverted "V" when viewed from front; dorsal surface of abdominal segment eight without dark spots
   Yellowstriped armyworm, *Spodoptera ornithogalli* (Guenée)

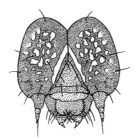

Head capsule of yellowstriped armyworm with inverted "V."

White bands form inverted "Y" when viewed from front; dorsal surface of abdominal segment eight with four dark spots
   Fall armyworm, *Spodoptera frugiperda* (J.E. Smith)

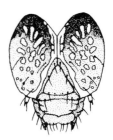

Head capsule of fall armyworm with inverted "Y."

15. Abdominal spiracles surrounded by black areas which do not unite to form a continuous black line connecting the spiracles; yellow or pink stripe below the spiracles
   Clover cutworm, *Scotogramma trifolii* (Hufnagel)

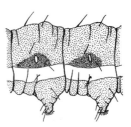

Dark pigment surrounding abdominal spiracles of clover cutworm.

Abdominal spiracles not surrounded by discontinuous black areas 16

16. Caterpillar with brick-red stripes dorsally, separated by a pale stripe
   Redbacked cutworm, *Euxoa ochrogaster* (Guenée)
   Caterpillar lacking brick-red stripes dorsally 17

17. Epicranium (sides of head) joined narrowly when viewed from above 18
   Epicranium (sides of head) joined broadly when viewed from above 23

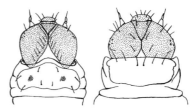

Sides of head joined narrowly, viewed from above.

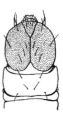

Sides of head joined broadly, viewed from above.

18. Dark markings on top of head restricted to submedial arc; prothoracic spiracle round or only slightly oval; abdomen generally uniform tan or gray
      Pale western cutworm, *Agrotis orthogonia* Morrison

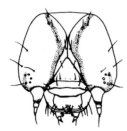

Pale western cutworm with dark marking on head restricted to sub-median arc.

    Dark markings on top of head widespread; abdomen usually dark or distinctly marked 19
19. Abdomen black (mature larva) or gray (immature), sometimes with broad brown stripe dorsally; skin with numerous coarse granules; prothoracic spiracle about twice as long as wide
      Black cutworm, *Agrotis ipsilon* (Hufnagel)
    Abdomen not black, usually indistinctly marked brown gray and tan, and often with a broad light stripe dorsally 20
20. Prothoracic spiracle oval, about 1.5 times as long as wide 21
    Prothoracic spiracle elongate, about twice as long as wide 22

Oval prothoracic spiracle.   Elongate prothoracic spiracle.

21. Caterpillar mostly gray in color, but with indistinct brown markings; lacking coarse skin texture
      Dingy cutworm, *Feltia jaculifera* (Guenée) and *F. subgothica* (Haworth)
    Caterpillar mostly yellowish-brown or sandy, but with indistinct brown markings; skin texture coarse, bearing conical projections
      Granulate cutworm, *Agrotis subterranea* (Fabricius)
22. Caterpillar mostly grayish-brown, but with a narrow wavy lateral line
      Darksided cutworm, *Euxoa messoria* (Harris)

Caterpillar generally yellowish-brown, but with a broad dark band laterally
      Army cutworm, *Euxoa auxiliaris* (Grote)
23. Top of head lacking reticulations (net-like markings); prothoracic plate dark centrally; abdomen lacking distinct stripes laterally
      Western bean cutworm, *Loxagrotis albicosta* (Smith)

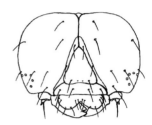

Top of western bean cutworm head, lacking reticulations.

    Top of head bearing reticulations 24
24. Abdominal segment eight bearing four raised brown or black spots dorsally 25

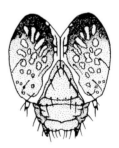

Fall armyworm head with dark pattern.

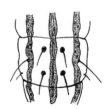

Eighth abdominal segment with four dark, raised spots.

    Abdominal segment eight lacking four brown or black spots dorsally 27
25. Microspines absent from dorsal surface of abdomen; reticulations on head create dark-colored pattern on light background
      Fall armyworm, *Spodoptera frugiperda* (J.E. Smith)
    Minute microspines present on dorsal surface of abdomen, imparting a rough feel; reticulations on head create light-colored pattern on dark background 26

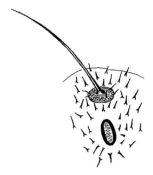

Minute microspines on surface of abdomen.

26. Small spines on the tubercles of the first, second, and eighth abdominal segments measure no more than one-fourth the height of the tubercle; a common vegetable pest, particularly on corn
Corn earworm, *Heliothis zea* (Boddie)

Small spines on tubercle of corn earworm measuring one-fourth the height of tubercle.

Small spines on the tubercles of the first, second, and eighth abdominal segments measure about half the height of the tubercle; not commonly a vegetable pest, and never on corn
Tobacco budworm, *Heliothis virescens* (Fabricius)

Small spines on tubercle of tobacco budworm measuring about one-half the height of tubercle.

27. Body color bronze, but with five distinct pale stripes (three when viewed from above), each about one-half as wide as the bronze area separating them; reticulations on head weak
Bronzed cutworm, *Nephelodes minians* (Guenée)
Body color and stripe pattern not as described above; reticulations on head distinct 28
28. Reticulations on head create light-colored pattern on dark background; body color variable but dark lateral stripe usually present; microspines present on abdomen
Corn earworm, *Heliothis zea* (Boddie)

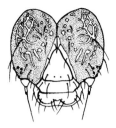

Corn earworm with light pattern on dark head capsule.

Reticulations on head create dark-colored pattern on light background; lateral stripe variable; microspines not present on abdomen 29
29. Abdominal color variable but usually with several distinct alternating dark and light stripes; spiracles dark brown or black throughout; submedial arcs bordering suture on head
Armyworm, *Pseudaletia unipunctata* (Haworth)

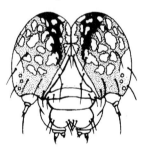

Armyworm head capsule with submedial arcs.

Beet armyworm head capsule lacking submedial arcs.

Abdominal color variable but usually lacking dark stripes; spiracles white or light brown and bordered by dark ring; suture not bordered by submedial arc; similar to clover cutworm but lacking dark areas near abdominal spiracles
Beet armyworm, *Spodoptera exigua* (Hübner)

Note: The armyworms and cutworms consist of a large and diverse number of caterpillars in the family Noctuidae. They are not easily differentiated, and nearly all keys, including this one, are based on the mature larva. Keys usually require high levels of magnification and examination of setal patterns and mouthparts; by avoiding these difficult-to-discern characters, this key sacrifices accuracy for ease-of-use. A more detailed key to noctuid larvae was provided by Stehr (1987); other detailed keys included by Crumb (1956) and Godfrey (1972).

KEY TO COMMON "CABBAGE WHITE" BUTTERFLIES

1. Upper surface of forewing with at least 3–4 (in male) or numerous (in female) dark spots centrally (do not consider pigmentation of the wing tip)

        Southern cabbageworm, *Pontia protodice* (Boisduval and LeConte)

   Upper surface of forewing with 0–2 black spots centrally 2

2. Largest of the "cabbage whites"; wingspan about 6–8 cm. May be white, tan or brown. Dark band along outer margin of forewing, or the band bearing deep indentations of white resulting in a scalloped effect. Central forewing with 0–2 spots.

        Southern white, *Ascia monuste* (Linnaeus)

   Forewing not marked with band along entire outer edge, or not deeply indented 3

3. A small butterfly (3–5 cm wingspan), usually entirely white; rarely with black spots on, or at the tip of, the forewing. Wing veins, and especially veins of hindwing when viewed from below, marked with dusky stripes

        Mustard white, *Pieris napi* (Linnaeus)

   A butterfly of medium size (4.5–6.5 cm wingspan) with black at tip of forewing, and usually with pronounced black spot(s) centrally. Veins of hindwing not distinctly darkened.

        Imported cabbageworm, *Pieris rapae* (Linnaeus)

KEY TO COMMON PIERID "CABBAGEWORMS"

1. Larva with black transverse bands crossing body (*Note*: this is not a typical pierid cabbageworm, but a pyralid that affects crucifers in the southeastern states)

        Cross-striped cabbageworm, *Evergestis rimosalis* (Guenée)

   Larva lacking transverse bands; larva either without pattern or with stripes running length of body 2

2. Larvae green and velvety in appearance 3

   Larva gray, bluish, or brownish, and not particularly velvet-like in appearance 4

3. Larva with thin yellow line down center of back, and a broken yellow line or series of yellow spots on each side

        Imported cabbageworm, *Pieris rapae* (Linnaeus)

   Larva without yellow line dorsally; stripe, if present, is green. Yellow on sides surrounding spiracles

        Mustard white, *Pieris napi* (Linnaeus)

4. Larva with four yellow stripes running length of body

        Southern cabbageworm, *Pontia protodice* (Boisduval & LeConte)

   Larva with five orange or yellow-green stripes running length of body

        Southern white, *Ascia monuste* (Linnaeus)

Note: Armyworms and cutworms, loopers, and other caterpillars also affect cabbage.

KEY TO COMMON LOOPERS AFFECTING VEGETABLES

1. Larva with 4–5 pairs of prolegs 2
   Larva with three pairs of prolegs 4
2. Larva with four pairs of prolegs 3
   Larva with five pairs of prolegs
   This is not a looper; see Armyworms and Cutworms
3. Feeding on legumes
   Green cloverworm, *Plathypea scabra* (Fabricius)
   Feeding on okra
   Okra caterpillar, *Anomis erosa* Hübner
4. Very small, nipple-like structures (vestigial prolegs) found on abdominal segments three and four 5
   Nipple-like structures absent from abdominal segments three and four 6

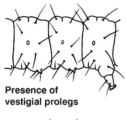

Presence of vestigial prolegs

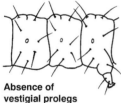

Absence of vestigial prolegs

Abdominal segments three and four with vestigial prolegs (*above*) or lacking vestigial prolegs (*below*).

5. Thoracic legs usually dark; microspines present on body, but only observable under very high magnification
   Soybean looper, *Pseudoplusia includens* (Walker)
   Thoracic legs not dark; microspines absent
   Cabbage looper, *Trichoplusia ni* (Hübner)

6. Body color predominantly brown; longitudinal stripe pattern continuing onto head capsule; large, dark transverse spots usually occur dorsally between the segments at about the mid-point of the body
   Striped grass looper, *Mocis latipes* Guenée
   Body color predominantly green; stripes not continuing onto head capsule; transverse spots absent 7
7. Microspines absent from abdominal segments 8
   Microspines present on abdominal segments 9
8. Found in eastern North America
   Plantain looper, *Autographa precationis* (Guenée)
   Found in western North America
   Alfalfa looper, *Autographa californica* (Speyer)
9. Head lacking black lateral bands, though spots may be present 10
   Head with black lateral bands extending through stemmata 11
10. Principally southern in distribution: California and Gulf Coast states; spiracles dark in color
    Bean leafskeletonizer, *Autoplusia egena* (Guenée)
    Principally northern in distribution; spiracles white, but with black rim
    Celery looper, *Anagrapha falcifera* (Kirby)
11. Thoracic legs generally black; lateral white line on abdomen is weak; dark bar on head is strong; tubercles above abdominal spiracles often black
    Bilobed looper; *Megalographa biloba* (Stephens)
    Thoracic legs pale; lateral line on abdomen bold; dark bar on head weak; abdominal tubercles not black
    Celery looper, *Anagrapha falcifera* (Kirby)

Note: This key uses readily observable characters, but lacks the precision of keys based on mouthparts and setal patterns. For greater resolution, consult Crumb (1956), Eichlin and Cunningham (1978), and LaFontaine and Poole (1991).

KEY TO COMMON STALK BORERS AFFECTING CORN

1. Larva white or whitish, often with distinct
   dark spots 2
   Larva light brown or dark brown, with bands,
   stripes or indistinct spots. 5
2. Larva bearing dark spots dorsally and laterally;
   summer form of *Diatraea*
   spp. 3
   Larva lacking distinct markings . . . overwintering
   form of *Diatraea* spp.

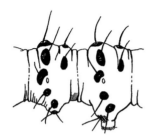

Lateral spots on southwestern corn borer embrace the spiracular openings.

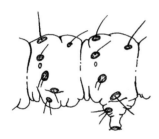

Lateral spots on other borers not embracing the spiracles.

3. Spot adjacent to spiracle on abdominal segments
   elongate and bean-shaped, partly embracing
   spiracle
     Southwestern corn borer, *Diatraea grandiosella*
     (Dyar)
   Spot adjacent to spiracle on abdominal segments
   more circular, not embracing spiracle 4
4. Dorsal spots on abdominal segments surrounded
   by light brown or pink pigmentation
     Sugarcane borer, *Diatraea saccharalis* (Fabricius)
   Dorsal spots on abdominal segments contrasting
   distinctly with background color, not surrounded
   by brown or pink pigmentation
     Southern cornstalk borer, *Diatraea crambidoides*
     (Grote)
5. Larva light brown with small, round, indistinct
   spots
     European corn borer, *Ostrinia nubilalis* (Hübner)
   Larva dark brown or gray-green with stripes
   or bands 6
6. Larva dark brown with white stripe dorsally and
   broken white stripe laterally
     Stalk borer, *Papaipema nebris* (Guenée)
   Larva greenish or gray with reddish bands (early
   instars) or white stripes (late instars)
     Lesser cornstalk borer, *Elasmopalpus lignosellus*
     (Zeller)

Note: This key is based on "typical" summer forms. Larval color changes as larvae near pupation or prepare for overwintering, and such larvae do not conform to the aforementioned descriptions. See Peterson (1948) and Stehr (1987) for more detailed keys.

GUIDE TO COMMON ADULT FLEA BEETLES
(Vegetable-feeding *Chaetocnema*, *Disonycha*, *Epitrix*, *Phyllotreta*, *Psylliodes*, and *Systena*)

Size moderate to large, usually 3.5 mm long or greater

Beetles 5–6 mm long (*Disonycha*)
    Spinach flea beetle [preferred hosts:
        Chenopodiaceae]
    Threespotted flea beetle [preferred hosts:
        Chenopodiaceae]
    Yellownecked flea beetle [preferred hosts:
        Chenopodiaceae]
Beetles 3.5–4.5 mm long (*Systena*)
    Elongate flea beetle [preferred hosts: variable]
    Palestriped flea beetle [preferred hosts: variable]
    Redheaded flea beetle [preferred hosts: variable]
    Smartweed flea beetle [preferred hosts: variable]

Size small, usually 3 mm long or less

Elytra covered with short hairs (*Epitrix*)
    Elytra color black
        Eggplant flea beetle [preferred hosts:
           Solanaceae]
        Potato flea beetle [preferred hosts: Solanaceae]
        Tuber flea beetle [preferred hosts: Solanaceae]
    Elytra color bronze, or with bronze luster
        Southern tobacco flea beetle [preferred hosts:
           Solanaceae]
        Tobacco flea beetle [preferred hosts: Solanaceae]
        Western potato flea beetle [preferred hosts:
           Solanaceae]
Elytra lacking layer of hairs
    Antennae with 10 segments (*Psylliodes*)

Hop flea beetle [preferred hosts: variable]
Antennae with 11 segments (*Phyllotreta*, *Chaetocnema*)
    Elytra color black, or blackish with yellow
    stripes or spots
        Elytra with yellow stripes or spots
           Horseradish flea beetle [preferred hosts:
               Cruciferae]
           Striped flea beetle [preferred hosts:
               Cruciferae]
           Western striped flea beetle [preferred
               hosts: Cruciferae]
           Zimmermann's flea beetle [preferred
               hosts: Cruciferae]
        Elytra lacking yellow stripes or spots
           Desert corn flea beetle [preferred hosts:
               Gramineae]
    Elytra color bronze, or with bronze, blue,
    or green luster
        Cabbage flea beetle [preferred hosts:
           Cruciferae]
        Corn flea beetle [preferred hosts:
           Gramineae]
        Crucifer flea beetle [preferred hosts:
           Cruciferae]
        Sweetpotato flea beetle [preferred hosts:
           Convolvulaceae]
        Toothed flea beetle [preferred hosts:
           Gramineae]
        Western black flea beetle [preferred hosts:
           Cruciferae]

Note: Identification of flea beetles can be achieved by using keys provided by Blatchley (1910), Chittenden (1927), Hatch (1971) and Smith (1970, 1985). An excellent source of information on flea beetle biology, including some minor pests not discussed here, was provided by Campbell *et al.* (1989).

KEY TO GENERA OF WIREWORMS COMMONLY AFFECTING VEGETABLES

1. Antennae conspicuous, with distal segments enlarged; suture on top of head is Y-shaped
   False wireworm, family Tenebrionidae
   Antennae not conspicuous, distal segments reduced; suture on top of head not Y-shaped
   Wireworm, family Elateridae 2
2. Terminal abdominal segment lacking a notch at the tip 3
   Terminal abdominal segment with small or pronounced notch at tip 4
3. Terminal abdominal segment somewhat flattened and scalloped
   Melanotus spp. (*M. communis, M. longulus oregonensis*)

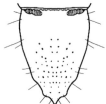

Terminal abdominal segment, Oregon wireworm.   Terminal abdominal segment, corn wireworm.

Terminal abdominal segment somewhat conical or bulbous, not flattened nor scalloped
   Agriotis spp. (*A. mancus*)

Terminal abdominal segment, wheat wireworm.

4. Notch at tip of abdomen wide, approximately one-third the width of the tip, and nearly closed; lateral margins and tip of abdomen bearing only dull or inconspicuous protuberances
   Limonius spp. (*L. agonus, L. californicus, L. canus*)

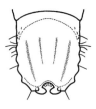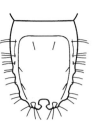

Terminal abdominal segment, sugarbeet wireworm (*left*), eastern field wireworm (*center*), Pacific Coast wireworm (*right*).

Notch at tip of abdomen not nearly closed, or if nearly closed then notch is minute; lateral margins and tip of abdomen often bearing sharply pointed protuberances 5
5. Notch at tip of abdomen small or "V"-shaped; found in southern United States
   Conoderus spp. (*C. falli, C. vespertinus, C. amplicollis*)

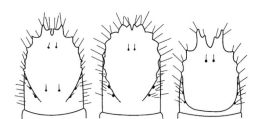

Terminal abdominal segment, southern potato (*left*), gulf (*center*), and tobacco wireworm (*right*).

Notch at tip of abdomen large, not "V"-shaped; found in northern United States and Canada
   Ctenicera spp. (*C. pruinina, C. aeripennis aeripennis, C. a. destructor, C. glauca*)

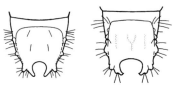

Terminal abdominal segment, Great Basin wireworm (*left*), and Puget Sound wireworm (*right*).

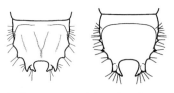

Terminal abdominal segment, prairie grain wireworm (*left*) and dryland wireworm (*right*).

Note: Keys for the identification of over 30 genera of wireworms were supplied by Becker and Dogger (1991). Wilkinson (1963) provided an excellent key to distinguish among pest wireworms found in British Columbia and adjacent areas.

KEY TO COMMON THRIPS AFFECTING VEGETABLES

1. Color grayish-black with front wings banded light and dark[a]

    Bean thrips, *Caliothrips fasciatus* (Pergande)

    Color not grayish-black, or lacking banded wings 2

2. Anterior angles of pronotum bearing stout hairs (anteroangular hairs) which are discernably larger than hairs located centrally on the pronotum 3

    Anterior angles of pronotum bearing stout hairs, which are not discernably larger than hairs found centrally on the pronotum 4

3. Anteroangular hairs nearly twice as long as the other stout hairs located along the anterior margin of the pronotum; body generally dark brown

    Tobacco thrips, *Frankliniella fusca* (Hinds)

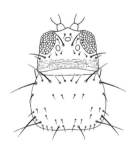

Head and thorax, tobacco thrips.

Anteroangular hairs about same size as other stout hairs located along anterior margin of pronotum; body yellow or brownish

    Western flower thrips, *Frankliniella occidentalis* (Pergande)

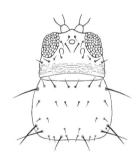

Head and thorax, western flower thrips.

4. Hairs on top of head originating within "triangle" formed by ocelli; body color yellowish-brown; found throughout North America and on many crops, but especially on onion and cabbage

    Onion thrips, *Thrips tabaci* Lindeman

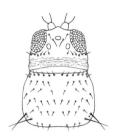

Head and thorax, onion thrips.

Hairs on top of head not originating within "triangle" formed by ocelli 5

5. Posterior margin of pronotum with stout hairs; body color yellow; hosts usually cucumbers, eggplant, and tomato; distribution limited to Hawaii, southern Florida, and the Caribbean region

    Melon thrips, *Thrips palmi* Karny

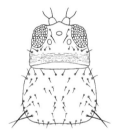

Head and thorax, melon thrips.

Posterior margin with uniformly small hairs; body color brown or pink; hosts usually corn and other grasses; found widely in North America

    Grass thrips, *Anaphothrips obscurus* (Müller)

[a]Thrips with three dark spots on each wing and a light body color are likely sixspotted thrips, *Scolothrips sexmaculatus* (Pergande). This species is predatory on phytophagous thrips, other small insects, and mites.

Note: More complete keys to thrips were supplied by Palmer *et al.* (1989), Oetting *et al.* (1993), and Mound and Marullo (1996).

KEY TO COMMON MITES AFFECTING VEGETABLES

1. Mite with only two pairs of legs; body elongate and tapering from anterior to posterior; body measuring only about 0.2 mm long; orange-yellow in color; causing bronzing of leaves and deformed tissue growth of solanaceous plants, damage beginning low in the plant and progressing upward
   Tomato russet mite, *Aculops lycopersici* (Massee)
   Mite with four pairs of legs 2

2. Mites moving rapidly in seemingly random directions; usually not abundant; not producing silk; often tan or light brown; adult body size 0.4 mm or larger
   These may be predatory mites in the family Phytoseiidae
   Mites rather sedentary or moving slowly; often abundant; sometimes producing silk; size variable 3

3. Body whitish and translucent; size variable; damage variable but not expressed as speckling; silk not apparent 4
   Body yellowish, greenish, or reddish, often with dark spots; body size usually 0.4–0.5 mm; feeding injury initially expressed as speckling, followed by bronzing of foliage; silk usually present on foliage 5

4. Small mites, adult females measuring only 0.2–0.3 mm long; feeding on foliage, primarily of solanaceous crops; foliage usually distorted and plant response variable, but speckling is not a common response to feeding

   Broad mite, *Polyphagotarsonemus latus* (Banks)
   Large mites, adult females measuring 0.5–0.8 mm or larger in length; feeding below-ground on bulbs and tubers or on injured, decaying tissue; largely a pest of onion and related crops
   Bulb mites, *Rhizoglyphus echinopus* (Fumouze and Robin) and *Rhizoglyphus robini* Claparede

5. During periods of feeding, body color reddish with dark lateral spots
   Tumid spider mite, *Tetranychus tumidus* Banks
   During periods of feeding, body color yellowish or greenish and usually with dark lateral spots 6

6. Body usually with two dark irregular spots present laterally, with dark spots covering less than one-half the length of the body and not usually extending to the posterior tip of the body
   Twospotted spider mite, *Tetranychus urticae* Koch
   Body usually with four dark spots present laterally, or spots extending to the posterior tip of the body 7

7. Feeding on broadleaf plants
   Strawberry spider mite, *Tetranychus turkestani* Ugarov and Nikolski
   Feeding on grasses, particularly corn
   Banks grass mite, *Oligonychus pratensis* (Banks)

Note: Mites are difficult to distinguish, even for experts. Final determination should be made by an authority following collection of a large sample of mites to include males of the species.

KEY TO ADULT SLUGS COMMONLY AFFECTING VEGETABLES

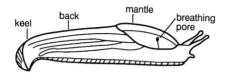

Diagram of typical slug.

1. Slug appearing to be hump-backed, bearing swollen area (mantle) immediately behind head 3
   Slug not bearing swelling behind head, mantle extending the length of the slug's body 2

2. Body blackish with a yellow stripe dorsally extending the length of the body
      Laevicaulis alte (Férussac)
   Body color brownish-gray mottled with black, and with dorsal stripe white and lateral stripes (if present) dark
      Leidyula floridana (Leidy)

3. Respiratory pore at lateral edge of mantle located anterior to mid-point of mantle; body without dorsal ridge or keel on body; body compact 4
   Respiratory pore at lateral edge of mantle located posterior to mid-point of mantle; body with dorsal ridge or keel along at least a portion of the body; bodyelongate 6

4. Larger slugs, usually 100–130 mm long; generally uniform black
      Black slug, *Arion ater rufus* (Linnaeus)
   Smaller slugs, usually 40 mm or less in length 5

5. Respiratory pore located below the dark band on the mantle; body length 30–40 mm; body mucus and sole (anterior region of ventral body surface) mucus colorless
      Banded slug, *Arion circumscriptus* Johnston

Respiratory pore located within the dark band on the mantle; body length 25–30 mm; body mucus yellow to orange, sole mucus colorless
      Garden slug, *Arion hortensis* Férussac

6. Dorsal ridge or keel extending from mantle to posterior end 7
   Dorsal ridge or keel not evident, or occurring along the posterior end of body only, not reaching mantle, or not apparent 8

7. Dorsal ridge or keel usually dark; body color variable but often yellowish brown; when viewed from below, periphery is indistinct in color from the central area; length often 45–55 mm but sometimes attaining 70 mm
      Greenhouse slug, *Milax gagates* (Draparnaud)
   Dorsal ridge or keel light, usually yellow or orange; body color variable but usually dark brown, gray, or black; when viewed from below, periphery is distinctly different in color from the central area; length 60–70 mm
      Subterranean slug, *Tandonia budapestensis* (Hazay)

8. Larger slugs, measuring 75–150 mm long 9
   Smaller slugs, measuring 25–60 mm long 10

9. Grayish or brownish with rows of darker spots or irregular stripes; mucus colorless
      Spotted slug, *Limax maximus* Linnaeus
   Brownish with irregular yellowish spots; mucus yellow
      Tawny garden slug, *Limax flavus* Linnaeus

10. Length 20–25 mm; color variable, ranging from gray to brown or brown-black, but with few or no distinct spots or markings; mucus colorless
      Marsh slug, *Deroceras laeve* (Müller)
    Length about 50 mm; color cream to gray or reddish-brown and usually well-marked with spots; mucus milky white
      Gray garden slug, *Deroceras reticulatum* (Müller)

KEY TO ADULT SNAILS COMMONLY AFFECTING VEGETABLES

1. Snail shell compact, flattened or oval, not substantially taller than maximum diameter 2
   Snail shell elongate, at least twice as tall as maximum diameter 9
2. Small snails, less than 25 mm in diameter 3
   Larger snails, exceeding 25 mm in diameter 6
3. Diameter of shell 12–16 mm in diameter; color light brown, often with a single spiral brown band; lip at shell opening turned back (reflected); presently restricted to Hawaii and the Gulf Coast region
       Bradybaena similaris (Rang)
   Diameter of shell greater than 19 mm or greater 4
4. Lip at shell opening reddish brown to almost black, and turned back (reflected); shell 22–23 mm in diameter; color variable, usually yellow, reddish, or brown, sometimes with narrow dark bands; known principally from northeastern North America and California
       Brown-lipped snail, *Cepaea nemoralis* (Linnaeus)
   Lip at shell opening white; shell 19–22 mm in diameter; color variable, usually ivory, yellow, light brown, or reddish 5
5. Lip at shell opening turned back (reflected); shell usually yellow with reddish brown bands; presently restricted to northeastern North America
       White-lipped snail, *Cepaea hortensis* (Müller)
   Lip at shell opening not turned back; shell color ivory yellow or white with highly variable brown banding pattern; presently restricted to California
       White garden snail, *Theba pisana* (Müller)
6. Shell whitish or gray, dark bands present or absent 7
   Shell yellowish or brown, distinct dark bands usually absent or indistinct 8
7. Shell 32–45 mm in diameter; color whitish gray, sometimes with light brown bands; distribution presently restricted to Michigan
       Roman snail, *Helix pomatia* Linnaeus
   Shell 28–35 mm in diameter; color entirely white, sometimes with dark brown bands; distribution throughout southern states
       Milk snail, *Otala lactea* (Müller)
8. Shell about 20–30 mm in diameter and 20 mm in height; brown bands absent; lip of shell not turned back
       Singing snail, *Helix aperta* Born
   Shell about 28–38 mm in diameter and 35 mm in height; brown bands usually present; lip of shell turned back
       Brown garden snail, *Helix* (or *Cryptomphalus*) *aspersa* Müller
9. Shell elongate, tapering gradually, and with tip absent from adult specimens; color pinkish brown; measuring up to 45 mm in height and 14 mm in diameter
       Decollate snail, *Rumina decollata* (Linnaeus)
   Shell tapering rapidly to a point; color yellowish or grayish with reddish brown transverse streaks; measuring up to 125 mm long and 60 mm in diameter
       Giant African snail, *Achatina fulica* Bowdich

Appendix B

Other Sources of Information about Vegetable Crop Pests

UNITED STATES

Alabama

Department of Entomology and Plant Pathology
Auburn University
301 Funchess Hall
Auburn University, AL 36849-5413
Tel: 205-844-5006
Fax: 205-844-5005
Web: http://www.ag.auburn.edu/dept/ent/misc/
    publications.html
    http://www.aces.edu/department/ipm/

Arizona

Department of Entomology
University of Arizona
Tucson, AZ 85721
Tel: 520-621-1151
Fax: 520-621-1150

Arkansas

Department of Entomology
University of Arkansas
Fayetteville, AR 72701
Tel: 501-575-2451
Fax: 501-575-2452
Web: http://ipm.uaex.edu/insects.htm

California

Department of Environmental Science, Policy and
    Management
Division of Insect Biology
University of California
Berkeley, CA 94720

Tel: 510-642-1603
Fax: 510-642-7428

Department of Entomology
University of California
Davis, CA 95616-8584
Tel: 916-752-0475
Fax: 916-752-1537
Web: http://ucipm.ucdavis.edu/

Department of Entomology
University of California
Riverside, CA 92521
Tel: 909-787-3718
Fax: 909-787-3086
Email: insects@ucracl.ucr.edu

Colorado

Department of Bioagricultural Sciences and Pest
    Management
Colorado State University
Fort Collins, CO 80523-1177
Tel: 970-491-5261
Fax: 970-491-3862
Email: bspm@lamar.colostate.edu
Web: http://www.colostate.edu/Depts/IPM/
    csuipm.html
    http://yuma.acns.colostate.edu/Depts/CoopExt/
    PUBS/INSECT/pubins.html

Connecticut

Department of Plant Science
University of Connecticut
438 Whitney Road Extension
Storrs, CT 06269-3043
Tel: 860-486-3438

Email: ipm@canr1.cag.uconn.edu
Web: http://www.lib.uconn.edu/CANR/ces/ipm/
  ipmveg.htm?33,24

Connecticut Agricultural Experiment Station
123 Huntington Street
P.O. Box 1106
New Haven, CT 06504
Tel: 203-974-8466
Fax: 203-974-8502
Web: http://www.curenet.org/memberprofiles/
  ctagricultural.html

Delaware

Department of Entomology and Applied Ecology
University of Delaware
Newark, DE 19717-1303
Tel: 302-831-2526
Fax: 302-831-3651
Web: http://bluehen.ags.udel.edu/deces/

District of Columbia

Department of Entomology
National Museum of Natural History, Room W308
Smithsonian Institution
Washington, DC 20560-0105
Tel: 202-357-2078
Fax: 202-786-2894

Florida

Department of Entomology and Nematology
University of Florida
P.O. Box 110620
Gainesville, FL 32611-0620
Tel: 352-392-1901
Fax: 352-392-0190
Email: entnem@gnv.ifas.ufl.edu
Web: http://www.ifas.ufl.edu/~insect/

Center for Studies in Entomology
Florida A and M University
Tallahassee, FL 32307-4100
Tel: 904-599-3912
Fax: 904-561-2221

Georgia

Department of Entomology
University of Georgia
413 Biological Sciences Building
Athens, GA 30602-2603
Tel: 706-542-2023

Fax: 706-542-2279
Web: http://www.ces.uga.edu/pubs/pubsubj.html

Department of Entomology
Coastal Plain Experiment Station
University of Georgia
Tifton, GA 31793
Tel: 912-386-3374
Fax: 912-386-3086

Hawaii

Department of Entomology
University of Hawaii at Manoa
3050 Maile Way, Gilmore 310
Honolulu, HI 96822
Tel: 808-956-7076
Fax: 808-956-2428

Hawaii Department of Agriculture
P.O. Box 22159
Honolulu, HI 96823-2159
Tel: 808-973-9560
Web: http://www.hawaiiag.org/212.htm

Idaho

Division of Entomology
University of Idaho
Moscow, ID 83844-2339
Tel: 208-885-7543
Fax: 208-885-7760
Web: http://info.ag.uidaho.edu/catalog/
  catalog_frm.htm

Illinois

Center for Economic Entomology
Illinois Natural History Survey
607 East Peabody Drive
Champaign, IL 61820
Tel: 217-333-6656
Fax: 217-333-4949

Center for Biodiversity
Illinois Natural History Survey
607 East Peabody Drive
Champaign, IL 61820
Tel: 217-333-6846
Fax: 217-333-4949

Department of Entomology
University of Illinois
320 Morrill Hall
505 South Goodwin Avenue
Urbana, IL 61801

Tel: 217-333-2910
Fax: 217-244-3499
Email: entwork@life.uiuc.edu
Web: http://www.life.uiuc.edu/entomology/home.
html
http://www.aces.uiuc.edu/~ipm/fruits/
fruits.html

Indiana

Department of Entomology
Purdue University
1158 Entomology Hall
West Lafayette, IN 47907-1158
Tel: 765-494-4554
Fax: 765-494-0535
Web: http://www.agcom.purdue.edu/AgCom/
Pubs/ent.htm

Iowa

Department of Entomology
Iowa State University
411 Science II
Ames, IA 50011-3222
Tel: 515-294-7400
Fax: 515-294-5957
Web: http://www.ipm.iastate.edu/ipm/iiin/

Kansas

Department of Entomology
University of Kansas
Lawrence, KS 66045
Tel: 785-864-7366
Fax: 785-864-5321

Department of Entomology
Kansas State University
123 Waters Hall
Manhattan, KS 66506-4004
Tel: 785-532-6154
Fax: 785-532-6232
Email: entomology@ksu.edu
Web: http://www.oznet.ksu.edu/entomology/

Kentucky

Department of Entomology
University of Kentucky
Lexington, KY 40546-0091
Tel: 606-257-7450
Fax: 606-323-1120
Web: http://www.uky.edu/Agriculture/
Entomology/entfacts/efhtmlst.htm

Louisiana

Department of Entomology
Louisiana State University
402 Life Sciences Building
Baton Rouge, LA 70803-1710
Tel: 504-388-1634
Fax: 504-388-1643
Web: http://www.lsu.edu/guests/wwwent2/

Maine

Department of Biological Sciences
Entomology Division
Deering Hall
University of Maine
Orono, ME 04469
Tel: 207-581-2957
Fax: 207-581-2999

Maryland

Systematic Entomology Laboratory
Building 046, BARC-West
Agricultural Research Service, USDA
10300 Baltimore Avenue
Beltsville, MD 20705
Tel: 301-504-5183
Fax: 301-504-6482

Department of Entomology
University of Maryland
4112 Plant Science Building
College Park, MD 20742-4454
Tel: 301-405-3911
Fax: 301-314-9290
Web: http://www.agnr.umd.edu/users/ipmnet./
ipmnet.htm
http://www.agnr.umd.edu/ces/pubs/online.
html

Massachusetts

Department of Entomology
University of Massachusetts
Fernald Hall
Amherst, MA 01003
Tel: 413-545-2283
Fax: 413-545-2115
Email: entomology@ent.umass.edu
Web: http://www.umass.edu/umext/programs/
agro/vegsmfr/index.html

Michigan

Department of Entomology
Michigan State University

243 Natural Science Building
East Lansing, MI 48824
Tel: 517-355-4662
Fax: 517-353-4354
Web: http://www.ent.msu.edu/dept/docs/
    ipmfacts.html
    http://insects.ummz.lsa.umich.edu/MES/notes/
    noteslist.html

Minnesota

Department of Entomology
University of Minnesota
219 Hodson Hall
1980 Folwell Avenue
St. Paul, MN 55108-6125
Tel: 612-624-3278
Fax: 612-625-5299
Email: entodept@gold.tc.umn.edu
Web: http://www3.extension.umn.edu/projects/
    yardandgarden/Ent.htm
    http://www3.extension.umn.edu/vegipm/intro/
    pestfact.htm

Mississippi

Department of Entomology and Plant Pathology
Mississippi State University
Box 9775
Mississippi State, MS 39762-9775
Tel: 601-325-2085
Fax: 601-325-8837
Web: http://www.msstate.edu/Entomology/
    ENTPLP.html

Missouri

Department of Entomology
University of Missouri
1-87 Agriculture Building
Columbia, MO 65211
Tel: 573-882-7894
Fax: 573-882-1469
Web: http://muextension.missouri.edu/xplor/
    agguides/pests/index.htm

Montana

Department of Entomology
Montana State University
Bozeman, MT 59717
Tel: 406-994-3860
Fax: 406-994-6029
Web: http://scarab.msu.montana.edu/ipm

Department of Entomology
University of Nebraska
Lincoln, NE 68583-0816
Tel: 402-472-2123
Fax: 402-472-4687
Web: http://ianrwww.unl.edu/pubs/insects/
    index.htm

New Hampshire

University of New Hampshire
State Extension, Taylor Hall
59 College Road
Durham, NH 03824-3587
Tel: 603-862-1520
Fax: 603-862-1585
Web: http://ceinfo.unh.edu/agpubs2.htm

New Jersey

Department of Entomology
Rutgers, The State University of New Jersey
Blake Hall
93 Lipman Drive
New Brunswick, NJ 08901-8524
Tel: 908-932-9459
Fax: 908-932-7229
Web: http://www.rce.rutgers.edu/ag/
    agpestcontrol/index.htm

New Mexico

Department of Entomology, Plant Pathology and
    Weed Science
New Mexico State University
Box 30003, Department 3BE
Las Cruces, NM 88003-0003
Tel: 505-646-3225
Fax: 505-646-8087
Web: http://www.cahe.nmsu.edu/resources/
    welcome.html

New York

Department of Entomology-Geneva
Cornell University
NYS Agricultural Experiment Station
Geneva, NY 14456
Tel: 315-781-2323
Fax: 315-787-2326
Web: http://www.nysaes.cornell.edu/ent/
    biocontrol/
    http://www.nysaes.cornell.edu:80/ipmnet/ny/
    vegetables/index.html

Department of Entomology
Cornell University
Ithaca, NY 14853
Tel: 607-255-3253
Fax: 607-255-0939
Web: http://www.cals.cornell.edu/dept/entom/

Entomology Concentration Leader
SUNY College of Environment Science and Forestry
133 Illick Hall
Syracuse, NY 13210-2788
Tel: 315-470-6742
Fax: 315-470-6934

North Carolina

Department of Entomology
North Carolina State University
Raleigh, NC 27695-7613
Tel: 919-515-8888
Fax: 919-515-7746
Web: http://www.ces.ncsu.edu/depts/ent/notes/
http://ipmwww.ncsu.edu/

North Carolina Pest Identification
Web: http://ipmwww.ncsu.edu/PEST_ID/pestid.
html

North Dakota

Department of Entomology
North Dakota State University
202 Hultz Hall
Box 5346, University Station
Fargo, ND 58105-5446
Tel: 701-237-7582
Fax: 701-237-8557
Web: http://www.ndsu.nodak.edu/entomology/

Ohio

Department of Entomology
Ohio State University
Columbus, OH 43210
Tel: 614-292-8209
Fax: 614-292-2180
Email: montano.3@osu.edu
Web: http://www.ag.ohio-state.edu/~ohioline/
b672/index.html

Department of Entomology
OARDC
Wooster, OH 44691
Tel: 216-263-3730
Fax: 216-263-3686

Oklahoma

Department of Entomology and Plant Pathology
Oklahoma State University
127 Noble Research Center
Stillwater, OK 74078
Tel: 405-774-5527
Fax: 405-744-6039
Web: http://www.ento.okstate.edu/factshts.htm

Oregon

Department of Entomology
Oregon State University
Cordley Hall, Room 2046
Corvallis, OR 97334-2709
Tel: 503-737-4733
Fax: 503-737-3643
Email: ent-off@bcc.orst.edu
Web: http://www.ent.orst.edu/entomology/

Pennsylvania

Department of Entomology
Pennsylvania State University
University Park, PA 16802
Tel: 814-865-1895
Fax: 814-865-3048

Puerto Rico

Crop Protection Department
University of Puerto Rico
Box 5000
Mayazuez, PR 00681-9030
Tel: 809-265-3859
Fax: 809-265-0860

Rhode Island

Department of Plant Sciences
University of Rhode Island
Kingston, RI 02881
Tel: 401-874-2924
Fax: 401-874-5296

South Carolina

Department of Entomology
Clemson University
114 Long Hall, Box 340365
Clemson, SC 29634
Tel: 864-656-3111
Fax: 864-656-5065
Web: http://dpr.clemson.edu/

South Dakota

Department of Plant Science
South Dakota State University
Box 2207-A
Brookings, SD 57007
Tel: 605-688-5123
Fax: 605-688-4602

Tennessee

Department of Entomology and Plant Pathology
University of Tennessee
Knoxville, TN 39701
Tel: 615-947-7135
Web: http://web.utk.edu/~extepp/fruitveg.htm

Texas

Department of Entomology
Texas A&M University
College Station, TX 77843-2475
Tel: 409-845-2516
Fax: 409-845-6305
Email: entomain@tamu.edu
Web: http://entowww.tamu.edu/
    http://insects.tamu.edu/extension/category.html

Department of Plant and Soil Science
Texas Tech University
Box 42122
Lubbock, TX 79409-2122
Tel: 806-742-2838
Fax: 806-742-0775

Utah

Insect Biology Division
Department of Biology
Utah State University
Logan, UT 84322-5305
Tel: 435-797-2515
Fax: 435-797-1575
Email: entpath@ext.usu.edu
Web: http://www.ext.usu.edu/ag/ipm/
    infactsh.htm

Vermont

Plant and Soil Science Department
University of Vermont
Hills Building
Burlington, VT 05405-0082
Tel: 802-656-2630
Fax: 802-656-4656

Virginia

Department of Entomology
Virginia Polytechnic Institute and State University
Mail Code 0319
Blacksburg, VA 24061-0319
Tel: 540-231-6341
Fax: 540-231-9131
Web: http://everest.ento.vt.edu/~idlab/vegpests/
    veg1.html

Washington

Department of Entomology
Washington State University
FSHN Building 166
Pullman, WA 99164-6382
Tel: 509-335-5505
Fax: 509-335-1009
Email: entom@wsu.edu
Web: http://IPM.wsu.edu/Archive.html#outline

West Virginia

Division of Plant and Soil Sciences
West Virginia University
P.O. Box 6108
Morgantown, WV 26506-6108
Tel: 304-293-4817 (Division)
Tel: 304-293-6023 (Entomology)
Fax: 304-293-3740

Wisconsin

Department of Entomology
University of Wisconsin
Madison, WI 53706
Tel: 608-262-3227
Fax: 608-262-3322
Web: http://entomology.wisc.edu/entomology/

Wyoming

Department of Renewable Resources
University of Wyoming
P.O. Box 3354, University Station
Laramie, WY 82071-3103
Tel: 307-766-3103
Fax: 307-766-3379
Web: http://www.sdvc.uwyo.edu/grasshopper/

CANADA

Alberta

Department of Biological Sciences
University of Alberta

CW 4-05 Biological Sciences Building
Edmonton, AB
Canada T6G 2E9
Tel: 403-492-3308
Fax: 403-492-9234

Alberta Agriculture, Food and Rural Development
10800 -97 Avenue
Edmonton, AB
Canada T5K 2B6
Tel: 780-427-2137
Fax: 780-422-6035
Web: http://www.agric.gov.ab.ca/navigation/pests/
   plantinsects/oilseed/col_index.html

British Columbia

Department of Biological Sciences
Simon Fraser University
Burnaby, BC
Canada V5A 1S6
Tel: 604-291-4475
Fax: 604-291-3496
Web: http://www.biol.sfu.ca/homepage.html

British Columbia Ministry of Environment, Lands and
   Parks
Web: http://pupux1.env.gov.bc.ca/~ipmis/ipmis.
   html

Manitoba

Department of Entomology
University of Manitoba
Winnipeg, MB
Canada R3T 2N2
Tel: 204-474-9257
Fax: 204-474-7628

Newfoundland

Department of Biology
Memorial University
St. Johns, NF
Canada A1B 3X9
Tel: 709-737-7497
Fax: 709-737-3018

Ontario

Southern Crop Protection and Food Research Centre
1391 Sanford Street
London, ON
Canada N5V 4T3
Tel: 519-457-1470

Fax: 519-457-3997
Web: http://res.agr.ca/lond/pmrc/pmrchome.html

Collections and Research
Canadian Museum of Nature
P.O. Box 3443, Station D
Ottawa, ON
Canada K1P 6P4
Tel: 613-364-4060
Fax: 613-364-4027

Eastern Cereal and Oilseed Research Centre
Biological Resources Program
Agriculture and Agri-food Canada
Ottawa, ON
Canada K1A 0C6
Tel: 613-759-1864
Fax: 613-759-1924
Web: http://res.agr.ca/ecorc/

Entomology Section
Centre for Biodiversity and Conservation Biology
Royal Ontario Museum
100 Queens Park
Toronto, ON
Canada M5S 2C6
Tel: 416-586-8059
Fax: 416-586-5553

Department of Zoology
University of Toronto
Toronto, ON
Canada M5S 3G5
Tel: 416-978-2084
Fax: 416-978-8532

Quebec

Department of Natural Resource Sciences
Faculty of Agricultural and Environmental Sciences
McGill University (Macdonald Campus)
21, 111 Lakeshore Road
Sainte-Anne de Bellevue, PQ
Canada H9X 3V9
Tel: 514-398-7890
Fax: 514-398-7990

Lyman Entomological Museum and Research
   Laboratory
Mcdonald Campus of McGill University
Sainte-Anne de Bellevue, PQ
Canada H9X 1C0
Tel: 514-398-7914
Fax: 514-398-7990
Email: lyman@nrs.mcgill.ca

Saskatchewan

Department of Biology
University of Saskatchewan
112 Science Place
Saskatoon, SK
Canada S7N 5E2
Tel: 306-966-4400
Fax: 306-966-4461

MISCELLANEOUS

Oregon State Multilink Vegetable Production and Pest
    Control Site
http://www.orst.edu/Dept/NWREC/veglink.html

Caterpillar Hostplants
http://www.nhm.ac.uk/entomology/hostplants/

Index of Insect Internet Resources
http://www.ent.iastate.edu/List/

Moths of the United States
http://www.npwrc.usgs.gov/resource/distr/lepid/
    moths/mothsusa.htm

Insecticide and Toxicology Information
http://ace.orst.edu/info/extoxnet/

Comprehensive Source of Pesticide Labels
http://www.edms.net/

Biological Control
http://www.nysaes.cornell.edu/ent/biocontrol/

Biological Control Virtual Information Center
http://ipmwww.ncsu.edu/biocontrol/
    biocontrol.html

Orthoptera Species File
http://viceroy.eeb.uconn.edu/orthoptera

Agriculture and Agri-Food Canada Website for Pest
    Management Research Centers
http://res.agr.ca/

National IPM Network
http://www.reeusda.gov/agsys/nipmn/index.htm

National IPM, Southeastern Server
http://ipmwww.ncsu.edu/

National IPM, Western Region
http://www.colostate.edu/Depts/IPM/index.html

National IPM, Northeastern Region
http://www.nysaes.cornell.edu/ipmnet/

National IPM, North Central Region
http://www.ipm.iastate.edu/ipm/~ipmn/

National IPM, Southern Region
http://ipm-www.ento.vt.edu:8000/nipmn/

Appendix C
Vegetable Plant Names

COMMON NAME, SCIENTIFIC NAME, AND PLANT FAMILY

Common Name	Scientific Name	Plant Family
Arrugula	*Eruca sativa*	Cruciferae
Artichoke, Globe	*Cynara scolymus*	Compositae
Artichoke, Jerusalem	*Helianthus tuberosus*	Compositae
Asparagus	*Asparagus officinalis*	Liliaceae
Bean, Broad	*Vicia faba*	Leguminosae
Bean, Lima	*Phaseolus limensis*	Leguminosae
Bean, Mung	*Phaseolus aureus*	Leguminosae
Bean, Snap	*Phaseolus vulgaris*	Leguminosae
Beet	*Beta vulgaris*	Chenopodiaceae
Bok Choy	*Brassica campestris*, var. *pekinensis*	Cruciferae
Broccoli	*Brassica oleracea*, var. *italica*	Cruciferae
Broccoli Raab	*Brassica campestris*, var. *ruvo*	Cruciferae
Brussels Sprout	*Brassica oleracea*, var. *gemmifera*	Cruciferae
Cabbage	*Brassica oleracea*, var. *capitata*	Cruciferae
Cabbage, Bok Choy	*Brassica campestris*, var. *pekinensis*	Cruciferae
Cabbage, Chinese	*Brassica campestris*, var. *chinensis*	Cruciferae
Calabaza	*Cucurbita moscata*	Cucurbitaceae
Cantaloupe	*Cucumis melo*	Cucurbitaceae
Cardoon	*Cynara cardunculus*	Compositae
Carrot	*Daucus carota*	Umbelliferae
Cauliflower	*Brassica oleracea*, var. *botrytis*	Cruciferae
Celeriac	*Apium graveolens*, var. *rapaceum*	Umbelliferae
Celery	*Apium graveolens*, var. *dulce*	Umbelliferae
Celtuce	*Lactuca sativa*, var. *asparagina*	Compositae
Chard	*Beta vulgaris*, var. *cicla*	Chenopodiaceae
Chayote	*Sechium edule*	Cucurbitaceae
Chervil, Salad	*Anthriscus cerefolium*	Umbelliferae
Chervil, Turnip-rooted	*Chaerophyllum bulbosum*	Umbelliferae
Chickpea	*Cicer arietinum*	Leguminosae
Chicory	*Chicorium intybus*	Compositae
Chili	*Capsicum annuum*	Solanaceae
Chive	*Allium schoenoprasum*	Alliaceae
Collard	*Brassica oleracea*, var. *acephala*	Cruciferae
Coriander	*Coriandrum sativum*	Umbelliferae

(Continues)

COMMON NAME, SCIENTIFIC NAME, AND PLANT FAMILY (Continued)

Common Name	Scientific Name	Plant Family
Corn, Sweet	*Zea mays*	Gramineae
Cowpea	*Vigna sinensis*	Leguminosae
Cucumber	*Cucumis sativus*	Cucurbitaceae
Daikon	*Raphanus sativus*	Cruciferae
Eggplant	*Solanum melongena*	Solanaceae
Endive	*Cichorium endivia*	Compositae
Escarole	*Cichorium endivia*	Compositae
Faba Bean	*Vicia faba*	Leguminosae
Fennel	*Foeniculum vulgare*	Umbelliferae
Garbanzo	*Cicer arietinum*	Leguminosae
Garlic	*Allium sativum*	Alliaceae
Garlic, Elephant	*Allium ampeloprasum*	Alliaceae
Gherkin	*Cucumis anguria*	Cucurbitaceae
Globe Artichoke	*Cynara scolymus*	Compositae
Horseradish	*Amoracia lapathifolia*	Cruciferae
Husk Tomato	*Pysalis pruinosa*	Solanaceae
Jerusalem Artichoke	*Helianthus tuberosus*	Compositae
Kale	*Brassica oleracea*, var. *cephala*	Cruciferae
Kale, Sea	*Crambe maritima*	Cruciferae
Kohlrabi	*Brassica oleracea*, var. *gongylodes*	Cruciferae
Leek	*Allium ampeolprasum*, var. *leek*	Alliaceae
Lentil	*Lepidium sativum*	Cruciferae
Lettuce, Head	*Lactuca sativa*, var. *capitata*	Compositae
Lettuce, Leaf	*Lactuca sativa*, var. *crispa*	Compositae
Lettuce, Romaine	*Lactuca sativa*, var. *longifolia*	Compositae
Mushroom	*Agaricus* sp.	Agaricaceae
Muskmelon	*Cucumis melo*	Cucurbitaceae
Mustard	*Brassica jundea*, var. *crispifolia*	Cruciferae
Okra	*Hibuscus esculentus*	Malvaceae
Onion	*Allium cepa*	Alliaceae
Parsley	*Petroselinum crispum*	Umbelliferae
Parsnip	*Pastinaca sativa*	Umbelliferae
Pea, Edible-podded	*Pisum sativum*	Leguminosae
Pea, Garden	*Pisum sativum*	Leguminosae
Pea, Pigeon	*Cajanus cajan*	Leguminosae
Pea, Southern	*Vigna sinensis*	Leguminosae
Pepper, Bell	*Capsicum annuum*	Solanaceae
Pepper, Chili	*Capsicum annuum*	Solanaceae
Potato	*Solanum tuberosum*	Solanaceae
Pumpkin	*Cucurbita* spp.	Cucurbitaceae
Purslane	*Portulaca oleracea*	Portulacaceae
Radicchio	*Cichorium intybus*	Compositae
Radish	*Raphanus sativus*	Cruciferae
Radish, Chinese	*Raphanus sativus*	Cruciferae
Rhubarb	*Rheum rhabarbarum*	Polygonaceae
Rutabaga	*Brassica napus*, var. *napobrassica*	Cruciferae
Salsify	*Tragopogon porrifolius*	Compositae
Sea Kale	*Crambe maritima*	Cruciferae

(Continues)

COMMON NAME, SCIENTIFIC NAME, AND PLANT FAMILY (*Continued*)

Common Name	Scientific Name	Plant Family
Shallot	*Allium cepa*, var. *aggregatum*	Alliaceae
Southern Pea	*Vigna sinensis*	Leguminosae
Spinach	*Spinacia oleracea*	Chenopodiaceae
Spinach, New Zealand	*Tetragonia tetragonioides*	Aizoceae
Squash, Summer	*Cucurbita pepo*	Cucurbitaceae
Squash, Winter	*Cucurbita maxima*	Cucurbitaceae
Squash, Winter	*Cucurbita moschata*	Cucurbitaceae
Sweet Potato	*Ipomoea batatas*	Convolvulaceae
Swiss Chard	*Beta vulgaris*, var. *cicla*	Chenopodiaceae
Tomatillo	*Physalis ixocarpa*	Solanaceae
Tomato	*Lycopersicon esculentum*	Solanaceae
Tomato, Husk	*Physalis pruinosa*	Solanaceae
Turnip	*Brassica rapa*, var. *rapifera*	Cruciferae
Watercress	*Nasturtium officinale*	Cruciferae
Watermelon	*Citrullus lanatus*	Cucurbitaceae
Yam	*Dioscorea spp.*	Dioscoreaceae

SCIENTIFIC NAME, COMMON NAME, AND PLANT FAMILY

Scientific Name	Common Name	Plant Family
Agaricus sp.	Mushroom	Agaricaceae
Allium ampeloprasum	Garlic, Elephant	Alliaceae
Allium ampeolprasum, var. *leek*	Leek	Alliaceae
Allium cepa	Onion	Alliaceae
Allium cepa, var. *aggregatum*	Shallot	Alliaceae
Allium sativum	Garlic	Alliaceae
Allium schoenoprasum	Chive	Alliaceae
Amoracia lapathifolia	Horseradish	Cruciferae
Anthriscus cerefolium	Chervil, Salad	Umbelliferae
Apium graveolens, var. *dulce*	Celery	Umbelliferae
Apium graveolens, var. *rapaceum*	Celeriac	Umbelliferae
Asparagus officinalis	Asparagus	Liliaceae
Beta vulgaris	Beet	Chenopodiaceae
Beta vulgaris, var. *cicla*	Chard	Chenopodiaceae
Beta vulgaris, var. *cicla*	Swiss Chard	Chenopodiaceae
Brassica campestris, var. *chinensis*	Cabbage, Chinese	Cruciferae
Brassica campestris, var. *pekinensis*	Bok Choy	Cruciferae
Brassica campestris, var. *pekinensis*	Cabbage, Bok Choy	Cruciferae
Brassica campestris, var. *ruvo*	Broccoli Raab	Cruciferae
Brassica jundea, var. *crispifolia*	Mustard	Cruciferae
Brassica napus, var. *napobrassica*	Rutabaga	Cruciferae
Brassica oleracea, var. *acephala*	Collard	Cruciferae
Brassica oleracea, var. *botrytis*	Cauliflower	Cruciferae
Brassica oleracea, var. *capitata*	Cabbage	Cruciferae
Brassica oleracea, var. *cephala*	Kale	Cruciferae

(*Continues*)

SCIENTIFIC NAME, COMMON NAME, AND PLANT FAMILY   (Continued)

Scientific Name	Common Name	Plant Family
Brassica oleracea, var. *gemmifera*	Brussels Sprout	Cruciferae
Brassica oleracea, var. *gongylodes*	Kohlrabi	Cruciferae
Brassica oleracea, var. *italica*	Broccoli	Cruciferae
Brassica rapa, var. *rapifera*	Turnip	Cruciferae
Cajanus cajan	Pea, Pigeon	Leguminosae
Capsicum annuum	Chili	Solanaceae
Capsicum annuum	Pepper, Bell	Solanaceae
Capsicum annuum	Pepper, Chili	Solanaceae
Chaerophyllum bulbosum	Chervil, Turnip-rooted	Umbelliferae
Chicorium intybus	Chicory	Compositae
Cicer arietinum	Chickpea	Leguminosae
Cicer arietinum	Garbanzo	Leguminosae
Cichorium endivia	Endive	Compositae
Cichorium endivia	Escarole	Compositae
Cichorium intybus	Radicchio	Compositae
Citrullus lanatus	Watermelon	Cucurbitaceae
Coriandrum sativum	Coriander	Umbelliferae
Crambe maritima	Kale, Sea	Cruciferae
Crambe maritima	Sea Kale	Cruciferae
Cucumis anguria	Gherkin	Cucurbitaceae
Cucumis melo	Cantaloupe	Cucurbitaceae
Cucumis melo	Muskmelon	Cucurbitaceae
Cucumis sativus	Cucumber	Cucurbitaceae
Cucurbita maxima	Squash, Winter	Cucurbitaceae
Cucurbita moscata	Calabaza	Cucurbitaceae
Cucurbita moschata	Squash, Winter	Cucurbitaceae
Cucurbita pepo	Squash, Summer	Cucurbitaceae
Cucurbita spp.	Pumpkin	Cucurbitaceae
Cynara cardunculus	Cardoon	Compositae
Cynara scolymus	Artichoke, Globe	Compositae
Cynara scolymus	Globe Artichoke	Compositae
Daucus carota	Carrot	Umbelliferae
Dioscorea spp.	Yam	Dioscoreaceae
Eruca sativa	Arrugula	Cruciferae
Foeniculum vulgare	Fennel	Umbelliferae
Helianthus tuberosus	Artichoke, Jerusalem	Compositae
Helianthus tuberosus	Jerusalem Artichoke	Compositae
Hibuscus esculentus	Okra	Malvaceae
Ipomoea batatas	Sweet Potato	Convolvulaceae
Lactuca sativa, var. *asparagina*	Celtuce	Compositae
Lactuca sativa, var. *capitata*	Lettuce, Head	Compositae
Lactuca sativa, var. *crispa*	Lettuce, Leaf	Compositae
Lactuca sativa, var. *longifolia*	Lettuce, Romaine	Compositae
Lepidium sativum	Lentil	Cruciferae
Lycopersicon esculentum	Tomato	Solanaceae
Nasturtium officinale	Watercress	Cruciferae
Pastinaca sativa	Parsnip	Umbelliferae
Petroselinum crispum	Parsley	Umbelliferae

(Continues)

SCIENTIFIC NAME, COMMON NAME, AND PLANT FAMILY   *(Continued)*

Scientific Name	Common Name	Plant Family
Phaseolus aureus	Bean, Mung	Leguminosae
Phaseolus limensis	Bean, Lima	Leguminosae
Phaseolus vulgaris	Bean, Snap	Leguminosae
Physalis ixocarpa	Tomatillo	Solanaceae
Physalis pruinosa	Tomato, Husk	Solanaceae
Pisum sativum	Pea, Edible-podded	Leguminosae
Pisum sativum	Pea, Garden	Leguminosae
Portulaca oleracea	Purslane	Portulacaceae
Pysalis pruinosa	Husk Tomato	Solanaceae
Raphanus sativus	Daikon	Cruciferae
Raphanus sativus	Radish	Cruciferae
Raphanus sativus	Radish, Chinese	Cruciferae
Rheum rhabarbarum	Rhubarb	Polygonaceae
Sechium edule	Chayote	Cucurbitaceae
Solanum melongena	Eggplant	Solanaceae
Solanum tuberosum	Potato	Solanaceae
Spinacia oleracea	Spinach	Chenopodiaceae
Tetragonia tetragonioides	Spinach, New Zealand	Aizoceae
Tragopogon porrifolius	Salsify	Compositae
Vicia faba	Bean, Broad	Leguminosae
Vicia faba	Faba Bean	Leguminosae
Vigna sinensis	Cowpea	Leguminosae
Vigna sinensis	Pea, Southern	Leguminosae
Vigna sinensis	Southern Pea	Leguminosae
Zea mays	Corn, Sweet	Gramineae

PLANT FAMILY, COMMON NAME, AND SCIENTIFIC NAME

Plant Family	Common Name	Scientific Name
Agaricaceae	Mushroom	*Agaricus* sp.
Aizoceae	Spinach, New Zealand	*Tetragonia tetragonioides*
Alliaceae	Chive	*Allium schoenoprasum*
Alliaceae	Garlic	*Allium sativum*
Alliaceae	Garlic, Elephant	*Allium ampeloprasum*
Alliaceae	Leek	*Allium ampeolprasum*, var. *leek*
Alliaceae	Onion	*Allium cepa*
Alliaceae	Shallot	*Allium cepa*, var. *aggregatum*
Chenopodiaceae	Beet	*Beta vulgaris*
Chenopodiaceae	Chard	*Beta vulgaris*, var. *cicla*
Chenopodiaceae	Spinach	*Spinacia oleracea*
Chenopodiaceae	Swiss Chard	*Beta vulgaris*, var. *cicla*
Compositae	Artichoke, Globe	*Cynara scolymus*
Compositae	Artichoke, Jerusalem	*Helianthus tuberosus*
Compositae	Cardoon	*Cynara cardunculus*

(Continues)

PLANT FAMILY, COMMON NAME, AND SCIENTIFIC NAME (*Continued*)

Plant Family	Common Name	Scientific Name
Compositae	Celtuce	*Lactuca sativa*, var. *asparagina*
Compositae	Chicory	*Chicorium intybus*
Compositae	Endive	*Cichorium endivia*
Compositae	Escarole	*Cichorium endivia*
Compositae	Globe Artichoke	*Cynara scolymus*
Compositae	Jerusalem Artichoke	*Helianthus tuberosus*
Compositae	Lettuce, Head	*Lactuca sativa*, var. *capitata*
Compositae	Lettuce, Leaf	*Lactuca sativa*, var. *crispa*
Compositae	Lettuce, Romaine	*Lactuca sativa*, var. *longifolia*
Compositae	Radicchio	*Cichorium intybus*
Compositae	Salsify	*Tragopogon porrifolius*
Convolvulaceae	Sweet Potato	*Ipomoea batatas*
Cruciferae	Arrugula	*Eruca sativa*
Cruciferae	Bok Choy	*Brassica campestris*, var. *pekinensis*
Cruciferae	Broccoli	*Brassica oleracea*, var. *italica*
Cruciferae	Broccoli Raab	*Brassica campestris*, var. *ruvo*
Cruciferae	Brussels Sprout	*Brassica oleracea*, var. *gemmifera*
Cruciferae	Cabbage	*Brassica oleracea*, var. *capitata*
Cruciferae	Cabbage, Bok Choy	*Brassica campestris*, var. *pekinensis*
Cruciferae	Cabbage, Chinese	*Brassica campestris*, var. *chinensis*
Cruciferae	Cauliflower	*Brassica oleracea*, var. *botrytis*
Cruciferae	Collard	*Brassica oleracea*, var. *acephala*
Cruciferae	Daikon	*Raphanus sativus*
Cruciferae	Horseradish	*Amoracia lapathifolia*
Cruciferae	Kale	*Brassica oleracea*, var. *cephala*
Cruciferae	Kale, Sea	*Crambe maritima*
Cruciferae	Kohlrabi	*Brassica oleracea*, var. *gongylodes*
Cruciferae	Lentil	*Lepidium sativum*
Cruciferae	Mustard	*Brassica jundea*, var. *crispifolia*
Cruciferae	Radish	*Raphanus sativus*
Cruciferae	Radish, Chinese	*Raphanus sativus*
Cruciferae	Rutabaga	*Brassica napus*, var. *napobrassica*
Cruciferae	Sea Kale	*Crambe maritima*
Cruciferae	Turnip	*Brassica rapa*, var. *rapifera*
Cruciferae	Watercress	*Nasturtium officinale*
Cucurbitaceae	Calabaza	*Cucurbita moscata*
Cucurbitaceae	Cantaloupe	*Cucumis melo*
Cucurbitaceae	Chayote	*Sechium edule*
Cucurbitaceae	Cucumber	*Cucumis sativus*
Cucurbitaceae	Gherkin	*Cucumis anguria*
Cucurbitaceae	Muskmelon	*Cucumis melo*
Cucurbitaceae	Pumpkin	*Cucurbita* spp.
Cucurbitaceae	Squash, Summer	*Cucurbita pepo*
Cucurbitaceae	Squash, Winter	*Cucurbita maxima*
Cucurbitaceae	Squash, Winter	*Cucurbita moschata*
Cucurbitaceae	Watermelon	*Citrullus lanatus*
Dioscoreaceae	Yam	*Dioscorea* spp.
Gramineae	Corn, Sweet	*Zea mays*

(*Continues*)

PLANT FAMILY, COMMON NAME, AND SCIENTIFIC NAME (*Continued*)

Plant Family	Common Name	Scientific Name
Leguminosae	Bean, Broad	*Vicia faba*
Leguminosae	Bean, Lima	*Phaseolus limensis*
Leguminosae	Bean, Mung	*Phaseolus aureus*
Leguminosae	Bean, Snap	*Phaseolus vulgaris*
Leguminosae	Chickpea	*Cicer arietinum*
Leguminosae	Cowpea	*Vigna sinensis*
Leguminosae	Faba Bean	*Vicia faba*
Leguminosae	Garbanzo	*Cicer arietinum*
Leguminosae	Pea, Edible-podded	*Pisum sativum*
Leguminosae	Pea, Garden	*Pisum sativum*
Leguminosae	Pea, Pigeon	*Cajanus cajan*
Leguminosae	Pea, Southern	*Vigna sinensis*
Leguminosae	Southern Pea	*Vigna sinensis*
Liliaceae	Asparagus	*Asparagus officinalis*
Malvaceae	Okra	*Hibiscus esculentus*
Polygonaceae	Rhubarb	*Rheum rhabarbarum*
Portulacaceae	Purslane	*Portulaca oleracea*
Solanaceae	Chili	*Capsicum annuum*
Solanaceae	Eggplant	*Solanum melongena*
Solanaceae	Husk Tomato	*Pysalis pruinosa*
Solanaceae	Pepper, Bell	*Capsicum annuum*
Solanaceae	Pepper, Chili	*Capsicum annuum*
Solanaceae	Potato	*Solanum tuberosum*
Solanaceae	Tomatillo	*Physalis ixocarpa*
Solanaceae	Tomato	*Lycopersicon esculentum*
Solanaceae	Tomato, Husk	*Physalis pruinosa*
Umbelliferae	Carrot	*Daucus carota*
Umbelliferae	Celeriac	*Apium graveolens,* var. *rapaceum*
Umbelliferae	Celery	*Apium graveolens,* var. *dulce*
Umbelliferae	Chervil, Salad	*Anthriscus cerefolium*
Umbelliferae	Chervil, Turnip-rooted	*Chaerophyllum bulbosum*
Umbelliferae	Coriander	*Coriandrum sativum*
Umbelliferae	Fennel	*Foeniculum vulgare*
Umbelliferae	Parsley	*Petroselinum crispum*
Umbelliferae	Parsnip	*Pastinaca sativa*

Appendix D
Journal Abbreviations and Journal Titles

Acarologia	*Acarologia*
Agric. Ecosyst. Environ.	*Agriculture, Ecosystems and Environment*
Am. Entomol.	*American Entomologist*
Am. Midl. Nat.	*American Midland Naturalist*
Am. Pot. J.	*American Potato Journal*
An. Soc. Entomol. Brasil	*Anais do Sociedade Entomologica do Brasil*
Ann. Agric. Fenn.	*Annales Agriculture Fenniae*
Ann. Appl. Biol.	*Annals of Applied Biology*
Ann. Entomol. Fennici	*Annales Entomologici Fennici*
Ann. Entomol. Soc. Am.	*Annals of the Entomological Society of America*
Ann. Entomol. Soc. Quebec	*Annals of the Entomological Society of Quebec*
Annu. Rev. Entomol.	*Annual Review of Entomology*
Appl. Agric. Res.	*Applied Agricultural Research*
Appl. Entomol. Zool.	*Applied Entomology and Zoology*
Aust. J. Zool.	*Australian Journal of Zoology*
Biocontrol Sci. Tech.	*Biocontrol Science and Technology*
Biol. Control	*Biological Control*
Bull. Brooklyn Entomol. Soc.	*Bulletin of the Brooklyn Entomological Society*
Bull. Entomol. Res.	*Bulletin of Entomological Research*
Bull. Entomol. Soc. Am.	*Bulletin of the Entomological Society of America*
Bull. Soc. Entomol. Egypte	*Bulletin of the Entomological Society of Egypt*
Bull. So. California Acad. Sci.	*Bulletin Southern California Academy of Sciences*
California Agric.	*California Agriculture*
Can. Entomol.	*Canadian Entomologist*
Can. J. Bot.	*Canadian Journal of Botany*
Can. J. Plant Sci.	*Canadian Journal of Plant Science*
Can. J. Res.	*Canadian Journal of Research*
Can. J. Zool.	*Canadian Journal of Zoology*
Clin. All.	*Clinical Allergy*
Coleop. Bull.	*The Coleopterists Bulletin*
Crop Prot.	*Crop Protection*
Crop Sci.	*Crop Science*
Ecol. Entomol.	*Ecological Entomology*
Ecology	*Ecology*
Entomol. Am.	*Entomologica Americana*
Entomol. Exp. Appl.	*Entomologia Experimentalis et Applicata*
Entomol. Mon. Mag.	*Entomologist's Monthly Magazine*
Entomol. News	*Entomological News*
Entomophaga	*Entomophaga*

Environ. Entomol.　　　　　　　Environmental Entomology
Evolution　　　　　　　　　　Evolution
Exp. Appl. Acarol.　　　　　　Experimental and Applied Acarology
Experientia　　　　　　　　　Experientia
FAO Plant Prot. Bull.　　　　　Food and Agriculture Organization of the United
　　　　　　　　　　　　　　　Nations, Plant Protection Bulletin

Fieldiana　　　　　　　　　　Fieldiana
Florida Entomol.　　　　　　　Florida Entomologist
Fund. Appl. Nematol.　　　　　Fundamentals of Applied Nematology
Great Lakes Entomol.　　　　　Great Lakes Entomologist
Hilgardia　　　　　　　　　　Hilgardia
HortScience　　　　　　　　　HortScience
Indian J. Agric. Sci.　　　　　Indian Journal of Agricultural Science
Indian J. Entomol.　　　　　　Indian Journal of Entomology
Insect Life　　　　　　　　　Insect Life
Insect Sci. Applic.　　　　　　Insect Science and its Application
Insecta Matsu.　　　　　　　　Insecta Matsumurana
Insecta Mun.　　　　　　　　Insecta Mundi
Integr. Pest Manage. Rev.　　　Integrated Pest Management Reviews
IOBC, WPRS Bull.　　　　　　International Organization of Biological Control,
　　　　　　　　　　　　　　　Western Palearctic Regional Section Bulletin

IPM Pract.　　　　　　　　　The IPM Practitioner
J. Agric. Entomol.　　　　　　Journal of Agricultural Entomology
J. Agric. Res.　　　　　　　　Journal of Agricultural Research
J. Agric. Sci.　　　　　　　　Journal of Agricultural Science
J. Agric. Univ. Puerto Rico　　Journal of Agriculture of the University of Puerto Rico
J. Am. Soc. Hort. Sci.　　　　Journal of the American Society for Horticultural Science
J. Appl. Biol.　　　　　　　　Journal of Applied Biology
J. Appl. Ecol.　　　　　　　　Journal of Applied Ecology
J. Aust. Entomol. Soc.　　　　Journal of the Australian Entomological Society
J. Biol. Sci. Res.　　　　　　Journal of Biological Science Research
J. Chem. Ecol.　　　　　　　Journal of Chemical Ecology
J. Econ. Entomol.　　　　　　Journal of Economic Entomology
J. Entomol. Sci.　　　　　　　Journal of Entomological Science
J. Entomol. Soc. Brit. Columbia　Journal of the Entomological Society of British
　　　　　　　　　　　　　　　Columbia
J. Ethol.　　　　　　　　　　Journal of Ethology
J. Georgia Entomol. Soc.　　　Journal of the Georgia Entomological Society
J. Insect Behav.　　　　　　　Journal of Insect Behavior
J. Insect Pathol.　　　　　　　Journal of Insect Pathology
J. Insect Physiol.　　　　　　Journal of Insect Physiology
J. Invertebr. Pathol.　　　　　Journal of Invertebrate Pathology
J. Kansas Entomol. Soc.　　　Journal of the Kansas Entomological Society
J. Rio Grande Valley Hort. Soc.　Journal Rio Grande Horticultural Society
J. Lepid. Soc.　　　　　　　　Journal of the Lepidopterists' Society
J. Nat. Hist.　　　　　　　　Journal of Natural History
J. Nematol.　　　　　　　　　Journal of Nematology
J. New York Entomol Soc.　　　Journal of the New York Entomological Society
J. Prod. Agric.　　　　　　　Journal of Production Agriculture
J. Range Manage.　　　　　　Journal of Range Management
J. Res. Lep.　　　　　　　　　Journal of Research on the Lepidoptera
J. Stored Prod. Res.　　　　　Journal of Stored Product Research
J. Zool.　　　　　　　　　　Journal of Zoology (London)
Manitoba Entomol.　　　　　　Manitoba Entomologist

Mem. Am. Entomol. Soc.	*Memoirs of the American Entomological Society*
Mem. Entomol. Inter.	*Memoirs on Entomology, International*
Mem. Entomol. Soc. Canada	*Memoirs of the Entomological Society of America*
Nature	*Nature*
Naturwissenschaften	*Naturwissenschaften*
Netherlands J. Zool.	*Netherlands Journal of Zoology*
New Zealand Entomol.	*New Zealand Entomologist*
Oecologia	*Oecologia*
Ohio J. Sci.	*Ohio Journal of Science*
Pan-Pac. Entomol.	*Pan-Pacific Entomologist*
Pedobiologia	*Pedobiologia*
Pest. Sci.	*Pesticide Science*
Physiol. Entomol.	*Physiological Entomology*
Phytoparasitica	*Phytoparasitica*
Phytopathology	*Phytopathology*
Phytoprotection	*Phytoprotection*
Plant Dis.	*Plant Disease*
Plant Dis. Rep.	*Plant Disease Reporter*
Plant Pathol.	*Plant Pathology*
Proc. Entomol. Soc. Brit. Columbia	*Proceedings of the Entomological Society of British Columbia*
Proc. Entomol. Soc. Ontario	*Proceedings of the Entomological Society of Ontario*
Proc. Entomol. Soc. Philadelphia	*Proceedings of the Entomological Society of Philadelphia*
Proc. Entomol. Soc. Washington	*Proceedings of the Entomological Society of Washington*
Proc. Florida State Hort. Soc.	*Proceedings of the Florida State Horticultural Society*
Proc. Hawaiian Entomol. Soc.	*Proceedings of the Hawaiian Entomological Society*
Proc. U.S. Nat. Mus.	*Proceedings of the United States National Museum*
Prot. Ecol.	*Protection Ecology*
Psyche	*Psyche*
Quaes. Entomol.	*Quaestiones entomologicae*
Quebec Soc. Prot. Plants	*Quebec Society for the Protection of Plants, Report*
Revista Bras. Entomol.	*Revista Brasileira de Entomologia*
Revista Bras. Zool.	*Revista Brasileira de Zoologia*
Rivista Instit. Colombiano Agropec.	*Rivista Instituto Colombiano Agropecuario*
Sci. Agric.	*Scientific Agriculture*
Sociobiology	*Sociobiology*
Soil Tillage Res.	*Soil and Tillage Research*
Southwestern Entomol.	*Southwestern Entomologist*
Syst. Entomol.	*Systematic Entomology*
Trans. R. Entomol. Soc.	*Transactions of the Royal Entomological Society*
Trends Agric. Sci.	*Trends in Agricultural Science*
Trop. Agric.	*Tropical Agriculture*
Trop. Pest Manage.	*Tropical Pest Management*
Trop. Sci.	*Tropical Science*
Univ. California Pub. Entomol.	*University of California Publications in Entomology*
Z. Angew. Entomol.	*Zeitschrift fur Angewandte Entomologie*

Glossary

Abdomen, the posterior of the three main body divisions of an insect.

Abiotic disease, a disease caused by factors other than pathogens (e.g., weather or chemicals.

Acaricide, a pesticide applied to manage mite populations, also called miticide.

Acid, having a pH of less than 7.

Action threshold, a level of pest abundance that stimulates action to protect plants from serious damage.

Adult, the sexually mature stage of an animal; usually the winged stage in insects; this stage does not molt.

Aestivation, a state of inactivity during the summer months.

Alate, bearing wings.

Alkaline, basic, having a pH greater than 7.

Allelopathy, the ability of a plant species to produce substances that are toxic to certain other plants.

Anal plate, the shield-like plate or dorsal covering on the terminal segment in immatures, usually dark in color.

Annual, a plant that normally completes its life cycle of seed germination, vegetative growth, reproduction, and death in a single growing season or year.

Antenna (pl., Antennae), the paired segmented sensory organs, borne one on each side of the head, commonly protruding foward and termed horns or feelers.

Anterior, in front; usually used to refer to the end of the body containing the head.

Apical, pertaining to the apex or outer end.

Apterous, lacking wings.

Arista, a large hair or bristle on the antenna of flies.

Axil, the upper angle where a leaf or twig joins the stem from which it grows.

Bacillus thuringiensis, a bacterium that causes disease in insects; formulations of the bacteria are used as insecticides, with different strains used for suppression of caterpillars, beetles, and flies.

Band, a transverse line, usually wide, crossing the body (often confused with "stripe," a term used to designate a longitudinal line running the length of the body).

Band application, an application in which an insecticide is applied in strips, usually to the bed or seed row.

Basal, at or pertaining to the base or point of attachment, or nearest the main body.

Beneficials, organisms that provide a benefit to crop production, especially natural enemies of pests and plant pollinators such as bees.

Biennial, a plant that completes its life cycle in more than one year and usually does not flower until the second growing season.

Binomial sampling, a sampling method that involves recording only the presence or absence of members of the population being sampled (such as an insect pest) on a sample unit (such as a leaf), rather than counting the numbers of individuals; presence-absence sampling.

Biological control, the action of parasites, predators, or pathogens in maintaining another organism's population density at a lower average level than would occur in their absence. Biological control may occur naturally in the fied or result from manipulation or introduction of biological control agents.

Biotic diease, disease caused by a pathogen, such as a bacterium, fungus, or virus.

Botanical, derived from plants or plant parts; often used to describe insecticides derived from plant.

Brachypterous, having short wings or elytra.

Bract, a modified leaf at the base of a flower.

Broadcast application, the application of a material such as an insecticide to the entire surface of a field.

Brood, all the individuals of a generation; insects that hatch at about the same time.

Bt, acronym for *Bacillus thuringiensis.*

C, centigrade; a unit of temperature on the centigrade thermometer.

Calibrate, to standardize or correct the measuring devices on instruments; to adjust nozzles on a spray apparatus.

Canker, a dead, discolored, often sunken area (lesion) on a root, trunk, stem, or branch.

Canopy, the leafy part of plants or trees.

Carcinogen, a substance or agent capable of causing cancer.

Carnivorous, feeding on animals.

Caterpillar, the larva of a butterfly, moth, sawfly, or scorpionfly.

Catkin, a cluster of flowers in the form of a spike.

Cauda, the pointed tip of the abdomen in aphids.

Caudal, pertaining to the anal end of the insect body.

Cercus, a process located laterally near the tip of the abdomen; often an important diagnostic character in identification.

Cervical shield, plate on the dorsal surface of caterpillars just behind the head; also known as thoracic plate.

Chelicerae, the pincer-like first pair of appendages in arachnids.

Chlorophyll, the green pigment of plants that captures, from sunlight, the energy necessary for photosynthesis.

Chlorosis, yellowing or bleaching of normal green plant tissue usually caused by the loss of chlorophyll.

Chlorinated hydrocarbons, a class of synthetic insecticides containing chlorine as one of the constituents; chlorinated hydrocarbon insecticides are typically very persistent and earlier were used widely for soil and seed treatments; also known as organochlorines.

Chorion, the outer layer of an insect egg.

Circulative virus, a virus that systemically infects its insect vector and usually is transmitted for the remainder of the vector's life.

Cladophyll, in plants, a branchlet that functions as a leaf.

Claw, a hollow, sharp organ, located at the tip of the tarsus (foot).

Clublike, a structure with a knob or swelling distally, usually used in reference to antennae.

cm, centimeter; a unit of length equivalent to 1/100 of a meter or 0.3937 inches.

Cocoon, a sheath, usually of silk, formed by an insect larva as a chamber for pupation.

Collophore, a tube-like structure located ventrally on the first abdominal segment of springtails (Collembola).

Companion planting, the practice of planting certain plant species often herbs on close association with crop plants to repel the practice of planting certain pests.

Complete metamorphosis, change in body form in which the insect displays the egg, larval, pupal and adult stages.

Corium, the thickened-basal region of the front wing in Hemiptera.

Cornicle, two tubular structures located on the posterior part of an aphid's abdomen.

Cortex, tissue between the phloem and the epidermis in roots and stems.

Costa, the basal segment of the leg, articulating with the body.

Cotyledon, a leaf formed within the seed and present on a seedling at germination; seed leaf.

Cover crops, cultivation of a second type of crop primarily to improve the production system for a primary crop (e.g., legumes maintained during the winter season to improve soil condition).

Crawler, the active first instar of a scale insect.

Crochets, minute hooks on the prolegs of caterpillars.

Cross resistance, in pest management, resistance of a pest population to a pesticide to which it has not been exposed that accompanies the development of resistance to a pesticide to which it has been exposed.

Crown, the basal region of a plant, usually at the juncture of the stem and roots.

Cultivar, an agricultural plant variety or strain developed for specific horticultural properties.

Dactyls, large finger-like tibial projections found on the front legs of mole crickets.

Degree-day, a unit combining temperature and time, and used to measure and predict growth of organisms; sometimes called "day-degrees."

Developmental threshold, the lowest temperature at which growth occur.

Diapause, a period of dormancy ("hiberation") in arthropods.

Disk, a soil cultivator made up of many circular blades that act to break up the soil into smaller aggregates.

Distal, pertaining to the part of an appendage furthest from the body.

Dormant, to become inactive during periods of cold weather.

Dorsal, referring to the upper surface.

Economic threshold, a level of pest abundance or damage at which the cost of control equals the crop value gained from instituting the control procedure.

Ectoparasite, a parasite that lives outside (on the surface) of the host.

Egg, the first free-living stage of most animals, contained within a chorion (shell).

Elytron (pl., Elytra), the thickened front wing of beetles, serving primarily for protection of the hind wings or flight wings.

Endoparasite, a parasite that lives inside the host.

Entomopathogenic, causing disease and/or death in insects.

Epicranium, the upper portion of the head.

Epidermis, the outermost layer of living cells on a plant or animal.

Extrafloral nectary, a nectar-secreting gland found outside the flower.

F, fahrenheit; a unit of temperature on the Fahrenheit thermometer.

Fallow, cultivated land allowed to remain free of crops during the normal growing season.

Femur (pl., Femora), a segment of the leg between the trochanter and tibia; one of the two largest portions of the insect leg, often expanded to enhance hopping.

Furcula, a structure found beneath the abdomen of springtails (Colembolla), attached apically, and used for leaping.

Frass, the solid fecal material produced by insects.

Fumigation, treatment that uses a pesticide applied in a gaseous form.

Fungicide, a pesticide used to manage growth of fungi.

Gaster, the swollen, terminal abdominal segments of ants behind the constriction or pedicel.

Genitalia, the modified abdominal segments used in copulation.

h, hour; a unit of time equivalent to 1/24 of a day.

Halter (pl., Halteres), the vestigial, modified hind wings of Diptera that function as balancing organs.

Hemelytra, the front wings in Hemiptera, in which the basal portion is thickened and the distal portion is membranous.

Herbivore, animal that consumes plant tissue.

Herbivory, consumption of plant material.

Honeydew, sugary anal secretions of aphids and some other insects.

Horticultural oil, petroleum or botanical oil used to control pests on plants.

Host, an organism that provides food and/or shelter for another organism.

Hyphae, the thread-like mycelial tissue of fungi.

Immature, the feeding stage of insects after birth but before adulthood, includes both larvae and nymphs.

Incomplete metamorphosis, change in body form in which the insect display the egg, nymphal, and adult stages.

Infection, the entry and establishment of a pathogen into a host.

Infestation, the presence of numerous pests in a field, on plants, or in the soil.

Inoculum, a small number of pests or pathogens that can, or does, lead to a greater abundance, and subsequent damage.

Inorganic, not containing the element carbon; usually derived from naturally occurring minerals.

Insecticide, a pesticide applied to manage insect populations.

Instar, the stage of larvae and nymphs between molts.

Integrated pest management, an approach to prevention of damage by pests that focuses on the long-term suppression of pest abundance through a combination of environmentally sound techniques such as biological control and resistant plant varieties, rather than depending solely on chemical insecticides; insecticides are applied only when needed rather than routinely; also known as "IPM."

Integument, the outer covering or cuticle of an insect.

Invertebrate, animals lacking and internal skeleton or "backbone," such as insects, mites, and worms.

IPM, see Integrated Pest Management.

Juvenile, the immature form of a nematode; the stages between the egg and adult form.

Larva (pl., Larvae), the growing stage of an insect with complete metamorphosis; the feeding stage between the egg and pupal stages; in mites the immature stages are also called larvae; larval insects usually bear little resemblance to the adult stage.

Lesion, a localized area of damage or infection, usually discolored or deformed.

Lodging, toppling of crop plants, often from wind or rain, and ultimately due to destruction of roots by insects.

Mandible, one of the pair of jaws in insects, attached to the head; usually stout and tooth-like in appearance.

Median, at or near the middle.

Meristem, the cells capable of division located at the growing point of a plant.

Mesothorax, the second or middle thoracic segment; the segment bearing the middle pair of legs; the segment bearing the first pair of wings.

Metamorphosis, a change in body form as the insect grows from the immature to the adult stage.

m, meter; a unit of length equivalent to 39.37 inches.

Microbial pesticide, microbial organisms that are applied like chemical pesticides for the suppression of pest abundance; various bacterial, fungal, viral, and microsporidian pathogens registered with the Environmental Protection Agency for use as a pest control agent.

Middorsal, refers to the middle of the upper region or "back" (as opposed to the subdorsal or slightly to the side or lateral).

Mineral oil, horticultural oil that is derived from petroleum.

Miticide, a pesticide used to control mites; also called "acaricide."

mm, millimeter; a unit of length equivalent to 1/10 of a centimeter or 0.0394 inches.

Molt, to shed or cast off the outer body covering, a necessary prerequisite to grow and attain the adult stage.

Monitoring, careful observation of pest abundance and damage; pest scouting.

Motile, active, able to move freely.

Mulch, a layer of material placed on the soil to retard growth of weeds; in commercial crop production plastic is often used, whereas in the home garden organic materials such as leaves, straw, wood chips, and pine needles are often used.

Mummy, the remains of an aphid whose insides have been consumed by a parasitoid.

Natural enemies, organisms normally killing arthropods that are considered to be pests without human intervention; examples include predatory insects or vertebrates, insect parasitoids, and microbial pathogens causing disease.

Neem, a botanical insecticide derived from the tropical tree *Azadirachta indica*; extracts of the foliage confer both feeding deterrent and toxic properties.

Neonate, recently hatched.

Nocturnal, active at night.

Non-persistent virus, a virus carried on the mouthparts of a vector, and usually lost after the insect feeds a few times; a styletborne virus.

Nymph, the immature stage of insects with incomplete metamorphosis; the nymph usually resembles the adult except for undeveloped wings.

Oil, see Horticultural Oil.

Omnivorous, having a broad diet, consisting of both plant and animal material.

Organic, chemically, this is a material that contains the elements carbon and hydrogen; agriculturally, this refers to plants cultured without synthetic fertilizers or pesticides.

Organophosphate, a class of synthetic insecticides derived from phosphoric acid, capable of disrupting neurotransmission in insects and vertebrates.

Oviparous, producing eggs.

Ovipositor, the structure, usually in the shape of a tube, that is used to deposit eggs.

Palps, small antenna-like sensory appendages attached to the mouth.

Parasite, an organism that obtains its food by feeding on the body of another organism, its host.

Parasitoid, a parasite that kills its host at about the time the parasite completes it development.

Parthenogenesis, development from an egg that is not fertilized.

Pathogen, a disease-causing microbial organism.

Pedicel, the constricted region of the abdomen in Hymenoptera; in ants the pedicel bears one or more upright lobes.

Perennial, a plant that lives at least three years and reproduces at least twice.

Persistent virus, a virus that passes through the body of the vector, and that usually persists for the remainer of the vector's life.

Pesticide, a material that kills pests; this term is often used to describe insecticides, which are pesticides that kill insects.

Petiole, stalk that connects the leaf to a stem; in Hymenoptera it is sometimes used in place of pedicel.

Pheromone, chemical substance secreted into the environment that affects the behavior or physiology of other members of the same species.

Phytophagous, feeding on plants or plant products.

Phytotoxicity, damage to a plant due to contact with a chemical toxin.

Plumose, feather-like in structure with a single thick stem and numerous parallel branches; usually used to describe antennae of Lepidoptera.

Pollinator, the agent of pollen transfer in plants, often bees.

Posterior, the hind region of the body, or referring to the end containing the anus.

Postplant, refers to treatments applied to a crop after planting.

Preplant, refers to treatments applied to a crop before planting.

Prepupa, an active but non-feeding stage of insects; the period immediately preceding the molt to the pupal stage.

Proleg, a fleshy, unsegmented leg found on the abdomen of caterpillars.

Pronotum, the upper or dorsal surface of the prothorax.

Prothoracic plate, equivalent to thoracic plate.

Prothorax, the most anterior of the three thoracic segments.

Proximal, pertaining to the part of an appendage closer to the body.

Pubescence, a covering of setae (hairs).

Pubescent, covered with hair-like structures (setae in insects, trichomes in plants).

Punctate, containing impressed points, punctures, or dimples.

Pupa, the non-feeding, immobile stage between the larval and adult stages in insects with complete metamorphosis; a stage where major reorganization of the body take place.

Pupate, to molt from the larval stage to the pupa.

Puparium, the hardened, thickened integument of the last instar larva of Diptera, in which the pupa is formed (plural, puparia).

Pyrethroids, synthetic insecticides that are structurally similar to the toxic components of pyrethrum; also called synthetic pyrethroids.

Pyrethrum, natural insecticide derived from certain plants in the genus *Chrysanthemum*; it is highly valued for its rapid effects on insects and low toxicity to mammals.

Reservoir, a site where organisms can survive, usually in relatively small numbers, and then invade or repopulate an area.

Resistant, tolerant of conditions that are deleterious to other strains of the same species; usually applied to plant tolerance of pest damage, or arthropod tolerance of pesticides.

Reticulations, a net-like structure, usually referring to the pigmented pattern on the head of Lepidoptera.

Rostrum, in weevils, the snout-like prolongation of the head containing the mouthparts distally; in Hemiptera, this sometimes refers to the beak or piercing-sucking mouthparts.

Rotation, in agriculture, purposeful alternation of crops grown on the same plot of land; in pest control, purposeful alternation of insecticides used to control a pest population.

Row covers, a covering, usually consisting of spun-bonded polyester, that is placed over crops to protect them from adverse weather or pests.

Sanitation, the practice of eliminating pests, or materials or sites that might harbor pests.

Saprophagous, feeding on dead or dying plant or animal tissue.

Scale, a modified, flattened seta on the surface of an insect.

Scientific name, a Latin or Latinized name given to all biological organisms and consisting of two parts—a genus and species; the scientific name often include the name of the individual(s) originally describing the species.

Scutellum, in Hemiptera, the triangular mesothoracic region between the base of the wings.

Segment, a major subdivision of the body or appendage, separated from other segments by areas of flexibility.

Selective pesticide, pesticides that are toxic principally to the target pest; having few adverse effects on non-target organisms.

Semicircular, like the half of a circle.

Senescence, the period of life after maturity.

Sequential sampling, a sampling protocol where the sampling continues only until a decision can be reached, rather than requiring a fixed number of samples to be taken.

Serrate, notched, like the teeth of a saw.

Sessile, immobile; incapable of moving.

Sex pheromone, pheromones that attract the opposite sex for mating; these are often used, in conjunction with traps, to monitor abundance of insects.

Sheath, a structure enclosing others.

Skeletonize, to remove the tissue of a leaf except for the veins, leaving a "skeleton."

Sooty mold, a dark-colored fungus growing on the honeydew secreted by insects, usually aphids or scales.

Spermatophore, a covering or capsule around the sperm.

Spindle-shaped, elongate-cylindrical, thicker in the middle and tapering to each end.

Spine, a large, stout seta or thorn-like process.

Spiracle, an external opening of the system of ducts used to transfer atmospheric gases into, and out of, the body of arthropods; they are commonly found along each side of the body.

Spore, a reproductive stage of fungi, usually somewhat resistant to adverse environmental conditions that is capable of growing into a new organism.

sq, square; a quantity multiplied by itself.

Stemmata, the small, simple eyes found on some insects, usually on the side of the head.

Stolon, a basal branch or stem that is inclined to root; a plant runner.

Stridulate, to produce a noise by rubbing together two surfaces.

Stripe, a line that runs horizontally or lengthwise on an insect (often confused with "band" or transverse line).

Stylet, the elongated needle-like portions of the piercing-sucking type of insect mouthparts.

Styletborne virus, see Non-persistent Virus.

Subdorsal, a region between the dorsal and lateral areas.

Submedial arc, pigmentation in the form of an arc occurring on the face or top of the head in caterpillars.

Subspiracular, the area immediately below the spiracles.

Supraspiracular, the area immediately above the spiracles.

Taproot, a large primary root growing downwards, and giving rise to numerous, smaller lateral roots.

Tarsus, the portion of the insect distal to the tibia, and often bearing claws; the foot.

Tegmen (pl., Tegmina), the thickened front wing of Orthoptera and related insect, and the thickened-basal portion of the front wings of Hemiptera.

Thoracic plate, shield-like dorsal covering or plate on the body segment immediately behind the head, usually dark in color; also known as cervical shield.

Thread-like, long, thin structure, approximately equal in diameter throughout; usually used in reference to the antennae.

Throrax, the second or middle of the three major body regions of insects, and the section bearing wings and jointed (true) legs.

Tibia, the section of the insect leg between the femur and the tarsus, usually one of the largest sections and ofen bearing spines or spurs.

Trap crop, a crop or portion of a crop that is intended to lure insects away from the main crop.

Trapezoidal, a four-sided figure in which two sides are parallel and two are not.

Trichomes, hairs or small spines on the surface of a plant.

Trochanter, a small section of the leg connecting the coxa and femur.

Tubercle, a small raised area or pimple; in caterpillars a hair often originates from these raised areas.

Tympanum, a membrane-covered cavity on the thorax, abdomen, or leg of insects that functions like an ear.

Variety, an identifiable strain of a species, usually bred for a particular horticultural purpose; also called a "cultivar".

Vascular system, the system of plant tissues that conducts water, minerals, and products of photosynthesis within the plant.

Vein, a tube running through the wings of insect, through which blood is pumped.

Ventral, the lower surface.

Virulence, the ability of a pathogen to infect a host and cause disease.

Viviparous, bearing living young, as opposed to eggs.

Vestigial, small or degenerate; the remains of a previously functioning organ.

Volunteer plants, the unexpected and undesired emergence of plants, usually self-seeded by the previous plants.

Whorl, the arrangement of leaves in a circle around the stem.

Wing, paired membranous flight organs of insects, originating at the mesothorax and metathorax; the number varies (0, 2, or 4) and they may or may not be functional.

Wing pad, the underdeveloped wings of nymphs.

References

Abdel–Gawaad, A. A. and A. Y. El–Shazli. 1971. Studies on the common cutworm *Agrotis ypsilon* Rott. I. Life cycle and habits. *Z. Angew. Entomol.* 68: 409–412.

Abdel–Malek, A., N. Z. Dimetry, S. El–Ziady, and F. M. El–Hawwary. 1982. Ecological studies on *Aphis craccivora* Koch. III. The role of day length as an environmental factor regulating development and form produced. *Z. Angew. Entomol.* 93: 238–243.

Abdul–Nasr, S. and A. M. Awadalla. 1957. External morphology and biology of the bean pod-borer, *Etiella zinckenella* Treit. *Bull. Soc. Entomol. Egypte* 41: 591–620.

Abdul–Sattar, A. A. and T. F. Watson. 1982. Effects of *Bacillus thuringiensis* var. *kurstaki* on tobacco budworm (Lepidoptera: Noctuidae) adult and egg stages. *J. Econ. Entomol.* 75: 596–598.

Abrew–Rodriguez, E. and M. Perez–Escobar. 1983. Oviposition preference of *Diaprepes abbreviatus* (Coleoptera: Curculionidae) on various ornamental plants. *J. Agric. Univ. Puerto Rico* 67: 117–120.

Adams, C. L. and R. B. Selander. 1979. The biology of blister beetles of the Vittata group of the genus *Epicauta* (Coleoptera, Meloidae). *Bull. Am. Mus. Nat. Hist.* 162. 266 pp.

Adams, C. T. 1983. Destruction of eggplants in Marion County, Florida by red imported-fire ants (Hymenoptera: Formicidae). *Florida Entomol.* 66: 518–520.

Adams, C. T. 1986. Agricultural and medical impact of the imported fire ants. Pages 48–57 *in* C. S. Lofgren and R. K. Vander Meer (eds.). *Fire Ants and Leaf-Cutting Ants, Biology and Management.* Westview Press, Boulder.

Adams, C. T., T. E. Summers, C. S. Lofgren, D. A. Focks, and J. C. Prewitt. 1981. Interrelationship of ants and the sugarcane borer in Florida sugarcane fields. *Environ. Entomol.* 10: 415–418.

Adams, J. A. 1949. The oriental beetle as a turf pest associated with the Japanese beetle in New York. *J. Econ. Entomol.* 42: 366–371.

Adams, J. R., R. L. Wallis, T. A. Wilcox, and R. M. Faust. 1968. A previously undescribed polyhedrosis of the zebra caterpillar, *Ceramica picta*. *J. Invertebr. Pathol.* 11: 45–58.

Adams, R. G. and L. M. Los. 1986. Monitoring adult corn flea beetles (Coleoptera: Chrysomelidae) in sweet corn fields with color sticky traps. *Environ. Entomol.* 15: 867–873.

Adams, R. G., R. A. Ashley, and M. J. Brennan. 1990. Row covers for excluding insect pests from broccoli and summer squash plantings. *J. Econ. Entomol.* 83: 948–954.

Adams, R. G., K. D. Murray, and L. M. Los. 1989. Effectiveness and selectivity of sex pheromone lures and traps for monitoring fall armyworm (Lepidoptera: Noctuidae) adults in Connecticut sweet corn. *J. Econ. Entomol.* 82: 285–290.

Adler, P. H. and C. R. L. Adler. 1988. Behavioral time budget for larvae of *Heliothis zea* (Lepidoptera: Noctuidae) on artificial diet. *Ann. Entomol. Soc. Am.* 81: 682–688.

Adler, P. H., M. B. Willey, and M. R. Bowen. 1991. Temporal oviposition patterns of *Heliothis zea* and *Spodoptera ornithogalli*. *Entomol. Exp. Appl.* 58: 159–164.

Adlerz, W. C. 1975. Natural control of three rindworm species and chemical control of the granulate cutworm, *Feltia subterranea* on watermelon. *Proc. Florida State Hort. Soc.* 88: 204–207.

Adlerz, W. C. and P. H. Everett. 1968. Aluminum foil and white polyethylene mulches to repel aphids and control watermelon mosaic. *J. Econ. Entomol.* 61: 1276–1279.

Afun, J. V. K., L. E. N. Jackai, and C. J. Hodgson. 1991. Calendar and monitored insecticide application for the control of cowpea pests. *Crop Prot.* 10: 363–370.

Ahmad, T. R., K. P. Pruess, and S. D. Kindler. 1984. Non-crop grasses as hosts for the chinch bug, *Blissus leucopterus* (Say) (Hemiptera: Lygaeidae). *J. Kansas Entomol. Soc.* 57: 17–20.

Ainslie, G. G. 1922. Webworms injurious to cereal and forage crops and their control. *USDA Farmers' Bull.* 1258. 16 pp.

Ainslie, G. G. 1923a. Striped sod webworm, *Crambus mutabilis* Clemens. *J. Agric. Res.* 24: 399–414.

Ainslie, G. G. 1923b. Silver-striped webworm, *Crambus praefectellus* Zincken. *J. Agric. Res.* 24: 415–425.

Ainslie, G. G. 1927. The larger sod webworm. *USDA Tech. Bull.* 31. 17 pp.

Airey, W. J. 1986. The influence of an alternative food on the effectiveness of propietary molluscicidal pellets against two species of slugs. *J. Mollusc. Stud.* 52: 206–213.

Akey, D. H. and S. D. Beck. 1971. Continuous rearing of the pea aphid, *Acyrthosiphon pisum*, on a holidic diet. *Ann. Entomol. Soc. Am.* 64: 353–356.

Alborn, H., H. Karlsson, L. Lundgren, P. Ruuth, and G. Stenhagen. 1985. Resistance in crop species of the genus *Brassica* to oviposition by the turnip root fly, *Hylemya floralis*. *Oikos* 44: 61–69.

Alcock, J. 1971. The behavior of a stinkbug, *Eustichus conspersus* Uhler (Hemiptera: Pentatomidae). *Psyche* 78: 215–228.

Alcock, J. and K. M. O'Neill. 1986. Density-dependent mating tactics in the Grey hairstreak, *Strymon melinus* (Lepidoptera: Lycaenidae). *J. Zool.* (A) 209: 105–113.

Aldrich, J. R., J. W. Neal, Jr., J. E. Oliver, and W. R. Lusby. 1991a. Chemistry vis-a-vis maternalism in lace bugs (Heteroptera: Tingidae): alarm pheromones and exudate defense in *Corythucha* and *Gargaphia* species. *J. Chem. Ecol.* 17: 2307–2322.

Aldrich, J. R., M. P. Hoffmann, J. P. Kochansky, W. R. Lusby, J. E. Eger, and J. A. Payne. 1991b. Identification and attractiveness of a major pheromone component for nearctic *Euschistus* spp. stink bugs (Heteroptera: Pentatomidae). *Environ. Entomol.* 20: 477–483.

Aldrich, J. R., W. R. Lusby, B. E. Marron, K. C. Nicolaou, M. P. Hoffmann, and L. T. Wilson. 1989. Pheromone blends of green stink bugs and possible parasitoid selection. *Naturwissenschaften* 76: 173–175.

Alexander, C. P. and G. W. Byers. 1981. Pages 153–190 *in* J. F. McAlpine, B. V. Peterson, G. E. Shewell, H. J. Teskey, J. R. Vockeroth, and D. M. Wood (eds.). *Manual of Nearctic Diptera*, Volume 1. *Agric. Canada Res. Branch Mono.* 27.

Alexander, R. D. 1957. The taxonomy of the field crickets of the eastern United States (Orthoptera: Gryllidae: Acheta). *Ann. Entomol. Soc. Am.* 50: 584–602.

Alexander, R. D. and G. H. Meral. 1967. Seasonal and daily chirping cycles in the northern spring and fall field crickets, *Gryllus veletis* and *G. pennsylvanicus. Ohio J. Sci.* 67: 200–209.

Al-Ghamdi, K. M., R. K. Stewart, and G. Boivin. 1995. Synchrony between populations of the tarnished plant bug, *Lygus lineolaris* (Palisot de Beauvois) (Hemiptera: Miridae), and its egg parasitoids in southwestern Quebec. *Can. Entomol.* 127: 457–472.

Ali, A. and M. J. Gaylor. 1992. Effects of temperature and larval diet on development of the beet armyworm (Lepidoptera: Noctuidae). *Environ. Entomol.* 21: 780–786.

All, J. N. 1988. Fall armyworm (Lepidoptera: Noctuidae) infestations in no-tillage cropping systems. *Florida Entomol.* 71: 268–272.

All, J. N. and R. N. Gallaher. 1977. Detrimental impact of no-tillage corn-cropping systems involving insecticides, hybrids, and irrigation on lesser cornstalk borer infestations. *J. Econ. Entomol.* 70: 361–365.

All, J. N., R. N. Gallaher, and M. D. Jellum. 1979. Influence of planting date, preplanting weed control, irrigation, and conservation tillage practices on efficacy of planting time insecticide applications for control of lesser cornstalk borer in field corn. *J. Econ. Entomol.* 72: 265–268.

All, J. N., A. Javid, and P. Guillebeau. 1986. Control of fall armyworm with insecticides in north Georgia sweetcorn. *Florida Entomol.* 69: 598–602.

All, J. N., J. D. Stancil, T. B. Johnson, and R. Gouger. 1996. Controlling fall armyworm (Lepidoptera: Noctuidae) infestations in whorl stage corn with genetically modified *Bacillus thuringiensis* formulations. *Florida Entomol.* 79: 311–317.

Allen, C. R., S. Demarais, and R. S. Lutz. 1994. Red imported fire ant impact on wildlife: an overview. *Texas J. Sci.* 46: 51–59.

Allen, W. R. and W. L. Askew. 1970. A simple technique for mass-rearing the onion maggot (Diptera: Anthomyiidae) on an artificial diet. *Can. Entomol.* 102: 1554–1558.

Alleyne, E. H. and F. O. Morrison. 1977a. The effects of insecticides on the lettuce root aphid, *Pemphigus bursarius* (L.). *Ann. Soc. Entomol. Quebec* 22: 36–39.

Alleyne, E. H. and F. O. Morrison. 1977b. The lettuce root aphid, *Pemphigus bursarius* (L.) (Homoptera: Aphidoidea) in Quebec, Canada. *Ann. Entomol. Soc. Quebec* 22: 171–180.

Alleyne, E. H. and F. O. Morrison. 1977c. The natural enemies of the lettuce root aphid, *Pemphigus bursarius* (L.) in Quebec, Canada. *Ann. Entomol. Soc. Quebec* 22: 181–187.

Alleyne, E. H. and F. O. Morrison. 1978a. Vertical distribution of overwintering soil apterae of the lettuce root aphid, *Pemphigus bursarius* (L.) in Quebec. *Ann. Soc. Entomol. Quebec* 23: 155–157.

Alleyne, E. H. and F. O. Morrison. 1978b. Resistance of lettuce varieties to attacks by the root apterae of *Pemphigus bursarius* (L.) *Ann. Soc. Entomol. Quebec* 23: 158–167.

Allsopp, P. G. 1980. The biology of false wireworms and their adults (soil-inhabiting Tenebrionidae) (Coleoptera): a review. *Bull. Entomol. Res.* 70: 343–379.

Allsopp, P. G., M. G. Klein, and E. L. McCoy. 1992. Effect of soil moisture and soil texture on oviposition by Japanese beetle and rose chafer (Coleoptera: Scarabaeidae). *J. Econ. Entomol.* 85: 2194–2200.

Allsopp, P. G., T. L. Ladd, Jr., and M. G. Klein. 1992. Sample sizes and distributions of Japanese beetles (Coleoptera: Scarabaeidae) captured in lure traps. *J. Econ. Entomol.* 85: 1797–1801.

Allyson, S. 1976. North American larvae of the genus *Loxostege* Hübner (Lepidoptera: Pyralidae: Pyraustinae). *Can. Entomol.* 108: 89–104.

Alm, S. R., M. C. Villani, and W. Roelofs. 1999. Oriental beetles (Coleoptera: Scarabaeidae): current distribution in the United States and optimization of monitoring traps. *J. Econ. Entomol.* 92: 931–935.

Alm, S. R., F. R. Hall, T. P. McGovern, and R. N. Williams. 1986. Attraction of *Glischrochilus quadrisignatus* (Coleoptera: Nitidulidae) to semiochemicals: butyl acetate and propyl propionate. *J. Econ. Entomol.* 79: 654–658.

Alm, S. R., R. N. Williams, T. P. McGovern, and F. R. Hall. 1989. Effective chemical structures, release methods, and trap heights for attracting *Glischrochilus quadrisignatus* (Coleoptera: Nitidulidae). *J. Econ. Entomol.* 82: 477–481.

Alm, S. R., T. Yeh, J. L. Hanula, and R. Georgis. 1992. Biological control of Japanese, oriental, and black turfgrass ataenius beetle (Coleoptera: Scarabaeidae) larvae with entomopathogenic nematodes (Nematoda: Steinernematidae, Heterorhabditidae). *J. Econ. Entomol.* 85: 1660–1665.

Alm, S. R., T. Yeh, M. L. Campo, C. G. Dawson, E. B. Jenkins, and A. E. Simeoni. 1994. Modified trap designs and heights for increased capture of Japanese beetle adults (Coleoptera: Scarabaeidae). *J. Econ. Entomol.* 87: 775–780.

Altieri, M. A. and S. R. Gliessman. 1983. Effects of plant diversity on the density and herbivory of the flea beetle, *Phyllotreta cruciferae* Goeze, in California collards (*Brassica oleracea*) cropping systems. *Crop Prot.* 2: 497–501.

Aluja, M. 1996. Future trends in fruit fly management. Pages 309–320 *in* B. A. McPheron and G. J. Steck (eds.) Fruit Fly Pests: a World Assessment of their Biology and Management. St. Lucie Press, Delray Beach, Florida.

Alvarado, L., D. B. Hogg, and J. L. Wedberg. 1989. Effects of corn and selected weed species on feeding behavior of the stalk borer, *Papaipema nebris* (Lepidoptera: Noctuidae). *Great Lakes Entomol.* 22: 65–69.

Alvarado–Rodriguez, B. 1987. Parasites and disease associated with larvae of beet armyworm *Spodoptera exigua* (Lepidoptera: Noctuidae), infesting processing tomatoes in Sinaloa, Mexico. *Florida Entomol.* 70: 444–449.

Alvarez, A. and G. Sanchez. 1981. Ciclo de vida y descripcion del gusano agrimensor, *Mocis latipes*. Rivista Instit. Colombiano Agropec. 16: 57–63.

Al–Zubaidi, F. S. and J. L. Capinera. 1983. Application of different nitrogen levels to the host plant and cannibalistic behavior of beet armyworm, *Spodoptera exigua* (Hübner) (Lepidoptera: Noctuidae). *Environ. Entomol.* 12: 1687–1689.

Al–Zubaidi, F. S. and J. L. Capinera. 1984. Utilization of food and nitrogen by the beet armyworm, *Spodoptera exigua* (Hübner) (Lepidoptera: Noctuidae), in relation to food type and dietary nitrogen levels. *Environ. Entomol.* 13: 1604–1608.

Al–Zubaidi, F. S. and J. L. Capinera. 1986. The effects of different types of food on the biology of beet armyworm, *Spodoptera exigua* (Hubn.) (Lepidoptera: Noctuidae). *J. Biol. Sci. Res.* 17: 153–160.

Ameen, A. O. and R. N. Story. 1997a. Fecundity and longevity of the yellowmargined leaf beetle (Coleoptera: Chrysomelidae) on crucifers. *J. Agric. Entomol.* 14: 157–162.

Ameen, A. O. and R. N. Story. 1997b. Feeding preferences of larval and adult *Microtheca ochroloma* (Coleoptera: Chrysomelidae) for crucifer foliage. *J. Agric. Entomol.* 14: 363–368.

Anderson, A. and J. A. Sharman. 1983. Effect of chlorfenvinphos and isofenphos on Carabidae and Staphylinidae (Col.) and their predation of eggs of *Delia floralis* Fallén (Diptera, Anthomyiidae) in field experiments. *Z. Angew. Entomol.* 95: 206–213.

Anderson, L. D. 1954. The tomato russet mite in the United States. *J. Econ. Entomol.* 47: 1001–1005.

Anderson, T. E., G. G. Kennedy, and R. E. Stinner. 1984. Distribution of the European corn borer, *Ostrinia nubilalis* (Hübner) (Lepidoptera: Pyralidae), as related to oviposition preference of the spring-colonizing generation in eastern North Carolina. *Environ. Entomol.* 13: 248–251.

Andow, D. A. 1990. Characterization of predation on egg masses of *Ostrinia nubilalis* (Lepidoptera: Pyralidae). *Ann. Entomol. Soc. Am.* 83: 482–486.

Andow, D. A., G. C. Klacan, D. Bach, and T. C. Leahy. 1995. Limitations of *Trichogramma nubilale* (Hymenoptera: Trichogrammatidae) as an inundative biological control of *Ostrinia nubilalis* (Lepidoptera: Crambidae). *Environ. Entomol.* 24: 1352–1357.

Andreadis, T. G. 1982a. Current status of imported and native parasites of the European corn borer (Lepidoptera: Pyralidae) in Connecticut. *J. Econ. Entomol.* 75: 626–629.

Andreadis, T. G. 1982b. Impact of *Nosema pyrausta* on field populations of *Macrocentrus grandii*, an introduced parasite of the European corn borer, *Ostrinia nubilalis*. *J. Invertebr. Pathol.* 39: 298–302.

Andreadis, T. G. 1984. Epizootiology of *Nosema pyrausta* in field populations of the European corn borer (Lepidoptera: Pyralidae). *Environ. Entomol.* 13: 882–887.

Andrews, G. L., F. A. Harris, P. P. Sikorowski, and R. E. McLaughlin. 1975. Evaluation of *Heliothis* nuclear polyhedrosis virus in a cottonseed oil bait for control of *Heliothis virescens* and *H. zea* on cotton. *J. Econ. Entomol.* 68: 87–90.

Angalet, G. W. and R. Fuester. 1977. The *Aphidius* parasites of the pea aphid *Acyrthosiphon pisum* in the eastern half of the United States. *Ann. Entomol. Soc. Am.* 70: 87–96.

Angalet, G. W. and N. A. Stevens. 1977. The natural enemies of *Brachycolus asparagi* in New Jersey and Delaware. *Environ. Entomol.* 6: 97–100.

Annan, I. B. and M. K. Bergman. 1988. Effects of the onespotted stink bug (Hemiptera: Pentatomidae) on growth and yield of corn. *J. Econ. Entomol.* 81: 649–653.

Annan, I. B., W. M. Tingey, and G. A. Schaefers. 1997. Population dynamics and clonal comparisons of cowpea aphid (Homoptera: Aphididae) on resistant and susceptible cowpea cultivars. *Environ. Entomol.* 26: 250–255.

Annis, B. and L. E. O'Keeffe. 1987. Influence of pea genotype on parasitization of the pea weevil, *Bruchus pisorum* (Coleoptera: Bruchidae) by *Eupteromalus leguminis* (Hymenoptera: Pteromalidae). *Environ. Entomol.* 16: 653–655.

Annis, B., G. Tamaki, and R. E. Berry. 1981. Seasonal occurrence of wild secondary hosts of the green peach aphid, *Myzus persicae* (Sulzer), in agricultural systems in the Yakima Valley. *Environ. Entomol.* 10: 307–312.

Antignus, Y., N. Mor, R. B. Joseph, M. Lapidot, and S. Cohen. 1996. Ultraviolet-absorbing plastic sheets protect crops from insect pests and from virus diseases vectored by insects. *Environ. Entomol.* 25: 919–924.

Apablaza, J. U., A. J. Keaster, and R. H. Ward. 1977. Orientation of corn-infesting species of wireworms toward baits in the laboratory. *Environ. Entomol.* 6: 715–718.

App, B. A. 1938. *Euxesta stigmatias* Loew, an otitid fly infesting ear corn in Puerto Rico. *J. Agric. Univ. Puerto Rico* 23: 181–187.

Appel, A. G. 1988. Water relations and desiccation tolerance of migrating garden millipedes (Diplopoda: Paradoxosomatidae). *Environ. Entomol.* 17: 463–466.

Apple, J. W., E. T. Walgenbach, and W. J. Knee. 1971. Thermal requirements for northern corn rootworm egg hatch. *J. Econ. Entomol.* 64: 853–856.

Apriyanto, D., J. D. Sedlacek, and L. H. Townsend. 1989a. Feeding activity of *Euschistus servus* and *E. variolarius* (Heteroptera: Pentatomidae) and damage to an early growth stage of corn. *J. Kansas Entomol. Soc.* 62: 392–399.

Apriyanto, D., L. H. Townsend, and J. D. Sedlacek. 1989b. Yield reduction from feeding by *Euschistus servus* and *E. Variolarius* (Heteroptera: Pentatomidae) on stage V2 corn. *J. Econ. Entomol.* 82: 445–448.

Arant, F. S. 1929. Biology and control of the southern corn rootworm. *Alabama Agric. Exp. Stn. Bull.* 230. 46 pp.

Arant, F. S. 1938. Life history and control of the cowpea curculio. *Alabama Agric. Exp. Stn. Bull.* 246. 34 pp.

Arant, F. S. 1942. Effectiveness of derris and cube in pickleworm control. *J. Econ. Entomol.* 35: 870–872.

Archer, T. L. and E. D. Bynum, Jr. 1994. Corn earworm (Lepidoptera: Noctuidae) biology on food corn on the High Plains. *Environ. Entomol.* 23: 343–348.

Archer, T. L., E. D. Bynum, Jr., and A. Knutson. 1983. Winter management of the southwestern corn borer (Lepidoptera: Pyralidae), using several cultural practices on different dates. *J. Econ. Entomol.* 76: 872–876.

Archer, T. L., G. L. Musick, and R. L. Murray. 1980. Influence of temperature and moisture on black cutworm (Lepidoptera: Noctuidae) development and reproduction. *Can. Entomol.* 112: 665–673.

Archer, T. L., B. R. Wiseman, and A. J. Bockholt. 1994. Factors affecting corn earworm (Lepidoptera: Noctuidae) resistance in food corn. *J. Agric. Entomol.* 11: 9–16.

Arita, L. H., S. C. Furutani, and J. J. Moniz. 1988. Preferential feeding by the Chinese rose beetle (Coleoptera: Scarabaeidae) on ethephon-treated plants. *J. Econ. Entomol.* 81: 1373–1376.

Armstrong, A. M. 1991. Field evaluations of pigeon pea genotypes for resistance against pod borers. *J. Agric. Univ. Puerto Rico* 75: 73–79.

Armstrong, A. M. 1994. Insecticides to combat damage by *Anthonomus eugenii* Cano in pepper var. Cubanelle in Puerto Rico. *J. Agric. Univ. Puerto Rico* 78: 23–31.

Arnaud, Jr., P. H. 1978. A host-parasite catalog of North American Tachinidae (Diptera). *USDA Misc. Publ.* 1319. 860 pp.

Arnett, R. H. 1968. The Beetles of the United States (A Manual for Identification). The American Entomological Institute. Ann Arbor, Mich. 1112 pp.

Arnett, Jr., R. H. 1985. American Insects, a Handbook of the Insects of America North of Mexico. Van Nostrand-Reinhold, New York. 850 pp.

Arnoldi, D., R. K. Stewart, and G. Boivin. 1991. Field survey and laboratory evaluation of the predator complex of *Lygus lineolaris* and *Lygocoris communis* (Hemiptera: Miridae) in apple orchards. *J. Econ. Entomol.* 84: 830–836.

Arnott, Jr., D. A. 1934. Observations on the flight of adults of the genus Crambus with special reference to the economic species. *Proc. Entomol. Soc. Ontario* 65: 98–107.

Arslan, A., P. M. Bessey, K. Matsuda, and N. F. Oebker. 1985. Physiological effects of psyllid (*Paratrioza cockerelli*) on potato. *Am. Pot. J.* 62: 9–22.

Ascerno, M. E., F. L. Pfleger, and H. F. Wilkins. 1981. Effect of root rot and *Rhizoglyphus robini* on greenhouse-forced Easter lily development. *Environ. Entomol.* 10: 947–949.

Ascerno, M. E., F. L. Pfleger, F. Morgan, and H. F. Wilkins. 1983. Relationship of *Rhyzoglyphus robini* (Acari: Acaridae) to root rot control in greenhouse-forced Easter lilies. *Environ. Entomol.* 12: 422–425.

Ashley, T. R. 1979. Classification and distribution of fall armyworm parasites. *Florida Entomol.* 62: 114–123.

Ashley, T. R., B. R. Wiseman, F. M. Davis, and K. L. Andrews. 1989. The fall armyworm: a bibliography. *Florida Entomol.* 72: 152–202.

Aslam, M. and R. J. Whitworth. 1988. Development of the southwestern corn borer *Diatraea grandiosella* Dyar, on corn and johnsongrass. *Southwestern Entomol.* 13: 191–197.

Asteraki, E. J. 1993. The potential of carabid beetles to control slugs in grass-clover swards. *Entomophaga* 38: 193–198.

Atyeo, W. T., G. T. Weekman, and D. E. Lawson. 1964. The identification of *Diabrotica* species by chorion sculpturing. *J. Kansas Entomol. Soc.* 37: 9–11.

Auclair, J. L. 1959. Life history, effects of temperature and relative humidity, and distribution of the Mexican bean beetle, *Epilachna varivestris* Mulsant (Coleoptera: Coccinellidae), in Quebec, with a review of the pertinent literature in North America. *Ann. Entomol. Soc. Quebec.* 5: 18–43.

Auclair, J. L. and P. N. Srivastava. 1977. Distinction de biotypes dans les populations du puceron du pois, *Acyrthosiphon pisum* (Harris), de l'Amerique du Nord. *Can. J. Zool.* 55: 983–989.

Auffenberg, K. and L. A. Stange. 1986. Snail-eating snails of Florida. *Florida Dept. Agric. Cons. Serv. Entomol. Circ.* 285. 4 pp.

Ayers, J. E., A. A. MacNab, R. C. Tetrault, and J. O. Yocum. 1979. The influence of selected insecticides on yield and the incidence of Stewart's wilt in sweet corn. *Plant Dis. Rep.* 63: 634–638.

Ayre, G. L. 1985. Cold tolerance of *Pseudaletia unipunctata* and *Peridroma saucia* (Lepidoptera: Noctuidae). *Can. Entomol.* 117: 1055–1060.

Ayre, G. L. 1990. The response of flax to different population densities of the red-backed cutworm, *Euxoa ochrogaster* (Gn.) (Lepidoptera: Noctuidae). *Can. Entomol.* 122: 21–28.

Ayre, G. L., W. J. Turnock, and D. L. Struble. 1982a. Spatial and temporal variability in sex attractant trap catches of *Scotogramma trifolii* (Lepidoptera: Noctuidae) in relation to its biology in Manitoba. *Can. Entomol.* 114: 145–154.

Ayre, G. L., W. J. Turnock, and D. L. Struble. 1982b. Spatial and temporal variability in sex attractant trap catches of *Euxoa messoria* and *E. ochrogaster* (Lepidoptera: Noctuidae) in relation to their biology in Manitoba. *Can. Entomol.* 114: 993–1001.

Ayre, G. L., W. J. Turnock, and D. L. Struble. 1983. Spatial and temporal variability in sex attractant catches of *Leucania commoides* and *Peridroma saucia* (Lepidoptera: Noctuidae) in relation to their biology in Manitoba. *Can. Entomol.* 115: 1573–1582.

Aziz, S. A. 1973. Toxicity of certain insecticide standards against the southern armyworm. *J. Econ. Entomol.* 66: 68–70.

Bach, C. E. and B. E. Tabashnik. 1990. Effects of nonhost plant neighbors on population densities and parasitism rates of the diamondback moth (Lepidoptera: Plutellidae). *Environ. Entomol.* 19: 987–994.

Back, E. A. 1922. Weevils in beans and peas. *USDA Farmers' Bull.* 1275. 35 pp.

Back, E. A. and C. E. Pemberton. 1914. The life history of the melon fly. *J. Agric. Res.* 3: 269–274.

Back, E. A. and C. E. Pemberton. 1918a. The melon fly. *USDA Bur. Entomol. Bull.* 643. 31 pp.

Back, E. A. and C. E. Pemberton. 1918b. The Mediterranean fruit fly in Hawaii. *USDA Bull.* 536. 119 pp.

Backus, E. A. and W. B. Hunter. 1989. Comparison of feeding behavior of the potato leafhopper, *Empoasca fabae* (Homoptera: Cicadellidae), on alfalfa and broad bean leaves. *Environ. Entomol.* 18: 473–480.

Bacon, O. G. 1960. Control of the potato tuberworm in potatoes. *J. Econ. Entomol.* 53: 868–871.

Bacon, O. G., J. N. Seiber, and G. G. Kennedy. 1976. Evaluation of survey trapping techniques for potato tuberworm moths with chemical baited traps. *J. Econ. Entomol.* 69: 569–572.

Badii, M. H. and J. A. McMurtry. 1984. Feeding behavior of some phytoseiid predators on the broad mite, *Polyphagotarsonemus latus* (Acari: Phytoseiidae, Tarsonemidae). *Entomophaga* 29: 49–53.

Bailey, C. G. 1976a. A quantitative study of consumption and utilization of various diets in the bertha armyworm, *Mamestra configurata* (Lepidoptera: Noctuidae). *Can. Entomol.* 108: 1319–1326.

Bailey, C. G. 1976b. Temperature effects on non-diapause development in *Mamestra configurata* (Lepidoptera: Noctuidae). *Can. Entomol.* 108: 1339–1344.

Bailey, C. G. and M. K. Mukerji. 1976. Feeding habits and food preferences of *Melanoplus bivittatus* and *M. femurrubrum* (Orthoptera: Acrididae). *Can. Entomol.* 108: 1207–1212.

Bailey, N. S. 1951. The Tingoidea of New England and their biology. *Entomol. Am.* 31: 1–140.

Bailey, S. F. 1933a. The biology of the bean thrips. *Hilgardia* 7: 467–522.

Bailey, S. F. 1933b. A contribution to the knowledge of the western flower thrips, *Frankliniella californica* (Moulton). *J. Econ. Entomol.* 26: 836–840.

Bailey, S. F. 1937. The bean thrips. *California Agric. Exp. Stn. Bull.* 609. 36 pp.

Bailey, S. F. and H. H. Keifer. 1943. The tomato russet mite, *Phyllocoptes destructor* Keifer: its present status. *J. Econ. Entomol.* 36: 706–712.

Bailey, W. C. and L. P. Pedigo. 1986. Damage and yield loss induced by stalk borer (Lepidoptera: Noctuidae) in field corn. *J. Econ. Entomol.* 79: 233–237.

Bailey, W. C., G. D. Buntin, and L. P. Pedigo. 1985. Phenology of the adult stalk borer, *Papaipema nebris* (Guenée), (Lepidoptera: Noctuidae) in Iowa. *Environ. Entomol.* 14: 267–271.

Bain, A. and L. LeSage. 1998. A late seventeenth century occurrence of *Phyllotreta striolata* (Coleoptera: Chrysomelidae) in North America. *Can. Entomol.* 130: 715–719.

Baines, D., R. Stewart, and G. Boivin. 1990. Consumption of carrot weevil (Coleoptera: Curculionidae) by five species of carabids (Coleoptera: Carabidae) abundant in carrot fields in southwestern Quebec. *Environ. Entomol.* 19: 1146–1149.

Baker, E. W. and D. M. Tuttle. 1994. *A Guide to the Spider Mites of the United States*. Indira Pub. House, West Bloomfield, Michigan. 347 pp.

Baker, G. L. and J. L. Capinera. 1997. Nematodes and nematomorphs as control agents of grasshoppers and locusts. *Mem. Entomol. Soc. Canada* 171: 157–211.

Baker, P. B. 1986. Responses by Japanese and oriental beetle grubs (Coleoptera: Scarabaeidae) to bendiocarb, chlorpyrifos, and isofenphos. *J. Econ. Entomol.* 79: 452–454.

Baker, P. B., A. M. Shelton, and J. T. Andaloro. 1982. Monitoring of diamondback moth (Lepidoptera: Yponomeutidae) in cabbage with pheromones. *J. Econ. Entomol.* 75: 1025–1028.

Baker, R., R. H. Herbert, and R. A. Lormer. 1982. Chemical components of the rectal gland secretions of male *Dacus cucurbitae*, the melon fly. *Experientia* 38:232–233.

Baker, W. A., W. G. Bradley, and C. A. Clark. 1949. Biological control of the European corn borer in the United States. *USDA Tech. Bull.* 983. 185 pp.

Bakker, F. M. and M. W. Sabelis. 1989. How larvae of *Thrips tabaci* reduce the attack success of phytoseiid predators. *Entomol. Exp. Appl.* 50:47–51.

Balciunas, J. and K. Knopf. 1977. Orientation, flight speeds, and tracks of three species of migrating butterflies. *Florida Entomol.* 60:37–39.

Balduf, W. V. 1931. The oviposition habits of *Feltia subgothica* Haw. (Noctuidae, Lep.). *Proc. Entomol. Soc. Washington* 33:81–88.

Baliddawa, C. W. 1984. Movement and feeding activity of adult pea leaf weevil, *Sitona lineatus* L. in an oat-broadbean diculture. *Insect Sci. Applic.* 5:33–39.

Ball, E. D., B. L. Boyden, and W. E. Stone. 1932. Some major celery insects in Florida. *Florida Agric. Exp. Stn. Bull.* 250. 22 pp.

Ball, E. D., J. A. Reeves, and W. E. Stone. 1935. Biological and ecological factors in the control of the celery leaf tier in Florida. *USDA Tech. Bull.* 463. 55 pp.

Banham, F. L. 1968. Thrips infesting the tips of asparagus spears. *J. Entomol. Soc. Brit. Columbia* 65:16–18.

Banks, C. J. 1954. A method for estimating populations and counting large numbers of *Aphis fabae* Scop. *Bull. Entomol. Res.* 45:751–756.

Banks, C. J. and E. D. M. Macaulay. 1964. The feeding, growth and reproduction of *Aphis fabae* Scop. on *Vicia faba* under experimental conditions. *Ann. Appl. Biol.* 53:229–242.

Banks, C. J., E. D. M. Macaulay, and J. Holman. 1968. Cannibalism and predation by aphids. *Nature* 218:491.

Banks, W. A., C. S. Lofgren, and D. P. Wojcik. 1978. A bibliography of imported fire ants and the chemicals and methods used for their control. *USDA, ARS-S-180.* 35 pp.

Banks, W. A., D. P. Jouvenaz, D. P. Wojcik, and C. S. Lofgren. 1985. Observations on fire ants, *Solenopsis* spp., in Mato Grosso, Brazil. *Sociobiology* 11:143–152.

Baranowski, R. M. and J. A. Slater. 1986. Coreidae of Florida (Hemiptera: Heteroptera). *Arthropods of Florida and Neighboring Land Areas*, Vol. 12. 82 pp.

Barber, G. W. 1937. Seasonal availability of food plants of two species of *Heliothis* in eastern Georgia. *J. Econ. Entomol.* 30:150–158.

Barber, G. W. 1942. Control of earworms in corn by birds. *J. Econ. Entomol.* 35:511–513.

Barber, G. W. 1944. Mineral oils, alone or combined with insecticides, for control of earworms in sweet corn. *USDA Tech. Bull.* 880. 83 pp.

Barber, G. W. and F. F. Dicke. 1937. The effectiveness of cultivation as a control for the corn earworm. *USDA Tech. Bull.* 561. 16 pp.

Barber, H. S. 1916. A review of North American tortoise beetles (Chrysomelidae: Cassidinae). *Proc. Entomol. Soc. Washington* 43:113–127.

Barber, H. S. 1935. The tobacco and solanum weevils of the genus *Trichobaris*. *USDA Misc. Publ.* 226. 28 pp.

Barbercheck, M. E. and H. K. Kaya. 1991. Competitive interactions between entomopathogenic nematodes and *Beauveria bassiana* (Deuteromycotina: Hyphomycetes) in soilborne larvae of *Spodoptera exigua* (Lepidoptera: Noctuidae). *Environ. Entomol.* 20:707–712.

Bardner, R. and K. E. Fletcher. 1979. Larvae of the pea and bean weevil, *Sitona lineatus*, and the yield of field beans. *J. Agric. Sci.* 92:109–112.

Bardner, R., K. E. Fletcher, and J. H. Stevenson. 1978. Pre-flowering and post-flowering insecticide applications to control *Aphis fabae* on field beans: their biological and economic effectiveness. *Ann. Appl. Biol.* 88:265–271.

Bari, M. A. and H. K. Kaya. 1984. Evaluation of the entomogenous nematode *Neoaplectana carpocapsae (Steinernema feltiae)* Weiser (Rhabditida: Steinernematidae) and the bacterium *Bacillus thuringiensis* Berliner var. *kurstaki* for suppression of the artichoke plume moth (Lepidoptera: Pterophoridae). *J. Econ. Entomol.* 77:225–229.

Bari, M. A. and W. H. Lange. 1980. Influence of temperature on the development, fecundity, and longevity of the artichoke plume moth. *Environ. Entomol.* 9:673–676.

Barker, G. M. and R. A. McGhie. 1984. The biology of introduced slugs (Pulmonata) in New Zealand 1. Introduction and notes on *Limax maximus*. *New Zealand Entomol.* 8:106–111.

Barlow, T. and L. H. Rolston. 1981. Types of host plant resistance to the sweetpotato weevil found in sweet potato roots. *J. Kansas Entomol. Soc.* 54:649–657.

Barlow, V. M., L. D. Godfrey, and R. F. Norris. 1999. Population dynamics of *Lygus hesperus* (Heteroptera: Miridae) on selected weeds in comparison with alfalfa. *J. Econ. Entomol.* 92:846–852.

Barnes, H. F. 1937a. The asparagus miner (*Melanagromyza simplex* H. Loew) (Agromyzidae; Diptera). *Ann. Appl. Biol.* 24:574–588.

Barnes, H. F. 1937b. Methods of investigating the bionomics of the common crane-fly, *Tipula paludosa* Meigen, together with some results. *Ann. Appl. Biol.* 24:356–368.

Barnes, H. F. and J. W. Weil. 1942. Baiting slugs using metaldehyde mixed with various substances. *Ann. Appl. Biol.* 29:56–68.

Barnes, H. F. and J. W. Weil. 1944. Slugs in gardens: their numbers, activities and distribution. *Part I. J. Anim. Ecol.* 13:140–175.

Barnes, M. M. 1970. Genesis of a pest: *Nysius raphanus* and *Sisymbrium irio* in vineyards. *J. Econ. Entomol.* 63:1462–1463.

Barratt, B. I. P., R. A. Byers, and D. L. Bierlein. 1994. Conservation tillage crop yields in relation to grey garden slug [*Deroceras reticulatum* (Müller)] (Mollusca: Agriolimacidae) density during establishment. *Crop Prot.* 13:49–52.

Barrett, C. F., P. H. Westdal, and H. P. Richardson. 1965. Biology of *Pachygonatopus minimus* Fenton (Hymenoptera: Dryinidae) a parasite of the six-spotted leafhopper, *Macrosteles fascifrons* (Stål), in Manitoba. *Can. Entomol.* 97:216–221.

Barrett, J. R., H. O. Deay, and J. G. Hartsock. 1971. Reduction in insect damage to cucumbers, tomatoes, and sweet corn through use of electric light traps. *J. Econ. Entomol.* 64:1241–1249.

Barrows, E. M. and M. E. Hooker. 1981. Parasitization of the Mexican bean beetle by *Pediobius foveolatus* in urban vegetable gardens. *Environ. Entomol.* 10:782–786.

Barry, R. M., J. H. Hatchett, and R. D. Jackson. 1974. Cage studies with predators of the cabbage looper, *Trichoplusia ni*, and corn earworm, *Heliothis zea*, in soybeans. *J. Georgia Entomol.* 9:71–78.

Bartels, D. W. and W. D. Hutchison. 1995. On-farm efficacy of aerially applied *Bacillus thuringiensis* for European corn borer (Lepidoptera: Pyralidae) and corn earworm (Lepidoptera: Noctuidae) control in sweet corn. *J. Econ. Entomol.* 88:380–386.

Bartelt, R. J., P. F. Dowd, and R. D. Platner. 1991. Aggregation pheromone of *Carpophilus lugubris*, new pest management tools for the nitidulid beetles. Pages 27–40 *in* P. A. Hedin (ed.). Naturally Occurring Pest Bioregulators. *Am. Chem. Soc. Symp. Ser.* 449.

Bartelt, R. J., R. S. Vetter, D. G. Carlson, and T. C. Baker. 1994. Responses to aggregation pheromones for five *Carpophilus* species (Coleoptera: Nitidulidae) in a California date garden. *Environ. Entomol.* 23:1534–1543.

Bartlett, B. R. 1968. Outbreaks of two-spotted spider mites and cotton aphids following pesticide treatment. I. Pest stimulation vs. natural enemy destruction as the cause of outbreaks. *J. Econ. Entomol.* 61:297–303.

Basinger, A. J. 1927. The eradication campaign against the white snail (*Helix pisana*) at LaJolla, California. *Bull. California Dept. Agric.* 16: 51–76.

Basinger, A. J. 1931. The European brown snail in California. *California Agric. Exp. Stn. Bull.* 515. 22 pp.

Basky, Z. 1984. Effect of reflective mulches on virus incidence in seed cucumbers. *Prot. Ecol.* 6: 57–61.

Batiste, W. C. and W. H. Olson. 1973. Laboratory evaluations of some solanaceous plants as possible hosts for tomato pinworm. *J. Econ. Entomol.* 66: 109–111.

Batiste, W. C., J. Joos, and R. C. King. 1970a. Studies on sources of the tomato pinworm attacking tomatoes in northern California. *J. Econ. Entomol.* 63: 1484–1486.

Batiste, W. C., R. C. King, and J. Joos. 1970b. Field and laboratory evaluations of insecticides for control of the tomato pinworm. *J. Econ. Entomol.* 63: 1479–1484.

Bauernfeind, R. J. and G. E. Wilde. 1993. Control of army cutworm (Lepidoptera: Noctuidae) affects wheat fields. *J. Econ. Entomol.* 86: 159–163.

Baur, M. E., H. K. Kaya, and B. E. Tabashnik. 1997. Efficacy of a dehydrated steinernematid nematode against black cutworm (Lepidoptera: Noctuidae) and diamondback moth (Lepidoptera: Plutellidae). *J. Econ. Entomol.* 90: 1200–1206.

Bautista, R. C. and R. F. L. Mau. 1994. Preferences and development of western flower thrips (Thysanoptera: Thripidae) on plant hosts of tomato spotted wilt tospovirus in Hawaii. *Environ. Entomol.* 23: 1501–1507.

Beach, R. M. and J. W. Todd. 1985. Parasitoids and pathogens of the soybean looper, *Pseudoplusia includens* (Walker), in south Georgia soybean. *J. Entomol. Sci.* 20: 318–323.

Beach, R. M. and J. W. Todd. 1988. Development, reproduction, and longevity of *Autographa biloba* (Lepidoptera: Noctuidae), with observations on laboratory adaptation. *Ann. Entomol. Soc. Am.* 81: 943–949.

Beard, R. L. 1935. Further observations on the squash bug in Connecticut. *Connecticut Agric. Exp. Stn. Bull.* 383: 338–339.

Beard, R. L. 1940. The biology of *Anasa tristis* De Geer with particular reference to the tachinid parasite, *Trichopoda pennipes* Fabr. *Connecticut Agric. Exp. Stn. Bull.* 440: 597–679.

Beavers, J. B. 1982. Biology of *Diaprepes abbreviatus* (Coleoptera: Curculionidae) reared on an artificial diet. *Florida Entomol.* 65: 263–269.

Beavers, J. B. and A. G. Selhime. 1978. Flight behavior and dispersal of *Diaprepes abbreviatus*. *Florida Entomol.* 61: 89–91.

Beavers, J. B., C. W. McCoy, and D. T. Kaplan. 1983. Natural enemies of subterranean *Diaprepes abbreviatus* (Coleoptera: Curculionidae) larvae in Florida. *Environ. Entomol.* 12: 840–843.

Beavers, J. B., S. A. Lovestrand, and A. G. Selhime. 1980. Establishment of the exotic parasite *Tetrastichus haitiensis* (Hym: Eulophidae) and recovery of a new Trichogramma (Hym: Trichogrammatidae) from root weevil egg masses in Florida. *Entomophaga* 25: 91–94.

Beavers, J. B., T. P. McGovern, and V. E. Adler. 1982. *Diaprepes abbreviatus*: laboratory and field behavioral and attractancy studies. *Environ. Entomol.* 11: 436–439.

Bechinski, E. J. and L. P. Pedigo. 1983. Green cloverworm (Lepidoptera: Noctuidae) population dynamics: pupal life table studies in Iowa soybeans. *Environ. Entomol.* 12: 656–661.

Bechinski, E. J., J. F. Bechinski, and L. P. Pedigo. 1983a. Survivorship of experimental green cloverworm (Lepidoptera: Noctuidae) pupal cohorts in soybeans. *Environ. Entomol.* 12: 662–668.

Bechinski, E. J., C. D. McNeal, and J. J. Gallian. 1989. Development of action thresholds for the sugarbeet root maggot (Diptera: Otitidae). *J. Econ. Entomol.* 82: 608–615.

Bechinski, E. J., G. D. Buntin, L. P. Pedigo, and H. G. Thorvilson. 1983b. Sequential count and decision plans for sampling green cloverworm (Lepidoptera: Noctuidae) larvae in soybean. *J. Econ. Entomol.* 76: 806–812.

Bechinski, E. J., D. O. Everson, C. D. McNeal, and J. J. Gallian. 1990. Forecasting peak seasonal capture of sugarbeet root maggot (Diptera: Otitidae) with sticky-stake traps in Idaho. *J. Econ. Entomol.* 83: 2078–2085.

Beck, S. D. 1987. Developmental and seasonal biology of *Ostrinia nubilalis*. Pages 59–96 in G. E. Russell (ed.). *Agricultural Zoology Reviews*, Vol. 2. Intercept, Wimborne, Dorset.

Beck, S. D. 1988. Thermoperiod and larval development of *Agrotis ipsilon* (Lepidoptera: Noctuidae). *Ann. Entomol. Soc. Am.* 81: 831–835.

Beckam, C. M. 1953. Biology and control of vegetable weevil. *Georgia Agric. Exp. Stn. Tech. Bull.* 2. 36 pp.

Becker, E. C. 1956. Revision of the nearctic species Agriotes (Coleoptera: Elateridae). *Can. Entomol.* 88 (Suppl. 1) 101 pp.

Becker, E. C. and J. R. Dogger. 1991. Elateridae (Elateroidea) (including Dicronychidae, Lissomidae). Pages 410–417 in F. W. Stehr (ed.) *Immature Insects*, Vol. 2. Kendall/Hunt Publishing, Dubuque, Iowa.

Beckham, C. M. and M. Dupree. 1952. Attractants for the green June beetle with notes on seasonal occurrence. *J. Econ. Entomol.* 45: 736–737.

Beirne, B. P. 1952. The nearctic species of Macrosteles (Homoptera: Cicadellidae). *Can. Entomol.* 84: 208–232.

Beirne, B. P. 1971. Pest insects of annual crop plants in Canada. I. Lepidoptera, II. Diptera, III. Coleoptera. *Mem. Entomol. Soc. Canada* 78. 124 pp.

Beirne, B. P. 1972. Pest insects of annual crop plants in Canada. IV. Hemiptera-Homoptera V. Orthoptera VI. Other groups. *Mem. Entomol. Soc. Canada* 85. 73 pp.

Belair, G. and G. Boivin. 1985. Susceptibility of the carrot weevil (Coleoptera: Curculionidae) to *Steinernema feltiae*, *S. bibionis*, and *Heterorhabditis*. *J. Nematol.* 17: 363–366.

Belair, G. and G. Boivin. 1995. Evaluation of *Steinernema carpocapsae* Weiser for control of carrot weevil adults, *Listronotus oregonensis* (LeConte) (Coleoptera: Curculionidae), in organically grown carrots. *Biocontrol Sci. Tech.* 5: 225–231.

Belcher, D. W. 1989. Influence of cropping systems on the number of wireworms (Coleoptera: Elateridae) collected in baits in Missouri cornfields. *J. Kansas Entomol. Soc.* 62: 590–592.

Bell, J. V. and R. J. Hamalle. 1971. A bacterium and dipterous parasite in wild populations of cowpea curculio larvae: effects of treatment with spores of *Metarrhizium anisopliae*. *J. Invertebr. Pathol.* 17: 256–259.

Bell, J. V., R. J. Hamalle, and J. A. Onsager. 1972. Mortality of larvae and pupae of the banded cucumber beetle in soil and sand following topical application of fungus spores. *J. Econ. Entomol.* 65: 605–606.

Bell, M. R. and J. L. Hayes. 1994. Areawide management of cotton bollworm and tobacco budworm (Lepidoptera: Noctuidae) through application of a nuclear polyhedrosis virus on early-season alternate hosts. *J. Econ. Entomol.* 87: 53–57.

Bell, M. R. and C. L. Romine. 1980. Tobacco budworm field evaluation of microbial control in cotton using *Bacillus thuringiensis* and a nuclear polyhedrosis virus with a feeding adjuvant. *J. Econ. Entomol.* 73: 427–430.

Bell, R. A. and F. G. Joachim. 1976. Techniques for rearing laboratory colonies of tobacco hornworms and pink bollworms. *Ann. Entomol. Soc. Am.* 69: 365–373.

Bellinger, R. G. and R. L. Pienkowski. 1989. Polymorphic development in relation to the life history of *Melanoplus femurrubrum* (Orthoptera: Acrididae). *Ann. Entomol. Soc. Am.* 82: 166–171.

Belloncik, S., C. Lavallee, and I. Quevillon. 1985. *Euxoa scandens* and *Euxoa messoria*. Pages 293–299 *in* P. Singh and R. F. Moore (eds.). *Handbook of Insect Rearing*. Vol 2. Elsevier, Amsterdam.

Bellows, T. S., T. M. Perring, R. J. Gill, and D. H. Headrick. 1994. Description of a species of *Bemisia* (Homoptera: Aleyrodidae). *Ann. Entomol. Soc. Am.* 87: 195–206.

Bender, D. A. and W. P. Morrison. 1989. Species composition and control of thrips in Texas High Plains onions. *J. Agric. Entomol.* 6: 257–263.

Benedict, J. H., E. S. Sachs, D. W. Altman, W. R. Deaton, R. J. Kohel, D. R. Ring, and S. A. Berberich. 1996. Field performance of cottons expressing transgenic CryIA insecticidal proteins for resistance to *Heliothis virescens* and *Helicoverpa zea* (Lepidoptera: Noctuidae). *J. Econ. Entomol.* 89: 230–238.

Bengtsson, M., G. Karg, P. A. Kirsch, J. Lofqvist, A. Sauer, and P. Witzgall. 1994. Mating disruption of pea moth *Cydia nigricana* F. (Lepidoptera: Tortricidae) by a repellent blend of sex pheromone and attraction inhibitors. *J. Chem. Ecol.* 20: 871–887.

Ben-Ze'ev, I. S. 1986. Notes on Entomophthorales (Zygomycotina) collected by T. Petch: II. *Erynia ellisiana* sp. nov., non *Erynia forficulae* (Giard.), comb. nov., pathogens of Forficulidae (Dermaptera). *Mycotaxon* 27: 263–269.

Berdegue, M., M. K. Harris, D. W. Riley, and B. Villalon. 1994. Host plant resistance on pepper to the pepper weevil, *Anthonomus eugenii* Cano. *Southwestern Entomol.* 19: 265–271.

Berenbaum, M. R., B. Moreno, and E. Green. 1992. Soldier bug predation on swallowtail caterpillars (Lepidoptera: Papilionidae): circumvention of defensive chemistry. *J. Insect Behav.* 5: 547–553.

Berger, E. W. 1920. The semitropical armyworm. State Plant Board of Florida *Quarterly Bull.* 4: 17–31.

Bergman, M. K. and F. T. Turpin. 1984. Impact of corn planting date on the population dynamics of corn rootworms (Coleoptera: Chrysomelidae). *Environ. Entomol.* 13: 898–901.

Beri, S. K. 1974. Biology of a leaf miner *Liriomyza brassicae* (Riley) (Diptera: Agromyzidae). *J. Nat. Hist.* 8: 143–151.

Berisford, Y. C. and C. H. Tsao. 1974. Field and laboratory observations of an entomogenous infection of the adult seedcorn maggot, *Hylemya platura* (Diptera: Anthomyiidae). *J. Georgia Entomol. Soc.* 9: 104–110.

Bernhardt, J. L. and J. R. Phillips. 1985. Identification of eggs of the bollworm, *Heliothis zea* (Boddie), and the tobacco budworm, *Heliothis virescens* (F.). *Southwestern Entomol.* 10: 236–238.

Bernhardt, J. L. and M. Shepard. 1978. Validation of a physiological day equation: development of the Mexican bean beetle on snap beans and soybeans. *Environ. Entomol.* 7: 131–135.

Berry, E. C. and R. P. Knake. 1987. Population suppression of black cutworm (Lepidoptera: Noctuidae) larvae with seed treatments. *J. Econ. Entomol.* 80: 921–924.

Berry, P. A. 1947. Investigations on the white-fringed beetle group in South America. *J. Econ. Entomol.* 40: 705–709.

Berry, R. E. 1972. Garden symphylan: reproduction and development in the laboratory. *J. Econ. Entomol.* 65: 1628–1632.

Berry, R. E. and H. H. Crowell. 1970. Effectiveness of Bay37289 as a transplant dip to control the garden symphylan in broccoli. *J. Econ. Entomol.* 63: 1718–1719.

Berry, R. E., J. Liu, and G. Reed. 1997. Comparison of endemic and exotic entomopathogenic nematode species for control of Colorado potato beetle (Coleoptera: Chrysomelidae). *J. Econ. Entomol.* 90: 1528–1533.

Bess, H. A. and F. H. Haramoto. 1961. Contributions to the biology and ecology of the oriental fruit fly, *Dacus dorsalis* Hendel (Diptera: Tephritidae), in Hawaii. *Hawaii Agric. Exp. Stn. Tech. Bull.* 44. 30 pp.

Bess, H. A., F. H. Haramoto, and A. D. Hinckley. 1963. Population studies of the oriental fruit fly, *Dacus dorsalis* Hendel (Diptera: Tephritidae). *Ecology* 44: 197–201.

Bessin, R. T. and T. E. Reagan. 1990. Fecundity of sugarcane borer (Lepidoptera: Pyralidae), as affected by larval development on gramineous host plants. *Environ. Entomol.* 19: 635–639.

Bessin, R. T. and T. E. Reagan. 1993. Cultivar resistance and arthropod predation of sugarcane borer (Lepidoptera: Pyralidae) affects incidence of deadhearts in Louisiana sugarcane. *J. Econ. Entomol.* 86: 929–932.

Bessin, R. T., E. B. Moser, and T. E. Reagan. 1990. Integration of control tactics for management of the sugarcane borer (Lepidoptera: Pyralidae) in Louisiana sugarcane. *J. Econ. Entomol.* 83: 1563–1569.

Best, R. L. and C. C. Beegle. 1977. Consumption of *Agrotis ipsilon* by several species of carabids found in Iowa. *Environ. Entomol.* 6: 532–534.

Bethke, J. A. and T. D. Paine. 1991. Screen hole size and barriers for exclusion of insect pests of glasshouse crops. *J. Entomol. Sci.* 26: 169–177.

Bethke, J. A., R. A. Redak, and T. D. Paine. 1994. Screens deny specific pests entry to greenhouses. *California Agric.* 48(3): 37–40.

Beyer, A. H. 1921. Garden flea-hopper in alfalfa and its control. *USDA Bull.* 964. 27 pp.

Beyer, A. H. 1922. The bean leafhopper and hopperburn with methods of control. *Florida Agric. Exp. Stn. Bull.* 164: 63–88.

Bharadwaj, R. K. 1966. Observations on the bionomics of *Euborellia annulipes* (Dermaptera: Labiduridae). *Ann. Entomol. Soc. Am.* 59: 441–450.

Bhatti, J. S. 1980. Species of the genus *Thrips* from India (Thysanoptera). *Syst. Entomol.* 5: 109–166.

Bhirud, K. M. and H. N. Pitre. 1972. Bioactivity of systemic insecticides in corn: relationship to leafhopper vector control and corn stunt disease incidence. *J. Econ. Entomol.* 65: 1134–1140.

Biever, K. D. and P. E. Boldt. 1971. Continuous laboratory rearing of the diamondback moth and related biological data. *Ann. Entomol. Soc. Am.* 64: 651–655.

Biever, K. D. and R. L. Chauvin. 1992. Suppression of the Colorado potato beetle (Coleoptera: Chrysomelidae) with augmentative releases of predaceous stinkbugs (Hemiptera: Pentatomidae). *J. Econ. Entomol.* 85: 720–726.

Bigger, J. H. 1928. Hibernation studies of *Colaspis brunnea* (Fab.). *J. Econ. Entomol.* 21: 268–273.

Bing, J. W., W. D. Guthrie, F. F. Dicke, and J. J. Obrycki. 1991. Seedling stage feeding by corn leaf aphid (Homoptera: Aphididae): influence on plant development in maize. *J. Econ. Entomol.* 84: 625–632.

Bintcliffe, E. J. B. and S. D. Wratten. 1980. Resistance in peas to the pea aphid, 1978. *Ann. Appl. Biol.* (Suppl.) 94: 52–53.

Biron, D. X. Langlet, G. Boivin, and E. Brunel. 1998. Expression of early and late-emerging phenotypes oin both diapausing and non-diapausing *Delia radicum* L. pupae. *Entomol. Exp. Appl.* 87: 119–124.

Bisabri-Ershadi, B. and L. E. Ehler. 1981. Natural biological control of western yellow-striped armyworm, *Spodoptera praefica* (Grote), in hay alfalfa in northern California. *Hilgardia* 49: 1–23.

Bishop, G. W. and J. W. Guthrie. 1964. Home gardens as a source of the green peach aphid and virus diseases in Idaho. *Am. Pot. J.* 41: 28–34.

Bjerke, J. M., A. W. Anderson, and T. P. Freeman. 1992. Morphology of the larval stages of *Tetanops myopaeformis* (Roder) (Diptera: Otitidae). *J. Kansas Entomol. Soc.* 65: 59–65.

Black, Jr., E. R., F. M. Davis, C. A. Henderson, and W. A. Douglas. 1970. The role of birds in reducing overwintering populations of the southwestern corn borer, *Diatraea grandiosella* (Lepidoptera: Crambidae), in Mississippi. *Ann. Entomol. Soc. Am.* 63: 701–706.

Blackman, R. L. and V. F. Eastop. 1984. *Aphids on the World's Crops: An Identification and Information Guide.* John Wiley & Sons, Chichester, England. 466 pp.

Blackmer, J. L. and G. W. Bishop. 1991. Population dynamics of *Rhopalosiphum padi* (Homoptera: Aphididae) in corn in relation to barley yellow dwarf epidemiology in southwestern Idaho. *Environ. Entomol.* 20: 166–173.

Blaine, W. D. and F. L. McEwen. 1984. Nutrition of the onion maggot, *Delia antiqua* (Diptera: Anthomyiidae). *Can. Entomol.* 116: 473–477.

Blakeley, P. E., L. A. Jacobson, and R. R. Förster. 1958. Rearing the army cutworm, *Chorizagrotis auxiliaris* (Grote) (Lepidoptera: Noctuidae), in the laboratory. *Can. Entomol.* 90: 301–302.

Blanchard, R. A. 1942. Hibernation of the corn earworm in the central and northeastern parts of the United States. *USDA Tech. Bull.* 838. 13 pp.

Blanchard, R. A. and C. B. Conger. 1932. Notes on *Prodenia praefica* Grote. *J. Econ. Entomol.* 25: 1059–1070.

Blank, R. H., D. S. Bell, and M. H. Olson. 1985. Insecticide baits for control of the black field cricket (*Teleogryllus commodus*). *New Zealand J. Exp. Agric.* 13: 263–269.

Blatchley, W. S. 1910. An Illustrated Descriptive Catalogue of the Coleoptera or Beetles Known to Occur in Indiana. The Nature Publishing Company, Indianapolis. 1385 pp.

Blatchley, W. S. 1920. Orthoptera of Northeastern America with Especial Reference to the Faunas of Indiana and Florida. The Nature Publishing Co., Indianapolis. 784 pp.

Blau, W. S. 1981. Life history variation in the black swallowtail butterfly. *Oecologia* 48: 116–122.

Blenk, R. G., R. J. Gouger, T. S. Gallo, L. K. Jordan, and E. Howell. 1985. *Agrotis ipsilon*. Pages 177–187 *in* P. Sing and R. F. Morse (eds.). *Handbook of Insect Rearing.* Vol. 2. Elsevier, Amsterdam.

Blickenstaff, C. C. 1979. History and biology of the western bean cutworm in southern Idaho, 1942–1977. *Idaho Agric. Exp. Stn. Bull.* 592. 23 pp.

Blickenstaff, C. C. and P. M. Jolley. 1982. Host plants of western bean cutworm. *Environ. Entomol.* 11: 421–425.

Blickenstaff, C. C. and R. E. Peckenpaugh. 1976. Sticky stake traps for monitoring fly populations of the sugarbeet root maggot and predicting maggot populations and damage ratings. *J. Am. Soc. Sugar Beet Technol.* 19: 112–117.

Blickenstaff, C. C. and R. E. Peckenpaugh. 1981. Insecticide tests for control of the western bean cutworm. *USDA, ARS. ARR-21.* 31 pp.

Blight, M. M. and L. J. Wadhams. 1987. Male-produced aggregation pheromone in pea and bean weevil, *Sitona lineatus* (L.). *J. Chem. Ecol.* 13: 733–739.

Blodgett, S. L., J. E. Carrel, and R. A. Higgins. 1991. Cantharidin content of blister beetles (Coleoptera: Meloidae) collected from Kansas alfalfa and implications for inducing cantharidiasis. *Environ. Entomol.* 20: 776–780.

Blower, J. G. (ed.). 1974. Myriapoda. *Symp. Zool. Soc. London* 32. 707 pp.

Blua, M. J., H. A. Yoshida, and N. C. Toscano. 1995. Ovipositional preference of two *Bemisia* species (Homoptera: Aleyrodidae). *Environ. Entomol.* 24: 88–93.

Bodenheimer, F. S. and E. Swirski. 1957. The Aphidoidea of the Middle East. Weizmann Sc. Press, Jerusalem. 378 pp.

Bodnaryk, R. P. 1978. Factors affecting diapause development and survival in the pupa of *Mamestra configurata* (Lepidoptera: Noctuidae). *Can. Entomol.* 110: 183–191.

Bodnaryk, R. P. and R. J. Lamb. 1991a. Mechanisms of resistance to the flea beetle, *Phyllotreta cruciferae* (Goeze), in mustard seedlings, *Sinapis alba* L. *Can. J. Plant Sci.* 71: 13–20.

Bodnaryk, R. P. and R. J. Lamb. 1991b. Influence of seed size in canola, *Brassica napus* L. and mustard, *Sinapis alba* L., on seedling resistance against flea beetles, *Phyllotreta cruciferae* (Goeze). *Can. J. Plant Sci.* 71: 397–404.

Boetel, M. A., D. D. Walgenbach, G. L. Hein, B. W. Fuller, and M. E. Gray. 1992. Oviposition site selection of the northern corn rootworm (Coleoptera: Chrysomelidae). *J. Econ. Entomol.* 85: 246–249.

Bohlen, P. J. and G. W. Barrett. 1990. Dispersal of the Japanese beetle (Coleoptera: Scarabaeidae) in strip-cropped soybean agroecosystems. *Environ. Entomol.* 19: 955–960.

Boiteau, G. 1984. Effect of planting date, plant spacing, and weed cover on populations of insects, arachnids, and entomophthoran fungi in potato fields. *Environ. Entomol.* 13: 751–756.

Boiteau, G. 1986. Native predators and the control of potato aphids. *Can. Entomol.* 118: 1177–1183.

Boiteau, G., J. R. Bradley, Jr., and J. W. Van Duyn. 1979. Bean leaf beetle: emergence patterns of adults from overwintering sites. *Environ. Entomol.* 8: 427–431.

Boiteau, G., J. R. Bradley, Jr., and J. W. Van Duyn. 1980. Bean leaf beetle: seasonal history of the overwintering population in eastern North Carolina. *J. Georgia Entomol. Soc.* 15: 138–151.

Boiteau, G., W. P. L. Osborn, and M. E. Drew. 1997. Residual activity of imidacloprid controlling Colorado potato beetle (Coleoptera: Chrysomelidae) and three species of potato colonizing aphids (Homoptera: Aphididae). *J. Econ. Entomol.* 90: 309–319.

Boiteau, G., G. C. Misener, R. P. Singh, and G. Bernard. 1992. Evaluation of a vacuum collector for insect pest control in potato. *Am. Pot. J.* 69: 157–166.

Boiteau, G., Y. Pelletier, G. C. Misener, and G. Bernard. 1994. Development and evaluation of a plastic trench barrier for protection of potato from walking adult Colorado potato beetles (Coleoptera: Chrysomelidae). *J. Econ. Entomol.* 87: 1325–1331.

Boivin, G. 1985. Evaluation of monitoring techniques for the carrot weevil, *Listronotus oregonensis* (Coleoptera: Curculionidae). *Can. Entomol.* 117: 927–933.

Boivin, G. 1987. Seasonal occurrence and geographical distribution of the carrot rust fly (Diptera: Psilidae) in Quebec. *Environ. Entomol.* 16: 503–506.

Boivin, G. 1988. Effects of carrot developmental stages on feeding and oviposition of carrot weevil, *Listronotus oregonensis* (Le Conte) (Coleoptera: Curculionidae). *Environ. Entomol.* 17: 330–336.

Boivin, G. 1993. Density dependence of *Anaphes sordidatus* (Hymenoptera: Mymaridae) parasitism on eggs of *Listronotus oregonensis* (Coleoptera: Curculionidae). *Oecologia* 93: 73–79.

Boivin, G. 1999. Integrated management for carrot weevil. *Integr. Pest Manage. Rev.* 4: 21–37.

Boivin, G. and G. Belair. 1989. Infectivity of two strains of *Steinernema feltiae* (Rhabditida: Steinernematidae) in relation to temperature, age, and sex of carrot weevil (Coleoptera: Curculionidae) adults. *J. Econ. Entomol.* 82: 762–765.

Boivin, G. and D. l. Benoit. 1987. Predicting onion maggot (Diptera: Anthomyiidae) flights in southwestern Quebec using degree-days and common weeds. *Phytoprotection* 68: 65–70.

Boivin, G., S. m. Cote, and J. R. Anciso. 1990. Egg parasitoid of a carrot weevil, *Listronotus texanus* (Stockton), in the lower Rio Grande Valley. *J. Rio Grande Valley Hort. Soc.* 43: 91–92.

Boivin, G., J.-P.R. LeBlanc, and J. A. Adams. 1991. Spatial dispersion and sequential sampling plan for the tarnished plant bug (Hemiptera: Miridae) on celery. *J. Econ. Entomol.* 84: 158–164.

Bolin, P. C., W. D. Hutchison, and D. W. Davis. 1996. Resistant hybrids and *Bacillus thuringiensis* for management of European corn borer (Lepidoptera: Pyralidae) in sweet corn. *J. Econ. Entomol.* 89: 82–91.

Boivin, G., R. K. Stewart, and I. Rivard. 1982. Sticky traps for monitoring phytophagous mirids (Hemiptera: Miridae) in an apple orchard in southwestern Quebec. *Environ. Entomol.* 11: 1067–1070.

Bolland, H. R., J. Gutierrez, and C. H. W. Flechtmann. 1998. World Catalog of the Spider Mite Family (Acari: Tetranychidae). Brill, Leiden. 392 pp.

Bomar, C. R., J. A. Lockwood, M. A. Pomerinke, and J. D. French. 1993. Multiyear evaluation of the effects of *Nosema locustae* (Microsporidia: Nosematidae) on rangeland grasshopper (Orthoptera: Acrididae) population density and natural biological controls. *Environ. Entomol.* 22: 489–497.

Bongers, W. 1970. Aspects of host-plant relationship of the Colorado beetle. Meded. Landbouwhogesch. *Wageningen* 70: 1–77.

Bonjour, E. L. and W. S. Fargo. 1989. Host effects on the survival and development of *Anasa tristis* (Heteroptera: Coreidae). *Environ. Entomol.* 18: 1083–1085.

Bonjour, E. L., W. S. Fargo, and P. E. Rensner. 1990. Ovipositional preference of squash bugs (Heteroptera: Coreidae) among cucurbits in Oklahoma. *J. Econ. Entomol.* 83: 943–947.

Bonjour, E. L., W. S. Fargo, A. A. Al-Obaidi, and M. E. Payton. 1993. Host effects on reproduction and adult longevity of squash bugs (Heteroptera: Coreidae). *Environ. Entomol.* 22: 1344–1348.

Borden, A. D., H. F. Madsen, and A. H. Retan. A stink bug, *Eustichus conspersus*, destructive to deciduous fruits in California. *J. Econ. Entomol.* 45: 254–257.

Bottrell, D. G., R. D. Brigham, and L. B. Jordan. 1973. Carrot beetle: pest status and bionomics on cultivated sunflower. *J. Econ. Entomol.* 66: 86–90.

Bottrell, D. G., R. D. Brigham, B. C. Clymer, and J. R. Cate, Jr. 1970. Evaluation of insecticides for controlling the carrot beetle in sunflower. Texas Agric. Exp. Stn. Prog. Rpt. 2831: 1–17.

Boucias, D. G. and G. L. Nordin. 1977. A granulosis virus from *Diacrisia virginica* (Lepidoptera: Arctiidae). *J. Invertebr. Pathol.* 30: 434–435.

Boucias, D. G. and G. L. Nordin. 1978. A cytoplasmic polyhedrosis virus isolated from the banded woolly bear, *Isia isabella. J. Invertebr. Pathol.* 31: 131–133.

Boyd, M. L. and G. L. Lentz. 1994. Seasonal incidence of the cabbage seedpod weevil (Coleoptera: Curculionidae) on rapeseed in west Tennessee. *Environ. Entomol.* 23: 900–905.

Bracken, G. K. 1988. Seasonal occurrence and infestation potential of cabbage maggot, *Delia radicum* (L.) (Diptera: Anthomyiidae), attacking rutabaga in Manitoba as determined by captures of females in water traps. *Can. Entomol.* 120: 609–614.

Bragg, D. E. 1994. Notes on parasitoids of *Platyptilia carduidactyla* (Riley) (Lepidoptera: Pterophoridae) in transition zone southeastern Washington. *Pan-Pac. Entomol.* 70: 322–323.

Brannon, L. W. 1937. Life-history studies of the squash beetle in Alabama. *Ann. Entomol. Soc. Am.* 30: 43–50.

Brannon, L. W. 1938. The sweetpotato leaf beetle. USDA Circ. 495. 9 pp.

Branson, T. F. and R. D. Johnson. 1973. Adult western corn rootworms: oviposition, fecundity, and longevity in the laboratory. *J. Econ. Entomol.* 66: 417–418.

Branson, T. F. and E. E. Ortman. 1967a. Host range of larvae of the northern corn rootworm (Coleoptera: Chrysomelidae). *J. Kansas Entomol. Soc.* 40: 412–414.

Branson, T. F. and E. E. Ortman. 1967b. Host range of larvae of the western corn rootworm. *J. Econ. Entomol.* 60: 201–203.

Branson, T. F. and E. E. Ortman. 1970. The host range of larvae of the western corn rootworm: further studies. *J. Econ. Entomol.* 63: 800–803.

Branson, T. F. and E. E. Ortman. 1971. Host range of larvae of the northern corn rootworm: further studies. *J. Kansas Entomol. Soc.* 44: 50–52.

Branson, T. F. and G. R. Sutter. 1985. Influence of population density of immatures on size, longevity, and fecundity of adult *Diabrotica virgifera* (Coleoptera: Chrysomelidae). *Environ. Entomol.* 14: 687–690.

Branson, T. F., J. J. Jackson, and G. R. Sutter. 1988. Improved method for rearing *Diabrotica virgifera virgifera* (Coleoptera: Chrysomelidae). *J. Econ. Entomol.* 81: 410–414.

Branson, T. F., J. Reyes R., and H. Valdes M. 1982. Field biology of Mexican corn rootworm, *Diabrotica virgifera zeae* (Coleoptera: Chrysomelidae), in central Mexico. *Environ. Entomol.* 11: 1078–1083.

Branson, T. F., G. R. Sutter, and J. R. Fisher. 1980. Plant response to stress induced by artificial infestations of western corn rootworm. *Environ. Entomol.* 9: 253–257.

Brazzel, J. R., L. D. Newsom, J. S. Roussel, C. Lincoln, F. J. Williams, and G. Barnes. 1953. Bollworm and tobacco budworm as cotton pests in Louisiana and Arkansas. Louisiana Agric. Exp. Stn. Tech. Bull. 482. 47 pp.

Breeland, S. G. 1958. Biological studies on the armyworm, *Pseudaletia unipunctata* (Haworth), in Tennessee (Lepidoptera: Noctuidae). *J. Tennessee Acad. Sci.* 33: 263–347.

Brewer, F. D. and W. A. Jones, Jr. 1985. Comparison of meridic and natural diets on the biology of *Nezara viridula* (Heteroptera: Pentatomidae) and eight other phytophagous Heteroptera. *Ann. Entomol. Soc. Am.* 78: 620–625.

Brewer, M. J. and R. N. Story. 1987. Larval spatial patterns and sequential sampling plan for pickleworm, *Diaphania nitidalis* (Stoll) (Lepidoptera: Pyralidae), on summer squash. *Environ. Entomol.* 16: 539–544.

Brewer, M. J. and J. T. Trumble. 1994. Beet armyworm resistance to fenvalerate and methomyl: resistance variation and insecticide synergism. *J. Agric. Entomol.* 11: 291–300.

Brewer, M. J., R. N. Story, and V. L. Wright. 1987. Development of summer squash seedlings damaged by striped and spotted cucumber beetles (Coleoptera: Chrysomelidae). *J. Econ. Entomol.* 80: 1004–1009.

Brewer, M. J., D. S. Schuster, J. T. Trumble, and B. Alvarado-Rodriguez. 1993. Tomato pinworm (Lepidoptera: Gelechiidae) resistance to fenvalerate from localities in Sinaloa, Mexico and California, USA. *Trop. Agric.* 70: 179–184.

Brewer, M. J., J. T. Trumble, B. Alvarado-Rodriguez, and W. E. Chaney. 1990. Beet armyworm (Lepidoptera: Noctuidae) adult and larval susceptibility to three insecticides in managed habitats and relationship to laboratory selection for resistance. *J. Econ. Entomol.* 83: 2136–2146.

Briese, D. T. and H. A. Mende. 1981. Differences in susceptibility to a granulosis virus between field populations of the potato moth, *Phthorimaea operculella* (Zeller) (Lepidoptera: Gelechiidae). *Bull. Entomol. Res.* 71: 11–18.

Brindley, T. A. and F. F. Dicke. 1963. Significant developments in European corn borer research. *Annu. Rev. Entomol.* 8: 155–176.

Brindley, T. A., J. C. Chamberlin, and F. G. Hinman. 1946. The pea weevil and methods for its control. USDA Farmers' Bull. 1971. 24 pp.

Brindley, T. A., A. N. Sparks, W. B. Showers, and W. D. Guthrie. 1975. Recent research advances on the European corn borer in North America. *Annu. Rev. Entomol.* 20: 221–239.

Brittain, W. H. and C. B. Gooderham. 1916. An insect enemy of the parsnip. *Can. Entomol.* 2: 37–41.

Britton, W. E. 1919. Insects attacking squash, cucumber, and allied plants in Connecticut. Connecticut Agric. Exp. Stn. Bull. 216: 33–51.

Broadbent, A. B. and T. H. A. Olthof. 1995. Foliar application of *Steinernema carpocapsae* (Rhabditida: Steinernematidae) to control *Liriomyza trifolii* (Diptera: Agromyzidae) larvae in chrysanthemums. *Environ. Entomol.* 24: 431–435.

Broadbent, L. 1954. The different distribution of two *Brassica* viruses in the plant and its influence on spread in the field. *Ann. Appl. Biol.* 41: 174–182.

Brodeur, J. and J. N. McNeil. 1994. Seasonal ecology of *Aphidius nigripes* (Hymenoptera: Aphididae), a parasitoid of *Macrosiphum euphorbiae* (Homoptera: Aphididae). *Environ. Entomol.* 23: 292–298.

Brooks, A. R. 1951. Identification of the root maggots (Diptera: Anthomyiidae) attacking cruciferous garden crops in Canada, with notes on biology and control. *Can. Entomol.* 83: 109–120.

Brooks, W. M. 1986. Comparative effects of *Nosema epilachnae* and *Nosema varivestis* on the Mexican bean beetle, *Epilachna varivestis*. *J. Invertebr. Pathol.* 48: 344–354.

Brooks, W. M., E. I. Hazard, and J. Becnel. 1985. Two new species of *Nosema* (Microsporida: Nosematidae) from the Mexican bean beetle *Epilachna varivestis* (Coleoptera: Coccinellidae). *J. Protozool.* 32: 525–535.

Brown, D. W. and R. A. Goyer. 1984. Comparative consumption rates for selected insect predators of two defoliating caterpillar species of soybean in Louisiana. *J. Georgia Entomol. Soc.* 19: 299–303.

Brown, E. A. and A. J. Keaster. 1983. Field response of *Melanotus depressus* and *M. verberans* (Coleoptera: Elateridae) to flight traps and to pheromone of tufted apple budmoth (*Platynota idaeusalis* Walker) (Lepidoptera: Tortricidae). *J. Kansas Entomol. Soc.* 56: 47–49.

Brown, E. A. and A. J. Keaster. 1986. Activity and dispersal of adult *Melanotus depressus* and *Melanotus verberans* (Coleoptera: Elateridae) in a Missouri cornfield. *J. Kansas Entomol. Soc.* 59: 127–132.

Brown, E. S. and C. F. Dewhurst. 1975. The genus *Spodoptera* (Lepidoptera, Noctuidae) in Africa and the Near East. *Bull. Entomol. Res.* 65: 221–262.

Brown, J. J., T. Jermy, and B. A. Butt. 1980. The influence of an alternate host plant on the fecundity of the Colorado potato beetle, *Leptinotarsa decemlineata* (Coleoptera: Chrysomelidae). *Ann. Entomol. Soc. Am.* 73: 197–199.

Brown, J. K., H. S. Costa, and F. Laemmlen. 1992. First report of whitefly-associated squash silverleaf disorder of *Cucurbita* in Arizona and of white streaking disorder of *Brassica* species in Arizona and California. *Plant Dis.* 76: 426.

Brown, J. K., D. R. Frolich, and R. C. Rosell. 1995. The sweetpotato/silverleaf whiteflies: biotypes of *Bemisia tabaci* Genn. or a species complex? *Annu. Rev. Entomol.* 40: 511–534.

Broza, M. and B. Sneh. 1994. *Bacillus thuringiensis* ssp. *kurstaki* as an effective control agent of lepidopteran pests in tomato fields in Israel. *J. Econ. Entomol.* 87: 923–928.

Brust, G. E. 1994. Natural enemies in straw-mulch reduce Colorado potato beetle populations and damage in potato. *Biol. Control* 4: 163–169.

Brust, G. E. and R. E. Foster. 1995. Semiochemical-based toxic baits for control of striped cucumber beetle (Coleoptera: Chrysomelidae) in cantaloupe. *J. Econ. Entomol.* 88: 112–116.

Brust, G. E. and R. E. Foster. 1999. New economic threshold for striped cucumber beetle (Coleoptera: Chrysomelidae) in cantaloupe in the midwest. *J. Econ. Entomol.* 92: 936–940.

Brust, G. E. and G. J. House. 1990. Influence of soil texture, soil moisture, organic cover, and weeds on oviposition preference of southern corn rootworm (Coleoptera: Chrysomelidae). *Environ. Entomol.* 19: 966–971.

Brust, G. E., R. E. Foster, and W. G. Buhler. 1996. Comparison of insecticide use programs for managing the striped cucumber

beetle (Coleoptera: Chrysomelidae) in muskmelon. *J. Econ. Entomol.* 89: 981–986.

Bruton, B. D., L. D. Chandler, and M. E. Miller. 1989. Relationships between pepper weevil and internal mold of sweet pepper. *Plant Dis.* 73: 170–173.

Bryan, D. E. and R. F. Smith. 1956. The *Frankliniella occidentalis* (Pergande) complex in California (Thysanoptera: Thripidae). *Univ. California Publ. Entomol.* 10: 359–410.

Bucher, G. E. and G. K. Bracken. 1976. The bertha armyworm, *Mamestra configurata* (Lepidoptera: Noctuidae). Artificial diet and rearing technique. *Can. Entomol.* 108: 1327–1338.

Bucher, G. E. and G. K. Bracken. 1977. The bertha armyworm, *Mamestra configurata* (Lepidoptera: Noctuidae). Sterilization of adults by exposure of prepupal-pupal stages to 25°. *Can. Entomol.* 109: 549–553.

Bucher, G. E. and G. K. Bracken. 1979. The bertha armyworm, *Mamestra configurata* (Lepidoptera: Noctuidae). An estimate of light and pheromone trap efficiency based on captures of newly emerged moths. *Can. Entomol.* 111: 977–984.

Bucher, G. E. and H. H. Cheng. 1970. Use of trap plants for attracting cutworm larvae. *Can. Entomol.* 102: 797–798.

Bucher, G. E. and H. H. Cheng. 1971. Mortality in larvae of *Euxoa messoria* (Lepidoptera: Noctuidae) collected from the tobacco area of Ontario. *Can. Entomol.* 103: 888–892.

Bucher, G. E. and W. J. Turnock. 1983. Dosage responses of the larval instars of the bertha armyworm, *Mamestra configurata* (Lepidoptera: Noctuidae), to a native nuclear polyhedrosis. *Can. Entomol.* 115: 341–349.

Buntin, G. D. and L. P. Pedigo. 1981. Dispersion and sequential sampling of green cloverworm eggs in soybeans. *Environ. Entomol.* 10: 980–985.

Buntin, G. D. and L. P. Pedigo. 1983. Seasonality of green cloverworm (Lepidoptera: Noctuidae) adults and an expanded hypothesis of population dynamics in Iowa. *Environ. Entomol.* 12: 1551–1558.

Buntin, G. D. and P. L. Raymer. 1994. Pest status of aphids and other insects in winter canola in Georgia. *J. Econ. Entomol.* 87: 1097–1104.

Buntin, G. D., L. P. Pedigo, and W. B. Showers. 1990. Temporal occurrence of the variegated cutworm (Lepidoptera: Noctuidae) adults in Iowa with evidence for migration. *Environ. Entomol.* 19: 603–608.

Buntin, G. D., J. N. All, D. V. McCracken, and W. L. Hargrove. 1994. Cover crop and nitrogen fertility effects on southern corn rootworm (Coleoptera: Chrysomelidae) damage in corn. *J. Econ. Entomol.* 87: 1683–1688.

Burbutis, P. P. and L. P. Kelsey. 1970. Pest status of the green cloverworm on lima beans in Delaware. *J. Econ. Entomol.* 63: 1956–1958.

Burbutis, P. P., N. Erwin, and L. R. Ertle. 1981. Reintroduction and establishment of *Lydella thompsoni* and notes on other parasites of the European corn borer in Delaware. *Environ. Entomol.* 10: 779–781.

Burch, J. B. 1960. Some snails and slugs of quarantine significance to the United States. USDA, ARS 82–1. 73 pp.

Burch, J. B. and T. A. Pearce. 1990. Terrestrial Gastropoda. Pages 201–309 *in* D. L. Dindal (ed.). *Soil Biology Guide*. Wiley & Sons, New York.

Burdette, R. C. 1935. The biology and control of the pepper maggot *Zonosemata electa* Say. Trypetidae. New Jersey Agric. Exp. Stn. Bull. 585. 26 pp.

Buren, W. F. 1972. Revisionary studies on the taxonomy of the imported fire ants. *J. Georgia Entomol. Soc.* 7: 1–26.

Buren, W. F., G. E. Allen, W. H. Whitcomb, F. E. Lennartz, and R. N. Williams. 1974. Zoogeography of the imported fire ants. *J. New York Entomol. Soc.* 82: 113–124.

Burgess, L. 1977. Flea beetles (Coleoptera: Chrysomelidae) attacking rape crops in the Canadian prairie provinces. *Can. Entomol.* 109: 21–32.

Burgess, L. 1980. Predation of adults of the flea beetle *Phyllotreta cruciferae* by lacewing larvae (Neuroptera: Chrysopidae). *Can. Entomol.* 112: 745–6.

Burgess, L. 1982. Occurrence of some flea beetle pests of parkland rapeseed crops in open prairie and forest in Saskatchewan (Coleoptera: Chrysomelidae). *Can. Entomol.* 114: 623–627.

Burgess, L. and C. F. Hinks. 1987. Predation of adults of the crucifer flea beetle, *Phyllotreta cruciferae* (Goeze), by the northern fall field cricket, *Gryllus pennsylvanicus* Burmeister (Orthoptera: Gryllidae). *Can. Entomol.* 119: 495–496.

Burgess, L. and D. T. Spurr. 1984. Distribution and abundance of overwintering flea beetles (Coleoptera: Chrysomelidae) in a grove of trees. *Environ. Entomol.* 13: 941–944.

Burgess, L. and H. H. Weegar. 1986. A method for rearing *Nysius ericae* (Hemiptera: Lygaeidae), the false chinchbug. *Can. Entomol.* 118: 1059–1061.

Burgess, L. and J. E. Wiens. 1976. Maintaining a colony of the striped flea beetle, *Phyllotreta striolata* (Coleoptera: Chrysomelidae), in the greenhouse. *Can. Entomol.* 108: 53–55.

Burgess, L. and J. E. Wiens. 1980. Dispensing allyl isothiocyanate as an attractant for trapping crucifer-feeding flea beetles. *Can. Entomol.* 112: 93–97.

Burgess, L., J. Dueck, and D. L. McKenzie. 1983. Insect vectors of the yeast *Nematospora coryli* in mustard, *Brassica juncea*, crops in southern Saskatchewan. *Can. Entomol.* 115: 25–30.

Burkness, E. C. and W. D. Hutchison. 1997. Development and validation of a binomial sequential sampling plan for striped cucumber beetle (Coleoptera: Chrysomelidae) in cucurbits. *J. Econ. Entomol.* 90: 1590–1594.

Burkness, E. C. and W. D. Hutchison. 1998. Development and validation of a fixed-precision sampling plan for estimating striped cucumber beetle (Coleoptera: Chrysomelidae) density in cucurbits. *Environ. Entomol.* 27: 178–183.

Burleigh, J. G. 1971. Parasites reared from the soybean looper in Louisiana 1968–1969. *J. Econ. Entomol.* 64: 1550–1551.

Burleigh, J. G. 1972. Population dynamics and biotic controls of the soybean looper in Louisiana. *Environ. Entomol.* 1: 290–294.

Burn, A. J. 1982. The role of predator searching efficiency in carrot fly egg loss. *Ann. Appl. Biol.* 101: 154–159.

Burn, A. J. 1984. Life tables for the carrot fly, *Psila rosae*. *J. Appl. Ecol.* 21: 891–902.

Burns, E. L. and P. E. A. Teal. 1989. Response of male potato stem borer moths, *Hydraecia micacea* (Esper) to conspecific females and synthetic pheromone blends in the laboratory and field. *J. Chem. Ecol.* 15: 1365–1378.

Burrage, R. H. 1963. Seasonal feeding of larvae of *Ctenicera destructor* and *Hypolithus bicolor* (Coleoptera: Elateridae) on potatoes placed in the field at weekly intervals. *Ann. Entomol. Soc. Am.* 56: 306–313.

Burris, E., A. M. Pavloff, B. R. Leonard, J. B. Graves, and G. Church. 1990. Evaluation of two procedures for monitoring populations of early season insect pests (Thysanoptera: Thripidae and Homoptera: Aphididae) in cotton under selected management strategies. *J. Econ. Entomol.* 83: 1064–1068.

Burstein, M. and D. Wool. 1993. Gall aphids do not select optimal galling sites (*Smynthurodes betae*; Pemphigidae). *Ecol. Entomol.* 18: 155–164.

Burton, R. L. 1970. A low-cost artificial diet for the corn earworm. *J. Econ. Entomol.* 63: 1969–1970.

Burton, R. L. and D. J. Schuster. 1986. Meridic diet for rearing tomato pinworm larvae (Lepidoptera: Gelechiidae). *J. Econ. Entomol.* 79: 1143–1146.

Burton, R. L., K. J. Starks, and D. C. Peters. 1980. The army cutworm. Oklahoma Agric. Exp. Stn. Bull. 739. 35 pp.

Busching, M. K. and F. T. Turpin. 1976. Oviposition preferences of black cutworm moths among various crop plants, weeds, and plant debris. *J. Econ. Entomol.* 69: 587–590.

Busching, M. K. and F. T. Turpin. 1977. Survival and development of black cutworm (*Agrotis ipsilon*) larvae on various species of crop plants and weeds. *Environ. Entomol.* 6: 63–65.

Buschman, L. L. and W. H. Whitcomb. 1980. Parasites of *Nezara viridula* (Hemiptera: Pentatomidae) and other Hemiptera in Florida. *Florida Entomol.* 63: 154–162.

Buschman, L. L., H. N. Pitre, and H. F. Hodges. 1981. Soybean cultural practices: effects on populations of green cloverworm, velvetbean caterpillar, loopers and *Heliothis complex*. *Environ. Entomol.* 10: 631–641.

Buschman, L. L., F. R. Lamm, P. E. Sloderbeck, and G. L. Dick. 1985. Chemigation in corn: effects of nonemulsifiable oils and sprinkler package on the efficacy of corn borer (Lepidoptera: Pyralidae) insecticides. *J. Econ. Entomol.* 78: 1331–1336.

Bush, G. L. 1965. The genus *Zonosemata*, with notes on the cytology of two species (Diptera: Tephritidae). Psyche 72: 307–323.

Bushing, R. W. and V. E. Burton. 1974. Leafhopper damage to silage corn in California. *J. Econ. Entomol.* 67: 656–658.

Butcher, F. G. 1936. Studies of seasonal occurrence of injuries to potato tubers in western New York. *J. Econ. Entomol.* 29: 486–490.

Butler, Jr., G. D. 1970. Temperature and the development of egg and nymphal stages of *Lygus desertus*. *J. Econ. Entomol.* 63: 1994–1995.

Butler, Jr., G. D. 1976. Bollworm: development in relation to temperature and larval food. *Environ. Entomol.* 5: 520–522.

Butler, Jr., G. D. and T. J. Henneberry. 1990a. Pest control on vegetables and cotton with household cooking oils and liquid detergents. *Southwestern Entomol.* 15: 123–131.

Butler, Jr., G. D., and T. J. Henneberry. 1990b. Cottonseed oil and Safer insecticidal soap: effects on cotton and vegetable pests and phytotoxicity. *Southwestern Entomol.* 15: 257–264.

Butler, Jr., G. D., T. J. Henneberry, and W. D. Hutchison. 1986. Biology, sampling and population dynamics of *Bemisia tabaci*. Pages 167–195 *in* G. E. Russell (ed.). *Agricultural Zoology Reviews*, Vol. 1. Intercept, Ponteland, Newcastle upon Tyne.

Butler, Jr., G. D., T. J. Henneberry, P. A. Stansly, and D. J. Schuster. 1993. Insecticidal effects of selected soaps, oils and detergents on the sweetpotato whitefly: (Homoptera: Aleyrodidae). *Florida Entomol.* 76: 161–167.

Butler, L. I., J. E. Halfhill, L. M. McDonough, and B. A. Butt. 1977. Sex attractant of the alfalfa looper *Autographa californica* and the celery looper *Anagrapha falcifera* (Lepidoptera: Noctuidae). *J. Chem. Ecol.* 3: 65–70.

Butler, L. I., L. M. McDonough, J. A. Onsager, and B. J. Landis. 1975. Sex pheromones of the Pacific Coast wireworm, *Limonius canus*. *Environ. Entomol.* 4: 229–230.

Buxton, G. M., D. B. Thomas, and R. C. Froeschner. 1983. Revision of the *Sayi*-group of *Chlorochroa* Stål (Hemiptera: Pentatomidae). Occ. Pap. Entomol. California Dept. Food Agric. 29. 33 pp.

Buxton, J. H. and D. S. Madge. 1974. Artificial incubation of eggs of the common earwig, *Forficula auricularia* (L.). *Entomol. Mon. Mag.* 110: 55–57.

Buxton, J. H. and D. S. Madge. 1976a. The food of the European earwig (*Forficula auricularia* L.) in hop gardens. *Entomol. Mon. Mag.* 112: 231–237.

Buxton, J. H. and D. S. Madge. 1976b. The evaluation of the European earwig (*Forficula auricularia*) as a predator of the damson-hop aphid (*Phorodon humuli*). I. Feeding experiments. *Entomol. Exp. Appl.* 19: 109–114.

Byers, G. W. 1973. A mating aggregation of *Nysius raphanus* (Hemiptera: Lygaeidae). *J. Kansas Entomol. Soc.* 46: 281–282.

Byers, J. R. and D. L. Struble. 1990. Identification of sex pheromones of two sibling species in dingy cutworm complex, *Feltia jaculifera* (Gn.) (Lepidoptera: Noctuidae). *J. Chem. Ecol.* 16: 2981–2992.

Byers, J. R., B. D. Hill, and G. B. Schaalje. 1992. Effect of inactivity associated with interstadial molts on short-term efficacy of insecticides for control of pale western cutworm (Lepidoptera: Noctuidae). *J. Econ. Entomol.* 85: 1146–1149.

Byers, J. R., D. S. Yu, and J. W. Jones. 1993. Parasitism of the army cutworm, *Euxoa auxiliaris* (Grt.) (Lepidoptera: Noctuidae), by *Copidosoma bakeri* (Howard) (Hymenoptera: Encyrtidae) and effect on crop damage. *Can. Entomol.* 125: 329–335.

Byers, J. R., D. L. Struble, C. E. Herle, G. C. Kozub, and J. D. LaFontaine. 1990. Electroantennographic responses differentiate sibling species of dingy cutworm complex, *Feltia jaculifera* (Gn.) (Lepidoptera: Noctuidae). *J. Chem. Ecol.* 16: 2969–2980.

Bykova, Y. P. 1984. Forecasting abundance of beet webworm on the basis of estimation of the ecological conditions for the formation of its hibernating pool (*Loxostege sticticalis* L., Lepidoptera, Pyralidae). *Entomol. Rev.* 63: 1–10.

Bynum, Jr., E. D. and T. L. Archer. 1987. Wireworm (Coleoptera: Elateridae) sampling for semiarid cropping systems. *J. Econ. Entomol.* 80: 164–168.

Cabanillas, H. E. and J. R. Raulston. 1994. Pathogenicity of *Steinernema riobravis* against corn earworm, *Helicoverpa zea* (Boddie). *Fund. Appl. Nematol.* 17: 219–223.

Cabanillas, H. E. and J. R. Raulston. 1995. Impact of *Steinernema* (Rhabditida: Steinernematidae) on the control of *Helicoverpa zea* (Lepidoptera: Noctuidae) in corn. *J. Econ. Entomol.* 88: 58–64.

Cabanillas, H. E. and J. R. Raulston. 1996. Evaluation of *Steinernema riobravis*, *S. carpocapsae*, and irrigation timing for the control of corn earworm, *Helicoverpa zea*. *J. Nematol.* 28: 75–82.

Cadogan, B. L. 1983. Biology and potential for increase of *Hellula phidilealis* (Lepidoptera: Pyralidae) in Barbados. *Environ. Entomol.* 12: 1805–1807.

Caffrey, D. J. and G. W. Barber. 1919. The grain bug. *USDA Bull.* 779. 35 pp.

Caffrey, D. J. and L. H. Worthley. 1927. A progress report on the investigations of the European corn borer. *USDA Bull.* 1476. 154 pp.

Cagle, L. R. 1956. Life history of the spider mite *Tetranychus atlanticus* McG. Virginia Agric. Exp. Stn. Tech. Bull. 12422 pp.

Cagle, L. R. and H. W. Jackson. 1947. Life history of the garden fleahopper. Virginia Agric. Exp. Stn. Tech. Bull. 107. 27 pp.

Calkins, C. O. and V. M. Kirk. 1975. Distribution of false wireworms (Coleoptera: Tenebrionidae) in relation to soil texture. *Environ. Entomol.* 4: 373–374.

Callahan, P. S. 1958. Behavior of the imago of the corn earworm, *Heliothis zea* (Boddie), with special reference to emergence and reproduction. *Ann. Entomol. Soc. of Am.* 51: 271–283.

Cameron, A. E. 1914. A contribution to a knowledge of the belladonna leaf-miner, *Pegomyia hyoscyami*, Panz., its life-history and biology. *Ann. Appl. Biol.* 1: 43–76.

Cammell, M. E. 1981. The black bean aphid, *Aphis fabae*. *Biologist* 28: 247–257.

Campbell, A. and M. Mackauer. 1975. Thermal constants for development of the pea aphid (Homoptera: Aphididae) and some of its parasites. *Can. Entomol.* 107: 419–423.

Campbell, C. D. and W. D. Hutchison. 1995. Rearing methods and demographic statistics for a subterranean morph of the sugarbeet root aphid (Homoptera: Aphididae). *Can. Entomol.* 127: 65–77.

Campbell, C. D., J. F. Walgenbach, and G. G. Kennedy. 1991. Effect of parasitoids on lepidopterous pests in insecticide-treated and untreated tomatoes in western North Carolina. *J. Econ. Entomol.* 84: 1662–1667.

Campbell, J. M., M. J. Sarazin, and D. B. Lyons. 1989. Canadian Beetles (Coleoptera) Injurious to Crops, Ornamentals, Stored Products, and Buildings. Publ. 1826. Agriculture Canada, Ottawa. 491 pp.

Campbell, R. E. 1920. The broad-bean weevil. USDA Bull. 807. 23 pp.

Campbell, R. E. 1924. Notes on injurious southwestern Tenebrionidae (Col.) Entomol. News 35: 1–7.

Campbell, R. E. 1927. The celery leaf-tyer, *Phlyctaenia rubicalis* Guen., in California (Lep.) *Pan-Pac. Entomol.* 4: 77–84.

Campbell, R. E. and V. Duran. 1929. Notes on the sugar-beet army worm in California. Mthly. Bull. California Dept. Agric. 18: 267–275.

Campbell, W. V. and C. H. Brett. 1966. Varietal resistance of beans to the Mexican bean beetle. *J. Econ. Entomol.* 59: 899–902.

Campbell, W. V., D. A. Mount, and B. S. Heming. 1971. Influence of organic matter content of soils on insecticidal control. *J. Econ. Entomol.* 64: 41–44.

Cancelado, R. E. and E. B. Radcliffe. 1979. Action thresholds for potato leafhopper on potatoes in Minnesota. *J. Econ. Entomol.* 72: 566–569.

Canerday, T. D. and F. S. Arant. 1966. The looper complex in Alabama (Lepidoptera, Plusiinae). *J. Econ. Entomol.* 59: 742–743.

Cantelo, W. W., M. Jacobson, and A. W. Hartstack. 1982. Moth trap performance: Jackson trap vs. Texas pheromone trap. *Southwestern Entomol.* 7: 212–215.

Capinera, J. L. 1974a. Damage to asparagus seedlings by *Brachycolus asparagi*. *J. Econ. Entomol.* 67: 447–448.

Capinera, J. L. 1974b. Biology of the asparagus beetles, *Crioceris asparagi* and *Crioceris duodecimpunctata*, in western Massachusetts. M. S. Thesis, University of Massachusetts, Amherst. 81 pp.

Capinera, J. L. 1976. Asparagus beetle defense behavior: adaptations for survival in dispersing and non-dispersing species. *Ann. Entomol. Soc. Am.* 69: 269–272.

Capinera, J. L. 1978a. Consumption of sugarbeet foliage by the pale-striped flea beetle. *J. Econ. Entomol.* 71: 301–303.

Capinera, J. L. 1978b. Consumption of sugarbeet foliage by the salt-marsh caterpillar. *J. Econ. Entomol.* 71: 661–663.

Capinera, J. L. 1978c. Variegated cutworm: consumption of sugarbeet foliage and development on sugarbeet. *J. Econ. Entomol.* 71: 978–980.

Capinera, J. L. 1979a. Zebra caterpillar (Lepidoptera: Noctuidae): foliage consumption and development of larvae on sugarbeet. *Can. Entomol.* 111: 905–909.

Capinera, J. L. 1979b. Parasitism of the palestriped flea beetle (Chrysomelidae) by an allantonematid nematode *Howardula* sp.: effect of parasitism on adult longevity and sugarbeet foliage consumption. *Ann. Entomol. Soc. Am.* 72: 348–349.

Capinera, J. L. 1980. Visual responses of some sugarbeet insects to sticky traps of various yellow and orange hues positioned at two heights. *Southwestern Entomol.* 5: 76–79.

Capinera, J. L. 1981. Some effects of infestation by bean aphid, *Aphis fabae* Scopoli, on carbohydrate and protein levels in sugarbeet plants, and procedures for estimating economic injury levels. *Z. Angew. Entomol.* 92: 374–384.

Capinera, J. L. 1986. Field key for identification of caterpillars found on field and vegetable crops in Colorado. *Colorado State Univ. Ext. Serv. Bull.* 535A. 13 pp.

Capinera, J. L. 1993a. Differentiation of nymphal instars in *Schistocerca americana* (Orthoptera: Acrididae). *Florida Entomol.* 76: 175–179.

Capinera, J. L. 1993b. Host-plant selection by *Schistocerca americana* (Orthoptera: Acrididae). *Environ. Entomol.* 22: 127–133.

Capinera, J. L. 1994. Pickleworm and melonworm. Pages 140–145 *in* D. Rosen, F. D. Bennett, and J. L. Capinera (eds.). Pest Management in the Subtropics: Biological Control—A Florida Perspective. Intercept, Andover, UK.

Capinera, J. L. and B. E. Hibbard. 1987. Bait formulations of chemical and microbial insecticides for suppression of crop-feeding grasshoppers. *J. Agric. Entomol.* 4: 337–344.

Capinera, J. L. and D. R. Horton. 1989. Geographic variation in effects of weather on grasshopper infestation. *Environ. Entomol.* 18: 8–14.

Capinera, J. L. and M. R. Kanost. 1979. Susceptibility of the zebra caterpillar *Ceramica picta* to *Autographa californica* nuclear polyhedrosis virus. *J. Econ. Entomol.* 72: 570–572.

Capinera, J. L. and J. H. Lilly. 1975a. Bionomics and biotic control of the asparagus beetle, *Crioceris asparagi*, in western Massachusetts. *Environ. Entomol.* 4: 93–96.

Capinera, J. L. and J. H. Lilly. 1975b. *Tetrastichus asparagi*, parasitoid of the asparagus beetle: some aspects of host-parasitoid interaction. *Ann. Entomol. Soc. Am.* 68: 595–596.

Capinera, J. L. and W. J. Roltsch. 1981. The predatory ant *Formica neoclara*: effect on within-field bean aphid distribution, and activity in relation to thermal conditions. *J. Kansas Entomol. Soc.* 54: 578–586.

Capinera, J. L. and R. A. Schaefer. 1983. Field identification of adult cutworms, armyworms, and similar crop pests collected from light traps in Colorado. Colorado State Univ. Ext. Serv. Bull. 514A. 24 pp.

Capinera, J. L. and T. S. Sechrist. 1982. Grasshoppers (Acrididae) of Colorado. Colorado State Exp. Stn. Bull. 584S. 161 pp.

Capinera, J. L. and M. R. Walmsley. 1978. Visual responses of some sugarbeet insects to sticky traps and water pan traps of various colors. *J. Econ. Entomol.* 71: 926–927.

Capinera, J. L., D. R. Gardner, and F. R. Stermitz. 1985. Cantharidin levels in blister beetles (Coleoptera: Meloidae) associated with alfalfa in Colorado. *J. Econ. Entomol.* 78: 1052–1055.

Capinera, J. L., S. E. Naranjo, and A. R. Renaud. 1981a. Alfalfa webworm: larval development in response to diet and temperature. *Southwest. Entomol.* 6: 10–16.

Capinera, J. L., A. R. Renaud, and S. E. Naranjo. 1981b. Alfalfa webworm: foliage consumption and host preference. *Southwestern. Entomol.* 6: 18–22.

Capinera, J. L., T. J. Weissling, and E. E. Schweizer. 1985. Compatibility of intercropping with mechanized agriculture: effects of strip intercropping of pinto beans and sweet corn on insect abundance in Colorado. *J. Econ. Entomol.* 78: 354–357.

Capinera, J. L., D. R. Horton, N. D. Epsky, and P. L. Chapman. 1987. Effects of plant density and late-season defoliation on yield of field beans. *Environ. Entomol.* 16: 274–280.

Capinera, J. L., D. Pelissier, G. S. Menout, and N. D. Epsky. 1988. Control of black cutworm, *Agrotis ipsilon* (Lepidoptera: Noctuidae), with entomogenous nematodes (Nematoda: Steinernematidae, Heterorhabditidae). *J. Invertebr. Pathol.* 52: 427–435.

Capps, H. W. 1946. Description of the larva of *Keiferia peniculo* Heinrich, with a key to the larvae of related species attacking eggplant, pepper, potato and tomato in the United States. *Ann. Entomol. Soc. Am.* 39: 561–563.

Capps, H. W. 1967. Review of some species of *Loxostege* Hübner and descriptions of new species (Lepidoptera, Pyraustidae: Pyraustinae). *Proc. U. S. Nat. Mus.* 120 (3561): 1–75.

Cardona, C., R. Gonzalez, and A. V. Schoonhoven. 1982. Evaluation of damage to common beans by larvae and adults of *Diabrotica balteata* and *Cerotoma facialis*. *J. Econ. Entomol.* 75: 324–327.

Cardona, C., C. E. Posso, J. Kornegay, J. Valor, and M. Serrano. 1989. Antibiosis effects of wild dry bean accessions on the Mexican

bean weevil and the bean weevil (Coleoptera: Bruchidae). *J. Econ. Entomol.* 82: 310–315.

Carey, J. R. and J. W. Bradley. 1982. Developmental rates, vital schedules, sex ratios, and life tables for *Tetranychus urticae, T. turkestani* and *T. pacificus* (Acarina: Tetranychidae) on cotton. Acarologia 23: 333–345.

Carlson, E. C. 1964. Effect of flower thrips on onion seed plants and a study of their control. *J. Econ. Entomol.* 57: 735–741.

Carner, G. R. and O. W. Barnett. 1975. A granulosis virus of the green cloverworm. *J. Invertebr. Pathol.* 25: 269–271.

Carroll, D. P. and S. C. Hoyt. 1984. Augmentation of European earwigs (Dermaptera: Forficulidae) for biological control of apple aphid (Homoptera: Aphididae) in an apple orchard. *J. Econ. Entomol.* 77: 738–740.

Carroll, D. P., J. T. S. Walker, and S. C. Hoyt. 1985. European earwigs (Dermaptera: Forficulidae) fail to control apple aphids on bearing apple trees and woolly apple aphids (Homoptera: Aphididae) in apple rootstock stool beds. *J. Econ. Entomol.* 78: 972–974.

Carruth, L. A. 1942. An investigation of the mineral oil treatment for corn earworm control. *J. Econ. Entomol.* 35: 227–233.

Carruthers, R. I., D. L. Haynes, and D. M. MacLeod. 1985. *Entomophthora muscae* (Entomophthorales: Entomophthoracae) mycosis in the onion fly, *Delia antiqua* (Diptera: Anthomyiidae). *J. Invertebr. Pathol.* 45: 81–93.

Carruthers, R. I., G. H. Whitfield, and D. L. Haynes. 1984. Sampling program for estimating plant damage caused by the onion maggot (Diptera: Anthomyiidae) on the regional and field levels. *J. Econ. Entomol.* 77: 1355–1363.

Carruthers, R. I., M. E. Ramos, T. S. Larkin, D. L. Hostetter, and R. S. Soper. 1997. The *Entomophaga grylli* (Fresenius) Batko species complex: its biology, ecology, and use for biological control of pest grasshoppers. *Me. Entomol. Soc. Canada* 171: 329–353.

Carter, M. and P. Feeny. 1985. Techniques for maintaining a culture of the black swallowtail butterfly, *Papilio polyxenes asterius* Stoll (Papilionidae). *J. Lepid. Soc.* 39: 125–133.

Carter, W. 1941. *Peregrinus maidis* (Ashm.) and the transmission of corn mosaic. *Ann. Entomol. Soc. Am.* 34: 551–556.

Cartwright, B., J. V. Edelson, and C. Chambers. 1987. Composite action thresholds for the control of lepidopterous pests on fresh-market cabbage in the lower Rio Grande Valley of Texas. *J. Econ. Entomol.* 80: 175–181.

Cartwright, B., J. C. Palumbo, and W. S. Fargo. 1990a. Influence of crop mulches and row covers on the population dynamics of the squash bug (Heteroptera: Coreidae) on summer squash. *J. Econ. Entomol.* 83: 1988–1993.

Cartwright, B., T. G. Teague, L. D. Chandler, J. V. Edelson, and G. Bentsen. 1990b. An action threshold for management of the pepper weevil (Coleoptera: Curculionidae) on bell peppers. *J. Econ. Entomol.* 83: 2003–2007.

Cartwright, O. L. 1929. The maize billbug in South Carolina. South Carolina Agric. Exp. Stn. Bull. 257. 35 pp.

Cartwright, O. L. 1934. The southern corn stalk borer in South Carolina. *South Carolina Agric. Exp. Stn. Bull.* 294. 32 pp.

Cary, L. R. 1902. The grass thrips. Maine Agric. Exp. Stn. Bull. 83. 128 pp.

Casanova, R. I. 1977. Notes on the life cycle of the bean leaf skeletonizer, *Syngrapha egena* (Guenée) (Lepidoptera: Noctuidae) in Puerto Rico. *J. Agric. Univ. Puerto Rico.* 61: 253–255.

Castineiras, A., R. M. Baranowski, and H. Glenn. 1996a. Temperature response of tow strains of *Ceranisus menes* (Hymenoptera: Eulophidae) reared on *Thrips palmi* (Thysanoptera: Thripidae). *Florida Entomol.* 79: 13–19.

Castineiras, A., R. M. Baranowski, and H. Glenn. 1997. Distribution of *Neoseiulus cucumeris* (Acarina: Phytoseiidae) and its prey,

Thrips palmi (Thysanoptera: Thripidae) within eggplants in south Florida. *Florida Entomol.* 80:211–217.

Castineiras, A., J. E. Pena, R. Duncan, and L. Osborne. 1996b. Potential of *Beauveria bassiana* and *Paecilomyces fumosoroseus* (Deuteromycotina: Hyphomycetes) as biological control agents of *Thrips palmi* (Thysanoptera: Thripidae). *Florida Entomol.* 79:458–461.

Castle, S. J., T. M. Perring, C. A. Farrar, and A. N. Kishaba. 1992. Field and laboratory transmission of watermelon mosaic virus 2 and zucchini yellow mosaic virus by various aphid species. *Phytopathology.* 82:235–240.

Castor, L. L., J. E. Ayers, A. A. MacNab, and R. A. Krause. 1975. Computerized forecasting system for Stewart's bacterial disease on corn. *Plant Dis. Rep.* 59:533–536.

Castro, V., C. Rivera, S. A. Isard, R. Gamez, J. Fletcher, and M. E. Irwin. 1992. The influence of weather and microclimate on *Dalbulus maidis* (Homoptera: Cicadellidae) flight activity and the incidence of diseases within maize and bean monocultures and bicultures in tropical America. *Ann. Appl. Biol.* 121:469–482.

Causey, N. B. 1943. Studies on the life history and the ecology of the hothouse millipede, *Orthomorpha gracilis* (C.L. Koch, 1847). *Am. Midl. Nat.* 29:670–682.

Cave, R. D. 1992. Inventory of parasitic organisms fo the striped grass looper, *Mocis latipes* (Lepidoptera: Noctuidae), in Honduras. *Florida Entomol.* 75:592–598.

Chalfant, R. B. 1973a. Chemical control of the southern green stinkbug, tomato fruitworm and potato aphid on vining tomatoes in southern Georgia. *J. Georgia Entomol. Soc.* 8:279–283.

Chalfant, R. B. 1973b. Cowpea curculio: control in southern Georgia. *J. Econ. Entomol.* 66:727–729.

Chalfant, R. B. 1975. A simplified technique for rearing the lesser cornstalk borer (Lepidoptera: Phycitidae). *J. Georgia Entomol. Soc.* 10:33–37.

Chalfant, R. B. and C. H. Brett. 1967. Interrelationship of cabbage variety, season, and insecticide on control of the cabbage looper and the imported cabbageworm. *J. Econ. Entomol.* 60:687–690.

Chalfant, R. B. and E. R. Mitchell. 1967. Some effects of food and substrate on oviposition of the spotted cucumber beetle. *J. Econ. Entomol.* 60:1010–1012.

Chalfant, R. B. and D. R. Seal. 1991. Biology and management of wireworms on sweet potato. Pages 303–326 *in* R. K. Jansson and K.V Raman (eds.). Sweet Potato Pest Management: a Global Perspective. Westview Press, Boulder, Colorado.

Chalfant, R. B. and J. R. Young. 1988. Cowpea curculio, *Chalcodermes aeneus* Boheman (Coleoptera: Curculionidae); insecticidal control on the southern pea in Georgia, 1980–1986. *Appl. Agric. Res.* 3:8–11.

Chalfant, R. B., S. A. Harmon, and L. Stacey. 1979a. Chemical control of the sweet potato flea beetle and southern potato wireworm on sweet potatoes in Georgia. *J. Georgia Entomol.* 14:354–358.

Chalfant, R. B., B. Mullinix, and S. S. Nilakhe. 1982. Southern peas: interrelationships among growth stage, insecticide applications, and yield in Georgia. *J. Econ. Entomol.* 75:405–409.

Chalfant, R. B., S. C. Phatak, and E. D. Threadgill. 1979b. Protection of direct-seeded tomatoes from early insect injury with systemic insecticides in Georgia. *J. Econ. Entomol.* 72:587–589.

Chalfant, R. B., E. F. Suber, and T. D. Canerday. 1972. Resistance of southern peas to cowpea curculio in the field. *J. Econ. Entomol.* 65:1679–1682.

Chalfant, R. B., K. Bondari, H. R. Sumner, and M. R. Hall. 1992a. Reduction of wireworm (Coleoptera: Elateridae) damage in sweet potato with insecticides applied by chemigation. *J. Econ. Entomol.* 86:123–130.

Chalfant, R. B., C. A. Jaworski, A. W. Johnson, and D. R. Summer. 1977. Reflective film mulches, millet barriers, and pesticides:

effects on watermelon mosaic virus, insects, nematodes, soilborne fungi, and yield of yellow summer squash. *J. Am. Soc. Hort. Sci.* 102:11–15.

Chalfant, R. B., M. R. Hall, A. W. Johnson, D. R. Seal, and K. Bondari. 1992b. Effects of application methods, timing, and rates of insecticides and nematicides on yield and control of wireworms (Coleoptera: Elateridae) and nematodes (Tylenchida: Heteroderidae) that affect sweet potato. *J. Econ. Entomol.* 85:878–887.

Chalfant, R. B., C. S. Creighton, G. L. Greene, E. R. Mitchell, J. M. Stanley, and J. C. Webb. 1974. Cabbage looper: populations in BL traps baited with sex pheromone in Florida, Georgia, and South Carolina. *J. Econ. Entomol.* 67:741–745.

Chamberlin, F. S. and J. N. Tenhet. 1926. The seasonal history and food habits of the tobacco budworm, *Heliothis virescens* Fab., in the southern tobacco-growing region. *J. Econ. Entomol.* 19:611–614.

Chamberlin, F. S. and H. H. Tippins. 1948. *Microtheca ochroloma*, an introduced pest of crucifers, found in Alabama. *J. Econ. Entomol.* 41:979–980.

Chamberlin, F. S., J. N. Tenhet, and A. G. Boving. 1924. Life-history studies of the tobacco flea-beetle in the southern cigar-wrapper district. *J. Agric. Res.* 29:575–584.

Chamberlin, J. R., J. W. Todd, R. J. Beshear, A. K. Culbreath, and J. W. Demski. 1992. Overwintering hosts and wingform of thrips, *Frankliniella* spp., in Georgia (Thysanoptera: Thripidae): implications for management of spotted wilt disease. *Environ. Entomol.* 21:121–128.

Chamberlin, R. V. and R. L. Hoffman. 1958. Checklist of the millipeds of North America. U. S. Nat. Mus. Bull. 212. 236 pp.

Chambers, R. J. 1986. Preliminary experiments on the potential of hoverflies (*Dipt.: Syrphidae*) for the control of aphids under glass. *Entomophaga* 31:197–204.

Chambers, R. J., S. Long, and N. L. Helyer. 1993. Effectiveness of *Orius laevigatus* (Hem.: Anthocoridae) for the control of *Frankliniella occidentalis* on cucumber and pepper in the UK. *Biocontrol Sci. Tech.* 3:295–307.

Chambers, W. S. and R. W. Sites. 1989. Overwintering thrips fauna in croplands of the Texas South Plains. *Southwestern Entomol.* 14:325–328.

Champlain, R. A. and G. D. Butler, Jr. 1967. Temperature effects on development of the egg and nymphal stages of *Lygus hesperus* (Hemiptera: Miridae). *Ann. Entomol. Soc. Am.* 60:519–521.

Chant, D. A. and J. H. McLeod. 1952. Effects of certain climatic factors on the daily abundance of the European earwig, *Forficula auricularia* L. (Dermaptera: Forficulidae), in Vancouver, British Columbia. *Can. Entomol.* 84:174–180.

Chapman, J. A., J. I. Romer, and J. Stark. 1955. Ladybird beetles and army cutworm adults as food for grizzly bears in Montana. *Ecology* 36:156–158.

Chapman, P. J. and G. E. Gould. 1929. Sweet potato sawfly. Virginia Truck Exp. Stn. Bull. 68:769–786.

Chapman, P. J. and S. E. Lienk. 1981. Flight periods of adults of cutworms, armyworms, loopers and others (family Noctuidae) injurious to vegetable and field crops. New York Agric. Exp. Stn. Search: Agric. Bull. 14. 43 pp.

Chapman, P. L. 1988. Local differences in host use by two populations of the Colorado potato beetle. *Ecology* 69:823–831.

Chapman, R. F., E. A. Bernays, and S. J. Simpson. 1981. Attraction and repulsion of the aphid, *Cavariella aegopodii*, by plant odors. *J. Chem. Ecol.* 7:881–888.

Chellemi, D. O., J. E. Funderburk, and D. W. Hall. 1994. Seasonal abundance of flower-inhabiting *Frankliniella* species (Thysanoptera: Thripidae) on wild plant species. *Environ. Entomol.* 23:337–342.

Cheng, H. H. 1971. Field studies on the chemical control of the dark-sided cutworm (Lepidoptera: Noctuidae) on tobacco in Ontario, with particular reference to Dursban. *Can. Entomol.* 103: 649–653.

Cheng, H. H. 1973a. Observations on the bionomics of the dark-sided cutworm, *Euxoa messoria* (Lepidoptera: Noctuidae), in Ontario. *Can. Entomol.* 105: 311–322.

Cheng, H. H. 1973b. Further field evaluation of insecticides for control of the dark-sided cutworm (Lepidoptera: Noctuidae) on tobacco in Ontario. *Can. Entomol.* 105: 1351–1357.

Cheng, H. H. 1973c. Laboratory and field tests with *Bacillus thuringiensis* against the dark-sided cutworm, *Euxoa messoria* (Lepidoptera: Noctuidae), on tobacco. *Can. Entomol.* 105: 941–945.

Cheng, H. H. 1977. Insect parasites of the dark-sided cutworm, *Euxoa messoria* (Lepidoptera: Noctuidae), in Ontario. *Can. Entomol.* 109: 137–142.

Cheng, H. H. 1980. Toxicity and persistence of pyrethroid insecticides as foliar sprays against dark-sided cutworm (Lepidoptera: Noctuidae) on tobacco in Ontario. *Can. Entomol.* 112: 451–456.

Cheng, H. H. 1981. Additional hymenopterous parasites newly recorded from the dark-sided cutworm, *Euxoa messoria* (Lepidoptera: Noctuidae), in Ontario. *Can. Entomol.* 113: 773–774.

Cheng, H. H. 1984. Residual toxicity of six pyrethroid and two organophosphorous insecticides on soil surface against dark-sided cutworm (Lepidoptera: Noctuidae) on tobacco in Ontario. *Can. Entomol.* 116: 11–17.

Cheng, H. H. and D. L. Struble. 1982. Field evaluation of blacklight and sex attractant traps for monitoring seasonal distribution of the darksided cutworm (Lepidoptera: Noctuidae) in Ontario. *Can. Entomol.* 114: 1121–1125.

Cherry, R. H. and D. G. Hall. 1986. Flight activity of *Melanotus communis* (Coleoptera: Elateridae) in Florida sugar cane fields. *J. Econ. Entomol.* 79: 626–628.

Cheshire, J. M. and T. J. Riley. 1988. Occurrence of a soil-inhabiting pest of row crops, *Melanotus similis* (Kirby) (Coleoptera: Elateridae), in forest habitats. *J. Entomol. Sci.* 23: 75–76.

Chiang, H. C. 1973. Bionomics of the northern and western corn rootworms. *Annu. Rev. Entomol.* 18: 47–72.

Chiang, H. C., L. K. French, and D. E. Rasmussen. 1980. Quantitative relationship between western corn rootworm population and corn yield. *J. Econ. Entomol.* 73: 665–666.

Chiang, H. C., D. Rasmussen, and R. Gorder. 1971. Survival of corn rootworm larvae under minimum tillage conditions. *J. Econ. Entomol.* 64: 1576–1577.

Childs, G., S. L. Poe, and M. J. Bassett. 1976. Response of the two-potted spider mite to *Phaseolus* cultivars. Proc. Florida State Hort. Soc. 89: 149–150.

Chillcott, J. G. 1959. The *Pegomyia hyoscyami* (spinach leaf miner) complex in North America (Diptera: Muscidae). *Can. Entomol.* 41: 167–170.

Chippendale, G. M. 1979. The southwestern corn borer, *Diatraea grandiosella*: case history of an invading insect. Missouri Agric. Exp. Stn. Res. Bull. 1031. 52 pp.

Chisholm, M. D., W. F. Steck, E. W. Underhill, and P. Palaniswamy. 1983. Field trapping of diamondback moth *Plutella xylostella* using an improved four-component sex attractant blend. *J. Chem. Ecol.* 9: 113–118.

Chittenden, F. H. 1899. Some insects injurious to garden and orchard crops. USDA Div. Entomol. Bull. 19 (N.S.). 99 pp.

Chittenden, F. H. 1900. Some insects injurious to garden crops. USDA Div. Entomol. Bull. 23 (N.S). 92 pp.

Chittenden, F. H. 1901. The fall army worm and variegated cutworm. USDA Div. Entomol. Bull. (N.S.) 29. 64 pp.

Chittenden, F. H. 1902. Some insects injurious to vegetable crops. USDA Div. Entomol. Bull. 33 (N.S.). 117 pp.

Chittenden, F. H. 1907. The Colorado potato beetle. USDA Bur. Entomol. Circ. 87. 15 pp.

Chittenden, F. H. 1908a. The common squash bug. USDA Bur. Entomol. Bull. 39. 5 pp.

Chittenden, F. H. 1908b. The harlequin cabbage bug. USDA Bur. Entomol. Circ. 103. 10 pp.

Chittenden, F. H. 1911a. The southern beet webworm. USDA Bur. Entomol. Bull. 109: 17–22.

Chittenden, F. H. 1911b. The Hawaiian beet webworm. USDA Bur. Entomol. Bull. 109: 1–14.

Chittenden, F. H. 1912a. Notes on the cucumber beetles. USDA Bur. Entomol. Bull. 82: 67–75.

Chittenden, F. H. 1912b. The yellow-necked flea-beetle. USDA Bur. Entomol. Bull. 82: 29–32.

Chittenden, F. H. 1912c. The parsnip leaf-miner. USDA Bur. Entomol. Bull. 82: 9–13.

Chittenden, F. H. 1912d. The celery caterpillar. USDA Bur. Entomol. Bull. 82: 20–24.

Chittenden, F. H. 1912e. The broad-bean weevil. USDA Bur. Entomol. Bull. 96: 59–82.

Chittenden, F. H. 1912f. The cowpea weevil. USDA Bur. Entomol. Bull. 96: 83–94.

Chittenden, F. H. 1913a. The spotted beet webworm. USDA Bur. Entomol. Bull. 127: 1–11.

Chittenden, F. H. 1913b. The Abutilon moth. USDA Bur. Entomol. Bull. 126. 10 pp.

Chittenden, F. H. 1916a. The common cabbage worm. USDA Farmers' Bull. 766. 16 pp.

Chittenden, F. H. 1916b. The rose-chafer: a destructive garden and vineyard pest. USDA Farmers' Bull. 721. 8 pp.

Chittenden, F. H. 1917. The asparagus beetles and their control. *USDA Farmers' Bull.* 837. 15 pp.

Chittenden, F. H. 1919. Control of the onion thrips. USDA Farmers' Bull. 1007. 16 pp.

Chittenden, F. H. 1923. The striped cucumber beetle and how to control it. USDA Farmers' Bull. 1322. 16 pp.

Chittenden, F. H. 1924. The argus tortoise beetle. *J. Agric. Res.* 27: 43–51.

Chittenden, F. H. 1927. The species of *Phyllotreta* north of Mexico. *Entomol. Am.* 8: 1–63.

Chittenden, F. H. and D. E. Fink. 1922. The green June beetle. USDA Bull. 891. 52 pp.

Chittenden, F. H. and N. F. Howard. 1917. The horse-radish flea-beetle: its life history and distribution. USDA Bur. Entomol. Bull. 535: 1–16.

Chittenden, F. H. and H. O. Marsh. 1912. The imported cabbage webworm. USDA Bur. Entomol. Bull. 109: 23–45.

Chittenden, F. H. and H. O. Marsh. 1920. The western cabbage flea-beetle. USDA Bur. Entomol. Bull. 902. 21 pp.

Chittenden, F. H. and H. M. Russell. 1910. The semitropical armyworm. USDA Bur. Entomol. Bull. 66: 53–70.

Cho, J. J., R. F. L. Mau, D. Gonsalves, and W. C. Mitchell. 1986. Reservoir weed hosts of tomato spotted wilt virus. *Plant Dis.* 70: 1014–1017.

Cho, J. J., W. C. Mitchell, R. F. L. Mau, and K. Sakimura. 1987. Epidemiology of tomato spotted wilt virus disease on crisphead lettuce in Hawaii. *Plant Dis.* 71: 505–508.

Cho, K., C. S. Eckel, J. F. Walgenbach, and G. C. Kennedy. 1995. Spatial distribution and sampling procedures for *Frankliniella* spp. (Thysanoptera: Thripidae) in staked tomato. *J. Econ. Entomol.* 88: 1658–1665.

Chow, Y. S., S. C. Chiu, and C. C. Chien. 1974. Demonstration of a sex pheromone of the diamondback moth (Lepidoptera: Plutellidae). *Ann. Entomol. Soc. Am.* 67: 510–512.

Christiansen, K. 1964. Bionomics of Collembola. *Annu. Rev. Entomol.* 9: 147–178.

Christiansen, K. A. 1990. Insecta: Collembola. Pages 965–995 *in* D.L. Dindal (ed.). *Soil Biology Guide.* John Wiley & Sons, New York.

Christiansen, K. and P. Bellinger. 1998. The Collembola of North America North of the Rio Grande: a Taxonomic Analysis. Grinnell College, Grinell, Iowa. 1520 pp.

Christie, J. R. 1930. Notes on larval nemas from insects. *J. Parasitol.* 16: 250–256.

Christie, R. D., J. T. Schulz, and N. C. Gudmestad. 1993. Potato flea beetle (Coleoptera: Chrysomelidae) evaluated as a possible vector of ring rot bacterium in potatoes. *J. Econ. Entomol.* 86: 1223–1227.

Chuman, T., P. L. Guss, R. E. Doolittle, J. R. McLaughlin, J. L. Krysan, J. M. Schalk, and J. H. Tumlinson. 1987. Identification of female-produced sex pheromone from banded cucumber beetle, *Diabrotica balteata* LeConte (Coleoptera: Chrysomelidae). *J. Chem. Ecol.* 13: 1601–1616.

Church, N. S. and G. H. Gerber. 1977a. The development and habits of *Linsleya sphaericollis* (Coleoptera: Meloidae). *Can. Entomol.* 109: 375–380.

Church, N. S. and G. H. Gerber. 1977b. Observations on the ontogeny and habits of *Lytta nuttalli, L. viridana,* and *L. cyanipennis* (Coleoptera: Meloidae): the adults and eggs. *Can. Entomol.* 109: 565–573.

Church, N. S. and R. W. Salt. 1952. Some effects of temperature on development and diapause in eggs of *Malanoplus bivittatus* (Say) (Orthoptera: Acrididae). *Can. J. Zool.* 30: 173–184.

Chyzik, R., I. Glazer, and M. Klein. 1996. Virulence and efficacy of different entomopathogenic nematode species against western flower thrips (*Frankliniella occidentalis*). *Phytoparasitica* 24: 103–110.

Clancy, D. W. and H. D. Pierce. 1966. Natural enemies of some lygus bugs. *J. Econ. Entomol.* 59: 853–858.

Clark, M. S., J. M. Luna, N. D. Stone, and R. R. Youngman. 1994. Generalist predator consumption of armyworm (Lepidoptera: Noctuidae) and effect of predator removal on damage in no-till corn. *Environ. Entomol.* 23: 617–622.

Clarke, R. G. and G. E. Wilde. 1970. Association of the green stink bug and the yeast-spot disease organism of soybeans. 1. Length of retention, effect of molting, isolation from feces and saliva. *J. Econ. Entomol.* 63: 200–204.

Clarke, W. T. 1901. The potato-worm in California (*Gelechia operculella,* Zeller). California Agric. Exp. Stn. Bull. 135. 30 pp.

Clausen, C. P. 1978. Introduced Parasites and Predators of Arthropod Pests and Weeds: A World Review. USDA Agric. Handbook 480. 545 pp.

Clegg, J. M. and C. A. Barlow. 1982. Escape behaviour of the pea aphid *Acyrthosiphon pisum* (Harris) in response to alarm pheromone and vibration. *Can. J. Zool.* 60: 2245–2252.

Clemens, B. 1862. North American micro-Lepidoptera. *Proc. Entomol. Soc., Philadelphia* 1: 147–151.

Clement, S. L., L. V. Kaster, W. B. Showers, and R. S. Schmidt. 1985. Seasonal changes in the reproductive condition of female black cutworm moths (Lepidoptera: Noctuidae). *J. Kansas Entomol. Soc.* 58: 62–68.

Cleveland, T. C. 1987. Predation by tarnished plant bugs (Heteroptera: Miridae) of *Heliothis* (Lepidoptera: Noctuidae) eggs and larvae. *Environ. Entomol.* 16: 37–40.

Cline, L. D. and D. H. Habeck. 1977. Reproductive biology of the granulate cutworm. *J. Georgia Entomol. Soc.* 12: 34–41.

Cloutier, C. and F. Bauduin. 1995. Biological control of the Colorado potato beetle *Leptinotarsa decemlineata* (Coleoptera: Chrysomelidae) in Quebec by augmentative releases of the two-spotted stinkbug *Perillus bioculatus* (Hemiptera: Pentatomidae). *Can. Entomol.* 127: 195–212.

Cloutier, C. and M. Mackauer. 1975. A modified technique for rearing of the pea aphid on artificial diets. *Entomol. Exp. Appl.* 18: 395–396.

Coaker, T. H. and D. A. Williams. 1963. The importance of some Carabidae and Staphylinidae as predators of the cabbage root fly, *Erioischia brassicae* (Bouché). *Entomol. Exp. Appl.* 6: 156–164.

Cobb, P. P. and M. H. Bass. 1975. Beet armyworm (Lepidoptera: Noctuidae): a laboratory rearing technique. *J. Georgia Entomol. Soc.* 10: 190–191.

Cock, M. J. W. 1986. *Bemisia tabaci* – a literature survey. CAB International, UK. 121 pp.

Cock, M. J. W. 1993. *Bemisia tabaci* – an update 1986–1992 on the cotton whitefly with an annotated bibliography. CAB International, UK. 78 pp.

Cockbain, A. J., S. M. Cook, and R. Bowen. 1975. Transmission of broad bean stain virus and Echtes Ackerbohnenmosaik-virus to field beans. *Ann. Appl. Biol.* 81: 331–339.

Cocke, Jr., J., J. W. Stewart, M. Morris, and W. Newton. 1994. Emergence patterns of Mexican corn rootworm, *Diabrotica virgifera* Zeae, adults in south and central Texas. *Southwestern Entomol.* 19: 347–354.

Cockerham, K. L. and O. T. Deen. 1936. Notes on life history, habits and distribution of *Heteroderes laurentii* Guer. *J. Econ. Entomol.* 29: 288–296.

Cockerham, K. L., O. T. Deen, M. B. Christian, and L. D. Newsom. 1954. The biology of the sweet-potato weevil. Louisiana Agric. Exp. Stn. Tech. Bull. 483. 30 pp.

Cockfield, S. D. 1992. Groundnut oil application and varietal resistance for control of *Callosobruchus maculatus* (F.) in cowpea grain in the Gambia. *Trop. Pest Manage.* 38: 268–270.

Cockfield, S. D. and D. A. Potter. 1984. Predation on sod webworm (Lepidoptera: Pyralidae) eggs as affected by chlorpyrifos application to Kentucky bluegrass turf. *J. Econ. Entomol.* 77: 1542–1544.

Cohen, S. 1981. Reducing the spread of aphid-transmitted viruses in peppers by coarse-net cover. *Phytoparasitica* 9: 69–76.

Cohen, S. and M. J. Berlinger. 1986. Transmission and cultural control of whitefly-borne viruses. *Agric. Ecosystems Environ.* 17: 89–97.

Cole, C. L. 1988. Stratification and survival of diapausing burrowing bugs. *Southwestern. Entomol.* 13: 243–246.

Cole, R. A., S. A. Rollason, A. Morgan, J. Gilchrist, S. A. Baker, and P. Springer. 1988. A rapid non-destructive technique for screening carrot seedlings for resistance to carrot fly damage. *Ann. Appl. Biol.* 112: 159–166.

Collantes, L. G., K. V. Raman, and F. H. Cisneros. 1986. Effect of six synthetic pyrethroids on two populations of potato tuber moth, *Phthorimaea operculella* (Zeller) (Lepidoptera: Gelechiidae), in Peru. *Crop. Prot.* 5: 355–357.

Collier, R. H. and S. Finch. 1983. Effects of intensity and duration of low temperatures in regulating diapause development of the cabbage root fly (*Delia radicum*). *Entomol. Exp. Appl.* 34: 193–200.

Collier, R. H. and S. Finch. 1990. Some factors affecting the efficiency of sticky board traps for capturing the carrot fly, *Psila rosae* (Diptera: Psilidae). *Bull. Entomol. Res.* 80: 153–158.

Collier, R. H., M. S. Elliott, and S. Finch. 1994. Development of the overwintering stages of the carrot fly, *Psila rosae* (Diptera: Psilidae). *Bull. Entomol. Res.* 84: 469–476.

Collins, R. D. and E. Grafius. 1986. Impact of the egg parasitoid, *Anaphes sordidatus* (Hymenoptera: Mymaridae), on the carrot weevil (Coleoptera: Curculionidae). *Environ. Entomol.* 15: 469–475.

Collins, W. E. 1956. On the biology and control of *Drosophila* on tomatoes for processing. *J. Econ. Entomol.* 49: 607–610.

Cone, W. W., S. Predki, and E. C. Klostermeyer. 1971b. Pheromone studies of the twospotted spider mite. 2. Behavioral response of males to quiescent deutonymphs. *J. Econ. Entomol.* 64: 379–382.

Cone, W. W., L. M. McDonough, J. C. Maitlen, and S. Burdajewicz. 1971a. Pheromone studies of the twospotted spider mite. 1. Evidence of a sex pheromone. *J. Econ. Entomol.* 64: 355–358.

Connell, W. A. 1956. Nitidulidae of Delaware. Delaware Agric. Exp. Stn. Tech. Bull. 318. 66 pp.

Conradi, A. F. 1906. Insects of the garden. Texas Agric. Exp. Stn. Bull. 89. 52 pp.

Cook, C. A. and J. J. Neal. 1999. Feeding behavior of larvae of *Anasa tristis* (Heteroptera: Coreidae) on pumpkin and cucumber. *Environ. Entomol.* 28: 173–177.

Cook, W. C. 1930. Field studies of the pale western cutworm (*Porosagrotis orthogonia* Morr.). Montana Agric. Exp. Stn. Bull. 225. 79 pp.

Cook, W. C. 1941. The beet leafhopper. USDA Farmers' Bull. 1886. 21 pp.

Cook, W. C. 1963. Ecology of the pea aphid in the Blue Mountain area of eastern Washington and Oregon. USDA Tech. Bull. 1287. 48 pp.

Cook, W. C. 1967. Life history, host plants, and migrations of the beet leafhopper in the western United States. USDA Tech. Bull. 1365. 122 pp.

Cooley, R. A. 1905. The white lined morning sphinx. Montana Agric. Exp. Stn. Bull. 62: 224–227.

Coop, L. B. and R. E. Berry. 1986. Reduction in variegated cutworm (Lepidoptera: Noctuidae) injury to peppermint by larval parasitoids. *J. Econ. Entomol.* 79: 1244–1248.

Coquillett, D. W. 1881. On the early stages of *Plusia precationis,* Guenée. *Can. Entomol.* 13: 21–23.

Cormier, D., A. B. Stevenson, and G. Boivin. 1996. Seasonal ecology and geographical distribution of *Anaphes listronoti* and *A. victus* (Hymenoptera: Myrmaridae), egg parasitoids of the carrot weevil (Coleoptera: Curculionidae) in central Ontario. *Environ. Entomol.* 25: 1376–1382.

Costa, H. S., J. K. Brown, and D. N. Byrne. 1991. Life history traits of the whitefly, *Bemisia tabaci* (Homoptera: Aleyrodidae) on six virus-infected or healthy plant species. *Environ. Entomol.* 20: 1102–1107.

Costa, H. S., M. J. Johnson, and D. E. Ullman. 1994. Row covers effect on sweetpotato whitefly (Homoptera: Aleyrodidae) densities, incidence of silverleaf, and crop yield in zucchini. *J. Econ. Entomol.* 87: 1616–1621.

Costa, H. S., D. E. Ullman, M. W. Johnson, and B. E. Tabashnik. 1993. Association between *Bemisia tabaci* density and reduced growth, yellowing, and stem blanching of lettuce and kai choy. *Plant Dis.* 77: 969–972.

Cottier, W. 1953. Aphids of New Zealand. N. Zea. Depart. Sci. and Ind. Res. Bull. 106. 382 pp.

Coudriet, D. L., A. N. Kishaba, J. D. McCreight, and G. W. Bohn. 1979. Varietal resistance in onions to thrips. *J. Econ. Entomol.* 72: 614–615.

Coudriet, D. L., D. E. Meyerdirk, N. Prabhaker, and A. N. Kishaba. 1986. Bionomics of sweetpotato whitefly (Homoptera: Aleyrodidae) on weed hosts in the Imperial Valley, California. *Environ. Entomol.* 15: 1179–1183.

Coudriet, D. L., N. Prabhaker, A. N. Kishaba, and D. E. Meyerdirk. 1985. Variation in developmental rate on different hosts and overwintering of the sweetpotato whitefly, *Bemisia tabaci* (Homoptera: Aleyrodidae). *Environ. Entomol.* 14: 516–519.

Coulson, J. C. 1962. The biology of *Tipula subnodicornis* Zetterstedt, with comparative observations on *Tipula paludosa* Meigen. *J. Anim. Ecol.* 31: 1–21.

Cowan, F. T. 1929. Life history, habits, and control of the Mormon cricket. USDA Tech. Bull. 161. 28 pp.

Cowan, F. T. and H. J. Shipman. 1940. Control of the Mormon cricket by the use of poisoned bait. USDA Cir. 575. 16 pp.

Cowart, F. F. 1953. Factors involved in forage quality for dairy cows. Georgia Exp. Stn. Tech. Bull. 3: 1–35.

Cowles, R. S. and M. G. Villani. 1994. Soil interactions with chemical insecticides and nematodes used for control of Japanese beetle (Coleoptera: Scarabaeidae) larvae. *J. Econ. Entomol.* 87: 1014–1021.

Cragg, J. B. and M. H. Vincent. 1952. The action of metaldehyde on the slug *Agriolimax reticulatus* (Müller). *Ann. Appl. Biol.* 39: 392–406.

Craig, C. H. 1963. The alfalfa plant bug, *Adelphocoris lineolatus* (Goeze) in northern Saskatchewan. *Can. Entomol.* 95: 6–13.

Cranshaw, W. S. 1993. An annotated bibliography of potato/tomato psyllid, *Paratrioza cockerelli* (Sulc) (Homoptera; Psyllidae). Colorado State Uni. Agric. Exp. Stn. Bull. TB93–5. 51 pp.

Creager, D. B. and F. J. Spruijt. 1935. The relation of certain fungi to larval development of *Eumerus tuberculatus* Rond. (Syrphidae, Diptera). *Ann. Entomol. Soc. Am.* 28: 425–437.

Credland, P. F. and A. W. Wright. 1990. Oviposition deterrents of *Callosobruchus maculatus* (Coleoptera: Bruchidae). *Physiol. Entomol.* 15: 285–298.

Creighton, C. S. and G. Fassuliotis. 1980. Seasonal population fluctuations of *Filipjevimermis leipsandra* and infectivity of juveniles on the banded cucumber beetle. *J. Econ. Entomol.* 73: 296–300.

Creighton, C. S. and G. Fassuliotis. 1982. Mass rearing a mermithid nematode, *Filipjevimermis leipsandra* (Mermithida: Mermithidae) on the banded cucumber beetle (Coleoptera: Chrysomelidae). *J. Econ. Entomol.* 75: 701–703.

Creighton, C. S. and G. Fassuliotis. 1983. Infectivity and suppression of the banded cucumber beetle (Coleoptera: Chrysomelidae) by the mermithid nematode *Filipjevimermis leipsandra* (Mermithida: Mermithidae). *J. Econ. Entomol.* 76: 615–618.

Creighton, C. S. and G. Fassuliotis. 1985. *Heterorhabditis* sp. (Nematoda: Heterorhabditidae): a nematode parasite isolated from the banded cucumber beetle *Diabrotica balteata. J. Nematol.* 17: 150–153.

Creighton, C. S. and T. L. McFadden. 1976. Field tests of insecticidal sprays and baits for control of tomato fruitworm on tomatoes. *J. Georgia Entomol. Soc.* 11: 101–105.

Creighton, C. S., W. S. Kinard, and N. Allen. 1961. Effectiveness of *Bacillus thuringiensis* and several chemical insecticides for control of budworms and hornworms on tobacco. *J. Econ. Entomol.* 54: 1112–1114.

Creighton, J. T. 1936. *Anomis erosa* Hubn. as an insect pest of cotton. *J. Econ. Entomol.* 29: 279–282.

Creighton, W. S. 1950. The ants of North America. Bull. Mus. Comp. Zool., Harvard Univ. 104: 1–585.

Crosby, C. R. and M. D. Leonard. 1914. The tarnished plant bug. New York (Cornell) *Agric. Exp. Stn. Bull.* 346: 459–526.

Cross, J. V. and P. Bassett. 1982. Damage to tomato and aubergine by broad mite, *Polyphagotarsonemus latus* (Banks). *Plant Pathol.* 31: 391–393.

Crumb, S. E. 1926. The bronzed cutworm (*Nephelodes emmedonia* Cramer) (Lepidoptera). *Proc. Entomol. Soc. Washington* 28: 201–207.

Crumb, S. E. 1929. Tobacco cutworms. *USDA Tech. Bull.* 88. 179 pp.

Crumb, S. E. 1956. The larvae of the Phalaenidae. USDA Tech. Bull. 1135. 356 pp.

Crumb, S. E., P. M. Eide, and A. E. Bonn. 1941. The European earwig. USDA Tech. Bull. 766. 76 pp.

Cruz, C. 1975. Observations on pod borer oviposition and infestation of pigeonpea varieties. *J. Agric. Univ. Puerto Rico* 59: 63–68.

Cuda, J. P. and H. R. Burke. 1985. Systematics of the larvae and pupae of three species of *Trichobaris* (Coleoptera: Curculionidae). *J. Kansas Entomol. Soc.* 58: 701–707.

Cuda, J. P. and H. R. Burke. 1986. Reproduction and development of the potato stalk borer, (Coleoptera: Curculionidae) with notes on field biology. *J. Econ. Entomol.* 79: 1548–1554.

Culliney, T. W. 1986. Predation on adult *Phyllotreta* flea beetles by *Podisus maculiventris* (Hemiptera: Pentatomidae) and *Nabicula americolimbata* (Hemiptera: Nabidae). *Can. Entomol.* 118: 731–732.

Culliney, T. W. 1990. Population performance of *Thrips palmi* (Thysanoptera: Thripidae) on cucumber infected with a mosaic virus. *Proc. Hawaiian Entomol. Soc.* 30: 85–89.

Cunningham, R. T. 1989. Parapheromones. Pages 221–230 *in* A.S. Robinson and G. Hooper (eds.). Fruit Flies: their Biology, Natural Enemies and Control, Vol. 3a. Elsevier, New York.

Cunningham, R. T. and L. F. Steiner. 1972. Field trial of cue-lure + naled on saturated fiberboard blocks for control of the melon fly by the male annihilation technique. *J. Econ. Entomol.* 65: 505–507.

Cunningham, R. T., L. F. Steiner, K. Ohinata, and G. J. Farias. 1970. Mortality of male melon flies and male Mediterranean fruit flies treated with aerial sprays of lure and naled formulated with a monoglyceride or silicaceous extender. *J. Econ. Entomol.* 63: 106–110.

Cusson, M., R. S. Vernon, and B. D. Roitberg. 1990. A sequential sampling plan for adult tuber flea beetles (*Epitrix tuberis* Gentner): dealing with "edge effects." *Can. Entomol.* 122: 537–546.

Cuthbert, Jr., F. P. 1968. Bionomics of a mermithid (Nematode) parasite of soil-inhibiting larvae of certain chrysomelids (Coleoptera). *J. Invertebr. Pathol.* 12: 283–287.

Cuthbert, Jr., F. P. and B. W. Davis, Jr. 1970. Resistance in sweetpotatoes to damage by soil insects. *J. Econ. Entomol.* 63: 360–363.

Cuthbert, F. P. and B. W. Davis. 1972. Factors contributing to cowpea curculio resistance in southern peas. *J. Econ. Entomol.* 65: 778–781.

Daane, K. M., K. S. Hagen, D. Gonzalez, and L. E. Caltagirone. 1995. European asparagus aphid. Pages 120–122 *in* J. R. Nechols, L. A. Andres, J. W. Beardsley, R. D. Goeden, and C. G. Jackson (eds.). Biological Control in the Western United States. Univ. of California Publ. 3361.

DaCosta, C. P. and C. M. Jones. 1971. Resistance in cucumber, *Cucumis sativus* L., to three species of cucumber beetles. *HortScience* 6: 340–342.

d'Aguilar, J. and J. Missionnier. 1957. Differences biologiques et morphologiques entre *Pegomyia betae* Curt. et *P. hyoscyami* Panz (Dipt. Muscidae). *Bull. Soc. Entomol. France* 62: 124–131.

Dahl, R. A. and D. L. Mahr. 1991. Light trap records of *Phyllophaga* (Coleoptera: Scarabaeidae) in Wisconsin, 1984–1987. *Great Lakes Entomol.* 24: 1–8.

Daigle, C. J., D. J. Boethel, and J. R. Fuxa. 1988. Parasitoids and pathogens of green cloverworm (Lepidoptera: Noctuidae) on an uncultivated spring host (vetch, *Vicia* spp.) and a cultivated summer host (soybean, *Glycine max*). Environ. Entomol. 17: 90–96.

Daigle, C. J., D. J. Boethel, and J. R. Fuxa. 1990. Parasitoids and pathogens of soybean looper and velvetbean caterpillar (Lepidoptera: Noctuidae) in soybeans in Louisiana. *Environ. Entomol.* 19: 746–752.

Dakin, Jr., M. E. 1985. A review of the *Melanoplus femur-rubrum* group (Orthoptera: Acrididae: Melanoplinae) in the southeastern United States. *Trans. Am. Entomol. Soc.* 111: 385–398.

Dakin, Jr., M. E. and K. L. Hays. 1970. A synopsis of Orthoptera (Sensu Lato) of Alabama. Alabama Agric. Exp. Stn. Bull. 404. 118 pp.

Damicone, J. P., W. J. Manning, and D. N. Ferro. 1987. Influence of management practices on severity of stem and crown rot, incidence of asparagus miner, and yield of asparagus grown from transplants. *Plant Dis.* 71: 81–84.

Daniels, N. E. 1977. Soil insect control in corn. Southwestern Entomol. 2: 127–131.

Daniels, N. E. 1978. Insecticidal and cultural control of the southwestern corn borer. *Southwestern Entomol.* 3: 308–314.

Danielson, S. D. and R. E. Berry. 1978. Redbacked cutworm: sequential sampling plans in peppermint. *J. Econ. Entomol.* 71: 323–328.

Daoust, R. A. and R. M. Pereira. 1986. Survival of *Beauveria bassiana* (Deuteromycetes: Moniliales) conidia on cadavers of cowpea pests stored outdoors and in laboratory in Brazil. *Environ. Entomol.* 15: 642–647.

Dapsis, L. J. and D. N. Ferro. 1983. Effectiveness of baited cone traps and colored sticky traps for monitoring adult cabbage maggots with notes on female ovarian development. *Entomol. Exp. Appl.* 33: 35–42.

Das, G. P. and K. V. Raman. 1994. Alternate hosts of the potato tuber moth, *Phthorimaea operculella* (Zeller). *Crop Prot.* 13: 83–86.

Daugherty, D. M. and C. H. Brett. 1966. Nitidulidae associated with sweet corn in North Carolina. North Carolina Agric. Exp. Stn. Bull. 171. 40 pp.

Davis, E. G., J. R. Horton, C. H. Gable, E. V. Walter, R. A. Blanchard, and C. Heinrich. 1933. The southwestern corn borer. *USDA Tech. Bull.* 388. 62 pp.

Davis, F. M. and P. P. Sikorowski. 1978. Susceptibility of the southwestern corn borer, *Diatraea grandiosella* Dyar (Lepidoptera: Pyralidae) to the baculovirus of *Autographa californica*. *J. Kansas Entomol. Soc.* 51: 11–13.

Davis, F. M., T. G. Bird, A. E. Knutson, and S.-S. Ng. 1986. Evaluation of trapping methods for southwestern corn borer (Lepidoptera: Pyralidae) using synthetic sex pheromone. *J. Econ. Entomol.* 79: 1385–1389.

Davis, F. M., W. P. Williams, S.-S. Ng, and G. W. Videla. 1991. Growth and survival of southwestern corn borer on whorl and reproductive stage plants of selected corn hybrids. *Southwestern Entomol.* 16: 144–154.

Davis, G. C. 1893. Insects injurious to celery. Michigan Agric. Exp. Stn. Bull. 102. 32 pp.

Davis, J. J. 1909. Biological studies on three species of Aphididae. *USDA Bur. Entomol. Tech. Series* 12: 123–168.

Davis, J. J. 1912. The alternanthera worm, *Hymenia perspectalis* Hubn. (*Spolodea perspectalis*). Report of the State Entomologist on the Noxious and Beneficial Insects of the State of Illinois 27: 103–106.

Davis, J. J. 1915. The pea aphis with relation to forage crops. USDA Bull. 276. 67 pp.

Davis, J. J. 1917. The corn root-aphis and methods of controlling it. USDA Farmers' Bull. 891. 12 pp.

Davis, J. J. 1922. Common white grubs. USDA Farmers' Bull. 940. 30 pp.

Davis, J. J. and P. Luginbill. 1921. The green June beetle or fig eater. North Carolina Agric. Exp. Stn. Bull. 242. 35 pp.

Davis, J. J. and A. F. Satterthwait. 1916a. The false cabbage aphid (*Aphis pseudobrassicae* Davis). Purdue Univ. Agric. Exp. Stn. Bull. 185: 915–939.

Davis, J. J. and A. F. Satterthwait. 1916b. Life-history studies of *Cirphis unipunctata*, the true army worm. *J. Agric. Res.* 6: 799–812.

Davis, P. M. and L. P. Pedigo. 1989. Analysis of spatial patterns and sequential count plans for stalk borer (Lepidoptera: Noctuidae). *Environ. Entomol.* 18: 504–509.

Davis, P. M. and L. P. Pedigo. 1990a. Evaluation of two management strategies for stalk borer, *Papaipema nebris*, in corn. *Crop Prot.* 9: 387–391.

Davis, P. M. and L. P. Pedigo. 1990b. Yield response of corn stands to stalk borer (Lepidoptera: Noctuidae) injury imposed during early development. *J. Econ. Entomol.* 83: 1582–1586.

Davis, P. M. and L. P. Pedigo. 1991. Injury profiles and yield responses of seedling corn attacked by stalk borer (Lepidoptera: Noctuidae). *J. Econ. Entomol.* 84: 294–299.

Davis, R. 1966. Biology of the leafhopper *Dalbulus maidis* at selected temperatures. *J. Econ. Entomol.* 59: 766.

Dawes, M. A., R. S. Saini, M. A. Mullen, J. H. Brower, and P. A. Loretan. 1987. Sensitivity of sweetpotato weevil (Coleoptera: Curculionidae) to gamma radiation. *J. Econ. Entomol.* 80: 142–146.

Dawkins, G., M. Luxton, and C. Bishop. 1985. Transmission of liquorice rot of carrots by slugs. *J. Mollusc. Stud.* 51: 83–85.

Dawkins, G., J. Hilsop, M. Luxton, and C. Bishop. 1986. Transmission of bacterial soft rot of potatoes by slugs. *J. Mollusc. Stud.* 52: 25–29.

Dawson, G. W., D. C. Griffiths, L. A. Merritt, A. Mudd, J. A. Pickett, L. J. Wadhams, and C. M. Woodcock. 1990. Aphid semiochemicals - a review - and recent advances on the sex pheromone. *J. Chem. Ecol.* 16: 3019–3030.

Day, A. and H. Crosby. 1972. Further field evaluation of insecticides for control of southern potato wireworms. *J. Econ. Entomol.* 65: 1164–1165.

Day, A. and W. J. Reid, Jr. 1969. Response of adult southern potato wireworms to light traps. *J. Econ. Entomol.* 62: 314–319.

Day, A., F. P. Cuthbert, Jr., and W. J. Reid, Jr. 1971. The southern potato wireworm, its biology and economic importance in coastal South Carolina. USDA Tech. Bull. 1443. 33 pp.

Day, A., J. M. Stanley, J. C. Webb, and J. G. Hartsock. 1973. Southern potato wireworms: light-trap catches of adults in an isolated agricultural area. *J. Econ. Entomol.* 66: 757–760.

Day, W. H. 1987. Biological control efforts against *Lygus* and *Adelphocoris* spp. infesting alfalfa in the United States, with notes on other associated Mirid species. Pages 20–39 *in* Hedlund, R. C., and H. M. Graham (eds.). Economic Importance and Biological Control of *Lygus* and *Adelphocoris* in North America. USDA Agric. Res. Serv., ARS-64. 95 pp.

Day, W. H. 1996. Evaluation of biological control of the tarnished plant bug (Hemiptera: Miridae) in alfalfa by the introduced parasite *Peristenus digoneutis* (Hymenoptera: Braconidae). *Environ. Entomol.* 25: 512–518.

Day, W. H. and L. B. Saunders. 1990. Abundance of the garden fleahopper (Hemiptera: Miridae) on alfalfa and parasitism by *Leiophron uniformis* (Gahan) (Hymenoptera: Braconidae). *J. Econ. Entomol.* 83: 101–106.

Day, W. H., R. C. Hedlund, L. B. Saunders, and D. Coutinot. 1990. Establishment of *Peristenus digoneutis* (Hymenoptera: Braconidae), a parasite of the tarnished plant bug (Hemiptera: Miridae), in the United States. *Environ. Entomol.* 19: 1528–1533.

Dean, D. E. and D. J. Schuster. 1995. *Bemisia argentifolii* (Homoptera: Aleyrodidae) and *Macrosiphum euphorbiae* (Homoptera: Aphididae) as prey for two species of Chrysopidae. *Environ. Entomol.* 24: 1562–1568.

Dean, G. J. W. 1973. Bionomics of aphids reared on cereals and some Gramineae. *Ann. Appl. Biol.* 73: 127–135.

Dean, T. W. 1985. Behavioral biology of the striped grass looper, *Mocis latipes* (Guenée), in north-central Florida. Unpublished Ph.D. dissertation, University of Florida, Gainesville. 122 pp.

DeBach, P. 1942. The introduction and recovery of *Trissolcus murgantiae* Ashm. in California. *J. Econ. Entomol.* 35: 787.

Debolt, J. W. 1982. Meridic diet for rearing successive generations of *Lygus hesperus*. *Ann. Entomol. Soc. Am.* 75: 119–122.

Debolt, J. W. and F. K. Wright. 1976. Beet armyworm: inherited sterility produced by irradiation of pupae or adults. *J. Econ. Entomol.* 69: 336–338.

Debolt, J. W., W. W. Wolf, T. J. Henneberry, and P. V. Vail. 1979. Evaluation of light traps and sex pheromone for control of cabbage looper and other lepidopterous insect pests of lettuce. USDA Tech. Bull. 1606. 39 pp.

Decker, G. C. 1931. The biology of the stalk borer *Papaipema nebris* (Gn.). Iowa Agric. Exp. Stn. Res. Bull. 143: 292–351.

de Coss–Romero, M. and J. E. Pena. 1998. Relationship of broad mite (Acara: Tarsonemidae) to host phenology and injury levels in *Capsicum annuum* L. *Florida Entomol.* 81: 515–526.

Decou, G. C. 1994. Biological control of the two-spotted spider mite (Acarina: Tetranychidae) on commercial strawberries in Florida with *Phytoseiulus persimilis* (Acarina: Phytoseiidae). *Florida Entomol.* 77: 33–41.

Decoursey, R. M. and C. O. Esselbaugh. 1962. Descriptions of the nymphal stages of some North American Pentatomidae (Hemiptera-Heteroptera). *Ann. Entomol. Soc. Am.* 55: 323–342.

Dedryver, C. A. 1978. Facteurs de limitation des populations d'*Aphis fabae* dans l'ouest de la France. III. Repartition et incidence des differentes especes d'*Entomophthora* dans les populations. *Entomophaga* 23: 137–151.

Deedat, Y. D. and C. R. Ellis. 1983. Damage caused by potato stem borer (Lepidoptera: Noctuidae) to field corn. *J. Econ. Entomol.* 76: 1055–1060.

Deedat, Y. D., C. R. Ellis, and J. Elmhurst. 1982. Control of potato stem borer, *Hydraecia micacea* (Lepidoptera: Noctuidae), in field corn. Proc. Entomol. Soc. Ontario 113: 43–51.

Deedat, Y. D., C. R. Ellis, and R. J. West. 1983. Life history of the potato stem borer (Lepidoptera: Noctuidae) in Ontario. *J. Econ. Entomol.* 76: 1033–1037.

Deheer, C. J. and D. W. Tallamy. 1991. Affinity of spotted cucumber beetle (Coleoptera: Chrysomelidae) larvae to cucurbitacins. *Environ. Entomol.* 20: 1173–1175.

Dejean, A., P. R. Ngnegueu, and T. Bourgoin. 1996. Trophobiosis between ants and *peregrinus maidis* (Hemiptera, Fulgoromorpha, Delphacidae). *Sociobiol.* 28: 111–120.

Dekle, G. W. 1976. Illustrated key to caterpillars on corn. Florida Dept. Agric. Cons. Serv. Div. Plant Ind. Bull. 4. 16 pp.

de Kort, C. A. D. 1990. Thirty-five years of diapause research with the Colorado potato beetle. Entomol. *Exp. Appl.* 56: 1–13.

Delate, K. M., J. K. Brecht, and J. A. Coffelt. 1990. Controlled atmosphere treatments for control of sweetpotato weevil (Coleoptera: Curculionidae) in stored tropical sweet potatoes. *J. Econ. Entomol.* 83: 461–465.

DeLong, D. M. 1931. A revision of the American species of *Empoasca* known to occur north of Mexico. USDA Tech. Bull. 231. 59 pp.

DeLong, D. M. 1938. Biological studies on the leafhopper *Empoasca fabae* as a bean pest. USDA Tech Bull. 618. 60 pp.

DeLong, D. M. 1971. The bionomics of leafhoppers. *Annu. Rev. Entomol.* 16: 179–210.

Demerec, M. (ed.). 1950. Biology of Drosophila. Wiley & Sons, New York. 632 pp.

den Ouden, H., J. Theunissen, and A. Heslinga. 1984. Protection of cabbage against oviposition of cabbage root fly *Delia brassicae* L. by controlled release of naphthalene. *Z. Angew. Entomol.* 97: 341–346.

DePew, L. J. 1980. Pale western cutworm: chemical control and effect on yield of winter wheat in Kansas. *J. Econ. Entomol.* 73: 138–140.

Derridj, S., H. Lefer, M. Augendre, and Y. Durand. 1988. Use of strips of *Zea mays* L. to trap European corn borer (*Ostrinia nubilalis* Hbn.) oviposition in maize fields. *Crop Prot.* 7: 177–182.

Dethier, V. G. 1988. The feeding behavior of a polyphagous caterpillar (*Diacrisia virginica*) in its natural habitat. *Can. J. Zool.* 66: 1280–1288.

Diawara, M. M., J. T. Trumble, M. L. Lacy, K. K. White, and W. G. Carson. 1996. Potential of somaclonal celeries for use in integrated pest management. *J. Econ. Entomol.* 89:218–223.

Dick, G. L. and L. L. Buschman. 1995. Seasonal occurrence of a fungal pathogen, *Neozygites adjarica* (Entomophthorales: Neozygitaceae), infecting banks grass mites, *Oligonychus pratensis*, and twospotted spider mites, *Tetranychus urticae* (Acari: Tetranychidae), in field corn. *J. Kansas Entomol. Soc.* 68:425–436.

Dickson, M. H. and C. J. Eckenrode. 1975. Variation in *Brassica oleracea* resistance to cabbage looper and imported cabbage worm in the greenhouse and field. *J. Econ. Entomol.* 68:757–760.

Dickson, M. H. and C. J. Eckenrode. 1980. Breeding for resistance in cabbage and cauliflower to cabbage looper, imported cabbageworm, and diamondback moth. *J. Am. Soc. Hort. Sci.* 105:782–785.

Dickson, M. H., A. M. Shelton, S. D. Eigenbrode, M. L. Vamosy, and M. Mora. 1990. Selection for resistance to diamondback moth (*Plutella xylostella*) in cabbage. *HortScience* 25:1643–1646.

Dilbeck, J. D., J. W. Todd, and T. Canerday. 1974. Pickleworm resistance in *Cucurbita. Florida Entomol.* 57:27–32.

Dills, L. E. and M. L. Odland. 1948. Cabbage caterpillar insecticide tests. *J. Econ. Entomol.* 41:948–950.

Dindonis, L. L. and J. R. Miller. 1980. Host-finding responses of onion and seedcorn flies to healthy and decomposing onions and several synthetic constituents of onion. *Environ. Entomol.* 9:467–472.

Dindonis, L. L. and J. R. Miller. 1981. Onion fly and little house fly host finding selectively mediated by decomposing onion and microbial volatiles. *J. Chem. Ecol.* 7:419–426.

Dingler, M. 1934. Uber unsere beiden spargelkater (*Crioceris duodecimpunctata* L. und *Cr. asparagi* L.). *Z. Angew. Entomol.* 21:415–442.

Ditman, L. P. 1937. Observations on poison baits for corn ear worm control. *J. Econ. Entomol.* 30:116–118.

Ditman, L. P. and E. N. Cory. 1931. The corn earworm biology and control. *Maryland Agric. Exp. Stn. Bull.* 328. 482 pp.

Ditman, L. P., E. N. Cory, and A. R. Buddington. 1936. The vinegar gnats or pomace flies—their relation to the canning of tomatoes. Maryland Agric. Exp. Stn. Bull. 400:91–111.

Dixon, A. F. G. 1971. The life-cycle and host preferences of the bird cherry-oat aphid, *Rhopalosiphum padi* L., and their bearing on the theories of host alternation in aphids. *Ann. Appl. Biol.* 68:135–147.

Dixon, A. F. G. and D. M. Glen. 1971. Morph determination in the bird cherry-oat aphid, *Rhopalosiphum padi* L. Ann. Appl. Biol. 68:11–21.

Dixon, A. F. G. and S. D. Wratten. 1971. Laboratory studies on aggregation, size and fecundity in the black bean aphid, *Aphis fabae. Scop. Bull. Entomol. Res.* 61:97–111.

Doane, E. N., B. L. Parker, and Y. Pivot. 1995. Method for mass rearing even-aged western flower thrips on beans. Pages 587–593 *in* B. L. Parker, M. Skinner, and T. Lewis (eds.). Thrips Biology and Management. Plenum Press, London.

Doane, J. F. 1963. Studies on oviposition and fecundity of *Ctenicera destructor* (Brown) (Coleoptera: Elateridae). *Can. Entomol.* 95:1145–1153.

Doane, J. F. 1969a. Effect of temperature on water absorption, development, and hatching in eggs of the prairie grain wireworm, *Ctenicera destructor. Ann. Entomol. Soc. Am.* 62:567–572.

Doane, J. F. 1969b. A method for separating the eggs of the prairie grain wireworm, *Ctenicera destructor*, from soil. *Can. Entomol.* 101:1002–1004.

Doane, J. F. 1977. Spatial pattern and density of *Ctenicera destructor* and *Hypolithus bicolor* (Coleoptera: Elateridae) in soil in spring wheat. *Can. Entomol.* 109:807–822.

Doane, J. F. 1981. Evaluation of a larval trap and baits for monitoring the seasonal activity of wireworms in Saskatchewan. *Environ. Entomol.* 10:335–342.

Doane, J. F. 1983. Attraction of the lesser bulb fly *Eumerus strigatus* (Diptera: Syrphidae) to decomposing oatmeal. *New Zealand Entomol.* 7:419.

Doane, J. F. and R. K. Chapman. 1962a. Oviposition preference of the cabbage maggot, *Hylemya brassicae* (Bouché). *J. Econ. Entomol.* 55:137–138.

Doane, J. F. and R. K. Chapman. 1962b. Control of root maggots on radish, turnip, and rutabaga in Wisconsin. *J. Econ. Entomol.* 55:160–164.

Doane, J. F., Y. W. Lee, J. Klingler, and N. D. Westcott. 1975. The orientation response of *Ctenicera destructor* and other wireworms (Coleoptera: Elateridae) to germinating grain and to carbon dioxide. *Can. Entomol.* 107:1233–1252.

Doederlein, T. A. and R. W. Sites. 1993. Host plant preferences of *Frankliniella occidentalis* and *Thrips tabaci* (Thysanoptera: Thripidae) for onions and associated weeds on the Southern High Plains. *J. Econ. Entomol.* 86:1706–1713.

Doherty, J. A. and J. D. Callos. 1991. Acoustic communication in the trilling field cricket, *Gryllus rubens* (Orthoptera: Gryllidae). *J. Insect Behav.* 4:67–82.

Domek, J. M. and D. T. Johnson. 1987. Evidence of a sex pheromone in the green June beetle, *Cotinis nitida* (Coleoptera: Scarabaeidae). *J Entomol. Sci.* 22:264–267.

Domek, J. M. and D. T. Johnson. 1988. Demonstration of semiochemically induced aggregation in the green June beetle, *Cotinis nitida* (L.) (Coleoptera: Scarabaeidae). *Environ. Entomol.* 17:147–149.

Domek, J. M. and D. T. Johnson. 1990. Inhibition of aggregation behavior in the green June beetle (Coleoptera: Scarabaeidae) by antibiotic treatment of food substrate. *Environ. Entomol.* 19:995–1000.

Domek, J. M. and D. T. Johnson. 1991. Effect of food and mating on longevity and egg production in the green June beetle (Coleoptera: Scarabaeidae). *J. Entomol. Sci.* 26:345–349.

Dominick, C. B. 1943. Life history of the tobacco flea beetle. Virginia Agric. Exp. Stn. Bull. 355. 39 pp.

Dominick, C. B. 1971. Collection of the tobacco flea beetle on colored panels. *J. Econ. Entomol.* 64:1575.

Dominick, C. B. and G. Wene. 1941. Notes on the hibernation of the tobacco flea beetle and on the parasite, *Microctonus epitricis* (Vier.). *J. Econ. Entomol.* 34:395–396.

Doncaster, J. P. 1956. The rice root aphid. *Bull. Entomol. Res.* 47:741–747.

Don–Pedro, K. N. 1989. Mode of action of fixed oils against eggs of *Callosobruchus maculatus* (F.). *Pest. Sci.* 26:107–115.

Dornan, A. P. and J. G. Stewart. 1995. Population dynamics of the European corn borer, *Ostrinia nubilalis* (Hübner) (Lepidoptera: Pyralidae) attacking potatoes in Prince Edward Island. *Can. Entomol.* 127:255–262.

Dornan, A. P., M. K. Sears, and J. G. Stewart. 1995. Evaluation of a binomial model for insecticide application to control lepidopterous pests in cabbage. *J. Econ. Entomol.* 88:302–306.

Doucette, C. F. 1947. Host plants of the cabbage seedpod weevil. *J. Econ. Entomol.* 40:838–840.

Douglas, W. A. 1947. The effect of husk extension and tightness on earworm damage to corn. *J. Econ. Entomol.* 40:661–664.

Douglass, J. R. and W. C. Cook. 1954. The beet leafhopper. USDA Circ. 942. 21 pp.

Dowd, P. F. and T. C. Nelsen. 1994. Seasonal variation of sap beetle (Coleoptera: Nitidulidae) populations in central Illinois cornfield-oak woodland habitat and potential influence of weather patterns. *Environ. Entomol.* 23:1215–1223.

Dowd, P. F., R. J. Bartelt, and D. T. Wicklow. 1992. Novel insect trap useful in capturing sap beetles (Coleoptera: Nitidulidae) and other flying insects. *J. Econ. Entomol.* 85: 772–778.

Dowd, P. F., D. E. Moore, F. E. Vega, M. R. McGuire, R. J. Bartelt, T. C. Nelsen, and D. A. Miller. 1995. Occurrence of a mermithid nematode parasite of *Carpophilus lugubris* (Coleoptera: Nitidulidae) in central Illinois. *Environ. Entomol.* 24: 1245–1251.

Dowell, R. V. 1990. Integrating biological control of whiteflies into crop management systems. Pages 315–335 *in* D. Gerling (ed.). Whiteflies: Their Bionomics, Pest Status and Management. Intercept, Andover, UK.

Downie, N. M. and R. H. Arnett, Jr. 1996. The Beetles of Northeastern North America. Vols. I and II. Sandhill Crane Press, Gainesville, Florida. 1721 pp.

Downing, A. S., C. G. Erickson, and M. J. Kraus. 1991. Field evaluation of entomopathogenic nematodes against citrus root weevils (Coleoptera: Curculionidae) in Florida citrus. *Florida Entomol.* 74: 584–586.

Dozier, H. L. 1917. The life-history of the okra or mallow caterpillar (*Cosmophila erosa* Hübner). *J. Econ. Entomol.* 10: 536–542.

Drake, C. J. 1920. The southern green stink-bug in Florida. Florida State Plant Bd. Quart. Bull. 4: 41–94.

Drake, C. J. and H. M. Harris. 1926. Insect enemies of melons and cucumbers in Iowa. Iowa Agric. Exp. Stn. Bull. 90. 12 pp.

Drake, C. J. and H. M. Harris. 1931. The pale-striped flea beetle, a pest of young seedling onions. *J. Econ. Entomol.* 24: 1132–1137.

Drake, C. J., G. C. Decker, and O. E. Tauber. 1945. Observation on oviposition and adult survival of some grasshoppers of economic importance. *Iowa State Coll. J. Sci.* 19: 207–223.

Drake, E. L. and F. L. McEwen. 1959. Pathology of a nuclear polyhedrosis of the cabbage looper, *Trichoplusia ni* (Hübner). *J. Insect Pathol.* 1: 281–293.

Drost, Y. C., J. C. van Lenteren, and H. J. W. van Roermund. 1998. Life-history parameters of different biotypes of *Bemisia tabaci* (Hemiptera: Aleyrodidae) in relation to temperature and host plant: a selective review. *Bull. Entomol. Res.* 88: 219–229.

Drummond, P. C. 1965. The terrestrial isopod crustaceans (Oniscoidea) of Florida. Unpublished M. S. Thesis, University of Florida, Gainesville. 57 pp.

Dufault, C. P. and T. H. Coaker. 1987. Biology and control of the carrot fly, *Psila rosae* (F.). Pages 97–134 *in* G. E. Russell (ed.). *Agr. Zool. Rev.*, Vol. 2. Intercept, Wimborne, Dorset.

Dufault, C. P. and M. K. Sears. 1982. Evaluation of insecticidal baits for control of the cabbage maggot, *Delia (Hylemya) brassicae* (Diptera: Anthomyiidae), on rutabagas. Proc. Entomol. Soc. Ontario 113: 7–15.

Duffus, S. R., J. D. Busacca, and R. B. Carlson. 1983. Evaluation of sampling methods for dingy cutworm larvae (Lepidoptera: Noctuidae). *J. Econ. Entomol.* 76: 1260–1261.

Dugas, A. L. 1938. The striped flea beetle. Louisiana Agric. Exp. Stn. Bull. 298: 25–27.

Duke, M. E. and E. P. Lampert. 1987. Sampling procedures for tobacco flea beetles (Coleoptera: Chrysomelidae) in flue-cured tobacco. *J. Econ. Entomol.* 80: 81–86.

Dunbar, D. M. and R. L. Beard. 1975. Present status of milky disease of Japanese and Oriental beetles in Connecticut. *J. Econ. Entomol.* 68: 453–457.

Dunn, J. A. 1959a. The survival in soil of apterae of the lettuce root aphid, *Pemphigus bursarius* (L.). *Ann. Appl. Biol.* 47: 766–771.

Dunn, J. A. 1959b. The biology of lettuce root aphid. *Ann. Appl. Biol.* 47: 475–491.

Dunn, J. A. 1960. The natural enemies of the lettuce root aphid, *Pemphigus bursarius* (L.). *Bull. Entomol. Res.* 51: 271–278.

Dunn, J. A. 1965. Studies on the aphid, *Cavariella aegopodii* Scop. I. On willow and carrot. *Ann. Appl. Biol.* 56: 429–438.

Dunn, J. A. 1970. The susceptibility of varieties of carrot to attack by the aphid, *Cavariella aegopodii* (Scop.) *Ann. Appl. Biol.* 66: 301–312.

Dunn, J. A. 1974. The influence of host plant on the production of sexuparae in the aphid *Pemphigus bursarius*. *Entomol. Exp. and Appl.* 17: 445–457.

Dunn, J. A. and D. P. H. Kempton. 1980a. Susceptibilities to attack by root aphid in varieties of lettuce. *Ann. Appl. Biol.* (Suppl.) 94: 56–57.

Dunn, J. A. and D. P. H. Kempton. 1980b. Susceptibilities to attack by top aphids in varieties of lettuce. *Ann. Appl. Biol.* (Suppl.) 94: 58–59.

Dunn, J. A. and J. Kirkley. 1966. Studies on the aphid, *Cavariella aegopodii* Scop. II. On secondary hosts other than carrot. *Ann. Appl. Biol.* 58: 213–217.

Dunn, P. H., I. M. Hall, and M. L. Snideman. 1964. Bioassay of *Bacillus thuringiensis*-based microbial insecticides. III. Continuous propagation of the salt-marsh caterpillar, *Estigmene acrea*. *J. Econ. Entomol.* 57: 374–377.

Dupnik, T. D. and J. A. Kamm. 1970. Development of an artificial diet for *Crambus trisectus*. *J. Econ. Entomol.* 63: 1578–1581.

Dupree, M. 1965. Observations on the life history of the lesser cornstalk borer. *J. Econ. Entomol.* 58: 1156–1157.

Dupree, M. 1970. Ultra-low-volume insecticide sprays for control of the cowpea curculio. *J. Georgia Entomol. Soc.* 5: 39–41.

Dupree, M. 1973. The effects of reflective aluminum foils and polyethylene films on two insect pests of summer squash. *J. Georgia Entomol. Soc.* 2: 146–148.

Dupree, M., T. L. Bissell, and C. M. Beckham. 1955. The pickleworm and its control. Georgia Agric. Exp. Stn. Bull. N. S. 5: 1–34.

Durant, J. A. 1973. Notes on factors influencing observed leafhopper (Homoptera: Cicadellidae) population densities on corn. *J. Georgia Entomol. Soc.* 8: 1–5.

Durant, J. A. 1974. Effectiveness of several insecticides against the maize billbug. *J. Georgia Entomol. Soc.* 9: 191–193.

Dyar, H. G. 1895. Preparatory stages of *Phlegethontius cingulata*. *Entomol. News* 6: 95–97.

Dyar, H. G. and C. Heinrich. 1927. The American moths of the genus *Diatraea* and allies. *Proc. U.S. Nat. Mus.* 71: 1–48.

East, D. A., J. V. Edelson, and B. Cartwright. 1989. Relative cabbage consumption by the cabbage looper (Lepidoptera: Noctuidae), beet armyworm (Lepidoptera: Noctuidae), and diamondback moth (Lepidoptera: Plutellidae). *J. Econ. Entomol.* 82: 1367–1369.

Eastop, V. F. 1981. The wild hosts of aphid pests. Pages 285–298 *in* J. M. Thresh (ed.). Pests, Pathogens and Vegetation. Pitman Advanced Publishing Program, Boston.

Eastop, V. F. 1983. The biology of the principal aphid virus vectors. Pages 115–132 *in* R. T. Plumb and J. M. Thresh (eds.). *Plant Virus Epidemiology*. Blackwell Sci., Oxford.

Ebbert, M. A. and L. R. Nault. 1994. Improved overwintering ability in *Dalbulus maidis* (Homoptera: Cicadellidae) vectors infected with *Spiroplasma kunkelii* (Mycoplasmatales: Spiroplasmataceae). *Environ. Entomol.* 23: 634–644.

Echendu, T. N. C. 1991. Ginger, cashew and neem as surface protectants of cowpeas against infestation and damage by *Callosobruchus maculatus* (Fab.). *Trop. Sci.* 31: 209–211.

Eckenrode, C. J. 1973. Foliar sprays for control of the aster leafhopper on carrots. *J. Econ. Entomol.* 66: 265–266.

Eckenrode, C. J. and R. K. Chapman. 1971a. Effect of various temperatures upon rate of development of the cabbage maggot under artificial conditions. *Ann. Entomol. Soc. Am.* 64: 1079–1083.

Eckenrode, C. J. and R. K. Chapman. 1971b. Observations on cabbage maggot activity under field conditions. *Ann. Entomol. Soc. Am.* 64:1226–1230.

Eckenrode, C. J. and R. K. Chapman. 1972. Seasonal adult cabbage maggot populations in the field in relation to thermal-unit accumulations. *Ann. Entomol. Soc. Am.* 65:151–156.

Eckenrode, C. J. and J. P. Nyrop. 1986. Impact of physical injury and commercial lifting on damage to onion bulbs by larvae of onion maggot (Diptera: Anthomyiidae). *J. Econ. Entomol.* 79:1606–1608.

Eckenrode, C. J., P. S. Robbins, and J. T. Andaloro. 1983. Variations in flight patterns of European corn borer (Lepidoptera: Pyralidae) in New York. *Environ. Entomol.* 12:393–396.

Eckenrode, C. J., E. V. Vea, and K. W. Stone. 1975. Population trends of onion maggots correlated with air thermal unit accumulations. *Environ. Entomol.* 4:785–789.

Eckenrode, C. J., N. L. Gauthier, D. Danielson, and D. R. Webb. 1973. Seedcorn maggot: seed treatments and granule furrow applications for protecting beans and sweet corn. *J. Econ. Entomol.* 66:1191–1194.

Economopoulos, A. P. 1989. Use of traps based on color and/or shape. Pages 315–327 *in* A. S. Robinson and G. Hooper (eds.). Fruit Flies: their Biology, Natural Enemies and Control, Vol. 3b. Elsevier, New York.

Eddy, C. O. and E. M. Livingstone. 1931. *Frankliniella fusca* Hinds (Thrips) on seedling cotton. South Carolina Agric. Exp. Stn. Bull. 271. 23 pp.

Eddy, C. O. and L. C. McAlister, Jr. 1927. The Mexican bean beetle. South Carolina Agric. Exp. Stn. Bull. 236. 38 pp.

Eddy, C. O. and W. C. Nettles. 1930. The bean leaf beetle. South Carolina Agric. Exp. Stn. Bull. 265. 25 pp.

Edelson, J. V. 1985a. A sampling method for estimating absolute numbers of thrips on onions. *Southwestern Entomol.* 10:103–106.

Edelson, J. V. 1985b. Biology of a carrot weevil, *Listronotus texanus* (Coleoptera: Curculionidae): range and seasonality of infestations. *J. Econ. Entomol.* 78:895–897.

Edelson, J. V. 1986. Biology and control of a carrot weevil in the lower Rio Grande Valley, Texas. *J. Rio Grande Valley Hort. Soc.* 39:79–82.

Edelson, J. V. and J. J. Magaro. 1988. Development of onion thrips, *Thrips tabaci* Lindeman as a function of temperature. *Southwestern Entomol.* 13:171–176.

Edelson, J. V., B. Cartwright, and T. A. Royer. 1986. Distribution and impact of *Thrips tabaci* (Thysanoptera: Thripidae) on onion. *J. Econ. Entomol.* 79:502–505.

Edelson, J. V., B. Cartwright, and T. A. Royer. 1989. Economics of controlling onion thrips (Thysanoptera: Thripidae) on onions with insecticides in south Texas. *J. Econ. Entomol.* 82:561–564.

Edelson, J. V., J. Trumble, and R. Story. 1988. Cabbage development and associated lepidopterous pest complex in the southern USA. *Crop Prot.* 7:396–402.

Edney, E. B. 1954. Woodlice and the land habitat. *Biol. Rev.* 29:185–219.

Edwards, C. A. 1958. The ecology of Symphyla. Part I. Populations. *Entomol. Exp. Appl.* 1:308–319.

Edwards, C. A. 1959. The ecology of Symphyla. Part II. Seasonal soil migrations. *Entomol. Exp. Appl.* 2:257–267.

Edwards, C. A. 1961. The ecology of Symphyla. Part III. Factors controlling soil distributions. *Entomol. Exp. Appl.* 4:239–256.

Edwards, M. A. and W. D. Seabrook. 1997. Evidence for an airborne sex pheromone in the Colorado potato beetle, *Leptinotarsa decemlineata*. *Can. Entomol.* 129:667–672.

Ehler, E. 1977a. Natural enemies of cabbage looper on cotton in the San Joaquin Valley. *Hilgardia* 45:72–105.

Ehler, E. 1977b. Parasitization of cabbage looper in California cotton. *Environ. Entomol.* 6:783–784.

Eichlin, T. D. and H. B. Cunningham. 1978. The Plusinae (Lepidoptera: Noctuidae) of America North of Mexico, emphasizing genitalic and larval morphology. USDA Tech. Bull. 1567. 122 pp.

Eichlin, T. D. and W. D. Duckworth. 1988. The Moths of America North of Mexico Including Greenland. Fascicle 5.1: Sesoidea, Sesiidae. Wedge Entomological Research Foundation, Washington, D. C. 176 pp.

Eichmann, R. D. 1943. Asparagus miner really not a pest. *J. Econ. Entomol.* 36:849–852.

Eigenbrode, S. D. and J. T. Trumble. 1994. Fruit-based tolerance to damage by beet armyworm (Lepidoptera: Noctuidae) in tomato. *Environ. Entomol.* 23:937–942.

Eigenbrode, S. D., K. A. Stoner, A. M. Shelton, and W. C. Kain. 1991. Characteristics of glossy leaf waxes associated with resistance to diamondback moth (Lepidoptera: Plutellidae) in *Brassica oleracea*. *J. Econ. Entomol.* 84:1609–1618.

Eigenbrode, S. D., C. White, M. Rohde and C. J. Simon. 1998. Behavior and effectiveness of adult *Hippodamia convergens* (Coleoptera: Coccinellidae) as a predator of *Acyrthosiphon pisum* on a glossy-wax mutant of *Pisum sativum*. *Environ. Entomol.* 91:902–909.

Eilenberg, J. 1987. Abnormal egg-laying behaviour of female carrot flies *(Psila rosae)* induced by the fungus *Entomophthora muscae*. *Entomol. Exp. Appl.* 43:61–65.

Eilenberg, J. and H. Philipsen. 1988. The occurrence of Entomophthorales on the carrot fly *(Psila rosae* F.) in the field during two successive seasons. *Entomophaga* 33:135–144.

Ekbom, B. S. 1987. Incidence counts for estimating densities of *Rhopalosiphum padi* (Homoptera: Aphididae). *J. Econ. Entomol.* 80:933–935.

El-Agamy, F. M. and K. F. Haynes. 1992. Susceptibility of the pea aphid (Homoptera: Aphididae) to an insecticide and a predator in the presence of synthetic aphid alarm pheromone. *J. Econ. Entomol.* 85:794–798.

Elberson, L. R., V. Borek, J. P. McCaffrey, and M. J. Morra. 1996. Toxicity of rapeseed meal-amended soil to wireworms, *Limonius californicus* (Coleoptera: Elateridae). *J. Agric. Entomol.* 13:323–330.

Elden, T. C. 1982. Evaluation of systemic insecticide applications for control of Mexican bean beetle on soybeans in Maryland. *J. Georgia Entomol. Soc.* 17:54–59.

Eliot, I. M. and C. G. Soule. 1921. Caterpillars and their Moths. The Century Company, New York. 302 pp.

Eller, F. J., R. J. Bartelt, B. S. Shasha, D. J. Schuster, D. G. Riley, P. A. Stansly, T. F. Mueller, K. D. Shuler, B. Johnson, J. H. Davis, and C. A. Sutherland. 1994. Aggregation pheromone for the pepper weevil, *Anthonomus eugenii* Cano (Coleoptera: Curculionidae): identification and field activity. *J. Chem. Ecol.* 20:1537–1555.

Elliott, C. and Poos, F. W. 1940. Seasonal development, insect vectors, and host range of bacterial wilt of sweet corn. *J. Agric. Res.* 60:645–686.

Elliott, H. J. and F. J. D. McDonald. 1976. Reproduction in a parthenogenetic aphid, *Aphis craccivora* Koch: embryology, ovarian development and fecundity of apterae and alatae. *Aust. J. Zool.* 24:49–63.

Elliott, N. C. and R. W. Kieckhefer. 1989. Effects of constant and fluctuating temperatures on immature development and age-specific life tables of *Rhopalosiphum padi* (L.) (Homoptera: Aphididae). *Can. Entomol.* 121:131–140.

Elliott, N. C., R. W. Kieckhefer, and D. D. Walgenbach. 1988. Effects of constant and fluctuating temperatures on developmental rates and demographic statistics for the corn leaf aphid (Homoptera: Aphididae). *J. Econ. Entomol.* 81:1383–1389.

Elliott, N. C., R. W. Kieckhefer, and D. D. Walgenbach. 1990. Binomial sequential sampling methods for cereal aphids in small grains. *J. Econ. Entomol.* 83: 1381–1387.

Elliott, W. M. 1980. Monitoring annual flight patterns of the potato aphid, *Macrosiphum euphorbiae* (Homoptera: Aphididae), in southern Ontario. *Can. Entomol.* 112: 963–968.

Elliott, W. M. 1981. The relationship of embryo counts and suction trap catches to populations dynamics of *Macrosiphum euphorbiae* (Homoptera: Aphididae) on tomatoes in Ontario. *Can. Entomol.* 113: 1113–1122.

Ellis, P. R., J. A. Hardman, R. A. Cole, and K. Phelps. 1987. The complementary effects of plant resistance and the choice of sowing and harvest times in reducing carrot fly *(Psila rosae)* damage to carrots. *Ann. Appl. Biol.* 111: 415–424.

Ellis, P. R., J. A. Hardman, P. Crisp, and A. G. Johnson. 1979. The influence of plant age on resistance of radish to cabbage root fly egg-laying. *Ann. Appl. Biol.* 93: 125–131.

Ellsbury, M. M. and M. W. Nielson. 1981. Comparative host plant range studies of the blue alfalfa aphid, *Acyrthosiphon kondoi* Shinji, and the pea aphid, *Acyrthosiphon pisum* (Harris) (Homoptera: Aphididae). USDA Tech. Bull. 1639. 14 pp.

Ellsbury, M. M., J. J. Jackson, W. D. Woodson, D. L. Beck, and K. A. Stange. 1996. Efficacy, application distribution, and concentration by stemflow of *Steinernema carpocapsae* (Rhabditida: Steinernematidae) suspensions applied with a lateral-move irrigation system for corn rootworm (Coleoptera: Chrysomelidae) control in maize. *J. Econ. Tnomol.* 89: 74–81.

Elmore, J. C. and A. F. Howland. 1943. Life history and control of the tomato pinworm. USDA Tech. Bull. 841. 30 pp.

Elmore, J. C., A. C. Davis, and R. E. Campbell. 1934. The pepper weevil. USDA Tech. Bull. 447. 25 pp.

Elmstrom, K. M., D. A. Andow, and W. W. Barclay. 1988. Flea beetle movement in a broccoli monoculture and diculture. *Environ. Entomol.* 17: 299–305.

Elnagar, S. and A. F. Murant. 1976. Relations of the semi-persistent viruses, parsnip yellow fleck and anthriscus yellows, with their vector, *Cavariella aegopodii. Ann. Appl. Biol.* 84: 153–167.

Elsey, K. D. and R. L. Rabb. 1967. Biology of the cabbage looper on tobacco in North Carolina. *J. Econ. Entomol.* 60: 1636–1639.

Elsey, K. D. and R. L. Rabb. 1970a. Analysis of the seasonal mortality of the cabbage looper in North Carolina. *Ann. Entomol. Soc. Am.* 63: 1597–1604.

Elsey, K. D. and R. L. Rabb. 1970b. Biology of *Voria ruralis* (Diptera: Tachinidae). *Ann. Entomol. Soc. Am.* 63: 216–222.

Elsey, K. D. 1977a. Parasitism of some economically important species of Chrysomelidae by nematodes of the genus *Howardula. J. Invertebr. Pathol.* 29: 384–385.

Elsey, K. D. 1977b. Dissemination of *Howardula* sp. nematodes by adult tobacco flea beetles (Coleoptera: Chrysomelidae). *Can. Entomol.* 109: 1283–1285.

Elsey, K. D. 1977c. *Howardula dominicki* n. sp. infesting the tobacco flea beetle in North Carolina. *J. Nematol.* 9: 338–342.

Elsey, K. D. 1979. *Howardula colaspidis* (Allantonematidae) n. sp., a new parasite of the grape colaspis (Coleoptera: Chrysomelidae). *Nematologica* 25: 54–61.

Elsey, K. D. 1980. Pickleworm: mortality on cucumbers in the field. *Environ. Entomol.* 9: 806–809.

Elsey, K. D. 1988. Cucumber beetle seasonality in coastal South Carolina. *Environ. Entomol.* 17: 496–502.

Elsey, K. D. 1993. Cold tolerance of the southern green stink bug (Heteroptera: Pentatomidae). *Environ. Entomol.* 22: 567–570.

Elsey, K. D. and J. M. Pitts. 1976. Parasitism of the tobacco flea beetle by a sphaerulariid nematode *Howardula* sp. *Environ. Entomol.* 5: 707–711.

Elsey, K. D., J. A. Klun, and M. Schwarz. 1989. Pickleworm sex pheromone: potential for use in cucumber pest management. *J. Agric. Entomol.* 6: 275–282.

Elsey, K. D., J. A. Klun, and M. Schwarz. 1991. Forecasting pickleworm (Lepidoptera: Pyralidae) larval infestations using sex pheromone traps. *J. Econ. Entomol.* 84: 1837–1841.

Elsey, K. D., T. L. McFadden, and R. B. Cuthbert. 1984. Improved rearing system for pickleworm and melonworm (Lepidoptera: Pyralidae). *J. Econ. Entomol.* 77: 1070–1072.

Elsey, K. D., J. E. Pena, and V. H. Waddill. 1985. Suitability of potential wild hosts of *Diaphania* species in southern Florida. *Florida Entomol.* 68: 682–686.

Eltoum, E. M. A. and R. E. Berry. 1985. Influence of garden symphylan (Symphyla: Scutigerellidae) root injury on physiological processes in snap beans. *Environ. Entomol.* 14: 408–412.

Engelken, L. K., W. B. Showers, and S. E. Taylor. 1990. Weed management to minimize black cutworm (Lepidoptera: Noctuidae) damage in no-till corn. *J. Econ. Entomol.* 83: 1058–1063.

Erlandson, M. A. 1990. Biological and biochemical comparison of *Mamestra configurata* and *Mamestra brassicae* nuclear polyhedrosis virus isolates pathogenic for the bertha armyworm, *Mamestra configurata* (Lepidoptera: Noctuidae). *J. Invertebr. Pathol.* 56: 47–56.

Esselbaugh, C. O. 1946. A study of the eggs of the Pentatomidae (Hemiptera). *Ann. Entomol. Soc. Am.* 39: 667–691.

Esselbaugh, C. O. 1948. Notes on the bionomics of some midwestern Pentatomidae. *Entomol. Am.* 28: 1–73.

Essig, E. O. 1934. Insects of Western North America. MacMillan, New York. 1035 pp.

Essig, E. O. 1938. Aphids feeding on celery in California. *Hilgardia* 11: 459–492.

Ester, A., S. B. Hofstede, P. S. R. Kosters, and C. P. DeMoel. 1994. Filmcoating of cauliflower seed (*Brassica oleracea* L. var. *botrytis* L.) with insecticides to control the cabbage root fly *(Delia radicum). Crop Prot.* 13: 14–19.

Etienne, J., J. Guyot, and X. van Waetermeulen. 1990. Effect of insecticides, predation, and precipitation on populations of *Thrips palmi* on aubergine (eggplant) in Guadeloupe. *Florida Entomol.* 73: 339–342.

Etzel, R. W. and F. L. Petitt. 1992. Association of *Verticillium lecanii* with population reduction of red rice root aphid (*Rhopalosiphum rufiabdominalis*) on aeroponically grown squash. *Florida Entomol.* 75: 605–606.

Evans, A., S. Wratten, C. Frampton, S. Causer, and M. Hamilton. 1997. Row covers: effects of wool and other materials on pest numbers, microclimate, and crop quality. *J. Econ. Entomol.* 90: 1661–1664.

Evans, K. A. and L. J. Allen-Williams. 1992. Electroantennogram responses of the cabbage seed weevil, *Ceutorhynchus assimilis*, to oilseed rape, *Brassica napus* ssp. *oleifera*, volatiles. *J. Chem. Ecol.* 18: 1641–1659.

Evans, W. G. and G. G. Gyrisco. 1956. Notes on the biology of the pea aphid. *J. Econ. Entomol.* 49: 878–879.

Eveleens, K. G., R. van den Bosch, and L. E. Ehler. 1973. Secondary outbreak induction of beet armyworm by experimental insecticide applications in cotton in California. *Environ. Entomol.* 2: 497–503.

Everts, K. L., H. F. Schwartz, N. D. Epsky, and J. L. Capinera. 1985. Effects of maggots and wounding on occurrence of *Fusarium* basal rot of onions in Colorado. *Plant Dis.* 69: 878–881.

Ewen, A. B. and M. K. Mukerji. 1980. Evaluation of *Nosema locustae* (Microsporidia) as a control agent of grasshopper populations in Saskatchewan. *J. Invertebr. Pathol.* 35: 295–303.

Ewing, H. E. 1914. The common red spider or spider mite. *Oregon Agric. Exp. Stn. Bull.* 121. 95 pp.

Eyer, J. R. 1922. The bionomics and control of the onion maggot. *Pennsylvania Agric. Exp. Stn. Bull.* 171. 27 pp.

Eyer, J. R. 1937. Physiology of psyllid yellows of potatoes. *J. Econ. Entomol.* 30: 891–898.

Ezzat, Y. M. 1958. Maconellicoccus hirsutus (Green), a new genus, with rediscription of the species. Bull. Entomol. Soc. Egypte 42: 377–383.

Facundo, H. T., A. Zhang, P. S. Robbins, S. R. Alm, C. E. Linn, Jr., M. G. Villani, and W. L. Roelofs. 1994. Sex pheromone responses of the oriental beetle (Coleoptera: Scarabaeidae). *Environ. Entomol.* 23: 1508–1515.

Fan, Y. and F. L. Petitt. 1994. Biological control of broad mite, *Polyphagotarsonemus latus* (Banks), by *Neoseiulus barkeri* Hughes on pepper. *Biol. Control* 4: 390–395.

Fargo, W. S., P. E. Rensner, E. L. Bonjour, and T. L. Wagner. 1988. Population dynamics in the squash bug (Heteroptera: Coreidae) – squash plant (Cucurbitales: Cucurbitaceae) system in Oklahoma. *J. Econ. Entomol.* 81: 1073–1079.

Faville, E. E. and P. J. Parrott. 1899. The potato-stalk weevil. Kansas Agric. *Exp. Stn. Bull.* 82: 1–12.

Federici, B. A. 1978. Baculovirus epizootic in a larval population of the clover cutworm, *Scotogramma trifolii*, in Southern California. *Environ. Entomol.* 7: 423–427.

Federici, B. A. 1980. Isolation of an iridovirus from two terrestrial isopods, the pill bug, *Armadillidium vulgare*, and the sow bug, *Porcellio dilatatus*. *J. Invertebr. Pathol.* 36: 373–381.

Federici, B. A. 1982. A new type of insect pathogen in larvae of the clover cutworm, *Scotogramma trifolii*. *J. Invertebr. Pathol.* 40: 41–54.

Federici, B. A. 1984. Diseases of terrestrial isopods. *Symp. Zool. Soc. London* 53: 233–245.

Feeny, P., K. L. Paauwe, and N. J. Demong. 1970. Flea beetles and mustard oils: host plant specificity of *Phyllotreta cruciferae* and *P. striolata* adults (Coleoptera: Chrysomelidae). *Ann. Entomol. Soc. Am.* 63: 832–841.

Feese, H. and G. Wilde. 1977. Factors affecting survival and reproduction of the Banks grass mite, *Oligonychus pratensis*. *Environ. Entomol.* 6: 53–56.

Felland, C. M. 1990. Habitat-specific parasitism of the stalk borer (Lepidoptera: Noctuidae) in northern Ohio. *Environ. Entomol.* 162–166.

Felland, C. M., L. A. Hull, D. A. J. Teulon, and E. A. Cameron. 1993. Overwintering of western flower thrips (Thysanoptera: Thripidae) in Pennsylvania. *Can. Entomol.* 125: 971–973.

Fenemore, P. G. 1980. Susceptibility of potato cultivars to potato tuber moth, *Phthorimaea operculella* Zell. (Lepidoptera: Gelechiidae). *New Zealand J. Agric. Res.* 23: 539–546.

Feng, M.-G., J. B. Johnson, and S. E. Halbert. 1992. Parasitoids (Hymenoptera: Aphidiidae and Aphelinidae) and their effect on aphid (Homoptera: Aphididae) populations in irrigated grain in southwestern Idaho. *Environ. Entomol.* 21: 1433–1440.

Feng, M.-G., J. B. Johnson, R. M. Nowierski, and S. E. Halbert. 1992. Population trends and biological aspects of cereal aphids (Homoptera: Aphididae), and their natural mortality factors on winter wheat in southwestern Idaho. *Pan–Pac. Entomol.* 68: 248–260.

Fenton, F. A. 1926. Observations on the biology of *Melanotus communis* and *Melanotus pilosus*. *J. Econ. Entomol.* 19: 502–504.

Fenton, F. A. and A. Hartzell. 1923. Bionomics and control of the potato leafhopper, *Empoasca mali* LeBaron. *Iowa Agric. Exp. Sta. Bull.* 78: 379–440.

Fereres, A., M. J. Blua, and T. M. Perring. 1992. Retention and transmission characteristics of zucchini yellow mosaic virus by *Aphis gossypii* and *Myzus persicae* (Homoptera: Aphididae). *J. Econ. Entomol.* 85: 759–765.

Fereres, A., P. Perez, C. Gemeno, and F. Ponz. 1993. Transmission of Spanish pepper- and potato-PVY isolates by aphid (Homoptera: Aphididae) vectors: epidemiological implications. *Environ. Entomol.* 22: 1260–1265.

Ferguson, A. W. and I. H. Williams. 1991. Deposition and longevity of oviposition-deterring pheromone in the cabbage seed weevil. *Physiol. Entomol.* 16: 27–33.

Ferguson, J. E., E. R. Metcalf, R. L. Metcalf, and A. M. Rhodes. 1983. Influence of cucurbitacin content of the cotyledons of Cucurbitaceae cultivars upon feeding behavior of Diabroticina beetles (Coleoptera: Chrysomelidae). *J. Econ. Entomol.* 76: 47–51.

Fernald, H. T. and W. E. Hinds. 1900. The grass thrips *Anaphothrips striata* (Osb.). Massachusetts Agric. Exp. Stn. Bull. 67: 1–12.

Ferro, D. N. and G. Fletcher-Howell. 1985. Controlling European corn borer (Lepidoptera: Pyralidae) on successively planted sweet corn in western Massachusetts. *J. Econ. Entomol.* 78: 902–907.

Ferro, D. N. and R. L. Gilbertson. 1982. Bionomics and population dynamics of the asparagus miner, *Ophiomyia simplex* (Loew), in western Massachusetts. *Environ. Entomol.* 11: 639–644.

Ferro, D. N. and G. J. Suchak. 1980. Assessment of visual traps for monitoring the asparagus miner, *Ophiomyia simplex*, Agromyzidae: Diptera. *Entomol. Exp. Appl.* 28: 177–182.

Ferro, D. N., J. D. MacKenzie, and D. C. Margolies. 1980. Effect of mineral oil and a systemic insecticide on field spread of aphid-borne maize dwarf mosaic virus in sweet corn. *J. Econ. Entomol.* 73: 730–735.

Ferro, D. N., B. J. Morzuch, and D. Margolies. 1983. Crop loss assessment of the Colorado potato beetle (Coleoptera: Chrysomelidae) on potatoes in western Massachusetts. *J. Econ. Entomol.* 76: 349–356.

Ferro, D. N., A. F. Tuttle, and D. C. Weber. 1991. Ovipositional and flight behavior of overwintered Colorado potato beetle (Coleoptera: Chrysomelidae). *Environ. Entomol.* 20: 1309–1314.

Ferro, D. N., J. A. Logan, R. H. Voss, and J. S. Elkington. 1985. Colorado potato beetle (Coleoptera: Chrysomelidae) temperature-dependent growth and feeding rates. *Environ. Entomol.* 14: 343–348.

Figueroa, W. and J. Roman. 1990a. Parasitism of entomophilic nematodes on the sugarcane rootstalk borer, *Diaprepes abbreviatus* (L.) (Coleoptera: Curculionidae), larvae. *J. Agric. Univ., Puerto Rico* 74: 197–202.

Figueroa, W. and J. Roman. 1990b. Boicontrol of the sugarcane rootstalk borer, *Diaprepes abbreviatus* (L.) (Coleoptera: Curculionidae), with entomophilic nematodes. *J. Agric. Univ., Puerto Rico* 74: 395–404.

Filinger, G. A. 1931. The garden symphylid, *Scutigerella immaculata* Newport. Ohio Agric. Exp. Stn. Bull. 486. 33 pp.

Finch, S. 1989. Ecological considerations in the management of *Delia* pest species in vegetable crops. *Annu. Rev. Entomol.* 34: 117–137.

Finch, S. 1991. Influence of trap surface on the numbers of insects caught in water traps in brassica crops. *Entomol. Exp. Appl.* 59: 169–173.

Finch, S. 1993. Integrated pest management of the cabbage root fly and the carrot fly. *Crop Prot.* 12: 423–430.

Finch, S. and T. H. Coaker. 1969. A method for the continuous rearing of the cabbage root fly *Erioischia brassicae* (Bch.) and some observations on its biology. *Bull. Entomol. Res.* 58: 619–627.

Finch, S. and R. H. Collier. 1985. Laboratory studies on aestivation in the cabbage root fly *(Delia radicum)*. *Entomol. Exp. Appl.* 38: 137–143.

Finch, S. and R. H. Collier. 1989. Effects of the angle of inclination of traps on the numbers of large Diptera caught on sticky boards in certain vegetable crops. *Entomol. Exp. Appl.* 52: 23–27.

Finch, S. and C. J. Eckenrode. 1985. Influence of unharvested, cull-pile, and volunteer onions on populations of onion maggot (Diptera: Anthomyiidae). *J. Econ. Entomol.* 78: 542–546.

Finch, S. and G. Skinner. 1976. The effect of plant density on populations of the cabbage root fly (*Erioischia brassicae* (Bch.)) and the cabbage stem weevil (*Ceutorhynchus quadridens* (Panz.)) on cauliflowers. *Bull. Entomol. Res.* 66: 113–123.

Finch, S. and G. Skinner. 1982. Trapping cabbage root flies in traps baited with plant extracts and with natural and synthetic isothiocyanates. *Entomol. Exp. Appl.* 31: 133–139.

Finch, S. and A. R. Thompson. 1992. Pests of cruciferous crops. Pages 87–138 in R. G. McKinlay (ed.). *Vegetable Crop Pests.* CRC Press, Boca Raton.

Finch, S., M. E. Cadoux, C. J. Eckenrode, and T. D. Spittler. 1986. Appraisal of current strategies for controlling onion maggot (Diptera: Anthomyiidae) in New York State. *J. Econ. Entomol.* 79: 736–740.

Fink, D. E. 1913. The asparagus miner and the twelve-spotted asparagus beetle. New York Agric. Exp. Stn. (Cornell) Bull. 331: 410–435.

Fink, D. E. 1915. The eggplant lace-bug. *USDA Bull.* 239. 7 pp.

Finke, M. D. and J. M. Scriber. 1988. Influence on larval growth of the eastern black swallowtail butterfly *Papilio polyxenes* (Lepidoptera: Papilionidae) of seasonal changes in nutritional parameters of Umbelliferae species. *Am. Midl. Nat.* 119: 45–62.

Finlayson, D. G., M. J. Brown, C. J. Campbell, A. T. S. Wilkinson, and I. H. Williams. 1972. Insecticides against tuber flea beetle on potatoes in British Columbia (Chrysomelidae: Coleoptera). *J. Entomol. Soc. Brit. Columbia* 69: 9–13.

Finney, C. L. 1964. The rearing of the variegated cutworm, *Peridroma saucia*, in the laboratory. *J. Econ. Entomol.* 57: 788–790.

Fisher, J. R. and M. K. Bergman. 1986. Field sampling of larvae and pupae. Pages 101–121 in J. L. Krysan and T. A. Miller (eds.). Methods for the Study of Pest *Diabrotica*. Springer-Verlag, Berlin New York.

Fisher, J. R. and L. E. O'Keeffe. 1979a. Seasonal migration and flight of the pea leaf weevil, *Sitona lineatus* (Coleoptera: Curculionidae) in northern Idaho and eastern Washington. *Entomol. Exp. Appl.* 26: 189–196.

Fisher, J. R. and L. E. O'Keeffe. 1979b. Host potential of some cultivated legumes for the pea leaf weevil, *Sitona lineatus* (Linnaeus) (Coleoptera: Curculionidae). *Pan-Pac. Entomol.* 55: 199–201.

Fisher, J. R. and L. E. O'Keeffe. 1979c. Food plants of the pea leaf weevil *Sitona lineatus* (Linnaeus) (Coleoptera: Curculionidae) in northern Idaho and eastern Washington. *Pan-Pac. Entomol.* 55: 202–207.

Fisher, J. R., J. J. Jackson, and A. C. Lew. 1994. Temperature and diapause development in the egg of *Diabrotica barberi* (Coleoptera: Chrysomelidae). *Environ. Entomol.* 23: 464–471.

Fisher, J. R., A. J. Keaster, and M. L. Fairchild. 1975. Seasonal vertical movement of wireworm larvae in Missouri: influence of soil temperature on the genera *Melanotus* Escholtz and *Conoderus* Escholtz. *Ann. Entomol. Soc. Am.* 68: 1071–1073.

Fisher, J. R., G. R. Sutter, and T. F. Branson. 1991. Influence of corn planting date on the survival and on some reproductive parameters of *Diabrotica virgifera virgifera* (Coleoptera: Chrysomelidae). *Environ. Entomol.* 20: 185–189.

Fisher, M. S. 1980. Superfamily Papilionoidea Latreille, 1809 (Parnassians, swallowtails). Pages 175–185 in C. D. Ferris and F. M. Brown (eds.). Butterflies of the Rocky Mountain States. Univ. of Oklahoma Press, Norman.

Fisher, T. W. and R. E. Orth. 1985. Observations of the snail *Rumina decollata* Linnaeus, 1758 (Stylommatophora: Subulinidae) with particular reference to its effectiveness in the biological control of *Helix aspersa* Müller, 1774 (Stylommatophora: Helicidae) in California. Dept. Entomol., Univ. of California, Riverside Occasional Paper #1. 111 pp.

Fitt, G. P. 1989. The ecology of *Heliothis* species in relation to agro-ecosystems. *Annu. Rev. Entomol.* 34: 17–52.

Fitzner, M. S., D. W. Hagstrum, D. A. Knauft, K. L. Buhr, and J. R. McLaughlin. 1985. Genotypic diversity in the suitability of cowpea (Rosales: Leguminosae) pods and seeds for cowpea weevil (Coleoptera: Bruchidae) oviposition and development. *J. Econ. Entomol.* 78: 806–810.

Flanders, K. L. and E. B. Radcliffe. 1989. Origins of potato leafhoppers (Homoptera: Cicadellidae) invading potato and snap bean in Minnesota. *Environ. Entomol.* 18: 1015–1024.

Flanders, K. L., L. W. Bledsoe, and C. R. Edwards. 1984. Effects of insecticides on *Pediobius foveolatus* (Hymenoptera: Eulophidae), a parasitoid of the Mexican bean beetle (Coleoptera: Coccinellidae). *Environ. Entomol.* 13: 902–906.

Flanders, K. L., E. B. Radcliffe, and D. W. Ragsdale. 1991. Potato leafroll virus spread in relation to densities of green peach aphid (Homoptera: Aphididae): implications for management thresholds for Minnesota seed potatoes. *J. Econ. Entomol.* 84: 1028–1036.

Fleischer, S. J. and M. J. Gaylor. 1987. Seasonal abundance of *Lygus lineolaris* (Heteroptera: Miridae) and selected predators in early season uncultivated hosts: implications for managing movement into cotton. *Environ. Entomol.* 16: 379–389.

Fleming, W. E. 1968. Biological control of the Japanese beetle. USDA Tech. Bull. 1383. 78 pp.

Fleming, W. E. 1972. Biology of the Japanese beetle. USDA Tech. Bull. 1449. 129 pp.

Fleming, W. E. 1976. Integrating control of the Japanese beetle – a historical review. USDA Tech. Bull. 1545. 65 pp.

Fleming, W. E., F. W. Metzger, and M. R. Osburn. 1934. Protecting plants in the home yard from injury by the Japanese beetle. USDA Circ. 326. 13 pp.

Fletcher, J. and A. Gibson. 1901. The life-history of the greenhouse leaf-tyer. *Can. Entomol.* 33: 140–144.

Flint, W. P., G. H. Dungan, and J. H. Bigger. 1935. Fighting the chinch bug on Illinois farms. Illinois Agric. Exp. Stn. Circ. 431. 16 pp.

Floyd, E. H. 1940. Investigations on the biology and control of the alfalfa caterpillar, *Colias eurytheme* Bois. Louisiana Agric. Exp. Stn. Bull. 323: 16–24.

Fluke, C. L. 1929. The known predacious and parasitic enemies of the pea aphid in North America. Wisconsin Agric. Exp. Stn. Bull. 93. 47 pp.

Flynn, J. L. and T. E. Reagan. 1984. Corn phenology in relation to natural and simulated infestations of the sugarcane borer (Lepidoptera: Pyralidae). *J. Econ. Entomol.* 77: 1524–1529.

Flynn, J. L., T. E. Reagan, and E. O. Ogunwolu. 1984. Establishment and damage of the sugarcane borer (Lepidoptera: Pyralidae) in corn as influenced by plant development. *J. Econ. Entomol.* 77: 691–697.

Folsom, J. W. 1933. The economic importance of Collembola. *J. Econ. Entomol.* 26: 934–939.

Folsom, J. W. and P. A. Woke. 1939. The field cricket in relation to the cotton plant in Louisiana. USDA Tech. Bull. 642. 28 pp.

Folwell, R. J., J. A. Gefre, S. M. Lutz, J. E. Halfhill, and G. Tamaki. 1990. Methods and costs of suppressing *Brachycorynella asparagi* Mordvilko (Homoptera: Aphididae). *Crop Prot.* 9: 259–264.

Foote, R. H. 1959. Notes on the genus *Euleia* Walker in North America. *J. Kansas Entomol. Soc.* 32: 145–150.

Foote, R. H., F. L. Blanc, and A. L. Norrbom. 1993. Handbook of the Fruit Flies (Diptera: Tephritidae) of America North of Mexico. Comstock, Ithaca. 571 pp.

Foott, W. H. 1963. The biology and control of the pepper maggot, *Zonosemata electa* (Say) (Diptera: Trypetidae) in southwestern Ontario. *Proc. Entomol. Soc. Ontario* 93: 75–81.

Foott, W. H. 1968a. The importance of *Solanum carolinense* L. as a host of the pepper maggot, *Zonosemata electa* (Say) (Diptera: Tephritidae) in southwestern Ontario. *Proc. Entomol. Soc. Ontario* 98: 16–17.

Foott, W. H. 1968b. Laboratory rearing of the pepper maggot, *Zonosemata electa* (Say) (Diptera: Tephritidae). *Proc. Entomol. Soc. Ontario* 98: 18–21.

Foott, W. H. 1973. Observations on Coccinellidae in corn fields in Essex County, Ontario. *Proc. Entomol. Soc. Ontario* 104: 16–21.

Foott, W. H. 1974. Effects of granular systemic insecticides on populations of the corn leaf aphid and yields of field corn in southwestern Ontario. *Proc. Entomol. Soc. Ontario* 105: 75–79.

Foott, W. H. 1975. Chemical control of the corn leaf aphid and effects on yields of field corn. *Proc. Entomol. Soc., Ontario* 106: 49–51.

Foott, W. H. 1977. Biology of the corn leaf aphid, *Rhopalosiphum maidis* (Homoptera: Aphididae), in southwestern Ontario. *Can. Entomol.* 109: 1129–1135.

Foott, W. H. 1978. Effects of granular systemic insecticides on populations of the corn leaf aphid and yields of field corn in southwestern Ontario. *Proc. Entomol. Soc. Ontario* 105: 75–79.

Foott, W. H. and J. E. Hybsky. 1976. Capture of *Glischrochilus quadrisignatus* (Coleoptera: Nitidulidae) in bait traps, 1970–74. *Can. Entomol.* 108: 837–839.

Foott, W. H. and P. R. Timmins. 1977. Biology of *Glischrochilus quadrisignatus* (Coleoptera: Nitidulidae) in southwestern Ontario. *Proc. Entomol. Soc. Ontario* 108: 37–44.

Foott, W. H. and P. R. Timmins. 1979. The rearing and biology of *Glischrochilus quadrisignatus* (Coleoptera: Nitidulidae) in the laboratory. *Can. Entomol.* 111: 1337–1344.

Forbes, A. R. and J. R. Mackenzie. 1982. The lettuce aphid, *Nasonovia ribisnigri* (Homoptera: aphididae) damaging lettuce crops in British Columbia. *J. Entomol. Soc. Brit. Columbia*, 79: 28–31.

Forbes, S. A. 1892. The northern corn root worm. Report of the State Entomologist on the Noxious and Beneficial Insects of the State of Illinois 18: 135–149.

Forbes, S. A. 1903. The colaspis root-worm. Report of the State Entomologist on the Noxious and Beneficial Insects of the State of Illinois 22: 145–149.

Forbes, S. A. 1904. The more important insect injuries to Indian corn. Illinois Agric. Exp. Stn. Bull. 95: 331–399.

Forbes, S. A. 1908. Habits and behavior of the corn-field ant, *Lasius niger americanus*. Illinois Agric. Exp. Stn. Bull. 131. 45 pp.

Forgash, A. J. 1981. Insecticide resistance of the Colorado potato beetle, Leptinotarsa decemlineata (Say). Pages 34–46 *in* Lashombe, J. H. and R. Casagrande (eds.). Advances in Potato Pest Management. Hutchinson Ross, Stroudsburg, Pennsylvania.

Forschler, B. T. and W. A. Gardner. 1990. A review of the scientific literature on the biology and distribution of the genus *Phyllophaga* (Coleoptera: Scarabaeidae) in the southeastern United States. *J. Entomol. Sci.* 25: 628–651.

Forschler, B. T. and W. A. Gardner. 1991. Flight activity and relative abundance of phytophagous Scarabaeidae attracted to blacklight traps in Georgia. *J. Agric. Entomol.* 8: 179–187.

Foster, D. E., W. B. Showers, W. H. Hendrix III, W. K. Wintersteen, and J. W. Bing. 1990. Effect of incorporation on the efficacy of selected pyrethroids for control of black cutworm (Lepidoptera: Noctuidae). *J. Econ. Entomol.* 83: 2073–2077.

Foster, J. E. and D. M. Daugherty. 1969. Isolation of the organism causing yeast-spot disease from the salivary system of the green stink bug. *J. Econ. Entomol.* 62: 424–427.

Foster, M. A. and W. G. Ruesink. 1984. Influence of flowering weeds associated with reduced tillage in corn on a black cutworm (Lepidoptera: Noctuidae) parasitoid, *Meteorus rubens* (Nees von Esenbeck). *Environ. Entomol.* 13: 664–668.

Foster, R. E. 1989. Strategies for protecting sweet corn ears from damage by fall armyworms (Lepidoptera: Noctuidae) in southern Florida. *Florida Entomol.* 72: 146–151.

Foster, R. E., J. J. Tollefson, and K. L. Steffey. 1982. Sequential sampling plans for adult corn rootworms (Coleoptera: Chrysomelidae). *J. Econ. Entomol.* 75: 791–793.

Foster, R. and B. Flood. (eds.). 1995. Vegetable Insect Management with Emphasis on the Midwest. Meister Publ. Co., Willoughby, Ohio. 206 pp.

Fournier, F., G. Boivin, and R. K. Stewart. 1994. Comparison of binomial and Iwao type sequential sampling plans for monitoring onion thrips (*Thrips tabaci*) (Thysanoptera: Thripidae) in onions. *Phytoprotection* 75: 69–78.

Fournier, F., G. Boivin, and R. K. Stewart. 1995. Effect of *Thrips tabaci* (Thysanoptera: Thripidae) on yellow onion yields and economic thresholds for its management. *J. Econ. Entomol.* 88: 1401–1407.

Fox, D. E. 1938. Occurrence of the beet leafhopper and associated insects on secondary plant successions in southern Idaho. USDA Tech. Bull. 607. 44 pp.

Fox, L. and B. J. Landis. 1973. Notes on the predaceous habits of the gray field slug, *Deroceras laeve*. *Environ. Entomol.* 2: 306–307.

Franclemont, J. G. 1980. "*Noctua c-nigrum*" in eastern North America, the description of two new species of *Xestia* Hübner (Lepidoptera: Noctuidae: Noctuinae). Proc. Entomol. Soc. Washington 82: 576–586.

Frank, J. H. 1971. Carabidae (Coleoptera) as predators of the red-backed cutworm (Lepidoptera: Noctuidae) in central Alberta. *Can. Entomol.* 103: 1039–1044.

Frank, J. H., J. P. Parkman, and F. D. Bennett. 1995. *Larra bicolor* (Hymenoptera: Sphecidae), a biological control agent of *Scapteriscus* mole crickets (Orthoptera: Gryllotalpidae), established in northern Florida. *Florida Entomol.* 78: 619–623.

Frank, J. H., T. J. Walker, and J. P. Parkman. 1996. The introduction, establishment, and spread of *Ormia depleta* in Florida. *Biol. Control* 6: 368–377.

Frank, R., H. E. Braun, G. Ritcey, F. L. McEwen, and G. J. Sirons. 1982. Pesticide residues in onions and carrots grown on organic soils, Ontario 1975 to 1980. *J. Econ. Entomol.* 75: 560–565.

Fransen, J. J. 1990. Natural enemies of whiteflies: fungi. Pages 187–210 *in* D. Gerling (ed.). Whiteflies: their Bionomics, Pest Status and Management. Intercept, UK.

Frantz, G., F. Parks, and H. C. Mellinger. 1995. Thrips population trends in peppers in southwest Florida. Pages 111–114 *in* B. L. Parker, M. Skinner, and T. Lewis (eds.). Thrips Biology Management. Plenum, New York.

Frazer, B. D. 1972. Life tables and intrinsic rates of increase of apterous black bean aphids and pea aphids, on broad bean (Homoptera: Aphididae). *Can. Entomol.* 104: 1717–1722.

Frazer, B. D., D. Raworth, and T. Gossard. 1976. Faba bean: low resistance to pea aphids, *Acyrthosiphon pisum* (Homoptera: Aphididae), in eleven cultivars. *Can. J. Plant Sci.* 56: 451–453.

Free, J. B., A.W. Ferguson, and S. Winfield. 1983. Effect of various levels of infestation by the seed weevil (*Ceutorhynchus assimilis* Payk.) on the seed yield of oil-seed rape (*Brassica napus* L.). *J. Agric. Sci.* 101: 589–596.

Frick, K. E. 1951. *Liriomyza langei*, a new species of leaf miner of economic importance in California. *Pan-Pac. Entomol.* 27: 81–88.

Friend, R. B. 1929. The Asiatic beetle in Connecticut. Connecticut Agric. Exp. Stn. Bull. 304. 664 pp.

Friend, R. B. 1931. The squash vine borer, *Melittia satyrinformis* Hübner. Connecticut Agric. Exp. Stn. Bull. 328: 587–608.

Friend, R. B. and N. Turner. 1931. The Mexican bean beetle in Connecticut. *Connecticut Agric. Exp. Stn. Bull.* 332. 108 pp.

Friend, W. G. and D. G. Harcourt. 1957. Influence of three types of soil on damage caused by the cabbage maggot, *Hylemya brassicae* (Bouché). *Quebec Soc. Prot. Plants* 38: 36–39.

Fritz, G. N. 1983. A technique for separating mole crickets from soil. *Florida Entomol.* 66: 360–362.

Froeschner, R. C. 1944. Contributions to a synopsis of the Hemiptera of Missouri, Part III. Lygaeidae, Pyrrhocoridae, Piesmidae, Tingididae, Enicocephalidae, Phymatidae, Ploiariidae, Reduviidae, Nabidae. *Am. Midl. Nat.* 31: 638–683.

Fronk, W. D. 1950. Cultural and biological control of the southern corn rootworm in peanuts. *J. Econ. Entomol.* 43: 22–24.

Frost, S. W. 1924. A study of the leaf-mining Diptera of North America. *New York* (Cornell) Agric. Exp. Stn. Mem. 78. 228 pp.

Frost, S. W. 1955. Cutworms of Pennsylvania. Pennsylvania Agric. Exp. Stn. Bull. 596. 29 pp.

Fuchs, T. W. and J. A. Harding. 1979. Seasonal abundance of the sugarcane borer, *Diatraea saccharalis*, on sugarcane and other hosts in the lower Rio Grande Valley of Texas. *Southwestern Entomol.* 4: 125–131.

Fuglie, K., H. B. Salah, M. Essamet, A. B. Temime, and A. Rahmouni. 1993. The development and adoption of integrated pest management of the potato tuber moth, *Phthorimaea operculella* (Zeller), in Tunisia. *Insect Sci. Applic.* 14: 501–509.

Fullaway, D. T. 1911. Insects attacking the sweet potato in Hawaii. Hawaii Agric. Exp. Stn. Bull. 22. 31 pp.

Fuller, B. W. and T. E. Reagan. 1988. Comparative predation of the sugarcane borer (Lepidoptera: Pyralidae) on sweet sorghum and sugarcane. *J. Econ. Entomol.* 81: 713–717.

Fulton, B. B. 1924. The European earwig. Oregon Agric. Exp. Stn. Bull. 207. 29 pp.

Fulton, B. B. 1942. The cabbage maggot in North Carolina. North Carolina Agric. Exp. Stn. Bull. 335. 24 pp.

Fulton, B. B. 1947. Biology and control of the pickleworm. North Carolina Agric. Exp. Stn. Bull. 85. 26 pp.

Funderburk, J. E. and L. P. Pedigo. 1983. Effects of actual and simulated seedcorn maggot (Diptera: Anthomyiidae) damage on soybean growth and yield. Environ. Entomol. 12: 323–330.

Funderburk, J. E., D. C. Herzog, and R. E. Lynch. 1987. Seasonal abundance of lesser cornstalk borer (Lepidoptera: Pyralidae) adults in soybean, peanut, corn, sorghum, and wheat in northern Florida. *J. Entomol. Sci.* 22: 159–168.

Funderburk, J. E., L. G. Higley, and L. P. Pedigo. 1984a. Seedcorn maggot (Diptera: Anthomyiidae) phenology in central Iowa and examination of a thermal-unit system to predict development under field conditions. *Environ. Entomol.* 13: 105–109.

Funderburk, J. E., D. C. Herzog, T. P. Mack, and R. E. Lynch. 1985. Sampling lesser cornstalk borer (Lepidoptera: Pyralidae) adults in several crops with reference to adult dispersion patterns. *Environ. Entomol.* 14: 452–458.

Funderburk, J. E., D. C. Herzog, T. P. Mack, and R. E. Lynch. 1986. Phenology and dispersion patterns of larval lesser cornstalk borers (Lepidoptera: Pyralidae) in grain sorghum in north Florida. *Environ. Entomol.* 15: 905–910.

Funderburk, J. E., D. G. Boucias, D. C. Herzog, R. K. Sprenkel, and R. E. Lynch. 1984b. Parasitoids and pathogens of larval lesser cornstalk borers (Lepidoptera: Pyralidae) in northern Florida. *Environ. Entomol.* 13: 1319–1323.

Furutani, S. C. and L. H. Arita. 1990. Effect of light exposure and carbohydrate content of snap bean leaves on Chinese rose beetle (Coleoptera: Scarabaeidae) feeding. *J. Econ. Entomol.* 83: 2022–2025.

Furutani, S. C., L. H. Arita, and J. K. Fujii. 1990. Relationship between simulated Chinese rose beetle (Coleoptera: Scarabaeidae) feeding and photosynthetic rate reduction. *Proc. Hawaiian Entomol. Soc.* 30: 97–104.

Fye, R. E. and W. C. McAda. 1972. Laboratory studies on the development, longevity, and fecundity of six lepidopterous pests of cotton in Arizona. USDA Tech. Bull. 1454. 73 pp.

Gammon, E. T. 1943. Helicid snails in California. Bull. California Dept. Agric. 32: 173–187.

Gardner, W. A. and R. Noblet. 1978. Effects of host age, route of infection, and quantity of inoculum on the susceptibility of *Heliothis virescens*, *Spodoptera eridania*, *S. frugiperda* to *Beauveria bassiana*. *J. Georgia Entomol. Soc.* 13: 214–222.

Gardner, W. A., R. Noblet, and R. D. Schwehr. 1984. The potential of microbial agents in managing populations of the fall armyworm (Lepidoptera: Noctuidae). *Florida Entomol.* 67: 325–332.

Garman, H. 1901. Enemies of cucumbers and related plants. Kentucky Agric. Exp. Stn. Bull. 91: 1–56.

Garman, H. 1917. Observations and experiments on the bean and pea weevils in Kentucky. Kentucky Agric. Exp. Stn. Bull. 213: 307–333.

Gaum, W. G., J. H. Giliomee, and K. L. Pringle. 1994. Life history and life tables of western flower thrips, *Frankliniella occidentalis* (Thysanoptera: Thripidae), on English cucumbers. *Bull. Entomol. Res.* 84: 219–224.

Gauthier, N. L., R. N. Hofmaster, and M. Semel. 1981. History of Colorado potato beetle control. Pages 13–33 *in* Lashombe, J. H. and R. Casagrande (eds.). Advances in Potato Pest Management. Hutchinson Ross, Stroudsburg, Pennsylvania.

Gauthier, N. L., P. A. Logan, L. A. Tewksbury, C. F. Hollingsworth, D. C. Weber, and R. G. Adams. 1991. Field bioassay of pheromone lures and trap designs for monitoring adult corn earworm (Lepidoptera: Noctuidae) in sweet corn in southern New England. *J. Econ Entomol.* 84: 1833–1836.

Gelernter, W. D., N. C. Toscano, K. Kido, and B. A. Federici. 1986. Comparison of a nuclear polyhedrosis virus and chemical insecticides for control of the beet armyworm (Lepidoptera: Noctuidae) on head lettuce. *J. Econ Entomol.* 79: 714–717.

Gentile, A. G. and A. K. Stoner. 1968a. Damage by larvae of the tobacco flea beetle to tomato seedlings. *J. Econ Entomol.* 61: 152–154.

Gentile, A. G. and A. K. Stoner. 1968b. Resistance in *Lycopersicon* spp. to the tobacco flea beetle. *J. Econ Entomol.* 61: 1347–1349.

Gentile, A. G. and A. K. Stoner. 1968c. Resistance in *Lycopersicon* and *Solanum* species to the potato aphid. *J. Econ Entomol.* 61: 1152–1154.

Gentile, A. G. and A. W. Vaughan. 1974. Control of the lettuce root aphid in Massachusetts. *J. Econ Entomol.* 67: 556.

Gentner, L. G. 1944. The black flea beetles of the genus *Epitrix* commonly identified as *cucumeris* (Harris) (Coleoptera: Chrysomelidae). *Proc. Entomol. Soc. Washington* 46: 137–149.

Gentry, C. R., W. A. Dickerson, Jr., T. J. Henneberry, and A. H. Baumhover. 1971. Pheromone-baited blacklight traps for controlling cabbage loopers on shade-grown tobacco in Florida. USDA, ARS. Prod. Res. Rpt. 133. 12 pp.

Gentry, C. R., F. R. Lawson, C. M. Knott, J. M. Stanley, and J. J. Lam, Jr. 1967. Control of hornworms by trapping with blacklight and stalk cutting in North Carolina. *J. Econ Entomol.* 60: 1437–1442.

Genung, W. G. 1959. Ecological and cultural factors affecting chemical control of subterranean cutworms in the Everglades. Proc. Florida State Hort. Soc. 72: 163–167.

Genung, W. G. 1970. Flooding experiments for control of wireworms attacking vegetable crops in the Everglades. *Florida Entomol.* 53: 55–63.

Genung, W. G. 1972. Seasonal occurrence of click beetles (Coleoptera: Elateridae) in the Everglades as measured by two types of traps. *Florida Entomol.* 55: 35–41.

Genung, W. G. and R. J. Allen, Jr. 1974. Bionomics and control of the striped grass looper *Mocis latipes* in the Everglades and adjacent areas (Lepidoptera: Noctuidae). University of Florida, Belle Glade AREC Res. Rep. EV-1974.

George, B. W. and A. M. Hintz. 1966. Immature stages of the western corn rootworm. *J. Econ Entomol.* 59: 1139–1142.

George, K. S. 1957. Preliminary investigations on the biology and ecology of the parasites and predators of *Brevicoryne brassicae* (L.) *Bull. Entomol. Res.* 48: 619–629.

George, K. S. 1962. Root nodule damage by larvae of *Sitona lineatus* and its effect on yield of green peas. *Plant Pathol.* 11: 172–176.

George, K. S., W. I.St.G. Light, and R. Gair. 1962. The effect of artificial defoliation of pea plants on the yield of shelled peas. *Plant Pathol.* 11: 73–80.

Gerber, G. H. 1976. Effects of feeding by adults of the red turnip beetle, *Entomoscelis americana* Brown (Coleoptera: Chrysomelidae), during late July and August on the yield of rapeseed (Cruciferae). *Manitoba Entomol.* 10: 31–35.

Gerber, G. H. 1981. Cold-hardiness in the eggs of the red turnip beetle, *Entomoscelis americana* (Coleoptera: Chrysomelidae). *Can. Entomol.* 113: 795–800.

Gerber, G. H. 1989. The red turnip beetle, *Entomoscelis americana* (Coleoptera: Chrysomelidae): distribution, temperature adaptations, and zoogeography. *Can. Entomol.* 121: 315–324.

Gerber, G. H. 1995. Fecundity of *Lygus lineolaris* (Heteroptera: Miridae). *Can. Entomol.* 127: 263–264.

Gerber, G. H. and R. J. Lamb. 1982. Phenology of egg hatching for the red turnip beetle, *Entomoscelis americana* (Coleoptera: Chrysomelidae). *Environ. Entomol.* 11: 1258–1263.

Gerber, G. H. and A. A. Obadofin. 1981a. The suitability of nine species of Cruciferae as hosts for the larvae of the red turnip beetle, *Entomoscelis americana* (Coleoptera: Chrysomelidae). *Can. Entomol.* 113: 407–413.

Gerber, G. H. and A. A. Obadofin. 1981b. Growth, development, and survival of the larvae of the red turnip beetle, *Entomoscelis americana* (Coleoptera: Chrysomelidae), on *Brassica campestris* and *B. napus* (Cruciferae). *Can. Entomol.* 113: 395–406.

Gerber, G. H. and C. E. Osgood. 1975. *Collops vittatus* (Coleoptera: Melyridae): a predator of flea beetle adults in rapeseed. *Manitoba Entomol.* 9: 61.

Gerber, G. H. and J. Walkof. 1992. Phenology and reproductive status of adult redbacked cutworms, *Euxoa ochrogaster* (Guenée) (Lepidoptera: Noctuidae), in southern Manitoba. *Can. Entomol.* 124: 541–551.

Gergerich, R. C., H. A. Scott, and J. P. Fulton. 1986. Evaluation of *Diabrotica* beetles as vectors of plant viruses. Pages 227–249 *in* J. L. Krysan and T. A. Miller (eds.). Methods for the Study of Pest *Diabrotica*. Springer-Verlag, Berlin, New York.

Gerling, D. 1990. *Natural enemies of whiteflies: predators and parasitoids.* Pages 147–185 *in* D. Gerling (ed.). Whiteflies: their Bionomics, Pest Status and Management. Intercept, UK.

Gerling, D. (ed.) 1990. *Whiteflies: their Bionomics, Pest Status and Management.* Intercept, UK. 348 pp.

Gerling, D., A. R. Horowitz, and J. Baumgaertner. 1986. Autecology of *Bemisia tabaci. Agric. Ecosystems Environ.* 17: 5–19.

Gerson, U. 1992. Biology and control of the broad mite, *Polyphagotarsonemus latus* (Banks) (Acari: Tarsonemidae). *Exp. Appl. Acarol.* 13: 163–178.

Gerson, U., S. Capua, and D. Thorens. 1983. Life history and life tables of *Rhizoglyphus robini* Claparede (Acari: Astigmata: Acaridae). *Acarologia* 24: 439–448.

Gerson, U., E. Cohen, and S. Capua. 1991. Bulb mite, *Rhizoglyphus robini* (Astigmata: Acaridae): as an experimental animal. *Exp. Appl. Acarol.* 12: 103–110.

Gerson, U., S. Yathom, and J. Katan. 1981. A demonstration of bulb mite control by solar heating of the soil. *Phytoparasitica* 9: 153–155.

Gerson, U., S. Yathom, S. Capua, and D. Thorens. 1985. *Rhizoglyphus robini* Claparede (Acari: Astigmata: Acaridae) as a soil mite. *Acarologia* 26: 371–380.

Gesell, S. G. and A. A. Hower, Jr. 1973. Garden symphylan: comparison of row and broadcast application of granular insecticides for control. *J. Econ Entomol.* 66: 822–823.

Getz, L. L. and L. F. Chichester. 1971. Introduced European slugs. *The Biol.* 53: 118–127.

Getzin, L. W. 1973. Supression of onion thrips with systemic-insecticide soil treatments. *J. Econ Entomol.* 66: 975–977.

Getzin, L. W. 1982. Seasonal activity and geographical distribution of the carrot rust fly (Diptera: Psilidae) in western Washington. *J. Econ Entomol.* 75: 1029–1033.

Getzin, L. W. 1983. Damage to inflorescence of cabbage seed plants by the pale legume bug (Heteroptera: Miridae). *J. Econ Entomol.* 76: 1083–1085.

Getzin, L. W. 1985. Chemical control of the springtail *Onychiurus pseudarmatus* (Collembola: Onychiuridae). *J. Econ Entomol.* 78: 1337–1340.

Getzin, L. W. and C. H. Shanks. 1964. Infection of the garden symphylan, *Scutigerella immaculata* (Newport), by *Entomophthora coronata* (Constantin) Kevorkian and *Metarrhizium anisopliae* (Metchnikoff) Sorokin. *J. Invertebr. Pathol.* 6: 542–543.

Gharib, A. H. and J. A. Wyman. 1991. Food consumption and survival of *Trichoplusia ni* (Lepidoptera: Noctuidae) larvae following intoxication by *Bacillus thuringiensis* var. *kurstaki* and Thuringiensin. *J. Econ Entomol.* 84: 436–439.

Ghidiu, G. M. and R. W. van Vranken. 1995. A modified carrot weevil (Coleoptera: Curculionidae) monitoring trap. *Florida Entomol.* 78: 627–630.

Gholson, L. E. and W. B. Showers. 1979. Feeding behavior of black cutworms on seedling corn and organic baits in the greenhouse. *Environ. Entomol.* 8: 552–557.

Gibson, R. W., J. A. Pickett, G. W. Dawson, A. D. Rice, and M. F. Stribley. 1984. Effects of aphid alarm pheromone derivatives and related compounds on non- and semi-persistent plant virus transmission by *Myzus persicae. Ann. Appl. Biol.* 104: 203–209.

Giebink, B. L., J. M. Scriber, and D. B. Hogg. 1985. Developmental rates of the hop vine borer and potato stem borer (Lepidoptera: Noctuidae): implications for insecticidal control. *J. Econ Entomol.* 78: 311–315.

Giebink, B. L., J. M. Scriber, and J. L. Wedberg. 1984. Biology and phenology of the hop-vine borer, *Hydraecia immanis* Guenée, and detection of the potato stem borer, *H. micacea* (Esper) (Lepidoptera: Noctuidae), in Wisconsin. *Environ. Entomol.* 13: 1216–1224.

Giebink, B. L., J. M. Scriber, and J. L. Wedberg. 1992. Suitability of selected broad-leaved weeds for survival and growth of two stalk-boring *Hydraecia* species (Lepidoptera: Noctuidae). *Great Lakes Entomol.* 25: 245–251.

Gilbertson, G. I. and W. R. Horsfall. 1940. Blister beetles and their control. *South Dakota Agric. Exp. Stn. Bull.* 340. 23 pp.

Gilbertson, R. L., W. J. Manning, and D. N. Ferro. 1985. Association of the asparagus miner with stem rot caused in asparagus by *Fusarium* species. *Phytopathology* 75: 1188–1191.

Gilbertson, R. L., W. M. Brown, Jr., E. G. ruppel, and J. L. Capinera. 1986. Association of corn stalk rot *Fusarium* spp. and western corn rootworm beetles in Colorado. *Phytopathology* 76: 1309–1314.

Gilboa, S. and H. Podoler. 1995. Presence-absence sequential sampling for potato tuberworm (Lepidoptera: Gelechiidae) on processing tomatoes: selection of sample sites according to predictable seasonal trends. *J. Econ Entomol.* 88: 1332–1336.

Gilkeson, L. A. and S. B. Hill. 1987. Release rates for control of green peach aphid (Homoptera: Aphididae) by the predatory midge *Aphidoletes aphidimyza* (Diptera: Cecidomyiidae) under winter greenhouse conditions. *J. Econ Entomol.* 80: 147–150.

Gillespie, D. R. and D. M. J. Quiring. 1990. Biological control of fungus gnats, Bradysia spp. (Diptera: Sciaridae), and western flower thrips, *Frankliniella occidentalis* (Pergande) (Thysanoptera: Thripidae), in greenhouses using a soil-dwelling predatory mite, *Geolaelaps* sp. nr. *Aculeifer* (Canestrini) (Acari: Laelapidae). *Can. Entomol.* 122: 975–983.

Gillespie, D. R. and D. M. J. Quiring. 1992. Flight behavior of the greenhouse whitefly, *Trialeurodes vaporariorum* (Westwood) (Homoptera: Aleyrodidae), in relation to yellow sticky traps. *Can. Entomol.* 124: 907–916.

Gillette, C. P. 1905. Beet worms and their remedies I. The beet webworm II. The beet army-worm. Colorado Agric. Exp. Stn. Bull. 98. 15 pp.

Gilmore, J. U. 1938. Observations on the hornworms attacking tobacco in Tennessee and Kentucky. *J. Econ Entomol.* 31: 706–712.

Gilstrap, F. E., T. J. Kring, and G. W. Brooks. 1984. Parasitism of aphids (Homoptera: Aphididae) associated with Texas sorghum. *Environ. Entomol.* 13: 1613–1617.

Girault, A. A. and A. H. Rosenfeld. 1907. Biological notes on the Colorado potato beetle, *Leptinotarsa decemlineata* (Say), with technical description of its stages. *Psyche* 14: 45–57.

Girling, D. J. (ed.) 1992. *Thrips palmi.* A Literature Survey with an Annotated Bibliography. International Institute of Biological Control, Silwood Park, Ascot, U.K. 37 pp.

Givovich, A., J. Weibull, and J. Pettersson. 1988. Cowpea aphid performance and behaviour on two resistant cowpea lines. *Entomol. Exp. Appl.* 49: 259–264.

Gladstone, S. M., A. de la Llana, R. Rios, and L. Lopez. 1994. Egg parasitoids of the corn leafhopper, *Dalbulus maidis* (DeLong and Wolcott) (Homoptera: Cicadellidae) in Nicaraguan maize. Proc. Entomol. Soc., Washington 96: 143–146.

Glass, E. H. 1943. Host plants of the tobacco flea beetle, *Epitrix parvula* F. *Virginia Agric. Exp. Stn. Bull.* 85. 22 pp.

Glen, R., K. M. King, and A. P. Arnason. 1943. The identification of wireworms of economic importance in Canada. *Can. J. Res.* (D) 21: 358–387.

Gliessman, S. R. and M. A. Altieri. 1982. Polyculture cropping has advantages. *California Agric.* 36: 14–16.

Goats, G. C. 1985. Effect of methiocarb, carbaryl, pirimiphos-methyl and polybutenes with deltamethrin upon woodlice on cucumber. Tests of Agrochemicals and Cultivars, Vol. 6, Suppl. to Ann. Appl. Biol. 106: 12–13.

Godan, D. 1983. *Pest Slugs and Snails.* Springer-Verlag, Berlin. 445 pp.

Godin, C. and G. Boivin. 1998a. Seasonal occurrence of lepidopterous pests of cruciferous crops in southwestern Quebec in relation to degree-day accumulations. *Can. Entomol.* 130: 173–185.

Godin, C. and G. Boivin. 1998b. Lepidopterous pests of *Brassica* crops and their parasitoids in southwestern Quebec. *Environ. Entomol.* 27: 1157–1165.

Godfrey, G. L. 1972. A review and reclassification of larvae in the subfamily Hadeninae (Lepidoptera, Noctuidae) of America north of Mexico. *USDA Tech. Bull.* 1450. 265 pp.

Godin, C. and G. Boivin. 1998c. Occurrence of *Cotesia rubecula* (Hymenoptera: Braconidae) in Quebec, 30 years after its introduction in North America. *Can. Entomol.* 130: 733–734.

Godfrey, L. D. and W. E. Chaney. 1995. Temporal and spatial distribution patterns of aphids (Homoptera: Aphididae) on celery. *J. Econ Entomol.* 88: 294–301.

Godfrey, L. D., L. J. Meinke, R. J. Wright, and G. L. Hein. 1995. Environmental and edaphic effects on western corn rootworm (Coleoptera: Chrysomelidae) overwintering egg survival. *J. Econ Entomol.* 88: 1445–1454.

Goettel, M. S. and B. J. R. Philogene. 1978a. Effects of photoperiod and temperature on the development of a univoltine population of the banded woollybear, *Pyrrharctia (Isia) isabella. J. Insect Physiol.* 24: 523–527.

Goettel, M. S. and B. J. R. Philogene. 1978b. Laboratory rearing of the banded woollybear, *Pyrrharctia (Isia) isabella* (Lepidoptera: Arctiidae), on different diets with notes on the biology of the species. *Can. Entomol.* 110: 1077–1086.

Goettel, M. S. and B. J. R. Philogene. 1979. Further studies on the biology of *Pyrrharctia (Isia) isabella* (Lepidoptera: Arctiidae) III. The relation between head capsule width and number of instars. *Can. Entomol.* 111: 323–326.

Goff, C. C. and A. N. Tissot. 1932. The melon aphid, *Aphis gossypii* Glover. Florida Agric. Exp. Stn. Bull. 252. 23 pp.

Goff, C. C. and J. W. Wilson. 1937. The pepper weevil. Florida Agric. Exp. Stn. Bull. 310. 12 pp.

Goh, K. S. and W. H. Lange. 1980a. Spatial distribution patterns and sampling plans for immature stages of the artichoke plume moth. *J. Econ Entomol.* 73: 113–116.

Goh, K. S. and W. H. Lange. 1980b. Life tables for the artichoke plume moth in California. *J. Econ Entomol.* 73: 153–158.

Gojmerac, W. L. 1956. Description of the sugar beet root maggot, *Tetanops myopaeformis* (von Roder), with observations on reproductive capacity. *Entomol. News* 67: 203–210.

Golden, K. L. and L. J. Meinke. 1991. Immature development, fecundity, longevity, and egg diapause of *Diabrotica longicornis* (Coleoptera: Chrysomelidae). *J. Kansas Entomol. Soc.* 64: 251–256.

Golino, D. A., G. N. Oldfield, and D. J. Gumpf. 1987. Transmission characteristics of the beet leafhopper transmitted virescence agent. *Phytopathology* 77: 954–957.

Golino, D. A., G. N. Oldfield, and D. J. Gumpf. 1989. Experimental hosts of the beet leafhopper-transmitted virescence agent. *Plant Dis.* 73: 850–854.

Gonzalez, D., R. Friesen, T. F. Leigh, T. Wilson, and M. Waggoner. 1995a. Naturally-occurring biological control: western flower thrips impact on spider mites in California cotton. Pages 317–323 in B. L. Parker, M. Skinner, and T. Lewis (eds.). *Thrips Biology and Management.* Plenum, London.

Gonzalez, D., K. S. Hagen, P. Stary, G. W. Bishop, D. W. Davis, and K. S. Pike. 1995b. Pea aphid and blue alfalfa aphid. Pages 129–135 in Nechols, J. R., L. A. Andres, J. W. Beardsley, R. D. Goeden, and C. G. Jackson (eds.). Biological Control in the Western United States. Univ. of California Publ. 3361.

Goodenough, J. L., A. E. Knutson, and F. M. Davis. 1989. Trap comparisons and behavioral observations for the male southwestern corn borer (Lepidoptera: Pyralidae). *J. Econ Entomol.* 82: 1460–1465.

Gooding, R. H., B. M. Rolseth, J. R. Byers, and C. E. Herle. 1992. Electrophoretic comparisons of pheromotypes of the dingy cutworm, *Feltia jaculifera* (Gn.) (Lepidoptera: Noctuidae). *Can. J. Zool.* 70: 79–86.

Gorder, N. K. N. and J. W. Mertins. 1984. Life history of the parsnip webworm, *Depressaria pastinacella* (Lepidoptera: Oecephoridae), in central Iowa. *Ann. Entomol. Soc. Am.* 77: 568–573.

Gordon, F. C. and D. A. Potter. 1986. Japanese beetle (Coleoptera: Scarabaeidae) traps: evaluation of single and multiple arrangements for reducing defoliation in urban landscape. *J. Econ Entomol.* 79: 1381–1384.

Gordon, R. and A. M. Armstrong. 1990. Biologia del picudo del pimiento, *Anthonomus eugenii*, Cano (Coleoptera: Curculionidae), en Puerto Rico. *J. Agric. Univ.*, Puerto Rico 74: 69–73.

Gould, G. E. 1931. *Pangaeus uhleri*, a pest of spinach. *J. Econ Entomol.* 24: 484–486.

Gould, G. E. 1948. Insect-problems in corn processing plants. *J. Econ Entomol.* 41: 774–778.

Gould, H. J. and C. W. Graham. 1977. The incidence of *Aphis fabae* Scop. on spring-sown field beans in south-east England and the efficiency of control measures. *Plant Pathol.* 26: 189–194.

Gould, J. R. and S. E. Naranjo. 1999. Distribution and sampling of *Bemisia argentifolii* (Homoptera: Aleyrodidae) and *Eretmocerus eremicus* (Hymenoptera: Aphelinidae) on cantaloupe vines. *J. Econ Entomol.* 92: 402–408.

Graf, J. E. 1914. A preliminary report on the sugar-beet wireworm. USDA Bur. Entomol. Bull. 123. 68 pp.

Graf, J. E. 1917. The potato tuber moth. *USDA Bull.* 427. 56 pp.

Grafius, E. and E. A. Morrow. 1982. Damage by the tarnished plant bug and alfalfa plant bug (Heteroptera: Miridae) to asparagus. *J. Econ Entomol.* 75: 882–884.

Grafius, E. and F. W. Warner. 1989. Predation by *Bembidion quadrimaculatum* (Coleoptera: Carabidae) on *Delia antiqua* (Diptera: Anthomyiidae). *Environ. Entomol.* 18: 1056–1059.

Graham, H. M. and O. T. Robertson. 1970. Host plants of *Heliothis virescens* and *H. zea* (Lepidoptera: Noctuidae) in the lower Rio Grande Valley, Texas. *Ann. Entomol. Soc. Am.* 63: 1261–1265.

Graham, H. M., N. S. Hernandez, Jr., and J. R. Llanes. 1972. The role of host plants in the dynamics of populations of *Heliothis* Spp. *Environ. Entomol.* 1: 424–431.

Graham, H. M., C. G. Jackson, and J. W. Debolt. 1986. *Lygus* spp. (Hemiptera: Miridae) and their parasites in agricultural areas of southern Arizona. *Environ. Entomol.* 15: 132–142.

Granados, R. R., J. S. Granados, K. Maramorosch, and J. Reinitz. 1968. Corn stunt virus: transmission by three cicadellid vectors. *J. Econ Entomol.* 61: 1282–1287.

Gratwick, M. (ed.). 1992. Crop Pests in the UK, Collected Edition of MAFF Leaflets. Chapman & Hall, London. 490 pp.

Gray, M. E. and J. J. Tollefson. 1988a. Survival of the western and northern corn rootworms (Coleoptera: Chrysomelidae) in different tillage systems throughout the growing season of corn. *J. Econ Entomol.* 81: 178–183.

Gray, M. E. and J. J. Tollefson. 1988b. Emergence of the western and northern corn rootworms (Coleoptera: Chrysomelidae) from four tillage systems. *J. Econ Entomol.* 81: 1398–1403.

Grayson, J. M. 1946. Life history and habits of *Strigoderma arboricola*. *J. Econ Entomol.* 39: 163–167.

Greene, G. L. 1970a. Head measurements and weights of the bean leaf roller, *Urbanus proteus* (Hesperiidae). *J. Lepid. Soc.* 24: 47–51.

Greene, G. L. 1970b. Hatching of bean leaf roller eggs as influenced by insect diet components. *J. Econ Entomol.* 63: 321.

Greene, G. L. 1971a. Instar distributions, natural populations, and biology of the bean leaf roller. *Florida Entomol.* 54: 213–219.

Greene, G. L. 1971b. Economic damage levels of bean leaf roller populations on snap beans. *J. Econ Entomol.* 64: 673–674.

Greenough, D. R. and L. L. Black. 1990. Aluminum-surfaced mulch: an approach to the control of tomato spotted wilt virus in solanaceous crops. *Plant Dis.* 74: 805–808.

Greenstone, M. H. 1995. Bollworm or budworm? Squashblot immunoassay distinguishes eggs of *Helicoverpa zea* and *Heliothis virescens* (Lepidoptera: Noctuidae). *J. Econ Entomol.* 88: 213–218.

Gregory, B. M., C. S. Barfield, and J. B. Chapin. 1988. Morphological differences between adult *Anticarsia gemmatalis* and *Mocis latipes* (Lepidoptera: Noctuidae). *Florida Entomol.* 71: 352–359.

Griffith, R. J. 1975. The West Indian sugarcane rootstalk borer weevil in Florida. *Proc. Florida State Hort. Soc.* 88: 87–90.

Grigarick, A. A. and W. H. Lange. 1962. Host relationships of the sugar-beet root aphid in California. *J. Econ Entomol.* 55: 760–764.

Grissell, E. E. 1981. *Edovum puttleri*, N. G., N. Sp. (Hymenoptera: Eulophidae), an egg parasite of the Colorado potato beetle (Chyrsomelidae). *Proc. Entomol. Soc. Washington.* 83: 790–796.

Gross, Jr., H. R. and R. Johnson. 1985. *Archytas marmoratus* (Diptera: Tachinidae): advances in large-scale rearing and associated biological studies. *J. Econ Entomol.* 78: 1350–1353.

Gross, Jr., H. R. and J. R. Young. 1977. Comparative development and fecundity of corn earworm reared on selected wild and cultivated early-season hosts common to the southeastern U. S. *Ann. Entomol. Soc. Am.* 70: 63–65.

Gross, Jr., H. R., S. D. Pair, and R. C. Layton. 1985. *Archytas marmoratus* (Diptera: Tachinidae): screened-cage performance of mechanically extracted maggots against larval populations of *Heliothis zea* and *Spodoptera frugiperda* (Lepidoptera: Noctuidae) on whorl and early tassel-stage corn. *J. Econ Entomol.* 78: 1354–1357.

Gross, Jr., H. R., B. R. Wiseman, and W. W. McMillian. 1976. Comparative suitability of whorl stages of sweet corn for establishment by larvae of the corn earworm. *Environ. Entomol.* 5: 955–958.

Gross, P. 1986. Life histories and geographic distributions of two leafminers, *Tildenia georgei* and *T. inconspicuella* (Lepidoptera: Gelechiidae), on solanaceous weeds. *Ann. Entomol. Soc. Am.* 79: 48–55.

Gross, P. and P. W. Price. 1988. Plant influences on parasitism of two leafminers: a test of enemy-free space. *Ecology* 69: 1506–1516.

Grundler, J. A., D. L. Hostetter, and A. J. Keaster. 1987. Laboratory evaluation of *Vairimorpha necatrix* (Microspora: Microsporidia) as a control agent for the black cutworm (Lepidoptera: Noctuidae). *Environ. Entomol.* 16: 1228–1230.

Guppy, J. C. 1961. Life history and behaviour of the armyworm, *Pseudaletia unipunctata* (Haw.) (Lepidoptera: Noctuidae), in eastern Ontario. *Can. Entomol.* 93: 1141–1153.

Guppy, J. C. 1969. Some effects of temperature on the immature stages of the armyworm, *Pseudaletia unipunctata* (Lepidoptera: Noctuidae), under controlled conditions. *Can. Entomol.* 101: 1320–1327.

Guppy, J. C. 1982. Effects of temperature and light intensity on nocturnal activity patterns of the northern June beetle, *Phyllophaga fusca*, and the common June beetle, *P. anxia* (Coleoptera: Scarabaeidae). *Can. Entomol.* 114: 1151–1157.

Guppy, J. C. and D. G. Harcourt. 1970. Spatial pattern of the immature stages and teneral adults of *Phyllophaga* spp. (Coleoptera: Scarabaeidae) in a permanent meadow. *Can. Entomol.* 102: 1354–1359.

Guppy, J. C. and D. G. Harcourt. 1973. A sampling plan for studies on the population dynamics of white grubs, *Phyllophaga* spp. (Coleoptera: Scarabaeidae). *Can. Entomol.* 105: 479–483.

Guppy, J. C. and C. D. F. Miller. 1970. Identification of cocoons and last-instar larval remains of some hymenopterous parasitoids of the armyworm, *Pseudaletia unipunctata*, in eastern Ontario. *Can. Entomol.* 102: 1320–1337.

Guss, P. L. 1976. The sex pheromone of the western corn rootworm (*Diabrotica virgifera*). *Environ. Entomol.* 5: 219–223.

Guss, P. L., T. F. Branson, and J. L. Krysan. 1976. Adaptation of a dry diet for adults of the western corn rootworm. *J. Econ Entomol.* 69: 503–505.

Guss, P. L., J. H. Tumlinson, P. E. Sonnet, and J. R. McLaughlin. 1983. Identification of a female-produced sex pheromone from the southern corn rootworm, *Diabrotica undecimpunctata howardi* Barber. *J. Chem. Ecol.* 8: 545.

Gustin, R. D. and G. E. Wilde. 1984. Cold hardiness of western corn rootworm (Coleoptera: Chrysomelidae) eggs from northern and southern zones of the corn belt. *J. Kansas Entomol. Soc.* 57: 722–725.

Gwynne, D. T. 1984. Sexual selection and sexual differences in Mormon crickets (Orthoptera: Tettigoniidae, *Anabrus simplex*). *Evolution* 38: 1011–1022.

Gwynne, D. T. 1993. Food quality controls sexual selection in Mormon crickets by altering male mating investment. *Ecology* 74: 1406–1413.

Gyrisco, A. A., D. Landman, A. C. York, B. J. Irwin, and E. J. Armbrust. 1978. The literature of arthropods associated with alfalfa. IV. A bibliography of the potato leafhopper. Illinois Agric. Exp. Stn. Spec. Publ. 51. 75 pp.

Habeck, D. H. 1964. Notes on the biology of the Chinese rose beetle, *Adoretus sinicus* Burmeister in Hawaii. Proc. Hawaiian Entomol. Soc. 18: 399–403.

Habeck, D. H. 1963. Description of immature stages of the Chinese rose beetle, *Adoretus sinicus* Burmeister (Coleoptera: Scarabaeidae). Proc. Hawaiian Entomol. Soc. 18: 251–258.

Habeck, D. H. 1976. Selective feeding on sweet potato varieties by southern armyworm. *Florida Entomol.* 59: 396.

Habib, M. E. M., L. M. Paleari, and M. E. C. Amaral. 1983. Effect of three larval diets on the development of the armyworm, *Spodoptera latifascia* Walker, 1856 (Noctuidae, Lepidoptera). *Revista Bras. Zool.* 1: 177–182.

Hackett, K. J., R. F. Whitcomb, T. B. Clark, R. B. Henegar, D. E. Lynn, A. G. Wagner, J. G. Tully, G. E. Gasparich, D. L. Rose, P. Carle, J. M. Bove, M. Konai, E. A. Clark, J. R. Adams, and D. L. Williamson. 1996. *Spiroplasma leptinotarsae* sp. nov., a mollicute uniquely adapted to its host, the Colorado potato beetle, *Leptinotarsa decemlineata* (Coleoptera: Chrysomelidae). *Intl. J. Syst. Bacteriol.* 46: 906–911.

Hagel, G. T. and R. O. Hampton. 1970. Dispersal of aphids and leafhoppers from red clover to red Mexican beans, and the spread of bean yellow mosaic by aphids. *J. Econ Entomol.* 63: 1057–1060.

Hagel, G. T. and B. J. Landis. 1967. Biology of the aster leafhopper, *Macrosteles fascifrons* (Homoptera: Cicadellidae), in eastern Washington, and some overwintering sources of aster yellows. *Ann. Entomol. Soc. Am.* 60: 591–595.

Hagel, G. T. and B. J. Landis. 1972. Chemical control of the two-spotted spider mite on field beans. *J. Econ Entomol.* 65: 775–778.

Hagel, G. T., D. W. Burke, and M. J. Silbernagel. 1981. Response of dry bean selections to field infestations of seedcorn maggot in central Washington. *J. Econ Entomol.* 74: 441–443.

Hagel, G. T., B. J. Landis, and M. C. Ahrens. 1973. Aster leafhopper: source of infestation, host plant preference, and dispersal. *J. Econ Entomol.* 66: 877–881.

Hagen, A. F. 1962. The biology and control of the western bean cutworm in dent corn in Nebraska. *J. Econ Entomol.* 55: 628–631.

Hagen, A. F. 1976. A fourteen-year summary of light trap catches of the western bean cutworm in Nebraska *Loxagrotis albicosta* (Smith) (Lepidoptera: Noctuidae). *J. Kansas Entomol. Soc.* 49: 537–540.

Halfhill, J. E. 1967. Mass propagation of pea aphids. *J. Econ Entomol.* 60: 298–299.

Halfhill, J. E. 1982. Evaluation of western yellowstriped armyworm (Lepidoptera: Noctuidae) as a pest of lentils. *J. Econ Entomol.* 75: 733–735.

Halfhill, J. E., J. A. Gefre, and G. Tamaki. 1984. Cultural practices inhibiting overwintering survival of *Brachycolus asparagi* Mordvilko (Homoptera: Aphididae). *J. Econ Entomol.* 77: 954–956.

Hall, D. G. 1982. A parasite, *Pristocera armifera* (Say), of the wireworm *Melanotus communis* (Gyll.) in south Florida. *Florida Entomol.* 65: 574.

Hall, D. G. 1985. Parasitoids of grasslooper prepupae and pupae in south Florida sugarcane. *Florida Entomol.* 68: 486–487.

Hall, D. G. 1986. Sampling for the sugarcane borer (Lepidoptera: Pyralidae) in sugarcane. *J. Econ Entomol.* 79: 813–816.

Hall, D. G. 1995. A revision to the bibliography of the sugarcane rootstalk borer weevil, *Diaprepes abbreviatus* (Coleoptera: Curculionidae). *Florida Entomol.* 78: 364–377.

Hall, I. M. 1957. Use of a polyhedrosis virus to control the cabbage looper on lettuce in California. *J. Econ Entomol.* 50: 551–553.

Hallock, H. C. 1932. Life history and control of the Asiatic garden beetle. USDA Circ. 24616 pp.

Hallock, H. C. 1934. The Asiatic garden beetle as a pest in vegetable gardens. *J. Econ Entomol.* 27: 476–481.

Hallock, H. C. 1935. Movements of larvae of the oriental beetle through soil. *J. New York Entomol. Soc.* 43: 413–425.

Hallock, H. C. 1936. Notes on biology and control of the Asiatic garden beetle. *J. Econ Entomol.* 29: 348–356.

Hamilton, C. C. and L. G. Gemmell. 1934. Some field tests showing the comparative efficiency of derris, pyrethrum and hellebore powders on different insects. *J. Econ Entomol.* 27: 446–453.

Hamilton, D. W., P. H. Schwartz, B. G. Townshend, and C. W. Jester. 1971. Traps reduce an isolated infestation of Japanese beetle. *J. Econ Entomol.* 64: 150–153.

Hamilton, G. C. and J. Lashomb. 1996. Comparison of conventional and biological control intensive pest management programs on eggplant in New Jersey. *Florida Entomol.* 79: 488–496.

Hamilton, G. C., J. H. Lashomb, S. Arpaia, R. Chianese, and M. Mayer. 1998. Sequential sampling plans for Colorado potato beetle (Coleoptera: Chrysomelidae) in eggplant. *Environ. Entomol.* 27: 33–38.

Hamm, J. J. and R. E. Lynch. 1982. Comparative susceptibility of the granulate cutworm, fall armyworm, and corn earworm to some entomopathogens. *J. Georgia Entomol. Soc.* 17: 363–369.

Hammond, R. B. 1985. Development and survival of the Mexican bean beetle, *Epilachna varivestis* Mulsant, on two host plants. *J. Kansas Entomol. Soc.* 57: 695–699.

Hammond, R. B. and R. L. Cooper. 1993. Interaction of planting times following the incorporation of a living, green cover crop and control measures on seedcorn maggot populations in soybean. *Crop Prot.* 12: 539–543.

Hammond, R. B., J. A. Smith, and T. Beck. 1996. Timing of molluscicide applications for reliable control in no-tillage field crops. *J. Econ Entomol.* 89: 1028–1032.

Hammond, R. B., T. Beck, J. A. Smith, R. Amos, J. Barker, R. Moore, H. Siegrist, D. Slates, and B. Ward. 1999. Slugs in conservation tillage corn and soybeans in the eastern corn belt. *J. Entomol. Sci.* 34: 467–478.

Hamon, N., L. Allen-Williams, J. B. Lee, and R. Bardner. 1984. Larval instar determination of the pea and bean weevil *Sitona lineatus* L. (Coleoptera: Curculionidae). *Entomol. Mon. Mag.* 120: 167–171.

Hamon, N., R. Bardner, L. Allen-Williams, and J. B. Lee. 1987. Flight periodicity and infestation size of *Sitona lineatus*. *Ann. Appl. Biol.* 111: 271–284.

Hamon, N., R. Bardner, L. Allen-Williams, and J. B. Lee. 1990. Carabid populations in field beans and their effect on the population dynamics of *Sitona lineatus* (L.). *Ann. Appl. Biol.* 117: 51–62.

Hammond, R. B. and B. R. Stinner. 1987. Seedcorn maggots (Diptera: Anthomyiidae) and slugs in conservation tillage systems in Ohio. *J. Econ Entomol.* 80: 680–684.

Hander, C. A., P. J. McLeod, and H. A. Scott. 1993. Incidence of aphids (Homoptera: Aphididae) and associated potyviruses in summer squash in Arkansas. *J. Entomol. Sci.* 28: 73–81.

Hanna, H. Y., R. N. Story, and A. J. Adams. 1987. Influence of cultivar, nitrogen, and frequency of insecticide application on vegetable leafminer (Diptera: Agromyzidae) population density and dispersion on snap beans. *J. Econ Entomol.* 80: 107–110.

Hanson, A. J. 1933. The potato flea beetles, *Epitrix cucumeris* and *Epitrix subcrinita*. Washington Agric. Exp. Stn. Bull. 280. 27 pp.

Hanson, A. J. and R. L. Webster. 1936. The pea moth *Laspeyresia nigricana* Steph. Washington Agric. Exp. Stn. Bull. 327. 22 pp.

Hanson, A. J., E. C. Carlson, E. P. Breakey, and R. L. Webster. 1948. Biology of the cabbage seedpod weevil in northwestern Washington. Washington Agric. Exp. Stn. Bull. 498. 15 pp.

Hanula, J. L. and T. G. Andreadis. 1988. Parasitic microorganisms of Japanese beetle (Coleoptera: Scarabaeidae) and associated scarab larvae in Connecticut soils. *Environ. Entomol.* 17: 709–714.

Hao, G. T. and H. N. Pitre. 1970. Relationship of vector numbers and age of corn plants at inoculation to severity of corn stunt disease. *J. Econ Entomol.* 63: 924–927.

Hara, A. H. and S. Matayoshi. 1990. Parasitoids and predators of insect pests on chrysanthemums in Hawaii. Proc. Hawaiian Entomol. Soc. 30: 53–58.

Hara, A. H., H. K. Kaya, R. Gaugler, L. M. Lebeck, and C. L. Mello. 1993. Entomopathogenic nematodes for biological control of the leafminer, *Liriomyza trifolii* (Dipt.: Agromyzidae). *Entomophaga* 38: 359–369.

Harcourt, D. G. 1955. Biology of the diamondback moth, *Plutella maculipennis* (Curt.) (Lepidoptera: Plutellidae), in eastern Ontario. Rpt. Quebec Soc. Prot. Plants. 37: 155–160.

Harcourt, D. G. 1957. Biology of the diamondback moth, *Plutella maculipennis* (Curt.) (Lepidoptera: Plutellidae), in Eastern Ontario. II. Life-history, behaviour, and host relationships. *Can. Entomol.* 89: 554–564.

Harcourt, D. G. 1960. Biology of the diamondback moth, *Plutella maculipennis* (Curt.) (Lepidoptera: Plutellidae), in eastern Ontario. III. Natural enemies. *Can. Entomol.* 92: 419–428.

Harcourt, D. G. 1961. Design of a sampling plan for studies on the population dynamics of the diamondback moth, *Plutella maculipennis* (Curt.) (Lepidoptera: Plutellidae). *Can. Entomol.* 93: 820–831.

Harcourt, D. G. 1962. Design of a sampling plan for studies on the population dynamics of the imported cabbageworm, *Pieris rapae* (L.) (Lepidoptera: Pieridae). *Can. Entomol.* 94: 849–859.

Harcourt, D. G. 1963a. Biology of cabbage caterpillars in eastern Ontario. Proc. Entomol. Soc. Ontario. 93: 61–75.

Harcourt, D. G. 1963b. Major mortality factors in the population dynamics of the diamondback moth, *Plutella maculipennis* (Curt.) (Lepidoptera: Plutellidae). Mem. Entomol. Soc. Canada. 32: 55–66.

Harcourt, D. G. 1964. Population dynamics of *Leptinotarsa decemlineata* (Say) in eastern Ontario II. Population and mortality estimation during six age intervals. *Can. Entomol.* 96: 1190–1198.

Harcourt, D. G. 1971. Population dynamics of *Leptinotarsa decemlineata* (Say) in eastern Ontario III. Major population processes. *Can. Entomol.* 103: 1049–1061.

Harding, J. A. 1961. Effect of migration, temperature, and precipitation on thrips infestations in south Texas. *J. Econ Entomol.* 54: 77–79.

Harding, J. A. 1976a. *Heliothis* spp.: seasonal occurrence, hosts and host importance in the lower Rio Grande Valley. *Environ. Entomol.* 5: 666–668.

Harding, J. A. 1976b. *Heliothis* spp.: parasitism and parasites plus host plants and parasites of the beet armyworm, diamondback moth and two tortricids in the lower Rio Grande Valley of Texas. *Environ. Entomol.* 5: 669–671.

Hardman, J. A. and P. R. Ellis. 1982. An investigation of the host range of the carrot fly. *Ann. Appl. Biol.* 100: 1–9.

Hardman, J. A., P. R. Ellis, and P. L. Saw. 1990. Further investigations of the host range of the carrot fly, *Psila rosae* (F.). *Ann. Appl. Biol.* 117: 495–506.

Hardwick, D. F. 1965a. The *ochrogaster* group of the genus *Euxoa* (Lepidoptera: Noctuidae), with description of a new species. *Can. Entomol.* 97: 673–678.

Hardwick, D. F. 1965b. The corn earworm complex. Entomol. Soc. Can. Mem. 40. 246 pp.

Hardy, D. E. 1949. Studies in Hawaiian fruit flies. Proc. Entomol. Soc. Washington 51: 181–205.

Hare, J. D. 1980. Impact of defoliation by the Colorado potato beetle on potato yields. *J. Econ Entomol.* 73: 369–373.

Hare, J. D. 1984. Suppression of the Colorado potato beetle, *Leptinotarsa decemlineata* (Say) (Coleoptera: Chrysomelidae), on solanaceous crops with a copper-based fungicide. *Environ. Entomol.* 13: 1010–1014.

Hare, J. D. 1990. Ecology and management of the Colorado potato beetle. *Annu. Rev. Entomol.* 35: 81–100.

Hare, J. D., P. A. Logan, and R. J. Wright. 1983. Suppression of Colorado potato beetle, *Leptinotarsa decemlineata* (Say), (Coleoptera: Chrysomelidaea) populations with antifeedant fungicides. *Environ. Entomol.* 12: 1470–1477.

Hargreaves, E. 1914. The life-history and habits of the greenhouse white fly (*Aleyrodes vaporariorum*). *Ann. Appl. Biol.* 1: 303–334.

Harman, G. E., C. J. Eckenrode, and D. R. Webb. 1978. Alteration of spermosphere ecosystems affecting oviposition by the bean seed fly and attack by soilborne fungi on germinating seeds. *Ann. Appl. Biol.* 90: 1–6.

Harper, A. M. 1962. Life history of the sugar-beet root maggot *Tetanops myopaeformis* (Roder) (Diptera: Otitidae) in southern Alberta. *Can. Entomol.* 94: 1334–1340.

Harper, A. M. 1963. Sugar-beet root aphid, *Pemphigus betae* Doane (Homoptera: Aphididae), in southern Alberta. *Can. Entomol.* 95: 863–873.

Harper, A. M. and T. P. Story. 1962. Reliability of trapping in determining the emergence period and sex ratio of the sugar-beet root maggot *Tetanops myopaeformis* (Roder) (Diptera). *Can. Entomol.* 94: 268–912.

Harper, A. M., B. D. Schaber, T. P. Story, and T. Entz. 1990. Effect of swathing and clear-cutting alfalfa on insect populations in southern Alberta. *J. Econ Entomol.* 83: 2050–2057.

Harper, J. D. 1970. Laboratory production of *Peridroma saucia* and its nuclear polyhedrosis virus. *J. Econ Entomol.* 63: 1633–1634.

Harper, J. D. 1971. Preliminary testing of a nuclear polyhedrosis virus to control the variegated cutworm on peppermint. *J. Econ Entomol.* 64: 1573–1574.

Harper, J. D. 1981. Citrus pulp bait insecticide formulations for control of velvetbean caterpillar and bean leaf beetle on soybean. *Florida Entomol.* 64: 538–540.

Harper, J. D. and G. R. Carner. 1973. Incidence of *Entomophthora* sp. and other natural control agents in populations of *Pseudoplusia includens* and *Trichoplusia ni. J. Invertebr. Pathol.* 22: 80–85.

Harrell, E. A., J. R. Young, and W. W. Hare. 1977. Insect control on late-planted sweet corn. *J. Econ Entomol.* 70: 129–131.

Harries, F. H. and J. R. Douglass. 1948. Bionomic studies on the beet leafhopper. *Ecol. Mono.* 18: 45–79.

Harrington, C. D. 1941. Influence of aphid resistance in peas upon aphid development, reproduction and longevity. *J. Agric. Res.* 62: 461–466.

Harris, C. R. 1976. Redbacked cutworm resistance to organochlorine insecticides. *J. Econ Entomol.* 69: 614–616.

Harris, C. R. and H. J. Svec. 1966. Mass rearing of the cabbage maggot under controlled environmental conditions, with observations on the biology of cyclodiene-susceptible and resistant strains. *J. Econ Entomol.* 59: 569–573.

Harris, C. R. And H. J. Svec. 1976. Onion maggot resistance to insecticides. *J. Econ Entomol.* 69: 617–620.

Harris, C. R., G. F. Manson, and J. H. Mazurek. 1962a. Development of insecticidal resistance by soil insects in Canada. *J. Econ Entomol.* 55: 777–780.

Harris, C. R., J. H. Mazurek, and G. V. White. 1962b. The life history of the black cutworm, *Agrotis ipsilon* (Hufnagel), under controlled conditions. *Can. Entomol.* 94: 1183–1187.

Harris, C. R., H. J. Svec, and J. A. Begg. 1966. Mass rearing of root maggots under controlled environmental conditions: seed-corn maggot, *Hylemya cilicrura*; bean seed fly, *H. liturata*; *Euxesta notata*; and *Chaetopsis* sp. *J. Econ Entomol.* 59: 407–410.

Harris, C. R., H. J. Svec, and R. A. Chapman. 1977. The effectiveness and persistence of some insecticides used for control of the variegated cutworm attacking tomatoes in southwestern Ontario. Proc. Entomol. Soc., Ontario 108: 63–68.

Harris, C. R., J. H. Tolman, and H. V. Svec. 1982. Onion maggot (Diptera: Anthomyiidae) resistance to some insecticides following selection with parathion or carbofuran. *Can. Entomol.* 114: 681–685.

Harris, C. R., S. A. Turnbull, and D. G. R. McLeod. 1985. Contact toxicity of twenty-one insecticides to adults of the carrot rust fly (Diptera: Psilidae). *Can. Entomol.* 117: 1025–1027.

Harris, C. R., H. J. Svec, S. A. Turnbull, and W. W. Sans. 1975a. Laboratory and field studies on the effectiveness of some insecticides in controlling the armyworm. *J. Econ Entomol.* 68: 513–516.

Harris, C. R., H. J. Svec, W. W. Sans, A. Hikichi, S. C. Phatak, R. Frank, and H. E. Braun. 1975b. Efficacy, phytotoxicity, and persistence of insecticides used as pre- and postplanting treatments for control of cutworms attacking vegetables in Ontario. Proc. Entomol. Soc., Ontario 105: 65–75.

Harris, E. J. and R. Y. Okamoto. 1991. A method for rearing *Biosteres arisanus* (Hymenoptera: Braconidae) in the laboratory. *J. Econ Entomol.* 84: 417–422.

Harris, V. E. and J. W. Todd. 1980. Duration of immature stages of the southern green stink bug, *Nezara viridula* (L.), with a comparative review of previous studies. *J. Georgia Entomol. Soc.* 15: 114–124.

Harris, V. E. and J. W. Todd. 1981. Rearing the southern green stink bug, *Nezara viridula*, with relevant aspects of its biology. *J. Georgia Entomol. Soc.* 16: 203–211.

Harris, V. E., J. W. Todd, J. C. Webb, and J. C. Benner. 1982. Acoustical and behavioral analysis of the songs of the southern green stink bug, *Nezara viridula. Ann. Entomol. Soc. Am.* 75: 234–249.

Harrison, F. P. 1962. Infestation of sweet corn by the dusky sap beetle, *Carpophilus lugubris. J. Econ Entomol.* 55: 922–925.

Harrison, F. P., R. A. Bean, and O. J. Qawiyy. 1980a. No-till culture of sweet corn in Maryland with reference to insect pests. *J. Econ Entomol.* 73: 363–365.

Harrison, F. P., L. P. Ditman, and W. E. Bickley. 1954. Habits of *Drosophila* with reference to animal excrement. *J. Econ Entomol.* 47: 935.

Harrison, M. D., J. W. Brewer, and L. D. Merrill. 1980b. Insect involvement in the transmission of bacterial pathogens. Pages 201–292 *in* K. F. Harris and K. Maramorosch (eds.). Vectors of Plant Pathogens. Academic Press, San Diego.

Harrison, P. K. and R. W. Brubaker. 1943. The relative abundance of cabbage caterpillars on cole crops grown under similar conditions. *J. Econ Entomol.* 36: 589–592.

Harrison, R. G. 1979. Flight polymorphism in the field cricket *Gryllus pennsylvanicus. Oecologia* 40: 125–132.

Hartley, G. G. 1990. Multicellular rearing methods for the beet armyworm, soybean looper, and velvetbean caterpillar (Lepidoptera: Noctuidae). *J. Entomol. Sci.* 25: 326–340.

Hartstack, A. W., J. A. Witz, and D. R. Buck. 1979. Moth traps for the tobacco budworm. *J. Econ Entomol.* 72: 519–522.

Hatch, M. H. 1949. Studies on the fauna of Pacific Northwest greenhouses (Isopoda, Coleoptera, Dermaptera, Orthoptera, Gastropoda). *J. New York Entomol. Soc.* 57: 141–165.

Hatch, M. H. 1971. The Beetles of the Pacific Northwest. Part V. Rhipiceroidea, Sternoxi, Phytophaga, Rhynchophora, and Lamellicornia. Univ. Washington Publ. Biol. 16. 662 pp.

Hatch, M. H. 1982. The Beetles of the Pacific Northwest. Part III. Pselaphidae and Diversicornia. Univ. Washington Publ. Biol. 16. 503 pp.

Hatchett, S. P. 1947. Biology of the Isopoda of Michigan. *Ecol. Mono.* 17: 47–79.

Hattori, M. and A. Sato. 1983. A technique for rearing the limabean pod borer, *Etiella zinckenella* Treitshke (Lepidoptera: Pyralidae) in large scale. *Appl. Entomol. Zool.* 18: 330–334.

Hausmann, S. M. and J. R. Miller. 1989. Ovipositional preference and larval survival of the onion maggot (Diptera: Anthomyiidae) as influenced by previous maggot feeding. *J. Econ Entomol.* 82: 426–429.

Havlickova, H. 1980. Causes of different feeding rates of pea leaf weevil *Sitona lineatus* on three pea cultivars. *Entomol. Exp. Appl.* 27: 287–292.

Havukkala, I. and M. Virtanen. 1984. Oviposition of single females of the cabbage root flies *Delia radicum* and *D. floralis* (Diptera: Anthomyiidae) in the laboratory. *Ann. Entomol. Fennici* 50: 81–84.

Havukkala, I. and M. Virtanen. 1985. Behavioural sequence of host selection and oviposition in the turnip root fly *Delia floralis* (Fall.) (Anthomyiidae). *Z. Angew. Entomol.* 100: 39–47.

Hawkins, C. D. B., M. I. Whitecross, and M. J. Aston. 1986. Long-term effects on cowpea plant growth of a short-term cowpea aphid infestation. *Can. J. Bot.* 64: 1727–1732.

Hawkins, J. H. 1936. The bionomics and control of wireworms in Maine. Maine Agric. Exp. Stn. Bull. 381. 146 pp.

Hawley, I. M. 1918. Insects injurious to the hop in New York, with special reference to the hop grub and the hop redbug. New York (Cornell) Agric. Exp. Stn. Mem. 15: 142–224.

Hawley, I. M. 1922a. Insects and other animal pests injurious to field beans in New York. New York (Cornell) Agric. Exp. Stn. Mem. 55: 942–1037.

Hawley, I. M. 1922b. The sugar-beet root-maggott (*Tetanops aldrichi* Hendel), a new pest of sugar-beets. *J. Econ Entomol.* 15: 388–391.

Hayakawa, D. L., E. Grafius, and F. W. Stehr. 1990. Effects of temperature on longevity, reproduction, and development of the asparagus aphid (Homoptera: Aphididae) and the parasitoid, *Diaeretiella rapae* (Hymenoptera: Braconidae). *Environ. Entomol.* 19: 890–897.

Hayes, J. L. and M. Bell. 1994. Evaluation of early-season baculovirus treatment for suppression of *Heliothis virescens* and *Helicoverpa zea* (Lepidoptera: Noctuidae) over a wide area. *J. Econ Entomol.* 87: 58–66.

Hayes, W. P. 1917. Studies on the life-history of *Ligyrus gibbosus* DeG. (Coleoptera). *J. Econ Entomol.* 10: 253–261.

Hayes, W. P. 1920. The maize billbug or elephant bug. Kansas Agric. Exp. Stn. Tech. Bull. 6. 27 pp.

Hayes, W. P. 1921. *Strigoderma arboricola* Fab. – its life cycle (Scarab. Coleop.). *Can. Entomol.* 53: 121–125.

Hayden, J. and E. Grafius. 1990. Activity of cyromazine on onion maggot larvae (Diptera: Anthomyiidae) in the soil. *J. Econ Entomol.* 83: 2398–2400.

Haynes, K. F., M. C. Birch, and J. A. Klun. 1981. Sex pheromone offers promise for control of artichoke plume moth. *California Agric.* 35:13–14.

Haynes, K. F., L. K. Gaston, M. M. Pope, and T. C. Baker. 1983. Rate and periodicity of pheromone release from individual female artichoke plume moths, *Platyptilia carduidactyla* (Lepidoptera: Pterophoridae). *Environ. Entomol.* 12:1597–1600.

Haynes, R. L. and C. M. Jones. 1975. Wilting and damage to cucumber by spotted and striped cucumber beetles. *HortScience* 10:265–266.

Hayslip, N. C. 1943. Notes on biological studies of mole crickets at Plant City, Florida. *Florida Entomol.* 26:33–46.

Hayslip, N. C., W. G. Genung, E. G. Kelsheimer, and J. W. Wilson. 1953. Insects attacking cabbage and other crucifers in Florida. Florida Agric. Exp. Stn. Bull. 534. 57 pp.

Hazan, A., U. Gerson, and A. S. Tahori. 1974. Life history and life tables of the carmine spider mite. *Acarologia* 15:414–440.

Headrick, D. H. and R. D. Goeden. 1996. Issues concerning the eradication or establishment and biological control of the Mediterranean fruit fly, *Ceratitis capitata* (Weidemann) (Diptera: Tephritidae), in California. *Biol. Control* 6:412–421.

Heath, R. R., N. D. Epsky, B. D. Dueben, and W. L. Meyer. 1996. Systems to monitor and suppress *Ceratitis capitata* (Diptera: Tephritidae) populations. *Florida Entomol.* 79:144–153.

Heath, R. R., J. A. Coffelt, P. E. Sonnet, F. I. Proshold, B. Dueben, and J. H. Tumlinson. 1986. Identification of sex pheromone produced by female sweetpotato weevil, *Cylas formicarius elegantulus* (Summers). *J. Chem. Ecol.* 12:1489–1503.

Heathcote, G. D. 1962. The suitability of some plant hosts for the development of the peach-potato aphid, *Myzus persicae* (Sulzer). *Entomol. Exp. Appl.* 5:114–118.

Heichel, G. H., D. C. Sands, and J. B. Kring. 1977. Seasonal patterns and reduction by carbofuran of Stewart's bacterial wilt of sweet corn. *Plant Dis. Rep.* 61:149–153.

Heie, O. E. 1979. Revision of the aphid genus *Nasonovia* Mordvilko, including *Kakimia* Hottes and Frison, with keys and descriptions of the species of the world (Homoptera: Aphididae). *Entomol. Scand. Suppl.* 9:1–105.

Heim, D. C., G. G. Kennedy, and J. W. Van Duyn. 1990. Survey of insecticide resistance among North Carolina Colorado potato beetle (Coleoptera: Chrysomelidae) populations. *J. Econ Entomol.* 83:1229–1235.

Hein, G. L. and J. J. Tollefson. 1984. Comparison of adult corn rootworm (Coleoptera: Chrysomelidae) trapping techniques as population estimators. *Environ. Entomol.* 13:266–271.

Hein, G. L., J. J. Tollefson, and P. N. Hinz. 1985. Design and cost considerations in the sampling of northern and western corn rootworm (Coleoptera: Chrysomelidae) eggs. *J. Econ Entomol.* 78:1495–1499.

Heinz, K. M. and W. E. Chaney. 1995. Sampling for *Liriomyza huidobrensis* (Diptera: Agromyzidae) larvae and damage in celery. *Environ. Entomol.* 24:204–211.

Heinz, K. M. and F. G. Zalom. 1995. Variation in trichome-based resistance to *Bemisia argentifolii* (Homoptera: Aleyrodidae) oviposition on tomato. *J. Econ Entomol.* 88:1494–1502.

Heinz, K. M., L. M. Heinz, and M. P. Parrella. 1996. Natural enemies of western flower thrips indigenous to California ornamentals. *IOBC, WPRS Bull.* 19:51–55.

Heinz, K. M., M. P. Parrella, and J. P. Newman. 1992. Time-efficient use of yellow sticky traps in monitoring insect populations. *J. Econ Entomol.* 85:2263–2269.

Helfer, J. R. 1972. *The Grasshoppers, Cockroaches, and their Allies.* Wm. C. Brown Company, Dubuque, Iowa. 359 pp.

Helm, C. G., M. R. Jeffords, S. L. Post, and M. Kogan. 1983. Spring feeding activity of overwintered bean leaf beetles (Coleoptera: Chrysomelidae) on non-leguminous hosts. *Environ. Entomol.* 12:321–322.

Henderson, C. A. and F. M. Davis. 1969. The southwestern corn borer and its control. Mississippi Agric. Exp. Stn. Bull. 773. 16 pp.

Henderson, C. F. and L. J. Padget. 1949. White-fringed beetles, distribution, survey, and control. USDA Entomol. Plant Quarantine Rpt. 779. 19 pp.

Henderson, I. (ed.) 1989. Slugs and snails in world agriculture. British Crop Protection Council Monogr. No. 41. 422 pp.

Henderson, I. F. (ed.) 1996. Slug and snail pests in agriculture. British Crop Protection Council Symposium Proc. No. 66. 450 pp.

Hendrichs, J. and M. A. Hendrichs. 1990. Mediterranean fruit fly (Diptera: Tephritidae) in nature: location and diel pattern of feeding and other activities on fruiting and nonfruiting hosts and nonhosts. *Ann. Entomol. Soc. Am.* 83:632–641.

Hendrickson, Jr., R. M., F. Gruber, G. Mailloux, and J. J. Drea. 1991. Parasite colonizations against *Crioceris asparagi* (L.) and *C. duodecimpunctata* (L.) (Coleoptera: Chrysomelidae) in North America from 1983 to 1988. Proc. Entomol. Soc., Washington 93:67–69.

Hendrix III, W. H. and W. B. Showers. 1990. Evaluation of differently colored bucket traps for black cutworm and armyworm (Lepidoptera: Noctuidae). *J. Econ Entomol.* 83:596–598.

Henne, R. C. 1970. Effect of five insecticides on populations of the six-spotted leafhopper and the incidence of aster yellows in carrots. *Can. J. Plant Sci.* 50:169–174.

Henneberry, T. J. 1994. Effects of temperature on tobacco budworm (Lepidoptera: Noctuidae) pupal diapause, initiation and final stage of movement of stemmatal eyespots and adult emergence. *Southwestern Entomol.* 19:329–333.

Henneberry, T. J. and A. N. Kishaba. 1967. Mating and oviposition of the cabbage looper in the laboratory. *J. Econ Entomol.* 60:692–696.

Henneberry, T. J., G. D. Butler, Jr., and D. L. Coudriet. 1993. Tobacco budworm (Lepidoptera: Noctuidae): effects of temperature and photoperiod on larval and pupal development, larval mortality and induction of pupal diapause. *Southwestern Entomol.* 18:269–279.

Henry, J. E. 1985. *Melanoplus* spp. Pages 451–464 *in* P. Singh and R. F. Moore (eds.). Handbook of Insect Rearing, Vol. II. Elsevier, Amsterdam/New York.

Henry, J. E. and J. A. Onsager. 1982. Experimental control of the Mormon cricket, *Anabrus simplex*, by *Nosema locustae* (Microspora: Microsporida), a protozoan parasite of grasshoppers (Ort.: Acrididae). *Entomophaga* 27:197–201.

Henry, T. J. 1983. The garden fleahopper genus *Halticus* (Hemiptera: Miridae): resurrection of an old name and key to species of the western hemisphere. Proc. Entomol. Soc., Washington 85:607–611.

Henry, T. J. and R. C. Froeschner (eds.). 1988. Catalog of the Heteroptera, or True Bugs, of Canada and the Continental United States. E. J. Brill, New York. 958 pp.

Heppner, J. B. 1975. The bean leaf roller, *Urbanus proteus*, and the related *Urbanus dorantes* in Florida (Lepidoptera: Hesperiidae). *J. Georgia Entomol. Soc.* 10:328–332.

Heppner, J. B. 1998. *Spodoptera* armyworms in Florida (Lepidoptera: Noctuidae). Florida Dept. Agric. Cons. Serv., Div. Plant Industry, Entomol. Circ. 390. 5pp.

Herrick, G. W. and J. W. Hungate. 1911. The cabbage aphis. New York (Cornell) Agric. Exp. Stn. Bull. 300:715–746.

Herzog, D. C. 1977. Bean leaf beetle: parasitism by *Celatoria diabroticae* (Shimer) and *Hyalomyodes triangulifer* (Loew). *J. Georgia Entomol. Soc.* 12:64–68.

Herzog, D. C. 1980. Sampling methods in soybean entomology. Pages 141–161 *in* Kogan, M. and D. C. Herzog (eds.). Sampling Methods in Soybean Entomology. Springer-Verlag, Berlin/New York.

Herzog, D. C., C. E. Eastman, and L. D. Newsom. 1974. Laboratory rearing of the bean leaf beetle. *J. Econ Entomol.* 67: 794–795.

Hesler, L. S. and G. R. Sutter. 1993. Effect of trap color, volatile attractants, and type of toxic bait dispenser on captures of adult corn rootworm beetles (Coleoptera: Chrysomelidae). *Environ. Entomol.* 22: 743–750.

Hetrick, L. A. 1947. The cowpea curculio: its life history and control. Virginia Agric. Exp. Stn. Bull. 409. 23 pp.

Hickman, J. M. and S. D. Wratten. 1996. Use of *Phacelia tanacetifolia* strips to enhance biological control of aphids by hoverfly larvae in cereal fields. *J. Econ Entomol.* 89: 832–840.

High, M. M. 1939. The vegetable weevil. USDA Circ. 530. 25 pp.

Highland, H. B. and P. F. Lummus. 1986. Use of light traps to monitor flight activity of the burrowing bug, *Pangaeus bilineatus* (Hemiptera: Cydnidae), and associated field infestations in peanuts. *J. Econ Entomol.* 79: 523–526.

Higley, L. G. and L. P. Pedigo. 1984. Seedcorn maggot (Diptera: Anthomyiidae) population biology and aestivation in central Iowa. *Environ. Entomol.* 13: 1436–1442.

Hildebrand, E. M. 1954. The generation time of the leafhopper *Baldulus maidis* in Texas. *Plant Dis. Rep.* 38: 572–573.

Highland, H. B. and J. E. Roberts. 1989. Oviposition of the stalk borer *Papaipema nebris* (Lepidoptera: Noctuidae) among various plants, and plant characteristics for ovipositional preference. *J. Entomol. Sci.* 24: 70–77.

Hill, B. D., J. R. Byers, and G. B. Schaalje. 1992. Crop protection from permethrin applied aerially to control pale western cutworm (Lepidoptera: Noctuidae). *J. Econ Entomol.* 85: 1387–1392.

Hill, C. C. 1925. Biological studies of the green clover worm. USDA Bull. 1336. 19 pp.

Hill, R. E. 1946. Influence of food plants on fecundity, larval development and abundance of the tuber flea beetle in Nebraska. Nebraska Agric. Exp. Stn. Bull. 143. 16 pp.

Hill. R. E. and W. J. Gary. 1979. Effects of the microsporidium, *Nosema pyrausta*, on field populations of European corn borers in Nebraska. *Environ. Entomol.* 8: 91–95.

Hill, R. E. and H. D. Tate. 1942. Life history and habits of the potato flea beetle in western Nebraska. *J. Econ Entomol.* 35: 879–884.

Hinds, W. E. 1904. Life history of the salt-marsh caterpillar (*Estigmene acrea* Dru.) at Victoria, Tex. USDA Div. Entomol. Bull. 44: 80–84.

Hinks, C. F. and J. R. Byers. 1976. Biosystematics of the genus *Euxoa* (Lepidoptera: Noctuidae) V. Rearing procedures, and life cycles of 36 species. *Can. Entomol.* 108: 1345–1357.

Hinsch, R. T., P. V. Vail, J. S. Tebbets, and D. F. Hoffmann. 1991. Live insects and other arthropods on California iceberg head and shredded lettuce. *Southwestern Entomol.* 16: 261–266.

Hirose, Y. 1991. Pest status and biological control of *Thrips palmi* in southeast Asia. Pages 57–60 *in* N. S. Talekar (ed.). Thrips in Southeast Asia. Asian Veg. Res. and Devt. Ctr., Taipei, Taiwan.

Hirose, Y., H. Kajita, M. Takagi, S. Okajima, B. Napompeth, and S. Buranapanichpan. 1993. Natural enemies of *Thrips palmi* and their effectiveness in the native habitat, Thailand. *Biol. Control* 3: 1–5.

Hobbs, H. A., L. L. Black, R. N. Story, R. A. Valverde, W. P. Bond, J. M. Gatti, Jr., D. O. Schaeffer, and R. R. Johnson. 1993. Transmission of tomato spotted wilt virus from pepper and three weed hosts by *Frankliniella fusca*. *Plant Dis.* 77: 797–799.

Hodges, R. W. 1971. The Moths of America North of Mexico including Greenland. Fascicle 21. Sphingoidea, Hawkmoths. E. W. Classey Ltd., London. 158 pp.

Hodson, W. E. H. 1927. The bionomics of the lesser bulb flies, *Eumerus strigatus*, Flyn., and *Eumerus tuberculatus*, Rond., in south-west England. *Bull. Entomol. Res.* 17: 373–384.

Hodson, W. E. H. 1928. The bionomics of the bulb mite, *Rhizoglyphus echinopus*, Fumouze & Robin. *Bull. Entomol. Res.* 19: 187–200.

Hoebeke, E. R. and A. G. Wheeler, Jr. 1985. *Sitona lineatus* (L.), the pea leaf weevil: first records in eastern North America (Coleoptera: Curculionidae). Proc. Entomol. Soc., Washington. 87: 216–220.

Hoerner, J. L. 1933. The alfalfa webworm (*Loxostege commixtalis* Walker). Colorado State Entomol. Circ. 58: 1–12.

Hoerner, J. L. 1948. The cutworm *Loxagrotis albicosta* on beans. *J. Econ Entomol.* 41: 631–635.

Hoerner, J. L. and C. P. Gillette. 1928. The potato flea beetle. Colorado Agric. Exp. Stn. Bull. 337. 19 pp.

Hoffman, R. L. 1990. Diplopoda. Pages 835–860 *in* D. L. Dindal (ed.). *Soil Biology Guide*. John Wiley & Sons, New York.

Hoffmann, C. H. 1936. Additional data on the biology and ecology of *Strigoderma arboricola* Fab. (Scarabaeidae-Coleoptera). Bull. Brooklyn Entomol. Soc. 31: 108–110.

Hoffmann, K. M. 1987. Earwigs (Dermaptera) of South Carolina, with a key to the eastern North American species and a checklist of the North American fauna. Proc. Entomol. Soc., Washington 89: 1–14.

Hoffmann, M. P., J. J. Kirkwyland, and P. M. Davis. 1995. Spatial distribution of adult corn flea beetles, *Chaetocnema pulicaria* (Coleoptera: Chrysomelidae), in sweet corn and development of a sampling plan. *J. Econ Entomol.* 88: 1324–1331.

Hoffmann, M. P., J. J. Kirkwyland, R. F. Smith, and R. F. Long. 1996a. Field tests with kairomone-baited traps for cucumber beetles and corn rootworms in cucurbits. *Environ. Entomol.* 25: 1173–1181.

Hoffmann, M. P., L. T. Wilson, F. G. Zalom, and R. J. Hilton. 1991a. Dynamic sequential sampling plan for *Helicoverpa zea* (Lepidoptera: Noctuidae) eggs in processing tomatoes: parasitism and temporal patterns. *Environ. Entomol.* 20: 1005–1012.

Hoffmann, M. P., N. A. Davidson, L. T. Wilson, L. E. Ehler, W. A. Jones, and F. G. Zalom. 1991b. Imported wasp helps control southern green stink bug. *California Agric.* 45: 20–22.

Hoffmann, M. P., J. P. Nyrop, J. J. Kirkwyland, D. M. Riggs, D. O. Gilrein, and D. D. Moyer. 1996b. Sequential sampling plan for scheduling control of lepidopteran pests of fresh market sweet corn. *J. Econ Entomol.* 89: 386–395.

Hofmaster, R. N. 1949. Biology and control of the potato tuberworm with special reference to eastern Virginia. Virginia Truck Exp. Stn. Bull. 111: 1826–1882.

Hofmaster, R. N. 1961. Seasonal abundance of the cabbage looper as related to light trap collections, precipitation, temperature and the incidence of a nuclear polyhedrosis virus. *J. Econ Entomol.* 54: 796–798.

Hogg, D. B. 1985. Potato leafhopper (Homoptera: Cicadellidae) immature development, life tables, and population dynamics under fluctuating temperature regimes. *Environ. Entomol.* 14: 349–355.

Holloway, T. E., W. E. Haley, U. C. Loftin, and C. Heinrich. 1928. The sugar-cane borer in the United States. USDA Tech. Bull. 41. 77 pp.

Hollingsworth, C. S. and C. A. Gatsonis. 1990. Sequential sampling plans for green peach aphid (Homoptera: Aphididae) on potato. *J. Econ Entomol.* 83: 1365–1369.

Hollingsworth, R. G., B. E. Tabashnik, D. E. Ullman, M. W. Johnson, and R. Messing. 1994. Resistance of *Aphis gossypii* (Homoptera: Aphididae) to insecticides in Hawaii: spatial patterns and relation to insecticide use. *J. Econ Entomol.* 87: 293–300.

Holt, J. and N. Birch. 1984. Taxonomy, evolution and domestication of *Vicia* in relation to aphid resistance. *Ann. Appl. Biol.* 105: 547–556.

Holtzer, T.O, T. M. Perring, and M. W. Johnson. 1984. Winter and spring distribution and density of Banks grass mite (Acari: Tetranychidae) in adjacent wheat and corn. *J. Kansas Entomol. Soc.* 57: 333–335.

Hooker, W. A. 1907. The tobacco thrips, a new and destructive enemy of shade-grown tobacco. USDA Bur. Entomol. Bull. 65. 23 pp.

Hopkin, S. P. 1997. Biology of the Springtails (Insecta: Collembola). Oxford Univ. Press, London. 330 pp.

Hopkin, S. P. and H. J. Read. 1992. The Biology of Millipedes. Oxford Univ. Press, London. 233 pp.

Horne, J. and P. Bailey. 1991. *Bruchus pisorum* L. (Coleoptera, Bruchidae) control by a knockdown pyrethroid in field peas. *Crop Prot.* 10: 53–56.

Horn, K. F., C. G. Wright, and M. H. Farrier. 1979. The lace bugs (Hemiptera: Tingidae) of North Carolina and their hosts. North Carolina Agric. Exp. Stn. Tech. Bull. 257. 22 pp.

Horne, P. A. 1993. Sampling for the potato moth (*Phthorimaea operculella*) and its parasitoids. *Aust. J. Exp. Agric.* 33: 91–96.

Horowitz, A. R., Z. Mendelson, and I. Ishaaya. 1997. Effect of abamectin mixed with mineral oil on the sweetpotato whitefly (Homoptera: Aleyrodidae). *J. Econ Entomol.* 90: 349–353.

Horsfall, J. L. 1924. Life history studies of *Myzus persicae* Sulzer. Pennsylvania Agric. Exp. Stn. Bull. 185. 16 pp.

Horsfall, J. L. 1943. Biology and control of common blister beetles in Arkansas. Arkansas Agric. Exp. Stn. Bull. 436. 55 pp.

Horsfall, J. L. and J. R. Eyer. 1921. Preliminary notes on control of millipedes under sash. *J. Econ Entomol.* 14: 269–272.

Horsfall, J. L. and F. A. Fenton. 1922. The onion thrips in Iowa. Iowa Agric. Exp. Stn. Bull. 205: 55–68.

Horsfall, W. R. 1941. Biology of the black blister beetle (Coleoptera: Meloidae). *Ann. Entomol. Soc. Am.* 34: 114–126.

Horton, D. R. and J. L. Capinera. 1987a. Seasonal and host plant effects on parasitism of Colorado potato beetle by *Myiopharus doryphorae* (Riley) (Diptera: Tachinidae). *Can. Entomol.* 119: 729–734.

Horton, D. R. and J. L. Capinera. 1987b. Effects of plant diversity, host density, and host size on population ecology of the Colorado potato beetle (Coleoptera: Chrysomelidae). *Environ. Entomol.* 16: 1019–1026.

Horton, D. R., J. L. Capinera, and P. L. Chapman. 1988. Local differences in host use by two populations of the Colorado potato beetle. *Ecology* 69: 823–831.

Hostetter, D. L. and B. Puttler. 1991. A new broad host spectrum nuclear polyhedrosis virus isolated from a celery looper, *Anagrapha falcifera* (Kirby), (Lepidoptera: Noctuidae). *Environ. Entomol.* 20: 1480–1488.

Hostetter, D. L., B. Puttler, A. H. McIntosh, and R. E. Pinnell. 1990. A nuclear polyhedrosis virus of the yellowstriped armyworm (Lepidoptera: Noctuidae). *Environ. Entomol.* 19: 1150–1154.

Hou, R. F. and M. A. Brooks. 1975. Continuous rearing of the aster leafhopper, *Macrosteles fascifrons*, on a chemically defined diet. *J. Insect Physiol.* 21: 1481–1483.

Hough-Goldstein, J. A. 1987. Tests of a spun polyester row cover as a barrier against seedcorn maggot (Diptera: Anthomyiidae) and cabbage pest infestations. *J. Econ Entomol.* 80: 768–772.

Hough-Goldstein, J. A. and K. A. Hess. 1984. Seedcorn maggot (Diptera: Anthomyiidae) infestation levels and effects on five crops. *Environ. Entomol.* 13: 962–965.

Hough-Goldstein, J. A. and D. McPherson. 1996. Comparison of *Perillus bioculatus* and *Podisus maculiventris* (Hemiptera: Pentatomi-

dae) as potential control agents of the Colorado potato beetle (Coleoptera: Chrysomelidae). *J. Econ Entomol.* 89: 1116–1123.

Hough-Goldstein, J. A. and J. M. Whalen. 1996. Relationship between crop rotation distance from previous potatoes and colonization and population density of Colorado potato beetle. *J. Agric. Entomol.* 13: 293–300.

Hough-Goldstein, J. A., G. E. Heimpel, H. E. Bechmann, and C. E. Mason. 1993. Arthropod natural enemies of the Colorado potato beetle. *Crop Prot.* 12: 324–334.

Houser, J. S. and W. V. Balduf. 1925. The striped cucumber beetle, *Diabrotica vittata* Fabr. Ohio Agric. Exp. Stn. Bull. 388: 241–364.

Houser, J. S., T. L. Guyton, and P. R. Lowry. 1917. The pink and green aphid of potato. Ohio Agric. Exp. Stn. Bull. 317: 61–88.

Howard, L. O. 1894. Damage by the American locust. Insect Life 7: 220–229.

Howard, N. F. and L. L. English. 1924. Studies of the Mexican bean beetle in the southeast. USDA Bull. 1243. 50 pp.

Howard, N. F. and B. J. Landis. 1936. Parasites and predators of the Mexican bean beetle in the United States. USDA Circ. 418. 12 pp.

Howard, R. J., J. A. Garland, and W. L. Seaman (eds). 1994. Diseases and Pests of Vegetable Crops in Canada. Entomological Society of Canada, Ottawa, Ontario. 554 pp.

Howe, R. W. and J. E. Currie. 1964. Some laboratory observations on the rates of development, mortality and oviposition of several species of Bruchidae breeding in stored pulses. *Bull. Entomol. Res.* 55: 437–477.

Howe, W. L. and A. M. Rhodes. 1973. Host relationships of the squash vine borer, *Melittia cucurbitae* with species of *Cucurbita*. *Ann. Entomol. Soc. Am.* 66: 266–269.

Howe, W. L., J. R. Sanborn, and A. M. Rhodes. 1976. Western corn rootworm adult and spotted cucumber beetle associations with *Cucurbita* and cucurbitacins. *Environ. Entomol.* 5: 1043–1048.

Howell, J. F. 1979. Phenology of the adult spotted cutworm in the Yakima Valley. *Environ. Entomol.* 8: 1065–1069.

Howell, W. E. and G. I. Mink. 1977. Role of aphids in the epidemiology of carrot virus diseases in central Washington. *Plant Dis. Rep.* 61: 841–844.

Howitt, A. J. 1959. Control of *Scutigerella immaculata* (Newport) in the Pacific Northwest with soil fumigants. *J. Econ Entomol.* 52: 678–683.

Howitt, A. J. and S. G. Cole. 1959. Chemical control of the carrot rust fly, *Psila rosae* (F.), in western Washington. *J. Econ Entomol.* 52: 963–966.

Howitt, A. J. and S. G. Cole. 1962. Chemical control of slugs affecting vegetables and strawberries in the Pacific Northwest. *J. Econ Entomol.* 55: 320–325.

Howitt, A. J., J. S. Waterhouse, and R. M. Bullock. 1959. The utility of field tests for evaluating insecticides against the garden symphylid. *J. Econ Entomol.* 52: 666–672.

Howlader, M. A. and G. H. Gerber. 1986. Calling behavior of the bertha armyworm, *Mamestra configurata* (Lepidoptera: Noctuidae). *Can. Entomol.* 118: 735–743.

Hoy, C. W. and D. W. Kretchman. 1991. Thrips (Thysanoptera: Thripidae) injury to cabbage cultivars in Ohio. *J. Econ Entomol.* 84: 971–977.

Hoy, C. W., S. E. Heady, and T. A. Koch. 1992. Species composition, phenology, and possible origins of leafhoppers (Cicadellidae) in Ohio vegetable crops. *J. Econ Entomol.* 85: 2336–2343.

Hoy, C. W., C. Jennison, A. M. Shelton, and J. T. Andaloro. 1983. Variable-intensity sampling: a new technique for decision making in cabbage pest management. *J. Econ Entomol.* 76: 139–143.

Hruska, A. J., S. M. Gladstone, and R. Obando. 1996. Epidemic roller coaster: maize stunt disease in Nicaragua. *Am. Entomol.* 42: 248–252.

Hsiao, T. H. 1988. Host specificity, seasonality and bionomics of Leptinotarsa beetles. Pages 581–599 in P. Jolivet, E. Petitpierre and T. H. Hsiao (eds.). Biology of Chrysomelidae. Kluwer Academic, Dordrecht/Norwell, MA.

Hsaio, T. H. and G. Fraenkel. 1968. Selection and specificity of the Colorado potato beetle for solanaceous and nonsolanaceous plants. Ann. Entomol. Soc. Am. 61: 493–503.

Hsing-Yeh, L., D. J. Gumpf, G. N. Oldfield, and E. C. Calavan. 1983. The relationship of Spiroplasma citri and Circulifer tenellus. Phytopathology 73: 585–590.

Huckett, H. C. 1934. Field tests on Long Island of derris as an insecticide for the control of cabbage worms. J. Econ Entomol. 27: 440–445.

Huckett, H. C. 1946. DDT and other new insecticides for control of cauliflower worms on Long Island. J. Econ Entomol. 39: 184–188.

Hudon, M. and E. J. LeRoux. 1986a. Biology and population dynamics of the European corn borer (Ostrinia nubilalis) with special reference to sweet corn in Quebec. I. Systematics, morphology, geographical distribution, host range, economic importance. Phytoprotection 67: 39–54.

Hudon, M. and E. J. LeRoux. 1986b. Biology and population dynamics of the European corn borer (Ostrinia nubilalis) with special reference to sweet corn in Quebec. II. Bionomics. Phytoprotection 67: 81–92.

Hudon, M. and E. J. LeRoux. 1986c. Biology and population dynamics of the European corn borer (Ostrinia nubilalis) with special reference to sweet corn in Quebec. III. Population dynamics and spatial distribution. Phytoprotection 67: 93–115.

Hudon, M., E. J. LeRoux, and D. G. Harcourt. 1989. Seventy years of European corn borer (Ostrinia nubilalis) research in North America. Pages 53–96 in G. E. Russell (ed.). Agricultural Zoology Reviews. Vol. 3. Intercept, Wimborne, Dorset, UK.

Hudson, A. 1982. Evidence for reproductive isolation between Xestia adela Franclemont and Xestia dolosa Franclemont (Lepidoptera: Noctuidae). Proc. Entomol. Soc., Washington 84: 775–780.

Hudson, A. 1983. Comparative studies of the development of two species of spotted cutworm Xestia adela and Xestia dolosa (Lepidoptera: Noctuidae), and identification of larvae by electrophoresis. Proc. Entomol. Soc., Washington 85: 612–618.

Hudson, W. G. 1987. Variability in development of Scapteriscus acletus (Orthoptera: Gryllotalpidae). Florida Entomol. 70: 403–404.

Hudson, W. G. 1988. Field sampling of mole crickets (Orthoptera: Gryllotalpidae: Scapteriscus): a comparison of techniques. Florida Entomol. 71: 214–216.

Hudson, W. G. 1989. Field sampling and population estimation of the tawny mole cricket (Orthoptera: Gryllotalpidae). Florida Entomol. 72: 337–343.

Huffaker, C. B. 1941. Egg parasites of the harlequin bug in North Carolina. J. Econ Entomol. 34: 117–118.

Huffaker, C. B., M. van de Vrie, and J. A. McMurray. 1969. The ecology of tetranychid mites and their natural control. Annu. Rev. Entomol. 14: 125–174.

Huffman, F. R. and J. A. Harding. 1980. Pitfall collected insects from various lower Rio Grande Valley habitats. Southwestern Entomol. 5: 33–46.

Hughes, A. M. 1961. The Mites of Stored Food. Ministry of Agriculture, Fisheries and Food Tech. Bull. 9. H. M. Stationery Office, London. 287 pp.

Hughes, J. H. 1943. The alfalfa plant bug: Adelphocoris lineolatus (Goeze) and other Miridae (Hemiptera) in relation to alfalfa-seed production in Minnesota. Minnesota Agric. Exp. Stn. Tech. Bull. 161. 80 pp.

Hughes, R. D. 1959. The natural mortality of Erioischia brassicae (Bouché) (Diptera, Anthomyiidae) during the egg stage of the first generation. J. Anim. Ecol. 28: 343–357.

Hughes, R. D. and B. Mitchell. 1960. The natural mortality of Erioischia brassicae (Bouché) (Dipt., Anthomyiidae): life tables and their interpretation. J. Anim. Ecol. 29: 359–374.

Hung, A. C. F., M. R. Barlin, and S. B. Vinson. 1977. Identification, distribution, and biology of fire ants in Texas. Texas Agric. Exp. Stn. Bull. B-1185. 24 pp.

Hunter, P. J. 1967. The effect of cultivations on slugs of arable land. Plant Pathol. 16: 153–156.

Hunter, P. J. 1968. Studies on slugs of arable ground II. Life cycles. Malacologia 6: 379–389.

Hunter, R. E. and T. F. Leigh. 1965. A laboratory life history of the conserpse stink bug, Euschistus conspersus (Hemiptera: Pentatomidae). Ann. Entomol. Soc. Am. 58: 648–649.

Hutchison, W. D. and D. B. Hogg. 1983. Cornicle length as a criterion for separating field-collected nymphal instars of the pea aphid, Acyrthosiphon pisum (Homoptera: Aphididae). Can. Entomol. 115: 1615–1619.

Hutchison, W. D. and D. B. Hogg. 1984. Demographic statistics for the pea aphid (Homoptera: Aphididae) in Wisconsin and a comparison with other populations. Environ. Entomol. 13: 1173–1181.

Hyslop, J. A. 1912a. The alfalfa looper. USDA Bur. Entomol. Bull. 95: 109–118.

Hyslop, J. A. 1912b. The legume pod moth. The legume pod maggot. USDA Bur. Entomol. Bull. 95: 89–108.

Hyslop, J. A. 1912c. The false wireworms of the Pacific Northwest. USDA Bur. Entomol. Bull. 95: 73–87.

Hyslop, J. A. 1915. Wireworms attacking cereal and forage crops. USDA Bull. 156. 34 pp.

Idris, A. B. and E. Grafius. 1995. Wildflowers as nectar sources for Diadegma insulare (Hymenoptera: Ichneumonidae), a parasitoid of diamondback moth (Lepidoptera: Yponomeutidae). Environ. Entomol. 24: 1726–1735.

Ignoffo, C. M. and C. Garcia. 1979. Susceptibility of larvae of the black cutworm to species of entomopathogenic bacteria, fungi, protozoa, and viruses. J. Econ Entomol. 72: 767–769.

Ignoffo, C. M., D. L. Hostetter, K. D. Biever, C. Garcia, G. D. Thomas, W. A. Dickerson, and R. Pinnell. 1978. Evaluation of an entomopathogenic bacterium, fungus, and virus for control of Heliothis zea on soybeans. J. Econ Entomol. 71: 165–168.

Inglis, G. D., D. L. Johnson, and M. S. Goettel. 1996. Effects of temperature and thermoregulation on mycosis by Beauveria bassiana in grasshoppers. Biol. Control 7: 131–139.

Ingram, J. W. and W. A. Douglas. 1932. Notes on the life history of the striped blister beetle in southern Louisiana. J. Econ Entomol. 25: 71–74.

Isely, D. 1927. The striped cucumber beetle. Arkansas Agric. Exp. Stn. Bull. 216. 31 pp.

Isely, D. 1929. The southern corn rootworm. Arkansas Agric. Exp. Stn. Bull. 232. 31 pp.

Isely, D. 1930. The biology of the bean leaf-beetle. Arkansas Agric. Exp. Stn. Bull. 248. 20 pp.

Isely, D. 1946. The cotton aphid. Arkansas Agric. Exp. Stn. Bull. 462. 29 pp.

Isman, M. B. 1993. Growth inhibitory and antifeedant effects of azadirachtin on six noctuids of regional economic importance. Pest. Sci. 38: 57–63.

Jackson, C. G. and H. M. Graham. 1983. Parasitism of four species of Lygus (Hemiptera: Miridae) by Anaphes ovijentatus (Hymenoptera: Mymaridae) and an evaluation of other possible hosts. Ann. Entomol. Soc. Am. 76: 772–775.

Jackson, C. G., G. D. Butler, Jr., and D. E. Bryan. 1969. Time required for development of Voria ruralis and its host, the cabbage looper, at different temperatures. J. Econ Entomol. 62: 69–70.

Jackson, C. G., D. E. Bryan, G. D. Butler, Jr., and R. Patana. 1970. Development, fecundity, and longevity of *Leschenaultia adusta*, a tachinid parasite of the salt-marsh caterpillar. *J. Econ Entomol.* 63: 1396–1397.

Jackson, D. J. 1920. Bionomics of weevils of the genus *Sitones* injurious to leguminous crops in Britain. *Ann. Appl. Biol.* 7: 269–298.

Jackson, D. M. and R. L. Campbell. 1975. Biology of the European crane fly, *Tipula paludosa* Meigen, in western Washington (Tipulidae; Diptera). Washington State Univ. Tech. Bull. 81. 23 pp.

Jackson, J. J. 1986. Rearing and handling of *Diabrotica virgifera* and *Diabrotica undecimpunctata howardi*. Pages 25–47 *in* J. L. Krysan and T. A. Miller (eds.). Methods for the Study of Pest *Diabrotica*. Springer-Verlag, Berlin/New York.

Jackson, J. J. 1996. Field performance of entomopathogenic nematodes for suppression of western corn rootworm (Coleoptera: Chrysomelidae). *J. Econ Entomol.* 89: 366–372.

Jackson, J. J. and N. C. Elliott. 1988. Temperature-dependent development of immature stages of the western corn rootworm, *Diabrotica virgifera virgifera* (Coleoptera: Chrysomelidae). *Environ. Entomol.* 17: 166–171.

Jackson, J. J. and L. S. Hesler. 1995. Placement and application rate of the nematode *Steinernema carpocapsae* (Rhabditidae: Steinernematidae) for suppression of the western corn rootworm (Coleoptera: Chrysomelidae). *J. Kansas Entomol. Soc.* 68: 461–467.

Jackson, J. J. and G. R. Sutter. 1985. Pathology of a granulosis virus in the army cutworm, *Euxoa auxiliaris* (Lepidoptera: Noctuidae). *J. Kansas Entomol. Soc.* 58: 353–355.

Jacob, D. and G. M. Chippendale. 1971. Growth and development of the southwestern corn borer, *Diatraea grandiosella*, on a meridic diet. *Ann. Entomol. Soc. Am.* 64: 485–488.

Jacobson, L. A. 1970. Laboratory ecology of the red-backed cutworm, *Euxoa ochrogaster* (Lepidoptera: Noctuidae). *Can. Entomol.* 102: 85–89.

Jacobson, L. A. 1971. The pale western cutworm, *Agrotis orthogonia* Morrison (Lepidoptera: Noctuidae): a review of research. *Quaes. Entomol.* 7: 414–436.

Jacobson, L. A. and P. E. Blakeley. 1959. Development and behavior of the army cutworm in the laboratory. *Ann. Entomol. Soc. Am.* 52: 100–105.

Jacques, R. L. and D. C. Peters. 1971. Biology of *Systena frontalis*, with special reference to corn. *J. Econ Entomol.* 64: 135–138.

Janes, M. J. and G. L. Greene. 1970. An unusual occurrence of loopers feeding on sweet corn ears in Florida. *J. Econ Entomol.* 63: 1334–1335.

Jansen, W. P. and R. Staples. 1970a. Transmission of cowpea mosaic virus by the Mexican bean beetle. *J. Econ Entomol.* 63: 1719–1720.

Jansen, W. P. and R. Staples. 1970b. Effect of cowpeas and soybeans as source or test plants of cowpea mosaic virus on vector efficiency and retention of infectivity of the bean leaf beetle and the spotted cucumber beetle. *Plant Dis. Rep.* 12: 1053–1054.

Jansson, R. K. 1991. Biological control of *Cylas* spp. Pages 169–201 *in* R. K. Jansson and K.V Raman (eds.). Sweet Potato Pest Management: A Global Perspective. Westview Press, Boulder, Colorado.

Jansson, R. K. and A. G. B. Hunsberger. 1991. Diel and ontogenetic patterns of oviposition in the sweetpotato weevil (Coleoptera: Curculionidae). *Environ. Entomol.* 20: 545–550.

Jansson, R. K. and S. H. Lecrone. 1989. Evaluation of food baits for pre-plant sampling of wireworms (Coleoptera: Elateridae) in potato fields in southern Florida. *Florida Entomol.* 72: 503–510.

Jansson, R. K. and S. H. Lecrone. 1991. Effects of summer cover crop management on wireworm (Coleoptera: Elateridae) abundance and damage to potato. *J. Econ. Entomol.* 84: 581–586.

Jansson, R. K. and Z. Smilowitz. 1985. Influence of nitrogen on population parameters of potato insects: abundance, development,

and damage of the Colorado potato beetle, *Leptinotarsa decemlineata* (Coleoptera: Chrysomelidae). *Environ. Entomol.* 14: 500–506.

Jansson, R. K. and Z. Smilowitz. 1986. Influence of nitrogen on population parameters of potato insects: abundance, population growth, and within-plant distribution of the green peach aphid, *Myzus persicae* (Homoptera: Aphididae). *Environ Entomol.* 15: 49–55.

Jansson, R. K., S. H. Lecrone, and R. Gaugler. 1993. Field efficacy and persistence of entomopathogenic nematodes (Rhabditida: Steinernematidae, Heterorhabditidae) for control of sweetpotato weevil (Coleoptera: Apionidae) in southern Florida. *J. Econ. Entomol.* 86: 1055–1063.

Jansson, R. K., L. J. Mason, and R. R. Heath. 1991a. Use of sex pheromone for monitoring and managing *Cylas formicarius*. Pages 97–138 *in* R. K. Jansson and K. V. Raman (eds.). Sweet Potato Pest Management: a Global Perspective. Westview Press, Boulder, Colorado.

Jansson, R. K., A. G. B. Hunsberger, S. H. Lecrone, and S. K. O'Hair. 1990a. Seasonal abundance, population growth, and within-plant distribution of sweet potato weevil (Coleoptera: Curculionidae) on sweet potato in southern Florida. *Environ Entomol.* 19: 313–321.

Jansson, R. K., S. H. Lecrone, R. R. Gaugler, and G. C. Smart, Jr. 1990b. Potential of entomopathogenic nematodes as biological control agents of sweetpotato weevil (Coleoptera: Curculionidae). *J. Econ. Entomol.* 83: 1818–1826.

Jansson, R. K., G. L. Leibee, C. A. Sanchez, and S. H. Lecrone. 1991b. Effects of nitrogen and foliar biomass on population parameters of cabbage insects. *Entomol. Exp. Appl.* 61: 7–16.

Jansson, R. K., F. I. Proshold, L. J. Mason, R. R. Heath, and S. H. Lecrone. 1990c. Monitoring sweetpotato weevil (Coleoptera: Curculionidae) with sex pheromone: effects of dosage and age of septa. *Trop. Pest Manage.* 36: 263–269.

Jaques, R. P. 1973. Tests on microbial and chemical insecticides for control of *Trichoplusia ni* (Lepidoptera: Noctuidae) and *Pieris rapae* (Lepidoptera: Pieridae) on cabbage. *Can. Entomol.* 105: 21–27.

Jaques, R. P. 1977. Field efficacy of viruses infectious to the cabbage looper and imported cabbageworm on late cabbage. *J. Econ. Entomol.* 70: 111–118.

Jaronski, S. T. and M. S. Goettel. 1997. Development of *Beauvaria bassiana* for control of grasshoppers and locusts. Mem. Entomol. Soc. Canada 171: 225–237.

Jarvis, J. L. and T. A. Brindley. 1965. Predicting moth flight and oviposition of European corn borer by the use of temperature accumulations. *J. Econ. Entomol.* 58: 300–302.

Jarvis, J. L. and W. D. Guthrie. 1987. Ecological studies of the European corn borer (Lepidoptera: Pyralidae) in Boone County, Iowa. *Environ Entomol.* 16: 50–58.

Javahery, M. 1990. Biology and ecological adaptation of the green stink bug (Hemiptera: Pentatomidae) in Quebec and Ontario. *Ann. Entomol. Soc. Am.* 83: 201–206.

Jaworska, M. and D. Ropek. 1994. Influence of host-plant on the susceptibility of *Sitona lineatus* L. (Col., Curculionidae) to *Steinernema carpocapsae* Weiser. *J. Invertebr. Pathol.* 64: 96–99.

Jedlinski, H. 1981. Rice root aphid, *Rhopalosiphum rufiabdominalis*, a vector of barley yellow dwarf virus in Illinois, and the disease complex. *Plant Dis.* 65: 975–978.

Jefferson, R. N., J. G. Bald, F. S. Morishita, and D. H. Close. 1956. Effect of Vapam on *Rhizoglyphus* mites and gladiolus soil diseases. *J. Econ. Entomol.* 49: 584–589.

Jeffords, M. R., C. G. Helm, and M. Kogan. 1983. Overwintering behavior and spring colonization of soybean by the bean leaf beetle (Coleoptera: Chrysomelidae) in Illinois. *Environ. Entomol.* 12: 1459–1463.

Jemal, A. and M. Hugh-Jones. 1993. A review of the red imported fire ant (*Solenopsis invicta* Buren) and its impact on plant, animal, and human health. *Prev. Vet. Med.* 17: 19–32.

Jensen, R. L., L. D. Newsom, and J. Gibbens. 1974. The soybean looper: effects of adult nutrition on oviposition, mating frequency, and longevity. *J. Econ. Entomol.* 67: 467–470.

Jensen, S. G. 1985. Laboratory transmission of maize chlorotic mottle virus by three species of corn rootworms. *Plant Dis.* 69: 864–868.

Jeppson, L. R., H. H. Keifer, and E. W. Baker. 1975. Mites Injurious to Economic Plants. Univ. California Press, Berkeley. 614 pp.

Jewett, H. H. 1926. The tobacco flea beetle. Kentucky Agric. Exp. Stn. Bull. 266: 51–69.

Jewett, H. H. 1929. Potato flea-beetles. Kentucky Agric. Exp. Stn. Bull. 297: 283–301.

Johannsen, O. A. 1913. Potato flea beetle. Maine Agric. Exp. Stn. Bull. 211: 37–56.

Johnson, A. W. 1974. *Bacillus thuringiensis* and tobacco budworm control on flue-cured tobacco. *J. Econ. Entomol.* 67: 755–759.

Johnson, D. L. 1997. Nosematidae and other protozoa as agents for control of grasshoppers and locusts: current status and prospects. *Mem. Entomol. Soc.* Canada 171: 375–389.

Johnson, D. L. and M. G. Dolinski. 1997. Attempts to increase the prevalence and severity of infection of grasshoppers with the entomopathogen *Nosema locustae* Canning (Microsporidia: Nosematidae) by repeated field application. Mem. Entomol. Soc. Canada 171: 391–400.

Johnson, D. L. and M. S. Goettel. 1993. Reduction of grasshopper populations following field application of the fungus *Beauvaria bassiana*. *Biocontrol Sci. Tech.* 3: 165–175.

Johnson, D. L. and J. E. Henry. 1987. Low rates of insecticides and *Nosema locustae* (Microsporidia: Nosematidae) on baits applied to roadsides for grasshopper (Orthoptera: Acrididae) control. *J. Econ. Entomol.* 80: 685–689.

Johnson, D. L. and E. Pavlikova. 1986. Reduction of consumption by grasshoppers (Orthoptera: Acrididae) infected with *Nosema locustae* Canning (Microsporidia: Nosematidae). *J. Invertebr. Pathol.* 48: 232–238.

Johnson, M. W. 1986. Population trends of a newly introduced species, *Thrips palmi* (Thysanoptera: Thripidae), on commercial watermelon plantings in Hawaii. *J. Econ. Entomol.* 79: 718–720.

Johnson, M. W. 1987. Parasitization of *Liriomyza* spp. (Diptera: Agromyzidae) infesting commercial watermelon plantings in Hawaii. *J. Econ. Entomol.* 80: 56–61.

Johnson, M. W. and D. M. Nafus. 1995. Melon thrips. Pages 79–80 *in* J. R. Nechols, L. A. Andres, J. W. Beardsley, R. D. Goeden, and C. G. Jackson (eds.). Biological Control in the Western United States. Univ. California Publ. 3361.

Johnson, M. W., E. R. Oatman, and J. A. Wyman. 1980a. Natural control of *Liriomyza sativae* (Dip.: Agromyzidae) in pole tomatoes in southern California. *Entomophaga* 25: 193–198.

Johnson, M. W., E. R. Oatman, J. A. Wyman, and R. A. van Steenwyk. 1980b. A technique for monitoring *Liriomyza sativae* in fresh market tomatoes. *J. Econ. Entomol.* 73: 552–555.

Johnson, M. W., L. C. Caprio, J. A. Coughlin, B. E. Tabashnik, J. A. Rosenheim, and S. C. Welter. 1992. Effect of *Trialeurodes vaporariorum* (Homoptera: Aleyrodidae) on yield of fresh market tomatoes. *J. Econ. Entomol.* 85: 2370–2376.

Johnson, P. M. and A. M. Ballinger. 1916. Life-history studies of the Colorado potato beetle. *J. Agric. Res.* 5: 917–925.

Johnson, R. R., L. L. Black, H. A. Hobbs, R. A. Valverde, R. N. Story, and W. P. Bond. 1995. Association of *Frankliniella fusca* and three winter weeds with tomato spotted wilt virus in Louisiana. *Plant Dis.* 79: 572–576.

Johnson, S. J. and J. W. Smith, Jr. 1981. Ecology of *Elasmopalpus lignosellus* parasite complex on peanuts in Texas. *Ann. Entomol. Soc. Am.* 74: 467–471.

Johnson, T. B. and L. C. Lewis. 1982a. Pathogenicity of two nuclear polyhedrosis viruses in the black cutworm, *Agrotis ipsilon* (Lepidoptera: Noctuidae). *Can. Entomol.* 114: 311–316.

Johnson, T. B. and L. C. Lewis. 1982b. Evaluation of *Rachiplusia ou* and *Autographa californica* nuclear polyhedrosis viruses in suppressing black cutworm damage to seedling corn in greenhouse and field. *J. Econ. Entomol.* 75: 401–404.

Johnson, T. B., F. T. Turpin, M. M. Schreiber, and D. R. Griffith. 1984. Effects of crop rotation, tillage, and weed management systems on black cutworm (Lepidoptera: Noctuidae) infestations in corn. *J. Econ. Entomol.* 77: 919–921.

Johnson, V. 1973. The female and host of *Triozocera mexicana* (Strepsiptera: Mengeidae). *Ann. Entomol. Soc. Am.* 66: 671–672.

Johnston, F. A. 1915. Asparagus-beetle egg parasite. *J. Agric. Res.* 4: 303–315.

Jones, C. G., T. A. Hess, D. W. Whitman, P. J. Silk, and M. S. Blum. 1987. Effects of diet breadth on autogenous chemical defense of a generalist grasshopper. *J. Chem. Ecol.* 13: 283–297.

Jones, D. W. 1917. The European earwig and its control. *USDA Bull.* 566. 12 pp.

Jones, D. and J. Granett. 1982. Feeding site preferences of seven lepidopteran pests of celery. *J. Econ. Entomol.* 75: 449–453.

Jones, H. A., S. F. Bailey, and S. L. Emsweller. 1935. Field studies of *Thrips tabaci* Lind. with especial reference to resistance in onions. *J. Econ. Entomol.* 28: 678–680.

Jones, M. P. and B. S. Heming. 1979. Effects of temperature and relative humidity on embryogenesis in eggs of *Mamestra configurata* (Walker) (Lepidoptera: Noctuidae). *Quaes. Entomol.* 15: 257–294.

Jones, T. H. 1916a. Notes on *Anasa andresii* Guer., an enemy of cucurbits. *J. Econ. Entomol.* 9: 431–434.

Jones, T. H. 1916b. The eggplant tortoise beetle. *USDA Bull.* 422. 8 pp.

Jones, T. H. 1918a. The southern green plant-bug. *USDA Bull.* 689. 27 pp.

Jones, T. H. 1918b. The granulated cutworm, an important enemy of vegetable crops in Louisiana. *USDA Bull.* 703: 7–14.

Jones, T. H. 1923. The eggplant leaf-miner, *Phthorimaea glochinella* Zeller. *J. Agric. Res.* 26: 567–570.

Jones, T. H. and M. P. Hassell. 1988. Patterns of parasitism by *Trybliographa rapae*, a cynipid parasitoid of the cabbage root fly, under laboratory and field conditions. *Ecol. Entomol.* 13: 309–317.

Jones, V. P. and R. D. Brown. 1983. Reproductive responses of the broad mite, *Polyphagotarsonemus latus* (Acari: Tarsonemidae), to constant temperature-humidity regimes. *Ann. Entomol. Soc. Am.* 76: 466–469.

Jones, Jr., W. A. and M. J. Sullivan. 1981. Overwintering habitats, spring emergence patterns, and winter mortality of some South Carolina Hemiptera. *Environ. Entomol.* 10: 409–414.

Jones, W. A. and M. J. Sullivan. 1982. Role of host plants in population dynamics of stink bug pests of soybean in South Carolina. *Environ. Entomol.* 11: 867–875.

Jones, W. A., B. M. Shepard, and M. J. Sullivan. 1996. Incidence of parasitism of pentatomid (Heteroptera) pests of soybean in South Carolina with a review of studies in other states. *J. Agric. Entomol.* 13: 243–263.

Jouvenaz, D. P. 1983. Natural enemies of fire ants. *Florida Entomol.* 66: 111–121.

Jouvenaz, D. P. 1986. Diseases of fire ants: problems and opportunities. Pages 327–338 *in* C. S. Lofgren and R. K. Vander Meer (eds.). Fire Ants and Leaf-Cutting Ants. Westview Press, Boulder.

Jubb, Jr., G. L. and T. F. Watson. 1971a. Development of the egg parasite *Telenomus utahensis* in two pentatomid hosts in relation to temperature and host age. *Ann. Entomol. Soc. Am.* 64: 202–205.

Jubb, Jr., G. L. and T. F. Watson. 1971b. Parasitization capabilities of the pentatomid egg parasite *Telenomus utahensis* (Hymenoptera: Scelionidae). *Ann. Entomol. Soc. Am.* 64: 452–456.

Judd, G. J. R. and J. H. Borden. 1989. Distant olfactory response of the onion fly, *Delia antiqua*, to host-plant odour in the field. *Physiol. Entomol.* 14: 429–441.

Judd, G. J. R. and R. S. Vernon. 1985. Seasonal activity of adult carrot rust flies, *Psila rosae* (Diptera: Psilidae), in the lower Fraser Valley, British Columbia. *Can. Entomol.* 117: 375–381.

Judd, G. J. R., R. S. Vernon, and J. H. Borden. 1985. Commercial implementation of a monitoring program for *Psila rosae* (F.) (Diptera: Psilidae) in southwestern British Columbia. *J. Econ. Entomol.* 78: 477–481.

Judd, G. J. R., G. H. Whitfield, and H. E. L. Maw. 1991. Temperature-dependent development and phenology of pepper maggots (Diptera: Tephritidae) associated with pepper and horsenettle. *Environ Entomol.* 20: 22–29.

Judge, F. D. 1968. Polymorphism in a subterranean aphid, *Pemphigus bursarius*. I. Factors affecting the development of sexuparae. *Ann. Entomol. Soc. Am.* 61: 819–827.

Jyani, D. B., N. C. Patel, R. C. Jhala, and J. R. Patel. 1995. Bioefficacy of neem and synthetic insecticides on serpentine leafminer (*Liriomyza trifolii*) (Diptera: Agromyzidae) infesting pea (*Pisum sativum*). *Indian J. Agric. Sci.* 65: 373–376.

Kaakeh, W. and J. D. Dutcher. 1993. Population parameters and probing behavior of cowpea aphid (Homoptera: Aphididae), on preferred and non-preferred host cover crops. *J. Entomol. Sci.* 28: 145–155.

Kabashima, J., D. K. Giles, and M. P. Parella. 1995. Electrostatic sprayers improve pesticide efficacy in greenhouses. *California Agric.* 49(4): 31–35.

Kabissa, J. and W. D. Fronk. 1986. Bean foliage consumption by Mexican bean beetle (Coleoptera: Coccinellidae) and its effect on yield. *J. Kansas Entomol. Soc.* 59: 275–279.

Kajita, H. 1986. Predation by *Amblyseius* spp. (Acarina: Phytoseiidae) and *Orius* sp. (Hemiptera: Anthocoridae) on *Thrips palmi* Karny (Thysanoptera: Thripidae). *Appl. Entomol. Zool.* 21: 482–484.

Kamau, A. W., J. M. Mueke, and B. M. Khaemba. 1992. Resistance of tomato varieties to the tomato russet mite, *Aculops lycopersici* (Massee) (Acarina: Eriophyidae). *Insect Sci. Applic.* 13: 351–356.

Kamm, J. A. 1971. Silvertop of bluegrass and bentgrass produced by *Anaphothrips obscurus*. *J. Econ. Entomol.* 64: 1385–1387.

Kamm, J. A. 1972. Thrips that affect production of grass seed in Oregon. *J. Econ. Entomol.* 65: 1050–1055.

Kamm, J. A. 1990. Biological observations of glassy cutworm (Lepidoptera: Noctuidae) in western Oregon. *Pan-Pac. Entomol.* 66: 66–70.

Kamm, J. A., L. M. McDonough, and R. D. Gustin. 1982. Armyworm (Lepidoptera: Noctuidae) sex pheromone: field tests. *Environ Entomol.* 11: 917–919.

Kanga, L. H. B., F. W. Plapp, Jr., B. F. McCutchen, R. D. Bagwell, and J. D. Lopez, Jr. 1996. Tolerance to cypermethrin and endosulfan in field populations of the bollworm (Lepidoptera: Noctuidae) from Texas. *J. Econ. Entomol.* 89: 583–589.

Kanga, L. H. B., F. W. Plapp, Jr., G. W. Elzen, M. L. Wall, and J. D. Lopez, Jr. 1995. Monitoring for resistance to organophosphorus, carbamate, and cyclodiene insecticides in tobacco budworm adults (Lepidoptera: Noctuidae). *J. Econ. Entomol.* 88: 1144–1149.

Kard, B. M. R. and F. P. Hain. 1990. Flight patterns and white grub population densities of three beetle species (Coleoptera: Scara-baeidae) in the mountains of northwestern North Carolina. *J. Entomol. Sci.* 25: 34–43.

Kard, B. M. R., F. P. Hain, and W. M. Brooks. 1988. Field suppression of three white grub species (Coleoptera: Scarabaeidae) by the entomogenous nematodes *Steinernema feltiae* and *Heterorhabditis heliothidis*. *J. Econ. Entomol.* 81: 1033–1039.

Kareiva, P. 1985. Finding and losing host plants by *Phyllotreta*: patch size and surrounding habitat. *Ecology* 66: 1809–1816.

Kareiva, P. and R. Sahakian. 1990. Tritrophic effects of a simple architectural mutation in pea plants. *Nature* 345: 433–434.

Kaster, L. V. and W. B. Showers. 1982. Evidence of spring immigration and autumn reproductive diapause of the adult black cutworm in Iowa. *Environ Entomol.* 11: 306–312.

Katsoyannos, B. I. 1989. Response to shape, size and color. Pages 307–324 in A. S. Robinson and G. Hooper (eds.). Fruit Flies: their Biology, Natural Enemies and Control, Vol. 3a. Elsevier, New York.

Kaufmann, T. 1968. A laboratory study of feeding habits of *Melanoplus differentialis* in Maryland (Orthoptera: Acrididae). *Ann. Entomol. Soc. Am.* 61: 173–180.

Kawada, K. and T. Murai. 1979. Apterous males and holocyclic reproduction of *Lipaphis erysimi* in Japan. *Entomol. Exp. Appl.* 26: 343–345.

Kawai, A. and C. Kitamura. 1987. Studies on population ecology of *Thrips palmi* Karny XV. Evaluation of effectiveness of control methods using a simulation model. *Appl. Entomol. Zool.* 22: 292–302.

Kawate, M. K. and J. A. Coughlin. 1995. Increased green onion yields associated with abamectin treatments for *Liriomyza sativae* (Diptera: Agromyzidae) and *Thrips tabaci* (Thysanoptera: Thripidae). *Proc. Hawaiian Entomol. Soc.* 32: 103–112.

Kaya, H. K. 1985. Susceptibility of early larval stages of *Pseudaletia unipunctata* and *Spodoptera exigua* (Lepidoptera: Noctuidae) to the entomogenous nematode *Steinernema feltiae* (Rhabditida: Steinernematidae). *J. Invertebr. Pathol.* 46: 58–62.

Kearns, R. S. and R. T. Yamamoto. 1981. Maternal behavior and alarm response in the eggplant lace bug, *Gargaphia solani* Heidemann (Tingidae: Heteroptera). *Psyche* 88: 215–230.

Keaster, A. J., M. A. Jackson, E. Levine, J. J. Tollefson, and F. T. Turpin. 1987. Monitoring adult *Melanotus* (Coleoptera: Elateridae) in the Midwest with the pheromone of tufted apple budmoth (*Platynota idaeusalis* Walker) (Lepidoptera: Tortricidae). *J. Kansas Entomol. Soc.* 60: 576–577.

Keep, E. and J. B. Briggs. 1971. A survey of *Ribes* species for aphid resistance. *Ann. Appl. Biol.* 68: 23–30.

Keever, D. W. and L. D. Cline. 1983. Effect of light trap height and light source on the capture of *Cathartus quadricollis* (Guerin-Meneville) (Coleoptera: Cucujidae) and *Callosobruchus maculatus* (F.) (Coleoptera: Bruchidae) in a warehouse. *J. Econ. Entomol.* 76: 1080–1082.

Keifer, H. H. 1946. A review of North American economic eriophyid mites. *J. Econ. Entomol.* 39: 563–570.

Kelleher, J. S. 1958. Life-history and ecology of *Hylemya planipalpis* (Stein) (Diptera: Anthomyiidae), a root maggot attacking radish in Manitoba. *Can. Entomol.* 90: 675–680.

Keller, J. E. and J. R. Miller. 1990. Onion fly oviposition as influenced by soil temperature. *Entomol. Exp. Appl.* 54: 37–42.

Kelly, E. O. G. and T. S. Wilson. 1918. Controlling the garden webworm in alfalfa fields. *USDA Farmers' Bull.* 944. 7 pp.

Kelly, J. W., P. H. Adler, D. R. Decoteau, and S. Lawrence. 1989. Colored reflective surfaces to control whitefly on poinsettia. *HortScience* 24: 1045.

Kelsheimer, E. G. 1949. Control of insect pests of cucumber and squash. Florida Agric. Exp. Stn. Bull. 465. 15 pp.

Kelsheimer, E. G., N. C. Hayslip, and J. W. Wilson. 1950. Control of budworms, earworms and other insects attacking sweet corn and green corn in Florida. Florida Agric. Exp. Stn. Bull. 466. 38 pp.

Kelton, L. A. 1975. The Lygus bugs (Genus *Lygus* Hahn) of North America (Heteroptera: Miridae). Mem. Entomol. Soc. Can. 95. 65 pp.

Kendall, D. M. and L. B. Bjostad. 1990. Phytohormone ecology: herbivory by *Thrips tabaci* induces greater ethylene production in intact onions than mechanical damage alone. *J. Chem. Ecol.* 16: 981–991.

Kendall, D. M. and J. L. Capinera. 1987. Susceptibility of onion growth stages to onion thrips (Thysanoptera: Thripidae) damage and mechanical defoliation. *Environ Entomol.* 16: 859–863.

Kendall, D. M. and J. L. Capinera. 1990. Geographic and temporal variation in the sex ratio of onion thrips. *Southwestern Entomol.* 15: 80–88.

Kendall, D. M., P. G. Kevan, and J. D. LaFontaine. 1981. Nocturnal flight activity of moths (Lepidoptera) in alpine tundra. *Can. Entomol.* 113: 607–614.

Kennedy, G. G. 1975. Trap design and other factors influencing capture of male potato tuberworm moths by virgin female baited traps. *J. Econ. Entomol.* 68: 305–308.

Kennedy, G. G. and M. F. Abou-Ghadir. 1979. Bionomics of the turnip aphid on two turnip cultivars. *J. Econ. Entomol.* 72: 754–757.

Kennedy, G. G. and E. R. Oatman. 1976. *Bacillus thuringiensis* and pirimicarb: selective insecticides for use in pest management on broccoli. *J. Econ. Entomol.* 69: 767–772.

Kennedy, J. S., M. F. Day, and V. F. Eastop. 1962. A Conspectus of Aphids as Vectors of Plant Viruses. Commonwealth Institute of Entomology, London. 114 pp.

Khaemba, B. M. and M. W. Ogenga-Latigo. 1985. Effects of the interaction of two levels of the black bean aphid, *Aphis fabae* Scopoli (Homoptera: Aphididae), and four stages of plant growth and development on the performance of the common bean, *Phaseolus vulgaris* L., under greenhouse conditions in Kenya. *Insect Sci. Applic.* 6: 645–648.

Khalsa, M. S., M. Kogan, and W. H. Luckmann. 1979. *Autographa precationis* in relation to soybean: life history, and food intake and utilization under controlled conditions. *Environ Entomol.* 8: 117–122.

Khan, A., W. M. Brooks, and H. Hirschmann. 1976. *Chromonema heliothidis* n. gen., n. sp. (Steinernematidae, Nematoda), a parasite of *Heliothis zea* (Noctuidae, Lepidoptera), and other insects. *J. Nematol.* 8: 159–168.

Khattat, A. R. and R. K. Stewart. 1975. Damage by tarnished plant bug to flowers and setting pods of green beans. *J. Econ. Entomol.* 68: 633–635.

Khattat, A. R. and R. K. Stewart. 1977. Development and survival of *Lygus lineolaris* exposed to different laboratory rearing conditions. *Ann. Entomol. Soc. Am.* 70: 274–278.

Khattat, A. R. and R. K. Stewart. 1980. Population fluctuations and interplant movements of *Lygus lineolaris*. *Ann. Entomol. Soc. Am.* 73: 282–287.

Khomyakova, V. O., E. P. Bykova, and V. S. Uzikhina. 1986. Effect of food and photoperiodic conditions on development of beet webworm *Loxostege sticticalis* L. (Lepidoptera, Pyralidae). *Entomol. Rev.* 65: 90–96.

Kieckhefer, R. W. 1984. Cereal aphid (Homoptera: Aphididae) preferences for and reproduction on some warm-season grasses. *Environ Entomol.* 13: 888–891.

Kieckhefer, R. W. and R. D. Gustin. 1967. Cereal aphids in South Dakota. I. Observations of autumnal bionomics. *Ann. Entomol. Soc. Am.* 60: 514–516.

Kim, T. H. and C. J. Eckenrode. 1983. Establishment of a laboratory colony and mass rearing of *Delia florilega* (Diptera: Anthomyiidae). *J. Econ. Entomol.* 76: 1467–1469.

Kim, T. H. and C. J. Eckenrode. 1984. Separation of *Delia florilega* from *D. platura* (Diptera: Anthomyiidae). *Ann. Entomol. Soc. Am.* 77: 414–416.

Kim, T. H. and C. J. Eckenrode. 1987. Bionomics of the bean seed maggot, *Delia florilega* (Diptera: Anthomyiidae), under controlled conditions. *Environ Entomol.* 16: 881–886.

Kim, T. H., C. J. Eckenrode, and M. H. Dickson. 1985. Resistance in beans to bean seed maggots (Diptera: Anthomyiidae). *J. Econ. Entomol.* 78: 133–137.

King, A. B. S. and J. L. Saunders. 1984. The Invertebrate Pests of Annual Food Crops in Central America. Overseas Development Administration, London. 166 pp.

King, E. G. and R. J. Coleman. 1989. Potential for biological control of *Heliothis* species. *Annu. Rev. Entomol.* 34: 53–75.

King, E. G. and G. G. Hartley. 1985. *Heliothis virescens*. Pages 323–328 *in* P. Singh and R. F. Moore (eds.). Handbook of Insect Rearing, Vol. II. Elsevier, New York.

King, E. G., J. Sanford, J. W. Smith, and D. F. Martin. 1981. Augmentative release of *Lixophaga diatraeae* (Dip.: Tachinidae) for suppression of early-season sugarcane borer populations in Louisiana. *Entomophaga* 26: 59–69.

King, E. W. 1969. Determination of the optimum rainfall conditions for spring emergence of the carrot beetle, *Bothynus gibbosus* (Coleoptera: Scarabaeidae). *Ann. Entomol. Soc. Am.* 62: 1336–1339.

King, E. W. 1981. Rates of feeding by four lepidopterous defoliators of soybeans. *J. Georgia Entomol. Soc.* 16: 283–288.

King, K. M. 1926. The red-backed cutworm and its control in the Prairie Provinces. Canada Dept. Agric. Pamphlet 69 (N.S.). 13 pp.

King, K. M. 1928a. Economic importance of wireworms and false wireworms in Saskatchewan. *Sci. Agric.* 8: 693–706.

King, K. M. 1928b. *Barathra configurata* Wlk., an armyworm with important potentialities on the northern prairies. *J. Econ. Entomol.* 21: 279–293.

King, K. M. and N. J. Atkinson. 1928. The biological control factors of the immature stages of *Euxoa ochrogaster* Gn. (Lepidoptera, Phalaenidae) in Saskatchewan. *Ann. Entomol. Soc. Am.* 21: 167–188.

Kinoshita, G. B., C. R. Harris, H. J. Svec, and F. L. McEwen. 1978. Laboratory and field studies on the chemical control of the crucifer flea beetle, *Phyllotreta cruciferae* (Coleoptera: Chrysomelidae) on cruciferous crops in Ontario. *Can. Entomol.* 110: 795–803.

Kinoshita, G. B., H. J. Svec, C. R. Harris, and F. L. McEwen. 1979. Biology of the crucifer flea beetle, *Phyllotreta cruciferae* (Coleoptera: Chrysomelidae), in southwestern Ontario. *Can. Entomol.* 111: 1395–1407.

Kinzer, R. E., S. Young, and R. R. Walton. 1972. Rearing and testing tobacco thrips in the laboratory to discover resistance in peanuts. *J. Econ. Entomol.* 65: 782–785.

Kirby, R. D. and J. E. Slosser. 1984. Composite economic threshold for three lepidopterous pests of cabbage. *J. Econ. Entomol.* 77: 725–733.

Kirfman, G. W., A. J. Keaster, and R. N. Story. 1986. An improved wireworm (Coleoptera: Elateridae) sampling technique for midwest cornfields. *J. Kansas Entomol. Soc.* 59: 37–41.

Kirk, M., S. R. Temple, C. G. Summers, and L. T. Wilson. 1991. Transmission efficiencies of field-collected aphid (Homoptera: Aphididae) vectors of beet yellows virus. *J. Econ. Entomol.* 84: 638–643.

Kirk, V. M. 1957a. Maize billbug control in South Carolina. South Carolina Agric. Exp. Stn. Bull. 452. 29 pp.

Kirk, V. M. 1957b. Preplanting treatment for billbug control on corn. *J. Econ. Entomol.* 50: 707–709.

Kirk, V. M. 1975. Biology of *Stenolophus* (= *Agnoderus*) *comma*, a ground beetle of cropland. *Ann. Entomol. Soc. Am.* 68:135–138.

Kirk, V. M. 1981. Base of corn stalks as oviposition sites for western and northern corn rootworms (Diabrotica: Coleoptera). *J. Kansas Entomol. Soc.* 54:255–262.

Kitch, L. W., G. Ntoukam, R. E. Shade, J. L. Wolfson, and L. L. Murdock. 1992. A solar heater for disinfesting stored cowpeas on subsistence farms. *J. Stored Prod. Res.* 28:261–267.

Kjaer-Pedersen, C. 1992. Flight behaviour of the cabbage seedpod weevil. *Entomol. Exp. Appl.* 62:61–66.

Klahn, S. A. 1987. Cantharidin in the natural history of the Meloidae (Coleoptera). Unpublished M. S. Thesis, Colorado State University, Ft. Collins. 136 pp.

Klein, M. G. and R. Georgis. 1992. Persistence of control of Japanese beetle (Coleoptera: Scarabaeidae) larvae with steinernematid and heterorhabditid nematodes. *J. Econ. Entomol.* 85:727–730.

Kloen, H. and M. A. Altieri. 1990. Effect of mustard (*Brassica hirta*) as a non-crop plant on competition and insect pests in broccoli (*Brassica oleracea*). *Crop Prot.* 9:90–96.

Klostermeyer, E. C. 1942. The life history and habits of the ringlegged earwig, *Euborellia annulipes* (Lucus) (Order Dermaptera). *J. Kansas Entomol. Soc.* 15:13–18.

Klostermeyer, L. E. 1985. Japanese beetle (Coleoptera: Scarabaeidae) traps: comparison of commercial and homemade traps. *J. Econ. Entomol.* 78:454–459.

Klun, J. A. and J. F. Robinson. 1971. European corn borer moth: sex attractant and sex attraction inhibitors. *Ann. Entomol. Soc. Am.* 64:1083–1086.

Klun, J. A., C. L. Tipton, and T. A. Brindley. 1967. 2,4-Dihydroxy-7-methoxy-1,4-benzoxazin-3-one (DIMBOA), an active agent in the resistance of maize to the European corn borer. *J. Econ. Entomol.* 60:1529–1533.

Klun, J. A., M. Schwarz, B. A. Leonhardt, and W. W. Cantelo. 1990. Sex pheromone of the female squash vine borer (Lepidoptera: Sesiidae). *J. Entomol. Sci.* 25:64–72.

Klun, J. A., C. C. Blickenstaff, M. Schwarz, B. A. Leonhardt, and J. R. Plimmer. 1983. Western bean cutworm, *Loxagrotis albicosta* (Lepidoptera: Noctuidae): female sex pheromone identification. *Environ Entomol.* 12:714–717.

Klun, J. A., K. F. Haynes, B. A. Bierl-Leonhardt, M. C. Birch, and J. R. Plimmer. 1981. Sex pheromone of the female artichoke plume moth, *Platyptilia carduidactyla. Environ Entomol.* 10:763–765.

Klun, J. A., B. A. Leonhardt, M. Schwartz, A. Day, and A. K. Raina. 1986. Female sex pheromone of the pickleworm. *J. Chem. Ecol.* 12:239–249.

Knight, H. H. 1941. The plant bugs, or Miridae, of Illinois. Illinois Nat. Hist. Surv. Bull. 22. 234 pp.

Knowles, C. O., D. D. Errampalli, and G. N. El–Sayed. 1988. Comparative toxicities of selected pesticides to bulb mite (Acari: Acaridae) and twospotted spider mite (Acari: Tetranychidae). *J. Econ. Entomol.* 81:1586–1591.

Knowlton, G. F. 1944. Pentatomidae eaten by Utah birds. *J. Econ. Entomol.* 37:118–119.

Knowlton, G. F. and M. J. Janes. 1931. Studies on the biology of *Paratrioza cockerelli* (Sulc). *Ann. Entomol. Soc. Am.* 24:283–291.

Knowlton, G. F., D. R. Maddock, and S. L. Wood. Insect food of the sagebrush swift. *J. Econ. Entomol.* 39:382–383.

Knutson, A. E. and F. E. Gilstrap. 1989a. Direct evaluation of natural enemies of the southwestern corn borer (Lepidoptera: Pyralidae) in Texas corn. *Environ Entomol.* 18:732–739.

Knutson, A. E. and F. E. Gilstrap. 1989b. Predators and parasites of the southwestern corn borer (Lepidoptera: Pyralidae) in Texas corn. *J. Kansas Entomol. Soc.* 62:511–520.

Knutson, A. E. and F. E. Gilstrap. 1990. Life tables and population dynamics of the southwestern corn borer (Lepidoptera: Pyralidae) in Texas corn. *Environ Entomol.* 19:684–696.

Knutson, A. E., G. B. Cronholm, and W. P. Morrison. 1982. Seasonal occurrence of southwestern corn borer adults in the Texas High Plains. *Southwestern Entomol.* 7:159–165.

Knutson, A. E., F. M. Davis, T. G. Bird, and W. P. Morrison. 1987. Monitoring southwestern corn borer, *Diatraea grandiosella* Dyar, with pheromone-baited traps. *Southwestern Entomol.* 12:65–71.

Knutson, A. E., J. A. Jackman, G. B. Cronholm, S.-S. Ng, F. M. Davis, and W. P. Morrison. 1989. Temperature-dependent model for predicting emergence of adult southwestern corn borer (Lepidoptera: Pyralidae) in Texas. *J. Econ. Entomol.* 82:1230–1236.

Knutson, H. 1944. Minnesota Phalaenidae (Noctuidae), the seasonal history and economic importance of the more common and destructive species. Minnesota Agric. Exp. Stn. Bull. 165. 128 pp.

Kochansky, J., R. T. Carde, J. Liebherr, and W. L. Roelofs. 1975. Sex pheromones of the European corn borer in New York. *J. Chem Ecol.* 1:225–231.

Kodet, R. T., M. W. Nielson, and R. O. Kuehl. 1982. Effect of temperature and photoperiod on the biology of blue alfalfa aphid, *Acyrthosiphon kondoi* Shinji. USDA Tech. Bull. 1660. 10 pp.

Kogan, J., D. K. Sell, R. E. Stinner, J. R. Bradley, Jr., and M. Kogan. 1978. The literature of arthropods associated with soybean. V. A bibliography of *Heliothis zea* (Boddie) and *H. virescens* (F.) (Lepidoptera: Noctuidae). International Soybean Program Series 17. 240 pp.

Kogan, M., G. P. Waldbauer, G. Boiteau, and C. E. Eastman. 1980. Sampling bean leaf beetles on soybean. Pages 201–236 *in* M. Kogan and D. C. Herzog (eds.). Sampling Methods in Soybean Entomology. Springer-Verlag, New York.

Koinzan, S. D. and K. P. Pruess. 1975. Effects of a wide-area application of ULV malathion on leafhoppers in alfalfa. *J. Econ. Entomol.* 68:267–268.

Kok, L. T. and T. J. McAvoy. 1989. Fall broccoli pests and their parasites in Virginia. *J. Entomol. Sci.* 24:258–265.

Kono, T. and C. S. Papp. 1977. Handbook of Agricultural Pests. Aphids, Thrips, Mites, Snails, and Slugs. California Dept. Agric. Div. Plant Indust., Sacramento. 205 pp.

Kostal, V. and S. Finch. 1994. Influence of background on host-plant selection and subsequent oviposition by the cabbage root fly (*Delia radicum*). *Entomol. Exp. Appl.* 70:153–163.

Koul, O. and M. B. Isman. 1991. Effects of azadirachtin on the dietary utilization and development of the variegated cutworm *Peridroma saucia*. *J. Insect Physiol.* 37:591–598.

Koyama, J., T. Teruya, and K. Tanaka. 1984. Eradication of the oriental fruit fly (Diptera: Tephritidae) from the Okinawa Islands by a male annihilation method. *J. Econ. Entomol.* 77:468–472.

Kozlowski, M. W., S. Lux, and J. Dmoch. 1983. Oviposition behaviour and pod marking in the cabbage seed weevil, *Ceutorhynchus assimilis*. *Entomol. Exp. Appl.* 34:277–282.

Kraemer, M. E., T. Mebrahtu, and M. Rangappa. 1994. Evaluation of vegetable soybean genotypes for resistance to Mexican bean beetle (Coleoptera: Coccinellidae). *J. Econ. Entomol.* 87:252–257.

Krasnoff, S. B. and D. D. Yager. 1988. Acoustic response to a pheromonal cue in the arctiid moth *Pyrrharctia isabella*. *Physiol. Entomol.* 13:433–440.

Krasnoff, S. B., L. B. Bjostad, and W. L. Roelofs. 1987. Quantitative and qualitative variation in male pheromones of *Phragmatobia fuliginosa* and *Pyrrharctia isabella* (Lepidoptera: Arctiidae). *J. Chem Ecol.* 13:807–822.

Kring, J. B. 1955. Biological separation of *Aphis gossypii* Glover and *Aphis sedi* Kaltenbach. *Ann. Entomol. Soc. Am.* 48:442–444.

Kring, J. B. 1958. Feeding behavior and DDT resistance of *Epitrix cucumeris* (Harris). *J. Econ. Entomol.* 51: 823–828.

Kring, J. B. 1959. The life cycle of the melon aphid, *Aphis gossypii* Glover, an example of facultative migration. *Ann. Entomol. Soc. Am.* 52: 284–286.

Kring, T. J. 1985. Key and diagnosis of the instars of the corn leaf aphid *Rhopalosiphum maidis* (Fitch). *Southwestern Entomol.* 10: 289–293.

Kring, T. J., J. R. Ruberson, D. C. Steinkraus, and D. A. Jacobson. 1993. Mortality of *Helicoverpa zea* (Lepidoptera: Noctuidae) pupae in ear-stage field corn. *Environ Entomol.* 22: 1338–1343.

Krombein, K. V., P. D. Hurd, Jr., D. R. Smith, and B. D. Burks. 1979. Catalog of Hymenoptera in America North of Mexico. Smithsonian Institution Press, Washington, D. C. 273 pp.

Kroschel, J., H. J. Kaack, E. Fritsch, and J. Huber. 1996. Biological control of the potato tuber moth (*Phthorimaea operculella* Zeller) in the Republic of Yemen using granulosis virus: propagation and effectiveness of the virus in field trials. *Biocontrol Sci. Tech.* 6: 217–226.

Krueger, S. R., J. R. Nechols, and W. A. Ramoska. 1992. Habitat distribution and infection rates of the fungal pathogen, *Beauveria bassiana* (Balsamo) Vuillemin in endemic populations of the chinch bug, *Blissus leucopterus* (Say) (Hemiptera: Lygaeidae) in Kansas. *J. Kansas Entomol. Soc.* 65: 115–124.

Krysan, J. L. 1986. Introduction: biology, distribution, and identification of pest *Diabrotica*. Pages 1–23 in J. L. Krysan and T. A. Miller (eds.). Methods for the Study of Pest *Diabrotica*. Springer-Verlag, New York.

Krysan, J. L. and T. F. Branson. 1983. Biology, ecology, and distribution of Diabrotica. Pages 144–150 *in* D. T. Gordon, J. K. Knoke, L. R. Nault, and R. M. Ritter (eds.). Proceedings International Maize Virus Disease Colloquium and Workshop, 2–6 August 1982. Ohio State Univ., Ohio Agric. Res. Devt. Ctr., Wooster. 266 pp.

Krysan, J. L., R. F. Smith, and P. L. Guss. 1983. *Diabrotica barberi* (Coleoptera: Chrysomelidae) elevated to species rank based on behavior, habitat choice, morphometrics, and geographical variation of color. *Ann. Entomol. Soc. Am.* 76: 197–204.

Krysan, J. L., R. F. Smith, T. F. Branson, and P. L. Guss. 1980. A new subspecies of *Diabrotica virgifera* (Coleoptera: Chrysomelidae): description, distribution, and sexual compatibility. *Ann. Entomol. Soc. Am.* 73: 123–130.

Kuba, H. and J. Koyama. 1985. Mating behavior of wild melon flies, *Dacus cucurbitae* Coquillett (Diptera: Tephritidae) in a field cage: courtship behavior. *Appl. Entomol. Zool.* 20: 365–372.

Kuba, H. and Y. Sokei. 1988. The production of pheromone clouds by spraying in the melon fly, *Dacus cucurbitae* Coquillett (Diptera: Tephritidae). *J. Ethol.* 6: 105–110.

Kuitert, L. C. and R. V. Connin. 1952. Biology of the American grasshopper in the southeastern United States. *Florida Entomol.* 35: 22–33.

Kundu, R. and A. F. G. Dixon. 1994. Feeding on their primary host by the return migrants of the host-alternating aphid, *Cavariella aegopodii*. *Ecol. Entomol.* 19: 83–86.

Ladd, Jr., T. L. 1984. Eugenol-related attractants for the northern corn rootworm (Coleoptera: Chrysomelidae). *J. Econ. Entomol.* 77: 339–341.

Ladd, Jr., T. L. and M. G. Klein. 1982. Japanese beetle (Coleoptera: Scarabaeidae): effect of trap height on captures. *J. Econ. Entomol.* 75: 746–747.

Ladd, Jr., T. L. and W. A. Rawlins. 1965. The effects of the feeding of the potato leafhopper on photosynthesis and respiration in the potato plant. *J. Econ. Entomol.* 58: 623–628.

LaFontaine, J. D. 1987. The Moths of America North of Mexico. Fascicle 27.2: Noctuoidea, Noctuidae (part). Wedge Entomological Research Foundation, Washington, D.C. 237 pp.

LaFontaine, J. D. and R. W. Poole. 1991. The Moths of America North of Mexico. Fasicle 25.1: Noctuoidea, Noctuidae (part). Wedge Entomological Research Foundation, Washington D.C. 182 pp.

Lafrance, J. 1967. The life history of *Agriotes mancus* (Say) (Coleoptera: Elateridae) in the organic soils in southwestern Quebec. *Phytoprotection* 48: 53–57.

Lafrance, J. and J. P. Perron. 1959. Notes on life-history of the onion maggot, *Hylemya antiqua* (Meig.) (Diptera: Anthomyiidae), in sandy and organic soils. *Can. Entomol.* 91: 633–638.

Laing, J. E. 1969. Life history and life table of *Tetranychus urticae* Koch. *Acarologia* 11: 32–42.

Lamb, R. J. 1975. Effects of dispersion, travel, and environmental heterogeneity on populations of the earwig *Forficula auricularia* L. *Can. J. Zool.* 53: 1855–1867.

Lamb, R. J. 1976. Parental behavior in the Dermaptera with special reference to *Forficula auricularia* (Dermaptera: Forficulidae). *Can. Entomol.* 108: 609–619.

Lamb, R. J. 1983. Phenology of flea beetle (Coleoptera: Chrysomelidae) flight in relation to their invasion of canola fields in Manitoba. *Can. Entomol.* 115: 1493–1502.

Lamb, R. J. and W. G. Wellington. 1974. Techniques for studying the behavior and ecology of the European earwig, *Forficula auricularia* (Dermaptera: Forficulidae). *Can. Entomol.* 106: 881–888.

Lamb, R. J. and W. G. Wellington. 1975. Life history and population characteristics of the European earwig, *Forficula auricularia* (Dermaptera: Forficulidae), at Vancouver, British Columbia. *Can. Entomol.* 107: 819–824.

Lamb, R. J., G. H. Gerber, and G. F. Atkinson. 1984. Comparison of developmental rate curves applied to egg hatching data of *Entomoscelis americana* Brown (Coleoptera: Chrysomelidae). *Environ Entomol.* 13: 868–872.

Lamb, R. J., W. J. Turnock, and H. N. Hayhoe. 1985. Winter survival and outbreaks of bertha armyworm, *Mamestra configurata* (Lepidoptera: Noctuidae), on canola. *Can. Entomol.* 117: 727–736.

Lambert, J. D. H., J. T. Arnason, A. Serratos, B. J. R. Philogene, and M. A. Faris. 1987. Role of intercropped red clover in inhibiting European corn borer (Lepidoptera: Pyralide) damage to corn in eastern Ontario. *J. Econ. Entomol.* 80: 1192–1196.

Lamp, W. O., M. J. Morris, and E. J. Armbrust. 1984. Suitability of common weed species as host plants for the potato leafhopper, *Empoasca fabae*. *Entomol. Exp. Appl.* 36: 125–131.

Lamp, W. O., G. R. Nielsen, and S. Danielson. 1994. Patterns among host plants of potato leafhopper, *Empoasca fabae* (Homoptera: Cicadellidae). *J. Kansas Entomol. Soc.* 67: 354–368.

Lampert, E. P., D. C. Cress, and D. L. Haynes. 1984. Temporal and spatial changes in abundance of the asparagus miner, *Ophiomyia simplex* (Loew) (Diptera: Agromyzidae), in Michigan. *Environ Entomol.* 13: 733–736.

Lampman, R. L. and R. L. Metcalf. 1987. Multicomponent kairomonal lures for southern and western corn rootworms (Coleoptera: Chrysomelidae: *Diabrotica* spp.). *J. Econ. Entomol.* 80: 1137–1142.

Lamson, Jr., G. H. 1922. The rose chafer as a cause of death of chickens. *Connecticut Agric. Exp. Stn. Bull.* 110: 117–134.

Lance, D. R. 1988. Potential of 8-methyl-2-decyl propanoate and plant-derived volatiles for attracting corn rootworm beetles (Coleoptera: Chrysomelidae) to toxic bait. *J. Econ. Entomol.* 81: 1359–1362.

Lance, D. R. and J. R. Fisher. 1987. Food quality of various plant tissues for adults of the northern corn rootworm (Coleoptera: Chrysomelidae). *J. Kansas Entomol. Soc.* 60: 462–466.

Lance, D. R. and G. R. Sutter. 1990. Field-cage and laboratory evaluations of semiochemical-based baits for managing western corn rootworm (Coleoptera: Chrysomelidae). *J. Econ. Entomol.* 83: 1085–1090.

Lanchester, H. P. 1946. Larval determination of six economic species of *Limonius* (Coleoptera: Elateridae). *Ann. Entomol. Soc. Am.* 39: 619–626.

Landis, B. J. and N. F. Howard. 1940. *Paradexodes epilachnae*, a tachinid parasite of the Mexican bean beetle. USDA Tech. Bull. 721. 32 pp.

Landis, B. J., D. M. Powell, and L. Fox. 1972. Overwintering and winter dispersal of the potato aphid in eastern Washington. *Environ Entomol.* 1: 68–71.

Landis, D. A., E. Levine, M. J. Haas, and V. Meints. 1992. Detection of prolonged diapause of northern corn rootworm in Michigan (Coleoptera: Chrysomelidae). *Great Lakes Entomol.* 25: 215–222.

Landolt, P. J. 1990. Trapping the green June beetle (Coleoptera: Scarabaeidae) with isopropanol. *Florida Entomol.* 73: 328–330.

Landolt, P. J. 1995. Attraction of *Mocis latipes* (Lepidoptera: Noctuidae) to sweet baits in traps. *Florida Entomol.* 73: 523–530.

Landolt, P. J. 2000. New chemical attractants for trapping *Lacanobia subjuncta, Mamestra configurata*, and *Xestia c-nigrum* (Lepidoptera: Noctuidae) (Lepidoptera: Noctuidae). *J. Econ. Entomol.* 93: 101–106.

Landolt, P. J. and R. R. Heath. 1989. Lure composition, component ratio, and dose for trapping male *Mocis latipes* (Lepidoptera: Noctuidae) with synthetic sex pheromone. *J. Econ. Entomol.* 82: 307–309.

Lane, M. C. 1931. The Great Basin wireworm in the Pacific Northwest. USDA Farmers' Bull. 1657. 8 pp.

Lange, C. E., C. M. MacVean, J. E. Henry, and D. A. Streett. 1995. *Heterovesicula cowani* N. G., N. Sp. (Heterovesiculidae N. Fam.), a microsporidian parasite of Mormon crickets, *Anabrus simplex* Haldeman, 1852 (Orthoptera: Tettigoniidae). *J. Euk. Microbiol.* 42: 552–558.

Lange, Jr., W. H. 1941. The artichoke plume moth and other pests injurious to the globe artichoke. *California Agric.* Exp. Stn. Bull. 653. 71 pp.

Lange, W. H. 1945. *Autographa egena* (Guen.) a periodic pest of beans in California. *Pan-Pac. Entomol.* 21: 13.

Lange, Jr., W. H. 1950. Biology and systematics of plume moths of the genus *Platyptilia* in California. *Hilgardia* 19: 561–668.

Lange, W. H., A. A. Grigarick, and E. C. Carlson. 1957. Serpentine leaf miner damage. *California Agric.* 11(3): 3–5.

Langford, G. S. 1930. Some factors relating to the feeding habits of grasshoppers. Colorado Agric. Exp. Stn. Bull. 354. 53 pp.

Langford, G. S. 1933. Observations on cultural practices for the control of the potato tuber worm, *Phthorimaea operculella* Zell. *J. Econ. Entomol.* 26: 135–137.

Langford, G. S. and E. N. Cory. 1932. Observations on the potato tuber moth. *J. Econ. Entomol.* 25: 625–635.

Langridge, W. H. R. 1983. Characterization of a cytoplasmic polyhedrosis virus from *Estigmene acrea* (Lepidoptera). *J. Invertebr. Pathol.* 42: 259–263.

Langston, R. L. and J. A. Powell. 1975. The earwigs of California (Order Dermaptera). Bull. California Insect Surv. Vol. 20. 25 pp.

Lanteri, A. A. and A. E. Marvaldi. 1995. *Graphognathus* Buchanan a new synonym of *Naupactus* Dejean and systematics of the *N. leucoloma* species group (Coleoptera: Curculionidae). Coleop. Bull. 49: 206–228.

Larew, H. G. and J. C. Locke. 1990. Repellency and toxicity of a horticultural oil against whiteflies on chrysanthemum. *HortScience* 25: 1406–1407.

Larsen, K. J., L. R. Nault, and G. Moya-Raygoza. 1992. Overwintering biology of *Dalbulus* leafhoppers (Homoptera: Cicadellidae): adult populations and drought hardiness. *Environ. Entomol.* 21: 566–577.

Larson, A. O. and C. K. Fisher. 1938. The bean weevil and the southern cowpea weevil in California. USDA Tech. Bull. 593. 71 pp.

Larson, A. O., T. A. Brindley, and F. G. Hinman. 1938. Biology of the pea weevil in the Pacific Northwest with suggestions for its control on seed peas. USDA Tech. Bull. 599. 48 pp.

Lasack, P. M. and L. P. Pedigo. 1986. Movement of stalk borer larvae (Lepidoptera: Noctuidae) from noncrop areas into corn. *J. Econ. Entomol.* 79: 1697–1702.

Lasack, J. A., W. C. Bailey, and L. P. Pedigo. 1987. Assessment of stalk borer (Lepidoptera: Noctuidae) population dynamics by using logistic development curves and partial life tables. *Environ. Entomol.* 16: 296–303.

Lasota, J. A. and L. T. Kok. 1986. Parasitism and utilization of imported cabbageworm pupae by *Pteromalus puparum* (Hymenoptera: Pteromalidae). *Environ. Entomol.* 15: 994–998.

Latheef, M. A. and D. G. Harcourt. 1974. The dynamics of *Leptinotarsa decemlineata* populations on tomato. *Entomol. Exp. Appl.* 17: 67–76.

Latheef, M. A. and R. D. Irwin. 1983. Seasonal abundance and parasitism of lepidopterous larvae on *Brassica* greens in Virginia. *J. Georgia Entomol. Soc.* 18: 164–168.

Latheef, M. A. and J. H. Ortiz. 1983a. The influence of companion herbs on egg distribution of the imported cabbageworm, *Pieris rapae* (Lepidoptera: Pieridae), on collards plants. *Can. Entomol.* 115: 1031–1038.

Latheef, M. A. and J. H. Ortiz. 1983b. Influence of companion plants on oviposition of imported cabbageworm, *Pieris rapae* (Lepidoptera: Pieridae), and cabbage looper, *Trichoplusia ni* (Lepidoptera: Noctuidae), on collards plants. *Can. Entomol.* 115: 1529–1531.

Latheef, M. A. and J. H. Ortiz. 1984. Influence of companion herbs on *Phyllotreta cruciferae* (Coleoptera: Chrysomelidae) on collards plants. *J. Econ. Entomol.* 77: 80–82.

Latheef, M. A., J. H. Ortiz, and A. Q. Sheikh. 1984. Influence of intercropping on *Phyllotreta cruciferae* (Coleoptera: Chrysomelidae) populations on collards plants. *J. Econ. Entomol.* 77: 1180–1184.

Latin, R. X. and G. L. Reed. 1985. Effect of root feeding by striped cucumber beetle larvae on the incidence and severity of Fusarium wilt of muskmelon. *Phytopathology* 75: 209–212.

Latta, R. 1939. Observations on the nature of bulb mite attack on Easter lilies. *J. Econ. Entomol.* 32: 125–128.

Latta, R. and F. R. Cole. 1933. A comparative study of the species of *Eumerus* known as the lesser bulb flies. California Dept. Agric. Mon. Bull. 22: 142–152.

Laub, C. A. and J. M. Luna. 1992. Winter cover crop suppression practices and natural enemies of armyworm (Lepidoptera: Noctuidae) in no-till corn. *Environ Entomol.* 21: 41–49.

Lavoipierre, M. M. J. 1940. *Hemitarsonemus latus* (Banks) (Acarina), a mite of economic importance new to South Africa. *J. Entomol. Soc. So. Africa.* 3: 116–123.

Lawson, D. D. and G. T. Weekman. 1966. A method of recovering eggs of the western corn rootworm from the soil. *J. Econ. Entomol.* 59: 657–659.

Lawson, F. R. 1959. The natural enemies of the hornworms on tobacco (Lepidoptera: Sphingidae). *Ann. Entomol. Soc. Am.* 52: 741–755.

Lawson, F. R., R. L. Rabb, F. E. Guthrie, and T. G. Bowery. 1961. Studies of an integrated control system for hornworms on tobacco. *J. Econ. Entomol.* 54: 93–97.

Launchbaugh, J. L. and C. E. Owensby. 1982. Transient depredation of early spring range; spotted cutworms [*Amathes c-nigrum* (L.)] as a possible cause. *J. Range Manage.* 35: 538–539.

Layland, J. K., M. Upton, and H. H. Brown. 1994. Monitoring and identification of *Thrips palmi* Karny (Thysanoptera: Thripidae). *J. Aust. Entomol. Soc.* 33: 169–173.

Leather, S. R. 1981. Factors affecting egg survival in the bird cherry-oat aphid, *Rhopalosiphum padi*. *Entomol. Exp. Appl.* 30: 197–199.

Leather, S. R. and A. F. G. Dixon. 1982. Secondary host preferences and reproductive activity of the bird cherry-oat aphid, *Rhopalosiphum padi*. *Ann. Appl. Biol.* 101: 219–228.

Leather, S. R., K. F. A. Walters, and A. F. G. Dixon. 1989. Factors determining the pest status of the bird cherry-oat aphid, *Rhopalosiphum padi* (L.) (Hemiptera: Aphididae), in Europe: a study and review. Bull. *Entomol. Res.* 79: 345–360.

Lederhouse, R. C., S. G. Codella, and P. J. Cowell. 1987. Diurnal predation on roosting butterflies during inclement weather: a substantial source of mortality in the black swallowtail, *Papilio polyxenes* (Lepidoptera: Papilionidae). *J. New York Entomol. Soc.* 95: 310–319.

Lee, J. and C. J. P. Upton. 1992. Relative efficiency of insecticide treatments in reducing yield loss from *Sitona* in faba beans. *Ann. Appl. Biol. (Suppl.)* 13: 6–7.

Lee, B. L. and M. H. Bass. 1969. Rearing technique for the granulate cutworm and some effects of temperature on its life cycle. *Ann. Entomol. Soc. Am.* 62: 1216–1217.

Lee, P. E. and A. G. Robinson. 1958. Studies on the six-spotted leafhopper, *Macrosteles fascifrons* (Stål.), and aster yellows in Manitoba. *Can. J. Plant Sci.* 38: 320–327.

Legg, D. E. and H. C. Chiang. 1984. European corn borer (Lepidoptera: Pyralidae) infestations: relating captures in pheromone and black-light traps in southern Minnesota cornfields. *J. Econ. Entomol.* 77: 1445–1448.

Lei, Z., T. A. Rutherford, and J. M. Webster. 1992. Heterorhabditid behavior in the presence of the cabbage maggot, *Delia radicum*, and its host plants. *J. Nematol.* 24: 9–15.

Leibee, G. L. 1984. Influence of temperature on development and fecundity of *Liriomyza trifolii* (Burgess) (Diptera: Agromyzidae) on celery. *Environ Entomol.* 13: 497–501.

Leibee, G. L. and K. E. Savage. 1992. Evaluation of selected insecticides for control of diamondback moth and cabbage looper in cabbage in central Florida with observations on insecticide resistance in the diamondback moth. *Florida Entomol.* 75: 585–591.

Leibee, G. L. and J. L. Capinera. 1995. Pesticide resistance in Florida insects limits management options. *Florida Entomol.* 78: 386–399.

Leiby, R. W. 1920. The larger corn stalk-borer in North Carolina. North Carolina Dept. Agric. Bull. 274, Vol. 41(13). 85 pp.

Leigh, T. F. 1961. Insecticidal susceptibility of *Nysius raphanus*, a pest of cotton. *J. Econ. Entomol.* 54: 120–122.

Leigh, T. F. and D. Gonzalez. 1976. Field cage evaluation of predators for control of *Lygus hesperus* Knight on cotton. *Environ Entomol.* 5: 948–952.

Lema, K-M. and S. L. Poe. 1979. Age specific mortality of *Liriomyza sativae* due to *Chrysonotomyia formosa* and parasitization by *Opius dimidiatus* and *Chrysonotomyia formosa*. *Environ Entomol.* 8: 935–937.

Lentz, G. L. and L. P. Pedigo. 1975. Population ecology of parasites of the green cloverworm in Iowa. *J. Econ. Entomol.* 68: 301–304.

Lentz, G. L., A. Y. Chambers, and R. M. Hayes. 1983. Effects of systemic insecticide-nematicides on midseason pest and predator populations in soybean. *J. Econ. Entomol.* 76: 836–840.

Leszczynski, B., W. W. Cone, and L. C. Wright. 1986. Changes in the sugar metabolism of asparagus plants infested by asparagus aphid, *Brachycorynella asparagi* and green peach aphid, *Myzus persicae*. *J. Agric. Entomol.* 3: 25–30.

Leonard, B. R., D. J. Boethel, A. N. Sparks, Jr., M. B. Layton, J. S. Mink, A. M. Pavloff, E. Burris, and J. B. Graves. 1990. Variations in response of soybean looper (Lepidoptera: Noctuidae) to selected insecticides in Louisiana. *J. Econ. Entomol.* 83: 27–34.

Leonard, D. E. 1966. Biosystematics of the "Leucopterus complex" of the genus *Blissus* (Heteroptera: Lygaeidae). Connecticut Agric. Exp. Stn. Bull. 677. 47 pp.

Leonard, D. E. 1968. A revision of the genus *Blissus* (Heteroptera: Lygaeidae) in eastern North America. *Ann. Entomol. Soc. Am.* 61: 239–250.

Letourneau, D. K. 1986. Associational resistance in squash monocultures and polycultures in tropical Mexico. *Environ Entomol.* 15: 285–292.

Letourneau, D. K. and M. A. Altieri. 1983. Abundance patterns of a predator, *Orius tristicolor* (Hemiptera: Anthocoridae), and its prey, *Frankliniella occidentalis* (Thysanoptera: Thripidae): habitat attraction in polycultures versus monocultures. *Environ Entomol.* 12: 1464–1469.

Leuck, D. B. 1966. Biology of the lesser cornstalk borer in south Georgia. *J. Econ. Entomol.* 59: 797–801.

Levine, E. 1983. Temperature requirements for development of the stalk borer, *Papaipema nebris* (Lepidoptera: Noctuidae). *Ann. Entomol. Soc. Am.* 76: 892–895.

Levine, E. 1985. Oviposition by the stalk borer, *Papaipema nebris* (Lepidoptera: Noctuidae), on weeds, plant debris, and cover crops. *J. Econ. Entomol.* 78: 65–68.

Levine, E. 1986a. Oviposition preference of *Hydraecia immanis* (Lepidoptera: Noctuidae) in cage tests. *J. Econ. Entomol.* 79: 1544–1547.

Levine, E. 1986b. Termination of diapause and postdiapause development in eggs of the stalk borer (Lepidoptera: Noctuidae). *Environ Entomol.* 15: 403–408.

Levine, E. 1989. Forecasting *Hydraecia immanis* (Lepidoptera: Noctuidae) moth phenology based on light trap catches and degree-day accumulations. *J. Econ. Entomol.* 82: 433–438.

Levine, E. 1993. Effect of tillage practices and weed management on survival of stalk borer (Lepidoptera: Noctuidae) eggs and larvae. *J. Econ. Entomol.* 86: 924–928.

Levine, E. and A. Felsot. 1985. Effectiveness of acephate and carbofuran seed treatments to control the black cutworm, *Agrotis ipsilon* (Lepidoptera: Noctuidae), on field corn. *J. Econ. Entomol.* 78: 1415–1420.

Levine, E. and M. E. Gray. 1994. Use of cucurbitacin vial traps to predict corn rootworm (Coleoptera: Chrysomelidae) larval injury in a subsequent crop of corn. *J. Entomol. Sci.* 29: 590–600.

Levine, E. and H. Oloumi-Sadeghi. 1991. Management of diabroticite rootworms in corn. *Annu. Rev. Entomol.* 36: 229–255.

Levine, E. and H. Oloumi-Sadeghi. 1992. Field evaluation of *Steinernema carpocapsae* (Rhabditida: Steinernematidae) against black cutworm (Lepidoptera: Noctuidae) larvae in field corn. *J. Environ. Sci.* 27: 427–435.

Levine, E., S. L. Clement, and D. A. McCartney. 1984. Effect of seedling injury by the stalk borer (Lepidoptera: Noctuidae) on regrowth and yield of corn. *J. Econ. Entomol.* 77: 167–170.

Levine, E., H. Oloumi-Sadeghi, and J. R. Fisher. 1992. Discovery of multiyear diapause in Illinois and South Dakota northern corn rootworm (Coleoptera: Chrysomelidae) eggs and incidence of the prolonged diapause trait in Illinois. *J. Econ. Entomol.* 85: 262–267.

Levine, E., S. L. Clement, W. L. Rubink, and D. A. McCartney. 1983. Regrowth of corn seedlings after injury at different growth stages by black cutworm, *Agrotis ipsilon* (Lepidoptera: Noctuidae) larvae. *J. Econ. Entomol.* 76: 389–391.

Levine, E., S. L. Clement, L. V. Kaster, A. J. Keaster, W. G. Ruesink, W. B. Showers, and F. T. Turpin. 1982. Black cutworm, *Agrotis ipsilon* (Lepidoptera: Noctuidae), pheromone trapping: a regional research effort. Bull. *Entomol. Soc. Am.* 28: 139–142.

Levins, R. A., S. L. Poe, R. C. Littell, and J. P. Jones. 1975. Effectiveness of a leafminer control program for Florida tomato production. *J. Econ. Entomol.* 68: 772–774.

Levy, R. and D. H. Habeck. 1976. Descriptions of the larvae of *Spodoptera sunia* and *S. latifascia* with a key to the mature *Spodoptera* larvae of the eastern United States (Lepidoptera: Noctuidae). *Ann. Entomol. Soc. Am.* 69: 585–588.

Lewis, L. C., R. D. Gunnarson, and J. E. Cossentine. 1982. Pathogenicity of *Vairimorpha necatrix* (Microsporidia: Nosematidae) against *Ostrinia nubilalis* (Lepidoptera: Pyralidae). *Can. Entomol.* 114: 599–603.

Lewis, P. A., R. L. Lampman, and R. L. Metcalf. 1990. Kairomonal attractants for *Acalymma vittatum* (Coleoptera: Chrysomelidae). *Environ Entomol.* 19: 8–14.

Lewis, T. and D. M. Sturgeon. 1978. Early warning of egg hatching in pea moth (*Cydia nigricana*). *Ann. Appl. Biol.* 88: 199–210.

Lewis, W. J. and J. R. Brazzel. 1968. A three-year study of parasites of the bollworm and the tobacco budworm in Mississippi. *J. Econ. Entomol.* 61: 673–676.

Lextrait, P., J.-C. Biemont, and J. Pouzat. 1995. Pheromone release by the two forms of *Callosobruchus maculatus* females: effects of age, temperature and host plant. *Physiol. Entomol.* 20: 309–317.

Li, S. Y., D. E. Henderson, and R. Feng. 1993. Larval parasitoid of the zebra caterpillar, *Melanchra picta* (Harris) (Lepidoptera: Noctuidae) on blueberry in British Columbia. *Can. Entomol.* 124: 405–406.

Liburd, O. E., R. A. Casagrande, and S. R. Alm. 1998. Evaluation of various color hydromulches and weed fabric on broccoli insect populations. *J. Econ. Entomol.* 91: 256–262.

Lilly, C. E. and G. A. Hobbs. 1956. Biology of the superb plant bug, *Adelphocoris superbus* (Uhl.) (Hemiptera: Miridae), in southern Alberta. *Can. Entomol.* 88: 118–125.

Lilly, C. E. and G. A. Hobbs. 1962. Effects of spring burning and insecticides on the superb plant bug, *Adelphocoris superbus* (Uhl.), and associated fauna in alfalfa seed fields. Can. J. Plant Sci. 42: 53–61.

Lim, K. P. and R. K. Stewart. 1976. Parasitism of the tarnished plant bug, *Lygus lineolaris* (Hemiptera: Miridae), by *Peristenus pallipes* and *P. pseudopallipes* (Hymenoptera: Braconidae). *Can. Entomol.* 108: 601–608.

Lim, K.-P., R. K. Stewart and W. N. Yule. 1980a. A historical review of the bionomics and control of *Phylllophaga anxia* (LeConte) (Coleoptera: Scarabaeidae), with special reference to Quebec. Ann. Entomol. Soc. Quebec 25: 163–178.

Lim, K. P. and W. N. Yule and C. R. Harris. 1980b. The toxicity of ten insecticides to third stage grubs of *Phyllophaga anxia* (LeConte) (Coleoptera: Scarabaeidae). *Phytoprotection* 61: 55–60.

Lim, K. P., R. K. Stewart and W. N. Yule. 1981b. Natural enemies of the common June beetle, *Phyllophaga anxia*, (Coleoptera: Scarabaeidae) in southern Quebec. Ann. Entomol. Soc. Quebec 26: 14–27.

Lim, K. P., W. N. Yule and R. K. Stewart. 1981a. Distribution and life history of *Phyllophaga anxia* (Coleoptera: Scarabaeidae) in southern Quebec. Ann. Entomol. Soc. Quebec 26: 100–112.

Lin, H., P. L. Phelan, and R. J. Bartelt. 1992. Synergism between synthetic food odors and the aggregation pheromone for attracting *Carpophilus lugubris* in the field (Coleoptera: Nitidulidae). *Environ Entomol.* 21: 156–159.

Lin, S. Y. H. and J. T. Trumble. 1985. Influence of temperature and tomato maturity on development and survival of *Keifera lycopersicella* (Lepidoptera: Gelechiidae). *Environ Entomol.* 14: 855–858.

Lindegren, J. E. and P. V. Vail. 1986. Susceptibility of Mediterranean fruit fly, melon fly, and oriental fruit fly (Diptera: Tephritidae) to the entomogenous nematode *Steinernema feltiae* in laboratory tests. *Environ Entomol.* 15: 465–468.

Lindegren, J. E., T. T. Wong, and D. O. McInnis. 1990. Response of Mediterranean fruit fly (Diptera: Tephritidae) to the entomogenous nematode *Steinernema feltiae* in field tests in Hawaii. *Environ Entomol.* 19: 383–386.

Lindsay, D. R. 1943. The biology and morphology of *Colaspis flavida* (Say). Unpublished Ph.D. Thesis, Iowa State College, Ames. 108 pp.

Linduska, J. J. 1979. Insecticides applied to the soil for control of wireworms on sweet potatoes in Maryland. *J. Econ. Entomol.* 72: 24–26.

Lingren, P. D., V. M. Bryant, Jr., J. R. Raulston, M. Pendleton, J. Westbrook, and G. D. Jones. 1993. Adult feeding host range and migratory activities of corn earworm, cabbage looper, and celery looper (Lepidoptera: Noctuidae) moths as evidenced by attached pollen. *J. Econ. Entomol.* 86: 1429–1439.

Linn, C. E., Jr., L. B. Bjostad, J. W. Du, and W. L. Roelofs. 1984. Redundancy in a chemical signal: behavioral responses of male *Trichoplusia ni* to a 6–component sex pheromone blend. *J. Chem Ecol.* 10: 1635–1658.

Linn, C. E., Jr., J. Du, A. Hammond, and W. L. Roelofs. 1987. Identification of unique pheromone components for soybean looper moth. *J. Chem Ecol.* 13: 1351–1360.

Liquido, N. J. 1991. Effect of ripeness and location of papaya fruits on the parasitization rates of oriental fruit fly and melon fruit fly (Diptera: Tephritidae) by braconid (Hymenoptera) parasitoids. *Environ. Entomol.* 20: 1732–1736.

Liquido, N. J. and T. Nishida. 1985. Variation in number of instars, longevity, and fecundity of *Cyrtorhinus lividipennis* Reuter (Hemiptera: Miridae). *Ann. Entomol. Soc. Am.* 78: 459–463.

List, G. M. 1921. The Mexican bean beetle. Colorado Agric. Exp. Stn. Bull. 271. 58 pp.

List, G. M. 1939a. The effect of temperature upon egg deposition, egg hatch and nymphal development of *Paratrioza cockerelli* (Sulc). *J. Econ. Entomol.* 32: 30–36.

List, G. M. 1939b. The potato and tomato psyllid and its control on tomatoes. Colorado Agric. Exp. Stn. Bull. 454. 33 pp.

Litsinger, J. A. and Ruhendi. 1984. Rice stubble and straw mulch suppression of preflowering insect pests of cowpeas sown after puddled rice. *Environ. Entomol.* 13: 500–514.

Liu, H. J., F. L. McEwen, and G. Ritcey. 1982. Forecasting events in the life cycle of the onion maggot, *Hylemya antiqua* (Diptera: Anthomyiidae): application to control schemes. *Environ. Entomol.* 11: 751–755.

Liu, M. Y. and C. N. Sun. 1984. Rearing diamondback moth (Lepidoptera: Yponomeutidae) on rape seedlings by a modification of the Koshihara and Yamada method. *J. Econ. Entomol.* 77: 1608–1609.

Liu, T. X. and P. A. Stansly. 1995. Toxicity and repellency of some biorational insecticides to *Bemisia argentifolii* on tomato plants. *Entomol. Exp. Appl.* 74: 137–143.

Liu, T.-X., R. D. Oetting, and G. D. Buntin. 1994. Temperature and diel catches of *Trialeurodes vaporariorum* and *Bemisia tabaci* (Homoptera: Aleyrodidae) adults on sticky traps in the greenhouse. *J. Entomol. Sci.* 29: 222–230.

Loan, C. C. 1965. Life cycle and development of *Leiophron pallipes* Curtis (Hymenoptera: Braconidae, Euphorinae) in five mirid hosts in the Belleville district. Proc. Entomol. Soc. Ont. 95: 115–121.

Loan, C. C. 1967a. Studies on the taxonomy and biology of the Euphorinae (Hymenoptera: Braconidae). I. Four new Canadian species of *Microtonus*. *Ann. Entomol. Soc. Am.* 60: 230–235.

Loan, C. C. 1967b. Studies on the taxonomy and biology of the Euphorinae (Hymenoptera: Braconidae). II. Host relations of six *Microctonus* species. *Ann. Entomol. Soc. Am.* 60: 236–240.

Loan, C. C. and S. R. Shaw. 1987. Euphorine parasites of *Lygus* and *Adelphocoris* (Hymenoptera: Braconidae and Heteroptera: Miridae). Pages 69–75 *in* Hedlund, R. C., and H. M. Graham (eds.). Economic Importance and Biological Control of *Lygus* and *Adelphocoris* in North America. USDA, Agric. Res. Serv., ARS-64. 95 pp.

Loebenstein, G. and B. Raccah. 1980. Control of non-persistently transmitted aphid-borne viruses. *Phytoparasitica* 8: 221–235.

Loera, J. and R. E. Lynch. 1987. Evaluation of pheromone traps for monitoring lesser cornstalk borer adults in beans. *Southwestern Entomol.* 12: 51–57.

Lofgren, C. S. 1986. The economic importance and control of imported fire ants in the United States. Pages 227–256 *in* S. B. Vinson (ed.). Economic Impact and Control of Social Insects. Praeger, New York.

Lofgren, C. S., W. A. Banks, and B. M. Glancey. 1975. Biology and control of imported fire ants. *Annu. Rev. Entomol.* 20: 1–30.

Logan, P. A., R. A. Casagrande, H. H. Faubert, and F. A. Drummond. 1985. Temperature-dependent development and feeding of immature Colorado potato beetles, *Leptinotarsa decemlineata* (Say) (Coleoptera: Chrysomelidae). *Environ. Entomol.* 14: 275–283.

Long, W. H. and S. D. Hensley. 1972. Insect pests of sugar cane. *Annu. Rev. Entomol.* 17: 149–176.

Loomans, A. J. M., T. Murai, J. P. N. F. van Heest, and J. C. van Lenteren. 1995. *Ceranisus menes* (Hymenoptera: Eulophidae) for control of western flower thrips: biology and behavior. Pages 263–268 *in* B. L. Parker, M. Skinner, and T. Lewis (eds.). Thrips Biology and Management. Plenum Press, London.

Lopez, Jr., J. D., J. L. Goodenough, and K. R. Beerwinkle. 1994. Comparison of two sex pheromone trap designs for monitoring corn earworm and tobacco budworm (Lepidoptera: Noctuidae). *J. Econ. Entomol.* 87: 793–801.

Lopez, Jr., J. D., T. N. Shaver, and J. L. Goodenough. 1990. Multispecies trapping of *Helicoverpa* (*Heliothis*) *zea*, *Spodoptera frugiperda*, *Pseudaletia unipuncta*, and *Agrotis ipsilon* (Lepidoptera: Noctuidae). *J. Chem Ecol.* 16: 3479–3491.

Losey, J. E., S. J. Fleischer, D. D. Calvin, W. L. Harkness, and T. Leahy. 1995. Evaluation of *Trichogramma nubilalis* and *Bacillus thuringiensis* in management of *Ostrinia nubilalis* (Lepidoptera: Pyralidae) in sweet corn. *Environ. Entomol.* 24: 436–445.

Losey, J. E., P. Z. Song, D. M. Schmidt, D. D. Calvin, and D. J. Liewehr. 1992. Larval parasitoids collected from overwintering European corn borer (Lepidoptera: Pyralidae) in Pennsylvania. *J. Kansas Entomol. Soc.* 65: 87–90.

Loughran, J. C. and D. W. Ragsdale. 1986a. Life cycle of the bean leaf beetle, *Cerotoma trifurcata* (Coleoptera: Chrysomelidae), in southern Minnesota. *Ann. Entomol. Soc. Am.* 79: 34–38.

Loughran, J. C. and D. W. Ragsdale. 1986b. *Medina* n. sp. (Diptera: Tachinidae): a new parasitoid of the bean leaf beetle, *Cerotoma trifurcata* (Coleoptera: Chrysomelidae). *J. Kansas Entomol. Soc.* 59: 468–473.

Lovell, O. H. 1932. The vegetable weevil *Listroderes obliquus*. *California Agric. Exp. Stn. Bull.* 546. 19 pp.

Lovett, A. L. and A. B. Black. 1920. The gray garden slug. *Oregon Agric. Exp. Stn. Bull.* 170. 43 pp.

Lowery, D. T. and M. K. Sears. 1986. Effect of exposure to the insecticide azinphosmethyl on reproduction of green peach aphid (Homoptera: Aphididae). *J. Econ. Entomol.* 79: 1534–1538.

Lowery, D. T., M. B. Isman, and N. L. Brard. 1993. Laboratory and field evaluation of neem for the control of aphids (Homoptera: Aphididae). *J. Econ. Entomol.* 86: 864–870.

Lowery, D. T., M. K. Sears, and C. S. Harmer. 1990. Control of turnip mosaic virus of rutabaga with applications of oil, whitewash, and insecticides. *J. Econ. Entomol.* 83: 2352–2356.

Lowry, V. K., J. W. Smith, Jr., and F. L. Mitchell. 1992. Life-fertility tables for *Frankliniella fusca* (Hinds) and *F. occidentalis* (Pergande)

(Thysanoptera: Thripidae) on peanut. *Ann. Entomol. Soc. Am.* 85: 744–754.

Lublinkhof, J. and D. E. Foster. 1977. Development and reproductive capacity of *Frankliniella occidentalis* (Thysanoptera: Thripidae) reared at three temperatures. *Kansas Entomol. Soc.* 50: 313–316.

Luckmann, W. H. 1963. Observations on the biology and control of *Glischrochilus quadrisignatus*. *J. Econ. Entomol.* 56: 681–686.

Luckmann, W. H., J. T. Shaw, D. W. Sherrod, and W. G. Ruesink. 1976. Developmental rate of the black cutworm. *J. Econ. Entomol.* 69: 386–388.

Ludwig, K. A. and R. E. Hill. 1975. Comparison of gut contents of adult western and northern corn rootworms in northeast Nebraska. *Environ Entomol.* 4: 435–438.

Luginbill, P. 1922a. The southern corn rootworm and farm practices to control it. USDA Farmers' Bull. 950. 10 pp.

Luginbill, P. 1922b. Bionomics of the chinch bug. USDA Bull. 1016. 14 pp.

Luginbill, P. 1928. The fall armyworm. USDA Tech. Bull. 34. 91 pp.

Luginbill, P. 1938. Control of common white grubs in cereal and forage crops. USDA Farmers' Bull. 1798. 19 pp.

Luginbill, P. and G. G. Ainslie. 1917. The lesser cornstalk borer. USDA Bull. 539. 27 pp.

Luginbill, Sr., P. and H. R. Painter. 1953. May beetles of the United States and Canada. USDA Tech. Bull. 1060. 102 pp.

Lund, R. D. and F. T. Turpin. 1977. Serological investigation of black cutworm larval consumption by ground beetles. *Ann. Entomol. Soc. Am.* 70: 322–324.

Lye, B.-H. and R. N. Story. 1988. Feeding preference of the southern green stink bug (Hemiptera: Pentatomidae) on tomato fruit. *J. Econ. Entomol.* 81: 522–526.

Lye, B.-H. and R. N. Story. 1989. Spatial dispersion and sequential sampling plan of the southern green stink bug (Hemiptera: Pentatomidae) on fresh market tomatoes. *Environ Entomol.* 18: 139–144.

Lye, B.-H., R. N. Story, and V. L. Wright. 1988a. Southern green stink bug (Hemiptera: Pentatomidae) damage to fresh market tomatoes. *J. Econ. Entomol.* 81: 189–194.

Lye, B.-H., R. N. Story, and V. L. Wright. 1988b. Damage threshold of the southern green stink bug, *Nezara viridula* (Hemiptera: Pentatomidae), on fresh market tomatoes. *J. Entomol. Sci.* 23: 366–373.

Lynch, R. E., J. W. Garner, and L. W. Morgan. 1984a. Influence of systemic insecticides on thrips damage and yield of Florunner peanuts in Georgia. *J. Agric. Entomol.* 1: 33–42.

Lynch, R. E., J. A. Klun, B. A. Leonhardt, M. Schwarz, and J. W. Garner. 1984b. Female sex pheromone of the lesser cornstalk borer, *Elasmopalpus lignosellus* (Lepidoptera: Pyralidae). *Environ. Entomol.* 13: 121–126.

MacFarlane, J. H. and A. J. Thorsteinson. 1980. Development and survival of the twostriped grasshopper, *Melanoplus bivittatus* (Say) (Orthoptera: Acrididae), on various single and multiple plant diets. *Acrida* 9: 63–76.

MacGillivray, M. E. and G. B. Anderson. 1958. Development of four species of aphids (Homoptera) on potato. *Can. Entomol.* 40: 148–155.

Mack, T. P. and C. B. Backman. 1984. Effects of temperature and adult age on the oviposition rate of *Elasmopalpus lignosellus* (Zeller), the lesser cornstalk borer. *Environ. Entomol.* 13: 966–969.

Mack, T. P. and Z. Smilowitz. 1980. The development of a green peach aphid natural enemy sampling procedure. *Environ. Entomol.* 9: 440–445.

Mack, T. P., D. P. Davis, and C. B. Backman. 1991. Predicting lesser cornstalk borer (Lepidoptera: Pyralidae) larval density from estimates of adult abundance in peanut fields. *J. Entomol. Sci.* 26: 223–230.

Mack, T. P., D. P. Davis, and R. E. Lynch. 1993. Development of a system to time scouting for the lesser cornstalk borer (Lepidoptera: Pyralidae) attacking peanuts in the southeastern United States. *J. Econ. Entomol.* 86: 164–173.

Mackauer, M. 1968. Insect parasites of the green peach aphid, *Myzus persicae* Sulz., and their control potential. Entomphaga 13: 91–106.

Mackauer, M. and T. Finlayson. 1967. The hymenopterous parasites (Hymenoptera: Aphididae et Aphelinidae) of the pea aphid in eastern North America. *Can. Entomol.* 99: 1051–1082.

Mackenzie, J. R. and R. S. Vernon. 1988. Sampling for distribution of the lettuce aphid, *Nasonovia ribisnigri* (Homoptera: Aphididae), in fields and within heads. *J. Entomol. Soc. Brit. Columbia* 85: 10–14.

Mackenzie, J. R., R. S. Vernon, and S. Y. Szeto. 1988. Efficacy and residues of foliar sprays against the lettuce aphid, *Nasonovia ribisnigri* (Homoptera: Aphididae), on crisphead lettuce. *J. Entomol. Soc. Brit. Columbia* 85: 3–9.

MacLean, G. F. and F. G. Butcher. 1934. Studies of milliped and gnat injuries to potato tubers. *J. Econ. Entomol.* 27: 106–108.

MacVean, C. M. 1987. Ecology and management of Mormon cricket, *Anabrus simplex* Haldeman. Pages 116–136 *in* J. L. Capinera (ed.). Integrated Pest Management on Rangeland: A Shortgrass Prairie Perspective. Westview Press, Boulder.

MacVean, C. M. and J. L. Capinera. 1991. Pathogenicity and transmission potential of *Nosema locustae* and *Vairimorpha* n. sp. (Protozoa: Microsporida) in Mormon crickets (*Anabrus simplex*; Orthoptera: Tettigoniidae): a laboratory evaluation. *J. Invertebr. Pathol.* 57: 23–36.

MacVean, C. M. and J. L. Capinera. 1992. Field evaluation of two microsporidian pathogens, an entomopathogenic nematode, and carbaryl for suppression of the Mormon cricket, *Anabrus simplex* Hald. (Orthoptera: Tettigoniidae). *Biol. Control* 2: 59–65.

MacVean, C. M., J. W. Brewer, and J. L. Capinera. 1982. Field tests of antidesiccants to extend the infection period of an entomogenous nematode, *Neoaplectana carpocapsae*, against the Colorado potato beetle. *J. Econ. Entomol.* 75: 97–101.

Madden, A. H. and F. S. Chamberlin. 1945. Biology of the tobacco hornworm in the southern cigar-tobacco district. USDA Tech. Bull. 896. 51 pp.

Maddock, D. R. and C. F. Fehn. 1958. Human ear invasion by adult scarabaeid beetles. *J. Econ. Entomol.* 51: 546–547.

Madueke, E.-D.N. and T. H. Coaker. 1984. Temperature requirements of the white fly *Trialeurodes vaporariorum* (Homoptera: Aleyrodidae) and its parasitoid *Encarsia formosa* (Hymenoptera: Aphelinidae). *Entomol. Gener.* 9: 149–154.

Mahr, S. E. R., J. A. Wyman, and R. K. Chapman. 1993. Variability in aster yellows infectivity of local populations of the aster leafhopper (Homoptera: Cicadellidae) in Wisconsin. *J. Econ. Entomol.* 86: 1522–1526.

Mahrt, G. G. and C. C. Blickenstaff. 1979. Host plants of the sugarbeet root maggot, *Tetanops myopaeformis*. *Ann. Entomol. Soc. Am.* 72: 627–631.

Mahrt, G. G., R. L. Stoltz, C. C. Blickenstaff, and T. O. Holtzer. 1987. Comparisons between blacklight and pheromone traps for monitoring the western bean cutworm (Lepidoptera: Noctuidae) in south central Idaho. *J. Econ. Entomol.* 80: 242–247.

Maiteki, G. A. and R. J. Lamb. 1985a. Growth stages of field peas sensitive to damage by the pea aphid, *Acyrthosiphon pisum* (Homoptera: Aphididae). *J. Econ. Entomol.* 78: 1442–1448.

Maiteki, G. A. and R. J. Lamb. 1985b. Spray timing and economic threshold for the pea aphid, *Acyrthosiphon pisum* (Homoptera: Aphididae), on field peas in Manitoba. *J. Econ. Entomol.* 78: 1449–1454.

Maiteki, G. A. and R. J. Lamb. 1987. Sequential decision plan for control of pea aphid, *Acyrthosiphon pisum* (Homoptera: Aphididae), on field peas in Manitoba. *J. Econ. Entomol.* 80: 605–607.

Majchrowicz, I., T. J. Poprawski, P. H. Robert, and N. K. Maniania. 1990. Effects of entomopathogenic and opportunistic fungi on *Delia antiqua* (Diptera: Anthomyiidae) at low relative humidity. *Environ. Entomol.* 19: 1163–1167.

Mani, M. 1989. A review of the pink mealybug—*Maconellicoccus hirsutus* (Green). *Insect Sci. Appl.* 10: 157–167.

Manson, D. C. M. 1972. A contribution to the study of the genus *Rhizoglyphus* Claparede, 1869 (Acarina: Acaridae). *Acarologia* 13: 621–650.

Manuwoto, S. and J. M. Scriber. 1982. Consumption and utilization of three maize genotypes by the southern armyworm. *J. Econ. Entomol.* 75: 163–167.

Maple, J. D. 1937. The biology of *Ooencyrtus johnsoni* (Howard), and the role of the egg shell in the respiration of certain encyrtid larvae (Hymenoptera). *Ann. Entomol. Soc. Am.* 30: 123–154.

Marco, S. 1986. Incidence of aphid-transmitted virus infections reduced by whitewash sprays on plants. *Phytopathology* 76: 1344–1348.

Marco, S. 1993. Incidence of nonpersistently transmitted viruses in pepper sprayed with whitewash, oil, and insecticide, alone or combined. *Plant Dis.* 77: 1119–1122.

Maredia, K. M. and J. A. Mihm. 1991. Sugarcane borer (Lepidoptera: Pyralidae) damage to maize at four plant growth stages. *Environ. Entomol.* 20: 1019–1023.

Marenco, R. J., R. E. Foster, and C. A. Sanchez. 1992. Sweet corn response to fall armyworm (Lepidoptera: Noctuidae) damage during vegetative growth. *J. Econ. Entomol.* 85: 1285–1292.

Margolies, D. C. 1987. 1987. Conditions eliciting aerial dispersal behavior in Banks grass mite, *Oligonychus pratensis* (Acari: Tetranychidae). *Environ. Entomol.* 16: 928–932.

Markham, P. G. 1994. Transmission of geminiviruses by *Bemisia tabaci*. *Pest. Sci.* 42: 123–128.

Marrone, P. G. and R. E. Stinner. 1983a. Effects of soil moisture and texture on oviposition preference of the bean leaf beetle, *Cerotoma trifurcata* (Förster) (Coleoptera: Coccinellidae). *Environ. Entomol.* 12: 426–428.

Marrone, P. G. and R. E. Stinner. 1983b. Effects of soil physical factors on egg survival of the bean leaf beetle, *Cerotoma trifurcata* (Förster) (Coleoptera: Chrysomelidae). *Environ. Entomol.* 12: 673–679.

Marrone, P. G. and R. E. Stinner. 1983c. Bean leaf beetle, *Cerotoma trifurcata* (Förster) (Coleoptera: Chrysomelidae): physical factors affecting larval movement in soil. *Environ. Entomol.* 12: 1283–1285.

Marrone, P. G. and R. E. Stinner. 1984. Influence of soil physical factors on survival and development of the larvae and pupae of the bean leaf beetle, *Cerotoma trifurcata* (Coleoptera: Chrysomelidae). *Can. Entomol.* 116: 1015–1023.

Marsh, H. O. 1911. The Hawaiian beet webworm. USDA Bur. Entomol. Bull. 109. 14 pp.

Marsh, H. O. 1912a. Biologic notes on species of *Diabrotica* in southern Texas. USDA Bur. Entomol. Bull. 82: 76–84.

Marsh, H. O. 1912b. Biologic and economic notes on the yellow-bear caterpillar. USDA Bur. Entomol. Bull. 82: 59–66.

Marsh, H. O. 1912c. The sugar-beet webworm. USDA Bur. Entomol. Bull. 109, Part VI: 57–70.

Marsh, H. O. 1913. The striped beet caterpillar. USDA Bur. Entomol. Bull. 127, Part II: 13–18.

Marsh, H. O. 1917. Life history of *Plutella maculipennis*, the diamondback moth. *J. Agric. Res.* 10: 1–10.

Marsh, P. M. 1977. Notes on the taxonomy ad nomenclature of *Aphidius* species (Hym.: Aphidiidae) parasitic on the pea aphid in North America. *Entomophaga* 22: 365–372.

Martel, P., G. Boivin, and J. Belcourt. 1986. Efficacy and persistence of different insecticides against the tarnished plant bug, *Lygus lineolaris* (Heteroptera: Miridae), on a season-long host plant, *Coronilla varia*. *J. Econ. Entomol.* 79: 721–725.

Martel, P., H. J. Svec, and C. R. Harris. 1975. Mass rearing of the carrot weevil, *Listronotus oregonensis* (Coleoptera: Curculionidae), under controlled environmental conditions. *Can. Entomol.* 107: 95–98.

Martel, P., H. J. Svec, and C. R. Harris. 1976. The life history of the carrot weevil, *Listronotus oregonensis* (Coleoptera: Curculionidae) under controlled conditions. *Can. Entomol.* 108: 931–934.

Martel, P., J. Belcourt, D. Choquette, and G. Boivin. 1986. Spatial dispersion and sequential sampling plan for the Colorado potato beetle (Coleoptera: Chrysomelidae). *J. Econ. Entomol.* 79: 414–417.

Martin, C. H. 1934. Notes on the larval feeding habits and the life history of *Eumerus tuberculatus* Rondani. Bull. Brooklyn Entomol. Soc. 29: 27–36.

Martin, J. H. 1987. An identification guide to common whitefly pest species of the world (Homoptera: Aleyrodidae). *Trop. Pest Manage.* 33: 298–322.

Martin, M. W. and P. E. Thomas. 1986. Levels, dependability, and usefulness of resistance to tomato curly top disease. *Plant Dis.* 70: 136–141.

Martin, P. B., P. D. Lingren, and G. L. Greene. 1976a. Relative abundance and host preferences of cabbage looper, soybean looper, tobacco budworm, and corn earworm on crops grown in northern Florida. *Environ. Entomol.* 5: 878–882.

Martin, P. B., P. D. Lingren, G. L. Greene, and A. H. Baumhover. 1981a. Seasonal occurrence of *Rachiplusia ou, Autographa biloba*, and associated entomophages in clover. *J. Georgia Entomol. Soc.* 16: 288–295.

Martin, P., P. D. Lingren, G. L. Greene, and E. E. Grissell. 1981b. The parasitoid complex of three noctuids (Lep.) in a northern Florida cropping system: seasonal occurrence, parasitization, alternate hosts, and influence of host-habitat. *Entomophaga* 26: 401–419.

Martin, P. B., P. D. Lingren, G. L. Greene, and R. L. Ridgway. 1976b. Parasitization of two species of Plusiinae and *Heliothis* spp. after releases of *Trichogramma pretiosum* in seven crops. *Environ. Entomol.* 5: 991–995.

Martin, W. D. and G. A. Herzog. 1987. Life history studies of the tobacco flea beetle, *Epitrix hirtipennis* (Melsheimer) (Coleoptera: Chrysomelidae). *J. Entomol. Sci.* 22: 237–244.

Martinson, T. E., J. P. Nyrop, and C. J. Eckenrode. 1988. Dispersal of the onion fly (Diptera: Anthomyiidae) and larval damage in rotated onion fields. *J. Econ. Entomol.* 81: 508–514.

Martinson, T. E., J. P. Nyrop, and C. J. Eckenrode. 1989. Long-range host-finding behavior and colonization of onion fields by *Delia antiqua* (Diptera: Anthomyiidae). *J. Econ. Entomol.* 82: 1111–1120.

Mason, G. A., B. E. Tabashnik, and M. W. Johnson. 1989. Effects of biological and operational factors on evolution of insecticide resistance in *Liriomyza* (Diptera: Agromyzidae). *J. Econ. Entomol.* 82: 369–373.

Mason, L. J. and T. P. Mack. 1984. Influence of temperature on oviposition and adult female longevity for the soybean looper, *Pseudoplusia includens* (Walker) (Lepidoptera: Noctuidae). *Environ Entomol.* 13: 379–383.

Mason, L. J., R. K. Jansson, and R. R. Heath. 1990. Sampling range of male sweetpotato weevils (*Cylas formicarius elegantulus*) (Summers) (Coleoptera: Curculionidae) to pheromone traps: influence of pheromone dosage and lure age. *J. Chem Ecol.* 16: 2493–2502.

Matheny, Jr., E. L. 1981. Contrasting feeding habits of pest mole cricket species. *J. Econ. Entomol.* 74: 444–445.

Matheny, E. L. and E. A. Heinrichs. 1971. Hatching of sod webworm eggs in relation to low and high temperatures. *Ann. Entomol. Soc. Am.* 64: 116–119.

Matheny, Jr., E. L., A. Tsedeke, and B. J. Smittle. 1981. Feeding response of mole cricket nymphs (Orthoptera: Gryllotalpidae: *Scapteriscus*) to radiolabeled grasses with, and without, alternative foods available. *J. Georgia Entomol. Soc.* 16: 492–495.

Matteson, J. W. 1966. Colonization and mass production of the false wireworm *Eleodes suturalis*. *J. Econ. Entomol.* 59: 26–27.

Matthews-Gehringer, D. and J. Hough-Goldstein. 1988. Physical barriers and cultural practices in cabbage maggot (Diptera: Anthomyiidae) management on broccoli and Chinese cabbage. *J. Econ. Entomol.* 81: 354–360.

Mattson, D. J., C. M. Gillin, S. A. Benson, and R. R. Knight. 1991. Bear feeding activity at alpine insect aggregation sites in the Yellowstone ecosystem. *Can. J. Zool.* 69: 2430–2435.

Maxson, A. C. 1948. Insects and Diseases of the Sugar Beet. Beet Sugar Development Foundation, Ft. Collins, Colo. 425 pp.

Mayer, D. F. and C. A. Johansen. 1978. Bionomics of *Meloe niger* Kirby (Coleoptera: Meloidae) a predator of the alkali bee, *Nomia melanderi* Cockerell (Hymenoptera: Halictidae). *Melanderia* 28: 1–22.

Mayer, D. F., J. D. Lunden, and L. Rathbone. 1987. Evaluation of insecticides for *Thrips tabaci* (Thysanoptera: Thripidae) and effects of thrips on bulb onions. *J. Econ. Entomol.* 80: 930–932.

Maynard, E. A. 1951. A Monograph of the Collembola or Springtail Insects of New York State. Comstock Pub. Co., Ithaca. 339 pp.

Mays, W. T. and L. T. Kok. 1997. Oviposition, development, and host preference of the cross-striped cabbageworm (Lepidoptera: Pyralidae). *Environ. Entomol.* 26: 1354–1360.

McAuslane, H. J., C. R. Ellis, and O. B. Allen. 1987. Sequential sampling of adult northern and western corn rootworms (Coleoptera: Chrysomelidae) in southern Ontario. *Can. Entomol.* 119: 577–585.

McAvoy, T. J. and L. T. Kok. 1992. Development, oviposition, and feeding of the cabbage webworm (Lepidoptera: Pyralidae). *Environ. Entomol.* 21: 527–533.

McAvoy, T. J. and J. C. Smith. 1979. Feeding and developmental rates of the Mexican bean beetle on soybeans. *J. Econ. Entomol.* 72: 835–836.

McCarthy, W. J., R. R. Granados, G. R. Sutter, and D. W. Roberts. 1975. Characterization of entomopox virions of the army cutworm, *Euxoa auxiliaris* (Lepidoptera: Noctuidae). *J. Invertebr. Pathol.* 25: 215–220.

McClanahan, R. J. 1963. Food preferences of the six-spotted leafhopper, *Macrosteles fascifrons* (Stål). Proc. Entomol. Soc. Ont. 93: 90–92.

McClanahan, R. J. 1975a. Tests of insecticides for asparagus beetle control and residues on the crop. Proc. Entomol. Soc. Ontario. 106: 52–55.

McClanahan, R. J. 1975b. Insecticides for control of the Colorado potato beetle (Coleoptera: Chrysomelidae). *Can. Entomol.* 107: 561–565.

McCloud, E. S., D. W. Tallamy, and F. T. Halaweish. 1995. Squash beetle trenching behavior: avoidance of cucurbitacin induction or mucilaginous plant sap? *Ecol. Entomol.* 20: 51–59.

McCoy, C. E. and T. A. Brindley. 1961. Biology of the four-spotted fungus beetle, *Glischrochilus q. quadrisignatus* and its effect on European corn borer populations. *J. Econ. Entomol.* 54: 713–717.

McCreight, J. D. and A. N. Kishaba. 1991. Reaction of *Cucurbit* species to squash leaf curl virus and sweet-potato whitefly. *J. Am. Soc. Hort. Sci.* 116: 137–141.

McCutcheon, G. S., S. G. Turnipseed, and M. J. Sullivan. 1990. Parasitization of lepidopterans as affected by nematicide-insecticide use in soybean. *J. Econ. Entomol.* 83: 1002–1007.

McDonald, S. 1979. Evaluation of insecticides for control of the army cutworm. *J. Econ. Entomol.* 72: 277–280.

McDonald, S. 1981a. Evaluation of organophosphorus and pyrethroid insecticides for control of the pale western cutworm. *J. Econ. Entomol.* 74: 45–48.

McDonald, S. 1981b. Laboratory evaluation of new insecticides for control of redbacked cutworm larvae. *J. Econ. Entomol.* 74: 593–596.

McEwen, F. L. and G. E. R. Hervey. 1960. Mass-rearing the cabbage looper, *Trichoplusia ni*, with notes on its biology in the laboratory. *Ann. Entomol. Soc. Am.* 53: 229–234.

McEwen, F. L., G. Ritcey, and H. J. Liu. 1984. Laboratory studies of radiation-induced sterility on the onion maggot, *Delia antiqua* (Diptera: Anthomyiidae). *Can. Entomol.* 116: 119–122.

McEwen, F. L., W. T. Schroeder, and A. C. Davis. 1957. Host range and transmission of the pea enation mosaic virus. *J. Econ. Entomol.* 50: 770–775.

McEwen, L. C. 1987. Function of insectivorous birds in a shortgrass IPM system. Pages 324–333 *in* J. L. Capinera (ed.). Integrated Pest Management on Rangeland, a Shortgrass Prairie Perspective. Westview Press, Boulder.

McGovern, T. P. and T. L. Ladd, Jr. 1990. Attractants for the northern corn rootworm (Coleoptera: Chrysomelidae): alkyl- and alkenylphenols. *J. Econ. Entomol.* 83: 1316–1320.

McHugh, Jr., J. J. and R. E. Foster. 1995. Reduction of diamondback moth (Lepidoptera: Plutellidae) infestation in head cabbage by overhead irrigation. *J. Econ. Entomol.* 88: 162–168.

McKenzie, C. L., B. Cartwright, M. E. Miller, and J. V. Edelson. 1993. Injury to onions by *Thrips tabaci* (Thysanoptera: Thripidae) and its role in the development of purple blotch. *Environ. Entomol.* 22: 1266–1277.

McKinlay, K. S. 1981. The importance of dry plant material in the diet of the grasshopper *Melanoplus sanguinipes* (Orthoptera: Acrididae). *Can. Entomol.* 113: 5–8.

McKinlay, R. G. 1985. Effect of undersowing potatoes with grass on potato aphid numbers. *Ann. Appl. Biol.* 106: 23–29.

McKinlay, R. G. 1992. Vegetable Crop Pests. CRC Press, Boca Raton. 406 pp.

McKinney, K. B. 1944. The cabbage looper as a pest of lettuce in the Southwest. USDA Tech. Bull. 846. 30 pp.

McLain, D. K., N. B. Marsh, J. R. Lopez, and J. A. Drawdy. 1990. Intravernal changes in the level of parasitization of the southern green stink bug (Hemiptera: Pentatomidae), by the featherlegged fly (Diptera: Tachinidae): host sex, mating status, and body size as correlated factors. *J. Entomol. Sci.* 25: 501–509.

McLaughlin, J. R., A. Q. Antonio, S. L. Poe, and D. R. Minnick. 1979. Sex pheromone biology of the adult tomato pinworm, *Keiferia lycopersicella* (Walsingham). *Florida Entomol.* 62: 35–41.

McLaughlin, J. R. and R. R. Heath. 1989. Field trapping and observations of male velvetbean caterpillar moths and trapping of *Mocis* spp. (Lepidoptera: Noctuidae: Catacolinae) with calibrated formulations of sex pheromone. *Environ. Entomol.* 18: 933–938.

McLaughlin, J. R., E. R. Mitchell, and P. Kirsch. 1994. Mating disruption of diamondback moth (Lepidoptera: Plutellidae) in cabbage: reduction of mating and suppression of larval populations. *J. Econ. Entomol.* 87: 1198–1204.

McLean, D. L. and M. G. Kinsey. 1961. A method for rearing the lettuce root aphid, *Pemphigus bursarius*. *J. Econ. Entomol.* 54: 1256–1257.

McLeod, D. G. R. and L. L. Gualtieri. 1992. Yellow pan water traps for monitoring the squash vine borer, *Melittia cucurbitae* (Lepi-

doptera: Sesiidae) in home gardens. Proc. Entomol. Soc. Ontario. 123: 133–135.

McLeod, D. G. R., J. W. Whistlecraft, and C. R. Harris. 1985. An improved rearing procedure for the carrot rust fly (Diptera: Psilidae) with observations on life history and conditions controlling diapause induction and termination. *Can. Entomol.* 117: 1017–1024.

McLeod, J. H. and D. A. Chant. 1952. Notes on the parasitism and food habits of the European earwig, *Forficula auricularia* L. (Dermaptera: Forficulidae). *Can. Entomol.* 84: 343–345.

McLeod, P. 1987. Effect of low temperature on *Myzus persicae* (Homoptera: Aphididae) on overwintering spinach. *Environ. Entomol.* 16: 796–801.

McLeod, P. 1991. Influence of temperature on translaminar and systemic toxicities of aphicides for green peach aphid (Homoptera: Aphididae) suppression on spinach. *J. Econ. Entomol.* 84: 1558–1561.

McLeod, P. J., S. Y. Young, and W. C. Yearian. 1982. Application of a baculovirus of *Pseudoplusia includens* to soybean: efficacy and seasonal persistence. *Environ. Entomol.* 11: 412–416.

McLeod, P. J., D. C. Steinkraus, J. C. Correll, and T. E. Morelock. 1998. Prevalence of *Erynia neoaphidis* (Entomophthorales: Entomophthoraceae) infections of green peach aphid (Homoptera: Aphididae) on spinach in the Arkansas River Valley. *Environ. Entomol.* 27: 796–800.

McMillian, W. W., B. R. Wiseman, and N. W. Widstrom. 1977. An evaluation of commercial sweet corn hybrids for damage by *Heliothis zea*. *J. Georgia Entomol.* 12: 75–79.

McMurtry, J. A., C. B. Huffaker, and M. van de Vrie. 1970. Tetranychid enemies: their biological characters and the impact of spray practices. *Hilgardia* 11: 331–390.

McNeil, J. N. 1987. The true armyworm, *Pseudaletia unipunctata*: a victim of the Pied Piper or a seasonal migrant? *Insect Sci. Appl.* 8: 591–597.

McPhail, M. 1937. Relation of time of day, temperature and evaporation to attractiveness of fermenting sugar solution to Mexican fruit fly. *J. Econ. Entomol.* 30: 793–799.

McPhail, M. 1939. Protein lures for fruit flies. *J. Econ. Entomol.* 32: 758–761.

McPherson, J. E. 1982. The Pentatomoidea (Hemiptera) of Northeastern North America with Emphasis on the Fauna of Illinois. Southern Illinois Univ. Press, Carbondale. 240 pp.

McPherson, R. M. and L. D. Newsom. 1984. Trap crops for control of stink bugs in soybean. *J. Georgia Entomol. Soc.* 19: 470–480.

McPherson, R. M., R. J. Beshear, and A. K. Culbreath. 1992. Seasonal abundance of thrips (Thysanoptera: suborders Terebrantia and Tubulifera) in Georgia flue-cured tobacco and impact of management practices on the incidence of tomato spotted wilt virus. *J. Entomol. Sci.* 27: 257–268.

McPherson, R. M., G. W. Zehnder, and J. C. Smith. 1988. Influence of cultivar, planting date, and row width on abundance of green cloverworms (Lepidoptera: Noctuidae) and green stink bugs (Heteroptera: Pentatomidae) in soybean. *J. Entomol. Sci.* 23: 305–313.

McSorley, R. and R. K. Jansson. 1991. Spatial patterns of *Cylas formicarius* in sweet potato fields and development of a sampling plan. Pages 157–167 *in* R. K. Jansson and K. V. Raman (eds.). Sweet Potato Pest Management: A Global Perspective. Westview Press, Boulder, Colorado.

McSorley, R. and V. H. Waddill. 1982. Partitioning yield loss on yellow squash into nematode and insect components. *J. Nematol.* 14: 110–118.

Mead, A. R. 1971a. Helicid land mollusks introduced into North America. *The Biol.* 53: 104–111.

Mead, A. R. 1971b. Status of *Achatina* and *Rumina* in the United States. *The Biol.* 53: 112–117.

Meade, A. B. and A. G. Peterson. 1967. Some factors influencing population growth of the aster leafhopper in Anoka County, Minnesota. *J. Econ. Entomol.* 60: 936–941.

Meade, D. L. and D. N. Byrne. 1991. The use of *Verticillium lecanii* against subimaginal instars of *Bemisia tabaci*. *J. Invertebr. Pathol.* 27: 296–298.

Meade, T. and J. D. Hare. 1991. Differential performance of beet armyworm and cabbage looper (Lepidoptera: Noctuidae) larvae on selected *Apium graveolens* cultivars. *Environ. Entomol.* 20: 1636–1644.

Meats, A. 1989. Abiotic mortality factors - temperature. Pages 229–239 *in* A. S. Robinson and G. Hooper (eds.) Fruit Flies: their Biology, Natural Enemies and Control, Vol. 3b. Elsevier, Amsterdam.

Medina–Gaud, S., E. Abreu, F. Gallardo, and R. A. Franqui. 1989. Natural enemies of the melonworm, *Diaphania hyalinata* L. (Lepidoptera: Pyralidae) in Puerto Rico. *J. Agric. Univ. Puerto Rico* 73: 313–320.

Medler, J. T. 1957. Migration of the potato leafhopper – a report on a cooperative study. *J. Econ. Entomol.* 50: 493–497.

Melander, A. L. and M. A. Yothers. 1917. The coulee cricket. Washington Agric. Exp. Stn. Bull. 137. 55 pp.

Mellors, W. K., A. Allegro, and A. M. Wilson. 1984. Temperature-dependent simulation of the effects of detrimental high temperatures on the survival of Mexican bean beetle eggs (Coleoptera: Coccinellidae). *Environ. Entomol.* 13: 86–94.

Mena–Covarrubias, J., F. A. Drummond, and D. L. Haynes. 1996. Population dynamics of the Colorado potato beetle (Coleoptera: Chrysomelidae) on horsenettle in Michigan. *Environ. Entomol.* 25: 68–77.

Mendoza, C. E. and D. C. Peters. 1964. Species differentiation among mature larvae of *Diabrotica undecimpunctata howardi, D. virgifera,* and *D. longicornis. J. Kansas Entomol. Soc.* 37: 123–125.

Messina, F. J. 1984. Influence of cowpea pod maturity on the oviposition choices and larval survival of a bruchid beetle *Callosobruchus maculatus. Entomol. Exp. Appl.* 35: 241–248.

Metcalf, R. L. and E. R. Metcalf. 1992. Plant Kairomones in Insect Chemical Ecology and Control. Chapman and Hall, New York. 168 pp.

Metcalf, R. L. and R. A. Metcalf. 1993. Destructive and useful Insects, their Habits and Control. Fifth Ed. McGraw-Hill, New York. 1074 pp.

Metcalf, R. L., J. E. Ferguson, R. Lampman, and J. F. Andersen. 1987. Dry cucurbitacin-containing baits for controlling diabroticite beetles (Coleoptera: Chrysomelidae). *J. Econ. Entomol.* 80: 870–875.

Metcalf, Z. P. 1917. Biological investigation of *Sphenophorus callosus* Oliv. North Carolina Agric. Exp. Stn. Tech. Bull. 13. 123 pp.

Metcalf, Z. P. and G. W. Underhill. 1919. The tobacco flea beetle. North Carolina Agric. Exp. Stn. Bull. 239. 47 pp.

Mendoza, C. E. and D. C. Peters. 1964. Species differentiation among mature larvae of *Diabrotica undecimpunctata howardi, D. virgifera,* and *D. longicornis. J. Kansas Entomol. Soc.* 37: 123–125.

Mertz, B. P., S. J. Fleischer, D. D. Calvin, and R. L. Ridgway. 1995. Field assessment of *Trichogramma brassicae* (Hymenoptera: Trichogrammatidae) and *Bacillus thuringiensis* for control of *Ostrinia nubilalis* (Lepidoptera: Pyralidae) in sweet corn. *J. Econ. Entomol.* 88: 1616–1625.

Meyerdirk, D. E. and N. A. Hessein. 1985. Population dynamics of the beet leafhopper, *Circulifer tenellus* (Baker), and associated *Empoasca* spp. (Homoptera: Cicadellidae) and their egg parasitoids on sugar beets in southern California. *J. Econ. Entomol.* 78: 346–353.

Meyerdirk, D. E. and M. S. Moratorio. 1987. *Circulifer tenellus* (Baker), the beet leafhopper (Homoptera: Cicadellidae): laboratory studies on fecundity and longevity. *Can. Entomol.* 119: 443–447.

Michelbacher, A. E. 1938. The biology of the garden centipede, *Scutigerella immaculata. Hilgardia* 11: 55–148.

Michelbacher, A. E. 1942. A synopsis of the genus *Scutigerella* (Symphyla: Scutigerellidae). *Ann. Entomol. Soc. Am.* 35: 267–288.

Michelbacher, A. E. and R. F. Smith. 1943. Some natural factors limiting the abundance of the alfalfa butterfly. *Hilgardia* 15: 369–397.

Michelbacher, A. E., G. F. MacLeod, and R. F. Smith. 1943. Control of *Diabrotica*, or western spotted cucumber beetle, in deciduous fruit orchards. *California Agric.* Exp. Stn. Bull. 681. 34 pp.

Michelbacher, A. E., W. W. Middlekrauff, and O. G. Bacon. 1952. Stink bug injury to tomatoes in California. *J. Econ. Entomol.* 45: 126.

Michelbacher, A. E., W. W. Middlekauff, O. G. Bacon, and J. E. Swift. 1955. Controlling melon insects and spider mites. *California Agric.* Exp. Stn. Bull. 749. 46 pp.

Michels, Jr., G. J. and R. W. Behle. 1989. Influence of temperature on reproduction, development, and intrinsic rate of increase of Russian wheat aphid, greenbug, and bird cherry-oat aphid (Homoptera: Aphididae). *J. Econ. Entomol.* 82: 439–444.

Michels, Jr., G. J. and C. C. Burkhardt. 1981. Economic threshold levels of the Mexican bean beetle on pinto beans in Wyoming. *J. Econ. Entomol.* 74: 5–6.

Michelsen, V. 1980. A revision of the beet leaf-miner complex, *Pegomya hyoscyami* s.lat. (Diptera: Anthomyiidae). *Entomol. Scand.* 11: 297–309.

Middlekauff, W. W. 1951. Field studies on the bionomics and control of the broad bean weevil. *J. Econ. Entomol.* 44: 240–243.

Milbrath, L. R., M. J. Weiss and B. J. Schatz. 1995. Influence of tillage system, planting date, and oilseed crucifers on flea beetle populations (Coleoptera: Chrysomelidae). *Can. Entomol.* 127: 289–293.

Miles, M. 1948. Field observations on the bean seed fly (seed corn maggot), *Chortophila cilicrura*, Rond., and *C. trichodactyla*, Rond. Bull. Entomol. Res. 38: 559–574.

Miles, M. 1951. Factors affecting the behaviour and activity of the cabbage root fly (*Erioischia brassicae* Bche). *Ann. Appl. Biol.* 38: 425–432.

Miles, M. 1952. Studies of British anthomyiid flies. III. Immature stages of *Delia cilicrura* (Rond.), *D. trichodactyla* (Rond.), *Erioischia brassicae* (Bch.), *E. floralis* (Fall.) and *Pegohylemyia fugax* (Mg.). *Bull. Entomol. Res.* 43: 83–90.

Miles, M. 1953. Studies of British anthomyiid flies. V. The onion fly, *Delia antiqua* (Mg.). *Bull. Entomol. Res.* 44: 583–588.

Miles, M. 1954a. Field studies on the influence of weather conditions on egg-laying by the cabbage root fly, *Erioischia brassicae* Bche. I. *Ann. Appl. Biol.* 40: 717–725.

Miles, M. 1954b. Field studies on the influence of weather on egg-laying by the cabbage root fly, *Erioischia brassicae* Bche. II. *Ann. Appl. Biol.* 41: 586–590.

Miles, M. 1955a. Studies of British anthomyiid flies. VI. The annual cycle of generations in some anthomyiid root flies. *Bull. Entomol. Res.* 46: 11–19.

Miles, M. 1955b. Studies of British anthomyiid flies. VII. The onion-fly complex. *Bull. Entomol. Res.* 46: 21–26.

Miles, M. 1956. Observations on the emergence periods, larval populations and parasitism of the cabbage root fly (*Erioschia brassicae* Bche.). *Ann. Appl. Biol.* 44: 492–498.

Millar, K. V. and M. B. Isman. 1988. The effects of a spunbonded polyester row cover on cauliflower yield loss caused by insects. *Can. Entomol.* 120: 45–47.

Miller, D. F. 1930. The effect of temperature, relative humidity and exposure to sunlight upon the Mexican bean beetle. *J. Econ. Entomol.* 23: 945–955.

Miller, J. C. 1977. Ecological relationships among parasites of *Spodoptera praefica*. *Environ. Entomol.* 6: 581–585.

Miller, J. R. and R. S. Cowles. 1990. Stimulo-deterrent diversion: a concept and its possible application to onion maggot control. *J. Chem Ecol.* 16: 3197–3212.

Miller, J. R. and B. K. Haarer. 1981. Yeast and corn hydrolysates and other nutritious materials as attractants for onion and seed flies. *J. Chem Ecol.* 7: 555–562.

Miller, K. V. and R. N. Williams. 1981. An annotated bibliography of the genus *Glischrochilus* Reitter (Coleoptera: Nitidulidae, Cryptarchinae). Ohio Agric. Res. Devt. Ctr. Res. Circ. 266. 65 pp.

Miller, L. A. and A. J. DeLyzer. 1960. A progress report on studies of biology and ecology of the six-spotted leafhopper, *Macrosteles fascifrons* (Stål.), in southwestern Ontario. Proc. Entomol. Soc. Ont. 90: 7–13.

Miller, L. A. and R. J. McClanahan. 1960. Life-history of the seed-corn maggot, *Hylemya cilicrura* (Rond.) and of *H. liturata* (Mg.) (Diptera: Anthomyiidae) in southwestern Ontario. *Can. Entomol.* 92: 210–221.

Milliken, F. B. 1916. The false chinch bug and measures for controlling it. *USDA Farmers' Bull.* 762: 1–4.

Milliken, F. B. 1918. *Nysius ericae*, the false chinch bug. *J. Agric. Res.* 13: 571–578.

Milliken, F. B. 1921. Results of work on blister beetles in Kansas. USDA Bull. 967. 26 pp.

Mills, H. B. 1934. A Monograph of the Collembola of Iowa. Collegiate Press, Ames, Iowa. 141 pp.

Milner, R. J. and G. G. Lutton. 1986. Dependence of *Verticillium lecanii* (Fungi: Hyphomycetes) on high humidities for infection and sporulation using *Myzus persicae* (Homoptera: Aphididae) as host. *Environ. Entomol.* 15: 380–382.

Miner, F. D. 1966. Biology and control of stink bugs on soybeans. Arkansas Agr. Exp. Stn. Bull. 708. 40 pp.

Minkenberg, O. P. J. M. 1988. Life history of the agromyzid fly *Liriomyza trifolii* on tomato at different temperatures. *Entomol. Exp. Appl.* 48: 73–84.

Minkenberg, O. P. J. M. and J. C. van Lenteren. 1986. The leafminers *Liriomyza bryoniae* and *L. trifolii* (Diptera: Agromyzidae), their parasites and host plants: a review. Wageningen Agric. Univ. Papers 86–2. 50 pp.

Mitchell, E. R. 1967. Life history of *Pseudoplusia includens* (Walker) (Lepidoptera: Noctuidae). *J. Georgia Entomol. Soc.* 2: 53–57.

Mitchell, E. R. 1973. Nocturnal activity of adults of three species of loopers, based on collections in pheromone traps. *Environ. Entomol.* 2: 1078–1080.

Mitchell, E. R. 1978. Relationship of planting date to damage by earworms in commercial sweet corn in north central Florida. *Florida Entomol.* 61: 251–255.

Mitchell, E. R. 1979. Migration by *Spodoptera exigua* and *S. frugiperda*, North American style. Pages 386–393 *in* Movement of Highly Mobile Insects: Concepts and Methodology in Research. North Carolina State Univ., Raleigh, North Carolina.

Mitchell, E. R. and J. H. Tumlinson. 1994. Response of *Spodoptera exigua* and *S. eridania* (Lepidoptera: Noctuidae) males to synthetic pheromone and *S. exigua* females. *Florida Entomol.* 77: 237–247.

Mitchell, E. R., M. Jacobson, and A. H. Baumhover. 1975. *Heliothis* spp.: disruption of pheromonal communication with (Z)-9-tetradecen-1-ol formate. *Environ. Entomol.* 4: 577–579.

Mitchell, E. R., H. Sugie, and J. H. Tumlinson. 1983. *Spodoptera exigua*: capture of feral males in traps baited with blends of pheromone components. *J. Chem Ecol.* 9: 95–104.

Mitchell, E. R., R. B. Chalfant, G. L. Greene, and C. S. Creighton. 1975. Soybean looper: populations in Florida, Georgia, and South Carolina, as determined with pheromone-baited BL traps. *J. Econ. Entomol.* 68: 747–750.

Mitchell, E. R., M. Kehat, F. C. Tingle, and J. R. McLaughlin. 1997. Suppression of mating by beet armyworm (Noctuidae: Lepidoptera) in cotton with pheromone. *J. Agric. Entomol.* 14: 17–28.

Mitchell, F. L. and J. R. Fuxa. 1987. Distribution, abundance, and sampling of fall armyworm (Lepidoptera: Noctuidae) in south-central Louisiana cornfields. *Environ. Entomol.* 16: 453–458.

Mittler, T. E., J. A. Tsitsipis, and J. E. Kleinjan. 1970. Utilization of dehydroascorbic acid and some related compounds by the aphid *Myzus persicae* feeding on an improved diet. *J. Insect Physiol.* 16: 2315–2326.

Miyatake, T., K. Kawasaki, T. Kohama, S. Moriya, and Y. Shimoji. 1995. Dispersal of male sweetpotato weevils (Coleoptera: Curculionidae) in fields with or without sweet potato plants. *Environ. Entomol.* 24: 1167–1174.

Mohamed, A. K. A., J. V. Bell, and P. P. Sikorowski. 1978. Field cage tests with *Nomuraea rileyi* against corn earworm larvae on sweet corn. *J. Econ. Entomol.* 71: 102–104.

Monti, L., B. Lalanne–Cassou, P. Lucas, C. Malosse, and J.–F. Silvain. 1995. Differences in sex pheromone communication systems of closely related species: *Spodoptera latifascia* (Walker) and *S. descoinsi* Lalanne–Cassou & Silvain (Lepidoptera: Noctuidae). *J. Chem Ecol.* 21: 641–660.

Moon, M. S. 1967. Phagostimulation of a monophagous aphid. Oikos 18: 96–101.

Moore, K. C. and M. A. Erlandson. 1988. Isolation of *Aspergillus parasiticus* Speare and *Beauvaria bassiana* (Bals.) Vuillemin from melanopline grasshoppers (Orthoptera: Acrididae) and demonstration of their pathogenicity in *Melanoplus sanquinipes* (Fabricius). *Can. Entomol.* 120: 989–991.

Moran, E. J. and C. Lyle. 1940. Observations on *Cirphis unipunctata* Haworth in Mississippi. *J. Econ. Entomol.* 33: 768–769.

Morgan, L. W. and J. C. French. 1971. Granulate cutworm control in peanuts in Georgia. *J. Econ. Entomol.* 64: 937–939.

Morgan, L. W. and J. W. Todd. 1975. Insecticidal baits for control of corn earworm on peanuts and soybeans and green cloverworm on soybeans. *J. Georgia Entomol. Soc.* 10: 18–25.

Morrill, W. L. 1975. Density and emergence of Japanese beetle and green June beetle adults from fescue sod in northern Georgia. *J. Georgia Entomol. Soc.* 10: 277–280.

Morris, O. N. 1986. Susceptibility of the bertha armyworm, *Mamestra configurata* (Lepidoptera: Noctuidae), to commercial formulations of *Bacillus thuringiensis* var. *kurstaki*. *Can. Entomol.* 118: 473–478.

Morris, O. N. 1987. Evaluation of the nematode, *Steinernema feltiae* Filipjev, for the control of the crucifer flea beetle, *Phyllotreta cruciferae* (Goeze) (Coleoptera: Chrysomelidae). *Can. Entomol.* 119: 95–102.

Morris, O. N. and V. Converse. 1991. Effectiveness of steinernematid and heterorhabditid nematodes against noctuid, pyralid, and geometrid species in soil. *Can. Entomol.* 123: 55–61.

Morris, R. F. 1958. Biology and control of the purple-backed cabbageworm in Newfoundland. *J. Econ. Entomol.* 3: 281–284.

Morrison, W. P., B. C. Pass, and C. S. Crawford. 1972. Effect of humidity on eggs of two populations of the bluegrass webworm. *Environ. Entomol.* 1: 218–221.

Morrison, W. P., D. E. Mock, J. D. Stone, and J. Whitworth. 1977. A bibliography of the southwestern corn borer, *Diatraea grandiosella* Dyar (Lepidoptera: Pyralidae). *Bull. Entomol. Soc. Am.* 23: 185–190.

Morrow, B. J., J. C. Pendland, and D. G. Boucias. 1987. *Entomophaga aulicae* in tomato hornworm, *Manduca quinquemaculata*. *J. Invertebr. Pathol.* 50: 330–332.

Morse, J. G., R. L. Metcalf, J. R. Carey, and R. V. Dowell. (eds.) 1995. Proceedings: The Medfly in California: Defining Critical Research. Univ. California, Riverside. 318 pp.

Moulton, M. E., R. A. Higgins, S. M. Welch, and F. L. Poston. 1992. Mortality of second-generation immatures of the southwestern corn borer (Lepidoptera: Pyralidae). J. Econ. Entomol. 85: 963–966.

Mound, L. A. and R. Marullo. 1996. The thrips of Central and South America: an introduction (Insecta: Thysanoptera). Mem. Entomol. Inter. 6: 1–487.

Muchmore, W. B. 1990. Terrestrial Isopoda. Pages 805–817 in D. L. Dindal (ed.). Soil Biology Guide. John Wiley & Sons, New York.

Mueller, A. J. and A. W. Haddox. 1980. Observations on seasonal development of bean leaf beetle, Cerotoma trifurcata (Förster) and incidence of bean pod mottle virus in Arkansas soybean. J. Georgia Entomol. Soc. 15: 398–403.

Mueller, A. J. and S. Kunnalaca. 1979. Parasites of green cloverworm on soybean in Arkansas. Environ. Entomol. 8: 376–379.

Mueller, T. F., L. H. M. Blommers, and P. J. M. Mols. 1988. Earwig (Forficula auricularia) predation on the woolly apple aphid, Eriosoma lanigerum. Entomol. Exp. Appl. 47: 145–152.

Mukerji, M. K. and D. G. Harcourt. 1970. Design of a sampling plan for studies on the population dynamics of the cabbage maggot, Hylemya brassicae (Diptera: Anthomyiidae). Can. Entomol. 102: 1513–1518.

Mukerji, M. K., A. B. Ewen, C. H. Craig, and R. J. Ford. 1981. Evaluation of insecticide-treated bran baits for grasshopper control in Saskatchewan (Orthoptera: Acrididae). Can. Entomol. 113: 705–710.

Mulkern, G. B., J. F. Anderson, and M. A. Brusven. 1962. Biology and Ecology of North Dakota grasshoppers I. Food habits and preferences of grasshoppers associated with alfalfa fields. North Dakota Agric. Exp. Stn. Res. Rep. 7. 26 pp.

Mulkern, G. B., D. R. Toczek, and M. A. Brusven. 1964. Biology and ecology of North Dakota grasshoppers II. Food habits and preference of grasshoppers associated with the Sand Hills Prairie. North Dakota Agric. Exp. Stn. Res. Rep. 11. 59 pp.

Mullen, M. A. 1981. Sweetpotato weevil, Cylas formicarius elegantulus (Summers): development, fecundity, and longevity. Ann. Entomol. Soc. Am. 74: 478–481.

Mullen, M. A., A. Jones, D. R. Paterson, and T. E. Boswell. 1985. Resistance in sweet potatoes to the sweetpotato weevil, Cylas formicarius elegantulus (Summers). J. Entomol. Sci. 20: 345–350.

Mullen, M. A., A. Jones, R. T. Arbogast, J. M. Schalk, D. R. Paterson, T. E. Boswell, and D. R. Earhart. 1980. Field selection of sweet potato lines and cultivars for resistance to the sweetpotato weevil. J. Econ. Entomol. 73: 288–290.

Munkvold, G. P., D. C. McGee, and A. Iles. 1996. Effects of imidacloprid seed treatment of corn on foliar feeding and Erwinia stewartii transmission by the corn flea beetle. Plant Dis. 80: 747–749.

Munroe, E. 1966. Revision of North American species of Udea Guenée (Lepidoptera: Pyralidae). Mem. Entomol. Soc. Canada 49. 57 pp.

Munroe, E. 1972. The Moths of America North of Mexico. Fascicle 13.1B; pages 137–250. Pyraloidea, Pyralidae (Part). E. W. Classey, Ltd., London.

Munroe, E. 1973. The Moths of America North of Mexico. Fascicle 13.1C; pages 253–304. Pyraloidea, Pyralidae (Part). E. W. Classey, Ltd., London.

Munroe, E. 1976. The Moths of America North of Mexico. Fascicle 13.2A and B; pages 1–78. Pyraloidea, Pyralidae (Part). E. W. Classey, Ltd., London.

Munyaneza, J. and J. E. McPherson. 1994. Comparative study of life histories, laboratory rearing, and immature stages of Euschistus servus and Euschistus variolarius (Hemiptera: Pentatomidae). Great Lakes Entomol. 26: 263–274.

Musick, G. J. and P. J. Suttle. 1973. Suppression of armyworm damage to no-tillage corn with granular carbofuran. J. Econ. Entomol. 66: 735–737.

Musick, G. J., H. C. Chiang, W. H. Luckmann, Z. B. Mayo, and F. T. Turpin. 1980. Impact of planting dates of field corn on beetle emergence and damage by the western and the northern corn rootworms in the corn belt. Ann. Entomol. Soc. Am. 73: 207–215.

Mussen, E. C. and H. C. Chiang. 1974. Development of the picnic beetle, Glischrochilus quadrisignatus (Say), at various temperatures. Environ. Entomol. 3: 1032–1034.

Nair, G. A. 1984. Breeding and population biology of the terrestrial isopod, Porcellio laevis (Latreille), in the Delhi region. Symp. Zool. Soc. London 53: 315–337.

Nair, K. S. S. and F. L. McEwen. 1973. The seed maggot complex, Hylemya (Delia) platura and H. (Delia) liturata (Diptera: Anthomyiidae), as primary pests of radish. Can. Entomol. 105: 445–447.

Namba, R. and S. Y. Higa. 1971. Host plant studies of the corn planthopper, Peregrinus maidis (Ashmead), in Hawaii. Proc. Hawaiian Entomol. Soc. 21: 105–108.

Namba, R. and E. S. Sylvester. 1981. Transmission of cauliflower mosaic virus by the green peach, turnip, cabbage, and pea aphids. J. Econ. Entomol. 74: 546–551.

Naranjo, S. E. 1996. Sampling Bemisia for research and pest management applications. Pages 206–224 in D. Gerling and D. Mayer (eds.). Bemisia 1995: Taxonomy, Damage Control and management. Intercept Press, Andover, United Kingdom

Naranjo, S. E. and A. J. Sawyer. 1987. Reproductive biology and survival of Diabrotica barberi (Coleoptera: Chrysomelidae): effect of temperature, food, and seasonal time of emergence. Ann. Entomol. Soc. Am. 80: 841–848.

Naranjo, S. E. and A. J. Sawyer. 1988. Impact of host plant phenology on the population dynamics and oviposition of northern corn rootworms, Diabrotica barberi (Coleoptera: Chrysomelidae), in field corn. Environ. Entomol. 17: 508–521.

Natwick, E. G. and F. G. Zalom. 1984. Surveying sweetpotato whitefly in the Imperial Valley. California Agric. 38(2): 11.

Nault, B. A. and G. G. Kennedy. 1996a. Timing insecticide applications for managing European corn borer (Lepidoptera: Pyralidae) infestations in potato. Crop Prot. 15: 465–471.

Nault, B. A. and G. G. Kennedy. 1996b. Sequential sampling plans for use in timing insecticide applications for control of European corn borer (Lepidoptera: Pyralidae) in potato. J. Econ. Entomol. 89: 1468–1476.

Nault, B. A. and G. G. Kennedy. 1996c. Evaluation of Colorado potato beetle (Coleoptera: Chrysomelidae) defoliation with concomitant European corn borer (Lepidoptera: Pyralidae) damage on potato yield. J. Econ. Entomol. 89: 475–480.

Nault, B. A. and G. G. Kennedy. 1999. Influence of foliar-applied Bacillus thuringiensis subsp. tenebrionis and an early potato harvest on abundance and overwinter survival of Colorado potato beetles (Coleoptera: Chrysomelidae) in North Carolina. J. Econ. Entomol. 92: 1165–1171.

Nault, L. R. 1983. Origins of leafhopper vectors of maize pathogens in Mesoamerica. Pages 75–82 in D. T. Gordon, J. K. Knoke, L. R. Nault, and R. M. Ritter (eds.). Proceedings International Maize Virus Disease Collquium and Workshop, 2–6 August 1982. Ohio State Univ., Ohio Agric. Res. Devt. Ctr., Wooster.

Nechols, J. R. 1987. Voltinism, seasonal reproduction, and diapause in the squash bug (Heteroptera: Coreidae) in Kansas. Environ. Entomol. 16: 269–273.

Nechols, J. R., J. L. Tracy, and E. A. Vogt. 1989. Comparative ecological studies of indigenous egg parasitoids (Hymenoptera:

Scelionidae: Encyrtidae) of the squash bug, *Anasa tristis* (Hemiptera: Coreidae). *J. Kansas Entomol. Soc.* 62:177–188.

Negron, J. F. and T. J. Riley. 1985. Effect of chinch bug (Heteroptera: Lygaeidae) feeding in seedling field corn. *J. Econ. Entomol.* 78:1370–1372.

Negron, J. F. and T. J. Riley. 1987. Southern green stink bug, *Nezara viridula* (Heteroptera: Pentatomidae), feeding in corn. *J. Econ. Entomol.* 80:666–669.

Negron, J. F. and T. J. Riley. 1990. Long-term effects of chinch bug (Hemiptera: Lygaeidae) feeding on corn. *J. Econ. Entomol.* 83:618–620.

Negron, J. F. and T. J. Riley. 1991. Seasonal migration and overwintering of the chinch bug (Hemiptera: Lygaeidae) in Louisiana. *J. Econ. Entomol.* 84:1681–1685.

Neilson, C. L. and D. G. Finlayson. 1953. Notes on the biology of the tuber flea beetle, *Epitrix tuberis* Gentner (Coleoptera: Chrysomelidae), in the interior of British Columbia. *Can. Entomol.* 85:31–32.

Neiswander, C. R. 1944. The ring-legged earwig, *Euborellia annulipes* (Lucas). Ohio Agric. Exp. Stn. Bull. 648. 14 pp.

Neuenschwander, P. and K. S. Hagen. 1980. Role of the predator *Hemerobius pacificus* in a non-insecticide treated artichoke field. *Environ. Entomol.* 9:492–495.

Neunzig, H. H. 1963. Wild host plants of the corn earworm and the tobacco budworm in eastern North Carolina. *J. Econ. Entomol.* 56:135–139.

Neunzig, H. H. 1964. The eggs and early-instar larvae of *Heliothis zea* and *Heliothis virescens* (Lepidoptera: Noctuidae). *Ann. Entomol. Soc. Am.* 57:98–102.

Neunzig, H. H. 1969. The biology of the tobacco budworm and the corn earworm in North Carolina with particular reference to tobacco as a host. North Carolina Agric. Exp. Stn. Tech. Bull. 196. 76 pp.

Newman, W. and D. Pimentel. 1974. Garden peas resistant to the pea aphid. *J. Econ. Entomol.* 67:365–367.

Newsom, L. D., J. S. Roussel, and C. E. Smith. 1953. The tobacco thrips, its seasonal history and status as a cotton pest. Louisiana Agric. Exp. Stn. Tech. Bull. 474. 36 pp.

Ng, S.–S., F. M. Davis, and J. C. Reese. 1993. Southwestern corn borer (Lepidoptera: Pyralidae) and fall armyworm (Lepidoptera: Noctuidae): comparative developmental biology and food consumption and utilization. *J. Econ. Entomol.* 86:394–400.

Ng, S.–S., F. M. Davis, and W. P. Williams. 1990. Ovipositional response of southwestern corn borer (Lepidoptera: Pyralidae) and fall armyworm (Lepidoptera: Noctuidae) to selected maize hybrids. *J. Econ. Entomol.* 83:1575–1577.

N'Guessan, K. F. and R. B. Chalfant. 1990. Dose response of the cowpea curculio (Coleoptera: Curculionidae) from different regions of Georgia to some currently used pyrethroid insecticides. *J. Entomol. Sci.* 25:219–222.

Nguyen, K. B. and G. C. Smart, Jr. 1992. Life cycle of *Steinernema scapterisci* Nguyen and Smart, 1990. *J. Nematol.* 24:160–169.

Ni, X., J. P. McCaffrey, R. L. Stoltz, and B. L. Harmon. 1990. Effects of postdiapause adult diet and temperature on oogenesis in the cabbage seedpod weevil (Coleoptera: Chrysomelidae). *J. Econ. Entomol.* 83:2246–2251.

Nickle, D. A. and J. L. Castner. 1984. Introduced species of mole crickets in the United States, Puerto Rico, and the Virgin Islands (Orthoptera: Gryllotalpidae). *Ann. Entomol. Soc. Am.* 77:450–465.

Nickle, D. A. and T. J. Walker. 1974. A morphological key to field crickets of southeastern United States (Orthoptera: Gryllidae: *Gryllus*). *Florida Entomol.* 57:8–12.

Nickle, W. R., W. J. Connick, Jr., and W. W. Cantelo. 1994. Effects of pesta-pelletized *Steinernema carpocapsae* (All) on western corn rootworms and Colorado potato beetles. *J. Nematol.* 26:249–250.

Nielsen, B. S. and T. S. Jensen. 1993. Spring dispersal of *Sitona lineatus*: the use of aggregation pheromone traps for monitoring. *Entomol. Exp. Appl.* 66:21–30.

Nielson, M. W. 1968. The leafhopper vectors of phytopathogenic viruses (Homoptera: Cicadellidae) taxonomy, biology, and virus transmission. USDA Tech. Bull. 1382. 386 pp.

Niemczyk, H. D. 1975. Status of insecticide resistance in Japanese beetle in Ohio, 1974. *J. Econ. Entomol.* 68:583–584.

Nihoul, P. 1993. Controlling glasshouse climate influences the interaction between tomato glandular trichome, spider mite and predatory mite. *Crop Prot.* 12:443–447.

Nilakhe, S. S., R. B. Chalfant, and S. V. Singh. 1981a. Field damage to lima beans by different stages of southern green stink bug. *J. Georgia Entomol. Soc.* 16:392–396.

Nilakhe, S. S., R. B. Chalfant, and S. V. Singh. 1981b. Damage to southern peas by different stages of the southern green stink bug. *J. Georgia Entomol. Soc.* 16:409–414.

Nishida, T. 1954. Further studies on the treatment of border vegetation for melon fly control. *J. Econ. Entomol.* 47:226–229.

Nishida, T. 1955. Natural enemies of the melon fly, *Dacus cucurbitae* Coq. in Hawaii. *J. Econ. Entomol.* 48:171–178.

Nishida, T. 1978. Management of the corn planthopper in Hawaii. *FAO Plant Prot. Bull.* 26:5–9.

Nishida, T. and H. A. Bess. 1957. Studies on the ecology and control of the melon fly *Dacus (Strumeta) cucurbitae* Coquillett (Diptera: Tephritidae). Hawaii Agric. Exp. Stn. Tech. Bull. 34. 44 pp.

Nitao, J. K. and M. R. Berenbaum. 1988. Laboratory rearing of the parsnip webworm, *Depressaria pastinacella* (Lepidoptera: Oecophoridae). *Ann. Entomol. Soc. Am.* 81:485–487.

Nolting, S. P. and F. L. Poston. 1982. Application of *Bacillus thuringiensis* through center-pivot irrigation systems for control of the southwestern corn borer and European corn borer (Lepidoptera: Pyralidae). *J. Econ. Entomol.* 75:1069–1073.

North, R. C. and A. M. Shelton. 1986a. Ecology of Thysanoptera within cabbage fields. *Environ. Entomol.* 15:520–526.

North, R. C. and A. M. Shelton. 1986b. Overwintering of the onion thrips, *Thrips tabaci* (Thysanoptera: Thripidae), in New York. *Environ. Entomol.* 15:695–699.

Norton, A. P. and S. C. Welter. 1996. Augmentation of the egg parasitoid *Anaphes iole* (Hymenoptera: Mymaridae) for *Lygus hesperus* (Heteroptera: Miridae) management in strawberries. *Environ. Entomol.* 25:1406–1414.

Novero, E. S., R. H. Painter, and C. V. Hall. 1962. Interrelations of the squash bug, *Anasa tristis*, and six varieties of squash (*Cucurbita* spp.). *J. Econ. Entomol.* 55:912–919.

Nuessly, G. S. and R. T. Nagata. 1995. Pepper varietal response to thrips feeding. Pages 115–118 *in* B. L. Parker, M. Skinner, and T. Lewis (eds.), Thrips Biology and Management. Plenum Press, New York.

Nuttycombe, J. W. 1930. Oviposition of the corn earworm moth in relation to nectar flow of some flowering plants. *J. Econ. Entomol.* 23:725–729.

Nwanze, K. F., E. Horber, and C. W. Pitts. 1975. Evidence for ovipositional preference of *Callosobruchus maculatus* for cowpea varieties. *Environ. Entomol.* 4:409–412.

Oatman, E. R. 1959. Host range studies of the melon leaf miner, *Liriomyza pictella* (Thomson) (Diptera: Agromyzidae). *Ann. Entomol. Soc. Am.* 52:739–741.

Oatman, E. R. 1966a. Parasitization of corn earworm eggs on sweet corn silk in southern California, with notes on larval infestations and predators. *J. Econ. Entomol.* 59:830–835.

Oatman, E. R. 1966b. An ecological study of cabbage looper and imported cabbageworm populations on cruciferous crops in southern California. *J. Econ. Entomol.* 59:1134–1139.

Oatman, E. R. 1970. Ecological studies of the tomato pinworm on tomato in southern California. *J. Econ. Entomol.* 63: 1531–1534.

Oatman, E. R. and A. E. Michelbacher. 1958. The melon leafminer, *Liriomyza pictella* (Thomson) (Diptera: Agromyzidae). *Ann. Entomol. Soc. Am.* 51: 557–566.

Oatman, E. R., and G. R. Platner. 1969. An ecological study of insect populations on cabbage in southern California. *Hilgardia* 40: 1–40.

Oatman, E. R. and G. R. Platner. 1971. Biological control of the tomato fruitworm, cabbage looper, and hornworms on processing tomatoes in southern California, using mass releases of *Trichogramma pretiosum*. *J. Econ. Entomol.* 64: 501–506.

Oatman, E. R. and G. R. Platner. 1972. An ecological study of lepidopterous pests affecting lettuce in coastal southern California. *Environ. Entomol.* 1: 202–204.

Oatman, E. R. and G. R. Platner. 1973. Parasitization of natural enemies attacking the cabbage aphid on cabbage in southern California. *Environ. Entomol.* 2: 365–367.

Oatman, E. R. and G. R. Platner. 1974. Parasitization of the potato tuberworm in southern California. *Environ. Entomol.* 3: 262–264.

Oatman, E. R. and G. R. Platner. 1989. Parasites of the potato tuberworm, tomato pinworm, and other, closely related gelechiids. Proc. Hawaiian Entomol. Soc. 29: 23–30.

Oatman, E. R., J. A. Wyman, and G. R. Platner. 1979. Seasonal occurrence and parasitization of the tomato pinworm on fresh market tomatoes in southern California. *Environ. Entomol.* 8: 661–664.

Oatman, E. R., I. M. Hall, K. Y. Arakawa, G. R. Platner, L. A. Bascom, and C. C. Beegle. 1970. Control of the corn earworm on sweet corn in southern California with a nuclear polyhedrosis virus and *Bacillus thuringiensis*. *J. Econ. Entomol.* 63: 415–421.

Obrycki, J. J., M. J. Tauber, and W. M. Tingey. 1983. Predator and parasitoid interaction with aphid-resistant potatoes to reduce aphid densities: a two-year field study. *J. Econ. Entomol.* 76: 456–462.

Oetting, R. D., R. J. Beshear, T.-X. Liu, S. K. Braman, and J. R. Baker. 1993. Biology and identification of thrips on greenhouse ornamentals. Georgia Agric. Exp. Stn. Res. Bull. 414. 20 pp.

Ofuya, T. I. and K. A. Bamigbola. 1991. Damage potential, growth and development of the seed beetle, *Callosobruchus maculatus* (Fabricius) (Coleoptera: Bruchidae), on some tropical legumes. *Trop. Agric.* 68: 33–36.

Ogenga-Latigo, M. W. and B. M. Khaemba. 1985. Some aspects of the biology of the black bean aphid *Aphis fabae* Scopoli reared on the common bean *Phaseolus vulgaris* L. *Insect Sci. Applic.* 6: 591–593.

Ogunwolu, E. O. and D.H Habeck. 1975. Comparative life-histories of three *Mocis* spp. in Florida (Lepidoptera: Noctuidae). *Florida Entomol.* 58: 97–103.

Ogunwolu, E. O. and D.H Habeck. 1979. Descriptions and keys to larvae and pupae of the grass loopers, *Mocis* spp., in Florida (Lepidoptera: Noctuidae). *Florida Entomol.* 62: 402–407.

O'Hayer, K. W., G. A. Schultz, C. E. Eastman, and J. Fletcher. 1984. Newly discovered plant hosts of *Spiroplasma citri*. *Plant Dis.* 68: 336–338.

Ohinata, K., M. Jacobson, R. M. Kobayashi, D. L. Chambers, M.S, Fujimoto, and H. H. Higa. 1982. Oriental fruit fly and melon fly: biological and chemical studies of smoke produced by males. *J. Environ. Sci. Health* (A) 17: 197–216.

Ohnesorge, B. and G. Rapp. 1986. Monitoring *Bemisia tabaci*: a review. *Agric. Ecosystems Environ.* 17: 21–27.

Okabe, K. and H. Amano. 1991. Penetration and population growth of the Robine bulb mite, *Rhizoglyphus robini* Claparede (Acari: Acaridae), on healthy and *Fusarium*-infected rakkyo bulbs. *Appl. Entomol. Zool.* 26: 129–136.

Okumura, G. T. 1962. Identification of lepidopterous larvae attacking cotton with illustrated key (primarily California species). California Dept. Agric. Bur. Entomol. Spec. Publ. 282. 80 pp.

Oliver, A. D. and J. B. Chapin. 1981. Biology and illustrated key for the identification of twenty species of economically important noctuid pests. Louisiana Agric. Exp. Stn. Bull. 733. 26 pp.

Oliver, A. D. and J. B. Chapin. 1983. Biology and distribution of the yellowmargined leaf beetle, *Microtheca ochroloma* Stål, with notes on *M. picea* (Guerin) (Coleoptera: Chrysomelidae) in Louisiana. *J. Georgia Entomol. Soc.* 18: 229–234.

Olmstead, K. L. and R. F. Denno. 1992. Cost of shield defence for tortoise beetles (Coleoptera: Chrysomelidae). *Ecol. Entomol.* 17: 237–243.

Olmstead, K. L. and R. F. Denno. 1993. Effectiveness of tortoise beetle larval shields against different predator species. *Ecology* 74: 1394–1405.

Olsen, C. E., K. S. Pike, L. Boydston, and D. Allison. 1993. Keys for identification of apterous viviparae and immatures of six small grain aphids (Homoptera: Aphididae). *J. Econ. Entomol.* 86: 137–148.

Olson, D. C. and R. W. Rings. 1969. Climbing responses of the spotted cutworm, *Amathes c-nigrum*. *Ann. Entomol. Soc. Am.* 62: 1403–1406.

O'Neal, J. and G. P. Markin. 1975a. The larval instars of the imported fire ant, *Solenopsis invicta* Buren. *J. Kansas Entomol. Soc.* 48: 141–151.

O'Neal, J. and G. P. Markin. 1975b. Brood development of the various castes of the imported fire ant, *Solenopsis invicta* Buren. *J. Kansas Entomol. Soc.* 48: 132–159.

O'Neill, R. V. and D. E. Reichle. 1970. Urban infestation by the millipede, *Oxidus gracillus* (Koch). *J. Tennessee Acad. Sci.* 45: 114–115.

Onillon, J. C. 1990. The use of natural enemies for the biological control of whiteflies. Pages 287–313 *in* D. Gerling (ed.). Whiteflies: their Bionomics, Pest Status and Management. Intercept, United Kingdom.

Onsager, J. A. 1975. Pacific Coast wireworm: relationship between injury and damage to potatoes. *J. Econ. Entomol.* 68: 203–204.

Onsager, J. A. and G. B. Hewitt. 1982. Rangeland grasshoppers: average longevity and daily rate or mortality among six species in nature. *Environ. Entomol.* 11: 127–133.

Onsager, J. A. and G. B. Mulkern. 1963. Identification of eggs and egg-pods of North Dakota grasshoppers (Orthoptera: Acrididae). North Dakota Agric. Exp. Stn. Bull. 446. 47 pp.

Opler, P. A. and G. O. Krizek. 1984. Butterflies East of the Great Plains, an Illustrated Natural History. The Johns Hopkins Univ. Press, Baltimore. 294 pp.

Oramas, D., J. Rodriguez, and A. L. Gonzales. 1990. Effect on yam (*Dioscorea rotundata* Poir) of soil spray and seed treatment with the nematicide-insecticide Oxamyl L, and soil treatments with Phenamiphos 15G. *J. Agric. Univ. Puerto Rico* 74: 103110.

O'Rourke, P. K., E. C. Burkness, and W. D. Hutchison. 1998. Development and validation of a fixed-precision sequential sampling plan for aster leafhopper (Homoptera: Cicadellidae) in carrot. *Environ. Entomol.* 27: 1463–1468.

Orozco-Santos, M., O. Perez-Zamora, and O. Lopez-Arriaga. 1995. Floating row cover and transparent mulch to reduce insect populations, virus diseases and increase yield in cantaloupe. *Florida Entomol.* 78: 493–501.

Orr, D. B., D. J. Boethel, and M. B. Layton. 1989. Effect of insecticide applications in soybeans on *Trissolcus basalis* (Hymenoptera: Scelionidae). *J. Econ. Entomol.* 82: 1078–1084.

Osborn, H. 1916. Studies of life histories of leafhoppers of Maine. Maine Agric. Exp. Stn. Bull. 248: 53–80.

Osborne, L. S. and Z. Landa. 1992. Biological control of whiteflies with entomopathogenic fungi. *Florida Entomol.* 75: 456–471.

Osborne, L. S. and F. L. Petitt. 1985. Insecticidal soap and the predatory mite, *Phytoseiulus persimilis* (Acari: Phytoseiidae), used in management of the twospotted spider mite (Acari: Tetranychidae) on greenhouse grown foliage plants. *J. Econ. Entomol.* 78: 687–691.

Ota, A. K. 1973. Wireworm damage to polyethylene tubing used in a drip irrigation system. *J. Econ. Entomol.* 66: 824–825.

Overholt, W. A. and J. W. Smith, Jr. 1989. *Pediobius furvus* parasitization of overwintering generation southwestern corn borer pupae. *Southwestern Entomol.* 14: 35–40.

Overholt, W. A., A. E. Knutson, J. W. Smith, Jr., and F. E. Gilstrap. 1990. Distribution and sampling of southwestern corn borer (Lepidoptera: Pyralidae) in preharvest corn. *J. Environ. Entomol.* 83: 1370–1375.

Own, O. S. and W. M. Brooks. 1986. Interactions of the parasite *Pediobius foveolatus* (Hymenoptera: Eulophidae) with two *Nosema* spp. (Microsporida: Nosematidae) of the Mexican bean beetle (Coleoptera: Coccinellidae). *Environ. Entomol.* 15: 32–39.

Ozaki, H. Y. and W. G. Genung. 1982. Insecticide evaluation for pepper weevil control. *Proc. Florida State Hort. Soc.* 95: 347–348.

Paddock, F. B. 1912. The sugar-beet web worm *Loxostege sticticalis* Linn. *J. Econ. Entomol.* 5: 436–443.

Paddock, F. B. 1915a. The harlequin cabbage-bug. *Texas Agric. Exp. Stn. Bull.* 179. 9 pp.

Paddock, F. B. 1915b. The turnip louse. *Texas Agric. Exp. Stn. Bull.* 180. 77 pp.

Paddock, F. B. 1918. Studies on the harlequin bug. *Texas Agric. Exp. Stn. Bull.* 227. 65 pp.

Paddock, F. B. and H. J. Reinhard. 1919. The cowpea weevil. *Texas Agric. Exp. Stn. Bull.* 256. 92 pp.

Pair, S. D. 1994. Japanese honeysuckle (Caprifoliaceae): newly discovered host of *Heliothis virescens* and *Helicoverpa zea* (Lepidoptera: Noctuidae). *Environ. Entomol.* 23: 906–911.

Pair, S. D. 1997. Evaluation of systemically treated squash trap plants and attracticidal baits for early-season control of striped and spotted cucumber beetles (Coleoptera: Chrysomelidae) and squash bug (Hemiptera: Coreidae) in cucurbit crops. *J. Econ. Entomol.* 90: 1307–1314.

Pair, S. D. and H. R. Gross, Jr. 1984. Field mortality of pupae of the fall armyworm, *Spodoptera frugiperda* (J.E. Smith), by predators and a newly discovered parasitoid, *Diapetimorpha introita*. *J. Georgia Entomol. Soc.* 19: 22–26.

Palaniswamy, P. and R. P. Bodnaryk. 1994. A wild *Brassica* from Sicily provides trichome-based resistance against flea beetles, *Phyllotreta cruciferae* (Goeze) (Coleoptera: Chrysomelidae). *Can. Entomol.* 126: 1119–1130.

Palaniswamy, P. and R. J. Lamb. 1992. Host preferences of the flea beetles *Phyllotreta cruciferae* and *P. striolata* (Coleoptera: Chrysomelidae) for crucifer seedlings. *J. Econ. Entomol.* 85: 743–752.

Palamiswamy, P. and E. W. Underhill. 1988. Mechanisms of orientation disruption by sex pheromone components in the redbacked cutworm, *Euxoa ochrogaster* (Guenée) (Lepidoptera: Noctuidae). *Environ. Entomol.* 17: 432–441.

Palaniswamy, P. and I. Wise. 1994. Effects of neem-based products on the number and feeding activity of a crucifer flea beetle, *Phyllotreta cruciferae* (Goeze) (Coleoptera: Chrysomelidae), on canola. *J. Agric. Entomol.* 11: 49–60.

Palaniswamy, P., E. W. Underhill, and M. D. Chisholm. 1984. Orientation disruption of *Euxoa messoria* (Lepidoptera: Noctuidae) males with synthetic sex attractant components: field and flight tunnel studies. *Environ. Entomol.* 13: 36–40.

Palaniswamy, P., E. W. Underhill, W. F. Steck, and M. D. Chisholm. 1983. Responses of male redbacked cutworm, *Euxoa ochrogaster* (Lepidoptera: Noctuidae), to sex pheromone components in a flight tunnel. *Environ. Entomol.* 12: 748–752.

Pallant, D. 1972. The food of the grey field slug, *Agriolimax reticulatus* (Müller), on grassland. *J. Anim. Ecol.* 41: 761–769.

Palmer, J. M., L. A. Mound, and G. J. du Heaume. 1989. CIE guides to insects of importance to man. 2. Thysanoptera. CAB International, Wallingford, United Kingdom. 73 pp.

Palmer, M. A. 1952. Aphids of the Rocky Mountain Region. Thomas Say Foundation, Vol. 5. 452 pp.

Palumbo, J. C. 1995. Developmental rate of *Liriomyza sativae* (Diptera: Agromyzidae) on lettuce as a function of temperature. *Southwestern Entomol.* 20: 461–465.

Palumbo, J. C. and D. L. Kerns. 1994. Effects of imidacloprid as a soil treatment on colonization of green peach aphid and marketability of lettuce. *Southwestern Entomol.* 19: 339–346.

Palumbo, J. C., W. S. Fargo, and E. L. Bonjour. 1991. Colonization and seasonal abundance of squash bugs (Heteroptera: Coreidae) on summer squash with varied planting dates in Oklahoma. *J. Econ. Entomol.* 84: 224–229.

Palumbo, J. C., A. Tonhasca, Jr., and D. N. Byrne. 1995. Evaluation of three sampling methods for estimating adult sweetpotato whitefly (Homoptera: Aleyrodidae) abundance on cantaloupes. *J. Econ. Entomol.* 88: 1393–1400.

Paris, O. H. 1963. The ecology of *Armadillidium vulgare* (Isopoda: Oniscoidea) in California grassland: food, enemies, and weather. *Ecol. Mono.* 33: 1–22.

Parish, H. E. 1934. Biology of *Euschistus variolarius* P. De B. (family Pentatomidae; order Hemiptera). *Ann. Entomol. Soc. Am.* 27: 50–54.

Parker, F. W. and N. M. Randolph. 1972. Mass rearing the chinch bug in the laboratory. *J. Econ. Entomol.* 65: 894–895.

Parker, H. L. 1951. Parasites of the lima-bean pod borer in Europe. USDA Tech. Bull. 1036. 28 pp.

Parker, J. R. 1915. An outbreak of the alfalfa looper. *J. Econ. Entomol.* 8: 286–291.

Parker, J. R. 1930. Some effects of temperature and moisture upon *Melanoplus mexicanus* Saussure and *Camnula pellucida* Scudder (Orthoptera). Montana Agric. Exp. Stn. Bull. 223. 132 pp.

Parker, J. R. 1939. Grasshoppers and their control. USDA Farmers' Bull. 1828. 37 pp.

Parker, J. R. and R. L. Shotwell. 1932. Devastation of a large area by the differential and the two-striped grasshoppers. *J. Econ. Entomol.* 25: 174–187.

Parker, J. R. and C. Wakeland. 1957. Grasshopper egg pods destroyed by larvae of bee flies, blister beetles, and ground beetles. USDA Tech. Bull. 1165. 29 pp.

Parker, J. R., R. C. Newton, and R. L. Shotwell. 1955. Observations on mass flights and other activities of the migratory grasshopper. USDA Tech. Bull. 1109. 46 pp.

Parker, J. R., A. L. Strand, and H. L. Seamans. 1921. Pale western cutworm (*Porosagrotis orthogonia* Morr.). *J. Agric. Res.* 22: 289–321.

Parker, W. B. 1910. The life history and control of the hop flea-beetle. USDA Bur. Entomol. Bull. 82: 33–58.

Parker, W. E. 1994. Evaluation of the use of food baits for detecting wireworms (*Agriotes* spp., Coleoptera: Elateridae) in fields intended for arable crop production. *Crop. Prot.* 13: 271–276.

Parkman, P. and M. Shepard. 1981. Foliage consumption by yellow-striped armyworm larvae after parasitization by *Euplectrus plathypenae*. *Florida Entomol.* 64: 192–194.

Parkman, J. P. and G. C. Smart, Jr. 1996. Entomopathogenic nematodes, a case study: introduction of *Steinernema scapterisci* in Florida. *Biocontrol Sci. Tech.* 6: 413–419.

Parkman, P., J. A. Dusky, and V. H. Waddill. 1989. Biological studies of *Liriomyza sativae* (Diptera: Agromyzidae) on castor bean. *Environ. Entomol.* 18: 768–772.

Parkman, J. P., J. H. Frank. K. B. Nguyen, and G. C. Smart, Jr. 1993b. Dispersal of *Steinernema scapterisci* (Rhabditida: Steinernematidae) after inoculative applications for mole cricket (Orthoptera: Gryllotalpidae) control in pastures. *Biol. Control* 3: 226–232.

Parkman, J. P., J. H. Frank, T. J. Walker, and D. J. Schuster. 1996. Classical biological control of *Scapteriscus* spp. (Orthoptera: Gryllotalpidae) in Florida. *Environ. Entomol.* 25: 1415–1420.

Parkman, J. P., W. G. Hudson, J. H. Frank, K. B. Nguyen, and G. C. Smart, Jr. 1993a. Establishment and persistence of *Steinernema scapterisci* (Rhabditida: Steinernematidae) in field populations of *Scapteriscus* spp. mole crickets (Orthoptera: Gryllotalpidae). *J. Entomol. Sci.* 28: 182–190.

Parmenter, R. R. and J. A. McMahon. 1988. Factors limiting populations of arid-land darkling beetles (Coleoptera: Tenebrionidae): predation by rodents. *Environ. Entomol.* 17: 280–286.

Parrella, M. P. 1987. Biology of *Liriomyza*. *Annu. Rev. Entomol.* 32: 201–224.

Parrella, M. P. and J. A. Bethke. 1984. Biological studies of *Liriomyza huidobrensis* (Diptera: Agromyzidae) on chrysanthemum, aster and pea. *J. Econ. Entomol.* 77: 342–345.

Parrella, M. P. and C. B. Kiel. 1984. Integrated pest management: the lesson of *Liriomyza*. *Bull. Entomol. Soc. Am.* 30: 22–25.

Parrella, M. P. and L. T. Kok. 1977. The development and reproduction of *Bedellia somnulentella* on hedge bindweed and sweet potato. *Ann. Entomol. Soc. Am.* 70: 925–928.

Parrella, M. P., K. L. Robb, and J. Bethke. 1983. Influence of selected host plants on the biology of *Liriomyza trifolii* (Diptera: Agromyzidae). *Ann. Entomol. Soc. Am.* 76: 112–115.

Parrella, M. P., V. P. Jones, R. R. Youngman, and L. M. Lebeck. 1985. Effect of leaf mining and leaf stippling of *Liriomyza* spp. on photosynthetic rates of chrysanthemum. *Ann. Entomol. Soc. Am.* 78: 90–93.

Parrella, M. P., J. T. Yost, K. M. Heinz, and G. W. Ferrentino. 1989. Mass rearing of *Diglyphus begini* (Hymenoptera: Eulophidae) for biological control of *Liriomyza trifolii* (Diptera: Agromyzidae). *J. Econ. Entomol.* 82: 420–425.

Parrella, M. P., T. S. Bellows, R. J. Gill, J. K. Brown, and K. M.Heinz. 1992. Sweet-potato whitefly: prospects for biological control. *California Agric.* 46(1): 25–26.

Parrella, M. P., T. D. Paine, J. A. Bethke, K. L. Robb, and J. Hall. 1991. Evaluation of *Encarsia formosa* (Hymenoptera: Aphelinidae) for biological control of sweet-potato whitefly (Homoptera: Aleyrodidae) on poinsettia. *Environ. Entomol.* 20: 713–719.

Pashley, D. P. 1988. Current status of fall armyworm host strains. *Florida Entomol.* 71: 227–234.

Passoa, S. 1991. Color identification of economically important *Spodoptera* larvae in Honduras (Lepidoptera: Noctuidae). *Insecta Mun.* 5: 185–196.

Patch, E. M. 1915. Pink and green aphid of potato (*Macrosiphum solanifolii*). Maine Agric. Exp. Stn. Bull. 242: 205–223.

Patch, E. M. 1924. The buckthorn aphid (*Aphis abbreviata* Patch). Maine Agric. Exp. Stn. Bull. 317: 29–52.

Patch, E. M. 1925. The melon aphid (*Aphis gossypii* Glover). Maine Agric. Exp. Stn. Bull. 326: 185–196.

Patch, E. M. 1928. The foxglove aphid on potato and other plants. Maine Agric. Exp. Stn. Bull. 346: 49–60.

Patel, V. C. and H. N. Pitre. 1971. Transmission of bean pod mottle virus to soybean by the striped blister beetle, *Epicauta vittata*. *Plant Dis. Rep.* 55: 628–629.

Patel, V. C. and H. N. Pitre. 1976. Transmission of bean pod mottle virus by bean leaf beetle and mechanical inoculation to soybeans at different stages of growth. *J. Georgia Entomol. Soc.* 11: 289–293.

Pathak, R. S. 1988. Genetics of resistance to aphid in cowpea. *Crop Sci.* 28: 474–476.

Patrock, R. J. and D. J. Schuster. 1992. Feeding, oviposition and development of the pepper weevil (*Anthonomus eugenii* Cano), on selected species of Solanaceae. *Trop. Pest Manage.* 38: 65–69.

Päts, P. and R. S. Vernon. 1999. Fences excluding cabbage maggot flies and tiger flies (Diptera: Anthomyiidae) from large plantings of radish. *Environ. Entomol.* 28: 1124–1129.

Patton, R. L. and G. A. Mail. 1935. The grain bug (*Chlorochroa sayii* Stål) in Montana. *J. Econ. Entomol.* 28: 906–913.

Pausch, R. D. 1979. Observations on the biology of the seed corn beetles, *Stenolophus comma* and *Stenolophus lecontei*. *Ann. Entomol. Soc. Am.* 72: 24–28.

Pausch, R. D. and L. M. Pausch. 1980. Observations on the biology of the slender seedcorn beetle, *Clivina impressifrons* (Coleoptera: Carabidae). *Great Lakes Entomol.* 13: 189–194.

Pavuk, D. M. and B. R. Stinner. 1991. Relationship between weed communities in corn and infestation and damage by the stalk borer (Lepidoptera: Noctuidae). *J. Entomol. Sci.* 26: 253–260.

Payne, C. C. 1981. The susceptibility of the pea moth, *Cydia nigricana*, to infection by the granulosis virus of the codling moth, *Cydia pomonella*. *J. Invertebr. Pathol.* 38: 71–77.

Payne, C. C., G. M. Tatchell, and C. F. Williams. 1981. The comparative susceptibilities of *Pieris brassicae* and *P. rapae* to a granulosis virus from *P. brassicae*. *J. Invertebr. Pathol.* 38: 273–280.

Pearson, G. A. 1995. Sesiid pheromone increases squash vine borer (Lepidoptera: Sesiidae) infestation. *Environ. Entomol.* 24: 1627–1632.

Peay, W. E., G. W. Beards, and A. A. Swenson. 1969. Field evaluations of soil and foliar insecticides for control of the sugarbeet root maggot. *J. Econ. Entomol.* 62: 1083–1087.

Pedigo, L. P. 1971. Ovipositional response of *Plathypena scabra* (Lepidoptera: Noctuidae) to selected surfaces. *Ann. Entomol. Soc. Am.* 64: 647–651.

Pedigo, L. P., E. J. Bechinski, and R. A. Higgins. 1983. Partial life tables of the green cloverworm (Lepidoptera: Noctuidae) in soybean and a hypothesis of population dynamics in Iowa. *Environ. Entomol.* 12: 186–195.

Pedigo, L. P., J. D. Stone, and G. L. Lentz. 1973. Biological synopsis of the green cloverworm in central Iowa. *J. Econ. Entomol.* 66: 665–673.

Pelletier, Y., C. D. McLeod, and G. Bernard. 1995. Description of sublethal injuries caused to the Colorado potato beetle (Coleoptera: Chrysomelidae) by propane flamer treatment. *J. Econ. Entomol.* 88: 1203–1205.

Pena, J. E. 1992. Predator-prey interactions between *Typhlodromalus peregrinus* and *Polyphagotarsonemus latus*: effects of alternative prey and other food resources. *Florida Entomol.* 75: 241–248.

Pena, J. E. 1988. Chemical control of broad mites (Acarina: Tarsonemidae) in limes (*Citrus latifolia*). *Proc. Florida State Hort. Soc.* 101: 247–249.

Pena, J. E. and R. C. Bullock. 1994. Effects of feeding of broad mite (Acari: Tarsonemidae) on vegetative plant growth. *Florida Entomol.* 77: 180–184.

Pena, J. E. and L. Osborne. 1996. Biological control of *Polyphagotarsonemus latus* (Acarina: Tarsonemidae) in greenhouses and field trials using introductions of predaceous mites (Acarina: Phytoseiidae). *Entomophaga* 41: 279–285.

Pena, J. E. and V. H. Waddill. 1981. Southern armyworm and black cutworm damage to cassava at different growth stages. *J. Econ. Entomol.* 74: 271–275.

Pena, J. E. and V. H. Waddill. 1983. Larval and egg parasitism of *Keiferia lycopersicella* (Walsingham) (Lepidoptera: Gelechiidae) in southern Florida tomato fields. *Environ. Entomol.* 12: 1322–1326.

Pena, J. E., R. M. Baranowski, and H. A. Denmark. 1989. Survey of predators of the broad mite in southern Florida. *Florida Entomol.* 72: 373–377.

Pena, J. E., L. S. Osborne, and and R. E. Duncan. 1996. Potential of fungi as biocontrol agents of *Polyphagotarsonemus latus* (Acari: Tarsonemidae). *Entomophaga* 41: 27–36.

Pena, J. E., V. H. Waddill, and K. D. Elsey. 1987a. Population dynamics of the pickleworm and the melonworm (Lepidoptera: Pyralidae) in Florida. *Environ. Entomol.* 16: 1057–1061.

Pena, J. E., V. H. Waddill, and K. D. Elsey. 1987b. Survey of native parasites of the pickleworm, *Diaphania nitidalis* Stoll, and melonworm, *Diaphania hyalinata* (L.) (Lepidoptera: Pyralidae), in southern and central Florida. *Environ. Entomol.* 16: 1062–1066.

Pena, J. E., K. Pohronezny, V. H. Waddill, and J. Stimac. 1986. Tomato pinworm (Lepidoptera: Gelechiidae) artificial infestation: effect on foliar and fruit injury of ground tomatoes. *J. Econ. Entomol.* 79: 957–960.

Pencoe, N. L. and P. B. Martin. 1981. Development and reproduction of fall armyworms on several wild grasses. *Environ. Entomol.* 10: 999–1002.

Peng, C. and M. J. Weiss. 1992. Evidence of an aggregation pheromone in the flea beetle, *Phyllotreta cruciferae* (Goeze) (Coleoptera: Chrysomelidae). *J. Chem Ecol.* 18: 875–884.

Peng, C. and R. N. Williams. 1991. Effect of trap design, trap height, and habitat on the capture of sap beetles (Coleoptera: Nitidulidae) using whole-wheat bread dough. *J. Econ. Entomol.* 84: 1515–1519.

Pepper, B. B. 1942. The carrot weevil, *Listronotus latiusculus* (Bohe), in New Jersey and its control. New Jersey Agric. Exp. Stn. Bull. 693. 20 pp.

Pepper, B. B. 1943. The relationship between cropping practices and injury by *Heliothis armigera* with especial reference to lima beans and tomatoes. *J. Econ. Entomol.* 36: 329–330.

Pepper, J. H. 1938. The effect of certain climactic factors on the distribution of the beet webworm (*Loxostege sticticalis* L.) in North America. *Ecology* 19: 565–571.

Pepper, J. H. and E. Hastings. 1941. Life history and control of the sugar-beet webworm *Loxostege sticticalis* (L.). Montana Agric. Exp. Stn. Bull. 389. 32 pp.

Perfecto, I. 1991. Ants (Hymenoptera: Formicidae) as natural control agents of pests in irrigated maize in Nicaragua. *J. Econ. Entomol.* 84: 65–70.

Perfecto, I. and A. Sediles. 1992. Vegetational diversity, ants (Hymenoptera: Formicidae), and herbivorous pests in a neotropical agroecosystem. *Environ. Entomol.* 21: 61–67.

Perkins, W. D. 1979. Laboratory rearing of the fall armyworm. *Florida Entomol.* 62: 87–91.

Perring, T. M. and C. A. Farrar. 1986. Historical perspective and current world status of the tomato russet mite (Acari: Eriophyidae). Entomol. Soc. Am. Misc. Publ. 63. 19 pp.

Perring, T. M., C. A. Farrar, and R. N. Royalty. 1987. Intraplant distribution and sampling of spider mites (Acari: Tetranychidae) on cantaloupe. *J. Econ. Entomol.* 80: 96–101.

Perring, T. M., C. A. Farrar, and N. C. Toscano. 1988. Relationships among tomato planting date, potato aphids (Homoptera: Aphididae), and natural enemies. *J. Econ. Entomol.* 81: 1107–1112.

Perring, T. M., C. A. Farrar, K. Mayberry, and M. J. Blua. 1992. Research reveals pattern of cucurbit virus spread. *California Agric.* 46(2): 35–39.

Perring, T. M., T. O. Holtzer, J. L. Toole, J. M. Norman, and G. L. Myers. 1984. Influences of temperature and humidity on preadult development of the Banks grass mite (Acari: Tetranychidae). *Environ. Entomol.* 13: 338–343.

Perron, J. P. 1971. Insect pests of carrots in organic soils of southwestern Quebec with special reference to the carrot weevil, *Listronotus oregonensis* (Coleoptera: Curculionidae). *Can. Entomol.* 103: 1441–1448.

Perron, J. P. 1972. Effects of some ecological factors on populations of the onion maggot, *Hylemya antiqua* (Meig.), under field conditions in southwestern Quebec. *Ann. Entomol. Soc. Quebec* 17: 29–45.

Perron, J. P. and J. Lafrance. 1961. Notes on the life-history of the onion maggot, *Hylema antiqua* (Meig.) (Diptera: Anthomyiidae) reared in field cages. *Can. Entomol.* 93: 101–106.

Persoons, C. J., C. van der Kraan, W. J. Nooijen, F. J. Ritter, S. Voerman, and T. C. Baker. 1981. Sex pheromone of the beet armyworm, *Spodoptera exigua*: isolation, identification and preliminary field evaluation. *Entomol. Exp. Appl.* 30: 98–99.

Persoons, C. J., S. Voerman, P. E. J. Verwiel, F. J. Ritter, W. J. Nooyen, and A. K. Minks. 1976. Sex pheromone of the potato tuberworm moth, *Phthorimaea operculella*: isolation, identification and field evaluation. *Entomol. Exp. Appl.* 20: 289–300.

Pesho, G. R., F. J. Muehlbauer, and W. H. Harberts. 1977. Resistance of pea introductions to the pea weevil. *J. Econ. Entomol.* 70: 30–33.

Peters, D. and G. Lebbink. 1973. The effect of oil on the transmission of pea enation mosaic virus during short inoculation probes. *Entomol. Exp. Appl.* 16: 185–190.

Peters, L. L. 1983. Chinch bug (Heteroptera: Lygaeidae) control with insecticides on wheat, field corn, and grain sorghum, 1981. *J. Econ. Entomol.* 76: 178–181.

Peterson, A. 1923. The pepper maggot, a new pest of peppers and eggplants, *Spilographa electa* Say. (Trypetidae). New Jersey Agric. Exp. Stn. Bull. 373. 22 pp.

Peterson, A. 1948. Larvae of Insects, an Introduction to Nearctic Species. Part I Lepidoptera and Plant Infesting Hymenoptera. Edwards Brothers, Inc., Ann Arbor, MI. 315 pp.

Peterson, A. 1957. Larvae of Insects, an Introduction to Nearctic Species. Part II Coleoptera, Diptera, Neuroptera, Siphonaptera, Mecoptera, Trichoptera. Edwards Brothers, Inc., Ann Arbor, MI. 416 pp.

Peterson, R. K. D., L. G. Higley, and W. C. Bailey. 1988. Phenology of the adult celery looper, *Syngrapha falcifera* (Lepidoptera: Noctuidae), in Iowa: evidence for migration. *Environ. Entomol.* 17: 679–684.

Peterson, R. K. D., L. G. Higley, and W. C. Bailey. 1990. Occurrence and relative abundance of *Papaipema* species (Lepidoptera: Noctuidae) in Iowa. *J. Kansas Entomol. Soc.* 63: 447–449.

Peterson, R. K. D., L. G. Higley, G. D. Buntin, and L. P. Pedigo. 1993. Flight activity and ovarian dynamics of the yellow woollybear, *Spilosoma virginica* (F.) (Lepidoptera: Arctiidae), in Iowa. *J. Kansas Entomol. Soc.* 66: 97–103.

Peterson, R. K. D., R. B. Smelser, T. H. Klubertanz, L. P. Pedigo, and W. C. Welbourn. 1992. Ectoparasitism of the bean leaf beetle (Coleoptera: Chrysomelidae) by *Trombidium hyperi* Vercammen-Grandjean, Van Driesche, and Gyrisco and *Trombidium newelli* Welbourn and Flessel (Acari: Trombidiidae). *J. Agric. Entomol.* 9: 99–107.

Petitt, F. L. and Z. Smilowitz. 1982. Green peach aphid feeding damage to potato in various plant growth stages. *J. Econ. Entomol.* 75: 431–435.

Petitt, F. L. and D. O. Wietlisbach. 1994. Laboratory rearing and life history of *Liriomyza sativae* (Diptera: Agromyzidae) on lima bean. *Environ. Entomol.* 23: 1416–1421.

Petralia, R. S. and S. B. Vinson. 1979. Developmental morphology of larvae and eggs of the imported fire ant, *Solenopsis invicta*. *Ann. Entomol. Soc. Am.* 72: 472–484.

Pfadt, R. E. 1949. Food plants as factors in the ecology of the lesser migratory grasshopper, *Melanoplus mexicanus* (Sauss.). Wyoming Agric. Exp. Stn. Bull. 290. 51 pp.

Pfadt, R. E. 1994a. Differential grasshopper, *Melanoplus differentialis* (Thomas), Species fact sheet *in* Field Guide to Common Western Grasshoppers. Wyoming Agric. Exp. Stn. Bull. 912. 4 pp.

Pfadt, R. E. 1994b. Migratory grasshopper, *Melanoplus sanguinipes* (Fabricius), Species fact sheet *in* Field Guide to Common Western Grasshoppers. Wyoming Agric. Exp. Stn. Bull. 912. 4 pp.

Pfadt, R. E. 1994c. Mormon cricket, *Anabrus simplex* Haldeman, Species fact sheet *in* Field Guide to Common Western Grasshoppers. Wyoming Agric. Exp. Stn. Bull. 912. 4 pp.

Pfadt, R. E. 1994d. Redlegged grasshopper, *Melanoplus femurrubrum* (De Geer), Species fact sheet *in* Field Guide to Common Western Grasshoppers. Wyoming Agric. Exp. Stn. Bull. 912. 4 pp.

Pfadt, R. E. 1994e. Twostriped grasshopper, *Melanoplus bivittatus* (Say), Species fact sheet *in* Field Guide to Common Western Grasshoppers. Wyoming Agric. Exp. Stn. Bull. 912. 4 pp.

Pfadt, R. E. and D. S. Smith. 1972. Net reproductive rate and capacity for increase of the migratory grasshopper, *Melanoplus sanguinipes sanguinipes* (F.). *Acrida* 1: 149–165.

Pfaffenberger, G. S. 1985. Description, differentiation, and biology of the four larval instars of *Acanthoscelides obtectus* (Say) (Coleoptera: Bruchidae). *Coleop. Bull.* 39: 239–256.

Phelan, P., M. E. Montgomery, and L. R. Nault. 1976. Orientation and locomotion of apterous aphids dislodged from their hosts by alarm pheromone. *Ann. Entomol. Soc. Am.* 69: 1153–1156.

Phelan, P., K. H. Norris, and J. F. Mason. 1996. Soil-management history and host preference by *Ostrinia nubilalis:* evidence for plant mineral balance mediating insect-plant interactions. *Environ. Entomol.* 1329–1336.

Phillips, K. A. and J. O. Howell. 1980. Biological studies on two *Euchistus* species in central Georgia apple orchards. *J. Georgia Entomol. Soc.* 15: 349–355.

Phillips, V. T. 1946. The biology and identification of trypetid larvae (Diptera: Trypetidae). *Mem. Am. Entomol. Soc.* 12: 1–161.

Phillips, W. J. 1909. The slender seed-corn ground-beetle. USDA Bull. 85: 13–28.

Phillips, W. J. 1914. Corn-leaf blotch miner. *J. Agric. Res.* 2: 15–31.

Phillips, W. J., G. W. Underhill, and F. W. Poos. 1921. The larger corn stalk-borer in Virginia. Virginia Agric. Exp. Stn. Bull. 22. 30 pp.

Pickel, C., R. C. Mount, F. G. Zalom, and L. T. Wilson. 1983. Monitoring aphids on brussels sprouts. *California Agric.* 37(3): 24–25.

Pickering, J. and A. P. Gutierrez. 1991. Differential impact of the pathogen *Pandora neoaphidis* (R. & H.) Humber (Zygomycetes: Entomophthorales) on the species composition of *Acyrthosiphon* aphids in alfalfa. *Can. Entomol.* 123: 315–320.

Pickett, C. H. and F. E. Gilstrap. 1984. Phytoseiidae (Acarina) associated with Banks grass mite infestations in Texas. *Southwestern Entomol.* 9: 125–133.

Pickford, R. J. J. 1984. Evaluation of soil treatment for control of *Thrips tabaci* on cucumbers. *Ann. Appl. Biol. (Suppl.)* 104: 18–19.

Pierce, W. D. 1907. Notes on the economic importance of sowbugs. USDA Bur. Entomol. Bull. 64: 15–22.

Pierce, W. D. 1915. Some sugar-cane and root-boring weevils of the West Indies. *J. Agric Res.* 4: 255–267.

Pierce, W. D. 1918. Weevils which affect irish potato, sweet potato, and yam. *J. Agric Res.* 12: 601–611.

Pike, K. S., L. Boydston, and D. Allison. 1990. Alate aphid viviparae associated with small grains in North America: a key and morphometric characterization. *J. Kansas Entomol. Soc.* 63: 559–602.

Pike, K. S., R. L. Rivers, and Z. B. Mayo. 1977. Geographical distribution of the known *Phyllophaga* and *Cyclocephala* species in the north central states. Nebraska Agric. Exp. Stn. Misc. Publ. 34. 13 pp.

Pimentel, D. 1961. Natural control of aphid populations on cole crops. *J. Econ. Entomol.* 54: 885–888.

Pinto, J. D. 1980. Behavior and taxonomy of the *Epicauta maculata* group. Univ. California Pub. Entomol. 89: 1–111.

Pinto, J. D. 1991. The taxonomy of North American *Epicauta* (Coleoptera: Meloidae), with a revision of the nominate subgenus and a survey of courtship behavior. Univ. California Pub. Entomol. Bull. 110. 372 pp.

Pinto, J. D. and R. B. Selander. 1970. The bionomics of blister beetles of the genus *Meloe* and a classification of the new world species. Illinois Biol. Monogr. 42. 222 pp.

Pitre, H. N. 1970. Notes on the life history of *Dalbulus maidis* on gama grass and plant susceptibility to the corn stunt disease agent. *J. Econ. Entomol.* 63: 1661–1662.

Pitre, H. N. 1989. Polymorphic forms of the bean leaf beetle, *Cerotoma trifurcata* (Förster) (Coleoptera: Chrysomelidae), in soybeans in Mississippi: transmission of bean pod mottle virus. *J. Entomol. Sci.* 24: 582–587.

Pitre, H. N. and D. B. Hogg. 1983. Development of the fall armyworm on cotton, soybean and corn. *J. Georgia Entomol.* Soc. 18: 187–194.

Pitre, Jr., H. N. and E. J. Kantack. 1962. Biology of the banded cucumber beetle, *Diabrotica balteata*, in Louisiana. *J. Econ. Entomol.* 55: 904–906.

Pivnick, K. A., R. J. Lamb, and D. Reed. 1992. Response of flea beetles, *Phyllotreta* spp., to mustard oils and nitriles in field trapping experiments. *J. Chem Ecol.* 18: 863–873.

Pivnick, K. A., D. W. Reed, J. G. Millar, and E. W. Underhill. 1991. Attraction of northern false chinch bug *Nysius niger* (Heteroptera: Lygaeidae) to mustard oils. *J. Chem Ecol.* 17: 931–941.

Platner, G. R. and E. R. Oatman. 1968. An improved technique for producing potato tuberworm eggs for mass production of natural enemies. *J. Econ. Entomol.* 61: 1054–1057.

Pletsch, D. J. 1947. The potato psyllid, *Paratrioza cockerelli* (Sulc), its biology and control. Montana Agric. Exp. Stn. Bull. 446. 95 pp.

Poe, S. L. 1976. Reinfestation of treated tomato fields by mole crickets. *Florida Entomol.* 59: 88.

Poinar, Jr., G. O. 1978. Generation polymorphism in *Neoaplectana glaseri* Steiner (Steinernematidae: Nematoda), redescribed from *Strigoderma arboricola* (Fab.) (Scarabaeidae: Coleoptera) in North Carolina. *Nematologica* 24: 105–114.

Poinar, Jr., G. O. 1979. Nematodes for Biological Control of Insects. CRC Press, Boca Raton. 277 pp.

Polaszek, A., G. A. Evans, and F. D. Bennett. 1992. *Encarsia* parasitoids of *Bemisia tabaci* (Hymenoptera: Aphelinidae, Homoptera: Aleyrodidae): a preliminary guide to identification. *Bull. Entomol. Res.* 82: 375–392.

Polivka, J. B. 1960. Effect of lime applications to soil on Japanese beetle larval populations. *J. Econ. Entomol.* 53: 476–477.

Pond, D. D. 1960. Life history studies of the armyworm, *Pseudaletia unipunctata* (Lepidoptera: Noctuidae), in New Brunswick. *Ann. Entomol. Soc. Am.* 53: 661–665.

Poos, F. W. 1932. Biology of the potato leafhopper, *Empoasca fabae* (Harris), and some closely related species of *Empoasca. J. Econ. Entomol.* 25: 639–646.

Poos, F. W. 1936. Certain insect vectors of *Aplanobacter stewarti. J. Agric Res.* 52: 585–608.

Poos, F. W. 1939. Host plants harboring *Aplanobacter stewarti* without showing external symptoms after inoculation by *Chaetocnema pulicaria. J. Econ. Entomol.* 32: 881–882.

Poos, F. W. 1951. Control of the garden webworm in alfalfa. USDA Leaflet 304: 1–4.

Poos, F. W. 1955. Studies of certain species of *Chaetocnema*. *J. Econ. Entomol.* 48: 555–563.

Poos, F. W. and C. M. Haenseler. 1931. Injury to varieties of eggplant by the potato leafhopper, *Empoasca fabae* (Harris). *J. Econ. Entomol.* 24: 890–892.

Poos, F. W. and H. S. Peters. 1927. The potato tuber worm. Virginia Truck Exp. Stn. Bull. 61: 597–630.

Poos, F. W. and F. F. Smith. 1931. A comparison of oviposition and nymphal development of *Empoasca fabae* (Harris) on different host plants. *J. Econ. Entomol.* 24: 361–371.

Poos, F. W. and N. H. Wheeler. 1943. Studies on host plants of the leafhoppers of the genus *Empoasca*. USDA Tech. Bull. 850. 51 pp.

Poprawski, T. J. and W. N. Yule. 1992. Acari associated with *Phyllophaga anxia* (Leconte) (Coleoptera: Scarabaeidae) in southern Quebec and eastern Ontario. *Can. Entomol.* 124: 397–403.

Poprawski, T. J., M. Marchal, and P.-H.Robert. 1985. Comparative susceptibility of *Otiorhynchus sulcatus* and *Sitona lineatus* (Coleoptera: Curculionidae) early stages to five entomopathogenic hyphomycetes. *Environ. Entomol.* 14: 247–253.

Port, C. M. and G. R. Port. 1986. The biology and behaviour of slugs in relation to crop damage and control. Pages 255–299 *in* G. E. Russell (ed.). Agricultural Zoology Reviews, Vol. 1. Intercept, Ponteland, Newcastle upon Tyne.

Porter, S. D. 1988. Impact of temperature on colony growth and developmental rates of the ant, *Solenopsis invicta*. *J. Insect Physiol.* 34: 1127–1133.

Porter, S. D. and W. R. Tschinkel. 1986. Adaptive value of nanitic workers in newly founded red imported-fire ant colonies (Hymenoptera: Formicidae). *Ann. Entomol. Soc. Am.* 79: 723–726.

Portillo, H. E., H. N. Pitre, D. H. Meckenstock, and K. L. Andrews. 1991. Langosta: a lepidopterous pest complex on sorghum and maize in Honduras. *Florida Entomol.* 74: 287–296.

Portillo, H. E., H. N. Pitre, D. H. Meckenstock, and K. L. Andrews. 1996a. Feeding preferences of neonates and late instar larvae of a lepidopterous pest complex (Lepidoptera: Noctuidae) on sorghum, maize, and noncrop vegetation in Honduras. *Environ. Entomol.* 25: 589–598.

Portillo, H. E., H. N. Pitre, D. H. Meckenstock, and K. L. Andrews. 1996b. Oviposition preference of *Spodoptera latifascia* (Lepidoptera: Noctuidae) for sorghum, maize and non-crop vegetation. *Florida Entomol.* 79: 552–562.

Poston, F. L., R. J. Whitworth, J. Loera, and H. B. Safford. 1979. Effects of substrate characteristics on the ovipositional behavior of the southwestern corn borer. *Ann. Entomol. Soc. Am.* 72: 47–50.

Poston, F. L., R. J. Whitworth, S. M. Welch, and J. Loera. 1983. Sampling southwestern corn borer populations in postharvest corn. *Environ. Entomol.* 12: 33–36.

Poswal, M. A., R. C. Berberet, and L. J. Young. 1990. Time-specific life tables for *Acyrthosiphon kondoi* (Homoptera: Aphididae) in first crop alfalfa in Oklahoma. *Environ. Entomol.* 19: 1001–1009.

Potts, M. J. and N. Gunadi. 1991. The influence of intercropping with *Allium* on some insect populations in potato (*Solanum tuberosum*). *Ann. Appl. Biol.* 119: 207–213.

Power, A. G. 1989. Influence of plant spacing and nitrogen fertilization in maize on *Dalbulus maidis* (Homoptera: Cicadellidae), vector of corn stunt. *Environ. Entomol.* 18: 494–498.

Powell, C. A. and P. J. Stoffella. 1993. Influence of endosulfan sprays and aluminum mulch on sweetpotato whitefly disorders of zucchini squash and tomatoes. *J. Prod. Agric.* 6: 118–121.

Powell, D. M. 1980. Control of the green peach aphid on potatoes with soil systemic insecticides: preplant broadcast and planting time furrow applications, 1973–77. *J. Econ. Entomol.* 73: 839–843.

Powell, D. M. and W. T. Mondor. 1976. Area control of the green peach aphid on peach and the reduction of potato leaf roll virus. *Am. Pot. J.* 53: 123–139.

Power, A. G., C. M. Rodriguez, and R. Gamez. 1992. Evaluation of two leafhopper sampling methods for predicting the incidence of a leafhopper-transmitted virus of maize. *J. Econ. Entomol.* 85: 411–415.

Prabhaker, N., D. L. Coudriet, A. N. Kishaba, and D. E. Meyerdirk. 1986. Laboratory evaluation of neem-seed extract against larvae of the cabbage looper and beet armyworm (Lepidoptera: Noctuidae). *J. Econ. Entomol.* 79: 39–41.

Prando, H. F. and F. Z. da Cruz. 1986. Aspectos da biologia de *Liriomyza huidobrensis* (Blanchard, 1926) (Diptera, Agromyzidae) em laboratorio. *An. Soc. Entomol., Brasil* 15: 77–88.

Prescott, H. W. 1960. Suppression of grasshoppers by nemestrinid parasites (Diptera). *Ann. Entomol. Soc. Am.* 53: 513–521.

Prescott, H. W. and M. M. Reeher. 1961. The pea weevil, an introduced pest of legumes in the Pacific Northwest. USDA Tech. Bull. 1233. 12 pp.

Price, J. F. and S. L. Poe. 1976. Response of *Liriomyza* (Diptera: Agromyzidae) and its parasites to stake and mulch culture of tomatoes. *Florida Entomol.* 59: 85–87.

Price, J. F. and D. J. Schuster. 1991. Effects of natural and synthetic insecticides on sweetpotato whitefly *Bemisia tabaci* (Homoptera: Aleyrodidae) and its hymenopterous parasitoids. *Florida Entomol.* 74: 60–68.

Price, J. F. and D. Taborsky. 1992. Movement of immature *Bemisia tabaci* (Homoptera: Aleyrodidae) on poinsettia leaves. *Florida Entomol.* 75: 151–153.

Prokopy, R. J., R. H. Collier, and S. Finch. 1983. Visual detection of host plants by cabbage root flies. *Entomol. Exp. Appl.* 34: 85–89.

Prostak, D. J. 1995. Corn earworm, *Helicoverpa zea* (Boddie). Pages 75–79 *in* R. G. Adams and J. C. Clark (eds.). Northeast Sweet Corn Production and Integrated Pest Management Manual. Coop. Ext. Serv., University of Connecticut, Storrs.

Pruess, K. P. 1961. Oviposition response of the army cutworm, *Chorizagrotis auxiliaris*, to different media. *J. Econ. Entomol.* 54: 273–277.

Pruess, K. P. 1963. Effects of food, temperature, and oviposition site on longevity and fecundity of the army cutworm. *J. Econ. Entomol.* 56: 219–221.

Pruess, K. P. 1967. Migration of the army cutworm, *Chorizagrotis auxiliaris* (Lepidoptera: Noctuidae). I. Evidence for a migration. *Ann. Entomol. Soc. Am.* 60: 910–920.

Pruess, K. P. 1974. Tarnished and alfalfa plant bugs in alfalfa: population suppression with ULV malathion. *J. Econ. Entomol.* 67: 525–528.

Pruess, K. P. and N. C. Pruess. 1971. Telescopic observation of the moon as a means for observing migration of the army cutworm, *Chorizagrotis auxiliaris* (Lepidoptera: Noctuidae). *Ecology* 52: 999–1007.

Prystupa, B. D., N. J. Holliday, and G. R. B. Webster. 1987. Molluscicide efficacy against the marsh slug, *Deroceras laeve* (Stylommatophora: Limacidae), on strawberries in Manitoba. *J. Econ. Entomol.* 80: 936–943.

Purcell, M., M. W. Johnson, L. M. Lebeck, and A. H. Hara. 1992. Biological control of *Helicoverpa zea* (Lepidoptera: Noctuidae) with *Steinernema carpocapsae* (Rhabditida: Steinernematidae) in corn used as a trap crop. *Environ. Entomol.* 21: 1441–1447.

Puttarudriah, M. 1953. The natural control of the alfalfa looper in central California. *J. Econ. Entomol.* 46: 723.

Puttler, B. and S. E. Thewke. 1970. Biology of *Microplitis feltiae* (Hymenoptera: Braconidae), a parasite of the black cutworm, *Agrotis ipsilon*. *Ann. Entomol. Soc. Am.* 63: 645–648.

Puttler, B. and S. E. Thewke. 1971. Field and laboratory observations of *Hexamermis arvalis* (Nematoda: Mermithidae) a parasite of cutworms. *Ann. Entomol. Soc. Am.* 64: 1102–1106.

Puttler, B., R. E. Sechrist, and D. M. Daughtery, Jr. 1973. *Hexamermis arvalis* parasitizing *Agrotis ipsilon* in corn and the origin of the pest infestation. *Environ. Entomol.* 2: 963–965.

Quaintance, A. L. 1898a. Three injurious insects. Bean leaf-roller. Corn delphax. Canna leaf-roller. Florida Agric. Exp. Stn. Bull. 45. 74 pp.

Quaintance, A. L. 1898b. The strawberry thrips and the onion thrips. Florida Agric. Exp. Stn. Bull. 46: 75–114.

Quaintance, A. L. 1901. The pickleworm. Georgia Agric. Exp. Stn. Bull. 54: 73–94.

Quaintance, A. L. and C. T. Brues. 1905. The cotton bollworm. USDA Bur. Entomol. Bull. 50. 155 pp.

Quate, L. W. and S. E. Thompson. 1967. Revision of click beetles of genus *Melanotus* in America north of Mexico (Coleoptera: Elateridae). Proc. U. S. Nat. Mus., Vol. 121, No. 3568. 84 pp.

Quintela, E. D., J. Fan, and C. W. McCoy. 1998. Development of *Diaprepes abbreviatus* (Coleoptera: Curculionidae) on artificial diet and citrus root substrates. *J. Econ. Entomol.* 91: 1173–1179.

Rabasse, J. M., C. A. Dedryver, J. Molinari, and J. P. Lafont. 1982. Facteurs de limitation des populations d'*Aphis fabae* dans l'ouest de la France. IV. Nouvelles donnees sur le deroulement des epizooties a entomophthoracees sur feverole de printemps. *Entomophaga* 27: 39–53.

Rabb, R. L. 1963. Biology of *Conoderus vespertinus* in the Piedmont section of North Carolina (Coleoptera: Elateridae). *Ann. Entomol. Soc. Am.* 56: 669–676.

Rabb, R. L. 1966. Diapause in *Protoparce sexta* (Lepidoptera: Sphingidae). *Ann. Entomol. Soc. Am.* 59: 160–165.

Rabb, R. L. 1969. Environmental manipulations as influencing populations of tobacco hornworms. Proc. Tall Timbers Conf. Ecol. Anim. Control Habitat Manage. 1: 175–191.

Radcliffe, E. B. 1982. Insect pests of potato. *Annu. Rev. Entomol.* 27: 173–204.

Radcliffe, E. B. and D. K. Barnes. 1970. Alfalfa plant bug injury and evidence of plant resistance in alfalfa. *J. Econ. Entomol.* 63: 1995–1996.

Radcliffe, E. B. and R. K. Chapman. 1966. Varietal resistance to insect attack in various cruciferous crops. *J. Econ. Entomol.* 59: 120–125.

Radke, S. G., A. W. Benton, and W. G. Yendol. 1973. Effect of temperature and light on the development of cowpea aphid, *Aphis craccivora* Koch. *Indian J. Entomol.* 35: 107–118.

Raina, A. K., P. S. Benepal, and A. Q. Sheikh. 1978. Evaluation of bean varieties for resistance to Mexican bean beetle. *J. Econ. Entomol.* 71: 313–314.

Raina, A. K., J. A. Klun, M. Schwarz, A. Day, B. A. Leonhardt, and L. W. Douglass. 1986. Female sex pheromone of the melonworm, *Diaphania hyalinata* (Lepidoptera: Pyralidae), and analysis of male responses to pheromone in a flight tunnel. *J. Chem Ecol.* 12: 229–237.

Raman, K. V. 1988. Control of potato tuber moth *Phthorimaea operculella* with sex pheromones in Peru. *Agric. Ecosys. Environ.* 21: 85–99.

Raman, K. V. and E. H. Alleyne. 1991. Biology and management of the West Indian sweet potato weevil, *Euscepes postfasciatus*. Pages 263–281 *in* R. K. Jansson and K.V Raman (eds.). Sweet Potato Pest Management: A Global Perspective. Westview Press, Boulder, Colorado.

Ramoska, W. A. 1984. The influence of relative humidity on *Beauveria bassiana* infectivity and replication in the chinch bug, *Blissus leucopterus*. *J. Invertebr. Pathol.* 43: 389–394.

Ramoska, W. A. and T. Todd. 1985. Variation in efficacy and viability of *Beauveria bassiana* in the chinch bug (Hemiptera: Lygaeidae) as a result of feeding activity on selected host plants. *Environ. Entomol.* 14: 146–148.

Ramsey, H. L. 1971. Garden symphylan populations in laboratory cultures. *J. Econ. Entomol.* 64: 657–660.

Rausher, M. D. 1984. Tradeoffs in performance on different hosts: evidence from within- and between-site variation in the beetle *Deloyala guttata*. *Evolution* 38: 582–595.

Rawlins, W. A. 1940. Biology and control of the wheat wireworm, *Agriotes mancus* Say. New York (Cornell) Agric. Exp. Stn. Bull. 738. 30 pp.

Rawlins, W. A. 1955. *Rhizoglyphus solani*, a pest of onions. *J. Econ. Entomol.* 48: 334.

Raworth, D. A., B. D. Frazer, N. Gilbert, and W. G. Wellington. 1984. Population dynamics of the cabbage aphid, *Brevicoryne brassicae* (Homoptera: Aphididae) at Vancouver, British Columbia. I. Sampling methods and population trends. *Can. Entomol.* 116: 861–870.

Reagan, T. E. 1986. Beneficial aspects of the imported fire ant: a field ecology approach. Pages 58–71 *in* C. S. Lofgren and R. K. Vander Meer (eds.). Fire Ants and leaf-cutting Ants, Biology and Management. Westview Press, Boulder, Colorado.

Redak, R. A., J. L. Capinera, and C. D. Bonham. 1992. Effects of sagebrush removal and herbivory by Mormon crickets (Orthoptera: Tettigoniidae) on understory plant biomass and cover. *Environ. Entomol.* 21: 94–102.

Redfern, R. E. 1967. Instars of southern armyworm determined by measurement of head capsule. *J. Econ. Entomol.* 60: 614–615.

Redfern, R. E. and J. R. Raulston. 1970. Improved rearing techniques for the southern armyworm. *J. Econ. Entomol.* 63: 296–297.

Reed, D. K., G. L. Reed, and C. S. Creighton. 1986. Introduction of entomogenous nematodes into trickle irrigation systems to control striped cucumber beetle (Coleoptera: Chrysomelidae). *J. Econ. Entomol.* 79: 1330–1333.

Reed, G. L., W. B. Showers, J. L. Huggans, and S. W. Carter. 1972. Improved procedures for mass rearing the European corn borer. *J. Econ. Entomol.* 65: 1472–1476.

Reed, H. E. and R. A. Byers. 1981. Flea beetles attacking forage kale: effect of carbofuran and tillage methods. *J. Econ. Entomol.* 74: 334–337.

Rees, N. E. 1973. Arthropod and nematode parasites, parasitoids, and predators of Acrididae in America North of Mexico. USDA Tech. Bull. 1460. 288 pp.

Rees, N. E. and J. A. Onsager. 1982. Influence of predators on the efficiency of the *Blaesoxipha* spp. parasites of the migratory grasshpper. *Environ. Entomol.* 11: 426–428.

Reese, J. C., L. M. English, T.-R. Yonke, and M. L. Fairchild. 1972. A method for rearing black cutworms. *J. Econ. Entomol.* 65: 1047–1050.

Rehn, J. A. G. and H. J. Grant, Jr. 1961. A monograph of the Orthoptera of North America (north of Mexico). Vol. 1. Monogr. No. 12, Acad. Nat. Sci. Phila. pp. 1–257.

Reid, J. C. and G. L. Greene. 1973. The soybean looper: pupal weight, development time, and consumption of soybean foliage. *Florida Entomol.* 56: 203–206.

Reid, Jr., W. J. 1940. Biology of the seed-corn maggot in the coastal plain of the south Atlantic states. USDA Tech. Bull. 723. 43 pp.

Reid, Jr., W. J. and C. O. Bare. 1952. Seasonal populations of cabbage caterpillars in the Charleston, S. C., area. *J. Econ. Entomol.* 45: 695–699.

Reidell, W. E. and G. R. Sutter. 1995. Soil moisture and survival of western corn rootworm larvae in field plots. *J. Kansas Entomol. Soc.* 68: 80–84.

Reinert, J. A. 1975. 1975. Life history of the striped grassworm, *Mocis latipes*. *Ann. Entomol. Soc. Am.* 68: 201–204.

Reinhard, H. J. 1923. The sweet potato weevil. Texas Agric. Exp. Stn. Bull. 308. 90 pp.

Reinhard, H. J. 1929. The cotton-square borer. Texas Agric. Exp. Stn. Bull. 401. 36 pp.

Reinink, K. and F. L. Dieleman. 1989. Resistance in lettuce to the leaf aphids *Macrosiphum euphorbiae* and *Uroleucon sonchi*. *Ann. Appl. Biol.* 115: 489–498.

Rennie, J. 1917. On the biology and economic significance of *Tipula paludosa*. *Ann. Appl. Biol.* 3: 116–137.

Rentz, D. C. and J. D. Birchim. 1968. Revisionary studies in the nearctic Decticinae. Mem. Pac. Coast Entomol. Soc., Vol. 3. 173 pp.

Rice, R. E. and F. E. Strong. 1962. Bionomics of the tomato russet mite, *Vasates lycopersici* (Massee). *Ann. Entomol. Soc. Am.* 55: 431–435.

Richards, O. W. 1940. The biology of the small white butterfly (*Pieris rapae*), with special reference to the factors controlling its abundance. *J. Anim. Ecol.* 9: 243–288.

Richards, W. R. 1960. A synopsis of the genus *Rhopalosiphum* in Canada (Homoptera: Aphididae). *Can. Entomol.* 92 (Suppl. 13) 51 pp.

Richman. D. B., W. F. Buren, and W. H. Whitcomb. 1983. Predatory arthropods attacking the eggs of *Diaprepes abbreviatus* (L.) (Coleoptera: Curculionidae) in Puerto Rico and Florida. *J. Georgia Entomol. Soc.* 18: 335–342.

Richman, D. B., R. C. Hemenway, Jr., and W. H. Whitcomb. 1980. Field cage evaluation of predators of the soybean looper, *Pseudoplusia includens* (Lepidoptera: Noctuidae). *Environ. Entomol.* 9: 315–317.

Richman, D. B., D. C. Lightfoot, C. A. Sutherland, and D. J. Ferguson. 1993. A manual of the grasshoppers of New Mexico (Orthoptera: Acrididae and Romaleidae). New Mexico State Univ. Ext. Serv. Handbook No. 7. 112 pp.

Ridgway, R. L. and G. G. Gyrisco. 1960. Effect of temperature on the rate of development of *Lygus lineolaris* (Hemiptera: Miridae). *Ann. Entomol. Soc. Am.* 53: 691–694.

Riggin, T. M., K. E. Espelie, B. R. Wiseman, and D. J. Isenhour. 1993. Distribution of fall armyworm (Lepidoptera: Noctuidae) parasitoids on five corn genotypes in south Georgia. *Florida Entomol.* 76: 292–302.

Riggin, T. M., B. R. Wiseman, D. J. Isenhour, and K. E. Espelie. 1992. Incidence of fall armyworm (Lepidoptera: Noctuidae) parasitoids on resistant and susceptible corn genotypes. *Environ. Entomol.* 21: 888–895.

Riley, C. V. 1870a. The flea-like negro-bug. Pages 33–37 *in* Second Annual Report on the Noxious, Beneficial and Other Insects, of the State of Missouri, Jefferson City.

Riley, C. V. 1870b. Insects infesting the sweet-potato. Tortoise beetles (Coleoptera, Cassidae). Pages 56–64 *in* Second Annual Report on the Noxious, Beneficial and Other Insects, of the State of Missouri, Jefferson City.

Riley, C. V. 1871. The common yellow bear—*Spilosoma virginica*, Fabr. (Lepidoptera, Arctiidae). Pages 68–69 *in* Third Annual Report on the Noxious, Beneficial and Other Insects, of the State of Missouri, Jefferson City.

Riley, C. V. 1873. The false chinch bug—*Nysius destructor*, N. Sp. (Subord. Heteroptera, Fam. Lygaeidae), a new enemy to the grape-vine, potato, cabbage, and many cruciferous plants. Pages 111–114 *in* Fifth Annual Report on the Noxious, Beneficial and other Insects, of the State of Missouri, Jefferson City.

Riley, C. V. 1884. Zimmermann's flea beetle. Pages 304–308 *in* Commissioner of Agriculture, Annu. Rpt. USDA for 1884.

Riley, C. V. 1888. The parsnip web-worm (*Depressaria heracliana* DeG.). *Insect Life* 1: 94–98.

Riley, C. V. 1890. The rose chafer (*Macrodactylus subspinosus*, Fabr.). *Insect Life* 2: 295–302.

Riley, D. G. and E. G. King. 1994. Biology and management of the pepper weevil Anthonomus eugenii Cano (Coleoptera: Curculoionidae): a review. *Trends Agric. Sci.* 2: 109–121.

Riley, D. G. and J. C. Palumbo. 1995a. Action thresholds for *Bemisia argentifolii* (Homoptera: Aleyrodidae) in cantaloupe. *J. Econ. Entomol.* 88: 1733–1738.

Riley, D. G. and J. C. Palumbo. 1995b. Interaction of silverleaf whitefly (Homoptera: Aleyrodidae) with cantaloupe yield. *J. Econ. Entomol.* 88: 1726–1732.

Riley, D. G. and D. J. Schuster. 1992. The occurrence of *Catolaccus hunteri*, a parasitoid of *Anthonomus eugenii*, in insecticide treated bell pepper. *Southwestern Entomol.* 17: 71–72.

Riley, D. G. and D. J. Schuster. 1994. Pepper weevil adult response to colored sticky traps in pepper fields. *Southwestern Entomol.* 19: 93–107.

Riley, D. G. and A. N. Sparks, Jr. 1993. Managing the sweetpotato whitefly in the lower Rio Grande Valley of Texas. Texas A & M Univ. Ext. Serv. B-5082. 12 pp.

Riley, D. G., D. J. Schuster, and C. S. Barfield. 1992a. Sampling and dispersion of pepper weevil (Coleoptera: Curculionidae) adults. *Environ. Entomol.* 21: 1013–1021.

Riley, D. G., D. J. Schuster, and C. S. Barfield. 1992b. Refined action threshold for pepper weevil adults (Coleoptera: Curculionidae) in bell peppers. *J. Econ. Entomol.* 85: 1919–1925.

Riley, E. G. 1986. Review of the tortoise beetle genera of the tribe Cassidini occurring in America north of Mexico (Coleoptera: Chrysomelidae: Cassidinae). *J. New York Entomol. Soc.* 94: 98–114.

Riley, T. I. 1983. Damage to soybean seeds and seedlings by larvae of the flea beetle *Systena frontalis* (Coleoptera: Chrysomelidae). *J. Georgia Entomol. Soc.* 18: 291–293.

Riley, T. J. and A. J. Keaster. 1979. Wireworms associated with corn: identification of larvae of nine species of *Melanotus* from the north central states. *Ann. Entomol. Soc. Am.* 72: 408–414.

Rings, R. W. 1977a. An illustrated field key to common cutworm, armyworm, and looper moths in the north central states. Ohio Agric. Res. Dev. Ctr. Res. Circ. 227. 60 pp.

Rings, R. W. 1977b. A pictorial field guide to the armyworms and cutworms attacking vegetables in the north central states. Ohio Agric. Res. Dev. Ctr. Res. Circ. 231. 36 pp.

Rings, R. W. and F. J. Arnold. 1974. An annotated bibliography of the glassy cutworm, *Crymodes devastator* (Brace). Ohio Agric. Res. Dev. Ctr. Res. Circ. 199. 45 pp.

Rings, R. W. and B. A. Johnson. 1977. An annotated bibliography of the spotted cutworm, *Amathes c-nigrum* (Linnaeus). Ohio Agric. Res. Dev. Ctr. Res. Circ. 235. 50 pp.

Rings, R. W. and E. W. Metzler. 1982. Two newly detected noctuids (*Hydraecia immanis* and *Hydraecia micacea*) of potential economic importance in Ohio. *Ohio J. Sci.* 82: 299–302.

Rings, R. W. and G. J. Musick. 1976. A pictorial field key to the armyworms and cutworms attacking corn in the north central states. Ohio Agric. Res. Devt. Ctr. Res. Circ. 221. 36 pp.

Rings, R. W., B. A. Baughman, and F. J. Arnold. 1974a. An annotated bibliography of the bronzed cutworm, *Nephelodes minians* Guenée. Ohio Agric. Res. Dev. Ctr. Res. Circ. 200. 36 pp.

Rings, R. W., B. A. Baughman, and F. J. Arnold. 1975a. An annotated bibliography of the dingy cutworm complex, *Feltia ducens* Walker and *Feltia subgothica* Haworth. Ohio Agric. Res. Dev. Ctr. Res. Circ. 202. 18 pp.

Rings, R. W., B. A. Johnson, and F. J. Arnold. 1975b. An annotated bibliography of the dark-sided cutworm, *Euxoa messoria* (Harris). Ohio Agric. Res. Dev. Ctr. Res. Circ. 205. 50 pp.

Rings, R. W., B. A. Johnson, and F. J. Arnold. 1976a. A worldwide, annotated bibliography of the variegated cutworm, *Peridroma saucia* Hübner. Ohio Agric. Res. Dev. Ctr. Res. Circ. 219. 126 pp.

Rings, R. W., B. A. Johnson, and F. J. Arnold. 1976b. Host range of the variegated cutworm on vegetables: a bibliography. *Bull. Entomol. Soc. Am.* 22:409–415.

Rings, R. W., F. J. Arnold, A. J. Keaster, and G. J. Musick. 1974b. A worldwide, annotated bibliography of the black cutworm, *Agrotis ipsilon* (Hufnagel). Ohio Agric. Res. Dev. Ctr. Res. Circ. 198. 106 pp.

Risch, S. 1981. Ants as important predators of rootworm eggs in the neotropics. *J. Econ. Entomol.* 74:88–90.

Ritcey, G., R. McGraw, and F. L. McEwen. 1982. Insect control on potatoes in Ontario from 1973 to1982. *Proc. Entomol. Soc. Ontario* 113:1–6.

Ritcey, G., F. L. McEwen, H. E. Braun, and R. Frank. 1991. Persistence and biological activity of residues of granular insecticides in organic soil and onions with furrow treatment for control of the onion maggot (Diptera: Anthomyiidae). *J. Econ. Entomol.* 84:1339–1343.

Ritcher, P. O. 1940. Kentucky white grubs. Kentucky Agric. Exp. Stn. Bull. 401. 157 pp.

Ritcher, P. O. 1966. White Grubs and their Allies, a Study of North American Scarabaeoid Larvae. Oregon State Univ. Press, Corvallis, OR. 219 pp.

Rivers, R. L., K. S. Pike, and Z. B. Mayo. 1977. Influence of insecticides and corn tillage systems on larval control of *Phyllophaga anxia*. *J. Econ. Entomol.* 70:794–796.

Roach, S. H. 1975. *Heliothis* spp.: larvae and associated parasites and diseases on wild host plants in the Pee Dee area of South Carolina. *Environ. Entomol.* 4:725–728.

Ro, T. H., G. E. Long, and H. H. Toba. 1998. Predicting phenology of green peach aphid (Homoptera: Aphididae) using degree-days. *Environ. Entomol.* 27:337–343.

Robert, Y. and J. M. Rabasse. 1977. Role ecologique de *Digitalis purpurea* dans la limitation naturelle des populations du puceron strie de la pomme de terre *Aulacorthum solani* par *Aphidius urticae* dans l'ouest de la France. *Entomophaga* 22:373–382.

Roberts, D. W., R. A. LeBrun, and M. Semel. 1981. Control of the Colorado potato beetle with fungi. Pages 119–137 *in* Lashombe, J. H. and R. Casagrande (eds.). Advances in Potato Pest Management. Hutchinson Ross, Stroudsburg, Pennsylvania.

Roberts, P. M. and J. N. All. 1993. Hazard for fall armyworm (Lepidoptera: Noctuidae) infestation of maize in double-cropping systems using sustainable agricultural practices. *Florida Entomol.* 76:276–283.

Robinson, A. G. and S.-J. Hsu. 1963. Host plant records and biology of aphids on cereal grains and grasses in Manitoba (Homoptera: Aphididae). *Can. Entomol.* 95:134–137.

Robinson, J. F., A. Day, R. Cuthbert, and E. V. Wann. 1979. The pickleworm: laboratory rearing and artificial infestation of cucumbers. *J. Econ. Entomol.* 72:305–307.

Rock, G. C. 1979. Relative toxicity of two synthetic pyrethroids to a predator *Amblyseius fallacis* and its prey *Tetranychus urticae*. *J. Econ. Entomol.* 72:293–294.

Rockburne, E. W. and J. D. Lafontaine. 1976. The cutworm moths of Ontario and Quebec. Can. Dept. Agric. Publ. 1593. 164 pp.

Rockwood, L. P. 1925. An outbreak of *Agrotis ypsilon* Rott. on overflow land in western Oregon. *J. Econ. Entomol.* 18:717–721.

Rockwood, L. P. and T. R. Chamberlin. 1943. The western spotted cucumber beetle as a pest of forage crops in the Pacific Northwest. *J. Econ. Entomol.* 36:837–842.

Rodriguez–del-Bosque, L. A., J. W. Smith, Jr., and H. W. Browning. 1990. Feeding and pupation sites of *Diatraea lineolata*, *D. saccharalis*, and *Eoreuma loftini* (Lepidoptera: Pyralidae) in relation to corn phenology. *J. Econ. Entomol.* 83:850–855.

Roe, R. M. 1981. A bibliography of the sugarcane borer, *Diatraea saccharalis* (Fabricius), 1887–1980. USDA, ARS. ARM-S 20. 101 pp.

Roe, R. M., A. M. Hammond, Jr., and T. C. Sparks. 1982. Growth of larval *Diatraea saccharalis* (Lepidoptera: Pyralidae) on an artificial diet and synchronization of the last larval stadium. *Ann. Entomol. Soc. Am.* 75:421–429.

Roelofs, W. L. and A. Comeau. 1970. Lepidopterous sex attractants discovered by field screening tests. *J. Econ. Entomol.* 63:969–974.

Roessler, Y. 1989. Insecticidal bait and cover sprays. Pages 329–336 *in* A. S. Robinson and G. Hooper (eds.). Fruit Flies: their Biology, Natural Enemies and Control, Vol. 3b. Elsevier, Amsterdam/New York.

Rogers, C. E. 1974. Bionomics of the carrot beetle in the Texas Rolling Plains. *Environ. Entomol.* 3:969–974.

Rogers, C. E. and G. R. Howell. 1973. Behavior of the carrot beetle on native and treated commercial sunflower. Texas Agric. Exp. Stn. Prog. Rpt. 3249:1–7.

Rogers, C. E. and O. G. Marti, Jr. 1993. Infestation dynamics and distribution of *Noctuidonema guyanense* (Nematoda: Aphelenchoididae) on adults of *Spodoptera frugiperda* and *Mocis latipes* (Lepidoptera: Noctuidae). *Florida Entomol.* 76:326–333.

Rogers, C. E., R. D. Kirby, and J. P. Hollingsworth. 1978. Nocturnal flight behavior of the carrot beetle. *Southwestern Entomol.* 3:271–275.

Rogers, C. E., A. M. Simmons, and O. G. Marti. 1991. *Noctuidonema guyanense*: an ectoparasitic nematode of fall armyworm adults in the tropical Americas. *Florida Entomol.* 74:246–257.

Rogers, C. E., T. E. Thompson, and M. J. Wellik. 1980. Survival of *Bothynus gibbosus* (Coleoptera: Scarabaeidae) on *Helianthus* species. *J. Kansas Entomol. Soc.* 53:490–494.

Roitberg, B. D. and J. H. Myers. 1978. Adaptation of alarm pheromone responses of the pea aphid *Acyrthosiphon pisum* (Harris). *Can. J. Zool.* 56:103–108.

Rollo, C. D. and C. R. Ellis. 1974. Sampling methods for the slugs, *Deroceras reticulatum* (Müller), *D. laeve* (Müller), and *Arion fasciatus* Nilsson in Ontario corn fields. Proc. Entomol. Soc. Ontario 105:89–95.

Rolston, L. H. 1955. The southwestern corn borer in Arkansas. Arkansas Agric. Exp. Stn. Bull. 553. 40 pp.

Rolston, L. H. and T. Barlow. 1980. Insecticide control of a white grub, *Phyllophaga ephilida* Say, (Coleoptera: Scarabaeidae) on sweet potato. *J. Georgia Entomol. Soc.* 15:445–449.

Rolston, L. H. and R. L. Kendrick. 1961. Biology of the brown stink bug, *Euschistus servus* Say. *J. Kansas Entomol. Soc.* 34:151–157.

Rolston, L. H. and P. Rouse. 1965. The biology and ecology of the grape colaspis, *Colaspis flavida* Say, in relation to rice production in the Arkansas Grand Prairie. Arkansas Agric. Exp. Stn. Bull. 694. 31 pp.

Rolston, L. H., T. Barlow, A. Jones, and T. Hernandez. 1981. Potential of host-plant resistance in sweet potato for control of a white grub, *Phyllophaga ephilida* Say (Coleoptera: Scarabaeidae). *J. Kansas Entomol. Soc.* 54:378–380.

Roltsch, W. J. and M. A. Mayse. 1984. Population studies of *Heliothis* spp. (Lepidoptera: Noctuidae) on tomato and corn in southeast Arkansas. *Environ. Entomol.* 13:292–299.

Romanow, L. R., J. W. Moyer, and G. G. Kennedy. 1986. Alteration of efficiencies of acquisition and inoculation of watermelon mosaic virus 2 by plant resistance to the virus and to an aphid vector. *Phytopathology* 76:1276–1281.

Rose, A. H., R. H. Silversides, and O. H. Lindquist. 1975. Migration flight by an aphid, *Rhopalosiphum maidis* (Hemiptera: Aphididae), and a noctuid, *Spodoptera frugiperda* (Lepidoptera: Noctuidae). *Can. Entomol.* 107: 567–576.

Rosenheim, J. A., S. C. Welter, M. W. Johnson, R. F. L. Mau, and L. R. Gusukuma–Minuto. 1990. Direct feeding damage on cucumber by mixed-species infestations of *Thrips palmi* and *Frankliniella occidentalis* (Thysanoptera: Thripidae). *J. Econ. Entomol.* 83: 1519–1525.

Rosenstiel, R. G. 1972. Control of young larvae of the rough strawberry root weevil with carbofuran. *J. Econ. Entomol.* 65: 881.

Rottger, U. 1979. Rearing the sugar beet miner *Pegomya betae*. *Entomol. Exp. Appl.* 25: 109–112.

Rowe, J. A. and G. F. Knowlton. 1935. Studies upon the morphology of *Paratrioza cockerelli* (Sulc). *J. Utah Acad. Sci.* 12: 233–237.

Rowley, W. A. and D. C. Peters. 1972. Scanning electron microscopy of the eggshell of four species of *Diabrotica* (Coleoptera: Chrysomelidae). *Ann. Entomol. Soc. Am.* 65: 1188–1191.

Royalty, R. N. and T. M. Perring. 1987. Comparative toxicity of acaracides to *Aculops lycopersici* and *Homeopronematus anconai* (Acari: Eriophyidae, Tydeidae). *J. Econ. Entomol.* 80: 348–351.

Royalty, R. N. and T. M. Perring. 1988. Morphological analysis of damage to tomato leaflets by tomato russet mite (Acari: Eriophyidae). *J. Econ. Entomol.* 81: 816–820.

Royer, L., J. Le Lannic, J. P. Nenon, and G. Boivin. 1998. Response of first-instar *Aleochara bilineata* larvae to the puparium morphology of its dipteran host. *Entomol. Exp. Appl.* 87: 217–220.

Royer, L., and G. Boivin. 1999. Infochemicals mediating the foraging behaviour of *Aleochara bilineata* Gyllenhal adults: sources of attractants. *Entomol. Exp. Appl.* 90: 199–205.

Ru, N. and R. B. Workman. 1979. Seasonal abundance and parasites of the imported cabbageworm, diamondback moth, and cabbage webworm in northeast Florida. *Florida Entomol.* 62: 68–69.

Ruberson, J. R., G. A. Herzog, W. R. Lambert, and W. J. Lewis. 1994. Management of the beet armyworm (Lepidoptera: Noctuidae) in cotton: role of natural enemies. *Florida Entomol.* 77: 440–453.

Ruesink, W. G. 1986. Egg sampling techniques. Pages 83–99 *in* J. L. Krysan and T. A. Miller (eds.). Methods for the Study of Pest *Diabrotica*. Springer-Verlag, Berlin/New York.

Rummel, D. R. and M. D. Arnold. 1989. Estimating thrips populations in cotton with conventional sampling and a plant washing technique. *Southwestern Entomol.* 14: 279–285.

Runner, G. A. 1914. The so-called tobacco wireworm in Virginia. USDA Bull. 78. 30 pp.

Rusoke, D. G. and T. Fatunla. 1987. Inheritance of pod seed resistance to the cow-pea seed beetle (*Callosbruchus maculatus* Fabr.). *J. Agric. Sci.* 108: 655–660.

Russell, C. E. 1981. Predation on the cowpea curculio by the red imported fire ant. *J. Georgia Entomol. Soc.* 16: 13–15.

Russell, H. M. 1912. The bean thrips. USDA Bull. 118. 49 pp.

Russell, H. M. and F. A. Johnston. 1912. The life history of *Tetrastichus asparagi* Crawf. *J. Econ. Entomol.* 5: 429–433.

Russell, W. R., P. J. McLeod, and T. L. Lavy. 1993. Acephate as a potential management tool of *Helicoverpa zea* (Lepidoptera: Noctuidae) in snap bean. *J. Econ. Entomol.* 86: 860–863.

Russin, J. S., D. B. Orr, M. B. Layton, and D. J. Boethel. 1988. Incidence of microorganisms in soybean seeds damaged by stink bug feeding. *Phytopathology* 78: 306–310.

Rust, M. K. and D. A. Reierson. 1977. Effectiveness of barrier toxicants against migrating millipedes. *J. Econ. Entomol.* 70: 477–479.

Rust, R. W. 1977. Evaluation of trap crop procedures for control of Mexican bean beetle in soybeans and lima beans. *J. Econ. Entomol.* 70: 630–632.

Saba, F. 1970. Host plant spectrum and temperature limitations of *Diabrotica balteata*. *Can. Entomol.* 102: 684–691.

Saba, F. 1974. Life history and population dynamics of *Tetranychus tumidus* in Florida (Acarina: Tetranychidae). *Florida Entomol.* 57: 4–63.

Sailer, R. I. 1954. Concerning *Pangaeus bilineatus* (Say) (Hemiptera: Pentatomidae, subfamily Cydninae). *J. Kans. Entomol. Soc.* 27: 41–44.

Sakimura, K. 1932. Life history of *Thrips tabaci* L. on *Emilia sagittata* and its host plant range in Hawaii. *J. Econ. Entomol.* 25: 884–891.

Sakimura, K. 1963. *Frankliniella fusca*, an additional vector for the tomato spotted wilt virus, with notes on *Thrips tabaci*, another vector. *Phytopathology* 53: 412–415.

Salas, J. and O. Mendoza. 1995. Biology of the sweetpotato whitefly (Homoptera: Aleyrodidae) on tomato. *Florida Entomol.* 78: 154–160.

Salas–Aguilar, J. and L. E. Ehler. 1977. Feeding habits of *Orius tristicolor*. *Ann. Entomol. Soc. Am.* 70: 60–62.

Salguero Navas, V. E., J. E. Funderburk, S. M. Olson, and R. J. Beshear. 1991a. Damage to tomato fruit by the western flower thrips (Thysanoptera: Thripidae). *J. Entomol. Sci.* 26: 436–442.

Salguero Navas, V. E., J. E. Funderburk, R. J. Beshear, S. M. Olson, and T. P. Mack. 1991b. Seasonal patterns of *Frankliniella* spp. (Thysanoptera: Thripidae) in tomato flowers. *J. Econ. Entomol.* 84: 1818–1822.

Salguero Navas, V. E., J. E. Funderburk, T. P. Mack, R. J. Beshear, and S. M. Olson. 1994. Aggregation indices and sample size curves for binomial sampling of flower-inhabiting *Frankliniella* species (Thysanoptera: Thripidae) on tomato. *J. Econ. Entomol.* 87: 1622–1626.

Sanborn, C. E. 1916. The alfalfa web worm. Oklahoma Agric. Exp. Stn. Bull. 109. 7 pp.

Sanborn, S. M., J. A. Wyman, and R. K. Chapman. 1982a. Threshold temperature and heat unit summations for seedcorn maggot development under controlled conditions. *Ann. Entomol. Soc. Am.* 75: 103–106.

Sanborn, S. M., J. A. Wyman, and R. K. Chapman. 1982b. Studies on the European corn borer in relation to its management on snap beans. *J. Econ. Entomol.* 75: 551–555.

Sanchez, N. E. and J. A. Onsager. 1988. Life history parameters in *Melanoplus sanguinipes* (F.) in two crested wheatgrass pastures. *Can. Entomol.* 120: 39–44.

Sanchez, N. E. and J. A. Onsager. 1994. Effects of dipterous parasitoids on reproduction of *Melanoplus sanguinipes* (Orthoptera: Acrididae). *J. Orth. Res.* 3: 65–68.

Sanderson, E. D. 1906. Report of miscellaneous cotton insects in Texas. USDA Bur. Entomol. Bull. 57. 63 pp.

Sands, D. C., G. H. Heichel, and J. B. Kring. 1979. Carbofuran reduction of Stewart's disease in corn inoculated with *Erwinia stewartii*. *Plant Dis. Rep.* 63: 631–633.

Sandstrom, J. 1994. High variation in host adaptation among clones of the pea aphid, *Acyrthosiphon pisum* on peas, *Pisum sativum*. *Entomol. Exp. Appl.* 71: 245–256.

Santiago–Alvarez, C. and B. A. Federici. 1978. Notes on the first-instar and two parasites of the clover cutworm, *Scotogramma trifolii* (Noctuidae; Hadeninae). *J. Res. Lep.* 17: 226–230.

Santiago–Alvarez, C., B. A. Federici, and J. J. Johnson. 1979. Clover cutworm, *Scotogramma trifolii*: a semidefined larval diet and colony maintenance. *Ann. Entomol. Soc. Am.* 72: 667–668.

Santos, G. P., G. W. Cosenza, and J. C. Albino. 1960. Biologia de *Spodoptera latifascia* (Walker, 1856) (Lepidoptera: Noctuidae) sobre folhas de eucalipto. *Revista Bras. Entomol.* 24: 153–155.

Sappington, T. W. and W. B. Showers. 1983. Comparison of three sampling methods for monitoring adult European corn borer (Lepidoptera: Pyralidae) population trends. *J. Econ. Entomol.* 76: 1291–1297.

Sappington, T. W. and O. R. Taylor. 1990. Genetic sources of pheromone variation in *Colias eurytheme* butterflies. *J. Chem Ecol.* 16: 2755–2770.

Satterthwait, A. F. 1919. How to control billbugs destructive to cereal and forage crops. USDA Farmers' Bull. 1003. 23 pp.

Satterthwait, A. F. 1933. Larval instars and feeding of the black cutworm, *Agrotis ypsilon* Rott. *J. Agric Res.* 46: 517–530.

Saunders, W. 1873. The *Isabella* tiger moth. *Pyrrharctia (Spilosoma) isabella. Can. Entomol.* 5: 75–77.

Saxena, R. C. 1975. Integrated approach for the control of *Thrips tabaci* Lind. *Indian J. Agric. Sci.* 45: 434–436.

Saynor, M. and D. S. Hill. 1977. Chemical control of onion fly, *Delia antiqua. Ann. Appl. Biol.* 85: 113–120.

Scales, A. L. 1973. Parasites of the tarnished plant bug in the Mississippi Delta. *Environ. Entomol.* 2: 304–306.

Schaaf, A. C. 1972. The parasitoid complex of *Euxoa ochrogaster* (Guenée) (Lepidoptera: Noctuidae). *Quaes. Entomol.* 8: 81–120.

Schaafsma, A. W., F. Meloche, and R. E. Pitblado. 1996. Effect of mowing corn stalks and tillage on overwintering mortality of European corn borer (Lepidoptera: Pyralidae) in field corn. *J. Econ. Entomol.* 89: 1587–1592.

Schaber, B. D., A. M. Harper, and T. Entz. 1990. Effect of swathing alfalfa for hay on insect dispersal. *J. Econ. Entomol.* 83: 2427–2433.

Schaefer, P. W., R. J. Dysart, R. V. Flanders, T. L. Burger, and K. Ikebe. 1983. Mexican bean beetle (Coleoptera: Coccinellidae) larval parasite *Pediobius foveolatus* (Hymenoptera: Eulophidae) from Japan: field release in the United States. *Environ. Entomol.* 12: 852–854.

Schalk, J. M. 1973. Chickpea resistance to *Callosobruchus maculatus* in Iran. *J. Econ. Entomol.* 66: 578–579.

Schalk, J. M. 1986. Rearing and handling of *Diabrotica balteata*. Pages 49–56 *in* J. L. Krysan and T. A. Miller (eds.). Methods for the Study of Pest *Diabrotica*. Springer-Verlag, Berlin/New York.

Schalk, J. M. and C. S. Creighton. 1989. Influence of sweet potato cultivars in combination with a biological control agent (Nematoda: *Heterorhabditis heliothidis*) on larval development of the banded cucumber beetle (Coleoptera: Chrysomelidae). *Environ. Entomol.* 18: 897–899.

Schalk, J. M. and R. L. Fery. 1982. Southern green stink bug and leaf-footed bug: effect on cowpea production. *J. Econ. Entomol.* 75: 72–75.

Schalk, J. M. and R. L. Fery. 1986. Resistance in cowpea to the southern green stink bug. *HortScience* 2: 1189–1190.

Schalk, J. M. and A. K. Stoner. 1979. Tomato production in Maryland: effect of different densities of larvae and adults of the Colorado potato beetle. *J. Econ. Entomol.* 72: 826–829.

Schalk, J. M., P. D. Dukes, A. Jones and Jarret. 1991. Evaluation of sweetpotato clones for soil insect damage. *HortScience* 26: 1548–1549.

Schalk, J. M., J. R. Bohac, P. D. Dukes, and W. R. Martin. 1993. Potential of non-chemical control strategies for reduction of soil insect damage in sweet-potato. *J. Am. Soc. Hort. Sci.* 118: 605–608.

Schalk, J. M., A. Jones, P. D. Dukes, and K. P. Burnham. 1992. Responses of soil insects to mixed and contiguous plantings of resistant and susceptible sweet-potato cultivars. *HortScience* 27: 1089–1091.

Scharff, D. K. 1954. The role of food plants and weather in the ecology of *Melanoplus mexicanus mexicanus* (Sauss.). *J. Econ. Entomol.* 47: 485–489.

Scheffrahn, R. H., R.-C. Hsu, N.-Y. Su, and J. P. Toth. 1987. Composition of adult exocrine gland secretions from three species of burrower bugs (Hemiptera: Cydnidae) in Florida. *J. Entomol. Sci.* 22: 367–370.

Schlinger, E. I. and J. C. Hall. 1960. Biological notes on Pacific coast aphid parasites, and lists of California parasites (Aphidiinae) and their aphid hosts (Hymenoptera: Braconidae). *Ann. Entomol. Soc. Am.* 53: 404–415.

Schneider, W. D., J. R. Miller, J. A. Breznak, and J. F. Fobes. 1983. Onion maggot, *Delia antiqua*, survival and development on onions in the presence and absence of microorganisms. *Entomol. Exp. Appl.* 33: 50–56.

Schoenbohm, R. B. and F. T. Turpin. 1977. Effect of parasitism by *Meteorus leviventris* on corn foliage consumption and corn seedling cutting by the black cutworm. *J. Econ. Entomol.* 70: 457–459.

Schoene, W. J. 1916. The cabbage maggot: its biology and control. New York (Geneva) Agric. Exp. Stn. Bull. 419: 99–160.

Schoene, W. J. and G. W. Underhill. 1933. Economic status of the green stinkbug with reference to the succession of its wild hosts. *J. Agric Res.* 46: 863–866.

Schoof, H. F. and J. W. Kilpatrick. 1958. House fly resistance to organophosphorous compounds in Arizona and Georgia. *J. Econ. Entomol.* 51: 546–547.

Schoonhoven, A. V., C. Cardona, and J. Valor. 1983. Resistance to the bean weevil and the Mexican bean weevil (Coleoptera: Bruchidae) in non-cultivated common bean accessions. *J. Econ. Entomol.* 76: 1255–1259.

Schoonhoven, A. V., J. Piedrahita, R. Valderrama, and G. Galvez. 1978. Biologia, dano y control del acaro tropical *Polyphagotarsonemus latus* (Banks) (Acarina; Tarsonemidae) en frijol. *Turrialba* 28: 77–80.

Schotzko, D. J. and L. E. O'Keeffe. 1986. Comparison of sweep net, D-vac, and absolute sampling for *Lygus hesperus* (Heteroptera: Miridae) in lentils. *J. Econ. Entomol.* 79: 224–228.

Schotzko, D. J. and L. E. O'Keeffe. 1988. Effects of food plants and duration of hibernal quiescence on reproductive capacity of pea leaf weevil (Coleoptera: Curculionidae). *J. Econ. Entomol.* 81: 490–496.

Schotzko, D. J. and L. E. O'Keeffe. 1989a. Comparison of sweep net, D-vac, and absolute sampling, and diel variation of sweep net sampling estimates in lentils for pea aphid (Homoptera: Aphididae), nabids (Hemiptera: Nabidae), lady beetles (Coleoptera: Coccinellidae), and lacewings (Neuroptera: Chrysopidae). *J. Econ. Entomol.* 82: 491–506.

Schotzko, D. J. and L. E. O'Keeffe. 1989b. *Lygus hesperus* distribution and sampling procedures in lentils. *Environ. Entomol.* 18: 308–314.

Schroeder, P. C., C. S. Ferguson, A. M. Shelton, W. T. Wilsey, M. P. Hoffmann, and C. Petzoldt. 1996. Greenhouse and field evaluations of entomopathogenic nematodes (Nematoda: Heterorhabditidae and Steinernematidae) for control of cabbage maggot (Diptera: Anthomyiidae) on cabbage. *J. Econ. Entomol.* 89: 1109–1115.

Schroeder, W. J. and J. B. Beavers. 1985. Semiochemicals and *Diaprepes abbreviatus* (Coloptera: Curculionidae) behavior: implications for survey. *Florida Entomol.* 68: 399–402.

Schroeder, W. J. and D. S. Green. 1983. *Diaprepes abbreviatus* (Coleoptera: Curculionidae); oil sprays as a regulatory treatment, effect on egg attachment. *J. Econ. Entomol.* 76: 1395–1396.

Schroeder, W. J., R. A. Hamlen, and J. B. Beavers. 1979. Survival of *Diaprepes abbreviatus* larvae on selected native and ornamental Florida plants. *Florida Entomol.* 62: 309–312.

Schuster, D. J. 1978. Tomato pinworm: chemical control on tomato seedlings for transplant. *J. Econ. Entomol.* 71: 195–196.

Schuster, D. J. 1982. Tomato pinworm: reduction of egg hatch with insecticides. *J. Econ. Entomol.* 75: 144–146.

Schuster, D. J. 1989. Development of tomato pinworm (Lepidoptera: Gelechiidae) on foliage of selected plant species. *Florida Entomol.* 72: 216–219.

Schuster, D. J. and R. L. Burton. 1982. Rearing the tomato pinworm (Lepidoptera: Gelechiidae) in the laboratory. *J. Econ. Entomol.* 75: 1164–1165.

Schuster, D. J. and P. H. Everett. 1982. Control of the beet armyworm and pepper weevil on pepper. Proc. Florida State Hort. Soc. 95: 349–351.

Schuster, D. J. and J. F. Price. 1992. Seedling feeding damage and preference of *Scapteriscus* spp. mole crickets (Orthoptera: Gryllotalpidae) associated with horticultural crops in west-central Florida. *Florida Entomol.* 75: 115–119.

Schuster, D. J., J. P. Jones, and P. H. Everett. 1976. Effect of leafminer control on tomato yield. Proc. Florida State Hort. Soc. 89: 154–156.

Schuster, D. J., J. B. Kring, and J. F. Price. 1991. Association of the sweet-potato whitefly with a silverleaf disorder of squash. *HortScience* 26: 155–156.

Schuster, D. J., J. P. Gilreath, R. A. Wharton, and P. R. Seymour. 1991. Agromyzidae (Diptera) leafminers and their parasitoids in weeds associated with tomato in Florida. *Environ. Entomol.* 20: 720–723.

Schuster, D. J., T. F. Mueller, J. B. Kring, and J. F. Price. 1990. Relationship of the sweet-potato whitefly to a new tomato fruit disorder in Florida. *HortScience* 25: 1618–1620.

Schwartz, H. F., N. D. Epsky, and J. L. Capinera. 1988. Onion transplant pink root and thrips contamination, and their control in Colorado. *Appl. Agric. Res.* 3: 71–74.

Schwartz, M. D. and R. G. Foottit. 1998. Revision of the nearctic species of the genus *Lygus* Hahn, with a review of the palaearctic species (Heteroptera: Miridae). Mem. Entomol. Inter. 10: 1–428.

Scott, D. B. 1964. The economic significance of Collembola in the Salinas Valley of California. *J. Econ. Entomol.* 57: 297–298.

Scott, D. R. 1970. Lygus bugs feeding on developing carrot seed: plant resistance to that feeding. *J. Econ. Entomol.* 63: 959–961.

Scott, H. G. 1961. Collembola: pictorial keys to the Nearctic genera. *Ann. Entomol. Soc. Am.* 54: 104–113.

Scott, J. A. 1986. The Butterflies of North America, a Natural History and Field Guide. Stanford Univ. Press, Stanford, CA. 583 pp.

Scott, L. B. and J. Milam. 1943. Isoamyl salicylate as an attractant for hornworm moths. *J. Econ. Entomol.* 36: 712–715.

Seal, D. R. 1994. Field studies in controlling melon thrips, *Thrips palmi* Karny (Thysanoptera: Thripidae) on vegetable crops using insecticides. Proc. Florida State Hort. Soc. 107: 159–162.

Seal, D. R. and R. M. Baranowski. 1992. Effectiveness of different insecticides for the control of melon thrips, *Thrips palmi* Karny (Thysanoptera: Thripidae) affecting vegetables in south Florida. Proc. Florida State Hort. Soc. 105: 315–319.

Seal, D. R. and R. B. Chalfant. 1994. Bionomics of *Conoderus rudis* (Coleoptera: Elateridae): newly reported pest of sweet potato. *J. Econ. Entomol.* 87: 802–809.

Seal, D. R. and R. K. Jansson. 1989. Biology and management of corn-silk fly, *Euxesta stigmatis* Loew (Diptera: Otitidae), on sweet corn in southern Florida. Proc. Florida State Hort. Soc. 102: 370–373.

Seal, D. R. and R. K. Jansson. 1993. Oviposition and development of *Euxesta stigmatis* (Diptera: Otitidae). *Environ. Entomol.* 22: 88–92.

Seal, D. R., R. M. Baranowski, and J. D. Bishop. 1993. Effectiveness of insecticides in controlling *Thrips palmi* Karny (Thysanoptera: Thripidae) on different vegetable crops in south Florida. Proc. Florida State Hort. Soc. 106: 228–233.

Seal, D. R., R. B. Chalfant, and M. R. Hall. 1992a. Effectiveness of different seed baits and baiting methods for wireworms (Coleoptera: Elateridae) in sweet-potato. *Environ. Entomol.* 21: 957–963.

Seal, D. R., R. B. Chalfant, and M. R. Hall. 1992b. Effects of cultural practices and rotational crops on abundance of wireworms (Coleoptera: Elateridae) affecting sweet-potato in Georgia. *Environ. Entomol.* 21: 969–974.

Seal, D. R., R. K. Jansson, and K. Bondari. 1995. Bionomics of *Euxesta stigmatis* (Diptera: Otitidae) on sweet corn. *Environ. Entomol.* 24: 917–922.

Seal, D. R., R. K. Jansson, and K. Bondari. 1996. Abundance and reproduction of *Euxesta stigmatis* (Diptera: Otitidae) on sweet corn in different environmental conditions. *Florida Entomol.* 79: 413–422.

Seal, D. R., R. McSorley, and R. B. Chalfant. 1992c. Seasonal abundance and spatial distribution of wireworms (Coleoptera: Elateridae) in Georgia sweet-potato fields. *J. Econ. Entomol.* 85: 1802–1808.

Sears, M. K. and G. Boiteau. 1989. Parasitism of Colorado potato beetle (Coleoptera: Chrysomelidae) eggs by *Edovum puttleri* (Hymenoptera: Eulophidae) on potato in eastern Canada. *J. Econ. Entomol.* 82: 803–810.

Sears, M. K. and C. P. Dufault. 1986. Flight activity and oviposition of the cabbage maggot, *Delia radicum* (Diptera: Anthomyiidae), in relation to damage to rutabagas. *J. Econ. Entomol.* 79: 54–58.

Seamans, H. L. 1928. Forecasting outbreaks of the army cutworm (*Chorizagrotis auxiliaris* Grote). Proc. Entomol. Soc. Ontario 58: 76–85.

Seamans, H. L. 1935. Forecasting outbreaks of the pale western cutworm (*Agrotis orthogonia* Morr.). *J. Econ. Entomol.* 28: 425–428.

Sechriest, R. E. 1972. Control of the garden symphylan in Illinois cornfields. *J. Econ. Entomol.* 65: 599–600.

Sechriest, R. E. and D. W. Sherrod. 1977. Pelleted bait for control of the black cutworm in corn. *J. Econ. Entomol.* 70: 699–700.

Sechriest, R. E., H. B. Petty, and D. E. Kuhlman. 1971. Toxicity of selected insecticides to *Clivina impressifrons*. *J. Econ. Entomol.* 64: 210–213.

Sedlacek, J. D. and L. H. Townsend. 1988. Impact of *Euschistus servus* and *E. variolarius* (Heteroptera: Pentatomidae) feeding on early growth stages of corn. *J. Econ. Entomol.* 81: 840–844.

Segarra–Carmona, A. E. and P. Barbosa. 1988. Relative susceptibility of *Crotalaria* spp. to attack by *Etiella zinckenella* in Puerto Rico. *J. Agric. Univ. Puerto Rico* 72: 147–152.

Segarra–Carmona, A. E. and A. Pantoja. 1988a. Evaluation of relative sampling methods for population estimation of the pepper weevil, *Anthonomus eugenii* Cano (Coleoptera: Curculionidae). *J. Agric. Univ. Puerto Rico* 72: 387–393.

Segarra–Carmona, A. E. and A. Pantoja. 1988b. Sequential sampling plan, yield loss components and economic thresholds for the pepper weevil, *Anthonomus eugenii* Cano (Coleoptera: Curculionidae). *J. Agric. Univ. Puerto Rico* 72: 375–385.

Sekul, A. A. and A. N. Sparks. 1976. Sex attractant of the fall armyworm moth. USDA Tech. Bull. 1542. 6 pp.

Selander, R. B. 1955. The blister beetle genus *Linsleya* (Coleoptera, Meloidae). *Am. Mus. Novit.* 1730: 1–30.

Selander, R. B. 1960. Bionomics, systematics, and phylogeny of *Lytta*, a genus of blister beetles (Coleoptera, Meloidae). Illinois Biol. Mono. 28. 295 pp.

Selander, R. B. 1981. Evidence for a third type of larval prey in blister beetles (Coleoptera: Meloidae). *J. Kansas Entomol. Soc.* 54: 757–783.

Selander, R. B. 1982. Further studies of predation of meloid eggs by meloid larvae (Coleoptera). *J. Kansas Entomol. Soc.* 55: 427–441.

Selman, C. L. and H. E. Barton. 1972. Seasonal trends in catches of moths of twelve harmful species in blacklight traps in northeast Arkansas. *J. Econ. Entomol.* 65: 1018–1021.

Selvan, S., P. S. Grewal, R. Gaugler, and M. Tomalak. 1994. Evaluation of steinernematid nematodes against *Popillia japonica* (Coleoptera: Scarabaeidae) larvae: species, strains, and rinse after application. *J. Econ. Entomol.* 87: 605–609.

Semel, M. 1956. Polyhedrosis wilt of cabbage looper on Long Island. *J. Econ. Entomol.* 49: 420–421.

Semtner, P. J. 1984a. Effect of early-season infestations of the tobacco flea beetle (Coleoptera: Chrysomelidae) on the growth and yield of flue-cured tobacco. *J. Econ. Entomol.* 77: 98–102.

Semtner, P. J. 1984b. Effect of transplanting date on the seasonal abundance of the tobacco flea beetle (Coleoptera: Chrysomelidae) on flue-cured tobacco. *J. Georgia Entomol. Soc.* 19: 49–55.

Semtner, P. J., M. Rasnake, and T. R. Terrill. 1980. Effect of host-plant nutrition on the occurrence of tobacco hornworms and tobacco flea beetles on different types of tobacco. *J. Econ. Entomol.* 73: 221–224.

Senanayake, D. G. and N. J. Holliday. 1988. Comparison of visual, sweep-net, and whole plant bag sampling methods for estimating insect pest populations on potato. *J. Econ. Entomol.* 81: 1113–1119.

Senanayake, D. G., S. F. Pernal, and N. J. Holliday. 1993. Yield responses of potatoes to defoliation by the potato flea beetle (Coleoptera: Chrysomelidae) in Manitoba. *J. Econ. Entomol.* 86: 1527–1533.

Setiawan, D. P. and D. W. Ragsdale. 1987. Use of aluminum-foil and oat-straw mulches for controlling aster leafhopper, *Macrosteles fascifrons* (Homoptera: Cicadellidae), and aster yellows in carrots. *Great Lakes Entomol.* 20: 103–109.

Severin, H. C. 1935. The common black field cricket, a serious pest in South Dakota. South Dakota Agric. Exp. Stn. Bull. 295. 51 pp.

Severin, H. H. P. 1930. Life-history of beet leafhopper, *Eutettix tenellus* (Baker) in California. Univ. of California Publ. Entomol. 5: 37–88.

Severin, H. H. P. 1933. Field observations on the beet leafhopper, *Eutettix tenellus*, in California. *Hilgardia* 7: 281–350.

Shands, W. A. and G. W. Simpson. 1971. Seasonal history of the buckthorn aphid and suitability of alder-leaved buckthorn as a primary host in northeastern Maine. Maine Agric. Exp. Stn. Tech. Bull. 51. 24 pp.

Shands, W. A. and G. W. Simpson. 1972. Effects of aluminum foil mulches upon abundance of aphids on, and yield of potatoes in northeastern Maine. *J. Econ. Entomol.* 65: 507–510.

Shands, W. A., I. M. Hall, and G. W. Simpson. 1962. Entomophthoraceous fungi attacking the potato aphid in northeastern Maine in 1969. *J. Econ. Entomol.* 55: 174–179.

Shands, W. A., G. W. Simpson, and H. J. Murphy. 1972. Effects of cultural methods for controlling aphids on potatoes in northeastern Maine. Maine Agric. Exp. Stn. Tech. Bull. 57. 31 pp.

Shands, W. A., G. W. Simpson, C. F. W. Muesebeck, and H. E. Wave. 1965. Parasites of potato-infesting aphids in northeastern Maine. Maine Agric. Exp. Stn. Bull. T19. 77 pp.

Shanks, Jr., C. H. 1966. Factors that affect reproduction of the garden symphylan, *Scutigerella immaculata*. *J. Econ. Entomol.* 59: 1403–1406.

Shannag, H. K. and J. L. Capinera. 1995. Evaluation of entomopathogenic nematode species for the control of melonworm (Lepidoptera: Pyralidae). *Environ. Entomol.* 24: 143–148.

Shannag, H. K., S. E. Webb, and J. L. Capinera. 1994. Entomopathogenic nematode effect on pickleworm (Lepidoptera: Pyralidae) under laboratory and field conditions. *J. Econ. Entomol.* 87: 1205–1212.

Shapiro, A. M. 1968. Laboratory feeding preferences of the banded woollybear, *Isia isabella*. *Ann. Entomol. Soc. Am.* 61: 1221–1224.

Shapiro, D. I, J. R. Cate, J. Pena, A. Hunsberger, and C. W. McCoy. 1999. Effects of temperature and host age on suppression of *Diaprepes abbreviatus* (Coleoptera: Curculionidae) by entomopathogenic nematodes. *J. Econ. Entomol.* 92: 1086–1092.

Sharma, R. K., A. Durazo, and K. S. Mayberry. 1980. Leafminer control increases summer squash yields. *California Agric.* 34(6): 21–22.

Sharma, R. K., J. M. Larrivee, and L. M. Theriault. 1976. Durees des stades larvaires chez le puceron du pois, *Acyrthosiphon pisum* (Harris) (Aphididae: Homoptera), sur les pois de la variete Perfection. *Ann. Entomol. Soc. Quebec* 21: 144–146.

Sharp, J. L. 1995. Mortality of sweet-potato weevil (Coleoptera: Apionidae) stages exposed to gamma irradiation. *J. Econ. Entomol.* 88: 688–692.

Shaw, M. E. and B. C. Kirkpatrick. 1993. The beet leafhopper-transmitted virescence agent causes tomato big bud disease in California. *Plant Dis.* 77: 290–295.

Shaw, M. W., R. M. Allan, E. A. Hunter, and I. K. Munro. 1993. A relationship between dry matter and damage by larvae of the turnip root fly *Delia floralis* in cultivars of the swedish turnip *Brassica napus*. *Entomol. Exp. Appl.* 67: 129–133.

Shaw, J. T., W. G. Ruesink, S. P. Briggs, and W. H. Luckmann. 1984. Monitoring populations of corn rootworm beetles (Coleoptera: Chrysomelidae) with a trap baited with cucurbitacins. *J. Econ. Entomol.* 77: 1495–1499.

Shean, B. and W. S. Cranshaw. 1991. Differential susceptibilities of green peach aphid (Homoptera: Aphididae) and two endoparasitoids (Hymenoptera: Encyrtidae and Braconidae) to pesticides. *J. Econ. Entomol.* 84: 844–850.

Shelton, A. M. and J. E. Hunter. 1985. Evaluation of the potential of the flea beetle *Phyllotreta cruciferae* to transmit *Xanthomonas campestris* pv. *campestris*, causal agent of black rot of crucifers. *Can. J. Plant Pathol.* 7: 308–310.

Shelton, A. M. and R. C. North. 1986. Species composition and phenology of Thysanoptera within field crops adjacent to cabbage fields. *Environ. Entomol.* 15: 513–519.

Shelton, A. M. and R. C. North. 1987. Injury and control of onion thrips (Thysanoptera: Thripidae) on edible podded peas. *J. Econ. Entomol.* 80: 1325–1330.

Shelton, A. M. and D. M. Soderlund. 1983. Varying susceptibility to methomyl and permethrin in widely separated cabbage looper (Lepidoptera: Noctuidae) populations within eastern North America. *J. Econ. Entomol.* 76: 987–989.

Shelton, A. M. and J. A. Wyman. 1979a. Potato tuberworm damage to potatoes under different irrigation and cultural practices. *J. Econ. Entomol.* 72: 261–264.

Shelton, A. M. and J. A. Wyman. 1979b. Time of tuber infestation and relationships between pheromone catches of adult moths, foliar larval populations, and tuber damage by the potato tuberworm. *J. Econ. Entomol.* 72: 599–601.

Shelton, A. M., R. F. Becker, and J. T. Andaloro. 1983a. Varietal resistance to onion thrips (Thysanoptera: Thripidae) in processing cabbage. *J. Econ. Entomol.* 76: 85–86.

Shelton, A. M., J. Theunissen, and C. W. Hoy. 1994. Efficiency of variable-intensity and sequential sampling for insect control decisions in cole crops in the Netherlands. *Entomol. Exp. Appl.* 70: 209–215.

Shelton, A. M., W. T. Wilsey, and M. A. Schmaedick. 1998. Management of onion thrips (Thysanoptera: Thripidae) on cabbage by using plant resistance and insecticides. *J. Econ. Entomol.* 91: 329–333.

Shelton, A. M., J. A. Wyman, and A. J. Mayor. 1981. Effects of commonly used insecticides on the potato tuberworm and its associated parasites and predators in potatoes. *J. Econ. Entomol.* 74: 303–308.

Shelton, A. M., M. K. Sears, J. A. Wyman, and T. C. Quick. 1983b. Comparison of action thresholds for lepidopterous larvae on fresh-market cabbage. *J. Econ. Entomol.* 76: 196–199.

Shelton, A. M., C. W. Hoy, R. C. North, M. H. Dickson, and J. Barnard. 1988. Analysis of resistance in cabbage varieties to damage by Lepidoptera and Thysanoptera. *J. Econ. Entomol.* 81: 634–640.

Shelton, A. M., J. P. Nyrop, R. C. North, C. Petzoldt, and R. Foster. 1987. Development and use of a dynamic sequential sampling program for onion thrips, *Thrips tabaci* (Thysanoptera: Thripidae), on onions. *J. Econ. Entomol.* 80: 1051–1056.

Shelton, A. M., J. R. Stamer, W. T. Wilsey, B. O. Stoyla, and J. T. Andaloro. 1982. Onion thrips (Thysanoptera: Thripidae) damage and contamination in sauerkraut. *J. Econ. Entomol.* 75: 492–494.

Shepard, B. M., K. D. Elsey, A. E. Muckenfuss, and H. D. Justo, Jr. 1994. Parasitism and predation on egg masses of the southern green stink bug, *Nezara viridula* (L.) (Heteroptera: Pentatomidae), in tomato, okra, cowpea, soybean, and wild radish. *J. Agric. Entomol.* 11: 375–381.

Shepard, M. 1972. Spatial patterns and overcrowding of the bean leaf roller, *Urbanus proteus* (Lepidoptera: Hesperiidae). *Ann. Entomol. Soc. Am.* 65: 1124–1125.

Shepard, M. 1973a. A sequential sampling plan tor treatment decisions on the cabbage looper on cabbage. *Environ. Entomol.* 2: 901–903.

Shepard, M. 1973b. Response of *Melanotus communis* (Coleoptera: Elateridae) larvae to soil temperature and moisture. *Can. Entomol.* 105: 577–580.

Shepard, M., G. R. Carner, and S. G. Turnipseed. 1977. Colonization and resurgence of insect pests of soybean in response to insecticides and field isolation. *Environ. Entomol.* 6: 501–506.

Sherman, M. and M. Tamashiro. 1954. The sweetpotato weevils in Hawaii, their biology and control. Univ. Hawaii Tech. Bull. 23. 36 pp.

Sherrod, D. W., J. T. Shaw, and W. H. Luckmann. 1979. Concepts on black cutworm field biology in Illinois. *Environ. Entomol.* 8: 191–195.

Shields, E. J. 1983. Development rate of variegated cutworm (Lepidoptera: Noctuidae). *Ann. Entomol. Soc. Am.* 76: 171–172.

Shields, E. J., D. I. Rouse, and J. A. Wyman. 1985. Variegated cutworm (Lepidoptera: Noctuidae): leaf-area consumption, feeding site preference, and economic injury level calculation for potatoes. *J. Econ. Entomol.* 78: 1095–1099.

Shipp, J. L. 1995. Monitoring of western flower thrips on glasshouse and vegetable crops. Pages 547–555 in B. L. Parker, M. Skinner, and T. Lewis (eds.). Thrips Biology and Management. Plenum Press, London.

Shipp, J. L. and G. H. Whitfield. 1991. Functional response of the predatory mite, *Amblyseius cucumeris* (Acari: Phytoseiidae), on western flower thrips, *Frankliniella occidentalis* (Thysanoptera: Thripidae). *Environ. Entomol.* 20: 694–699.

Shirck, F. H. 1951. Hibernation of onion thrips in southern Idaho. *J. Econ. Entomol.* 44: 1020–1021.

Shorey, H. H. 1963. The biology of *Trichoplusia ni* (Lepidoptera: Noctuidae). II. Factors affecting adult fecundity and longevity. *Ann. Entomol. Soc. Am.* 56: 476–480.

Shorey, H. H. and L. D. Anderson. 1960. Biology and control of the morning-glory leaf miner, *Bedellia somnulentella*, on sweet potatoes. *J. Econ. Entomol.* 53: 1119–1122.

Shorey, H. H. and R. L. Hale. 1965. Mass-rearing of the larvae of nine noctuid species on a simple artificial medium. *J. Econ. Entomol.* 58: 522–524.

Shorey, H. H., L. A. Andres, and R. L. Hale. 1962. The biology of *Trichoplusia ni* (Lepidoptera: Noctuidae). I. Life history and behavior. *Ann. Entomol. Soc. Am.* 55: 591–597.

Shotwell, R. L. 1930. A study of the lesser migratory grasshopper. USDA Tech. Bull 190. 34 pp.

Shotwell, R. L. 1941. Life histories and habits of some grasshoppers of economic importance on the Great Plains. USDA Tech. Bull. 774. 48 pp.

Shotwell, R. L. 1942. Evaluation of baits and bait ingredients used in grasshopper control. USDA Tech. Bull. 793. 51 pp.

Short, D. E. and R. J. Luedtke. 1970. Larval migration of the western corn rootworm. *J. Econ. Entomol.* 63: 325–326.

Shour, M. H. and T. C. Sparks. 1981. Biology of the soybean looper, *Pseudoplusia includens:* characterization of last-stage larvae. *Ann. Entomol. Soc. Am.* 74: 531–535.

Showers, W. B., E. C. Berry, and L. von Kaster. 1980. Management of second-generation European corn borer by controlling moths outside the cornfield. *J. Econ. Entomol.* 73: 88–91.

Showers, W. B., L. V. Kaster, and P. G. Mulder. 1983. Corn seedling growth stage and black cutworm (Lepidoptera: Noctuidae) damage. *Environ. Entomol.* 12: 241–244.

Showers, W. B., G. L. Reed, and H. Oloumi-Sadeghi. 1974. European corn borer: attraction of males to synthetic lure and to females of different strains. *Environ. Entomol.* 3: 51–58.

Showers, W. B., R. B. Smelser, A. J. Keaster, F. Whitford, J. F. Robinson, J. D. Lopez, and S. E. Taylor. 1989. Recapture of marked black cutworm (Lepidoptera: Noctuidae) males after long-range transport. *Environ. Entomol.* 18: 447–458.

Showers, W. B., A. J. Keaster, J. R. Raulston, W. H. Hendrix III, M. E. Derrick, M. D. McCorcle, J. F. Robinson, M. O. Way, M. J. Wallendorf, and J. L. Goodenough. 1993. Mechanism of southward migration of a noctuid moth [*Agrotis ipsilon* (Hufnagel)]: a complete migrant. *Ecology* 74: 2303–2314.

Shull, W. E. 1933. An investigation of the *Lygus* species which are pests of beans (Hemiptera, Miridae). Idaho Agric. Exp. Stn. Res. Bull. 11. 42 pp.

Siebert, M. W. 1975. Candidates for the biological control of *Solanum elaeagnifolium* Cav. (Solanaceae) in South Africa. 1. Laboratory studies on the biology of *Gratiana lutescens* (Boh.) and *Gratiana pallidula* (Boh.) (Coleoptera: Cassididae). *J. Entomol. Soc. South. Africa* 38: 297–304.

Siegfried, B. D. and C. A. Mullin. 1990. Effects of alternative host plants on longevity, oviposition, and emergence of western and northern corn rootworms (Coleoptera: Chrysomelidae). *Environ. Entomol.* 19: 474–480.

Silberglied, R. E. and O. R. Taylor, Jr. 1978. Ultraviolet reflection and its behavioral role in the courtship of the sulfur butterflies *Colias eurytheme* and *C. philodice* (Lepidoptera, Pieridae). *Behav. Ecol. Sociobiol.* 3: 203–243.

Silbernagel, M. J. and A. M. Jafri. 1974. Temperature effects on curly top resistance in *Phaseolus vulgaris*. *Phytopathol.* 64: 825–827.

Sim, R. J. 1934. Characters useful in distinguishing larvae of *Popillia japonica* and other introduced Scarabaeidae from native species. USDA Circ. 334. 20 pp.

Simigrai, M. and R. E. Berry. 1974. Resistance in broccoli to the garden symphylan. *J. Econ. Entomol.* 67: 371–373.

Simmons, A. M. 1994. Oviposition on vegetables by *Bemisia tabaci* (Homoptera: Aleyrodidae): temporal and leaf surface factors. *Environ. Entomol.* 23: 381–389.

Simmons, A. M. 1999. Nymphal survival and movement of crawlers of *Bemisia argentifolii* (Homoptera: Aleyrodidae) on leaf surfaces of selected vegetables. *Environ. Entomol.* 28: 212–216.

Simmons, A. M. and C. E. Rogers. 1996. Ectoparasitic acugutturid nematodes of adult Lepidoptera. *J. Nematol.* 28: 1–7.

Simmons, A. M. and K. V. Yeargan. 1988a. Development and survivorship of the green stink bug, *Acrosternum hilare* (Hemiptera: Pentatomidae) on soybean. *Environ. Entomol.* 17: 527–532.

Simmons, A. M. and K. V. Yeargan. 1988b. Feeding frequency and feeding duration of the green stink bug (Hemiptera: Pentatomidae) on soybean. *J. Econ. Entomol.* 81: 812–815.

Simmons, C. L., L. P. Pedigo, and M. E. Rice. Evaluation of seven sampling techniques for wireworms (Coloptera: Elateridae). *Environ. Entomol.* 27: 1062–1068.

Simonet, D. E. and B. L. Davenport. 1981. Temperature requirements for development and oviposition of the carrot weevil. *Ann. Entomol. Soc. Am.* 74: 312–315.

Simonet, D. E. and R. L. Pienkowski. 1980. Temperature effect on development and morphometrics of the potato leafhopper. *Environ. Entomol.* 9: 798–800.

Simonet, D. E., S. L. Clement, W. L. Rubink, and R. W. Rings. 1981. Temperature requirements for development and oviposition of *Peridroma saucia* (Lepidoptera: Noctuidae). *Can. Entomol.* 113: 891–897.

Simser, D. 1992. Field application of entomopathogenic nematodes for control of *Delia radicum* in collards. *J. Nematol.* 24: 374–378.

Singh, P. and R. F. Moore (eds.). 1985. Handbook of Insect Rearing Vol. II. Elsevier, Amsterdam/New York. 514 pp.

Singh, S. R., R. A. Luse, K. Leuschner, and D. Nangju. 1978. Groundnut oil treatment for the control of *Callosobruchus maculatus* (F.) during cowpea storage. *J. Stored Prod. Res.* 14: 77–80.

Sinha, S. N., A. K. Chakrabarti, N. P. Agnihotri, H. K. Jain, and V. T. Gajbhiye. 1984. Efficacy and residual toxicity of some systemic granular insecticides against *Thrips tabaci* on onion. *Trop. Pest. Manage.* 30: 32–35.

Sirur, G. M. and C. A. Barlow. 1984. Effects of pea aphids (Homoptera: Aphididae) on the nitrogen fixing activity of bacteria in the root nodules of pea plants. *J. Econ. Entomol.* 77: 606–611.

Sites, R. W., W. S. Chambers, and B. J. Nichols. 1992. Diel periodicity of thrips (Thysanoptera: Thripidae) dispersion and the occurrence of *Frankliniella williamsi* on onions. *J. Econ. Entomol.* 85: 100–105.

Sivapragasam, A. and A. M. Abdul Aziz. 1992. Cabbage webworm on crucifers in Malaysia. Pages 75–80 *in* N. S. Talekar (ed.) Diamondback Moth and Other Crucifer Pests. Asian Vegetable Research and Development Center, Taipei, Taiwan.

Slater, J. A. and R. M. Baranowski. 1978. How to Know the True Bugs (Hemiptera-Heteroptera). Brown, Dubuque., IA. 256 pp.

Slaymaker, P. H. and N. P. Tugwell. 1982. Low-labor method for rearing the tarnished plant bug (Hemiptera: Miridae). *J. Econ. Entomol.* 75: 487–488.

Slaymaker, P. H. and N. P. Tugwell. 1984. Inexpensive female-baited trap for the tarnished plant bug (Hemiptera: Miridae). *J. Econ. Entomol.* 77: 1062–1063.

Sloderbeck, P. E. and K. V. Yeargan. 1983a. Green cloverworm (Lepidoptera: Noctuidae) populations in conventional and double-crop, no-till soybeans. *J. Econ. Entomol.* 76: 785–791.

Sloderbeck, P. E. and K. V. Yeargan. 1983b. Comparison of *Nabis americoferus* and *Nabis roseipennis* (Hemiptera: Nabidae) as predators of the green cloverworm (Lepidoptera: Noctuidae). *Environ. Entomol.* 12: 161–165.

Slosser, E., W. E. Pinchak, and D. R. Rummel. 1989. A review of known and potential factors affecting the population dynamics of the cotton aphid. *Southwestern Entomol.* 14: 302–313.

Smart, L. E., M. M. Blight, J. A. Pickett, and B. J. Pye. 1994. Development of field strategies incorporating semiochemicals for the control of the pea and bean weevil, *Sitona lineatus* L. *Crop Prot.* 13: 127–135.

Smelser, R. B. and L. P. Pedigo. 1991. Phenology of *Cerotoma trifurcata* on soybean and alfalfa in central Iowa. *Environ. Entomol.* 20: 514–519.

Smelser, R. B., W. B. Showers, R. H. Shaw, and S. E. Taylor. 1991. Atmospheric trajectory analysis to project long-range migration of black cutworm (Lepidoptera: Noctuidae) adults. *J. Econ. Entomol.* 84: 879–885.

Smith, A. M. 1992. Modeling the development and survival of eggs of pea weevil (Coleoptera: Bruchidae). *Environ. Entomol.* 21: 314–321.

Smith, A. M. and G. Hepworth. 1992. Sampling statistics and a sampling plan for eggs of pea weevil (Coleoptera: Bruchidae). *J. Econ. Entomol.* 85: 1791–1796.

Smith, C. E. and N. Allen. 1932. The migratory habit of the spotted cucumber beetle. *J. Econ. Entomol.* 25: 53–57.

Smith, C. E. and R. W. Brubaker. 1938. Observations on cabbage worm populations at Baton Rouge, La. *J. Econ. Entomol.* 31: 697–700.

Smith, D. B. and M. K. Sears. 1982. Evidence for dispersal of diamondback moth, *Plutella xylostella* (Lepidoptera: Plutellidae), into southern Ontario. *Proc. Entomol. Soc. Ontario* 113: 21–27.

Smith, D. B., A. J. Keaster, J. M. Cheshire, Jr., and R. H. Ward. 1981. Self-propelled soil sampler: evaluation for efficiency in obtaining estimates of wireworm populations in Missouri cornfields. *J. Econ. Entomol.* 74: 625–629.

Smith, D. S. 1966. Fecundity and oviposition in the grasshoppers *Melanoplus sanguinipes* (F.) and *Melanoplus bivittatus* (Say). *Can. Entomol.* 98: 617–621.

Smith, E. H. 1970. Taxonomic revision of the genus *Systena* Chevrolat (Coleoptera: Chrysomelidae, Alticinae). Unpublished M.S. Thesis, Purdue University. 179 pp.

Smith, E. H. 1985. Revision of the genus *Phyllotreta* Chevrolat of America north of Mexico Part I. The maculate species (Coleoptera: Chrysomelidae, Alticinae). *Fieldiana* 28: 1–168.

Smith, F. F. and A. L. Boswell. 1970. New baits and attractants for slugs. *J. Econ. Entomol.* 63: 1919–1922.

Smith, H. A., J. L. Capinera, J. E. Pena, and B. Linbo-Terhaar. 1994. Parasitism of pickleworm and melonworm (Lepidoptera: Pyralidae) by *Cardiochiles diaphaniae* (Hymenoptera: Braconidae). *Environ. Entomol.* 23: 1283–1293.

Smith, J. B. 1891. The rose-chafer, or "rosebug." New Jersey Agric. Exp. Stn. Bull. 82. 40 pp.

Smith, J. B. 1893. Insects injurious to cucurbs. New Jersey Agric. Exp. Stn. Bull. 94. 40 pp.

Smith, J. B. 1897. The harlequin cabbage bug and the melon plant louse. New Jersey Agric. Exp. Stn. Bull. 121. 14 pp.

Smith, J. B. 1910. Insects injurious to sweet potatoes in New Jersey. New Jersey Agric. Exp. Stn. Bull. 229. 16 pp.

Smith, Jr., J. W. and S. J. Johnson. 1989. Natural mortality of the lesser cornstalk borer (Lepidoptera: Pyralidae) in a peanut agroecosystem. *Environ. Entomol.* 18: 69–77.

Smith, Jr., J. W. and J. T. Pitts. 1974. Pest status of *Pangaeus bilineatus* attacking peanuts in Texas. *J. Econ. Entomol.* 67: 111–113.

Smith, Jr., J. W., S. J. Johnson, and R. L. Sams. 1981. Spatial distribution of lesser cornstalk borer eggs in peanuts. *Environ. Entomol.* 10: 192–193.

Smith, L. B. 1919. The life history and biology of the pink and green aphid (*Macrosiphum solanifolii* Ashmead). Virginia Truck Exp. Stn. Bull. 27: 27–79.

Smith, M. T., G. Wilde, and R. Mize. 1981. Chinch bug: damage and effects of host plant and photoperiod. *Environ. Entomol.* 10: 122–124.

Smith, R. C. and W. W. Franklin. 1954. The garden webworm—*Loxostege similalis* Guen. as an alfalfa pest in Kansas. *J. Kansas Entomol. Soc.* 27: 27–39.

Smith, R. F. and A. E. Michelbacher. 1949. The development and behavior of populations of *Diabrotica 11-punctata* in foothill areas of California. *Ann. Entomol. Soc. Am.* 42: 497–510.

Smith, R. I. 1910. Insect enemies of cantaloupes, cucumbers and related plants. North Carolina Agric. Exp. Stn. Bull. 205. 40 pp.

Smith, R. I. 1911. Two important cantaloupe pests. North Carolina Agric. Exp. Stn. Bull. 214: 101–146.

Smith, R. W. 1965. A field population of *Melanoplus sanguinipes* (Fab.) (Orthoptera: Acrididae) and its parasites. *Can. J. Zool.* 43: 129–201.

Smith, S. A. and T. G. Taylor. 1996. Production cost for selected vegetables in Florida. Florida Coop. Ext. Serv. Circ. 1176. 65 pp.

Smitley, D. R. 1996. Incidence of *Popillia japonica* (Coleoptera: Scarabaeidae) and other scarab larvae in nursery fields. *J. Econ. Entomol.* 89: 1262–1266.

Smits, P. H., M. van de Vrie, and J. M. Vlak. 1987. Nuclear polyhedrosis virus for control of *Spodoptera exigua* larvae on glasshouse crops. *Entomol. Exp. Appl.* 43: 73–80.

Snodgrass, G. L. and Y. H. Fayad. 1991. Euphorine (Hymenoptera: Braconidae) parasitism of the tarnished plant bug (Heteroptera: Miridae) in the areas of Washington County, Mississippi disturbed and undisturbed by agricultural production. *J. Entomol. Sci.* 26: 350–356.

Snodgrass, G. L. and E. A. Stadelbacher. 1994. Population levels of tarnished plant bugs (Heteroptera: Miridae) and beneficial arthropods following early season treatments of *Geranium dissectum* for control of bollworms and tobacco budworms (Lepidoptera: Noctuidae). *Environ. Entomol.* 23: 1091–1096.

Snow, J. W. and P. S. Callahan. 1968. Biological and morphological studies of the granulate cutworm, *Feltia subterranea* (F.) in Georgia and Louisiana. Georgia Agric. Exp. Stn. Bull. 42. 23 pp.

Snow, S. J. 1925. Observations on the cutworm, *Euxoa auxiliaris* Grote, and its principal parasites. *J. Econ. Entomol.* 18: 602–609.

Sohati, P. H., G. Boivin, and R. K. Stewart. 1992. Parasitism of *Lygus lineolaris* eggs on *Coronilla varia*, *Solanum tuberosum*, and three host weeds in southeastern Quebec. *Entomophaga* 37: 515–523.

Sohati, P. H., R. K. Stewart, and G. Boivin. 1989. Egg parasitoids of the tarnished plant bug, *Lygus lineolaris* (P. de B.) (Hemiptera: Miridae), in Quebec. *Can. Entomol.* 121: 1127–1128.

Solomon, J. D. 1988. Observations on two *Papaipema* borers (Lepidoptera: Noctuidae) as little known pests of intensively cultured hardwood trees. *J. Entomol. Sci.* 23: 77–82.

Son, K.–C., R. F. Severson, and S. J. Kays. 1991. Pre- and postharvest changes in sweet-potato root surface chemicals modulating insect resistance. *HortScience* 26: 1514–1516.

Soni, S. K. 1976. Effect of temperature and photoperiod on diapause induction in *Erioischia brassicae* (Bch.) (Diptera, Anthomyiidae) under controlled conditions. *Bull. Entomol. Res.* 66: 125–131.

Soo Hoo, C. R., D. L. Coudriet, and P. V. Vail. 1984. *Trichoplusia ni* (Lepidoptera: Noctuidae) larval development on wild and cultivated plants. *Environ. Entomol.* 13: 843–846.

Soper, R. S. 1981. Role of entomophthoran fungi in aphid control for potato integrated pest management. Pages 153–177 *in* Lashombe, J. H. and R. Casagrande (eds.). Advances in Potato Pest Management. Hutchinson Ross, Stroudsburg, Pennsylvania.

Sorensen, K. A. and J. R. Baker. 1983. Insect and related pests of vegetables: some important, common, and potential pests in the southeastern United States. North Carolina Agric. Ext. Serv. AG-295. 173 pp.

Sorenson, C. J. and L. Cutler. 1954. The superb plant bug, *Adelphocoris superbus* (Uhler): its life history and its relation to seed development in alfalfa. Utah Agric. Exp. Stn. Bull. 370. 20 pp.

Sorenson, C. J. and H. F. Thornley. 1941. The pale western cutworm (*Agrotis orthogonia* Morrison). Utah Agric. Exp. Stn. Bull. 297. 23 pp.

Soroka, J. J. and P. A. MacKay. 1991. Antibiosis and antixenosis to pea aphid (Homoptera: Aphididae) in cultivars of field peas. *J. Econ. Entomol.* 84: 1951–1956.

Soroka, J. J. and M. K. Pritchard. 1987. Effects of flea beetle feeding on transplanted and direct-seeded broccoli. *Can. J. Plant Sci.* 67: 549–557.

Sosa, Jr., O. 1990. Oviposition preference by the sugarcane borer (Lepidoptera: Pyralidae). *J. Econ. Entomol.* 83: 866–868.

Soteres, K. M., R. C. Berberet, and R. W. McNew. 1984. Parasites of larval *Euxoa auxiliaris* (Grote) and *Peridroma saucia* (Hübner) (Lepidoptera: Noctuidae) in alfalfa fields of Oklahoma. *J. Kansas Entomol. Soc.* 57: 63–68.

Soule, C. 1896. Life history of *Deilephila lineata*. Psyche 7: 458–459.

Sparks, A. N. 1979. A review of the biology of the fall armyworm. *Florida Entomol.* 62: 82–87.

Sparks, A. N., H. C. Chiang, C. A. Triplehorn, W. D. Guthrie, and T. A. Brindley. 1967. Some factors influencing populations of the European corn borer, *Ostrinia nubilalis* (Hübner) in the north central states: resistance of corn, time of planting and weather conditions Part II, 1958–1962. Iowa Agric. Exp. Stn. Res. Bull. 559. 103 pp.

Spencer, K. A. 1981. A revisionary study of the leaf-mining flies (Agromyzidae) of California. Univ. California Spec. Publ. 3273. 489 pp.

Spencer, K. A. and G. C. Steyskal. 1986. Manual of the Agromyzidae (Diptera) of the United States. USDA, ARS Agric. Handbook 638. 478 pp.

Spike, B. P., G. E. Wilde, T. W. Mize, R. J. Wright, and S. D. Danielson. 1994. Bibliography of the chinch bug, *Blissus leucopterus leucopterus* (Say) (Heteroptera: Lygaeidae) since 1888. *J. Kansas Entomol. Soc.* 67: 116–125.

Srikanth, J. and N. H. Lakkundi. 1988. A method for estimating populations of *Aphis craccivora* Koch on cowpea. *Trop. Pest Manage.* 34: 335–337.

Srinivasan, K. and P. N. Krishna Moorthy. 1992. Development and adoption of integrated pest management for major pests of cabbage using Indian mustard as a trap crop. Pages 511–521 *in* N. S. Talekar (ed.) Diamondback Moth and Other Crucifer Pests. Asian Vegetable Research and Development Center, Taipei, Taiwan.

Stadelbacher, E. A. 1981. Role of early-season wild and naturalized host plants in the buildup of the F_1 generation of *Heliothis zea* and *H. virescens* in the Delta of Mississippi. *Environ. Entomol.* 10: 766–770.

Stahl, C. F. 1920. Studies on the life history and habits of the beet leafhopper. *J. Agric. Res.* 20: 245–252.

Stam, P. A., L. D. Newsom, and E. N. Lambremont. 1987. Predation and food as factors affecting survival of *Nezara viridula* (L.) (Hemiptera: Pentatomidae) in a soybean ecosystem. *Environ. Entomol.* 16: 1211–1216.

Stange, L. A. 1978. The slugs of Florida (Gastropoda: Pulmonata). Florida Dept. Agric. Cons. Serv. Entomol. Circ. 197. 4 pp.

Stanley, W. W. 1936. Studies of the ecology and control of cutworms in Tennessee. Tennessee Agric. Exp. Stn. Bull. 159. 16 pp.

Stannard, Jr., L. J. and S. M. Vaishampayan. 1971. *Ovacarus clivinae*, new genus and species (Acarina: Podapolipidae), and endoparasite of the slender seedcorn beetle. *Ann. Entomol. Soc. Am.* 64: 887–890.

Stark, J. D., R. I. Vargas, and R. K. Thalman. 1991. Diversity and abundance of oriental fruit fly parasitoids (Hymenoptera: Braconidae) in guava orchards in Kauai, Hawaii. *J. Econ. Entomol.* 84: 1460–1467.

Starratt, A. N. and D. G. R. McLeod. 1982. Monitoring fall armyworm, *Spodoptera frugiperda* (Lepidoptera: Noctuidae), moth populations in southwestern Ontario with sex pheromone traps. *Can. Entomol.* 114: 545–549.

Steck, W., M. D. Chisholm, and E. W. Underhill. 1980a. Optimized conditions for sex attractant trapping of male redbacked cutworm moths, *Euxoa ochrogaster* (Guenée). *J. Chem Ecol.* 6: 585–591.

Steck, W., E. W. Underhill, B. K. Bailey, and M. D. Chisholm. 1977. A sex attractant for male moths of the glassy cutworm *Crymodes devastator* (Brace): a mixture of Z-11-hexadecen-1-yl acetate, Z-11-hexadecenal and Z-7-dodecen-1-yl acetate. *Environ. Entomol.* 6: 270–273.

Steck, W., E. W. Underhill, M. D. Chisholm, C. C. Peters, H. G. Philip, and A. P. Arthur. 1979. Sex pheromone traps in population monitoring of adults of the bertha armyworm, *Mamestra configurata* (Lepidoptera: Noctuidae). *Can. Entomol.* 111: 91–95.

Steck, W. F., E. W. Underhill, B. K. Bailey, and M. D. Chisholm. 1980b. Improved sex attractant blend for adult males of the glassy cutworm, *Crymodes devastator* (Lepidoptera: Noctuidae). *Can. Entomol.* 112: 751–752.

Steck, W. F., E. W. Underhill, B. K. Bailey, and M. D. Chisholm. 1982. A four-component sex attractant for male moths of the armyworm, *Pseudaletia unipuncta*. *Entomol. Exp. Appl.* 32: 302–304.

Steck, W. F., E. W. Underhill, M. D. Chisholm, and H. S. Gerber. 1979. Sex attractant for male alfalfa looper moths, *Autographa californica*. *Environ. Entomol.* 8: 373–375.

Steffey, K. L., M. E. Gray, and D. E. Kuhlman. 1992. Extent of corn rootworm (Coleoptera: Chrysomelidae) larval damage in corn after soybeans: search for the expression of the prolonged diapause trait in Illinois. *J. Econ. Entomol.* 85: 268–275.

Stegmaier, C. E. 1966a. Host plants and parasites of *Liriomyza munda* in Florida (Diptera: Agromyzidae). *Florida Entomol.* 49: 81–86.

Stegmaier, C. E. 1966b. Host plants and parasites of *Liriomyza trifolii* in Florida (Diptera: Agromyzidae). *Florida Entomol.* 49: 75–80.

Stegmaier, C. E. 1967. Host plants of *Liriomyza brassicae*, with records of their parasites from south Florida (Diptera: Agromyzidae). *Florida Entomol.* 50: 257–261.

Stehr, F. W. (ed.). 1987. Immature Insects, Vol. 1. Kendall/Hunt Publ., Dubuque. 754 pp.

Stehr, F. W. (ed.). 1991. Immature Insects, Vol. 2. Kendall/Hunt Publ., Dubuque. 975 pp.

Steiner, L. F., E. J. Harris, W. C. Mitchell, M. S. Fujimoto, and L. D. Christenson. 1965a. Melon fly eradication by overflooding with sterile flies. *J. Econ. Entomol.* 58: 519–522.

Steiner, L. F., W. C. Mitchell, E. J. Harris, T. T. Kozuma, and M. S. Fujimoto. 1965b. Oriental fruit fly eradication by male annihilation. *J. Econ. Entomol.* 58: 961–964.

Steiner, L. F., W. G. Hart, E. J. Harris, R. T. Cunningham, K. Ohinata, and D. C. Kamakahi. 1970. Eradication of the oriental fruit fly from the Mariana Islands by the methods of male annihilation and sterile insect release. *J. Econ. Entomol.* 63: 131–135.

Steiner, W. W. M., D. J. Voegtlin, and M. E. Irwin. 1985. Genetic differentiation and its bearing on migration in North American populations of the corn leaf aphid, *Rhopalosiphum maidis* (Fitch) (Homoptera: Aphididae). *Ann. Entomol. Soc. Am.* 78: 518–525.

Steinhaus, E. A. 1957. New records of insect-virus diseases. *Hilgardia* 26: 417–430.

Steinhaus, E. A. and J. P. Dineen. 1960. Observations on the role of stress in a granulosis of the variegated cutworm. *J. Insect Pathol.* 2: 55–65.

Steinkraus, D. C., A. J. Mueller, and R. A. Humber. 1993a. *Furia virescens* (Thaxter) Humber (Zygomycetes: Entomophthoraceae) infections in the armyworm, *Pseudaletia unipunctata* (Haworth) (Lepidoptera: Noctuidae) in Arkansas with notes on other natural enemies. *J. Entomol. Sci.* 28: 376–386.

Steinkraus, D. C., G. O. Boys, T. J. Kring, and J. R. Ruberson. 1993b. Pathogenicity of the facultative parasite, *Chroniodiplogaster aerivora* (Cobb) (Rhabditida: Diplogasteridae) to corn earworm (*Helicoverpa zea* [Boddie]) (Lepidoptera: Noctuidae). *J. Invertebr. Pathol.* 61: 308–312.

Stephenson, J. W. 1975. Laboratory observations on the effect of soil compaction on slug damage to winter wheat. *Plant Pathol.* 24: 9–11.

Stephenson, J. W. and L. V. Knutson. 1966. A resume of recent studies of invertebrates associated with slugs. *J. Econ. Entomol.* 59: 356–360.

Stevens, L. M., A. L. Steinhauer, and J. R. Coulson. 1975a. Suppression of Mexican bean beetle on soybeans with annual inoculative releases of *Pediobius foveolatus*. *Environ. Entomol.* 4: 947–952.

Stevens, L. M., A. L. Steinhauer, and T. C. Elden. 1975b. Laboratory rearing of the Mexican bean beetle and the parasite, *Pediobius foveolatus*, with emphasis on parasite longevity and host-parasite ratios. *Environ. Entomol.* 4: 953–957.

Stevenson, A. B. 1976a. Seasonal history of the carrot weevil, *Listronotus oregonensis* (Coleoptera: Curculionidae) in the Holland Marsh, Ontario. Proc. Entomol. Soc., Ontario 107: 71–78.

Stevenson, A. B. 1976b. Carrot rust fly: chemical control of first generation larvae in organic soil in Ontario. *J. Econ. Entomol.* 69: 282–284.

Stevenson, A. B. 1981. Development of the carrot rust fly, *Psila rosae* (Diptera: Psilidae), relative to temperature in the laboratory. *Can. Entomol.* 113: 569–574.

Stevenson, A. B. 1983. Seasonal occurrence of carrot rust fly (Diptera: Psilidae) adults in Ontario and its relation to cumulative degree-days. *Environ. Entomol.* 12: 1020–1025.

Stevenson, A. B. 1986. Relationship between temperature and development of the carrot weevil, *Listronotus oregonensis* (LeConte) (Coleoptera: Curculionidae), in the laboratory. *Can. Entomol.* 118: 1287–1290.

Stevenson, A. B. and E. S. Barszcz. 1991. Influence of prepupation environment on diapause induction in the carrot rust fly, *Psila rosae* (Fab.) (Diptera: Psilidae). *Can. Entomol.* 123: 41–53.

Stevenson, A. B. and G. Boivin. 1990. Interaction of temperature and photoperiod in control of reproductive diapause in the carrot weevil (Coleoptera: Curculionidae). *Environ. Entomol.* 19: 836–841.

Stewart, J. G. 1994. Monitoring adult European corn borer (Lepidoptera: Pyralidae) in potatoes on Prince Edward Island. *Environ. Entomol.* 23: 1124–1128.

Stewart, J. G. and L. S. Thompson. 1989. The spatial distribution of spring and summer populations of adult potato flea beetles, *Epitrix cucumeris* (Harris) (Coleoptera: Chrysomelidae), on small plots of potatoes. *Can. Entomol.* 121: 1097–1101.

Stewart, J. K., Y. Aharoni, P. L. Hartsell, and D. K. Young. 1980. Acetaldehyde fumigation at reduced pressures to control the green peach aphid on wrapped and packed head lettuce. *J. Econ. Entomol.* 73: 149–152.

Stewart, J. W., C. Cole, and P. Lummus. 1989. Winter survey of thrips (Thysanoptera: Thripidae) from certain suspected and confirmed hosts of tomato spotted-wilt virus in south Texas. *J. Entomol. Sci.* 24: 392–401.

Stewart, J. W., J. Cocke, Jr., J. Taylor, and R. Williams. 1995. Mexican corn rootworm emergence from corn fields where sorghum was grown the previous season. *Southwestern Entomol.* 20: 229–230.

Stewart, P. A. and A. P. Baker, Jr. 1970. Rate of growth of larval tobacco hornworms reared on tobacco leaves and on artificial diet. *J. Econ. Entomol.* 63: 535–536.

Stewart, R. K. and A. R. Khattat. 1980. Pest status and economic thresholds of the tarnished plant bug, *Lygus lineolaris* (Hemiptera

(Heteroptera): Miridae), on green beans in Quebec. *Can. Entomol.* 112: 301–305.

Stewart, R. K. and H. Khoury. 1976. The biology of *Lygus lineolaris* (Palisot de Beauvois) (Hemiptera: Miridae) in Quebec. *Ann. Entomol. Soc., Quebec* 21: 52–63.

Steyskal, G. C. 1987. Otitidae. Pages 799–808 *in* J. F. McAlpine (ed.). Manual of Nearctic Diptera, Vol. 2. Agric. Canada Res. Branch Mono. 28.

Stinner, B. R., D. A. McCartney, and W. L. Rubink. 1984. Some observations on ecology of the stalk borer (*Papaipema nebris* (Gn.): Noctuidae) in no-tillage corn agroecosystems. *J. Georgia Entomol. Soc.* 19: 229–234.

Stockton, W. D. 1963. New species of *Hyperodes* Jekel and a key to the nearctic species of the genus (Coleptera: Curculionidae). *Bull. So. California Acad. Sci.* 62: 140–149.

Stoetzel, M. B. 1990. Aphids (Homoptera: Aphididae) colonizing leaves of asparagus in the United States. *J. Econ. Entomol.* 83: 1994–2002.

Stoetzel, M. B., G. L. Miller, P. J. O'Brien, and J. B. Graves. 1996. Aphids (Homoptera: Aphididae) colonizing cotton in the United States. *Florida Entomol.* 79: 193–205.

Stoltz, R. L. and R. L. Förster. 1984. Reduction of pea leaf roll of peas (*Pisum sativum*) with systemic insecticides to control the pea aphid (Homoptera: Aphididae) vector. *J. Econ. Entomol.* 77: 1537–1541.

Stoltz, R. L. and C. D. McNeal, Jr. 1982. Assessment of insect emigration from alfalfa hay to bean fields. *Environ. Entomol.* 11: 578–580.

Stone, J. D. and L. P. Pedigo. 1972. Development and economic-injury level of the green cloverworm on soybean in Iowa. *J. Econ. Entomol.* 65: 197–201.

Stone, M. W. 1941. Life history of the sugar-beet wireworm in southern California. USDA Tech. Bull. 744. 88 pp.

Stone, M. W. 1965. Biology and control of the lima-bean pod borer in southern California. USDA Tech. Bull. 1321. 46 pp.

Stone, M. W. and J. Wilcox. 1979. The gulf wireworm in California (Coleoptera: Elateridae). *Pan-Pac. Entomol.* 55: 235–238.

Stone, M. W. and J. Wilcox. 1983. Light trap collections of three introduced *Conoderus* species (Coleoptera: Elateridae) in southern California. *Pan-Pac. Entomol.* 58: 202–205.

Stone, T. B. and S. R. Sims. 1993. Geographic susceptibility of *Heliothis virescens* and *Helicoverpa zea* (Lepidoptera: Noctuidae) to *Bacillus thuringiensis*. *J. Econ. Entomol.* 86: 989–994.

Stone, W. E., B. L. Boyden, C. B. Wisecup, and E. C. Tatman. 1932. Control of the celery leaf-tier in Florida. Florida Agric. Exp. Stn. Bull. 251. 23 pp.

Stoner, K. A. 1990. Glossy leaf wax and plant resistance to insects in *Brassica oleracea* under natural infestation. *Environ. Entomol.* 19: 730–739.

Stoner, K. A. 1992. Density of imported cabbageworms (Lepidoptera: Pieridae), cabbage aphids (Homoptera: Aphididae), and flea beetles (Coleoptera: Chrysomelidae) on glossy and trichome-bearing lines of *Brassica oleracea*. *J. Econ. Entomol.* 85: 1023–1030.

Stoner, K. A. and A. M. Shelton. 1988. Influence of variety on abundance and within-plant distribution of onion thrips (Thysanoptera: Thripidae) on cabbage. *J. Econ. Entomol.* 81: 1190–1195.

Stoner, K. A., F. J. Ferrandino, M. P. N. Gent, W. H. Elmer, and J. A. Lamondia. 1996. Effects of straw mulch, spent mushroom compost, and fumigation on the density of Colorado potato beetles (Coleoptera: Chrysomelidae) in potatoes. *J. Econ. Entomol.* 89: 1267–1280.

Storey, C. L. 1978. Mortality of cowpea weevil in a low-oxygen atmosphere. *J. Econ. Entomol.* 71: 833–834.

Storms, W. W., C. Berry, and W. Withee. 1981. Miller moth asthma. *Clin. All.* 11: 55–59.

Story, R. N. and A. J. Keaster. 1982a. The overwintering biology of the black cutworm, *Agrotis ipsilon*, in field cages (Lepidoptera: Noctuidae). *J. Kansas Entomol. Soc.* 55: 621–624.

Story, R. N. and A. J. Keaster. 1982b. Temporal and spatial distribution of black cutworms in midwest field crops. *Environ. Entomol.* 11: 1019–1022.

Story, R. N. and A. J. Keaster. 1983. Modified larval bait trap for sampling black cutworm (Lepidoptera: Noctuidae) populations in field corn. *J. Econ. Entomol.* 76: 662–666.

Story, R. N., F. J. Sundstrom, and E. G. Riley. 1983. Influence of sweet corn cultivar, planting date, and insecticide on corn earworm damage. *J. Georgia Entomol. Soc.* 18: 350–353.

Story, R. N., A. J. Keaster, W. B. Showers, and J. T. Shaw. 1984. Survey and phenology of cutworms (Lepidoptera: Noctuidae) infesting field corn in the midwest. *J. Econ. Entomol.* 77: 491–494.

Stracener, C. L. 1931. Economic importance of the salt-marsh caterpillar (*Estigmene acraea* Drury) in Louisiana. *J. Econ. Entomol.* 24: 835–838.

Strand, M. R. 1990. Characterization of larval development in *Pseudoplusia includens* (Lepidoptera: Noctuidae). *Ann. Entomol. Soc. Am.* 83: 538–544.

Straub, R. W. 1982. Occurrence of four aphid vectors of maize dwarf mosaic virus in southeastern New York. *J. Econ. Entomol.* 75: 156–158.

Straub, R. W. 1983. Minimization of insecticide treatment for first-generation European corn borer (Lepidoptera: Pyralidae) control in sweet corn. *J. Econ. Entomol.* 76: 345–348.

Straub, R. W. 1984. Maize dwarf mosaic virus: symptomatology and yield reactions of susceptible and resistant sweet corns. *Environ. Entomol.* 13: 318–323.

Straub, R. W. and C. W. Boothroyd. 1980. Relationship of corn leaf aphid and maize dwarf mosaic disease to sweet corn yields in southeastern New York. *J. Econ. Entomol.* 73: 92–95.

Straub, R. W. and A. C. Davis. 1978. Onion maggot: evaluation of insecticides for protection of onions in muck soils. *J. Econ. Entomol.* 71: 684–686.

Straub, R. W. and P. C. Huth. 1976. Correlations between phenological events and European corn borer activity. *Environ. Entomol.* 5: 1079–1082.

Streett, D. A., S. A. Woods, and M. A. Erlandson. 1997. Entomopoxviruses of grasshoppers and locusts: biology and biological control potential. *Mem. Entomol. Soc. Canada* 171: 115–130.

Strong, F. E. and J. W. Apple. 1958. Studies on the thermal constants and seasonal occurrence of the seed-corn maggot in Wisconsin. *J. Econ. Entomol.* 51: 704–707.

Stroyan, H. L. G. 1963. The British Species of *Dysaphis* Borner (*Sappaphis* auctt. nec Mats.), Part 2. H. M. Stationery Office, London. 119 pp.

Struble, D. L. 1981a. A four-component pheromone blend for optimum attraction of redbacked cutworm males, *Euxoa ochrogaster* (Guenée). *J. Chem Ecol.* 7: 615–625.

Struble, D. L. 1981b. Modification of the attractant blend for adult males of the army cutworm, *Euxoa auxiliaris* (Grote), and the development of an alternate three-component attractant blend for this species. *Environ. Entomol.* 10: 167–170.

Struble, D. L. and J. R. Byers. 1987. Identification of sex pheromone components of darksided cutworm, *Euxoa messoria*, and modification of sex attractant blend for adult males. *J. Chem Ecol.* 13: 1187–1199.

Struble, D. L. and L. A. Jacobson. 1970. A sex pheromone in the redbacked cutworm. *J. Econ. Entomol.* 63: 841–844.

Struble, D. L. and C. E. Lilly. 1977. An attractant for the beet webworm, *Loxostege sticticalis* (Lepidoptera: Pyralidae). *Can. Entomol.* 109: 261–266.

Struble, D. L. and G. E. Swailes. 1977a. Sex attractant for clover cutworm, *Scotogramma trifolii*: field tests with various ratios of Z-11-hexadecen-1-yl acetate and Z-11-hexadecen-1-ol, and with various quantities of attractant of two types of carriers. *Can. Entomol.* 109: 369–373.

Struble, D. L. and G. E. Swailes. 1977b. A sex attractant for the adult males of the army cutworm, *Euxoa auxiliaris*: a mixture of Z-5-tetradecen-1-yl acetate and E-7-tetradecen-1-yl acetate. *Environ. Entomol.* 6: 719–724.

Struble, D. L. and G. E. Swailes. 1978. A sex attractant for adult males of the pale western cutworm, *Agrotis orthogonia* (Lepidoptera: Noctuidae). *Can. Entomol.* 110: 769–773.

Struble, D. L., G. L. Ayre, and J. R. Byers. 1984. Minor sex-pheromone components of *Mamestra configurata* (Lepidoptera: Noctuidae) and improved blends for attraction of male moths. *Can. Entomol.* 116: 103–105.

Struble, D. L., G. E. Swailes, and G. L. Ayre. 1977. A sex attractant for males of the darksided cutworm, *Euxoa messoria* (Lepdioptera: Noctuidae). *Can. Entomol.* 109: 975–980.

Struble, D. L., G. E. Swailes, W. F. Steck, E. W. Underhill, and M. D. Chisolm. 1976. A sex attractant for the adult males of variegated cutworm, *Peridroma saucia. Environ. Entomol.* 5: 988–990.

Stuart, J., G. E. Wilde, and J. H. Hatchett. 1985. Chinch-bug (Heteroptera: Lygaeidae) reproduction, development, and feeding preference on various wheat cultivars and genetics sources. *Environ. Entomol.* 14: 539–543.

Sturtevant, A. H. 1921. The North American species of Drosophila. Carnegie Inst. of Washington, Washington. 150 pp.

Su, H. C. F. 1976. Toxicity of a chemical component of lemon oil to cowpea weevils. *J. Georgia Entomol. Soc.* 11: 297–301.

Su, H. C. F. 1978. Laboratory study on toxic effect of black pepper varieties to three species of stored-product insects. *J. Georgia Entomol. Soc.* 13: 269–274.

Su, H. C. F. 1991. Toxicity and repellency of chenopodium oil to four species of stored-product insects. *J. Entomol. Sci.* 26: 178–182.

Sudbrink, Jr., D. L. and J. F. Grant. 1995. Wild host plants of *Helicoverpa virescens* (Lepidoptera: Noctuidae) in eastern Tennessee. *Environ. Entomol.* 24: 1080–1085.

Sudbrink, Jr., D. L., T. P. Mack, and G. W. Zehnder. 1998. Alternate host plants of cowpea curculio, (Coleoptera: Curculionidae) in Alabama. *Florida Entomol.* 81: 373–383.

Suett, D. L., A. A. Jukes, and K. Phelps. 1993. Stability of accelerated degradation of soil-applied insecticides: laboratory behaviour of aldicarb and carbofuran in relation to their efficacy against cabbage root fly *(Delia radicum)* in previously treated field soils. *Crop Prot.* 12: 431–441.

Sullivan, D. J. and R. van den Bosch. 1971. Field ecology of the primary parasites and hyperparasites of the potato aphid, *Macrosiphum euphorbiae*, in the east San Francisco Bay area. *Ann. Entomol. Soc. Am.* 64: 389–394.

Sullivan, M. J. and C. H. Brett. 1974. Resistance of commercial crucifers to the harlequin bug in the coastal plain of North Carolina. *J. Econ. Entomol.* 67: 262–264.

Summers, C. G. and A. S. Newton. 1989. Economic significance of sugarbeet root aphid, *Pemphigus populivenae* Fitch (Homoptera: Aphididae) in California. *Appl. Agric. Res.* 4: 162–167.

Sumner, H. R., R. B. Chalfant, and D. Cochran. 1991. Influence of chemigation parameters on fall armyworm control in field corn. *Florida Entomol.* 74: 280–287.

Sutherland, D. W. S. 1966. Biological investigations of *Trichoplusia ni* (Hübner) and other Lepidoptera damaging cruciferous crops on Long Island, New York. New York (Cornell) Agric. Exp. Stn. Mem. 399. 99 pp.

Sutherland, J. A. 1986. Damage by *Cylas formicarius* Fab. to sweet-potato vines and tubers, and the effect of infestations on total yield in Papua New Guinea. *Trop. Pest Manage.* 32: 316–323.

Sutter, G. R. 1972. A pox virus of the army cutworm. *J. Invertebr. Pathol.* 19: 375–382.

Sutter, G. R. 1973. A nonoccluded virus of the army cutworm. *J. Invertebr. Pathol.* 21: 62–70.

Sutter, G. R. and E. Miller. 1972. Rearing the army cutworm on an artificial diet. *J. Econ. Entomol.* 65: 717–718.

Sutter, G. R., E. Miller, and C. O. Calkins. 1972. Rearing the pale western cutworm on artificial diet. *J. Econ. Entomol.* 65: 1470–1471.

Sutton, S. 1972. Woodlice. Pergamon Press, Oxford. 143 pp.

Sutton, S. L., M. Hassall, R. Willows, R. C. Davis, A. Grundy, and K. D. Sunderland. 1984. Life histories of terrestrial isopods: a study of intra- and interspecific variation. Symp. Zool. Soc., London 53: 269–294.

Swailes, G. E. 1958. Periods of flight and oviposition of the cabbage maggot, *Hylemya brassicae* (Bouché) (Diptera: Anthomyiidae), in southern Alberta. *Can. Entomol.* 90: 434–435.

Swailes, G. E. 1959. Resistance in rutabagas to the cabbage maggot, *Hylemya brassicae* (Bouché) (Diptera: Anthomyiidae). *Can. Entomol.* 91: 700–703.

Swailes, G. E. and D. L. Struble. 1979. Variation in catches with sex attractant of the clover cutworm, *Scotogramma trifolii*, and army cutworm, *Euxoa auxiliaris* (Lepidoptera: Noctuidae), due to trap location. *Can. Entomol.* 111: 11–14.

Swain, R. B. 1944. Nature and extent of Mormon cricket damage to crop and range plants. USDA Tech. Bull. 866. 44 pp.

Sweetman, H. L. 1926. Results of life history studies of *Diabrotica 12-punctata* Fabr. (Chrysomelidae, Coleoptera). *J. Econ. Entomol.* 19: 484–490.

Sweetman, H. L. and H. T. Fernald. 1930. Ecological studies of the Mexican bean beetle. Massachusetts Agric. Exp. Stn. Bull. 261. 32 pp.

Swenk, M. H. 1925. The chinch bug and its control. Nebraska Agric. Exp. Stn. Circ. 28. 34 pp.

Swenson, A. A. and W. E. Peay. 1969. Color and natural products attracting the adult sugarbeet root maggot in southcentral Idaho. *J. Econ. Entomol.* 62: 910–912.

Symondson, W. O. C. 1994. The potential of *Abax parallelepipedus* (Col.: Carabidae) for mass breeding as a biological control agent against slugs. *Entomophaga* 39: 323–333.

Tabashnik, B. E., N. L. Cushing, N. Finson, and M. W. Johnson. 1990. Field development of resistance to *Bacillus thuringiensis* in diamondback moth (Lepidoptera: Plutellidae). *J. Econ. Entomol.* 83: 1671–1676.

Tahvanainen, J. 1983. The relationship between flea beetles and their cruciferous host plants: the role of plant and habitat characteristics. Oikos 40: 433–437.

Takahashi, R. 1965. Some new and little-known Aphididae from Japan (Homoptera). *Insecta Matsu.* 28: 19–61.

Takara, J. and T. Nishida. 1983. Spatial distribution of the migrants of the corn delphacid, *Peregrinus maidis* (Ashmead) (Homoptera: Delphacidae) in cornfields. Proc. Hawaiian Entomol. Soc. 24: 327–333.

Talekar, N. S. 1982. Effects of a sweet-potato weevil (Coleoptera: Curculionidae) infestation on sweet-potato root yields. *J. Econ. Entomol.* 75: 1042–1044.

Talekar, N. S. 1987. Resistance in sweet potato to sweetpotato weevil. *Insect Sci. Applic.* 8: 819–823.

Talekar, N. S. and K. W. Cheng. 1987. Nature of damage and sources of resistance to sweet-potato vine borer (Lepidoptera: Pyralidae) in sweet potato. *J. Econ. Entomol.* 80: 788–791.

Talekar, N. S. and C. P. Lin. 1992. Characterization of *Callosobruchus chinensis* (Coleoptera: Bruchidae) resistance in mungbean. *J. Econ. Entomol.* 85: 1150–1153.

Talekar, N. S. and C. P. Lin. 1994. Characterization of resistance to limabean pod borer (Lepidoptera: Pyrallidae) in soybean. *J. Econ. Entomol.* 87: 821–825.

Talekar, N. S. and Y. H. Lin. 1981. Two sources with differing modes of resistance to *Callosobruchus chinensis* in mungbean. *J. Econ. Entomol.* 74: 639–642.

Talekar, N. S. and G. V. Pollard. 1991. Vine borers of sweet potato. Pages 327–339 *in* R. K. Jansson and K.V Raman (eds.). Sweet Potato Pest Management: A Global Perspective. Westview Press, Boulder, Colorado.

Talekar, N. S., H. C. Yang, S. T. Lee, B. S. Chen, and L. Y. Sun (eds.). 1985. Annotated Bibliography of Diamondback Moth. Asian Vegetable Research and Development Center, Taipei, Taiwan. 469 pp.

Tallamy, D. W. 1985. "Egg dumping" in lace bugs (*Gargaphia solani*, Hemiptera: Tingidae). *Behav. Ecol. Sociobiol.* 17: 357–362.

Tamaki, G. 1975. Weeds in orchards as important alternate sources of green peach aphids in late spring. *Environ. Entomol.* 4: 958–960.

Tamaki, G. and W. W. Allen. 1969. Competition and other factors influencing the population dynamics of *Aphis gossypii* and *Macrosiphoniella sanborni* on greenhouse chrysanthemums. *Hilgardia* 39: 447–505.

Tamaki, G. and B. A. Butt. 1977. Biology of the false celery leaftier and damage to sugarbeets. *Environ. Entomol.* 6: 35–38.

Tamaki, G. and B. A. Butt. 1978. Impact of *Perillus bioculatus* on the Colorado potato beetle and plant damage. USDA Tech. Bull. 1581. 11 pp.

Tamaki, G. and L. Fox. 1982. Weed species hosting viruliferous green peach aphids, vector of beet western yellows virus. *Environ. Entomol.* 11: 115–117.

Tamaki, G. and J. E. Halfhill. 1968. Bands on peach trees as shelters for predators of the green peach aphid. *J. Econ. Entomol.* 61: 707–711.

Tamaki, G., B. Annis, and M. Weiss. 1981. Response of natural enemies to the green peach aphid in different plant cultures. *Environ. Entomol.* 10: 375–378.

Tamaki, G., R. L. Chauvin, and A. K. Burditt, Jr. 1983a. Field evaluation of *Doryphorophaga doryphorae* (Diptera: Tachinidae), a parasite, and its host the Colorado potato beetle (Coleoptera: Chrysomelidae). *Environ. Entomol.* 12: 386–389.

Tamaki, G., L. Fox, and P. Featherston. 1982. Laboratory biology of the dusky sap beetle and field interaction with the corn earworm in ears of sweet corn. *J. Entomol. Soc. Brit. Columbia* 79: 3–8.

Tamaki, G., J. A. Gefre, and J. E. Halfhill. 1983b. Biology of morphs of *Brachycolus asparagi* Mordvilko (Homoptera: Aphididae). *Environ. Entomol.* 12: 1120–1124.

Tamaki, G., H. R. Moffitt, and J. E. Turner. 1975. The influence of perennial weeds on the abundance of the redbacked cutworm on aspargus. *Environ. Entomol.* 4: 274–276.

Tamaki, G., R. E. Weeks, and B. J. Landis. 1972. Biology of the zebra caterpillar, *Ceramica picta* (Lepidoptera: Noctuidae). *Pan-Pac. Entomol.* 48: 208–214.

Tamaki, G., B. Annis, L. Fox, R. K. Gupta, and A. Meszleny. 1982. Comparison of yellow holocyclic and green anholocyclic strains of *Myzus persicae* (Sulzer): low temperature adaptability. *Environ. Entomol.* 11: 231–233.

Tan, F. M. and C. R. Ward. 1977. Laboratory studies on the biology of the Banks grass mite. *Ann. Entomol. Soc. Am.* 70: 534–536.

Tanada, Y. 1959. Synergism between two viruses of the armyworm, *Pseudaletia unipunctata* (Haworth) (Lepidoptera, Noctuidae). *J. Insect Pathol.* 1: 215–231.

Tanada, Y. 1961. The epizootiology of virus diseases in field populations of the armyworm, *Pseudaletia unipunctata* (Haworth). *J. Insect Pathol.* 3: 310–323.

Tanaka, N., L. F. Steiner, K. Ohinata, and O. Kamoto. 1969. Low-cost larval rearing medium for mass production of Oriental and Mediterranean fruit flies. *Econ. Entomol.* 62: 967–968.

Tappan, W. B. 1970. *Nysius raphanus* attacking tobacco in Florida and Georgia. *J. Econ. Entomol.* 63: 658–660.

Tappan, W. B. 1986. Relationship of sampling time to tobacco thrips (Thysanoptera: Thripidae) numbers in peanut foliage buds and flowers. *J. Econ. Entomol.* 79: 1359–1363.

Tappan, W. B. and D. W. Gorbet. 1981. Economics of tobacco thrips control with systemic pesticides on Florunner peanuts in Florida. *J. Econ. Entomol.* 74: 283–286.

Tauber, M. J. and C. A. Toschi. 1965. Bionomics of *Euleia fratria* (Loew) (Diptera: Tephritidae) I. Life history and mating behavior. *Can. J. Zool.* 43: 369–379.

Tawfik, M. F. S., K. T. Awadallah, and F. F. Shalaby. 1976. The life history of *Bedellia somnulentella* Zell. (Lepidoptera: Lyonetiidae). *Bull. Soc. Entomol. Egypte* 60: 25–33.

Taylor, E. A. 1954. Parasitization of the salt-marsh caterpillar in Arizona. *J. Econ. Entomol.* 47: 525–530.

Taylor, Jr., O. R., J. W. Grula, and J. L. Hayes. 1981. Artificial diets and continuous rearing methods for the sulfur butterflies *Colias eurytheme* and *C. Philodice* (Pieridae). *J. Lepid. Soc.* 35: 281–289.

Taylor, P. S. and E. J. Shields. 1995. Phenology of *Empoasca fabae* (Harris) Homoptera: Cicadellidae) in its overwintering area and proposed seasonal phenology. *Environ. Entomol.* 24: 1096–1108.

Taylor, P. S., J. L. Hayes, and E. J. Shields. 1993. Demonstration of pine feeding by *Empoasca fabae* (Harris)(Homoptera: Cicadellidae) using an elemental marker. *J. Kansas Entomol. Soc.* 66: 250–252.

Taylor, R. A. J. and D. Reling. 1986. Preferred wind direction of long distance leafhopper (*Empoasca fabae*) migrants and its relevance to the return migration of small insects. *J. Anim. Ecol.* 55: 1103–1114.

Taylor, R. G. and D. G. Harcourt. 1975. The distributional pattern of *Crioceris asparagi* (L.) (Coleoptera: Chrysomelidae) on asparagus. *Proc. Entomol. Soc. Ontario.* 105: 22–28.

Taylor, R. G. and D. G. Harcourt. 1978. Effects of temperature on developmental rate of the immature stages of *Crioceris asparagi* (Coleoptera: Chrysomelidae). *Can. Entomol.* 110: 57–62.

Taylor, T. A. and V. M. Stern. 1971. Host-preference studies with the egg parasite *Trichogramma semifumatum* (Hymenoptera: Trichogrammatidae). *Ann. Entomol. Soc. Am.* 64: 1381–1390.

Teal, P. E. A., R. J. West, and J. E. Laing. 1983. Identification of a blend of sex pheromone components of the potato stem borer (Lepidoptera: Noctuidae) for monitoring adults. *Proc. Entomol. Soc. Ontario* 114: 15–19.

Teerling, C. R., H. D. Pierce, Jr., J. H. Borden, and D. R. Gillespie. 1993. Identification and bioactivity of alarm pheromone in the western flower thrips, *Frankliniella occidentalis*. *J. Chem Ecol.* 19: 681–697.

Temerak, S. A., D. G. Boucias, and W. H. Whitcomb. 1984. A singly-embedded nuclear polyhedrosis virus and entomophagous insects associated with populations of the bean leafroller *Urbanus proteus* L. (Lepid., Hesperiidae). *Z. Angew. Entomol.* 97: 187–191.

Teulon, D. A. J. 1992. Laboratory technique for rearing western flower thrips (Thysanoptera: Thripidae). *J. Econ. Entomol.* 85: 895–899.

Teulon, D. A. J., D. R. Penman, and P. M. J. Ramakers. 1993. Volatile chemicals for thrips (Thysanoptera: Thripidae) host-finding and applications for thrips pest management. *J. Econ. Entomol.* 86: 1405–1415.

Theunissen, J. and H. Legutowska. 1992. Observers' bias in the assessment of pest and disease symptoms in leek. *Entomol. Exp. Appl.* 64:101–109.

Thomas, C. A. 1936. The tomato pin worm. Pennsylvania Agric. Exp. Stn. Bull. 337. 15 pp.

Thomas, C. A. 1940. The biology and control of wireworms. Pennsylvania Agric. Exp. Stn. Bull. 392. 90 pp.

Thomas, P. E. 1972. Mode of expression of host preference by *Curculifer tenellus*, the vector of curly top virus. *J. Econ. Entomol.* 65:119–123.

Thomas, P. E. and R. K. Boll. 1977. Effect of host preference on transmission of curly top virus to tomato by the beet leafhopper. *Phytopathology* 67:903–905.

Thomas, P. E. and M. W. Martin. 1971. Vector preference, a factor of resistance to curly top virus in certain tomato cultivars. *Phytopathology* 61:1257–1260.

Thomas, W. A. 1928. The Porto Rican mole cricket. USDA Farmers' Bull. 1561. 8 pp.

Thome, C. R., M. E. Smith, and J. A. Mihm. 1992. Leaf feeding resistance to multiple insect species in a maize diallel. *Crop Sci.* 32:1460–1463.

Thompson, D. C., J. L. Capinera, and S. D. Pilcher. 1987. Comparison of an aerial water-pan pheromone trap with traditional trapping techniques for the European corn borer (Lepidoptera: Pyralidae). *Environ. Entomol.* 16:154–158.

Thompson, J. N. and P. W. Price. 1977. Plant plasticity, phenology, and herbivore dispersion: wild parsnip and the parsnip webworm. *Ecology* 58:1112–1119.

Thompson, L. S. 1967. Reduction of lettuce yellows with systemic insecticides. *J. Econ. Entomol.* 60:716–718.

Thompson, L. S. and J. B. Sanderson. 1977. Pea moth control in field peas with insecticides and the effect on crop yield. *J. Econ. Entomol.* 70:518–520.

Thorne, G. 1940. The hairworm, *Gordius robustus* Leidy, as a parasite of the Mormon cricket, *Anabrus simplex* Haldeman. *J. Washington Acad. Sci.* 30:219–231.

Thorvilson, H. G. and L. P. Pedigo. 1984. Epidemiology of *Nomuraea rileyi* (Fungi: Deuteromycotina) in *Plathypena scabra* (Lepidoptera: Noctuidae) populations from Iowa soybeans. *Environ. Entomol.* 13:1491–1497.

Thorvilson, H. G., L. P. Pedigo, and L. C. Lewis. 1985a. *Plathypena scabra* (F.) (Lepidoptera: Noctuidae) populations and the incidence of natural enemies in four soybean tillage systems. *J. Econ. Entomol.* 78:213–218.

Thorvilson, H. G., L. P. Pedigo, and L. C. Lewis. 1985b. The potential of alfalfa fields as early-season nurseries for natural enemies of *Plathypena scabra* (F.) (Lepidoptera: Noctuidae). *J. Kansas Entomol. Soc.* 58:597–604.

Throne, J. E. and C. J. Eckenrode. 1985. Emergence patterns of the seedcorn maggot, *Delia platura* (Diptera: Anthomyiidae). *Environ. Entomol.* 14:182–186.

Throne, J. E. and C. J. Eckenrode. 1986. Development rates for the seed maggots *Delia platura* and *D. florilega* (Diptera: Anthomyiidae). *Environ. Entomol.* 15:1022–1027.

Thurston, G. S. and W. N. Yule. 1990. Control of larval northern corn rootworm (*Diabrotica barberi*) with two steinernematid nematode species. *J. Nematol.* 22:127–131.

Thurston, R., K. J. Starks, and G. M. Boush. 1956. Control of green June beetle adults with insecticides. *J. Econ. Entomol.* 49:828–830.

Timper, P., H. K. Kaya, and R. Gaugler. 1988. Dispersal of the entomogenous nematode *Steinernema feltiae* (Rhabditida: Steinernematidae) by infected adult insects. *Environ. Entomol.* 17:546–550.

Tingey, W. M. and J. E. Laubengayer. 1981. Defense against the green peach aphid and potato leafhopper by glandular trichomes of *Solanum berthaultii*. *J. Econ. Entomol.* 74:721–725.

Tingey, W. M. and S. L. Sinden. 1982. Glandular pubescence, glycoalkaloid composition, and resistance to the green peach aphid, potato leafhopper, and potato fleabeetle in *Solanum berthaultii*. *Am. Pot. J.* 59:95–106.

Tingle, F. C. and E. R. Mitchell. 1977. Seasonal populations of armyworms and loopers at Hastings, Florida. *Florida Entomol.* 60:115–122.

Tingle, F. C., T. R. Ashley, and E. R. Mitchell. 1978. Parasites of *Spodoptera exigua*, *S. eridania* (Lep.: Noctuidae) and *Herpetogramma bipunctalis* (Lep.: Pyralidae) collected from *Amaranthus hybridus* in field corn. *Entomophaga* 23:343–347.

Tingle, F. C., E. R. Mitchell, and J. R. McLaughlin. 1994. Lepidopterous pests of cotton and their parasitoids in a double-cropping environment. *Florida Entomol.* 77:334–341.

Tipping, C., K. B. Nguyen, J. E. Funderburk, and G. C. Smart, Jr. 1998. *Thripenema fuscum* n. sp. (Tylenchida: Allantonematidae), a parasite of the tobacco thrips, *Frankliniella fusca* (Thysanoptera). *J. Nematol.* 30:232–236.

Tippins, H. H. 1982. A review of information on the lesser cornstalk borer *Elasmopalpus lignosellus* (Zeller). Georgia Agric. Exp. Stn. Spec. Publ. 17. 65 pp.

Toapanta, M., J. Funderburk, S. Webb, D. Chellemi, and J. Tsai. 1996. Abundance of *Frankliniella* spp. (Thysanoptera: Thripidae) on winter and spring host plants. *Environ. Entomol.* 25:793–800.

Toba, H. H. 1985a. Damage to potato tubers by false wireworm larvae. *J. Agric. Entomol.* 2:13–19.

Toba, H. H. 1985b. Lateral movement of sugarbeet wireworm larvae in soil. *J. Agric. Entomol.* 2:248–255.

Toba, H. H. 1987. Treatment regimens for insecticidal control of wireworms on potato. *J. Agric. Entomol.* 4:207–212.

Toba, H. H. and D. M. Powell. 1986. Soil application of insecticides for controlling three insect pests of potatoes in Washington. *J. Agric. Entomol.* 3:87–99.

Toba, H. H. and J. E. Turner. 1979. Chemical control of wireworms on potatoes. *J. Econ. Entomol.* 72:636–641.

Toba, H. H. and J. E. Turner. 1981a. Comparison of four methods of applying insecticides for control of wireworms on potatoes. *J. Econ. Entomol.* 74:259–265.

Toba, H. H. and J. E. Turner. 1981b. Wireworm injury to potatoes in relation to tuber weight. *J. Econ. Entomol.* 74:514–516.

Toba, H. H. and J. E. Turner. 1983. Evaluation of baiting techniques for sampling wireworms (Coleoptera: Elateridae) infesting wheat in Washington. *J. Econ. Entomol.* 76:850–855.

Toba, H. H., A. N. Kishaba, R. Pangaldan, and P. V. Vail. 1973. Temperature and the development of the cabbage looper. *Ann. Entomol. Soc. Am.* 66:965–974.

Toba, H. H., J. E. Lindegren, J. E. Turner, and P. V. Vail. 1983. Susceptibility of the Colorado potato beetle and the sugarbeet wireworm to *Steinernema feltiae* and *S. glaseri*. *J. Nematol.* 15:597–601.

Todd, E. L. and R. W. Poole. 1980. Keys and illustrations for the armyworm moths of the noctuid genus *Spodoptera* Guenée from the Western Hemisphere. *Ann. Entomol. Soc. Am.* 73:722–738.

Todd, J. L., L. V. Madden, and L. R. Nault. 1991. Comparative growth and spatial distribution of *Dalbulus* leafhopper populations (Homoptera: Cicadellidae) in relation to maize phenology. *Environ. Entomol.* 20:556–564.

Todd, J. W. 1989. Ecology and behavior of *Nezara viridula*. *Annu. Rev. Entomol.* 34:273–292.

Todd, J. W. and D. C. Herzog. 1980. Sampling phytophagous Pentatomidae on soybean. Pages 438–478 *in* M. Kogan and D. C.

Herzog (eds.). Sampling Methods in Soybean Entomology. Springer-Verlag, Berlin/New York.

Todd, J. W. and W. J. Lewis. 1976. Incidence and oviposition patterns of *Trichopoda pennipes* (F.), a parasite of the southern green stink bug, *Nezara viridula* (L.). *J. Georgia Entomol. Soc.* 11: 50–54.

Todd, J. W. and F. W. Schumann. 1988. Combination of insecticide applications with trap crops of early maturing soybean and southern peas for population management of *Nezara viridula* in soybean (Hemiptera: Pentatomidae). *J. Entomol. Sci.* 23: 192–199.

Tollefson, J. J. 1986. Field sampling of adult populations. Pages 123–146 *in* J. L. Krysan and T. A. Miller (eds.). Methods for the Study of Pest *Diabrotica*. Springer-Verlag, Berlin/New York.

Tollefson, J. J. 1990. Comparison of adult and egg sampling for predicting subsequent populations of western and northern corn rootworms (Coleoptera: Chrysomelidae). *J. Econ. Entomol.* 83: 574–579.

Tomlin, A. D., J. J. Miller, C. R. Harris, and J. H. Tolman. 1985. Arthropod parasitoids and predators of the onion maggot (Diptera: Anthomyiidae) in southwestern Ontario. *J. Econ. Entomol.* 78: 975–981.

Tompkins, G. J., J. J. Linduska, J. M. Young, and E. M. Dougherty. 1986. Effectiveness of microbial and chemical insecticides for controlling cabbage looper (Lepidoptera: Noctuidae) and imported cabbageworm (Lepidoptera: Pieridae) on collards in Maryland. *J. Econ. Entomol.* 79: 497–501.

Tonhasca, Jr., A. and B. R. Stinner. 1991. Effects of strip intercropping and no-tillage on some pests and beneficial invertebrates of corn in Ohio. *Environ. Entomol.* 20: 1251–1258.

Tonhasca, Jr., A., J. C. Palumbo, and D. N. Byrne. 1994. Distribution patterns of *Bemisia tabaci* (Homoptera: Aleyrodidae) in cantaloupe fields in Arizona. *Environ. Entomol.* 23: 949–954.

Toscano, N. C. and V. M. Stern. 1976. Development and reproduction of Euschistus consperus at different temperatures. *Ann. Entomol. Soc. Am.* 69: 839–840.

Toscano, N. C. and V. M Stern. 1980. Seasonal reproductive condition of *Euschistus consperus*. Ann . Entomol. Soc. Am. 73: 85–88.

Toscano, N. C., R. R. Youngman, E. R. Oatman, P. A. Phillips, M. Jiminez, and F. Munoz. 1987. Implementation of integrated pest management program for fresh market tomatoes. *Appl. Agric. Res.* 1: 315–324.

Toth, M., C. Lofstedt, B. S. Hansson, G. Szocs, and A. I. Farag. 1989. Identification of four components from the female sex pheromone of the lima-bean pod borer, *Etiella zinckenella*. *Entomol. Exp. Appl.* 51: 107–112.

Tourneur, J.–C. and J. Gingras. 1992. Egg laying in a northeastern North American (Montreal, Quebec) population of *Forficula auricularia* L. (Dermaptera: Forficulidae). *Can. Entomol.* 124: 1055–1061.

Townsend, M. L., D. C. Steinkraus, and D. T. Johnson. 1994. Mortality response of green June beetle (Coleoptera: Scarabaeidae) to four species of entomopathogenic nematodes. *J. Entomol. Sci.* 29: 268–275.

Tracy, J. L. and J. R. Nechols. 1987. Comparisons between the squash bug egg parasitoids *Ooencyrtus anasae* and *O.* sp. (Hymenoptera: Encyrtidae): development, survival, and sex ratio in relation to temperature. *Environ. Entomol.* 16: 1324–1329.

Tracy, J. L. and J. R. Nechols. 1988. Comparison of thermal responses, reproductive biologies, and population growth potentials of the squash bug egg parasitoids *Ooencyrtus anasae* and *O.* sp. (Hymenoptera: Encyrtidae). *Environ. Entomol.* 17: 636–643.

Tran, B., J. Darquenne, and J. Huignard. 1993. Changes in responsiveness to factors inducing diapause termination in *Bruchus rufimanus* (Boh.) (Coleoptera: Bruchidae). *J. Insect Physiol.* 39: 769–774.

Traynier, R. M. M. 1975. Field and laboratory experiments on the site of oviposition by the potato moth *Phthorimaea operculella* (Zell.) (Lepidoptera, Gelechiidae). *Bull. Entomol. Res.* 65: 391–398.

Treat, T. L. and J. E. Halfhill. 1973. Rearing alfalfa loopers and celery loopers on an artificial diet. *J. Econ. Entomol.* 66: 569–570.

Trichilo, P. J. and T. F. Leigh. 1988. Influence of resource quality on the reproductive fitness of flower thrips (Thysanoptera: Thripidae). *Ann. Entomol. Soc. Am.* 81: 64–70.

Triplehorn, B. W. and L. R. Nault. 1985. Phylogenic classification of the genus *Dalbulus* (Homoptera: Cicadellidae), and notes on the phylogeny of the Macrostelini. *Ann. Entomol. Soc. Am.* 78: 291–315.

Trivedi, T. P. and D. Rajagopal. 1992. Distribution, biology, ecology and management of potato tuber moth, *Phthorimaea operculella* (Zeller) (Lepidoptera: Gelechiidae): a review. *Trop Pest Manage.* 38: 279–285.

Trottier, M. R., O. N. Morris, and H. T. Dulmage. 1988. Susceptibility of the bertha armyworm, *Mamestra configurata* (Lepidoptera, Noctuidae), to sixty-one strains from ten varieties of *Bacillus thuringiensis*. *J. Invertebr. Pathol.* 51: 242–249.

Trumble, J. T. 1982a. Temporal occurrence, sampling, and within-field distribution of aphids on broccoli in coastal California. *J. Econ. Entomol.* 75: 378–382.

Trumble, J. T. 1982b. Within-plant distribution and sampling of aphids (Homoptera: Aphididae) on broccoli in southern California. *J. Econ. Entomol.* 75: 587–592.

Trumble, J. T. and T. C. Baker. 1984. Flight phenology and pheromone trapping of *Spodoptera exigua* (Hübner) (Lepidoptera: Noctuidae) in southern coastal California. *Environ. Entomol.* 13: 1278–1282.

Trumble, J. T., H. Nakakihara, and G. W. Zehnder. 1982a. Comparisons of traps and visual searches of foliage for monitoring aphid (Heteroptera: Aphididae) population density in broccoli. *J. Econ. Entomol.* 75: 853–856.

Trumble, J. T., H. Nakakihara, and W. Carson. 1982b. Monitoring aphid infestations on broccoli. *California Agric.* 36(6): 15–16.

Tryon, Jr., E. H., The striped earwig, and ant predators of sugarcane rootstalk borer, in Florida citrus. *Florida Entomol.* 69: 336–343.

Tsai, J. H. 1988. Bionomics of *Dalbulus maidis* (DeLong and Wolcott), a vector of mollicutes and virus (Homoptera: Cicadellidae). Pages 209–221 *in* K. Maramorosch and S. P. Raychaudhuri (eds.). Mycoplasma Disease of Crops, Basic and Applied Aspects. Springer-Verlag, Berlin/New York.

Tsai, J. H. 1996. Development and oviposition of *Peregrinus maidis* (Homoptera: Delphacidae) on various host plants. *Florida Entomol.* 79: 19–26.

Tsai, J. H. and Y.–H. Liu. 1998. Effect of temperature on development, survivorship, and reproduction of rice root aphid (Homoptera: Aphididae). *Environ. Entomol.* 27: 662–666.

Tsai, J. H. and K. Wang. 1996. Development and reproduction of *Bemisia argentifolii* (Homoptera: Aleyrodidae) on five host plants. *Environ. Entomol.* 25: 810–816.

Tsai, J. H. and S. W. Wilson. 1986. Biology of *Peregrinus maidis* with descriptions of immature stages (Homoptera: Delphacidae). *Ann. Entomol. Soc. Am.* 79: 395–401.

Tsai, J. H. and T. A. Zitter. 1982. Characteristics of maize stripe virus transmission by the corn delphacid. *J. Econ. Entomol.* 75: 397–400.

Tsai, J. H., B. Steinberg, and B. W. Falk. 1990. Effectiveness and residual effects of seven insecticides on *Dalbulus maidis* (Homoptera: Cicadellidae) and *Peregrinus maidis* (Homoptera: Delphacidae). *J. Entomol. Sci.* 25: 106–111.

Tsai, J. H., B. Yue, S. E. Webb, J. E. Funderburk, and H. T. Hsu. 1995. Effects of host plant and temperature on growth and reproduction of *Thrips palmi* (Thysanoptera: Thripidae). *Environ. Entomol.* 24: 1598–1603.

Tschinkel, W. R. 1992. Brood raiding and the population dynamics of founding and incipient colonies of the fire ant, *Solenopsis invicta*. *Ecol. Entomol.* 17: 179–188.

Tsitsipis, J. A. and T. E. Mittler. 1976. Development, growth, reproduction, and survival of apterous virginoparae of *Aphis fabae* at different temperatures. *Entomol. Exp. Appl.* 19: 1–10.

Tu, C. M. and C. R. Harris. 1988. A description of the development and pathogenicity of *Entomophthora muscae* (Cohn) in the onion maggot, *Delia antiqua* (Meigen). *Agric. Ecosystems Environ.* 20: 143–146.

Tukahirwa, E. M. and T. H. Coaker. 1982. Effect of mixed cropping on some insect pests of brassicas; reduced *Brevicoryne brassicae* infestations and influences on epigeal predators and the disturbance of ovipositon behaviour in *Delia brassicae*. *Entomol. Exp. Appl.* 32: 129–140.

Tulisalo, U. and M. Markkula. 1970. Resistance of pea to the pea weevil *Sitona lineatus* (L.) (Col., Curculionidae). *Ann. Agric. Fenn.* 9: 139–141.

Tumlinson, J. H., E. R. Mitchell, S. M. Browner, and D. A. Lindquist. 1972. A sex pheromone for the soybean looper. *Environ. Entomol.* 1: 466–468.

Tumlinson, J. H., M. G. Klein, R. E. Doolittle, T. L. Ladd, Jr., and A. T. Proveaux. 1977. Identification of the female Japanese beetle sex pheromone: inhibition of male response by an enantiomer. *Science* 197: 789–792.

Tumlinson, J. H., D. E. Hendricks, E. R. Mitchell, R. E. Doolittle, and M. M. Brennan. 1975. Isolation, identification and synthesis of the sex pheromone of the tobacco budworm. *J. Chem Ecol.* 1: 203–214.

Tumlinson, J. H., E. R. Mitchelll, P. E. A. Teal, R. R. Heath, and L. J. Mengelkoch. 1986. Sex pheromone of fall armyworm, *Spodoptera frugiperda* (Smith). Identification of components critical to attraction in the field. *J. Chem Ecol.* 12: 1909–1926.

Turgeon, J.J, J. N. McNeil, and W. L. Roelofs. 1983. Responsiveness of *Pseudaletia unipunctata* males to the female sex pheromone. *Physiol. Entomol.* 8: 339–344.

Turner, C. E., R. W. Pemberton, and S. S. Rosenthal. 1987. Host range and new host records for the plume moth *Platyptilia carduidactyla* (Lepidoptera: Pterophoridae) from California thistles (Asteraceae). Proc. Entomol. Soc., Washington 89: 132–136.

Turner, N. 1945. Some fundamental aspects of control of the European corn borer. Connecticut Agric. Exp. Stn. Bull. 495. 43 pp.

Turnock, W. J. 1987. Predicting larval abundance of the bertha armyworm, *Mamestra configurata* Wlk., in Manitoba from catches of male moths in sex attractant traps. *Can. Entomol.* 119: 167–178.

Turnock, W. J. 1988. Density, parasitism, and disease incidence of larvae of the bertha armyworm, *Mamestra configurata* Walker (Lepidoptera: Noctuidae), in Manitoba, 1973–1986. *Can. Entomol.* 120: 401–413.

Turnock, W. J. and R. J. Bilodeau. 1984. Survival of pupae of *Mamestra configurata* (Lepidoptera: Noctuidae) and two of its parasites in untilled and tilled soil. *Can. Entomol.* 116: 257–267.

Turnock, W. J. and R. J. Bilodeau. 1985. A comparison of three methods of examining the density of larvae of the bertha armyworm, *Mamestra configurata*, in fields of canola (*Brassica* spp.). *Can. Entomol.* 117: 1065–1066.

Turnock, W. J. and S. A. Turnbull. 1994. The development of resistance to insecticides by the crucifer flea beetle, *Phyllotreta cruciferae* (Goeze). *Can. Entomol.* 126: 1369–1375.

Turnock, W. J., R. J. Lamb, and R. J. Bilodeau. 1987. Abundance, winter survival, and spring emergence of flea beetles (Coleoptera: Chrysomelidae) in a Manitoba grove. *Can. Entomol.* 119: 419–426.

Tuskes, P. M. and M. D. Atkins. 1973. Effect of temperature on occurrence of color phases in the alfalfa caterpillar (Lepidoptera: Pieridae). *Environ. Entomol.* 2: 619–622.

Tuthill, L. D. 1943. The psyllids of America north of Mexico (Psyllidae: Homoptera). *Iowa State Coll. J. Sci.* 17: 443–660.

Ueckert, D. N. and R. M. Hansen. 1970. Seasonal dry-weight composition in diets of Mormon crickets. *J. Econ. Entomol.* 63: 96–98.

Umesh, K. C., J. Valencia, C. Hurley, W. D. Gubler, and B. W. Falk. 1995. Stylet oil provides limited control of aphid-transmitted viruses in melons. *California Agric.* 49(3): 22–24.

Underhill, E. W., W. F. Steck, and M. D. Chisholm. 1977. A sex pheromone mixture for the bertha armyworm moth, *Mamestra configurata*: (Z)-9-tetradecen-1-ol acetate and (Z)-11-hexadecen-1-ol acetate. *Can. Entomol.* 109: 1335–1340.

Underhill, G. W. 1923. The squash lady-bird beetle. Virginia Agric. Exp. Stn. Bull. 232. 24 pp.

Underhill, G. W. 1928. Life history and control of the pale-striped and banded flea beetles. Virginia Agric. Exp. Stn. Bull. 264. 20 pp.

Underhill, G. W. 1934. The green stinkbug. Virginia Agric. Exp. Stn. Bull. 294. 26 pp.

Utida, S. 1972. Density dependent polymorphism in the adult of *Callosobruchus maculatus* (Coleoptera, Bruchidae). *J. Stored Prod. Res.* 8: 111–126.

Uvah, I. I. I. and T. H. Coaker. 1984. Effect of mixed cropping on some insect pests of carrots and onions. *Entomol. Exp. Appl.* 36: 159–167.

Vail, K. M., L. T. Kok, and T. J. McAvoy. 1991. Cultivar preferences of lepidopterous pests of broccoli. *Crop Prot.* 10: 199–204.

Vail, P. V., T. J. Henneberry, and R. Pengalden. 1967a. An artificial diet for rearing the salt marsh caterpillar, *Estigmene acrea* (Lepidoptera: Arctiidae), with notes on the biology of the species. *Ann. Entomol. Soc. Am.* 60: 134–138.

Vail, P. V., R. E. Seay, and J. DeBolt. 1980. Microbial and chemical control of the cabbage looper on fall lettuce. *J. Econ. Entomol.* 73: 72–75.

Vail, P. V., D. L. Jay, F. D. Stewart, A. J. Martinez, and H. T. Dulmage. 1978. Comparative susceptibility of *Heliothis virescens* and *H. zea* to the nuclear polyhedrosis virus isolated from *Autographa californica*. *J. Econ. Entomol.* 71: 293–296.

Vail, P. V., M. W. Stone, J. C. Maitlen, D. A. George, and L. I. Butler. 1967b. Performance of insecticides against cabbage and green peach aphids on leafy vegetables and persistence of residues during cool weather. *J. Econ. Entomol.* 60: 537–541.

Valenzuela, H. R. and D. R. Bienz. 1989. Asparagus aphid feeding and freezing damage asparagus plants. *J. Am. Soc. Hort. Sci.* 114: 578–581.

Valles, S. M., and J. L. Capinera. 1992. Periodicity of attraction of adult melonworm, *Diaphania hyalinata*. *Florida Entomol.* 75: 390–392.

Valles, S. M. and J. L. Capinera. 1993. Response of larvae of the southern armyworm, *Spodoptera eridania* (Cramer) (Lepidoptera: Noctuidae), to selected botanical insecticides and soap. *J. Agric. Entomol.* 10: 145–153.

Valles, S. M., J. L. Capinera, and P. E. A. Teal. 1991. Evaluation of pheromone trap design, height, and efficiency for capture of male *Diaphania nitidalis* (Lepidoptera: Pyralidae) in a field cage. *Environ. Entomol.* 20: 1274–1278.

Valles, S. M., R. R. Heath, and J. L. Capinera. 1992. Production and release of sex pheromone by *Diaphania nitidalis* (Lepidoptera: Pyralidae): periodicity, age, and density effects. *Ann. Entomol. Soc. Am.* 85: 731–735.

van Alphen, J. J. M. 1980. Aspects of the foraging behaviour of *Tetrastichus asparagi* Crawford and *Tetrastichus* spec. (Eulophidae), gregarious egg parasitoids of the asparagus beetles *Crioceris asparagi* L. and *C. duodecimpunctata* L. (Chrysomelidae). I. Host species selection, host-stage selection and host discrimination. *Netherlands J. Zool.* 30: 307–325.

van Dam, W. and G. Wilde. 1977. Biology of the bean leafroller *Urbanus proteus* (Lepidoptera: Hesperidae). *J. Kansas Entomol. Soc.* 50: 157–160.

van den Bosch, R. and R. F. Smith. 1955. A taxonomic and distributional study of the species of *Prodenia* occurring in California. *Pan-Pac. Entomol.* 31: 21–28.

Vander meer, R. K. 1988. Behavioral and biochemical variation in the fire ant, *Solenopsis invicta*. Pages 223–255 *in* R. L. Jeanne (ed.). Interindividual Behavioral Variability in Social Insects. Westview Press, Boulder.

van de Vrie, M., J. A. McMurtry, and C. B. Huffaker. 1972. Biology, ecology, and pest status, and host-plant relations of tetranychids. *Hilgardia* 41: 343–432.

van Emden, H. F. 1966. Studies on the relations of insect and host plant. III. A comparison of the reproduction of *Brevicoryne brassicae* and *Myzus persicae* (Hemiptera: Aphididae) on Brussels sprout plants supplied with different rates of nitrogen and potassium. *Entomol. Exp. Appl.* 9: 444–460.

van Emden, H. F., V. F. Eastop, R. D. Hughes, and M. J. Way. 1969. The ecology of *Myzus persicae*. *Annu. Rev. Entomol.* 14: 197–270.

van Houten, Y. M. and P. van Stratum. 1995. Control of western flower thrips on sweet pepper in winter with *Amblyseius cucumeris* (Oudemans) and *A. degenerans* Berlese. Pages 245–248 *in* B. L. Parker, M. Skinner, and T. Lewis (eds.). Thrips Biology and Management. Plenum Press, London.

van Keymeulen, M., L. Hertveldt, and C. Pelerents. 1981. Methods for improving both the quantitative and qualitative aspects of rearing *Delia brassicae* for sterile release programmes. *Entomol. Exp. Appl.* 30: 231–240.

van Lenteren, J. C., H. J. W. van Roermund, and S. Sutterlin. 1996. Biological control of greenhouse whitefly (*Trialeurodes vaporariorum*) with the parasitoid *Encarsa formosa*: how does it work? *Biol. Control* 6: 1–10.

Van Name, W. G. 1936. The American land and fresh-water isopod Crustacea. *Bull. Am. Mus. Nat. Hist.* 71: 1–535.

Van Name, W. G. 1940. A supplement to the American land and fresh-water isopod Crustacea. *Bull. Am. Mus. Nat. Hist.* 77: 109–142.

Van Name, W. G. 1942. A second supplement to the American land and fresh-water isopod Crustacea. *Bull. Am. Mus. Nat. Hist.* 80: 299–329.

van Rijn, P. C. J., C. Mollema, and G. M. Steenhuis-Broers. 1995. Comparative life history studies of *Frankliniella occidentalis* and *Thrips tabaci* (Thysanoptera: Thripidae) on cucumber. *Bull. Entomol. Res.* 85: 285–297.

van Roermund, H. J. W. and J. C. van Lenteren. 1992. The parasite-host relationship between *Encarsia formosa* (Hymenoptera: Aphelinidae) and *Trialeurodes vaporariorum* (Homoptera: Aleyrodidae) XXXIV. Life history parameters of the greenhouse whitefly, *Trialeurodes vaporariorum* as a function of host plant and temperature. Wageningen Agric. Univ. Papers 92–3. 147 pp.

van Steenwyk, R. A. and E. R. Oatman. 1983. Mating disruption of tomato pinworm (Lepidoptera: Gelechiidae) as measured by pheromone trap, foliage, and fruit infestation. *J. Econ. Entomol.* 76: 80–84.

van Steenwyk, R. A. and N. C. Toscano. 1981. Relationship between lepidopterous larval density and damage in celery and celery plant growth analysis. *J. Econ. Entomol.* 74: 287–290.

van Steenwyk, R. A., E. R. Oatman, N. C. Toscano, and J. A. Wyman. 1983. Pheromone traps to time tomato pinworm control. *California Agric.* 37: 22–24.

VanWoerkom, G. J., F. T. Turpin, and J. R. Barrett, Jr. 1983. Wind effect on western corn rootworm (Coleoptera: Chrysomelidae). *Environ. Entomol.* 12: 196–200.

Vargas, R. I., D. Miyashita, and T. Nishida. 1984. Life history and demographic parameters of three laboratory-reared tephritids (Diptera: Tephritidae). *Ann. Entomol. Soc. Am.* 77: 651–656.

Vargas, R. I., J. D. Stark, and T. Nishida. 1989. Abundance, distribution, and dispersion indices of the oriental fruit fly and melon fly (Diptera: Tephritidae) on Kauai, Hawaiian Islands. *J. Econ. Entomol.* 82: 1609–1615.

Vargas, R. I., J. D. Stark, and T. Nishida. 1990. Population dynamics, habitat preference, and seasonal distribution patterns of oriental fruit fly and melon fly (Diptera: Tephritidae) in an agricultural area. *Environ. Entomol.* 19: 1820–1828.

Vargas, R. I., J. D. Stark, R. J. Prokopy, and T. A. Green. 1991. Response of oriental fruit fly (Diptera: Tephritidae) and associated parasitoids (Hymenoptera: Braconidae) to different-color spheres. *J. Econ. Entomol.* 84: 1503–1507.

Vargas, R. I., J. D. Stark, G. K. Uchida, and M. Purcell. 1993. Opiine parasitoids (Hymenoptera: Braconidae) of oriental fruit fly (Diptera: Tephritidae) on Kauai Island, Hawaii: islandwide relative abundance and parasitism rates in wild and orchard guava habitats. *Environ. Entomol.* 22: 246–253.

Vea, E. V. and C. J. Eckenrode. 1976. Resistance to seedcorn maggot in snap beans. *J. Econ. Entomol.* 5: 735–737.

Veazey, J. N., C. A. R. Kay, T. J. Walker, and W. H. Whitcomb. 1976. Seasonal abundance, sex ratio, and macroptery of field crickets in northern Florida. *Ann. Entomol. Soc. Am.* 69: 374–380.

Vega, F. E. and P. Barbosa. 1990. *Gonatopus bartletti* Olmi (Hymenoptera: Dryinidae) in Mexico: a previously unreported parasitoid of the corn leafhopper *Dalbulus maidis* (DeLong and Wolcott) and the Mexican corn leafhopper *Dalbulus elimatus* (Ball) (Homoptera: Cicadellidae). Proc. Entomol. Soc. Washington 92: 461–464.

Vega, F. E., P. Barbosa, and A. P. Panduro. 1990. An adjustable water-pan trap for simultaneous sampling of insects at different heights. *Florida Entomol.* 73: 656–660.

Vega, F. E., P. Barbosa, and A. P. Panduro. 1991. *Eudorylas* (*Metadorylas*) sp. (Diptera: Pipunculidae): a previously unreported parasitoid of *Dalbulus maidis* (DeLong and Wolcott) and *Dalbulus elimatus* (Ball) (Homoptera: Cicadellidae). *Can. Entomol.* 123: 241–242.

Velasco, L. R. I. and G. H. Walter. 1992. Availability of different host plant species and changing abundance of the polyphagous bug *Nezara viridula* (Hemiptera: Pentatomidae). *Environ. Entomol.* 21: 751–759.

Verma, J. S. 1955. Biological studies to explain the failure of *Cyrtorhinus mundulus* (Breddin) as an egg-predator of *Peregrinus maidis* (Ashmead) in Hawaii. Proc. Hawaiian Entomol. Soc. 15: 623–634.

Vernon, R. S. and J. H. Borden. 1979. *Hylemya antiqua* (Meigen): longevity and oviposition in the laboratory. *J. Entomol. Soc. Brit. Columbia* 76: 12–16.

Vernon, R. S. and J. H. Borden. 1983. Spectral specific discrimination by *Hylemya antiqua* (Meigen) (Diptera: Anthomyiidae) and other vegetable-infesting species. *Environ. Entomol.* 12: 650–655.

Vernon, R. S. and D. R. Gillespie. 1990. Spectral responsiveness of *Frankliniella occidentalis* (Thysanoptera: Thripidae) determined by trap catches in greenhouses. *Environ. Entomol.* 19: 1229–1241.

Vernon, R. S. and D. R. Gillespie. 1995. Influence of trap shape, size, and background color on captures of *Frankliniella occidentalis* (Thysanoptera: Thripidae) in a cucumber greenhouse. *J. Econ. Entomol.* 88: 288–293.

Vernon, R. S. and R. Houtman. 1983. Evaluation of sprayed and granular aphicides against the European asparagus aphid,

Brachycolus asparagi (Homoptera: Aphididae), in British Columbia. *J. Entomol. Soc. Brit. Columbia.* 80: 3–9.

Vernon, R. S. and J. R. MacKenzie. 1991a. Granular insecticides against overwintered tuber flea beetle, *Epitrix tuberis* Gentner (Coleoptera: Chrysomelidae), on potato. *Can. Entomol.* 123: 333–343.

Vernon, R. S. and J. R. MacKenzie. 1991b. Evaluation of foliar sprays against the tuber flea beetle, *Epitrix tuberis* Gentner (Coleoptera: Chrysomelidae), on potato. *Can. Entomol.* 123: 321–331.

Vernon, R. S. and D. R. Thomson. 1991. Overwintering of tuber flea beetles, *Epitrix tuberis* Gentner (Coleoptera: Chrysomelidae), in potato fields. *Can. Entomol.* 121: 239–240.

Vernon, R. S. and D. Thomson. 1993. Effects of soil type and moisture on emergence of tuber flea beetles, *Epitrix tuberis* (Coleoptera: Chrysomelidae) from potato fields. *J. Entomol. Soc. Br., Columbia* 90: 3–10.

Vernon, R. S., J. R. MacKenzie, and D. L. Bartel. 1990. Monitoring tuber flea beetle, *Epitrix tuberis* Gentner (Coleoptera: Chrysomelidae) on potato: parameters affecting the accuracy of visual sampling. *Can. Entomol.* 122: 525–535.

Vernon, R. S., J. W. Hall, G. J. R. Judd, and D. L. Bartel. 1989. Improved monitoring program for *Delia antiqua* (Diptera: Anthomyiidae). *J. Econ. Entomol.* 82: 251–258.

Viana, P. A. and E. F. da Costa. 1995. Efeito da umidade do solo sobre o dano da lagarta elasmo, *Elasmopalpus lignosellus* (Zeller) na cultura do milho. *An. Soc. Entomol. Brasil* 24: 209–214.

Vickery, R. A. 1924. The striped grass looper, *Mocis repanda* Fab., in Texas. *J. Econ. Entomol.* 17: 401–406.

Vickery, R. A. 1929. Studies on the fall armyworm in the Gulf coast district of Texas. USDA Tech. Bull. 138. 63 pp.

Vickery, V. R. and D. K. M. Kevan. 1985. The Grasshoppers, Crickets, and Related Insects of Canada and Adjacent Regions. The insects and Arachnids of Canada, Part 14. Agric. Canada Pub. 1777. 918 pp.

Villani, M. G., R. J. Wright, and P. B. Baker. 1988. Differential susceptibility of Japanese beetle, oriental beetle, and European chafer (Coleoptera: Scarabaeidae) larvae to five soil insecticides. *J. Econ. Entomol.* 81: 785–788.

Villanueva B., J. R. and F. E. Strong. 1964. Laboratory studies on the biology of *Rhopalosiphum padi* (Homoptera: Aphididae). *Ann. Entomol. Soc. Am.* 57: 609–613.

Vinal, S. C. and D. J. Caffrey. 1919. The European corn borer and its control. Massachusetts Agric. Exp. Stn. Bull. 189. 71 pp.

Vincent, C. and P. LaChance. 1993. Evaluation of a tractor-propelled vacuum device for management of tarnished plant bug (Heteroptera: Miridae) populations in strawberry plantations. *Environ. Entomol.* 22: 1103–1107.

Vincent, C. and R. K. Stewart. 1983. Crucifer-feeding flea beetle dispersal and statistics of directional data. *Environ. Entomol.* 12: 1380–1383.

Vinson, S. B. 1997. Invasion of the red imported fire ant (Hymenoptera: Formicidae). *Am. Entomol.* 43: 23–39.

Vinson, S. B. and L. Greenberg. 1986. The biology, physiology, and ecology of imported fire ants. Pages 193–226 *in* S. B. Vinson (ed.). Economic Impact and Control of Social Insects. Praeger, New York.

Voegtlin, D. 1984. Notes on *Hyadaphis foeniculi* and redescription of *Hyadaphis tataricae* (Homoptera: Aphididae). *Great Lakes Entomol.* 17: 55–67.

Vogt, E. A. and J. R. Nechols. 1991. Diel activity patterns of the squash bug egg parasitoid *Gryon pennsylvanicum* (Hymenoptera: Scelionidae). *Ann. Entomol. Soc. Am.* 84: 303–308.

Vogt, E. A. and J. R. Nechols. 1993. Responses of the squash bug (Hemiptera: Coreidae) and its egg parasitoid, *Gryon pennsylvanicum* (Hymenoptera: Scelionidae) to three *Cucurbita* cultivars. *Environ. Entomol.* 22: 238–245.

von Arx, R., O. Roux, and J. Baumgartner. 1990. Tuber infestation by potato tubermoth, *Phthorimaea operculella* (Zeller), at potato harvest in relation to farmers' practices. *Agric. Ecosys. Environ.* 31: 277–292.

von Stryk, F. G. and W. H. Foott. 1976. Residue levels in tomato products processed from fruit in contact with insecticide treated hampers. *Can. Entomol.* 108: 989–991.

Voss, R. H. and D. N. Ferro. 1990. Phenology of flight and walking by Colorado potato beetle (Coleoptera: Chrysomelidae) adults in western Massachusetts. *Environ. Entomol.* 19: 117–122.

Waddill, V. H. and R. A. Conover. 1978. Resistance of white-fleshed sweet potato cultivars to the sweet-potato weevil. *HortScience* 13: 476–477.

Waddill, V., K. Pohronezny, R. McSorley, and H. H. Bryan. 1984. Effect of manual defoliation on pole bean yield. *J. Econ. Entomol.* 77: 1019–1023.

Wade, J. S. 1921. Notes on ecology of injurious Tenebrionidae (Col.). *Entomol. News* 32: 1–6.

Wafa, A. K., F. M. El-Borolossy, and A. A. Khattab. 1969. Biological and ecological studies on Spodoptera exigua HB. *Bull. Soc. Entomol. Egypte* 53: 533–549.

Wakamura, S. and M. Takai. 1992. Control of the beet armyworm in open fields with sex pheromone. Pages 115–125 *in* N. S. Talekar (ed.) Diamondback Moth and other Crucifer Pests. Asian Research and Development Center, Taipei, Taiwan.

Wakeland, C. 1926. False wireworms injurious to dry-farmed wheat and a method of combatting them. Idaho Agric. Exp. Stn. Bull. 6. 52 pp.

Wakeland, C. 1959. Mormon crickets in North America. USDA Tech. Bull. 1202. 77 pp.

Wakeland, C. 1961. The replacement of one grasshopper species by another. USDA Prod. Res. Rep. 42. 9 pp.

Wakisaka, S., R. Tsukuda, and F. Nakasuji. 1992. Effects of natural enemies, rainfall, temperature and host plants on survival and reproduction of the diamondback moth. Pages 15–26 *in* N. S. Talekar (ed.) Diamondback Moth and Other Crucifer Pests. Asian Vegetable Research and Development Center, Taipei, Taiwan.

Walgenbach, J. F. 1994. Distribution of parasitized and nonparasitized potato aphid (Homoptera: Aphididae) on staked tomato. *Environ. Entomol.* 23: 795–804.

Walgenbach, J. F. 1997. Effect of potato aphid (Homoptera: Aphididae) on yield, quality, and economics of staked-tomato production. *J. Econ. Entomol.* 90: 996–1004.

Walgenbach, J. F. and E. A. Estes. 1992. Economics of insecticide use on staked tomatoes in western North Carolina. *J. Econ. Entomol.* 85: 888–894.

Walgenbach, J. F. and J. A. Wyman. 1984. Colorado potato beetle (Coleoptera: Chrysomelidae) development in relation to temperature in Wisconsin. *Ann. Entomol. Soc. Am.* 77: 604–609.

Walkden, H. H. 1937. Notes on the life history of the bronzed cutworm in Kansas. *J. Kansas Entomol. Soc.* 10: 52–58.

Walkden, H. H. 1950. Cutworms, armyworms, and related species attacking cereal and forage crops in the central Great Plains. USDA Circ. 849. 52 pp.

Walker, G. P., L. V. Madden, and D. E. Simonet. 1984a. Spatial dispersion and sequential sampling of the potato aphid, *Macrosiphum euphorbiae* (Homoptera: Aphididae), on processing-tomatoes in Ohio. *Can. Entomol.* 116: 1069–1075.

Walker, G. P., L. R. Nault, and D. E. Simonet. 1984b. Natural mortality factors acting on potato aphid (*Macrosiphum euphorbiae*) populations in processing-tomato fields in Ohio. *Environ. Entomol.* 13: 724–732.

Walker, H. G. and L. D. Anderson. 1933. Report on the control of the harlequin bug, *Murgantia histrionica* Hahn, with notes on the severity of an outbreak of this insect in 1932. *J. Econ. Entomol.* 26:129–135.

Walker, H. G. and L. D. Anderson. 1934. Notes on the use of Derris and Pyrethrum dusts for the control of certain insects attacking cruciferous crops. *J. Econ. Entomol.* 27:388–393.

Walker, H. G. and L. D. Anderson. 1939. The Hawaiian beet webworm and its control. Virginia Truck Exp. Stn. Bull. 103:1649–1659.

Walker, K. A., T. H. Jones, and R. D. Fell. 1993. Pheromonal basis of aggregation in European earwig, *Forficula auricularia* L. (Dermaptera: Forficulidae). *J. Chem Ecol.* 19:2029–2038.

Walker, T. J. 1982. Sound traps for sampling mole cricket flights (Orthoptera: Gryllotalpidae: *Scapteriscus*). *Florida Entomol.* 65:105–110.

Walker, T. J. (ed.). 1984. Mole crickets in Florida. Florida Agric. Exp. Stn. Bull. 846. 54 pp.

Walker, T. J. 1987. Wing dimorphism in *Gryllus rubens* (Orthoptera: Gryllidae). *Ann. Entomol. Soc. Am.* 80:547–560.

Walker, T. J. and D. Ngo. 1982. Mole crickets and pasture grasses: damage by *Scapteriscus vicinus*, but not by *S. acletus* (Orthoptera: Gryllotalpidae). *Florida Entomol.* 65:300–306.

Walker, T. J. and J. M. Sivinski. 1986. Wing dimorphism in field crickets (Orthoptera: Gryllidae: *Gryllus*). *Ann. Entomol. Soc. Am.* 79:84–90.

Walker, T. J., J. A. Reinert, and D. J. Schuster. 1983. Geographical variation in flights of the mole cricket, *Scapteriscus* spp. (Orthoptera: Gryllotalpidae). *Ann. Entomol. Soc. Am.* 76:507–517.

Wall, C. 1988. Application of sex attractants for monitoring the pea moth, *Cydia nigricana* (F.) (Lepidoptera: Tortricidae). *J. Chem Ecol.* 14:1857–1866.

Wall, C., D. G. Garthwaite, J. A. Blood Smyth, and A. Sherwood. 1987. The efficacy of sex-attractant monitoring for the pea moth, *Cydia nigricana*, in England, 1980–1985. *Ann. Appl. Biol.* 110:223–229.

Wallen, V. R., H. R. Jackson, and S. W. MacDiarmid. 1976. Remote sensing of corn aphid infestation, 1974 (Hemiptera: Aphididae). *Can. Entomol.* 108:751–754.

Wallis, R. L. 1946. Seasonal occurrence of the potato psyllid in the North Platte Valley. *J. Econ. Entomol.* 39:689–694.

Wallis, R. L. 1955. Ecological studies on the potato psyllid as a pest of potatoes. USDA Tech. Bull. 1107. 25 pp.

Wallis, R. L. 1962. Spring migration of the six-spotted leafhopper in the western Great Plains. *J. Econ. Entomol.* 55:871–874.

Walters, T. W. and C. J. Eckenrode. 1996. Integrated pest management of the onion maggot (Diptera: Anthomyiidae). *J. Econ. Entomol.* 89:1582–1586.

Walton, R. R. and G. A. Bieberdorf. 1948. Seasonal history of the southwestern corn borer, *Diatraea grandiosella* Dyar, in Oklahoma; and experiments on methods of control. Oklahoma Agric. Exp. Stn. Tech. Bull. T-32. 23 pp.

Walton, W. R. and C. M. Packard. 1947. The armyworm and its control. USDA Farmers' Bull. 1850. 10 pp.

Wang, B., D. N. Ferro, and D. W. Hosmer. 1997. Importance of plant size, distribution of egg masses, and weather conditions on egg parasitism of the European corn borer, *Ostrinia nubilalis* by *Trichogramma ostriniae* in sweet corn. *Entomol. Exp. Appl.* 83:337–345.

Wang, K., J. H. Tsai, and N. A. Harrison. 1997. Influence of temperature on development, survivorship, and reproduction of buckthorn aphid (Homoptera: Aphididae). *Ann. Entomol. Soc. Am.* 90:62–68.

Wang, R. Y., R. C. Gergerich, and K. S. Kim. 1994. The relationship between feeding and virus retention time in beetle transmission of plant viruses. *Phytopathology* 84:995–998.

Warburg, M. R. 1993. Evolutionary Biology of Land Isopods. Springer-Verlag, Berlin/New York. 159 pp.

Ward, A. and S. Morse. 1995. Partial application of insecticide to broad bean (*Vicia fabae*) as a means of controlling bean aphid (*Aphis fabae*) and bean weevil (*Sitona lineatus*). *Ann. Appl. Biol.* 127:239–249.

Ward, A. G. and B. C. Pass. 1969. Rearing sod webworms on artificial diets. *J. Econ. Entomol.* 62:510–511.

Ward, R. H. and A. J. Keaster. 1977. Wireworm baiting: use of solar energy to enhance early detection of *Melanotus depressus*, *M. verberans*, and *Aeolus mellillus*. *J. Econ. Entomol.* 70:403–406.

Warner, R. E. 1975. New synonyms, key, and distribution of *Graphognathus*, whitefringed beetles (Coleoptera: Curculionidae), in North America. USDA Coop. Econ. Ins. Rpt. 25:855–860.

Warren, A. D. 1990. Predation of five species of Noctuidae at ultraviolet light by the western yellowjacket (Hymenoptera: Vespidae). *J. Lepid. Soc.* 44:32.

Wasserman, S. S. 1981. Host-induced oviposition preferences and oviposition markers in the cowpea weevil, *Callosobruchus maculatus*. *Ann. Entomol. Soc. Am.* 74:242–245.

Waterhouse, J. S. 1968. Studies on the garden symphylan, *Scutigerella immaculata* (Symphyla: Scutigerellidae). *Can. Entomol.* 100:172–178.

Waterhouse, J. S. 1969. An evaluation of a new predaceous centipede *Lamyctes* sp., on the garden symphylan *Scutigerella immaculata*. *Can. Entomol.* 101:1081–1083.

Waterhouse, J. S. 1970. Distribution of the garden symphylan, *Scutigerella immaculata*, in the United States: a 15–year survey. *J. Econ. Entomol.* 63:390–394.

Watson, J. R. 1941. Migrations and food preferences of the lubberly locust. *Florida Entomol.* 24:40–42.

Watson, J. R. 1944. *Herse cingulatus* Fab. as an armyworm. *Florida Entomol.* 27:58.

Watson, J. R. and H. E. Bratley. 1940. Preliminary report on lubberly locust control. *Florida Entomol.* 23:7–10.

Watson, M. T. and B. W. Falk. 1994. Ecological and epidemiological factors affecting carrot motley dwarf development in carrots grown in the Salinas Valley of California. *Plant Dis.* 78:477–481.

Watts, J. G. 1934. A comparison of the life cycles of *Frankliniella tritici* (Fitch), *F. fusca* (Hinds) and *Thrips tabaci* Lind. (Thysanoptera-Thripidae) in South Carolina. *J. Econ. Entomol.* 27:1158–1159.

Wave, H. E. 1964. Effect of bait-trap color on attractancy to *Drosophila melanogaster*. *J. Econ. Entomol.* 57:295–296.

Wave, H. E., W. A. Shands, and G. W. Simpson. 1965. Biology of the foxglove aphid in the northeastern United States. USDA Tech. Bull. 1338. 40 pp.

Way, M. J. and M. E. Cammell. 1982. The distribution and abundance of the spindle tree, *Euonymus europaeus*, in southern England with particular reference to forecasting infestations of the black bean aphid, *Aphis fabae*. *J. Appl. Ecol.* 19:929–940.

Way, M. J. and G. Murdie. 1965. An example of varietal variations in resistance of Brussels sprouts. *Ann. Appl. Biol.* 56:326–328.

Way, M. J., M. E. Cammell, L. R. Taylor, and I. P. Woiwod. 1981. The use of egg counts and suction trap samples to forecast the infestation of spring-sown field beans, *Vicia faba*, by the black bean aphid, *Aphis fabae*. *Ann. Appl. Biol.* 98:21–34.

Way, M. J., M. E. Cammell, D. V. Alford, H. J. Gould, C. W. Graham, A. Lane, W. I. St. G. Light, J. M. Raymer, G. D. Heathcote, K. E. Fletcher, and K. Seal. 1977. Use of forecasting in chemical control of black bean aphid, *Aphis fabae* Scop., on spring-sown field beans, *Vicia faba* L. *Plant Pathol.* 26: 1–7.

Wearing, C. H. 1972. Responses of *Myzus persicae* and *Brevicoryne brassicae* to leaf age and water stress in Brussels sprouts grown in pots. *Entomol. Exp. Appl.* 15: 61–80.

Webb, R. E. and F. F. Smith. 1973. Incidence of reflective mulches on infestations of *Liriomyza munda* in snap bean foliage. *J. Econ. Entomol.* 66: 539–540.

Webb, R. E., F. F. Smith, and A. L. Boswell. 1970. In-furrow applications of systemic insecticides for control of Mexican bean beetle. *J. Econ. Entomol.* 63: 1220–1223.

Webb, S. E. 1995. Damage to watermelon seedlings caused by *Frankliniella fusca* (Thysanoptera: Thripidae). *Florida Entomol.* 78: 178–179.

Webb, S. E. and S. B. Linda. 1992. Evaluation of spunbounded polyethylene row covers as a method of excluding insects and viruses affecting fall-grown squash in Florida. *J. Econ. Entomol.* 85: 2344–2352.

Webb, S. E. and A. M. Shelton. 1988. Laboratory rearing of the imported cabbageworm. New York Agric. Exp. Stn. Food Life Sci. Bull. 122. 6 pp.

Weber, D. C., F. A. Drummond, and D. N. Ferro. 1995. Recruitment of Colorado potato beetles (Coleoptera: Chrysomelidae) to solanaceous hosts in the field. *Environ. Entomol.* 24: 608–622.

Webster, F. M. 1889. Some studies of the development of *Lixus concavus*, Say, and *L. macer*, Leconte. *Entomol. Am.* 5: 11–16.

Webster, F. M. 1907. The chinch bug. USDA Bur. Entomol. Bull. 69. 95 pp.

Webster, R. L. and D. Stoner. 1914. The eggs and nymphal stages of the dusky leaf bug *Calocoris rapidus* Say. *J. New York Entomol. Soc.* 22: 229–232.

Weibull, J. H. W. 1993. Bird cherry-oat aphid (Homoptera: Aphididae) performance on annual and perennial temperate-region grasses. *Environ. Entomol.* 22: 149–153.

Weigel, C. A. and R. H. Nelson. 1936. Heat treatments for control of bulb mite on tuberose. *J. Econ. Entomol.* 29: 744–749.

Weigel, C. A., B. M. Broadbent, A. Busck, and C. Heinrich. 1924. The greenhouse leaf-tyer, *Phlyctaenia rubigalis* (Guenée). *J. Agric Res.* 29: 137–158.

Weinberg, H. L. and W. H. Lange. 1980. Developmental rate and lower temperature threshold of the tomato pinworm. *Environ. Entomol.* 9: 245–246.

Weintraub, P. G. and A. R. Horowitz. 1995. The newest leafminer pest in Israel, *Liriomyza huidobrensis*. *Phytoparasitica* 23: 177–184.

Weintraub, P. G. and A. R. Horowitz. 1996. Spatial and diel activity of the pea leafminer (Diptera: Agromyzidae) in potatoes, *Solanum tuberosum*. *Environ. Entomol.* 25: 722–726.

Weintraub, P. G. and A. R. Horowitz. 1997. Systemic effects of a neem insecticide on *Liriomyza huidobrensis* larvae. *Phytoparasitica* 25: 283–289.

Weisenborn, W. D., J. T. Trumble, and E. R. Oatman. 1990. Economic comparison of insecticide treatment programs for managing tomato pinworm (Lepidoptera: Gelechiidae) on fall tomatoes. *J. Econ. Entomol.* 83: 212–216.

Weiss, H. B. 1912. Notes on *Lixus concavus*. *J. Econ. Entomol.* 5: 434–436.

Weiss, M. J. and Z. B. Mayo. 1983. Potential of corn rootworm (Coleoptera: Chrysomelidae) larval counts to estimate larval populations to make control decisions. *J. Econ. Entomol.* 76: 158–161.

Weiss, M. J., Z. B. Mayo, and J. P. Newton. 1983. Influence of irrigation practices on the spatial distribution of corn rootworm (Coleoptera: Chrysomelide) eggs in the soil. *Environ. Entomol.* 12: 1293–1295.

Weiss, M. J., B. C. Schatz, J. C. Gardner, and B. A. Nead. 1994. Flea beetle (Coleoptera: Chrysomelidae) populations and crop yield in field pea and oilseed rape intercrops. *Environ. Entomol.* 23: 654–658.

Weissling, T. J. and L. J. Meinke. 1991a. Potential of starch encapsulated semiochemical-insecticide formulations for adult corn rootworm (Coleoptera: Chrysomelidae) control. *J. Econ. Entomol.* 84: 601–609.

Weissling, T. J. and L. J. Meinke. 1991b. Semiochemical-insecticide bait replacement and vertical distribution of corn rootworm (Coleoptera: Chrysomelidae) adults: implications for management. *Environ. Entomol.* 20: 945–952.

Weisz, R., Z. Smilowitz, and S. Fleischer. 1996. Evaluating risk of Colorado potato beetle (Coleoptera: Chrysomelidae) infestation as a function of migratory distance. *J. Econ. Entomol.* 89: 435–441.

Wellik, M. J., J. E. Slosser, and R. D. Kirby. 1979. Evaluation of procedures for sampling *Heliothis zea* and *Keiferia lycopersicella* on tomatoes. *J. Econ. Entomol.* 72: 777–780.

Welter, S. C., J. A. Rosenheim, M. W. Johnson, R. F. L. Mau, and L. R. Gusukuma-Minuto. 1990. Effects of *Thrips palmi* and western flower thrips (Thysanoptera: Thripidae) on the yield, growth, and carbon allocation pattern in cucumbers. *J. Econ. Entomol.* 83: 2092–2101.

Welty, C. 1995. Monitoring and control of European corn borer, *Ostrinia nubilalis* (Lepidoptera: Pyralidae), on bell peppers in Ohio. *J. Agric. Entomol.* 12: 145–161.

Wene, G. P. 1958. Control of *Nysius raphanus* Howard attacking vegetables. *J. Econ. Entomol.* 51: 250–251.

Werner, F. G. 1945. A revision of the genus Epicauta in America north of Mexico (Coleoptera, Meloidae). *Bull. Mus. Comp. Zool.* 95: 421–517.

Werner, F. G., W. R. Enns, and F. H. Parker. 1966. The Meloidae of Arizona. Arizona Agric. Exp. Stn. Tech. Bull. 175. 96 pp.

West, R. J., J. E. Laing, and S. A. Marshall. 1983. Parasites of the potato stem borer, *Hydraecia micacea* (Lepidoptera: Noctuidae). Proc. Entomol. Soc., Ontario 114: 69–82.

West, R. J., J. E. Laing, and P. E. A. Teal. 1985. Method for rearing the potato stem borer, *Hydraecia micacea* Esper (Lepidoptera: Noctuidae), in the laboratory. *J. Econ. Entomol.* 78: 219–221.

Westbrook, J. K., W. W. Wolf, P. D. Lingren, J. R. Raulston, J. D. Lopez, Jr., J. H. Matis, R. S. Eyster, J. F. Esquivel, and P. G. Schleider. 1997. Early-season migratory flights of corn earworm (Lepidoptera: Noctuidae). *Environ. Entomol.* 26: 12–20.

Westdal, P. H. and W. Romanow. 1972. Observations on the biology of the flea beetle, *Phyllotreta cruciferae* (Coleoptera: Chrysomelidae). *Manitoba Entomol.* 6: 35–45.

Westdal, P. H., C. F. Barrett, and H. P. Richardson. 1961. The six-spotted leafhopper, *Macrosteles fascifrons* (Stål.) and aster yellows in Manitoba. *Can J. Plant Sci.* 41: 320–331.

Weston, P. A. and J. R. Miller. 1989. Ovipositional responses of seed-corn maggot, *Delia platura* (Diptera: Anthomyiidae), to developmental stages of lima bean. *Ann. Entomol. Soc. Am.* 82: 387–392.

Whalon, M. E. and B. L. Parker. 1978. Immunological identification of tarnished plant bug predators. *Ann. Entomol. Soc. Am.* 71: 453–456.

Wharton, R. A. and F. E. Gilstrap. 1983. Key to and status of opiine braconid (Hymenoptera) parasitoids used in biological control of *Ceratitis* and *Dacus s. l.* (Diptera: Tephritidae). *Ann. Entomol. Soc. Am.* 76: 721–742.

Wheeler, E. W. 1923. Some braconids parasitic on aphids and their life-history. (Hym.). *Ann. Entomol. Soc. Am.* 16: 1–29.

Wheeler, G. C. and J. Wheeler. 1990. Insecta: Hymenoptera Formicidae. Pages 1277–1294 *in* D. L. Dindal (ed.). Soil Biology Guide. Wiley & Sons, New York.

Wheeler, M. R. 1981a. The Drosophilidae: a taxonomic overview. Pages 1–97 *in* M. Ashburner, H. L. Carson, and J. N. Thompson, Jr. (eds.). The Genetics and Biology of Drosophila. Academic Press, New York.

Wheeler, M. R. 1981b. Geographical survey of Drosophilidae: Nearctic species. Pages 99–121 *in* M. Ashburner, H. L. Carson, and J. N. Thompson, Jr. (eds.). The Genetics and Biology of Drosophila. Academic Press, New York.

Wheeler, M. R. 1987. Drosophilidae. Pages 1011–1018 *in* J. F. McAlpine (ed.). Manual of Nearctic Diptera, Vol. 2. Agric. Canada Res. Branch Mono. 28.

Whelan, D. B. 1935. A key to the Nebraska cutworms and armyworms that attack corn. Univ. Nebraska Agric. Exp. Stn. Res. Bull. 81. 27 pp.

Whitcomb, W. D. 1938. The carrot rust fly. Massachusetts Agric. Exp. Stn. Bull. 352. 36 pp.

Whitcomb, W. D. 1953. The biology and control of *Lygus campestris* L. on celery. Massachusetts Agric. Exp. Stn. Bull. 473. 15 pp.

Whitcomb, W. D. 1965. The carrot weevil in Massachusetts: biology and control. Massachusetts Agric. Exp. Stn. Bull. 550. 30 pp.

Whitcomb, W. D., T. D. Gowan, and W. F. Buren. 1982. Predators of *Diaprepes abbreviatus* (Coleoptera: Curculionidae) larvae. *Florida Entomol.* 65: 150–158.

White, A. J., S. D. Wratten, N. A. Berry, and U. Weigmann. 1995. Habitat manipulation to enhance biological control of *Brassica* pests by hover flies (Diptera: Syrphidae). *J. Econ. Entomol.* 88: 1171–1176.

White, I. M. and M. M. Elson-Harris. 1992. Fruit Flies of Economic Significance: their Identification and Bionomics. CAB International, United Kingdom. 601 pp.

White, R. E. and H. S. Barber. 1974. Nomenclature and definition of the tobacco flea beetle, *Epitrix hirtipennis* (Melsh.), and of *E. fasciata* Blatchley, (Coleoptera: Chrysomelidae). Proc. Entomol. Soc. Washington 76: 397–400.

White, W. H. and L. W. Brannon. 1939. The harlequin bug and its control. USDA Farmer's Bull. 1712. 10 pp.

Whitfield, G. H. 1984. Temperature threshold and degree-day accumulation required for development of postdiapause sugarbeet root maggots (Diptera: Otitidae). *Environ. Entomol.* 13: 1431–1435.

Whitfield, G. H. and B. Grace. 1985. Cold hardiness and overwintering survival of the sugarbeet root maggot (Diptera: Otitidae) in southern Alberta. *Ann. Entomol. Soc. Am.* 78: 501–505.

Whitford, F. and W. B. Showers. 1984. Olfactory and visual response by black cutworm larvae (Lepidoptera: Noctuidae) in locating a bait trap. *Environ. Entomol.* 13: 1269–1273.

Whitman, D. W., C. G. Jones, and M. S. Blum. 1992. Defensive secretion production in lubber grasshoppers (Orthoptera: Romaleidae): influence of age, sex, diet, and discharge frequency. *Ann. Entomol. Soc. Am.* 85: 96–102.

Whitman, D. W., J. P. J. Billen, D. Alsop, and M. S. Blum. 1991. Anatomy, ultrastructure, and functional morphology of the metathoracic tracheal defensive glands of the grasshopper *Romalea guttata*. *Can. J. Zool.* 69: 2100–2108.

Whitmarsh, R. D. 1917. The green soldier bug *Nezara hilaris* Say. Ohio Agric. Exp. Stn. Bull. 310: 519–552.

Whittle, T. A. and R. L. Burton. 1980. Development of the southwestern corn borer on three artificial diets. *Southwestern Entomol.* 5: 257–260.

Whitworth, R. J. and F. L. Poston. 1979. A thermal-unit accumulation system for the southwestern corn borer. *Ann. Entomol. Soc. Am.* 72: 253–255.

Whitworth, R. J., F. L. Poston, S. M. Welch, and D. Calvin. 1984. Quantification of southwestern corn borer feeding and its impact on corn yield. *Southwestern Entomol.* 9: 308–318.

Wierenga, J. M., D. L. Norris, and M. E. Whalon. 1996. Stage-specific mortality of Colorado potato beetle (Coleoptera: Chrysomelidae) feeding on transgenic potatoes. *J. Econ. Entomol.* 89: 1047–1052.

Wilborn, R. and J. Ellington. 1984. The effect of temperature and photoperiod on the coloration of the *Lygus hesperus, desertinus* and *lineolaris. Southwestern Entomol.* 9: 187–197.

Wilcox, J. 1926. The lesser bulb fly, *Eumerus strigatus* Fallén, in Oregon. *J. Econ. Entomol.* 19: 762–772.

Wilde, G. E. 1968. A laboratory method for continuously rearing the green stink bug. *J. Econ. Entomol.* 61: 1763–1764.

Wilde, G. E. 1969. Photoperiodism in relation to development and reproduction in the green stink bug. *J. Econ. Entomol.* 62: 629–630.

Wilde, G. E. 1971. Temperature effect on development of western corn rootworm eggs. *J. Kansas Entomol. Soc.* 44: 185–186.

Wilde, G. E. and J. Morgan. 1978. Chinch bug on sorghum: chemical control, economic injury levels, plant resistance. *J. Econ. Entomol.* 71: 908–910.

Wilde, G. E., O. Russ, and T. W. Mize. 1986. Tillage, cropping, and insecticide use practice: effects on efficacy of planting time treatments for controlling greenbug (Homoptera: Aphididae) and chinch bug (Heteroptera: Lygaeidae) in seedling sorghum. *J. Econ. Entomol.* 79: 1364–1365.

Wildermuth, V. L. 1914. The alfalfa caterpillar. USDA Bull. 124. 90 pp.

Wildermuth, V. L. 1917. The desert corn flea-beetle. USDA Bull. 436. 23 pp.

Wildermuth, V. L. 1920. The alfalfa caterpillar. USDA Farmers' Bull. 1094. 16 pp.

Wildermuth, V. L. and E. V. Walter. 1932. Biology and control of the corn leaf aphid with special reference to the southwestern states. USDA Tech. Bull. 306. 21 pp.

Wilding, N. 1975. *Entomophthora* species infecting pea aphid. Trans. R. Entomol. Soc. 127: 171–183.

Wilding, N. 1981. The effect of introducing aphid-pathogenic Entomophthoraceae into field populations of *Aphis fabae*. *Ann. Appl. Biol.* 99: 11–23.

Wilding, N. and J. N. Perry. 1980. Studies on *Entomophthora* in populations of *Aphis fabae* on field beans. *Ann. Appl. Biol.* 94: 367–378.

Wildman, T. E. and W. W. Cone. 1986. Drip chemigation of asparagus with disulfoton: *Brachycorynella asparagi* (Homoptera: Aphididae) control and disulfoton degradation. *J. Econ. Entomol.* 79: 1617–1620.

Wildman, T. E. and W. W. Cone. 1988. Control of *Brachycorynella asparagi* (Homoptera: Aphididae) in irrigated asparagus with granular systemic insecticides, and disulfoton degradation in asparagus fern. *J. Econ. Entomol.* 81: 1196–1202.

Wiesenborn, W. D. and J. T. Trumble. 1988. Optimal oviposition by the corn earworm (Lepidoptera: Noctuidae) on whorl-stage sweet corn. *Environ. Entomol.* 17: 722–726.

Wilkinson, A. T. 1963. Wireworms of cultivated land in British Columbia. Proc. Entomol. Soc. Brit., Columbia 60: 3–17.

Wilkinson, A. T. S. and H. R. MacCarthy. 1967. The marsh crane fly, *Tipula paludosa* Mg., a new pest in British Columbia (Diptera: Tipulidae). J. Entomol. Soc. Brit., Columbia 64: 29–34.

Wilkinson, A. T. S., D. G. Finlayson, and C. J. Campbell. 1977. Soil incorporation of insecticides for control of wireworms in potato land in British Columbia. *J. Econ. Entomol.* 70: 755–758.

Williams, D. F., C. S. Lofgren, and A. Lemire. 1980. A simple diet for rearing laboratory colonies of the red imported fire ant. *J. Econ. Entomol.* 73: 176–177.

Williams, D. J. 1986. The identity and distribution of the genus *Maconellicoccus* Ezzat (Hemiptera: Pseudococcidae) in Africa. Bull. Entomol. Res. 76: 351–357.

Williams, D. J. 1996. A brief account of the hibiscus mealybug *Maconellicoccus hirsutus* (Hemiptera: Pseudococcidae), a pest of agriculture and horticulture, with descriptions of two related species from southern Asia. *Bull. Entomol. Res.* 86: 617–628.

Williams, F. X. 1931. Handbook of the Insects and other Invertebrates of Hawaiian Sugar Cane Fields. Hawaiian Sugar Planters' Assoc., Honolulu. 400 pp.

Williams, III, L., D. J. Schotzko, and J. P. McCaffrey. 1992. Geostatistical description of the spatial distribution of *Limonius californicus* (Coleoptera: Elateridae) wireworms in the northwestern United States, with comments on sampling. *Environ. Entomol.* 21: 983–995.

Williams, R. N. and K. V. Miller. 1982. Field assay to determine attractiveness of various aromatic compounds to rose chafer adults. *J. Econ. Entomol.* 75: 196–198.

Williams, R. N., T. P. McGovern, M. G. Klein, and D. S. Fickle. 1990. Rose chafer (Coleoptera: Scarabaeidae): improved attractants for adults. *J. Econ. Entomol.* 83: 111–116.

Williams, R. N., D. S. Fickle, M. Kehat, D. Blumberg, and M. G. Klein. 1983. Bibliography of the genus *Carpophilus* Stephens (Coleoptera: Nitidulidae). Ohio Agric. Res. Devt Ctr Res. Circ. 278. 95 pp.

Williamson, C. and M. B. von Wechmar. 1995. The effects of two viruses on the metamorphosis, fecundity, and longevity of the green stinkbug, *Nezara viridula*. *J. Invertebr. Pathol.* 65: 174–178.

Willson, H. R. and J. B. Eisley. 1992. Effects of tillage and prior crop on the incidence of five key pests on Ohio corn. *J. Econ. Entomol.* 85: 853–859.

Willson, H. R., M. Semel, M. Tebcherany, D. J. Prostak, and A. S. Hill. 1981. Evaluation of sex attractant and blacklight traps for monitoring black cutworm and variegated cutworm. *J. Econ. Entomol.* 74: 517–519.

Wilson, H. F., W. B. Hull, and A. S. Srivastava. 1947. A comparison of rotenone, DDT, and benzene hexachloride for pea aphid control. *J. Econ. Entomol.* 40: 101–103.

Wilson, H. F., R. C. Pickett, and L. G. Gentner. 1919. The common cabbage worm in Wisconsin. Univ. Wisconsin Agric. Exp. Stn. Res. Bull. 45. 35 pp.

Wilson, J. W. 1932. Notes on the biology of *Laphygma exigua* Huebner. *Florida Entomol.* 16: 33–39.

Wilson, J. W. 1933. The biology of parasites and predators of *Laphygma exigua* Huebner reared during the season of 1932. *Florida Entomol.* 17: 1–15.

Wilson, J. W. 1934. The asparagus caterpillar: its life history and control. Florida Agric. Exp. Stn. Bull. 271: 1–26.

Wilson, L. T., C. Pickel, R. C. Mount, and F. G. Zalom. 1983a. Presence-absence sequential sampling for cabbage aphid and green peach aphid (Homoptera: Aphididae) on Brussels sprouts. *J. Econ. Entomol.* 76: 476–479.

Wilson, L. T., F. G. Zalom, R. Smith, and M. P. Hoffmann. 1983b. Monitoring for fruit damage in processing tomatoes: use of a dynamic sequential sampling plan. *Environ. Entomol.* 12: 835–839.

Wilson, M. J., D. M. Glen, and S. K. George. 1993a. The rhabditid nematode *Phasmarhabditis hermaphrodita* as a potential biological control agent for slugs. *Biocontrol Sci. Tech.* 3: 503–511.

Wilson, M. J., D. M. Glen, S. K. George, and R. C. Butler. 1993b. Mass cultivation and storage of the rhabditid nematode *Phasmarhabditis hermaphrodita*, a biocontrol agent for slugs. *Biocontrol Sci. Tech.* 3: 513–521.

Wilson, M. J., D. M. Glen, S. K. George, and L. A. Hughes. 1995. Biocontrol of slugs in protected lettuce using the rhabditid nematode *Phasmarhabditis hermaphrodita*. *Biocontrol Sci. Tech.* 5: 233–242.

Wilson, T. H. and T. A. Cooley. 1972. A chalcidoid planidium and an entomophilic nematode associated with the western flower thrips. *Ann. Entomol. Soc. Am.* 65: 414–418.

Wilson, W. A. 1971. Nematode occurrence in Ontario earwigs (Nematoda: Dermaptera). *Can. Entomol.* 103: 1045–1048.

Winewriter, S. A. and T. J. Walker. 1988. Group and individual rearing of field crickets (Orthoptera: Gryllidae). *Entomol. News* 99: 53–62.

Wipfli, M. S., S. S. Peterson, D. B. Hogg, and J. L. Wedberg. 1992. Dispersion patterns and optimum sample size analyses for three plant bug (Heteroptera: Miridae) species associated with birdsfoot trefoil in Wisconsin. *Environ. Entomol.* 21: 1248–1252.

Wisecup, C. B. and N. C. Hayslip. 1943. Control of mole crickets by use of poisoned baits. USDA Leaflet 237. 6 pp.

Wiseman, B. R. and N. W. Widstrom. 1986. Mechanisms of resistance in 'Zapalote Chico' corn silks to fall armyworm (Lepidoptera: Noctuidae) larvae. *J. Econ. Entomol.* 79: 1390–1393.

Wiseman, B. R., W. W. McMillian, and N. W. Widstrom. 1976. Wireworm resistance among corn inbreds. *J. Georgia Entomol. Soc.* 11: 58–59.

Wiseman, B. R., W. P. Williams, and F. M. Davis. 1981. Fall armyworm: resistance mechanisms in selected corns. *J. Econ. Entomol.* 74: 622–624.

Wishart, G. 1957. Surveys of parasites of *Hylemya* spp. (Diptera: Anthomyiidae) that attack cruciferous crops in Canada. *Can. Entomol.* 79: 450–454.

Witkowski, J. F. and G. W. Echtenkamp. 1996. Influence of planting date and insecticide on the bean leaf beetle (Coleoptera: Chrysomelidae) abundance and damage in Nebraska soybean. *J. Econ. Entomol.* 89: 189–196.

Wittenborn, G. and W. Olkowski. 2000. Potato aphid monitoring and biocontrol in processing tomatoes. *IPM Pract.* 22 (3): 1–7.

Wojcik, D. P. 1986a. Observations on the biology and ecology of fire ants in Brazil. Pages 88–103 *in* C. S. Lofgren and R. K. Vander Meer (eds.). Fire Ants and Leaf-Cutting Ants. Westview Press, Boulder.

Wojcik, D. P. 1986b. Bibliography of imported fire ants and their control: second supplement. *Florida Entomol.* 69: 394–415.

Wojcik, D. P. and C. S. Lofgren. 1982. Bibliography of imported fire ants and their control: first supplement. *Bull. Entomol. Soc. Am.* 28: 269–276.

Wolcott, G. N. 1936. The life history of "*Diaprepes abbreviatus*" L., at Rio Piedras, Puerto Rico. *J. Agric. Univ. Puerto Rico* 20: 883–914.

Wolcott, G. N. 1948. The insects of Puerto Rico. *J. Agric. Univ. Puerto Rico* 32: 417–748.

Wolf, R. A., L. P. Pedigo, R. H. Shaw, and L. D. Newsom. 1987. Migration/transport of the green cloverworm, *Plathypena scabra* (F.) (Lepidoptera: Noctuidae), into Iowa as determined by synoptic-scale weather patterns. *Environ. Entomol.* 16: 1169–1174.

Wolfenbarger, D. and J. P. Sleesman. 1961. Resistance to the Mexican bean beetle in several bean genera and species. *J. Econ. Entomol.* 54: 1018–1022.

Wolfenbarger, D. A. 1967. Seasonal abundance and damage estimates of cabbage looper larvae and two aphid species. *J. Econ. Entomol.* 60: 277–279.

Wolfenbarger, D. A. and Wolfenbarger, D. O. 1966. Tomato yields and leaf miner infestations and a sequential sampling plan for determining need for control treatments. *J. Econ. Entomol.* 59: 279–283.

Wolfenbarger, D. O. 1940. Relative prevalence of potato flea beetle injuries in fields adjoining uncultivated areas. *Ann. Entomol. Soc. Am.* 33: 391–394.

Wolfenbarger, D. O. 1954. Potato yields associated with control of aphids and the serpentine leaf miner. *Florida Entomol.* 37: 7–12.

Wolfenbarger, D. O. 1972. Effects of temperatures on mortality of green peach aphids on potatoes treated with ethyl-methyl parathion. *J. Econ. Entomol.* 65: 881–882.

Wolfenbarger, D. O. and W. D. Moore. 1968. Insect abundances on tomatoes and squash mulched with aluminum and plastic sheetings. *J. Econ. Entomol.* 61: 34–36.

Wolfenbarger, D. O., J. B. Nosky, and R. L. McGarr. 1970. Toxicity of petroleum oils to larvae and eggs of some cotton insects and persistence of these oils on cotton foliage. *J. Econ. Entomol.* 63: 1765–1767.

Wong, T. T. Y. and M. M. Ramadan. 1987. Parasitization of the Mediterranean and Oriental fruit flies (Diptera: Tephritidae) in the Kula area of Maui, Hawaii. *J. Econ. Entomol.* 80: 77–80.

Wong, T. T. Y., J. B. Beavers, R. A. Sutton, and P. A. Norman. 1975. Field tests of insecticides for control of adult *Diaprepes abbreviatus* on citrus. *J. Econ. Entomol.* 68: 119–121.

Wong, T. T. Y., M. M. Ramadan, J. C. Herr, and D. O. McInnis. 1992. Suppression of a Mediterranean fruit fly (Diptera: Tephritidae) population with concurrent parasitoid and sterile fly releases in Kula, Maui, Hawaii. *J. Econ. Entomol.* 85: 1671–1681.

Wood, E. A. and K. J. Starks. 1972. Damage to sorghum by a lygaeid bug, *Nysius raphanus. J. Econ. Entomol.* 65: 1507–1508.

Wood, J. R., S. L. Poe, and N. C. Leppla. 1979. Winter survival of fall armyworm pupae in Florida. *Eviron. Entomol.* 8: 249–252.

Woodruff, R. E. 1974. A South American leaf beetle pest of crucifers in Florida. Florida Dept. Agric. and Consumer Serv. Entomol. Circ. 148. 2 pp.

Woodside, A. M. 1946. Life history studies of *Euschistus servus* and *E. tristigmus. J. Econ. Entomol.* 39: 161–163.

Woodson, W. D. and J. V. Edelson. 1988. Developmental rate as a function of temperature in a carrot weevil, *Listronotus texanus* (Coleoptera: Curculionidae). *Ann. Entomol. Soc. Am.* 81: 252–254.

Woodson, W. D. and W. S. Fargo. 1991. Interactions of temperature and squash bug density (Hemiptera: Coreidae) on growth of seedling squash. *J. Econ. Entomol.* 84: 886–890.

Woodson, W. D. and W. S. Fargo. 1992. Interactions of plant size and squash bug (Hemiptera: Coreidae) density on growth of seedling squash. *J. Econ. Entomol.* 85: 2365–2369.

Woodson, W. D. and J. J. Jackson. 1996. Developmental rate as a function of temperature in northern corn rootworm (Coleoptera: Chrysomelidae). *Ann. Entomol. Soc. Am.* 89: 226–230.

Woodson, W. D., J. V. Edelson, and T. A. Royer. 1989. Control of a carrot weevil, *Listronotus texanus* (Coleoptera: Curculionidae): timing pesticide applications and response to selected pesticides. *J. Econ. Entomol.* 82: 209–212.

Woodson, W. D., J. J. Jackson, and M. M. Ellsbury. 1996. Northern corn rootworm (Coleoptera: Chrysomelidae) temperature requirements for egg development. *Ann. Entomol. Soc. Am.* 89: 898–903.

Workman, Jr., R. B. 1958. The biology of the onion maggot, *Hylemya antiqua* (Meigen), under field and greenhouse conditions. Unpublished Ph.D. dissertation, Oregon State Univ. Press, Corvallis, OR. 82 pp.

Workman, R. B., R. B. Chalfant, and D. J. Schuster. 1980. Management of the cabbage looper and diamondback moth on cabbage by using two damage thresholds and five insecticide treatments. *J. Econ. Entomol.* 73: 757–758.

Worsham, E. L. and W. V. Reed. 1912. The mole cricket (*Scapteriscus didactylus* Latr.). Georgia Agric. Exp. Stn. Bull. 101. 263 pp.

Wozniak, C. A., G. A. Smith, D. T. Kaplan, W. J. Schroeder, and L. G. Campbell. 1993. Mortality and aberrant development of the sugarbeet root maggot (Diptera: Otitidae) after exposure to Steinernematid nematodes. *Biol. Control* 3: 221–225.

Wright, Jr., D. P. 1972. Rearing *Eleodes suturalis* without diapause. *J. Econ. Etnomol.* 65: 1731.

Wright, D. W. and Q. A. Geering. 1948. The biology and control of the pea moth, *Laspayresia nigricana*, Steph. *Bull. Etnomol. Res.* 39: 57–84.

Wright, D. W., Q. A. Geering, and D. G. Ashby. 1947. The insect parasites of the carrot fly, *Psila rosae*, Fab. *Bull. Entomol. Res.* 37: 507–529.

Wright, J. M. and G. C. Decker. 1958. Laboratory studies of the life cycle of the carrot weevil. *J. Econ. Entomol.* 51: 37–39.

Wright, L. C. and W. W. Cone. 1983a. Extraction from foliage and within-plant distribution for sampling of *Brachycolus asparagi* (Homoptera: Aphididae) on asparagus. *J. Econ. Entomol.* 76: 801–805.

Wright, L. C. and W. W. Cone. 1983b. Barriers for control of cutworm (Lepidoptera: Noctuidae) damage to Concord grape buds. *J. Econ. Entomol.* 76: 1175–1177.

Wright, L. C. and W. W. Cone. 1986. Sampling plan for *Brachycorynella asparagi* (Homoptera: Aphididae) in mature asparagus fields. *J. Econ. Entomol.* 79: 817–821.

Wright, L. C. and W. W. Cone. 1988a. Population dynamics of *Brachycorynella asparagi* (Homoptera: Aphididae) on undisturbed asparagus in Washington state. *Environ. Entomol.* 17: 878–886.

Wright, L. C. and W. W. Cone. 1988b. Population statistics for the asparagus aphid, *Brachycorynella asparagi* (Homoptera: Aphididae), on different ages of asparagus foliage. *Environ. Entomol.* 17: 699–703.

Wright, R. J. and S. D. Danielson. 1992. First report of the chinch bug (Heteroptera: Lygaeidae) egg parasitoid *Eumicrosoma beneficum* Gahan (Hymenoptera: Scelionidae) in Nebraska. *J. Kansas Entomol. Soc.* 65: 346–348.

Wright, R. J., J. F. Witkowski, G. Echtenkamp, and R. Georgis. 1993. Efficacy and persistence of *Steinernema carpocapsae* (Rhabditida: Steinernematidae) applied through a center-pivot irrigation system against larval corn rootworms (Coleoptera: Chrysomelidae). *J. Econ. Entomol.* 86: 1348–1354.

Wukasch, R. T. and M. K. Sears. 1981. Damage to asparagus by tarnished plant bugs, *Lygus lineolaris,* and alfalfa plant bugs, *Adelphocoris lineolatus* (Heteroptera: Miridae). Proc. Entomol. Soc. Ontario 112: 49–51.

Wylie, H. G. 1979. Observations of distribution, seasonal life history, and abundance of flea beetles (Coleoptera: Chrysomelidae) that infest rape crops in Manitoba. *Can. Entomol.* 111: 1345–1353.

Wylie, H. G. 1982. An effect of parasitism by *Microctonus vittatae* (Hymenoptera: Braconidae) on emergence of *Phyllotreta cruciferae* and *Phyllotreta striolata* (Coleoptera: Chrysomelidae) from overwintering sites. *Can. Entomol.* 114: 727–732.

Wylie, H. G. 1984. Oviposition and survival of three nearctic Euphorine braconids in crucifer-infesting flea beetles (Coleoptera: Chrysomelidae). *Can. Entomol.* 116: 1–4.

Wylie, H. G. and G. E. Bucher. 1977. The bertha armyworm, *Mamestra configurata* (Lepidoptera: Noctuidae). Mortality of immature stages on the rape crop, 1972–1975. *Can. Entomol.* 109: 823–837.

Wylie, H. G. and C. Loan. 1984. Five nearctic and one introduced euphorine species (Hymenoptera: Braconidae) that parasitize adults of crucifer-infesting flea beetles (Coleoptera: Chrysomelidae). *Can. Entomol.* 116: 235–246.

Wyman, J. A. 1979. Effect of trap design and sex attractant release rates on tomato pinworm catches. *J. Econ. Entomol.* 72: 865–868.

Wyman, J. A. 1992. Management approaches for cruciferous insect pests in central North America. Pages 503–509 *in* N. S. Talekar (ed.) Diamondback Moth and Other Crucifer Pests. Asian Vegetable Research and Development Center, Taipei, Taiwan.

Wyman, J. A., J. L. Libby, and R. K. Chapman. 1976. The role of seed-corn beetles in predation of cabbage maggot immature stages. *Environ. Entomol.* 5: 259–263.

Wyman, J. A., J. L. Libby, and R. K. Chapman. 1977. Cabbage maggot management aided by predictions of adult emergence. *J. Econ. Entomol.* 70: 327–331.

Wyman, J. A., N. C. Toscano, K. Kido, H. Johnson, and K. S. Mayberry. 1979. Effects of mulching on the spread of aphid-transmitted watermelon mosaic virus to summer squash. *J. Econ. Entomol.* 72: 139–143.

Wymore, F. H. 1931. The garden centipede. California Agric. Exp. Stn. Bull. 518. 22 pp.

Wynne, J. W., A. J. Keaster, K. O. Gerhardt, and G. F. Krause. 1991. Plant species identified as food sources for adult black cutworm (Lepidoptera: Noctuidae) in northwestern Missouri. *J. Kansas Entomol. Soc.* 64: 381–387.

Yamamoto, R. T. 1968. Mass rearing of the tobacco hornworm. I. Egg production. *J. Econ. Entomol.* 61: 170–174.

Yamamoto, R. T. 1969. Mass rearing of the tobacco hornworm. II. Larval rearing and pupation. *J. Econ. Entomol.* 62: 1427–1431.

Yamamoto, R. T. 1972. Flight behavior of *Manduca* moths as a factor in hostplant site selection. *Israel J. Entomol.* 7: 59–61.

Yang, S. L. and F. W. Zettler. 1975. Effects of alarm pheromones on aphid probing behavior and virus transmission efficiency. *Plant Dis. Rep.* 59: 902–905.

Yano, K., T. Miyake, and V. F. Eastop. 1983. The biology and economic importance of rice aphids (Hemiptera: Aphididae): a review. *Bull. Entomol. Res.* 73: 539–566.

Yathom, S., M. J. Berlinger, R. Dahan, and S. Voerman. 1979. Pheromone-baited traps as an aid in studying the phenology of the potato tuber moth, *Phthorimaea operculella* (Zell.), in Israel. *Phytoparasitica* 7: 195–197.

Yeargan, K. V. 1979. Parasitism and predation of stink bug eggs in soybean and alfalfa fields. *Environ. Entomol.* 8: 715–719.

Yokomi, R. K., K. A. Hoelmer, and L. S. Osborne. 1990. Relationships between the sweetpotato whitefly and the squash silverleaf disorder. *Phytopathology* 80: 895–900.

Yosef, R. and D. W. Whitman. 1992. Predator exaptations and defensive adaptations in evolutionary balance: no defense is perfect. *Evol. Ecol.* 6: 527–536.

Yoshida, H. A. and N. C. Toscano. 1994. Comparative effects of selected natural insecticides on *Heliothis virescens* (Lepidoptera: Noctuidae) larvae. *J. Econ. Entomol.* 87: 305–310.

Yoshiyasu, Y. 1975. Systematic study of the subfamily Pyraustinae of Japan (Lepidoptera: Pyralidae) I. Descriptions of a new species and two unrecorded species. *J. Fac. Agric. Kyushu Univ.* 19: 187–195.

Young, A. G. 1990. Assessment of slug activity using bran-baited traps. *Crop Prot.* 9: 355–358.

Young, A. G., G. R. Port, and D. B. Green. 1993. Development of a forecast of slug activity: validation of models to predict slug activity from meteorological conditions. *Crop Prot.* 12: 232–236.

Young, A. M. 1985. Natural history notes on *Astraptes* and *Urbanus* (Hesperiidae) in Costa Rica. *J. Lepid. Soc.* 39: 215–223.

Young, D. K. 1984a. Cantharidin and insects: an historical review. *Great Lakes Entomol.* 17: 187–194.

Young, D. K. 1984b. Field records and observations of insects associated with cantharidin. *Great Lakes Entomol.* 17: 195–199.

Young, H. C., B. A. App, J. B. Gill, and H. S. Hollingsworth. 1950. White-fringed beetles and how to combat them. USDA Circ. 850. 15 pp.

Young, J. R. 1979. Fall armyworm: control with insecticides. *Florida Entomol.* 62: 130–133.

Young, O. P. 1986. Host plants of the tarnished plant bug, *Lygus lineolaris* (Heteroptera: Miridae). *Ann. Entomol. Soc. Am.* 79: 747–762.

Young, O. P. 1995. Ground-surface activity of *Cotinis nitida* (L.) (Coleoptera: Scarabaeidae: Cetoniinae) larvae in an old-field habitat. *Coleop. Bull.* 49: 229–233.

Young, S. Y. 1990. Effect of nuclear polyhedrosis virus infection in *Spodoptera ornithogalli* larvae on post larval stages and dissemination by adults. *J. Invertebr. Pathol.* 55: 69–75.

Young, S. Y. and R. W. McNew. 1994. Persistence and efficacy of four nuclear polyhedrosis viruses for corn earworm (Lepidoptera: Noctuidae) on heading grain sorghum. *J. Entomol. Sci.* 29: 370–380.

Young, W. R. and J. A. Sifuentes. 1959. Biological and control studies on *Estigmene acrea* (Drury), a pest of corn in the Yaqui Valley, Sonora, Mexico. *J. Econ. Entomol.* 52: 1109–1111.

Youssef, K. H., S. M. Hammad, and A. R. Donia. 1973. Studies on biology of the cabbage webworm *Hellula undalis* F. (Lepidoptera, Pyraustidae). *Z. Angew. Entomol.* 74: 1–6.

Yudin, L. S., J. J. Cho, and W. C. Mitchell. 1986. Host range of western flower thrips, *Frankliniella occidentalis* (Thysanoptera: Thripidae), with special reference to *Leucaena glauca*. *Environ. Entomol.* 15: 1292–1295.

Yudin, L. S., W. C. Mitchell, and J. J. Cho. 1987. Color preference of thrips (Thysanoptera: Thripidae) with reference to aphids (Homoptera: Aphididae) and leafminers in Hawaiian lettuce farms. *J. Econ. Entomol.* 80: 51–55.

Yudin, L. S., B. E. Tabashnik, W. C. Mitchell, and J. J. Cho. 1991. Effects of mechanical barriers on distribution of thrips (Thysanoptera: Thripidae) in lettuce. *J. Econ. Entomol.* 84: 136–139.

Zacharuk, R. Y. 1962a. Distribution, habits, and development of *Ctenicera destructor* (Brown) in western Canada, with notes on the related species *C. aeripennis* (Kby.) (Coleoptera: Elateridae). *Can. J. Zool.* 40: 539–552.

Zacharuk, R. Y. 1962b. Seasonal behavior of larvae of *Ctenicera* spp. and other wireworms (Coleoptera: Elateridae), in relation to temperature, moisture, food, and gravity. *Can. J. Zool.* 40: 697–718.

Zacharuk, R. Y. 1963. Comparative food preferences of soil-, sand-, and wood-inhabiting wireworms (Coleoptera, Elateridae). *J. Entomol. Res.* 54: 161–165.

Zalom, F. G. 1981a. Effects of aluminum mulch on fecundity of apterous *Myzus persicae* on head lettuce in a field planting. *Entomol. Exp. Appl.* 30: 227–230.

Zalom, F. G. 1981b. The influence of reflective mulches and lettuce types on the incidence of aster yellows and abundance of its vector, *Macrosteles fascifrons* (Homoptera: Cicadellidae), in Minnesota. *Great Lakes Entomol.* 14: 145–150.

Zalom, F. G. and A. Jones. 1994. Insect fragments in processed tomatoes. *J. Econ. Entomol.* 87: 181–186.

Zalom, F. G. and C. Pickel. 1985. Damage by the cabbage maggot, *Hylemya (Delia) brassicae* (Diptera: Anthomyiidae), to brussels sprouts. *J. Econ. Entomol.* 78: 1251–1253.

Zalom, F. G., C. Castane, and R. Gabarra. 1995. Selection of some winter-spring vegetable crop hosts by *Bemisia argentifolii* (Homoptera: Aleyrodidae). *J. Econ. Entomol.* 88: 70–76.

Zalom, F. G., J. M. Smilanick, and L. E. Ehler. Fruit damage by stink bugs (Hemiptera: Pentatomidae) in bush-type tomatoes. *J. Econ. Entomol.* 90: 1300–1306.

Zalom, F. G., L. T. Wilson, and M. P. Hoffmann. 1986a. Impact of feeding by tomato fruitworm, *Heliothis zea* (Boddie) (Lepidoptera: Noctuidae), and beet armyworm, *Spodoptera exigua* (Hübner) (Lepidoptera: Noctuidae), on processing tomato fruit quality. *J. Econ. Entomol.* 79: 822–826.

Zalom, F. G., L. T. Wilson, and R. Smith. 1983. Oviposition patterns by several lepidopterous pests on processing tomatoes in California. *Environ. Entomol.* 12: 1133–1137.

Zalom, F. G., J. Kitzmiller, L. T. Wilson, and P. Gutierrez. 1986b. Observation of tomato russet mite (Acari: Eriophyidae) damage symptoms in relation to tomato plant development. *J. Econ. Entomol.* 79: 940–942.

Zangerl, A. R. and M. R. Berenbaum. 1992. Oviposition patterns and hostplant suitability: parsnip webworms and wild parsnip. *Am. Midl. Nat.* 128: 292–298.

Zangerl, A. R. and M. R. Berenbaum. 1993. Plant chemistry, insect adaptations to plant chemistry, and host plant utilization patterns. *Ecology* 74: 47–54.

Zehnder, G. W. 1986. Timing of insecticides for control of Colorado potato beetle (Coleoptera: Chrysomelidae) in eastern Virginia based on differential susceptibility of life stages. *J. Econ. Entomol.* 79: 851–856.

Zehnder, G. W. 1997. Population dynamics of whitefringed beetle (Coleoptera: Curculionidae) on sweet potato in Alabama. *Environ. Entomol.* 26: 727–735.

Zehnder, G. W. and J. Hough-Goldstein. 1990. Colorado potato beetle (Coleoptera: Chrysomelidae) population development and effects on yield of potatoes with and without straw mulch. *J. Eco. Entomol.* 83: 1982–1987.

Zehnder, G. and J. Speese III. 1987. Assessment of color response and flight activity of *Leptinotarsa decemlineata* (Say) (Coleoptera: Chrysomelidae) using window flight traps. *Environ. Entomol.* 16: 1199–1202.

Zehnder, G. W. and J. T. Trumble. 1984. Spatial and diel activity of *Liriomyza* species (Diptera: Agromyzidae) in fresh market tomatoes. *Environ. Entomol.* 13: 1411–1416.

Zehnder, G. W. and J. T. Trumble. 1985. Sequential sampling plans with fixed levels of precision for *Liriomyza* species (Diptera: Agromyzidae) in fresh market tomatoes. *J. Econ. Entomol.* 78: 138–142.

Zehnder, G. and J. D. Warthen. 1988. Feeding inhibition and mortality effects of neem-seed extract on the Colorado potato beetle. *J. Econ. Entomol.* 81: 1040–1044.

Zehnder, G., A. M. Vencill, and J. Speese III. 1995. Action thresholds based on plant defoliation for management of Colorado potato beetle (Coleoptera: Chrysomelidae) in potato. *J. Econ. Entomol.* 88: 155–161.

Zeiss, M. R. and L. P. Pedigo. 1996. Timing of food plant availability: effect on survival and oviposition of the bean leaf beetle (Coleoptera: Chrysomelidae). *Environ. Entomol.* 25: 295–302.

Zepp, D. B. and A. J. Keaster. 1977. Effects of corn plant densities on the girdling behavior of the southwestern corn borer. *J. Econ. Entomol.* 70: 678–680.

Zhang, A., H. T. Facundo, P. S. Robbins, C. E. Linn, Jr., J. L. Hanula, M. G. Villani, and W. L. Roelofs. 1994. Identification and synthesis of female sex pheromone of oriental beetle, *Anomala orientalis* (Coleoptera: Scarabaeidae). *J. Chem Ecol.* 20: 2415–2427.

Zhao, G., W. Liu, J. M. Brown, and C. O. Knowles. 1995. Insecticide resistance in field and laboratory strains of western flower thrips (Thysanoptera: Thripidae). *J. Econ. Entomol.* 88: 1164–1170.

Zimmerman, E. C. 1978. Insects of Hawaii. Vol. 9. Microlepidoptera, Part I. University Press of Hawaii, Honolulu. 881 pp.

Zuk, M. 1987. The effects of gregarine parasites on longevity, weight loss, fecundity and developmental time in the field crickets *Gryllus veletis* and *G. pennsylvanicus*. *Ecol. Entomol.* 12: 349–354.

Zurlini, G. and A. S. Robinson. 1978. Onion conditioning pertaining to larval preference, survival and rate of development in *Delia* (= *Hylemya*) *antiqua*. *Entomol. Exp. Appl.* 23: 279–286.

Index